International Seminar on Applied Physics, Optoelectronics and Photonics (APOP 2016)

MATEC Web of Conferences Volume 61 (2016)

Shanghai, China
28 – 29 May 2016

Editors:

Samir Ahmed Hamouda
Mohsen Mirzaei
Zhongqin Yu

ISBN: 978-1-5108-2599-4

Printed from e-media with permission by:

Curran Associates, Inc.
57 Morehouse Lane
Red Hook, NY 12571

Some format issues inherent in the e-media version may also appear in this print version.

Printed by Curran Associates, Inc. (2016)

For additional information, please contact EDP Sciences – Web of Conferences
at the address below.

EDP Sciences – Web of Conferences
17, Avenue du Hoggar
Parc d'Activité de Courtabœuf
BP 112
F-91944 Les Ulis Cedex A
France

Phone: +33 (0) 1 69 18 75 75
Fax: +33 (0) 1 69 28 84 91

contact-edps@webofconferences.org

Additional copies of this publication are available from:

Curran Associates, Inc.
57 Morehouse Lane
Red Hook, NY 12571 USA
Phone: 845-758-0400
Fax: 845-758-2633
Email: curran@proceedings.com
Web: www.proceedings.com

TABLE OF CONTENTS

Aims & scope

MATEC Web of Conferences is an open-access journal devoted to the publishing of conference proceedings dealing with Materials science, Engineering and Chemistry.

The journal aims to make conference proceedings visible, citable and widely usable by the scientific community. It offers outstanding services for the publication and dissemination of contents in order to ensure long-term archiving and maximum exposure of scientific events.

Coverage includes all engineering disciplines: civil, naval, mechanical, chemical, and electrical engineering as well as nanotechnology and metrology. The journal also deals with the physical-chemical characterization of materials, their implementation, their resistance in their environment, etc. Others subdisciplines of chemistry, such as analytical chemistry, petrochemistry, organic chemistry, etc., and even pharmacology, are also welcome.

Why publish with MATEC Web of Conferences?

- **Open access:** conference papers available worldwide.
- **Fast publication:** less than 3 months.
- **Total flexibility:** no page restrictions, free unlimited use of colour in the online version, any kind of scientific document welcomed (traditional articles, posters, abstracts and slideshows).
- **High visibility:** all documents indexed in main databases and easily found online.
- **Fair author rights:** CC-BY Creative Commons license.
- **Range of additional publishing options:** printed proceedings, CD-ROMs, copy-editing, etc.

Contact: Isaline Augusto
matec@webofconferences.org

ISSN (electronic edition): 2261-236X

www.matec-conferences.org

edp open Open Access Publishing from EDP Sciences

edp sciences

MATEC Web of Conferences

Statement of Peer review

In submitting conference proceedings to MATEC Web of Conferences, I certify to the Publisher that I adhere to the **Policy on Publishing Integrity** of the journal in order to safeguard good scientific practice in publishing.

1. All articles have been subjected to peer review administered by the proceedings editors.
2. Reviews have been conducted by expert referees, who have been requested to provide unbiased and constructive comments aimed, whenever possible, at improving the work.
3. Proceedings editors have taken all reasonable steps to ensure the quality of the materials they publish and their decision to accept or reject a paper for publication has been based only on the merits of the work and the relevance to the journal.

Title, date and place of the conference

The International Seminar on Applied Physics, Optoelectronics and Photonics (APOP 2016)

May 28-29, 2016, Shanghai, China.

Proceedings editor(s):

Prof. Samir Ahmed Hamouda Dr.Mohsen Mirzaei Prof.Zhongqin Yu

Date and editor's signature

2016/4/17

17, avenue du Hoggar - PA de Courtabœuf – BP 112 - 91944 Les Ulis Cedex A (France)
Tél. : 33 (0)1 69 18 75 75 - Fax : 33(0)1 69 07 45 17 – www.edpsciences.org

The field investigation on thermal comfort of tent in early autumn of Tianjin

Hao Yang [1,a] ,Yingni Jiang [2,b] ,Jingbo Liu [3,c] Zhengguo Gong [4,d]

Architecture Engineering Department, Logistics University of PAP, Tian'jin, china
[a] E-mail address:849253712@qq.com, [b] E-mail address: jyingni@sina.com
[c] E-mail address:568818102@qq.com, [d] E-mail address:782101251@qq.com

Abstract: In a university campus in Tianjin, four tents were set up to investigate the thermal environment and thermal comfort. Both the field investigation and questionnaires were adopted in this experiment. Two hundred people were investigated, and two hundred questionnaires were gotten. The results show that the thermal comfort zone of officers and soldiers is 24℃ to 28℃ in early fall, it is a wide range. There is a big error between the PMV index and the actual survey results, PMV calculation index is not accurate in tents environment. The results will have a significant effect on improving the thermal comfort of tents..

1. Introduction

In this paper, through test the thermal comfort conditions of tent, analysis the influence of factors on the thermal comfort degree, such as temperature, relative humidity, wind speed, etc. the questionnaire survey is carried out in the tent, and the actual environment parameters were measured. Combining with the parameter and questionnaire to find the best appropriate temperature range inside the tent, and compare the PMV with the thermal sensation MTS [1], which find the difference between the two factors. The research has reference meaning in improving the thermal environment in tents.

2. Research CONTENTS AND METHODS

2.1. Survey site

The survey site is located in a certain college of Tianjin, at the address of Tianjin dong li nearing the sea. The air flow rate is large outside. There are four tents which are far away and have no effect on each other.

2.2. The investigation of the time and weather conditions

The investigation was established from September 12, 2015 to September 22, 2015. The temperature and relative humidity recorder were used to measure the data throughout the day. It was carried out every half our. On September 15, for example, the temperature and relative humidity had a wide range. The temperature ranged from 16 to 36 degrees. The distribution of the temperature had twenty degrees. And the relative humidity ranged from 23% to 70%, the average of the relative humidity was 45%. The change of temperature and humidity are shown in figure1.

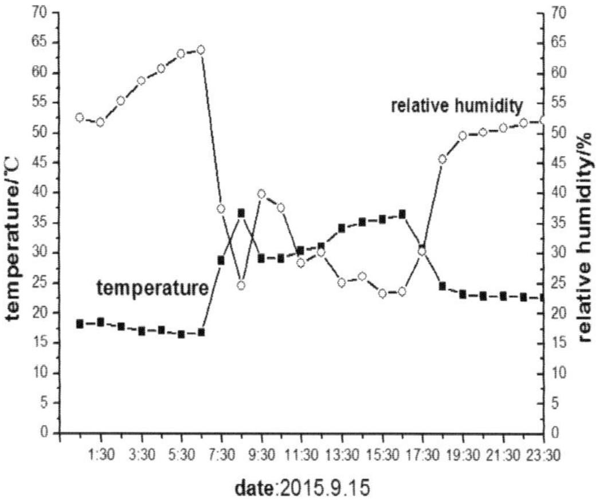

Figure 1: the change of temperature and humidity in whole day during the investigation

3. Research Analysis and methods

3.1. The environment parameters in tent and the results of the questionnaires

The thermal environment parameters in tent and the results of the questionnaires are shown in table 1, including the temperature, the relative humidity, the black ball temperature, and the air velocity, PMV (The predicted average index of thermal sensation vote), TSV (thermal sensation vote) and DPV (the air velocity sensation vote)[2].

Table 1: the questionnaires about the thermal environment parameters in tent

	the average value	standard deviation	maximum value	minimum value
air temperature	27.40	2.74	36.40	22.60
relative humidity	45.00	0.06	57.00	37.00
air velocity	1.80	2.43	8.20	0.06
mean radiant temperature	29.97	3.14	37.50	25.40
the black ball temperature	27.80	2.02	32.31	23.20
PMV	1.60	0.57	2.80	0.80
TSV	1.12	0.52	2.50	-1.00
DPV	-1.00	1.20	2.00	-3.00

During the experiment, each test was regarded as a condition. A total of thirty questionnaires and parameter test were conducted. The average temperature of each test is shown in figure2 and figure 3.

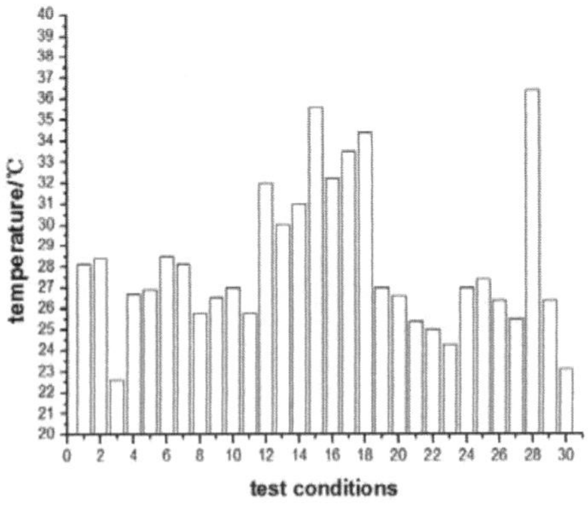

Figure 2: The distribution of average temperature on each test in tent

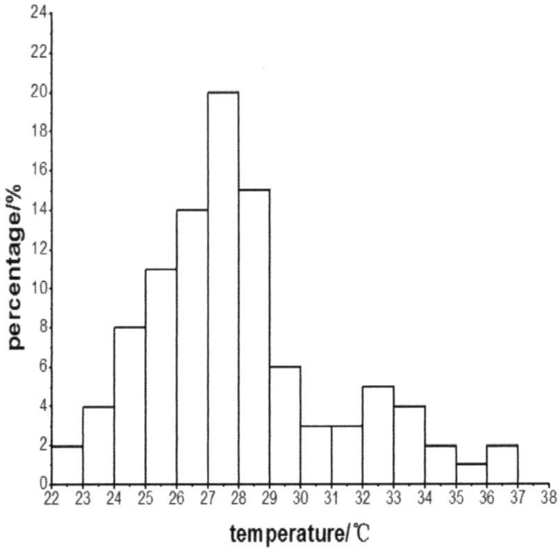

Figure 3: frequency distribution of the average temperature

There are thirty groups of test in total. The average temperature of each group was calculated. As you can see in figure 2, figure 3. There is a wide variety among the average temperature. Under the conditions of the experiment, the average temperature in tent is distributed from 22 to 37 degrees. The highest temperature can reach to 36.4 degree, and the lowest temperature is 22.6 degree. The temperature in the tent is high on the whole. Most of the temperature range from 24 to 29 degrees, it accounts for sixty-eight percent of the whole samples.

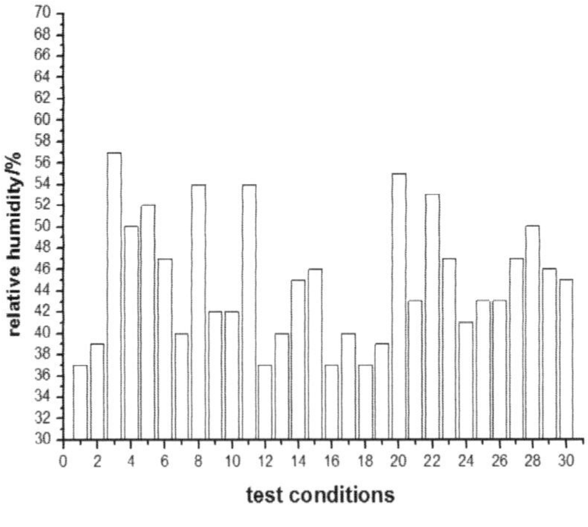

Figure 4: The distribution of the relative humidity on each test in tent

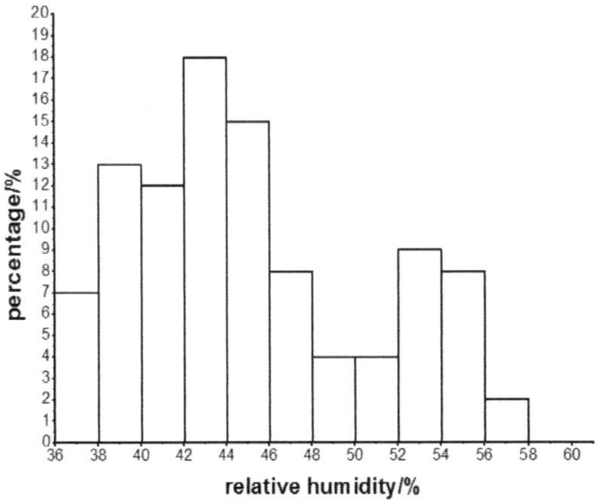

Figure 5: frequency distribution of the relative humidity

As you can see in figure 4, figure 5, under the conditions of the experiment, the relative humidity in tent is distributed from thirty-seven percent to fifty-seven percent. There is little variation on the relative humidity. The average humidity is forty-five percent, the relative humidity outside is also forty-five percent. Therefore the relative humidity is satisfied with normal result.

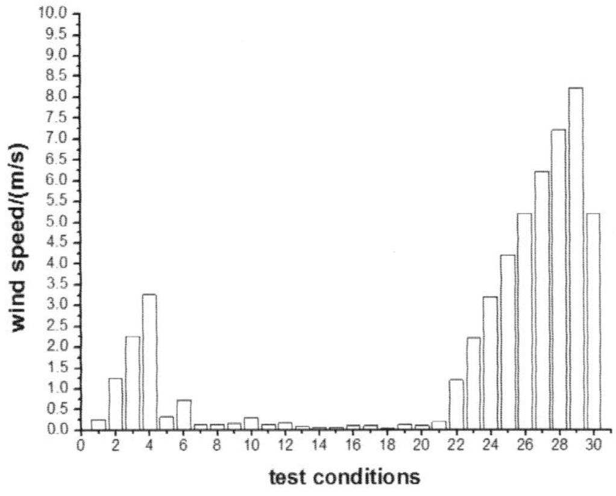

Figure 6: The distribution of the wind speed on each test in tent

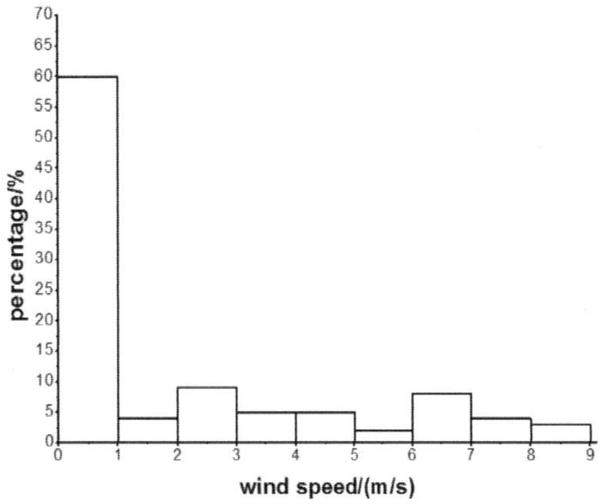

Figure 7: frequency distribution of the wind speed

As you can see in figure 6, figure 7, there is large variation on the air velocity in tent, range from 0.06 meters per second to 8.20 meters per second. Frequency distribution of the wind speed is polarized. Sixty percent of the wind speed is less than 1 meter per second, it illustrates that the air velocity is small on the whole during the experiment. But the result also have 15 percent of wind speed is more than 6 meters per second, it illustrates that the wind speed has a large variation.

3.2. The effect of temperature on the thermal comfort

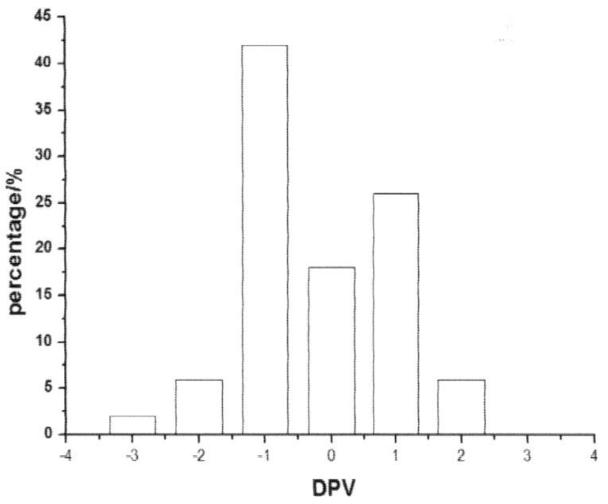

Figure 8: frequency distribution of the DPV

As you can see in figure 8, the reason why forty-two percent of the people feel a little stuffy in tent is that sixty percent of the wind speed is less than 1 meter per second when we have tested before. Meanwhile, according to the previous data, we can know that the wind speed is changing rapidly during the experiment, and the wind speed is polarized. So it gives an explanation on the consequence that twenty-six percent of people feel that there exist breeze, and eight percent of people feel a draft.

 DOI: 10.1051/matecconf/20166101001

4. Discussing

4.1. The comparison between the measured results and PMV

The PMV index represents the thermal sensation of most people, and it can predict the thermal sensation of people in the changing environment. The index is mainly used to provide a reference value of the thermal comfort. The PMV value can be divided into seven phases. When the PMV value is +3, it means that the thermal comfort is the hottest. And When the PMV value is -3[3], it means the thermal comfort is the coldest. The value is 0, which means the thermal comfort is moderate. According to the ISO7730, the calculation equation of PMV is shown as follows:

$$PMV=[0.303e^{-0.036M}+0.028]\{M-W-3.05*10^{-3}[5733-6.99(M-W)-P_a]-0.42[(M-W)-58.15]-1.7*10^{-5}M(5867-P_a)-0.0014M(34-t_a)-3.96*10^{-8}f_{cl}[(t_{cl}+273)^4-(\bar{t}_r+273)^4]-f_{cl}h_c(t_{cl}-t_a)\}$$ [4]

According to compare the calculated mean thermal sensation index PMV with the measured mean thermal sensation index MTS, the consequence is shown in figure 9.

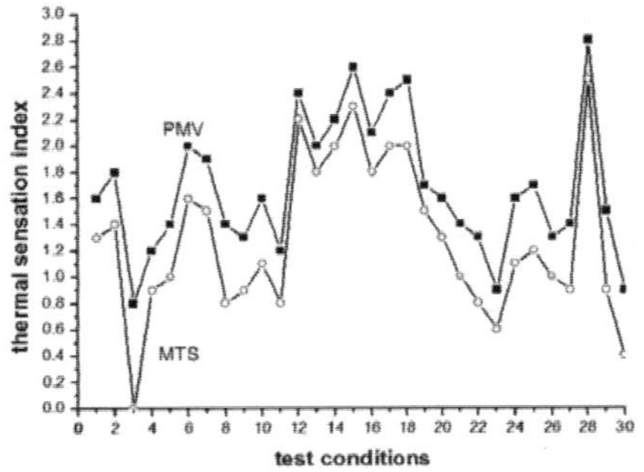

Figure 9: the value of PMV and MTS on each condition

According to figure 10, the PMV index in the ISO7730 exist a big error when it predicts the thermal environment in tent. The calculated result of the index underestimates the tolerance of the officers and soldiers. They have lived in tents for a long time, so they have some adaptability to the thermal environment of the tent. Therefore it is not accurate to use the PMV index to predict the officers and soldiers' comfort condition. The ordinary people feel a little hot in certain situation while the officers and soldiers feel it is comfortable. So a new evaluation model should be used to predict the thermal environment in tent.

5. Conclusions

(1) During the experiment, the range of parameters is t_a=22.6℃～36.4℃,φ=37%～57%, $\bar{t}_r$=25.4℃～37.5℃, v=0.06m/s～8.20m/s, the temperature and wind speed have a large variation, but there is little variation on the relative humidity, the radiant temperature is high.

(2) The wind speed has a significant effect on the thermal comfort in tent, and there is sixty percent of the wind speed that is less than 1 meter per second. Therefore forty-two percent of the people feel a little stuffy in tent. The reason why twenty-six percent of people feel that there exist breeze, and eight percent of people feel a draft is that many people are tend to sit near the window and door when they are tested, and another reason is that the wind speed has a great change during the experiment.

(3)The PMV index in ISO7730 is mainly used to predict the air conditioning environment. It will have some errors when it is used in non-air-conditioned environment. The result is different from the measured one, because the PMV index doesn't consider the officers and soldiers' adaptation to the tent. The ordinary people feel a little hot in certain situation while the officers and soldiers feel it is comfortable. Therefore, the interval value of the comfortable temperature is large.

References

1. Yingxin ,Zhu. Building environment(third edition).Beijing: China architecture,2010.
2. Qian,Wu. Study on the relationship between indoor air velocity and human comfort and physiological stress [D]. Chongqing university,2005
3. Fangjiu,Li,Ning,Lv. Measurement and evaluation of thermal environment in winter in a college classroom in Guangzhou[J]. Fujianarchitecture,2012,08:104-106.
4. Huaqiu,Tao, Li,Li. Thermal environment test and thermal comfort prediction of Xiamen university classroom in winter[J].HVAC,2012,04:72-75.

Bayesian denoising of white light interference signal in rough surface measurement

Zou Wendong[a], Zhu Lihua, Wang Weihong, Chen Chungeng
Nanchang Hangkong University, Key Laboratory of Nondestructive Test (Ministry of Education), Nanchang, Jiangxi, 330063)

Abstract—The measurement and quantitative analysis of three-dimensional microtopography of failure surface has important values for understanding the initiation and development of failure process. In this paper, we used scanning white-light interferometry (SWLI) for the characterization of metallic failure surface, but for it's large height variation, non-homogeneous and weak spectral reflectivity, the detected interferometric signals would be severely disturbed by the noises, consequently causing large recontruction error. For this reason, we used a novel denoising method based on Bayesian estimation to process the extremely weak white light interference signal, theoretical and algorithmic derivations were given out as well. Finally, the effect of signals denoising was evaluated and a quantitative comparison with other typical approaches showed that the Bayesian estimation method had obvious advantage in the signal-to-noise ratio improvement and made much more reduction in the mean squared error, along with a better smoothness and stability.

1 Introduction

Characteristic information on the material's surface is of great value which provides a stationary solution for deep study of it's formation processes and mechanical behaviors [1]. Generally, scanning electron microscopy (SEM) [2] is used to get the microstructure on the surface of an object because of high precision and large depth of field. However, in a number of specific applications, the uneven distribution of reflectivity on the object's surface causes the serious loss of fringe visibility and it happens frequently that the reflectivity on some of the regional surface is extremely low, so that the detection becomes much more difficult, leading to a lower signal-to-noise ratio of the interference signal and a larger variation range of the optical path difference. Therefore, the denoising processing with the testing signals on the object's surface plays a positive role, which can not only enhance the signal contrast, but highlight the characteristics of the useful signal also, so as to have a great improvement on the signal-to-noise ratio and make a big reduction on the mean squared error. According to the relevant literature [3-5], the study on the problem of signal detection mainly adopts such testing theories as stochastic resonance, time domain averaging, sampling integral, spectrum analysis, chaos theory, short-time Fourier and wavelet transform. This paper aims to discuss the method of signal detection based on Bayesian estimation [6], which is in combination with preprocessing and postprocessing, to make estimation of the white light interference signal collected, and thus improve the accuracy of the estimated parameter. Bayesian estimation, on the whole, is served as a method of model parameter estimation, whose basic idea is to set up a model equation, and then calculate the distribution of posterior probability, determining the maximum probability point of signal parameters, finally drawing a more accurate experimental result. That how to establish the model equation of white light interference signal on the object's surface, placed as the key component of Bayesian estimation, is the most difficult to design, and it would cause great error of signal reconstruction if unsuitably modeling.

In this article, scanning white-light interferometry is applied to the measurement of microcosmic morphology on the object's surface. What is more, MATLAB as a tool for simulation is in full use to detect the white light interference signal in a single pixel on the surface of objects through the processing with the sequences of white-light interferograms. Moreover, a new denoising method based on Bayesian estimation is proposed to cope with the white light interference signal in the interest of suppressing the noise in the interference signal, improving the signal-to-noise ratio and diminishing the mean squared error, which is conductive to providing available information and accurate judgment for further study on the failure mechanism and the invalidation formation.

2. Theory

2.1 *Rough surface measurement based on scanning white-light interferometry*

Scanning white-light interferometry [7] is a very important method in optical measurement, which eliminates the error of phase ambiguities and reduces the limitation of measurement range on one hand and possesses such characters as simple and small measuring structure, large

[a] Corresponding author: 18979106189@189.cn

 DOI: 10.1051/matecconf/20166101002

measuring range and high precision on the other hand, in contrast to the way using laser interferometry. Consequently, it is applicable for the signal detection on the complex measured surface, which has been widely used in those areas, such as optical fiber sensing technology, optical fiber dispersion measurement, non-contact measurement of surface topography, film thickness measurement and interference localization, so having a certain application value. To analyze the microstructure features on the surface of the objects, this article puts forward to adopt a structure of Linnik interference system [8] on the foundation of scanning white-light interferometry, among which the white-light interferogram is captured by the digital CCD plane array and the spatial orientation scanning is driven by the piezoelectric ceramic (PZT) [9]. The outcome of the experiment indicates that this detection device is provided with the transverse optical resolution of 1.18 μm, the longitudinal resolution of 10 nm and the maximum measuring range reaching to 120 μm.

In practice, however, due to the impact caused by the surface properties of beam splitter, the white light interferometer works with an optical path difference not being equal to zero, and thus the interference fringes of each wavelength will be stagger gradually with the increase of optical path difference and interference series, which makes the contrast of fringes decrease little by little and ultimately the interference fringes will disappear completely to a certain degree. As the figure 1 shows, it is a typical ideal white light interference signal, while the actual white light interference signal in a single pixel on the 30CrMnSiA alloy fracture surface, with the contrast of fringes approximately 0.04, is depicted in figure 2, whose curve presents a severe deformation and fluctuation with the zero-order stripe no longer being the center and bilateral symmetry. And a frame of white-light interferogram with its pixel array of 400×400 on the surface of the test sample is shown in figure 3.Through comparing with the ideal white light interference signal, it is not difficult to reveal that the actual white light interference signal is severely lost within the intense noise. In consequence, the process of de-noising the white light interference detection signal on the surface of the objects is particularly significant and worthy of further analysis [10-11].

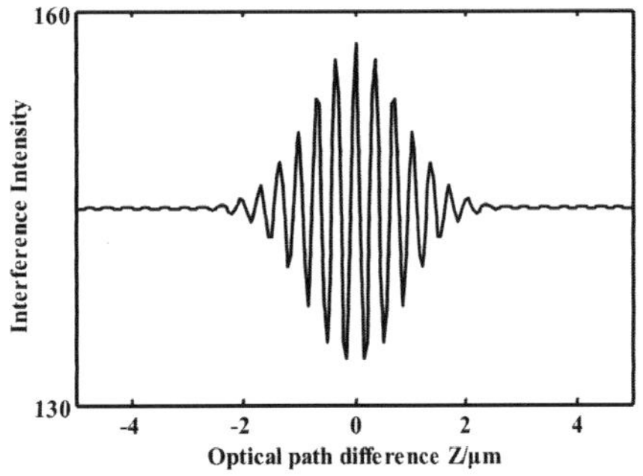

Figure 1. Typical ideal white light interference signal in a single pixel along scanning axis.

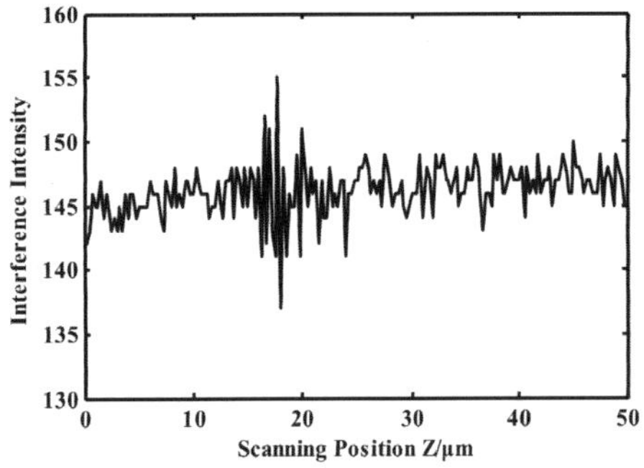

Figure 2. Actual white light interference signal in a single pixel on the metal surface along scanning position.

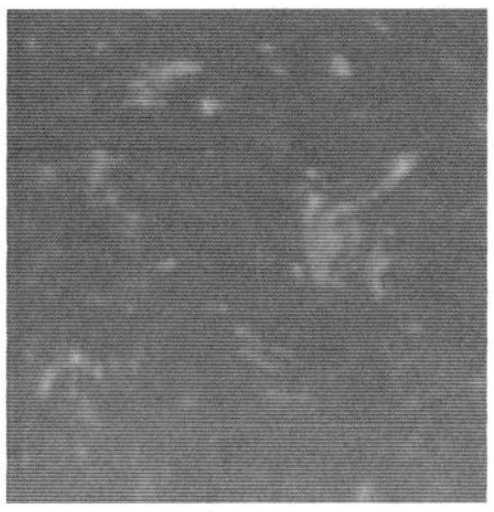

Figure 3. A frame of white-light interferogram on the metal surface.

2.2. Bayesian surface estimation

In this paper, the method on the base of Bayesian estimation, associated with preprocessing and postprocessing, is used to eliminate the noise in the interference signal on the surface, in allusion to the core idea of making the parameter estimation of the white light interference signal that has been collected already and accordingly improving its signal-to-noise ratio and reducing its mean squared error, which contributes to make the waveform of the interference signal more genuine, stable and reliable.

Assume a set U of pixel sites in a single white-light interferogram and a pixel index $i \in U$. In addition, make assumptions that the white light interference signal for each pixel on the measured surface is y^i and the possible denoised white light interference signal for each pixel is to be x^i correspondingly. Given as the mode of the conditional probability density $P(x|y)$, we make a posterior estimation to look for the denoised white light interference signal x. For each pixel i, according to the Bayesian paradigm, it can be expressed by

$$\mathrm{P}(x^i \mid y^i) \sim f(y^i \mid x^i) P(x^i) \tag{1}$$

$\mathrm{P}(x^i|y^i)$ is the a posteriori probability of a possible denoised white light interference signal value x^i given the collected data of white light interference signal y^i. The likelihood function $f(y^i|x^i)$ depicts the probability of the observations y^i given a possible value of white light interference signal x^i. The a priori probability $P(x^i)$ carries the prior knowledge on the surface.

Make δ represent the union of a pixel and its immediate neighborhood. And as a matter of convenience, we index the elements of δ with 0, ..., n, in which δ_0 denotes the measured pixel of the white-light interferograms. Then the formula (1) can be written as:

$$\mathrm{P}(x^{\delta} \mid y^{\delta}) \sim f(y^{\delta} \mid x^{\delta}) P(x^{\delta}) \quad (2)$$

where, x^{δ} and y^{δ} are combinations of the possible denoised values and all signal observations in δ, respectively.

In order to get the a posteriori probability of the possible denoised interference signal for the measured pixel, the likelihoods $f(y^i|x^i)$ at neighboring pixels in δ are assumed to be distributed independently and identically. And thus the formula (2) can be converted into:

$$\mathrm{P}(x^{\delta} \mid y^{\delta}) \sim \prod_{i=\delta_0}^{i=\delta_n} f(y^i \mid x^i) P(x^{\delta}) \quad (3)$$

In the application for this article, we need to look for the a priori probability $P(x^{\delta})$, which helps to express the prior knowledge on the object's surface. Thereby, we may consider:

$$P(x^{\delta}) = \begin{cases} 1/2\varepsilon & if\ x^{\delta_1}, \ldots, x^{\delta_n} \in [-\varepsilon + x^{\delta_0},\ x^{\delta_0} + \varepsilon] \\ 0 & otherwise \end{cases} \quad (4)$$

where ε is a constant and the possible denoised values x^{δ} is assumed to follow uniform distribution.

Plugging the a priori probability into the formula (3), we can write:

$$\mathrm{P}(x^{\delta} \mid y^{\delta}) \sim \prod_{i=\delta_0}^{i=\delta_n} \sum_{x=-\varepsilon+x^{\delta_0}}^{x=x^{\delta_0}+\varepsilon} f(y^i \mid x) / 2\varepsilon \quad (5)$$

By using the risk minimization Bayesian estimation to obtain the possible values, an equation will be given by

$$x^{\delta} = \arg\max_{x^{\delta}} [\mathrm{P}(x^{\delta} \mid y^{\delta})] \quad (6a)$$

$$= \arg\max_{x^{\delta}} [\prod_{i=\delta_0}^{i=\delta_n} \sum_{x=-\varepsilon+x^{\delta_0}}^{x=x^{\delta_0}+\varepsilon} f(y^i \mid x) / 2\varepsilon] \quad (6b)$$

Suppose that the noise, which has nothing to do with the detecting interference signal on the object's surface, is considered as Gaussian white noise with the mean of zero and the variance of $\sigma_n^{\ 2}$ and then we will obtain a computable formula as follows

$$x^{\delta_0} = \arg\max_{x^{\delta_0}} [\prod_{i=\delta_0}^{i=\delta_n} \sum_{x=-\varepsilon+x^{\delta_0}}^{x=x^{\delta_0}+\varepsilon} \frac{1}{\sqrt{2\pi}\sigma_n} \exp[-\frac{(y^i - x)^2}{2\sigma_n^{\ 2}}] / 2\varepsilon] \quad (7)$$

Consider that the possible denoised values x^i at the neighboring pixels in δ are closely equal to the measured pixel value, that is, the formula (7) is equivalent to

$$x^{\delta_0} = \arg\max_{x^{\delta_0}} [\prod_{i=\delta_0}^{i=\delta_n} \exp[-\frac{(y^i - x^{\delta_0})^2}{2\sigma_n^{\ 2}}]] \quad (8a)$$

$$= \arg\min_{x^{\delta_0}} [\prod_{i=\delta_0}^{i=\delta_n} (y^i - x^{\delta_0})^2] \quad (8b)$$

Aiming to attain the possible value x^{δ_0}, the algorithm offered above thus far depends on the following parameters: n, δ. With these determined parameters, our final execution is tested to be practical and thus accomplishes our anticipative goal. Without doubt, for other kinds of noise modeling, the idea in this paper is suitable as well.

3. Numerical analysis and results

3.1 Denoising process of white light interference signal using Bayesian estimation

Taking an application example, this paper which built upon the foundation of scanning white-light interferometry

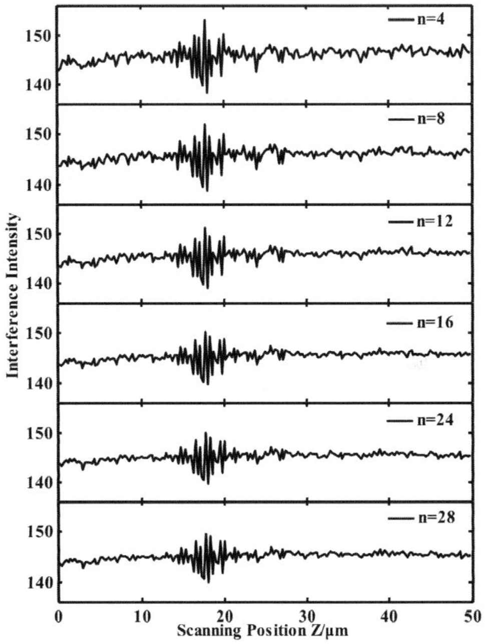

Figure 4. The denoised interference signals with the choice of different *n* based on Bayesian estimation applied to the white light interference signal on the 30CrMnSiA alloy fracture surface.

is involved to acquire the white light interference signal on the 30CrMnSiA alloy fracture surface. And we obtains the denoised white light interference signal from the calculable algorithm in the formula (8b), noting that the measured pixel is U(381,209) for all simulations, and its accuracy can be further increased with the choice of the suitable value n. As is illustrated in figure 4, it pictures the denoised interference signal by choosing different n with all other factors being equal, and the hisgrams and normal curves of white light interference signals are showed in figure 5.

For more detailed and direct display of the denois- ing performance, as tabulated in table 1, it shows the results of SNR(signal-to-noise ratio) and MSE(mean squared error), from which it is obvious that the performances of measured white light interference signal reduce significantly with the increase of the value of neighborhood pixels n.

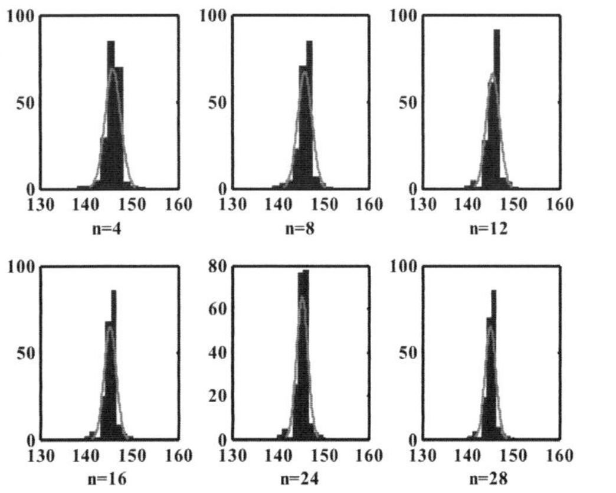

Figure 5. Histograms and normal curves of white light interference signals with different value of parameter *n*.

Above all, we are able to screen an intuitive $n = 4$ that better matches the Bayesian denoising algorithm.

Table 1. SNR and MSE results with the choice of different *n* based on Bayesian estimation for the white light interference signal on the metal surface.

n / Results	*n*=4	*n*=8	*n*=12	*n*=20	*n*=24	*n*=28
PSNR (dB)	48.404	45.683	45.103	43.876	43.223	42.924
MSE	0.939	1.757	2.008	2.666	3.096	3.316

3.2 White light interference signal denoising with different approaches

With a view to verify the superiority of the Bayesian denoising method, as is visualized in figure 6, different denoising approaches are employed to deal with the white light interference signal on the surface of 30CrMnSiA fracture, which provides a direct comparison among the different methods. Note that a scale of 3×3 pixels is chosen for the simulation and throughout the Bayesian denoising, n =4 is used which in other words means that δ includes the 4 nearest neighbors of the measured pixel .

Meanwhile, as presented in table 2, the results of SNR and MSE are used to have a good visual effect on the denoising quality of interference signal on the metal surface, and the hisgrams and normal curves of white light interference signal described in figure 7 are showed to graphically summarize and display the distri-
bution of the signal dataset which works for the more precise analysis of signal details and useful informations.

Correspondingly, the white light interference signal spectrums described in figure 8 are showed to graphically summarize and display the frequency distribution of the signal dataset which works for the more precise analysis of signal details and useful informations, with the amplitude of three main spectrum peaks from left to right listed in table 3.

Accompanying figure 5, figure 6, table 2 and table 3, some helpful conclusions are drawn:

1. The denoising processing with the detecting white light interference signal on the surface of metal not merely make the signal waveform more smooth and stable, but also evidently wipe off the outliers in the interference signal,

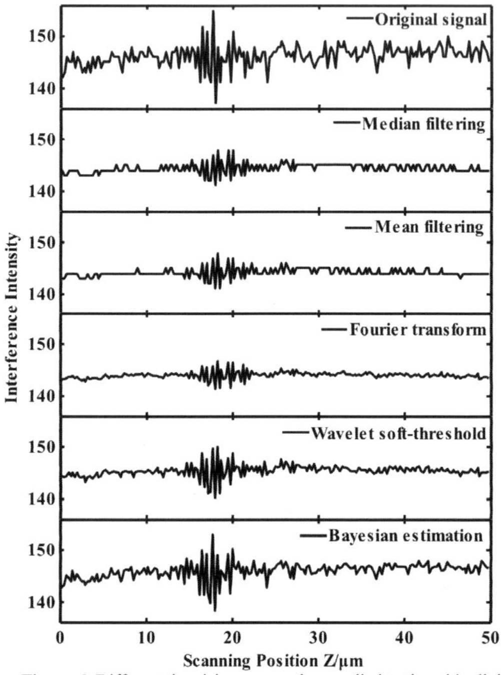

Figure 6. Different denoising approaches applied to the white light interference signal on the metal surface. From top to bottom:Median filtering, Mean filtering, Fourier transform, Wavelet soft-threshold and Bayesian estimation.

Table 2. SNR and MSE results of different denoising methods for the white light interference signal on the surface of metal.

Method / Results	Median filtering	Mean filtering	Fourier transform	Wavelet soft-threshold	Bayesian estimation
PSNR (dB)	40.182	39.649	39.207	43.247	48.404
MSE	6.235	7.050	7.805	3.079	0.939

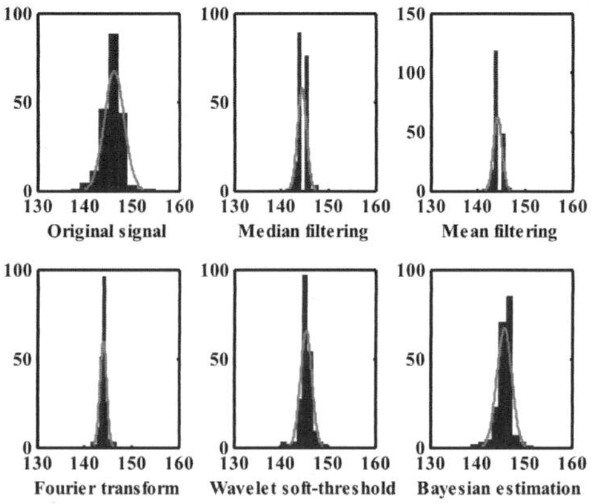

Figure 7. Histograms and normal curves of white light interference signals with different denoising approaches.

Table 3. Peak amplitude of the white light interference signal.

Method / Peak amplitude	Original signal	Medin filtering	Mean filtering	Fourier transform	Wavelet soft-threshold	Bayesian estimation
①	10.28	10.27	10.27	10.27	10.28	10.28
②	4.390	3.638	3.567	3.462	3.918	4.177
③	4.390	3.638	3.567	3.462	3.918	4.177

whereas our method can conserve much more detailed information and useful informations occurred in the edge. By seriously analyzing and researching in figure 5, we could observe clearly that the position with maximum modulation of white light interference signal that is also the position with zero light path difference is located at the scanning position Z=17.75 μm, which helps us detect the surface height at that tested pixel and realize the surface height construction from the white-light interferograms.

2. By comparing and analyzing, we are left in no doubt that our method has more or less improvements than other denoising algorithms, which leads to a better signal-to-noise ratio and a smaller mean squared error. The difference between the maximum and minimum in PSNR is close to 9.2 dB, while it is about 6.9 in MSE.

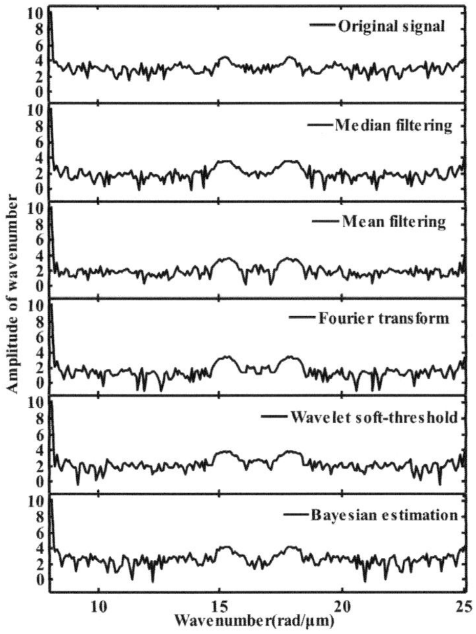

Figure 8. Spectrums of white light interference signals with different denoising approaches.

3. From the results of spectrum data, it is easily seen that the amplitude of denoised signals by the method of median filtering, mean filtering and Fourier transform decreases obviously. And compared with wavelet transform, the spectrum signal processed with bayesian estimation is most similar to the original shape, that is, our method greatly saves the spectrum characteristics of the original signal. Among these methods, the denoising effect of Fourier transform is relatively poorer, which seriously affects the frequency resolution of weak white light interference signals.

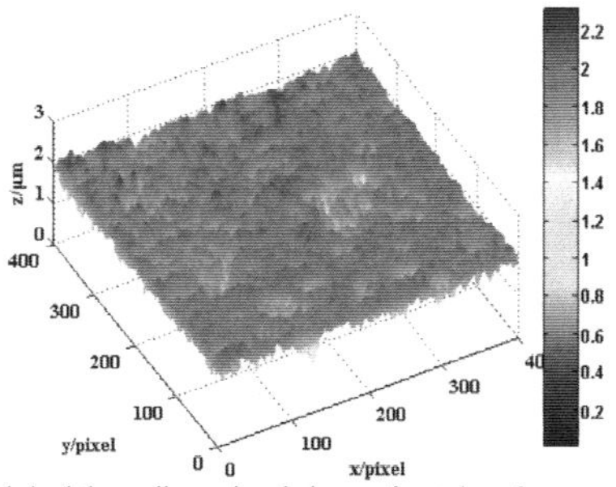

Figure 9. Original three-dimensional shape of metal surface

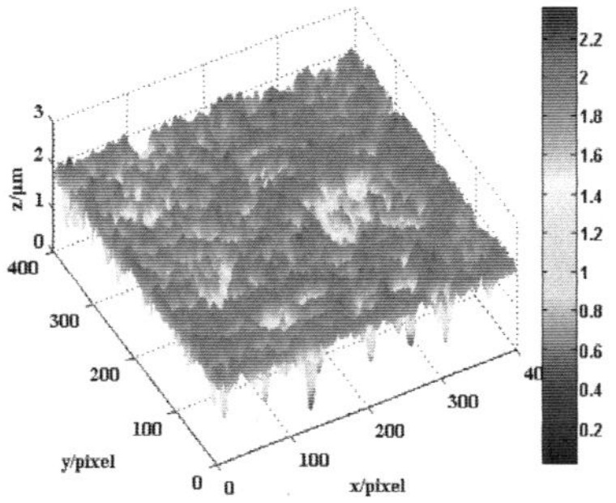

Figure 10. Three-dimensional shape of metal surface denoised by Bayesian estimation

For purpose of further studying the three-dimensional shape on the metal surface, the reconstruction image is shown in figure 10 through the use of the algorithm proposed, in which way we are able to have access to more comprehensive information such as quantitative depth, width and direction, etc. we can clearly observe the characteristic information on the metal surface.

4. Conclusion

In this paper, a new methodology for signal denoising is proposed, while the white light interference signal in a single pixel is extracted from the interferogram sequences that are obtained by the scanning white-light interferometry. And this kind of method applied to the detecting white light interference signal on the surface is based on Bayesian estimation where the noise is supposed to be distributed following a Gaussian distribution with the mean of zero. In order to challenge the algorithms based on Bayesian estimation, we choose different neighbor pixels with the choice of a more suitable variance of noise that better matches the Bayesian estimation. In addition, we make a quantitative comparison between our method and the other denoising approaches by the means of MATLAB implementation and the simulation results show that all four approaches are able to evidently suppress noise and decimate outliers, but that the Bayesian estimation could preserve more detail features with a continuous histogram distribution, and lead to a better PSNR by about 5.2 dB to 9.2 dB, which corresponds to a smaller MSE from about 2.2

to 6.9. Furthermore, through the way of three-dimensional shape reconstruction on the metal surface, we can not only have access to more comprehensive information of the surface, but also possess an available ways to validate the feasibility of signal detection by adopting the denoising method described in this article, with the analysis result revealing that the reconstruction effect is relatively satisfying and obvious [14]. Finally, associated with all the fruit of this research, we can draw a conclusion that the method based on Bayesian estimation represents an efficient mean for denoising the real signals on the surface.

Acknowledgments

This work is supported by National Natural Science Foundation of China under grant No. 51241005, The Industry Project of Science & Technology Pillar Program of Jiangxi Province, China (Grant No.20122BBE500040), Natural Science Foundation of Jiangxi Province, China (Grant No. 2008GZW0011, No.20122BAB202010).

References

1. Azevedo C R F, Marques E R. *Eng. Fail. Anal.*, **17**, (2010):286-300.
2. Cao Peng, Sun Libo, Shao Yuehua. Mod. Manu. Tech. and Equi., **1**, (2010): 24-26.
3. Zhang Haibo, Ye Xiaohui. Mod. Elec. Tech., **32**, 4(2009):124-126.
4. Wu Qiang, Zhang Fulai, He Zhuqing, et al. Research of denoising of low signal-to-noise ratio signal[J]. Elec. Pub. House, **4**, 1 (2009):310-311.
5. Li Chongsheng. Nature and Science, **3**,2 (2005):59-64.
6. Hou Jianwen, Tian Jinwen. Compu. Eng., **32**, 11 (2006):13-15.
7. Yang Guangxue, Zhang Yu, Kang Shouqiang. Foreign Elec. Meas. Tech., **27**, 11 (2008):20-22.
8. Zou Wendong, Du Nan, Xiao Huirong, et al. Bejing:CSMTE, (2005):983-986.
9. Wu Yanjie, Liu Ju, Meng Xianfeng, et al. Rare Metal Materials and Eng., **37**, s1 (2008):580-583.
10. Alfaouri M, Daqrouq K. ECG. *Science Publi.*, **5**, 3 (2008):276-281.
11. Same A, Oukhellou L, Come E, Aknin P. *Advances in Data Anal. and Classification*, **1**, 1(2007):39-51.
12. Hissmann M, Hamprecht F A. *Optica*l *Engineering*, **44**, 1(2005).
13. Gosteva A, Haiml M, Paschotta R, Keller U. Noise-related resolution limit of dispersion measurements with white-light interferometers[J]. *Opt. Society of America*, **22**, 9(2005):1868-1874.
14. Drost R J, Singer A C. *J Opt Soc Am A Opt Image Sci Vis*, **21**, 10(2004):1855-1868.
15. Achim A, Bezerianos A, and Tsakalides P. IEEE Transactions on Medical Imaging, **20** (2001): 772-83.
16. Abramovich F, Sapatinas T and Silverman B. J Royal Statist Society B, **60** (1998): 725-49.
17. Pižurica A, Philips W, Lemahieu I, Acheroy M. IEEE Trans Image Proc, **11** (2002): 545-57.

DOI: 10.1051/matecconf/20166101003

The Recursive Forward Dynamics of Flexible Mechanical Systems

Shao Bing[1,a], Yuan Entao[1]

[1] School of Mechanical, Shanghai Dianji University, Shanghai 201306, China

Abstract. Lie groups and Lie algebras are used to study the recursive dynamics of flexible multi-body systems. First the adjoint transformations and adjoint operators of Lie groups and Lie algebras are discussed. Then the generalized mass matrix of flexible mechanical systems is built on the basis of modal vector. And then the inverse and forward dynamics of flexible mechanical systems are constructed. These two recursive formulas are of high efficient. Finally a four-bar model with flexible body is simulated with above method. The simulation results show that with the method can be solved quickly and efficiently.

1 Introduction

The mechanical systems are becoming larger in scale and more complex. Many kinds of flexible material are used in engineering as members of mechanical systems. The assumption of rigid body dynamics has been unable to explain the dynamics of such systems. It is necessary to consider a wide range of sports deformed parts and components [1]. Kinetic analysis of the elastic deformation is not assumed to affect multiple body systems rigid body motion which is mainly confined to the low-speed linear elastic range. Rigid mechanical systems coupled nonlinear problem has become increasingly prominent. The dynamics of flexible multi-body systems have been rapidly developed [2].

There is a variety of flexible multi-body modeling methods. A motion can be expressed as the synthesis of rigid body motion and the relative flexibility of motion according to flexible multi-body system theory. The classic formula of flexible multi-body systems can be got if the additional relatively flexible movement is built in the body reference frame. Singh establish the equations of tree topology flexible multi-body system dynamics [3]. Shabana use linear and nonlinear finite element method combination of multi-body dynamics theory to establish a flexible body dynamics model [4]. However, these methods have a problem in dealing with the efficiency and integration method [5].

On the basis of the above methods, this paper describes the dynamics with Lie groups and Lie algebras. It establishes the dynamic equations by recursive methods. It leads to a recursive dynamics of flexible mechanical systems.

2 Recursive dynamics of flexible mechanical systems

2.1 Adjoint transformations of Lie groups and Lie algebras

A Lie group is a group and also it is a differentiable manifold expressed by *G*. The tangent space of a Lie group's element is called the Lie algebra for that group. The Lie algebra forms a vector space with a bilinear map which is called the Lie bracket [8].

The rotation matrix is an orthogonal matrix in three dimensional Euclidean space and it is an element of the special Orthogonal group *SO(3)*. The homogeneous transformation is the Lie group referred to as the special Euclidean group *SE(3)*. The Lie algebra of *SO(3)* is *so(3)* which consists of the set of skew symmetric matrices on $R^{3\times 3}$. Similar with it, the Lie algebra of *SE(3)* is *se(3)*. The relation between a Lie group *SO(3)* or *SE(3)* and the associated Lie algebra *so(3)* or *se(3)* is the matrix exponential.

A linear mapping on the Lie algebra is an element of the Lie group. This is called the Adjoint map on *SE(3)* and is denoted by *Ad*, where $Ad_G(h) = GhG^{-1}$, $G \in SE(3)$ and $h \in se(3)$. Note that *h* is the homogeneous matrix.

$$Ad_G(h) = \begin{bmatrix} R & 0 \\ [p]R & R \end{bmatrix} \begin{bmatrix} \omega_h \\ v_h \end{bmatrix} \quad (1)$$

The linear mapping on the dual space *se(3)* is a dual Adjoint operator denoted by Ad^*. For h^*:

$$Ad_G^*(h^*) = \begin{bmatrix} R^T & R^T[p]^T \\ 0 & R^T \end{bmatrix} \begin{bmatrix} \mathcal{M} \\ \mathcal{F} \end{bmatrix} \quad (2)$$

The Lie algebra can also be used as a linear mapping on itself denoted by *ad*. It has the form of $ad_g(h) = [g,h] = gh - hg$, where $g, h \in se(3)$.

[a] Corresponding author: shaob@sdju.edu.cn

$$ad_g(h) = \begin{bmatrix} [\omega_g] & 0 \\ [v_g] & [\omega_g] \end{bmatrix} \begin{bmatrix} \omega_h \\ v_h \end{bmatrix} \quad (3)$$

Similarly, the dual operator ad^* is given by $ad^*_g(h^*) = [g, h^*]$ and:

$$ad^*_g(h^*) = \begin{bmatrix} [\omega_g]^T & [v_g]^T \\ 0 & [\omega_g]^T \end{bmatrix} \begin{bmatrix} \mathcal{M} \\ \mathcal{F} \end{bmatrix} \quad (4)$$

In mutibody dynamics, the Adjoint operator Ad maps the spatial velocity screws from one reference frame to body frame and the dual Adjoint operator Ad^* maps the spatial force screws. The mapping ad and ad^* respectively define the standard cross product operation of the 6-vector velocity screws and force screws.

2.2 Generalized mass matrix of flexible mechanics

Deformation vectors are shown in Figure 1. OXYZ is inertial reference and oxyz is body-fixed reference. The vector of m point in deformable body is u_0 in body-fixed reference and u in inertial reference. The deformation displacement vector according to flexible is u_f.

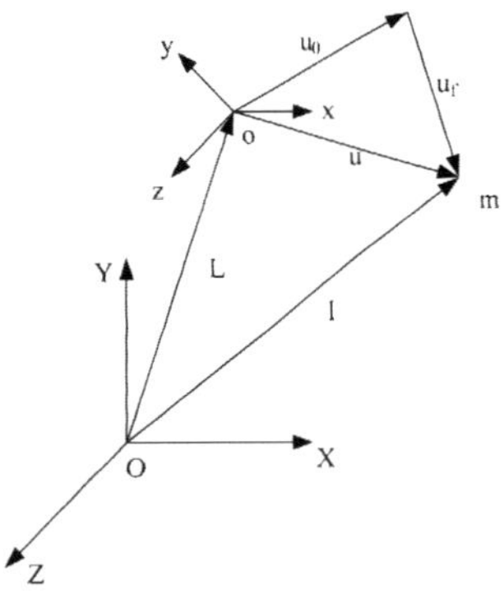

Figure 1. Deformation vectors in inertial reference system and the body-fixed reference

$$u = u_0 + u_f \quad (5)$$

$$u_f = \sum_{i=1}^{n_f} \psi(u_0) q_f(t) = \Psi^T q_f \quad (6)$$

Where $\Psi^T = [\psi_1, \psi_2, \cdots, \psi_{n_f}]$, $q_f^T = [q_{f1}, q_{f2}, \cdots, q_{fn_f}]$, and q_{fi} $(i = 1, 2, \cdots, n_f)$ are the i-th order deformation mode vectors and generalized coordinates respectively. n_f is the order of deformation mode. The vector of point m in the inertial coordinate system can be expressed as:

$$l = L + Ru = L + R(u_0 + \psi^T q_f) \quad (7)$$

R is the transformation matrix of body-fixed reference and inertial reference coordinate system. Take derivative of formula 7.

$$\dot{l} = \dot{L} + \dot{R}u + R\psi^T \dot{q}_f \quad (8)$$

$$\dot{R}u = \sum_{i=1}^{n_\theta} \frac{\partial R}{\partial \theta_i} u\dot{\theta}_i = [\frac{\partial R}{\partial \theta_1} u, \frac{\partial R}{\partial \theta_2} u, \cdots, \frac{\partial R}{\partial \theta_{n_\theta}} u]\dot{\Theta} = N\dot{\Theta} \quad (9)$$

Where $N \equiv [\frac{\partial R}{\partial \theta_1} u, \frac{\partial R}{\partial \theta_2} u, \cdots, \frac{\partial R}{\partial \theta_{n_\theta}} u]$. Equation 8 can be rewrited as follows.

$$\dot{l} = [E_3, N, R\Psi^T] \begin{bmatrix} \dot{L} \\ \dot{\Theta} \\ \dot{q}_f \end{bmatrix} \quad (10)$$

The kinetic energy of an object can be expressed as follows.

$$T = \frac{1}{2}\int_\Omega \dot{r}^T \dot{r} dm = \frac{1}{2}\dot{q}^T M \dot{q} \quad (11)$$

Where generalized mass matrix is M.

$$M \equiv \begin{bmatrix} M_{RR} & M_{R\theta} & M_{Rf} \\ M_{R\theta}^T & M_{\theta\theta} & M_{\theta f} \\ M_{Rf}^T & M_{\theta f}^T & M_{ff} \end{bmatrix}$$

Generalized coordinates is $\dot{q}$.

$$\dot{q} \equiv \begin{bmatrix} \dot{L} \\ \dot{\Theta} \\ \dot{q}_f \end{bmatrix}$$

$M_{RR} = \int_\Omega E_3 dm$, $M_{\theta\theta} = \int_\Omega N^T N dm$, $M_{ff} = \int_\Omega \Psi\Psi^T dm$

$M_{R\theta} = \int_\Omega N dm$, $M_{Rf} = R\int_\Omega \Psi^T dm$,

$M_{\theta f} = \int_\Omega N^T R \Psi^T dm$

Matrix subscript f and R represent flexible and rigid module. M_{ff} represents flexible and flexible coupling module. M_{Rf} represents flexible and rigid coupling module. M_{RR} is rigid space inertia. If the object is a rigid body, the generalized inertia matrix is $M \equiv \begin{bmatrix} M_{RR} & M_{R\theta} \\ M_{R\theta}^T & M_{\theta\theta} \end{bmatrix}$.

2.3 Inverse recursive dynamics of flexible mechanics

Similar with rigid body dynamics, we can simply expand each symbol to coordinate containing modal diformation using the finite element method to express the deformation of the flexible object.

$$
\begin{aligned}
&\textit{Giving} \quad V_0^m = 0, \dot{V}_0^m = 0 \\
&\textit{for } i = 1 \to n \\
&\mathbf{V}^m{}_i = Ad^m{}_{\mathbf{T}^{-1}_{i-1,i}} (\mathbf{V}^m{}_{i-1}) + \mathbf{S}^m{}_i \dot{q}^m \\
&\dot{\mathbf{V}}_i^m = Ad^m{}_{\mathbf{T}^{-1}_{i-1,i}} (\dot{\mathbf{V}}^m_{i-1}) + \mathbf{S}^m{}_i \ddot{q}_i^m + \mathbf{a}^m{}_i \\
&a^m{}_i = ad_{V_i^m} (S^m{}_i \dot{q}_i^m)
\end{aligned} \tag{12}
$$

$$
\begin{aligned}
&\textit{Giving} \quad F^m{}_n = 0 \\
&\textit{for } k = 1 \to n \\
&\mathbf{F}^m{}_i = Ad^{m*}_{\mathbf{T}^{-1}_{i,i+1}} (\mathbf{F}^m{}_{i+1}) + \mathbf{J}^m{}_i \dot{\mathbf{V}}_i^m - ad^{m*}_{\mathbf{V}_i} (\mathbf{J}^m{}_i \dot{\mathbf{V}}_i^m) + K_i^m \vartheta_i \\
&\boldsymbol{\tau}_i = \mathbf{S}_i^{m^{\mathrm{T}}} \mathbf{F}^m{}_i
\end{aligned} \tag{13}
$$

Where the superscript m represents a modal symbol.Velocity is synthesis of modal velocity and rigid body velocity $V^m{}_i \equiv \begin{pmatrix} \dot{\eta}_i \\ V_i \end{pmatrix}$. The generalized coordinate and velocity of link k is ϑ_i and $\dot{q}_i{}^m$. $\vartheta_i \equiv \begin{pmatrix} \eta_i \\ q_i \end{pmatrix} \in \mathbb{R}^{N(k)}$, $\dot{q}_i{}^m \equiv \begin{pmatrix} \dot{\eta}_i \\ \beta_i \end{pmatrix} \in \mathbb{R}^{N(k)}$.

Modal spatial displacement vector on the node j_i^{th} is defined as as $\Pi^j(i) \in R^{6 \times n_m(k)}$ and the modal matrix of link i is defined as $\Pi(i) \in R^{6n_s(i) \times n_m(i)}$.

$$
\Pi^j(i) \equiv \left[\Pi_1^j(i), ..., \Pi^j_{n_m(i)} \right] \quad \Pi(i) \equiv col\{\Pi^j(i)\} \tag{14}
$$

The Adjoint operator $Ad^m{}_{T^{-1}_{i-1,i}}$ of flexible mechanical system is the synthesis of original adjoint operator rigidity and modal matrix.

$$
Ad^m{}_{\mathbf{T}^{-1}_{i-1,i}} \equiv \begin{pmatrix} 0 & \left[\Pi^t(i+1)\right]^* Ad_{\mathbf{T}^{-1}_{i-1,i}} \phi(t_{k+1}, k) \\ 0 & Ad_{\mathbf{T}^{-1}_{i-1,i}} \end{pmatrix} \in \mathbb{R}^{\bar{N}(i+1) \times \bar{N}(i)} \tag{15}
$$

The modal spatial force of link i is $F_i^m \in \mathbb{R}^{\bar{N}(i)}$.

$$
F_i^m \equiv \begin{pmatrix} \Pi^*(i) F_i^s \\ F_i \end{pmatrix} \tag{16}
$$

And the modal stiffness matrix.

$$
K_i^m = \begin{pmatrix} \Pi^*(i) K_i^s \Pi(i) & 0 \\ 0 & 0 \end{pmatrix} \in R^{\bar{N}(i) \times \bar{N}(i)} \tag{17}
$$

2.4 Forward recursive dynamics of flexible mechanics

Calculation of articulated body inertia

$$
\begin{aligned}
&\textit{Giving} \quad \hat{J}^m_{n+1} = 0, z^m{}_{n+1} = 0 \\
&\textit{for } i = n \to 1 \\
&\hat{J}_i^m = J^m{}_i + Ad^{m*}_{T^{-1}_{i,i+1}} \left(I - \frac{\hat{J}^m_{i+1} S^m{}_{i+1} S^{mT}_{i+1}}{S^{mT}_{i+1} \hat{J}^m_{i+1} S^m{}_{i+1}}\right) \hat{J}^m_{i+1} Ad^m{}_{T^{-1}_{i,i+1}} \\
&b^m{}_i = -ad^{m*}_{V_i} (J^m{}_i V^m{}_i) \\
&B^m{}_i = Ad^{m*}_{T^{-1}_{i,i+1}} \left[z^m{}_{i+1} + \frac{\hat{J}^m_{i+1} S^m{}_{i+1} \left(\tau_{i+1} - S^{mT}_{i+1} \left(z^m{}_{i+1}\right)\right)}{S^{mT}_{i+1} \hat{J}^m_{i+1} S^m{}_{i+1}} \right] + b^m{}_i \\
&z^m{}_i = \hat{J}_i^m a^m{}_i + B^m{}_i
\end{aligned} \tag{18}
$$

Calculation of accerlation

$$
\begin{aligned}
&\textit{Giving} \quad \dot{V}_0^m = 0 \\
&\textit{for } i = 1 \to n \\
&\ddot{q}_i^m = \frac{\tau_i - S_i^{mT} \left[\hat{J}_i^m Ad^m{}_{T^{-1}_{i-1,i}} \left(\dot{V}^m_{i-1}\right) + z^m{}_i \right]}{S_i^{mT} \hat{J}_i^m S^m{}_i} \\
&\dot{V}_i^m = Ad^m{}_{T^{-1}_{i-1,i}} \left(\dot{V}^m_{i-1}\right) + S^m{}_i \ddot{q}_i^m + a^m{}_i
\end{aligned} \tag{19}
$$

3 Examples

In a Four-bar mechanism, link 2 is made of Aluminum which is considered as flexible body. The bar 3 and 4 are rigid. The parameters of link 2 are: the mass is 0.57kg, the length is 0.75m, the density is 7600kg/m3, the Young's modulus is 200GPa. The parameters of link 3 and 4 are the same of: the mass are 0.5kg and 9.028kg, the moment inertia are [5.0e-5 0 0;0 5.5e-3 0;0 0 5.5e-3] and [1.8e-3 0 0;0 0.678 0;0 0 0.678].

Figure 2 to 5 give the simulation results of position, velocity and acceleration of link 3. It can be seen that the acceleration of link 3 appears tremble and the flexible impact the position and velocity little.

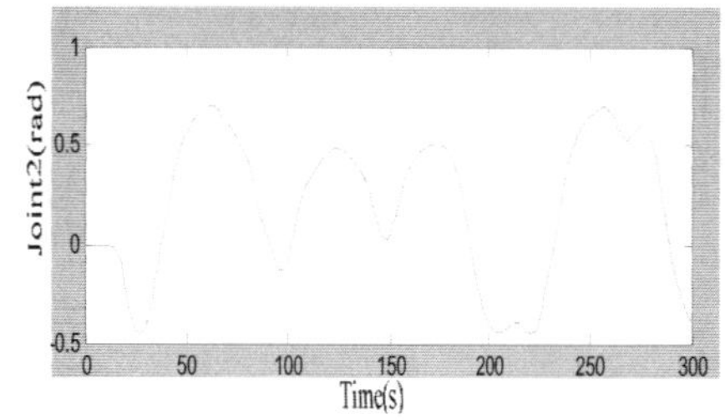

Figure 2. The body position of link 3

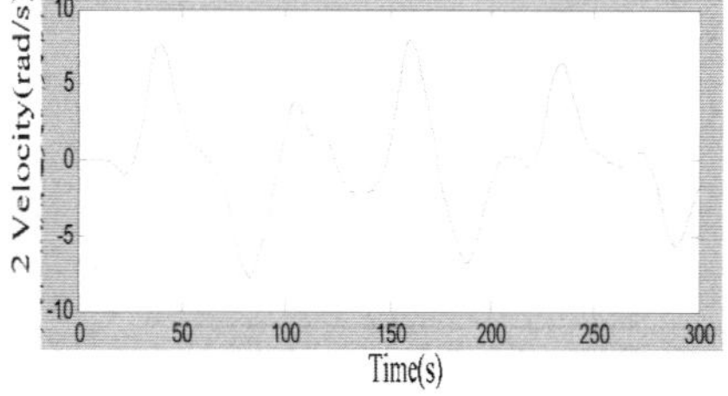

Figure 3. The body velocity of link 3

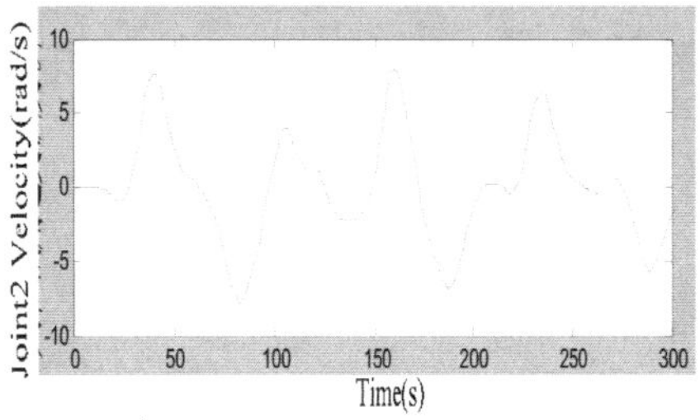

Figure 4. The body velocity of link 3

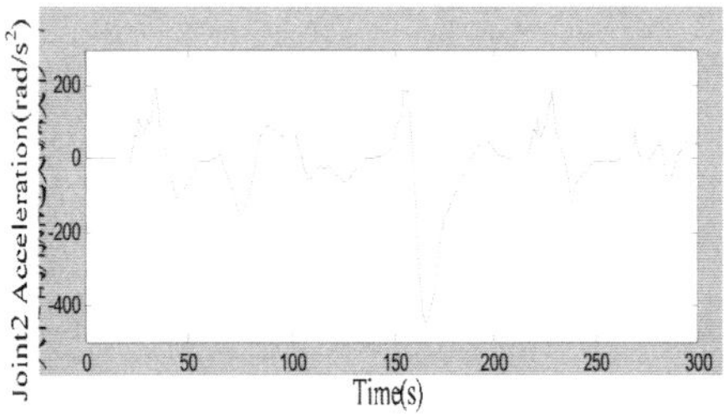

Figure 5. The body acceleration of link 3

4 Conclusions

The recursive dynamics of flexible multi-body systems are discussed. Also Lie groups and Lie algebras are used to model this problem. A four-bar model with flexible body is simulated with above method. The simulation results show that with the method can be solved quickly and efficiently.

From the above derivation, Lie groups and Lie algebras method can solve the flexible problems with higher efficiency. The finite element model is also a recursive method. In the calculation process almost all matrices have joined the modal matrix, which greatly increased the matrix order.

Acknowledgment

This research is partially supported by Young Teacher Training Project Funded Plan of Colleges of Shanghai (project number:ZZSDJ12005) and Scientific Research Project Start-up Fund of Shanghai Dianji University (project number: 11c404).

References

1. Fisk, Len A. The impact of space on society: Past, present and future. SpacePolicy, **24**, 175 (2008)
2. Smith, Charladen. Free-Flyer Capture -- New Robotic Challenges from the International Space Station. Journal of the British Interplanetary Society, , **60**, 339 (2007)
3. Ko Ayusawa. Minimal-set of Inertial Parameters Identification of Legged Robots Using Base-link Dynamics. The Japanese robot academic journals, , **27**, 1066 (2009)
4. Khail W, Kleinfinger J F. A new geometric notation for open and closed loop robot. *The IEEE International Conference on Robotics & Automation*, 1174 (1986)
5. Murostu Y, Senda K. Parameter Identification of Unknown Object Handled by Free Flying Space Robot. Journal of Guidance Control and Dynamics, **17**, 488 (1994)
6. Sujan V A, Dubowsky S. An Optimal Information Method for Mobile Manipulator Dynamic Parameter Identification. IEEE/ASME Transactionson Mechatronics, **2**, 215 (2005)
7. Katoh R, Sakon H, Yamamoto T. A Control Method for Space Manipulator Mounted on a Fre-Flying Robot by Using Parameter Identication. *The Asian-Pacic Vibration Conf*, 1205 (1993)
8. Abiko S, Yoshida K. On-line Parameter Identification of a Payload Handled by Flexible Based Manipulator. *2004 IEEE/RSJ International Conference on Intelligent Robots and Systems, Sendal,* 2930 (2004)
9. Guo Qi, Hong Bingrong. Two arms four degrees of freedom parameter identification of unknown object space robot. Robot, **25**, 512 (2005)

Dynamics of Complex Mechanical Systems Based on Lie Groups and Lie algebra

Shao Bing[1,a], Yuan Entao[1]

[1] *School of Mechanical, Shanghai Dianji University, Shanghai 201306, China*

Abstract. Adjoint transformations and adjoint operators of Lie groups and Lie algebra are discussed in this paper to study the recursive dynamics of complex systems. The dynamics based on Lie groups and Lie algebra of tree topology systems is then developed. Efficient dynamics algorithms for closed-loop systems and flexible systems can be developed. Simulation results show that the dynamic problems can be solved quickly and efficiently using the method. This study provides a foundation for real-time control based on dynamics.

1 Introduction

With the development of modern science and technology, there has been an engineering object composed by a large number of objects, such as vehicles, robots, aerospace vehicles. These complex mechanical systems contact with each other in a variety of manners. Multi-body system dynamics is the large-scale complex systems researching the forward, backward and mixed problem. Inverse dynamics can be applied to optimize the mechanical design and control [1]. By calculating the inverse kinematics, the supporting force and moments can be obtained. This complete information used in mechanical design can be more accurate than the traditional static design method. It can optimize the design and reduce the quality of the system and energy consumption. At the same time, the multi-body system inverse dynamics are applied to control. Forward dynamics in modern engineering technology is mainly used for virtual simulation which can save huge costs of design, manufacture and test.

The rapid development of large-scale systems based on the dynamics and efficiency of real-time control requires a substantial increase in dynamics [2]. Based on the need of high efficiency, recursive dynamical algorithms are obtained. Computational dynamics recursive algorithm is O (N), where N is the degrees of freedom, namely freedom increases linearly with the amount of calculation. For the inverse dynamics problem, first developed in the O (N) algorithm is Newton-Euler (referred to as NE) formula [3]. Rodriguez Guillermo implemented O(N) multi-body system dynamics spatial operator algebra (SOA) in reconfigurable spacecraft ground control [4]. This method for large complex machinery systems is becoming increasingly important. Dong Longlei et al established dynamics equation based on Kane dynamics with finite segment modeling method of rigid - flexible coupling multi-body system [5]. Ding Xilun analyzed Lagrange dynamics with Lie algebra to describe two spatial dynamics of a flexible rod robot and compared with ADAMS simulation results [6].

In summary, the rigid system of coupled nonlinear problems appearing in modern machinery and a variety of constraints are required to develop an efficient recursive algorithm. Based on the above methods, this paper dynamics developed the recursive dynamics based on Lie groups and Lie algebras which can be applied to the complex mechanical system. The kinetic equations by recursive methods can achieve efficient real-time dynamic simulation and control. Some examples show the correctness and efficiency of the method.

2 Recursive dynamics of Lie groups and Lie algebra

2.1 Adjoint transformations of Lie groups and Lie algebra

A Lie group is a group and also it is a differentiable manifold expressed by G. The tangent space of a Lie group's element is called the Lie algebra for that group. The Lie algebra forms a vector space with a bilinear map which is called the Lie bracket [7].

The rotation matrix is an orthogonal matrix in three dimensional Euclidean space and it is an element of the special Orthogonal group *SO(3)*. The homogeneous transformation is the Lie group referred to as the special Euclidean group *SE(3)*. The Lie algebra of *SO(3)* is *so(3)* which consists of the set of skew symmetric matrices on $R^{3\times 3}$. Similar with it, the Lie algebra of *SE(3)* is *se(3)*. The relation between a Lie group *SO(3)* or *SE(3)* and the associated Lie algebra *so(3)* or *se(3)* is the matrix exponential.

A linear mapping on the Lie algebra is an element of the Lie group. This is called the Adjoint map on *SE(3)* and is denoted by *Ad*, where $Ad_G(h) = GhG^{-1}$, $G \in SE(3)$ and $h \in se(3)$. Note that h is the homogeneous matrix.

[a] Corresponding author: shaob@sdju.edu.cn

DOI: 10.1051/matecconf/20166101004

$$Ad_G(h)=\begin{bmatrix} R & 0 \\ [p]R & R \end{bmatrix}\begin{bmatrix} \omega_h \\ v_h \end{bmatrix} \tag{1}$$

The linear mapping on the dual space *se(3)* is a dual Adjoint operator denoted by Ad^*. For h^*:

$$Ad_G^*(h^*)=\begin{bmatrix} R^T & R^T[p]^T \\ 0 & R^T \end{bmatrix}\begin{bmatrix} \mathcal{M} \\ \mathcal{F} \end{bmatrix} \tag{2}$$

The Lie algebra can also be used as a linear mapping on itself denoted by *ad*. It has the form of $ad_g(h)=[g,h]=gh-hg$, where $g,h\in se(3)$.

$$ad_g(h)=\begin{bmatrix} [\omega_g] & 0 \\ [v_g] & [\omega_g] \end{bmatrix}\begin{bmatrix} \omega_h \\ v_h \end{bmatrix} \tag{3}$$

Similarly, the dual operator ad^* is given by $ad_g^*(h^*)=[g,h^*]$ and:

$$ad_g^*(h^*)=\begin{bmatrix} [\omega_g]^T & [v_g]^T \\ 0 & [\omega_g]^T \end{bmatrix}\begin{bmatrix} \mathcal{M} \\ \mathcal{F} \end{bmatrix} \tag{4}$$

In multi-body dynamics, the Adjoint operator *Ad* maps the spatial velocity screws from one reference frame to body frame and the dual Adjoint operator Ad^* maps the spatial force screws. The mapping *ad* and ad^* respectively define the standard cross product operation of the 6-vector velocity screws and force screws.

2.2 Dynamics of tree topology

2.2.1 Outward recursion of spatial velocity

Given V_0, outward recursion to compute the velocities in depth first manner:

$$\boldsymbol{T}_{P,i}=\boldsymbol{M}_i e^{S_i q_i} \tag{5}$$

$$\boldsymbol{V}_i=Ad_{\boldsymbol{T}_{P,i}^{-1}}(\boldsymbol{V}_P)+\boldsymbol{S}_i\dot{q}_i \tag{6}$$

$$\boldsymbol{a}_i=ad_{V_i}(\boldsymbol{S}_i\dot{q}_i) \tag{7}$$

$$\boldsymbol{b}_i=-ad_{V_i}^*(\boldsymbol{J}_i\boldsymbol{V}_i) \tag{8}$$

Where *P* denotes the index of body *i* 's parent body.

The principle of inward recursive is based on breadth-first which begins from the farthest chain endings. Inward recursive calculate all J and B of links. It allows any number of external forces applied to the branch chain link.

2.2.2 Inward recursion of $\hat{\boldsymbol{J}}_i$ and $\boldsymbol{B}_i$

Inward recursion to compute the articulated body inertia and articulated body zero force in reversed breadth first manner.

$$\hat{\boldsymbol{J}}_i=\boldsymbol{J}_i+\sum_{j\in C}Ad_{T_{i,j}^{-1}}^*\hat{\boldsymbol{J}}_j^+Ad_{T_{i,j}^{-1}} \tag{9}$$

$$\boldsymbol{B}_i=\boldsymbol{b}_i+\sum_{j\in C}Ad_{T_{i,j}^{-1}}^*\boldsymbol{z}_j^+ \tag{10}$$

$$\boldsymbol{z}_i=\hat{\boldsymbol{J}}_i\boldsymbol{a}_i+\boldsymbol{B}_i+\sum_j Ad_{T_{i,j}^{-1}}^*\boldsymbol{F}_{i,j}^{'} \tag{11}$$

$$\hat{\boldsymbol{J}}_i^+=\left[\boldsymbol{I}-\frac{\hat{\boldsymbol{J}}_i\boldsymbol{S}_i\boldsymbol{S}_i^T}{\boldsymbol{S}_i^T\hat{\boldsymbol{J}}_i\boldsymbol{S}_i}\right]\hat{\boldsymbol{J}}_i \tag{12}$$

$$\boldsymbol{z}_i^+=\left[\boldsymbol{z}_i+\frac{\hat{\boldsymbol{J}}_i\boldsymbol{S}_i(\tau_i-\boldsymbol{S}_i^T\boldsymbol{z}_i)}{\boldsymbol{S}_i^T\hat{\boldsymbol{J}}_i\boldsymbol{S}_i}\right] \tag{13}$$

Where C denotes the set of child bodies.

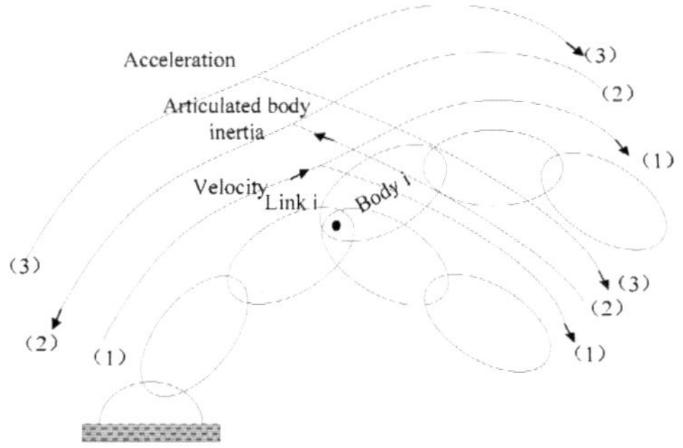

Figure 1. Forward dynamics of tree topology system

2.2.3Outward recursion of acceleration

Given $\dot{V}_0$, outward recursion to compute the acceleration in depth first manner:

$$\dot{\boldsymbol{V}}_i=\hat{\boldsymbol{J}}_iAd_{T_{P,i}^{-1}}(\dot{\boldsymbol{V}}_P)+\boldsymbol{S}_i\ddot{q}_i+\boldsymbol{a}_i \tag{14}$$

$$\ddot{q}_i=\frac{\tau_i-\boldsymbol{S}_i^T\left[\hat{\boldsymbol{J}}_iAd_{T_{P,i}^{-1}}(\dot{\boldsymbol{V}}_P)+\boldsymbol{z}_i\right]}{\boldsymbol{S}_i^T\hat{\boldsymbol{J}}_i\boldsymbol{S}_i} \tag{15}$$

For the complex mechanical systems with closed loop constraints, recursive dynamics method can be constructed by a series of natural orthogonal complement matrix, such that the closed loop system dynamic equations relating the generalized inertia matrix can be clearly shown, and then the generalized inertia matrix recursive decomposition kinetics can be obtained. As for the flexible lightweight mechanical system, a motion can be expressed as relatively flexible synthetic rigid body motion and movement. Using the finite element method to express the deformation of the flexible object, simply expand each symbol to represent rigid body containing modal distortion coordinate with the previous similar recursive dynamics.

3. Example

Each link of the six-bar press are made of steel, having a density of 7.8×10-6(kg/mm3), Young's modulus of 2.07×105(N/mm2), Poisson's ratio of 0.29. Parameters of each link are shown in Table 1.

Link 4 is the input with the speed of 1.47 (rad/s). The simulation curve is shown in figure 2 to figure 5.

Table 1. Parameters of the six-bar press

Parameters	Mass(kg)	Center of mass (mm)	Inertia (kg · mm^2)	Initial (deg)
Link 2	6.28	[181.1, 452.7, 0.0]	[5.3×10^5, 5.3×10^5, 1039.8]	340.9
Link 3	9.19	[1103.4, -143.8, 0.0]	[1.4×10^6, 1.4×10^6, 7.6×10^4]	316.9
Link 4	1.69	[1280.0, 251.6, 0.0]	[1.1×10^4, 1.0×10^4, 275.0]	0
Link 5	5.997	[1209.4, -1220.9, 0.0]	[4.6×10^5, 4.6×10^5, 992.7]	353.2
Slider	62.41	[1400.0, -1645.0, 0.0]	[4.2×10^5, 4.2×10^5, 4.2×10^5]	

As can be seen from the figure, work area of six-link press is longer than ordinary crank press. It can be achieved full capacity in the entire working area, so the six-link press is particularly suitable for deep drawing stretch craft work. Because of the smooth operation of the slider, operating noise is greatly reduced. When the slider reaches the maximum acceleration, the torque of each joint almost simultaneously reached the maximum. Visible dynamics simulation results obtained using the process can be drawn throughout the run all the links and the joints of the forces.

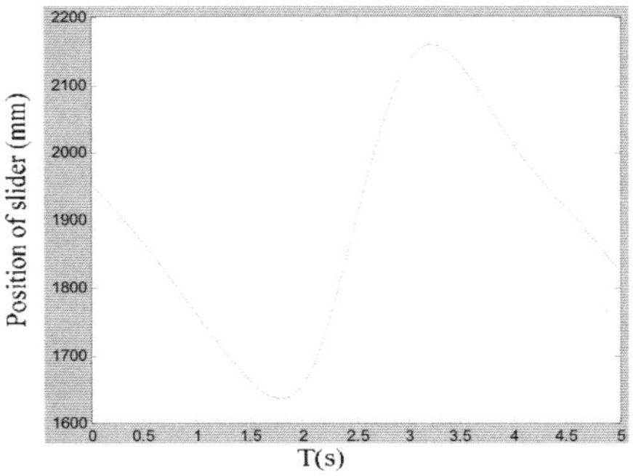

Figure 2. Position of slider

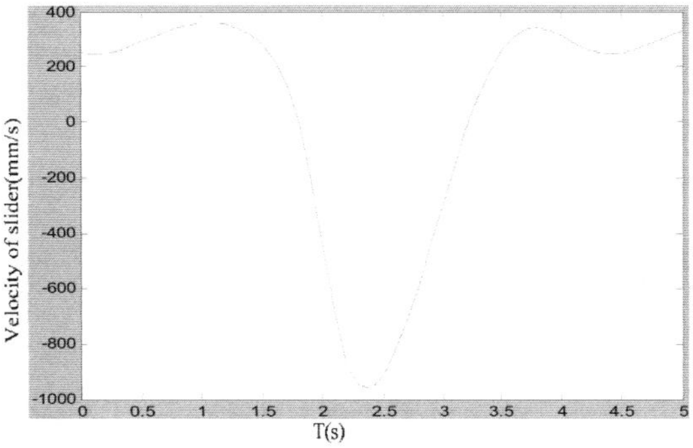

Figure 3. Velocity of slider

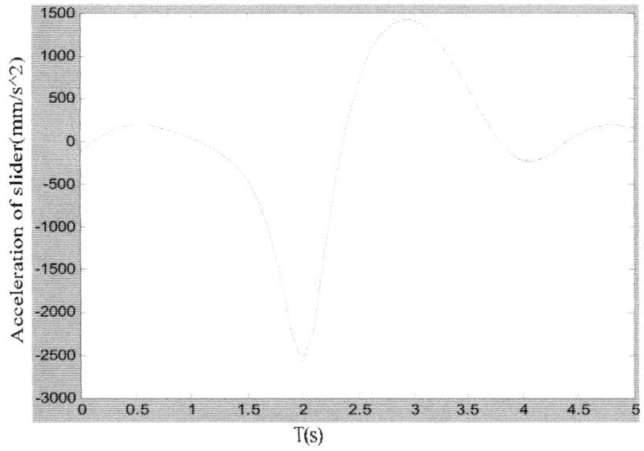

Figur 4. Acceleration of slider

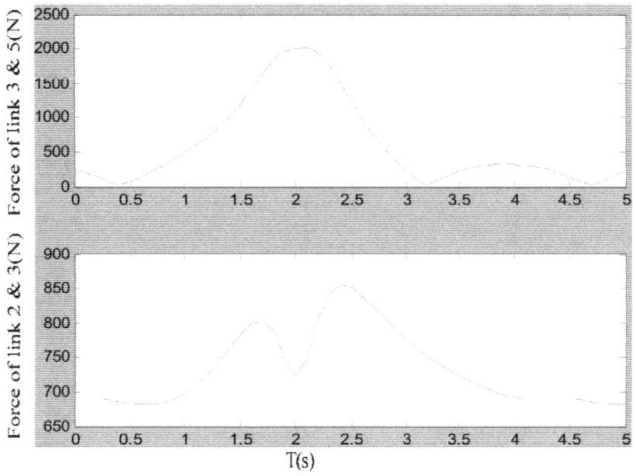

Figure 5. Force of link 2 & 3 and link 3 & 5

DOI: 10.1051/matecconf/20166101004

4. Conclusions

In this paper, Li Lie algebra theory is used to describe complex mechanical system. This efficient recursive dynamics is not only suitable for the chain, tree topology recursive mechanical system, and can be further extended to the closed-loop and flexible system. Its complexity is O (N). This concise, clear and efficient algorithm is very important for real-time dynamics. Virtual simulation positive dynamics achieved can save huge laboratory fees. Examples of use of the above kinetic method are validated.

Acknowledgment

This research is partially supported by Young Teacher Training Project Funded Plan of Colleges of Shanghai (project number:ZZSDJ12005) and Scientific Research Project Start-up Fund of Shanghai Dianji University (project number: 11c404).

References

1. Liu ZY, Hong JZ. Current Situation and Prospect of system dynamics of flexible multi-body. Computational Mechanics, **25**, 411 (2008)
2. Hwang Y L. Recursive Newton-Euler formulation for flexible dynamic manufacturing analysis of open loop robotic systems. International Journal of Advanced Manufacturing Technology, **29**, 598 (2006)
3. Feather stone R. Efficient factorization of the joint space inertia matrix for branched kinematical trees. International Journal of Robotics Research, **24**, 487 (2005)
4. Mukherjee R M , Jain A. Parallelization of the spatial operator algebra for dynamics of multi-body systems. *8th .World Congress on Computational Mechanics .Venice , Italy*, 269 (2008)
5. Dong LL, Yan GR, Yu JJ. Finite Segment Modeling of the flexible member with a center of rotation of the rigid body. Xi'an Jiaotong University, **35**, 977 (2001)
6. Ding HL, Liu Y. Lie group with Lie algebra analysis robot has a flexible deformable rod space dynamics. Mechanical Engineering, **43**, 184 (2007)
7. Park F C , Bobrow J E, Ploen S R. A lie group formulation of robot dynamics. International Journal of Robotics Research, **14**, 609 (1995)
8. Sohl G A, Bobrow J E .A recursive multibody dynamics and sensitivity algorithm for branched kinematical chains. Journal of Dynamic Systems, Measurement , and Control, **123**, 391 (2001)

 DOI: 10.1051/matecconf/20166101005

Dynamic Analysis of Power System with Photovoltaic Generation

Chia-Yu Hsu [1,a], Pei-Hwa Huang [1] and Kai-Sheng Yang [1]

[1] *Department of Electrical Engineering, National Taiwan Ocean University, Taiwan*

Abstract. With more penetration of solar power plants, photovoltaic generation will be exerting more influences on the power system. The main purpose of this paper is to study the dynamic characteristics of the power system with photovoltaic generation. First, power flow and bus voltage variations of the system with photovoltaic generation are calculated. Then according to the related interconnection criteria, the voltage variation and cable loading are inspected to determine whether the system complies with relevant specifications. Finally, the dynamic responses of the system are analyzed. This paper focuses on a sample power system with photovoltaic generation. The system dynamic characteristics are examined when three-phase faults occur, and the system voltage variations subject to illumination change are investigated. Results show that photovoltaic generation has less impact on system because the capacity of the photovoltaic generation is not high. The illumination change does affect both voltage and frequency of the system, and the system will return to a normal operating state after a period of time.

1 Introduction

Because of some challenges such as conventional energy sources consumption, pollution, global climate change and security of energy supply, significant efforts have been made to develop renewable energy sources such as solar energy, wind energy, and hydraulic energy. As a result, worldwide, there has been a move to significantly increase the utilization of environment friendly renewable resources and to increase the reliability of power supplied to the customers, and if possible, reduce losses in transmission and distribution systems [1]. One way of generating electricity from renewable sources is to use photoelectric cell that convert the energy of light directly into electricity [2].

The latest technological advancements in solar energy conversion and the increased support from government and private institutions have led to increased photovoltaic generation in recent years. Thus photovoltaic (PV) generation exert more influences on the system and the regulation of parallel operation for the photovoltaic generation has been becoming more and more stringent [3-6]. The main purpose of this paper is to study the analysis of the dynamic characteristics problem of power system with photovoltaic generation.

The major research work is focus on the detailed investigation of system dynamic characteristics with a sample power system as study cases. The system dynamic characteristics are examined when three-phase faults occur, and the system voltage variations subject to illumination change are investigated.

2 Overview of photovoltaic generation

Photovoltaic generation is a use of light energy into power, also known as photovoltaic cells. Photovoltaic power generation employs solar panels composed of a number of solar cells containing a photovoltaic material. Most of the solar cells are semiconductor material. Accordance with the chemical formula can be divided into silicon semiconductor and compound semiconductor. The main practical application of the silicon material battery, can be divided into stacking (Bulk Type) and Thin Film (Thin Film Type). Materials presently used for photovoltaic include monocrystalline silicon, polycrystalline silicon, amorphous silicon and cadmium telluride [7]. As there are many kinds of the production of materials, the manufacture of solar cells with conversion efficiency are also different.

3 Dynamic model of photovoltaic generation

3.1. Mathematical modelling of photovoltaic generation

A photovoltaic cell is based on the physical phenomenon called "photovoltaic effect". Figure 1 shows the equivalent circuit model of a PV module [7, 8]. This circuit model is composed of an ideal current source, a diode connected in parallel with the current source and a series resistor.

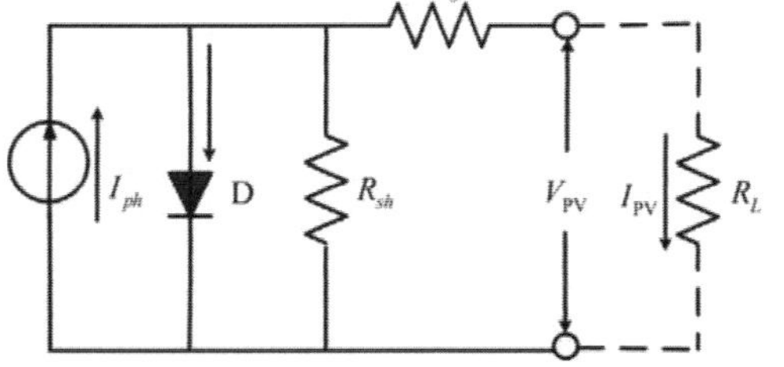

Figure 1. The equivalent circuit of PV cell model.

[a] Corresponding author: 20053003@mail.ntou.edu.tw

In Figure 1 the output terminal current I_{PV} may be described by (1)

$$I_{PV} = I_{ph} - I_D = I_{ph} - I_{sat}[\exp(\frac{qV_{pv}}{AKT}) - 1] \qquad (1)$$

Where I_{ph} is the short circuit current of module under a given solar irradiance (A); I_{PV} is the terminal current of a module (A); V_{PV} is the terminal voltage of a module. q is the coulomb constant(1.6×10^{-19}C); K is the Boltzmann's constant(1.38×10^{-23} J/°K); T is the cell temperature(°K); A is the ideal constant; I_{sat} is the diode saturation current (A).

From (2) the output power of PV module P_{PV} can be represented by

$$P_{PV} = I_{PV}V_{PV} = I_{ph}V_{PV} - I_{sat}V_{PV}(\exp(\frac{qV_{PV}}{AKT}) - 1) \qquad (2)$$

3.2 Integration of dynamic photovoltaic model

The Solar PV Unit dynamic model was developed to simulate performance of a photovoltaic (PV) plant connected to the grid via a power converter. The model is largely based on the generic type 4 wind model, WT4, with the added ability to simulate output changes due to solar irradiation. The model is then incorporated with the power converter/generator module and electrical control module of Type 4 wind plant available in the PSS/E library [9]. The reason for using the Type 4 wind power model is PV plants are connected to the grid using the same technology used by Type 4 wind farm. From the point of DC-link to the grid connection, both PV and Type 4 wind technology use similar control and inverter technology to inject power to the grid [9, 10].

Figure 2 shows the PV Generic Model comprises the following modules:

- PVGU: power converter/generator module
- PVEU: electrical control module
- PANEL: linearized model of a panel's output curve
- IRRAD: linearized solar irradiance profile

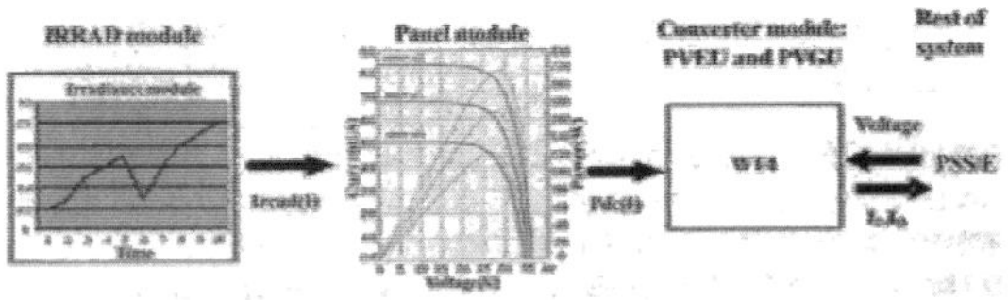

Figure 2. PV model connectivity diagram.

The modules are conventionally designated as wind modules with PVGU as the generator/converter module, PVEU as the electrical control module, PANEL as the mechanical module, and IRRAD as the pitch module. The power converter/generator module calculates the current injection to the grid based on filtered active and reactive power commands from the electrical control module as shown in Figure 3 and Figure 4 [9]. The reactive control calculates the reactive current command for the various control options, which could be any of the following:

- Remote bus voltage control
- Power factor control
- Reactive power control

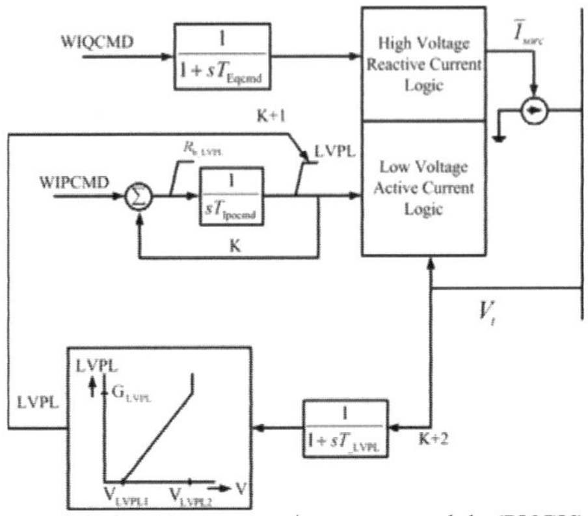

Figure 3. Power converter/generator module (PVGU).

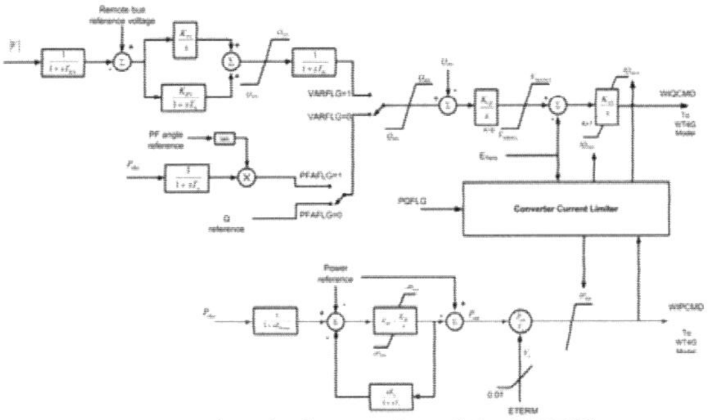

Figure 4. Electrical control module (PVEU).

4 System dynamics

4.1 Study system

In this paper, both system A and system B are taken as the study system of which both the investigation of incorporating PV generation on system dynamic characteristics and the effect of the changes in solar illumination. The single-line diagrams are shown in Figure 5 and Figure6.

The PV generations in the system A is divided into four groups, and use a step-up transformers 22.8/161kV up to 161 kV. The PV generation in the system B is integrated into the 161kV bus through a substation.

DOI: 10.1051/matecconf/20166101005

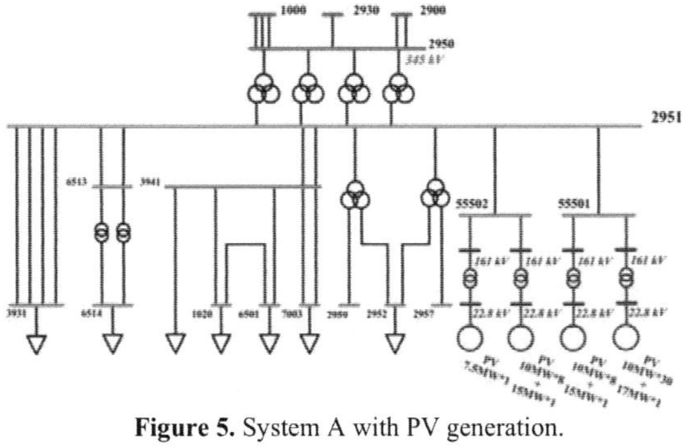

Figure 5. System A with PV generation.

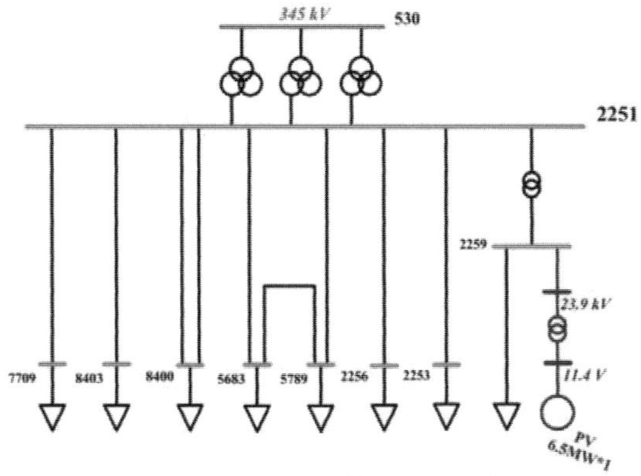

Figure 6. System B with PV generation.

4.2 Simulation result

When the disturbance applied to the system, the solar irradiation is set to constant ($1000\text{W}/\text{m}^2$). The simulation work to be conducted starts with a three-phase short circuit fault set to occur at bus 2951 and cleared after 4 cycles. Figure 7 to Figure 10 show the responses of PV generation terminal voltage, active power, and the reactive power, respectively. From Figure 7 to Figure 9, due to that the a three-phase short circuit fault set to occur at 161kV bus 2951, so the voltage response of bus 2951 drops more severe than another buses during a grid fault. Figure 10 shows the simulation results for the active power and reactive power of one PV generation. It can be observed that the PV generation provides the necessary reactive power compensation during system fault transient.

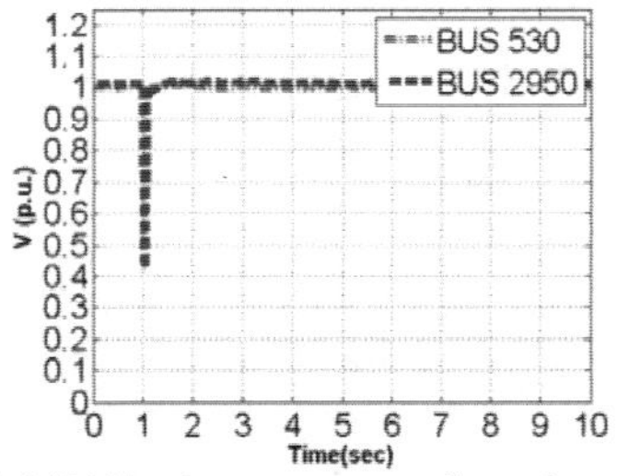

Figure 7. 345 kV voltage responses of sample power system with PV generation.

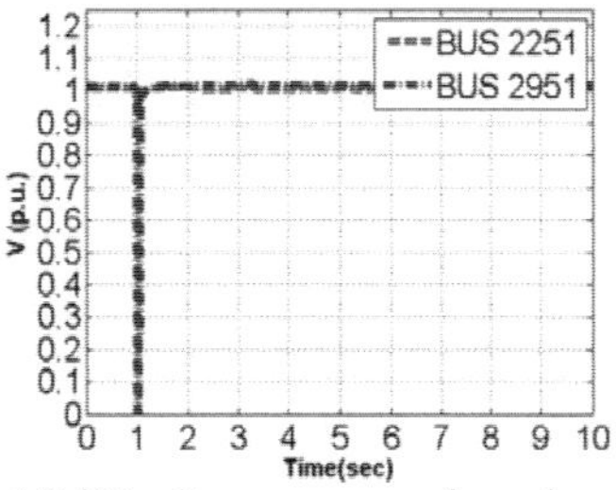

Figure 8. 161 kV voltage responses of sample power system with PV generation.

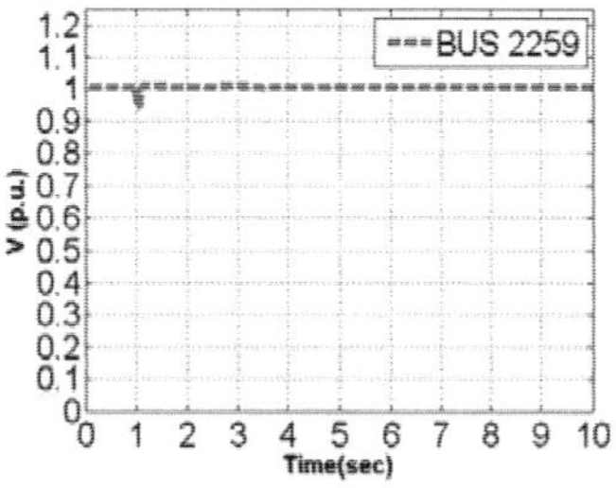

Figure 9. 23.9kV voltage responses of sample power system with PV generation.

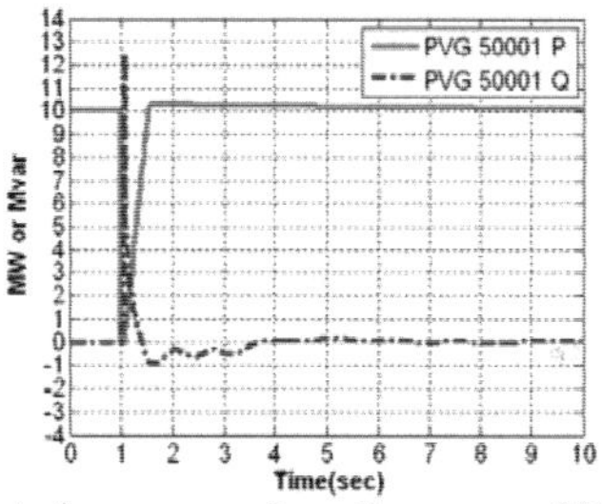

Figure 10. Active power and reactive power of PV generation.

PV generation is characterized as an intermittent source of energy as it is dependent on the environmental conditions. In this case, dynamic analysis is performed on the power system with PV generation under different illumination. The illumination is dropped from $1000\text{W}/\text{m}^2$ to $0\text{W}/\text{m}^2$ in 10 minutes and restored the illumination to $1000\text{W}/\text{m}^2$ in 5, 10, 30, 60 minutes, respectively. Figure 11 shows the active power responses are changed due to the changes in illumination. It is found from Figure 12 and Figure 17 that the illumination change does affect both voltage and frequency of the system.

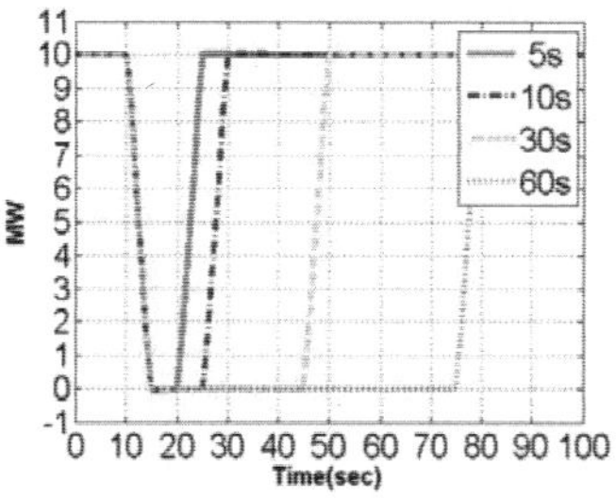

Figure 11. Active power for changes of illumination at PV generation.

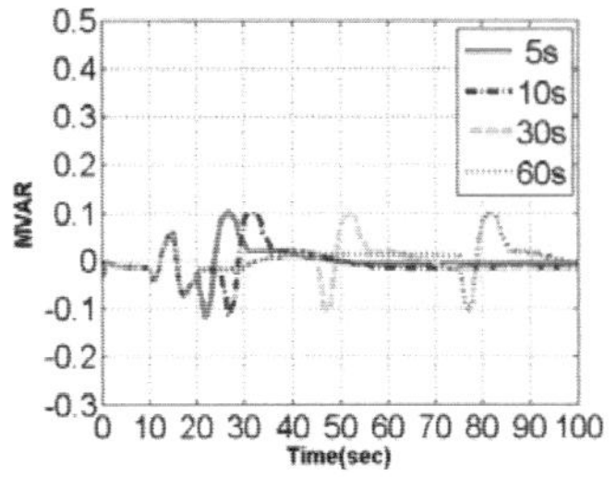

Figure 12. Reactive power for changes of illumination at PV generation.

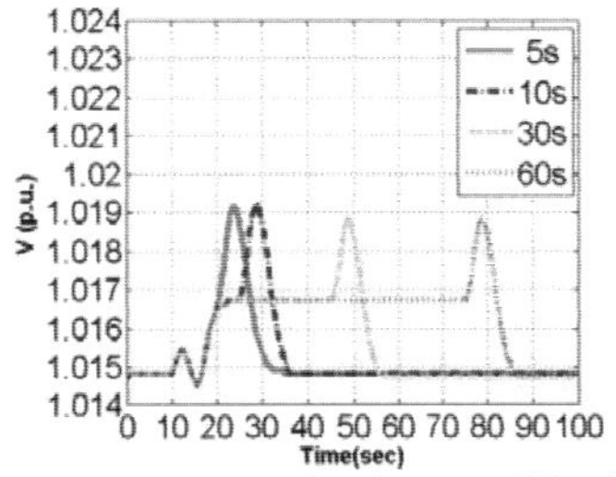

Figure 13. Voltage responses for changes of illumination at PV generation.

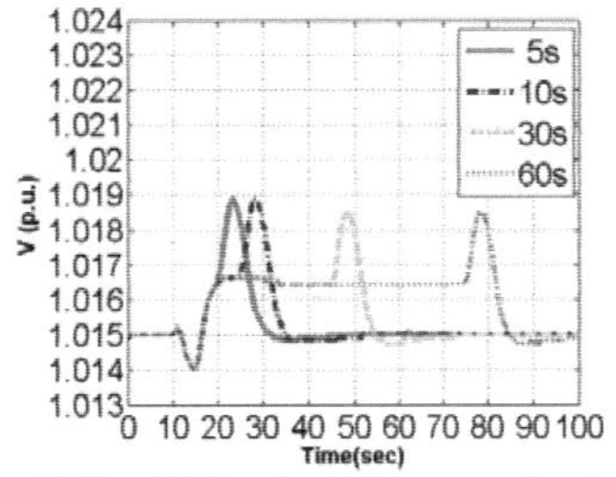

Figure 14. Bus 2951 voltage responses for changes of illumination.

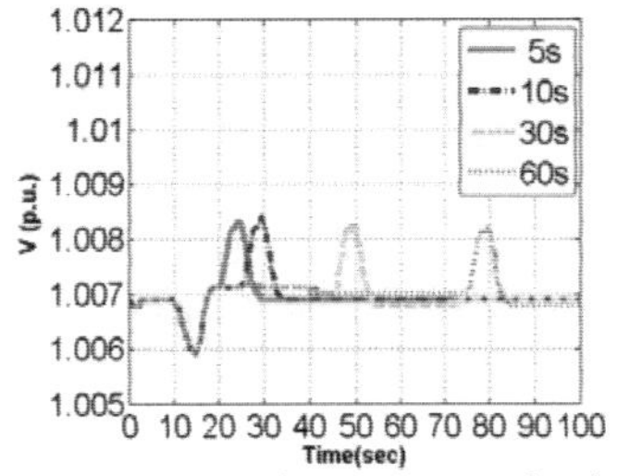

Figure 15. Bus 2259 voltage responses for changes of illumination.

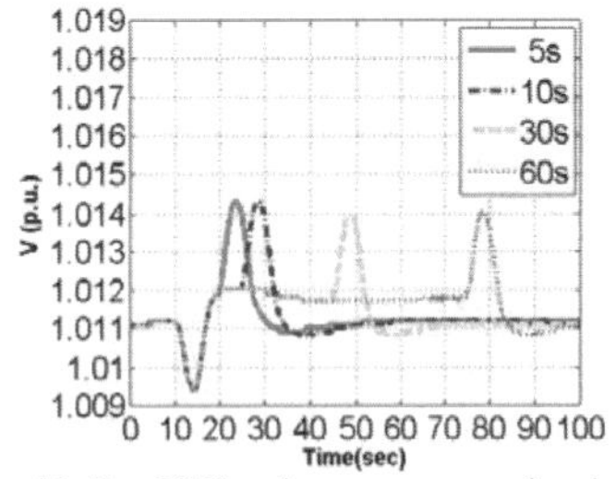

Figure 16. Bus 2950 voltage responses for changes of illumination.

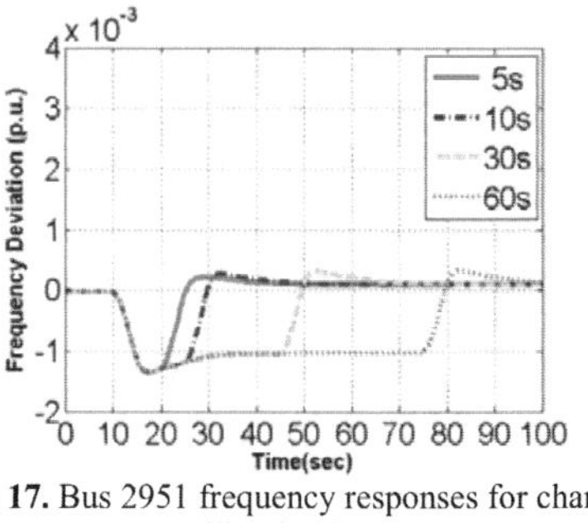

Figure 17. Bus 2951 frequency responses for changes of illumination.

5 Conclusion

The main purpose of this paper is to study the investigation of system dynamic characteristics with PV generation. The dynamic characteristics of the system are studied by simulating a three phase fault, and by changing the illumination. The demonstrated results show that PV generation provides the necessary reactive power compensation during system fault transient. The responses obtained from the studies indicate that the illumination change does affect both voltage and frequency of the system, and the system will return to a normal operating state after a period of time. The study results will be helpful to the practices of operation for the PV generation system

Acknowledgements

This work was supported in part by Ministry of Science and Technology under Grant MOST 104-2221-E-019-025 and MOST 105-3113-E-194-003.

References

1. G. Boyle, *Renewable Energy* (Oxford, 2004)
2. R. Foster, M. Ghassemi, and A. Cota, *Solar energy: Renewable Energy and the Environment* (CRC Press, 2010)
3. B. Tamimi, C. Canizares, and K. Bhattacharya, IEEE T. Sustain. Energ., **4**, 680-688 (2013)
4. A. R. Aarathi and M. V. Jayan, in *Proceedings of 2014 International Conference on Advances in Green Energy*, 26-32 (2014)
5. W. De Soto, S. A. Klein, and W. A. Beckman, Sol. Energy, **80,** 77-88 (2006)
6. Y. B. Wang, C. S. Wu, H. Liao, and H. H. Xu, in *Proceedings of IEEE International Conference on Industrial Technology*, 1-6 (2008)
7. S. A. Kalogirou, *Solar Energy Engineering: Processes and Systems* (Elsevier, 2014)
8. R. A. Messenger and J. Ventre, *Photovoltaic Systems Engineering* (CRC Press, 2004)
9. Siemens Power Technology International, *PSS/E Version 33.4 Program Operation Manual* (2013)
10. T. Alquthami, H. Ravindra, M. O. Faruque, M. Steurer, and T. Baldwin, in *Proceedings of IEEE North American Power Symposium*, 1-8 (2010)

Design of Film Digitizer Equipment and Film Weld DefectRecognition Technology in Pressure Vessel

Du Weidong[1], Wang Peng[2,a], Li Xiaobin[2], Lv Zhigang[2], Zhang Feng[2]

[1]Shandong Special Equipment Inspection Institute, Ji Nan, China
[2]School of Electronic and Information Engineering, Xi' an Technological University, Xi An, China

Abstract: According to the characteristics of image defects, this paper integrated use of image preprocessing, image segmentation algorithms, expert systems and other methods, presented a film weld defect identification method for non-destructive testing in the pressure vessel weld defect recognition, and to achieve in MATLAB. This paper also described the expert system's basic principles, structure, establish methods and identification strategy. In the light of the weld defect characteristics from different levels of recognition, through continuous testing to improve the film defect system library and to further enhance the defect recognition rate.
Keywords: Film Digitizer; Pressure vessel; Weld; Defect identification

1 Introduction

Industrial ray detection (RT) is an important method of non-destructive testing industry. In the internal quality testing of pressure vessel welds and castings, X-ray detection is an important non- destructive testing method[1], use the results to determine the type of weld defects, the location and size, thus, to judge whether the welding qualified. The usual X-ray film weld assessment is in a artificial way, putting the film on the fluorescent lamp to determine the welding quality, by experienced staff. The main shortcoming is this artificial weld evaluation full of uncertainty, evaluation results vary from person to person according to the level of employee's business, lead to the assessment results lack of accuracy and standardization. Another key point is that assessment operation process always is hard to grasp, using standard methods, cumbersome manual and a few inefficient ways to assessment the welding[2], therefore, traditional ways are starves for innovation. With the rapidly development of image processing technology and artificial intelligence technology, we can use computer to improve the assessment efficiency and accuracy, reduce the labor intensity and ensure the consistency of evaluation results, also can improve the recognition stability and liberate the workforce[3].

This paper summarizes existing research, proposed the weld defect recognition technology of the film pressure vessel.

2 Film digitizer equipment design

2.1 Equipment components and operating mechanism

The system uses machine vision approach to design a complete set of film digitization system, uses cold light source through the radiation film, in addition, the adjustable brightness are more than 1000001x, and take the 12bit industrial line matrix CCD camera as the image scan and capture device[4]. In order to shield the ambient light and cut the CCD noise, using a sealed enclosure loaded up all the devices, only opens a small window between the light source and CCD. CCD camera communicates with the computer through the USB port. With its own configured of the optical lens of Nikon AF 50mm. Equipment components shown in Figure 1.

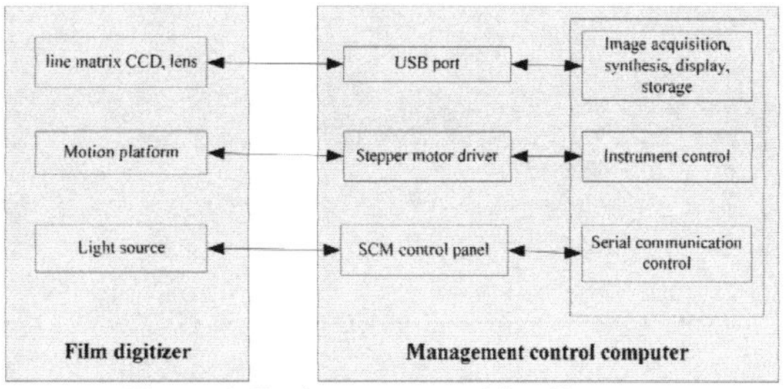

Figure 1. Equipment composition diagram.

Equipment using linear industrial CCD camera to capture the image, solve the shortcomings of array CCD's slow scanning speed, low resolution and the digitization of long film; Using the uniform LED line light source as the background light source to solve the common light source heat, insufficient strength, uneven light intensity and other issues, also can control the light intensity by the the gear switch on the chassis; Using the vertical film transport mechanism witch adopt two step motor drives the left driving wheel and the right driving wheel to control the negative transfer mechanism and the

[a] Corresponding author: wang_peng_1@126.com

clamping plate movement ,respectively. We can measure the location of film through the photoelectric position sensor, gather the running speed by the signal of circular grating sensor and control the speed of the stepper motor by movement control card. Thus, we can realize the synchronous control of image scanning and ensure the accuracy of the linear CCD scanning. The film recycling box is installed at the lower part of the back end of the case, so when the film scanning is over, the film will fall into the recovery box,automatically.

2.2 Information contained in digital film

Digitized film mainly consists of three areas: the weld zone, date area and code region. Weld defects present in the weld area. Defect feature and defect size obtained through image processing, according to the structure, size, surface morphology, etc. Combined with national standards and expertise to identify and classify the weld defects, ultimately. Defect characteristics are divided into boundary characteristics and regional gray feature. The boundary shape of defects is just a reflection of the type of defect, and the internal region of defects contains a large number of closely related with the defect type information. Based on the pressure vessel fusion welding butt joints quality grading requirements, weld defects can be divided into the following five categories: crack, incomplete penetration, lack of fusion, strip defects, circular defects. According to the nature and number of weld defects, quality grading are divided into four, I class quality is the best, IV class quality is the worst[5].

The instrument will transmit the digital film picture to computer, then, we can carry out the image enhancement, image denoising, image correction by computer software witch embeded out image preprocessing algorithm and defect recognition algorithm. Finally, we can realize the complete remote evaluation system.

3 Defect type recognition

Expert system based on image characteristics to identify parameters and the structure of expert systems is in the ongoing process of identifying logic until they have to meet the actual requirements of the solution. The system uses forward reasoning strategy, starting from the image input processing, forward reasoning, to the target state[6], the specific steps are as follows, shown in Figure 2.

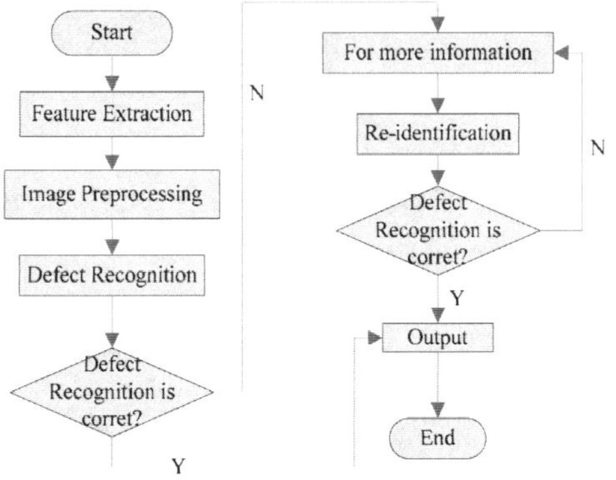

Figure 2. Defect type recognition flow chart.

The film weld image information's obtaining includes extraction of image features, image pre-processing and double-precision transform, and received the binary image[7]; Inference of the user input information and knowledge base helps to match the conditions of the various rules and regulations to the conclusions being stored in an integrated database for the initial identification. The results obtained defect category; If the results confirm the identification of weld defects, then, terminate the expert system knowledge-based identification of reasoning and output the results to identify; If the weld defect identification is not correct, system will give question to the user again, up to obtain more information about the characteristics of film defect feature and other information; Using reasoning inference results, again, to match with the information of expert system knowledge base for the deep identification, until the defect is correctly identified, it will output the reasoning results and defects type[8].

4 Results

Through film digitizer to obtain the original grayscale image film, as shown in Figure 3(a), and then process the original image by various image preprocessing method. Such as double of grayscale images, using im2double functions convert the original image to double precision image, then double-precision image result shown in Figure 3(b).

The nest image-processing operation are cut and split the double-precision image. Image segmentation is the key step of weld defect identify. The various parts of the image region will noticeable and obvious by pre-processing and with that use threshold-based method to achieve image segmentation. Film images will eventually be divided into three regions: the weld zone, data area, coding region, three regions of the image shown in Figure 3(c) (d) (e).

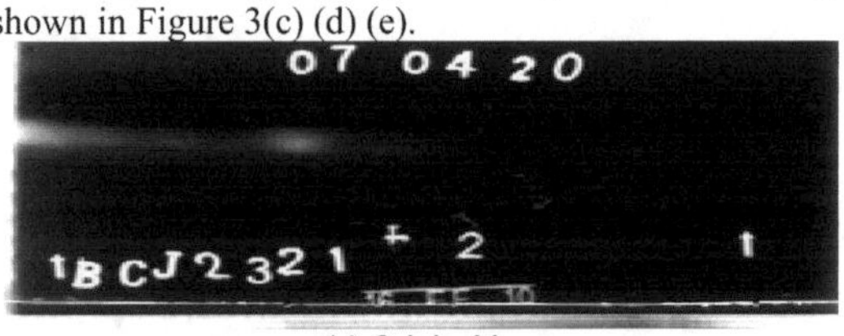

(a) Original image

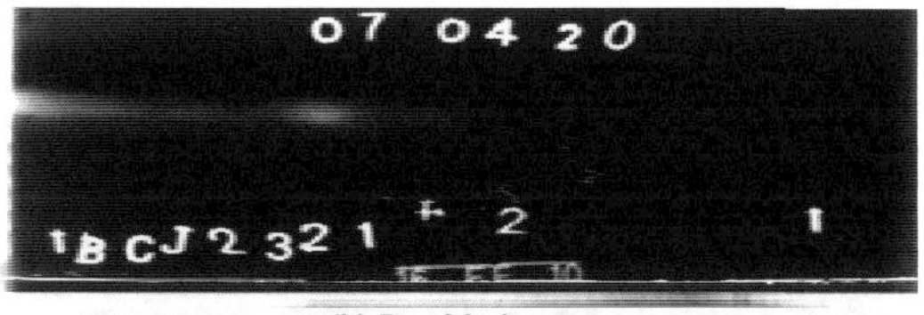

(b) Double image

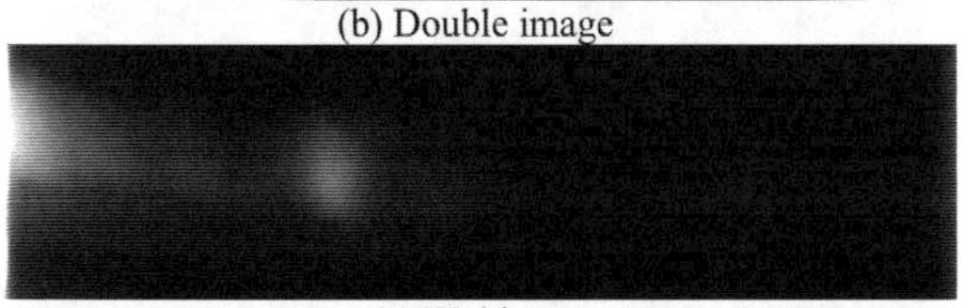

(c) Weld area

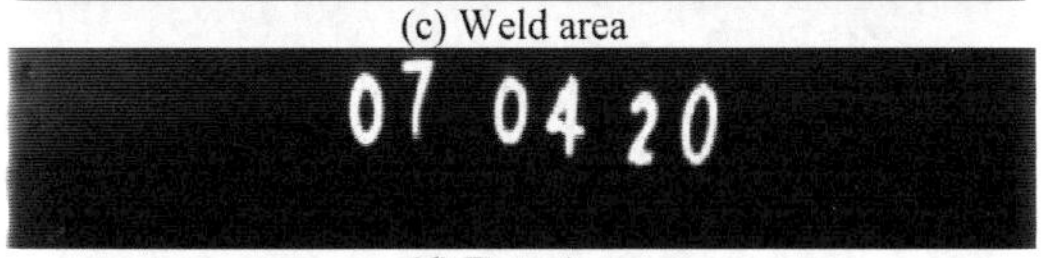

(d) Date Area

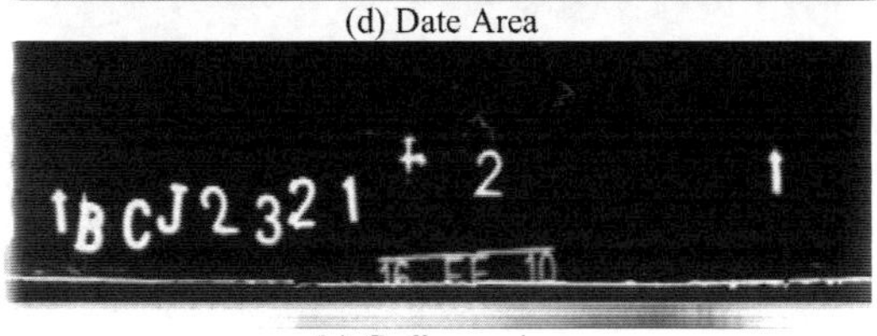

(e) Coding region

Figure 3. Film image.

And then we will implement the deep image processing of the split image parts, take the focus on the weld area. After equalize the weld area images, we can get the high contrast image shown in Figure 4(a). And then take the processing of image threshold on Figure 4(a) , we obtain the Figure 4(b). Finally, corrode the image and invert the pixels, at the same time, connecting the edges noise and the background to remove noise, then we get the image as shown in Figure 4(c).

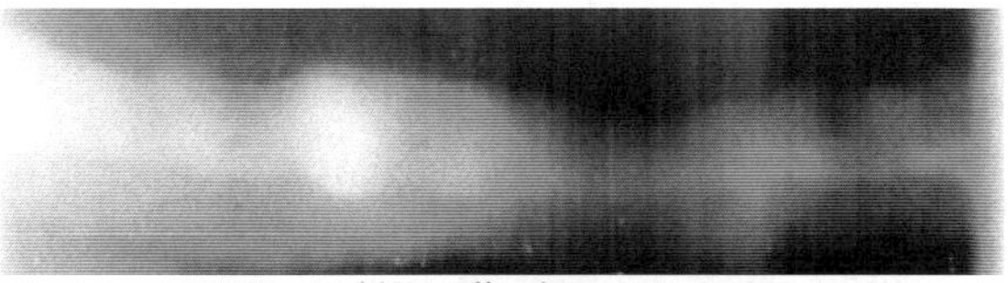

(a)Equalization

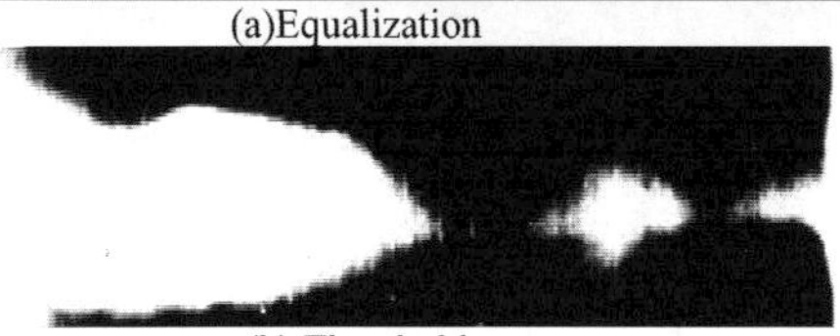

(b) Threshold

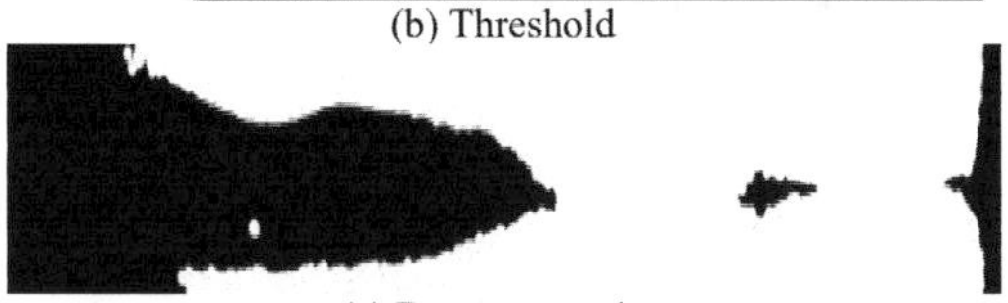

(c) Deep processing

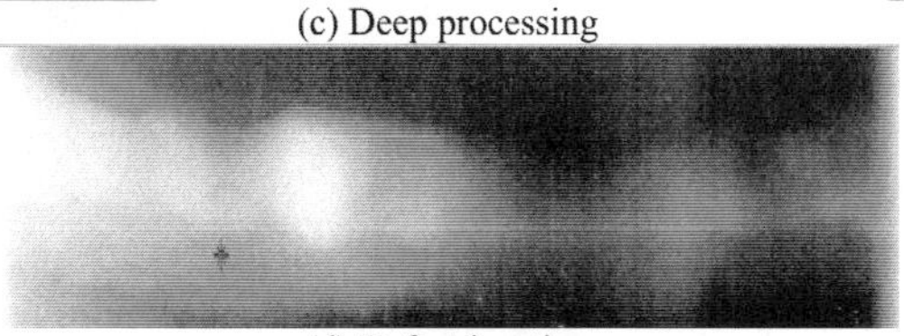

(d) Defect location

Figure 4. The results of weld area.

For the weld area images, first step is regional mark, mark the background area of 1; then, judge the dimension value is greater than 1, if the mark value is greater than 1, indicating a defect exists; finally, calculate the center defect region, for defect location. Positioning results shown in Figure 4(d). The red points are the weld defects. The conclusion is that the film's defect is non-fusion defect type.

5 Conclusion

In this paper, we proposed the expert system based film pressure vessel weld defect recognition technology in MATLAB for simulation and implementation. By a large number of repeated tests and constantly enrich and improve the expert system to improve the system' s ability to identify defects, experimental results show that the system identification of typical defects is better than traditional ways. But for some cross defects, the effect is not very satisfactory. The training with small sample size of typical film defects is also the cause of low recognition rate. In order to improve the recognition rate, we can combine the fuzzy neural network with expert system, thus further improve the recognition rate.

In pressure vessel X-ray detection in our industry, the results of this research both meet the current efficient and accurate detection of requirements, also meet the digital needs for the analysis of X-ray detection and identification of defects. The promotion of the technology can achieve good economic returns, and improve the standardization, traceability of detection, which full of great significant.

Acknowledgements

This work was supported by the 2011 Special research projects in shaanxi province education department (Project name: Intelligent analysis technical research film defects based on the expert system in pressure vessel NDT, Project number: 11JK0921),and transverse project of special equipment inspection institute in Shandong province (Project name: Digital film and defects in intelligent analysis technology research and promotion in the X-ray testing) ,and 2013 industrial research projects in shaanxi province science and technology department(Project name: X ray film digitizer and weld image defect detection and recognition system in nondestructive testing, Project number: 2013K07-15).

References

1. Kang Hongwei, Su Zhenwei, Ma Jie. Nondestructive Inspection.*Research on digital image processing algorithms x-ray weld defects*. **30**, 23 (2006)
2. Ren Dahai, You Zheng,Sun Changku,Ye Shenghua. Journal Of Tsinghua University. *Automatic inspection techniques by real-time radiography for weld defects*. **41**, 19 (2001)

3. Zhou Wei, Wang Chengxun. Transactions of the China Welding Institution. *Research and Application of automatic recognition system to weld defects*. **1**, 45 (1992)
4. Gao Lan, Song Qingguo, Wang Yanqiong, Hua Qing, Li Jun. Transportation Science & Technology. *Image Defect Recognition of Ship Welding Line Processing System Based on Machine Vision*. **6**, 105 (2007)
5. Zhang Tianpeng, Li Wei. Nondestructive Testing. *A Description of Standard JB/T 4730 — 2005 Nondestructive Testing of Pressure Equipments: Radiographic Testing(II)*. **28**, 33 (2006)
6. Zhang Xiaoguang, Li Yu, Xu Jianjian. Computer Engineering. *Weld Seam Defects Distinguishing Expert System Based 0n Neural Network*. **29**, 22 (2003)
7. Yin Ying, Mao Jian, Su Zhenwei. Nondestructive Testing. *PCA-Based Defect Enhancement and Segmentation for X-Ray Image of Welds*. **32**, 678 (2010)
8. Wang Yong, Guo Hui. Computer Aided Drafting, Design and Manufacturing. *Application of Support Vector Machine in Weld Defect Detection and Recognition of X-ray Images*. **24**, 22 (2014)

The Research on Transient Burning Rate of Solid Propellant by Digital Image Processing

Xin Peng[1,a], Bing Liu[2]

[1]*Photoelectronics Academy of National University of Defense Technology, 410005ChangSha, China*
[2]*Aerospace Academy of National University of Defense Technology, 410005ChangSha, China*

Abstract. In order to obtain the burn rate of the solid propellant that is the important parameter of transient burning, the new method named digital image processing is presented. In the article , the principle of digital image processing is analysed; The burning face of the sample in the each time is located according the image and the coordinates of the burning face is obtained. In experiment the transient burn rate is measured by digital image processing and the accuracy is acceptable.

1 Introduction

The transient burn rate is the important parameter in the research of the burning character, so the measure of the transient burn is very important[1]. At present, the target line method and the acoustic emission method is used to measure burn rate in the static state. The two method can not obtain the burn rate at each time and variation of burn rate in the process of the burning.

The digital image processing method can gather the image of the whole burning process and obtain the transient burn rate at frame per second. Moreover, the method can mend the impact of burn rate due to the unusual burning and can not disturb the experiment because of the non-contact.

2 The principle and flow path of the digital image processing

2.1 The method principle of the digital image processing

According to the technique of digital image process gather the image in the process of the burning and read the curve of burning face. The gray level of each pixel in the image can reflect the light intensity of corresponding position in the solid propellant. The image of the burning face contain the brighter dot, so can judge the critical curve and define the curve as transient burning position. Then, convert the critical curve coordinate to the physical coordinate and calculate the transient burn rate. The relationship between the two coordinates is as follow[2]:

$$\begin{bmatrix} x \\ y \end{bmatrix} = k \begin{bmatrix} u \\ v \end{bmatrix} \quad (1)$$

(x, y) is the physical coordinate, (u, v) is the critical coordinate, k is the similarity coefficient.

2.2 The flow path of the digital image processing

[a] Corresponding author:Xin Peng: 540057112@qq.com

2.1.1 Gray

The image of the burning process is RGB picture. Firstly, we change it to gray image. Because the most flame in the experiment is red, so we increase the weight of red gallery, the conversion relation is:

$$f = 0.8\times R + 0.1\times G + 0.1\times B \quad (2)$$

R, G, B is the three quantities of red .

2.1.2 Adjust the uniformity of light

Because the burning chamber is closed and the light on the solid propellant is uniformity, it is necessary to adjust the uniformity of light.

2.1.3 Gauss high pass filter

The Gauss high pass filter is used to get signal of the high frequency light and the flame, in order to locate the position of the burning face accurately.

2.1.4 Bilinear expansion

The burning face of the solid propellant is concave. The most slope of curve is 45° and 135° [3]. In order to link the light mass of the burning face smoothly, the bilinear expansion is used.

2.1.5 Binary

In order to get the burning profile easy, it is necessary to translate gray image into binary image.

2.1.6Connection profile

The profile of the binary image is the discrete light. The discrete light profile is connected line by line.

3 The experiment of transient burn rate measure

3.1 The device of the experiment

The device of the experiment of transient burn rate measure include burning chamber(Figure 1), CCD camera, light, fire device(Figure 2). The propellant fire from the head face. The process of the burning is recorded by the camera. The transient burning rate is measured according to the distance of two frames of the burning face. At the same way, A group transient burning rate is measured. In other words, transient burning rates are measured at time interval of frame of per second.

The experimental sample(Figure 3) is a rich fuel solid propellant strip(10cm*10cm*60cm). The sample is fixed on the fire device, then the sample is put in the burning room. The parallel light is set at the behind cameral in order to heightened the definition of the sample. The sample must parallel the ground in the process of burning. The face of sample must perpendicular to the optical axis in the process experiment.

Figure 1. Burning chamber.

Figure 2. Fire device.

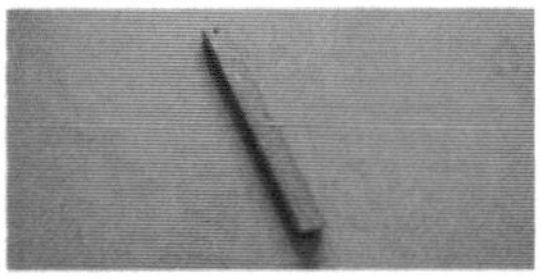

Figure 3. Sample.

3.2 The result of the digital imagine processing

The figure(4) to the figure(9) is the digital imagine processing result of t1, the figure(10) to figure(11) is the result of t2. The figure(8) to figure(9) is the effect picture of the location curve and the burning face curve. We can the green curve which is the location curve is good agreement with the burning face curve. the pixel coordinates of the burning face curve approximated to the pixel coordinates of the green curve. We can get the horizontal coordinate of the green curve and average it. Then change it to physical coordinate that is the distance of burning face to the bottom face. We can the distance of two green curve between the adjacent two frame image. The time is the intervals per frame. The transient burning rate equal to the distance divided by the time.

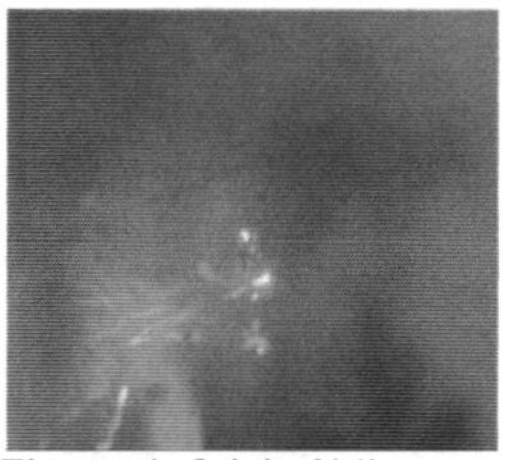

Figure 4. Original(t1)

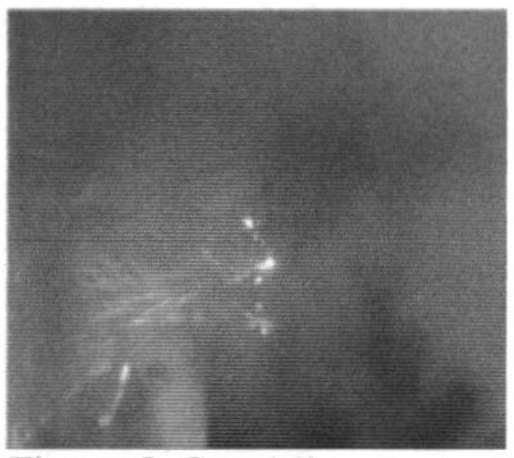

Figure 5. Gray(t1)

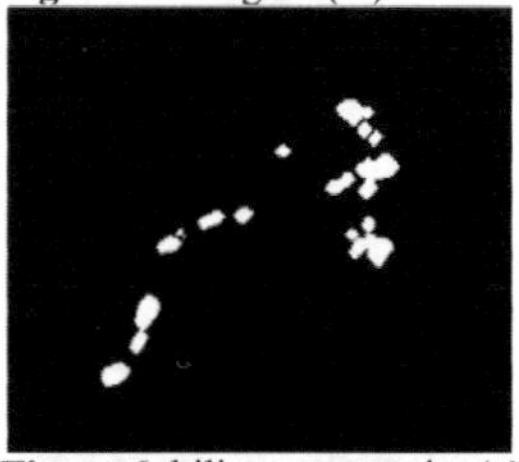

Figure 6. bilinear expansion(t1).

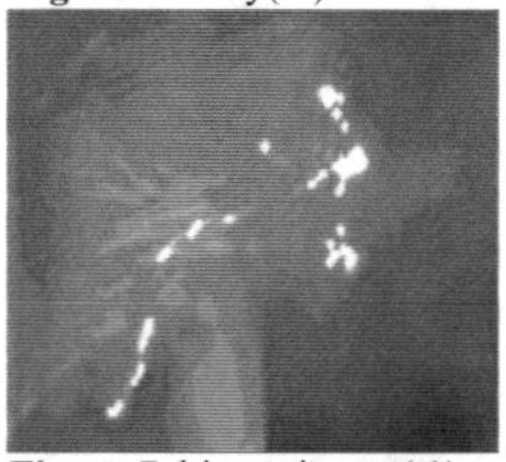

Figure 7. binary image(t1).

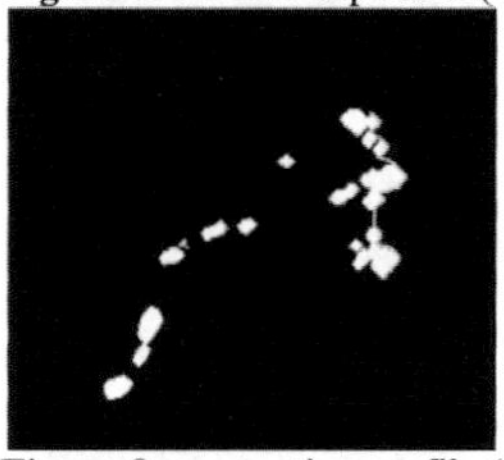

Figure 8. connection profile (t1).

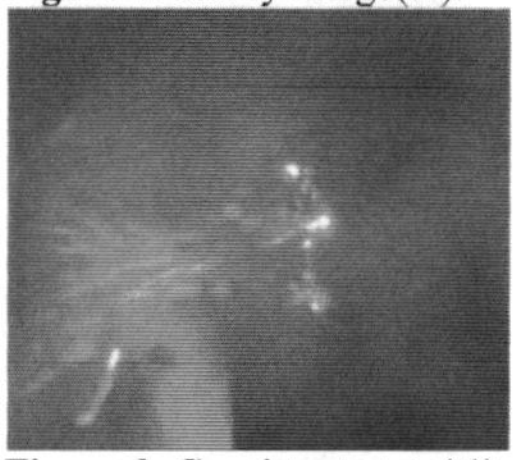

Figure 9. Consistent state(t1).

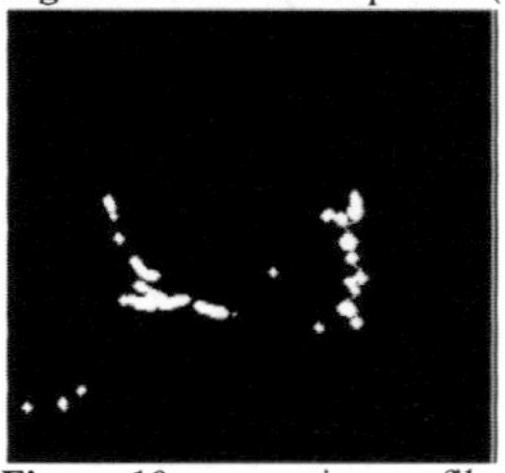

Figure 10. connection profile (t2) .

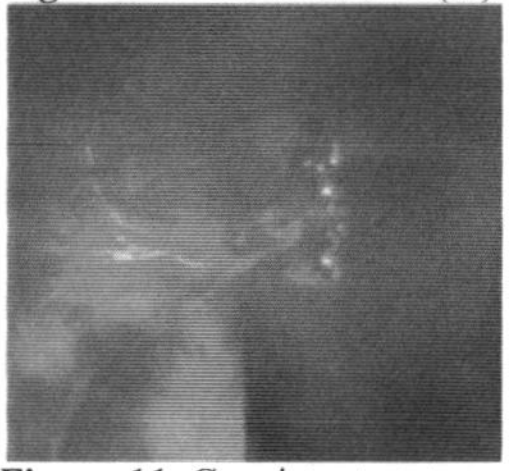

Figure 11. Consistent state (t2).

The averaged rate is equal that the length of the sample strip divide the time of total burning process. Compared the transient burning rate by the digital image processing to the averaged rate, there is small deviation. It is acceptable that the digital image processing is accurate method. The experimental parameters is show in the table(1) and the experimental result is show in table(2).

Table 1. Experimental parameters.

Element	Parameter
Pressure	2Mpa
Frame per second	30

Similarity coefficient	22.857
Average burning rate	20.6mm/s

Table 2. Experimental result.

	Pixel location	Transient burning rate	Deviation
T1	494.3786	21.8964 mm/s	6.3%
T2	511.4157	22.1198 mm/s	7.4%
T3	528.0257	21.5753 mm/s	4.7%
T4	545.2781	22.4050 mm/s	8.7%

4 Conclusion

We can get the transient burning rate by the digital image processing. The method is verified by experiment. Although the transient burning rate is not continuous, it can reflect the the burning character well.

References

1. Wei Zha, Hui Zhu, Ding You Fang, A Study on Mesurement of Fuel rich Propellant Combustion Property at Low Pressure, Energetic Materials, **3**,118 (1999)
2. Qi Feng Yu, Yang Shang. *Videometrics Principles and Researcher,* (2008)
3. Yu Q, Ding X, Lei Z, Optical measurement system for bridge displacement andvibration, *The 3rd Int.Conf. On Experimental Mechanics SPIE,* (2001)

Analysis of geometric factors for an X-ray fluorescence analyzer designed via the Monte Carlo method

LIU Hefan[1,a], TIAN Hong[1], ZHOU Laidong[1], ZHOU Zihang[1], HUANG Fengxia[1], LU Chengwei[1], ZENG Guoqiang[2] and HU Xiang[1]

[1]*Chengdu Research Academy of Environmental Sciences, Chengdu 610072, China*

[2]*Chengdu University of Technology, Sichuan, Chengdu, 610059, China*

Abstract. This paper presents a theoretical study on the design of an X-ray fluorescence (XRF) analyzer. The relative geometric positions of the analyzer's source, detector and specimen are emphasized, which impacts the analyte's characteristic X-ray fluorescence counts. The theoretical formula for the X-ray fluorescence intensity was derived. The geometric factors (angle and distance) were simulated using Monte Carlo Neutron-Particle Transport Code MCNP5. The Cu's X-ray characteristic peak counts were calculated. These Monte Carlo model calculations analyzed two types of geometry changes. The best geometric positions for the XRF analyzer had the incident angle for the excitation source (α) equivalent to the exit angle for the specimen's characteristic X-ray (β). The maximal characteristic X-ray peak counts were obtained when α and β were orthogonal, and the minimum counts were obtained when parallel. To increase the fluorescence counts, the source and detector should be set as close to the specimen as possible. This method and these conclusions can provide technical guidance for designing XRF analyzers.
Key words: Geometric factors; X-ray fluorescence analyzer; Monte Carlo method; MCNP code

1 Instruction

X-ray fluorescence (XRF) analyzers are one of the most important analytical instrument types for element analyses [1-6]. The analyte's characteristic X-ray fluorescence counts relate not only to its physical and chemical properties but also the geometric positions of the source, specimen and detector [7-9]. The XRF analyzer's geometric setup can change the detector's excitation area, detection area and X-ray absorption by the surrounding medium. The best geometric conditions yield the maximal counts, highest analytical accuracy, and minimal measuring time. Many methods are used to find the optimal geometric conditions when designing XRF analyzers. The Monte Carlo Neutron-Particle Transport Code MCNP5, produced by Los Alamos National Laboratory, is important software for nuclear analysis including simulating and calculating XRF spectroscopy [10]. With relative errors below 5%, the spectral intensity accuracy predicted by the MCNP5 code is approximately 95%, and the prediction accuracy ranges from 90% to 97% for unknown specimen compositions [11]. Several models were established in this paper using the MCNP5 code for XRF analyzers. The best geometric conditions were obtained by varying the geometric parameters for the XRF analyzer design via the Monte Carlo method.

2 Theoretical calculations

The basic hypotheses follow: the specimen surface is smooth; the elements in the analyte are evenly distributed; the excitation source is monoenergetic; the incident X-ray beam is parallel to the analyte's characteristic X-ray beam. According to the Beer-Lambert Law, the original incident X-ray fluorescence intensity of the specimen is:

$$I_0' = I_0 \exp[-\mu(S,\lambda_0)\rho x \cos\alpha] \quad (1)$$

In Eq. (1), I_0 is the incident X-ray fluorescence intensity in primary-sectional units. $\mu(S,\lambda_0)$ is the combined mass absorption coefficient. S is the analyte. λ_0 is the incident X-ray wavelength. ρ is the specimen's density. α is the incident angle between the incident X-ray beam and specimen surface.

If i elements are uniformly distributed throughout the specimen, C_i is its content, and I_x is the analyte's characteristic X-ray fluorescence intensity. Therefore, the characteristic X-ray intensity from a dx layer is as follows:

$$dI_x = I_0' C_i \mu(i,\lambda_i)\rho \cos\alpha E_i dx \quad (2)$$

In Eq. (2), element i is excited by incident X-rays of wavelength λ_i ; $\mu(i,\lambda_i)$ is the absorption coefficient quality; E_i is the probability for characteristic X-ray emission when excited by an incident X-ray with wavelength λ_i .

[a] Corresponding author: liuhefan@126.com

Characteristic X-rays are released uniformly from all directions, and only recorded within the solid angle of the detector. β is the exit angle for the X-ray characteristic beam from the specimen's surface. The crossing distance for the characteristic X-ray is $x\cos\beta$ based on the direction for the dx layer from the specimen to the detector. Therefore, the characteristic X-ray fluorescence intensity is as follows:

$$dI_i = \frac{d\Omega}{4\pi}\exp[-\mu(S,\lambda_i)\rho x\cos\beta]dI_x \quad (3)$$

Inserting Eq. (1) and Eq. (2) into Eq. (3) yields the following:

$$dI_i = \frac{d\Omega}{4\pi}I_0 C_i E_i \mu(i,\lambda_0)\rho\cos\alpha\exp[-\mu(S,\lambda_0)\rho x\cos\alpha - \mu(S,\lambda_i)\rho x\cos\beta]dx \quad (4)$$

Integrating Eq. (4) over the thicknesses, x, from $x = 0$ to $x \to \infty$ yields:

$$I_i = \frac{d\Omega}{4\pi}\frac{I_0 C_i E_i \mu(i,\lambda_0)\rho\cos\alpha}{\mu(S,\lambda_0)\rho\cos\alpha + \mu(S,\lambda_i)\rho\cos\beta} \quad (5)$$

In Eq. (5), I_i and C_i relate to not only the specimen's density, type, mass absorption coefficient, etc. but also the geometric layout for the XRF analyzer, such as the incidence angle, exit angle, detector's solid angle, and so on.

3 Monte Carlo model descriptions

3.1 Basic model

Many models to determine the influence of different geometric factors and XRF fluorescence counts have been established in this paper. The basic model is shown in Fig. 1.

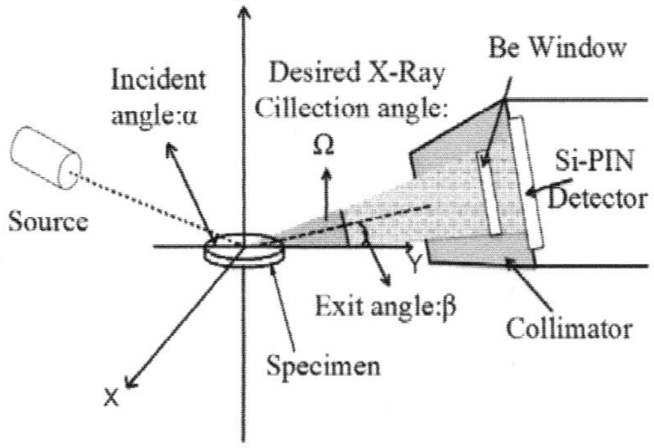

Fig. 1 The geometric layout for designing an X-ray fluorescence measurement system via the Monte Carlo method

In Fig. 1, the model's origin is centered in the specimen's plane. The Y-coordinate is parallel to the specimen, and the Z-coordinate is vertical to the specimen. This model is described in detail in Section 2.2 and the Section 2.3.

3.2 Angle descriptions

There are two methods for describing the angle with a Monte Carlo model. The distance between the excitation source and specimen ($H_{S\text{-}S}$) is 20 mm, so only the distance between the detector and specimen ($H_{D\text{-}S}$) varies.

(a) The value of α and β are simultaneously changed from 0° to 90°, $\alpha = \beta$;

(b) Set β=45° while changing α from 0° to 180°.

Because the incident X-ray is primarily a diverging cone, the excitation source has a scattering angle. For instance, the scattering angle for an X-ray tube, TUB00083, produced by Moxtek Inc. is 23°. Therefore, α and β in Fig. 1 are averages.

2.3 Distance descriptions

There are three ways to vary the distance using these Monte Carlo models.

(a) When $H_{D\text{-}S}$=20 mm and α=β=45°, $H_{S\text{-}S}$ is changed from 0.1 mm to 10^4 mm;

(b) When $H_{S\text{-}S}$=20 mm and α=β=45°, $H_{D\text{-}S}$ is changed from 0.1 mm to 100 mm;

(c) When $H_{S\text{-}S}$=$H_{D\text{-}S}$=20 mm and α=β=45°, the distance between the excitation source and detector ($H_{S\text{-}D}$) is changed from 0.001 mm to 100 mm.

4 Discussion

4.1 Impact of the angle

4.1.1 Simultaneously changing α and β from 0° to 90° (α=β)

Under these conditions, the characteristic X-ray peak counts for CuK_α and CuK_β are nearly constant, as shown in Fig. 2.

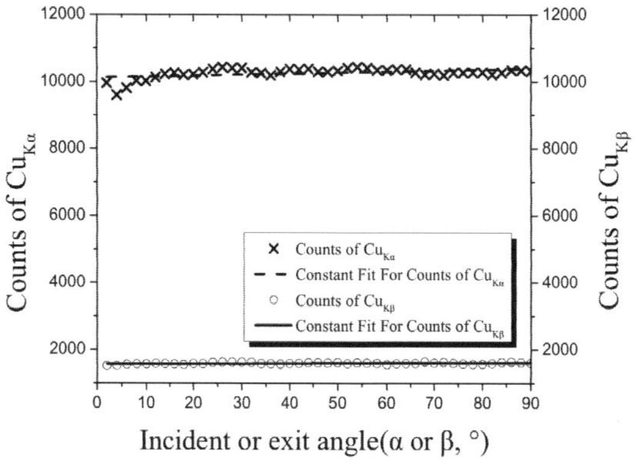

Fig. 2 Relationship between Cu's characteristic X-ray peak counts and the angle from the Monte Carlo calculations ($H_{S\text{-}S}$=$H_{D\text{-}S}$=20 mm, α=β)

In Fig. 2, changing α and β has no effect on the measurement results. This conclusion can guide XRF analyzer design.

4.1.2 β=45° while changing α from 0° to 180° (α≠β)

Under these conditions, the relationship between the angle and characteristic X-ray peak counts for CuK_{α} and CuK_{β} is shown in Fig. 3.

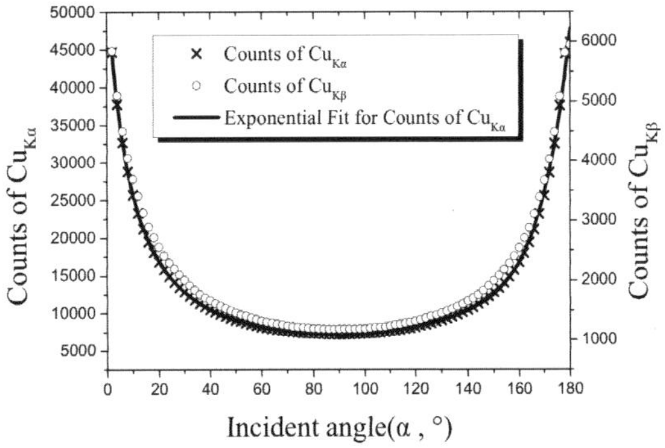

Fig. 3 Relationship between the Cu's characteristic X-ray peak counts and angle from the Monte Carlo calculations ($H_{S\text{-}S}$=$H_{D\text{-}S}$=20 mm, β=45°, $\alpha \neq \beta$)

In Fig. 3, when α and β are orthogonal, the characteristic X-ray peak counts for CuK_{α} and CuK_{β} are maximized; the counts become minimum when they are parallel.

This result occurs because the excitation source and detector are close and even overlap when α and β are orthogonal. In this case, α and β are at nearly a right-angle, i.e., $\sin\alpha \approx \sin\beta \approx 1$. The probability for the characteristic X-rays to enter the detection window is maximized. Interference due to incident X-rays from the source is minimized. The probability characteristic X-rays are excited by the incident X-rays is maximized. Therefore, the X-ray characteristic peak counts for CuK_{α} and CuK_{β} are maximal. Similarly, the incident X-ray counts recorded by the detector are maximized when α and β are parallel. The specimen's fluorescence probability decreases in the detector's solid angle. The probability the characteristic X-rays are excited by the source decreases. The background counts from scattering increases. In this case, multiple scatterings from the incident and characteristic X-rays further reduce the characteristic X-ray peak counts. Therefore, the characteristic X-ray peak counts for CuK_{α} and CuK_{β} are minimized.

These results confirm the reasons center or cyclic excitation sources are used for XRF analyzers in practice.

4.2 Impact of distance

4.2.1 Distance between the specimen and excitation source ($H_{S\text{-}S}$)

In Fig. 4, $H_{D\text{-}S}$ remains constant, $H_{S\text{-}S}$ is below 20 mm, and the characteristic X-ray peak counts for CuK_{α} and CuK_{β} remain maximal.

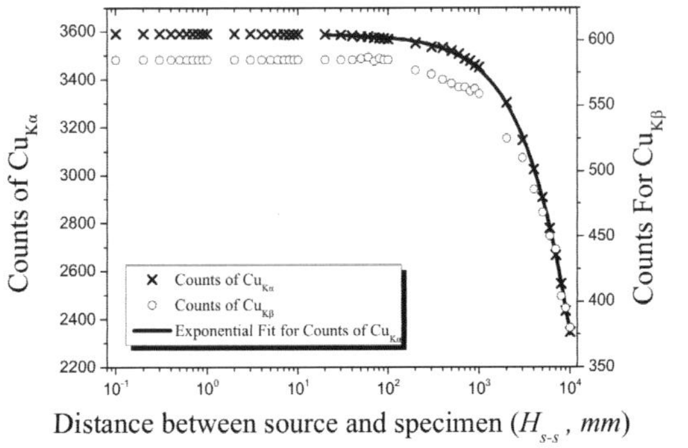

Fig. 4 Relationship between Cu's characteristic X-ray peak counts and $H_{S\text{-}S}$ in the Monte Carlo calculations ($H_{D\text{-}S}$=20 mm, α=β=45°)

Increasing $H_{S\text{-}S}$ decreases the incident X-ray irradiation at the specimen surface. The probability the characteristic X-rays are excited by the source also decreases, which reduces Cu's characteristic X-ray peak counts. In Fig. 4, increasing $H_{S\text{-}S}$ further (≥20 mm) exponentially decreases the counts.

4.2.2 Distance between the detector and specimen ($H_{D\text{-}S}$)

Holding $H_{S\text{-}S}$ constant while increasing $H_{D\text{-}S}$ (＜10 mm) exponentially decreased the characteristic X-ray peak counts for CuK_{α} and CuK_{β}. Increasing $H_{D\text{-}S}$ further (≥10 mm) reduces the characteristic X-ray peak counts to essentially nothing as shown in Fig. 5.

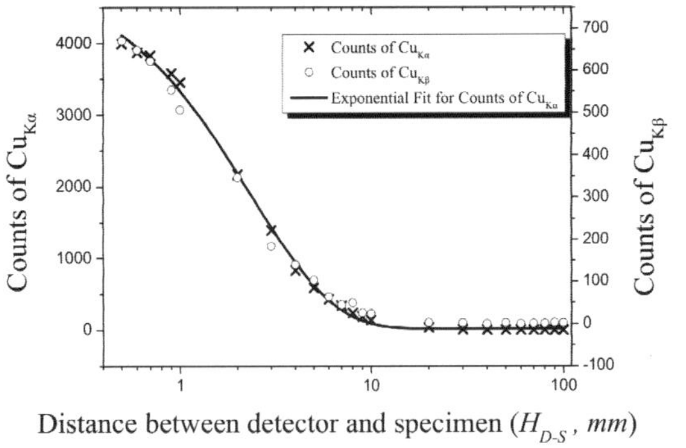

Fig. 5 Relationship between the Cu's X-ray characteristic peak counts and $H_{D\text{-}S}$ in Monte Carlo calculations ($H_{S\text{-}S}$=20 mm, α=β=45°)

Increasing *the* $H_{D\text{-}S}$ decreases the detector's solid angle and Cu's characteristic X-ray peak counts. When $H_{D\text{-}S}$ is above 10 mm, the characteristic X-rays are almost outside the solid angle entirely. Therefore, the counts are almost zero.

Based on Fig. 4 and Fig. 5, the detector and specimen should be as close as possible in the XRF analyzer design to increase the fluorescence counts.

4.2.3 Distance between the excitation source and detector (H_{S-D})

The specimen center is the origin for the coordinate system (in Fig. 1). In this case, the characteristic X-ray peak counts for CuK_α are almost zero when H_{S-D} is below 5 mm. Increasing H_{S-D} also increases the counts. When H_{S-D}=40 mm, the counts are maximal. Increasing H_{S-D} further (>40 mm) exponentially decreases the counts.

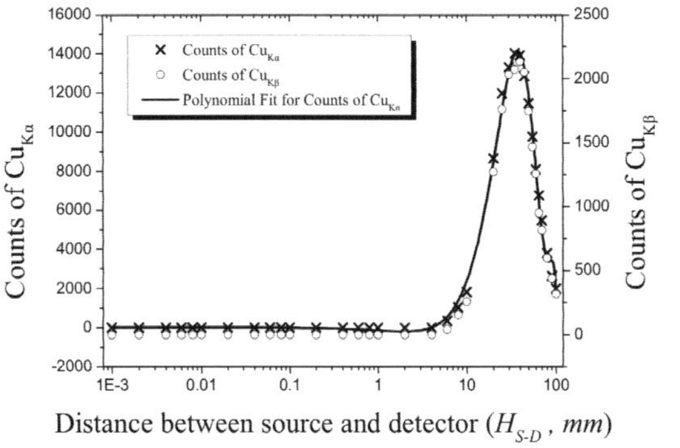

Fig. 6 Relationship between Cu's X-ray characteristic peak counts and H_{S-D} from the Monte Carlo calculations (H_{S-S}=H_{D-S}=20 mm, α=β=45°)

Extremely small or large H_{S-D} values cause the characteristic X-rays to mostly fall outside the detector's solid angle. Therefore, there was essentially no characteristic X-ray peak. There must be a preferred distance for the geometric conditions. In Fig. 6, H_{S-D}=40 mm was the best distance for this paper's model. However, H_{S-S}=H_{D-S}=20 mm when H_{S-D}=40 mm, the relationship between α and β is exactly orthogonal. This result is congruent with the conclusions in Section 3.1.2.

5 Conclusions

This paper studies the geometric effects for an XRF analyzer designed via the Monte Carlo method. The theoretical formula for the X-ray fluorescence intensity was derived. XRF analysis models were established using the MCNP5 Code. The results were as follows: when α and β are simultaneously changed and α=β, there is no effect on the measured results. There is non-linear relationship between α and the Cu's X-ray characteristic peak counts. The counts are maximized when α and β are orthogonal and minimized when α and β are parallel. When H_{D-S}=20 mm, α=β=45°, the X-ray characteristic fluorescence counts remain constant (H_{S-S} < 20 mm). Increasing H_{S-S} (≥20 mm) exponentially decreases the counts. When H_{S-S}=20 mm, α=β=45°, and H_{D-S} increases (< 10 mm), the characteristic X-ray peak counts for CuK_α and CuK_β decrease exponentially. Increasing H_{D-S} further (≥10 mm) reduces the characteristic X-ray peak counts to nearly zero. This method and these conclusions can provide technical guidance for designing X-ray fluorescence analyzers.

Acknowledgements

The presented results receive funding from the National High Technology Research and Development Program of China under contract number 2012AA061803. It is also supported by the Special Funds of the National Natural Science Foundation of China (Grant No. 41474159). This work is partly supported by the Opening Fund of Provincial Key Lab of Applied Nuclear Techniques in Geosciences (gnzds2014002, gnzds2014006).

References

1. Alireza Mazoochi, et al., NUCL INSTRUM METH A. **763** (2014) 538.
2. D. Hampai, et al., Spectrochimica Acta Part B: Atomic Spectroscopy. 101 (2014) 114.
3. P. Oberta, R. Mokso, NUCL INSTRUM METH A. **729** (2013) 85.
4. Graeme M. Hansford, NUCL INSTRUM METH A. **690** (2012) 117.
5. LU Anxiang, et al., SPECTROSC SPECT ANAL **30** (2010) 2848.
6. E G San Miguel, et al., APPL RADIAT ISOTOPES **61** (2004) 361.
7. Tektronix Corporation 2007 TEKVISA Programmer Manual Version1.1071-1101-00.
8. Will J.M., Duncan A.R 2008 Understanding XRF Spectrometry PT-**1** 7602.
9. Ge Liangquan, Zhang Ye, NUCLEAR TECHNIQUES **18** (1995) 332.
10. Sharshar T., et al., APPL RADIAT ISOTOPES **48** (1997) 695.
11. László Vincze, et al., Spectrochim Acta. **50B** (1993) 127.

Experiment Research on Shading to Improve the Thermal Environment of Tents

Jingbo Liu[1,a] , Yingni Jiang[2], Zhengguo Gong[3] and Hao Yang[4]

[1,2,3,4]Architecture Engineering Department, Logistics University of PAP. Tianjin, China

Abstract. Different black agricultural shading-nets were erected on two 93 type military cotton tents to improve the thermal environment of tents in this paper. Indoor temperature, RH (relative air humidity) and interior wall temperature were tested. The experimental results show that shading-nets a effect on the improvement of thermal environment and 6 needled filters shading-net is better than 3 needled filters shading-net. The conclusion of this paper provides a basis for the improvement of the thermal environment and energy saving design of the tent.

1 Introduction

The tent is army camping equipment support, but there is a big flaw: thermal resistance and thermal inertia are small and the internal environment is hot, humid and poorly ventilated, seriously affect the rest of personnel[1]. There are a lot of researches on the improvement of tent thermal environment, but few scholars have applied the sun-shading technology to the tent[2-5]. In this work, the authers erected black shading-nets on tents, tested and analyzed the characteristics of the thermal parameters. The research on shading provides data reference and theoretical basis for upgrading the tents.

2 Test object

The experiment object is 93 type military cotton tent. Size: Long 4.4m, wide 4.6m, top high 2.57m, edge high 1.4m[6]. All the doors and windows were opened. two types of shading-nets were selected and erected on the top of the tent. One is "3 needled filters shading-net" (Net A), which has up to 75% shading rate and the tent under Net A were named Tent A. The other is "6 needled filters shading-net" (Net B), which has up to 96% shading rate and the tent under Net A were named Tent B. The two shading-nets were erected about 1m higher than the top of tents.

3 Experiment scheme

The experiment was done in Tianjin On September 8, 2015 at 8:30 ~ 16:00. That day the weather was from sunny to cloudy and the air temperature was from 20 degrees to 29 degrees. Tianjin is located at north part of northern China plain, has a sub-humid warm temperate continental monsoon four distinct seasons. Winter of Tianjin is cold and dry, while summer is hot and wet.

Mercurial thermometers, psychrometer and portable infrared thermometer were used to test indoor temperature, relative humidity and interior wall temperature.Twenty-five mercurial thermometers were put on 5 layers to test indoor air temperature and there were 5 thermometers in each layers representative position. The height of layers was 0.1m, 0.5m, 1.0m, 1.5m, 2.0m, respectively. A psychrometer was put at tent center and another was put outside where there

Corresponding author: Jingbo Liu, 568818102@qq.com

were no sunshine to test the different RHs. An infrared thermometer were used to test 6 interior wall temperature, these are north wall, east wall, west wall, top surface and ground surface. All the data were recorded every half an hour in the test.

4 Experimental data analysis

4.1 Indoor air temperature

Indoor temperature of Tent A is shown in Figure 1, indoor temperature of Tent B is shown in Figure 2.

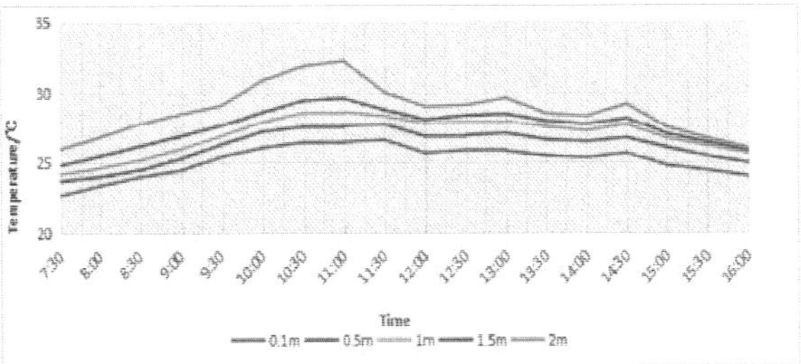

Figure 1. Indoor temperature of Tent A.

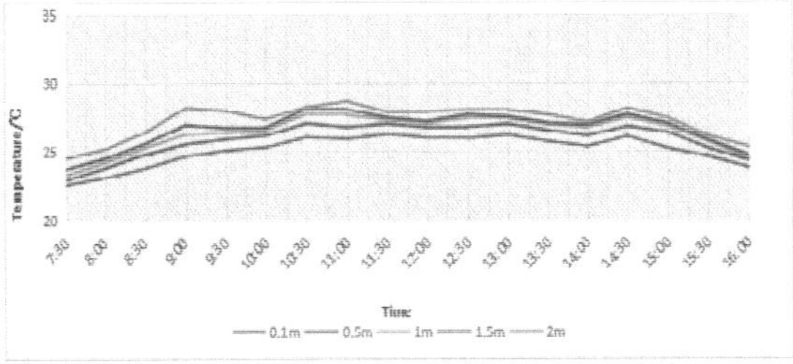

Figure 2. Indoor temperature of Tent B.

4.1.1 Analysis

It can be seen from Figure 1 that the indoor temperature rose at first, then dropped, and the highest temperature appears at 11:00. The reason is that the weather was clear before 11:00, the sunlight radiation transfer energy to Net A and tent, and the temperature was rising. After 11:00, it changed from sunny to cloudy, which caused the temperature to drop. While, there was little change in indoor temperature in Figure 2, the temperature was rising at 7:30~9:00 while it was dropped at 14:30~16:00.

The 2m layer was near to the top, its temperature rose faster than other layers. The highest temperature of Tent A was 32.2degrees, appeared at 11:00, 2.6 degrees higher than 1.5m layer and 3.0 degrees higher than outside. While, the 2.0m layer temperature of Tent B was lower than Tent A. The reason is that the gaps of Net A is larger than Net B. and the sunlight reach the top of Tent A through Net A. More and more heat accumulated under the top, which cause the temperature to rise. The results shows that the shading effect of Net B is better than Net A, and there is little sunlight reach Tent B.

The temperature difference between the 2m layer and 0.1 layer in Tent A is larger than Tent B. The maximum temperature difference was about 5.8degrees at 11:00. And the change of temperature was great in the entire test period. The influence of temperature cannot be ignored in reasonable breaks in Tent A. The temperature difference between adjacent layers was about 0.5degrees. What calls for special attention is that the temperature between 1.0m and 1.5m was closer to each other, especially at 11:30~14:30. The main reason is that there are 4 windows in tent at a same height, where there was the air convection between indoor and outside.

4.1.2 Discussion

The results indicated that the indoor temperature changes greatly between day and night. Because the small thermal resistance tarp, the indoor thermal environment is easy to be affected by different conditions: the top layer temperature was affected by the sunlight, the bottom layer was affected by the ground, the middle layer was affected air convection from windows. People in tent cannot have a good rest and sleep well.

Compared with Net A, Net B resists more sunlight and has a better shading effect, which result in the lower level of solar radiation and a more comfortable thermal environment. It is suggest that the improvement of thermal environment should be reasonable. When shading tent with nets, Net B can be selected as the best option.

The air in tent cannot have a strong convection with outside by the 4 same windows, which result in a stuffy environment. The present result suggests a means to improve the thermal environment that is to adjust a suitable size and height, especially in summer.

4.2 Air relative humidity

Air relative humidity is shown in figure 3.

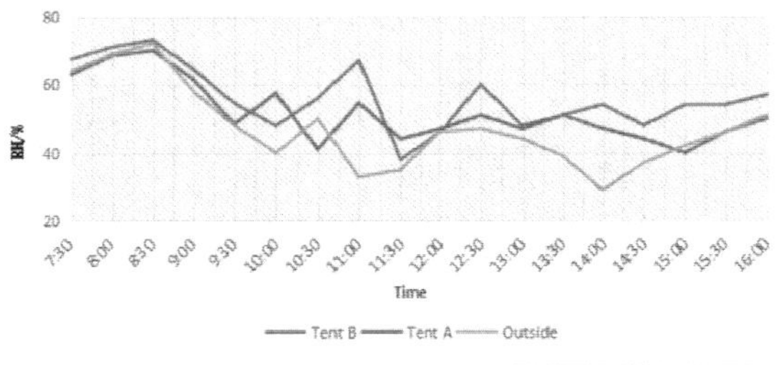

Figure 3. Air relative humidity.

4.2.1 Analysis

The result shows that the outside air RH was fit to the single peak curve, namely low at noon while high at morning and night. Show the opposite trend with the outside temperature. The main reason is the higher the drying temperature is, the lower RH value is at the same water holding capacity.

The RH of Tent A changed a lot in Figure 3. It was was high at 7:30~8:30, the reason is that the dew is easily to evaporate back into the atmosphere in the morning. The RH was nearly stable at 8:30~11:30, because of weaken air convection, the water holding capacity was stable. And the RH increase with the temperature dropping after 11:00.

The RH of Tent B was moderated so that it is always less than the variation of Tent A, and often much less. And because of weaken air convection, the indoor RH was higher than outside. The highest RH was 73%,the lowest RH was 38% and the mean RH was 56%.

There are some similarities of the curves above: low at noon while high at morning and night. Because of lower shading rate, Tent A is easier to receive solar radiation. Compared with Net B, the RH of Net A is especially sensitive to temperature changes with the relatively constant of water vapor content.

4.2.2 Discussion

The results show a high RH in tent. The main factor was the ground, where the dew on the ground and moisture in the soil evaporated with the temperature rising. To reduce the RH in tent, trips is necessary to be lay on the ground.

The air convention is weaken. Adjust the height and size of windows can reduce the RH. Whereas the higher the ventilation rate, the lower the indoor relative humidity and the higher the warm respiratory comfort.

The indoor RH can be influenced by shading rate of nets. Net B can easily shade more sunlight, and Net B is a better choice of tent shading.

4.3 Interior wall temperature

The interior wall temperatures of the Tent A are shown in Figure 4. The interior wall temperatures of the Tent A are shown in Figure 5.

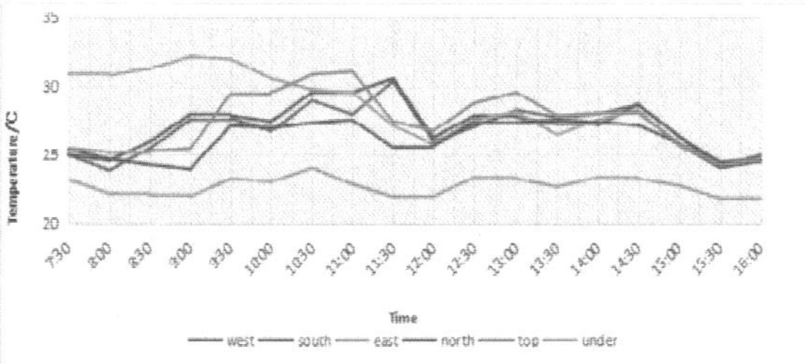

Figure 4. Interior wall temperatures of Tent A.

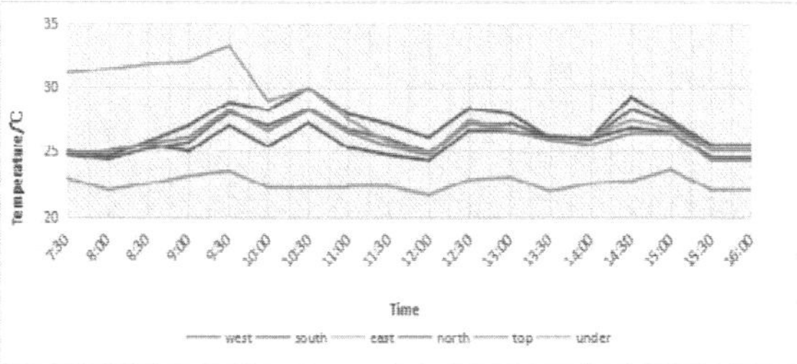

Figure 5. Interior wall temperatures of Tent B.

4.3.1 Analysis

The main reason for interior wall temperature rising was sunlight. You can see from Figure 3 and Figure 4 that the interior wall temperatures of Tent B is similar to Tent A. It's important to note, however, the curves of Tent B are more stable than Tent A.

The east wall temperature was obviously higher than other wall at 7:30~10:00 due to sunlight. And because of cloudy weather at afternoon, the west wall temperature was not as high as the east wall. The south wall temperature was higher than the north wall. The main reason for that is sunlight can reach the south wall and can not north. The ground temperature was stably 23 degrees. Because of larger gap of Net A, the roof temperature of Tent A is higher than Tent B at 9:30~11:00. After 14:30, the weather was cloudy and only a little sunlight shone on nets and tents, which make walls the near temperature.

4.3.1 Discussion

The walls of tent are greatly affected by the radiation of the sun. It shows that as the change of the solar azimuth, the same wall absorbed different solar radiation, the temperature changes. And as different walls absorbed different solar radiation at a same time, temperatures are different. The main reason is that the tent envelope thermal inertia is lower than fixed structures and can be easily affected by solar radiation.

The shading-nets can screen out much of the solar radiation and then the tent wall temperature dropped. Especially, Net B is rather better than Net A in shading.

The top of the tent can be shaded from the sun, but the east and west wall shading effect is limited. It is suggest that enhance the shading in east wall and the west wall and reduce the amount of radiation. Move the shading-net south in the northern area, which can reduce the amount of radiation in the south.

5 Conclusion

The tent thermal environment is greatly influenced by the outside, and the convection ability of indoor air and outside is weak. The erection of shading-net can effectively improve the indoor thermal environment. It is recommended that choosing the material with high shading rate as shading-net like Net B, and increasing the net area to the east, west and south, so that the radiation of sunlight could be reduced. Moreover, people can adjust the windows position and size to raise the capacity of air convection inside and outside the tent, and laying a cloth on the ground has the effect on damp-proof, reducing the relative humidity inside the tent.

References

1. S.Hu, Q.Meng, C.Wang. Experimental on the Improvement of the Thermal Environment of Tent. Journal of Logistical Engineering University: **02**, 81-83+87(2007).
2. T.Wang, E.Long, Q.Yuan, H.Chen. Research on Thermal Insulation Effect of Double-roof Tent. Building Science: **12**, 59-63(2010).
3. J.Ya, Y.Jin, S.Chen, P.Zeng, H,Wang. Improved Design of Disaster Relief Tent. Science & Technology: **17**, 13+80(2014).
4. R.Xu, G.Du. Development of Field Inflatable Medical Tent. Chinese Medical Equipment Journa: **14**, 85-87(2013).
5. W.Wang, L.LIu, G.Liu. Research on Army Green Heat Reflective Coatings for Tents. Paint & Coatings Industry: **9**, 22-24(2008).
6. C.Su. Basic situation. Tianjin County Yearbook. (Tianjin Academy of Social Sciences Press, 35-36(2014))

Thermodynamic Analysis of Stirling Heat Pump based on Thermoeconomic Optimization Criteria

Liya Xu[a]

College of Information and Electronic Engineering, Liming Vocational University, Quanzhou, 362000, People's Republic of China

Abstract: A connectionist investigation of irreversible Stirling heat pump cycles that includes both internal and external irreversibilities together finite heat capacities of external reservoirs was carried out. The thermoeconomic optimization for Stirling heat pump is reported. The heating load per unit total cost for the heat pump is proposed as objective functions for the optimization. The optimum performance parameters which maximize the objective functions are investigated. Since the optimization technique consists of both investment and energy consumption costs, the obtained results are more general and realistic.

1 Introduction

Throughout thermodynamics cycle approaches of refrigerators and heat pump systems, Stirling cycle is one of the important cycle models with regeneration. According to the theory of classical thermodynamics, the Stirling cycle may possess the condition of perfect regeneration through the use of a reversible regenerator and its theoretical efficiency can approach that of a reversible Carnot cycle. Recently huge amounts of attentions have been made by various Scientists on the addressed cycles. Curzon and Ahlborn were introduced the concept of finite time thermodynamics[1] by performing a new work on Carnot heat engine where as they high lighted the effectiveness of addressed engine that operates at maximum power is calculated with their introduced [$\eta_m = 1 - \sqrt{T_L / T_H}$]. It should be noted that the output of their proposed formula is always lower than the routine formula of Carnot [$\eta_C = 1 - T_L / T_H$] while closed more appropriate with the actual efficiencies of operating systems given by them. Wu has been effectively implemented the theorem of finite time thermodynamics to indicate the optimal performance of the various cycles such as Carnot, Stirling heat pump and Brayton systems for various situations[2,3]. Nowadays, optimum performance parameters of the Stirling refrigerators and Stirling heat pumps have been investigated systematically by performing finite time thermodynamics as a function of operating conditions [4–9]. In this paper, we have presented a general thermoeconomic optimization analysis of irreversible Stirling heat pump with imperfect regenerator and with finite heat capacity of external reservoirs and maximize the thermoeconomic function [10-12].

2 Stirling heat pump model

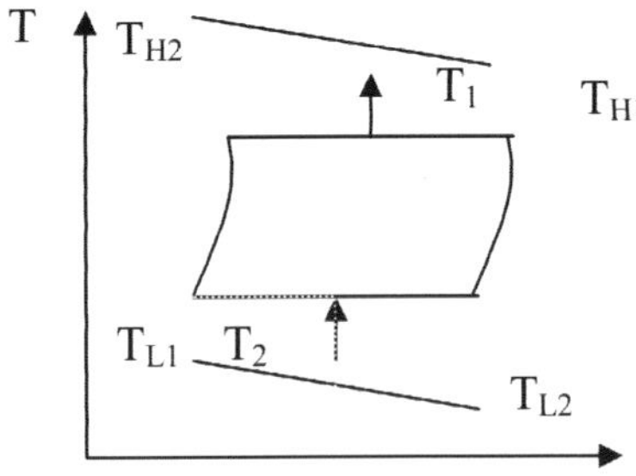

Figure 1. Schematic diagram of the Stirling heat pump

As depicted in Fig. 1, the cycle comprises two isochoric and two isothermal processes when the addressed working fluids of the cycle are the ideal gas. The mentioned cycle proximate the expansion stroke of an actual cycle as an isothermal process 1–2 while an irreversible isothermal heat added from a heat source of

[a] Corresponding author: *13860716115@163.com*

finite heat capacity at temperature T_c that the temperature changes in range of T_{L1} to T_{L2}. Addition of the heat from the regenerator to the addressed working fluid was considered as an isochoric process 2–3. The compression stroke was considered as isothermal processes 3–4 with an irreversible heat rejection to the heat sink of finite heat capacity at temperature T_h that temperature changes in range of T_{H1} to T_{H2}: Lastly, rejection of the heat from the working fluid that comes to the regenerator was considered as an isochoric process 4–1. As expressed previously, the heat transfer processes 1–2 and 3–4 in an actual cycle must happen throughout finite time. This needs the addressed heat processes would be preferred in the finite temperature difference and consequently, specified as stand hardly irreversible. In addition, the change of entropy through process 3–4 is greater than the change of entropy through process 1–2. So, for each cycle, some heat losses entire the regenerator, as a perfect regeneration needs unlimited regeneration area or time. Therefore, it is necessary to study an actual regenerator. Involving all the above mentioned aspects, the addressed heat pumps approaches turn irreversible. As mentioned previously, the external irreversibility is caused by difference of limited temperature between the external reservoirs and the cycle and on the other hand, regenerative loss.

By defining Q_1 and Q_2 that represent the value of heat gained at temperature T_1 from the source and the heat deliver at temperature T_2 to the sink, respectively, then [5,6]:

$$\begin{aligned} Q_1 &= U_H A_H \Delta T_H t_1 \\ &= C_H (T_{H1} - T_{H2}) t_1 = nRT_1 \ln(V_2 / V_1) \end{aligned} \tag{1}$$

$$\begin{aligned} Q_2 &= U_L A_L \Delta T_L t_2 \\ &= C_L (T_{L2} - T_{L1}) t_2 = nRT_2 \ln(V_2 / V_1) \end{aligned} \tag{2}$$

Where $U_H A_H$ and $U_L A_L$ are the heat transfer coefficient and area products, C_H and C_L are the heat capacitance rates of the source and sink reservoirs, respectively, and are defined as

$$\Delta T_H = \frac{(T_{H1} - T_1) - (T_{H2} - T_1)}{\ln[(T_{H1} - T_1)/(T_{H2} - T_1)]} \tag{3}$$

$$\Delta T_L = \frac{(T_2 - T_{L1}) - (T_2 - T_{L2})}{\ln[(T_2 - T_{L1})/(T_2 - T_{L2})]} \tag{4}$$

It is also consider the finite heat transfer through the regenerator. Let ΔQ_R is the regenerative heat loss per cycle during the two regeneration processes, which is proportional to the temperature difference of the working fluid as given by

$$\Delta Q_R = nC_p (1 - \eta_R)(T_1 - T_2) \tag{5}$$

$$\tau = \gamma(t_1 + t_2) \tag{6}$$

Where n is the number of moles, C_p the molar specific heat of the working fluid and η_R the effectiveness of the regenerator. The time of regenerative processes were considered and γ is ratio of the time per cycle to the temperature difference of the working fluid. From Eqs. (1) - (2) we can obtained

$$Q_1 = C_H \varepsilon_H (T_{H1} - T_H) t_1 = C_H (T_{H1} - T_{H2}) t_1 \tag{7}$$

$$Q_2 = C_L \varepsilon_L (T_L - T_{L1}) t_2 = C_L (T_{L2} - T_{L1}) t_2 \tag{8}$$

Where ε_H and ε_L are the effectiveness of the hot and cold side of heat exchangers, respectively and are given as $\varepsilon_H = 1 - e^{-\frac{U_H A_H}{C_H}}$ and $\varepsilon_L = 1 - e^{-\frac{U_L A_L}{C_L}}$.

From Eqs. (1) - (8), we can obtain the power input, heating load and the coefficient of performance are, respectively.

$$R_H = \frac{T_1}{\dfrac{\gamma T_H}{C_H \varepsilon_H (T_{H1} - T_H)} + \dfrac{\gamma T_2}{C_L \varepsilon_L (T_L - T_{L1})}} \tag{9}$$

$$COP_H = \frac{T_1}{T_1 + a(T_1 - T_2)} \tag{10}$$

Where $a = C_p (1 - \eta_R) / R \ln(V_2 / V_1)$. The fundamental optimum relations of the cycle are obtained as

$$R_H = \frac{KT_H COP_H [1 - (a + COP_H) COP_C^{-1}]}{\gamma (COP_H + a)(COP_H + a - 1)} \tag{11}$$

Using the above equations, one can obtain the expressions of the important performance parameters of the Stirling heat pump.

3 Thermoeconomic optimization

The objective function to be optimized for the heat pump is defined as the heating load per unit total cost, i.e., [9-11]

$$F = R_H / (C_i + C_e) \tag{12}$$

In which C_i and C_e indicate annual investment and energy consumption costs, respectively. For the investment cost, we may consider the investment costs of the main system components, which are the heat exchangers and the compressor together with its prime mover. In general, the investment cost of the heat exchangers is assumed to be proportional to the total heat-transfer area. On the other hand, the investment cost of the compressor and its driver is assumed to be proportional to its compression capacity. Thus, the annual investment cost of the system can be given as

$$C_i = a_1(A_H + A_L) \quad (13)$$

Where the proportionality coefficient for the investment cost of the heat exchangers, a, is equal to the capital recovery factor times investment cost per unit heat transfer area and its dimension is ncu /(year m2). The proportionality coefficient for the investment cost of the compressor and its driver, b, is equal to the capital recovery factor times investment cost per unit power and its dimensionis ncu /(year kW).The unit ncu stands for the national currency unit. The initial investment cost is converted to equivalent yearly payments using the capital recovery factor [9-11]. The annual energy consumption cost is proportional to the heat rate input, i.e.

$$C_e = bP \quad (14)$$

Where the coefficient, b ,is equal to the annual operation hours times price per unit energy and its dimension is ncu/(yeark W).Substituting Eqs.(1) to (5) and (9),(10) into Eq.(12), we get

$$F = R_H / [a_1(A_H + A_L) + bP] \quad (15)$$

Using Eqs. (1)-(11) and （15） We obtain

$$F=\{\frac{T_1}{\frac{\gamma T_H}{C_H\varepsilon_H(T_{H1}-T_H)}+\frac{\gamma T_2}{C_L\varepsilon_L(T_L-T_{L1})}}\}/\{a_1(A_H+A_L) + b[\frac{T_1-T_2}{\frac{\gamma T_H}{C_H\varepsilon_H(T_{H1}-T_H)}+\frac{\gamma T_L}{C_L\varepsilon_L(T_L-T_{L1})}}]\} \quad (16)$$

The objective function given in Eq.(16) can be plotted F/K [$K = C_H\varepsilon_H C_L\varepsilon_L / \gamma(\sqrt{C_H\varepsilon_H} + \sqrt{C_L\varepsilon_L})^2$] with respect to the coefficient of performance of the pump given in Eq.(10), the economical parameter, as shown in Fig.2. As can be seen from the figure, there exists a COP_F which maximizes the objective function.

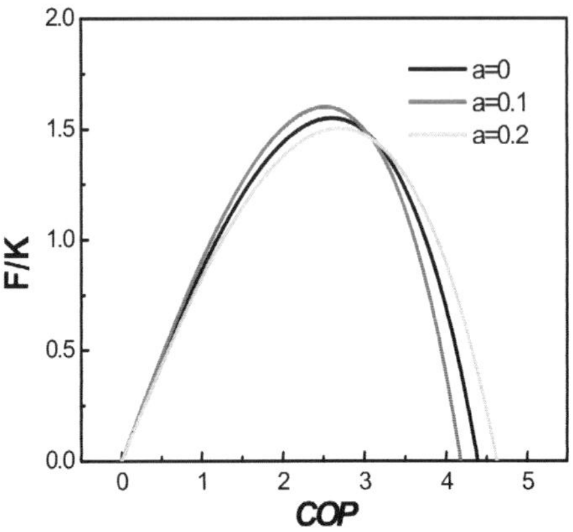

Figure 2 The alteration of the objective function for the heat pump with respect to COP (T_{H2} = 450K, T_H =420K, T_L =250K , T_{L2} =200K , a_1=0.5,b=1, $V_2/V_1 = 1.2$, $A_H = A_L$ =1m^2).

4 Conclusions

In this study, Finite time thermodynamics approach has been carried out to indicate input power, the heating load and the coefficient of performance of the Stirling heat pump. The thermoeconomic optimization for Stirling heat pump is reported. The heating load per unit total cost for the heat pump is proposed as objective functions for the optimization. The optimum performance parameters which maximize the objective functions are investigated.

Acknowledgements

This work was supported by the project of Science and Technology Research of Quanzhou, Fujian Provice of China (No.2009G10).

References

1. Curzon F L, Ahlborn B. Amer J Phys, Vol. 43(1975), p.19.
2. Wu C. Int J Ambient Energy, Vol. 14(1993), p.25.
3. Wu C.Energy Convers Manage, Vol. 36(1995), p.7.
4. [4] Chen L, Wu C, Sun F. Energy Convers Manage, Vol. 39(1998), p.113.
5. Chen L, Wu C, Sun F. Energy Convers Manage, Vol. 39(1998), p.445.
6. Chen L, Wu C, Sun F. Energy Convers Manage , Vol. 39(1998), p.117.
7. Chen L, Xiaoqin Z, Sun F, Wu C. Appl Energy, Vol. 84(2007), p.78.

8. Bi Y, Chen L, Sun F. Appl Energy, Vol. 85(2008), p.607.
9. Chen J. Energy Convers Manage, Vol. 39(1998),p.1255.
10. Kodal A,Sahin B,Yilmaz T. Energ Convers Manage, Vol. 41(2000), p.607.
11. Kodal A,Sahin B,Oktem A S. Energ Convers Manage, Vol. 41(2000),p.1989.
12. Sahin B,Kodal A,Koyun A. Energ Convers Manage, Vol. 42(2001),p.451.

MATEC Web of Conferences **61**, 01011 (2016) DOI: 10.1051/matecconf/20166101011
APOP2016

Energy Transports in Toda Lattices with Quasiperiodic On-site Potentials

Zhenjun Zhang[1,a], Jing Kang[1] and Chunmei Tang[1]

[1]*College of Science, Hohai University, 210098 Nanjing, Jiangsu, People's Republic of China*

Abstract. We study numerically the process of energy transports in Toda lattices with quasiperiodic on-site potentials. The total energy was initially equidistributed the 10% of lowest frequency linear modes. For Toda model without on-site potentials, only several new low frequency modes are excited, but the energy equipartition is not achieved. When the quasiperiodic on-site potentials are added, the energy transfers continuously to the high frequency modes and finally evolves towards energy equipartition. We further study the equipartition time T_{eq} as a function of energy density ε for different nonlinear parameters α and different strengths of on-site potentials δ. In the thermodynamic limit, the dependence of T_{eq} on ε is found to display a power law behaviour, that is, $T_{eq} \propto \varepsilon^{a}$. The exponenet a is found to be -2.03 and is independent of the values of α and δ.

1 Introduction

The study of energy transfers in nonlinear lattices has a long history but is still far from been completed. The first numerical calculation was performed by Fermi, Pasta, and Ulam (FPU) with the initial energy in a lowest frequency mode [1]. It was expected the nonlinearity could cause the energy transfers to the high frequency modes, resulting in energy equipartition. However, energy equipartition was not observed. On the contrary, the energy remained concentrated only on some modes. Since then, extensive studies on the energy transport behaviors in nonlinear lattices have been carried out [2-26]. For the FPU model, there is a general agreement that two well separated time scales are present for weak nonlinearity [15, 20-22, 24]. That is, in a relative short time scale, a kind of metastable state is appeared. In this state, in addition to the initial excited states, only several new low frequency modes are excited, but still far from energy equipartition. This is the state observed in the original FPU paper [1]. Such a state, however, is only stationary. On a much longer time scale, the high frequency modes take part in the energy sharing. Finally, the energy equipartition is achieved. The dependence of the equipartition time T_{eq} on the energy density ε in the thermodynamic limit was recently found to grow as $\varepsilon^{-9/4}$ when the initially frequency modes arbitrarily close to zero are excited [21]. Another extensively studied model is the classical lattice φ^4 model. The equipartition time in this model is found to be an exponential law with 1/ε [5].

On the other hand, the energy transport behaviors of Toda model have been found very similar to the FPU model in the short time scale [21, 22, 24]. But the difference is also obvious. Energy equipartition was not achieved in Toda model, which indicates that the Toda model is integrable and nonergodic [22].

[a] Corresponding author: hi_zhangzhenjun@sina.com

In this paper, we are devoted to the energy transport problem in Toda model with the effect of quasiperiodic on-site potentials. The purpose of this paper is to clarify two questions. Firstly, whether can the energy equipartition being achieved with the effect of quasiperiodic on-site potentials. Secondly, if it dose, what is the scaling law for the equipartition time T_{eq} as a function of the energy density ε in the thermodynamic limit. This paper is organized as follows. In Sec. 2, we outline the model, the calculated physical quantities, and the numerical approach. Numerical results are presented in Sec. 3. We summarize this work in Sec. 4.

2 Model, physical quantities, and numerical method

The Hamiltonian of the Toda model with on-site potentials is

$$H=\sum_{n=1}^{N}\frac{p_n^2}{2}+\sum_{n=0}^{N}V_T\left(q_{n+1}-q_n\right)+\frac{1}{2}\sum_{n=1}^{N}V_n q_n^2, \quad (1)$$

where V_T has the form

$$V_T\left(q_{n+1}-q_n\right)=\frac{1}{4}\alpha^{-2}\cdot\left[e^{2\alpha\left(q_{n+1}-q_n\right)}-1-2\alpha\left(q_{n+1}-q_n\right)\right]. \quad (2)$$

p_n and q_n are momentum and the displacement with respect to the equilibrium position of *n*th particle. *N* is the particle number of the system. α is the nonlinear parameter. V_n represents the on-site potentials. For Toda model, the on-site potentials for all *n* are zero. For quasiperiodic systems, on-site potentials are arranged in quasiperiodic sequences. The quasiperiodic systems represent an intermediate situation between periodic and random systems. One of the simplest prototypes of quasiperiodic systems is the Fibonacci lattice [27, 28].It has been rigorously proved [29] that the Fibonacci lattice has a singular continuous spectrum and critical wave functions, which means that the eigenstates are neither localized nor extended. In this paper, V_n takes two values V_A and V_B that are arranged according to Fibonacci sequences. The Fibonacci sequences are defined by the binary substitution rule

$$A \to AB, B \to A, \quad (3)$$

starting with letter A. To generate a quasiperiodic lattice we arrange the on-site potentials such that V_A (V_B) stands for A (B) in the Fibonacci sequences, with V_A=1+δ, V_B=1-δ, where δ represents the strength of on-site potentials. In this paper, we consider the case of fixed boundaries, i.e., q_0=q_{N+1}=0. In the linear case (α=0), the model of particles can be put in the form of N independent normal modes and is therefore integrable and nonergodic.

For α=0, Eq. (1) with $q_n = A_n e^{i\omega t}$ is reduced to the linear eignvalues problem with

$$-A_{n+1} - A_{n-1} + (2+V_n)A_n = \lambda A_n. \quad (4)$$

The eigenstates A_n^k are the linear modes. The frequencies $\omega_k = \sqrt{\lambda_k}$. To each mode k one can associate a harmonic energy

$$E_k = \frac{1}{2}\left(P_k^2 + \omega_k^2 Q_k^2\right), \quad (5)$$

and a phase φ_k is defined by

$$P_k = \sqrt{2E_k}\cos(\varphi_k), \; Q_k = \sqrt{\frac{2E_k}{\omega_k^2}}\sin(\varphi_k) \quad (6)$$

Energy equipartition means that the average of E_k up to time T, namely

$$\bar{E}_k(T) = \frac{1}{T}\int_0^T E_k(P(t), Q(t))dt \quad (7)$$

for large T converges to the energy density $\varepsilon = E/N$, E denoting the total energy of the system.

On the basis of the $\bar{E}_k$'s, one should then introduce a paremeter providing a distance to equipartition. A frequency used parameter is the "effective number of degrees of freedom"

$$\xi(t) = \frac{e^{\eta(t)}}{N}, \quad (8)$$

where η is the "epectral entropy"

$$\eta(t) = -\sum_{k=1}^{N} F_k(t)\log F_k(t), F_k(t) = \frac{\bar{E}_k(t)}{\sum_{j=1}^{N}\bar{E}_j(t)}. \quad (9)$$

Equipartition denotes to $\xi = 1$.

In this paper, the energy was initially equidistributed the 10% of lowest frequency normal modes, with random initial phases φ_k, uniformly distributed in the interval [0, 2π]. We use a $SABA_2C$ symplectic algorithm [30] to integrate the equation of motion derived by the Hamiltonian (1). The typical integration time step is Δt = 0.1. The corresponding relative error in energy conservation, when all modes are excited and do contribute to the total energy, is around 10^{-5} . A further decrease of time step by one order of magnitude does not affect the result. In our numerical calculations, we generate a very long but finite Fibonacci chain with size M (M>>N) according to the substitution rule mentioned above. Then we select a chain with N consecutive on-site potentials, with the starting position chosen randomly. Because of the nonuniform nature of Fibonacci lattices, the $\xi(t)$ differs slightly form chain to chain. For this reason, we present the results of the quantities defined above after first averaging them over ten different selections. We use $\langle \cdot \rangle$ to denote the averaging results. Averages on more than ten different selections are also calculated. The variation on the result is negligible.

3 Numerical results

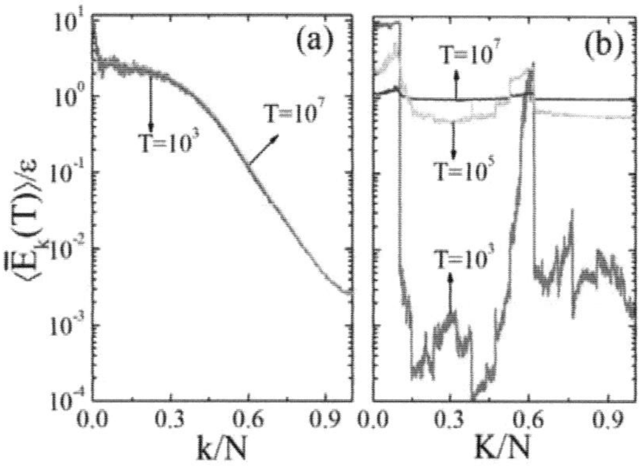

Figure 1. (a) The averaged energy spectrum at different times T for Toda model without on-site potentials for N=1024, α=1.0, ε=0.01. (b) The averaged energy spectrum at different times T for Toda model with quasiperiodic on-site potentials for N=1024, α=1.0, ε=0.01, and δ=0.5.

Figure 1(a) shows the results of $\langle \bar{E}_k(T) \rangle / \varepsilon$ versus k/N for Toda model without on-site potentials for N=1024, α=1.0, ε=0.01 at different times T. Quite soon, already at T=10^3, a well defined profile is formed, in which only some low frequency modes effectively take part in energy sharing. The energies of the remaining ones decay exponentially with k/N. Since then, the energy profile keeps its form nearly unchanged. Energy equipartition can not be achieved for Toda model without on-site potentials. It proves that the Toda model is integrable and nonergodic. Figure 1(b) shows the results of $\langle \bar{E}_k(T) \rangle / \varepsilon$ versus k/N for Toda model with quasiperiodic on-site potentials for N=1024, α=1.0, ε=0.01, and δ=0.5 at different times T. It can be seen that the energy of the initial excited modes decreases continuously with time. Meanwhile, the energy of other frequency modes grows continuously with time. Finally, the energy of all modes approaches to the global equipartition. It indicates that the integrable property is destroyed and the energy equipartition is therefore can be achieved with the effect of quasiperiodic on-site potentials.

DOI: 10.1051/matecconf/20166101011

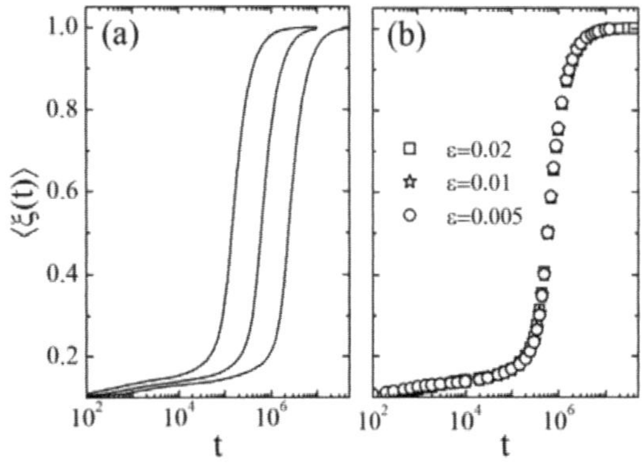

Figure 2. (a) The dependence of <ξ(t)> on time t for N=1024, δ=0.5, α=1.0 and different values of ε in semi-log scale. The curves from left to right at $<\xi(t)>=0.5$are for ε=0.02, 0.01, 0.005, respectively. (b) The results of $<\xi(t)>$ for different ε are horizontally translated, in such a way the coincide at $<\xi(t)>=0.5$ (the one at ε=0.01 is kept fixed).

In order to obtain the overall process of energy transport, we study the properties of <ξ(t)>. The results of <ξ(t)> at different ε with N=1024, α=1.0 and δ=0.5 in semi-log scale are shown in Fig. 2(a). It is seen that on a sufficiently large time scale, all values of <ξ(t)> grow from 0 to 1, with very similar sigmoidal profiles. It illustrates that the energy equipartition is sure to achieve on a sufficiently large time scale. When decreasing the value of ε, the time required to reach equipartition increases. Now, we turn to the definition of the equipartition time T_{eq}. We choose the time when <ξ(t)> reaches a value ξ_{eq} as the equipartition time T_{eq}. In this paper, we choose ξ_{eq}=0.5, as has been done in Ref. [21]. Obviously, the choice of ξ_{eq} is very artificial. We find that T_{eq} is quite insensitive to the choice of ξ_{eq}. This can be seen in Fig. 2(b) where the sigmoidal profiles from Fig. 2(a) are horizontally shifted to coincide at <ξ(t)>=0.5.

In the following, we study the equipartition time T_{eq} as a function of energy density ε for different nonlinear parameters α and different strengths of on-site potentials δ. Figure 3 shows the results of T_{eq} vs. ε in log-log scale, for different values of N, namely N=128, 256, 512, 1024, 2048, at fixed α=1.0, δ=0.5. In the explored range of ε, T_{eq} is found to be nearly independent of N when N is large. N=512 seems large enough that the thermodynamics limit is practically achieved. Moreover, T_{eq} versus ε in the thermodynamics limit displays a power law behavior, that is,

$$T_{eq} \propto \varepsilon^{a}. \tag{10}$$

The computed value of the exponent is a = -2.03, which is a little different from that of the FPU model. It illuminates that more time is required to approach equipartition when ε is decreased. It is known that the energy transfer from the initial excited modes to other modes is induced by the effect of nonlinearity. Less time is needed to approach equipartition when the strength of nonlinearity is increased. When increasing ε, both the energies in linear and nonlinear parts are raised. While the linear terms are quadratic, the nonlinear terms are greater than quadratic. Therefore, the proportion of the energy in nonlinear parts increases i.e., the strength of nonlinearity increases as ε is increased. Therefore, less time is needed to approach equipartition for larger ε and the exponent of ε is thereby less than zero. Besides, we also study the dependencies of T_{eq} on ε for different values of α. The results of T_{eq} as functions of ε in log-log scale, with N=1024, δ=0.5 and different values of α are shown in Fig. 4. It is seen that T_{eq} versus ε for different α all display power law behaviors. Moreover, the exponent of ε is nearly independent of the values of α. For fixed ε, it is found that more time is required to approach energy equipartition when α is decreased. The reason is that α is the nonlinear parameter. The strength of nonlinearity increases as the value of α is increased. Therefore, less time is needed to approach equipartition for larger α.

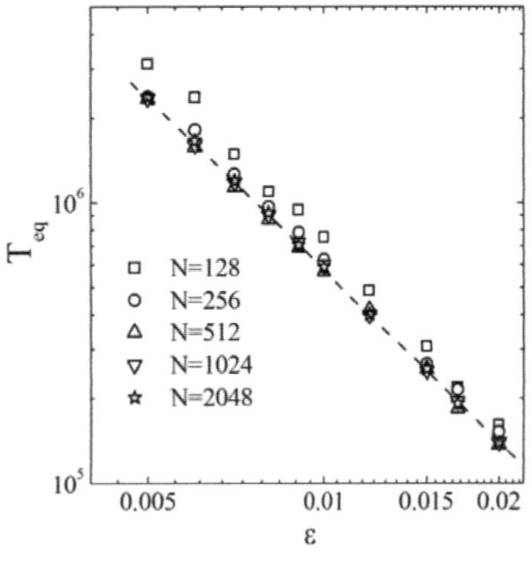

Figure 3. The equipartition time T_{eq} as a function of ε in log-log scale, with N=128, 256, 512, 1024, 2048 and fixed α=1.0, δ=0.5. The long dashed line has a slope of -2.03.

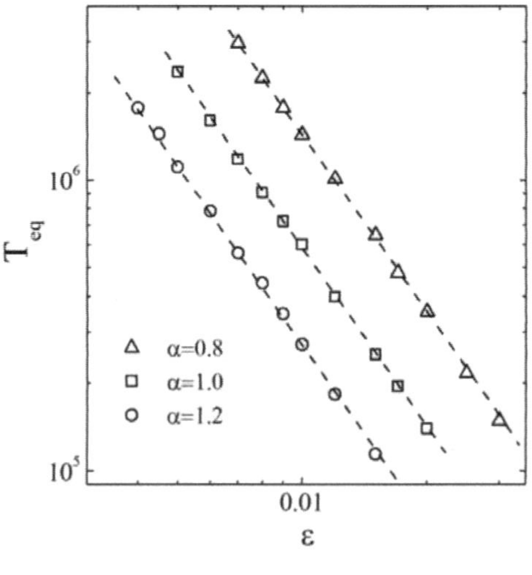

Figure 4. The equipartition time T_{eq} as functions of ε in log-log scale, with N=1024, δ=0.5 and different values of α. The slope of the long dashed lines is -2.03.

Finally, we study the dependencies of T_{eq} on ε for different strengths of on-site potentials in the thermodynamic limit. The results of the equipartition time T_{eq} as functions of ε in log-log scale, with N=1024, α=1.0 and different values of δ are shown in Fig. 5. It is seen that T_{eq} versus ε for different δ all display power law behaviors. Moreover, the exponent of ε is nearly independent of the values of δ. For fixed ε, less time is required to approach energy equipartition when the strength of on-site potentials is increased. It indicates that the strong on-site potentials can accelerate the process to energy equipartition.

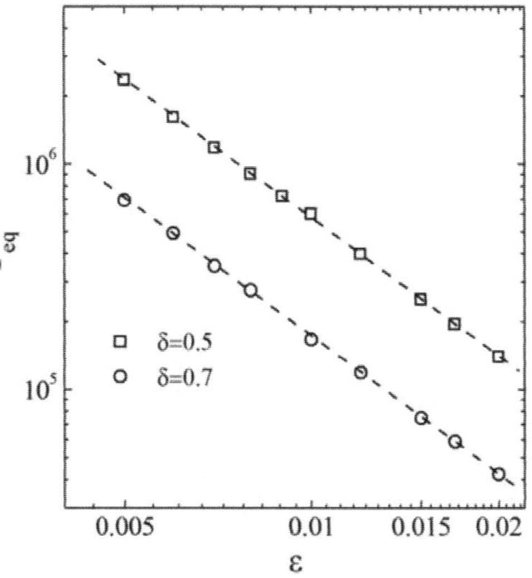

Figure 5. The equipartition time T_{eq} as functions of ε in log-log scale, with N=1024, α=1.0 and different values of δ. The slope of the long dashed lines is -2.03.

4 Conclusions

In this paper, we considered the energy transports and time scale to equipartition in Toda lattices with quasiperiodic on-site potentials. At first, the energy was equidistributed the 10% of lowest frequency modes. For Toda model without on-site potentials, only several new low frequency modes are excited. The energies of the remaining ones decay exponentially with k/N. Energy equipartition can not be achieved. It proves that the Toda model is integrable and nonergodic. When the quasiperiodic on-site potentials are added, the energy transfers continuously to the high frequency modes and finally evolves towards energy equipartition. It indicates that the integrable property is destroyed with the effect of quasiperiodic on-site potentials. For fixed α and δ, the dependence of T_{eq} on ε in the thermodynamics limit is $T_{eq} \propto \varepsilon^{a}$. The exponent a is found to be -2.03, which turns out to be independent of the values of α and δ and is a little different from that of the FPU model. For fixed ε, it is found that more time is required to approach energy equipartition when α is decreased. The reason is that α is the nonlinear parameter. The strength of nonlinearity increases as the value of α is increased. Therefore, less time is needed to approach equipartition for larger α. Besides, less time is required to approach energy equipartition when the strength of on-site potentials is increased, which indicates that the strong on-site potentials can accelerate the process to energy equipartition.

Acknowledgments

This work is supported by the National Science Foundation of China (Grant No. 11305045), the Special Foundation for theoretical physics Research Program of China (Grant No. 1134711), and the Fundamental Research Funds for the Central Universities of China (Grant Nos. 2013B00414, 2013B00314).

References

1. E. Fermi, J. Pasta, and S. Ulam, *in Collected Papers of E. Fermi,* edited by E. Segré (University of Chicago, Chicago, 1965), Vol. 2, p. 978.
2. N. J. Zabusky, and M. D. Kruskal, Phys. Rev. Lett. **15**, 240 (1965).
3. R. Livi, M. Pettini, S. Ruffo, M. Sparpaglione, and A.Vulpiani, Phys. Rev. A **28**, 3544 (1983).
4. R. Livi, M. Pettini, S. Ruffo, M. Sparpaglione, and A.Vulpiani, Phys. Rev. A **31**, 1039 (1985).
5. M. Pettini and M. Landolfi, Phys. Rev. A **41**, 768 (1990).
6. M. Pettini and M. Cerruti-Sola, Phys. Rev. A **44**, 975 (1991).
7. H. Kantz, R. Livi, and S. Ruffo, J. Stat. Phys. **76**, 627 (1994).
8. J. De Luca, A. J. Lichtenberg, and S. Ruffo, Phys. Rev. E **51**, 2877 (1995).
9. L. Casetti, M. Cerruti-Sola, M. Pettini, and E. D. G. Cohen, Phys. Rev. E **55**, 6566 (1997).
10. J. De Luca, A. J. Lichtenberg, and S. Ruffo, Phys. Rev. E **60**, 3781 (1999).
11. K. Ullman, A. J. Lichtenberg, and G. Corse, Phys. Rev. E **61**, 2471 (2000).
12. A. Ponno, L. Galgani, and F. Guerra, Phys. Rev. E **61**, 7081(2000).
13. P. Villain and M. Lewenstein, Phys. Rev. A **62**, 043601(2000).
14. J. De Luca, A. J. Lichtenberg, Phys. Rev. E **66**, 026206(2002).
15. L. Berchialla, A. Giorgilli, and S. Paleari, Phys. Lett. A **321**, 167 (2004).
16. D. K. Campbell, P. Rosenau, and G. M. Zaslavsky, Chaos **15**, 015101(2005).
17. G. P. Berman, F. M. Izrailev, Chaos **15**, 015104 (2005).
18. A. J. Lichtenberg, V. V. Mirnov, and C. Day, Chaos, **15**, 015109 (2005).
19. A. Carati, L. Galgani, A. Giorgilli, and S. Paleari, Phys. Rev. E **76** 022104 (2007).
20. G. Benettin, R. Livi, and A. Ponno, J. Stat. Phys. **135**, 873 (2009).

21. G. Benettin and A. Ponno, J. Stat. Phys. **144**, 793 (2011).
22. A. Ponno, H. Christodoulidi, Ch. Skokos, and S. Flach, Chaos **21**, 43127 (2011).
23. T. Genta, A. Giorgilli, S. Paleari, and T. Penati, Phys. Lett. A **376**, 2038 (2012).
24. G. Benettin, H. Christodoulidi, and A. Ponno, J. Stat. Phys. **152**, 195 (2013).
25. A. Maiocchi, D. Bambusi, and A. Carati, J. Stat. Phys. **155**, 300 (2014).
26. H. J. Matsuyama and T. Konishi, Phys. Rev. E **92**, 022917 (2015).
27. M. Kohmoto, L. P. Kadanoff, and C. Tang, Phys. Rev. Lett. **50**, 1870 (1983).
28. S. Ostlund, R. Pandit, D. Rand, H. J. Schellnhuber, and E. D. Siggia, Phys. Rev. Lett. **50**, 1873 (1983).
29. A. Sütö, Commun. Math. Phys. **111**, 409 (1987).
30. Ch. Skokos and E. Gerlach, Phys. Rev. E **82**, 036704 (2010)

MATEC Web of Conferences **61**, 01012 (2016) DOI: 10.1051/matecconf/20166101012
APOP2016

States and properties of metallic systems at a threshold breakdown of the through holes under power laser action (part 2). *Susceptibility*

Kalashnikov E[1a]., Kantor M[2]., Bugayev A[3].

[1] *Moscow Region State University, Moscow, Russia*

[2] *Ioffe Physical Technical Institute, RAS, S.-Petersburg, Russia*

[3] *Applied Research Center Old Dominion University, VA 23606 USA*

Abstract. Threshold breakdown of the through holes by power laser radiation of metallic foils is considered as response of metallic system to laser radiation. Binding experimentally determined response to the absolute temperature scale allows to determine the value of the imaginary part of the generalized susceptibility depending on temperature, the critical temperature of the transition "liquid metal - gas", states of the electronic subsystems at this temperature, and the reflectance coefficient values. *Key words:* Response, imaginary part of the generalized susceptibility, plasm frequency, frequency of electron-ion interaction, critical temperature of second order "liquid metal-gas" phase transition, reflectance coefficient.

1 Introduction

One of measuring in an experiment on the threshold holes of metallic foils under act of powerful laser radiation is a diameter of threshold exit hole [1]. – The diameter is absent if the influence of external fields of laser radiation is absent or is not enough. In other words, minimum diameter of threshold exit hole is, not that another, as response of the system to an external action [2]. A size of diameter of exit hole, in the created conditions, is fluctuating value under act of the external field. Our case, just gives possibility to connect properties of fluctuating magnitude with values characterizing behavior of body under external action [2]. Formally, response d is expressed through susceptibility operator α [2]:

$$d = \hat{\alpha} f \qquad (1)$$

here f - generalized force is time function

.

2 Model

The falling on a target the electromagnetic field of laser radiation interacts with the electronic subsystem of metal and enlarges in region of this interaction a chemical potential μ_e of electrons. (Pondermotive potential [3] $V = (q^2 /4c \cdot m \cdot \omega^2) \cdot E^2$ is about $10^{-6} \div 10^{-4} eV$ at created conditions $\lambda = 694$ nm and intensity $J \sim 10^6 \div 10^{10}\ W/cm^2$) Therefore electrons will begin to move into place, where their chemical potential is smaller or there (what is the same) where their number (density) is less. Into a metal, their density remains unchanged. Therefore, electrons begin the motion into place, where them less - from a metal to the interface of "metal-gas". Such motion, in particular, arises up, when a solid is primary quickly heated to the very high temperatures [4]. Kinetic energy of motion of electrons and ions at these densities (densities of metals in solid and liquid states) turns out comparable with coulomb energy of their interaction. Electrons are concentrating at a "gas-metal" interface. It leads to original polarization at a "interface" which is expressed as difference of chemical potentials:

$$(\mu_{e\_interface} - \mu_{e\_volume}) = e \cdot (\varphi_{interface} - \varphi_{volume}) \qquad (2)$$

here $\mu_{e\_interface}$ and μ_{e_volume} is chemical potentials of electrons on interface and in volume. Therefore, the arising up here electric field of E ($\partial\varphi/\partial r = (1/e) \cdot \partial\mu_e /\partial r = - E$; $\mu_e = \varepsilon_{Fermi}$, for cooper $\varepsilon_{Fermi} = 7\ eV$ so that the order of magnitude of electric-field tension corresponds to $- E \sim 7 \cdot 10^8 V/cm^2$) drags out positive ions from a warmed-up volume. Accelerated ions, passing through the "gas-metal" interface and, due to comparableness of kinetic energy of motion and coulomb energy of interaction, capture electrons, and abandon a metal.

[a] ekevkalashnikov1@gmail.com

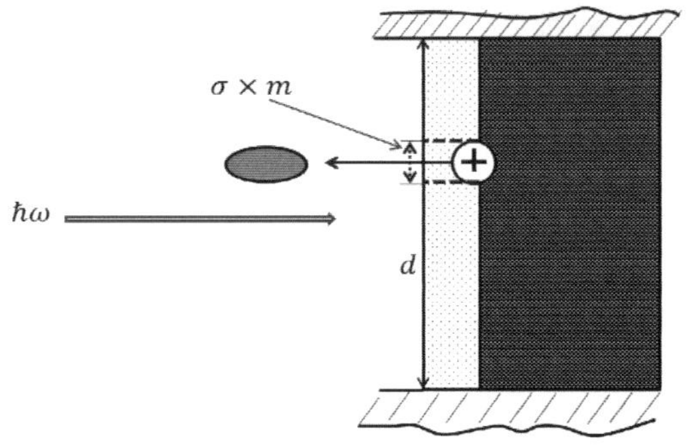

Fig. 1. Creation (annihilation) of element of interface, σ, on leaving of ion from the volume of metal

Therefore, an ion, passing through an interface, participates in formation of surface, increasing (or reduction) its area, fig.1 Therefore, an ion, passing through an interface and transforming itself into atom, is the same fluctuation part of surface which is the response of the system on the external field *f*, (appendix of A).. Therefore, the elemental act of displacement (crossing of interface) can be considered as an act of moving away (joining) of element of interface area per time. This element of area, σ, is approximately equal to the "area of surface» of shade of atom (appendix of A). Elemental area, **σ,** per time is connected with mass, *m*, of atom. It is therefore possible to talk about moving away (formation) per time of area in the interface, by the characterized composition $N \cdot \sigma \cdot m$, where *N* is an amount of atoms in this region. Then value *z* measureable in an experiment on the threshold breakdown of the through holes is connected to the mean value of *d* (appendix of A)

$$N \cdot \sigma \cdot m = \pi \cdot (d/2)^2 \cdot m = (\pi/4) \cdot (z \cdot \sqrt{m})^2 \quad (3)$$

Effective diameter of the threshold exit hole, $d = z \cdot \sqrt{m}$, is the response of the system on external action *f*. As a threshold diameter of *z* develops in time, then, in general case, *d* is the function of time of *t* :

$$d = d\,(t) \quad (3a)$$

(It is possible to construct generalized force *f* conjugated to the chosen response *d* [2]. In our case $f \sim (v \cdot \varepsilon / \omega)^{1/2}$. Here $\varepsilon = (1/c) \cdot (\sqrt{\xi \cdot \mu}) \cdot E \cdot H$ – electromagnetic energy density. $E \sim$ and $H \sim e^{+j\omega t}$ or $e^{-j\omega t}$ are tensions of the monochromatic field with cyclic frequency ω; *v* is volume accepting ε).

Dropping details of calculations [2] and taking into account that external perturbation is monochromatic, we will write down connection between the response *d (t)* of the system and external perturbation *f* in Fourier components:

$$d_\omega = \alpha(\omega) \cdot f_\omega \quad (4)$$

Here $\alpha(\omega)$ is generalized susceptibility. Symbol ω means Fourier component of value, for example:

$$d_\omega = \int_{-\infty}^{+\infty} d(t) e^{j\omega t} dt$$

Then, fluctuation-dissipative theorem [2] allows to write down Fourier component of square of diameter (areas) (3a) nascent hole in a kind:

$$\langle d^2 \rangle_\omega = 2 \cdot \hbar\alpha_2 \cdot \left[\frac{1}{2} + \frac{1}{e^{\hbar\omega/kT} - 1}\right] \quad (5)$$

here α_2 is imaginary part of generalized susceptibility (1) and (4):

$$\alpha(\omega) = \alpha_1 + j\alpha_2$$

$\hbar, k, \pi$ are constants Plank, Boltzmann and 3,14., T is an absolute temperature. The important feature $\langle d^2 \rangle_\omega$ consists of that it, practically, is composition of area of the threshold hole, as functions of temperature, on a delta function $\delta\,(\omega)$ (appendix of B).

3 Solution and Results

3.1 *Determination of critical temperature*

Inverting expression (5), we will find the temperature of the electron-ion system depending on the "diameter" of the threshold hole

$$T = \frac{\frac{\hbar\omega}{k}}{\ln\left[\frac{1 + \hbar\alpha_2/d^2}{1 - \hbar\alpha_2/d^2}\right]} \quad (6)$$

A feature of expression (6) is in what it decomposes on the "bunch" of lines - dependences of imaginary part of generalized susceptibility for different diameters versus temperature, рис.2. The curves of this bunch intersect nowhere, and each is begun with the minimum of real value.. Each of curves corresponds only to one certain

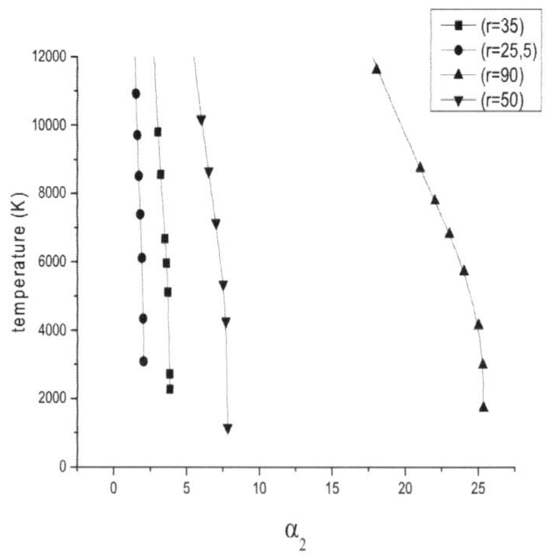

Fig.2. The temperature of the electron-ion system depending on α_2 is presented as a bunch of lines for every diameter. There are lines (for four diameters) of from a complete bunch.

diameter of exit hole. The analysis of behavior of these curves allows to find the real value of critical temperature (within the framework of the chosen approaching), and also, values of imaginary part of generalized susceptibility, α_2, at different temperatures. . Research of each of lines of bunch, fig. 2, finds out on them inflection points, fig. 3. An inflection point on dependence of $dT/d\alpha_2$ indicate on that, how energy only of one quantum, $\hbar\omega$, is adopted by the system at temperature of T, equaled to critical temperature of T_{cr}.

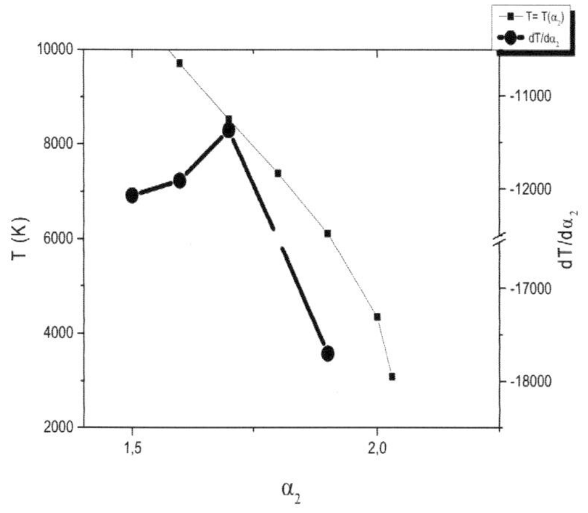

- **Figure 3.** Inflection point on a derivative $dT/d\alpha_2$. Squares designate dependence of T on α_2 for d ~ 25,5 μm. Mugs are the first derivative $dT/d\alpha_2$

Value, reciprocal to $dT/d\alpha_2$, i.e. $\frac{d\alpha_2}{dT} \sim \left(\frac{\hbar\omega}{kT}\right)\cdot\frac{1}{T}$, is analogue of heat capacity at permanent pressure [5]. A characteristic peak or inflection point on dependence of T versus α_2 determines the value of critical temperature, рис.3. . Thus, a critical temperature for a copper is equal to $T_{cr} = 6522K$

3.1.1 State of the electron-ion system

Imaginary part of susceptibility, α_2, is fully determined by the state of electronic subsystem and its interaction with an ionic subsystem.

Behavior of the two subsystems depends on the obtained energy amount at interaction with laser field, which, in turn, depends on a reflectivity (absorptions) coefficient. Now, knowing dependence (6) and fig. 2, 3, it is possible to specify behavior of the electron-ion system at critical temperature and near it.

Within the framework of Drude approaching [6-9] it is possible to determine the reflectivity of metals [7, 8]. The temperature range of research of reflectivities for metals is bounded enough, both in a theoretical and in experimental plan [10-12].

To calculate reflection coefficient we take into account our experimental data of response and with using (6).

Within the framework of Drude approximation an dielectric permittivity ε (ω) is expressed through external field frequency, ω, plasma frequency ω_p and electron-ion collisions frequency of ν [3, 6-13]:

$$\varepsilon(\omega) = 1 - \frac{\omega_p^2}{\omega\cdot(\omega + j\nu)} = \varepsilon_1(\omega) + j\varepsilon_2(\omega) \qquad (7)$$

(within the framework of Drude model, a value ν depending on a context is also mentioned as mean time of relaxation of electronic distribution [11] or ν^{-1}, and as time of relaxation of electrons at interaction both with electrons and with ions [7]).

From another side, permittivity ε(ω) and susceptibility α(ω) are connected by relationships [6,7]:

$$\varepsilon(\omega) = 1 + 4\pi\cdot\alpha(\omega) = 1 + 4\pi(\alpha_1 + j\alpha_2) \qquad (8)$$

Therefore, from (6) and (7) we find that ε_1 and ε_2:

$$\begin{cases} \varepsilon_1 = 1 + 4\pi\cdot\alpha_1(\omega) \\ \varepsilon_2 = 4\pi\cdot\alpha_2(\omega) \end{cases} \qquad (9)$$

Coming back to (7) we find:

$$\begin{cases} \varepsilon_1 = 1 - \omega_p^2/(\omega^2 + \nu^2) \\ \varepsilon_2 = \omega_p^2\cdot\nu/\omega(\omega^2 + \nu^2) \end{cases} \qquad (7a)$$

Here ω is frequency of the external field of radiation (radiations of laser), for a ruby laser $27\cdot10^{14}$ s^{-1}, ω_p is plasma frequency; ν is frequency of electron-ion interaction.

In the threshold breakdown of the through hole conditions of, [1], temperatures of experiment achieve an about a few thousand kelvin, so that $\hbar\omega \sim kT$.

At that rate $|\varepsilon_1| \sim |\varepsilon_2|$ are realized. Applications and calculations of conductivity and reflectivity coefficient connected by approaching $|\varepsilon_1| << |\varepsilon_2|$, [13], are not carried out. Electron-ion frequency ν is expressed from second equation of (7a) system:

$$\nu^2 - \frac{\omega_p^2}{\varepsilon_2\omega}\cdot\nu + \omega^2 = 0 \qquad (10)$$

$$\nu_{(+),(-)} = \frac{\omega_p^2}{2\varepsilon_2\cdot\omega}\left(1 \pm \sqrt{1 - 4\cdot\left(\frac{\omega}{\omega_p}\right)^4\cdot(\varepsilon_2)^2}\right) \qquad (10a)$$

The equation shows that frequency of electron-ion interaction has two roots. These two roots testify two different states of electronic subsystem

First of roots (+ before a root) talks that the electrons of conductivity with the increase of temperature more often collide with ions (mean free time of these electrons diminishes sharply). The system is kept in metallic state.

The second root talks that frequency of collisions falls. And it is possible (at temperatures below than critical, when the volume, occupied by the system also below, than at a critical temperature) only at reduction of number of electrons participating in conductivity. Reduction of number of electrons of conductivity talks that they (electrons) are localized on ions (or original ionic bubbles). Thus, a metal passes to the dielectric state.

. Because these both roots are equal in rights and coexist at identical temperatures, then it remains to suppose that we are at the coexistence of two states: liquid metallic and dielectric Thus there is the only condition - a critical temperature, pressure and volume, is a critical state of substance (or critical point) at which these two roots coincide:

$$\nu_{(+)} = \nu_{(-)} \tag{11}$$

Equality (11) means that subduplicate expression is equal to the zero. The condition (11) is reduced to the zero of subduplicate expression in (10a):

$$\omega_p^4 - (2\varepsilon_2\omega^2)^2 = 0$$

or

$$\varepsilon_2 = \frac{1}{2}\cdot\left(\frac{\omega_p}{\omega}\right)^2 \tag{12}$$

The value ε_2 is found from imaginary part of susceptibility α_2 (9), which corresponds to critical temperature, fig.3. From (10a) subject to condition (11), taking into account (12), the electron-ion colliding frequency at critical temperature is found $\nu_{(+)} = \nu_{(-)} = \nu$:

$$\nu = \frac{\omega_p^2}{\varepsilon_2\cdot\omega} = \omega \tag{13}$$

Such situation implies that external perturbation at a critical temperature distributes electrons and ions so, that frequency of their interaction becomes equal frequency of external field. Under act of the field of radiation of ruby laser ($\omega = 27\cdot10^{14}\,s^{-1}$) on a metal, at a critical temperature, this frequency equals to $\nu = 27\cdot10^{14}\,s^{-1}$

3.2 *Temperature behavior of parameters of the electron-ion system*

3.2.1 Reflectivity coefficient

In order to determine reflectivity (appendix of C) and its temperature dependence it is necessary to know temperature behavior of permittivity, plasma frequency, frequency of electro-ion interaction. These values are bound by the standard set of relationships (7) - (9), which include imaginary part of susceptibility α (ω). Therefore experimental dependence of temperature versus imaginary part of susceptibility, (fig.2, fig.3), allows to find temperature dependences ε_2, ω_p (fig. 4), ν (fig. 5) and reflectivity coefficient(fig. 6) (appendix of C).

*3.2.2 **Plasma frequency***

Temperature behavior of plasma frequency, ω_p plays an important role at consideration of reflectivity. ω_p (рис.4) is determined from next considerations.- It is considered that copper atoms are singly ionized (valence of Z = 1) in from melting temperature up to the critical one. Then we accept $\omega_p = 4\pi\cdot n\cdot^2/m = 31,80795\cdot10^{-8}\cdot n$, where $n = 6{,}022\cdot10^{23}\cdot Z\cdot\rho / A$; A = 64. Temperature dependence of copper density,ρ, is adopted from [16]. The measured values of metal density are taken into account up to the temperature of boiling only, inclusive. The value ω_p was determined from (12) at the critical temperature T_{cr} = 8522K, got from our results, fig.3, preliminary defining ε_2 on α_2. In addition, experiment determined imaginary part of susceptibility and, accordingly, to the permittivity, and the value of critical temperature do not allow to fluctuate of plasma frequency, due to correlation (10).

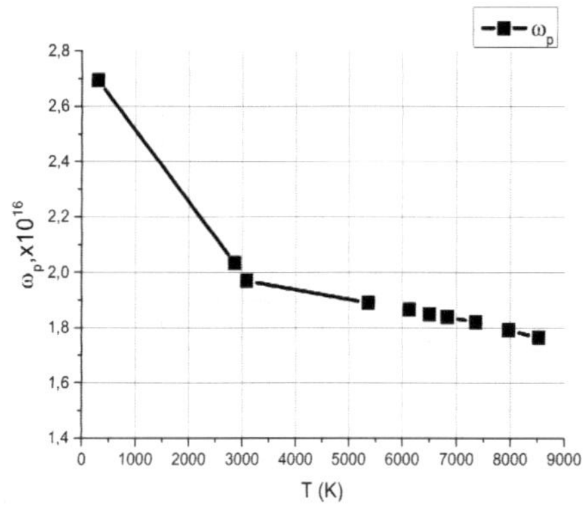

Fig.4.Temperature behavior of plasma frequency ω_p.

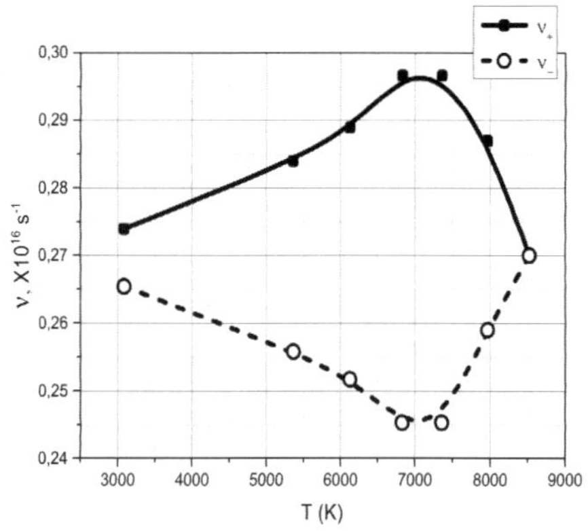

Fig. 5. Frequency ν is temperature dependence of electron-ion interaction .Upper curve corresponds ν(+) - металлической составляющей электрон-ионной системы, ν(-) - corresponds to "dielectric" part of the electron-ion system.

3.2.3 Electron-ion interaction frequency

Electron-ion interaction frequency ν consists of two branches ν(+) and ν(-), fig.6. First of the branches corresponds to "metallic» part, and second - to "dielectric" part of the electron-ion system at the examined conditions. ν(+) illustrates the growth of frequency of electron-ion collisions, and ν(-) is reduction of frequency of электрон-ионных collisions due to localization of electrons by the capture of them on ionic bubbles. It leads to growth of dielectric constituent of a metal.

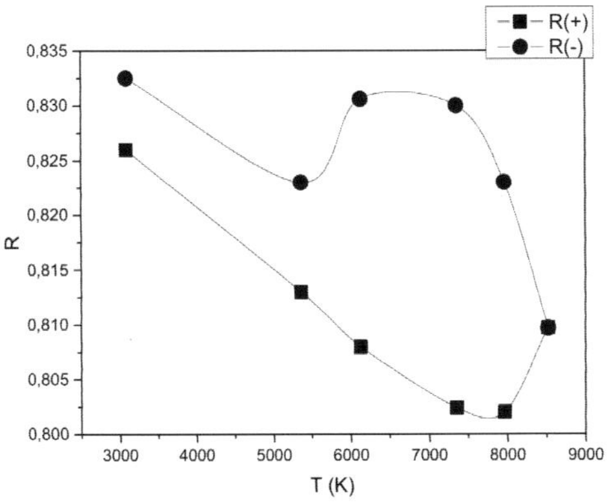

Fig.6. Reflectivity coefficients R (+) and R (-) for metallic and dielectric constituents of the electron-ion system

3.2.4 Critical temperature

In literature, there were a few different values of critical temperatures for a copper. A critical temperature was measured by the method of explosion of thin wire by a large current [14]: T_{cr} = 8390 K. In [15] a theoretical value T_{cr} = 7800K is got from the calculations of modelling a phase diagram. But in such calculations, if reflectivity (absorptions) were used, then for such temperatures they would be got only by extrapolation on high temperatures, as a direct measuring of these coefficients at high temperatures, higher 3000K in literature were not met. In [16] it was got T_{cr} = 9000 K but by extrapolation of density of the gas state of copper from the boiling temperature

3.2.5 Reflectivity coefficient

R calculated in accordance with the appendix C. Fig.6 demonstrates complication of behavior of reflectivity in the interval of from boiling up to the critical temperature of phase transition of liquid metal to gas state. Dielectric constituent part of reflectivity $R_{(-)}$ plays substantial role. - In a liquid metal near-by the critical temperature of $R_{(-)}$ is substantially larger than reflection of $R_{(+)}$ caused by a metallic constituent. Quite possible, that such situation is caused not simply by appearance of the constrained localized states of electrons on ions or ionic bubbles, but also by origin of overlapping between such constrained states [22, 23] leading to generalization of the localized states. In such regions electrons work again as in a metal. Herein, presumably, and essence of the critical phenomenon at what distinction between a liquid metal and gas disappears.

4 Conclusion

The idea of threshold breakdown of the through holes and application of fluctuation-dissipative theorem allowed to find the value of critical temperature, and also temperature dependences of the generalized susceptibility of the system. This combination opens the prospects of more thin and detailed high temperature measuring and possibilities of management by the electron-ion system at high temperatures

5 Appendix A

To estimate the mean (quantum mechanical) value of *d* from (1) and (3), we will take into account next moments: (*1*). At the densities of liquid metal and temperatures in thousands of Kelvin, coulomb energy of interaction of electrons and ions is comparable with their kinetic energy of motion [4]. In these conditions a liquid metal is the electron-ion system, being mixture of electron-ion gas (constrained electron-ion pairs) and electron-ion liquid. Such state is substantially heterogeneous and non-ideal. The last properties suppose a presence, both disconnected electrons and ions, providing the metallic state ("zone of conductivity" and level of Fermi in her), and appearing electron-ion pairs (atomic metallic gas) that can decompose in next moments. And at approaching of the liquid metallic state to the critical condition (to the critical temperature) frequency of origin of such virtual pairs grows (см.for example, рис.5,behaviorv(-))...... (*2*). In such electron-ion system, electrons can be considered, as electrons of the strongly excited state of atom at which peripheral electron of atom appear far remote from a positive ion [17] This situation is equivalent to what to electron being at long distance from a nucleus may attribute the large main quantum number *n* (and orbital number $l = 0$. So the electron corresponds to excited *S*-state). Motion of such electron (or such ion in the field of electron) can be considered as mutual motion in the coulomb field of atomic remain (ion) with the effective charge of $Z = 1$ [17]: electric-field tension arising up between the charges at that rate ~ $E_{n,l,m} \cdot (a \cdot e)^{-1}$, ($E_{n,l,m}$ is own energy of electron, *e* is a electron charge, *a* is a Bhor radius).................... ... (*3*). The conduction band in a copper appears as a result of hybridization 4s and 3d shells of atom [18]. So that the wave function of electron in the conductiion band can be decomposed on ball functions with certain quantum numbers (n, l, m) [17, 18, 19] including those that correspond to the states of electrons participating in interaction with the external field, Part of them participates in dispersion of the external field, another part – in absorption (in optical transitions). For example transitions $3d^{10}5s - 3d^{10}4p$, $3d^{10}6s - 3d^{10}4p$, $3d^{10}5,6,7d - 3d^{10}4p$ [20], with the corresponding to them values of energy $\sim 1{,}8;\ 2{,}9;\ 3{,}3\ eV$ correspond to the optical transitions.

Taking into account these facts, (1) - (3), we can do next estimations

(*a*). Remoteness of electron from a nucleus in approaching of hydrogen-like atom [17, 19] is determined by quantum mechanical mean distance of electron in an atom

$$\langle r\rangle_{n,l} = \frac{a}{2Z}\cdot[3n^2 - l\cdot(l+1)] \sim d/2$$

Then distance of 5S -state (at Z = 1, look higher, point (2)) equals to $\langle r\rangle_{5,l=0} = 19{,}838\cdot10^{-8}cm$, i.e., an electron is remote from a nucleus on mean distance $19{,}838\cdot10^{-8}cm$ (These distances are less at taking into account Thomas-Fermi approaching [21]) Tension of electric-field of such electron operating on rest atom (ion) equaled to $E_{n,l,m}/(e\cdot\langle r\rangle_{5,0})$ is the same, what $\frac{1}{e}\cdot\frac{\partial\mu_E}{\partial r} \approx 3{,}5\cdot10^7\,V/cm = -E$.

(*b*). The electromagnetic field of laser radiation falling on a target is changing slower than time of electron-electron (τ_{e-e}) interaction (period of oscillation for a ruby laser $\tau = 23\cdot10^{-16}s$ and $\omega_p^{-1}\sim\tau_{e-e} = 2.32\cdot10^{-16}s$, where ω_p is plasm frequency). It implies that an electronic subsystem will have time to be tuned under external influenceю Thus electrical fields of interaction between charged particles will look like slowly changing. Atom of copper in exited state is considered in hydrogen-like atom approaching. Character (potential) of electric field allows to split up of wave function of atom $\Psi(x_a, y_a, z_a; X, Y, Z) - \varphi_e(x,y,z)\cdot\Phi(X,Y,Z)$ in the adiabatic approaching on the wave function of electron $\varphi_e(x,y,z)$ (here *a* -нумерует electrons in an atom; (x,y,z) are coordinates of relative motion of electron, (X,Y,Z) coordinates of centre-of-mass of atom) and wave function of ion of atom $\Phi_n(X,Y,Z)$ so, that equalization of Schrodinger for the ion of atom will be written down, as

$$i\hbar\frac{\partial}{\partial t}\Phi_n = -\frac{\hbar^2}{2M}\Delta_{X,Y,Z}\Phi_n + [\mathcal{U}(X,Y,Z) + E_{n,l,m}(X,Y,Z)]\Phi_n$$

(A1)

Because de Broglie wavelength of ion of atom is very small a wave function for him is searched in a kind [17]

$$\Phi_n = A\cdot exp\left(-{}^{iS}/_{\hbar}\right)$$

(A2)

Where S, is an action [17], equals to

$$S = H\cdot t - P_X\cdot X - P_Y\cdot Y - P_Z\cdot Z$$

(A3)

Substitution of (A2) in (A1) with an account (A3) leads to hamiltonian

$$H = \vec{P}^2/_{2M} + \mathcal{U}(X,Y,Z) + E_{n,l,m}(X,Y,Z)$$

(A4) ,

It corresponds to classic behavior of ion of atom of copper, in the external field created by surroundings of ion, $\mathcal{U}(X,Y,Z)$ and by field, $E_{n,l,m}(X,Y,Z)$, created by remote valency electron.

We consider that $E_{n,l,m}(X,Y,Z)$ at moving of centre-mass does not change. The external field for the ion of $\mathcal{U}(X,Y,Z)$ is expressed through the difference of chemical potentials of electrons in the "interface" and in a "volume"

$$\mathcal{U}(X,Y,Z) = (\mu_{e\_interface} - \mu_{e\_volume}) = e\cdot(\varphi_{interface} - \varphi_{volume})$$

For simplicity it is accepted the vector of tension of electric-field is directed on a normal to the "interface", fig.1. Then equations of motion of centre-of-mass of remain of copper atom take the following form:

$$M\frac{d^2X}{dt^2} = 0$$
$$M\frac{d^2Y}{dt^2} = 0$$
$$M\frac{d^2Z}{dt^2} = -\frac{\partial}{\partial Z}\left(\mathcal{U}(X,Y,Z) + E_{n,l,m}(X,Y,Z) + \frac{q^2}{4m\omega\omega^2}J\right)$$

Value $|V| \ll |\mathcal{U}(X,Y,Z)|$, where $V = \frac{q^2}{4cm\omega^2}E^2$.

DOI: 10.1051/matecconf/20166101012

It is accepted the initial value of speed of remain in the moment of $t = 0$ is equal to the zero $\left(\frac{dX}{dt}, \frac{dY}{dt}, \frac{dZ}{dt}\right) = 0$. From it is followed, the conserved values are $X - X_0, Y - Y_0 \sim \langle r \rangle_{n,l}$ and $(X - X_0) \cdot (Y - Y_0)$. In other words, an area $(X - X_0) \cdot (Y - Y_0) = \sigma$, is equal to the area formed by the projection of atom on normal to moving of atom of plane, is conserved. Values of $M \cdot \sigma$ and $N \cdot M \cdot \sigma$ also are the invariants of motion during the process of transition of remain of atom from one position in other. Distance overcame by the remain of atom is

$$Z - Z_0 - -\frac{1}{2M}\left[\frac{\partial}{\partial Z}\left(u(X,Y,Z) + E_{n,l,m}(X,Y,Z)\right)\right] \cdot \tau^2$$

If to accept, that

$$Z - Z_0 = \langle r \rangle_{4,l=1} - \langle r \rangle_{5,l=0} = |7{,}671 \cdot 10^{-8}|\ cm$$

then the remain of atom will overcome this distance after $\tau \sim 10^{-14}s$, before will go across from the state $|n = 5, l = 0\rangle$ in $\langle n = 4, l = 1|$

6 Appendix B

Calculation of spectral density of square of diameter of the threshold breakdown of the through hole, $\langle d^2 \rangle_\omega$

Formation of hole is the temporal phenomenon. And the diameter of minimum exit hole depends on a diameter in the moment of time previous to the breakdown. So that it is possible to talk about temporal correlation function

$$\langle d(t) \cdot d(t') \rangle \qquad \text{(B1)},$$

. in which t' is time moment of through breakdown of minimum exit hole by the diameter of d. $t = 0$ is the preceding moment of time is chosen as point of reference. It means an exit hole arises up at $t' - 0 = \tau$. Spectral density of square of diameter of the through breakdown of minimum exit hole with respect to [2] is written down:

$$\langle d^2 \rangle_\omega = \int_{-\infty}^{+\infty} \langle d(0) \cdot d(\tau) \rangle e^{j\omega\tau} dt \qquad \text{(B2)}$$

It is assumed that diameter during τ is changed poorly the d (τ) is decomposed into row:

$$d(\tau) = d(0) + \dot{d} \cdot \tau \qquad \text{(B3)}$$

(B2) is rewritten with taking into account (B3)..

$$\langle d^2 \rangle_\omega = \int_{-\infty}^{+\infty} \langle d(0) \cdot (d(0) + \dot{d} \cdot \tau) \rangle e^{j\omega\tau} dt =$$

$$= \int_{-\infty}^{+\infty} \langle d(0) \cdot d(0) \rangle e^{j\omega\tau} dt + \int_{-\infty}^{+\infty} \langle d(0) \cdot \dot{d} \cdot \tau) \rangle e^{j\omega\tau} dt =$$

$$= \langle d(0) \cdot d(0) \rangle \int_{-\infty}^{+\infty} e^{j\omega\tau} dt + \tau \cdot \int_{-\infty}^{+\infty} \langle d(0) \cdot \dot{d}(0) \rangle e^{j\omega\tau} dt = \qquad \text{(B4)}$$

the one should notice

$$\int_{-\infty}^{+\infty} e^{j\omega\tau} dt - \delta(\omega) \qquad \text{(B5)}$$

$$\int_{-\infty}^{+\infty} \langle d(0) \cdot \dot{d}(0) \rangle e^{j\omega\tau} dt = \langle d(0) \cdot \dot{d}(0) \rangle \qquad \text{(B6)}$$

Continuing equality (B4) and taking into account relations (B5) and (B6), then (B4) :

$$= \langle d(0) \cdot d(0) \rangle \cdot \delta(\omega) + \tau \cdot \langle d(0) \cdot \dot{d}(0) \rangle \qquad \text{(B7)}$$

It is assumed that change of diameter $\dot{d}$ versus time and size of diameter d in one and the same moment of time are not correlated. Then the second member in (B7) equals to the zero. So finally:

$$\langle d^2 \rangle_\omega = d^2(0) \cdot \delta(\omega)$$

7 Appendix C

Reflectivity coefficient (relation of Frenel of electromagnetic radiation for the normally falling on a interface)

$$R = \left|\frac{\sqrt{\varepsilon}-1}{\sqrt{\varepsilon}+1}\right|^2 = \left|\frac{n-j\cdot k-1}{n+j\cdot k+1}\right|^2 = \frac{(n-1)^2+k^2}{(n+1)^2+k^2}$$

$$\varepsilon^{\frac{1}{2}} = n - j\cdot k$$

$$n = \left\{\frac{1}{2}\cdot\left[\varepsilon_1 + (\varepsilon_1^2 + \varepsilon_2^2)^{\frac{1}{2}}\right]\right\}^{\frac{1}{2}}$$

$$k = \left\{\frac{1}{2}\cdot(\varepsilon_1^2 + \varepsilon_2^2)^{\frac{1}{2}} - \varepsilon_1\right\}^{\frac{1}{2}}$$

$$\varepsilon = \varepsilon_1 + j\cdot\varepsilon_2 = 1 - \frac{\omega_p^2}{\omega^2+\nu^2} + j\cdot\frac{\omega_p^2\nu}{(\omega^2+\nu^2)\cdot\omega} =$$

$$= 1 - \frac{\omega_p^2\cdot\omega - j\omega_p^2\cdot\nu}{(\omega^2+\nu^2)\cdot\omega} = 1 - \frac{\omega_p^2\cdot(\omega - j\nu)}{(\omega + j\nu)\cdot(\omega - j\nu)\cdot\omega}$$

$$= 1 - \frac{\omega_p^2}{\omega\cdot(\omega + j\nu)}$$

$$\varepsilon_2 = 4\pi\cdot\alpha_2$$

References

1. E.V Kalashnikov, M. Yu Kantor, A.A Bugayev,D.V Kouprienko, A.V. Chikiryaka. Part 1
2. L.D. Landau and E. M Lifshitz, Statistical Physics, Elsevier .
3. D Bauer. Theory of intense laser-matter interaction. *Max-Plank-Institute für Kernphysik, Heidelberg, Germany*. 2006, 106 p.
4. Ya.B. Zel'dovich., Yu.P. Raizer. Physics of Shock Waves and High-Temperature Hydrodynamic Phenomena. DOVER PUBLICATIONS, INC. Mineola, NY, 2002
5. H.E Stanley. Introduction to phase transitions and critical phenomena. Clarendon Press. Oxford 1971
6. M.S. Dresselhaus. Optical Properties of Solids. Solid State Physics. Part II, 194 (1985).
7. D Arovas. Lecture Notes on Condensed Matter Physics. Departm.of Physics. University of California, San Diego. 2010, 179 p.
8. O.V Vaculenco., V.S Severin. Semiconductor Physics, Quantum Electronics and Optoelectronics, ***10***, 55, (2007).
9. K Fleischer., S Chandola., Th Hermann., N. Esser. , W. Richter. , J.F. McGilp. Phys. Rev. B. 195432, (2006).
10. William T Walter. Change in reflectivity of Metals under intense Laser Radiation. Polytechnic Institute of New York. 1-35 (1981.).
11. Kikuo Ujihara. Journ. Appl. Phys.***43***, 2376 (1972).
12. M. Schmid, S. Zehnder., P. Schwaller., B. Neuenschwander.,M.Held., U. Hunziker., J.Zürcher Advanced Laser Technologies (ALT 12), 2-6 September, 2012, Thun, Switzerland. 8 pages.
13. G.G Gladush., I. Smurov. Physics of laser materials Processing. (Theory and epxperiment).Springer-Verlag Berlin Heidelberg 2011
14. V.E Fortov., A.N. Driemin., A.B. Leont'ev. Teplophys. High Temper. ***13***, 1072 (1975).
15. D Autique., G. Clair, D. L'Hermite, V Alexiades., A. Bogaerts. , B. Rethfeld. Journ. Appl.Physics. ***114***, 023301 (2013).
16. J.A Cahill., A.D. Kirshenbaum. Journ. Phys. Chemistry.***66***,1080 (1962).
17. L. D. Landau. and E. M Lifshitz, Quantum Mechanics, Elsevier
18. **J.** Friedel ,: in The Physics of Metals, ed. J. Ziman, Cambridge Univ. (1969)
19. Hans A. Bethe. Quantum Mechanics of One- and Two-Electron Atoms Cornell University,
20. C.L. Lewis., G.P. Jackson., K.D. Stephen, V. Majidi., F.L King. Spectrochimica Acta. Part B.***56***, 481 (2001).
21. L Spruch. Rev. Mod. Phys.***63***, 151 (1991).
22. A.A Likal'ter. Teplophys..High Temper. ***25***, 424 (1987).
23. A.A. Likal'ter. Uspehy-Phys. ***170***, 831 (2000).

Identifying 1Method of Meat Containing Excessive Moisture Based on hyperspectral and SVM Multi-Information Fusion

GUO Peiyuan, XU Jingjing，XU Pan，DONG Xiaodong，LIU Yanfang
XING suxia, SUN mei

College of Computer and Information Engineering, Beijing Technology and Business University, Beijing, China,Email:ggppyy@126.com
Project 61473009 supported by NSFC

Abstract: In this paper, a quick and accurate detection method which can identify whether the meat contain excessive moisture is mentioned. By using near-infrared spectroscopy measurement and SVM Multi-Information Fusion, the meat moisture content model has been established. In order to improve the accuracy of NIR measurement predicted model and to reduce the measurement sensitivity, utilizing image information and the PH value data as the parameters of the meat moisture content model. The study concluded that the theory and method can be further extended to the detection of other related meat agricultural products.

1 Introduction

In recent years, food safety issues have attracted more and more attention, at the same time, the event "water-injected meat" which shocked consumers. Water-injected meat is the low-quality meat that illegal in pork and beef. The national standard meat moisture limit as follows: if the pork, beef, chicken, water content is more than 77%, can be judged as the meat injection. In the present stage, method of detecting water-injected meat has two main kinds: sensory detection (i.e. through view color) and physicochemical detection. Sensory evaluation methods can be used to quickly determine the meat, but the accuracy is poor; Physical and chemical detection method can obtain accurate results, but the chemical experiment needs for a long time.

With the development of information technology, NIR spectrum detection, computer vision, multi-information data fusion detection technology used in the food safety which can provide a rapid, effective and practical non-destructive testing methods for meat quality. so this paper use the moisture content of the meat as a main parameters，use the color value and the pH value of the meat as the auxiliary parameter parameters to identify the water-injected meat ， to extract the characteristics of these three data and carry on the multiple information fusion.

1.1 The Model of NIR for Water Content in Meat

NIR technology is the physical measurement techniques that it utilizes chemical properties in the near infrared region to rapidly determine content and property of one or more chemical components in the sample [1]. When using near infrared spectroscopy for determining the content of ingredients in the meat, usually chopped the meat and stir till to uniform mixture, so that it can accurately determine the content. This study uses FoodScan NIR Spectrometers of the FOSS Company for measuring meat spectrogram.

FoodScan NIR Spectrometers (shown in Fig.1) is a fast, accurate and easy to use instrument for analysing all stages of meat production - from checking incoming raw material to final product control. It is pre-calibrated to analyse all key parameters including fat analysis, moisture analysis, protein analysis, salt analysis and collagen analysis delivering results in just 50 seconds.

In order to eliminate the interference of environmental light to lighting equipment, the illumination system adopts the closed box with 400 mm x 400 mm square at the bottom. Box is made of white aluminum-plastic plate, the inner wall are white, the top two adjustable halogen light source. At the top of the box for the camera hole, placed the digital camera. Right below of the camera hole is the elevating pallet for carrying the measured meat. The illumination value of the whole measured meat can be adjusted by adjusting the height of the tray. The structure diagram of light box is shown in Fig.2.

Figure 1. FoodScan NIR Spectrometers

Figure 2. Image acquisition system

[a] Corresponding author: ggppyy@126.com

1.2 Spectral Data Acquisition and Preprocessing

In the experimental environment at room temperature 25°C, the sample set include 30 group normal meat samples and 30 group water-injected meat samples. Spectra collected and then completed each sample with a control experiment, which uses physicochemical method for measuring the moisture content of meat. In the MATLAB7.11.0 (R2010b), generated data file, viewed the sample group 1-40 as the training set and the sample group 41-60 as the test set for the predictive mode. After analysis and detection, moisture content in water-injected samples is more than 77%, in line with national standards, which can be used as water-injected meat samples for experimental research. The statistical result of the experimental data was shown in Table 1.

Table1: The moisture of training and testing set for the NIR mode

After repeated experiments, studies have adopted the multiple scatter correction (MSC) and centralized method to do pretreatment for the spectral data.

Multiplicative scatter correction (MSC), not only can effectively remove the uneven distribution of the sample particles and the particle size generated by scattering effects in the near-infrared spectrum, but also can modify a path length error in the measurement of the transmission spectrum.

Specific calculation method as follows:
(1) Calculate the average spectrum of the spectra of the samples collected:

$$\bar{x} = \frac{\sum_{i=1}^{n} x_i}{n} \quad (1)$$

(2) Calculate regressively for the above average spectrum:

$$x_i = a_i \bar{x} + b_i \quad (2)$$

(3) Make multiplicative scatter correction (MSC) calculation for each spectrum:

$$x_{i,MSN} = \frac{x_i - b_i}{m_i} \quad (3)$$

The mean spectrum centralized method is that the average spectrum of near infrared spectroscopy subtract the average spectrum of the calibration set and make the calibrating spectral matrix X column average to be zero; so the modeling is more suitable for the meat water content.

Specific calculation method as follows:

① Calculate the average spectrum of the calibration set of samples:

$$\bar{x}_j = \frac{\sum_{i=1}^{n} x_{ij}}{n} \quad (4)$$

In this structure: n is the number of calibration samples, $j = 1, 2, ..., m$.

② Calculate the mean center spectrum $x_i(1 \times m)$ of the unknown sample spectra $x_{i,mean}$:

$$x_{i,mean} = x_i - \bar{x} \quad (5)$$

1.3 BiPLS Method for Prediction Model

Backward interval partial least squares (BiPLS) method for prediction model In MATLAB7.11.0 (R2010b), the spectral information of training set sample is shown in Fig.3.

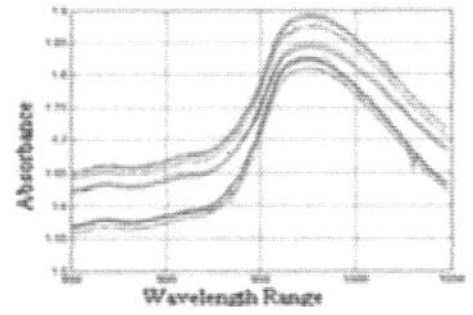

Figure 3. Spectral of training set samples

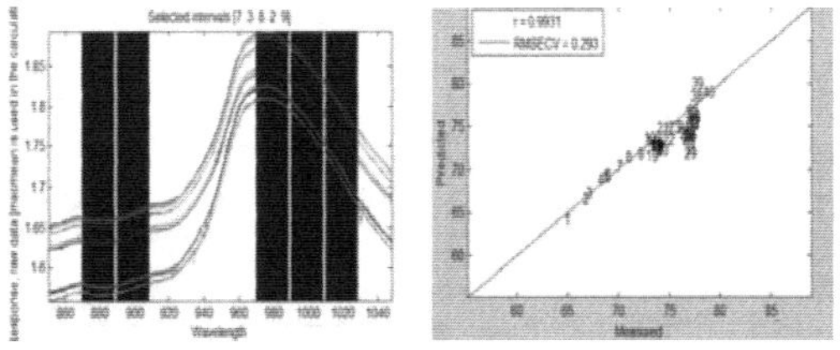

Figure 4. Select the best subintervals of the model

According to the minimum root mean squared error correction (RMSEC) principle of BiPLS algorithm (shown in Fig.4). choose 5 subintervals, including number 2, 3, 7, 8, 9, to establish PLS model. The root mean square error of cross validation (RMSECV) of the PLS models, RC, RP, root mean square error of prediction (RMSEP) were 0.9821, 0.442, 0.9985, 0.318. The number of variables which the model optimization needed is reduced by half and the prediction performance is good. In order to improve the accuracy of model, make a comprehensive evaluation to the meat, in this paper, by adding auxiliary variables pH value and meat color information for multi-data fusion detection, optimize the prediction model which has been established.

2 The Extraction of Meat Color Information

2.1 The Image Pretreatment and Color value Extraction

When collected images in the light box and would extract the color information of the image processing. Identifying the color of the meat with the computer vision technology is the use of the RGB color and HSI color model. The color characteristic value adopted in this study is the mean and standard deviation of the R, G, B and H, S, I component. Wherein the mean value of the color indicates the basic color of the meat, the standard deviation of the color indicates the uniformity of the meat color. In the MATLAB image processing toolbox, using the following procedures can obtain the R, G, B component of image：
R=photo (:,:,1); G= photo (:,:,2); B= photo(:,:,2);
According to the RGB to HSI conversion equation, In the MATLAB can extract H, S, I component:

$$H = \arccos\{\frac{[(R-G)+(R-B)]}{[2(R-G)^2+(R-B)(G-B)]^{1/2}}\} \tag{6}$$

$$S = 1 - \frac{3}{(R+G+B)}[\min(R,G,B)] \tag{7}$$

$$I = \frac{1}{3}(R+G+B) \tag{8}$$

By experiment and analysis, there are some differences in the color between the normal meat samples and the water-injected meat samples（shown in Fig.5 and Fig.6）; The normal meat sample is cherry red and with rich color, the color of water-injected meat samples is light pink because it injected a large number of moisture.

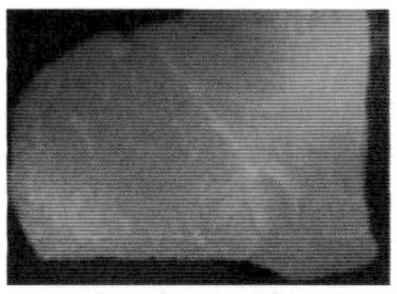

Figure 5. for normal meat

Figure 6. excessive moisture meat

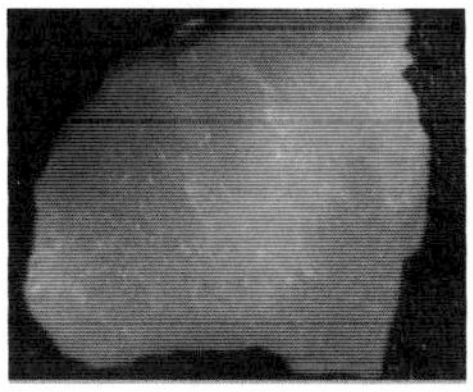

Figure 7.pH value of the meat

Therefore, the color characteristic values (R, G, B and H, S, I) of water-injected meat samples, can be used as one of the basis to judge whether it is water-injected meat.

2.2 Measurement of PH Value of Meat

The pH value is one of the important quality the meat, which influences the color, water-lockup, and the quality of the meat [3]. This paper adopts ray magnetic PHS-3D pH meter to measure the pH value of the meat, As shown in the figure.7.

before collecting the data need to break the meat to pieces. The electrode was inserted into the sample, record the value after mixing evenly, and repeat this operation until the record of pH value is stable. The statistical results of experimental data collected in this study are shown in Table 1.

Table 1: Measurement results of pH

PH value	Samples	Maximum	Minimum	Variance
Training set	40	5.66	5.32	0.01
Testing set	20	5.70	5.31	0.01

There is obvious difference in pH value between the normal meat and the water-injected meat, is the meat PH measurements shown in Fig.8. The pH of normal meat (pork through muscle) is generally greater than 5.4. Whereas the water-injected meat due to injection of a large number of moisture, make further loss of water-soluble substances (such as amino acids) in the meat and then lower the pH value. As a result, according to the trend of changes in the pH value can identify whether the meat is water-injected meat.

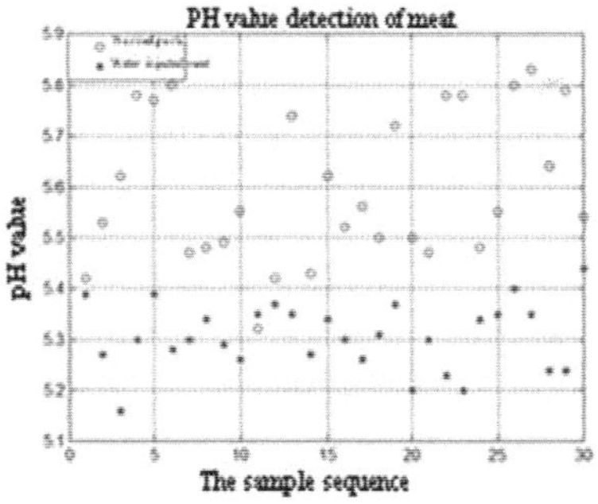

Figure 8. Measurement results of pH

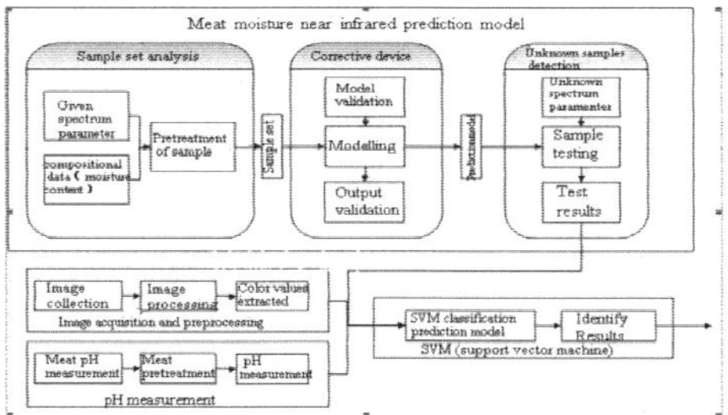

Figure 9. Support Vector Machine (SVM) prediction model

3 The Establishment of Multi-information Fusion

3.1 Support Vector Machine (SVM) Prediction Model

Although the use of near infrared spectroscopy can be used to predict the moisture content of meat, this method is not the original method and need constant maintenance. In order to guarantee the accuracy, this paper adopts the method of the multi-information fusion, using the moisture content of meat detected by NIR, the mean color value (H, I, S, R, G, B), the standard deviation obtained by the image acquisition system and the pH value measured as input features, adopting the SVM algorithm for multi-information fusion, to establish a water-injected identification model. The overall structure of this study is shown in Fig.9.

As shown in the figure.6, the moisture infrared predictive model added the structure of correction for improving the detection accuracy; and then the moisture value and the value of color and the pH value as input values of SVM classification prediction model.

$$f(x;w,b) = sign(w^T x + b) \tag{9}$$

$$G(x,z) = \varphi^T(x)\varphi(z) = \sum_{j=0}^{M} \varphi_j(x)\varphi_j(z) \tag{10}$$

$$\left| y_i(w^T x_i + b) \right| \geq 1, i = 1,2,...,k \tag{11}$$

In summary, using the mean and standard deviation of meat color value, pH value and the meat water content, a total of 14 characteristic component as input features for the support vector machine (SVM) classification prediction model, and using the 60 groups samples collected by experiment as sample set, which include 40 groups samples as the training set and 20 groups samples as the testing set; Namely, a total of 60 samples, each sample contains 14 characteristic component. Due to the choice of the integral operator kernel functions directly affect the performance of the classifier, so use a variety of the integral operator kernel function to establish the model and compare the results.

3.2 Support Vector Machine (SVM) Prediction Results

In the MATLAB7.11.0 (R2010b), generated data.mat file by used the collected data, then viewed The sample group 1-40 as the training set and the sample group 41-60 as the test set for the predictive mode. The lib-svm toolbox [5] is adopted to establish the SVM classifier prediction model, and select the parameter c=1, g=4; the result is shown in Fig.10. According to statistics, the results Accuracy = 95%, can meet the research requirements, and realize the fast discrimination whether the meat is the water-injected meat.

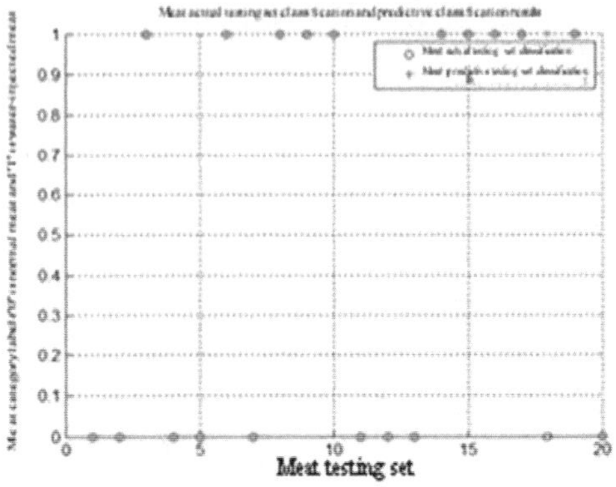

Figure 10. Support Vector Machine (SVM) prediction results

4 Conclusion

In this paper, used the NIR technique to measure the moisture content of the meat; In order to improve the NIR measurement precision and accuracy and overcome the meat moisture measurement is affected , the color value and the pH value of the meat as the auxiliary parameter was adopted, and then established and optimized the multi-information fusion model. As experiments shown, the method used in this paper compared to the other two common methods this study (shown in Table 2), can be in a relatively short period of time to make quick and accurate judgment whether the meat is water-injected meat.

Table 2: The comparison results of three detection methods

Detection methods	Sensory evaluation method	Physical and chemical detection method	Method in this paper
Advantage	Convenient and simple	Accurate	Convenient, simple and accurate
Disadvantage	Completely relies on the experience and with unilateral results	Operation is complex, time consuming, destructive,	Equipment support
Time	About 2-3 minutes	Take a long time	About 4-5 minutes

consequently, the method can realize rapid and accurate detection of water-injected meat. The health and quarantine departments and related meat manufacturers can use this fast, accurate, practical detection technology to detect the water-injected meat, eliminate unqualified meat in the first time and ensure the food safety.

The main author, Guo Peiyuan (1958 -), professor and Ph.D. is mainly engaged in researching of intelligent detection technology of agricultural products and multiple data fusion, embedded technology.

Acknowledgement

This work is supported by National Natural Science Foundation (NNSF) of China under Grant 61473009.

The main author, Guo Peiyuan (1958 -), professor and Ph.D. is mainly engaged in researching of intelligent detection technology of agricultural products and multiple data fusion, embedded technology.

References

1. Sun,Mei, Fu,Yan , Xu Ranran etc. Non-destructive testing of fruit quality based on hyperspectral imaging technology. Journal of Food Technology Science,31(2):67-72,2013.
2. Guo, Peiyuan. Bi Song, Chen, Tianhua, Xu, Guannan, Liu, Xing The fresh classification of pork detection based on multi-data fusion, 29th Chinese Control Conference 2010(7); 213-316
3. Zhao Jiewen, Zhang Haidong, Liu Muhua. Non-destructive Determination of Sugar Content in Apples Using Near Infrared Diffuse reflectance [J]. Transaction of the CSAE, 2010, 21 (3):163-165.
4. Andrews B S, Huntchison S, J. A. Journal of Muscle Foods, 2007, 18:
5. Shi Feng, Wang Xiaoshuan,YuLei et al. MATLAB Neural Network Analysis of 30 Cases[M]. Beijing: Beijing University of Aeronautics and Astronautics Press, 2011.
6. Guo Peiyuan, Chen Yan, Xuan Hong, et al. The Device on Intellectual Detection Techniques to Freshness of Pork. China, 101144780A [P]. 2008,03,19.
7. Zou Xiaobo, Zhao Jiewen, Xia Rong. Near infrared Determination of Sugar Content in Apples Based on Multi-resolution Decmpositiion and Interval Partial Least Square（IPLS）Method[J]. Transactions of the Chinese Society for Agricultural Machinery, 2010,37(6)79-82.
8. Li Xi. Research and Soft Sensing Applications of Partial Least-squares Regression [D]. Dalian: Dalian University of Technology. 2011.
9. Chen Bing, Li Shan, Wu Baohai et al. Multistage Manufacturing Quality Analysis and Forecasting on Dimension Association and Partial Least Square Regression. CIMS, 2010, 15(2):389-398.
10. Peiyuan Guo , Yan, Fu , Fang, Yuan , Man, Bao Research on intellectual detection and classification of pork freshness based on SOM neural network, Mechatronics and Automation 2011.8:432-435
11. Fu yan, Guo Peiyuan, Sun mei Study on Meat Quality Detection Techniques, Bioinformatics and Biomedical Engineering 2012.5:35-37
12. Panigrahi S, Balasubramanian S, GuH. Neural-network-integrated electronic nose for identification of spoiled beef[J]. LWT-Food Science andTechnology, 2012, 39: 135-145.
13. [13] Fu,Yan,Guo,Peiyuan, Sun,Mei,etc. Meat Quality Detection Technology Review. Journal of Beijing Technology and Business University (Natural Science),30(1):69-71,2012

A Nonlinear Blind Source Separation Method Based On Radial Basis Function and Quantum Genetic Algorithm

Pidong Wang[1,2], Ruoyun Tang[1,3], Di Zhao[1] , Hongyi Li[1,3,*], and Jiaxin Chen[4]

[1]*LMIB, School of Mathematics and Systems Science, Beihang University, Beijing 100191, China*
[2]*School of Reliability and Systems Engineering, Beihang University, Beijing 100191, China*
[3]*School of Software Engineering, Beihang University, Beijing 100191, China*
[4]*School of Computer Science and Engineering, Beihang University, Beijing 100191, China*
**Corresponding author (E-mail:Hongyili_buaa@163.com)*

Abstract. Blind source separation is a hot topic in signal processing. Most existing works focus on dealing with linear combined signals, while in practice we always encounter with nonlinear mixed signals. To address the problem of nonlinear source separation, in this paper we propose a novel algorithm using radial basis function neutral network, optimized by multi-universe parallel quantum genetic algorithm. Experiments show the efficiency of the proposed method.

Keywords. EMI signal, Non-linear blind source separation, Radial basis function, Quantum genetic algorithm

1 Introduction

In practice, the nonlinear mixed signals are widespread in blind source separation(BSS), which traditional methods based on linear mixture assumption like ICA are unable to deal with. According to [1,10-12], supposing that x, y are two independent random variables, and f, g are two nonlinear functions, then f(x) and g(y) are independent, too. This fact means that the nonlinear function of source signals could possibly be recovered with the independence assumption of source signals. [2] discusses the existence and uniqueness of solutions to nonlinear ICA and concludes that the solution exists but more assumptions should be added to confirm the uniqueness of solution.

Recently, a method utilizing the radial basis function neutral network (RBFNN) and chaotic self-adaptive particle swarm optimization was introduced in [3]. RBFNN does not deed to determine the mixture model. Theoretically, it is unsupervised and can approximate an arbitrary function [4- 5]. However, RBFNN always converges to a local minimum by adopting general optimization method, such as the gradient descent algorithm. Optimization algorithm searching for global minimum such as Genetic Algorithm [9-12] also doesn't perform well due to large amounts of control parameters, complex calculation and being sensitive to initial values. Quantum genetic algorithm (QGA) combine the features of quantum probability amplitude, superposition, polymorphism and group optimization, which could effectively improve the efficiency. Moreover, the multi-universe parallel quantum genetic algorithm (MPQGA) [6] integrates the concept of subgroup with QGA. We utilize the RBFNN to process signals, and employ fourth-order joint cumulants as independent criterion for separation to optimize parameters.

The rest of this article is as follows: in the second section presents the preliminaries. In section 3, we describe our proposed method in great detail. Then, in section 4, experiments results on synthetic data are provided to demonstrate the performance of the proposed method. Finally, we draw conclusions in the last section.

2 Preliminaries

In this section, we will briefly introduce the nonlinear mixed model, radial basis function neural network, and the multi-universe parallel quantum genetic algorithm.

2.1 Nonlinear mixed model

Nonlinear mixing model for nonlinear BSS could be expressed as x(t)=f(s(t)), where s(t) means the source signal, x(t) is the mixed signal, and $f(\cdot)$ is a unknown nonlinear function.

Supposing $f(\cdot)$ is reversible, and its inverse function exists, denoted by $g(\cdot,\theta)$, where θ is the undetermined parameter. Thus, we obtain the following

$$y(t) = g(x(t),\theta) = g(f(s(t)),\theta) = s(t) \qquad (1)$$

[a] Corresponding author: author@e-mail.org

where $y(t)$ could regarded as an estimation for the source signal $s(t)$.

It could be inferred that if $g(\cdot,\theta)$ could be approximated, $y(t)$ could be obtained from the observed signal. And RBFNN could approximate an unknown function, making it suitable for nonlinear BSS.

2.2 Radial basis function neural net

Broomhead put forward Radial basis Function Neural Network(RBFNN) in 1988 [7]. RBFNN is a feed-forward neural networks with ability of local approximation.

Fig. 1 shows the structure of an N-dimensional input, N-dimensional output network model. It consists of 3 layers; an input layer with N nodes, a hidden layer with one M nodes, and an output layer with N nodes. Nodes in hidden layer are called radial basis neurons and have local responses to the input layer. Nodes in output layer are called linear neurons and conduct linear accumulation on the output from hidden layer.

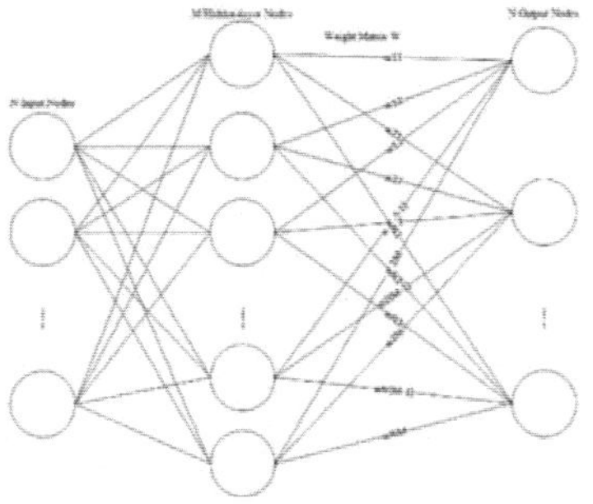

Figure. 1. The topology structure of a typical RBFN

RBFNN is usually utilized to approximate an unknown function $Q: R^n \to R^n$, which could be expressed as $U(X) = WK(X,P)$, where $W = [w_{ij}]$ is an $N \times N$ weight matrix of the output layer, $K(X,P)$ is the core function vector of this network composed of partly corresponding functions, and $X = [x_1, x_2, \Lambda, x_N]^{\mathrm{T}}$ is the input vector.

In this paper, we choose the Gaussian kernel as

$$K(X,P)=[1,\exp(-(X-\mu_1)^T(X-\mu_1)/\sigma_1^2),\exp(-(X-\mu_M)^T(X-\mu_M)/\sigma_M^2)]^T$$

, where $X = [x_1, x_2, \Lambda, x_N]^{\mathrm{T}}$ is the input vector, and , μ_i, σ_i are the center and width parameter of the i-th neuron. $P = (\mu_1, \Lambda, \mu_M, \sigma_1, \Lambda, \sigma_M)$ is the vector parameter for the kernel. And the first component of $K(X,P)$ is set to 1.

2.3 Fourth-order joint cumulant

For a zero-mean random variable, its fourth-order joint cumulant is

$$cum(y_1^2, y_2^2) = E\{y_1^2 y_2^2\} - E\{y_1^2\}E\{y_2^2\} - 2(E\{y_1 y_2\})^2 \quad (2)$$

According to [8], kurtosis is not appropriate for nonlinear problems, then the fourth-order cumulant is proposed as a criterion of independence. It tends to be 0, when the corresponding variables tend to be separated. Therefore it could be utilized for objective function of RBFNN.

2.4 Multi-universe parallel quantum genetic algorithm

It is difficult to obtain global optimal solution in the field of NP-hard problem like nonlinear BSS by using the gradient descent method. Multi-universe parallel quantum genetic algorithm (MPQGA) [6] is a kind of improvement for quantum genetic algorithm. This algorithm groups the individuals into several independent sub-groups, which are called as universes, according to their topology structure. Multi-state gene quantum coding is used to express individuals, which is called qubit coding. Quantum revolving door and quantum NOT gate are used to perform evolution and mutation operation on individuals in the same universe. Best immigration and quantum cross operation are performed between universes. More concrete content about MPQGA such as multiple universe parallel topology, Qubit coding of Chromosome and Information Exchanging Strategy are referred in [18].

3 Preliminaries A nonlinear BSS method based on radial basis function and quantum generic algorithm

In this paper, we use the topology structure as shown in Fig. 2.

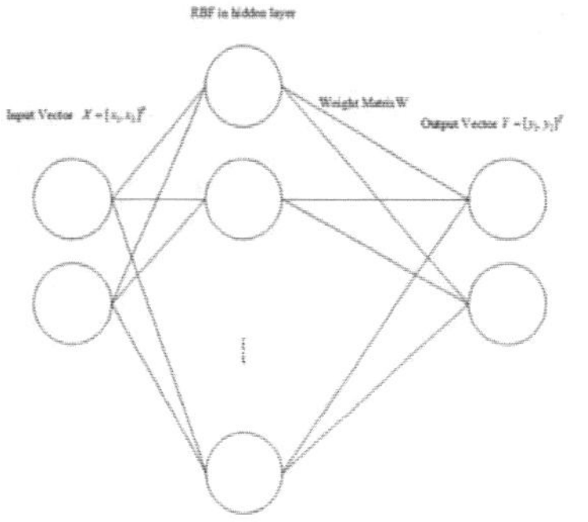

Figure. 2 Selected topology structure

The overall process of our proposed nonlinear BSS model based on RF and MPQGA is shown in Fig. 3. Assuming that nonlinear BSS problem is positive definite, input vector is consist of two mixed signals. After the preprocessing of centralization and whitening, mixed signal vector becomes $X = [x_1, x_2, \Lambda, x_N]^{\mathrm{T}}$, where $x_i = [x_{i,1}, x_{i,2}, \Lambda, x_{i,t}]^{\mathrm{T}}$ means a mixed signal with temporal structure, and $x_{i,t}$ means the value of signal x_i at time *t*.

The cluster step utilizes mixed signal at time t, i.e. $x_{:,t}=[x_{1,t},x_{2,t},\Lambda,x_{N,t}]^{\mathrm{T}}$, to get an M-dimensional clustering center μ, where the width parameter corresponding to the i-th center μ_i is set to mean value of the distances from the three nearest cluster centers. Then the hidden layer nodes of RBFNN, the nuclear function vector could be computed as follows:

$$K(X,P)=[1,\exp(-(X-\mu_1)^T(X-\mu_1)/\sigma_1^2),\exp(-(X-\mu_M)^T(X-\mu_M)/\sigma_M^2)]^{\mathrm{T}} \quad (3)$$

where X vector variable will accept the input vector of RBFNN as value.

After this step, we start to optimize RBFNN optimized by MPQGA iteratively. The input vector is the mixed signal $X=[x_1,x_2,\Lambda,x_N]^{\mathrm{T}}$. Then output vector of hidden layer nodes is:

$$[1,\exp(-(x-\mu_1)^T(x-\mu_1)/\sigma_1^2),\exp(-(x-\mu_M)^T(x-\mu_M)/\sigma_M^2)]^{\mathrm{T}} \quad (4)$$

And the output vector $y=[y_1,y_2,\Lambda,y_N]^{\mathrm{T}}$ could be computed through $N\times(M+1)$ dimension weight matrix $W=[w_{ij}]$ from hidden layer to output layer. After that, we calculate the objective function $fitness(y_1,y_2)=1/|cum\{y_1^2,y_2^2\}||$ to estimate independence of signal separation, which is the feedback for RBF neural network to iterate again. After reaching the iterative target, $y=[y_1,y_2,\Lambda,y_N]^{\mathrm{T}}$ will be regarded as the output vector of the RBF neural network and the estimation of source signal.

More details about Multi-universe parallel quantum genetic algorithm (MPQGA) to optimize the weight matrix $W=[w_{ij}]$ could refer to the [18].

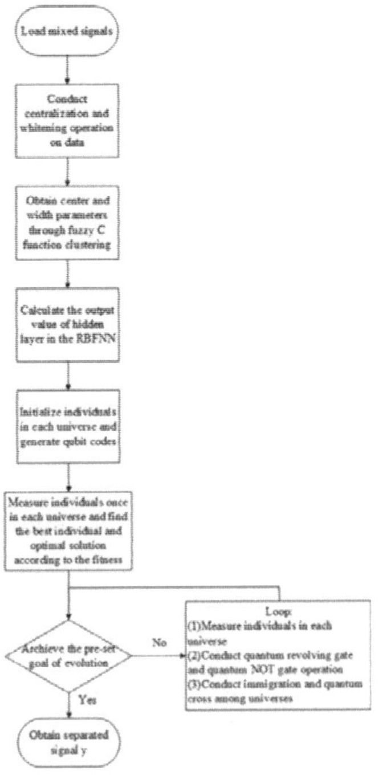

Figure. 3 Flow chart of the proposed method

4 Experimental results and analysis

Simulation experiment is done using Matlab2013a. The source signals are $s_1(t)=0.5(1+\sin(7\pi t))\cos t$ and $s_2(t)=\sin(22\pi t)$. The mixing function is $f(t)=s^3$ and the mixed signal vector is $x(t)=A_2 f(A_1 s(t))$, where A_1 and A_2 are random mixing matrix and $s(t)=[s_1(t),s_2(t)]^{\mathrm{T}}$ is source signal vector.

4.1 Fourth-order joint cumulant

For a zero-mean random variable, its fourth-order joint cumulant is

$$cum(y_1^2,y_2^2)=E\{y_1^2y_2^2\}-E\{y_1^2\}E\{y_2^2\}-2(E\{y_1y_2\})^2 \quad .(5)$$

According to [8], kurtosis is not appropriate for nonlinear problems, then the fourth-order cumulant is proposed as a criterion of independence. It tends to be 0, when the corresponding variables tend to be separated. Therefore it could be utilized for objective function of RBFNN.

4.2 Source signals and mixed signals

The figure of source signals and mixed signals is as follows.

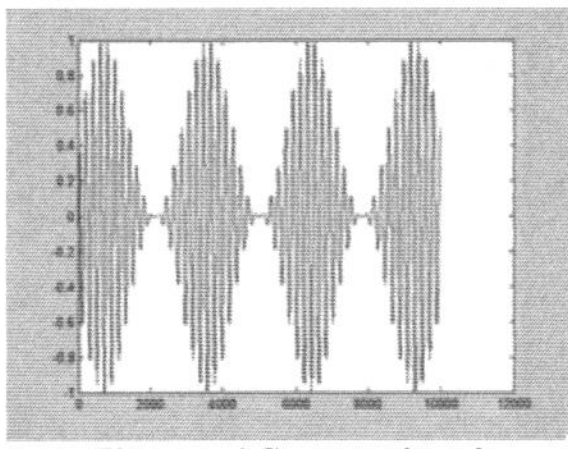

Figure. 4 Source signal $s_1(t)$

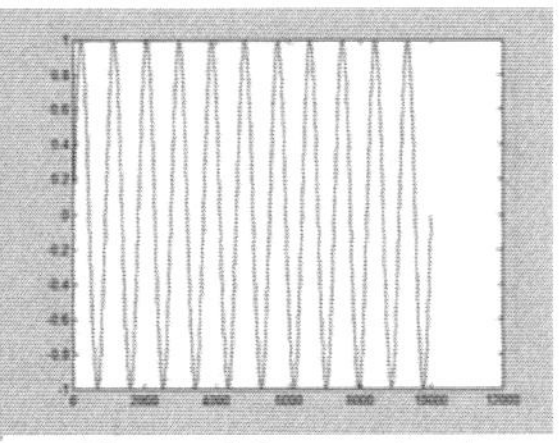

Figure. 5 Source signal $s_2(t)$

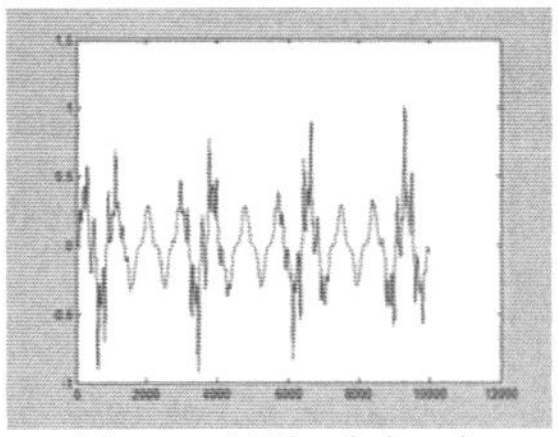

Figure. 6 Mixed signal $x_1(t)$

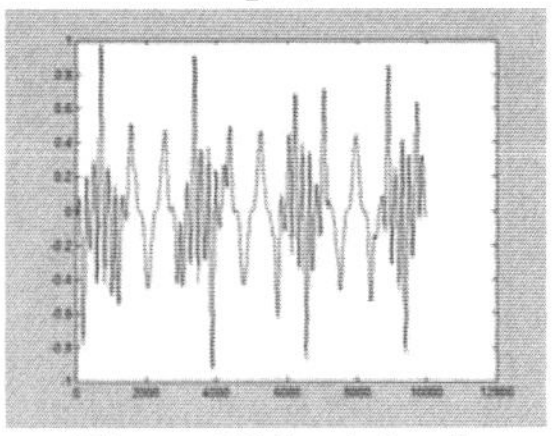

Figure. 7 Mixed signal $x_2(t)$

4.3 Experiment Results

4.3.1 Evaluate function

Similar coefficient is used to measure similarity between source signals and estimated signals for solving sequential uncertainty of output vector. The similarity coefficient between the source signal s_i and estimated signal y_j is defined as $\rho_{ij}=|s_i\cdot y_j|/\sqrt{s_i^2\cdot y_j^2}$.

The value tends to 1 when source signals and estimated signals tends to similar. The uncertainty could be eliminated by choosing the corresponding source signal and estimated signal with high similarity coefficient.

4.3.2 Experimental result

After 20000 generation, the result is like the following.

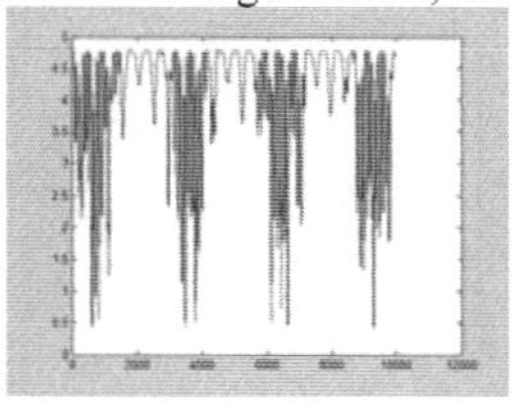

Figure. 8 Estimated signal y_1

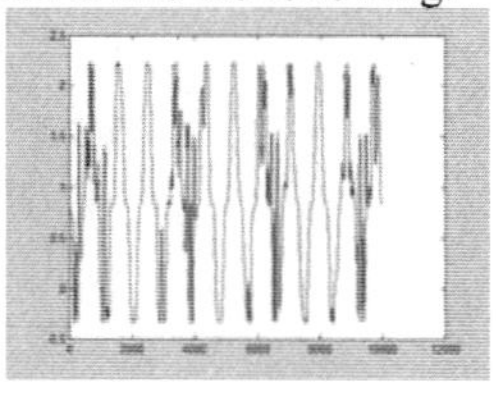

Figure. 9 Estimated signal y_2

Table 1 shows the similarity coefficients between each source signal and estimated signal is called original table of similarity coefficients.

The first row of Table 1 presents two estimated signals output by RBFNN optimized by MPQGA. And the first column presents two source signals.

Table 1. The similarity coefficients between each source signal and estimated signal

ρ_{ij}	y_1	y_2
s_1	0.0362	0.2270
s_2	0.0089	0.8898

Table 2. The similarity coefficients between the source signal and its corresponding estimated signal

$s_i\_y_j$	$s_1\_y_1$	$s_2\_y_2$
ρ_{ij}	0.0362	0.8898

The first row of Table 2 presents the chosen source signal and its corresponding estimated signal, e.g. $s_1\_y_1$ means the estimated signal, y_1 is the estimated signal of source signal s_1. The similarity between them is 0.0362. As the two tables shows, our method performs well on source signal s_2 while bad on s_1, due to the disadvantages of GA and search space fixed to [-1,2].

4.3.3 Comparison experiment to RBFNN optimized by natural gradient algorithm

For comparison, RBFNN optimized by natural gradient algorithm (NGA) is experimented for same parameters. Results are as the following.

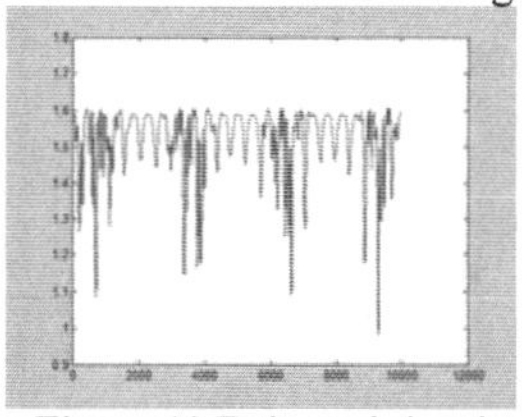

Figure.10 Estimated signal y_1

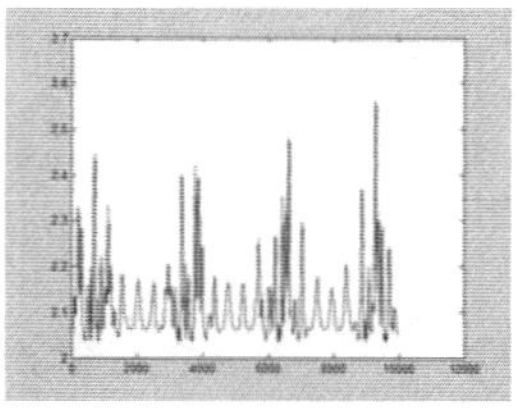

Figure. 11 Estimated signal y_2

Table 3 Original table

ρ_{ij}	y_1	y_2
s_1	0.5360	0.0226
s_2	0.6304	0.3572

Table 4 Corresponding table

$s_i\_y_j$	$s_1\_y_1$	$s_2\_y_2$
ρ_{ij}	0.5360	0.3572

The comparison table of the two methods is as the following.

Table 5 Comparisons between MPQGA and NGA

$s_i\_y_j$	$s_1\_y_1$	$s_2\_y_2$	$s_1\_y_1$	$s_2\_y_2$
Metho d	MPQG A	MPQG A	NGA	NGA
ρ_{ij}	0.0362	0.8898	0.5360	0.3572

The first row of figure indicates estimated signals computed from methods the second row indicates, such as $s_1\_y_1$ and MPQGA in type means the corresponding result s_1 and y_1 from the method optimized by MPQGA.

It could be seen that effect of our method is much better than the natural gradient one on s_2, though both low on s_1.

4.3.4 Comparison experiment to RBFNN optimized by MPQGA emphasizing search space

We found a drawback in experiment above about the same size of search space for each gene, caused by fixed explanation to codes of MPQGA, which leads to unbalance between size of search space and precision of search yielding local optimum. Experiment in 4.2.2 set the search space [-1,2] and precision 0.0001 to emphasize precision. A comparison experiment is done to show the difference between emphasizing precision and emphasizing search space, based on same parameters except for the explanation to codes. Results are as the following.

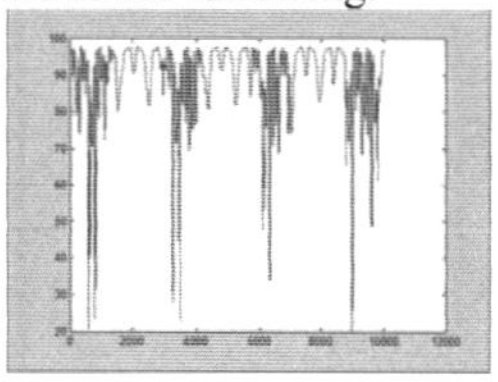

Figure.12 Estimated signal y_1

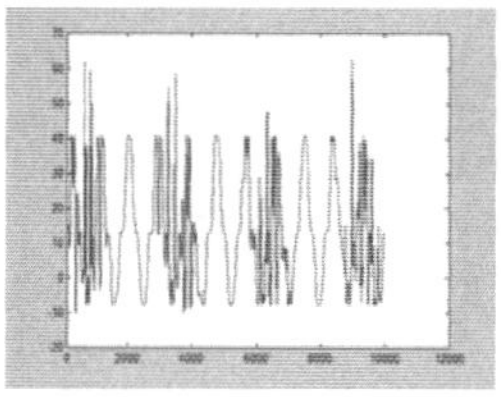

Figure.13 Estimated signal y_2

Table 6 Original table

ρ_{ij}	y_1	y_2
s_1	0.335	0.107
s_2	0.781	0.212

Table 7 Corresponding table

$s_i\_y_j$	$s_1\_y_1$	$s_2\_y_2$
ρ_{ij}	0.1078	0.7818

Compare to the experiment emphasizing search precision before,It could be seen that the MPQGA emphasizing search space is more powerful on the estimation of 1 s , but the similarity coefficient is still low because GA cannot balance search space and search precision adaptively.

Table 8 Comparisons between methods emphasizing precision and space

$s_i\_y_j$	$s_1\_y_1$	$s_2\_y_2$	$s_1\_y_1$	$s_2\_y_2$
Metho d	Precisio n	Precisio n	Space	Space
ρ_{ij}	0.0362	0.8898	0.1078	0.7818

5 Conclusion

Appropriate result is obtained in our experiment and shows our method's effectiveness. However, our method has high time complexity, and requires a large amount of sample data, which would be considered in future.

Acknowledgement

This work was supported by the National Natural Science Foundation of China (No.61379001) and the Key Project of Beihang University (No. 201412) ("Exploration and Practice of Innovation Ability Cultivation in Mathematical Education").

References

1. A. Papoulis, S. U. Pillai. Probability, random variables, and stochastic processes. Tata McGraw-Hill Education, (2002)
2. A. Hyvärinen A, P. Pajunen. Nonlinear independent component analysis: Existence and uniqueness results. Neural Networks, **12(3)**, 429-439 (1999)
3. M. Li, X. Huang, H. Liu, B. Liu, Y. Wu. Prediction of the gas solubility in polymers by a radial basis function neural network based on chaotic self-adaptive particle swarm optimization and a clustering method. Journal of Applied Polymer Science,**130(5)**, 3825-3832 (2013)
4. X. M. Wang. Nonlinear Blind Source Separation using GA optimized RBF-ICA and its Application to the Image Noise Removal. International Conference on Biotechnology, Chemical and Materials Engineering, (2011)
5. T. Kurihara, K. Jinno. Analysis of convergence property of PSO and its application to nonlinear blind source separation. IEEE Congress on Evolutionary, (2013)
6. J. Yang, Z. Zhuang, L. Shi. Multi-universe parallel quantum genetic algorithm. ACTA ELECTRONICA SINICA., **32(6)**, 923-928 (2004)
7. D. S. Broomhead, D.Lowe. Multivariate functional interpolation and adaptive networks. Complex System, **2**, 321-355 (1998)
8. J. A. Yang, Y. Zou, Z. Q. Zhuang. Nonlinear blind source separation algorithm using multi-universe parallel quantum genetic algorithm. Journal of Electronics and Information Technology, 2004, 26(8): 1210-1217.
9. G. Li, K. H. Lee, K. S. Leung, Genetic algorithm based on independent component analysis for global optimization. Parallel Problem Solving from Nature-PPSN IX, Springer Berlin Heidelberg, Iceland, 172-181 (2006)
10. H. Li, C. Wang, D. Zhao. An Improved EMD and Its Applications to Find the Basis Functions of EMI Signals. Mathematical Problems in Engineering, (2015)
11. H. Li, Z. Song, D. Zhao, P. Wang, J. Chen. A Single Channel EMI Signal Separation Method Based on Directly-mean Empirical Mode Decomposition. Journal of Infromation & Computation Science, **12(17)**, 6333-6340 (2015)
12. D. Zhao, H. Li, On the computation of inverses and determinants of a kind of special matrices. Applied Mathematics and Computation, **250**, 721-726 (2015)

MATEC Web of Conferences **61**, 01015 (2016) DOI: 10.1051/matecconf/20166101015
APOP2016

States and Properties of Metallic Systems at a Threshold Breakdown of the Through Holes Under Power Laser Action. Threshold Breakdown of Through Holes in Metal Foils by Powerful Laser Radiation (Part 1)

Kalashnikov E[1a], Kantor M[2]. Bugayev A[3]., Kouprienko D[2]., Chikiryaka A[2].

[1] Moscow Region State University, Moscow, Russia
[2] Ioffe Physical Technical Institute, RAS, S.-Petersburg, Russia
[3] Applied Research Center Old Dominion University, VA 23606 USA

Abstract. Energy of threshold breakdown of through holes in metal foils of different thicknesses by powerful laser radiation is investigated experimentally. Properties of foil matter at "liquid metal-gas" phase transition are revealed. One of controlled parameters of the threshold breakdown is the outlet on the shady side of the target. The threshold breakdown hole is outlet of hole for a given foil thickness when further decrease of energy is not able to create.

1 Introduction

Affecting of powerful laser irradiation on metallic (dielectric and semiconductor) targets leads to formation of craters and holes, moving away of substance from a target - to the ablation [1-12]. The special role are here playing the different thermodynamic states - "solid, liquid, gas, plasma" in which substance consistently goes across.

To control such possible states and transition between them it is necessary to know the amount of energy absorbed by a substance. As a rule, it, last, presents the greatest problem in the tasks of interaction of powerful laser radiation with a substance because as for this purpose it is necessary to know at least an absorption coefficient (which, in turn, depends on a temperature and, accordingly, the state of substance). Practically, the measuring of the temperature of the states of substance during process of experiment is not succeeded. At the best, a model of thermal problem is considered [2-4, 7, 10, 12] with respect to surface and volume temperature distribution at taking into account brought energy. After that, the temperature distribution is compared with experimental results. At low intensities, when perturbation of electronic subsystem is small, such "thermal approach" allows to analyze more complex problems [13]. Also, characteristics, which show that a substance is in a critical state (there is "liquid metal-gas" transition at the critical temperature), are also absent. Nevertheless, at certain cases it is possible to specify characteristics showing closeness to the critical temperature [9] and even to fix the borders of spinodal in a transition "liquid-gas" [3].

Aim of this part of the work is to find experimental characteristics of substance at interaction with laser radiation, which allow to define the presence of "liquid metal-gas" second order phase transition.

The work is devoted to experimental research of laser-induced breakdown threshold holes in metallic foils under act of single-pulse with minimum energy, which would be enough to provide minimum possible hole through in foil of the set thickness. Such statement of problem defined requirements to the radiative device – pulse duration and power of radiation.

2 Scheme of experiment

A ruby laser served as source of irradiation with the wavelength $\lambda = 694\ nm$ in the mode of free generation. An impulse duration was fixed: $\tau = 1ms$. The beam of laser was focused by a short-focus lens (with focal distance of $F = 20\ mm$) on the surface of metallic target in a spot by the diameter of $f = 50\ \mu m$. The laser radiation power density in experiments was varied within the bounds of $10^6 \div 10^8\ W/cm^2$, that, as is generally known [8,10], lies within bounds of transparency of evaporant. To register upraise of hole a detector in a camera-obscure was placed behind a target. Radiation power was registered by another photodetector. The detectors worked in the mode of synchronization with a laser. A delay time was measured during laser-induced breakdown threshold holes. Power of radiation was fixed and registered. Variation of this power (for maintenance of unchanging temporal structure of pulse) carry out by the graduated color filters. Laser-induced threshold holes of foils were studied in the atmosphere of air at normal pressure and room temperature.

Metallic foils were prepared from a copper (with 99,999-weight % of main component), stainless steel, niobium and aluminum. The prepared holes of foils were

[a] ekevkalashnikov1@gmail.com

studied under an optical and electron microscopes. In order to calculate mass of holes a longitudinal (along the axis of holes) micro sections were prepared.

3 Results

Results of experiments are presented on fig.1 - 6. At the small thicknesses of foils, *Fig.1a.* The threshold holes have a cap-like form, by a large diameter turned to the falling radiation. At the increase of thickness of foil the form of the threshold holes is changing and takes

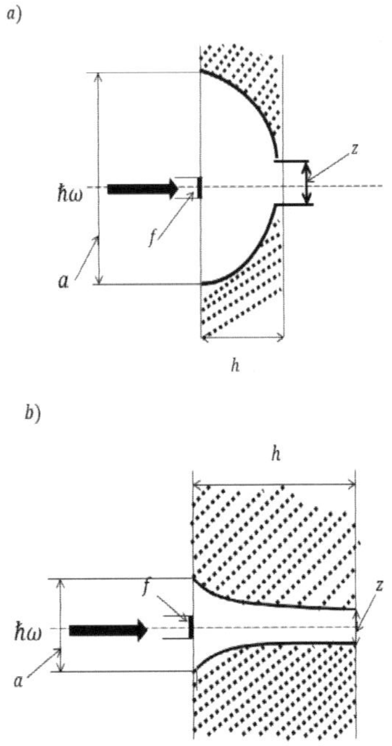

Fig.1. Form of the through threshold hole (along the axis of hole) for thin (a) and thick (b) foils; *a* and *z* is a entrance and outlet diameter of hole, accordingly, *h* is a target thickness, *f* is a focal spot

shape of the prelate truncated cone with hyperbola-like line formative, *Fig.1b.* - In thin foils the energy of radiation is spent on formation of large surface of input hole, substantially exceeding the size of focal spot. In thick foils, the size of input hole exceeds the size of focal spot not very. Size z of output dimeter (a shadow side) of threshold holes is different too. Fig.2 demonstrates the values of diameter, z, of conical holes measured (on a shadow side) on an exit.

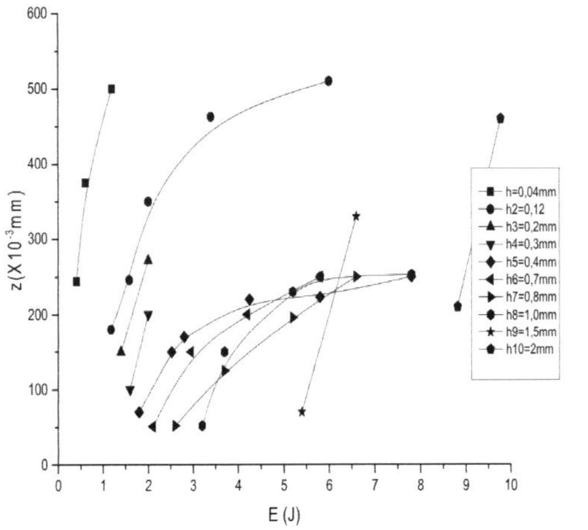

Fig.2. Outlet diameter of through hole vs energy of falling radiations for the different thicknesses of *h* foils.

Dependence of $z = z\ (E)$ on fig.2 is similar to a form dependence of entrance diameter of the deep craters with respect to energy of falling radiation [1, 8, 10]. But a main feature of these dependences (*Fig.2*) consists of minima of $z(E)$ of exit holes. So each minimum exit hole (threshold hole for each thick of foil) corresponds to threshold energy. Subsequent decreasing of energy does not lead to creation of hole through, although a craters and deep small holes appear.

Diameter of exit hole, z, at minimum possible energy, below which hole does not push (*Fig.2*) through is designated the threshold diameter of Ф corresponding to threshold energy E_p. Circumflex of threshold diameters of $z\ (E = E_p) = \Phi$ against energies of thresholdings resulted on *Fig.3.*

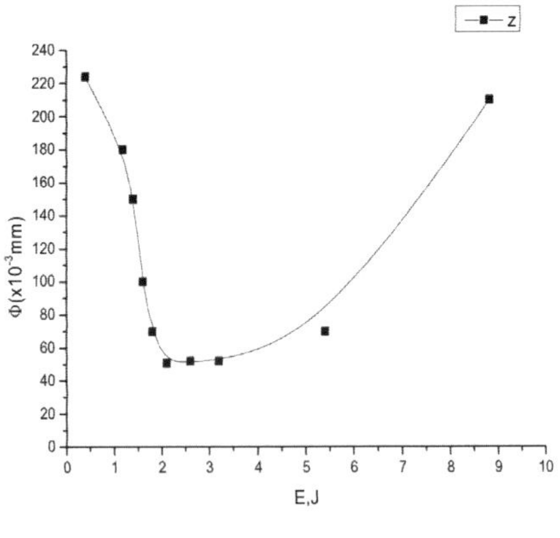

Fig.3. Dependence of threshold diameter of output diameter Φ on threshold energy of E_p for a copper.

On fig.4 dependence over of threshold diameter is brought on the thickness of foil, $\Phi = \Phi\ (h)$ for a copper and stainless steel, having different to heat conductivity.

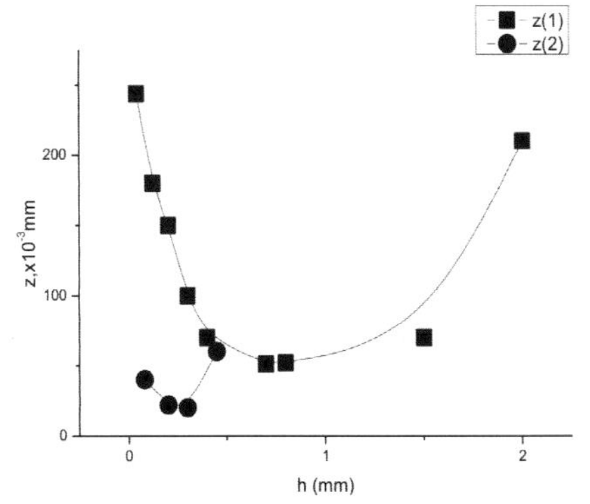

Fig.4. Diameter of the threshold hole depending on the thickness of foil: for a copper, z (1) = Ф1, and for stainless steel, z (2) = Ф2.

Finally, relation of threshold energy, E_p , absorbed by threshold holes with respect to their masses, E_p/m, versus falling energy of radiation of E has pic-like behavior. If it could be possible to express E as absolute temperature, T, the relation $(E_p/m)/E$, would be similar to heat capacity of C_p. Behavior of C_p vs temperature at second order phase transition has λ-like character [14]. So, the relation, $(E_p/m)/E$, could be served as the index of second order phase transition of "*liquid metal-gas*" one.

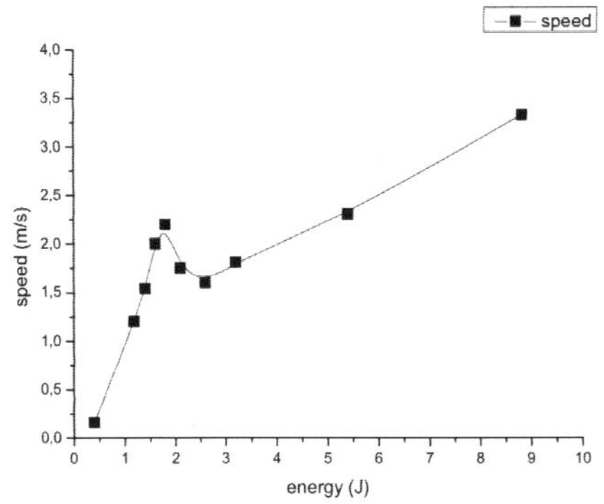

Fig.5. Speed of moving of "gas-liquid-solid" sandwich vs threshold energy of breakdown.

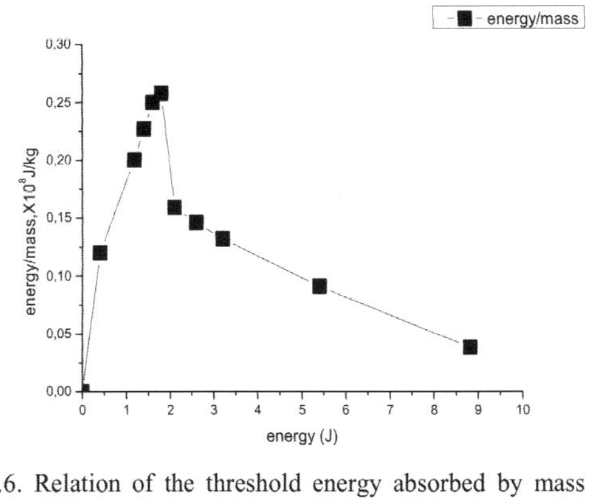

Fig.6. Relation of the threshold energy absorbed by mass of threshold hole to falling on a target of energy of radiation.

4 Discussion

Stream of radiation $10^8 \div 10^{10}$ W/cm^2 provides, as is generally known, mode of the developed evaporation. Mode of heat conductivity here, though eliminated, but determines the that minimum thickness of layer of $l \sim (\chi\tau)^{1/2}$, that has time to transform into a liquid and warmed up in times of action of impulse $\tau \sim 10^{-3}$ s to the high temperature of evaporation of T (here χ is a heat conductivity). However, at the sharp focusing of ray [8, 10], radius of the exposed to rays ground: $r_o \leq l$. Thermal losses become substantial and, in final analysis, determine the terms of forming of the through opening. It appears thus, that $l \sim h$ and condition of semi-infinite model $(h \geq 3\ (\chi\tau)^{1/2})$, [8, 10]) is not supported. In that case, using the traditional analysis of heat conductivity problem, we carry out the quality analysis of dispersion of falling energy. The appearing temperature drop,ΔT, between the lighted and shadow sides of target and between the center of the lighted spot and its periphery causes moving of front of moving away of substance "along" the (negative) gradient of temperature. Thermal losses through a surface normal to the falling ray (through front) equal to $P_h = \lambda{\cdot}\pi{\cdot}r^2{\cdot}\Delta T/h$ and the same through flank surface (truncated cone, approximately, cylinder) equal to $P_r = \lambda{\cdot}\pi{\cdot}r^2{\cdot}\Delta T/r$. The relation of these components

$$R = P_h / P_r = r^2/h^2$$

connects radius, r, with height, h, of threshold hole. It means that

(a) if $R \ll 1$, then all energy is spent on formation of hole.

(b) if $R \gg 1$, that all energy of impulse is spent on formation of area of surface turned of to the falling radiation.

Radius of hole arising up during of laser impulse with energy near its threshold value can be estimated through next relation

$$r^2 = \lambda{\cdot}\tau/\rho{\cdot}c$$

where λ is heat conductivity, ρ – density, c – heat capacity. In that case, relation of R can be written as $R = \lambda{\cdot}\tau/2\rho{\cdot}c{\cdot}h^2$ and boundary between cases (a) and (b), $R = 1$, determines suitable thickness of foil

$$h = (\lambda{\cdot}\tau/2\rho{\cdot}c)^{1/2}$$

As a result, the thickness of copper foil equals to $h_{Cu} = 0{,}25\sqrt{\tau}\ mm$ and the same for stainless steel (st_st) foil is $h_{st\_st} = 0{,}09\ \sqrt{\tau}\ mm$. It leads to $h_{Cu} / h_{st\_st} \approx 3$. This situation is well confirmed by an experiment, fig. 4: relationship of thicknesses h_{Cu} / h_{st_st} with the minimum through threshold holes is the same. Such situation supposes that process of motion of some complex formation of «gas-liquid-solid» through thickness of foil are fully controlled by heat conductivity of solid. . It means that creation of liquid metal on solid with next transformation into gas state takes place during the process. In this way, sandwich-like formation (solid-liquid-saturated gas) moves through thick of foil. It is impotent to notice a speed of the formation,

$V = V(E)$, has specific zigzag at max of $(E_p/m)/E$. It means that after the achievement of values of energies at that a peak appears in dependence of $(E_p/m)/E$ fig.6, there is a critical condition corresponding to the transition "liquid metal-gas"[14]. A liquid metal achieved at a temperature equal or near to the critical temperature. Therefore, that "front" of "liquid-gas" begins to move in an opposite side. A liquid, quickly broadening, loses the border of division "liquid-gas" and arrives at a critical condition.

Two «slowest» processes control heating, phase transformations and motion of such sandwich. First of them is due electron-phonon interaction and is characterized by time of changing lattice temperature $\tau_l \sim 10^{-10}s$ [15]. Second "slowest" process is connected with compression and decompression by the way of joining of atom to arising condensate (or to the exit from an arising up compression) of atoms. This process is characterized by time of order $\tau_2 = (a/v) \sim 10^{-14}$ s , where a is interatomic distance, v is speed of sound. It means that originated liquid will reach equilibrium state at certain temperature faster then will be cooled due to electron-phonon interaction. By the way, it is possible to suppose that at each value of threshold energy E_p , of laser field and at certain thickness, h, of foil in threshold hole, a certain equilibrium temperature will be set and exit threshold hole z will correspond to equilibrium temperature $z = z(T)$. Dependences of z = z (h) for investigated metals are similar. It means processes taking place at the threshold holes of foils on different metals are similar too.

Thus, it is possible to suppose that all set of descriptions of threshold holes at minimum possible (threshold) energy in an impulse largely caused by processes taking place at the critical and near to them parameters of the state of substance

5 Conclusion

A laser-induced threshold holes of foils finds out new characteristics of metals at its reaction on influence of laser radiation:

1) diameter of exit threshold hole, Φ = Φ(Ep), depending on energy of falling radiations Ep for the different thicknesses of h foils has the clearly expressed minimum. The characteristic, Φ = Φ(Ep), on the form remains unchanging for substances with the different heat conductivity- from badly conducting materials to the metals with good heat conductivity

2) relation (Ep/m)/E is similar to heat capacity of Cp at phase transition of the second order "liquid metal-gas". It shows extremum behavior versus energy and corresponds to a minimum of description = (Ep). So (Ep/m)/E could be served as the index of second order phase transition of "liquid metal-gas" one.

3) energy absorbed by threshold hole volume is fully controlled

References

1. J. Hamazaki, R. Morita, K. Chujo, Y. Koyashi Y, S. Tanda, T. Omatsu, OPTICS EXPRESS. 18, 2144 (2010).
2. D. Autrique, Z. Chen, V. Alexiades, A. Bogaerts, B. Rethefeld, http://www.math.utk.edu/~vasili/va/descr/laser/78. Autr.SantaFe_may12.preprt.pdf.
3. D. Autique, G. Clair, D. L'Hermite, V. Alexiades, A. Bogaerts, B. Rethfeld, Journ. Appl.Physics. 114, 023301 (2013).
4. O. Benavides, L. de la Cruz May, Gil A. Flores, PIERS Proc. Stockholm, Sweden. 12-15, 1468 (2013)
5. N.M. Bulgakova., A.V. Bulgakov, Appl. Phys. A 73, 199 (2001).
6. J.Meijer, Journ. Mater. Proces. Techn, 149, 2 (2004).
7. N. Kumar., S. Dash, A.K. Tyagi., B. Raj, Prama – J.Phys. 71, 529 (2008).
8. J. F. Ready. Industrial Applications of Lasers, Academic Press, New York.
9. A.A. Bugayev, M.C. Gupta., M. El-Bandrawy. Opt. Las. Engin. 44, 797 (2006)
10. G.G. Gladush., I. Smurov. Physics of laser materials Processing. (Theory and experiments).Springer- Verlag Berlin Heidelberg 2011.
11. A.A. Bugayev, E.V. Kalashnikov, M.Yu. Kantor., A.L. Stankevich . Phys.Solid.State, 34, 801 (1992).
12. S.I . Anisimov, B.S. Luk'yanchuk., Physics-Usekhi 45 (3), 293 (2002)
13. A.G. Ambrok., E.V. Kalashnikov., S.A. Kukushkin a) Journ. Appl. Physics, 91, 4961 (2002)
14. H.E. Stanley. Introduction to phase transitions and critical phenomena. Clarendon Press. Oxford 1971.
15. A.A. Samokhin. Proc. A.M. Prokhorov GPI RAS (Moscow: Nauka, 13, 3 (1988)).

Study of properties of solutions for quasilinear parabolic systems

Yunzhu Gao, Qiu Meng and Yingjia Guo
Department of Mathematics and Statistics, Beihua University, Jilin City, PR China

Abstract. In this paper, we concern with degenerate and quasilinear parabolic systems not in divergence form with null Dirichlet boundary conditions and positive initial conditions. The local existence and uniqueness of classical solution are proved. Moreover, It will be proved that all solutions exist globally with homogeneous Dirichlet boundary condition.

1 Introduction

In this paper, we consider the following degenerate and quasilinear parabolic systems not in divergence form:

$$\begin{cases} u_{it} = u_i^{p_i}(\Delta u_i + a_i u_{i+1}), i = 1,2,\cdots,m, u_{m+1} := u_1 \\ u_i(x,0) = u_{i0}(x), i = 1,2,\cdots,m, x \in \Omega, \\ u_i(x,t) = 0, i = 1,2,\cdots,m, x \in \partial\Omega, t > 0, \end{cases} \quad (1)$$

where a_i, p_i,are positive constants, $\Omega \in R^N$ is a bounded domain with smooth boundary $\partial\Omega$,The initial data $u_{i0}(x)$, $i = 1,2,\cdots,m$, satisfy

$$\begin{cases} u_{i0}(x) \in C^1(\bar{\Omega}), u_{i0}(x) > 0, i = 1,2,\cdots,m, in\ \Omega, \\ u_i(x,t) = 0, \dfrac{\partial u_{i0}}{\partial \eta} < 0, i = 1,2,\cdots,m, in\ \partial\Omega, \end{cases} \quad (2)$$

here η is the outward normal vector on $\partial\Omega$. The first aim of this paper is to prove the local existence and uniqueness of positive classical solutions. The first aim of this paper is to prove the local existence and uniqueness of positive classical solution of (1) exist globally if and only if $\Pi_{i=1}^m a_i \le \lambda_1^m$, where λ_1 is the first eigenvalue of $-\Delta$ in Ω with homogeneous Dirichlet boundary condition.

Throughout this paper we say that $(u_1, u_2, \cdots, u_m)$ is a classical solution of a initial and boundary value problem if $(u_1, u_2, \cdots, u_m) \in C^{2,1}(\Omega \times (0,T)) \cap C(\bar{\Omega} \times [0,T))$ for some $T: 0 \le T < +\infty$, and $(u_1, u_2, \cdots, u_m)$ satisfies the differential equations in $\Omega \times (0,T)$ and the initial and boundary conditions continuously of this problem.

This system can be used to describe the development of multiple groups in the dynamics of biological groups where u_i , $i = 1,2,\cdots,m$ are the densities of the different groups.

Our paper is organized as follows. In Section 2 we prove the local existence, uniqueness and comparison principle of solution to (1). In Section 3 we will prove that the solutions of (1) blows up in finite time if $\Pi_{i=1}^m a_i \ge \lambda_1^m$ and finally we prove that if $\Pi_{i=1}^m a_i \le \lambda_1^m$ then the solution of (1) exists globally.

2 Existence, uniqueness of solutions

In this section, we shall first consider local existence of system (1). Since $u_i, i = 1,2,\cdots,m$ on the boundary $\partial\Omega$, the equation of (1) is not strictly parabolic type. The standard parabolic theory [1,2] cannot be used directly to prove the local existence of solution to problem (1). To overcome this difficulty, we will use the standard approximate method, see [3], For $\varepsilon > 0$, considering the following approximate problem

$$\begin{cases} u_{i\varepsilon t} = f_{i\varepsilon}(u_{i\varepsilon})(\Delta u_{i\varepsilon} + a_i u_{i+1\varepsilon}, i = 1,2,\cdots,m, x \in \Omega, t > 0, \\ u_{i\varepsilon}(x,0) = u_{i0}(x) + \varepsilon, i = 1,2,\cdots,m, x \in \Omega, \\ u_{i\varepsilon}(x,t) = \varepsilon, i = 1,2,\cdots,m, x \in \partial\Omega, t > 0 \end{cases} \quad (3)$$

Where $u_{m+1} := u_1$, $f_{i\varepsilon}, i = 1,2,\cdots,m$ are smooth function, $f_{i\varepsilon} \ge (\frac{\varepsilon}{2})^{p_i}$, $i = 1,2,\cdots,m$

$$f_{i\varepsilon} = \begin{cases} u_i^{p_i} & , u_i \ge \varepsilon \\ (\frac{\varepsilon}{2})^{p_i} & , u_i < \varepsilon \end{cases}$$

The standard parabolic theory shows that (3) admits a unique maximal defined classical solution

$(u_{1\varepsilon},u_{2\varepsilon},\cdots,u_{m\varepsilon})$ which is defined on $[0,T(\varepsilon))$ here . $0<T(\varepsilon)\le\infty$ .The maximum principle implies that $u_{i\varepsilon}\ge\varepsilon$, $i=1,2,\cdots,m$ which gives $f_{i\varepsilon}(u_{i\varepsilon})\ge u_{i\varepsilon}^{pi}$ and hence $(u_{1\varepsilon},u_{2\varepsilon},\cdots,u_{m\varepsilon})$ the following problem

$$\begin{cases} u_{i\varepsilon t}=u_{i\varepsilon}^{pi}(\Delta u_{i\varepsilon}+a_i u_{i+1\varepsilon}),\ x\in\Omega,\ 0<t<T(\varepsilon),\\ u_{i\varepsilon}(x,0)=u_{i0}(x)+\varepsilon,\ x\in\Omega,\\ u_{i\varepsilon}(x,t)=\varepsilon,\ x\in\partial\Omega,\ 0<t<T(\varepsilon).\end{cases}\tag{4}$$

Where $i=1,2,\cdots,m$, $u_{m+1}\ge u_1$. The fact that $u_{i\varepsilon}\ge\varepsilon$ shows that problem (4) is not degenerate. Since problem (4) is quasimonotone increasing in $(u_{1\varepsilon},u_{2\varepsilon},\cdots,u_{m\varepsilon})$, applying the comparison principle for parabolic systems (.[1, Chapter IV]) we have the following two lemmas.

Lemma 2.1 Assume that positive functions $u_i\in C(\overline{\Omega}\times[0,T(\varepsilon))\cap C^{2,1}(\Omega\times(0,T(\varepsilon)))$ $,i=1,2,\cdots,m$. If $(u_1,u_2,\cdots,u_m)$ is a lower (or upper) solution of (4). Then $(u_1,u_2,\cdots,u_m)\le(\ge)(u_{1\varepsilon},u_{2\varepsilon},\cdots,u_{m\varepsilon})$ on $\overline{\Omega}\times[0,T(\varepsilon))$.

Lemma 2.2 Let $\varepsilon_1<\varepsilon_2$ then $T(\varepsilon_1)<T(\varepsilon_2)$ and $(u_{1\varepsilon_1},u_{2\varepsilon_1},\cdots,u_{m\varepsilon_1})<(u_{1\varepsilon_2},u_{2\varepsilon_2},\cdots,u_{m\varepsilon_2})$ on $\overline{\Omega}\times[0,T(\varepsilon_2))$.

It follows from Lemma 2.2 that there exist $T:0<T\le\infty$ and $(u_1(x,t),u_2(x,t),\cdots u_m(x,t))$ which is defined on $\overline{\Omega}\times[0,T)$ such that $T(\varepsilon)$, $T=\sup_{\varepsilon>o}T(\varepsilon)$ and $(u_{1\varepsilon}(x,t),u_{2\varepsilon}(x,t),\cdots u_{m\varepsilon}(x,t))\to$ $(u_1(x,t),u_2(x,t),\cdots u_m(x,t))$ as $\varepsilon\to 0$.

To discuss the positivity and regularity of $(u_1(x,t),u_2(x,t),\cdots u_m(x,t))$, we should give some estimate of the lower bound of $(u_1(x,t),u_2(x,t),\cdots u_m(x,t))$. Let λ_1 and $\varphi(x)>0(x\in\Omega)$ be respectively the first eigenvalue and the corresponding eigenfunction of the following eigenvalue problem

$$-\Delta\varphi=\lambda\varphi \text{ in } \Omega,\quad \varphi=0 \text{ on } \partial\Omega.$$

and think that $\max_{\overline{\Omega}}\varphi(x)=1$. Then $\lambda_1>0$ and $\frac{\partial\varphi}{\partial\eta}<0$ on $\partial\Omega$.

By the assumptions on $(u_{10},u_{20},\cdots u_{m0})$ we see that there exist positive constants k_i ,$i=1,2,\cdots,m$ such that

$$\begin{cases} u_{i0}(x)\ge k_i\varphi(x),\ i=1,2,\cdots m,\ x\in\overline{\Omega},\\ \frac{\lambda_1}{a_i}\le\frac{k_{i+1}}{k_i},\ i=1,2,\cdots m,k_{m+1}:=k_m,\ if \prod_{i=1}^{m}a_i\ge\lambda_1^m,\\ \frac{\lambda_1}{a_i}>\frac{k_{i+1}}{k_i},\ i=1,2,\cdots m,k_{m+1}:=k_m,\ if \prod_{i=1}^{m}a_i<\lambda_1^m,\end{cases}\tag{5}$$

Lemma 2.3 Let $\varepsilon<1$, $(u_{1\varepsilon},u_{2\varepsilon},\cdots u_{m\varepsilon})$ be the solution of (4), and positive constants $k_1,k_2,\cdots k_m$ satisfy (5). We have the followingestimates:

(i) If $\prod_{i=1}^{m}a_i\ge\lambda_1^m$, then for $i=1,2,\cdots,m$,

$$u_{i\varepsilon}(x,t)a_i\ge k_i\varphi(x)+\varepsilon,\quad\forall(x,t)\in\overline{\Omega}\times[0,T(\varepsilon)),\tag{6}$$

(ii) If $\prod_{i=1}^{m}a_i\le\lambda_1^m$, then for $i=1,2,\cdots,m$,

$$u_{i\varepsilon}(x,t)a_i\ge k_i\varphi(x)e^{-\mu t}+\varepsilon,\quad\forall(x,t)\in\overline{\Omega}\times[0,T(\varepsilon))\tag{7}$$

, where

$$\mu\ge\max\left\{(1+k_i)^{p_i}(\lambda_1-a_i\frac{k_{i+1}}{k_i}),i=1,2,\cdots,m,k_{m+1}:=k_1\right\}$$

.

Proof: (i) When $\prod_{i=1}^{m}a_i\ge\lambda_1^m$, set $\underline{u}_{i\varepsilon}=k_i\varphi(x)+\varepsilon,i=1,2,\cdots,m$, In view of (5) , for $u_{m+1}:=u_1$ and, $k_{m+1}:=k_1$ we have,

$$\underline{u}_{i\varepsilon}^{p_i}(\Delta\underline{u}_{i\varepsilon}+a_i\underline{u}_{i+1\varepsilon})=(k_i\varphi+\varepsilon)^{p_i}(-\lambda_1k_i\varphi+a_i(k_{i+1}\varphi+\varepsilon))$$
$$>(k_i\varphi+\varepsilon)^{p_i}(a_ik_{i+1}-\lambda_ik_i)\varphi$$
$$\ge 0=\underline{u}_{i\varepsilon t},\quad(x,t)\in\Omega\times(0,T(\varepsilon)),$$

,

$$\underline{u}_{i\varepsilon}(x,t)=\varepsilon,\quad(x,t)\in\partial\Omega\times(0,T(\varepsilon))$$
$$\underline{u}_{i\varepsilon}(x,0)\le\underline{u}_{i\varepsilon}(x,0),\quad x\in\overline{\Omega}.$$

This shows that $(\underline{u}_{1\varepsilon},\underline{u}_{2\varepsilon},\cdots\underline{u}_{m\varepsilon})$ is a lower solution of (4). Lemma 2.1 implies that (6) holds.

When $\prod_{i=1}^{m}a_i<\lambda_1^m$, set $\underline{u}_{i\varepsilon}(x,t)=k_i\varphi(x)e^{-\mu t}+\varepsilon,i=1,2,\cdots,m$ Then from (5) and (7), for $u_{m+1}:=u_1$ and $k_{m+1}:=k_1$ we have,

$$\underline{u}_{i\varepsilon t}=-\mu k_i\varphi e^{-\mu t},$$
$$\underline{u}_{i\varepsilon}^{p_i}(\Delta u_{i\varepsilon}+a_i\underline{u}_{i+1\varepsilon})$$
$$=(k_i\varphi e^{-\mu t}+\varepsilon)^{p_i}(-k_i\lambda_1\varphi e^{-\mu t}+a_ik_{i+1}\varphi e^{-\mu t}+a_i\varepsilon)$$
$$>(k_i\varphi e^{-\mu t}+\varepsilon)^{p_i}(a_ik_{i+1}-k_i\lambda_1)\varphi e^{-\mu t}$$
$$\ge(\varepsilon+k_i)^{p_i}(a_ik_{i+1}-k_i\lambda_1)\varphi e^{-\mu t}$$
$$>(1+k_i)^{p_i}(a_ik_{i+1}-k_i\lambda_1)\varphi e^{-\mu t}$$
$$\ge-\mu k_i\varphi e^{-\mu t}=\underline{u}_{i\varepsilon t},\quad(x,t)\in\Omega\times(0,T(\Omega))$$

$$\underline{u}_{i\varepsilon}(x,t)=\varepsilon, \quad (x,t)\in\partial\Omega\times(0,T(\varepsilon)),$$

$$\underline{u}_{i\varepsilon}(x,0)\le u_{i\varepsilon}(x,0), \quad x\in\overline{\Omega}.$$

The same reason as that of (i) shows that (6) holds. The proof is completed.

Theorem 2.1 Problem (1) has a positive classical solution $(u_1,u_2,\cdots u_m)\in[C^{2+\beta,1+\beta/2}(\Omega\times(0,T))\cap C(\overline{\Omega}\times[0,T))]^m$ for some $\beta:0<\beta<1$, and $T:0<T\le\infty$.

Proof: Lemma 2.3 shows that $(u_1,u_2,\cdots u_m)$, the limit of $(u_{1\varepsilon},u_{2\varepsilon},\cdots u_{m\varepsilon})$, is positive in $\Omega\times(0,T)$. In view of Lemma 2.2 and Lemma 2.3, the classical results of local Schauder estimates (see[4]) imply that $u_{i\varepsilon}\in C^{2+\alpha,(1+\alpha)/2}(\Omega\times(0,T(\varepsilon))$ for some $\alpha>0$. Furthermore, for any $\Omega_0\subset\Omega$ and $0<\tau<T_0<T$, we have

$(u_{1\varepsilon},u_{2\varepsilon},\cdots u_{m\varepsilon})\to(u_1,u_2,\cdots u_m)$ in $[C^{2+\beta,1+\beta/2}(\overline{\Omega}_0\times[\tau,T_0])]^m$, $0<\beta<\alpha$, $\varepsilon\to0^+$.

Therefore, $(u_1,u_2,\cdots u_m)$ belongs to $C^{2+\beta,1+\beta/2}(\Omega\times(0,T(\varepsilon))$ and satisfies the differential equations of (1) in $\Omega\times(0,T)$. Fix $\varepsilon_0:0<\varepsilon_0<1$. For any $\Omega_0\subset\Omega$ and $0<\varepsilon<\varepsilon_0$, From Lemma 2.1, we have $\underline{u}_{i\varepsilon}(x,t)<u_{i\varepsilon_0}(x,t), i=1,2,\cdots,m$ on $\overline{\Omega}_0\times[0,T(\varepsilon_0)/2]$, in view of Lemma 2.3, the L^p theory and embedding theorem show that the $C^{\alpha,\alpha/2}(\overline{\Omega}_0\times[0,T(\varepsilon_0)/2])$ norm of $u_{i\varepsilon}, i=1,2,\cdots,m$ is uniformly bounded above for all $\varepsilon<\varepsilon_0$. Hence $(u_{1\varepsilon},u_{2\varepsilon},\cdots u_{m\varepsilon})\to(u_1,u_2,\cdots u_m)$ in $[C^{\beta,\beta/2}(\overline{\Omega}_0\times[0,T(\varepsilon_0)/2])]^m$, $0<\beta<\alpha$, $\varepsilon\to0^+$, which implies that $u_i\in C(\Omega\times[0,T]), i-1,2,\cdots,m$. Similar to the arguments of [4,5,6] we can prove that $(u_1,u_2,\cdots u_m)$ is continuous on $\partial\Omega\times(0,T)$. Using the initial and boundary conditions of (4) we see that $(u_1,u_2,\cdots u_m)$ satisfies the initial and boundary conditions of (1), i.e. $(u_1,u_2,\cdots u_m)\in[C^{2+\beta,1+\beta/2}(\Omega\times(0,T)\cap C[0,T))]^m$ is a classical solution of (1). The proof is completed.

Next, we will prove the uniqueness of positive classical solution of (1).

Theorem 2.2 Problem (1) has a unique positive classical solution $(u_1,u_2,\cdots u_m)\in[C^{2+\beta,1+\beta/2}(\Omega\times(0,T)\cap C(\overline{\Omega\times}[0,T))]^m$, for some $\beta:0<\beta<1$ and $T:0<T\le\infty$. Moreover, if $T<\infty$ then $\overline{\lim_{t\to T^-}}\max_{\Omega}u(\cdot,t)=\infty$

Proof: Suppose that $(v_1,v_2,\cdots v_m)$ is also a positive classical solution of (1). By Lemma 2.1 we have that $v_i<u_{i\varepsilon}, i=1,2,\cdots,m$ and hence $v_i<u_i, i=1,2,\cdots,m$. For $\delta>0$ sufficiently small, denote $\Omega_\delta=\{x\in\Omega|\,dist(x,\partial\Omega)\ge\delta\}$. Let λ_δ and $\varphi_\delta(x)>0$ be respectively the first eigenvalue and the corresponding eigenfunction of $-\Delta$ in Ω_δ with homogeneous boundary condition, and think that $\max_{\Omega_\delta}\varphi_\delta=1$. Multiplying the equations of (1) $\varphi_\delta/u_i^{p_i}, i=1,2,\cdots,m$ respectively, and integrating the results by parts we have

$$\int_{\Omega_\delta}\xi_i(u_i)\varphi_\delta dx=\int_{\Omega_\delta}\xi_i(u_{i0})\varphi_\delta dx-\int_0^t\int_{\partial\Omega_\delta}u_i\frac{\partial\varphi_\delta}{\partial\eta}dS_xdt$$
$$+a_i\int_0^t\int_{\Omega_\delta}u_{i+1}\varphi_\delta dxdt-\lambda_\delta\int_0^t\int_{\Omega_\delta}u_i\varphi_\delta dxdt, \tag{8}$$

where $\xi_i(u_i)=\begin{cases}\dfrac{u_i^{1-p_i}}{1-p_i}, if\ p_i\ne1,\\ Inu_i, if\ p_i=1,\end{cases}$ $i=1,2,\cdots,m$,

$u_{m+1}:=u_1$.

Similarly

$$\int_{\Omega_\delta}\xi_i(v_i)\varphi_\delta dx=\int_{\Omega_\delta}\xi_i(u_{i0})\varphi_\delta dx-\int_0^t\int_{\partial\Omega_\delta}v_i\frac{\partial\varphi_\delta}{\partial\eta}dS_xdt$$
$$+a_i\int_0^t\int_{\Omega_\delta}v_{i+1}\varphi_\delta dxdt-\lambda_\delta\int_0^t\int_{\Omega_\delta}v_i\varphi_\delta dxdt, \tag{9}$$

where $i=1,2,\cdots,m, v_{m+1}:=v_1$. Then (8) subtract (9) respectively, we have

$$\int_{\Omega_\delta}[\xi_i(u_i)-\xi(v_i)]\varphi_\delta dx=-\int_0^t\int_{\partial\Omega_\delta}(u_i-v_i)\frac{\partial\varphi_\delta}{\partial\eta}dS_xdt$$
$$+a_i\int_0^t\int_{\Omega_\delta}(u_{i+1}-v_{i+1})\varphi_\delta dxdt$$
$$-\lambda_\delta\int_0^t\int_{\Omega_\delta}(u_i-v_i)\varphi_\delta dxdt. \tag{10}$$

For any given $T_0:0<T_0<T$, denot $M=\max_{\overline{\Omega}\times[0,T_0]}(\sum_{i=1}^m u_i)$. Applying $v_i<u_i, i=1,2,\cdots,m$, we get that

$$\int_{\Omega_\delta}[\xi_i(u_i)-\xi_i(v_i)]\varphi_\delta dx$$
$$=\int_{\Omega_\delta}(\int_0^1\xi_i'(v_i+s(u_i-v_i))ds)(u_i-v_i)\varphi_\delta dx$$

$$= \int_{\Omega_\delta} (\int_0^1 (v_i + s(u_i - v_i))^{-p_i} ds)(u_i - v_i)\varphi_\delta dx$$
$$\geq M^{-p_i} \int_{\Omega_\delta} (u_i - v_i)\varphi_\delta dx . \quad (11)$$

From (10) and (11) it follows that

$$\int_{\Omega_\delta} [\sum_{i=1}^m (u_i - v_i)]\varphi_\delta dx$$
$$\leq -M_1 \int_0^t \int_{\partial\Omega_\delta} \sum_{i=1}^m (u_i - v_i)\frac{\partial \varphi_\delta}{\partial \eta} dS_x dt$$
$$+ M_2 \int_{\Omega_\delta} [\sum_{i=1}^m (u_i - v_i)]\varphi_\delta dx dt ,$$

where $M_1 = \max\{M^{p_i}, i = 1,2,\cdots,m\}$, $M_2 = \sum_{i=1}^m M^{p_i}(a_i + \lambda_\delta)$. Applying Gronwall's lemma and then taking the limit as $\delta \to 0$ it follows that $u_i \equiv v_i, i = 1,2,\cdots,m$. The proof is completed.

Finally we will give the comparison principle without proof for Problem (1).

Theorem 2.3 Assume that $(\bar{u}_{10}(x),\bar{u}_{20}(x),\cdots,\bar{u}_{m0}(x))$ and $(\underline{u}_{10}(x),\underline{u}_{20}(x),\cdots,\underline{u}_{m0}(x))$ are continuous positive functions in Ω and satisfy (2). Let $(\bar{u}_1(x),\bar{u}_2(x),\cdots,\bar{u}_m(x))$ and $(\underline{u}_1(x),\underline{u}_2(x),\cdots,\underline{u}_m(x))$ be the positivesolutions of (1) with initial data $(\bar{u}_{10}(x),\bar{u}_{20}(x),\cdots,\bar{u}_{m0}(x))$ and $(\underline{u}_{10}(x),\underline{u}_{20}(x),\cdots,\underline{u}_{m0}(x))$, respectively. Denote by $\bar{T}$ and $\underline{T}$ their maximal existence time, If $(\bar{u}_{10}(x),\bar{u}_{20}(x),\cdots,\bar{u}_{m0}(x)) \geq (\underline{u}_{10}(x),\underline{u}_{20}(x),\cdots,\underline{u}_{m0}(x))$ in Ω, then $\bar{T} \leq \underline{T}$ and

$$(\bar{u}_1(x),\bar{u}_2(x),\cdots,\bar{u}_m(x)) \geq (\underline{u}_1(x),\underline{u}_2(x),\cdots,\underline{u}_m(x)),$$
$$\forall (x,t) \in \overline{\Omega} \times [0,\bar{T}) .$$

Acknowledgements

This work was financially supported by the National Natural Science Foundation of Tianyuan Youth Fund Project (11526035).

References

1. O.A. Ladyzenskaja, V.A. Solonnikov, N.N. Ural'ceva, Linear and Quasilinear Equations of Parabolic Type, Amer. Math. society, Providence, RI, 1968..
2. C.V. Pao, Nonlinear Parabolic and Elliptic Equations, Plenum, New York, 1992. J. Rauch. Discontinuous semilinear differential equations and multiple valued maps, Proc. Amer. Math. Soc. 64 (1977) 277-282.
3. S. Wang, M.X. Wang, C.H. Xie, A nonlinear degenerate diffusion equation not in divergence form, Z. Angew. Math. Phys. 51 (2000) 149-159.
4. A. Friedman, J.B. McLeod, Blow up of solutions of nonlinear degenerate parabolic equations, Arch. Rational Mech. Anal. 96 (1987) 55-80.
5. M. Wiegner, Blow-up for solutions of some degenerate parabolic equations, Differential Integral Equations 7 (1994) 1641-1647.
6. Y. Gao, W. Gao Existence and blow-up of solutions for a porous medium equation with nonlocal boundary condition. Applicable Analysis， 90(5)(2011)， 799-809.

MATEC Web of Conferences **61**, 01017 (2016) DOI: 10.1051/matecconf/20166101017
APOP2016

Exploration of Digital Frequency Band System

Xiaoping[1a] Xu, Anqi Wang[2] and Weiqi Wang[3]

[123] *Electronic information and control engineering, Beijing university of technology,China*

Abstract. Digital modulation techniques is a very important method of realizing modern communication.The showing up of Digital modulation techniques promoted the development of modern communication.. Research on digital communications modulation theory, provide an effective modulation, compressed signal transmission band, increase channel multiplexing efficiency, improve signal transmission reliability, has important significance in practice.

1 Introduction

Which will mainly explore research is the basic principle of digital band transmission system, using VHDL hardware description language based, MATLAB as an auxiliary tool, developed each module modem programming, and digital band transmission teach development system developed exploration of modulation and demodulation of the digital band systems research.

First, the digital band system into amplitude shift keying (ASK), frequency shift keying (FSK), phase shift keying (PSK), differential phase shift keying (DPSK), and were done for ASK, FSP, PSK, DPSK research. Each system debugging and simulation, thereby drawn from the analysis of digital band modulation process, results and characteristics.

2 Amplitude shift keying (ASK)

Digital amplitude modulation is to use digital baseband signal to control the amplitude of the sinusoidal carrier, the signal amplitude of the carrier varies with changes in the baseband signal changes, the most basic digital modulation amplitude modulation techniques include binary amplitude modulation and monitoring ary amplitude shift keying modulation . Fundamentals derivation omitted

Waveform simulation and waveform analysis：

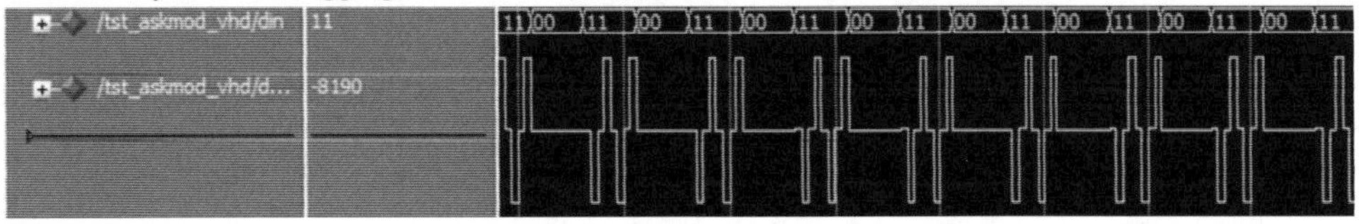

Figure 1 Simulation waveform modulation signal 2ASK

Corresponding author: xuxiaoping@bjut.edu.cn

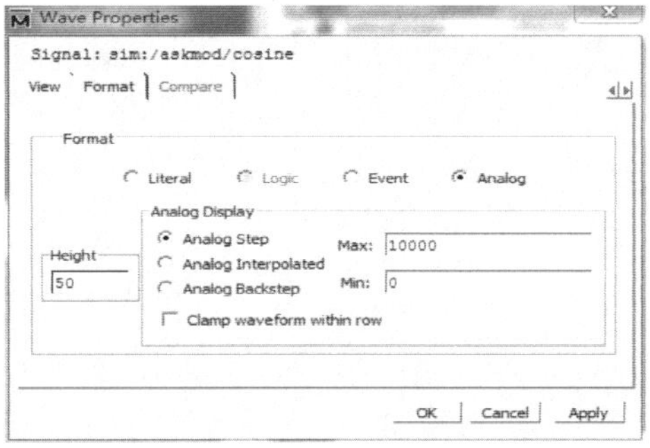

Figure 2 Waveform characteristics setting Figure

2ASK is based on the value of the input pin to adjust the amplitude of the output signal of the cosine signal can be seen from the figure, the output signal having a value of zero and a maximum value in both cases.

4ASK Simulation of the modulation signal waveform :

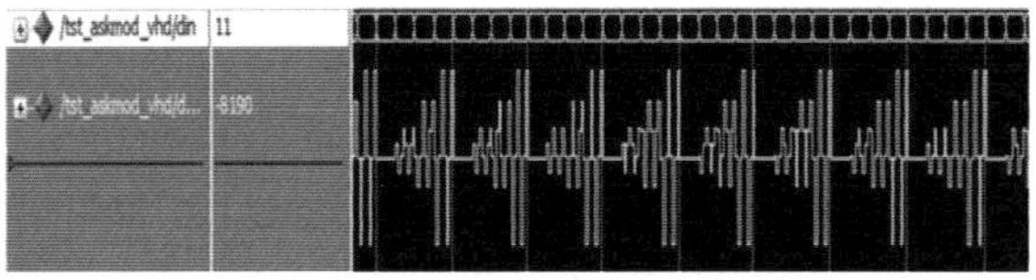

Figure 3 Simulation of the modulation signal waveform

When 4ASK modulation, the input signal is: 00,01,10,11, the amplitude of the output signal is shown, there are four, maximum, minimum, and the middle two values 0.3281max and 0.6563max (approximately 1 / 3max and 2 / 3max).

Demodulation section: 2ASK demodulated signal (without shaping filter) waveform shown in Figure 4, the input signal 2ASK rectification and low-pass filtering

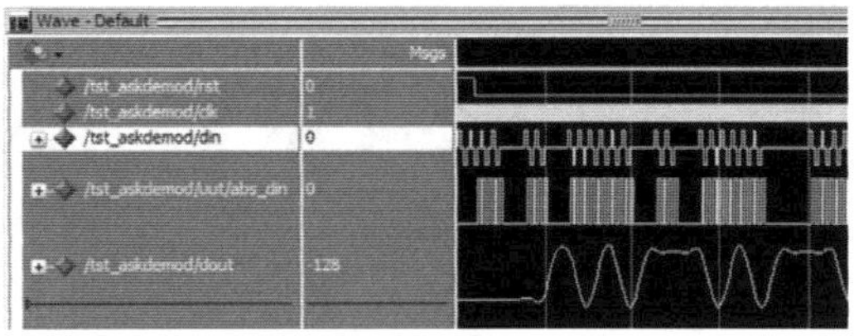

Figure 4 2ASK demodulated signal waveform simulation (without shaping filter)

2ASK demodulated signal (shaping filter) simulation waveforms in Figure 5, after the forming of 2ASK signal filtering, rectifying and low-pass filtering.

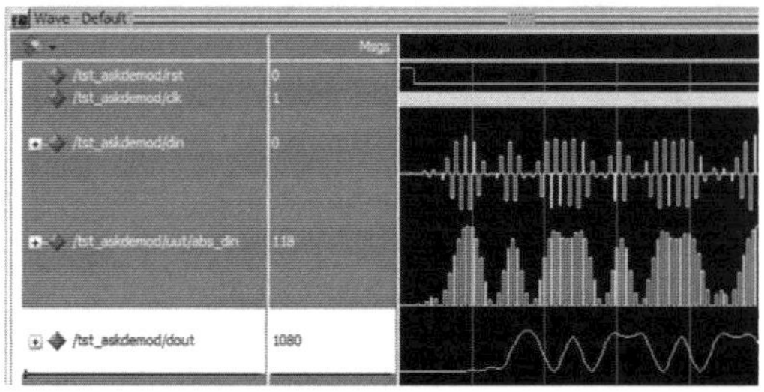

Figure 5 2ASK demodulated signal (shaping filter) simulation waveforms

4ASK demodulation signal (without shaping filter) simulation waveforms in Figure 6, the input signal to 4ASK rectification and low-pass filtering

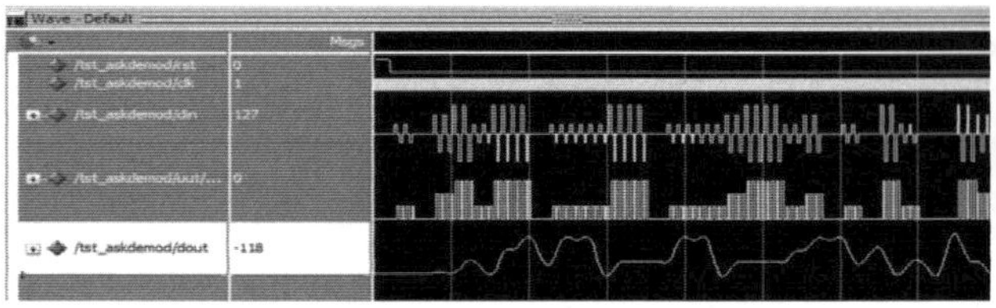

Figure 6 4ASK demodulation signal (without shaping filter) simulation waveforms

4ASK demodulation signal (shaping filter) simulation waveforms in Figure 7, after the forming of 4ASK signal filtering, rectifying and low-pass filtering.

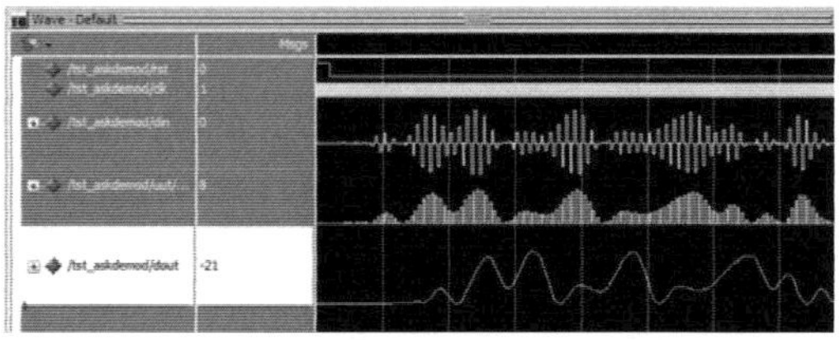

Figure 7 4ASK demodulation signal (shaping filter) simulation waveforms

The results of the simulation results with matlab (Figure 8) were compared and found consistent simulation waveforms, the program can prove ASK signal is demodulated.

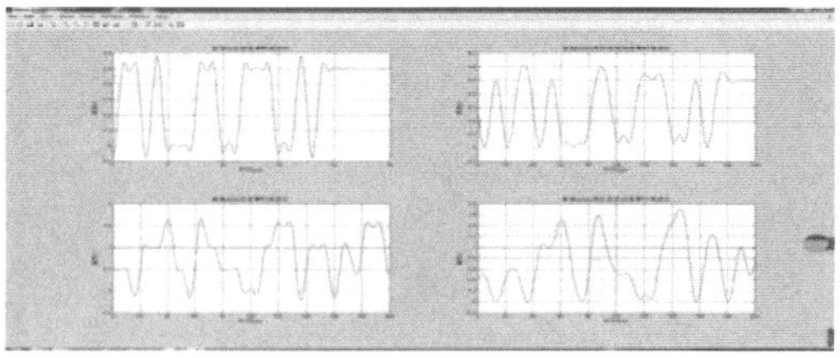

Figure 8 Mat lab for 2ASK, 4ASK demodulation simulation waveform shaping filter.

Frequency shift keying (FSK)

Fundamentals derivation omitted

DOI: 10.1051/matecconf/20166101017

Continuous Phase FSK systems design (CPFSK)

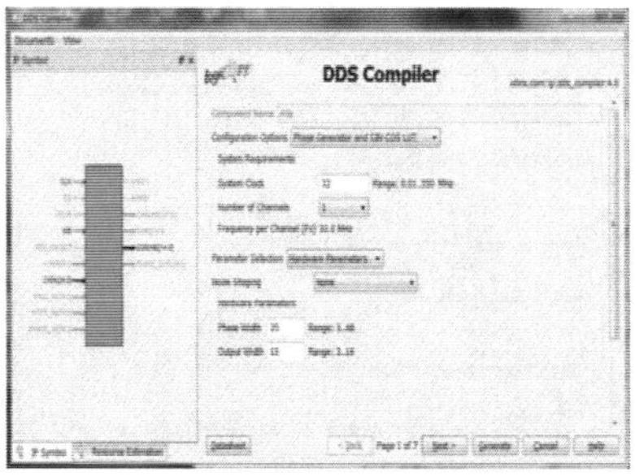

Figure 9 A part of DDS parameter settings

Figure Demodulation aspects of using non-coherent demodulation, design model design as shown in Figure 10.

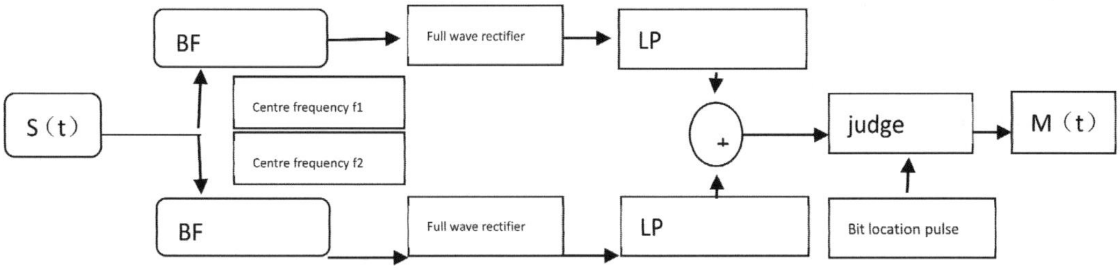

Figure 10 Non-coherent demodulation of the design model

Waveform simulation and waveform analysis

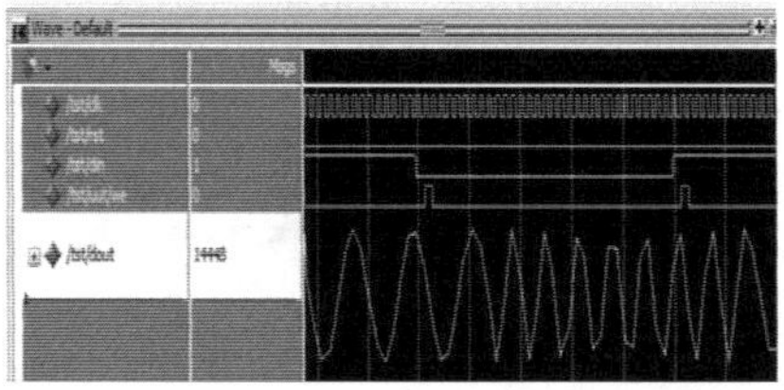

Figure 11 Continuous phase FSK modulation waveform simulation

Use DDS to produce a continuous phase frequency signal, high-frequency serial data clock detection Edge to DDS frequency control port write word after the transition data, the data enable signal is 1, initialization frequency.

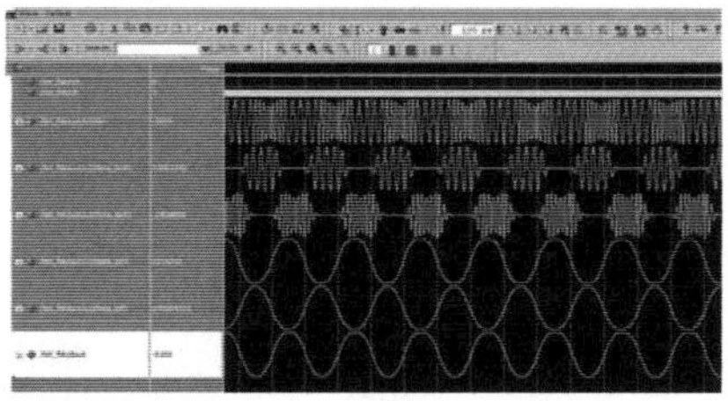

Figure 12 Continuous phase FSK modulation

waveform simulation

Modulated output as the demodulation input, through non-coherent demodulation get 12 simulation results. According to the simulation results, see the output signal maintains the continuity of the signal phase.

A phase shift keying (PSK)

Fundamentals derivation omitted

Four-phase relative phase shift keying (QPSK) system design

It's selected representative DQPSK, four phase relative phase shift keying modulation scheme uses a phase modulation method, primarily shaping filter, serial-parallel conversion, together with a differential coding, bipolar transform.

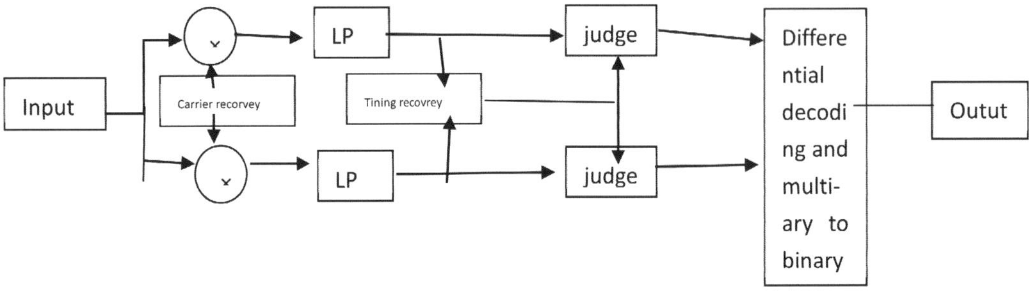

Figure 13 Block Diagram DQPSK coherent demodulation

Waveform simulation and waveform analysis

QPSK modulation waveform shown in Figure 14

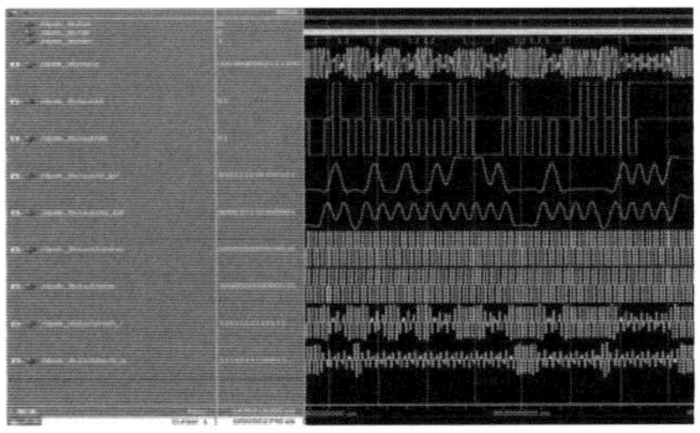

Figure 14 QPSK modulation waveform

QPSK demodulator:First carried out by costas carrier synchronization, demodulating the baseband signal output

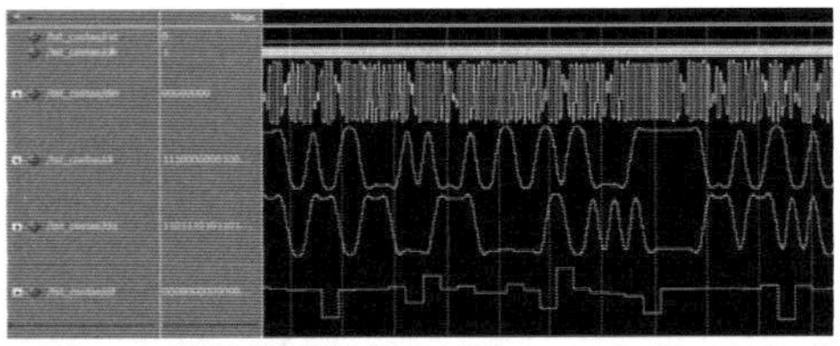

Figure 15 DQPSK demodulator polarity costas loop simulation waveforms

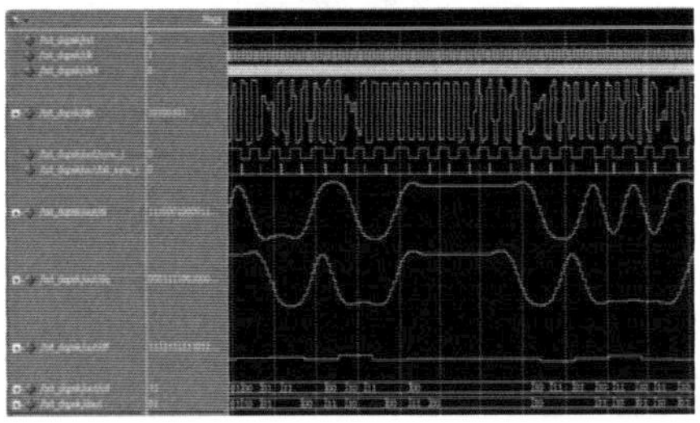

Figure 16 Polar costas loop when no deviation waveform diagram Convergence

Figure 15 were treated through various waveforms generated, three modules were converted code pattern (serial to parallel converter, differential coding, bipolar change), shaping filter, nco module (generated in the same direction orthogonal to the carrier signal and multiplier), a multiplier module, the final in-phase, quadrature branch combined, the output DQPSK signal.

By Costas loop to achieve a coherent carrier extraction, and generates orthogonal to the branch with the baseband waveform.Figure 15 by polar costas demodulated baseband waveform ring

As can be seen from FIG. 16, the data after run to 20,000, after the frequency difference loop convergence jitter significantly reduced, a difference also corresponds to the decrease, increase demodulation efficiency.

Then through BitSync for bit synchronization, were added to the bit timing judgment and differential decoding, then the individual modules are connected through the component is instantiated, it can make the whole demodulation 25

Through simulation overall map view, then loop stability data before and after the DQPSK demodulator exactly.

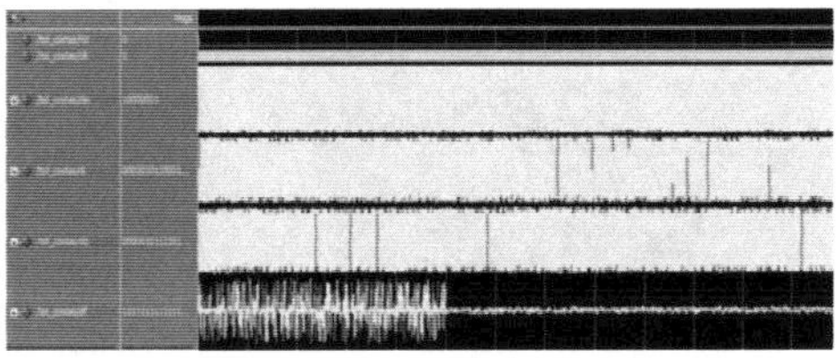

Figure 17 QPSK demodulation system simulation wavefor

Summary

Through this research and exploration, according to the bandwidth of each modulation system to analyze their anti-noise performance Specifically, look at the binary digital modulation contrast, the rate for data dissemination from PSK> FSK> ASK. ASK worst noise immunity, anti-fading capability is not strong, which generally adapt to constant use in the reference channel, FSK strong noise immunity, easy to implement, PSK in the cable channel to achieve high-speed transmission, but the requirements of a stable receiver the reference phase is used to distinguish between the various phases,

DOI: 10.1051/matecconf/20166101017

the best noise immunity, also strong resistance to decay, to implement more complex. The multi-band compared with the binary system improves bandwidth efficiency, but the complexity of the realization of the increase, at the same information rate, multi-band signal element duration wider than the binary, widening the symbol width will increase signal element can also reduce the impact of energy due to inter-symbol interference caused by the channel characteristics of.

References

1. Junnian Wang,Jinfeng Wu. Based on Matlab / Simulink communication principle visual dynamic simulation [J]. Gansu Agricultural University 2003(6).
2. Zhixing Zhang， matlab program design and application [M]， Tsinghua University Press,2004

Modulation Spliced Transform Binaural Cue Coding Algorithm-based Encoder

Xiaoping Xu , Jie Rong[a] and Anqi Wang

Electronic Information and Control Engineering,Beijng university of technology, China

Abstract. Perceptual parameters ICLD, ICTD and ICC has a very important practical significance, so the use of these parameters, the researchers propose parameters based on FFT frequency domain binaural cue coding method to achieve a better stereo encoding quality. Compared with the FFT, modulation spliced transform (Modulated Lapped Transform, MLT) has an energy concentration, facilitating the removal of the characteristics of inter-blocking effect. Therefore, this article will be applied to binaural cue coding MLT transform them, to achieve a binaural cue coding method based MLT transform

1 Implementation of the encoder

Left channel signal and right channel signal input, each frame of MLT transform signal independently, the time-domain signal into a frequency domain MLT coefficients; extract relevant information left channel signal and right channel signal in the frequency domain i.e. between spatial parameters, and use the extracted spatial parameters will do the next right-channel signal is mixed to mono MLT coefficient; and finally the mono signal in the frequency domain inverse MLT transform into the time domain, after windowing, and then through the stack then adding the processed output. At the same time, spatial parameter information obtained by coding a frequency domain is formed independently of the secondary side stereo coding information.After the encoded bit stream is formed in two parts, one containing the inter-channel stereo correlation parameters (ICDL, ICTL and ICC), the other part is a single channel signal after mixing.

2 Time - frequency transform

In the sub-frame processing voice and audio signals usually small error algorithm will bring inter-block effect, seriously affecting the quality of the signal processing, so as to avoid this effect, can be introduced into the language MLT transform audio signals processing them, when used for the conversion of audio signals in the frequency domain.Splicing modulation signal conversion can be accurately reconstructed, while both DCT (Discrete Cosine Transform, DCT) with similar energy concentration characteristics. MLT transform sampling precision, good signal reconstruction effect, a linear transformation.When modulation spliced transform, signal processing between frames will be half of the adjacent overlapping. If the input signal of 16kHz, select the frame length is 20ms, the actual processing frame length is 640 sampling points, and each performs a MLT transform the number of samples needed to get to 1280, so every time you need to change the modulation spliced two frames signal. After the MLT transform the current frame is completed, when the next frame is modulated spliced transformation, retention time domain information of

Corresponding author: 376760445@qq.com

 DOI: 10.1051/mateconf/20166101018

the frame as the first half of the time domain of information processing, while 640 samples will be used later as a signal for one frame the second half of temporal information processing frame. By analogy, then transform modulation overlapping coefficient can be obtained when each time-frequency transform.Forward modulating expression of spliced transformation:

$$X(k) = \sum_{n=0}^{2M-1} x(n)P(n,k) \quad (1.1)$$

Where, k = 0,1, ... M-1, M is the frame length, P (n, k) too - domain aliasing filter offset basic functions by cosine modulation of smooth window to get this function, such as a specific expression formula (1.2) :

$$P(n,k) = h(n)\sqrt{\frac{2}{M}}\cos\left(\frac{(2n+M+1)(2k+1)\pi}{4M}\right) \quad (1.2)$$

Wherein the window function h (n) of the low-pass half-band filter. IMLT need to use the current frame X (k) and the previous frame Xp (k), in order to restore the original signal becomes x (n), the formula (1.3) shown in the expression for the MLT inverse transform equation.

$$x(n) = \sum_{k=0}^{M-1} (X(k)P(n,k) + X^{p}(k)P(n+M,k)) \quad (1.3)$$

Usually used as formula (1.4) shown in the window function as a window to allow the signal reconstruction operation to achieve the best results.

$$h(n) = \sin\left(\frac{(2n+1)\pi}{4M}\right) \quad (1.4)$$

MLT operation for ultra-wideband speech input audio signal according to the following process of implementation:Let x (n) for each input signal MLT transform recent 1280 speech signal samples, each containing 640 samples of each of the previous and current frames.Where: x (0) is the first sampling point, 0≤n≤1279.After conversion output 640 MLT transform coefficients smlt (m), where: 0≤m <640MLT transform is:

$$s_{mlt}(m) = \sum_{n=0}^{1279}\sqrt{\frac{2}{640}}\sin\left(\frac{\pi}{1280}(n+0.5)\right)\cos\left(\frac{\pi}{640}(n-319.5)(m+0.5)\right)x(n)$$

(1.5)

MLT transform decomposed into a window function, overlapping, accumulate and type Ⅳ discrete cosine transform. Window functions overlap, accumulating detailed process is:

$$w(n) = \sin\left(\frac{\pi}{1280}(n+0.5)\right) \quad 0 \; I \; n \; I \; 640 \qquad (1.6)$$

Thus, we get v (n) be able to use type Ⅳ discrete cosine transform MLT transform operations to achieve the original signal. MLT transform can be rewritten in the form below:

$$s_{mlt}(m) = \sum_{n=0}^{639}\sqrt{\frac{2}{640}}\cos\left(\frac{\pi}{640}(n+0.5)(m+0.5)\right)v(n) \qquad (1.7)$$

The current practice, FFT fast algorithm has been applied to the DCT by adding a fast algorithm can effectively reduce the complexity of the DCT.After the MLT transform changes made above, in programming practices can be considerably reduced when the amount of computation to reduce the complexity of the algorithm.

3 Subband Division

During the encoding process the audio signal needs to subband division. MLT spectral coefficients subband division process using a uniform way division, in this paper the frequency domain signal lines per frame value N is 640, the 640 spectral coefficients B = 20 is divided into sub-bands, such as sub-band b, where A0 = 0, sub-band border demarcation as shown in Table 1, where b is the sub-band number, kb b represents the initial sub-band frequency points.

Table.1 Sub-band boundary comparison

b	0	1	2	3	4	5	6
k_b	0	32	64	96	128	160	192
b	7	8	9	10	11	12	13
k_b	224	256	288	320	352	384	416
b	14	15	16	17	18	19	20
k_b	448	480	512	544	576	608	639

4 Mixing down

In order to allow the input signal contains the original audio signal information will be taken down mixed multi-channel signal. Between simple superposition channel signal such that the signal can be weak or enlarge the content, and causing signal and the signal power of each channel are not equal, larger or smaller.

Hybrid technology is usually taken down in order to avoid the occurrence of this problem, this technology and signal proper balance, so that the signal power is approximately equal to the signal power of each channel.

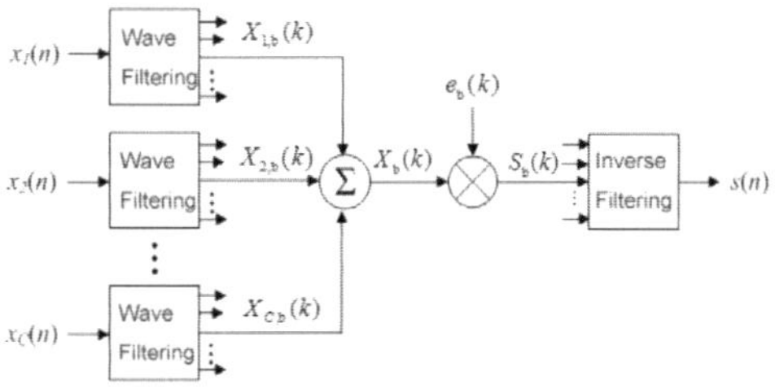

Figure 1 Down multi-channel signal mixing process schematic

As shown in the figure is down mixed multi-channel signal process. Downmix signal forming process, the first input audio signal xc (n) (1≤c≤C) by filtering decomposed into a plurality of sub-band signals Xc, b (k) (Xcb (k) for the MLT coefficients, b is the sub-the subscript, 0≤b≤B), followed by the corresponding sub-band signal for each channel signal obtained by adding Xb (k), and then multiplied by the weighting factor eb (k) generate Sb (k), the last of the B Sb (k) (for the MLT coefficients) of inverse filtering generation and signal s (n), where C is the number of channels, B is divided into sub-band number.The following two-channel signal will be deduced as an example the case of stereo weighting factor eb (k) of the expression:Dual-channel short-term power of the b sub-band within the mixed-signal estimation formula (1.8) below.

$$\begin{aligned} P_b &= \sum_{k=A_{b-1}}^{A_b-1} |S_b(k)|^2 \\ &= \sum_{k=A_{b-1}}^{A_b-1} |e_b(k)(X_{1,b}(k)+X_{2,b}(k))|^2 \\ &= e_b^{\,2}(k) \sum_{k=A_{b-1}}^{A_b-1} |X_{1,b}(k)+X_{2,b}(k)|^2 \end{aligned} \quad (1.8)$$

Short-term power estimate of the left and right channels and formula (1.9) below.

$$\begin{aligned} P_b' &= P_{1,b} + P_{2,b} \\ &= \sum_{k=A_{b-1}}^{A_b-1} |X_{1,b}(k)|^2 + \sum_{k=A_{b-1}}^{A_b-1} |X_{2,b}(k)|^2 \end{aligned} \quad (1.9)$$

Mixing process need to satisfy Pb = Pb ', can be derived eb (k) of the formula as in formula (1.10) below.

$$e_b(k) = \sqrt{\frac{\sum_{k=A_{b-1}}^{A_b-1} |X_{1,b}(k)|^2 + \sum_{k=A_{b-1}}^{A_b-1} |X_{2,b}(k)|^2}{\sum_{k=A_{b-1}}^{A_b-1} |X_{1,b}(k)+X_{2,b}(k)|^2}} \quad (1.10)$$

Left and right channel signals can be omitted in the subscript sub b, formula (1.11).

$$e_b(k) = \sqrt{\frac{\sum_{k=A_{b-1}}^{A_b-1} |X_1(k)|^2 + \sum_{k=A_{b-1}}^{A_b-1} |X_2(k)|^2}{\sum_{k=A_{b-1}}^{A_b-1} |X_1(k)+X_2(k)|^2}} \quad (1.11)$$

References

1. Ping Ping, Yicheng .Song Digital communications, Electronic Industry Press, 2008. 562
2. Xuanpeng Li. Based on spatial perception of stereo coding information [J] Southeast University, 2006,(04).

MATEC Web of Conferences **61**, 01019 (2016) DOI: 10.1051/matecconf/20166101019
APOP2016

Simulation of Frequency Hopping Communication System Based On MATLAB

Xiaoping Xu[1] Weiqi Wang[2] Lan Yi[3] Jie Rong[4] Anqi Wang[5]
[1, 2.3.4.5] Electronic Information and Control Engineering,Beijng university of technology, China

Abstract. In today's information age, how to carry out the accurate information communication is a crucial issue in the field of communication, which is widely used in civil and military fields. Frequency hopping communication system is a typical spectrum communication system, it is in military communication, mobile communication, computer wireless data transmission and wireless LAN and other fields have a very wide range of applications, has become the current short wave communication is one of the important development directions. In this paper, we introduce a design of frequency hopping communication system based on Matlab.Through the frequency hopping communication system the simulation research and theoretical analysis, to achieve the desired results.

1 Background

Spread spectrum communication has very strong anti-interference ability and security, first appeared in the military field. In recent years, spread spectrum communication is also widely used in the field of civil. Spread spectrum communication has two ways,they are the direct sequence spread spectrum (DSSS) and frequency hopping spread spectrum (FHSS).

2 The Working Principle and Characteristics of Spread Spectrum Communication

In the information theory, the channel capacity formula of the channel capacity formula is $C=w\log_2(1+SNR)$

This is the Shannon's capacity formula. C is the channel capacity (limit speed data transmission)W is the signal bandwidth and SNR is the signal to noise ratio.

The formula shows that the signal to noise ratio and the signal bandwidth can be the same as the transmission rate can be changed. We can improve the signal bandwidth to reduce the signal to noise ratio requirements, Thus enhancing the ability to resist interference. By expanding the signal bandwidth, the ability to resist interference is also improved.

As mentioned above, spread spectrum communication through the expansion of signal bandwidth, in exchange for a strong anti-interference ability, which is the biggest characteristic of the spread spectrum communication. In addition to the signal spectrum expansion, its power in the entire extended frequency potential, but also become sparse and close to the uniform distribution, which increases the difficulty of the third party capture. Who do not know the synchronization of spread spectrum signal of third-party recipients, spread spectrum signal as buried in the noise as it is difficult to detect. In addition, as well as to improve the time resolution of spread spectrum communication and multi access etc.. So spread spectrum communication technology has become a hot spot of research.

One of the important parameters of spread spectrum communication is spread spectrum gain, which reflects the system's ability to resist interference. It is a measure to improve the SNR.

$$G=SNR_{out}/SNR_{in}=R_c/R_d=B_c/B_d$$

DOI: 10.1051/matecconf/20166101019

R_s is the transmission rate of spread spectrum code, R_d is the transmission rate of information data, B_s is the bandwidth of spread spectrum code, B_d is the bandwidth of information data.

3 Frequency Hopping Communication System

Frequency hopping is to control the frequency synthesizer by using the pseudo code sequence to form the frequency synthesizer. The transmitted information code and the pseudo random sequence mode two (or the waveform multiplication) constitute the hopping instruction (i.e., the frequency hopping pattern), and by which the transmission frequency is randomly selected. The simplified block diagram of the frequency hopping communication system is shown in Figure 1. After modulation, the sender's information code sequence and the pseudo random sequence are controlled by different frequency hopping pattern. At the receiving end, the signal and noise received by the filter is sent to the mixer. The receiver's local oscillator signal is a frequency hopping signal, the same as the transmission end is the same, the frequency of the two frequency synthesizer is relatively, but the corresponding frequency has a frequency difference, just as the receiver of the intermediate frequency. As long as the sender and receiver of the pseudo-random code synchronization, it can make both the frequency hopping frequency synthesizer and the frequency synthesizer to generate the frequency synchronization, after mixing, you can get a constant frequency signal, and then the signal demodulation, you can recover the transmitted information. In the case of interference signal, the frequency of the frequency synthesizer is not related to the frequency of the frequency synthesizer. Therefore, it can not enter into the intermediate frequency channel, which can not interfere with the formation of the frequency hopping system.

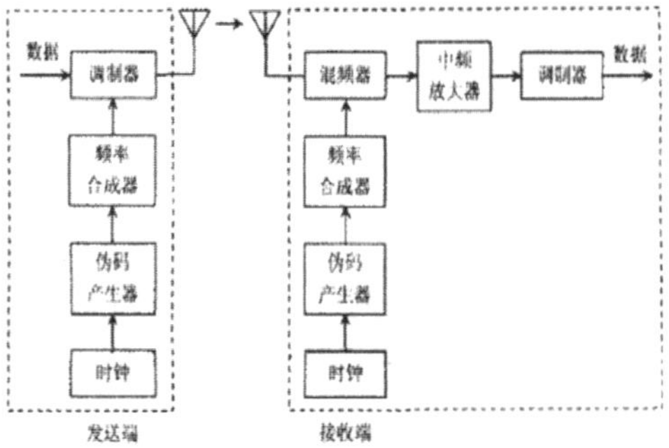

Figure 1 Frequency hopping communication system model

4 Simulation Platform of MATLAB System

MATLAB is a kind of mathematical application software, after many years of development, the development of the communication system, including the number of tools, become the most widely used in scientific research and engineering software. Simulink is a visual simulation tool in MATLAB, just click and drag the mouse operation to complete the simulation, it is simple and clear, and the user can immediately see the results of the simulation system. In addition, MATLab provides a program language based on matrix calculation, and it can also be shown by the function of writing the function of the combination of source program and graphics rendering.

Part in the MATLAB simulation procedures can be broadly divided into the majority of 6, respectively, the main function, sending module, receiver module, AWNG channel, Walsh function and error counters.

In this paper, a simple simulation is made on the source code of a frequency hopping communication, and the time domain waveform and spectrum of the signal are analyzed.

Procedures are mainly composed of the following steps: first, through the rand function to generate a data signal carrier and a fixed frequency carrier signal signal, and then the two BPSK modulation, the modulation output is modulated signal BPSK. Frequency signal spread is generated by the stitch function. We choose 6 kinds of frequency (we consider 6 kinds of jump frequency to meet the requirements of the working process and the effect of frequency hopping). Finally, signal BPSK and modulated signal spread frequency hopping modulation, frequency hopping spread spectrum signal.

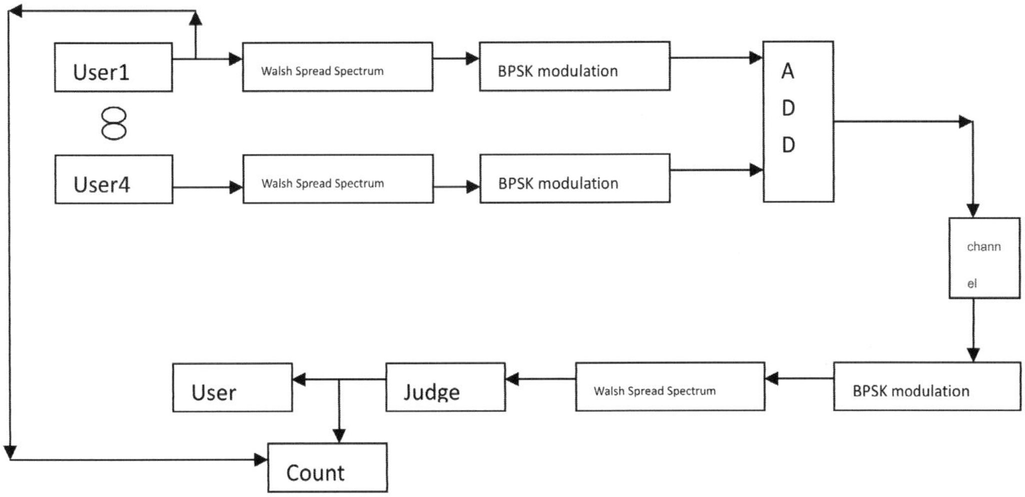

Figure 2 block diagram of simulation system

5 Simulation Results

Signal in frequency hopping spread spectrum before and after the waveform, the use of MATLAB simulation, as shown in figure 2: The original bit sequence is a randomly generated binary sequence as the data signal to be transmitted. After BPSK modulation, the BPSK modulated signal is obtained. Sread signal with 6 frequencies is a randomly generated FM signal.There are six kinds of frequency in the picture, frequency hopped spread spectrum signal is the output signal after frequency hopping modulation.The bandwidth is broadened, and the frequency components are increased and the frequency components are increased, so that the receiver is not aware of the original data.

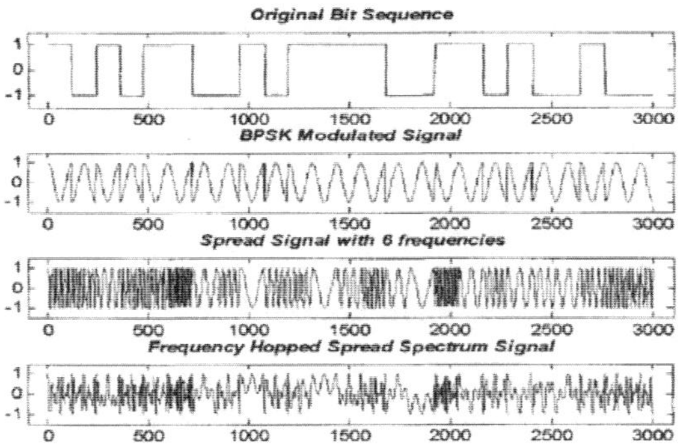

Figure 3 Frequency hopping communication signal

6 Conclusions

Frequency hopping communication is widely used in military and civilian fields because of its anti - jamming. At present, the popular Bluetooth technology and wireless LAN technology have the figure of frequency hopping technology. The modulation of transmitter in frequency hopping communication system through the MATLAB simulation analysis the principle of frequency hopping communication sender's working process and spread spectrum.

Building the frequency hopping system and simulation, and the error rate analysis. The theory basis of communication theory is based on the good advantages of frequency hopping communication:

1) considering the bandwidth and signal to noise ratio in order to find the optimal parameters, the maximum utilization of the anti-jamming performance is given;

2) the secrecy and non-distortion characteristics of frequency hopping communication.

3) frequency hopping communication uses a variety of patterns or clocks, so that the frequency resource sharing within a bandwidth to improve frequency utilization.

References

1. Junnian Wang,Jinfeng Wu. Based on Matlab / Simulink communication principle visual dynamic simulation [J]. Gansu Agricultural University 2003(6).
2. Ping Ping,Yicheng .Song Digital communications,Electronic Industry Press, 2008. 562
3. Zhenyu Li. Research on spread spectrum communications received coded modulation and related technologies [J] NORTHERN JIAOTONG UNIVERSITY newspaper, 1995,(04)

A wireless Transient Electromagnetic Data Acquisition System based on Virtual Instrument

Yue Li and Xuming Zeng [a]

Faculty of Mechanical and Electronic Information, China University of Geosciences, Wuhan, Hubei, China

Abstract. Electromagnetic survey equipment is widely used in the geophysical field, such as mineral exploration and underground oil detection. To meet the demand of high processing speed and flexible maneuverability, this paper proposes a wireless transient electromagnetic system combined virtual instrument and Wi-Fi technology. Transient electromagnetic receiver and transmitter, Wi-Fi communication links and human operation terminal comprise the whole acquisition system. With the implementation of software programmed by using LabVIEW, the system can process in real-time. This system replaces the old dedicated hardware with compact and flexible device with the same functionality and superior performance.

1 Introduction

Transient electromagnetic method (TEM) is one of the main method which is widely used in mineral exploration, aviation and other fields [1]. At present, Jilin University, Central South University and Chongqing University designed electromagnetic detection equipment which have the slow processing speed, single function, cable transmission and other defects, which cannot meet the actual measurement of the diversity, versatile needs [2-5]. To meet the demand of high processing speed, flexible maneuverability and remote monitoring, a method based on virtual instrument and Wi-Fi technology has been proposed in the transient electromagnetic data acquisition system.

Since the virtual instrument is created by computer technology, it has been became the combination of computer technology and instrument control, test and measurement technology [6]. It is more flexible by establishing the graphical "virtual" panel to realize the instrument's function for measurement, control, transformation, analysis, display [7], etc.

Wireless communication offers lower wiring costs, simple data transfer, and remote monitoring for electromagnetic data acquisition equipment, since the measuring point spatial position is not fixed. Applying virtual instrument and Wi-Fi communication, coupled with computer system's high speed, high precision processing, which can realize data collected under different field environment in real-time display, processing, preservation and analysis of complex algorithms, greatly alleviate the microcomputer in the instrument [8-9].

2 System design

2.1 The main structure of the system

The whole transient electromagnetic data acquisition system is mainly composed of the transmitter and receiver, Wi-Fi communication links and human operation terminal. The transmitter and receiver are used to transmit and receive signals to the underground. The communication between the transmitter and receiver through the communication links, containing the serial and wireless modes. Human operation terminal run the software and user control interface to interact with the transient electromagnetic data receiver via Wi-Fi communication links.

The main structure of this system is shown in figure 1. TEM transmitter consists of control module, the power drive module and system state detection module. All parts are under the unified dispatch of the control module, which can realize a variety of electromagnetic signal waveform synthesis, signal power, large current emission, and fault protection alarm functions.

[a] Corresponding author: zhangh_sun@foxmail.com

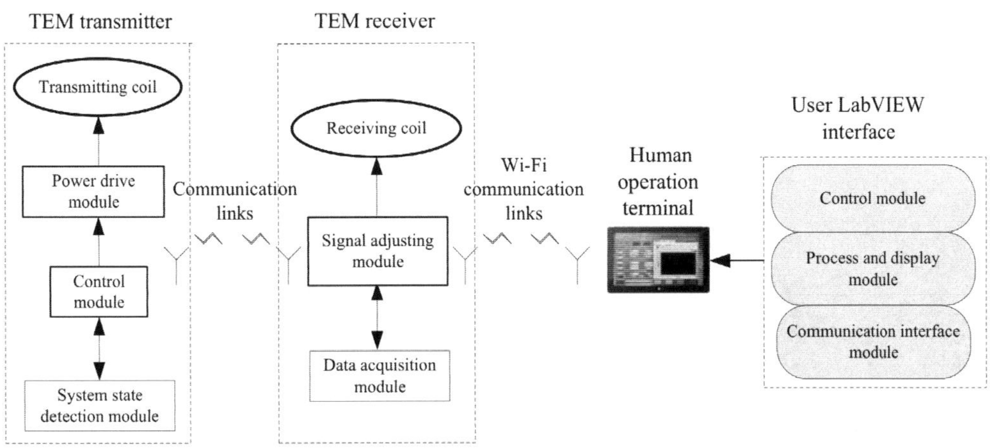

Figure 1. The main structure of the system.

TEM receiver includes a receiving coil, signal adjusting module, data acquisition module. Through receiving coil, the electromagnetic signal is carried into data acquisition from underground. It is essential to note that the signal acquired by the receiving coil need to be adjusted before entering into the data acquisition module, which may be too weak to be distinguished and contain lots of electromagnetic noise. Thus amplification and filtering are used in signal adjusting module.

For a desktop computer, it can use the computer's ethernet port for connecting to a Wi-Fi router. Also, if it is a newer laptop computer, such as a windows tablet, it very likely features built-in Wi-Fi capabilities. Based on data acquisition device and computer, the transient electromagnetic data acquisition system user interface is developed by using virtual instrument technology. User interface software is programmed on basis of LabVIEW platform and its usable software packages, including control module, data display module and data processing module, can realize the parameters of the transient electromagnetic data acquisition system setting and electromagnetic data in-real time processing and display.

2.2 The flow chart of the system

The transient electromagnetic data acquisition system based on virtual instrument takes on an orderly structure as figure 2 shows.

At the beginning, it is necessary to build Wi-Fi network connection. After the network connection is established, the computer user interface will be able to send commands to control the working status of the transmitter and receiver. When press the start button, the system starts monitoring. Computer user interface open TCP connection and send data to initial the whole hardware properties. TEM transmitter will transmit the electromagnetic exploration signal waveform and TEM receiver collect transient electromagnetic data. After collected, it will send the digital signal to upper user interface panel. According to the user's choice, the upper computer will display, process or storage the data. If the user does nothing, the procedure will stop monitoring and TCP connection will close. Then the whole system is waiting for next task.

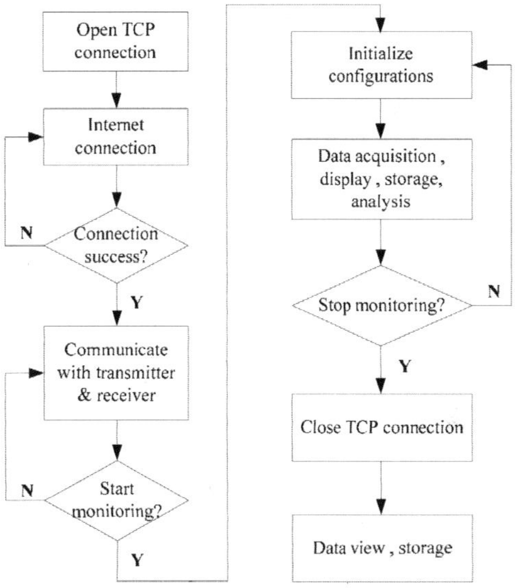

Figure 2. The flow chart of the system.

3 Virtual instrument interface design

Just the virtual instrument software makes the data acquisition device and computer become an integrated system including communication, data processing and display. We use LabVIEW platform written user in-terface programs. Generally program can be categorized into driver and application program. The driver can directly operate the register of data acquisition device hardware. The benefit is that the users are faced only with more understandable interface. The application program is used to complete display, process, storage and reproduce of the data.

User interface software is composed of three parts: control interface module, data process and display module, communication interface module. Control interface module sends commands to control the equipment and set the parameters of the system. Data process and display module realize data process, dis-play, storage and reproduce. Communication interface module build the Wi-Fi communication channel.

3.1. Control module

This module is designed for user to control the hardware. Figure 3 shows the control panel of the user interface.

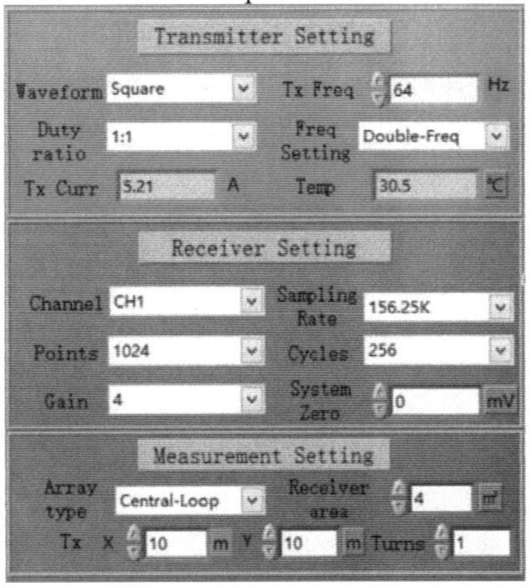

Figure 3. The control panel of the user interface.

Before the whole system starting to collect data, measurement parameters are required for the transmitter and receiver settings. In the middle of the user interface panel, the user-defined parameters can be configured. Drop-down menu contains parameters to choose, according to the needs, the user can select the appropriate parameters. The user-configured transmitter parameters include waveform selection, transmitting frequency and duty ratio, frequency setting, etc. As to receiver parameters, it is needed to configure the parameters of channel selection, sampling, cycle times, signal gain, the system zero. For channel selection, the channel number of the channel used for sample need to be determined. At most three channels can be selected. For the sample, the users can define sample frequency and the number of sample points per channel. For the cycle times, signal gain, it can enhance the signal amplitude and reduce noise. After all parameters finished setting, they all form in a unified command and then sent to the transmitter and receiver.

3.2 Process and display module

Data process and display module can easily display and process transient electromagnetic data in real-time, at the same time can finish to storage and view historical data.

After all parameters are determined and the system acquisition task is started, the electromagnetic acquisition equipment begins collecting data automatically. The original collected data from the communication interface are 16 bits integer number. Data process and display module calculates transient electromagnetic data into real voltage value. The waveform of voltage is displayed on the right of the user interface panel, then press the "save" button to save the results and parameter settings into text file with the test time. Our design is unique in that we display data in three ways, including one-dimensional arrays, two-dimensional waveform, called MATLAB logarithmic coordinate. When we need to analysis the historical data, it is convenient to view the settings and waveforms just choose the right file.

3.3 Communication interface module

According to the actual needs of communication with the transient electromagnetic data acquisition systems, communication interface module supports wireless communication modes, including Bluetooth and Wi-Fi mode. Bluetooth can realize the short-range wireless communications between user interface and the device, which is suitable for the test the performance of the device in lab environment. Using Wi-Fi mode can realize the data transportation over a long distance in the field [10], it is the resolution of processing collected data in real-time.

The wireless communication function of Wi-Fi mode is mainly to achieve by LabVIEW comes with TCP functions and VI. Run LabVIEW, create a new VI, added a block diagram "Open TCP Connection" function, and defined IP address, remote port and other terminals. The IP address and port number configured here must the same with the setting of the TEM receiver. User interface communicate with transient electromagnetic equipment via this channel. If TCP network connect success, the "successful connection" indicator will light. Figure 4 shows the TCP parameter setting of the user interface.

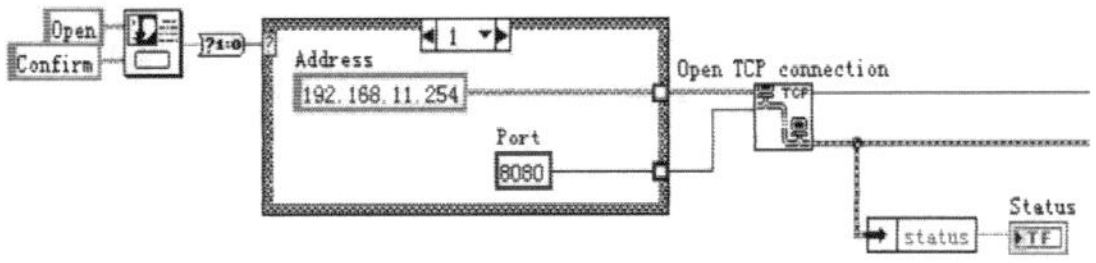

Figure 4. TCP parameter setting.

3.4 Analysis module

In order to obtain the stratigraphic information of the survey area, the data collected by the receiver is mainly use for data inversion. The analysis module call MATLAB Script node to inversion, which use improved damping least squares fitting algorithm. The input data of VI node is the response value after de-noising. After obtained inversion formation parameters by the inversion algorithm, using a forward inversion can get inversion of transient response value. In order to compare the effectiveness of the inversion algorithm, the analysis module output original voltage and inversion formation response curve. Figure 5 shows the part diagram of data inversion analysis module.

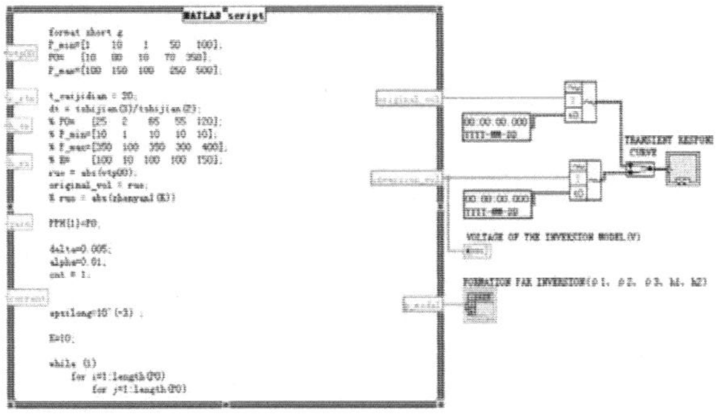

Figure 5. Diagram of data inversion analysis module.

4 Result

In order to verify the performance of this design, we do the test in the lab environment.

Once initialized, the electromagnetic acquisition equipment and user interface panel turned into the default settings. If the users are not familiar with the procedure of the whole transient electromagnetic data acquisition system, "help" button will gives user a brief introduction of how to operate the system, which makes the system easy to use.

4.1 Parameter settings and waveform display

At the beginning of the procedure, it is necessary to initialize hardware properties such as waveform selection, duty ratio, channel selection, sampling, etc. For the case of transmitter, the waveform is set as square wave, the duty ratio is 1 to 1, and the frequency mode is selected to double frequency. For the case of the data acquisition, the channel is chose channel one, the sampling frequency is set as 156.25K/s, acquisition points is 1024, the current system adjustment is zero. After entering the acquisition procedure, the measurement parameters such as device type and magnetic torque for receiving are all need to be configured firstly. The rest of configuration parameters are kept default. Then the acquisition system begins acquiring signal from channel one. The acquisition procedure would quit if there is no synchronization signal within 30 seconds.

Figure 6 shows hardware parameter settings and the waveform data. As is shown in the figure, the horizontal axis represents the acquisition time, the unit is millisecond. The longitudinal axis represent the actual voltage value, which is limited to between 0 and 1 after logarithmic coordinate transformation.

4.2 View results

In order to compare the different results in different fields, the previously saved data can be shown in figure 7 to view the data and parameters.

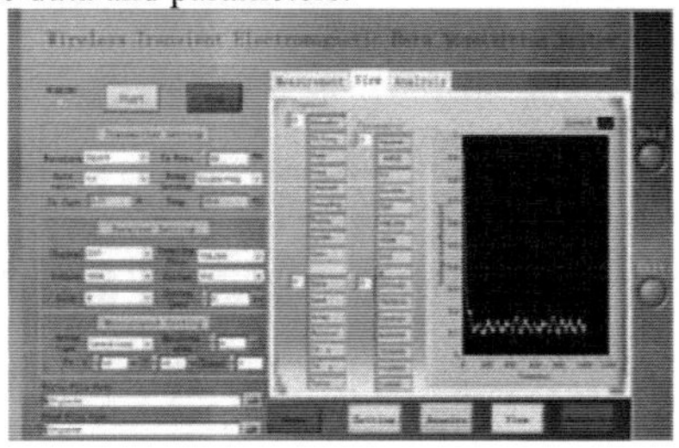

Figure 7. The view results of previously saved data file.

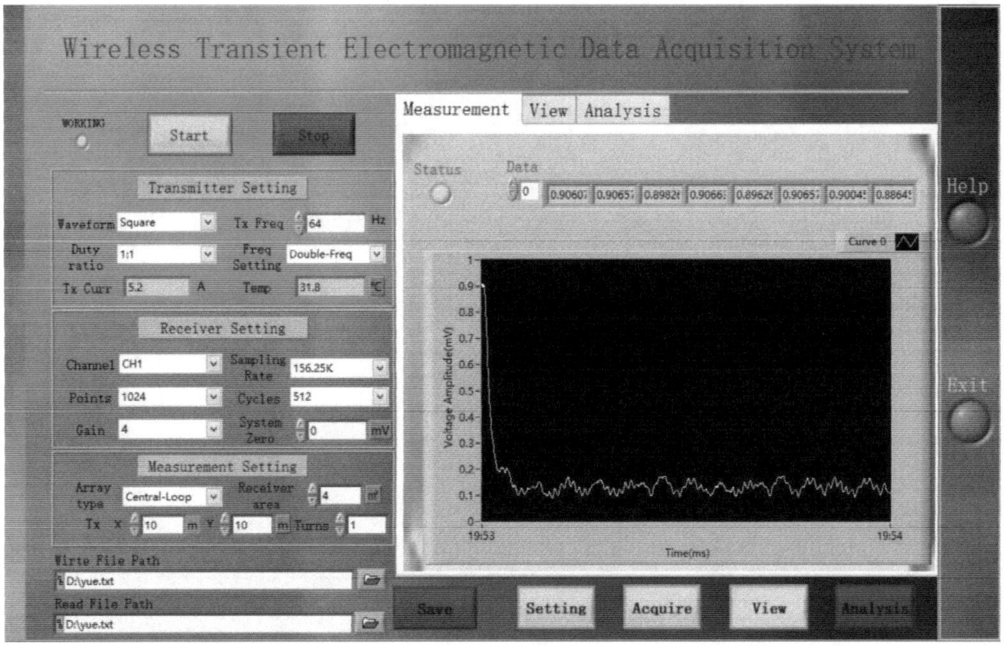

Figure 6. The result of the whole user interface.

Waveform		TX Freq		Duty ratio		Freq setting		Channel		Sampling Rate		Points		Cycles		Gain		System-zero
Square		64HZ		1:1		Double-Freq		CH1		156.25k		1024		512		4		0.0000
Save time		Save time		Array type		Receiver area		TX x		TX y		Turns						
2015/6/20		19:58:32		Central-Loop		4.000000		10.0000		10.0000		1.0000						
0.906	0.907	0.898	0.907	0.896	0.907	0.900	0.886	0.847	0.805	0.767	0.724	0.689	0.650	0.611	0.570	0.530	0.495	0.463
0.436	0.411	0.384	0.356	0.336	0.320	0.309	0.299	0.287	0.276	0.265	0.254	0.244	0.236	0.228	0.220	0.209	0.201	0.197
0.196	0.195	0.193	0.191	0.190	0.190	0.193	0.199	0.204	0.206	0.204	0.202	0.203	0.203	0.201	0.197	0.192	0.188	0.187
0.190	0.196	0.201	0.202	0.200	0.196	0.194	0.193	0.190	0.184	0.174	0.165	0.156	0.151	0.149	0.147	0.143	0.138	0.132
0.134	0.135	0.137	0.137	0.134	0.129	0.125	0.122	0.119	0.114	0.107	0.102	0.099	0.099	0.102	0.106	0.112	0.118	0.122
0.124	0.126	0.127	0.125	0.120	0.114	0.109	0.105	0.103	0.103	0.106	0.111	0.115	0.118	0.122	0.126	0.128	0.127	0.125
0.123	0.122	0.122	0.123	0.126	0.129	0.130	0.131	0.133	0.135	0.135	0.133	0.130	0.128	0.127	0.127	0.128	0.129	0.133
0.135	0.138	0.142	0.146	0.148	0.150	0.151	0.152	0.152	0.149	0.144	0.139	0.132	0.128	0.120	0.114	0.110	0.107	0.106
0.106	0.108	0.107	0.106	0.103	0.103	0.100	0.100	0.098	0.101	0.104	0.108	0.111	0.113	0.114	0.111	0.106	0.100	0.096
0.092	0.089	0.089	0.094	0.101	0.109	0.117	0.122	0.125	0.126	0.124	0.119	0.113	0.108	0.103	0.102	0.102	0.100	0.105
0.110	0.115	0.122	0.130	0.138	0.144	0.145	0.144	0.141	0.138	0.135	0.131	0.128	0.128	0.130	0.134	0.140	0.147	0.155
0.161	0.165	0.167	0.168	0.168	0.162	0.153	0.144	0.138	0.134	0.133	0.135	0.139	0.143	0.146	0.149	0.152	0.153	0.152
0.149	0.143	0.132	0.125	0.117	0.113	0.112	0.112	0.116	0.120	0.125	0.129	0.131	0.131	0.127	0.119	0.112	0.106	0.102
0.100	0.100	0.101	0.102	0.106	0.113	0.121	0.127	0.127	0.124	0.116	0.114	0.110	0.106	0.105	0.105	0.105	0.105	0.109
0.115	0.121	0.126	0.128	0.128	0.130	0.134	0.137	0.139	0.141	0.141	0.142	0.143	0.146	0.148	0.149	0.148	0.147	0.147
0.148	0.149	0.152	0.154	0.153	0.151	0.153	0.157	0.159	0.159	0.158	0.156	0.154	0.151	0.149	0.147	0.143	0.137	0.129
0.125	0.124	0.125	0.126	0.129	0.131	0.133	0.134	0.134	0.132	0.128	0.121	0.113	0.107	0.104	0.104	0.108	0.113	0.117
0.120	0.123	0.125	0.128	0.127	0.122	0.115	0.109	0.104	0.101	0.099	0.098	0.099	0.100	0.104	0.110	0.118	0.124	0.125
0.122	0.119	0.115	0.112	0.110	0.109	0.108	0.110	0.115	0.122	0.131	0.140	0.146	0.148	0.152	0.149	0.143	0.139	0.132
0.131	0.128	0.127	0.129	0.134	0.141	0.148	0.154	0.157	0.157	0.154	0.151	0.146	0.141	0.135	0.130	0.127	0.126	0.130
0.135	0.140	0.142	0.141	0.141	0.142	0.141	0.138	0.136	0.136	0.136	0.137	0.140	0.143	0.146	0.147	0.145	0.144	0.146
0.147	0.148	0.146	0.143	0.138	0.132	0.134	0.134	0.133	0.132	0.132	0.132	0.131	0.129	0.125	0.120	0.113	0.106	0.100

Figure 8. The example of storage file.

4.3 Data storage

In consideration of further analysis, it is necessary to store the transient electromagnetic data to a text file. Not only the collected electromagnetic data, but also the acquisition parameter configurations and time are all packed into the text file.

Figure 8 shows a text file of data storage. The example of storage file contains data and parameters.

Through the interface of select previously saved data file, it is easy to view the older parameters and data, which facilitates the experimental comparison.

5 Conclusion

A wireless transient electromagnetic data acquisition system is developed based on virtual instrument and Wi-Fi technology. The system combined the electromagnetic equipment and virtual instrument technology has the advantage of flexible maneuverability and real-time processing. Transient electromagnetic receiver and transmitter, Wi-Fi communication links and human operation terminal comprise the whole system, which is controlled by virtual instrument panel. The collected data can be easily stored, displayed and analysis in the computer. Furthermore, we have made the whole user interface into an application with the suffix of .Exe, which indicates that any computer can run the interface just use the .Exe.

Human operation terminal using the portable terminal, such as tablets, which facilitates fast moving when working in the field. We hope that the paper will help the process of research and development in this field, and we also welcome comments and suggestions for improvement.

Acknowledgement

This paper is sponsored by the Fundamental Research Funds of the Institute of Geophysical and Geochemical Exploration CAGS (GN: WHS201305); the Teaching Laboratory Opening Funds, China University of Geosciences, Wuhan (GN: SKJ2015220); the Key New Products and New Technology Research and Development Funds of Hubei Province of China (GN: 2012BAA06007).

References

1. Defu Cheng, Jun Lin, Shengbao Yu, etc, *Study on TEM weak signal detection techniques*, J. Jilin. U. Techno. Ed, **20**, 2(2002)
2. Zhong Wang, Yue Huang, Jun Lin, etc, *Development of high speed and large dynamic range transient electromagnetic receiver*, Chin. J. Sci. Instrum, **27**, 7(2006)
3. Qiyun Jiang, *Study on the key technology of wide field electromagnetic sounding instrument*(CSU, Chang Sha, 2010)
4. Jing Pei, Qiyun Jiang, *Wide-field electromagnetic receiver acquisition design based on AD7762*, PG, **25**, 3 (2010)
5. Zhihong Fu, Junli Zhao, Luowei Zhou, etc, *WTEM fast turn-off transient electromagnetic detection system*, Chin. J. Sci. Instrum, **29**, 5 (2008)
6. C.J. Kalkman, *LabVIEW: A software system for data acquisition, data analysis, and instrument control*, JCMC, **11**, 1 (2005)
7. Hu Liu, Shili Lu, Cangwen Jiao, *Virtual instrument techniques and their application to development of electromagnetic imaging system*, UG, **21**, 6 (2005)
8. Jie Xu, Benyue Lou, Jianying Zhang, *Collection and pretreatment of transient electromagnetic data based on LabVIEW*, Ind. Mine. Autom, **12**, 6 (2011)
9. Tao Lv, *Study on the key technology of wide field electromagnetic sounding instrument*(J. Jilin. U, Ji Lin, 2004)
10. L.W. Couch, *Digital & Analog communication systems* (Prentice Hall, Upper Saddle River, 2007)

Numerical simulation for Human Quiet Standing System

Caihong Wang [1,a], Jian Xu[2]

[1] Shanghai Publishing and Printing College, Shanghai, China
[2] School of Aerospace Engineering and Applied Mechanics, Tongji University, Shanghai, China

Abstract. Human quiet standing system was investigated with a 2-joint (ankle and hip) sagittal model of body movement with perturbation noise. In the case of the noise perturbations, the influence on the stability of time delay is discussed numerically in this paper. Analytical analyses on the mean of the station response reveal that the presence of time delay leads to periodic vibration in parameter space, and the mean value increases as the reflexes delay increasing. But the mean square of the stationary response varies with not only the noise but also the delay. We find the stochastic factor has no affection on the stability of human, but it will make the amplify oscillations diffusion, which can be seen from the shape change of the figure on the stationary joint probability density and the stationary probability density. All these conclusions are shown by the numerical simulation. In conclusion, the influence of reflexes delay and perturbation on the stability and quality of stance has been studied by successfully combined with the biomechanical concepts and tools..

1 Introduction

A large numbers of publication work [1-4] at the equilibrium stability of upright standing were based on the simple biomechanical model of the human body as single inverted pendulum. The multi-joint properties of the body were rarely taken into account in the previous studies. Furthermore, the delay was not taken into account in the feedback loop [5, 6]. However, the existence of delays essentially complicates postural and requires the strict adjustment of the feedback gains to maintain upright [7]. Also, the human nervous system operates in a very noisy environment and hence the noise effects on human postural. The noise perturbations have been considered into the single-link biomechanical model of the human body, but not in the double-link biomechanical model. In the present study, we take into account both delay in the feedback loop and stochastic perturbations in the double-joint body configuration. In 2009, a spatial double inverted pendulum moving in a three dimensional space was considered to be a model of human standing postural control by Xinjilefu [8]. Postural control was possible by on-line minimization of the system Lagrangian have been showed. The influence of reflexes delay and perturbations on the stability and quality of stance has been not studied by numerical method.

As one can see in [9], how the delay effect on the human posture during stance in the double pendulum with perturbation have been discussed. Is it consistent with [9]? Motivated by the above findings, the purpose of this study is to show the reflex delay and noise in the double pendulum system in numerical simulation.

This Letter is organized as follows. In Section 2, the effects of time delay and additive noise on dynamics of system (1) are discussed in detail by numerical Simulation. Finally, conclusions are drawn.

2 Numerical simulation

The equations of human quiet standing have been built in [9].

$$\begin{cases} (\frac{1}{3}m_1+m_2)l_1^2\ddot{\theta}_1+\frac{1}{2}m_2l_1l_2\ddot{\theta}_2\cos(\theta_1-\theta_2)+\frac{1}{2}m_2l_1l_2\dot{\theta}_2^2\sin(\theta_1-\theta_2) \\ -\mu_2(\dot{\theta}_2-\dot{\theta}_1)-(\frac{1}{2}m_1+m_2)gl_1\sin\theta_1+\alpha_1\theta_1+\mu_1\dot{\theta}_1-\alpha_2(\theta_2-\theta_1) \\ =\beta f(\theta_1(t-\tau))+\gamma\eta(t), \\ \frac{1}{2}m_2l_1l_2\ddot{\theta}_1\cos(\theta_1-\theta_2)+\frac{1}{3}m_2l_2^2\ddot{\theta}_2-\frac{1}{2}m_2l_1l_2\dot{\theta}_1^2\sin(\theta_1-\theta_2) \\ -\frac{1}{2}m_2gl_2\sin\theta_2+\alpha_2(\theta_2-\theta_1)+\mu_2(\dot{\theta}_2-\dot{\theta}_1)=0. \end{cases} \quad (1)$$

In [9], in the case of the combined effects of the noise, the influence on the stability of the reflexes delay will be firstly discussed. The stationary response of Eq. (1) will be calculated to study the influence. Because of the system is infinite dimensional stochastic system, it is difficult to calculate the accurate stationary response. Here, the approximate stationary response of stochastic oscillatory near critical delay of deterministic system is derived employing multi-scale method to find the relation with the delay and noise.

The influence of the reflexes delay and perturbation on the stability is found. To verify adequately the obtained results of the influences on the human during standing and explain the mechanism of human maintain upright stance, the stationary probability density function figures and time history plane portrait are analyzed as following. All the parameter values are the same as [9] taken.

[a] Corresponding author: wch042000@126.com

 DOI: 10.1051/matecconf/20166101021

Fig. 1 and Fig. 2 show the shape of the joint stationary probability density and the stationary probability density varies with the different reflex delay. For the human reflex delay $\tau = 0.14 < \tau_c = 0.184$, the shape of the stationary joint probability density $p(\theta_1, \dot{\theta}_1)$ and $p(\theta_1, \theta_2)$ have a single-peak bell shaped in Fig. 1(1). In Fig. 2(1), the function $p(\theta_1)$ have only a single maximum at the point $(0,0)$ too. It shows the probability when human at upright position is maximum. The human during standing are always controlled to the upright equilibrium position even if we are perturbed by the noise, as is shown in Fig. 5(1) and Fig. 6(1). The random trajectories of human movement perturbations noise are concentrated in the vicinity of point $(0,0)$. The equilibrium point $(0,0)$ is stable, which is natural to name this type of stochastic attractor by stochastic equilibrium point.

We can see that in this regime the additive noise plays the dominant role in amplifying the oscillations. It reflects mechanism of human maintain upright stance. However, if human reflexes delay $\tau = 0.20$ is beyond the critical delay, the shape of the joint stationary probability density $p(\theta_1, \dot{\theta}_1)$ and $p(\theta_1, \theta_2)$ become a crater-like form but not single-peak bell shaped form. Its center locates at the equilibrium point $(0,0)$ in Fig.4(b). In Fig. 2(2), the function $p(\theta_1)$ possesses a maximum at the points of deterministic cycle and a single minimum at the point $(0,0)$. It shows that the probability is small when human during standing is close to upright position. Furthermore, the probability density is smaller as the human stand is more close to upright position. At this moment, as above section stated human will start to periodically vibration. The random trajectories of human quiet standing system (1) are concentrated in the vicinity of deterministic cycle, as is shown in Fig.4(2) and Fig.5(2). The equilibrium point $(0,0)$ is unstable and a stochastic limit cycle appeared, as is shown in Fig.6. This phenomenon varies with reflexes delay is be defined as phenomenological bifurcation in random vibration theory. Fig. 2 and Fig. 3 shows the variation of the parameter γ of noise perturbation doesn't change the location of the extremal points of the related probability density function. In other words, noise perturbation has no influence on the stability of human. An increase of γ just only results into a growth of the dispersion of the stochastic human quiet standing system around the deterministic attractors (i.e. around the stable point $(0,0)$ for $\tau < \tau_c$ and around the stable limit cycle for $\tau > \tau_c$). Note that the deterministic bifurcation value $\tau_c = 0.184$ under additive Gaussian noise perturbations is not changed

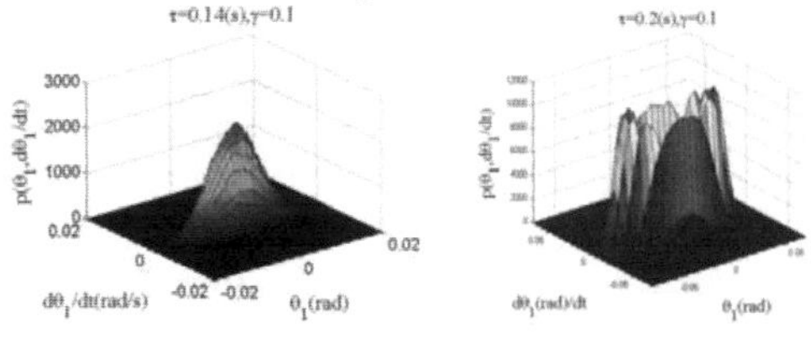

(a-1) (a-2)

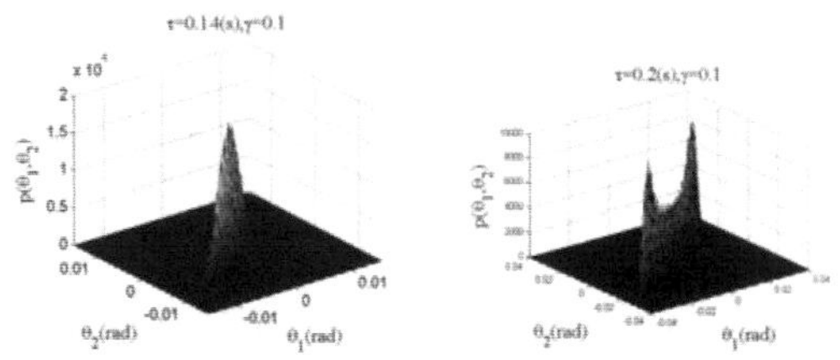

(b-1) (b-2)

Fig.1. The stationary joint probability density of human standing system (1) for $\gamma = 0.1$, (a-1), (b-1) $\tau = 0.14$, (a-2), (b-2) $\tau = 0.20$, the other parameters are set as Tab.1, (a): The stationary joint probability density $p(\theta_1, \dot{\theta}_1)$ the angular displacement and velocity in the ankle; (b): the stationary joint probability density $p(\theta_1, \theta_2)$ of the angular displacement in the ankle and in the hip

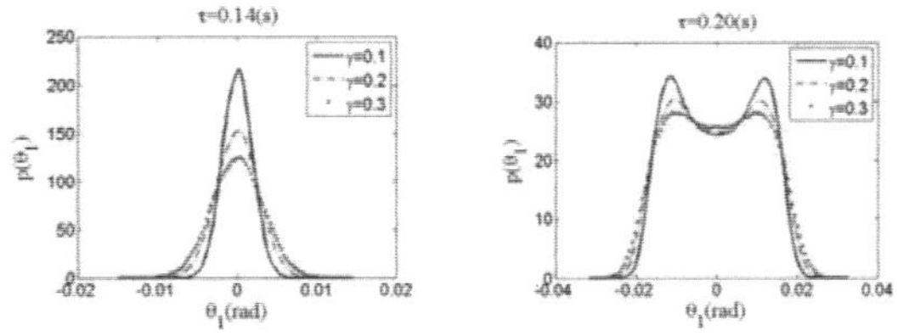

(a-1) (a-2)

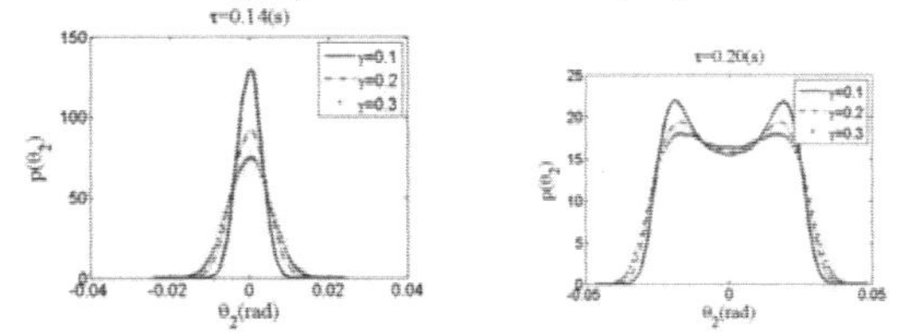

(b-1) (b-2)

Fig. 2. (a): the stationary probability density $p(\theta_1)$ of the angular displacement in the ankle; (b): the stationary probability density $p(\theta_2)$ of the angular displacement in the hip for $\gamma = 0.1$ (solid), $\gamma = 0.2$ (dashed), $\gamma = 0.3$ (dotted), (a-1),(b-1): $\tau = 0.14$, (a-2), (b-2): $\tau = 0.20$.

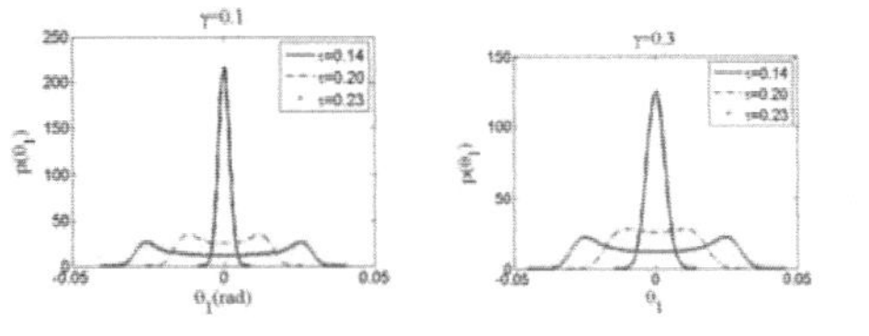

(a-1) (a-2)

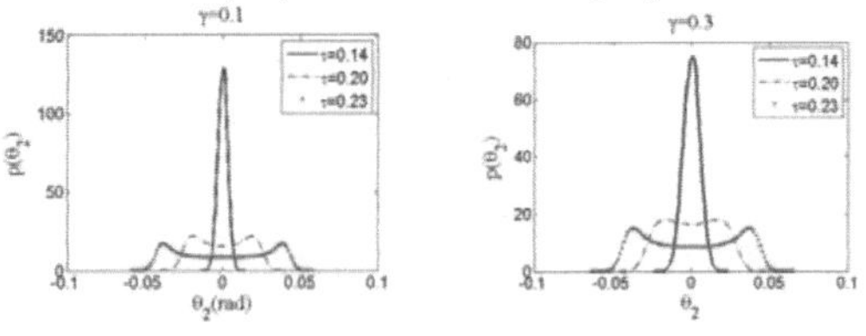

(b-1) (b-2)

Fig. 3. (a): the stationary probability density $p(\theta_1)$ of the angular displacement in the ankle; (b): the stationary probability density $p(\theta_2)$ of the angular displacement in the hip for $\tau = 0.14$ (solid), $\tau = 0.20$ (dashed), $\tau = 0.23$ (dotted). (a-1) (b-1): $\gamma = 0.10$, (a-2) (b-2): $\gamma = 0.30$.

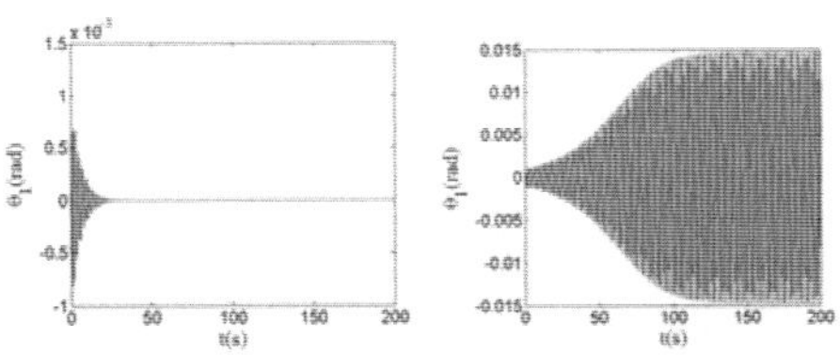

(a-1) (a-2)

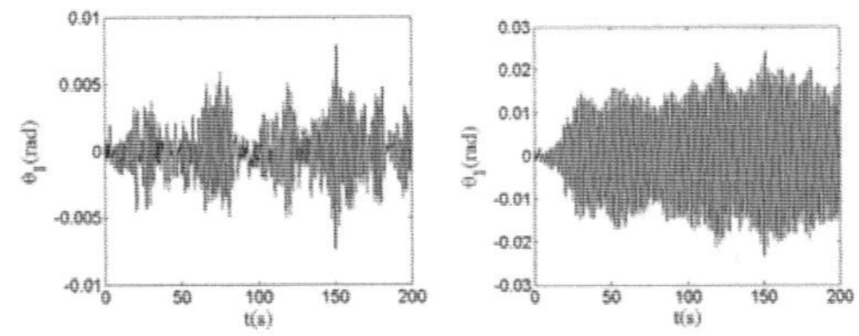

(b-1) (b-2)

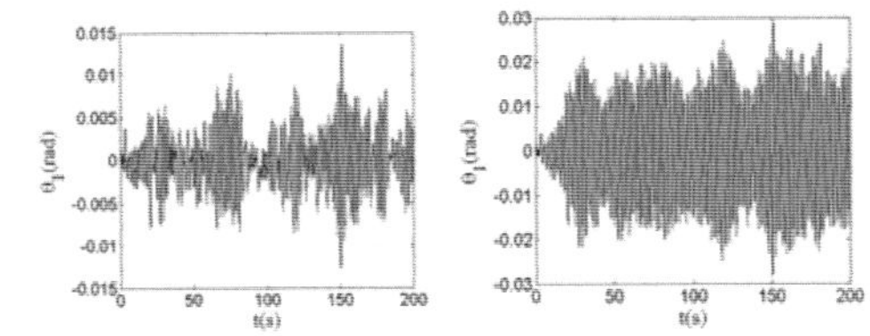

(c-1) (c-2)

Fig. 4. Time history plane portrait in the ankle for (a-1): $\tau = 0.14$, $\gamma = 0$; (a-2): $\tau = 0.20$, $\gamma = 0$; (b-1): $\tau = 0.14$, $\gamma = 0.1$; (b-2): $\tau = 0.20$, $\gamma = 0.1$; (c-1): $\tau = 0.14$, $\gamma = 0.3$; (c-2): $\tau = 0.20$, $\gamma = 0.3$.

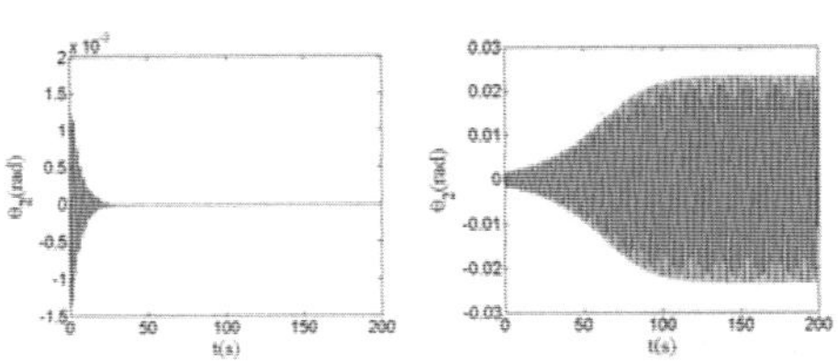

(a-1) (a-2)

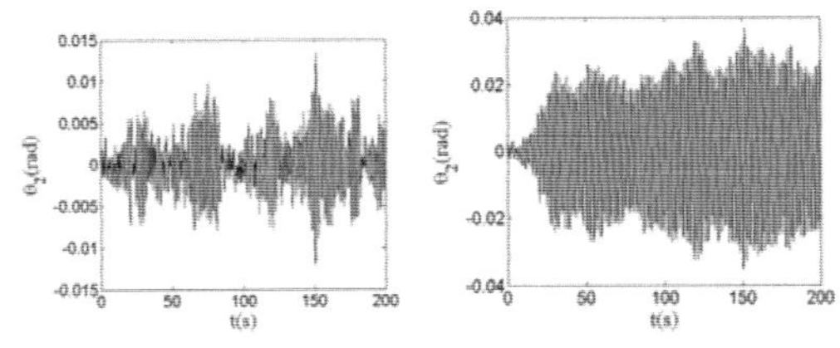

(b-2) (b-2)

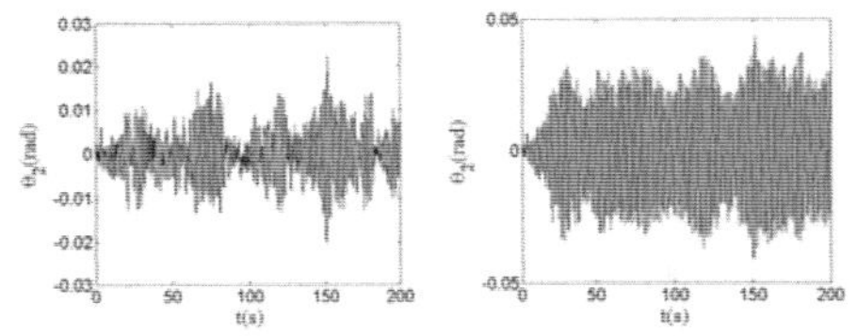

(c-1) (c-2)

Fig. 5. Time history plane portrait in the ankle for (a-1): $\tau = 0.14$, $\gamma = 0$; (a-2): $\tau = 0.20$, $\gamma = 0$; (b-1): $\tau = 0.14$, $\gamma = 0.1$; (b-2): $\tau = 0.20$, $\gamma = 0.1$; (c-1): $\tau = 0.14$, $\gamma = 0.3$; (c-2): $\tau = 0.20$, $\gamma = 0.3$.

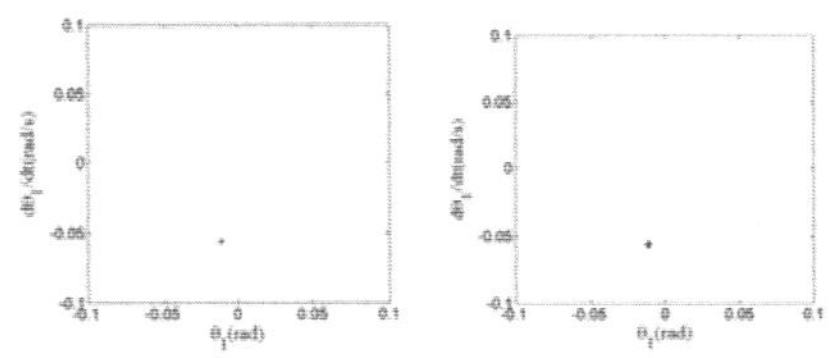

(a) (b)

Fig.6. Poincare mapping section when (a): $\tau = 0.20$, $\gamma = 0$; (b): $\tau = 0.20$, $\gamma = 0.1$.

3 Conclusions

In this paper, in the case of the combined effects of the noise, the influence on the stability of time delay is discussed by numerical simulation. The results show human reflexes delay still has a great influence on stability in human quiet standing. The human will be still stable when reflexes delay is small in the case under perturbation noise. While reflex delay exceeds the critical time delay, human will start to periodic vibration. And the first of stationary response will increase as human reflexes delay increasing. This influence of delay on the human movement is same as [9]. By analysis the shape change of the figure on the stationary probability density, we can also find the stochastic factor has no affection on the stability of human, but it makes the amplify oscillations diffusion, as is shown in Fig.6. All these results are particularly important in the clinical realm.

Acknowledgments

This research is supported by the state key program of national natural science of china 11032009, shanghai leading academic discipline project B302, and PhD start-up fund of shanghai publishing and printing college 95-A1030114203.

References

1. W. Yao, P. Yu, and C. Essex, Delayed stochastic differential model for quiet standing, PHYSICAL REVIEW E, VOLUME 63, 021902
2. Peterka RJ, Sensorimotor integration in human postural control, J Neurophysiol, 85: 1097-1118, 2002.
3. M. Ben Rabha, M.F. Boujmil, M. Saadoun, B. Bessaïs, Eur. Phys. J. Appl. Phys. (to be published)
4. Winter DA, Patla AE, Rietdyk S, Ishac M, Ankle muscle stiffness in the control of balance during quiet standing. J Neurophysiol 85:2630–2633, 2001.
5. Barin K, Evaluation of a generalized model of human postural dynamics and control in the sagittal plane. Biol Cybern 61:37–50, 1989.
6. Park S, Horak FB, Kuo AD, Postural feedback responses scale with biomechanical constraints in human standing. Exp Brain Res,154:417–427, 2004.
7. Rack PMH, Limitation of sensory feedback in control of posture and movement. In: The nervous system. Motor control. Bethesda, MD: Am. Physiol. Soc. section. 1, vol. II: 229–256, 1981.
8. Xinjilefu, V. Hayward, and H. Michalska. "Stabilization of the Spatial Double Inverted Pendulum Using Stochastic Programming Seen as a Model of Standing Postural Control," Proc. 9th IEEE-RAS Int. Conf. on Humanoid Robots, pp: 367-372, 2009.
9. Caihong Wang , Effects of Time Delay and Noise on Asymptotic Stability in the Double Inverted Pendulum for Human Quiet Standing,, MME-2015, 2015.

FTIR and ESEM Analysis of soil moisture Microscopic conservation feature with liquid membrane

Jian Gu[1,2], Liang Hao[2,3], Pengfei Huang[1,2], Na Tong[4], Xue Li[1,2],Guanghua Yin[1,a]

[1]*Institute of Applied Ecology, Chinese Academy of Sciences, Shenyang, 110016,China*

[2]*University of Chinese Academy of Science, Beijing, 100039,China*

[3]*Institute of Policy and Management, Chinese Academy of Sciences, Beijing, 100190,China*

[4]*Shenyang Institute of Engineering, Shenyang, 110136,China.*

Abstract. Liquid membrane applied in soil has a good soil moisture conservation and evaporation suppression performance. Application of spectrum analysis technology to understand its structure and micro morphological characteristics will be help to reveal the soil moisture conservation mechanism of liquid membrane. In this paper, we used the three types of liquid membrane as the research object based on the laboratory preparation, with infrared spectrometer (FTIR) and environmental scanning electron microscope (ESEM) as the means, analysis the soil moisture mechanism of liquid membrane.The results showed that the -OH between the CMC and PVA generated intermolecular hydrogen bonds, the formation of hydrogen bonds between molecules of the two components strengthened the two-phase's compatibility, increasing the liquid membrane's effective groups and forming a dense mesh structure.ESEM observation showed that liquid membrane can effectively cementing soil particles, generating the soil-membrane structure, reducing soil moisture to evaporate, improve soil moisture conservation performance.
keywords: liquid membrane; microscopic mechanism; FTIR;ESEM

1 Introduction

Water shortage is the key factor to restrict the agricultural sustainable development in arid and semiarid regions, plastic mulching technique is an effective way to solve the problem [1]. Since 1979, our country imported polyethylene mulch from Japan, its rapid development has made significant yield effect. However, polyethylene--mulch is a kind of macromolecular hydrocarbon, which is insoluble in water and not readily degrade in the soil, it has brought serious white pollution to the environment because of wide and long-term using in the agricultural production[2].Therefore, developing a environmentally friendly material to replace polyethylene mulch has become an urgent need for the development of ecological agriculture. Liquid membrane is a new type of covering products with organic polymer as the main material, which is the recent development and application of agricultural production[3].The product sprayed on the surface can effectively cementing soil particles, generating the soil-membrane structure, reducing soil moisture to evaporate, improve soil moisture conservation performance[4].Research shows that[5-7],Liquid membrane is showing superiority in the economy (reduce the amount), environment (security clearances dose), soil conservation (improved soil structure), irrigation efficiency (increase infiltration / reduce runoff), etc. However, the current researches on soil moisture conservation performance for liquid membrane mainly adopt the method of field experiment fixed-point observation, don't sets forth the mechanism of soil moisture conservation from the micro structure of liquid membrane[10-13], the research is not deep enough. The successful application of modern test technology Such as X-ray diffraction, infrared spectrum, differential thermal analysis, electron microscope, nuclear magnetic resonance (NMR) and electron spin resonance, etc, in the analysis of soil related components and structural analysis provides a way for the research on the microstructure of liquid membrane.

In view of this, this article try to infrared spectrometer (FT-IR) and environmental scanning electron microscope (ESEM) as the means, analysis the soil moisture mechanism of liquid membrane, try to reveal the soil moisture conservaton mechanism of liquid membrane from spectroscopy, so as to provide reference for the similar research and using for reference.

2 Experimental part

2.1 Preparation of liquid membrane

Liquid membrane is based on polyvinyl alcohol (PVA) and sodium carboxymethyl cellulose (CMC) as the main raw materials, crosslinking agents, defoamers, glycerin and water as the secondary raw materials, three types of liquid membrane were prepared and labeled as 1 (PVA : CMC=3:1), 2 (PVA : CMC=1:1) and 3 (PVA : CMC=3:1). Preparation methods are as follows:

[a] Corresponding author: ygh006@163.com

The pure PVA powder and pure CMC powder were mixed in deionized water according to three ratios, bathing for 3 h at 80 ℃ in the autoclave water, at the same time, constant stirring, the speed is 100r/min, the quality of the solution is obtained by 8%. After completely dissolved, add other auxiliary materials, 0.5h reaction, and the liquid membrane is prepared after cooling.

2.2 Test methods

2.2.1 Structure determination

The liquid membrane and PVA and CMC liquid flows into the watch glass, film casting, placed on the bench at room temperature for three days, and then alternate. Use the Nicolet 6700 Fourier infrared spectrometer (FTIR) produced by the Thermo Fisher company for functional characterization of liquid membrane， measuring the absorption spectrum of liquid membrane and structural analysis.

2.2.2 soil-liquid membrane microscopic moisture morphometric

The three kinds of liquid membrane were dissolved in water at a concentration of 1:20,using 300kg /hm2of liquid.After thorough stirring, the solution was sprayed on the surface of the soil. An identical amount of water was sprayed as a control. The soil was sampledtwice a month for four monthsafter spraying.A QuantaTM 250 ESEM produced by American FEI Company was used to determine and analyze the micromorphology of the liquid membrane-treated soil.

3 Results

3.1. Structure Determination

From the results of Figure 1, it can be seen that the PVA chain is flexible, while the CMC molecular chain contains a large number of ring structures.Although there is a big difference between the two groups of macromolecular chain structure, both of the PVA and the CMC molecules have a hydroxyl group, intermolecular hydrogen bonds between them can improve the compatibility between PVA and CMC.

In pure PVA film, 3251cm^{-1} is stretching vibration peak of -OH, 2906cm^{-1} is -CH stretching vibration of, 1085cm^{-1} is symmetrical and asymmetrical C-O absorption peak;Stretching vibration of pure CMC film -OH appears in 3356cm^{-1},2941cm^{-1} is -CH stretching vibration peak, 1600cm^{-1} is -COONa groups C = O symmetric and asymmetric vibration absorption peaks.When the CMC and PVA blended -OH on PVA molecular chains will bind with ionized carboxylate, compared to Na^+, - OH has the functions of providing electronic, so C = O infrared absorption peak moves to the high number of wave.For liquid film, the position of the OH peak was changed, and it appeared in the 3303cm^{-1}, 3265cm^{-1} and 3200cm^{-1}also shows CMC -OH and polyvinyl alcohol (PVA) -OH also produce intermolecular hydrogen bonds, the formation of intermolecular hydrogen bonds between the two components of molecular strengthen the compatibility between two phase, so that the liquid membrane formed a dense mesh structure, favorable for water and soil moisture conservation.

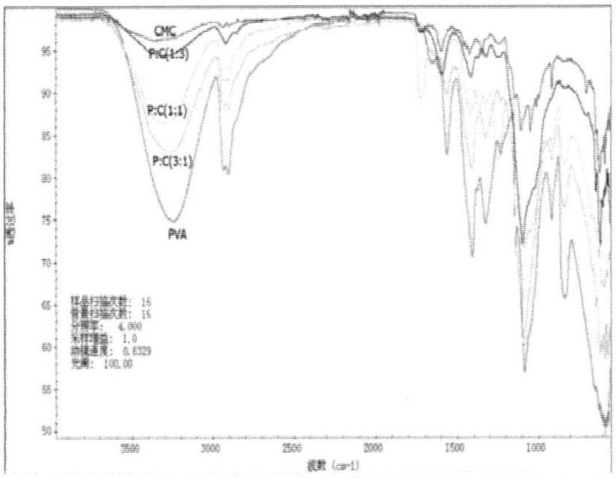

Figure1. FT-IR spactra of liquid membrane

3.2Soil membrane structure Micromorphological

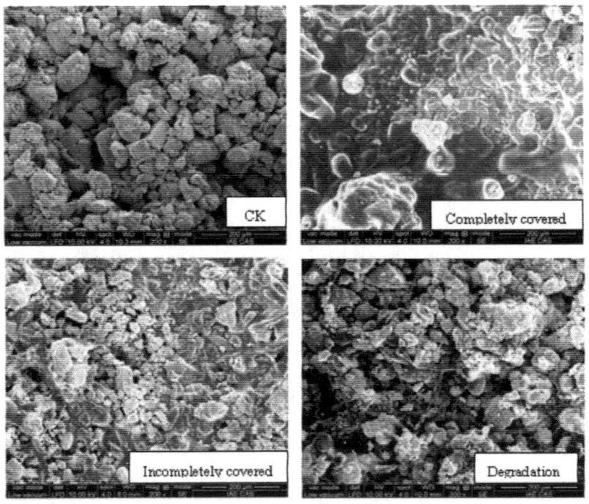

Figure 2. The soil surface micromorphology with different treatments

Figure 2 shows that the control group and the soil surface covered with liquid membrane have completely different structural characteristics.

The figure in upper left corner is the control, the soil surface layer of soil particle accumulation is compact, small pore, see more small particles in the larger pore filling, the formation of soil crust.The surface of the soil covering with liquid membrane is different,pictured top right completely covered is an ideal way to cover,liquid membrane completely wrapped in the soil particle surface, and can be connected to each other to form a closed membrane structure of soil;Pictured left partially covered, the soil surface is not completely covered, large pores, poor soil moisture results; the lower right picture shows the degradation effect after four months,most of the liquid membrane covering the soil surface degrade, only remains connection between part soil particles, Basically lost soil moisture conservation performance.

3.3 Three types of liquid membrane covering effect

Figure 3 shows the effect of the liquid membrane covering for half of a moon,Wherein the upper left corner picture shows the effect of the 2nd liquid membrane, film is completely wrapped in the soil surface, good sealing effect;top right photo shows the 2nd liquid membrane coverage effect, which is not complete, there are more pores, poor sealing effect; The lower left corner figure shows the effect of the 2nd liquid membrane,the material in the picture is less, more common filamentous material, good connection effect, but the effect is not ideal.The main cause of this phenomenon is that the different components of the liquid membrane.The viscosity of sodium carboxymethyl cellulose solution is relatively high, and the mobility is poor;while the polyvinyl alcohol solution flow better ductility.Therefore, the cover effect of 1nd liquid membrane which containing most polytene is best,no. 2 and no. 3, liquid membrane covering effect is poorer.

Figure 3. Three types of liquid membrane coverage of the soil surface micromorphology

3.4 The connection of the liquid membrane with soil particles

Figure4 shows the combination method of liquid membrane and soil particles.The hydrophilic group of the liquid membrane can be formed by hydrogen bonding with many substances, which can produce a bond between the two particles,a molecule can bond a few particles at the same time usually.As shown in the figure,under the effect of flocculant sticky, large particles surface adsorb a lot of fine particles in the soil. There are a lot of connection between the floc particles,forming a lattice-like structure.In the soil, the pore is large and the connectivity is good. The sticky substance formed after dissolution similar to the hydrogen bonding can connect two or more soil particles to form a stable structure.The "key" that plays a role of sticky flocculant can be found everywhere in the soil surface, which has a strong adhesive force. Its role as the the chemical bond in molecular structure, can make the soil particles contact no longer point to point between each other, but the connecting way of lattice frame, make the soil structure more stable and lattice frame of pore in soil not easy to damage. Liquid membrane can cover soil particles,their component filaments can bond soil particles.The improvement of soil physical and chemical properties and the accumulation of organic matter promote the formation of soil structure and enhanced the surface stability.

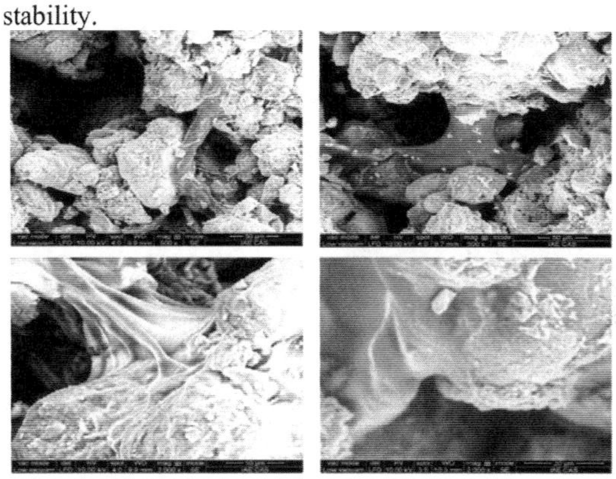

Figure 4. the connection of the liquid membrane with soil particles

4 conclusion

(1) The -OH in CMC which is the main component of the liquid membrane and the –OH in PVA can produce intermolecular hydrogen bonds, the formation of hydrogen bonds between the two components of molecular enhanced the compatibility between the two phase,and reached the purpose of increasing the effective group in liquid membrane .and the formation of dense network.

(2) Liquid membrane can effectively bond soil particles, forming a structure of the soil and liquid membrane and reduce evaporation, increase soil moisture conservation performance.In comparison with the micro morphology of different time, it was found that lliquid membrane degradation happens, will not cause pollution of soil environment.

(3)FTIR and ESEM can effectively reveal the soil moisture conservation mechanism of liquid membrane. Through the observation of the soil and liquid membrane in different periods, the degradation of liquid film can be inferred, which can provide theoretical basis for the popularization and application of liquid membrane.

Acknowledgements

The authors would like to express their sincere thanks to the editor and anonymous reviewer's comments and suggestions for the improvement of this paper.
This research was supported by the Special Fund for Agro-scientific Research in the Public Interest (201303125), the National Major Technology Program (2013BAD05B07; 2011BAD16B12; 2012BAD04B03) and the science and technology research projects of Liaoning province (2015103018).

References

1. Wang X., Xu G.B., Ren Z.G., Zhang Z.J., Jian Y.F., Zhang Y.M.Chinese J. Eco-Agr., **15(1)**,78-81(2007).
2. Yan C.R., Mei X.R., He W.Q., Zheng S.H. T. CSAE, **22(11)** , 269-272(2006).
3. Vlad S, Vlad A, Oprea S. Eur. Polym. J. **38(4)**, 829-835(2002).
4. Cascaval C, Rosu D, Rosu L. Polym. Test. **22(1)**, 45-49(2003).
5. Zhang J., Yao Y.Q., Lu J.J., Wang Y.H., Li J.H., Wang C.H. Chinese J. Soil Sci. **36(4)**,638-640(2005).
6. Yin G.H., Tong N., Hao L., Gu J., Liu Z.X. J. Soil Water Conserv. **26(4)**, 204-207(2012).
7. Li C.X.,Zhou X.G., Qiang X.M., Guo D.D., Liu Z.G., Zhang J.P. Maize Sci., **18(3)**, 108-112 (2010).
8. Lentz, R.D. and R.E. Sojka. Soil Sci. **158**, 274–282(1994).
9. G.F.Fanta,E.B. Encyclopaedia Polym. Sci. Techno.,**2(1)**, 665 (Wiley-Interscience,NY,1977).
10. Yin, G., L. Hao, Y. Zhao, J. Gu and Z. Liu. Chinese J. Eco. **32(6)**, 1405-1411 (2013).
11. Wang C.H.,Zhang J.,Li J.h.,Ding Z.Q.,Wang Y.H.,Yao Y.Q.,Lu J.J. Anhui Agric. Sci. **34(14)**, 3289-3290(2006).
12. Li, Y., Z. Wang, J. Zhang, X. Zheng, W. Li and G. Jiang. Aca. Agric. Boreali-Occidentalis Sin. **23(12)**, 110-115(2014).
13. Zhang, H., J. Li, Z. Jia, T. Zhang, X. Hou and P. Zhang. Agricul. Res. Arid Areas **30(2)** , 93-100(2012).

Data loss for PLC of nonlinear systems Iterative Learning Control Algorithm

Yinjun Zhang[1] Yinhui Li[1] Jianhuan Su[2] Zhengjie Lu[2]
[1] Aeronautics and Astronautics Engineering Institute, Air Force Engineering University, Xi'an 710038, Shanxi Province,
[2] School of physics and electrical engineering Hechi university Yizhou GuangXi China

Abstract- When we use power line as data carrier, due to the complexity of the PLC network environment, data packet loss frequently, so the paper deal with the iterative learning control for a class of nonlinear systems with measurement dropouts in the PLC, and studies the P-type iterative learning control algorithm convergence issues, the data packet loss is described as a stochastic Bernoulli process, on this basis we given convergence conditions for the P-type iterative learning control algorithm. The theoretically analysis is supported by the simulation of a numerical example; the convergence of ILC can be guaranteed when some output measurements are missing.

1 Introduction

ILC (Iterative learning control) is a Human-simulated intelligent control method, these cyclical and repetitive tasks can be achieved fully tracked through continuous learning within finite time interval. Since 1984, Japanese scholars Arimoto proposed the method, ILC has been a hot issue at home and abroad, and has achieved rich theoretical results and application results [2-6]. Robustness is an important research aspect for iterative learning control and is also applied to solve the real factory problem. Due to the theoretical study of iterative learning control algorithm has the stringent system conditions which is the repeatability but the real system is difficult to meet these conditions, so robust study of ILC focused on non-strict repeat the conditions, such as the non-repetitive initial conditions [7], non-repetitive desired trajectory [8], non-repetitive disturbances and noise [9-10], non-repetitive uncertain time-delay [11-13] and so on..

At present, we use net to control in real factory control system, such as DCS, FCS and remote control based on Internet. Net control is a closed-loop system, although net control system, compared with the traditional control system, is low cost, easy installation and maintenance, system flexibility, ease of troubleshooting etc., but net control brings data delay, data dropout and other problems[14-15]. However, it is less applied to the net network with ILC, currently a small number of studies have mainly focus on the linear systems. In the literature [16-17], the lifting method is used to examine data loss of linear discrete systems and the design of robust convergence and algorithm of iterative learning control (ILC). Given that most working systems are nonlinear and the lifting method of generations of learning is not suitable for nonlinear discrete systems, the same method cannot be applied to nonlinear systems. Thus, the present paper is aimed at adopting a new approach to explore the convergence problem of ILC of nonlinear systems in case of data loss in the power line communication.

2 Problem descriptions

Consider the PLC MIMO non liner discrete time system:

$$\begin{cases} x_k(t+1) = f(x_k(t)) + B(x_k(t))u_k(t) \\ y_k(t) = g(x_k(t)) + D(x_k(t))u_k(t) \end{cases} \quad (1)$$

Here, $x_k(t) \in \Re^n$, $u_k(t) \in \Re^m$, $y_k(t) \in \Re^r$ are state vector, control input and output vector respectively. For the MIMO system, we give below suppose 1.

Suppose 1 non-liner Function

$$f(\cdot):\Re^n \times [0,N] \to \Re^n, g(\cdot):\Re^n \times [0,N] \to \Re^r,$$

$$B(\cdot):\Re^n \times [0,N] \to \Re^{n\times m}, D(\cdot):\Re^n \times [0,N] \to \Re^{r\times m}$$

the state vector x meets Lipschitz conditions, that is for all $t \in [0, N]$ and any of $(x_1(t), x_2(t))$ represent bounded constant k_F, k_B, k_G, k_D meet

$$\|f(x_1(t)) - f(x_2(t))\| \le k_F \|x_1(t) - x_2(t)\|$$

$$\|B(x_1(t)) - B(x_2(t))\| \le k_E \|x_1(t) - x_2(t)\|$$

$$\|g(x_1(t)) - g(x_2(t))\| \le k_G \|x_1(t) - x_2(t)\|$$

$$\|D(x_1(t)) - D(x_2(t))\| \le k_D \|x_1(t) - x_2(t)\|$$

where $\|\cdot\|$ is Euclidean norm.

Suppose 2 Initial condition of system satisfies

$x_k(0)=x_d(0)$, where $x_k(0)$ is the initial value of the state variable with k times, $x_d(0)$ is the initial value of the desired state.

Suppose 3 For the desired tracking trajectory $y_d(t)$, there are desired input $u_d(t)$ and $x_d(t)$ that meet formula (2)

$$\begin{cases} x_d(t+1) = f(x_d(t)) + B(x_d(t))u_d(t) \\ y_d(t) = g(x_d(t)) + D(x_d(t))u_d(t) \end{cases} \quad (2)$$

For the MIMO non-liner system (1), consider P-type learning control algorithm:

$$u_{k+1}(t) = u_k(t) + \gamma e_k(t) \quad (3)$$

γ is learning gain matrix, $e_k(t)=y_d(t)-y_k(t)$ is system tracking error.

We assume output data $y_k(t)$ that is lost and each component of the output vector has same characteristics of loss. So the P-type ILC can be described as

$$u_{k+1}(t) = u_k(t) + \gamma\Theta e_k(t)$$

Where $\Theta = \begin{bmatrix} \eta_1(t) & \cdots & 0 \\ \vdots & \ddots & \vdots \\ 0 & \cdots & \eta_r(t) \end{bmatrix}_{r\times r}$, $\eta(t)$ is Bernoulli sequence, namely $\eta(t) \in \{0,1\}$, but $\eta(t)$ =0 or =1 is random, if $\eta(t)$ =0 indicates that the data is lost, $\eta(t)$ =1 means that no data is lost. Due to the randomness of data loss, we know $\eta(t)$ and $u_k(t),x_k(t),y_k(t)$ are unconnected. Assumption $P\{\eta(t)=1\}=E\{\eta(t)\}=\bar{\eta}$, $P\{\}$ is the probability, $E\{\cdot\}$ is desired factor. So $\bar{\eta}$ is the success rate of data transmission and satisfies $0 \le \bar{\eta} \le 1$ .Matrix Θ meet $P\{\eta_i(t)=1\} = E\{\eta_i(t)\} = \bar{\eta}\ i = 1,2,\cdots r.$

Before we give the convergence theorem, Let us make the following marks firstly:

$\|\cdot\|_\infty$ is infinite norm, we define vector x as $\|x\|_\infty = \max\limits_{1\le i\le n}\left|\sum_{j=n}^{n} x_j\right|$, where x_j denotes the j-th element of the vector z. for the $n\times n$ matrix M, we define $\|M\|_\infty = \max\limits_{1\le i\le n}\left|\sum_{j=n}^{n} m_{i,j}\right|$, $m_{i,j}$ denotes the element of matrix.

3 Convergence analyses

According to the below theorem we give the main result in the paper.

Theorem 1 For MIMO nonlinear system(1) to meet the suppose 1-3, using the formula(3) of ILC, when the system exits the output data dropout, if you choose the learning gain Γ for all t and k satisfy $\|I - \bar{\eta}\gamma D(x_k(t))\|_\infty < 1$, for all $t \in [0,N]$, we can get $\lim\limits_{k\to\infty} E\{y_k(t)\} \to y_d(t)$.

Proof. By the formula(1) and suppose3,we know

$$\begin{cases} y_d(t) = c(x_d(t)) + d(x_d(t))u_d(t) & (4) \\ y_k(t) = c(x_k(t)) + d(x_k(t))u_k(t) & (5) \end{cases}$$

Eq.(4) minus Eq.(5), we get $e_k(t)$.

$$\begin{aligned} e_k(t) &= y_d(t) - y_k(t) = \\ &(g(x_d(t)) + D(x_d(t))u_d(t) - \\ &(g(x_k(t)) + D(x_k(t))u_k(t) = \\ &g(x_d(t)) - g(x_k(t)) + D(x_d(t))u_d(t) - \\ &D(x_k(t))u_k(t) = \\ &g(x_d(t)) - g(x_k(t)) + D(x_k(t))du_k(t) \\ &+[D(x_d(t)) - D(x_k(t))]u_d(t) \end{aligned}$$

where $\delta u_k(t) = u_d(t) - u_k(t)$ according to the algorithm 3, we get following Eq.

$$\begin{aligned} \delta u_{k+1}(t) &= u_d(t) - u_{k+1}(t) = \\ &u_d(t) - u_k(t) - \gamma\eta(t)e_k(t), \end{aligned}$$

Namely

$$\begin{aligned} \delta u_{k+1}(t) &= \delta u_k(t) - \gamma\eta(t)e_k(t) = \\ &= [1 - \gamma\eta(t)d(x_k(t))]\delta u_k(t) \\ &- \gamma\eta(t)[c(x_d(t)) - c(x_k(t))] - \\ &\gamma\eta(t)[d(x_d(t)) - d(x_k(t))]u_d(t) \end{aligned} \quad (6)$$

According to the suppose 1 and meet Lipschitz, we can get the below Eq. by Eq.(6).

$$\begin{aligned} |\delta u_{k+1}(t)| &\le |1 - \gamma\eta(t)d(x_k(t))||\delta u_k(t)| + \\ &k_c|\gamma\eta(t)||x_d(t) - x_k(t)| \cdot \\ &k_d|\gamma\eta(t)||x_d(t) - x_k(t)||u_d(t)| = \\ &|1 - \gamma\eta(t)d(x_d(t))||\delta u_k(t)| + k_1|\delta x_k(t)| \end{aligned} \quad (7)$$

Where $\delta x_k(t) = x_d(t) - x_k(t)$,

$k_1 = k_c|\gamma\eta(t)||k_d\gamma\eta(t)||u_d(t)|$.

According to the Eq.(1)and suppose 3, we get

$$\begin{aligned} \delta x_k(t) &= x_d(t) - x_k(t) = \\ &f(x_d(t-1)) + b(x_d(t-1))u_d(t-1) - \\ &f(x_k(t-1)) + b(x_k(t-1))u_k(t-1), \end{aligned} \quad (8)$$

Then we get inequality (9)

$$|\delta x_k(t)| \le (k_f + k_b|u_d(t-1)|)|\delta x_k(t-1)| + |b(x_k(t-1))||\delta u_k(t-1)|$$

namely $|\delta u_k(t)| \le k_2|\delta u_k(t-1)| + k_3|\delta u_k(t-1)|$ (9)

Where $k_2 = |b(x_k(t-1))|$, $k_3 = k_f + k_b|u_d(t-1)|$.

As $|\delta u_k(t)|$ =0,so we expand right-hand member of Eq.(9) successively.

$$|\delta u_k(t)| \le \sum_{j=1}^{t} k_2 k_3^{t-j} |\delta u_k(j-1)| \quad (10)$$

Substitution Eq.(10) into Eq.(7),we get below Eq.(11).

$$|\delta u_{k+1}(t)| \le |1-\gamma\eta(t)d(x_k(t))||\delta u_k(t)| + k_1|\delta x_k(t)| \le |1-\gamma\eta(t)d(x_k(t))||\delta u_k(t)| + k_1\sum_{j=1}^{t} k_2 k_3^{t-j}|\delta u_k(j-1)| = |1-\gamma\eta(t)d(x_k(t))||\delta u_k(t)| + \sum_{i=0}^{t-1} K_{1,i}|\delta u_k(i)| \quad (11)$$

here $K_{1,i} = K_1 K_2 K_3^{t-i-1}$, because Eq.(11) includes random factors $\eta(t)$, considering the $\eta(t)$ and $u_k(t)$ are unconnected. The two ends of the Eq.(11) take mathematic expectation, then we get Eq.(12)

$$E\{|\delta u_{k+1}(t)|\} \le |1-\gamma\bar{\eta}d(x_k(t))|E\{|\delta u_k(t)|\} + \sum_{i=0}^{t-1}\bar{K}_{1,i}E\{|\delta u_k(i)|\} \quad (12)$$

where $\bar{K}_{1,i} = E\{K_{1,i}\} =$

$$E\left\{\sum_{i=0}^{t-1} k_c|\gamma\eta(t)|(1+k_d|u_d(t)|)k_2k_3^{t-i-1}\right\} \sum_{i=0}^{t-1} k_c(|\gamma\bar{\eta}| + k_d|\gamma\bar{\eta}||u_d(t)|)k_2k_3^{t-i-1}$$

We take Eq.(12) expand from t=10 to t=N successively, then we get Eq.(13)

$$E\{|\delta u_{k+1}(0)|\} \le |1-\gamma\bar{\eta}d(x_k(0))|E\{|\delta u_k(0)|\},$$
$$E\{|\delta u_{k+1}(1)|\} \le |1-\gamma\bar{\eta}d(x_k(1))|E\{|\delta u_k(1)|\} + \sum_{i=1}^{N-1}\bar{K}_{1,i}E\{|\delta u_k(i)|\} \quad (13)$$

We rewrite inequality(13) then get $V_{k+1} \le A_k V_k$

where: $V_k = \begin{bmatrix} E\{|\delta u_k(0)|\} \\ \vdots \\ E\{|\delta u_k(N)|\} \end{bmatrix}$,

$$A_k = \begin{bmatrix} |1-\gamma\bar{\eta}d(x_k(0))| & 0 & \cdots & 0 \\ \bar{K}_{1,0} & |1-\gamma\bar{\eta}d(x_k(1))| & \cdots & \vdots \\ \vdots & \vdots & \ddots & 0 \\ \bar{K}_{1,0} & \cdots & \bar{K}_{1,N-1} & |1-\gamma\bar{\eta}d(x_k(N))| \end{bmatrix}$$

So, according inequality (13), if spectral radius of matrix A_k meet $\rho(A_k) < 1$, then V_k follow iteration axis asymptotic convergence. When $k \to \infty$,for all $t = 0,\cdots,N$, V_k meet $E\{u_k(t)\} \to u_d(t)$ namely $E\{u_k(t)\} \to y_d(t)$ by suppose 3, as A_k is lower triangular matrix, their characteristic root is $|1-\gamma\bar{\eta}d(x_k(0))|$, $|1-\gamma\bar{\eta}d(x_k(1))|$, ... $|1-\gamma\bar{\eta}d(x_k(N))|$, when $\rho(A_k)<1$, all t and k meet $|1-\gamma\bar{\eta}d(x_k(k))|<1$. QED.

4 Simulation study

Consider to the below non-liner system

$$\begin{cases} x_{1,k}(t+1) = \sin(x_{1,k}(t)) + 2x_{2,k}(t)\sin t \\ x_{2,k}(t+1) = 2x_{1,k}(t) + \cos(x_{2,k}(t)) + [1+0.8e^{-t}\sin(x_{1,k}(t))]u_k(t) \\ y_k(t) = 0.2\sin(x_{1,k}(t)) + 0.1x_{2,k}(t) + u_k(t) \end{cases} \quad (14)$$

In the paper, we suppose that mathematic expectation output of the system is Eq.(15)

$$\bar{y}_d(t) = \sin(\pi t/100), t \in [0,500] \quad (15)$$

Initial conditions set $x_{1,k}(t) = x_{2,k}(t) = 0, u_0(k) = 0$ in the simulation system. In both cases output data lose in the system. The first case is $\bar{\eta} = 0.8$, 20% output data dropout, the second case is $\bar{\eta} = 0.6$, 40% output data dropout. We choose P-type ILC: $u_{k+1}(t) = u_k(t) + 0.5e_k(t)$

In the non-liner system (14), $d(x_k(t)) = 1$,we looked at convergence condition of theorem 1 in two data lose cases, when $\bar{\eta} = 0.8$, $|1-\gamma\bar{\eta}d(x_k(k))| = 0.6 < 1$, algorithm convergence.

Simulation results are figure 1 and figure 2. Fig.1 is maximum tracking error with different iteration time, obviously, when there is 20% measure output data lose, system tracing error still converges to 0. Fig.2 gives system output with different iteration. When $\bar{\eta} = 0.6$,

$$\left|1 - \gamma\bar{\eta} d(x_k(k))\right| = 0.7 < 1$$

, algorithm convergence remains the same, show fig.3 and fig.4

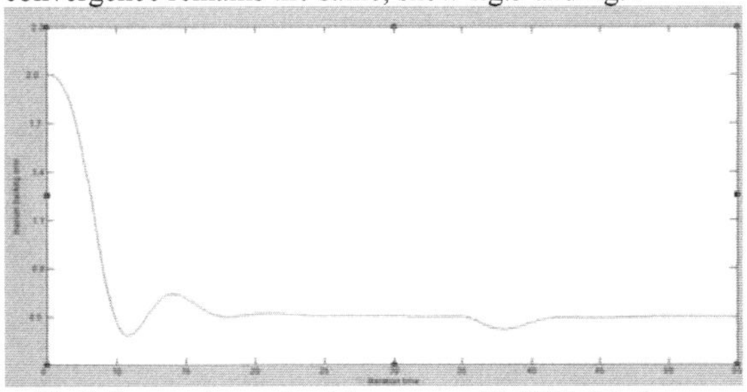

Fig.1 the maximum tracking error with 20% data lose

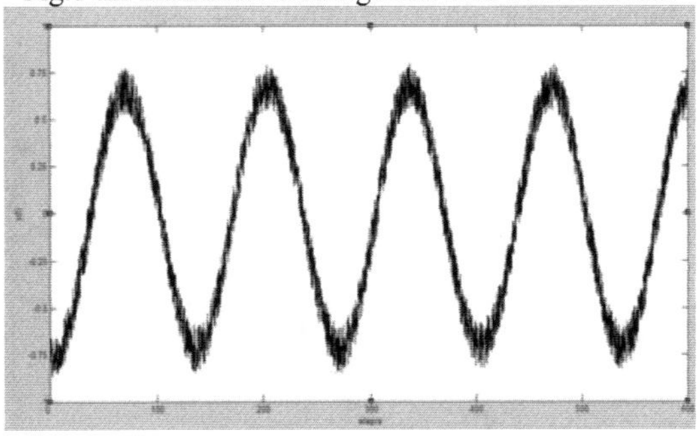

a) The fifth times iterations output

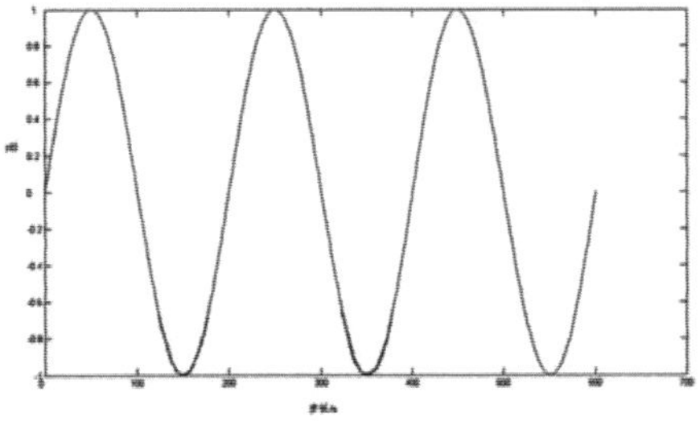

b) The twenty-fifth times iterations output

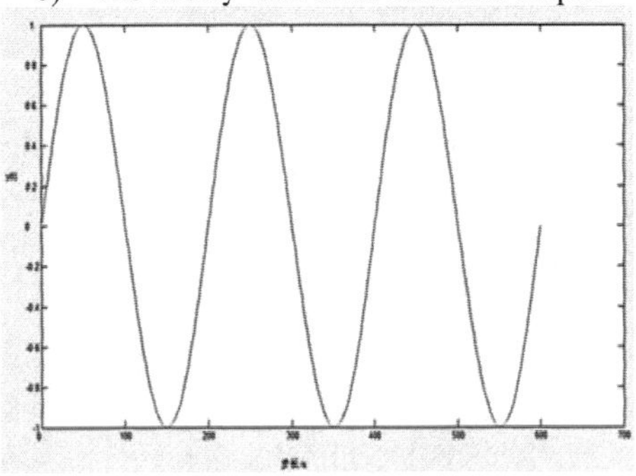

c) The thirty-fifth times iteration output

Fig.2 The system output profiles of different iterations for 20% lose

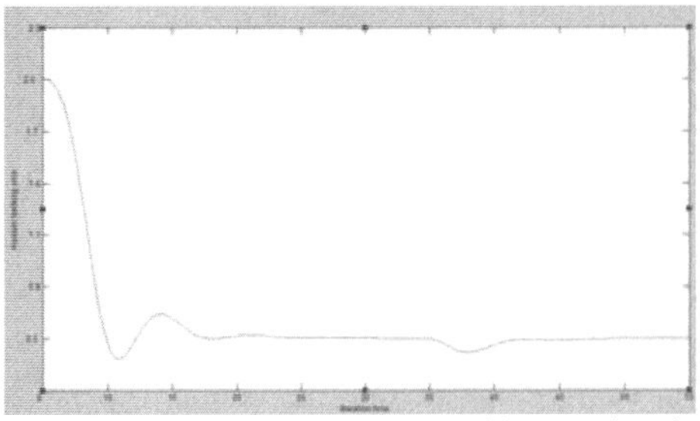

Fig.3 the maximum tracking error with 40% data lose

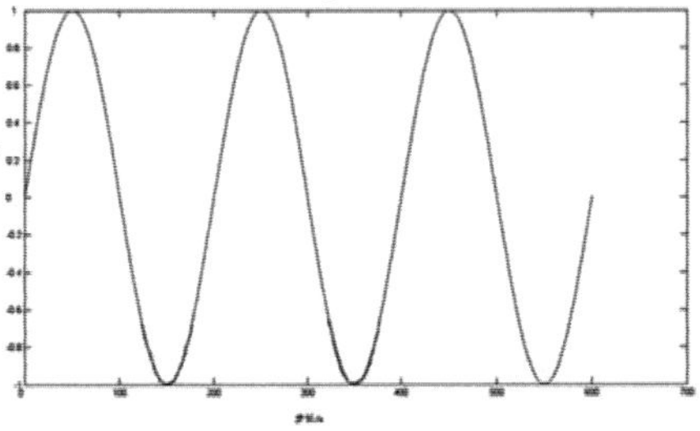

a) The thirty-fifth times iteration output

Fig.4 The system output profiles of different iterations for 40% lose

5 Conclusions

We research the presence of nonlinear system iteration learning control algorithm output measurement data loss robustness convergence problems in the paper and give the convergence condition. Theoretically proved that the convergence of the algorithm and the theoretical results through simulation. The results show that presence of certain nonlinear system when output data is lost, iterative learning control algorithm can guarantee the convergence of the tracking error, but the convergence rate increases will slow down with the degree of data loss.

References

1. ARIMOTO S, KAWAMURA S,MIYAZAKI F. Bettering operation of robotics [1]. Journal of Robotic System, 1984, 1 (2): 123 – 140.
2. XU J X. Analysis of iterative learning control for a class of nonlinear discrete-time system [J]. Automatic, 1997, 33(1 0): 1905 -1907
3. XU J X. Linear and Nonlinear Iterative Learning Control [M] Berlin: Springer Verlag,
4. 2003
5. AHN H S, CHEN Y Q, MOORE K L. Iterative learning control: brief survey and categorization [J]. IEEE Transactions on Systems, Man, and Cybernetics, Part C: Applicatiol1s and Reviews, 2007, 37(6) 1099 -1121.
   CHI R H, HOU Z S, XU J X. Adaptive ILC for a class of discrete time systems with iteration-varying trajectory and random initial condition[J]. Automatic, 2008, 44(8): 2207 - 2213
6. CHI Ronghu, HOU Zhongsheng, SUI Shulin. Non-parameter adaptive iterative learning control for the freeway traffic ramp metering[J]. Control Theory & Applications, 2008, 25(6): 1011 - 1015.

7. SUN M X, WANG D W. Initial shift issues on discrete-time iterative learning control with system relative degree [J]. IEEE Transactions on Automatic Control, 2003, 48(1): 144-148
8. YIN C K, XU J X, HOU Z S. A high-order internal model based iterative learning control scheme for nonlinear systems with time iteration-varying parameters [J]. IEEE Transactions on Automatic Control, 2010, 55(11)2665 - 2670
9. NORRLOF M, GUNNARSSON S. Disturbance aspects of iterative learning control [J]. Engineering Applicatiol1s of Artificial Intelligel1ce, 2001 , 14(1): 87 - 94.
10. SAAB SS. On a discrete-time stochastic learning control algorithm [J]. IEEE Transactions on Automatic Control, 2001, 46(8): 1333-1336.
11. MENG D, JIA Y, DU J, et al. Feedback approach to design fast iterative learning controller for a class of time-delay systems [J]. IET Control Theory and Applications, 2009, 3(2): 225 - 238.
12. CHEN W S, ZHANG L. Adaptive iterative learning control for nonlinearly parameterized systems with unknown time-varying delays[J]. lntemational Joumal of Control, Automation, and Systems, 2010, 8(2): 177 - 186
13. CHEN Weisheng, WANG Yuanliang, LI Junmin. Adaptive learning control for nonlinearly parameterized systems with periodically time-varying delay [J]. Acta Automatica Sinica, 2008, 34(1 2): 1556- 1560.)
14. YANG T C. Networked control system: a brief survey [J]. IEE Proceedings Control Theory and Applications, 2006, 153(4): 403-412
15. Li Hongbo, SUN Zengqi, SUN Fuchun. Networked control systems: an overview of state-of-the-art and the prospect in future research [J]. Control Theory & Applications, 2010, 27(2): 238 - 243.
16. AHN H S, CHEN Y Q, MOORE K L. Discrete time intermittent iterative learning control with independent data dropouts [C] //Proceedings of the 17th lFAC World Congress. Korea: IFAC, 2008: 12442-12447.
17. BU X H, HOU Z S. Stability of iterative learning control with data Dropout via asynchronous dynamical system[J].International Journal of Automation and Computing, 2011 ,8(1): 29 - 36.

Supported by KY2015LX339, 2015ER04, XJ2015QN001

DOI: 10.1051/matecconf/20166101024

Dynamic Analysis of a Micro Drilling System by the Modified Transfer Matrix Method

An-Chen Lee[1,a] and Danh-Tuyen Nguyen[2]

[1,2] Department of Mechanical Engineering, National Chiao Tung University, 1001 Ta Hsueh Road, Hsinchu 30010, Taiwan

Abstract. Normally, breakage is the main cause of failure of a micro drilling system, of which coupling effect of lateral and torsional displacements plays the most important role. This research developed a modified transfer matrix method to investigate the dynamics of a micro drilling system focusing on displacements induced by not only lateral, shear, torsional deformations, but also the coupling effect between lateral and torsional vibrations. The micro drilling system is modeled as pre-twisted rotating Timoshenko beam elements with a continuous-system concept. Each element includes components of linear bearings, axial drilling force, gyroscopic moments, and eccentricity. The overall transfer matrix of the system is developed to determine the state vector of the system including displacements and forces. Finally, a numerical example was presented, where the lateral and torsional displacements and critical speed of the micro drilling system at the micro drill's tip were presented.

1 Introduction

Nowadays, with the development of science and technology, there is an increasing demand for producing micro holes in some devices, for example, cameras, mobile phones, computers, medical instruments, micro dies and molds, fuel injection nozzles, watches, bearings, especially printed circuit boards (PCB). In methods to manufacture micro holes now, such as using micro mechanical machining or micro drilling, electron beam machining, electrical discharge machining, and laser machining, the micro drilling has higher accuracy, greater productivity, and more economical efficiency, compared with other approaches [1-3]. Hence, there are more and more researchers making great efforts to improve the ability of micro drilling to meet that demand. To achieve that purpose, it is necessary to understand the dynamic characteristics of micro drilling process [4, 5]. The transverse displacement affects hole's accuracy, and exhibits the maximum displacements if its operating speed is equal to a critical one. In the past, several researches on dynamics of micro drilling were published. To name few, Yang studied transverse displacement of a micro drill tip, but a lot of dynamic properties of the micro drilling system on the drilling process are not considered, such as eccentricity, gyroscopic moments, etc. [6]. Ehmann modeled micro drills as long twisted beams including effects of transverse shear, rotary inertia and gyroscopic moments during drilling. Models for the critical speeds and critical bucking loads were developed by the finite element method. However, the beams did not include the effect of eccentricity of the drilling system and the torsional displacements were ignored [7]. Zhaojun modeled a micro-drill-spindle system as Timoshenko beam elements. Besides considerations of transverse shear, rotary inertia and gyroscopic moments similar to Ehmann's models, the element models also included the effects of eccentricity of the spindle-clamp-drill system, axial drilling force, and bearings. Nevertheless, the element's torsional displacement was also ignored [8]. Ozdoganlar modeled the flute part of a micro drill, and other parts as a beam including torsional displacement and one-dimensional beams, respectively. However, the models did not consider the effect of eccentricity of the drilling system [9, 10].

In this research, the underlying micro drilling system includes a micro drill grasped in a spindle, which is supported by identical linear bearings. It is modeled as pre-twisted rotating Timoshenko beam elements with a continuous-system concept. Each element includes effects of linear bearings, axial drilling force, gyroscopic moments, and eccentricity. Its orientation is described by Euler angles. Then the kinetic, potential energies, and work done by non-conservative forces are developed, and the Hamilton's principle is applied to obtain its motion equations, as well as natural boundary conditions with respect to the fixed coordinates. The harmonic balance method is used to solve those equations. The state vector of the system containing displacements and forces is developed by the transfer matrix method (TMM). The most significant advantage of this method is that it does not require the storage and manipulation of large system arrays. Therefore, there is less demand on computer memory. After developing local transfer matrices of system elements and bearings, the overall transfer matrix of the system is developed to determine the state vector. Finally, a numerical example was presented and the results of the dynamic analysis showed the effects of

[a] Corresponding author: aclee@mail.nctu.edu.tw

external forces, bearing coefficients, and eccentricity to the displacements and critical speeds of the micro drilling system at the micro drill tip.

2 Motion Equations of the Drilling System Element

The structure of the micro drilling system can be simplified as shown in Fig.1. The micro drill E is grasped in the clamp C-D that is connected to the spindle A by a small shaft B. The system is supported by two identical linear bearings, and applied by an axial force F_a, and a torque T during drilling the work-piece W. The whole system's dimensions and parameters are shown in table 1 [8, 11].

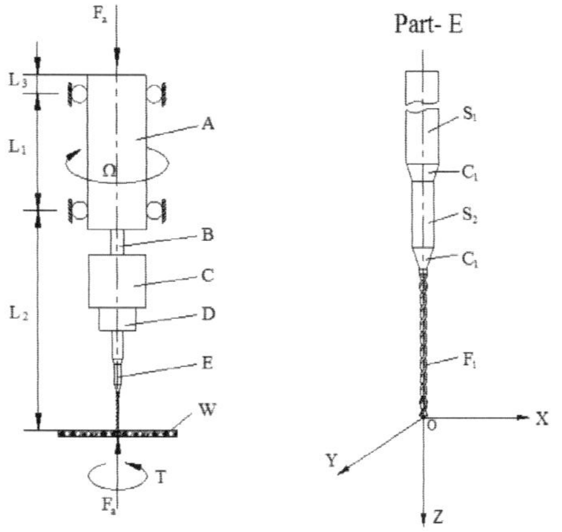

(a) Structure of micro drilling system (b) Micro drill

Figure 1. Micro drilling system

Because the axial displacement is much smaller than the other displacements, it is neglected in this research [12-14]. The micro drill and spindle of micro drilling system are modeled pre-twisted rotating Timoshenko beam elements as shown in Fig. 2.

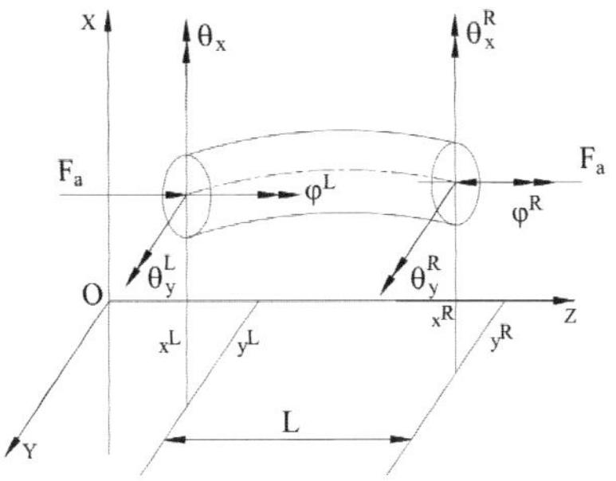

Figure 2. Pre-twisted rotating Timoshenko beam element

The motion equations of the pre-twisted rotating Timoshenko beam element are developed by applying the extended Hamilton's principle. According to this principle, for a rotating element with kinetic energy E_k, potential energy E_p, virtual work W done by the non-conservative force, its governing equation motions can be obtained by the following equation.

$$\delta\int_{t_1}^{t_2}\left(E_k - E_p + W\right)dt = 0 \tag{1}$$

The kinetic energy E_k includes translational and rotational ones. To make it simple for developing the rotational kinetic energy, the Euler angles (ϕ,θ,ψ) as shown in Fig. 3(a) are used. They were introduced by Leonhard Euler to describe orientations of rotating elements and explained via the following steps: (1) rotate the XYZ initial coordinate system, parallel to the fixed coordinate system, about the Z axis by an angle ϕ to lead to the $X'Y'Z'$ coordinate system, (2) rotate $X'Y'Z'$ about the X' axis by an angle θ to lead to the $U'V'W'$ coordinate system, (3) rotate $U'V'W'$ about the W' axis by an angle ψ to produce the UVW principal coordinate system. As the rotating element is deflected in position and orientation, the angular displacements as shown in Fig. 3(b) are the projections of angle θ, measuring counterclockwise from the Z fixed axis to the W spin axis, onto YZ and XZ planes, thus

$$\begin{cases}\theta_x = \theta\cos\phi \\ \theta_y = \theta\sin\phi\end{cases} \tag{2}$$

Furthermore, the spin angle about the W axis is obtained as $\Phi = \phi + \psi$ from the geometric configuration of a rotating element with a very small oblique angle θ. Applying the coordinate transformation formula, the components of the angular velocities in directions of the principal coordinates are obtained [11, 15].

$$\begin{cases}\omega_u = \dot{\theta}\cos\psi + \dot{\phi}\sin\theta\sin\psi \\ \omega_v = -\dot{\theta}\sin\psi + \dot{\phi}\sin\theta\cos\psi \\ \omega_w = \dot{\psi} + \dot{\phi}\cos\theta\end{cases} \tag{3}$$

Take the mass center C of the cross-section of the rotating element as the origin of the UVW principal coordinate system. Its kinetic energy E_k is calculated in the fixed coordinate system as follows:

$$E_k = \frac{1}{2}\rho\int_0^L\left[A\left(\dot{x}_c^2 + \dot{y}_c^2\right)\frac{1}{2}\left(I_u\omega_u^2 + I_v\omega_v^2 + I_p\omega_w^2\right)\right]dZ \tag{4}$$

where A is the cross sectional area, ρ is mass density. I_u, and I_v are area principal moments of inertia, I_p is area polar moment of inertia. The $\omega_u,\omega_v,\omega_w$ are angular velocities about principal axes U,V,W, respectively.

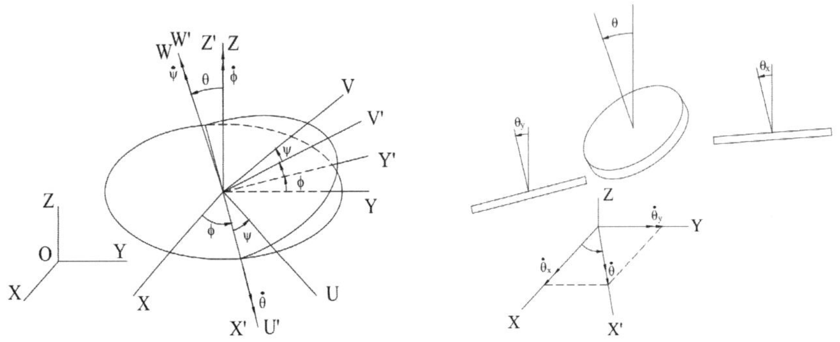

(a) Euler angles of a rotating element (b) Element projection

Figure 3. Orientation of a rotating element

Table 1. Structure dimensions of the micro drilling system

Name of shafts		Shaft symbol	Diameter (mm)	Length (mm)	Number of divided elements
Spindle		A	22	60	3
Shaft		B	6.3	9	1
Up clamp		C	22.5	19	1
Down clamp		D	14	8	1
	Shank	S_1	3	10	1
	Cone 1	C_1		4.1	5
Micro drill	Cylinder	S_2	0.8	2.6	1
	Cone 2	C_2		0.933	5
	Flute	F_1	0.3	5	20
Positions of bearings (mm)		L_1=11.25, L_3=37.5			

The Fig 4 shows the whirling orbit of a rotating element which is in asymmetric, eccentric form. From the geometric relations, the coordinates of the mass center C can be calculated as follows:

$$\begin{bmatrix} x_c \\ y_c \end{bmatrix} = \begin{bmatrix} x \\ y \end{bmatrix} + \begin{bmatrix} e_x \\ e_y \end{bmatrix} \tag{5}$$

$$\begin{bmatrix} e_x \\ e_y \end{bmatrix} = \begin{bmatrix} \cos(\Omega t+\varphi+\beta) & \text{-}\sin(\Omega t+\varphi+\beta) \\ \sin(\Omega t+\varphi+\beta) & \cos(\Omega t+\varphi+\beta) \end{bmatrix} \begin{bmatrix} e_u \\ e_v \end{bmatrix} \tag{6}$$

where, Ω is rotating speed , φ is torsional angle, β is the initial twist angle between principle axes U , and X axis, t is time, e_u , e_v are eccentricities corresponding to U, V coordinates, respectively. Substituting Eq. 6 into Eq. 5, and differentiating it yields,

$$\begin{aligned} \dot{x}_c = \dot{x} - e_v\left(\Omega+\dot{\varphi}\right)\cos\left(\Omega t+\varphi+\beta\right) \\ - e_u\left(\Omega+\dot{\varphi}\right)\sin\left(\Omega t+\varphi+\beta\right) \end{aligned} \tag{7}$$

$$\begin{aligned} \dot{y}_c = \dot{y} + e_u\left(\Omega+\dot{\varphi}\right)\cos\left(\Omega t+\varphi+\beta\right) \\ - e_v\left(\Omega+\dot{\varphi}\right)\sin\left(\Omega t+\varphi+\beta\right) \end{aligned} \tag{8}$$

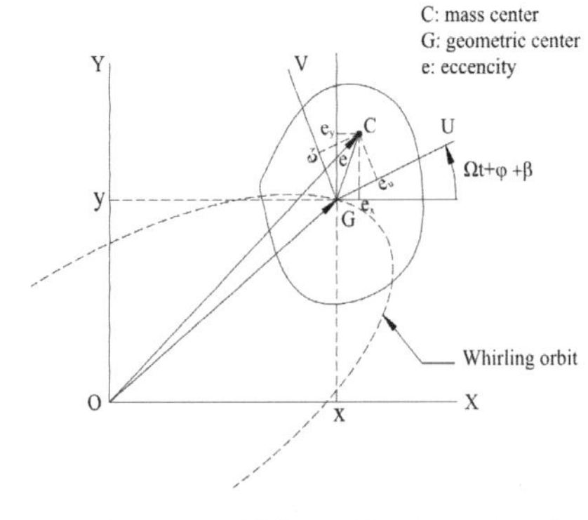

Figure 4. Whirling orbit of a rotating element

Replacing Eqs. 2, 3, 7, and 8, and $\Phi = \Omega t + \varphi + \beta$ into Eq. 4, one gets

$$
\begin{aligned}
E_k = \frac{1}{2}\rho\int_0^L \{ & A\big[\dot{x}^2 - 2\dot{x}e_v(\Omega+\dot{\varphi})\cos(\Omega t+\varphi+\beta) \\
& -2\dot{x}e_u(\Omega+\dot{\varphi})\sin(\Omega t+\varphi+\beta) \\
& +\dot{y}^2 + 2\dot{y}e_u(\Omega+\dot{\varphi})\cos(\Omega t+\varphi+\beta) \\
& -2\dot{y}e_v(\Omega+\dot{\varphi})\sin(\Omega t+\varphi+\beta) \\
& +(\Omega+\dot{\varphi})^2 e^2\big] + I_p(\Omega+\dot{\varphi})^2 \\
& +I_p(\Omega+\dot{\varphi})\left(\dot{\theta}_x\theta_y - \dot{\theta}_y\theta_x\right) + I\left(\dot{\theta}_x^2+\dot{\theta}_y^2\right) \\
& +2\Delta\dot{\theta}_x\dot{\theta}_y\sin 2(\Omega t+\varphi+\beta) \\
& +\Delta\left(\dot{\theta}_x^2-\dot{\theta}_y^2\right)\cos 2(\Omega t+\varphi+\beta)\}\,dZ
\end{aligned}
\tag{9}
$$

where, $I = \frac{1}{2}(I_u + I_v)$, $\Delta = \frac{1}{2}(I_u - I_v)$

Assuming the rotating element as shown in Fig 5 is acted by an axial force F_a , its potential energy E_p . According to the bending and shear deformations can be described as in the following:

$$
\begin{aligned}
E_p = \frac{1}{2}\int_0^L \Big[& E(I+\Delta\cos 2(\Omega t+\varphi+\beta))\left(\theta_x^{'}\right)^2 \\
& +E(I-\Delta\cos 2(\Omega t+\varphi+\beta))\left(\theta_y^{'}\right)^2 \\
& +2E\Delta\theta_x^{'}\theta_y^{'}\sin 2(\Omega t+\varphi+\beta) \\
& +k_s GA\left((x^{'}-\theta_y)^2+(y^{'}+\theta_x)^2\right) \\
& +GI_p\left(\varphi^{'}\right)^2 - F_a\left((x^{'})^2+(y^{'})^2\right)\Big]dZ
\end{aligned}
\tag{10}
$$

From Fig. 5, the work done by external force is

$$
\begin{aligned}
W = & V_x^R x + M_y^R\theta_y + V_y^R y + M_x^R\theta_x + T^R\varphi \\
& -\left(V_x^L x + M_y^L\theta_y + V_y^L y + M_x^L\theta_x + T^L\varphi\right)
\end{aligned}
\tag{11}
$$

Substituting Eqs. 9, 10, and 11 into Eq. 1, one gets motion equations of the rotating elements and natural boundary conditions as follows:

$$
\begin{aligned}
& \rho A\big[-\ddot{x} + \ddot{\varphi}e_v\cos(\Omega t+\varphi+\beta) + \ddot{\varphi}e_u\sin(\Omega t+\varphi+\beta) \\
& -(\Omega+\dot{\varphi})^2 e_v\sin(\Omega t+\varphi+\beta) + (\Omega+\dot{\varphi})^2 e_u\cos(\Omega t+\varphi+\beta)\big] \\
& +k_s GA\left(x^{''}-\theta_y^{'}\right) - F_a x^{''} = 0
\end{aligned}
\tag{12}
$$

$$
\begin{aligned}
& \rho A\big[-\ddot{y} - \ddot{\varphi}e_u\cos(\Omega t+\varphi+\beta) + \ddot{\varphi}e_v\sin(\Omega t+\varphi+\beta) \\
& +(\Omega+\dot{\varphi})^2 e_u\sin(\Omega t+\varphi+\beta) + (\Omega+\dot{\varphi})^2 e_v\cos(\Omega t+\varphi+\beta)\big] \\
& +k_s GA\left(y^{''}+\theta_x^{'}\right) - F_a y^{''} = 0
\end{aligned}
\tag{13}
$$

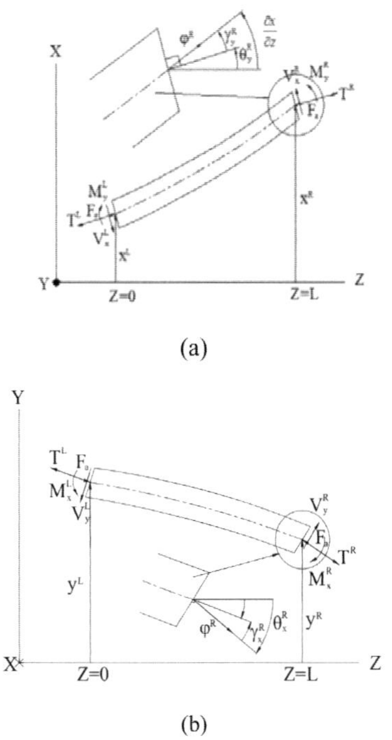

Figure 5. Forces, moments, and torques applying on a rotating element

$$
\begin{aligned}
& \rho I\ddot{\theta}_x + \frac{1}{2}\rho I_p\ddot{\varphi}\theta_y + \rho I_p(\Omega+\dot{\varphi})\dot{\theta}_y - EI\theta_x^{''} + k_s GA\left(\theta_x + y^{'}\right) \\
& +\rho\Delta\big[\ddot{\theta}_y\sin 2(\Omega t+\varphi+\beta) + 2\dot{\theta}_y(\Omega+\dot{\varphi})\cos 2(\Omega t+\varphi+\beta) \\
& +\ddot{\theta}_x\cos 2(\Omega t+\varphi+\beta) - 2\dot{\theta}_x(\Omega+\dot{\varphi})\sin 2(\Omega t+\varphi+\beta)\big] \\
& +E\Delta\big[-\theta_y^{''}\sin 2(\Omega t+\varphi+\beta) - 2\theta_y^{'}\varphi^{'}\cos 2(\Omega t+\varphi+\beta) \\
& -\theta_x^{''}\cos 2(\Omega t+\varphi+\beta) + 2\theta_x^{'}\varphi^{'}\sin 2(\Omega t+\varphi+\beta)\big] = 0
\end{aligned}
\tag{14}
$$

$$
\begin{aligned}
& \rho I\ddot{\theta}_y - \frac{1}{2}\rho I_p\ddot{\varphi}\theta_x - \rho I_p(\Omega+\dot{\varphi})\dot{\theta}_x - EI\theta_y^{''} + k_s GA\left(\theta_y - x^{'}\right) \\
& +\rho\Delta\big[\ddot{\theta}_x\sin 2(\Omega t+\varphi+\beta) + 2\dot{\theta}_x(\Omega+\dot{\varphi})\cos 2(\Omega t+\varphi+\beta) \\
& -\ddot{\theta}_y\cos 2(\Omega t+\varphi+\beta) \ +2\dot{\theta}_y(\Omega+\dot{\varphi})\sin 2(\Omega t+\varphi+\beta)\big] \\
& +E\Delta\big[-\theta_x^{''}\sin 2(\Omega t+\varphi+\beta) - 2\theta_x^{'}\varphi^{'}\cos 2(\Omega t+\varphi+\beta) \\
& +\theta_y^{''}\cos 2(\Omega t+\varphi+\beta) - 2\theta_y^{'}\varphi^{'}\sin 2(\Omega t+\varphi+\beta)\big] = 0
\end{aligned}
\tag{15}
$$

$$
\begin{aligned}
& \rho I_p\ddot{\varphi} + \frac{1}{2}\rho I_p\ddot{\theta}_x\theta_y - \frac{1}{2}\rho I_p\ddot{\theta}_y\theta_x + \rho A\big[-\ddot{x}e_v\cos(\Omega t+\varphi+\beta) \\
& -\ddot{x}e_u\sin(\Omega t+\varphi+\beta) + \ddot{y}e_u\cos(\Omega t+\varphi+\beta) \\
& -\ddot{y}e_v\sin(\Omega t+\varphi+\beta) + e^2\ddot{\varphi}\big] - GI_p\varphi^{''} \\
& +\rho\Delta\Big[-2\dot{\theta}_x\dot{\theta}_y\cos 2(\Omega t+\varphi+\beta) + \left(\dot{\theta}_x\right)^2\sin 2(\Omega t+\varphi+\beta) \\
& -\left(\dot{\theta}_y\right)^2\sin 2(\Omega t+\varphi+\beta)\Big] + E\Delta\Big[2\theta_x^{'}\theta_y^{'}\cos 2(\Omega t+\varphi+\beta) \\
& -\left(\theta_x^{'}\right)^2\sin 2(\Omega t+\varphi+\beta) \ +\left(\theta_y^{'}\right)^2\sin 2(\Omega t+\varphi+\beta)\Big] = 0
\end{aligned}
\tag{16}
$$

Note that the φ and x appeared in Eqs. 12, 15, and 16, so as φ and y appeared in Eqs. 13, 14, and 16. So there are the coupling effect between torsional displacement φ

and lateral displacements (x, y) . $(\Omega+\dot{\varphi})\dot{\theta}_y$,and $\dot{\theta}_x(\Omega+\dot{\varphi})$ in Eqs. 14, and 15, respectively present the gyroscopic effects.

Natural boundary conditions are

$$\begin{cases} V_x^R + k_s GA(\theta_y - x') + F_a \cdot x' = 0 \\ V_x^R - k_s GA(\theta_x + y') + F_a \cdot y' = 0 \\ T^R - GI_p\varphi' = 0 \\ M_x^R - EI\theta_x' - E\Delta\theta_x' \cos 2(\Omega t+\varphi+\beta) \\ -E\Delta\theta_y' \cos 2(\Omega t+\varphi+\beta) = 0 \\ M_y^R - EI\theta_y' - E\Delta\theta_y' \cos 2(\Omega t+\varphi+\beta) \\ -E\Delta\theta_x' \cos 2(\Omega t+\varphi+\beta) = 0 \end{cases} \text{, at } Z = L \tag{17}$$

$$\begin{cases} V_x^R + k_s GA(\theta_y - x') + F_a \cdot x' = 0 \\ V_x^R - k_s GA(\theta_x + y') + F_a \cdot y' = 0 \\ T^R - GI_p\varphi' = 0 \\ M_x^R - EI\theta_x' - E\Delta\theta_x' \cos 2(\Omega t+\varphi+\beta) \\ -E\Delta\theta_y' \cos 2(\Omega t+\varphi+\beta) = 0 \\ M_y^R - EI\theta_y' - E\Delta\theta_y' \cos 2(\Omega t+\varphi+\beta) \\ -E\Delta\theta_x' \cos 2(\Omega t+\varphi+\beta) = 0 \end{cases} \text{, at } Z = 0 \tag{18}$$

3 Transfer Matrix of the Micro Drilling System

The most important work of the transfer matrix method is to calculate the overall transfer matrix of the system. Based on it, the state vector including displacements, shear forces, moments, and torques at any position of the system can be evaluated. It is the multiplication of element transfer matrices that are developed by solving their motion equations.

The transfer matrix of pre-twisted rotating Timoshenko beam element is obtained by solving Eqs. 12-16. The solution process is presented below. Assuming the steady-state solutions of Eqs. 12-16 can be expressed in series form as

$$\begin{cases} x(Z,t) = x_0(Z) + \sum_{i=1}^{n} x_{ic}(Z)\cos i\Omega t + x_{is}(Z)\sin i\Omega t \\ y(Z,t) = y_0(Z) + \sum_{i=1}^{n} y_{ic}(Z)\cos i\Omega t + y_{is}(Z)\sin i\Omega t \\ \theta_x(Z,t) = \theta_{x,0}(Z) + \sum_{i=1}^{n} \theta_{x,ic}(Z)\cos i\Omega t + \theta_{x,is}(Z)\sin i\Omega t \\ \theta_y(Z,t) = \theta_{y,0}(Z) + \sum_{i=1}^{n} \theta_{y,ic}(Z)\cos i\Omega t + \theta_{y,is}(Z)\sin i\Omega t \\ \varphi(Z,t) = \varphi_0(Z) + \sum_{i=1}^{n} \varphi_{ic}(Z)\cos i\Omega t + \varphi_{is}(Z)\sin i\Omega t \end{cases} \tag{19}$$

where,
$x_0(Z)$, $y_0(Z)$, $\theta_{x,0}(Z)$, $\theta_{y,0}(Z)$, $\varphi_0(Z)$, $x_{ic}(Z)$, $x_{is}(Z)$, $y_{ic}(Z)$, $y_{is}(Z)$, $\theta_{x,ic}(Z)$, $\theta_{x,is}(Z)$, $\theta_{y,ic}(Z)$, $\theta_{y,is}(Z)$, $\varphi_{ic}(Z)$, and $\varphi_{is}(Z)$ are the mode functions of the relative 0th order and $n\times$ harmonic whirl with respect to the static coordinate system of the Timoshenko beam element.

$$\begin{aligned} \mathbf{X(Z)} = [& x_0(Z)\ x_{1c}(Z) \cdots x_{nc}(Z)\ x_{1s}(Z) \cdots x_{ns}(Z) \\ & y_0(Z)\ y_{1c}(Z) \cdots y_{nc}(Z)\ y_{1s}(Z) \cdots y_{ns}(Z) \\ & \theta_{x,0}(Z)\ \theta_{x,1c}(Z) \cdots \theta_{x,nc}(Z)\ \theta_{x,1s}(Z) \cdots \theta_{x,ns}(Z) \\ & \theta_{y,0}(Z)\ \theta_{y,1c}(Z) \cdots \theta_{y,nc}(Z)\ \theta_{y,1s}(Z) \cdots \theta_{y,ns}(Z) \\ & \varphi_0(Z)\ \varphi_{1c}(Z) \cdots \varphi_{nc}(Z)\ \varphi_{1s}(Z) \cdots \varphi_{ns}(Z)]^T \end{aligned} \tag{20}$$

The size of $\mathbf{X(Z)}$ is $(10n+5)\times 1$. Replacing Eq. 19 into Eqs. 12-16, ignoring nonlinear terms, and equating the same harmonic term, the following matrix equation can be obtained.

$$\mathbf{E_2 X''(Z) + E_1 X'(Z) + E_0 X(Z) = F_0} \tag{21}$$

where, $\mathbf{X''(Z)}$, $\mathbf{X'(Z)}$ are second and first order derivatives of $\mathbf{X(Z)}$ with respect to Z coordinate, respectively. $\mathbf{E_2}, \mathbf{E_1}$, and $\mathbf{E_0}$ are matrices with size of $(10n+5)\times(10n+5)$. The size of matrix $\mathbf{F_0}$ is $(10n+5)\times 1$. The general solutions of Eq .21 are the sum of homogeneous and particular solutions.

$$\mathbf{X(Z)} = \mathbf{X(Z)}^h + \mathbf{X(Z)}^p \tag{22}$$

The homogeneous solution $\mathbf{X(Z)}^h$ is [11]

$$\mathbf{X(Z)}^h = \sum_{i=1}^{k} C_i \mathbf{X}_i^h e^{\lambda_i Z} \tag{23}$$

where, λ_i is eigen-value, $\mathbf{X}_i^h$ is eigen-vector corresponding to λ_i , which are solutions of Eq. 24; C_i is an undetermined constant.

$$\left\{ \lambda \begin{bmatrix} \mathbf{0} & \mathbf{E_2} \\ \mathbf{E_2} & \mathbf{E_1} \end{bmatrix}_{k\times k} - \begin{bmatrix} \mathbf{E_2} & \mathbf{0} \\ \mathbf{0} & \mathbf{E_0} \end{bmatrix}_{k\times k} \right\} \begin{bmatrix} \lambda \mathbf{X}^h \\ \mathbf{X}^h \end{bmatrix}_{k\times 1} = \begin{bmatrix} \mathbf{0} \\ \mathbf{0} \end{bmatrix}_{k\times 1} \tag{24}$$

where, $k = 20n + 10$.

The particular solution is

$$\begin{aligned} \mathbf{X(Z)}^h = [& x_0^p\ x_{1c}^p \cdots x_{nc}^p\ x_{1s}^p \cdots x_{ns}^p\ y_0^p\ y_{1c}^p \cdots y_{nc}^p\ y_{1s}^p \cdots y_{ns}^p \\ & \theta_{x,0}^p\ \theta_{x,1c}^p \cdots \theta_{x,nc}^p\ \theta_{x,1s}^p \cdots \theta_{x,ns}^p\ \theta_{y,0}^p\ \theta_{y,1c}^p \cdots \theta_{y,nc}^p \\ & \theta_{y,1s}^p \cdots \theta_{y,ns}^p \varphi_0^p\ \varphi_{1c}^p \cdots \varphi_{nc}^p\ \varphi_{1s}^p \cdots \varphi_{ns}^p] \end{aligned} \tag{25}$$

where

$x_{1s}^p = -y_{1c}^p = e_v \cos\beta + e_u \sin\beta,\ x_{1c}^p = y_{1s}^p = e_v \sin\beta - e_u \cos\beta,$

$y_{2c}^p = x_{2s}^p = -\dfrac{g\Delta \cos 2\beta}{4\Omega^2 I},\ y_{2s}^p = x_{2c}^p = -\dfrac{\Delta g \sin 2\beta}{4\Omega^2 I},$

$\theta_{x,2c}^p = \theta_{y,2s}^p = \dfrac{\Delta\rho g \cos 2\beta}{k_s GI} Z,\ \theta_{x,2s}^p = -\theta_{y,2c}^p = \dfrac{\Delta\rho g \sin 2\beta}{k_s GI} Z,$

other elements = 0.

Substituting Eqs. 25, and 23 into Eq. 22, one gets

$$\mathbf{X}(\mathbf{Z}) = \left[G(Z)\right]\begin{bmatrix}\mathbf{C}\\1\end{bmatrix} \tag{26}$$

where $\left[G(Z)\right]$ is the matrix function of Z with the size of $(10n+5)\times(20n+11)$, $\mathbf{C} = \left[C_1\ C_2\ \cdots C_{20n+10}\right]^T$, where, C_i the undetermined constant in Eq. 23. Hence the displacement state vectors at two ends of the drill flute elements are

$$\mathbf{X}^R = \mathbf{X}(Z=L) = \left[G_L\right]\begin{bmatrix}\mathbf{C}\\1\end{bmatrix} \tag{27}$$

$$\mathbf{X}^L = \mathbf{X}(Z=0) = \left[G_0\right]\begin{bmatrix}\mathbf{C}\\1\end{bmatrix} \tag{28}$$

where $\left[G_L\right] = \left[G(Z=L)\right],\ \left[G_0\right] = \left[G(Z=0)\right]$.

Assuming the solutions of Eqs. 17, and 18 are described in Fourier series form as

$$\begin{cases} V_x(Z,t) = V_{x,0}(Z) + \sum_{i=1}^{n} V_{x,ic}(Z)\cos i\Omega t + V_{x,is}(Z)\sin i\Omega t \\ V_y(Z,t) = V_{y,0}(Z) + \sum_{i=1}^{n} V_{y,ic}(Z)\cos i\Omega t + V_{y,is}(Z)\sin i\Omega t \\ M_x(Z,t) = M_{x,0}(Z) + \sum_{i=1}^{n} M_{x,ic}(Z)\cos i\Omega t + M_{x,is}(Z)\sin i\Omega t \\ M_y(Z,t) = M_{y,0}(Z) + \sum_{i=1}^{n} V_{y,ic}(Z)\cos i\Omega t + M_{y,is}(Z)\sin i\Omega t \\ T(Z,t) = T_0(Z) + \sum_{i=1}^{n} T_{ic}(Z)\cos i\Omega t + T_{is}(Z)\sin i\Omega t \end{cases} \tag{29}$$

The mode function vector of the general force $\mathbf{F}(Z)$ is defined as below

$$\begin{aligned}\mathbf{F}(Z) = [&V_{x,0}(Z)\ V_{x,1c}(Z) \cdots V_{x,nc}(Z)\ V_{x,1s}(Z) \cdots V_{x,ns}(Z) \\ &V_{y,0}(Z)\ V_{y,1c}(Z) \cdots V_{y,nc}(Z)\ V_{y,1s}(Z) \cdots V_{y,ns}(Z) \\ &M_{x,0}(Z)\ M_{x,1c}(Z) \cdots M_{x,nc}(Z)\ M_{x,1s}(Z) \cdots M_{x,ns}(Z) \\ &M_{y,0}(Z)\ M_{y,1c}(Z) \cdots M_{y,nc}(Z)\ M_{y,1s}(Z) \cdots M_{y,ns}(Z) \\ &T_0(Z)\ T_{1c}(Z) \cdots T_{nc}(Z)\ T_{1c}(Z) \cdots T_{nc}(Z)]\end{aligned} \tag{30}$$

Substituting Eqs. 19, 27, 29, and 30 into Eq. 17, and equating the coefficients of the same harmonic term, one obtained the vector of general force general force $\mathbf{F}(Z)$ at the right end of the Timoshenko beam element.

$$\mathbf{F}^R = \left[H_L\right]\begin{bmatrix}\mathbf{C}\\1\end{bmatrix} \tag{31}$$

Similarly, Substituting Eqs. 19, 28, 29, and 30 into Eq. 18, the vector $\mathbf{F}(Z)$ at the left end is

$$\mathbf{F}^L = \left[H_0\right]\begin{bmatrix}\mathbf{C}\\1\end{bmatrix} \tag{32}$$

where $\left[H_L\right],\ \left[H_0\right]$ are matrices with the size of $(10n+5)\times(20n+11)$.

Combining Eqs. 27, and 31, one gets

$$\begin{bmatrix}\mathbf{X}^R\\ \mathbf{F}^R\\ 1\end{bmatrix} = \begin{bmatrix}G_L\\ H_L\\ M_U\end{bmatrix}\begin{bmatrix}\mathbf{C}\\1\end{bmatrix} \tag{33}$$

where $M_U = \left[0\ 0\ \cdots 1\right]_{1\times(20n+11)}$.

Let $\left[\mathbf{S}^R\right] = \begin{bmatrix}\mathbf{X}^R\\ \mathbf{F}^R\end{bmatrix}$, $\left[M_L\right] = \begin{bmatrix}G_L\\ H_L\\ M_U\end{bmatrix}$, and inserting them into Eq. 33 yields

$$\begin{bmatrix}\mathbf{S}^R\\1\end{bmatrix} = \left[M_L\right]\begin{bmatrix}\mathbf{C}\\1\end{bmatrix} \tag{34}$$

Similarly, combining Eqs 28, and 32, one gets

$$\begin{bmatrix}\mathbf{S}^L\\1\end{bmatrix} = \left[M_0\right]\begin{bmatrix}\mathbf{C}\\1\end{bmatrix} \tag{35}$$

where, $\left[M_L\right],\ \left[M_0\right]$ are matrices with the size of $(20n+11)\times(20n+11)$. From Eqs. 34, and 35, it yields

$$\begin{bmatrix}\mathbf{S}^R\\1\end{bmatrix} = \left[M_L\right]\begin{bmatrix}\mathbf{C}\\1\end{bmatrix} = \left[M_L\right]\left[M_0\right]^{-1}\begin{bmatrix}\mathbf{S}^L\\1\end{bmatrix} = \left[T^s\right]\begin{bmatrix}\mathbf{S}^L\\1\end{bmatrix} \tag{36}$$

where $\left[M_0\right]^{-1}$ is the inverse matrix of matrix $\left[M_0\right]$, $\left[T^s\right] = \left[M_L\right]\left[M_0\right]^{-1}$ is called as transfer matrix of the pre-twisted rotating Timoshenko beam element. From it, the relation of the state vector $\mathbf{S}$ including displacements, shear forces, moments, and torques at two element ends is developed. So far, transfer matrices of spindle and micro drill elements are derived from the transfer matrix of the of the pre-twisted rotating Timoshenko beam element. Transfer matrix of the bearing will be considered next.

The bearings supporting the micro drilling system are considered to be linear and isotropic. The bearing as shown in Fig. 6 creates forces F_x^b, F_y^b, bending moments

M_x^b, M_y^b, and torque T^b applying on the spindle. They are calculated by the following Eq. 37.

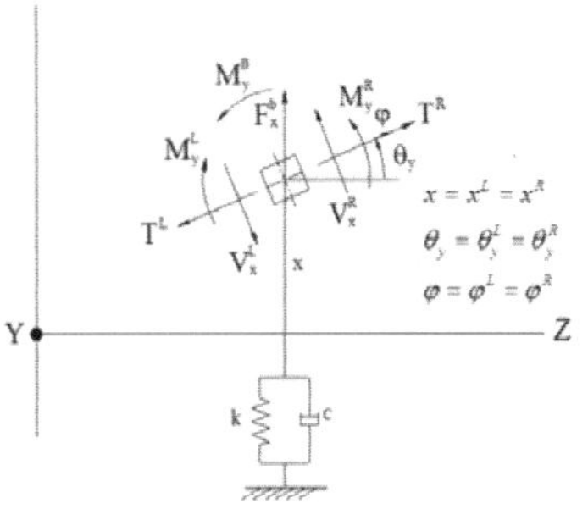

(a)

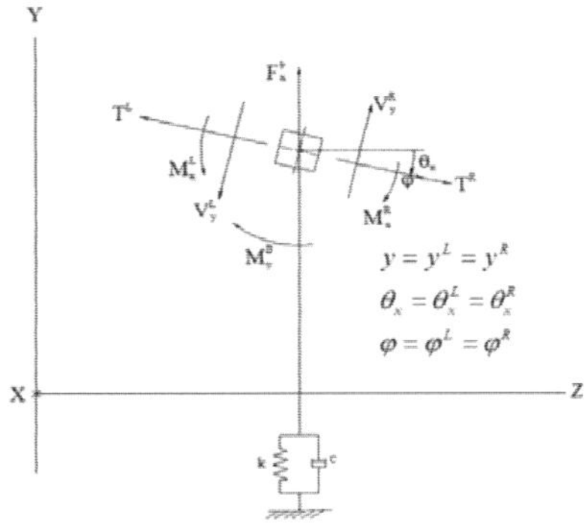

(b)

Figure 6. Forces, moments, and torques applying on a bearing node

$$\begin{cases}\begin{bmatrix}F_x^b\\F_y^b\end{bmatrix}=-\begin{bmatrix}K_{xx}&K_{xy}\\K_{yx}&K_{yy}\end{bmatrix}\begin{bmatrix}x\\y\end{bmatrix}-\begin{bmatrix}C_{xx}&C_{xy}\\C_{yx}&C_{yy}\end{bmatrix}\begin{bmatrix}\dot{x}\\\dot{y}\end{bmatrix}\\ \begin{bmatrix}M_x^b\\M_y^b\end{bmatrix}=-\begin{bmatrix}K_{\theta xx}&K_{\theta xy}\\K_{\theta yx}&K_{\theta yy}\end{bmatrix}\begin{bmatrix}\theta_x\\\theta_y\end{bmatrix}-\begin{bmatrix}C_{\theta xx}&C_{\theta xy}\\C_{\theta yx}&C_{\theta yy}\end{bmatrix}\begin{bmatrix}\dot{\theta}_x\\\dot{\theta}_y\end{bmatrix}\\ T^b=-K_\varphi\varphi-C_\varphi\dot{\varphi}\end{cases} \tag{37}$$

The equilibrium equations of the force, bending moment and torque acting on the bearing node as shown in Fig. 6 are

$$\begin{cases}\begin{bmatrix}V_x^R\\V_y^R\end{bmatrix}=\begin{bmatrix}V_x^L\\V_y^L\end{bmatrix}-\begin{bmatrix}F_x^b\\F_y^b\end{bmatrix}\\ \quad=\begin{bmatrix}V_x^L\\V_y^L\end{bmatrix}+\begin{bmatrix}K_{xx}&K_{xy}\\K_{yx}&K_{yy}\end{bmatrix}\begin{bmatrix}x\\y\end{bmatrix}+\begin{bmatrix}C_{xx}&C_{xy}\\C_{yx}&C_{yy}\end{bmatrix}\begin{bmatrix}\dot{x}\\\dot{y}\end{bmatrix}\\ \begin{bmatrix}M_x^R\\M_y^R\end{bmatrix}=\begin{bmatrix}M_x^L\\M_y^L\end{bmatrix}-\begin{bmatrix}M_x^b\\M_y^b\end{bmatrix}\\ \quad=\begin{bmatrix}M_x^L\\M_y^L\end{bmatrix}+\begin{bmatrix}K_{\theta xx}&K_{\theta xy}\\K_{\theta yx}&K_{\theta yy}\end{bmatrix}\begin{bmatrix}\theta_x\\\theta_y\end{bmatrix}+\begin{bmatrix}C_{\theta xx}&C_{\theta xy}\\C_{\theta yx}&C_{\theta yy}\end{bmatrix}\begin{bmatrix}\dot{\theta}_x\\\dot{\theta}_y\end{bmatrix}\\ T^R=T^L-T^b=T^L+K_\varphi\varphi+C_\varphi\dot{\varphi}\end{cases} \tag{38}$$

Replacing Eqs. 19, and 29 into Eq. 38, and equating the coefficients of the same harmonic term, one obtained

$$\begin{bmatrix}\mathbf{S}^R\\1\end{bmatrix}=\begin{bmatrix}T^b\end{bmatrix}_{(20n+11)\times(20n+11)}\begin{bmatrix}\mathbf{S}^L\\1\end{bmatrix} \tag{39}$$

where the state vector $\mathbf{S}$ contains all coefficients of Eqs. 19, and 29. The $\left[T^b\right]$ is the transfer matrix of the bearing. Then the overall transfer matrix of the whole system is the multiplication of the transfer matrices of all elements from the tip of the micro drill to its end.

According to the modified transfer matrix method, the state vectors at two ends of the micro drilling system are described as the following.

$$\begin{bmatrix}\mathbf{X}^R\\\mathbf{F}^R\\1\end{bmatrix}=\left[U\right]\begin{bmatrix}\mathbf{X}^L\\\mathbf{F}^L\\1\end{bmatrix} \tag{40}$$

where $\left[U\right]$ is the total transfer matrix of the system, and it is the multiplication of transfer matrices of elements. From Fig. 1, and Table 1, one gets

$$[U]=[T]_A[T]_B[T]_C[T]_D[T]_{S_1}[T]_{C_1}[T]_{S_2}[T]_{C_2}[T]_{F_1} \tag{41}$$

where $[T]_A$, $[T]_B$, $[T]_C$, $[T]_D$, $[T]_{S_1}$, $[T]_{C_1}$, $[T]_{S_2}$, $[T]_{C_2}$, and $[T]_{F_1}$ are the transfer matrices of parts A, B, C, D, S_1, C_1, S_2, C_2, and F_1, respectively.

The micro drill flute F_1 shown in Fig 7(a) is divided into small elements, so its transfer matrix $[T]_{F_1}$ is the multiplication of transfer matrices of these elements. The transfer matrix of each flute element is derived from the transfer matrix of the of the pre-twisted rotating Timoshenko beam element. To calculate it, it is necessary to evaluate the area moments of inertia (I_u, I_v) about its principal axes of the flute cross-section as well as its polar moments of inertia I_p, and the twist angle β between U principal axis and x-axis. Because the drill flute is formed by sweeping helically a flute cross-section, every drill flute element has the same area moments of inertia. Therefore, to be simple, I_u, I_v, and I_p are calculated at the flute cross-section perpendicular to the drill axis at the drill tip O as shown in Fig 7. Developing the profile of the cross section in Fig 7(a) is referred to the reference [16, 17]. Its moments of inertia I_u, I_v, and I_p are evaluated by Solidworks software, and its principal axes U, V is coincident to the axes x and y of the fixed coordinate system, respectively [18]. The twist angle β of the cross section of the flute element with coordinate Z_i as shown in Fig 7(b) is calculated by the following equation: $\beta = Z_i \dfrac{\tan(h_0)}{R}$, where R, h_0 are radius, helical angle of the micro drill, respectively.

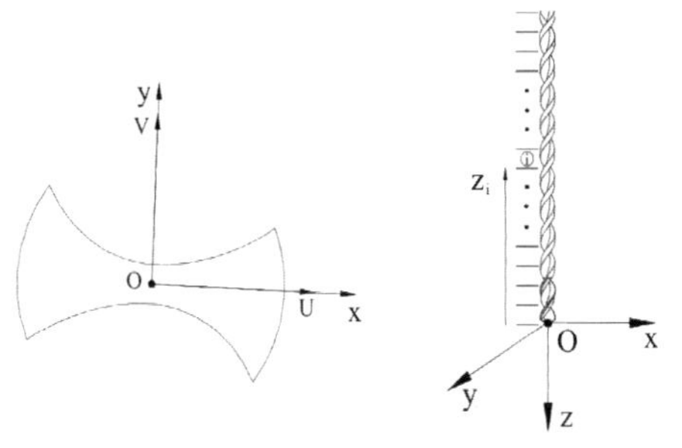

(a) Flute elements (b) Flute cross- section

Figure 7. Micro drill flute

To be simple, the conical part of the micro drill is considered as the sum of small cylinders as shown in Fig. 8. Hence, the transfer matrix of the conical part is the multiplication of transfer matrices of small cylinders. The area principal moments of inertia (I_u, I_v), area polar moment of inertia I_p, area A, and initial twist angle β of a cylinder are calculated as follows: $I = I_u = I_v = \dfrac{\pi r^4}{4}, I_p = \dfrac{\pi r^4}{2}, A = \pi r^2, \beta = 0,$ where r is the radius of the cross section.

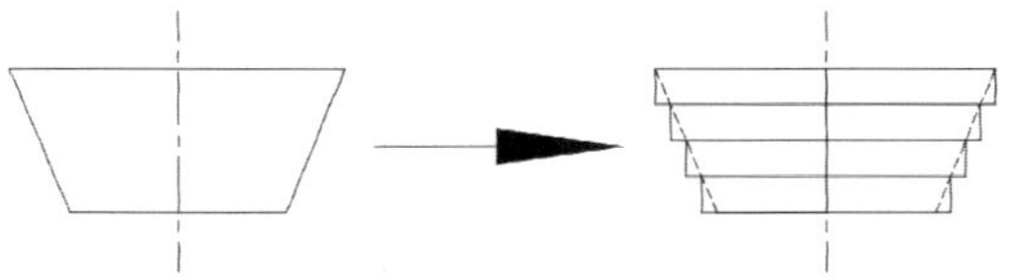

Figure 8. Approximation of conical part

The transfer matrix of spindle part A including two bearings, and three cylinder portions is

$$[T]_A = [T]_{L3}[T]^B[T]_{A2}[T]^B[T]_{L3} \tag{42}$$

4 Dynamic analysis of the micro drilling system

The vibration at the micro drill tip has the most important impact to the quality of drilled holes, and the cutting life time of the micro drill. Therefore, the dynamic analysis is carried out at the drill tip in this research. Namely, the state vectors containing displacements (x, y, θ_x, θ_y, φ), and the critical speeds of the system are evaluated.

After calculating all transfer matrices of system elements, the overall transfer matrix $[U]$ is calculated. Eq. 40 can be rewritten as Eq. 43

$$\begin{bmatrix} \mathbf{X}^R \\ \mathbf{F}^R \\ 1 \end{bmatrix} = \begin{bmatrix} U_{11} & U_{12} & \mathbf{u}_1 \\ U_{21} & U_{22} & \mathbf{u}_2 \\ \mathbf{0} & \mathbf{0} & 1 \end{bmatrix} \begin{bmatrix} \mathbf{X}^L \\ \mathbf{F}^L \\ 1 \end{bmatrix} \tag{43}$$

where U_{11}, U_{12}, U_{21}, U_{22} are matrices with the size of $(10n+5)\times(10n+5)$. $\mathbf{u}_1$, $\mathbf{u}_2$ are vectors with the size of $(10n+5)\times 1$. The $\mathbf{0}$ is zero vector with the size of $1\times(10n+5)$. Because there is no external force applying on the left end of the drilling system, one obtains $\mathbf{F}^L = [0\ 0\ \cdots 0]_{(10n+5)\times 1}$. Only torque T is the external force applying on the right end or drill tip of the system, i.e., $\mathbf{F}^R = [0\ 0\ \cdots 0\ -T\ 0 \cdots 0]_{(10n+5)\times 1}$,where $\mathbf{F}^R(8n+4,1) = -T$. From Eq. 43, one obtains

$$\mathbf{F}^R = U_{21}\mathbf{X}^L + U_{21}\mathbf{F}^L + \mathbf{u}_2 = U_{21}\mathbf{X}^L + \mathbf{u}_2$$

$$\Rightarrow \mathbf{X}^L = U_{21}^{-1}\left(\mathbf{F}^R - \mathbf{u}_2\right)$$

$$\mathbf{X}^R = U_{11}\mathbf{X}^L + U_{12}\mathbf{F}^L + \mathbf{u}_1 = U_{11}\mathbf{X}^L + \mathbf{u}_1$$

The vector $\mathbf{X}^R$ containing the displacements (x, y, θ_x, θ_y, φ) at the drill tip are calculated, i.e., the lateral and torsional displacements at the drill tip are evaluated.

Because lateral displacements including x and y, and torsional displacement φ play the most important role to the quality of the micro drill, they were analyzed in this section. The amplitude of the lateral (bending) displacement is calculated as below.

$$\text{Lateral amplitude} = \text{maximum value of } \sqrt{x^2 + y^2} \tag{44}$$

where x and y are evaluated by Eq. 19.

We chose order of whirling at the drill tip to be n=2. With the given parameters in Table 2, the lateral and torsional displacements are obtained. According to Fig. 9 (a, b), the lateral and torsional amplitudes is $5.39\cdot 10^{-8} m$ and $7.4101769\cdot 10^{-7} rad$ under 30000 rpm, respectively.

 DOI: 10.1051/matecconf/20166101024

Table 2. Parameters of the micro drilling system

Micro drill	Value		The material properties			
			Young's modulus E (N/m2)	Shear modulus G (N/m2)	Mass density ρ (Kg/m3)	Shear coefficient k
Radius R	$1.5\cdot10^{-4}m$					
Helical angle h_0	30^0	Flute part	650 × 109	280 × 109	15250	0.835
Semi point angle ρ_1	65^0	Other parts	200 × 109	90 × 109	7800	0.85

Bearing parameters

Bearing stiffness (N/m × 105)	Bearing damping (N/m × 103)
$K_{xx}=K_{yy}=1.7513$, $K_{xy}=K_{yx}=K_{\theta xx}=K_{\theta xy}=K_{\theta yy}=K_{\theta yx}=0$, $K_\varphi=0.3$	$C_{xx}=C_{yy}=1.7513$, $C_{xy}=C_{yx}=C_{\theta xx}=C_{\theta xy}=C_{\theta yy}=C_{\theta yx}=0$, $K_\varphi=0.001$

Eccentricity

$e_u = k_{eu}\cdot r$, $e_v = k_{ev}\cdot r$, where $k_{eu} = 1\cdot10^{-5}$, $k_{ev} = -9\cdot10^{-6}$, r is radius of each cross section.

Flute: $e_u = k_{euf}\cdot R$, $e_v = k_{evf}\cdot R$, , where $k_{euf} = 1\cdot10^{-5}$, $k_{evf} = -9\cdot10^{-6}, R = 1.5\cdot10^{-4}m$

$\Omega = 30000$ rpm , $T = 10^{-5}\,Nmm$, F_a=$7\cdot10^{-6}$N

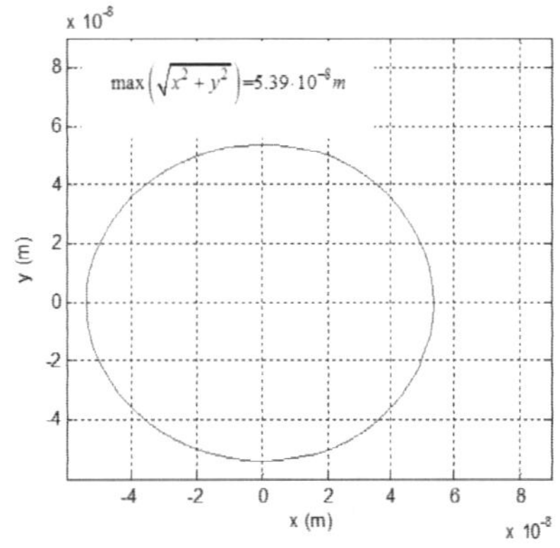

(a) Lateral displacement

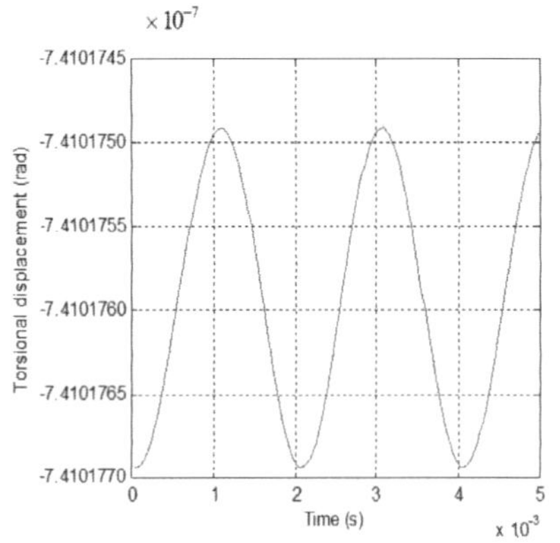

(b) Torsional displacement

Figure 9. Effects of external forces to displacements

Fig. 10 shows the lateral and torsional amplitudes. It shows that the lateral and torsional displacements are under resonance at rotating speed of 83080 rpm. Fig 11 shows the lateral and torsional displacement at the critical speed.

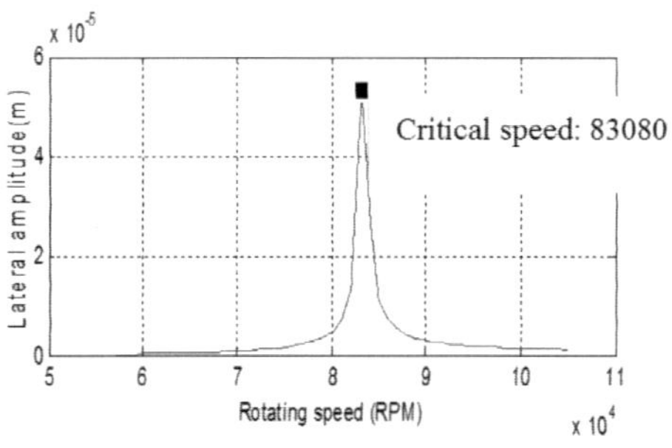

(a) Lateral amplitude

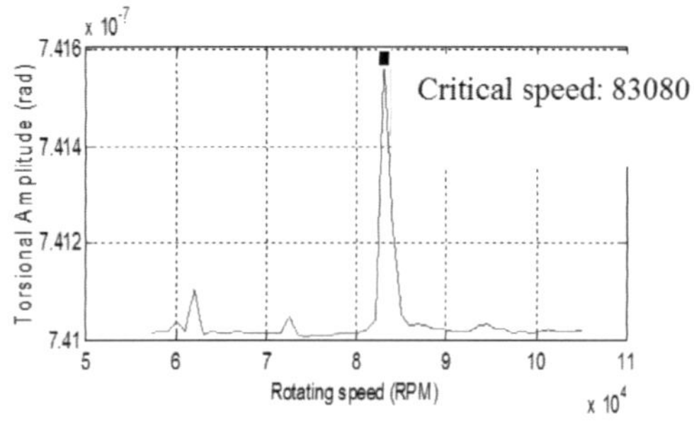

(b) Torsional amplitude

Figure 10. Effects of external forces to displacement amplitudes

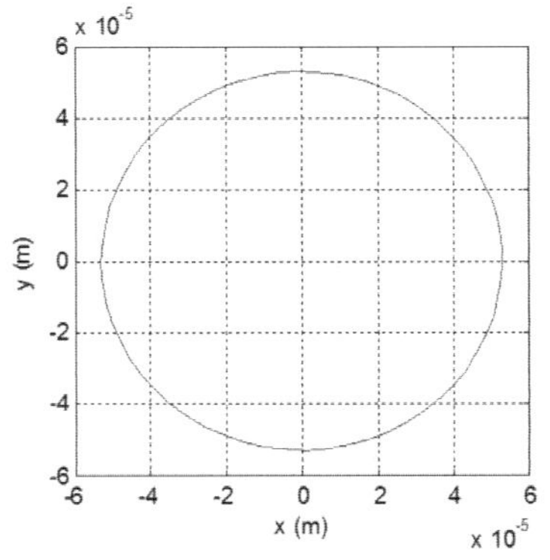

(a) Lateral displacement

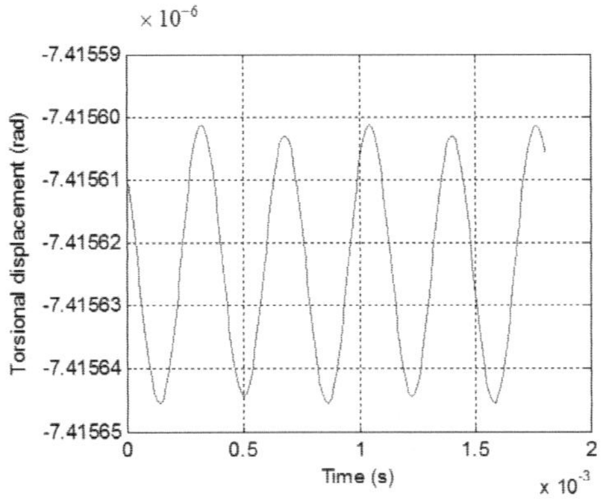

(b) Torsional displacement

Figure 11. Displacements at critical speed

5 Conclusions

The dynamic analysis of the micro drilling system was implemented by the new approach, modified transfer matrix, in this research. The micro drilling system is modeled as pre-twisted rotating Timoshenko beam elements using a continuous-system concept. Each element could include linear bearings, axial drilling force, gyroscopic moments, and eccentricity, especially coupling effect between lateral and torsional displacements. Therefore, the method has higher accuracy in evaluating the displacements and critical speeds of the system. The numerical example about effects of external forces, eccentricity, and bearings on lateral and torsional displacements and critical speeds of the system at the micro drill tip was presented in this research.

References

1. Y. Zhaojun, L. Wei, C. Yanhong, and W. Lijiang, Study for increasing micro-drill reliability by vibrating drilling, Reliab. Eng. Syst. Safe. **61,** 5 (1998)
2. J. Chae, S. S. Park, and T. Freiheit, Investigation of micro-cutting operations, Int. J. Mach. Tool. Manu. **46**, 20(2006)
3. T. Masuzawa, State of the Art of Micromachining, CIRP Annals - Manufacturing Technology, **49**, 16(2000)
4. C. K. Yu, C. H. Chang, Y. P. Ping, Y. K. Tien, and H. B. Wun, Natural properties in a micro drill cutting into bones, Life Sci. J. **6**, 6(2009)
5. B.-W. Huang, The Drilling Vibration Behavior of a Twisted Microdrill, J. Manuf. Sci. Eng. **126**,8 (2005)
6. Z. Yang, Q. Tan, and L. Wang, Principle of precision micro-drilling with axial vibration of low frequency, Int. J. Prod. Res. **40**, 7 (2002)
7. Y. Gong, K. F. Ehmann, and C. Lin, Analysis of dynamic characteristics of micro-drills, J. Mater. Process. Technol. **141**, 9(2003)
8. P. Yongchen, T. Qingchang, and Y. Zhaojun, A study of dynamic stresses in micro-drills under high-speed machining, Int. J. Mach. Tool. Manu. **46**,9 (2006)
9. S. Filiz and O. Burak Ozdoganlar, A model for bending, torsional, and axial vibrations of micro- and macro-drills including actual drill geometry—part I: model development and numerical solution, J. Manuf. Sci. Eng.**132**,(2010)
10. [10]S. Filiz and O. B. Ozdoganlar, A model for bending, torsional, and axial vibrations of micro- and macro-drills including actual drill geometry—part II: model validation and application, J. Manuf. Sci. Eng. **132**,(2010)
11. S.-C. Hsieh, J.-H. Chen, and A.-C. Lee, A modified transfer matrix method for the coupling lateral and torsional vibrations of symmetric rotor-bearing systems, J. Sound Vib. **289**, 10(2006)
12. H. P. Lee, Dynamic response of a rotating timoshenko shaft subject to axial forces and moving loads, J. Sound Vib. **181**, 9(1995)
13. D.-M. Ku, Finite element analysis of whirl speeds for rotor-bearing systems with internal damping, Mech. Syst. Sig. Process., **12,** (1998)
14. Q. H. Qin and C. X. Mao, Coupled torsional-flexural vibration of shaft systems in mechanical engineering—I. Finite element model, Comput. Struct. **58**, 9(1996)
15. Y. Kang, Y.-P. Shih, and A.-C. Lee, Investigation on the steady-state responses of asymmetric rotors, J. Sound Vib. **114**, 7(1992)
16. C. Lin, S. Kang, and K. Ehmann, "Planar micro-drill point design and grinding methods," Trans. North Am. Manuf. Res. Inst. SME, **20**, 7(1992)
17. S.-C. Hsieh, J.-H. Chen, and A.-C. Lee, A modified transfer matrix method for the coupled lateral and torsional vibrations of asymmetric rotor-bearing systems, J. Sound Vib. **312**, 9(2008)
18. A.-C. Lee, D.-T. Nguyen, and G.-T. Wu, Analyses of a new four-facet drill, Int. J. Adv. Manuf. Technol.**75**,14(2014)

MATEC Web of Conferences **61**, 01025 (2016) DOI: 10.1051/matecconf/20166101025
APOP2016

Optimization of the Excitation signal for a novel piezoelectric micro-jet

Kai Li[1,a], Jun-kao Liu[1], Wei-shan Chen[1] and Lu Zhang[2]

[1] State Key Laboratory of Robotics and System, Harbin Institute of Technology, Harbin 150001, China
[2] AVIC, Harbin Dongan Engine Corporation LTD, Harbin 150001, China

Abstract. Fourier series expansions of different excitation signals for a novel piezoelectric micro-jet is deducted in this paper. By comparison and analysis of the Fourier series expansions, we find that the motivational effect of square wave is preferable than the sine wave which is better than triangle wave. And by analysing the Fourier series expansion of trapezoidal wave, it indicates that the ratio of the waveform parameters to the period should be studied rather than the numerical value of the parameters for improving the injection performance. The numerical simulations and experiments are carried out, and the results are consistent with theoretical calculations.

1 Introduction

The piezoelectric micro-jet, which can achieve drop-on-demand injection, is wildly used in various industrial fields [1-4]. The trapezoidal pulse wave is usually chosen as the excitation pulse waveform [5-7], however, there is no research results explaining the reasons for not choosing a triangular wave or sine wave instead. People mainly focuses on analysing the influences of trapezoidal wave parameters on the injection performance through experiments and simulations [8-11], or researching on the new method for controlling of the trapezoidal wave to reduce the residual vibration by using the self-sensing capability of the piezoelectric vibrator [12-14].

In order to obtain the optimal excitation signal which is applied on the piezoelectric micro-jet for achieving better injection performance, several typical waveforms are analysed depend on theoretical derivation in this paper. By analysing the derived results, the preferable excitation signal is given. Then, the analysis results are verified by simulations and experiments.

2 Theoretical analysis

The typical excitation waveform for piezoelectric micro-jet mainly include sine wave, triangle wave, square wave and trapezoidal wave. It's hard to find out which is preferred just by comparing the waveform functions or the shapes of the wave. As all functions (periodic function satisfies the Dirichelet conditions) can be expanded into a form of trigonometric functions by Fourier series expansion theory, the preferable waveform for better injection can be obtained by analysing the Fourier series expansions of different waveforms with the same voltage amplitude (V_m) and period (T).

The sine wave is shown in Fig.1, and its function expression with time as the variable is shown in Eq.(1).

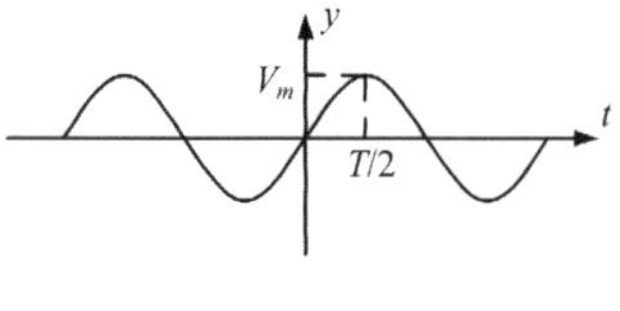

Figure 1. Function expression of sine wave.

$$f_{\text{sine}}(t) = V_m \sin(\frac{2\pi}{T}t) \quad (1)$$

The triangle wave is shown in Fig.2, and its Fourier series expansion with time as the variable is shown in Eq.(2).

[a] Corresponding author: sdcxlikai@126.com

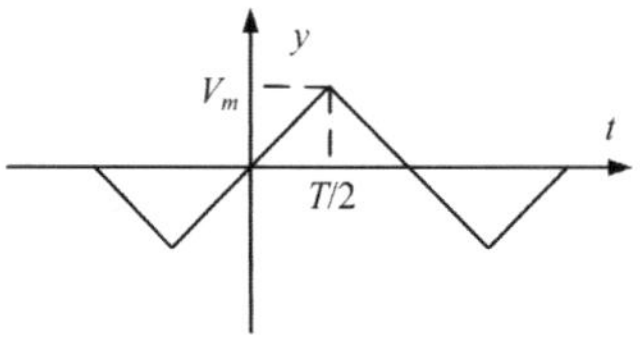

Figure 2. Fourier series expansion of triangle wave.

$$f_{\text{triangle}}(t)=\frac{8}{\pi^2}V_m\sum_{n=1}^{\infty}[\frac{1}{n^2}\sin(\frac{n\pi}{2})\sin(\frac{2n\pi}{T}t)] \quad (2)$$

The square wave is shown in Fig.3, and its Fourier series expansion with time as the variable is shown in Eq.(3).

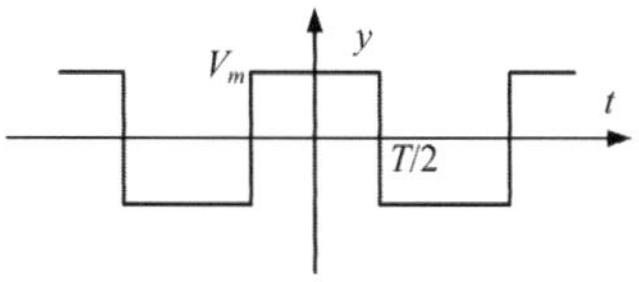

Figure 3. Fourier series expansion of square wave.

$$f_{\text{square}}(t)=\frac{4}{\pi}V_m\sum_{n=1}^{\infty}[\frac{1}{n}\sin(\frac{n\pi}{2})\cos(\frac{2n\pi}{T}t)] \quad (3)$$

The trapezoidal wave (T_r=T_f) is shown in Fig.4, and its Fourier series expansion with time as the variable is shown in Eq.(4), where T_r is rising time, T_h is holding time, T_f is falling time and T_s is spacing time, T_r+T_h+T_f+T_s =T.

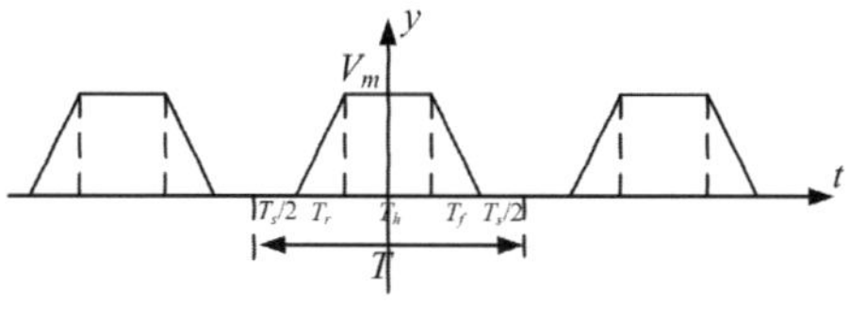

Figure 4. Fourier series expansion of trapezoidal wave (T_r=T_f).

$$f_{\text{trapezoidal}\,(T_r=T_f)}(t)=\frac{T_h+T_f}{T}V_m+\frac{T}{\pi^2T_f}V_m\sum_{n=1}^{\infty}[\frac{1}{n^2}(\cos\frac{T_h}{T}n\pi-\cos\frac{T_h+2T_f}{T}n\pi)\cos\frac{2n\pi}{T}t] \quad (4)$$

The trapezoidal wave (T_r≠T_f) is shown in Fig.5, and its Fourier series expansion with time as the variable is shown in Eq.(5), where a_n' and b_n' are shown in appendix.

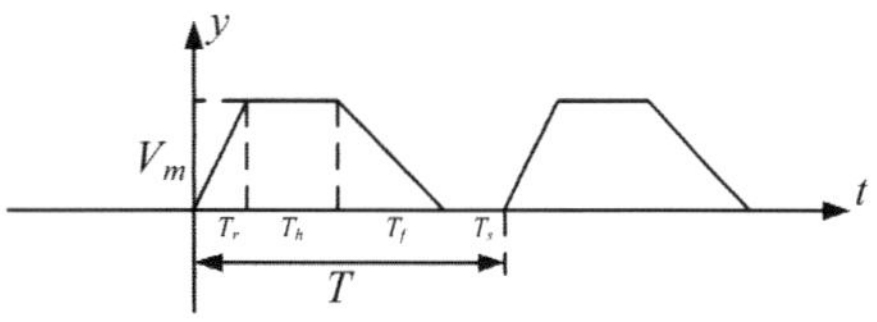

Figure 5. Fourier series expansion of trapezoidal wave (T_r≠T_f).

$$f_{\text{trapezoidal}\,(T_r\neq T_f)}(t)=\frac{T_r+2T_h+T_f}{2T}V_m+\frac{T}{\pi^2T_rT_f}V_m\sum_{n=1}^{\infty}\frac{1}{n^2}(a_n'\cos\frac{2n\pi}{T}t+b_n'\sin\frac{2n\pi}{T}t) \quad (5)$$

Thus, the fundamental components (n=1) of these excitation signals are shown as follows:

$$f^1_{\text{triangle}}(t)=\frac{8}{\pi^2}V_m\sin(\frac{2\pi}{T}t) \quad (6)$$

$$f^1_{\text{square}}(t)=\frac{4}{\pi}V_m\cos(\frac{2\pi}{T}t) \quad (7)$$

$$f^1_{\text{trapezoidal}\,(T_r=T_f)}(t)=\frac{T_h+T_f}{T}V_m+\frac{T}{\pi^2T_f}V_m[\cos(\frac{T_h}{T}\pi)-\cos(\frac{T_h+2T_f}{T}\pi)]\cos(\frac{2\pi}{T}t) \quad (8)$$

$$f^1_{\text{trapezoidal}\,(T_r\neq T_f)}(t)=\frac{T_r+2T_h+T_f}{2T}V_m+\frac{T}{\pi^2T_rT_f}V_m[a_n'\cos\frac{2\pi}{T}t+b_n'\sin\frac{2\pi}{T}t] \quad (9)$$

By comparing Eq.(1), (6), (7), (8) and (9), it demonstrate that the motivational effect of square wave is preferable than the sine wave which is better than triangle wave, as $8/\pi^2<1<4/\pi$ and the greater the excitation voltage amplitude, the better the performance of the injection. Meanwhile, due to the complex expressions of the Fourier series expansion of the trapezoidal wave, it is not easy to be compared with other waveforms. However, the square wave can be treated as a special form of the trapezoidal (T_r=T_f=0), the trapezoidal wave may be more effective as well than other waveforms. And by analysing Eq.(8) and (9), it indicates that the ratio of the waveform parameters to the period should be studied rather than the numerical value of the parameters for improving the injection performance. In order to prove the correction of the analysis above, numerical analysis of a novel piezoelectric micro-jet is carried out next.

3 Numerical analysis

In this paper, we use the novel piezoelectric micro-jet which is designed for bearing lubrication by our group, for more details, please refer to the references[8].The structure of the novel piezoelectric micro-jet is shown in Fig.6, and all the simulation parameter settings are in agreement with the references[8].

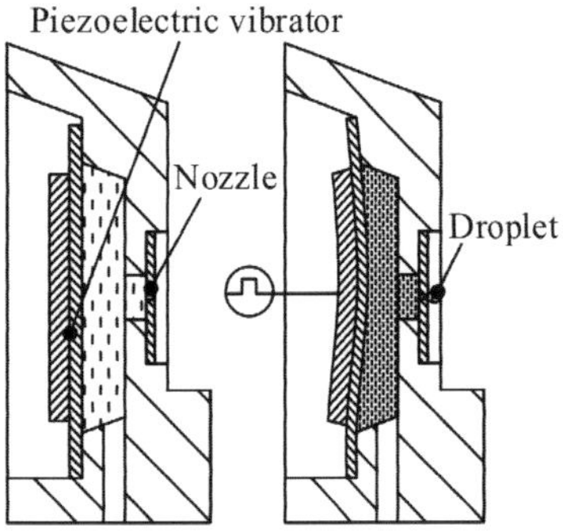

Figure 6. The structure of the micro-jet.[8]

We select the sine wave, trapezoidal wave (T_r : T_h : T_f=1:1:1), triangular wave and square wave as the excitation signals with the same voltage amplitude (V_m=100V) and period (T=55.77us). The displacement and velocity of one of the particles near the nozzle are studied by a finite element analysis. The curves of the vibration displacement of the same particle with different excitation signals is shown in Fig.7.

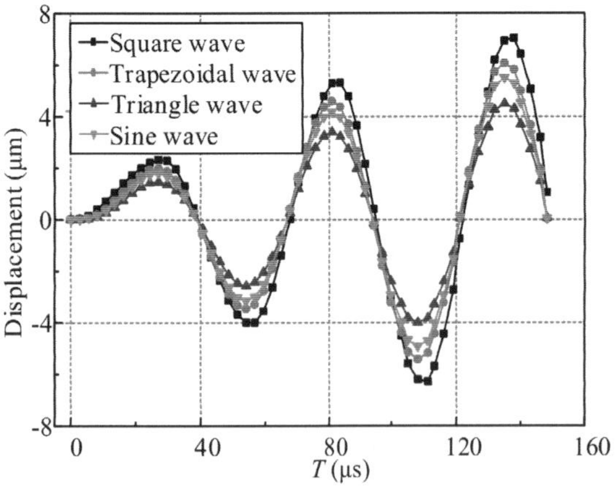

Figure 7 The he vibration displacement of the same particle with different excitation signals.

As we can see that the vibration displacement of the particle under the excitation of square wave is larger than that of trapezoidal wave or sine wave which are greater than that of triangle wave. This is consistent with the theoretical calculation. The curves of the vibration velocity of the same particle with different excitation signals is shown in Fig.8.

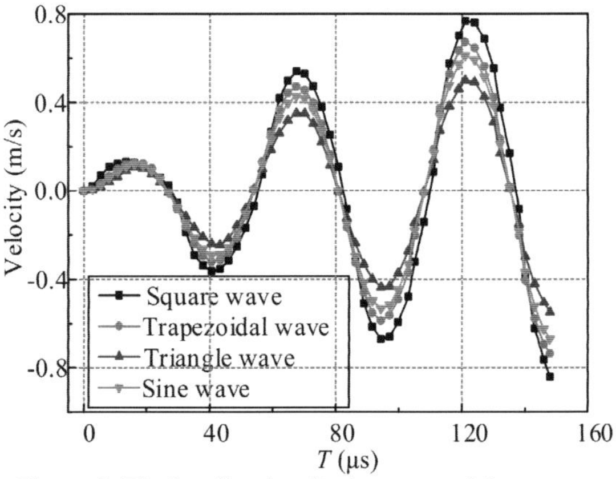

Figure 8. The he vibration displacement of the same particle with different excitation signals.

It can be seen that when the piezoelectric micro-jet is applied different excitation signals with the same voltage amplitude (V_m=100V) and period (T=55.77us), the velocity of the particle under the excitation of triangle wave is less than that of trapezoidal wave or sin wave which are inferior to that of square wave. This is also consistent with the theoretical calculation. The theoretical and numerical simulation results are verified by experiments, as described below.

4 Experimental analysis

An experimental research is carried out by using the experimental method as shown in references[8], and the properties of injection lubricating oil and the experimental equipment are the same as what is described in references[8]. In order to show the influences of the excitation signals obviously, only the square wave which obtain the best incentive effect and the sine wave which obtain the worst effect are selected in the experiments.

The injection status of the micro-jet under different excitation signals with same voltage amplitudes and period, are shown in Fig.9.

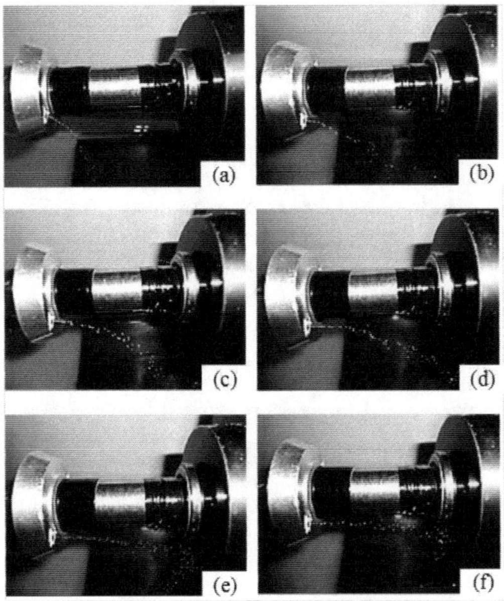

Figure 9. Injection status of the micro-jet under different excitation signals :(a) sine wave, and V_m=60V; (b) square wave, and V_m=60V; (c) sine wave, and V_m=80V; (d) square wave, and V_m=80V; (e) sine wave, and V_m=100V; (f) square wave, and V_m=100V.

As we can see from Fig.9, whether the excitation signal is sine wave or square wave, the injection intensity of the micro-jet increases along with the increase of the voltage amplitude. Meanwhile, the injection status of the micro-jet under the excitation of square wave is more intense than that of sine wave when the voltage amplitude of the excitation signal keeps the same. The experimental results are in agreement with the simulations and theoretical calculations.

Another evaluation of the piezoelectric micro-jet is the mass of the droplet, the mass of the droplet varies with different voltage amplitude of excitation signals are shown in Fig.10. It can be seen that the droplet mass increase gradually with the increase of the voltage amplitude for each excitation signals. And the mass of the droplet ejected out of the micro-jet under the square wave is larger than that of sine wave when the voltage amplitude keep the same. This also verifies the correctness of simulations and theoretical calculations.

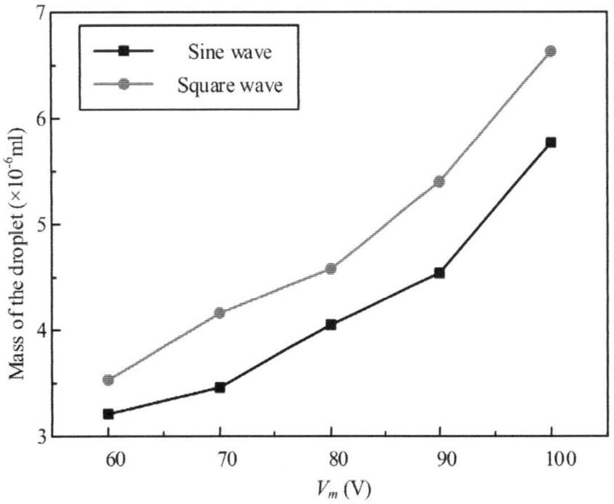

Figure 10. The change curves of the droplet mass along with the voltage amplitude of excitation signals.

5 Conclusions

A new method for analysing the excitation signals of the piezoelectric micro-jet is proposed in this paper. The theoretical calculations prove that the square wave or trapezoidal wave are more effectively than sine wave or square wave by the discussions of the Fourier series expansions of excitation signals. Simulations and experiments are carried out to verify the correctness of the theoretical calculations, and the results are in agreement with the theoretical calculations. Meanwhile, the ratio of the waveform parameters of the trapezoidal wave to the period should be studied rather than the numerical value of the parameters for improving the injection performance according to the analysis of the Fourier series expansion of the trapezoidal wave.

Appendix

$$a_n' = T_f\cos^2(\frac{T_r}{T}n\pi) - 2T_r\cos^2(\frac{T_r}{T}n\pi) - 2T_r\cos^2(\frac{T_h}{T}n\pi)$$
$$-T_r\cos^2(\frac{T_f}{T}n\pi)$$
$$+4T_r\cos^2(\frac{T_r}{T}n\pi)\cos^2(\frac{T_h}{T}n\pi)\sin^2(\frac{T_f}{T}n\pi)$$
$$+2T_r\cos^2(\frac{T_r}{T}n\pi)\cos^2(\frac{T_f}{T}n\pi)$$
$$+2T_r\cos^2(\frac{T_h}{T}n\pi)\cos^2(\frac{T_f}{T}n\pi)$$
$$-\frac{1}{2}T_r\sin(\frac{2T_r}{T}n\pi)\sin(\frac{2T_f}{T}n\pi)$$
$$-T_r\sin(\frac{2T_r}{T}n\pi)\sin(\frac{2T_h}{T}n\pi)$$
$$-\frac{1}{2}T_r\sin(\frac{2T_h}{T}n\pi)\sin(\frac{2T_f}{T}n\pi)$$
$$+T_r\sin(\frac{2T_r}{T}n\pi)\sin(\frac{2T_h}{T}n\pi)\cos^2(\frac{T_f}{T}n\pi)$$
$$+T_r\sin(\frac{2T_h}{T}n\pi)\sin(\frac{2T_f}{T}n\pi)\cos^2(\frac{T_r}{T}n\pi)$$
$$+T_r\sin(\frac{2T_r}{T}n\pi)\sin(\frac{2T_f}{T}n\pi)\cos^2(\frac{T_h}{T}n\pi) + T_r - T_f$$

$$b_n' = 2T_r\sin(\frac{2T_r}{T}n\pi)\cos^2(\frac{T_h}{T}n\pi)$$
$$+2T_r\sin(\frac{2T_h}{T}n\pi)\cos^2(\frac{T_r}{T}n\pi)$$
$$+T_r\sin(\frac{2T_r}{T}n\pi)\cos^2(\frac{T_f}{T}n\pi)$$
$$+T_r\sin(\frac{2T_f}{T}n\pi)\cos^2(\frac{T_r}{T}n\pi)$$
$$+T_r\sin(\frac{2T_h}{T}n\pi)\cos^2(\frac{T_f}{T}n\pi)$$
$$+T_r\sin(\frac{2T_f}{T}n\pi)\cos^2(\frac{T_h}{T}n\pi)$$
$$-2T_r\sin(\frac{2T_r}{T}n\pi)\cos^2(\frac{T_h}{T}n\pi)\cos^2(\frac{T_f}{T}n\pi)$$
$$-2T_r\sin(\frac{2T_h}{T}n\pi)\cos^2(\frac{T_r}{T}n\pi)\cos^2(\frac{T_f}{T}n\pi)$$
$$-2T_r\sin(\frac{2T_f}{T}n\pi)\cos^2(\frac{T_r}{T}n\pi)\cos^2(\frac{T_h}{T}n\pi)$$
$$+\frac{1}{2}T_f\sin(\frac{2T_r}{T}n\pi) - T_r\sin(\frac{2T_r}{T}n\pi)$$
$$-T_r\sin(\frac{2T_h}{T}n\pi) - \frac{1}{2}T_r\sin(\frac{2T_f}{T}n\pi)$$
$$+\frac{1}{2}T_r\sin(\frac{2T_r}{T}n\pi)\sin(\frac{2T_h}{T}n\pi)\sin(\frac{2T_f}{T}n\pi)$$

References

1. M Alomari,F H Mohamed,A W Basit,S Gaisford. Int. J. Pharm. , **494** 568-577 (2015)
2. J Xu,D Attinger. J. Micromech. Microeng. , **18** 10 (2008)
3. I Van Driessche,J Feys,S C Hopkins,P Lommens,X Granados,B A Glowacki,S Ricart,B Holzapfel,M Vilardell,A Kirchner, *et al.* Superconductor Science & Technology, **25** 12 (2012)
4. S Rahmati,S F Shirazi,H Baghayeri. Rapid Prototyping Journal, **15** 187-191 (2009)
5. J Brunahl,A M Grishin. Sensors and Actuators a-Physical, **101** 371-382 (2002)
6. M-H Tsai,W-S Hwang. Mater. Trans. , **49** 331-338 (2008)
7. H J Lin,H C Wu,T R Shan,W S Hwang. Mater. Trans. , **47** 375-382 (2006)
8. K Li,J Liu,W Chen,L Ye,L Zhang. Applied Sciences, **6** 38 (2016)
9. S Kim,J Sung,M H Lee. J. Therm. Sci. , **22** 29-35 (2013)
10. T M Liou,C Y Chan,K C Shih. Microfluid. Nanofluid. , **8** 575-586 (2010)
11. W J Shin,Y S Jeong,K Choi,W G Shin. Aerosol Sci. Technol. , **49** 1256-1262 (2015)
12. K-S Kwon,김우식. Transactions of the Korean Society for Noise and Vibration Engineering, **17** 333-341 (2007)
13. A A Khalate,X Bombois,G Scorletti,R Babuska,S Koekebakker,W De Zeeuw. Journal of Microelectromechanical Systems, **21** 1365-1374 (2012)
14. K S Kwon,W Kim. Sensors and Actuators a-Physical, **140** 75-83 (2007)

Strategies of the Game on the Use of BOT Model for the Eco-tourism Enterprises in the Infrastructure Development of Ecological Tourism

Yang Peitao, Wang Shuai and Zhou Ting

Dept. of Graduates, Central South University Forestry and Technical, Changsha Hunan, 410004

Abstract: Based on BOT model in the eco-tourism infrastructure and from the perspective of the Eco-tourism Enterprises, this article analyzes the game of the concession periods, the game of the costs and the game between the two to achieve the Maximization of Eco-tourism Enterprises' profits by using some related theories and methods.

Key Words: Eco-tourism infrastructure, Eco-tourism Enterprises ,BOT, Game

Introduction

With the rapid development of economy, the Great Western Development and the implementation of the strategy of central China, the development of eco-tourism industry (Lachapelle, P. (1998),Teece D. 1997,Kumaraswamy et al. 2001)is bound to become a new economic growth point, the investment demand gap of eco-tourism facilities will increase, relying solely on government investment can not meet the huge demand of the growing eco-tourism facilities. Tourism enterprises actively participate in the construction and operation of the eco-tourism facilities (Michael and Charles 2003)will certainly become an important way to solve the investment demand gap of the eco-tourism facilities, but the best path that tourism enterprises participate in the construction and operation of the eco-tourist facilities is to introduce the BOT model. The BOT model (YANG H and MENG Q 2002,Shengfeng L et al. 2008,Xinmin Z 2010)is that government departments will firstly transfer the concession of the eco-tourism facilities to tourism enterprises in a certain period of time, then Tourism enterprises are responsible for the construction and operation of the project, and charge fees to repay the debt or profit according to the specific way, finally Tourism enterprises will transfer the project of the eco-tourism facilities to the government departments with free or pre-agreed price until the expiration of the ecological tourism projects. Seen that the determination of the duration of the concession of the construction projects of the eco-tourism facilities will be the core issue that affects the revenue between tourism enterprises and government departments. From the perspective of tourism enterprises, For the BOT model in the construction projects of eco-tourism facilities, Using the correlation theories and method of game, the article analyses the game of the concession period, the cost and the relationship between them.

1. The game of The concession duration when eco-tourism enterprises make application of BOT model in the construction of tourist facilities.

The most important game between tourism enterprises and government departments in the construction projects of the eco-tourism facilities is the concession of the game, Including the scope of the powers of the concession, the concession period, financing, operation mode, toll pricing and more. Tourism enterprises hope to extend the concession period of the project of the eco-tourism facilities to achieve the maximization of the profits. However, government departments always try to shorten the concession period to maximize the effectiveness of eco-tourism facilities. Assumed that T_A is the minimum acceptable concession period to achieve the expected profit of the tourism enterprises and TB is the concession period to achieve the maximum expected utility allowed by the government departments, so T_A-T_B is the respective interests of the game space between Tourism enterprises and government departments. Considering the subsidy of the government departments, actually increasing the direct value of eco-tourism facility, the corresponding concession boundaries in tourism enterprises will be left to the T_C, then the game space has increased from T_A-T_B to T_C-T_B. Figure 1 describes the best strategy for tourism enterprises that make the application of BOT model in the construction of eco-tourism facilities to take the game in respect of a concession period with government departments. It can be seen from Figure 1 that only if the respective benefits or profits accepted by both tourism enterprises and Government departments of the game is the best value, both sides get the best TY of the concession period. In addition, tourism enterprises not only have increased the willingness to invest and have guaranteed the minimum level of debt repayment, but also have improved the ability for guarding against risks in order to ensure that the construction period and the quality of the eco-tourism facilities.

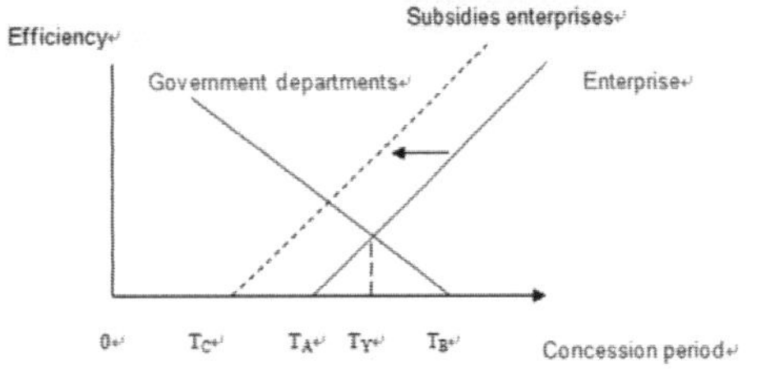

Figure 1 the game of the concession period in the application of BOT model in the eco-tourism infrastructure of the tourism enterprises

2 Cost game of the enterprises in the application of BOT model of the Eco-tourism facility construction

In the Investment in eco-tourism facilities, it is assumed that all content can be quantified in the concession agreement and are reflected in the total profits of BOT projects, profits can only be allocated between tourism enterprises and Government departments, without regard to other profit distribution scheme. A period of the application of BOT model of the Eco-tourism facility construction is N years, the economic life is Te years , the total cost of the Eco-tourism enterprises (CT)can be divided into the construction cost (Cn) and maintenance cost (CV), in which CN is a constant, Cv is equal to KC-a,(k>0,a>0),r is for annual interest rate. that is

$$C_T = C_N + C_V = \sum_{i=1}^{N} c_i\left(\frac{1}{1+r}\right)_{i-1} + \sum_{i=N+1}^{T} c_i\left(\frac{1}{1+r}\right)_{i-1}$$

In General, the maintenance costs of the eco-tourism facility CV will decrease with the increase of the construction costs CN, because the Higher the construction costs are ,the better the quality of the BOT projects of the eco-tourism infrastructure is, the maintenance costs in the later are the lower. On the contrary, the less the construction costs of the tourism enterprises in the construction of eco-tourism facility are ,the higher the maintenance costs in the construction of eco-tourism facility construction are , the longer the privilege period is ,the more the total costs of the investment will be .This is against the goals the tourism enterprises to maximize profits.

Because the tourism enterprises apply BOT to ecotourism facilities construction in order to reap the maximum profits at the minimum cost within the period of the concession in the project. They must make clear the relation between the total costs of the investment and the BOT concession for eco-tourism facility construction projects during the game. The game relation between the total investment costs, maintenance costs and concession period for BOT eco-tourism facility construction project can be described in Figure 2.

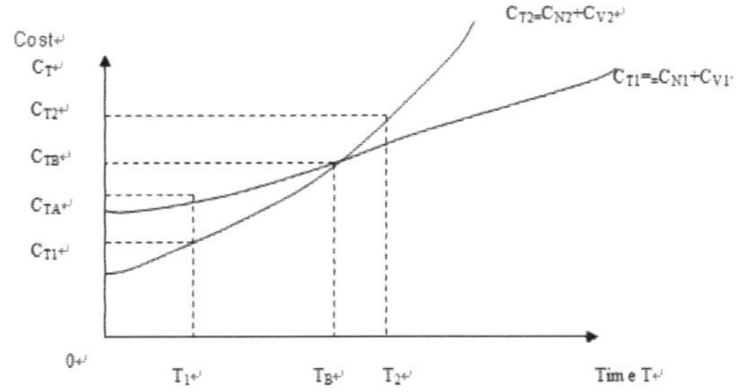

Figure 2 the costs game in the application of BOT model in the eco-tourism infrastructure of the tourism enterprises

From Figure (2) it can be seen when the concession period is less than TB .the tourism enterprises will choose the cost CT2 and the total cost is less. If the concession period is equal to TB, tourism enterprises will choose the cost CT1 or CT2, and both of the total costs are equal. But when selecting CT1 tourism enterprises have higher construction costs and lower maintenance costs, selecting the CT2 tourism enterprises have higher maintenance costs and lower construction costs; If the duration of the concession cost is greater than TB and tourism enterprises selects the cost CT1, the total cost is smaller..

It appears that the concession period in the BOT projects of the eco-tourism facilities is too short, the tourism enterprises in the BOT projects will choose a lower construction cost, which will affect the quality of the eco-tourism facility construction, So that the high maintenance costs of government departments will be very high and the income of government departments will be reduced after the ecological tourist facilities are transferred from the tourism enterprises over to the government. But if the concession of the BOT projects in the construction of the ecological tourism infrastructure, the income of the tourism enterprises will increase and the income of the government departments will decrease, in order to reduce the government initiative in introducing BOT mode in the construction of eco-tourism facilities. Therefore, the tourism enterprises should determine their own best cost strategy in the particular concession period in the BOT projects of the ecological tourism infrastructure not only to achieve their maximum profit, and must give the dual attention to the optimal of the government income, thus achieving the win-win pattern of tourism enterprises and Government departments in the BOT model of the eco-tourism facilities.

3 The dynamic game between the cost and the concession in the construction of the ecological tourism facilities where tourism enterprises use the BOT mode.

Due to considering only the direct benefits of the construction of ecological tourism facilities for the tourism enterprises and government departments and not taking into account the indirect usefulness of the eco-tourism facility construction, assuming that the total benefits of the eco-tourism facility construction as a whole are a constant R, including the direct benefits R2 to the tourism Enterprises and the direct gains R1 to government departments, the establishments of the decision models of the tourism enterprise are as follows:

$$\underset{T}{MAX}\int_0^T (R_1 - C_N - Cv)dt \qquad (1)$$

$$\text{s.t}\int_T^{Te} (R - R_1 - Cv)dt \ge \varpi \qquad (2)$$

Formula (1) refers to the function of the Maximum utility for the tourism enterprises when they choose the concession period T, Formula (2) is the participation constraint in the utility function of the tourism enterprises , namely the expected utility that government departments accept the contract with the tourism enterprises cannot be less than the maximum expected utility or the opportunity cost ϖ hat government departments don't accept the contract with the tourism enterprises. According to formula (1) and formula (2), a Lagrange function is constructed and the Lagrange Constant is λ, here is:

$$L(T, \lambda) = \int_0^T (R_1 - CN - Cv)dt + \lambda \left[\text{s.t.} \int_T^{Te} (R - R_1 - Cv)dt - \varpi\right] \qquad (3)$$

For the Lagrange function, the optimizations to the formula (3) are:

$$(\lambda+1)R_1+(\lambda-1)KC^{-a}-(\lambda R+C_N)=0 \qquad (4)$$

$$(Te-T)(R-R_1-KC^{-a})-\varpi=0 \qquad (5)$$

According to the analyses of the formula (4) and the formula (5), the optimal decision of

The concession period for the tourism enterprises to choose the BOT model in the ecological construction of tourism infrastructure is :

$$T=Te \quad \frac{\varpi}{R-R1-KC^{-a}} \qquad (6)$$

The optimization decision-making of construction cost of tourism enterprises is CN,

$$CN=(aKT)^{1/(a+1)}=\left[\frac{aKT_e(R_2-KC^{-a})-aK\varpi}{R_2-KC^{-a}}\right]^{1/(a+1)}$$

(7)

Thus, it is concluded that The optimization decision-making of construction cost of tourism enterprises is CN when the concession period of the tourism enterprises in the ecological construction of tourism facilities is T.

4 Conclusions

The concession period of Ecological construction of tourism facilities using the BOT model is very important, the Length of the concession directly determines the interests of tourism enterprises and Government departments, but also influences the cost and quality on the ecological construction of tourism infrastructure. By using the game theory, This paper examines the decision model of the concession period when the tourism enterprises use the BOT model in the ecological construction of tourism infrastructure, and analyses the game relationship between the construction cost and the concession period of the construction of eco-tourism facilities. Namely concession period is short, the tourism enterprise will choose the smallest construction costs to decline the quality of the construction of eco-tourism facilities and increase the cost of the maintenance. According to the income of government departments, the tourism enterprises should choose the best concession period to obtain the maximum benefit, ensuring the quality and technology of the BOT projects in the construction of eco-tourism facilities. It is a good theoretical significance and practical value that the systematic theory and methods on the BOT model in the construction of eco-tourism facilities are established.

Acknowledgements

Authors would like to thank professor Xiaoping LIAO at Central South University Forestry and Technical who provides language help.

Reference

1. Teece D.,Pisano G. &Shuen A.DynamicCapabilities and Strategic Management[M].Strategic Management Journal, 1997,18(7):509-533.
2. Lachapelle, P. Managing sanitation in protected areas: problems and challenges in Sagarmatha (Mt. Everest) National Park, Nepal. [J].*Himalayan Research Bulletin*, 1998:18, 53–57.
3. Kumaraswamy MM,zhang XQ.Govermental role in BOT-led infrastructure development [J].international journal of project management,2001,19:195-205
4. Michael J G， Charles Y J C. Valuation techniques for infrastructure investment decision [J]. Construction Management and Economics， 2003， 22： 373-383.
5. Xinmin Z. Effects of Operating BOT Model in Rural Eco-tourism on the New Countryside Construction [J]. Reformation and Strategy,2010,11:102-104
6. Shengfeng L. ,Daopin C. Lingyan H. The Risks of Applying BOT Model into the West Tourist Sites Exploitation [J]. Business research ,2008,2:14-18
7. Darrin Grimsey ,Mervyn K.Lewis. Evaluating the risk of public private partnerships for infrastructure projects [J].International Journal of Prject Management , 2002,20:107~118.
8. YANG H, MENG Q. A note on highway pricing and capacity choice in a road network under a build-operate-transfer scheme [J]. Transportation Research Part A, 2002, 36:659-663.
9. FREDI L, GUSTAVO K. South Israel 100 millionm3/y seawater desalination facility: build, operate and transfer (BOT) project [J]. Desalination,2003,156:29-37.
10. SMITH Nigel, ZHAO Hao, ZHU Yi-ran. The Huai-bei power plant and its implications for the Chinese BOT market [J]. International J of Project Management, 2004, 22:407-413

Study on Comprehensive Evaluation Index System for Education E-government Performance of College and University

Lijuan Ye, Fang Yu and Jiaming Zhong

Xiangnan University, 423000 Chenzhou Hunan,China

Abstract: First, it sets forth the connotation of education E-government performance evaluation in college and university ,Then it illustrates its design principles as directive, complete, scientific, feasible, and development. It also puts forward the construction process, including dispersion, convergence, test, and verification. Meanwhile, it determines the framework for this index system, including four primary indexes, which are information infrastructure, application system, income and benefit, insurance system, as well as eighteen secondary indexes, namely possession quality of computer products per capita, construction of office and business application system, hardware, software, personnel training, and views and talent environment, etc. By making use of analytic hierarchy process, it determines the weight of primary index and the secondary index, thus constructing a relatively complete comprehensive evaluation index system for education E-government performance of college and university.

1 Introduction

The construction of education E-government is an important part of education information and the national E-government affairs, and it is an important approach of advancing the educational modernization. To transform the working style and method, as well as to further improve the work quality and efficiency, current universities and colleges are advancing the e-government construction now, and with some results. But in all, it still has some problems, such as large input while small output, high investment while low efficiency. Therefore, e-government construction in colleges and universities is far from adapting to the requirements of educational reform and development in China. International practices and studies have shown that the e-government project is highly risky and extremely possible to fail, so many research institutions, consulting companies, and scholars in western countries have studied on the e-government performance evaluation. The study on e-government performance evaluation can be divided into five international models, namely: performance focusing on government websites, such as American Brown University, Accenture Consulting Company; technical index of infrastructures, such as IBM Company; comprehensive index system of software and hardware, such as the United Nations and American Administrative Society; network performance in the whole society, such as International Development Center for Harvard University; basic evaluation standards, such as OECD (Organization for Economic Cooperation and Development) [1-4].

Some domestic research institutions have evaluated the portal websites of central government and local governments, and many experts and scholars have conducted a lot of researches on the e-government evaluation index system and methods in line with the characteristics of domestic e-government construction. Zhang Chengfu, Tang Jun (2005) put forward that the e-government performance can be divided into three levels of output, result, and influence [5]. Yang Yunfei, Bai Qing (2004) established an e-government evaluation index system including five items, which are "electronic concentration, electronic safety, electronic management, electronic service, and electronic decision" [6]. Peng Xizheng (2005) maintained that the main factors of e-government performance evaluation are: evaluation index system, index evaluation standard, index weight, comprehensive scoring methods, and data collection method [7]. Ma Lianjie, Hu Xinli, Zhang Xiaolian et. al. (2005) designed a set of e-government project evaluation index from the perspective of cost benefit [8]. Liu Teng, Liu Ronghui, and Zhao Peng (2004) established internal and external index systems, as well as six subsystems [9]. Wang Xiezhou, Chen Yanhong (2004) established the evaluation index system for e-government performance from the quantity, quality, utilization, and development potential of e-government information resource construction [10].

In all, the e-government performance evaluation in China is still at the initial stage, and the construction of scientific e-government performance evaluation system is at the exploratory period.

2 Concept of Education E-government Performance Evaluation

2.1 Performance

Performance is the work expression and business result of the organization or its subsystems (department, procedure, work teams and employees), which reflects the following meanings: first, it is the expression of subsystem during the work process; second, it is the business result of subsystems.

2.2 Education E-government Performance in College and University

According to the meaning of "performance" discussed in the previous part, the education e-government performance in college and university can be summarized in two aspects: process and result, namely the e-government operation process and result for school education. The operation process is to improve the school targets through effectively allocating information resources; the operation result is the contribution degree of realizing enterprise targets by making use of education e-government results.

Therefore, education e-government performance is not a result reflection. It is not only the relevance between the function of final system (the provided product and service amount) and results (the result that can effectively satisfy the expected missions and targets); but also contains the strategic implementation, management control, and comprehensive ability of project management during the implementation process of education e-government. Wherein, it includes the efficiency of operation and the ability of sustainable development.

In terms of the education e-government evaluation in college and university, the determination of e-government level can be transformed from "focus on input" to "focus on performance", which has important policy guidance and practical value. The evaluation on "performance" can help us lead the education information to an efficient, competitive, and sustainable development direction; it can help the education information cooperate with the overall strategy of colleges and universities; help school leaders to correctly recognize and implement the information in colleges and universities, thus achieving practical efficiency and avoiding waste.

2.3 Education E-government Performance Evaluation in College and University

E-government performance evaluation refers to the comprehensive assessment and evaluation for the construction, operation, management process, and results of college or university e-government by making use of certain verification methods, qualitative index and evaluation standards. In this way, it can promote the enhancement of its e-government management and application level.

3 Design Principles of Index System

3.1 Scientific Principle

The index system should reflect science, and it must abide by the laws and characteristics of education e-government. Meanwhile, it should accurately reflect the basic features of evaluation objects, as well as the actual level and effect during the process with the attitude of being practical and realistic. The science of index includes coincidence, independence, objectiveness, and coordination.

3.2 Guiding Principle

Evaluation index system plays a guiding role in enhancing education E-government performance. Therefore, when we are establishing index system, we should observe the selection of index factors, the connotation of factors, as well as the determination of evaluation standard. Meanwhile, we should emphasize the characteristics of education e-government and the main factors that influence the education e-government. It should reflect the basic requirements of education e-government, and also make all the factors the self-checking dimensions in the school, as well as the targets can be achieved in recent period. Meanwhile, through the evaluation factors of the evaluation index system and the weight distribution, it should guide all the teachers and students to reach the targets of education e-government.

3.3 Feasible Principle

The education E-government involves wide areas, while the index system cannot involve all aspects. Therefore, we need to analyze carefully, and master the main contradiction. We need to ensure the integrity of the index system, and simplify the index factors to the largest extent, to make it easier to be operated, but not being away from the actual situation of the school. Each evaluation index should be concrete and specific in content, independent in connotation and extension, distinct in wording, and clear in semantic meaning. It is the regulation for action and operation, and the data collection is simple, and with stable data source, as well as standard data and consistent range. Therefore, it is convenient for evaluators to observe and measure. To sum up, formulating index system can measure the e-government performance of education, and provide convenience to the implementation during evaluation process.

3.4 Open and Adaptive Principle

The education e-government is a dynamic development course, so the E-government performance evaluation of education should be continuously improved and consummated along the course of development. Therefore, we should appropriately revise the established evaluation index system after applying for certain period in accordance with the development and requirement of the situation, so as to enhance the consistency between the E-government evaluation of education and the E-government target of education, and also improve the efficiency of the E-government evaluation in education.

3.5 Systematic Principle

The evaluation index system should satisfy the completeness of index. That is to say, the formulated e-government performance evaluation index system of education should not neglect any important information, but comprehensively, systematically, and naturally reflect or cover all the factors of e-government performance in education. If the evaluation index system is more comprehensive and systematic, it can reflect the nature of evaluation objects better, and the evaluation result will be more accurate. However, it should be remembered that we cannot get away from reality for over pursuing completeness. When designing the evaluation index system, we should determine the factors that can best reflect the attributions off evaluation objects and best satisfy the requirements of evaluation targets as indexes in line with the characteristics of evaluation objects and the specific evaluation targets. On the other hand, some unimportant and non-essential evaluation factors can be neglected or abandoned. In this way, we can better combine completeness and focus.

4 Construction Process of Evaluation Index System

4.1 Target Decomposition

First, the evaluation targets can be decomposed. The general targets can be divided into several sub targets (primary index). According to the complexity of sub targets, some of the sub targets can be further decomposed to several sub sub targets (secondary index). Then, we should carefully analyze all the sub sub targets (sub targets) and all the involved factors (tertiary index), and then list them out without any omission.

4.2 Classification and Induction

Through induction and deduction, all the factors should be selected and refined. By getting rid of the nonessential and unimportant factors, while merging the same types of factors, and adjusting contradictory factors or the factors with causal-effect relationship, it will form the primary index framework.

4.3 Comparison and Synthesis

Through comprehensively applying literature research method, comparative research method, interview method, induction and deduction method, we should comparatively analyze and refer to the evaluation indexes generated through above analysis with the existing evaluation indexes, thus obtaining more comprehensive, objective, and accurate evaluation indexes. Therefore, the primary index framework will be shaped.

4.4 Inspection and Revision

Select appropriate evaluation target to conduct test in small scale. Meanwhile, revise the evaluation index in accordance with the test results.

4.5 Verification and Perfection

By making use of Delphi method, and through rounds of questionnaire surveys and expert consultation, the index framework will be determined. Then, the investigated factors can be further analyzed and selected. Through expert judgment, the investigated factors will be sorted, to determine the key investigated factors.

4.6 Weight Determination

By making use of subjective weighting method and objective weighting method, the index weight may be determined, and the complete index system may be obtained at last.

5 Dimensional Analysis for Evaluation Index System

The following part will specifically analyze the four primary indexes to analyze e-government performance evaluation of education, which mainly collects data and judge the value from the following four evaluation dimensions.

(1) Information infrastructure

Information infrastructure includes the per capita possession amount of computer products, the proportion of computer network, the network performance level, information safety facilities, and other secondary indexes.

(2) Application system

The application system includes construction of office and business application system, construction of government information sources and basic data, construction of government information disclosure, construction of portal website, authorized software, and other secondary indexes.

(3) Cost and income

The cost and income includes the secondary indexes like hardware, software, personnel training, and operation input, enhancement of business quality and efficiency, reduction rate of administrative expense, improvement of scientific decisions, satisfactions from teachers, students, and the society.

(4) Safeguard system

The safeguard system includes the secondary indexes like views and talent environment, information planning and implementation level, standard application, as well as technical support and security safeguard.

6 Methods to Obtain and Process Index Data

6.1 Methods to Obtain Index Data

(1) Questionnaire

According to specific evaluation index, we should design effective questionnaire, to obtain evaluation index data.

(2) Expert evaluation

Based on experts' experience and knowledge, the qualitative index data can be obtained.

(3) Collective discussion and personal interviews

Collective discussion and personal interviews

mainly invite related responsible persons from the evaluated objects, as well as the personnel from construction and operation management departments to make evaluation.

(4) Archives consultation

According to the data filled in the questionnaire, we should refer to the archives from the evaluated organizations.

(5) On-site verification

According to the data filled in the questionnaire, we should verify the authenticity of the data on the site.

(6) User evaluation

6.2 Data Processing

Data can be divided into quantitative data and qualitative data, so we should often scientifically process data when applying the application model.

(1) Qualitative index processing

Qualitative data processing usually adopts questionnaire method and field inspection. In order to avoid errors caused by subjective judgment, and increase the accuracy of qualitative indicators, we can use membership value assignment method, and classify qualitative indexes into several classes, which are 1-10 respectively.

(2) Quantitative index processing

Because contents and dimensions of each qualitative index are different, these indexes should undergo dimensionless processing, and transfer the original quantitative index value to evaluation value.

7 Index Weight Determination through AHP Method

Index weight is an important component of index system, which decides the importance degree of each index in the whole evaluation system, and also determines that scientific and reasonable weight is a vital item in comprehensive evaluation. This paper adopts the analytic hierarchy process (AHP) on the basis of Delphi Method to determine the weight. AHP was put forward by the famous American operational scientist T. L. Satty in the 1970s, and it was a systematic and hierarchical analysis method by combining qualitative and quantitative factors. The basic thinking of AHP is to decompose the factors relating to decision into targets, standards, and programs. Based on this, we can conduct qualitative and quantitative analysis.

7.1 Establish Hierarchy Frame Model

The key of AHP lies in establishing hierarchical model. In order to solve modeling problems, T.L.Satty divided the factors included in the problems into several layers, which may be generally classified as the highest layer, the middle layer, and the lowest layer. According to the principle of AHP and the above analysis for evaluation index system, we can obtain the hierarchical model of evaluation shown.

7.2 Construct Judgment Matrix

After establishing hierarchical structural model, it should point out the relative importance of all the factors in each layer regarding to the related factors on the previous layer, and then mark the value. Then, it can construct the judgment matrix, which is a key step of AHP. In order to get the value of each element in the matrix, the experts should first compare all the factors and then give out the judgment value. Then, according to the nine-level scales , the judgment result will be quantified, thus forming the comparative judgment matrix. Through applying Delphi method, we should organize experts to conduct questionnaire method. And through collecting the contents of the investigation tables from experts, we have obtained five judgment matrixes (for limited length of the paper, they are not listed out).

7.3 Single Hierarchical Arrangement

Single hierarchical arrangement is the weight from all the indexes of a certain layer to the relative importance of indexes at the previous layer. The commonest calculating methods include square root method, sum, characteristic root method, and least square method. However, the least square method is adopted in this paper, and the specific calculation steps are shown as follows:

Providing:

$$z = \sum_{i=1}^{n}\sum_{j=1}^{n}[\ln a_{ij} - \ln(\frac{\overline{w}_i}{\overline{w}_j})]^2 \qquad (1)$$

According to the fair and just principles, several judges (or several group judges) are selected to take part in the evaluation, so the factors of the judgment matrix a_{ij} have several values. Therefore, the formula (1) can be expanded as:

$$z = \sum_{i=1}^{n}\sum_{j=1}^{n}\sum_{k=1}^{b}(\ln a_{ijk} - \ln \overline{w}_i + \ln \overline{w}_j)^2 \qquad (2)$$

Wherein, b is the constant, namely the number (or group number) of the judges.

Obtain the partial derivative of $\overline{w}_i (i = 1,2,\cdots,n)$ for both sides of the formula (2), which can be simplified as:

$$n \ln \overline{w}_i - \sum_{i=1}^{n} \ln \omega_i = \frac{1}{b}\sum_{j=1}^{n}\sum_{k=1}^{b} \ln a_{ijk} \qquad (3)$$

From formula (3), we can get:

$$\overline{w}_i = (\prod_{k=1}^{b}\prod_{j=1}^{n}\omega_j)^{\frac{1}{bn}}(\prod_{k=1}^{b}\prod_{j=1}^{n}a_{ijk})^{\frac{1}{bn}} \qquad (4)$$

At last, normalize the vector $\overline{W} = (\overline{W_1}, \overline{W_2}, \cdots, \overline{W_n})$.

$$w_i = \frac{\overline{w}_i}{\sum_{i=1}^{n} \overline{w}_i} = \frac{(\prod_{k=1}^{b}\prod_{j=1}^{n} a_{ijk})^{\frac{1}{bn}}}{\sum_{i=1}^{n}(\prod_{k=1}^{b}\prod_{j=1}^{n} a_{ijk})^{\frac{1}{bn}}} \qquad (5)$$

Then, $W = (W_1, W_2, \cdots, W_n)^T$ is the characteristic vector.

Therefore, we can obtain the following weight of primary index and secondary index, which is shown as follows.

$$\omega_1 = 0.356, \omega_2 = 0.236, \omega_3 = 0.215, \omega_4 = 0.193$$

$$\omega_{11} = 0.4235, \omega_{12} = 0.2271, \omega_{13} = 0.2271, \omega_{14} = 0.1223$$

$$\omega_{21} = 0.491, \omega_{22} = 0.232, \omega_{23} = 0.093, \omega_{24} = 0.138, \omega_{25} = 0.046$$

$$\omega_{31} = 0.263, \omega_{32} = 0.475, \omega_{33} = 0.058, \omega_{34} = 0.093, \omega_{35} = 0.111$$

$$\omega_{41} = 0.055, \omega_{42} = 0.565, \omega_{43} = 0.117, \omega_{44} = 0.263$$

In this way, we can obtain a complete evaluation index system, which is shown as Table 1.

8 Future Research Direction

Education e-government performance evaluation is a very complicated issue, especially for the determination of the evaluation index and the determination of each index weight. The education e-government performance evaluation index system of college and university with different types and in different areas is different. However, AHP can make our thinking process mathematical, modeling, systematic, and standardized, so making decisions through AHP can greatly enhance the efficiency of the decision making. Nevertheless, during the process of applying AHP, no matter establishing hierarchical structure or constructing judgment matrix, people's subjective judgment, selection, and preference has great influence on the result. Therefore, it is very subjective to make decisions by making use of AHP. In other words, how to overcome the subjective influence, how to make use of group judgment, and how to combine AHP with other decision making methods are the future research directions.

Table 1 Education E-government Performance Evaluation Index System in College and University

Primary Index		Secondary Index	
Index Item	Weight	Index Item	Weight
Information Infrastructure	0.356	per capita possession amount of computer product	0.4235
		computer network proportion	0.2271
		network performance level	0.2271
		information safety insurance	0.1223
Application System	0.236	construction of office and business application system	0.491
		construction of government information resource and basic data	0.232
		construction of government information public	0.093
		portal website construction	0.138
		authorized software	0.046
Cost and Income	0.215	software, hardware, personnel training, and operation input	0.263
		office quality and efficiency enhancement	0.475
		reduction ratio of administrative expense	0.058
		enhancement of scientific decision level	0.093
		satisfaction degree of teachers, students, and the society	0.111
Safety System	0.193	views and talent environment	0.055
		information planning and implementation level	0.565
		standard application	0.117
		technical support and security safeguard	0.263

Acknowledgment

This study is supported by the construct program of the key discipline in human province, Hunan Philosophy and social science program(NO: 13YBB205), Hunan science and technology program(NO:2014SK3229) , Hunan province ordinary university teaching reform project(xiangjiangtong[2013]NO:223),China.

References

1. OECD.www.oecd.org.(2001)
2. UN.www.un.org.8(2003)
3. Tang Jun. J.E-Government,24(2005)
4. LeiZhanbo,JiangXiaofang. J. Intelligence, 12(2006)
5. Zhang Chengfu, Tang Jun. J. E-government, 2005,(24).
6. Yang Yunfei, Bai Qinghua. J. Comp. Appl. Soft, 2004,(8).
Evaluation(Publishing House of Electronics Industry,2005)
7. Peng Xizheng. J. Info. Cons., 10(2005)
8. Ma Lianjie, Hu Xinli, Zhang Xiaolian, et. al. J. Hubei Social Science,11(2005)
9. Liu Tenghong, Liu Ronghui, Zhao Peng. J.Journal of Wuhan University of Technology (Information and Management Engineering),3(2004)
10. Wang Xiezhou, Chen Yanhong. J. Archives Science Study,6(2004)
11. Zhijuan Liu,Jiaming Zhong.J.IERI Procedia,2(2012)
12. Du Dong, Pang Qinghua. *Modern Comprehensive Evaluation Methods and Case Selection* (Tsinghua University Press,2006)
13. ZhangXiaoming. J.Terminology Standardization & Information Technology, 2(2005)
14. An Hong,Gao Xuedong,Zeng Dehua,etc.J.Transactions of Beijing Institute of Technology,**27**,8(2007)
15. Zhang Haofeng,Kong Fanshi. *Educational Information*

Analysis of Present State of the Digital Power Meter Field Calibration Technology

XU Yi-Hui[1], LI Zhen-hua[1]

[1]*Hubei Key Laboratory of Cascaded Hydropower Stations Operation & Control, Three Gorges University, Yichang 443002, Hubei Province, China*

Abstract: As one of the most important measurement equipment of the intelligent substation, digital electric energy meter fundamentally resolved the technical bottleneck of traditional substation. During the operation process, measurement accuracy and reliability are related to the safe and stable operation of power system. Based on the literatures related to the digital electric energy meter, this paper introduced the concept of digital meter, analyzed the error sources, and summarized its development status and the existing calibration method. The paper studied the calibration system based on IEC61850-9 and provided an accurate and safe application method for digital electric energy meter.

1 Introduction

Compared with the traditional electric energy meter, working principle and interface mode of digital electric energy meter have been changed fundamentally. Nowadays, most people have inherited the ideas of traditional instrument measurement, calibration parameters, calibration method, provision of technical indicators and so on[1], which has demanded new requirements for the calibration of the digital power meter.

As an electric energy transmission and consumption metering device, electric energy meter has gone through more than one hundred years of history. Its development has experienced induction type (mechanical), pulse type (electromechanical) electric energy meter and electronic energy meter etc[2]. In 1880, the world's first electric energy meter was born. Through continuous improvement, in nineteenth century the inductive power meter manufacturing theory was proposed and had since been widely used for its advantages of simple manufacturing, good reliability and low cost etc. With the updated requirements of the power management modernization, pulse electric energy meter came to the market with the accuracy of 2.0, 1.0 and highest 0.5. In 1976, electronic power meter prototype appeared in Japan. Following the development of electronic devices, especially the electric energy metering chip, the performance of electronic energy meter was improved greatly, and gradually replaced the ordinary induction meter.

Nowadays, with the advancements of intelligent substation[3] digitized standards, its corresponding technology, and the continuous development of the digital and network measurement mode, digital multifunctional electric meter based on IEC61850 has been widely used and become the development trend of the smart grid. Because there is no single standard of digital electric energy meter calibration, 347 digital substations all over the country that use digital electric energy meter are operating without complete test or the measuring accuracy cannot be guaranteed. This paper provided a variety of ideas for realizing high precision in the current verification technology.

2 Overview of Digital Electric Energy Meter

Digital electric energy meter uses digital interface, optical fiber interface, data interface etc. and the high speed optical fiber Ethernet are used in the physical layer and link layer. The electric current and voltage signals are digitized by digital substation fore-end electromagnetic or electronic transformer, and the digital signal is transmitted to the merging unit (equivalent to the analog signal processing and A/D converter) by the optical fiber. The merging unit outputs normative digital signal frame which is based on the IEC61850 standard. Digital power meter can figure out electric energy parameter by receiving this digital signal frame. The digital electric energy meter is a pure digital signal processing equipment or an IT equipment. It requires that the equipment is reliable, the algorithm is scientific and the calculation error is minimized.

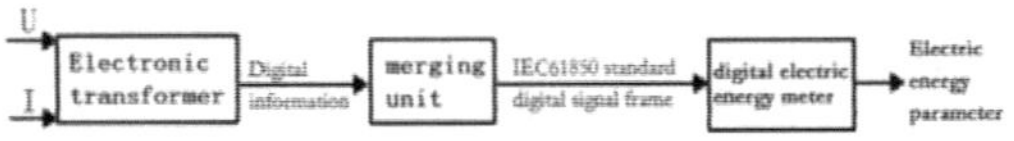

Fig1. Digital substation electrical energy measurement principle diagram

The digital substation electric energy metering device is comprised with an electronic type voltage, current transformer, combining unit (which plays a role of protocol conversion and time synchronization), digital electric energy meter and total station sampling synchronous clock as well as the connecting cable[4]. Compared with the traditional electric energy meter, digital substation metering device's current and voltage digital signal should guarantee the synchronization to ensure the accurate measurement. So it is necessary to design a new calibration device to replace the traditional electric energy meter calibration instrument. The device must have the following basic functions: first of all, it should be equipped with an optical fiber Ethernet interface; secondly, the link layer needs to have the ability to adapt to different standards; and thirdly, it should have the function of calculating the power. Most impotent thing is to have the function of preserving historical data for follow-up, and the detection system should also be able to receive the meter pulse signal to do some error analysis.

Considering the measurement structure principle, digital electric energy meter can be divided into two categories[5]: digital electric energy meter based on single chip microcomputer (MCU) and digital signal processor (DSP). DSP chip is 32 bit register with a very high clock rate and very rich signal processing optimized instruction set with its operational speed on "instruction level". The other category is based on the special electric energy metering chip, which has a strong anti-interference capability and good measurement accuracy. But the development cost is high and its flexibility is low because it is too dependent on the multi functions of the electric energy metering chip. The hardware frame of digital electric energy meter is more stable and the data processing is the key to realize the wide application of the digital electric energy meter. In other words, how to reform the electric energy meter and realize the high precision measurement relies on how to overcome the difficulty of electric energy parameter measurement algorithm.

Referring to the problems that occurred during the application of digital electric energy meter, research shows that the following points should be considered for high-precision: the measurement accuracy of the digital electric energy meter in the multi unit data packets, the measurement accuracy in different sampling ratio, the measurement accuracy at 0.01% Packet-loss Ratio, and the digital electric energy meter should keep the same level of accuracy when change the voltage and current specifications.

3 Analysis of electric energy meter error formation

3.1 Error formation of traditional electric energy meter

The error of the traditional measurement system is mainly from aspects such as Holzer component, shunt circuit, voltage divider circuit, TV and TA. The error characteristics curve of A/D preamplifier and A/D converter under different temperature, different load or power factors are nonlinear[6]. For example, Figure 2 is the error characteristic curve of the inductive electric energy meter under voltage change, in which mu=U/Ue. The relative error γ u represents the voltage influence, The induction watt-hour meter is composed of the electromagnetic element, the rotating element and the brake component. In reality, the turntable is affected by the driving torque and braking torque that is proportional to the load power. It is also affected by the inhibition torque, parasitic torque, friction torque and torque compensation. The influence of the turntable displacement, will directly affect the basic error of the electric energy meter.

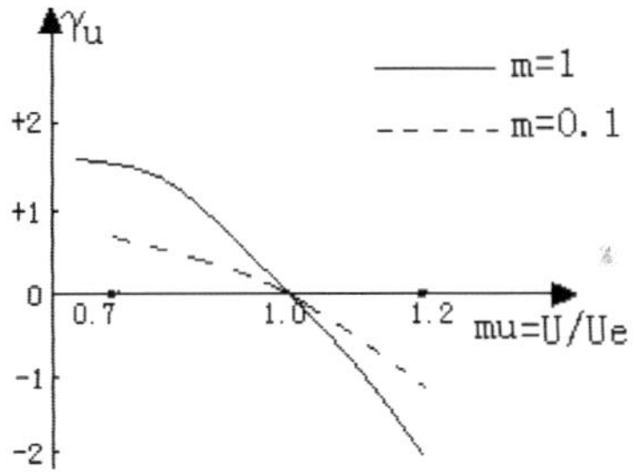

Figure 2. Voltage error characteristic diagram of induction type electric energy meter

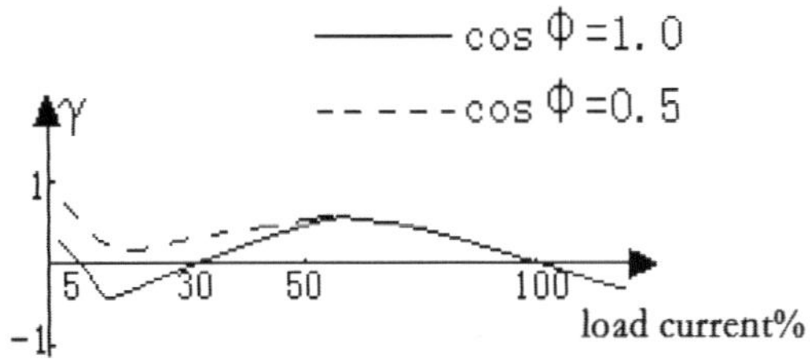

Figure 3. Load characteristic curve diagram of induction type electric energy meter

In the same way, the error characteristic curve of the induction watt-hour meter is nonlinear when load changes. In Figure 3, the electric energy meter working characteristic is unstable when the load current is less than 5%, The influences are from the friction torque, self-induced dynamic torque, the current magnet nonlinear error and nonlinear error changes when the load current is from 5 to 100 percent. Due to its different principle of electronic energy meter [7], the load characteristic curve is a straight line as shown in

Figure 4.

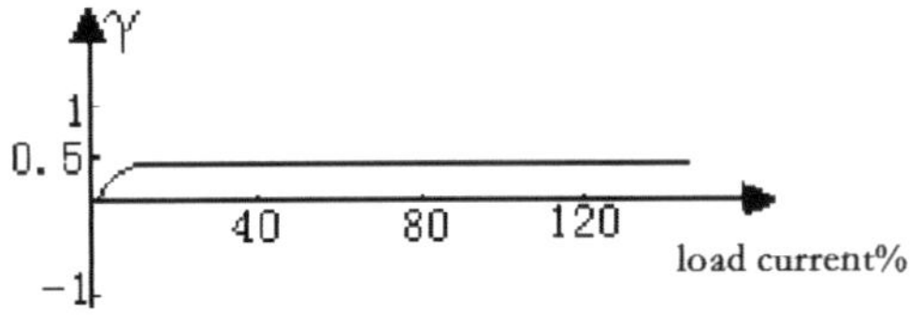

Fig4. Load characteristic curve diagram of Electronic energy meter

3.2 Error formation of the digital electric energy meter

In the digital electric energy meter, there is no start, running concept and overload limit. Small current or no current state does not affect the electrical energy measurement. Different from the traditional analog electronic energy meter, there is no impacting factor such as signal transformation, A/D conversion, sampling and synchronization that caused by the equipment itself. Changes in voltage, frequency fluctuations, three-phase imbalance, noise and other external signal impact also do not influence the digital electric energy meter. In theory, the digital electric energy meter is just a mathematical operation with no error. The actual error may be caused by three factors. The first is the abnormal phenomena due to the loss of frames and packet during the process of sending the packet, which will affect the meter's work function and reliability. It will not change the internal program (algorithm) so that the measurement performance is not impacted. Second, is the error caused by the difference between the electric energy pulse output and the actual electric energy accumulation[8]. Lastly, is the error caused by the floating point operation or the truncation error[9] which is the computer system inherent error. But in the digital substation, the sampling data error caused by the limit of the IEC61850 protocol is more suitable to be considered as the error of electronic transformers. In fact, in the field of electronic transformer, it is one part of the error and it is not reasonable to repeat the calculation. As for the tremble during the process of analog signal harmonic, noise and digital signal transmission, it is the error factor of mutual inductance and the combination unit. Even being disposed by the electric energy meter, it still belongs to the category of the algorithm error.

4 Research status and methods of electric energy meter

Overseas researches on electric energy meter testing technology have started very early, and a lot of relevant technical solutions for the measurement device state detection have been proposed. But there is no any related international or western standards to follow. In the field of digital electric energy meter, there still lacks of authoritative research results.

Domestic calibration technology is relatively weak. It was not until the mid-1980s, China began to develop field calibration instrument. The rapid expansion of market demand greatly stimulated the enthusiasm of domestic field calibration instrument research. In a short span of less than twenty years, China's field calibration apparatus has experienced three stages of development. The first stage is mechanical pointer type instrument, the second stage is digital instrument, and the third stage is intelligent instrument. With the continuous development of the digital process, many research institutions, universities and manufacturers have launched the exploration and research on the measurement system of digital electric energy. For different concepts and applications, a number of different digital electric energy meter calibration methods are proposed: digital power source and standard digital electric energy meter method, standard digital power source method, and analog power source and analog standard electric energy meter method.

4.1Digital power source and standard digital electric energy meter method

Principles of digital power source and standard digital electric energy meter method are shown in Figure 5. The digital power source and network equipment should be stable and reliable. The accuracy grades of standard digital electric energy meter should be two steps higher than the verified digital electric energy meter. It also requires that the standard digital electric energy meter must pass the energy value.

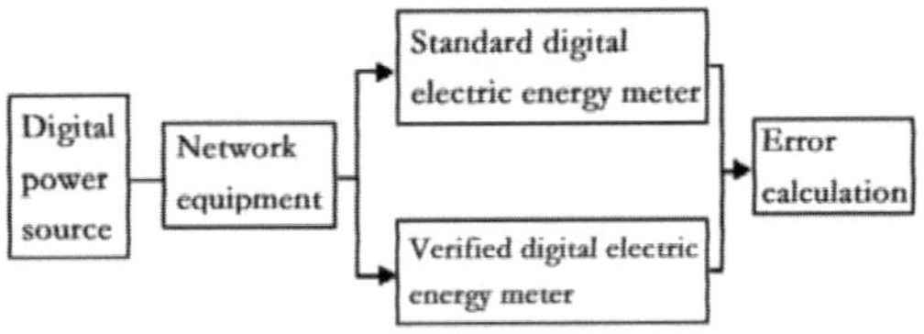

Figure 5. Principle block diagram of digital power source and standard digital power meter method

A number of scientific institutions and universities, such as Chinese Academy of Metrology, have proposed calibration methods based on the principles above. State Grid Hubei Electric Power Research Institute's Shen Li[7] et al. have designed a scheme of test device. It is able to simulate and test the meter accuracy when network packet is abnormal which provides a new design idea for the laboratory test of the digital electric energy meter. Zhejiang Electric Power Test Research Institute developed a digital electric energy meter detection device based on the digital power source and standard digital electric energy meter method[10]. In its unique design, the digital channels can be configured, primary and secondary pulse constants are optional, the IEC61850 related parameters can be configured; uses both of 220V and110V AC auxiliary power supply, and with single mode, multi-mode etc. This device can solve the problem of standard digital power source output with only one way, and also solve the problem of a network interface that cannot be connected to both single and multimode optical converters.

4.2 Standard digital power source method

Jiangsu Electric Power Research Institute, Nanjing Automation Co. Ltd, China Institute of Metrology and several institutions put forward the so-called standard digital power source method. Different from the China Institute of Metrology which obtain the standard electric energy value through the way of receiving line with IEC61850-9-1/2 communication protocol standards packet and then unpacking, the Nanjing Automation Co. Ltd obtains the standard electric energy value through direct calculation.

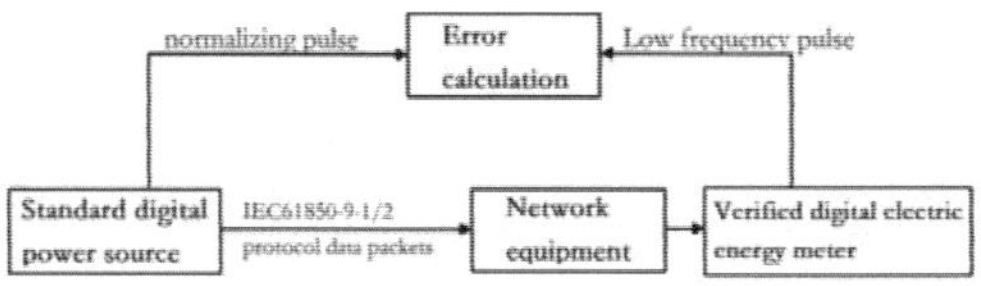

Figure 6. Principle block diagram of standard digital power method

4.3 Analog power source and analog standard electric energy meter method

Guangdong Electric Power Scientific Research Institute and other institutions have proposed a method based on the analog power source. The principle is to use three-phase power source output analog voltage and current signal directly to the standard analog meter, and after sampling and communication protocol conversion, analog voltage and current signal are provided to the verified digital electric energy meter and standard digital electric energy meter. The error of the digital electric energy meter can be obtained by comparing the electric energy value of standard digital electric energy meter and the verified electric energy meter.

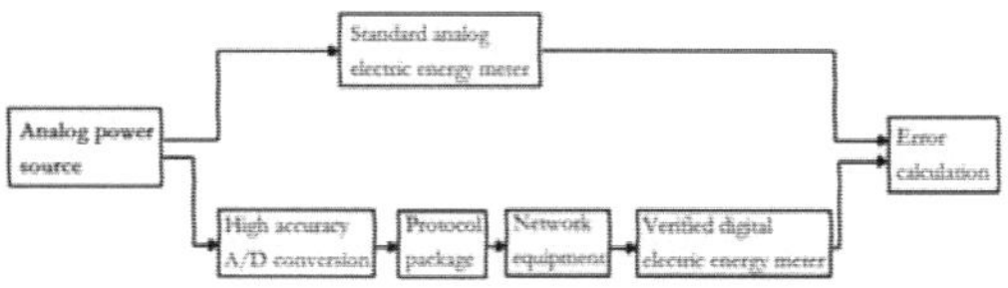

Figure 7. Principle block diagram of analog power source and standard analog power meter method

The first two methods are using the digital frequency direct synthesis technology to generate a sine wave signal. This method can set up the output sine wave amplitude, phase, frequency and channel number. Standard digital power source method can also test the error that caused by the error rate and the algorithm of the digital electric energy meter. But the calculated current and voltage values generated by the standard digital power source are not directly traceable to the standard analog values. The third method have better flexibility since it generates digital signal through high precision A/D conversion equipment acquisition of analog power source voltage current signal and then packaged.

4.4 Other related research results

In research on the field environment for test technology to simulate the merging unit, Wasion Group digital electric energy meter checking device first proposed the merging unit measurement technology, which can simulate the maximum three units in on-site inspection environment, support maximum of 50 channel sampling data and GMRP networking[11], and realize multi-fiber access port digital electric energy meter calibration[12].

On the electric energy meter error characteristic and the digital electric energy measurement traceability method[13], the Sichuan Electric Power Research Institute and the Tsinghua University have been working together to construct a kind of "input type digital standard power meter" which can output two types of digital parameters.

Zhejiang Han Pu Electric Power Technology Co. Ltd has designed a digital power meter calibration device based on embedded platform[14]. The device can realize the function of analog voltage and current collection, digital power output, and all kinds of electric energy calculation, display, storage, etc. It can also provide the function of MMS communication to complete the calibration of the electric energy meter.

Sichuan Electric Power Science Research Institute[15] developed a digital power meter with DSP as the core to support the IEC61850 protocol. It was an electric energy meter with integration of the meter and its power source realized the test of positive reverse and active reactive power. Based on their design, experts and scholars are able to continue to develop and optimize the power meter test function.

5 Research of digital power meter calibration system based on IEC61850 protocol

In recent years, based on the electronic transformer's good linear relationship with the measurement value, the wide measurement range, and the magnetic saturation is not a problem, output sampling data packets can be done with standard format. While the traditional power meter cannot receive the digital signal, the digital electric meter is coming into play.

In this paper, a portable digital electric energy meter calibrator is tested and verified in the laboratory and on-the-spot. It can effectively solve the problem of field calibration of digital electric energy meter with high accuracy and efficiency. The calibrator includes hardware and software platform. The main function of the hardware platform is that through the way of real-time receiving, to convert the standard channel merging unit's digital output and the electric energy meter pulse output into the corresponding electrical signal, and transmitted them to the software platform for analysis and calculation through the signal processing unit frame.

The software platform for digital power meter is shown in Figure 8.

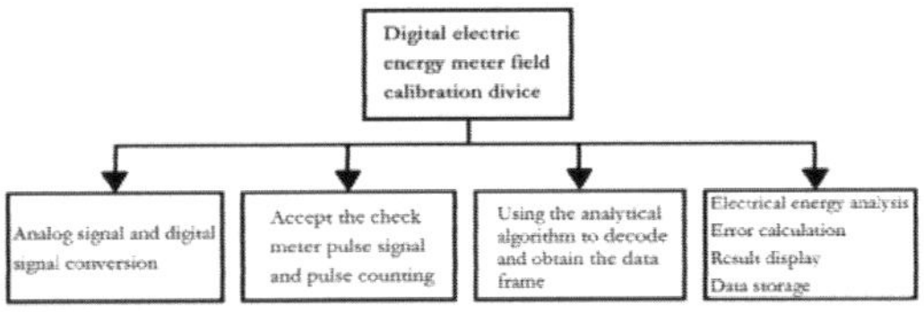

Figure 8. Digital electric energy meter field calibration software platform

Its main functions are summarized here. First, is the data frame decoding of the merging unit's digital output, which can obtain it's voltage and current signal and then perform the standard channel power calculation. The data frame of the merging unit output is generally in the 61850-9-2 IEC frame format. The software platform needs to receive the data of the merging unit in real time so that it can carry out the frame decoding accurately; Second, is to receive the pulse of the verified meter and analyze the corresponding electric energy. Finally, the software platform will compare the standard channel and the electric energy meter calibration data to get the error value and the results will be displayed in real time which can be stored as needed.

There are two key points of this research. First, the data acquisition and transmission system of the field digital electric energy meter is divided into the standard channel merging unit data acquisition[16] and the verified electric energy meter pulse output acquisition. Because it is compatible with the current mainstream synchronous mode and can achieve higher synchronization accuracy, the synchronization error between the standard and the check channel can be controlled within 0.01%. Second[17], based on high precision and multifunction clock synchronization method research, it can realize clock synchronization by taking advantage of the substation clock, and keep high precision synchronization by using its high frequency crystal oscillator clock. It is of great significance to the promotion of the digital electric energy meter technology development and the power grid's intelligent level.

Through the research of this project, we can develop a portable calibration device that can carry out real-time verification of digital electric energy meter, not only guarantee the instrument's accuracy and reliability but also improve the accuracy of the current channel and the voltage channel to the 0.05 levels. This work has an important theoretical and practical significance in regulating the economic benefit evaluation index of electric power industry and promoting the standardization and intelligence of the entire network.

6 Concluding remarks

The new generation of intelligent substation is in the direction of large-scale and practical applications. Meanwhile, ensuring the accuracy of the digital electric energy meter has great economic and social benefits to the substation construction. Study the existing research results is essential to the science and development in this area. . This paper provided a variety of design ideas for experts and scholars through the analysis on the digital power meter error form factor. The work summarized and analyzed the development status of existing methods and introduced digital electric energy meter calibration system based on the IEC61850-9 protocol.

Acknowledgement

This work was supported by Natural Science Fund of Hubei Province (grant number 2015CFB219) and Foundation of Education Department of Hubei Province (grant number B2015245).

References

1. M. Xiaoxing, D. Meimei, Z. Jian, Probe andAnalysis of the Traceability Method[J]. *Electrical Measurement & Instrumentation* **51**,23 (2014)
2. T. Min, Development of Electric Energy Meter[J]. *Electrical Theory and New Technology*, (2004)
3. W. Wei, F. Jianbo, Application of IEC61850 Standard in Digital Substation in 220kV Three Township[J]. *Power Technology*, **29**,3 (2012)
4. Z. Qiuyan, Analysis and Counter measures of Measurement Problems in the Construction of Digital Substation[J]. *Guizhou Electric Power Technology*, **14**,1 (2011)
5. H. Xiaobo, Research on the DigitalMulti-Function Power Meter Based on IEC61850[D]. *China University of Petroleum*, (2010)
6. W. Wenhua, F. Xiaoping, S. Xiaoxian, Comparison and Analysis of the Traditional Electric Energy Meter and Digital Electric Energy Meter[J]. *Zhejiang Electric Power*, (2011)
7. Xu Xinrui. Electronic Meter Characteristics and Selection[J]. *Shanghai Electric Power*, (2006)
8. Shen Li, Gan Yiyi, et al. Research and Application of Digital Electric Energy Meter[J]. *Electrical Measurement & Instrumentation*, **51**,24 (2014)

9. L. Qian, Z. ShuHan, et al. Digital Electric Energy Metering System of Field Verification Technology Research[J]. *Electrical Measurement & Instrumentation*, **47**,538 (2010)
10. Yao Li, Li Shaoteng, Lu Chunguang. The Digital Electric Energy Meter Detection Device Design [J]. (2010)
11. Cui Hongqi, LIAN Zhonglin, et al. GMRP Multicast Function Test[J]. *Journal of Changchun University of Science and Technology* (NATURAL SCIENCE EDITION)
12. Zhou Jianbo, Guo Lihuang, Research on theinterface of digital power meter sampling data[J]. *Electrical Measurement & Instrumentation*, **51**, 11(2013)
13. Lin Guoying, ZHOU Shangli, Sun Weiming, et al. Digital Substation Metering Device Research on Traceability[J]. *Purcell Power Automation*, **87**,(2010)
14. Lu Fengjie, Shen Tujiequn, et al. Research and Design of Digital Electric Energy Meter Calibration Device Based on IEC61850 Protocol[J]. *Electrical Measurement & Instrumentation*, **52**, 6 (2015)
15. Yang Huayun, AI Bing, et al. The Developmentof Digital Electric Energy Meter[J]. *Electrical Measurement & Instrumentation*, **50**, 570 (2013)
16. Mao Lijun. Analysis of the Causes of Failure ofthe Smart Meter Data Acquisition[J]. *Power Technology*, **30**, 6 (2013)
17. Cheng Zetao, Mao Yuanjun et al. Clock Synchronization Scheme of Smart Substation Based on Phase Locked Loop Technology [J]. *Power Technology*, **32** ,5 (2015)

MATEC Web of Conferences **61**, 02004 (2016) DOI: 10.1051/matecconf/20166102004
APOP2016

Analysis on the Conversion Mechanism of Electromagnetic Energy to Continuous Movement

Dan Xia[a], Dianyou Song and Zhiyong Wang

School of Electronic Information Engineering, Tianjin Key Laboratory of Film Electronic and Communication Device, Tianjin University of Technology, Tianjin 300384, China

Abstract. The interaction between current slow wave and moving conductive medium is analyzed, and correspondingly a novel conversion mechanism of electromagnetic energy to continuous movement is proposed. The theoretical analysis and its experimental validation have shown the feasibility of the promising RF and microwave motors.

1 Introduction

The interconversion of electric and mechanical energies is always a fundamental problem in energy conversion. Motors and generators are its representative devices. It is well known that almost all motors and generators convert energy from one form to another through the action of a magnetic field. Usually the magnetic field is confined in a magnetic circuit, and the rotating magnetic field in the gap is created by currents from multiphase electric circuit system; both magnetic and electric circuits are operating at a very low frequency, and therefore can be considered independent of each other. The progresses of science and technology have introduced new demands in energy conversion, and one is related to converting RF/microwave energy to motion or kinetic energy. As early as in the sixties of last century, so called parametric motor or microwave motor [1-2] appeared. That motor was essentially a DC motor with a rectenna, which inserted inside the waveguide to pick up and rectify the microwave energy into DC energy then feed it to the motor coil [3]. Without direct conversion way, this indirect scheme has been continued until now in electric vehicles [4] and flights [5-6] etc. Nowadays, Solar Power Station/Satellite is a sustainable energy source to help meet the growing energy needs of developed and developing nations. One of the most important technologies for its realization is Wireless Power Transmission (WPT) [7]. The more using of WPT for the next generation will make the conversion mechanism from electromagnetic (EM) energy to other energies much more important. Unfortunately, the above mentioned fundamentals in electrical machinery become invalid at RF and microwave, because electric and magnetic fields are strongly coupled to each other, magnetic and electric circuits are mixed together, and RF and microwave multiphase system, in addition to the rotating magnetic or electric fields hardly can be feasible.

Own to the application demands, the explorations of the direct conversion mechanism from EM energy to kinetic energy have never ceased. Physical scientists had found that EM force, referred to radiation pressure in optical frequency band, can manipulate micro-objects [8] and even make micro-objects rotating [9], but probably it is still not big enough with the viewpoint of electric machinery. The effort has continued so far to enhance EM forces [10]; as a part of this effort, here we will report our research on the conversion mechanism of EM energy (RF/microwave) to continuous movement, which may find variant applications in fields of Power and Energy Engineering.

2 Conversion mechanisms

The conversion mechanism of EM energy to continuous movement is based on the interaction between current slow wave and moving conductive medium. To analyze the interaction, we consider a simplified planar model of a motor driven by EM wave as shown in Figure 1. EM fields are excited by a surface current density in the stator at $z = 0$, i.e.

$$\boldsymbol{J} = \boldsymbol{y}_0 J \exp[\mathrm{j}(\omega t - \alpha x)] \quad (1a)$$

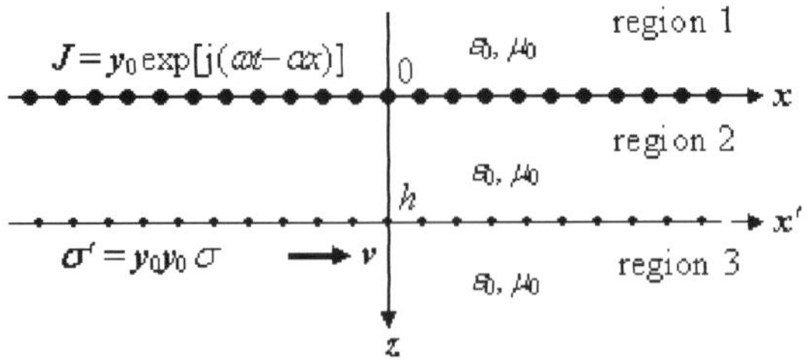

Figure 1. Current wave over a moving plate with dyad conductivity.

in which J, ω and α are the amplitude, the angular frequency and the propagation constant in x-direction, respectively. For a slow wave, α is greater than the propagation constant $k_0 = (\omega^2 \varepsilon_0 \mu_0)^{1/2}$ in the air, and therefore the phase velocity in x-direction

[a] Corresponding author: xiadan@tjut.edu.cn, waiting831@163.com

$$v_{\mathrm{p}}=\frac{\omega}{\alpha} \tag{1b}$$

is less than the light speed $c = \omega/k_0$ [11]. The rotor at $z = h$ is a very thin conductive plate which is moving at the velocity $\boldsymbol{v} = \boldsymbol{x}_0 v$, and its dyad surface conductivity represented in moving frame is

$$\boldsymbol{\sigma}' = \boldsymbol{y}_0\boldsymbol{y}_0\sigma \tag{2}$$

All other regions are stationary free space with permittivity ε_0 and permeability μ_0.

Electrical fields of slow TE (y-polarized) wave in regions $i = 1, 2$, and 3 are expressed as

$$\boldsymbol{E}_i=\boldsymbol{y}_0 V_i(z)\,\mathrm{e}^{\mathrm{j}(\omega t-\alpha x)} \tag{3a}$$

with

$$V_1(z)=b_1\exp(\mathrm{j}k_z z) \tag{3b}$$

$$V_2(z)=a_2\exp(-\mathrm{j}k_z z)+b_2\exp(\mathrm{j}k_z z) \tag{3c}$$

$$V_3(z)=a_3\exp(-\mathrm{j}k_z z) \tag{3d}$$

where b_1, a_2, b_2, and a_3 are constants to be determined. The propagation constant in z-direction is

$$k_z=\sqrt{\frac{\omega^2}{c^2}-\alpha^2}=-\mathrm{j}\gamma \tag{4a}$$

in which

$$\gamma=\sqrt{\alpha^2-\frac{\omega^2}{c^2}}=\omega\sqrt{\frac{1}{v_{\mathrm{p}}^2}-\frac{1}{c^2}} \tag{4b}$$

is real.

By use of Maxwell's equation $\nabla\times\boldsymbol{E} = -\mu_0\partial\boldsymbol{H}/\partial t$, the corresponding H-fields can be derived as

$$\boldsymbol{H}_i=\left[-\boldsymbol{x}_0 I_i(z)+\boldsymbol{z}_0\frac{\alpha}{\omega\mu_0}V_i(z)\right]\mathrm{e}^{\mathrm{j}(\omega t-\alpha x)} \tag{5a}$$

with

$$I_1(z)=-Yb_1\exp(\mathrm{j}k_z z) \tag{5b}$$

$$I_2(z)=Y\left[a_2\exp(-\mathrm{j}k_z z)-b_2\exp(\mathrm{j}k_z z)\right] \tag{5c}$$

$$I_3(z)=Ya_3\exp(-\mathrm{j}k_z z) \tag{5d}$$

and mode admittance

$$Y=\frac{k_z}{\omega\mu_0}=-\mathrm{j}\frac{\gamma}{\omega\mu_0}=-\mathrm{j}\frac{1}{\mu_0}\sqrt{\frac{1}{v_{\mathrm{p}}^2}-\frac{1}{c^2}} \tag{6}$$

The boundary conditions at the interfaces $z = 0$ and $z = h$ are

$$\boldsymbol{z}_0\times(\boldsymbol{E}_2-\boldsymbol{E}_1)\big|_{z=0}=0 \tag{7a}$$

$$\boldsymbol{z}_0\times(\boldsymbol{H}_2-\boldsymbol{H}_1)\big|_{z=0}=\boldsymbol{J} \tag{7b}$$

$$\boldsymbol{z}_0\times(\boldsymbol{E}_3-\boldsymbol{E}_2)\big|_{z=h}=0 \tag{7c}$$

$$\boldsymbol{z}_0\times(\boldsymbol{H}_3-\boldsymbol{H}_2)\big|_{z=h}=\boldsymbol{J}_{\mathrm{e}} \tag{7d}$$

where $\boldsymbol{J}_{\mathrm{e}}$ is the equivalent surface current density on the moving conductive plate. To count the contribution of the moving conductive plate, it is necessary to express the fields in moving frame. In the case of v much smaller than c, Galilean transformations $x' = x-vt$, $y' = y$, $z' = z$, and $t' = t$ can be applied [12]. The fields in moving frame are

$$\boldsymbol{E}_i'=\boldsymbol{E}_i+\boldsymbol{v}\times\mu_0\boldsymbol{H}_i \tag{8a}$$

$$\boldsymbol{H}_i'=\boldsymbol{H}_i-\boldsymbol{v}\times\varepsilon_0\boldsymbol{E}_i \tag{8b}$$

The conductive current can then obtain that

$$\boldsymbol{J}_{\mathrm{e}}'=\boldsymbol{\sigma}'\cdot(\boldsymbol{E}_3'+\boldsymbol{E}_2')\big|_{z'=h} \tag{9}$$

Both tangential (parallel to $\boldsymbol{v}$) component of H-field and current density are invariant for Galilean transformations. Therefore, substituting the fields in stationary frame into Eqs. (7a) ~ (7c) and those in moving frame into Eq. (7d), one can solve out the four unknown constants b_1, a_2, b_2, and a_3. Finally, we have

$$V_3(h)=-\frac{J\exp(-\gamma h)}{2\sigma\left(1-\dfrac{v}{v_{\mathrm{p}}}\right)-\mathrm{j}\dfrac{2}{\mu_0}\sqrt{\dfrac{1}{v_{\mathrm{p}}^2}-\dfrac{1}{c^2}}} \tag{10}$$

Nothing but the EM force density will make the rotor moving. It can be derived by use of $\boldsymbol{F} = \boldsymbol{J}_{\mathrm{e}}'\times\boldsymbol{B}_3' = \boldsymbol{J}_{\mathrm{e}}'\times\mu_0\boldsymbol{H}_3'$ and represented in terms of $V_3(h)$.

3 Performances

Suppose that the rotor is only possible to move in x-direction. Thus the x-component of EM force density F_x is most concerned, and it is related to the z-component of flux density

$$B_{3z}'=\left(\frac{1}{v_{\mathrm{p}}}-\frac{v}{c^2}\right)V_3(h)\,\mathrm{e}^{\mathrm{j}(\omega t-\alpha x)} \tag{11a}$$

and the y-component of current density

$$J_{\mathrm{e}}'=2\sigma\left(1-\frac{v}{v_{\mathrm{p}}}\right)V_3(h)\,\mathrm{e}^{\mathrm{j}(\omega t-\alpha x)} \tag{11b}$$

To keep continuous movement of rotor in x-direction, it is required that the effective (average) force density in x-direction $\langle F_x\rangle$ is non-zero. $\langle F_x\rangle$ can be obtained by integrating the EM force density over space period $2\pi/\alpha$ and time period $2\pi/\omega$, respectively, and the result is

$$\langle F_x\rangle=\frac{1}{2}\mathrm{Re}\left(J_{\mathrm{e}}'B_{3z}'^{*}\right) \tag{12}$$

where * denotes complex conjugate. Substituting Eq. (11) into Eq. (12), it follows that

$$\langle F_x\rangle=\frac{J^2\sigma\left(1-\dfrac{v}{v_{\mathrm{p}}}\right)\left(\dfrac{1}{v_{\mathrm{p}}}-\dfrac{v}{c^2}\right)}{4\sigma^2\left(1-\dfrac{v}{v_{\mathrm{p}}}\right)^2+\dfrac{1}{\mu_0^2}\left(\dfrac{1}{v_{\mathrm{p}}^2}-\dfrac{1}{c^2}\right)}\mathrm{e}^{-2\gamma h} \tag{13}$$

Letting $v = 0$ in Eq. (13), the starting force is expressed as

$$\langle F_x \rangle_s = \frac{1}{4} \frac{J^2 \sigma \dfrac{1}{v_p}}{\sigma^2 + \dfrac{1}{\mu_0^2}\left(\dfrac{1}{v_p^2} - \dfrac{1}{c^2}\right)} e^{-2\gamma h} \quad (14)$$

It is worth noting that the starting force is depended on the factors: v_p^{-1}, $\sigma[\sigma^2+(v_p^{-2}-c^{-2})/\mu_0^2]^{-1} \approx \sigma^{-1}$ and $J^2\exp(-2\gamma h)$. It implies that in order to obtain the larger starting force, we must make the phase velocity of EM wave, the surface conductivity and the air gap smaller, and in turn the amplitude of current source J larger. Taken typical parameters of phase velocity in x-direction $v_p = 0.01c$ and unit current $J = 1$, the force-speed curves with different values of surface conductivity and relative thickness of air gap h/λ, $\lambda = c/f$ is the wavelength in the gap, are calculated by using Eq. (13) and shown in Figure 2.

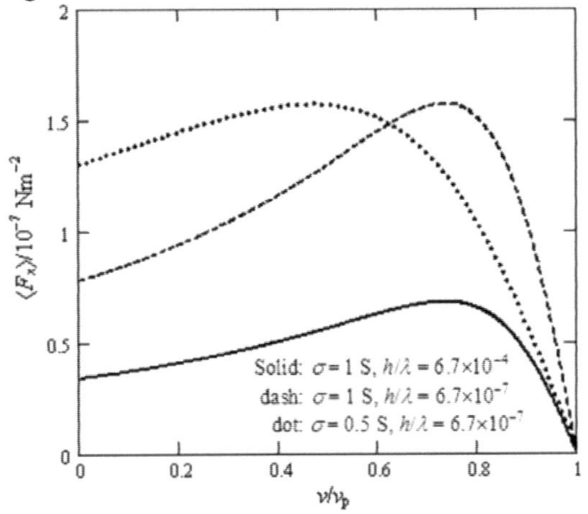

Figure 2. Force-speed curves.

It is obvious from the curves that the force is larger for smaller relative thickness of air gap. The other behaviors of the model are somewhat similar to the traditional induction motor. At synchronous speed $v = v_p$, the force is always vanished; at certain speed that is near but different from synchronous speed v_p, the force gets the maximum value, and less conductivity σ leads to larger starting force and bigger speed slip (v_p-v) for the maximum force.

4 Experimental validation

The experiment has two principal aims: to test if the above conversion mechanism really works and to estimate the starting force under the excitation. The prototype used in the experiment is shown in Figure 3. Here the moving part (rotor) is a conductive film on a disk-shaped substrate (Figure 3 right), while a circular array of slots are etched on the film to make the film anisotropic with the conductivity as defined by Eq. (2). The stationary part (Figure 3 left) can be regarded as a converter from high frequency voltage to slow wave, and it is designed with phase velocity $v_p \approx 0.01c$ and wave impedance $Z_0 \approx 75\Omega$.

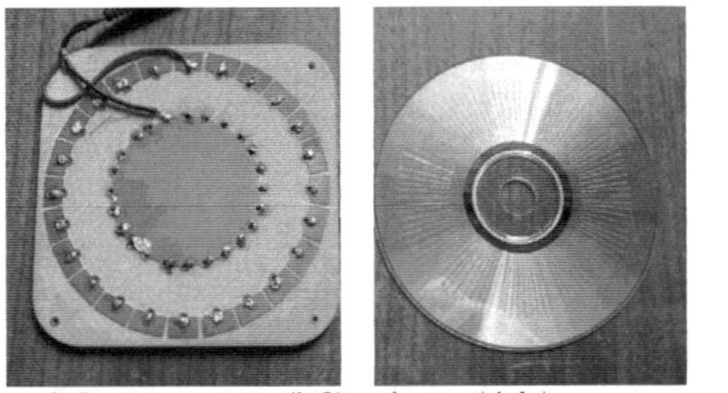

Figure 3. Prototype stator (left) and rotor (right).

In the experiment, the moving part is assembled as a torsion pendulum [13] so that the possible rotation can be observed and the toque can be estimated when it is in balance after rotating. The voltage-to-wave converter is excited by a high frequency signal generator at the starting port and matched with a 75Ω load at the end port. The travelling slow wave as defined by Eq. (3) is incident on the surface of the rotor and the EM force density in x-direction $\langle F_x \rangle$ makes it moving. Its movement can be observed in full frequency range from 0.150 to 1 GHz of the signal generator. The measured deviation angles θ of the torsion pendulum at different frequencies f are listed in Table 1.

Table 1. Deviation angles θ at different frequencies.

f/MHz	150	266	290	340	440
θ/degree	−15	−11.3	15	33.3	37.5
f/MHz	540	650	750	850	1000
θ/degree	37.5	37.5	37.5	56.3	41.3

The deviation angle can roughly reflect the amplitude of the starting force. The positive and negative angles in the table tell us an unusual fact that the forces can have different directions in different frequency bands. This interesting fact can be understood by invoking the particular frequency characteristics of the voltage-to-wave converter. Taking the stray parameters into account, the converter can be modeled as a composite right/left-handed transmission line [14], whose unitary cell and phase velocity-to-frequency curve [15] are shown in Figure 4. Here L_R and C_R are stray parameters, and L_L and C_L are real parameters. One can figure out from the curve that the phase velocity can change its direction near the resonant frequency, so can the starting force in the experiment [16].

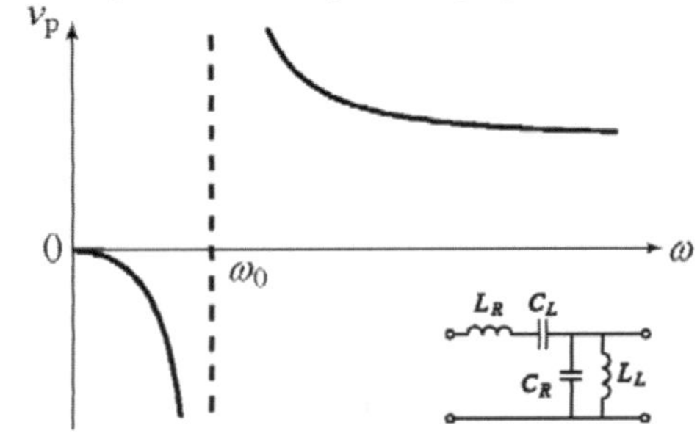

Figure 4. Unitary cell (lower right) and phase velocity-to-frequency curve of composite right/left-handed transmission line.

5 Conclusions

The proposed conversion mechanism of EM energy to continuous movement in this paper is based on the interaction between current slow wave and moving conductive medium. Invoking the novel exciting scheme of slow EM wave, the model of motor is successfully to operate at RF/microwave frequencies. To the best of our knowledge, it is the first time to make a macroscopic object into continuous moving state directly by RF/microwave EM force. Although the experiment of prototype is rather primary, the analysis based on rigorous EM theory and its experimental validation have provided sound base for further study on practical device of such a kind of motors, which may find a series of potential applications in the fields of Power and Energy Engineering, such as directly wireless energy storage by flywheel and wireless power distribution to mobile users. Moreover, the applications in other fields are also possible.

References

1. H.E. Stockman, Proceedings of the IEEE, **51**, 1253-1254 (1963)
2. R.C. Garnier, K. Ishii, Proceedings of the IEEE, **52**, 1380-1381 (1964)
3. R.W. Steinhoff, T.K. Ishii, IEEE T. Microw. Theory, **14**, 438-438 (1966)
4. A. Oida, H. Nakashima, J. Miyasaka, K. Ohdoi, H. Matsumoto, N. Shinohara, J. Terramechanics, **44**, 329-338 (2007)
5. J.J. Schlesak, A. Alden, T. Ohno, IEEE MTT-S, **1**, 283-286 (1988)
6. E. Shimane, S. Komatsu, K. Komurasaki, Y. Arakawa, 46th AIAA, 2008-1149
7. S.S. Mohammed, K. Ramasamy, T. Shanmuganantham, IJCA, **13**, 100-103 (2007)
8. A. Ashkin, IEEE J. Sel. Top. Quant., **6**, 841-856 (2000)
9. E. Higurashi, O. Ohguchi, T. Tamamura, H. Ukita, R. Sawada, J. Appl. Phys., **82**, 2773-2779 (1997)
10. A. Mizrahi, M. Horowitz, L. Schächter, Phys. Rev. A, **78**, 023802 (2008)
11. D. Xia, T.L. Dong, Chinese Sci. Bull., **55**, 3875-3879 (2010)
12. J.A. Kong, *Electromagnetic Wave Theory* (EMW Publishing, Cambridge, Massachusetts, 2000)
13. V.B. Braginsky, A.B. Manukin, G.M. Keiser, Phys. Today, **31**, 195-196 (1978)
14. T. Günel, AEÜ, **61**, 459-462 (2007)
15. C. Caloz, T. Itoh, *Electromagnetic Metamaterials: Transmission Line Theory and Microwave Applications: The Engineering Approach* (John Wiley & Sons, Inc., Hoboken, New Jersey, 2006)
16. D. Xia, R.X. Yao, P. Li, G. Feng, T.L. Dong, APPEEC 2011

Energy Characterization of End-Device in ZigBee-based Wireless Sensor Network

Yikui Zhang[1], Xiangbin Xin[1] and Youmeng Li[1,a]

[1] School of Computer Software, Tianjin University, Yaguan Road 135#, Jinnan District, Tianjin 300350, China

Abstract. In this paper, a flow of ZigBee end-device energy characterization is introduced. The variation of the energy and power consumption of wireless communication through ZigBee end-device is studied. The remainder of this article details the methods used to determine energy and power overheads of the end-device in four parts: start-up stage, basic energy expenditure, data-sending consumption and polling consumption. The impacts of software parameters like polling rate and data size on the energy consumption are analysed. Also, models of the power and energy are extracted. Our methods allow estimating the energy consumption of the ZigBee end-device when running a Z-Stack application on a specific hardware platform.

1 Introduction

ZigBee Wireless Technology is emerging as the leading method for implementing low-cost, low-data rate, short-range wireless networks with extended battery life for use in creating the "Internet of Things" [1]. In recent years, ZigBee wireless sensor networks are becoming widely applied in a large number of fields including health, agriculture, industrial and geology, military, home and emergency management. Most of these networks are composed of multiple sensor nodes, which are able to collect specific data like temperature, humidity, etc.

In ZigBee networks, there are three kinds of nodes: coordinator, router and end-device. End-Device can connect to coordinator and router but cannot support routing [2]. It is a ZigBee end-device that joins a network and normally turns off its receiver when idle. This requires its parent to hold messages until the end-device polls for its messages. In most cases, an end-device is equipped with some sensors for collecting data and is powered by batteries. So energy consumption of end-device is a vital aspect for software and hardware designers to consider. Accurate energy characterization can make it easy to estimate battery life according to its capacity, or to choose optimum battery capacity according to designed life cycle. In our work, we proposed a method and flow of characterizing energy and power consumption of end-device in ZigBee network, taking both hardware factor and software factor into consideration.

2 Related works

In order to characterize energy consumption and establish energy models on embedded system, several studies have proposed their approaches.

Dhouib et al. [3] presented a multi-layer power and energy estimation approach for embedded applications and operating systems. According to them, energy can be summed over the hyper period of the real-time system. They estimated energy and power consumption of standalone tasks and also added energy overheads of embedded operating system services such as timer interrupt, IPC and peripheral devices accesses.

Xia et al. [4] propose a quantitative approach, in which EOS and applications are executed on two cooperated simulators, to estimate and analyse the energy consumption of embedded operating system (EOS). They used a cycle-accurate micro-architecture power model to estimate instruction execution energy.

Waseem [5] discussed and compared currently used energy efficient mobile operating system. And then he summarized and discussed recent energy efficient techniques used to reduce the power consumption of mobile devices.

Baynes et al. [6] described their simulation environment and the modelling of embedded systems with SimBed, which measures the execution behaviour and power consumption of the RTOS and embedded applications. The authors compared three different RTOSs: μ COS, Echidna and NOS. However, their models are not deduced from measurements on actual hardware platform.

3 Energy characterization and estimation flow

Basically, the energy consumed by an end-device in a certain time duration T after it's powered up can be divided into four parts: start-up stage, basic energy expenditure, consumption and polling consumption.

[a] Corresponding author: liyoumeng@tju.edu.cn

DOI: 10.1051/mateconf/20166102005

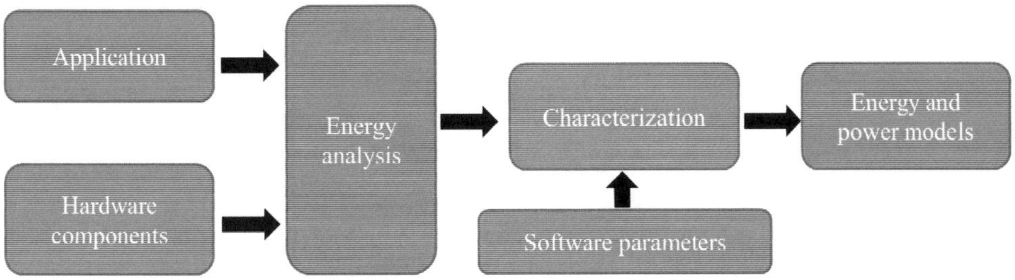

Figure 1. The methodology of end-device energy characterization.

Total energy can be given by Equation (1):

$$E_T = E_0 + E_b(T) + E_d(T) + E_p(T) \quad (1)$$

The proposed method consists in extracting an energy model of the end-device power overhead. The inputs are the generic application and the hardware platform. As showed in Figure 1, to characterize the energy consumed by the end-device, it's necessary to analyse the application running on the board. In the energy analysis step, a set of software parameters, which influence the energy consumption, are varied and then energy profiles are traced. The energy traces obtained are able to characterize the energy overhead of the end-device and then to model the power and energy consumption. After extracting the energy models, we determine the energy and power laws, as showed in Figure 2.

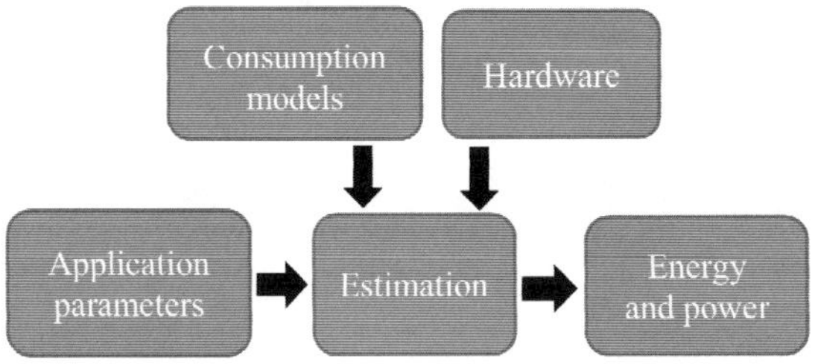

Figure 2. Estimation of the end-device energy consumption.

In Equation (1), T represents the total working time of the end-device. E_0 represents the energy consumption in the start-up stage, which is from being powered to successfully adding to the ZigBee network. In this stage, an end-device has to finish a few necessary activities, including hardware initialization, OS abstract layer initialization, ZigBee network discovery, network-adding request, and acquiring IP address. For a certain hardware platform, completing this stage takes a proximately constant time T_0. So T has no effect on E_0 when T is bigger than T_0 and we can make a presumption of $T >> T_0$.

$E_b(T)$ represents the basic energy consumption of the end-device from time T_0 to T and it's the energy consumed by core circuit and peripheral circuit to maintain basic operations except sending data. To characterize the basic energy consumption, we need to stop the end-device from sending data and polling messages.

$E_d(T)$ represents the extra energy for sending data to the ZigBee coordinator during the time from T_0 to T. An end-device is usually required to send fixed size of data to the coordinator periodically.

This energy is divided into two parts: data transmission and state switch.

$E_p(T)$ represents the energy for polling from time T_0 to T. End-Devices in the Z-Stack sample applications are setup initially with automatic polling for messages enabled. Three different polling options are supported, each controlled by a time delay parameter. If the end-device never needs to receive messages once it has joined the network, all three polling rates should be set to zero.

To characterize the energy overhead of the end-device, we should analyse the four parts and extract their energy models separately. In the next section, we build models for the four parts with the help of current and voltage data samples acquired by power monitor.

4 End-device power and energy models

4.1 Start-up stage

The trace of electric current change in start-up is showed in Figure 3(a). The last peak in Figure 3(a) indicates that the end-device has successfully joint into the ZigBee network and acquired its IP address. We sampled the data 10 times and calculated the energy for each, shown in Figure 4. As the calculated numbers has slight fluctuations, we consider E_0 a constant value and define it as the mean value of these samples.

Finally, E_0 equals 671.0121 mW•s and T_0 equals 3.27s.

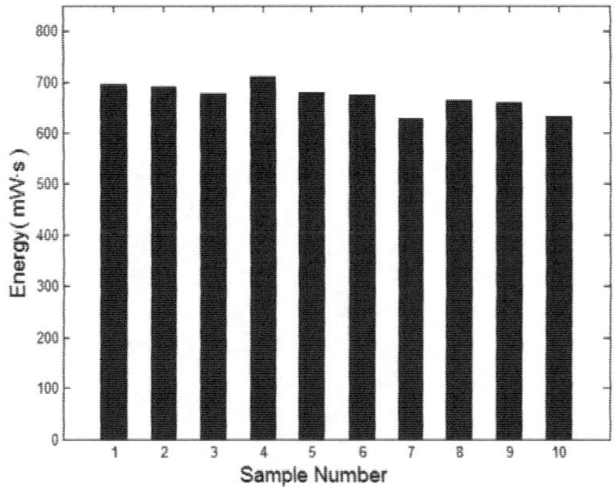

Figure 4. Ten energy samples in start-up stage.

4.2 Basic energy consumption

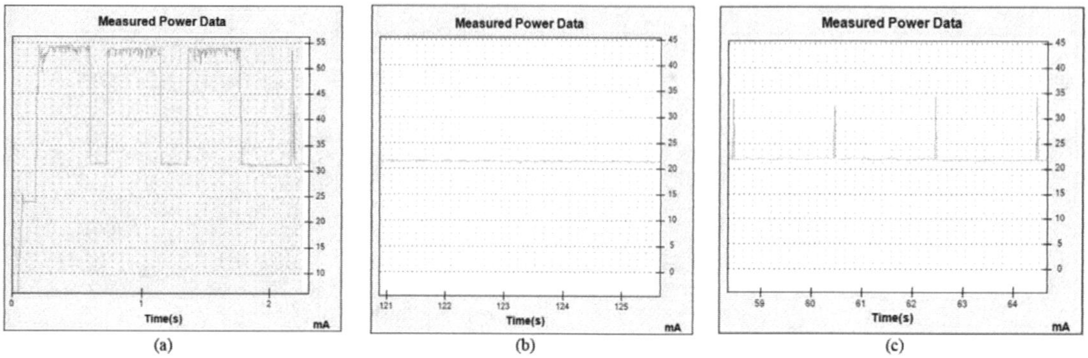

Figure 3. Traces of measured real-time current.

As is shown in Figure 3(b), when an end-device gets a steady state without data transmission and polling, its current-changing trace is similar to a horizontal line. So we define $E_b(T)$ by the Equation (2):

$$E_b(T) = P_b(T - T_0) \quad (2)$$

In Equation (2), P_b represents basic power and it's also a constant value and calculated following the Equation (3):

$$P_b = \sum_{i=1}^{N} U_i I_i / N \quad (3)$$

Where N is the sample size, U_i and I_i are respectively the voltage and current value of a sample point. We obtained about 25 thousand sample points within 5 seconds. Finally, P_b equals 106.8335 mW.

4.3 Data-sending energy consumption

Mostly, an end-device needs to report data to coordinator regularly when it's working. Our work here is to find out the relation between energy consumption and data size for one sending activity. As showed in Figure 3(c), each peak represents a sending activity and the sending process is extremely short. Energy consumed in this process can be calculated through integration. As this energy is consumed in two ways: changing device state and emitting electromagnetic waves through an antenna. The first part is stable and the second part is varied by the data size. We can define the Equation (4):

$$e_d = kS + e_0 \quad (4)$$

Where e_0 represents the energy consumed for changing device state, S represents the data size and k is the coefficient.

Figure 5 depicts the variation of measured data-sending energy consumption with data size. According to the fitting curve, k equals 0.0110 and e_0 equals 0.8573.

Meantime, $E_d(T)$ is varied by both S and time interval T_{si} , as showed in Equation (5):

$$E_d(T) = (kS + e_0) * (T - T_0) / T_{si} \quad (5)$$

4.4 Polling energy consumption

If an end-device needs to receive messages from other nodes, polling-rate must be set non-zero. The energy consumed for polling during a period of time is only varied by polling time interval T_{pi}. So $E_p(T)$ can be calculated by Equation (6):

$$E_p(T) = \begin{cases} e_p * \frac{T-T_0}{T_{pi}}, & T_{pi} \neq 0 \\ 0, & T_{pi} = 0 \end{cases} \quad (6)$$

Where e_p represents the energy consumed for polling once. e_p can be calculated through integration and it is assigned with the mean value of 100 sample, equaling 0.7032 mW•s.

From all the above, the total energy consumed by an end-device can be estimated when the following four parameters are given: working time T, data size S, sending time interval T_{si} and polling interval T_{pi}.

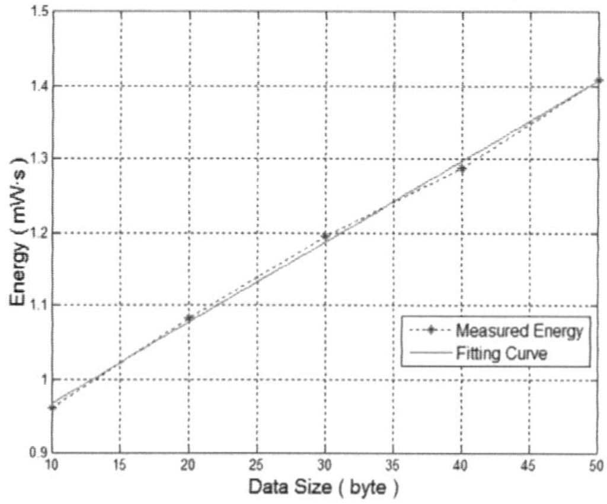

Figure 5. Energy variation as a function of data size.

5 Experiment

5.1 Hardware platform

The target hardware platform is the CC2530EB board from TI (Texas Instruments) [7]. It is equipped with an industry-standard enhanced 8051 processor as its MCU (Micro Controller Unit) [8]. The environment we build application on is ZigBee protocol stack (Z-Stack) from Texas Instruments. IEEE standard 802.15.4 defines physical (PHY) layer and media access control (MAC) layer [9]. Z-Stack Home is TI's ZigBee Home Automation (ZHA) compliant protocol stack for the CC2630, CC2530 and CC2538 System-on-Chip [10]. As shown is Figure 6, We use Mobile Device Power Monitor hardware and Power Tool software form Monsoon Solutions, Inc. to measure the power dissipation of an end-device. The Power Monitor is connected to a computer through a USB cable and an end-device is powered by the Power Monitor through a USB cable. The raw data file we get from Power Monitor is in the format of CSV, containing large amount of number pairs -- count time (s) and electric current (mA). These discrete data points can be processed with Matlab software through trapezoid integration to calculate the gross of energy consumed.

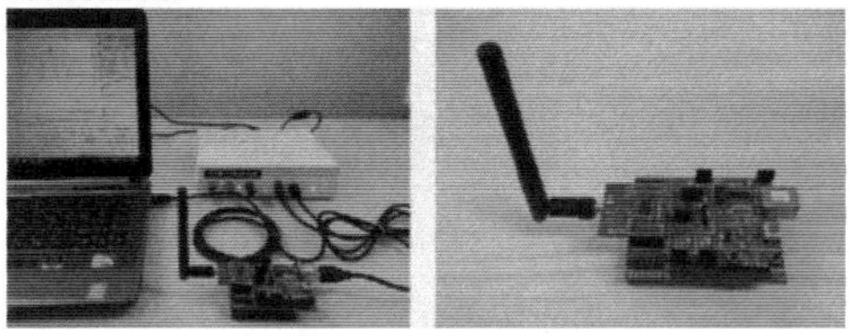

Figure 6. The Measurement Framework.

5.2 Experiment design and results

To verify the energy and power model, four experiments with variation of four parameters are performed on the hardware platform that is introduced above. With the Power Monitor successfully connected to the computer and IAR embedded workbench installed and open, we started the experiments. Initially, we set the polling off and make the end-device send 15 bytes data to the coordinator, which had already setup a ZigBee network, every each second. After 20 minutes, we turned off the end-device and export raw data file in the format of CSV. Then the file is simply processed by deleting its first title line so that it can be read by Matlab. Finally, the total energy consumed was calculated through integration. The application were compiled and loaded repeatedly with varied parameter combination. These groups of parameters and experimental results are shown in Table 1.

Table 1. Experimental results

No.	T (s)	T_{pi} (s)	T_{si} (s)	S (byte)	Estimated (mW·s)	Measured (mW·s)	Error (%)
1	1200	0	1	15	129745.2836	130203.6488	0.35
2	1200	1	2	25	130040.9358	130354.7847	0.24
3	1200	2	3	35	129438.2027	129023.0945	0.32
4	1200	3	4	45	129206.9646	129432.4539	0.17

According to Table 1, we note that the average estimation error is 0.27%. Actually, this statistics can't strongly prove the correctness of all our models. Because the basic energy consumption accounts for a huge proportion of total number. So we have to exclude basic energy to validate the other energy models, which is shown in Table 2.

we excluded basic energy and calculated the overall energy of the other three parts, because their values are in the same order of magnitude. Finally, according to Table 2, the average estimation error equals 3.20%, which proves that our models work well in this situation.

Table 2. Experimental results without basic energy

No.	T (s)	T_{pi} (s)	T_{si} (s)	S (byte)	Estimated (mW·s)	Measured (mW·s)	Error (%)
1	1200	0	1	15	1894.4292	1987.0564	4.67
2	1200	1	2	25	2190.0813	2135.1457	2.51
3	1200	2	3	35	1587.3483	1622.1621	2.19
4	1200	3	4	45	1356.1101	1402.6938	3.44

5.3 Deviation analysis

In our experiments, there are two main deviation sources: the Power Monitor and end-device itself. The Power Monitor we use is able to measure the current accurately, with a dual range, self-calibrating and integrating system used [11]. But it may still drop samples when running. The end-device may lose packets during the transfer process, especially when there exist WIFI or Bluetooth networks in the lab environment, which would surely increase the energy consumption as a result.

6 Conclusion

We have divided energy consumption of an end-device into four parts and proposed energy and power models of each part. The models are based on measurements on the hardware platform CC2530EB board and allow the characterization of the energy overhead of the end-device. Experiments show that these models can precisely characterize the energy overhead. For this reason, we plan, in the future work, to analyse more details of data-sending process and basic energy consumption. Apparently, these future works of this project will focus on optimizing the energy consumption of the end-device.

Acknowledgements

Our work was supported by Tianjin University-Qinghai Nationalities University Special Independent Fund, with the project number: 2015XZC-0038.

References

1. YF Zhong, YJ Liu. ZigBee wireless sensor networks [M]. BeUing University of posts and telecommunications press, 2011.
2. Abhishek Kumar, Sandeep Gupta, "Study on ZigBee Technology", IEEE, 3rd International Conference on

Electronics Computer Technology (ICECT), Volume **6**, Pages: 297 – 301, 8-10 April 2011.
3. Saadia Dhouib, Eric Senn, Jean-Philippe Diguet and Johann Laurent, Modelling and Estimating the Energy Consumption of Embedded Applications and Operating Systems, ISIC-2009 - 12th International Symposium on Integrated Circuits, Proceedings, p 457-461, 2009.
4. Zhao Xia, Guo Yao, Lei Zhi-Yong and Chen Xiang-qun, Estimation and analysis of embedded operating system energy consumption, Acta Electronica Sinica, v **36**, n 2, p 209-15, Feb. 2007.
5. Muhammad Waseem, Energy efficient mobile operating systems, International Journal of Advanced Networking and Applications, v **5**, n 1, 1812-17, July-Aug. 2013.
6. K Baynes, C Collins, E Fiterman, B Ganesh, P Kohout, C Smit, T Zhang, BL Jacob, The performance and energy consumption of embedded real-time operating systems. IEEE Trans. Comput. **52**, pp. 1454–1469 (2003)
7. Texas Instruments. CC2530 Datasheet [EB/OL]. http://www.ti.com.cn/cn/lit/ds/symlink/cc2530.pdf
8. Second Generation System-on-Chip Solution for 2.4 GHz IEEE 802.15.4 / RF4CE / ZigBee(2015), http://www.ti.com/product/CC2530/description
9. ZigBee Alliance. "ZigBee Overview". September 2003.
10. ZigBee protocol stack, http://www.ti.com.cn/tool/cn/z-stack#descriptionArea
11. Power Monitor End User Manual, http://msoon.github.io/powermonitor/PowerTool/doc/Power Monitor Manual.pdf

MATEC Web of Conferences **61**, 02006 (2016) DOI: 10.1051/matecconf/20166102006
APOP2016

A Course Scheduling Algorithm Based on Self-Adaptive Constrained Particle Swarm*

Cui Wei[1,a] and Long Xiaohong[1]

[1] *Luzhou Vocational & Technical College, 646005 Luzhou, Sichuan,China,*

Abstract. The automatic course scheduling in college is a complex constraint optimization problem. A kind of self-adaptive constraint particle swarm algorithm is proposed to avoid premature convergence of particle swarm optimization algorithm. A self-adaptive criterion is applied in different phases of algorithm execution to realize dynamic and adaptive switch between attractors and repellors during the algorithm's running process, and inertia weight will change dynamically and adaptively with the number of iterations. The algorithm performance can be further improved with a higher convergence speed and stronger robustness.

1 Introduction

The course scheduling in college is a very important and rather complex work in teaching management. Its main function is summarizing course application forms of various teaching institutes and departments and then formulating class timetables for the whole college according to the professional talent cultivation plans so as to enable the whole teaching to proceed in a planned and ordered way. The course scheduling system must ensure that no conflicts occur to teachers, students and classrooms in the course arrangement and meet the constraint conditions such as the teachers' requirements and classroom resources.

Therefore, the course scheduling is a constrained, multi-objective combinational optimization problem. How to make use of various existing techniques to realize the automation, scientification and rationalization of formulating class timetables and improve the resource utilization rate as well as the satisfaction of teachers and students to class timetables is one of burning questions in college teaching management work at present. It is of great practical significance to present a self-adaptive algorithm for the course scheduling problem that can meet constraints and reduce calculated amount.

To reduce the time needed to judge constraints in the operational process of the course scheduling system, this paper presents a new method of solving constrained optimization problems, self-adaptive constrained particle swarm algorithm by overcoming the limitation of particle swarm optimization algorithm. The results show that the improved algorithm has stronger optimizing capability and efficiency to solve constrained optimization problems.

2 Particle Swarm Optimization

The particle swarm optimization is a new evolutionary algorithm that has been developed in recent years. It is a kind of evolutionary computation technique, derived from the research on the birds preying behavior, was proposed by Dr. Eberhart and Dr. Kennedy in 1995. This algorithm is a simplified model which was first enlightened by the regularity of birds' swarm activities and then was built with swarm intelligence. By making use of the information sharing of individuals in the swarm, the algorithm makes the movement of the whole swarm produce an evolutionary process from disorder to order in the problem solution space and then obtains the optimal solution.

The course scheduling problem mainly involves the five elements such as teacher, course, class, time and classroom, among which teacher, course and class are closely related. The five elements involved in the course scheduling problem are set as follows:

Lesson Set: $L = \{L_i\}, i = 1,2,3 \cdots N_l$

Class Set: $C = \{C_i\}, i = 1,2,3 \cdots N_c$

Professor Set: $P = \{P_i\}, i = 1,2,3 \cdots N_p$

Time Set: $T = \{T_i\}, i = 1,2,3 \cdots N_t$

Room Set: $R = \{R_i\}, i = 1,2,3 \cdots N_r$

The Cartesian product of five elements constitutes the solution space of the course scheduling problem, and course scheduling is to find a solution that meets various constraints in this solutions space.

[a] Corresponding author: cccpmig29@gmail.com

3 The Key Problems In Self-Adaptive Constrained Particle Swarm Optimization

3.1 Constrained Swarm Coding

The constrained swarm coding is the key of self-adaptive constrained particle swarm optimization, and the coding quality determines the algorithm performance. In this algorithm, the constrained particle coding adopts decimal coding, whose structure is shown as Table 1.

Table 1. Constrained Particle Coding Table

Constrained Particle	digit
Teaching Building	2
Classroom	4
Class	6
Professor	6
Course	5
Time	2

For example, the particle with the code "04010411010100005410001 51" indicates such a course scheduling information: teaching building 04, classroom 0104, class 110101, the teacher with job number 000054 has course 10001 (appointed as Office Automation in our college) at the first two classes on Friday morning (time code is 51).

Every particle shows a possible course scheduling result, and the merit of particles can be evaluated according to the fitness function value.

3.2 Fitness Function

Let x be a feasible class timetable and X be the set of all feasible class timetables, and we build the fitness function as follows:

$$\max(f(x)) \qquad x \in X$$

$$f(x) = \omega_1 \times f_1 + \omega_2 \times f_2 + \omega_3 \times f_3 \quad (1)$$

Among them, x is a feasible class timetable, X is the set constituted by all feasible class timetables, ω_1、ω_2、ω_3 are the corresponding weights, which meet $\omega_1 + \omega_2 + \omega_3 = 1$. f_1、f_2、f_3 are fitness functions which meet course dispersion degree, time utilization ratio and classroom utilization ratio respectively.

4 SELF-ADAPTIVE CONSTRAINED PARTICLE SWARM OPTIMIZATION

4.1 The Description of Self-Adaptive Constrained Particle Swarm Optimization

Several methods have been proposed to overcome the premature convergence of continuous particle swarm optimization algorithm, yet at present there are not rather effective methods to solve the premature convergence problem of discrete particle swarm optimization algorithm. This paper has proposed a self-adaptive constrained particle swarm optimization algorithm based on constrained discrete particle swarm. To overcome the premature convergence of the algorithm, a criterion of controlling swarm diversity has been presented to measure the diversity of the current swarm, and a repulsive process has been introduced on the basis of the original algorithm to increase swarm diversity. When the swarm diversity is below the specified threshold, the repulsive process will be executed to make particles deviate from the worst solution of individuals and the optimal solution of the current swarm and search the unreached area until the swarm diversity exceeds the specified threshold, and then the attractive process will be executed to make individuals move towards the optional solution of the current swarm as well as individuals. To further improve the algorithm performance, a self-adaptive strategy has been defined in this paper to make the inertia weight change dynamically and adaptively with the iterations in the different executing phases of the algorithm, and several local search techniques have been introduced to speed up the convergence and improve the precision of the solution.

Thus, the two velocity updating formulas and position updating formulas of particles in the adaptive constrained particle optimization algorithm are defined as follows:

The velocity updating formulas:

Attractors:

$$x_i^{(t+1)} = \omega \times x_i^{(t)} + \phi_1(q_{i,b} - q_{i,c}) + \phi_2(g_b - q_{i,c}) \quad (2)$$

Repellors:

$$\boldsymbol{x}_i^{(t+1)} = \omega \times \boldsymbol{x}_i^{(t)} + \phi_1(\boldsymbol{q}_{i,c} - \boldsymbol{q}_{i,w}) + \phi_2(\boldsymbol{q}_{n,c} - \boldsymbol{g}_b) \quad (3)$$

Among them, ω denotes the inertia weight that is used to control the influence which the previous velocity has on the current velocity ϕ_1, ϕ_2 denote the learning coefficients, whose values are random numbers between 0 and 1.

Position updating formulas:

$$\boldsymbol{q}_i^{(t+1)} = \boldsymbol{q}_i^{(t)} + \boldsymbol{x}_i^{(t+1)} \quad (4)$$

It can be seen that, as for attractors, particles share the information of the current better solution and attract each other; as for repellors, particles share the information of the worst solution of individuals and the optimal solution of current swarm and move in the direction of deviating from these positions and search the potential optimal solution area that has not attained to increase the swarm diversity.

The discrete particle swarm optimization algorithm to solve the course scheduling problem proceeds as follows:

1) Initialize the swarm size M, ω, c_1, c_2, let the loop iterations k=0;

2) Randomly initialize the positions of particles in the swarm, calculating the fitness values of the initial particles, and sorting the fitness values in an ascendant manner;
3) Let $pb_i^{(t)} = X_i^{(t)}; gb^{(t)}$ is set to the position of the optimal particle in the initial swarm;
4) Judge whether the algorithm convergence criterion is met; if it is met, then output $gb^{(t)}$, and obtain the optimal class timetable, thus the execution of the algorithm ends; otherwise execute step (5);
5) Do the following operations on all the particles in the swarm:

Update the position of each particle according the formula 1) and formula 2); calculate the fitness value of the newly generated particle; if the fitness value of particles is better than $pb_i^{(t)}$, let $pb_i^{(t)} = x_i^{(t)}$; if the fitness value of the particle is better than $gb^{(t)}$, let $gb^{(t)} = x_i^{(t)}$;

6) Let k=k+1, and jump to step 4).

4.2 Experimental Result and Analysis

To verify the effectiveness of the self-adaptive constrained particle swarm optimization algorithm presented in this paper, the actual course scheduling from some college are chosen as the experimental data. The experimental environment is as follows: Intel(R)Pentium(R)D, CPU 3.00 GHz, 2.99 GHz, 4.00 GB primary memory, Windows Server 2003, and Matlab7.0

4.2.1 Experimental Data

This paper takes the course scheduling in some semester from some college as the experimental data, including 292 teachers, 163 classes and 865 courses. The courses, classes and teaching times per week of each teacher are referred to the actual workload.

4.2.2 Experimental Result

The experiment generates 292 particles altogether, each of which is a two-dimensional matrix;

w=0.9, c1=0.5, c2 =0.5; the maximum iteration times MaxDT =10. The analysis of the experimental result shows that the class timetables generated by the algorithm in this paper have no conflicts and all of them are arranged in the confined scope. The newly generated class timetables not only meet the conclusive condition but also appear more reasonable and humanized than the existing class timetables.

5 Conclusion

This paper has introduced a solution for the course scheduling problem in colleges by using the self-adaptive constrained particle swarm optimization algorithm. This solution introduces a repulsive process to prevent the swarm premature convergence, using attractive and repulsive processes to evolve the swarm constantly. To improve the algorithm performance, a strategy of dynamically and adaptively changing inertia weight has been proposed and local search techniques have been introduced.

The key point of the follow-up study is to further expand the experimental scope and introduce other intelligent algorithms to further improve the course scheduling system for the solution of the course scheduling problem in college.

Acknowledgement

* This paper is one of the research achievements of Scientific research project of Sichuan Provincial Department of Education (16ZB0457, 15ZB0405) and Luzhou Science and Technology Innovation Talent Cultivation Plan (2015-R-55(16/16),2014-R-52(18/20)).

References

1. Liao Caicai, Xi Guang, Xu Jianzhong. An Improved PSO Algorithm for Solution of Constraint Optimization Problem and its Application [J] Journal of Engineering Thermophysics, 2010.01,32-35
2. Zhang Huawei, Yang Kai. Research on Timetabling Problem Based on Particle Swarm Optimization [J]. Journal of Zhengzhou University of Light Industry (Natural Science), 2010.12,49-52
3. Yu Chengmin, Jiang Hua, Li Huan, Jia Baoxian. Research on Timetabling Problem Based on Discrete Particle Swarm Optimization [J]. Microcomputer Applications, 2010.07,67-71
4. Zhang Changsheng, Sun Jigui, Ouyang Dantong. A Self-Adaptive Discrete Particle Swarm Optimization Algorithm [J]. Chinese Journal of Electronics, 2009.02,299-304

A Narrative Approach to Detect the Vehicles using color, texture and edge based techniques

Ravi Gaurav[1,a] , Shubham Kumar[1] , Venkatesan S[2] and Ramesh Babu D.R.[2]

[1] Dept of CSE ,Dayanand Sagar College of Engineering, Bangalore,India
[2] Professor, Dept of CSE ,Dayanand Sagar College of Engineering, Bangalore,India

Abstract. Vehicle recognition is the chief stride in observing the speeding vehicles in a thruway. The feature arrangements caught by a stationary camera demonstrate to us that there's a requirement for a vehicle location calculation which handles sudden light change furthermore the situations where the closer view converges away from plain sight. This paper gives us a study of different foundation subtraction systems that are utilized for recognizing the vehicles effectively. Vehicles proceeding onward street are of significance on the grounds that issues like movement blockage, monetary waste, sticking on the underpasses and over-extensions (if the vehicle going through is not of the passable size) are connected with them.*Index Terms*—Vehicle Detection, video sequences, foreground, background, MATLAB, RGB conversion.

I Introduction

Hsu-Yung Cheng [1] utilized a pixel wise classification model for vehicle detection using Dynamic Bayesian Network (DBN). In numerous picture handling applications, closer view recognition is a vital step. Fitting recognition and ID of these frontal area protests just will prompt other handling like following. With a specific end goal to screen the vehicles in the expressway, we need to think seriously about the changing climate conditions and closer view protests that may converge away from plain sight later. The camera's perspectives will be unique in relation to distinctive positions of the camera which should likewise be considered in the recognition process. There is front perspective, back perspective, side view and top perspective. The stature at which the camera is put will likewise influence the identification's yield calculations. The camera details, the casing rate and range or separation secured by this altered camera likewise ought to be considered for proficient vehicle discovery. Also, as the camera will be altered and is inclined to part of aggravations to commotions, there will be clamor in the features which ought to be uprooted for effective vehicle recognition. A few techniques have been proposed for vehicle identification. Vehicle identification should be possible utilizing foundation subtraction calculations and vehicle model based methods. Every calculation has its own focal points and detriments. Be that as it may, these routines attempt to recognize vehicles vigorously and productively. Utilizing any location calculation first the foundation and the closer view are isolated. Progressively vehicle identification, first the vehicles' shadow must be uprooted keeping in mind the end goal to stay away from errors in the recognition process. This paper gives a study on the current calculations that work on beating the issues said before. Area (II) gives a review on the strategy which utilizes both shading and surface for vehicle discovery. Segment (III) depicts a strategy that uses shading change. Extracting information from the image, calculating certain metrics based on this information and comparing the values of these metrics top predetermined values[10].

2 COLOR AND TEXTURE BASED VEHICLE DETECTION

Shading and composition are two most critical qualities in picture preparing application. Visual shading difference is utilized to channel data present in every shading segment. (Tremeau et al. [2]) and to recognize among comparative dim scale intensities (Barilla-Perez & Spann [2]). There have been both shading based and composition based vehicle divisions. Be that as it may, reconciliation of these both will yield us better results in vehicle location. The street's shade is homogenous and they have the same surface all through. You can without much of a stretch identify vehicles making utilization of this. The vehicle's shade will demonstrate an enormous variety when contrasted and the street's shade. In any case, issue comes when the vehicle's shade is like the street's shade which can be understood by considering the composition. The initial phase in this system would be making a foundation

[a] Corresponding author: Ravi Gaurav, gaurav.ravi9@gmail.com

model. A foundation model can be made utilizing recursive or non-recursive systems. Since clamor is included, you can utilize middle channel. The contrasts between the present casing and the foundation can be figured utilizing the elements of composition and the L* intensities part of L*u*v* shading space. This system depends on a change location procedure that will consolidate both the power and surface contrasts between the present edge and the past edge. This strategies lives up to expectations useful for static foundations where there is no much distinction in the foundation's surface. Composition is viewed as a rich wellspring of data about the nature and three-dimensional state of items (Petrou (2006)). It comprises of complex visual examples and subpatterns with particular properties, for example, brilliance, shading and size (Materka &Strzelecki(1998)). There exist diverse methodologies for composition demonstrating, for example, measurable models, auxiliary models what's more, models taking into account changes. Materka & Strzelecki (1998) recommended that change systems allow the variety of spatial determination to speak to compositions at diverse scales. In this sense, Wavelet Transform is a feasible strategy to get time-recurrence limitation investigation and to recognize stationary and non-stationary compositions.

2.1 Color And Edge Based Vehicle Detection

Luo-wei Tsai, Jun-wei hsieh, Kuo-button fan et al.[3] proposed a methodology for identifying vehicles utilizing shading and edges. Dissimilar to the vast majority of the customary routines that uses the movement components to recognize vehicles, this technique presents another shading change model to locate the conceivable vehicle hopefuls. As the same vehicle can have distinctive hues in diverse climate and enlightenment conditions, vehicle identification taking into account hues weren't supported much. This proposed system, paying little respect to the fluctuating lighting conditions, is fit for distinguishing the vehicle pixels from the foundation. To begin with, every one of the info's shades pixel are anticipated on a shading space. At that point utilizing Bayesian system, vehicle hues can be recognized from the foundation. A vehicle will have distinctive size furthermore diverse introduction so a few vehicle theories are created from every pixel. The theories can be checked utilizing three critical components which are corners, edges and coefficients of wavelet changes. This system can be utilized to vigorously identify vehicles if there should arise an occurrence of static foundations.

2.2 ALGORITHM

A. Read/capture image

The picture is initially obtained from a live feature food or a current picture can be stacked from the memory. We shall consider that the obtained picture is in RGB group which is a genuine nature position for a picture [4], [5]. In MATLAB, thecaptured or imported RGB picture is three dimensional and every pixel is spoken to by a component of a network whosesize compares to the picture's measure.

B. Converting RGB image to Black and White

This procedure is done in two stages. The RGB picture is initially changed over to a two dimensional grayscale picture. The grayscale picture is only a lattice that holds the luminance (Y) estimations of the picture [6]. We get the luminance estimations of the picture by consolidating the RGB qualities utilizing the NTSC standard comparison (1) that duplicates the essential hues (red, green and blue) with coefficients in light of the affectability of our eyes to these hues as appeared: $Y = 0.3 * R + 0.59 * G + 0.11 * B$ (1)
The luminance picture is then changed over to high contrast (twofold) picture by a procedure called thresholding [7], [8]. A limit is set and the luminance of every pixel is contrasted and this edge. All values that are more noteworthy than this edge are supplanted with an intelligent one (white) and the qualities underneath this edge are supplanted by a legitimate zero (dark). The edge can be ascertained either by deciding the luminance estimations of pixels that relate to question areas in an example picture (i.e., by machine preparing) and afterward averaging these qualities using so as to bring about the limit, or a calculation that assesses the picture's histogram and amplifies the difference of power in the middle of items and foundations.

C Recognize edge of objects

The picture is currently a two dimensional cluster with double components. Limits of the items are perceived by first setting a solitary pixel on the article foundation interface as a beginning stage and moving in a clockwise or counter clockwise course and scanning for other item pixels. The pixels may be looked either slantingly (in 8-associated pixels) or edge-nearby pixels (in 4-joined pixels). By chasing for article pixels in a settled heading, the object's limit can be perceived [9].

D Colour recognition

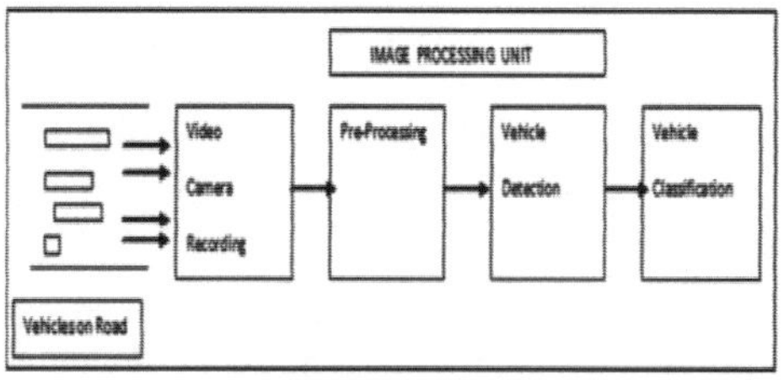

The items may be further assessed for their colour. This is finished by averaging the RGB content inside of the objects' limits. For instance an article with a generally high normal estimation of R over its surface, may have a shade of red. In MATLAB, the in pixel order can be utilized to gain the RGB data in a 1 X 3 grid over the whole protest surface which can be found the middle **Fig1**

E. Vehicle classification

Isolation of Objects by Blob Analysis and Calculation of Convex Hull: The vehicle arrangement begins with separating every item and reshaping it into a close polygon shape with the goal that it mirrors the genuine vehicle's measurements from an aeronautical camera. The Convex Hull of every item is figured and the articles are reshaped into close polygon shape and more organized than they were before this stride.

F: Post result of vehicle classification in MATLAB

The work has been executed utilizing MATLAB rendition 7.6.0 (R2008a). There are initially 12 vehicles: 2 little, 4 medium and 6 major. In the wake of trimming, the top little auto is disposed of as an after effect of touching the picture's limit close to the medium auto on the right half of the street. Consequently, the aggregate number of remaining vehicles is 9: 1 little, 2 medium and 5 major vehicles which are appropriately distinguished and demonstrated utilizing red, green and blue rectangular boxes around them for little, medium and enormous vehicles separately as shown in the below figure:

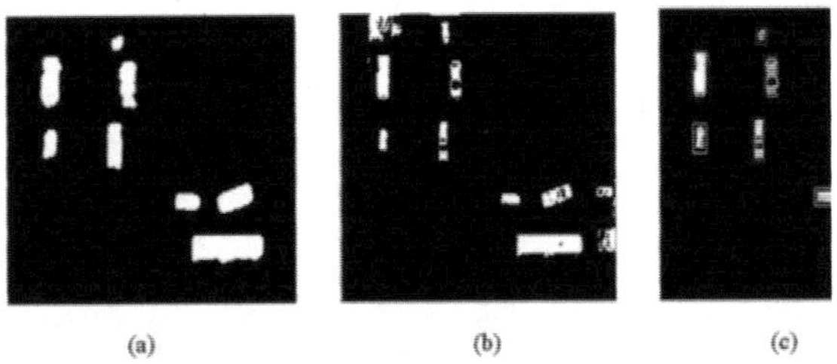

(a) (b) (c)

Fig2. : Output (a) Pre-processing; (b) Vehicle Detection Algorithm; (c) Vehicle Classification

G:PRE-PROCESSING TECHINIQUE:

1.Histogram equalization: is a technique for adjusting image intensities to enhance contrast. Let f be a given image represented as a mr by mc matrix of integer pixel intensities ranging. from 0 to L − 1. L is the number of possible intensity values, often 256[17].

Histogram evening out is a procedure for modifying picture intensities to upgrade contrast. Let f be a given picture spoke to as a mr by mc grid of whole number pixel intensities running from 0 to L − 1. L is the quantity of conceivable power values, regularly 256. Let p signify the standardized histogram of f with a container for every conceivable power.

value of. Utilizing the standards of added substance blending ,countless can henceforth be seen. fig1 shows Designing steps of the algorithm.

$$p_n = \frac{\text{number of pixels with intensity } n}{\text{total number of pixels}} \quad n = 0, 1, \ldots, L - 1.$$

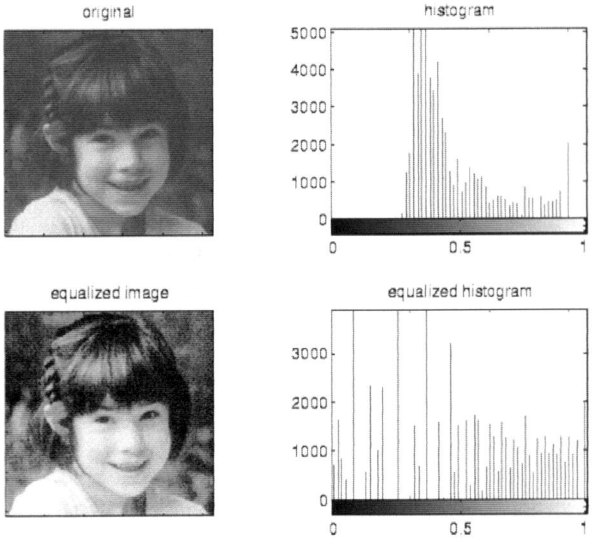

2. Digital filtering:

In picture preparing channels are for the most part used to stifle either the high frequencies in the picture, i.e. smoothing the picture, or the low frequencies, i.e. upgrading or identifying edges in the picture. A picture can be sifted either in the recurrence or in the spatial space. The main includes changing the picture into the recurrence space, increasing it with the recurrence channel capacity and re-changing the outcome into the spatial area. The channel capacity is formed to lessen a few frequencies and upgrade others. For instance, a straightforward lowpass capacity is 1 for frequencies littler than the cut-off recurrence and 0 for all others. The comparing procedure in the spatial space is to convolve the information picture f(i,j) with the channel capacity h(i,j). This can be composed as

$$g(i,j) = h(i,j) \odot f(i,j)$$

H: Input to the Pre-processing Phase:

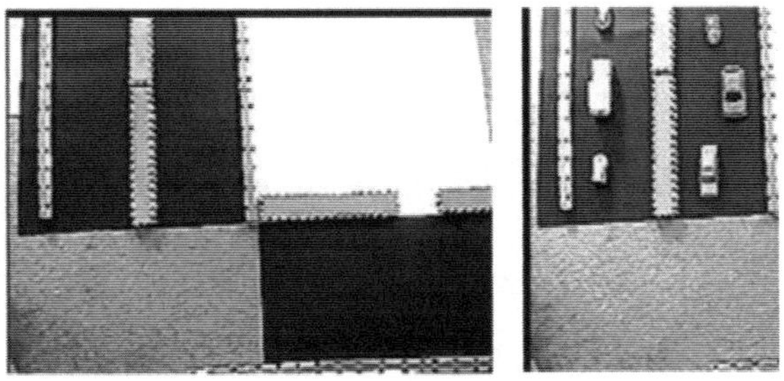

Fig: Reference and Current Image

I: Using Dual-Tree Complex Wavelet Transform (DTCWT)

As a structure, the Dual-Tree Complex Wavelet Transform (DTCWT) conquers a few challenges present in other Wavelet Transform techniques (Selesnick et al. (2005)) and offers the accompanying points of interest:

• Nearly move invariance. At the point when the change is connected, the outcomes appear to be somewhat influenced by interpretations in the first info

• Directional selectivity. A subsequent arrangement of 6 diverse positive and negative bearings: 15◦,45◦ , 75◦ , - 15◦ , - 45◦ and - 75◦.

• Moderate repetition. The repetition presented by Complex Wavelets is 2d:1 for d Dimensions (Kingsbury (1999))

K: Application of vehicle detection

(a) Automated Car Wash Detection: A computerized auto wash framework needs to stay in a low-control utilization standby mode until a vehicle is recognized entering the auto wash.

Description: Previous innovation obliged the utilization of an inductive circle framework inserted into the ground. Introducing an inductive circle framework obliges cutting into an extensive segment of black-top or cement. After some time these inductive circles break and must be replaced. Using a Sure Cross M-GAGE remote sensor hub disposes of the requirement for a huge inductive circle. The gadget may be mounted over the ground or subterranean, obliging just a six to eight inch width opening. Vehicle vicinity is distinguished by a M-GAGE Node that initiates the computerized auto wash framework to power up and triggers the section entryway.

(b)Indicating Traffic Levels in Vehicle Queues: Toll stalls on high-activity streets can rapidly go down amid top go times, bringing on cerebral pains for drivers and toll corner representatives. It is useful for specialists to know to what extent autos are holding up in every path so they can evaluate activity levels and distinguish moderate paths that may oblige additional consideration.

Description: Either M-Gage Vehicle Detection Sensors or R-Gage Radar Sensors can be utilized to screen an assigned checkpoint of every path and focus the length of time of time that every vehicle is ceased by then.

L: Post Processing and Enhancement

In post processing stage we enhance the detection and per-forms the labelling of connected components to get vehicle objects. Size and aspect ratio constraints are applied again after morphological operations in the post processing to eliminate unwanted objects i.e., no vehicle objects. nonvehicle objects.

Here we utilize haar wavelet based future extraction [11] to upgrade the location on vehicles characterized by SVM. In all cases, characterization is performed utilizing GMM. Wavelets are basically a kind of multi determination capacity rough guess that take into account the progressive deterioration of a sign or picture. They have been connected effectively to different issues including article location [12] [13], face acknowledgment [14] and picture recovery [15]. Diverse reasons make the components separated utilizing Haar wavelets appealing for vehicle discovery. Some of them are they frame a conservative representation, they encode edge data which is an imperative element for vehicle recognition, they catch data from various determination levels furthermore there exist quick calculations for registering these elements [13] [12]. A few reasons make these elements appealing for vehicle identification. To begin with, they xxform a reduced representation. Second, they encode edge data, an imperative element for vehicle location. Third, they catch data from various determination levels. At last, there exist quick calculations, particularly on account of Haar wavelets, for processing these components.

1: Flowchart showing the various step of processing of algorithm

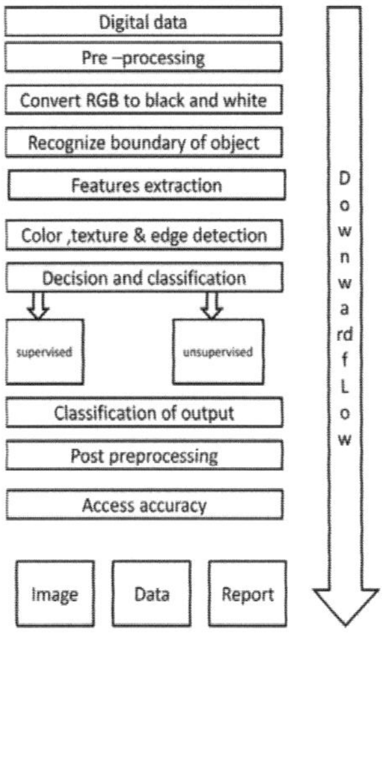

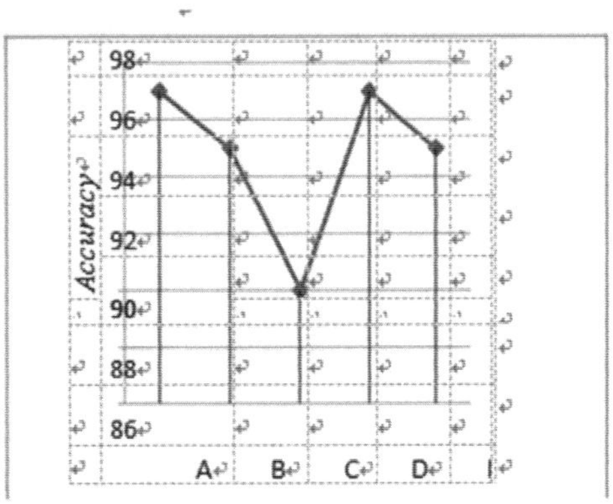

Methods ◆ Accuracy

Figure shown above diagrammatically representation of detection accuracy

Since method A and B is best among the all detection procedure.

Table -1

Name of vehicle detection technique used	ACCURACY RATE	False alarm
Color and texture based vehicle detection (A)	97%	Low
Color and edge based vehicle detection (B)	95%	Low
Block matching algorithm (C)	90%	High
Improved adative Gaussian mixture model (D)	94%	High
Combining motion and detection and background substraction (E)	94%	High

Conclusion

This paper has given techniques on the various methods that available for vehicle detection in the day time with the image sequence obtained from a stationary CCTV camera at different views. Table 1 gives the accuracy in detection accuracy produced by applying each of the five methods. We have developed an algorithmic approach to vehicle detection. This not only reduces the complexity of the system but enhances its use in the areas which are too difficult to be detected by normal means. This algorithm can be applied on. The calculation talked about in this paper is a basic, yet viable system for examining the shapes and shades of articles. The idea of perceiving shapes on the premise of effectively possible. In our examination, the calculation was utilized to perceive the states of all cells and the Extent parameter of typical cancer cells was noted. As malignancy cells contrast from solid cells fit as a fiddle and additionally shading, their Extent values and RGB substance don't coordinate those of cancer cells and can in this way be effortlessly distinguished and evaluated. The applications are not confined to the bio-restorative field but rather can reach out to any field that obliges order of distinctive articles in view of their physical appearance. This may incorporate quality assessment in sequential construction systems, counterfeit automated insight, PC vision, acknowledgment of vehicles at toll corners or movement signals for activity status determination, and so on.

References

1. Hsu-Yung Cheng, Chih-Chia Weng, and Yi-Ying Chen, —Vehicle Detection in Aerial Surveillance Using Dynamic Bayesian Networks‖, IEEE Trans. Image Process., vol. 21, no. 4, pp. 2152–2159, Apr. 2012.
2. Barilla-Perez, M. E. (2008). Colour-based Texture Image Segmentation,PhD thesis, University of Birmingham.
3. Koller D, Weber J. Huang T. Malik J. Ogasawara G. Rao B. Russell S.Towards robust automatic traffic scene analysis in real-time. inProceedings of the 33rd IEEE Conference on Decision and Control (Cat. No.94CH3460-3). IEEE. Part vol.4, 1994. 1994.
4. G. Wyszecki and W. S. Styles, "Color Science: Concepts and Methods, Quantitative Data and Formulae" (2nd edition New York: Wiley,1982).
5. R. S. Berns, "Principles of Color Technology" (3rd edition New York: Wiley, 2000).
6. J. L. Vincent, "Morphological Grayscale Reconstruction in Image Analysis: Applications and Efficient Algorithms", IEEE Transactions on Image Processing, vol. 2, pp. 176–201, 1993.
7. M. Sezgin and B. Sankur, "Survey over Image Thresholding Techniques and Quantitative Performance Evaluation", Journal of Electronic Imaging, vol. 13, no. 1, pp. 146–168, Jan. 2004.
8. N. R. Pal and D. Bhandari, "On Object Background Classification", International Journal Syst. Science, vol. 23, no. 11, pp. 1903–1920, Nov.1992.
9. F. Meyer, "Color image segmentation", Proceedings of 4th International Conference on Image Processing, pp. 523–548, 1992.
10. link.springer.com/chapter/10.1007%2F978-3-642-22456-0_4
11. Shalinee Patel, Pinal Trivedi, and Vrundali Gandhi, "2D Basic Shape Detection Using Region Properties", International Journal of Engineering Research & Technology, vol. 2, no. 5, pp. 1147-1153, May 2013.
12. Zehang Sun, George Bebis and Ronald Miller, —Quantized Wavelet Features and Support Vector Machines for On-Road Vehicle Detection,‖ Computer Vision Laboratory, Department of Computer Science, University of Nevada.

[a] Corresponding author: Ravi Gaurav, gaurav.ravi9@gmail.com

DOI: 10.1051/matecconf/20166102007

13. C. Papageorgiou and T. Poggio, —A trainable system for object detection," International Journal of Computer Vision, vol. 38, no. 1, pp. 15-33, 2000.
14. H. Schneiderman, A statistical approach to 3D object detection applied to faces and cars. CMU-RI-TR-00-06, 2000.
15. G. Garcia, G. Zikos, and G. Tziritas, —Wavelet packet analysis for face recognition," Image and Vision Computing, vol. 18, pp. 289-297, 2000.
16. C. Jacobs, A. Finkelstein and D. Salesin, —Fast multiresolution image querying," Proceedings of SIGGRAPH, pp. 277-286, 1995.
17. R. C. Gonzalez and R. E.Woods, Digital Image Processing, Third Edition, 2008.

A Narrative Methodology to Recognize Iris Patterns By Extracting Features Using Gabor Filters and Wavelets

Shristi Jha[1,a], Sindhu B.M.[1,b] and S.Venkatesan[2,c]

[1] Pre-final Year, Department of Computer Science & Engineering, Dayananda Sagar College of Engineering, Bangalore, India.
[2] Professor, Department of Computer Science & Engineering, Dayananda Sagar College of Engineering, Bangalore, India.
[a]shristij97@gmail.com, [b]sindhub1995@gmail.com, [c]selvamvenkatesan@gmail.com

Abstract: Iris pattern Recognition is an automated method of biometric identification that uses mathematical pattern-Recognition techniques on images of one or both of the irises of an individual's eyes, whose complex random patterns are unique, stable, and can be seen from some distance. Iris recognition uses video camera technology with subtle near infrared illumination to acquire images of the detail-rich, intricate structures of the iris which are visible externally. In this narrative research paper the input image is captured and the success of the iris recognition depends on the quality of the image so the captured image is subjected to the preliminary image preprocessing techniques like localization, segmentation, normalization and noise detection followed by texture and edge feature extraction by using Gabor filters and wavelets then the processed image is matched with templates stored in the database to detect the Iris Patterns.

Keywords: Gabor filters, Near Infrared, Gabor wavelet transform, Hough Transform, Iris patterns, Bit patterns, Hamming Distance

1 Introduction

Iris patterns are unique to every individual on earth" this concept was first proposed by Frank Burch in 1939. John Daugman working at university of Cambridge, England, developed and patented the first algorithms to perform iris recognition in 1991. These algorithms became widely licensed through a series of companies: IriScan (a start-up founded by Flom, Safir, and Daugman), Iridian, Sarnoff, Sensar, LG-Iris, Panasonic, Oki, BI2, IrisGuard, Unisys, Sagem, Enschede, Securimetrics and L-1, now owned by French company Morpho.Just as every human being has different fingerprints, so does the minute architecture of the iris exhibit variations in every subject examined. Its features represent a series of variable factors whose conceivable permutations and combinations are almost infinite. Iris recognition is an automated method of biometric identification that uses mathematical pattern-recognition techniques on video images of one or both of the irises of an individual's eyes, whose complex random patterns are unique, stable, and can be seen from some distance. Iris recognition uses video camera technology with subtle near infrared illumination to acquire images of the detail-rich, intricate structures of the iris which are visible externally. A key advantage of iris recognition,

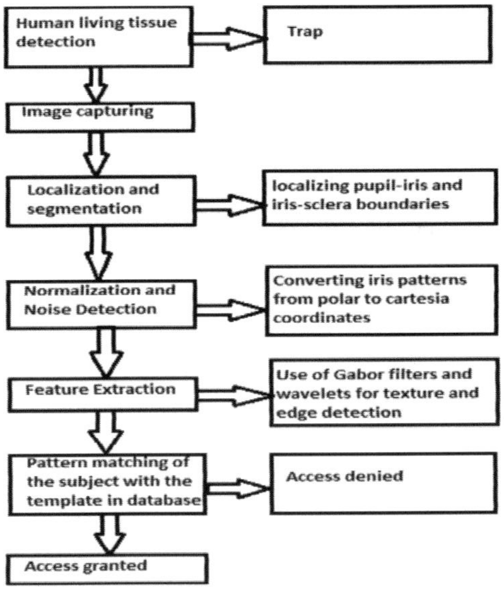

Figure 1. Work flow diagram

besides its speed of matching and its extreme resistance to false matches is the stability of the iris as an internal and protected, yet externally visible organ of the eye. The extraction of iris pattern has following steps: Image capturing, image segmentation, normalization, feature extraction and matching. Daugman [4] used multi scale quadrature wavelets to extract texture phase structure information of the iris to generate a 2048 bit iris code and compared the difference between iris representations by computing their hamming distance via the XOR operator.

Digital templates encoded from these patterns by mathematical and statistical algorithms allow the identification of an individual. Databases of enrolled templates are searched by matcher engines at speeds measured in the millions of templates per second per (single-core) CPU, and with remarkably low false match rates.

2. Image Capturing

Capturing an image of an iris is the initial stage of an iris based recognition system. Success rate of other recognition stages depends on quality of images taken. Images available in CASIA database do not have reflections in pupil and iris areas because NIR (700-900nm) is used for imaging. The majority of persons worldwide have "dark brown eyes", the dominant phenotype of the human population, revealing less visible texture in the Visible Wave length band but appearing richly structured, in the Near Infrared band(NIR). Using the NIR spectrum also enables the blocking of corneal specular reflections from a bright ambient environment, by allowing only those NIR wavelengths from the narrow-band illuminator back into the iris camera.

Figure 2.Image capturing

2. Localization and Segmentation :

Pupil is darkly pigmented compared to iris. Hence the use of canny edge Recognition results in determining points in iris pupil boundary. After determining the edge points by the use of circular Hogue transform the Centre and radius of iris circles are obtained. One of the important challenges in segmentation is detecting the boundary of iris and sclera. Illumination intensity distinction between iris and sclera is very low at the border; there are other edge points in the eye where illumination intensity is much more than that of the boundary of iris and sclera. As a result, edge Recognition algorithms which are able to detect outer iris edges identify those points as edge. Therefore, in order to detect iris outer boundary, these points have to be identified and eliminated. In this paper, available boundaries are initially enhanced and then extra edge points are identified and eliminated. At the end, through circular Hough transform, outer iris boundary is obtained. In order to enhance iris outer boundary edges, canny edge Recognition is performed on eye image in preprocessing stage.

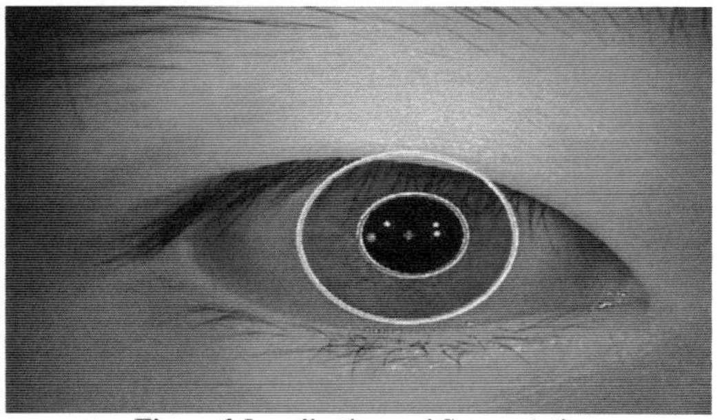

Figure 3.Localization and Segmentation

By performing such edge Recognition, a matrix is obtained with the same dimensions as of the image itself which its elements are high in areas where there is a definite boundary and the elements are low in areas where there is no perfectly definite boundary, such as iris outer boundary. Boundary between iris and eyelids can be differentiated as two lines. To localize them we use linear Hough transform[4]. Initially eyelid boundary should be detected by using canny edge Recognition. This method localizes eyelids with high precision by multiplying of 2.76 in matrix of pixel values of iris image and intensifying light in eye image, the edges are enhanced. The main disadvantage of this method is sclera boundary not being circular, is the result of angled imaging, and in these cases some information is lost.

3. Normalization and Noise Detection

Normalization is based on transforming iris areas from polar to Cartesian coordinates. This involves converting circular area into rectangle of equal area. Iris area is obtained as a strip with regard to iris boundary and pupillary center. 128 pupils centered circles are taken from iris pupil boundary and the pixels located on this circles are mapped into a rectangle. Pupil contracts or dilates with surrounding illumination level, so iris is not of the same size always. Hence choosing these 128 perfect circles normalize iris in terms of a rectangles of fixed sizes.

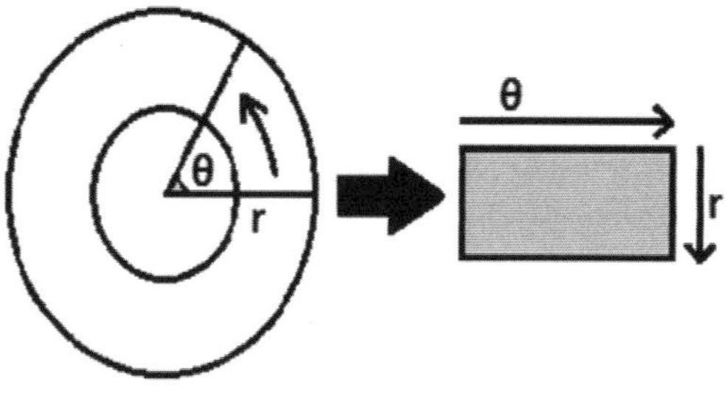

Figure 4.Normalization

The set of pixels containing only the iris, normalized by a rubber-sheet model[3] to compensate for pupil dilation or constriction, is then analyzed to extract a bit pattern encoding the information needed to compare two iris images. An iris code is a 512 byte record. Any noises such as salt and pepper or cases where dark areas appear too dark are processed to get better image quality. Amplitude information is discarded so that the template is unaffected by light intensity.

Table 1: Accuracy rate of iris boundary localization for three databases [1]

Database	Accuracy rate in pupil boundary localizatio n (%)	Accuracy rate in sclera boundary localization (%)	Accuracy rate in eyelids boundary localization (%)
CASIA Iris Image (ver. 1.0)	98.13	99	98
CASIA Iris Interval (ver.3.0)	99.73	99.19	98.63

4. Feature extraction:

Gabor filters are used to extract features. In image processing, a Gabor filter, named after Dennis Gabor, is a linear filter used for edge Recognition [11]. Gabor filters can serve as excellent band-pass filters for one-dimensional signals Frequency and orientation representations of Gabor filters are similar to those of the human visual system, and they have been found to be particularly appropriate for texture representation and discrimination. In the spatial domain, a 2D Gabor filter is a Gaussian kernel function modulated by a sinusoidal plane wave. Simple cells in the visual cortex of mammalian brains can be modeled by Gabor functions. Thus, image analysis with Gabor filters is thought to be similar to perception in the human visual system. Gabor wavelets are wavelets invented by Dennis Gabor using complex functions constructed to serve as a basis for Fourier transforms in information theory applications. Here is the formula of a complex Gabor function in space domain:

$$g(x, y) = s(x, y)\, w_r(x, y) \qquad (1)$$

Where s(x, y) is a complex sinusoid, known as the carrier, and $w_r(x, y)$ is a 2-D Gaussian-shaped function, known as the envelope.

They are also closely related to Gabor filters; the important property of the wavelet is that it minimizes the product of its standard deviations in the time and frequency domain. Put another way, the uncertainty in information carried by this wavelet is minimized. Gabor filters are directly related to Gabor wavelets, since they can be designed for a number of dilations and rotations. Usually, a filter bank consisting of Gabor filters with various scales and rotations is created. The filters are convolved with the signal, resulting in a so-called Gabor space. The Gabor space is very useful in image processing applications such as optical character recognition, iris recognition and recognition. The result of Gabor wavelet transform is a set of complex numbers that carry local amplitude and phase information about the iris pattern.

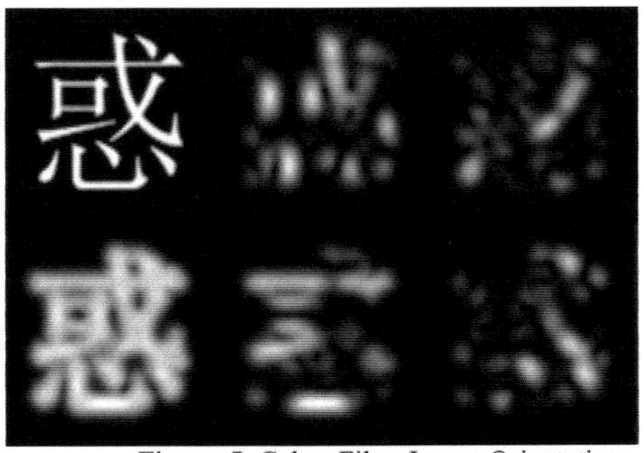

Figure 5. Gabor Filter Image Orientation

Figure 5 above shows the demonstration of Gabor filter: i.e, shown at 4 orientations at 0°, 45°, 90° and 135. The original image and superposition of all 4 images is shown on left.

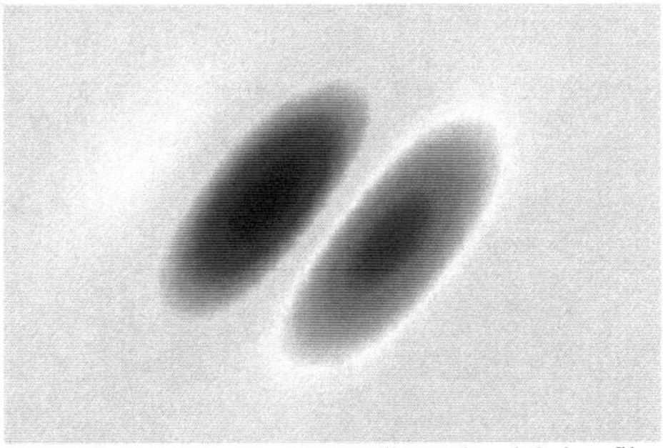

Figure 6. Image of Two dimensional Gabor filters

5. Matching

Most amplitude information is discarded, and the 2048 bits representing an iris pattern consist of phase information. Discarding the amplitude information ensures that the template remains largely unaffected by

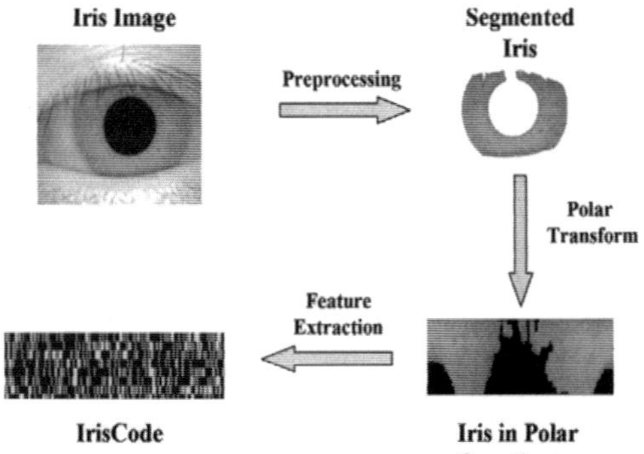

Figure 7.steps in iris recognition process

the changes in light illumination or camera gain and contributes to the long-term usability of the biometric template. For identification (one-to-many template matching) or verification (one-to-one template matching) a template created by imaging an iris is compared to stored template(s) in a database. If the Hamming distance is below the decision threshold, a positive identification has effectively been made (If a Hamming distance threshold of 0.26 is used, meaning that up to 26% of the bits in two Iris Codes are allowed to disagree due to imaging noise, reflections, etc., while still declaring them to be a match). Multi scale quadrature wavelets are used to extract texture, phase structure information of the iris to generate a 2048 bit iris code and compared with the template housed in database. The difference between iris representations is done by computing their hamming distance using the XOR operator. Table below gives truth table for the XOR. Considering the nth bit of r and s, we can know if they are similar or not:

Table 2: XOR truth table

r[n] s[n]	d(r[n] , s[n]) – distance between digits	Similarity check
0 0	0	similar
0 1	1	dissimilar
1 0	1	dissimilar
1 1	0	similar

The distance is calculated as the number of bits of value 1 in the XOR result of the given two binary strings. This is the definition of the Hamming distance. For instance, if we consider two 8-bits vectors R and S. The distance between them will be dI = I (R ⊕ S) where ⊕ is the XOR operator and 'I' is the function to calculate number of bits whose value is 1 in the binary XOR result. The mathematical algorithms allow the unique identification of a person by encoding digital templates from the iris patterns.

6. Future Enhancement

Iris scanners can be tricked using high resolution images of iris in place of a real eye. Iris scanning is susceptible to difference in light intensity and illumination. The remedy for this could be using spectral analysis instead of merely monochromatic cameras to distinguish iris tissue from other material; observing the characteristic natural movement of an eyeball testing for retinal retro reflection (red-eye effect) or for reflections from the eye's four optical surfaces (front and back of both cornea and lens) to verify their presence, position and shape. Some defense against the use of fake eyes and irises include changing lighting during the identification such that the pupillary reflex can be verified and the iris image be recorded at different pupil diameters.

A 2004 report from a German federal office stated that no live tissue Recognition technology was used for iris scanning, hence the use of sensors to detect live tissues or any of the above suggested methods could be used. The modern cataract surgery shows that it can change iris texture in such a way that iris pattern recognition is no longer feasible or the chance of falsely rejected cases is increased. Hence people with such surgeries must enroll a new request for iris scanning to government so that their database can be changed as early as possible. Consumption of alcohol may to lead to large dilation or contraction of the pupil, hence it would be advisable for the subject to restrain from alcohol at the time of verification.

The use of contact lenses is a major issue:

The verification rate with images from sensor and from cross-sensor is high and accurate in case of no contact lenses[10]. In case of soft lenses when it comes to sensor 47% of performance degradation is noted. In the case of matching images with no contact lens and images with textured contact lens, the verification rate is below 40% in all cases. The same situation holds for matching images with a soft lens against images with a textured lens. Hence it would be advisable to use a live tissue scanner to inform the subject to take the test without lenses.

Table 3: analyzing pattern recognition under lenses

Lens type	Sensor	Cross-sensor
No lens	Accurate results	Accurate results
Soft lenses	Accurate results	47% performance degradation
Textured lens	Not very accurate	Accuracy below 40%

7. Applications

Some of these include:

Millions of persons around the world have been enrolled in iris recognition systems for convenience purposes such as passport-free automated border-crossings. Aadhaar ID program deployment began in 2011 in India, wherein the government is enrolling the iris patterns of all citizens for the Aadhaar scheme for entitlements distribution, run by the Unique Identification Authority of India (UIDAI).SRI International Sarnoff has been developing an "Iris on the Move" system for U.S. Government clients, capable of identifying 30 people per minute. Most recently, they have specialized in a product where drivers can be identified without needing to leave their vehicle. Since 2011, Google uses iris scanners to control access to their datacenters. Today, all of the UAE's land, air and sea ports of entry are equipped with iris recognition systems.

8. Conclusion

Iris Pattern Recognition has been done using the Gabor filters after the captured image is subjected to the image preprocessing techniques. After the result analysis it is found that through this novel methodology of extracting the features with respect to textures Recognition and discrimination with multiple orientation of the Gabor filter has given the best results in terms of Recognition rates of iris patterns. For identification of one to many template matching and for verification one to one template matching is preferred.

References

1. J.canny:a computational approach to edge Recognition IEEE transaction and pattern analysis and machine intelligence Vol 8(1986)[8] L. Flom and A. safir iris recognition system. U.S. patent no, 4641394 (1987)
2. R . P. Wildes, "Iris Recognition: An Emerging Biometric Technology", Proceeding of the IEEE, vol. 85, (1999), pp 1348-1363
3. J. Daugman, "How iris recognition works" ,IEEE Transactions on Circuits and Systems for Video Technology, vol. 14, pp. 21–30, (2002).
4. J. Daugman " How iris recognition works". Proceedings of 2002 International Conference on Image Processing, Vol. 1, (2002).
5. Chinese Academy of Sciences – Institute of Automation.Databas of 756 Greyscale Eye Images. http://www.sinobiometrics.com Version 1.0,(2003)
6. J. Daugman, "The importance of being random: statistical principles of iris recognition'', Pattern Recognition 36 (2003) The Journal of the Pattern Recognition Society. Published by Elsevier Science Ltd.pp279 –291
7. J .Huang ,H. Wechsler, visual rputines for eye loalisation using learning and evolution, IEEE Trrans .Evol.Comput.(4)(1)(2000)73-82
8. K. W. Bowyer, K. Hollingsworth and P. J. Flynn, "Image understanding for iris biometrics: A survey", Elsevier computer vision and Image Understanding, vol.110, (2008), pp. 281-307.
9. Z. Wei, X. Qiu, Z. Sun, and T. Tan, "Counterfeit iris Recognition based on texture analysis," in Proceedings of the 18th International Conference on Pattern Recognition, (2008). [10] Segmentation Techniques for Iris Recognition System, International Journal of Scientific & Engineering Res V vol 2, Issue 4, April-2011 , ISSN 2229-5518 IJSER © @ 2011 BY Surjeet Singh, Kulbir Singh.
10. Mahmoud Mahlouji and Ali Noruzi IJCSI International Journal of Computer Science Issues, Vol. 9, Issue 1, No 3, January[(2012)
11. PrateekVerma , MaheedharDubey , Praveen Verma, SomakBasu, International Journal of Emerging Technology and Advanced Engineering Website: www.ijetae.com (ISSN 2250-2459, Volume 2, Issue 6, June (2012)
12. SajidaKalsoom, Sheikh Ziauddin, COMSATS Institute of Information Technology Islamabad, Pakistan, The Fourth International Conferences on Pervasive Patterns and Applications, (2013).
13. DakshaYadav, Student Member, IEEE, NamanKohli, Student Member, IEEE, James S. Doyle, Jr., Student Member, IEEE, Richa Singh, Member, IEEE, MayankVatsa, Member, IEEE, Kevin W. Bowyer, Fellow, IEEE. SUBMITTED TO IEEE TRANSACTIONS ON IFS, OCTOBER (2013).
14. M. Vatsa, R. Singh and A. Noore, "Improving Iris recognition performance using segmentation, QualityEnhancement", Match Score Fusion, and Indexing, IEEE Transactions on Systems, Man, and Cybernetics B,Cybern., vol. 38, no. 4, pp

Research on the Forecast Model of Total Viable Count on Bacon Based on Hyper spectral Imaging Technique

Junhua Zhao, Peiyuan Guo, Pan Xu, Shuo Liu and Kuncheng Yang

School of Computer and Information Engineering, Beijing Technology and Business University, Beijing 100048, China

Abstract: The total viable count (TVC) in bacon overweight can cause serious damage to human health .In order to find a rapid and nondestructive method of TVC, hyper spectral imaging technique was applied to quantitatively analysis of TVC on bacon .Comprehensively comparing the pretreatment method of multiple scattering, derivative method and so on ,finally the multiple scattering for pretreatment was used. And the interval optimization method of least squares model was set up to predict, and get a good prediction results .The correlation coefficient of the calibration and predictions respectively was 0.808 and 0.808 , interactive authentication root mean square error was 0.115 and 0.198 respectively. Therefore, hyper spectral imaging technique combining iPLS can be used for the rapid detection of TVC on bacon.

Key words: bacon; total viable count; hyper spectral imaging; interval optimization least-squares

Introduction

Chinese bacon is an important heritage of world food culture, however, research shows: due to the environmental health conditions of the slaughter process is not up to standard ,bacon will produce a large number of microbial residues which is not conductive to long-term preservation of bacon ,and harmful metabolites will be increased accordingly that will result in a decline in quality of bacon . So there is an urgent need for a new real-time ,fast ,accurate detection of bacon quality.

The total viable count (TVC) is a very important microbiological indexes to evaluate the quality of the bacon health .Most of the traditional detection methods are based on the method of microscope ,the method of plate counting , etc.[5].But most of them in detection method are low efficient ,have long time-consuming cycle ,and have the disadvantage of breaking meat and so on .

As a new detection technology, hyper spectral imaging technology [6] set of optoelectronics, computer science and other fields of cutting-edge technology in a body and conventional spectroscopy and imaging techniques also be combined together. The superiority of hyper spectral imaging technology [7] is reflected in the following aspects: Have a high spectral resolution, the resolution can be got NM level, and number of bands up to hundreds, so multiple narrow spectrum obtained can be approximated as a continuous; Atlas unity--hyper spectral image contains both spectral information inside the substances, but also image information outside the substances make it more accurate.

This experiment takes bacon as the research object and detects the total number of bacteria on the surface of the bacon. In 900nm-1700nm wavelength range of hyper spectral imaging system, acquire image and spectral information of the sample, determine the best modeling characteristic wavelengths, and then establish the prediction model based on interval partial lcast squarcs to analyzc and compare .Through the above steps ,we can achieve quality and safety detection about bacteria on the surface of bacon.

Project supported by the National Natural Science Foundation of China (61473009), Beijing Natural Science Foundation of China (4122020), Project funding for the promotion of graduate research capacity in 2015
Biography: Peiyuan Guo (1958-), Beijing Technology and Business University, Professor, doctor, Email: ggppyy@126.com
Junhua Zhao (1989-), Beijing Technology and Business University, master,Email: 990558865@qq.com

DOI: 10.1051/matecconf/20166102009

1 Material and Methods

1.1 Materials

The bacon in Leshan, Sichuan was taken as the experimental object. The collected sample is divided into 4 cm × 4 cm × 2 cm (length × width × thickness) ,and 40 samples were collected. In order to accelerate the process of bacon corruption , expose it to the 35 ℃ storage space of moisture ,air turbidity to be tested .Two meat samples taken randomly every 5d were tested during the experiment--one piece as the calibration set samples, another piece as the prediction set sample until all the samples out.

1.2 Instrument

Hyper spectral imaging data were acquired by GaiaSorter hyper spectral sorter systems from Beijing Zhuo Li Han Kuang Instrument Co., Ltd as shown in Figure 1.

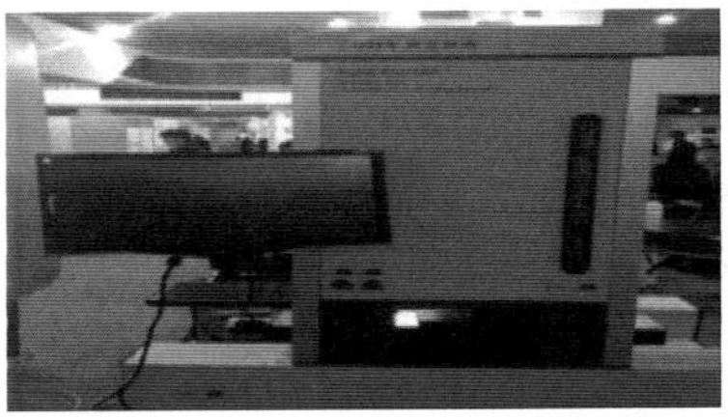

Figure 1. Spectrum analyzer system of sisuCHEMA.

1.3 The acquisition of spectral values and detection of total bacterial count

1.3.1 The acquisition of hyper spectral image

In order to overcome the influence of the image noise and dark current in the wavelengths of weak light intensity distribution, first you need to scan standard white to acquire all white calibration image Dw with the reflectivity of 1, then cover the lid of the camera to acquire all black calibration image Dd with the reflectivity of 0. After over a black and white correction, then the spectral image of sample Ds can be acquired. It can be based on Formula 1 to get reflectivity R of the interested pixel region C or band I of the relative spectral image of sample after a black and white corrected.

$$R(ci) = \frac{D_s(ci) - D_d(ci)}{D_w(ci) - D_d(ci)} \qquad (1)$$

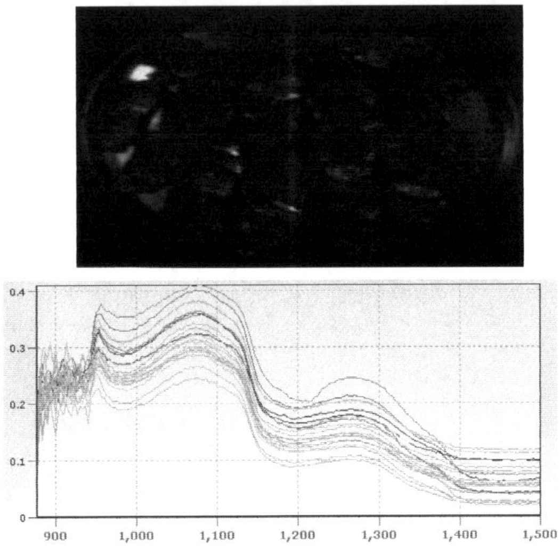

Figure 2. The original spectral on bacon.

The data analysis of hyper spectral image used ENVI software platform. To the hyper spectral images of sample corrected, select a region of interest on the surface and calculate the average reflectance spectroscopy .According to this method sequentially obtain a full-band spectral reflectance curves of the original of sample which are shown in Figure 2.

1.3.2 Determination microbes on the surface of the sample

The microbial detection experiments were carried out immediately after obtaining the spectral images of all the samples. Total number of colonies: In accordance with the GB/10278-1997" the total number of colonies of Chinese bacon" to determinate. Using the method of plate counting, select 2 – 3 sample dilutions of suitable dilution ,and take 1mL empty diluent and then place it in a sterile petri dish. Cool the 15mL medium of plate counting , flip it until the agar solidified ,and then cultivate it for 48h in the condition of 36℃. The total numbers of colonies gathered are shown in Figure 3, and the data was used as the standard reference data.

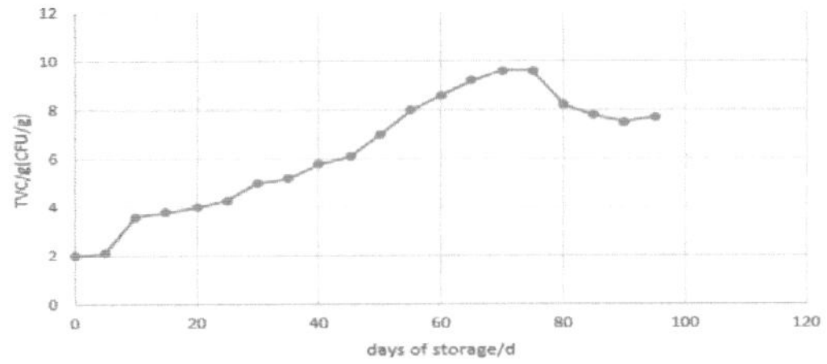

Figure 3. Bacon bacterial colonies.

2 Results and Analysis

2.1 Spectral data preprocessing

In this study, the original spectra were pretreated by using differential pretreatment, vector normalization, multiple scatter correction and the combination of the methods, etc. Finally make a model of the bacon, and use correlation coefficient R and root mean square error of cross- validation (RMSEVC) as 2 evaluation index of the model. While R-value is closer to 1, and RMSEVC value is smaller, the performance of the model is better. It can be drawn that within 900-1700nm wavelength range, the best spectral preprocessing method for modeling is multivariate scatter correction (MSC). This is mainly because MSC not only can effectively eliminate the scattering influence produced by the sample particles spotty and uneven distribution, but also can correct the error of optical path length of hyper spectral scanning sample. The model adopting MSC and standardized preprocessing method is optimal in which R is 0.934, the RMSECV is 0.186.

Using the above preprocessing method to preprocess hyper spectral data of sample of calibration set of the bacon collected, the spectrum are shown in Figure 4.

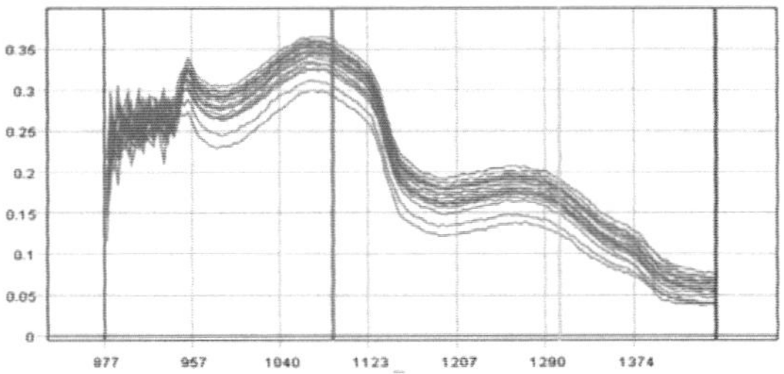

Figure 4. The pretreatment of spectrum.

2.2 Establishment of forecasting model of partial least squares for interval optimization(iPLS)

The method of partial least square is to eliminate redundant information and noise produced during the process of modeling, so decompose and change not only the matrix X of high spectral, but also the matrix Y of bacterial content of samples of the bacon to be detected. So at the same time, the effect of matrix Y on X is considered. The method of PLS makes the direct relationship between main components of *X* and *Y*. In addition, it can reduce the space of data dimension letting the high-dimensional data into low dimensional, so as to solve the nonlinear problems.

Due to the high spectrum carries the information of the whole structure of the material , in the process of analyzing and modeling of the spectral data , the phenomenon of the loss of target information in the spectra was avoided by using full spectrum data to establish the model. But when the band range of the spectrum is very large, it is easy to cause the information redundancy, the model complexity, and the reduction of accuracy of the model if still make model using the data of full spectrum. In the study, in order to avoid the above problems, the forecasting model is established by using the method of regression analysis of the interval partial least squares.

Use the method of regression analysis of interval partial least squares to establish prediction model .Full spectrum is divided equally into 10,11,12,13...20 interval ,and because the value of n does not have the corresponding theoretical formula ,the classification is on the basis of experience. PLS regression analysis was carried out in every sub interval divided, and the accuracy of model was evaluated by RMSECV. The smaller the value of the RMSECV, the better prediction ability of the model. In the case the high spectrum is divided into different numbers of sub regions. When the spectrum is divided equally into 12 regions, the number of the factor of principal components is determined for 10, and the analysis model of the total number of bacteria in bacon is optimal. At this point, the best range of local model is 8. The scatter plot of the actual and predicted value of the TVC on bacon surface got by using the hyper spectra processed as the input of partial least squares model are shown in Figure 5. The correlation coefficient r was 0.808 and RMSECV was 0.115.

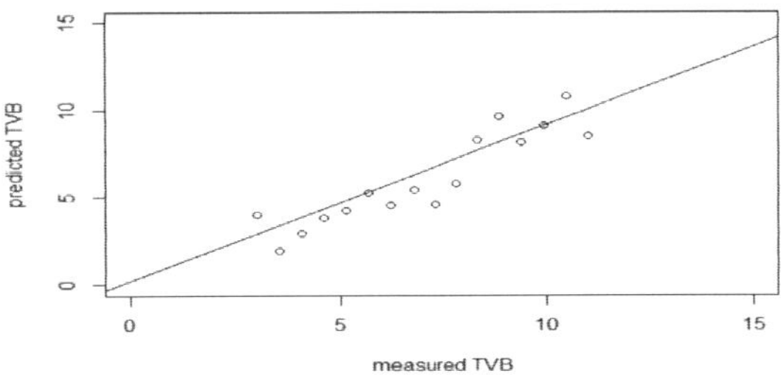

Figure 5. The scatter plot of the actual and predicted value of the TVC on bacon surface.

Enter the sample of bacon of validation set into the model of analysis prediction of the total number of bacteria in the bacon created by the model of regression analysis of interval partial least squares, compare the result of outputs of model and the true value obtained through the national standard method of measurement , and the results are shown in Figure 6. Blue represents the real value, and red represents the predicted value.

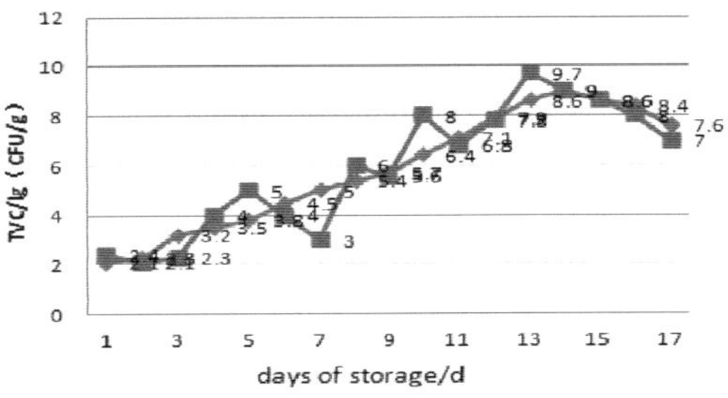

Figure 6. The comparison of the actual and predicted value of the TVC on bacon surface.

From the predicted results of the model, we can learn that correlation coefficient of test results by the detection model of content of the total number of bacteria on the surface of bacon established by modeling method of interval partial least squares and actual results is 0.798, and the error of root mean square of cross validation is 0.198.

3 Conclusion

In order to avoid the long time and the destruction of the sample of the traditional detection of the bacteria, the experiment adopted nondestructive testing by hyper spectra to the content of bacterial on the surface of bacon, and achieved the expected effect. In this paper, create a predictive model by using the hyper spectral technology combined with MSC + standardized calibration and interval partial least squares. But later need to increase the number of the samples of prediction set and calibration set , and hope to get better results.

References

1. H.H. Wang, D.Y. Liu, et al, Journal of meat research .*The situation and development trend of bacon industry present in China.* J, **27 (9)** : 36 to 40(2013)
2. W. Z. Sun, C. Cui, et al, Journal of food science and technology. *The research on influential factors of acid on Cantonese sausage storage* . J.(2007)
3. L. G. Zhou, L. Xiao, et al, Journal of meat research , *Near infrared spectroscopy rapidly detect bacon acid value and peroxide value.* J, **26 (3)** : 30-33(2012)
4. L. L. Zhao, L. D. Zhang, et al, Spectroscopy and spectral analysis. *The research on bacon quality by near infrared spectrum* . J, **27 (1)** : 46-49(2007)
5. N. N. Li, G.H. Zhou, et al, Journal of food industry science and technology. *The residues of pyrene in bacon by chromatography-fluorescence* . J, **34 (1)** : 319-323(2013)
6. L. L. Tao, X.J. Yang, et al, Spectroscopy and spectral analysis. *The research progress on the chemical composition by near infrared spectrum in poultry.* J, **33 (11)** : 3002-3009(2013)
7. J. H. Wu, Beijing: China agricultural university. *The forecast of fresh beef quality by hyper spectral.* D. (2010)
8. K. W. Liu, F. Cheng, et al, Spectroscopy and spectral analysis. *The detection on fat ,protein and moisture content in fresh pork by visible/near-infrared spectrum.* J, **29(1)**:102-105(2009)
9. [9] R. D. Dong, X. C Wang, et al, Spectroscopy and spectral analysis. *The detection nondestructively and rapidly on the content of phosphate in by infrared spectrum.* J,

33(6):1542-1546(2013)

10. [10] X. B. Zhou, X. L. Zhang, et al, Journal of agricultural engineering. *The detection on the content of cucumber leaf chlorophyll based on hyper spectral image*. J, **30(13)**:169-175(2014)

11. [11] Y. Y. Li, L. L. Zhang, Y. K. Pen, et al, Sensing for Agriculture and Food Quality and Safety . *Hyper spectral Imaging Technique for Determination of Pork Freshness Attributes.*, 1-9(2011)

12. [12] M. Kamruzzaman, et al, Analytica Chimica Acta . *Prediction of Some Quality Attributes of Lamb Meat Using Near-Infrared Hyper spectral Imaging and Multivariate Analysis* , **714**:57-67(2012)

13. [13] J. Zhao, Y. K. Pen, et al, Journal of agricultural engineering. T*he evaluation on beef tender degree Based on hyper spectral image texture feature*. J, **31(7)**:279-286(2015)

14. [14] D. Barbin, et al, Analytica Chimica Acta. *Predicting Quality and Sensory Attributes of Pork Using Near-Infrared Hyper spectral Imaging*, **719**:30-43(2012)

Segmentation of MRI Volume Data Based on Clustering Method

Dongsheng Ji[1,2,a], Peng Chen , Chi Zhang[3],

[1]*Gansu Radio & TV University, Lanzhou, 730000, China,*
[2]*School of information Science & Engineering, Lanzhou University, 730000, Lanzhou, China,*
[3]*State Gride Gansu Electric Power Company, Lanzhou,730000, China.*

Abstract. Here we analyze the difficulties of segmentation without tag line of left ventricle MR images, and propose an algorithm for automatic segmentation of left ventricle (LV) internal and external profiles. Herein, we propose an Incomplete K-means and Category Optimization (IKCO) method. Initially, using Hough transformation to automatically locate initial contour of the LV, the algorithm uses a simple approach to complete data subsampling and initial center determination. Next, according to the clustering rules, the proposed algorithm finishes MR image segmentation. Finally, the algorithm uses a category optimization method to improve segmentation results. Experiments show that the algorithm provides good segmentation results.

1 Introduction

Nuclear magnetic resonance (NMR) imaging technology can provide anatomical information of the heart including different fault plane image information. Compared with other medical imaging modalities, such as tomography x-ray radiography, echocardiography, the cardiac muscle tissue has a higher resolution in MR imaging. As it is a non-invasive technique, MR imaging poses no harm to the human body [1]. NMR technique can be used to effectively prevent and diagnose heart disease. As the left ventricle (LV) pumps oxygenated blood from the lungs to the rest of the body, LV plays an important role in cardiac function, and is the focus of our research program. Left ventricular surface reconstruction is based on inside and outside LV MR image contour segmentation, and canprovide border restrictions on myocardial motion analysis. Therefore, quick and accurate segmentation of LVprofiles is of great significance [2, 3].

Mainstream methods of left ventricular myocardial segmentation are based on edge [4, 5]. For example, Snake model, Level Set model, and Geometric active contour model are all improved models, but the mastoid muscle is often present and interferes with the MR image, due to local gradient maximum regions, weak edges, artefacts and other phenomena, resulting in a lot of difficulties in image segmentation methods based on edge. To solve these problems, an image segmentation model based on image priori knowledge was introduced to improve the robustness of the model. Thus it appears that newer methods are based on prior knowledge constraints and regional approaches. Recent studies [5,6] introduced the physiological cardiac structure constraint model to control the relative position of the LV internal and external contours, and introduced regional information to provide a category of information for image segmentation. Other studies such as Paragios' paper[7] have developed algorithms on the basis of geometric active contour model, which combined the edge of the image and the regional information and physiological structure constraints. Split left ventricular internal and external contour.

Paragios method uses Gaussian mixture model fitting the histogram to obtain regional information; however, EM algorithm for Gaussian mixture model has some dependence on the initial parameter estimates; therefore it may be stuck in local maxima. Paragios' method also has some limitations, for example, the use of GVF (Gradient Vector Flow) [8,9] instead of gradient field guide evolution to the edge of the curve, which can expand the model capture range. [10] The performance of GVF will be affected for uneven greyscale images or noisy images[11, 12]. At this point, the evolution of the curve from weak border into another area causes border leaks or local maxima in the gradient within the region and stops at the edge of the isolated region, and it cannot move to the real border. In addition, the curve evolution formula in Paragios method is complex, and, in order to assure its stability, requires special numerical methods, so, the numerical computational complexity is higher. Through the analysis of the various methods stated above, in this paper, from the view of simplicity and practicality, we put forward an interior and exterior contour automatic segmentation method of the LV.

In this paper, proposed automatic segmentation method of left ventricular internal and external contours. Short axis image of left ventricle myocardial inner and outer profiles approximate rounded, firstly, using Hough transform to automatically locate initial contour of the left ventricle; Then, based on geometric active contour model, K-means clustering algorithm to be applied to a coarse segmentation of objects in an image, provides regional information and cardiac Physiology structure

[a] jids8012@163.com

constraint, establish coupled curve evolution equations of myocardium interior and exterior contour, to realize the automatic segmentation of left ventricle.

2 Hough transform algorithm to describe MR left ventricular image edge

The extraction of features from shape of MR image is one of the principal responsibilities of an effective features descriptor. In this work, we proposed a contour descriptor, which characterizes local and global features extraction of left ventricular internal and external contours. Local features are extracted using Hough transform for describing the contour of the MR left ventricular. The descriptors are explained in the subsequent subsections.
Hough transform is an effective technique for detecting lines from an MR image, which is widely used in computer vision and pattern recognition. Hough transform maps a line in the spatial domain to a point in the Hough parametric domain, The effectiveness of the Hough transform based local features that provide small variations in histograms for intra-class and large variations for interclass is demonstrated.[13, 14,15].

3 Incomplete K-means clustering algorithm

3.1 K-means clustering algorithm

Clustering is a certain similarity criteria to divide the data into groups having similar properties of subclasses, and internal similarities greater than the similarity between such. Measurement of distance or similarity is the basis of clustering algorithms.

K-means clustering algorithm is one of the most common iterative operator with adjustment K-means center of mass[7]. K-means is a clustering method based on squared error, is also very famous hard clustering algorithm. The algorithm is simple and fast, and as with other clustering algorithms, K-means is an iterative optimization process.

Assuming that the sample set $X=\{x_1,x_2,...,x_N\}$ are divided into K classes, first, choose an initial separation for K classes, and calculate the mean vector μ of these classes, and then the remaining each sample by Euclidean distance is assigned to the closest one.

$$d(x_i,u)=|x_i-u_i|. \qquad (1)$$

Recalculated mean vector is assigned to each class as the new class of centers. Repeat this process until the mean vector μ converges.

Central idea of K-means clustering algorithm is to minimize the total distance in class. All methods are re-adjusted the cluster centers and the allocation of the sample to the nearest centroid in the same category:

$$E=\sum_{j=1}^{k}\sum_{x_i\in w_j}\|x_i-\mu_j\| \qquad (2)$$

Algorithm steps are as follows:

Step1: Set the maximum number of iterations and the minimum allowable deviation threshold, initialize all class center $\mu_j^{(0)}$;

Step2: Distribute x_i to the nearest centroid $\mu_j^{(n)}$ represented clustering;

Step3: Calculate the new center of mass $\mu_j^{(n+1)}$ and $E^{(n+1)}$;

Step4: Repeat Step 2 and Step 3, until K is less than the threshold, or when they reach the maximum number of iterations.

K-means belongs to unsupervised learning methods. In the application of image segmentation, generally the target image is divided into different classes according to the gray value or combining histogram for image clustering segmentation. Image pixel clustering in the feature space is found in the spatial characteristics of similar pixel point group. According to the different methods of selecting initial cluster centers, sample allocation rules, centroid calculation method, and stopping criterion is different, as there are several different versions of the algorithm. When the K-means algorithm for clustering, the initial cluster centers are generally randomly generated, the number of clusters are also manually set. Several methods of choosing clustering center are possible: (1) select special samples, (2) select K samples; (3) arrange all samples according to the distance, choose distance equal sample; (4) select maximum distance between samples. Many new K-means algorithms are presented to automatically determine the number of clusters and the initial cluster centers.

In this paper, the traditional K-means clustering algorithm is put forward with a few improvements: First, use the sample space streamlined ways to reduce a single iteration time-consumption; secondly, use a simple method to determine the initial clustering center; an incomplete clustering method is proposed to further improve the efficiency of clustering; and finally, the classification method to optimize the clustering results to get rid of some small area in cluster results.

3.1 Streamline the sample space

At present, there are some main sample cutting strategies [8]: One is based on image pixels, image blocking, with a mean or median of block as the new samples. Another strategy is based on image feature, which have the same characteristics of samples (pixels) into one class. The sample cutting strategy of this paper is to improve in the second method. We initially block feature space, divide pixels by their characteristics into a block of cells, and, the cell's value to reflect such characteristics of the number of pixels. Compared with traditional K-means clustering in the process of the results of all samples for processing, this algorithm can significantly improve the efficiency of clustering.

3.2 Determine clustering center

There are many optimization methods in the choice of K-means cluster centers which deal with image segmentation issues and achieve better results. These methods include: LINDE-BUZO-GRAY[9], (LBG), max-min distance-based (MMX) [10], Density-based methods (DEN) [11], and Subset Farthest First (SFF) [12]. In this paper, we give the method to determine the initial center in the previous sample based on streamline, the maximum minimum distance method is introduced into density as a measure, the initial clustering center K choose sample K features of highest density sample, then, iteration to determine the initial clustering center.

Problem: Data set $X=(x_1,x_2,...,x_n)$ is divided into classes, making the error sum E of squares of the clustering is minimal.

Define 1: The data object $x=(x^1,x^2,...,x^p)$ and $y=(y^1,y^2,...,y^p)$, the distance between them is:

$$d(x,y)=\sqrt{(x^1-y^1)^2+(x^2-y^2)^2+...+(x^p-y^p)^2} \quad (3)$$

3.3 Incomplete clustering

The traditional K-means clustering method in the process of cluster is to compute the distance between sample and each cluster center. As most of the sample is very obvious, this can be a way to quickly get it belongs to class, avoid to calculate the distance to each cluster center, and to shrink the clustering of time. Proposed algorithms will cluster near the center of the sample quickly to this clustering. The main idea is to proceed from the cluster center; cluster center R neighborhood of samples to the belonged class, it did not appear in any sample clustering center in the neighborhood, iteration to the clustering center no longer changes it due to recent class.

Define 2: Assume a point q in m-dimensional space $Z=(z_1,z_2,...,z_m)$, S is am-Korekata, V is a multi-dimensional region, if the following conditions are met:

(1) q is the center of S.

(2) R is a positive integer.

(3) Value area for $[q[i]-R,q[i]+R]$;

(4) V is a non-empty cell of S does not contain a q.

So, V is for q field (q,R), expressed in $T(q,R)$.

For the sample x, if it is located within the R field of its cluster center, it is attributed to m-centric class. The size of the field will become the important factors influencing the clustering effect. Determining ideal field radius is very complicated; as a result, this paper puts forward three kinds of approximate methods to determine the domain size R. When ith cluster, clustering center for $\phi^i=\{\varphi_1^i,\varphi_2^i,...,\varphi_k^i\}$, jth clustering radius from the center for $R^{i,j}$, Three methods of computation formula is as follows:

1)Adjacency minimum distance:

$$R^{i,j}=\frac{1}{2}\cdot\min(\left\|\varphi_m^i-\varphi_j^i\right\|^2);1\le m,j\le k,m\ne j \quad (4)$$

2) Adjacent average:

$$R^{i,j}=\frac{1}{2}\cdot\min(\left\|\varphi_m^i-\varphi_j^i\right\|^2);1\le m,j\le k,m\ne j \quad (5)$$

3) Global average:

$$R^{i,j}=\frac{1}{2C_k^2}\sum_{1\le m,j\le k,m\ne j}^{C_k^2}\left\|\varphi_m^i-\varphi_j^i\right\|^2 \quad (6)$$

The above are three ways to determine the domain radius R, it shows experiments on more than one image to get R.

3.4 Clustering optimization

Using "incomplete" K-means clustering method for image segmentation [13], clustering is based on features, and completely ignores the space information of image, therefore the clustering results have many small areas. This paper presents an optimization method of the classification result, this approach combines regional information, small areas on the clustering segmentation has been removed, and obtains good effect. The main idea is to use the method of hash $R\times R$ neighborhood of the statistical sample x that belongs to the range of sample number, again will belong to class x belongs to sample number compared with a threshold, if greater than threshold do not perform any operation; otherwise the sample x to neighborhood sample belongs to class most of the class.

Due to traverse the entire optimization process is based on the classification results, only in a single step to statistics class predetermined belongs to the neighborhood, so the entire complexity of clustering algorithm approach to O(n). Compared to other more complex clustering algorithms, this algorithm can greatly improve the efficiency of clustering optimization.

4 Left ventricle initial positioning of the inside and outside profile by Hough transform

In order to overcome the initial contour effects on geometric active contour segmentation results, Hough transformation [15] is used to estimate the initial internal and external contours. Hough transformation is a method of detecting the resolution defined shape of the image [14]. Hough transformation can be based on the prior knowledge of the approximate model to search the exact shape of the objects not known.

Using left ventricular internal and external contours prior shape similar to the circle, we adopted Hough transformation to estimate LV internal and external contours. In the process of LV systolic, when the LV in contour approximates a circle because of the influence of the mastoid muscle, there may be undetectable left heart interior contour phenomenon. In the course of left ventricular systolic, when the inner contour of the left ventricle does not approximate a circle, due to the effect of mastoid muscle, the image may appear less than left ventricular contour detection. We used Hough transformation in detecting left ventricular contour and then center of the outline of MR imaging parameters to estimate the radius of inner contour of the initial LV.

Left ventricular contour edges and greyscale consistency is better than the outer contour, so, left heart inner contour segmentation dependence on the initial contour is much smaller and does not need to be as accurate as outer contour. The initial outline to more accurately locate the real edge is close and the ease curve evolution converges to the correct contour.

Hough transformation can identify partial deformation or partially occluded objects, but, it has a feature of the large amount of calculation [16,17]. To reduce computational complexity, improve the accuracy of applying prior knowledge, the long axis and short axis plane intersection point must be within the ventricle, node in the left ventricular internal and external contours around the center point. Centered on the intersection a range of window (the window size by MR imaging parameters, can be obtained from MR DICOM image file header), calculate the Hough transformation in this window.

Hough transformation makes use of round transformation formulas for $(x-a)^2+(y-b)^2=\rho^2$, parameter space is (ρ,a,b).

This initial contour algorithm:

Step 1: According to MR imaging parameters to estimate the range $[R_{\min}, R_{\max}]$ of outdoor profile radius of left heart;

Step2: Hough transformation for circle radius in the $[R_{\min}, R_{\max}]$, record the number of Hough transformation to detect under different radius of circle;

Step3: Take minimum radius as left ventricular outer contour radius (If $n_r > 1$ the center take close to the longest axis and minor axis plane intersection center averages);

Step 4: Taking outer contour center combined by MR imaging parameters estimation within the outline of the radius, initialize the left ventricle contour.

Manually set $[R_{\min}, R_{\max}]$, using the loop method or gradient descent algorithm to get Hough transformation position left ventricular initial contour.

The effectiveness of the Hough transformation is based on local features that provide small variations for intra-class and large variations for interclass is demonstrated in Fig.1. Sobel operator is applied to extract the left ventricular contour as reference, with the Hough transformation algorithm initialization profiles

Hough transformation strong anti-blocking methods in this section to MR short-axis images are tagged as effective, as they showed the effectiveness of the Hough transformation based on local features that provide small variations in histograms for intra-class and large variations for interclass as demonstrated in Fig.2. Sobel operator is applied to extract the left ventricular contour as reference, with the Hough transformation algorithm initialization profiles.

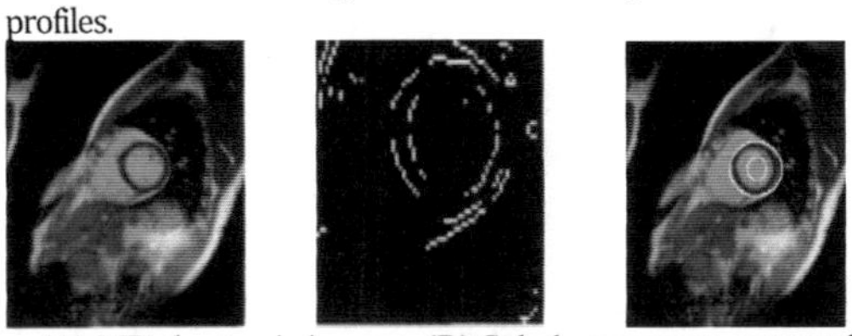

(A) MR short-axis images. (B) Sobel operator to extract the edge. (C) Initializes the internal and external contours.

Figure 1. Hough transformation position initial contour.

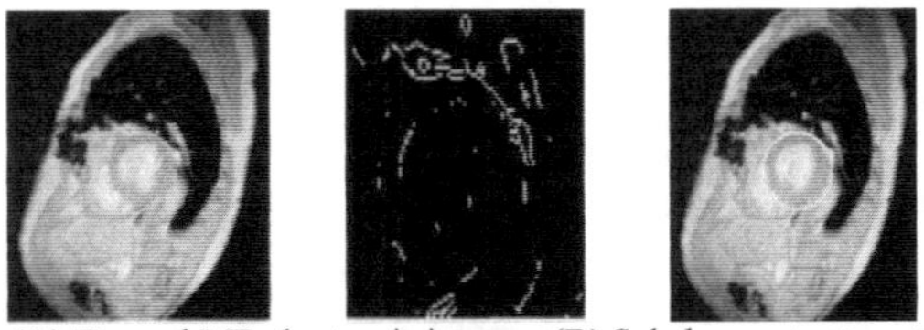

(A) Tagged MR short-axis images. (B) Sobel operator to extract the edge. (C) Initializes the internal and external contours.

Figure 2. Hough transformation position Tagged MR initial contour.

5 Analysis of the left ventricle MR images

Regional information provides target category information for segmentation model that is mainly clustering method based on statistical models (Gaussian mixture model) or other clustering methods based on distance to perform target classification.

Paragios obtained regional information by Gaussian mixture model fitting histogram [18] and the K-means algorithms were used to classify image target, and obtain regional information. According to left ventricular intimal area, myocardial region and background region, Paragios' method put MR image into three peak images. With three Gaussian distribution fitted histogram, then, follow the Gaussian function obtained by fitting the calculated probability of category area.

Different MRI also varies and the majority of hospital MR images are as shown in Fig.3. Inconsistent greyscale of the image and the histogram are not an ideal distribution of the three peaks, and EM algorithms for Gaussian mixture model estimation of initial parameters have certain dependencies.

Fig.4 shows the comparison of two methods of classification and three Gaussian distribution (green, red and blank).The classification results show that, although the K-means algorithm to sort results; however, due to myocardial grey scale similar to the background, left ventricular myocardium and the background are divided into different categories. This causes the outer contour of the left ventricle segmentation errors as external contour segmentation without using region information is used in this method, instead of using cardiac physiology constraints as prior information. If there are any deviation estimates of initial value and true value, EM algorithm might stall at some local maximum, so incomplete K-means algorithm are better than the algorithm of classification.

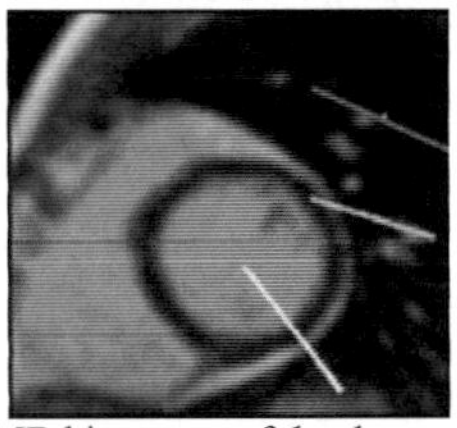

(A) MR histogram of the three peaks.

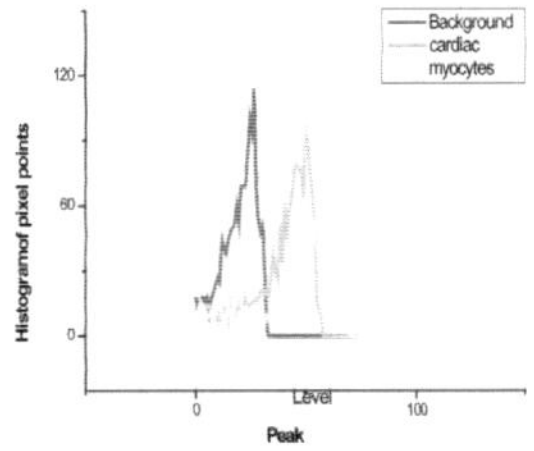

(B) The histogram corresponds to three peaks.

Figure 3. Ideal MR histogram of the three peaks.

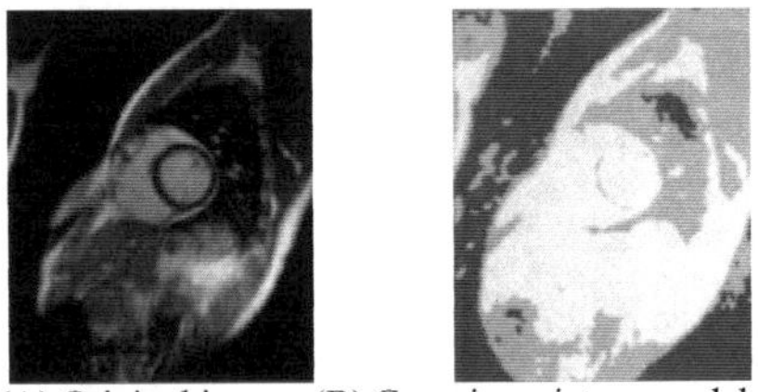

(A) Original image. (B) Gaussian mixture model classification results.

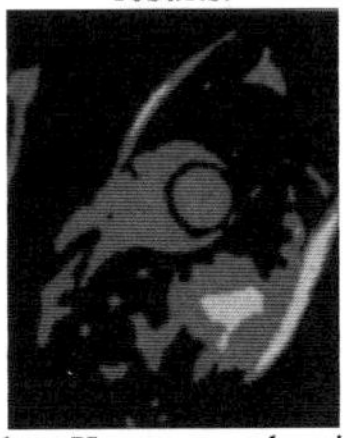

(C) Incomplete K - means algorithm results.

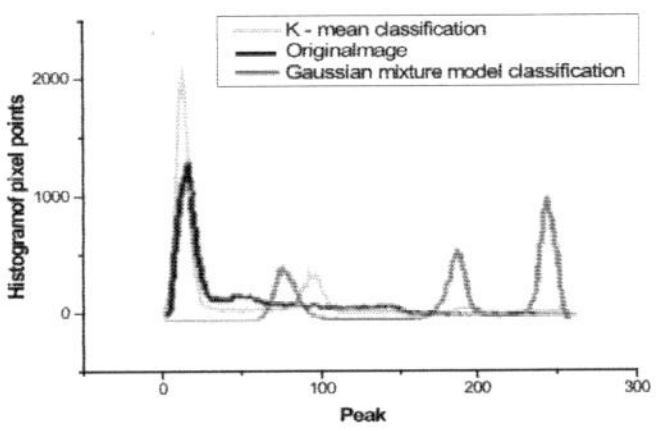

(D) Classification model fitting histogram results.

Figure 4. MR images of left ventricle classification results.

In order to compare the validity and robustness of the algorithm, the methods in this chapter are compared with geometric active contour model and its improved model Fig.5 shows that, in the case of the same iterative number, (α=0.4β=0.3γ=0.5$m=3, M=4, \delta=1$, time step $t=0.1$, number of iterations $n=350$).The method in this paper effectively partitioned the left ventricular internal and external contours, while other methods of segmentation results deviate from true profile (Fig.5E). It can be seen by comparing the role of various additions, contour area information; curve will swell when the curve evolves to left heart interior contour internal region; when the curve evolves beyond the inner contour, the curve will shrink. This acts as an adaptive balloon force. Left ventricular myocardial physiological structure constraint item control the relative position of the inner and outer contour evolution curve, to play the role of shape constraints.

Segmentation results compared with professional doctors hand-split, which is the gold standard for medical image segmentation. Comparing the results with manual segmentation results, using mean absolute distance (MAD) to measure the differences between the two, P is the algorithm to get the left ventricular contour lines, h for the hand-split result. MAD expressions are as follows:

$$e(P,H)=\frac{1}{2}\left\{\frac{1}{n}\sum_{i=1}^{n}d(p_i,H)+\frac{1}{m}\sum_{i=1}^{m}d(h_i,P)\right\} \quad (7)$$

$P=(p_1,p_2,...,p_n)$ and $H=(h_1,h_2,...,h_n)$ are the coordinates point on the contour, $d(p_i,H)$ is the nearest distance of point p_i to the H h coordinates:

$$d(p_i,H)=\min_j \left\|h_j-p_i\right\| \quad (8)$$

The methods in this paper and ASM model are presented in Table 1, in Fig.6 shows compare each method parameters average ,variance and maximum, and the MAD measurement results of 30 images, wherein the algorithm can split the left ventricle inner and outer contour and have better effect. The time complexity is close to that in geometric active contour model, and average of 100-200 iterations to converge.

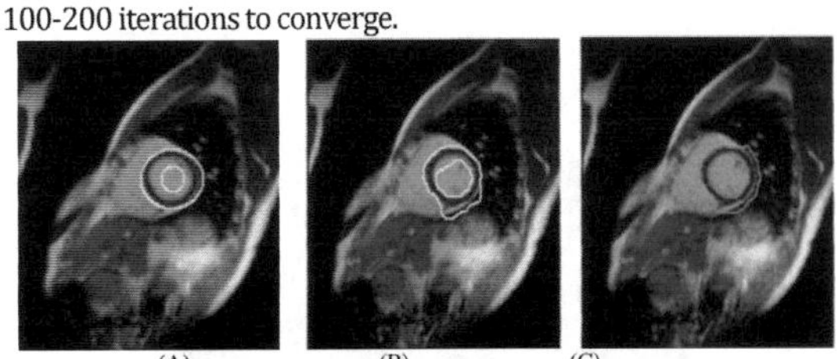

(A) (B) (C)

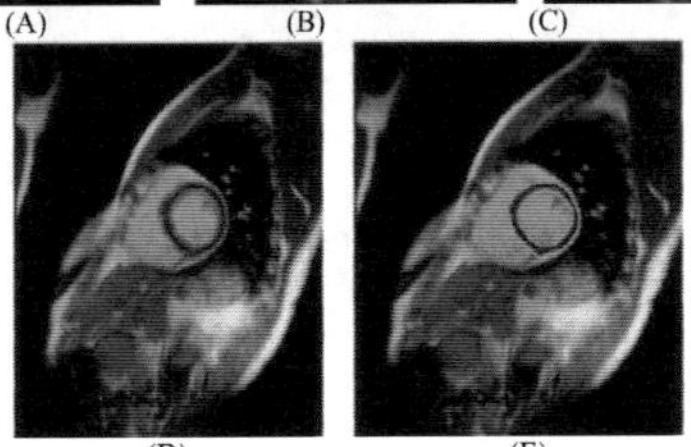

(D) (E)

(A) Hough locating initial contour. (B) Geometric active contour segmentation results. (C) Segmentation results of combining regional information. (D) Geometric active contour segmentation results after adding profiles regional information. (E) Incomplete K-means algorithm.

Figure 5. Comparisons of left ventricle segmentation result.

Table.1. MAD evaluation results with methods (pixels)

Method	Average	Variance	Maximum
Hybrid AAM	1.06	0.5	1.77
Shape particle filtering	1.10	0.3	1.90
Multi-view AAM	1.40	0.7	1.93
ASM	1.70	1.1	3.51
Extended AAM	1.06	0.48	1.34
Left ventricular external profile(IKCO)	0.80	0.11	1.05
Left ventricular internal profile(IKCO)	0.84	0.10	1.03

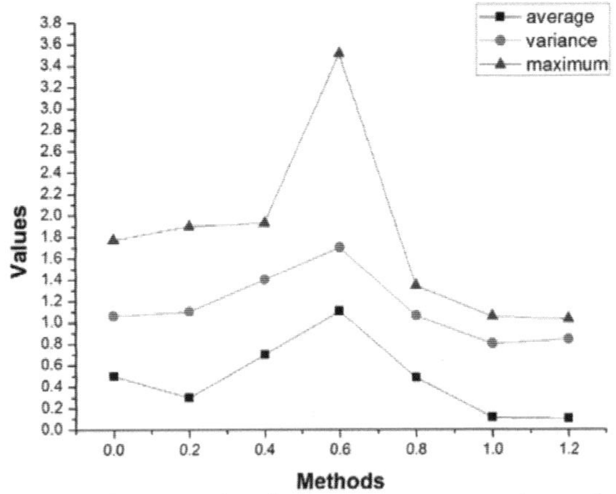

Figure 6. Compare each method parameters average ,variance and maximum.

Results from evaluation: the methods in this paper achieve high levels of accuracy, the left ventricular internal and external contours with an average error of less than 1 pixel, the literature of between the segmentation accuracy is 1-2 pixels and good stability than other methods in this chapter, and the segmentation error is the standard deviation of less than 0.2.

The comparison of left ventricular internal and external contours MAD value shows that the methods in this paper achieve high levels of accuracy, the left ventricular internal and external contours with an average error of less than 1 pixel, and the segmentation error is the standard deviation of less than 0.2.

6 Conclusion

MR image left ventricle inside and outside contour segmentation is the basis of left ventricular surface reconstruction, it can also provide boundary restrictions on myocardial motion analysis, at the same time provide carrier for left ventricular. Mainstream methods of left ventricular myocardial segmentation are image segmentation methods which are edge-driven, such as Snake model, Level Set model, Geometric active contour model and improved models. However, the mastoid muscle is often presents interference in the MR image resulting in local gradient maximum regions, weak edges and artifacts and other phenomena, thus results in many difficulties in image segmentation methods based on edge.

Hough transformation algorithm is robust, not sensitive to data incompleteness or noise, and can identify deformation or partially occluded object. To use this method to estimate the initial profile, the initial contours can be made more accurately and positioned in the vicinity of edge; curve evolution converges to the correct outline to overcome the influence of initial contour segmentation of geometric active contour model. Segmentation of left ventricle profiles, edge-based image segmentation models are often less than ideal results, so, a lot of prior knowledge need to be introduced to improve the robustness of the model. Shape constraint based on prior knowledge and information of the target area with more myocardial physiological constraints can be used to control the relative position of the left ventricular internal and external contours, and regional information can provide category information for image segmentation.

Incomplete K-means clustering algorithm is a clustering method based on squared error, and is simple and fast clustering method. While MR image grey is not consistent and histogram of image region is not according to left ventricular internal area, myocardial region and background region form ideal three peak distribution, therefore, the incomplete K-means algorithm for classification is better than the Gaussian mixture model. This algorithm is simple and practical, and our experimental results show that the algorithm is highly accurate.

Acknowledgments

This paper is supported by Science and Technology Plan of GanSu Province (No.1308RJZA266) and Gansu Radio & TV university youth fund project(NO.QN201501). This support is gratefully acknowledged.

References

1. A.F. Frangi, W.J. Niessen, M.A.Viergever. Three-dimensional Modeling for Functional Analysis of Cardiac Images: A Review. IEEE Transactions on medical imaging, 2001, 20(1):2-5.
2. American Heart Association. Heart and Stroke Statistical Update,1998.Website: www. Americanheart.org.
3. LiPei-hua, ZhangTian-wen. Review on active contour model (Snake model) [J]. Journal of Software, 2000, 11(6): 751-757.
4. C.L.Poh, R.I.Kitriey, R.B.K. Shrestha. Visualisation of cardiac wall motion using MR images. Computers in Cardiology, 2005, 17-19.
5. N.Paragios.A variational approach for the segmentation of the left ventricle in MR cardiac image analysis. International Journal of Computer Vision, 2002, 50(3): 345-362.
6. X.Zeng, L.Staib, R.Schukz, and J. Duncan. Volumetric Layer Segmentation Using Coupled Surfaces Propagation. Proceedings of the IEEE Conference on Computer Vision and Pattem Recognition, Santa Barbara,USA,23-25 Jun.1998,708-715.
7. Singh C, Bhatia N, Kaur A. Hough transform based fast skew detection and accurate skew correction methods. Pattern Recognit 2008;41:3528–46.
8. Gonzalez RC, Woods RE. Digital image processing. Prentice Hall; 2008.
9. Canny J. A computational approach for edge detection. IEEE Trans Pattern
10. E. Baeker, A. Jain. A clustering Performance measure based on fuzzy set decomposition. IEEE Transactions on Pattern Analysis and Machine Intelligenee, 1981,3(l):66-75
11. GOLDBERG N. Colour image quantization for high resolution graphics display [J]. Image and Vision Computing, 1991,9(5) : 303 -312.
12. MIGNOTTE M. Segmentation by fusion of histogram-based k-means clusters in different color spaces [J].IEEE Transactions on Image Processing, 2008,17(5) : 780 -787.
13. LINDE Y, BUZO A, GRAY R. An algorithm for vector quantizer design [J]. IEEE Transactions on Communications, 1980, 28(1) : 84 - 95.
14. AL-DAOUD M, ROBERTS S .New methods for the initialisation of clusters [J]. Pattern Recognition Letters, 1996, 17(5) : 451 - 455.
15. TURNBULL D, ELKAN C. Fast recognition of musical genres using RBF networks [J]. IEEE Transactions on Knowledge and Data Engineering, 2005, 17(4) : 580 -584.
16. MARTIN D, FOWLKES C, TAL D, et al. A database of human segmented natural images and its application to evaluating segmentation algorithms and measuring ecological statistics [C] / / Proceedings of 2001 International Conference on Computer Vision. Washing-ton, DC: IEEE Computer Society, 2001: 416 -423.
17. K.P. PhiliP. Automatic detection of myocardial contours in cine computed topographic images. PhD thesis, University of Iowa,1991
18. Da-meng DAI, De-jun MU, "A Fast Approach to K-means Clustering for Time Series Based on Symbolic Representation" , International Journal of Advancements in Computing Technology, vol. 4, no. 5, pp. 233-239, 2012.

Texture Features Based on Area Ternary Pattern for Classification Collagen Second Harmonic Generation Images of Skin Scar

Gaoqiang Liu, Mingyu Liu, Jichun Li, Encai Zhang, Kun Zhang, and Guannan Chen [a]

Key Laboratory of Optoelectronic Science and Technology for Medicine, Ministry of Education

Fujian Normal University

Fuzhou, 350007, China

Abstract. We proposed a method that based on area ternary pattern (ATP) operator to extract texture features of collagen second harmonic generation (SHG) images of skin scar. The proposed method includes texture feature extraction in collagen SHG images of skin scar, and classification with a support vector machine. By comparing with other texture feature extractive methods relating to the accuracy with specificity and sensibility, the method was demonstrated to obtain higher accuracy and better performance. It is credible that the proposed method improves clinical diagnosis of scar types and educes the development of new curative methods for dermatology and plastic surgery.

1 Introduction

Most wound on the skin healings result in scars especially after deep dermal injuries, mainly including normal and pathologic scars.[1] What scar may bring about pain, deformity, or limiting joint mobility is an inconvenient presence. We need distinguish different methods to aim at specific scars because different scars need to be treated in different ways. Nowadays, clinical examination and biopsy, which is complicated, are the main way to diagnose the types of scar.[2] Thus, clinicians and patients desire a simple and effective method for characterizing scar tissues. Image analysis of collagen fibers permit us to achieve the aim, since it has special and unique characteristics. What texture analysis about quantitative measures are hot research methods now can provide a novel technique for quantitative analysis.

The virtue of texture feature that is a kind of visual characteristics reflects the homogeneity of an image rather than depending on colors and brightness.[3] Appearance, arrangement and structure of image belong to image texture. Nowadays, there is not widely accepted texture extraction method that is fit for all types of medical images, because the texture information of image is complex.[4] Therefore, researching more accurate texture analysis method is instant and difficult task in current medical image processing researches.

In this paper, we discussed the texture feature extraction of collagen SHG images in human skin scars. The new local operator combined with line-likeness and phase congruency to represent texture information of medical images. In addition, we make the comparisons using these texture parameters and classification of scar images to evidence the effectiveness of proposed method.

2 Materials and methods

2.1 Images acquisition

[a] Corresponding author:edado@fjnu.edu.cn

We obtained these ripen skin scar specimens from our patients undergoing cesarean section located at the abdomen. Informed consent was achieved from every patient who participated in the study.

The normal and hypertrophic scar samples were collected during surgery at the abdomen from 4 subjects aged 29 to 55 years old. The samples were snap-frozen in liquid nitrogen (-196 $^{\circ}C$) to store right now after being severed from patients. We chose the center of the severed tissue as scar region for examination. SHG images were achieved using a commercial laser scanning microscopic imaging system equipped with a mode-locked femtosecond Ti: sapphire laser operating at 810 nm. The acquisition of the SHG image with 512×513 pixels was about 1.57s. In this study, one SHG image of collagen was acquired from each normal scar and skin tissue section.

2.2 Procedure of the proposed method

Fig. 1 shows the procedure of the proposed method, which included four steps as follows. Firstly, we acquired the original collagen SHG image from the surgery at the abdomen. Secondly, using area ternary pattern operator in original image acquire ATP image. Thirdly, The texture features including line-likeness and phase congruency of ATP image were extracted. At last, the texture features were classified by support vector machine (SVM). The detailed description could be found in the following sections.

2.3 Texture extraction

LTP operator has many advantages including robustness to illumination, rotation invariance and computational simplicity.[5] However, domain of LTP operator is limited in 3×3 neighborhoods, so texture information Reserved is restricted.

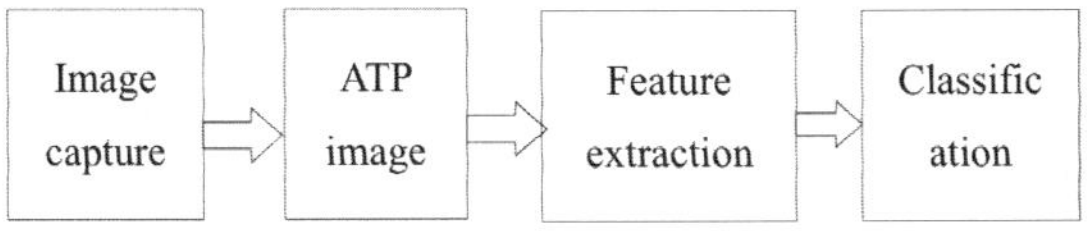

Figure 1. Procedure of the proposed method

We proposed a new local operator area ternary pattern (ATP) that is inspired from LTP operator. Extracting the line-likeness and phase congruency of ATP image is expedient to quantify texture features within the scar collagen SHG images. ATP is devised to inherited the advantages and extend function neighborhood to enhance effectiveness. Fig. 2 illuminates the procedure of ATP operator. The function neighborhood of ATP's center pixel g_c that is gray value of original image is $k\times k$ neighborhood, where $k=2x+1, x\in[1,2,3,4]$. As for a center pixel g_c, we could achieve the mean g_k of each $k\times k$ neighborhood. For decreasing the computational complex, we assume binary code instead of ternary code and disintegrate it into upper and lower patterns. In the end, It will be a pixel value in ATP image after upper and lower patterns perform a bitwise OR operation in decimal.The ATP code could be acquired in Equation 1.

$$ATP=\sum_{x=1}^{4} s(g_k-g_c)2^p, s(x)=\begin{cases}1, & x>t\\ 0, & -t\le x\le t, k=2x+1\\ -1, & x<-t\end{cases} \quad (1)$$

Where t is theshold that is the mean of g_k. In addition, for maintaining the uniform of code, we selected the ATP's code that only consist of at most one 0-1 and one 1-0 conversion when perceived as a rounded bit string. Due to the uncertainty of binary code of uniform pattern, we could not ensure correspondent decimal. The minimum value is defined as rotation invariance pattern to solve the problem.

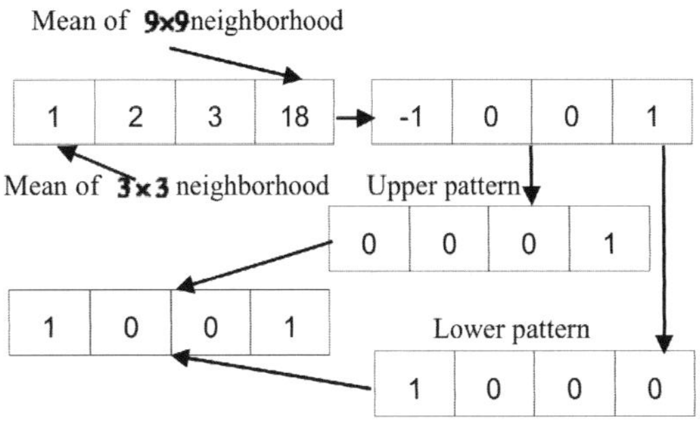

Figure 2. The procedure of ATP operator (value of center pixel is 8)

Line-likeness is a visual perception of textural features to represent lines, which is regarded as the number of equal for the direction of target edge and neighbor edge.[6] As a measure of line-likeness, co-occurrences in the same direction are weighted by +1 and vertical direction by -1. The line-likeness was calculated as follows:

$$F_{lin} = \sum_{i}^{n}\sum_{j}^{n} P(i,j)\cos[(i-j)\frac{2\pi}{n}] / \sum_{i}^{n}\sum_{j}^{n} P(i,j) \quad (2)$$

where $P(i,j)$ is the $n \times n$ local direction co-occurrence matrix of points at distance. The distance function d_4 was using four in our experiments.

Phase congruency (PC) is a dimensionless measure for the importance of a structure individually of the signal amplitude.[7] The relevant technique to calculate PC used in this paper is effective to combine with ATP. The phase congruency idea is rooted in the supposition that features are observed in image points, where Fourier constituents of the signal are maximally in phase. What the PC is expressed as the summation over orientation o and scale n was calculated as follows:

$$PC(x,y) = \frac{\sum_{o}\sum_{n} W(x) \lfloor A_{no}(x,y)\Delta\phi_{no}(x,y) - T \rfloor}{\sum_{o}\sum_{n} A_{no}(x,y) + \varepsilon} \quad (3)$$

$$\Delta\phi_{no}(x,y) = \cos(\phi_{no}(x,y) - \overline{\phi}_{o}(x,y)) - |\sin(\phi_{no}(x,y) - \overline{\phi}_{o}(x,y))| \quad (4)$$

$$W(x) = \frac{1}{1 + e^{r(c - s(x))}} \quad (5)$$

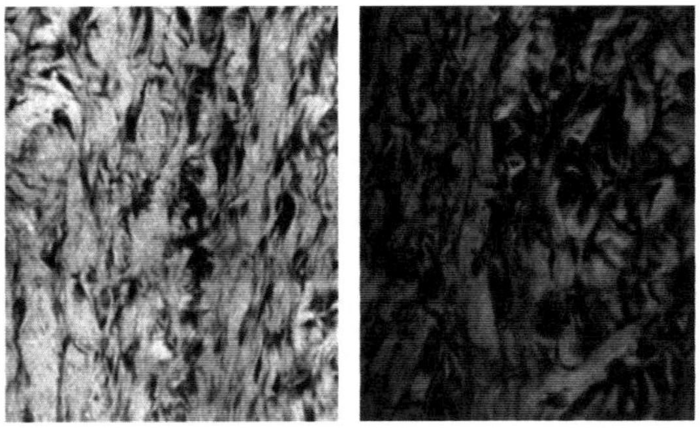

(a) (b)

Figure 3. Collagen SHG images in human (a) normal and (b) hypertrophic scar. Images are 512×513 pixels

$$s(x) = \frac{1}{N}\left(\frac{\sum_{n} A_{n}(x)}{A_{\max}(x) + \varepsilon}\right) \quad (6)$$

where $\lfloor \ \rfloor$ denotes that the enclosed quantity is equal to itself when its value is positive, otherwise it is zero. T is a additional item for noise and ε is a compensate item for avoiding division by zero. $\Delta\phi_{no}(x,y)$ is a sensitive measure of phase deviation. $\phi_{no}(x,y)$ is phase angle and $\overline{\phi}_{o}(x,y)$ is the mean phase angle at orientation o. $W(x)$ is a weighting function including threshold c of filter response and control factor r of sharpness of the threshold. $s(x)$ is frequency response spread function.

2.4 SVM

Support Vector Machine (SVM) is a mighty tool to binary classification in the field of pattern recognition and machine vision.[8] Because SVM have powerful classification performance and train samples speedily, we focused on SVM in this work. The SVM algorithm that is stemmed from the notion of the structural risk minimization is outstanding to learn non-linear or linear class boundaries.

The SVM algorithm functions as input labelled data from two groups, and generates new labelled or unlabeled data by an outputs model for classifying. As with many pattern methods, the classifier has two main processes that is training and testing. In the course of training, the training set consists of training data and label. The input data is applied to shine upon higher dimensional space in SVM by radial basis function (RBF) kernel. A hyper plane linear classifier is utilized in the higher dimensional space by pattern vectors. The RBF is shown as follows:

$$k(x_i, x_j) = \exp(-\gamma \|x_i - x_j\|^2), \gamma > 0 \quad (7)$$

where x_i and x_j are training vectors, γ is parameter of kernel.

3. Results and discussion

The collagen SHG images in human normal and hypertrophic scar tissues are revealed in Fig. 3. It is difficult to distinguish normal scar (see Fig. 3(a)) and hypertrophic scar (see Fig. 3(b)) in unaided eye because of similarity of fibroblasts and deposition of collagen fibers. So as to quantitatively depict the scars, we extracted texture features in normal and hypertrophic scar collagen SHG images. The proposed method that extracted the line-likeness and phase congruency of ATP images was suggested to describe the texture features in normal and hypertrophic scar collagen SHG images.

We compared our algorithm based on ATP operator with LTP operator and gray level co-occurrence matrix (GLCM) to reflect the validity. GLCM, which is extensively used for texture feature extraction, is based on gray-level statistical models in an image. GLCM could provide four texture features including homogeneity, entropy, contrast and correlation in grayscale images. Comparison of LTP and GLCM with ATP is shown in Table 1. It is obvious that accuracy rate required by our algorithm based on ATP is significantly better than other methods. The recognition effects of LTP operator and GLCM separately are worse in relation to ATP.

Table 1. Accuracy rates of different texture feature

Type	GLCM(%)	LTP(%)	ATP(%)
normal	66.67	46.67	80.00
hypertrophic	80.00	86.67	86.67
overall	73.33	66.67	83.33

We evaluated the performance of proposed method that line-likeness and phase congruency of ATP image are extracted using leaving one out. From the result of experiment, ATP operator have higher sensibility and lower specificity then other methods. In summary, we believe the ATP operator provide a effective method to achieve a relatively good performance for scar diagnosis.

4 Conclusion

In this work, a new method that extracted the line-likeness and phase congruency of ATP image in collagen SHG images from human normal and hypertrophic scar was proposed. Extracting the line-likeness and phase congruency of ATP image was applied to classify the images by SVM classifier. The proposed method about ATP operator was demonstrated to have highly advantage after comparing with other texture feature extraction methods. The proposed method belongs to image technology that is a noninvasive method for the scar diagnosis, which could be aroused by the development of new therapies and has highly application potential in the studies of some other domain.

We are focusing on more effective approaches to extract heterogeneous texture features of human scars. We believe the higher accuracy classifier of texture features will emerge in the near future.

Acknowledgement

This work is supported by the National Natural Science Foundation of China (Nos. 81101110), the Science and Technology Project of Fujian Province (Nos.2015J01300), and Fujian Normal University foundation for excellent young teachers (No. fjsdjk2012058).

References

1. X. Zhu, S. Zhuo, L. Zheng, K. Lu, X. Jiang, J. Chen, *et al.*, J BIOPHOTONICS, vol. **3**(2010).
2. K. Tan, D. McGrouther, A. Day, C. Milner, and A. Bayat, J EUR ACAD DERMATOL, vol. **25**(2011).
3. B. Li and M. Q.-H. Meng, IMAGE VISION COMPUT, vol. **27**(2009).
4. R. Ambekar, T.-Y. Lau, M. Walsh, R. Bhargava, and K. C. Toussaint, BIOMED OPT EXPRESS, vol. **3**(2012).
5. X. Tan and B. Triggs, *Analysis and Modeling of Faces and Gestures*(ed: Springer, 2007)
6. T. Majtner and D. Svoboda, *3D Imaging, Modeling, Processing, Visualization and Transmission* (3DIMPVT, 2012)
7. L. Zhang, L. Zhang, D. Zhang, and Z. Guo, PATTERN RECOGN, vol. **45**(2012).
8. M. Pal and G. M. Foody, IEEE T GEOSCI REMOTE, vol. **48**(2010).

Orthogonal Discriminant Diversity and Similarity Preserving Projection for Face Recognition

Yu'e Lin [1,a], Fengrui Zhu[1], Kui Lu[1], Guangyu Xu[1], Xingzhu Liang[1]

[1]*School of Computer Science and Engineering, Anhui University of Science and Technology, Huainan, 232001, China*

Abstract. Feature extraction is a crucial step for face recognition. In this paper, based on supervised local structure and diversity projection (SLSDP), a new feature extraction method called orthogonal discriminant diversity and similarity preserving projection (ODDSPP) is proposed for face recognition. ODDSPP defines two parameterless weighted matrices by taking into account the class label information and local structure. Thus ODDSPP could utilize both the diversity and similarity information of the data simultaneously for dimensionality reduction. Moreover, the proposed algorithm is able to extract the orthogonal discriminant vectors in the feature space and does not suffer from the small sample size problem, which is desirable for many pattern analysis applications. Experimental results on the ORL and AR databases show the effectiveness of the proposed method.

1 Introduction

Over the past few decades, a large volume of dimensionality reduction methods for face recognition have been proposed. The most well-know dimensionality reduction methods are principal component analysis (PCA)[1] and linear discriminant analysis (LDA)[2]. Recent studies have shown that large volumes of high dimensional data possibly reside on a nonlinear sub-manifold. However, both PCA and LDA fail to discover the underlying manifold structure, due to the fact that they aim only to preserve the global structures of the samples. The most representative manifold learning methods include Laplacian Eigenmap (LE)[3], Isomap [4], and Locally linear embedding (LLE)[5]. These nonlinear methods yield impressive results on some benchmark artificial datasets. However, they are difficult to find low dimensional embeddings of new test samples. Locality preserving projection (LPP)[6] is the linear approximation of LE and can compute the low dimensional embeddings for new test points. Motivated by LPP, many local discriminant approaches[7-11] have been developed for image classification, among which the most prevalent ones include margin fisher analysis (MFA)[7], locality preserving discriminant projection (LPDP)[8], local fisher discriminant analysis (LFDA)[9], local discriminant embedding(LDE)[10], locality sensitive discriminant analysis (LSDA)[11]. These methods are simple to compute and easy to operate. However, one common limitation of the above mentioned manifold-based learning algorithms is that they only pay attention on the local similarity information and neglect the diversity information of the samples. Impairing the diversity information of the samples may arise the over learning problem. In order to overcome the over learning problem, Gao et al. considered on the diversity weighted information between points and proposed the supervised local structure and diversity projection (SLSDP)[12]. SLSDP, which explicitly considers the variation of the values among nearby data, overcomes the over learning problem successfully. However, there are still some limitations existed in SLSDP. Firstly, the diversity matrix is constructed without the guide of the class label information that is very important for recognition problem. Secondly, the neighborhood size has an impact on the learning of the manifold and is usually set in advance to the same value for all samples. Finally, similar with LDA, SLSDP also suffers from the small sample size problem (SSS) when dealing with high-dimensional data recognition task, such as face recognition.

In this paper, we present a new method called orthogonal discriminant diversity and similarity preserving projection (ODDSPP), which can directly overcome the SSS problem and at the same time derive all the orthogonal optimal discriminant vectors. The rest of this paper is organized as follows. Section 2 briefly gives an overview of related works about SLSDP. In Section 3, the dimensionality reduction method of ODDSPP is presented. In Section 4, the performance of ODDSPP is experimentally evaluated on the ORL and AR databases. At last, a brief conclusion is offered in Section 5.

2 Related work on SLSDP

The aim of SLSDP is to find an optimal projection that can preserve the similarity and the diversity information at the same time. Let S denote the local weighted matrix with elements characterizing the similarity of two close data points with same labels and B denote the local weighted matrix with elements characterizing the diversity of two close data points.

[a] Corresponding author: yelin@aust.edu.cn

The elements of the local similarity weighted matrix S are defined as follows:

$$S_{ij}=\begin{cases}\exp(\frac{-\|x_i-x_j\|^2}{t}) & if\ x_i\in N_k(x_j)\ or\ x_j\in N_k(x_i)\ and\ l_i=l_j\\ 0 & ,\text{otherwise}\end{cases} \quad (1)$$

Where $N_k(x_i)$ denotes the set of k nearest neighbors of x_i, l_i and l_j denote the class label of data x_i and x_j, respectively. $t\in(0,+\infty)$ is a suitable constant.

The elements of the diversity weighted matrix B are defined as follows:

$$B_{ij}=\begin{cases}\exp(\frac{-b}{\|x_i-x_j\|^2}) & if\ x_i\in N_k(x_j)\ or\ x_j\in N_k(x_i)\\ 0 & ,\text{otherwise}\end{cases} \quad (2)$$

Where $b\in(0,+\infty)$ is a suitable constant. B_{ij} measures the contribution of x_i relative to x_j to the diversity information.

Using the similarity matrix S, the local similarity scatter S_L can be expressed as:

$$\begin{aligned}S_L&=\frac{1}{2}\sum_{i=1}^{N}\sum_{j=1}^{N}S_{ij}(x_i-x_j)(x_i-x_j)^T\\&=X(D-S)X^T\\&=XLX^T\end{aligned} \quad (3)$$

Where D is a diagonal matrix whose elements on diagonal are column sum of S, i.e. $D_{ii}=\sum_j S_{ij}$.

Using the local diversity weighted matrix B, the local diversity scatter S_D can be expressed as:

$$\begin{aligned}S_D&=\frac{1}{2}\sum_{i=1}^{N}\sum_{j=1}^{N}B_{ij}(x_i-x_j)(x_i-x_j)^T\\&=X(Q-B)X^T\\&=XL_BX^T\end{aligned} \quad (4)$$

Where Q is a diagonal matrix whose elements on diagonal are column sum of B, i.e. $Q_{ii}=\sum_j B_{ij}$.

The objective function of SLSDP can be expressed as follows:

$$J(W)=\min_{W^TW=I}\frac{\text{trace}(W^TS_LW)}{\text{trace}(W^TS_DW)} \quad (5)$$

3 Orthogonal Discriminant Diversity and Similarity Preserving Projection

In this section, we introduce the ODDSPP method, which can preserve both the similarity and diversity information and at the same time derive all the optimal orthogonal discriminant vectors in the low-dimensional space. We firstly redefine the discriminating similarity matrix $\overline{S}$ and discriminating diversity matrix $\overline{B}$, respectively. The elements of the discriminating similarity matrix $\overline{S}$ are defined as follows:

$$\overline{S}_{ij}=\begin{cases}\exp(\frac{-\|x_i-x_j\|^2}{t}) & if\ \|x_i-x_j\|\le d_i\ and\ l_i=l_j\\ 0 & ,\text{otherwise}\end{cases} \quad (6)$$

Where d_i denotes average distance between samples with the same label. $\overline{S}$ denotes the case that x_i and x_j have the same label .

The elements of the discriminating diversity matrix $\overline{B}$ are defined as follows:

$$\overline{B}_{ij}=\begin{cases}\exp(\frac{-b}{\|x_i-x_j\|^2})\left(1-\exp(\frac{-b}{\|x_i-x_j\|^2})\right) & if\ \|x_i-x_j\|\le d_i\ and\ l_i=l_j\\ \exp(\frac{-b}{\|x_i-x_j\|^2})\left(1+\exp(\frac{-b}{\|x_i-x_j\|^2})\right) & if\ \|x_i-x_j\|\le d\ and\ l_i\ne l_j\\ 0 & ,\text{otherwise}\end{cases} \quad (7)$$

Where d_i denotes average distance between samples of the same label and d denotes average distance between all samples.

From the Eq.6 and Eq.7, it is obvious to find that the discriminating similarity matrix and discriminating diversity matrix integrate both the local structure and class label information. The size of the local structure is parameterless, which is favorable for face recognition.

Then using the similarity matrix $\overline{S}$ and diversity matrix $\overline{B}$, the local similarity scatter $\overline{S}_L$ and local diversity scatter $\overline{S}_D$ can be expressed as:

$$\begin{aligned}\overline{S}_L&=\frac{1}{2}\sum_{i=1}^{N}\sum_{j=1}^{N}\overline{S}_{ij}(x_i-x_j)(x_i-x_j)^T\\&=X(\overline{D}-\overline{S})X^T\\&=X\overline{L}X^T\end{aligned} \quad (8)$$

and

$$\begin{aligned}\overline{S}_D&=\frac{1}{2}\sum_{i=1}^{N}\sum_{j=1}^{N}B_{ij}(x_i-x_j)(x_i-x_j)^T\\&=X(\overline{Q}-\overline{B})X^T\\&=X\overline{L}_BX^T\end{aligned} \quad (9)$$

Then, the objective function of ODDSPP is written as follows:

$$J(W)=\min_{W^TW=I}\frac{\mathrm{trace}(W^T\overline{S}_LW)}{\mathrm{trace}(W^T\overline{S}_DW)} \quad (10)$$

In order to avoid the SSS problem, the objective function of ODDSPP can be rewritten as:

$$J(W)=\min_{W^TW=I}\mathrm{trace}\,(W^T\overline{S}_LW-W^T\overline{S}_DW) \quad (11)$$

Owing to the dimensionality of the sample very high, firstly we need to calculate the total-scatter matrix S_t eigenvectors $u_1,u_2,\cdots,u_r$ corresponding to the first r non-zero eigen-values. The total-scatter matrix S_t is defined as:

$$S_t=X(I-M)(I-M)^TX^T=H_tH_t^T \quad (12)$$

Where I is the $N\times N$ identity matrix and M is a $N\times N$ matrix with all terms equal to $1/N$.

Denote $r=\mathrm{rank}(H_t)$, $H_t=U\Sigma E$ and $\Sigma=\begin{pmatrix}\Sigma_r & 0\\ 0 & 0\end{pmatrix}$.

We know that

$S_t=U\Sigma E^TE\Sigma U^T=U\Sigma\Sigma U^T=U\begin{pmatrix}\Sigma_m^2 & 0\\ 0 & 0\end{pmatrix}U^T$.Wh ere $U=(U_1,U_2)$, $U_1\in R^{n\times r}$, $U_2\in R^{n\times(N-r)}$.

Denote $V=U_1^T$.Then, Eq. 11 can be rewritten as:

$$J(A)=\min_{A^TA=I}\mathrm{trace}(A^T\overline{S}_L^VA-A^T\overline{S}_D^VA) \quad (13)$$

Where $\overline{S}_L^V=V^T\overline{S}_LV$ and $\overline{S}_D^V=V^T\overline{S}_DV$. Finally, we apply the Lagrange multipliers to the Eq.13 and set the derivative with respect to A to zero. The projection vector a that minimizes the Eq.13 is given by the minimum eigenvalue solution to the generalized eigenvalue problem

$$(\overline{S}_L^V-\overline{S}_D^V)a=\lambda a \quad (14)$$

Then we can obtain the eigenvectors $a_1,a_2,\cdots,a_k$ corresponding to the first k smallest eigen-values.

The procedure of computing ODDSPP can be summarized in the following steps:

Step 1: Constructing the discriminating similarity matrix $\overline{S}$ and the discriminating diversity matrix $\overline{B}$ according t o Eq.6 and Eq.7.

Step 2: Calculating the discriminating similarity scatter $\overline{S}_L$ and discriminating diversity scatter $\overline{S}_B$ according to Eq.8 and Eq. 9.

Step 3:Compute S_t according to Eq.12 and solve S_t by g eneralized eigenvalue decomposition, then we can obtain the basis vectors $V=U_1$;

Step4: Compute $\overline{S}_L^V=V^T\overline{S}_LV$ and $\overline{S}_D^V=V^T\overline{S}_DV$, then solve Eq.14 by generalized eigenvalue decomposition. Finally, we can obtain the optimal orthogonal transformation matrix $W=VA$.

4 Experimental Results

In order to test performance of our proposed method, the ORL and AR database are used. The ORL face database is composed of 40 distinct subjects. Each subject has 10 images under different expression and different views. Fig. 1 shows the sample images of one person in the ORL database. The AR face database contains over 4000 color face images from 126 people, including frontal views of faces with different facial expressions, lighting conditions, and occlusions. In the experiments, we choose a subset of AR database. The subset of AR database contains 100 individuals and 10 images for person. Fig. 2 shows the sample images of one person in the AR database. In the experiments, the nearest distance classifier is adopted. All methods are compared on the same training sets and testing sets. In each round, k(3-7) images are randomly selected from the database. For each k, 10 tests are performed and these results are averaged. The recognition rates of UDDPP, MFA and SLSDP are listed in Table 1 and 2 .

Fig. 1. Images of one person in the ORL database

Fig. 2. Images of one person in the AR database

Table 1. The comparison of recognition rates(%) on the ORL database

methods	3	4	5	6	7
UDDPP	90.36	93.25	94.75	96.25	96.67
MFA	90.17	92.50	94.20	95.63	96.50
SLSDP	90.00	92.08	93.00	93.75	95.83

Table 2. The comparison of recognition rates(%) on the AR database

methods	3	4	5	6	7
UDDPP	92.57	94.82	95.40	95.75	96.33
MFA	91.82	94.15	94.50	95.25	95.97
SLSDP	91.71	93.66	94.00	94.50	95.66

From Table 1 and 2, we find that ODDSPP is the most efficient dimensionality reduction method, and is much more efficient than SLSDP and MFA. The one reason may be that the ODDSPP not only overcomes the small sample size problems but also exploits the orthogonal features, which is very suitable for face recognition. The other reason may be that our method can deal with the similarity and diversity information simultaneously. Furthermore, the strong structure preserving and discriminating ability make the proposed method more suitable for the recognition tasks.

5 Conclusions

In this paper, we presented a novel algorithm which is based on SLSDP, namely ODDSPP, for dimensionality reduction. Two contributions were made in this paper. (1)Combined with local structure information and the class information, ODDSPP redefines the parameterless weighted matrices, then can prevent the main geometric structure of the data and has more discriminating power. (2)Using a difference-based optimization objective function, the optimized transformation matrix can be computed by solving an eigen-equation in our method. Thus ODDSPP does not suffer from the SSS problem. These merits make our method more robust and suitable for classification tasks. In further research, we will focus on the incremental learning methods. The incremental feature set could be utilized by manifold learning to improve its recognition performance for real-time classification.

Acknowledgement

This article is supported by the Nature Science Foundation of China (No. 61471004) , the Key Project of Higher Education Natural Science Foundation of Anhui Province (No. KJ2016A203) and the Master and Doctor Foundation of Anhui University Of Science and Technology(No. 2010yb026).

References

1. M Turk, A Pentland. Eigenfaces for recognition, Journal of Cognitive Neuroscience. 3(1) (1991) 71 - 86.
2. P.N. Belhumeur, J.P. Hespanha, D.J. Kiregman. Eigenfaces vs. fisherfaces: recognition using class specific linear projection.IEEE Transactions on Pattern Analysis and Machine Intelligence.19(7) (1997) 711 - 720
3. M. Belkin, P. Niyogi. Laplacian eigenmaps for dimensionality reduction and data representation, Neural Comput. 15 (6) (2003) 1373 - 1396
4. J. Tenenbaum, V. de Silva, J. Langford. A global geometric framework for nonlinear dimensionality reduction, Science.290 (5050) (2000) 2319 - 2323
5. S.T. Roweis, L.K. Saul. Nonlinear dimensionality reduction by locally linear embedding, Science. 290 (5050) (2000) 2323 - 2326
6. X.F. He, S.C. Yan, Y.X. Hu, et al. Face recognition using Laplacianfaces, IEEE Transactions on Pattern Analysis and Machine Intelligence.27(3) (2005) 328 - 340
7. D. Xu, S. Yan, D. Tao, et al. Marginal Fisher Analysis and Its Variants for Human Gait Recognition and Content-Based Image Retrieval. IEEE Transactions on Image processing. 16(11)(2007) 2811 - 2821
8. J. Gui, W. Jia, L. Zhu, et al. Locality preserving discriminant projection for face and palmprint recognition , Neurocomputing . 73(13) (2010) 2696-2707.
9. M. Sugiyama. Dimensionality reduction of multimodal labeled data by local Fisher discriminant analysis, Journal of Machine Learning Research. 8(5) (2007) 1027 - 1061.
10. H.T. Chen, H.W. Chang, T.L. Liu. Local discriminant embedding and its variants,in: Proceedings of the IEEE Computer Visual and Pattern Recognition (CVPR).2005, pp. 846 - 853.
11. D. Cai, X. He, K. Zhou, et al. Locality sensitive discriminant analysis, in:Proceedings of the 20th International Joint Conference Artificial Intelligence(IJCAI). 2007, pp. 708 - 713.
12. Q.X. Gao, D.Y. Xie, H. Xiu, et al. Supervised feature extraction based on information fusion of local structure and diversity information, Acta Autom. Sin.36(8)(2010)1108 - 1114.

A Method of MPPT Control Based on Power Variable Step-size in Photovoltaic Converter System

Xu Hui-xiang[1], Li Nian-qiang[2]
[1]School of Information Science and Engineering,University of Jinan, jinan 250022,China
[2]School of Information Science and Engineering,University of Jinan, jinan 250022,China

Abstract: Since the disadvantage of traditional MPPT algorithms of variable step-size, proposed power tracking based on variable step-size with the advantage method of the constant-voltage and the perturb-observe(P&O)[1-3]. The control strategy modify the problem of voltage fluctuation caused by perturb-observe method, at the same time, introducing the advantage of constant-voltage method and simplify the circuit topology. With the theoretical derivation, control the output power of photovoltaic modules to change the duty cycle of main switch. Achieve the maximum power stabilization output, reduce the volatility of energy loss effectively, and improve the inversion efficiency[3,4]. Given the result of experimental test based theoretical derivation and the curve of MPPT when the prototype work.

Key Words: MPPT, variable step-size, constant-voltage, P&O, inversion efficiency, circuit topology

1 Introduction

The maximum power point tracking(MPPT) is the important element of photovoltaic converter system. Its rationale is that control the duty cycle of main power switch by the acquisition value of voltage and current to achieve the maximum power point (MPP), and stability output. The control algorithm can be divided roughly into three methods: constant-voltage methods, perturb-observe methods, and incremental conductance[5]. The reference[4] proposed a method of variable step-size incremental conductance based on Boost topology, used $|dp/du|$ as the coefficient of variable step-size. But this method need collect current value, increase the complicated structure of circuit topology. The reference[6] proposed a prediction-based variable step perturbation and observation method to avoid misjudge. But this method still need to disturbance voltage.

In this paper, through the analysis of traditional MPPT algorithms of variable step-size and the advantage and disadvantage methods of the constant-voltage and perturb-observe, proposed a method of power variable step-size combining constant-voltage method and perturb-observe method, on the premise of no voltage disturbance, without input current collect input current to control the output power from photovoltaic modules with power variable step-size. Just collect the voltage value, simplify the circuit topology.

2 The analysis of advantage and disadvantage of three methods

The output curve of photovoltaic panels with the same temperature different light intensity and the same light intensity different temperature has shown in figure-1[5]. We can found that the MPP distribute near the dotted line in the picture.

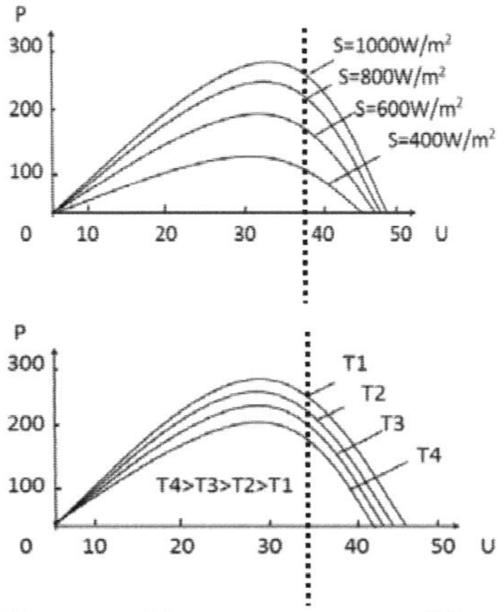

Fig.1 The curve of the same temperature different light intensity and the same light intensity different temperature

The thought control of constant-voltage method is fixed the input voltage on the dotted line to achieve the MPP, with low control precision, low efficiency, and can not tracking in real time[2,7].

Compared to constant-voltage method, the method of perturb-observe is more widely used. Judgment the change of power before and after and disturbance voltage to achieve MPPT[5,8]. But the disturbance is a kind of interference will occur voltage fluctuation, and reduce the converter efficiency.

The traditional variable step-size incremental conductance achieve MPPT through two ways: P-U derivation and power variation rate . The former must collect current value for the accurate calculation will complicate the circuit topology. The latter need to judgment and determine multiple stages, its will increase the complicate of algorithm with a complexity circuit topology.

3 A Method Of Mppt Control Based On Power Variable Step-Size

3.1 Control principle

Aim at the issue of complicate circuit based on $|dp/du|$ as variable step-size, and the problem of the constant-voltage and the perturb-observe, proposed a method of MPPT control based on power variable step-size, without current collect, and no voltage disturbance, computing power increment(Δp) by the value of voltage, control output power of photovoltaic to achieve MPPT. The figure-2 shown the curve of P-V. According to the method of constant-voltage, set the range of MPP, which is usually about 0.8 times of open-circuit voltage. Instead of $|dp/du|$. used Δp as the parameter of variable step-size, closer it gets to the MPP, smaller the step-size. When the Δp is zero, collect voltage and judgment whether within the 0.8 times rage of open-circuit voltage to avoid misjudgment. Change the duty cycle of the main switch by control output power of photovoltaic, accordingly change the input voltage, and adjust the output load. When get the MPP, because of the power constantly, the input voltage will not change, reducing the voltage fluctuation.

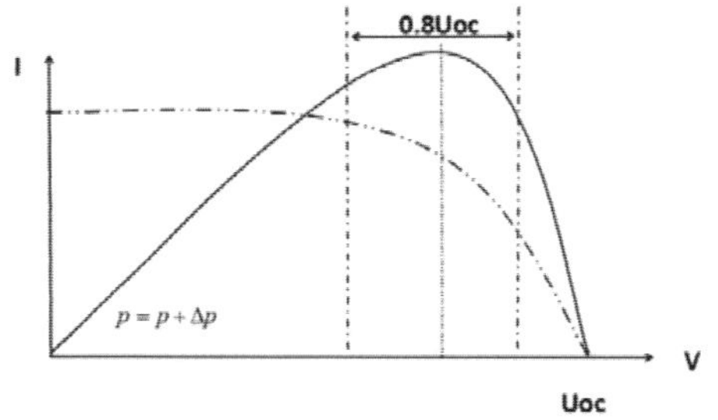

Fig.2 The curve of P-V

3.2 Theoretical Derivation

The figure-3 shown the equivalent output circuit of the photovoltaic[9]. The *Cin* is a large capacity electrolyte capacitor use for balance the instantaneous power of DC side and Ac side. According to the law of conservation of energy, the output energy of photovoltaic (Epv) is equal to the sum of the input energy of converter(Ei) and the energy(Ec) stored in Cin :

$$Ec = Epv - Ei \tag{1}$$

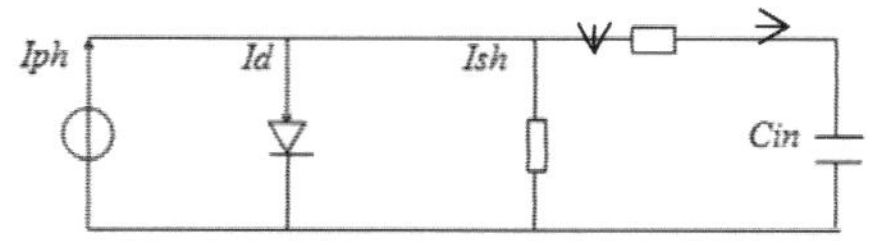

Fig.3 equivalent output circuit of the photovoltaic

Let Tp as a power cycle, Ppv is the average output power of photovoltaic, P_i is the input power of converter, in a power cycle, the above formula can be written as:

$$\frac{1}{2}C_{in}U_{t+Tp}^2 - \frac{1}{2}C_{in}U_t^2 = P_{pv}Tp - \int_t^{t+Tp} p_i dt$$

(2)

In the formula, U_t is the capacitor voltage in t moment, U_{t+tp} is the capacitor voltage after a power cycle.

Set t_0 to t_1, t_1 to t_2, the average output power was P_1, P_2, and the electrolytic capacitor voltage at the time of t_0, t_1 and t_2 are respectively U_0, U_1 and U_2. Take $t_1 = t_0 + T_p$, $t_2 = t_1 + T_p$ into the formula 2:

$$\frac{1}{2}C_{in}U_1^2 - \frac{1}{2}C_{in}U_0^2 = P_1T_p - \int_{t_0}^{t_1} p_i dt \quad (3)$$

$$\frac{1}{2}C_{in}U_2^2 - \frac{1}{2}C_{in}U_1^2 = P_2T_p - \int_{t_1}^{t_2} p_i dt \quad (4)$$

Because the grid energy is equal in two consecutive power cycles, so

$$\int_{t_0}^{t_1} p_i dt = \int_{t_1}^{t_2} p_i dt \quad (5)$$

subtract (3) from (4):

$$\Delta p = \Delta P_{pv} = P_2 - P_1 = \frac{Cin}{2Tp}(U_2^2 - 2U_1^2 + U_0^2) \quad (6)$$

$\frac{Cin}{2Tp}$ is a constant, Δp is power variation, take $\Delta p = U_2^2 - 2U_1^2 + U_0^2$ as the parameter of variable step-size. The calculation formula of duty cycle D and switch cycle Ts is[10]:

$$D = \frac{Ug}{N \bullet Upv + Ug} \quad (7)$$

$$Ts = Lp \bullet Pop(\frac{\sin \varphi_G}{Upv \bullet D})^2 \quad (8)$$

In formula (8), the Pop is the output peak power of converter and the Upv is the input voltage of converter, control the Ts and Upv by change the value of Pop, adjusting the duty cycle of switch and matching output impedance. The MPPT control flow chart of the method shown in figure-4.

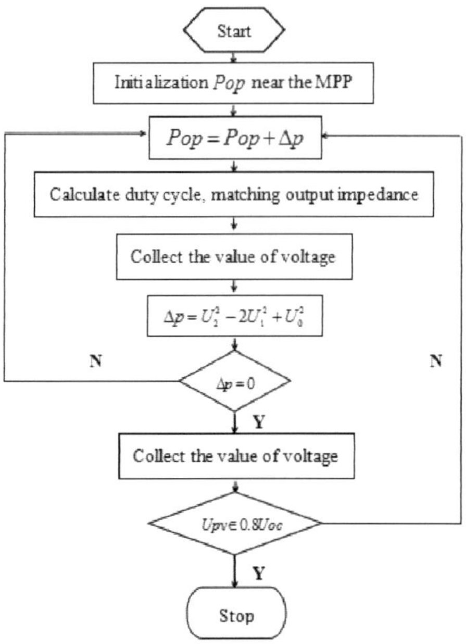

Fig.4 The MPPT control flow chart

4 Prototype testing and experimental curves

In this paper, the converter rated capacity is 230VA, input DC 30V, output AC 220V, the main power MOSFET used IRF4321, and the parasitic capacitance Coss = 390 pF, the primary induction of the main power switches is 5.4 μH, the leak inductance is 0.2 ηH, the ratio of turns is 1:6, the clamp capacitance Ca = 3.3 nF, the leak inductance recovery resistance R1 = 100K, the clamp diode used ES1D, the switch tube in secondary side used 17N80, and the commutation diode used C2D05120.

The experiment waveform are shown in figure-5. Figure-5(a) shown the Simulation of light intensity and temperature change curve in a day[9,11-12]. The figure-5(b) shown the Prototype maximum power point tracking in real time and the efficiency of MPPT. We can clearly see that, the power point stable near the MPP and the efficiency achieve 99.3%. The figure-6(c) shown the grid voltage waveform with the peak value is about 330V, and very smooth. The output converter efficiency can reach 96.2%.

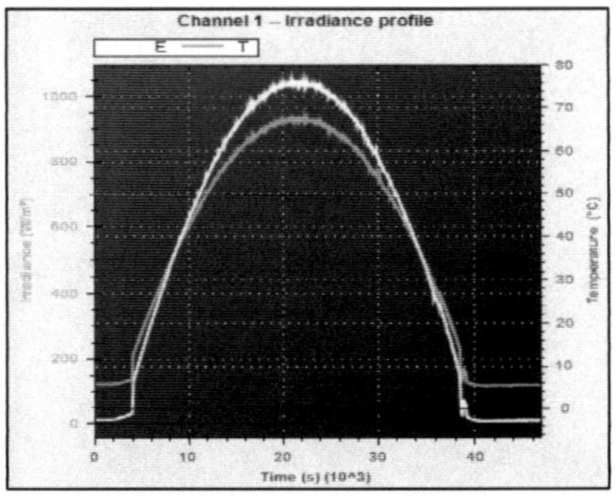

(a) The curve of light intensity and temperature

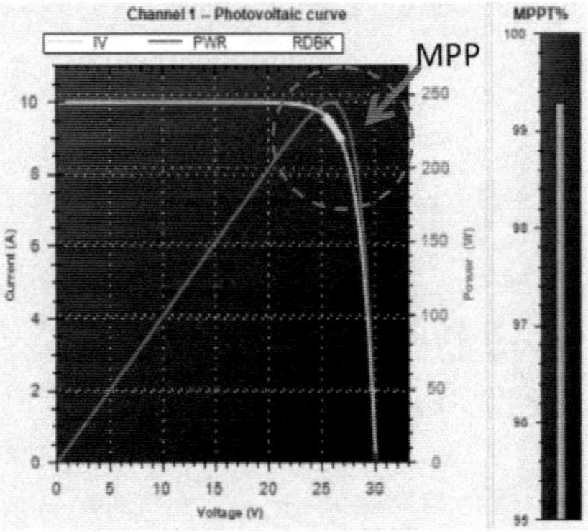

(b) The curve of MPPT and the efficiency

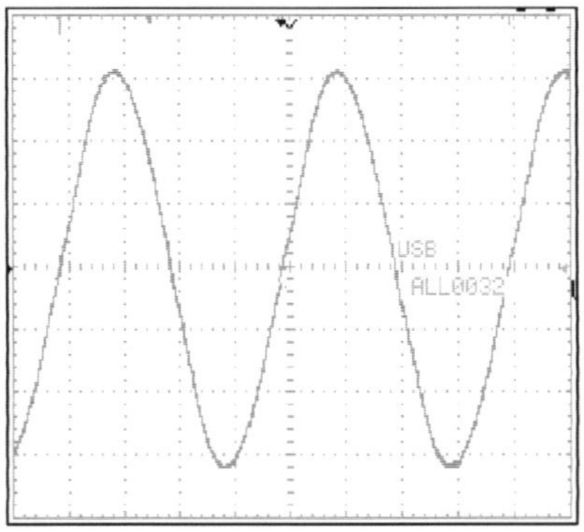

(t=10ms/lattice, peak value=100V/lattice)

(c) The converter AC output voltage

Fig. 5 Prototype testing and experimental curves

5 Conclusion

The Method of MPPT Control Based on Power Variable Step-size combining the advantage of constant-voltage and the perturb-observe, remove the circuit topology of current acquisition, simplify the circuit topology, reduce the algorithm requirements to the bottom of hardware. Improved the issue of voltage fluctuation caused by the method of perturb-observe, and the complicate of circuit topology and algorithm in traditional variable step-size incremental conductance.

References

1. WANG Ya-nan, YANG Xu-hong, WANG Jun-cheng, etc. One Novel Variable Step-size Strategy of MPPT for Photovoltaic System[J]ELECTRIC DRIVE,2015(01):54-57
2. XIONG Yuansheng, YU Li1, XU Jianming. MPPT control of photovoltaic generation system combining constant-voltage method with perturb-observe method[J]. Electric Power Automation Equipment,2009(06):85-88
3. Xue Yang, Wang Sha. Fuzzy Control Based on P&O Applied In Photovoltaic Maximum Power Tracking[J]. ACTA ENERGIAE SOLARIS SINICA,2014(09):1622-1626
4. LAI Dong-sheng, YANG Ping. Variable Step-size Incremental Conductance Used in Maximum Power Point Tracking Control of PV Power System[J]. Power Electronics,2012(3):40-42
5. XU Peng-wei, LIU Fei, LIU Bang-yin, DUAN Shan-xu. Analysis, Comparison and Improvement of Several MPPT Methods for PV System[J]. Power Electronics,2007(05):3-5
6. LIU Zeng-huan, LI Jie. Application of power prediction-based variable step perturbation and observation method in PV MPPT[J]. Manufacturing Automation,2015(02):2-4
7. YU Yun-jun, ZHANG Xiang, WANG Shi-sheng, ZHANG Yan-fei. Review of incremental conductance method for photovoltaic power generation MPPT[J]. Chinese Journal of Power Sources,2014(12):2440-2444
8. Wu Junjuan, Jiang Yida, Wang Qiang, Sun Xiaofeng. An Improved MPPT Control Algorithm of PV System[J]. ACTA ENERGIAE SOLARIS SINICA,2012(03):478-482
9. Yang Yongheng Zhou Keliang. Photovoltaic Cell Modeling and MPPT Control Strategies[J]. Transactions of China Electrotechnical Society, 2011(12):40-45

10. GU Junyin1, WU Hongfei2, CHEN Guocheng1, XING Yan 2. Soft-switching Grid-connected PV Inverter With Interleaved Flyback Topology[J]. Proceedings of the CSEE, 2011(12):40-45
11. Wu Dazhong, Wang Xiaowei. A Photovoltaic MPPT Fuzzy Controlling Algorithm[J]. ACTA ENERGIAE SOLARIS SINICA,2011(06):808-813
12. SUN Hang, DU Haijiang, JI Yingxu, YANG Bo. Photovoltaic distributed MPPT mechanism analysis and simulation study[J]. Power System Protection and Control,2015(01):48-54

DC link current simulation of voltage source inverter with random space vector pulse width modulation

Guoqiang Chen and Jianli Kang

School of Mechanical and Power Engineering, Henan Polytechnic University, Jiaozuo 454003, China

Abstract. Aiming at analysis complexity, a simulation model is built and presented to analyze and demonstrate the characteristics of the direct current (DC) link current of the three-phase two-level inverter with the random space vector pulse width modulation (SVPWM) strategy. The developing procedure and key subsystems of the simulation model are given in detail. Several experiments are done using the simulation model. The results verify the efficiency and convenience of the simulation model and show that the random SVPWM scheme, especially the random switching frequency scheme, can efficiently suppress the harmonic peaks of the DC link current.

1 Introduction

The voltage source inverter (VSI) has been widely in adjustable frequency drives (AFDs) in all kinds of fields [1]. For example, the drive system of the electric vehicle in the drive mode can be simplified in Figure 1. The battery provides the direct current (DC) power, the DC link transmits the DC voltage to the inverter, and the ride-through capacitors are connected to the DC link cables to store the harmonic energy. The inverter changes DC to alternating current (AC) to feed the load, the AC motor [2, 3]. The inverter uses semiconductor switches to make the load terminal connect to the positive or negative rail of the DC link. The semiconductor switch includes the insulated gate bipolar transistor (IGBT), the insulated gate commutated thyristor (IGCT), and the injection enhanced gate transistor (IEGT), the gate turn-off thyristor (GTO), and so on. The three-phase two-level inverter shown in Figure 1 is mainly composed of 6 switches. The switches are always fully controllable and can be turned on and off by the third leads, the gate leads T_1, T_2, T_3, T_4, T_5 and T_6 in Figure 1.

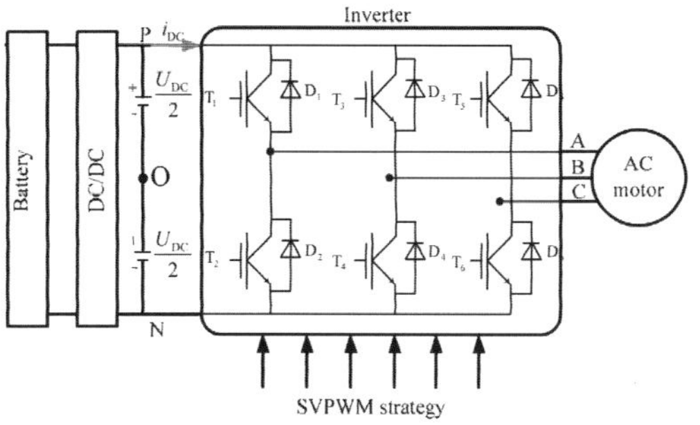

Figure 1. Drive system of electric vehicle in drive mode.

The six gate lead control signals are generated using the space vector pulse width modulation (SVPWM) strategy [4,5]. The SVPWM strategy uses the specific algorithm to compute the on and off state duration time of the six switches and generates the PWM gate control signals, so it is very convenient to implement in the digital control system. The input current i_{DC} of the inverter consists of high frequency rectangular stream of pulses with changeable widths and magnitudes [6]. The current pulses of the DC link current are phase current segments that flow through the DC link cable. The widths and magnitudes of the DC link current pulses highly depend on the AC motor speed and load torque, the SVPWM switching frequency and the SVPWM strategy. Besides the DC component, there are plenty of harmonics in the DC link current. The DC component that flows through the battery is corresponding to the output power of the AC motor. And the AC harmonics flow through the ride-through capacitor. The harmonic currents have significant influences on the working life of the capacitors and electromagnetic radiation [7]. The issues on how to compute the harmonic spectrum and the capacitor current RMS value of the DC link current and determine the capacitor volume have gained plenty of study. J.W. Kolar and S.D. Round [7] derived the analytical expression for the current stress on the DC-link capacitor. T.D. Nguyen et al. [8] proposed a new PWM strategy called Uni-DCPWM that can effectively reduce the RMS current value of the DC link capacitor. N. Rouhana et al. [6] presented a method to analyze the DC-link current harmonic components for the Uni-DC PWM and the traditional SVPWM strategy using the double Fourier integral formulation. J. Hobraiche et al. [9] proposed an optimal strategy that uses selective harmonic elimination PWM to control the DC-link current harmonics. X. Pei et al. [10] discussed analysis and calculation of the DC link current and voltage

[a] Corresponding author: jz97cgq@sina.com

ripples for a three-phase inverter with unbalanced load. The above study is focused on the deterministic SVPWM strategy. The volt-second balance principle that the SVPWM strategy adopts results in plenty of harmonics in the output waveforms of the inverter besides the required fundamental or first harmonic. The phenomenon is most serious in the deterministic SVPWM strategy because the harmonics around the multiples of the switching frequency have large magnitudes. The random strategy has been invented to suppress the peak harmonics and reduce the serious undesirable effects through spreading the harmonics continuously to a wideband area [11-19]. The DC link current presents tremendous difference in the random SVPWM strategy. The theoretical analysis and the analytical expression of the DC link current characteristic are always complicated and tedious. Therefore a simulation model is built and the characteristic of the DC link current with the random SVPWM strategy is discussed based on the simulation results in the paper.

2 Principle of random SVPWM

The topology of the inverter discussed in the paper is shown in Figure 1[2,4,5]. The DC link voltage is U_{DC}. The inverter has 8 permissible states that are corresponding to 8 basic space vectors as shown in Figure 2. An arbitrary command/reference voltage vector $\vec{U}_s$ with the vector amplitude U_o and the phase angle θ inside the hexagon region shown in Fig. 2 can be generated by two adjacent active vectors (for example, $\vec{U}_1$ and $\vec{U}_2$ in the first sextant) and the zero vectors. The on-state duration time T_1, T_2 and T_0 of the three basic vectors are determined by the identical volt-second balance at the periodical time interval/switching period T_s using Equation (1). There are large numbers of vector operation modes that satisfy this equation.

$$T_s\vec{U}_s = T_1\vec{U}_1 + T_2\vec{U}_2 \tag{1}$$

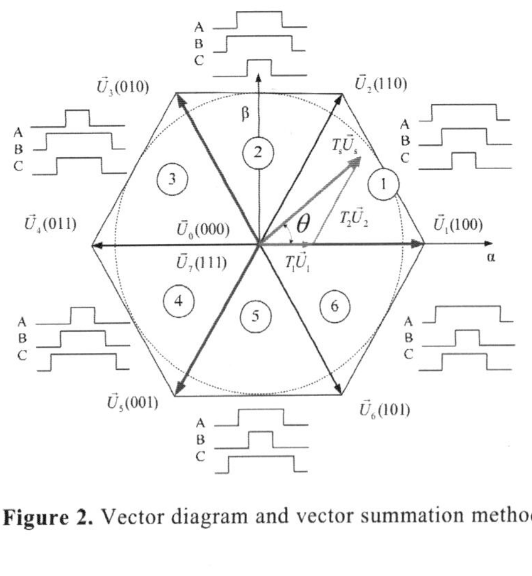

Figure 2. Vector diagram and vector summation method.

For the commonly used 7-segment SVPWM pattern operation of the inverter, the PWM waveforms in the 6 sextants are shown in Figure 2. The vector sequences in the 6 sextants are shown in Table 1.

Table 1. Vector sequences of 7-segment SVPWM pattern.

Sextant No.	Vector sequence
1	$\vec{U}_0(000)\to\vec{U}_1(100)\to\vec{U}_2(110)\to\vec{U}_7(111)$ $\to\vec{U}_2(110)\to\vec{U}_1(100)\to\vec{U}_0(000)$
2	$\vec{U}_0(000)\to\vec{U}_3(010)\to\vec{U}_2(110)\to\vec{U}_7(111)$ $\to\vec{U}_2(110)\to\vec{U}_3(010)\to\vec{U}_0(000)$
3	$\vec{U}_0(000)\to\vec{U}_3(010)\to\vec{U}_4(011)\to\vec{U}_7(111)$ $\to\vec{U}_4(011)\to\vec{U}_3(010)\to\vec{U}_0(000)$
4	$\vec{U}_0(000)\to\vec{U}_5(001)\to\vec{U}_4(011)\to\vec{U}_7(111)$ $\to\vec{U}_4(011)\to\vec{U}_5(001)\to\vec{U}_0(000)$
5	$\vec{U}_0(000)\to\vec{U}_5(001)\to\vec{U}_6(101)\to\vec{U}_7(111)$ $\to\vec{U}_6(101)\to\vec{U}_5(001)\to\vec{U}_0(000)$
6	$\vec{U}_0(000)\to\vec{U}_1(100)\to\vec{U}_6(101)\to\vec{U}_7(111)$ $\to\vec{U}_6(101)\to\vec{U}_1(100)\to\vec{U}_0(000)$

The duration time for the two basic active vectors in the first sextant is

$$\begin{cases} T_1 = \dfrac{\sqrt{3}}{2} M T_s \sin(\pi/3 - \theta) \\ T_2 = \dfrac{\sqrt{3}}{2} M T_s \sin\theta \end{cases} \tag{2}$$

where the modulation index M is always given by

$$M = \frac{U_o}{U_{DC}/2} \tag{3}$$

The total duration time of the two zero vectors (T_{00} for $\vec{U}_0$ and T_{07} for $\vec{U}_7$) is

$$\begin{aligned} T_0 &= T_s - T_1 - T_2 = T_{00} + T_{07} \\ &= T_s\left(1 - \frac{\sqrt{3}}{2} M \sin(\pi/3 + \theta)\right) \quad (0 \le \theta \le \pi/3) \end{aligned} \tag{4}$$

There are three single factor random SVPWM schemes as follows [18, 19].

(1) Random switching frequency SVPWM (RSFSVPWM): the switching frequency or the switching period T_s is randomized in an interval.

(2) Random zero-vector distribution SVPWM (RZDSVPWM): the zero vector duration time distribution ratio between the zero vectors $\vec{U}_0$ and $\vec{U}_7$ is controlled by a random variable R_0. The duration time T_{00} for $\vec{U}_0$ and T_{07} for $\vec{U}_7$ is

$$\begin{cases} T_{00} = R_0 T_0 \\ T_{07} = (1 - R_0) T_0 \end{cases} \quad (0 \le R_0 \le 1) \tag{5}$$

(3) Random pulse position SVPWM (RPPSVPWM): the pulse positions of the three phases are controlled by three independent variables R_1, R_2 and R_3. The duration time is shown in Figure 3 and expresses as

$$\begin{cases} t_1 = R_1 T_{00} \\ t_2 = R_2 T_1 \\ t_3 = R_3 T_2 \\ t_4 = T_{07} \\ t_5 = (1 - R_3) T_2 \\ t_6 = (1 - R_2) T_1 \\ t_7 = (1 - R_1) T_{00} \end{cases} \quad \begin{pmatrix} 0 \le R_1 \le 1 \\ 0 \le R_2 \le 1 \\ 0 \le R_3 \le 1 \end{pmatrix} \qquad (6)$$

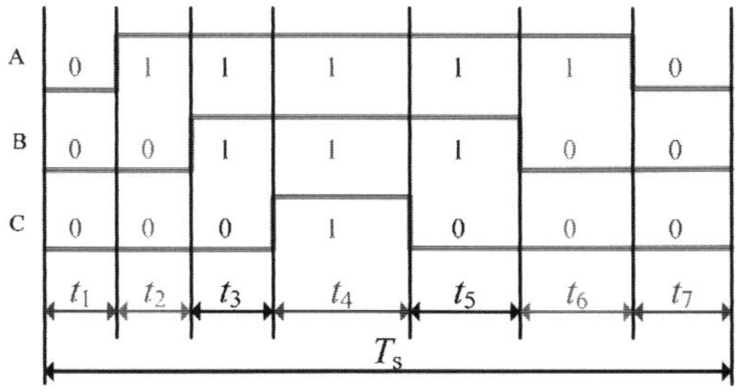

Figure 3. 7-segment SVPWM pattern in the first sextant.

3 Simulation model

A simulation model is built in MATLAB/Simulink combined with Plecs based on the control block shown in Figure 4 [20, 21]. The proportional–integral (PI) controller is used for the speed controller and current controllers. The 2-Level Converter (Ideal three-phase two-level converter.) block and PMSM (Permanent Magnet Synchronous Machine) block are used directly.

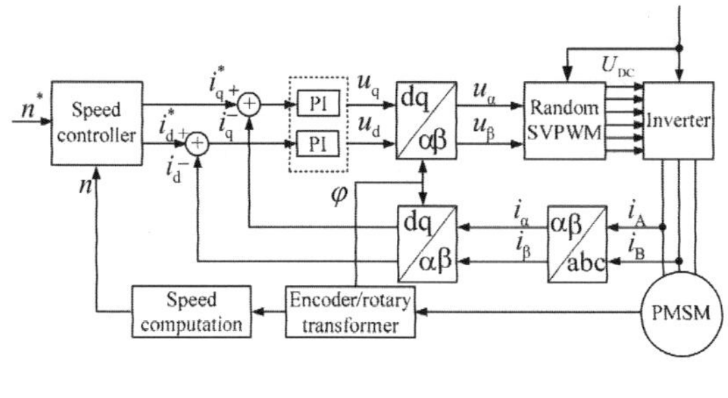

Figure 4. Control block diagram of an electric vehicle.

The random SVPWM subsystem is the key to the simulation and is built using Embedded MATLAB Function block. The flow chart of the random SVPWM subsystem is shown in Figure 5. One random variable is needed in the RZDSVPWM and RSFSVPWM schemes, and three random variables are needed in the RPPSVPWM scheme. The random variable is presented by the random values. The random values that represent the random variables in Equations (5) and (6) should be held constant in the one switching period. Global or persistent variables can be used to store the random values. Persistent variables differ from global variables in that persistent variables are known only to the function in which they are declared. So the persistent variable is more convenient here than the global variable because the random value is only used in one function. The random values are generated at the first simulation step (controlled by the simulation solver) and hold constant during the whole switching period.

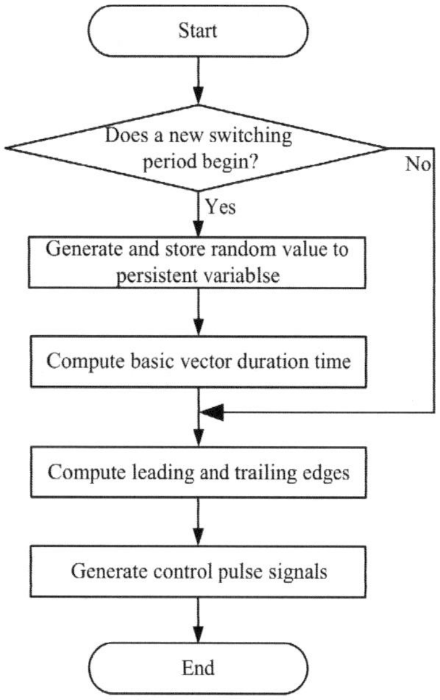

Figure 5. Flow chart of random SVPWM subsystem.

Random functions in MATLAB can be directly used to generate the random values that obey the specified distribution. For example, random (NAME, A, B) or random (NAME, A, B,C) returns an array of random numbers chosen from a two- or three-parameter probability distribution with parameter values A, B (and C). rand(N) returns an N-by-N matrix containing pseudorandom values drawn from the standard uniform distribution in the open interval (0,1).

4 Simulation analysis and discussion

The simulation experiments are done using the built simulation model described in Section 3. The DC link voltage is 300V. The number of the pole pairs of PMSM is 4, the stator resistance is 0.0113 Ω , the d-axis and q-axis inductance of the combined stator leakage and magnetizing inductance [Ld, Lq] is [0.000175 H, 0.000284 H], and the flux induced by magnet is 0.08424 Wb. The specified reference speed is 1200rad/s. All the four random variables in Equations (5) and (6) obey to the standard uniform distribution in the interval (0,1), and the random values are generated using MATLAB function rand. The switching frequency is 15000Hz for the deterministic SVPWM scheme, and the interval (10000Hz, 20000Hz) with the uniform distribution for

DOI: 10.1051/matecconf/20166102014

the RSFSVPWM scheme. The simulation results are shown in Figures 6 to 8.

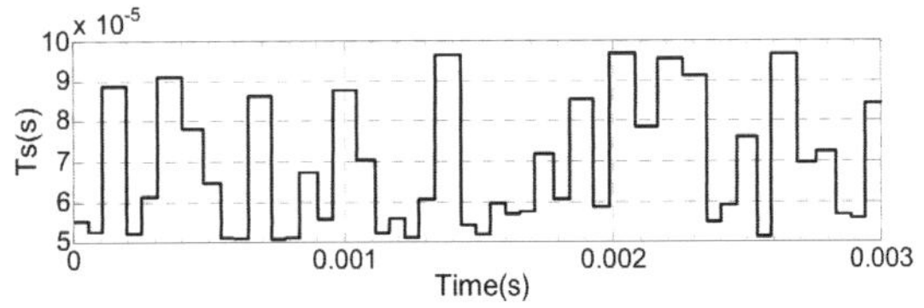

Figure 6. Switching period of RSFSVPWM.

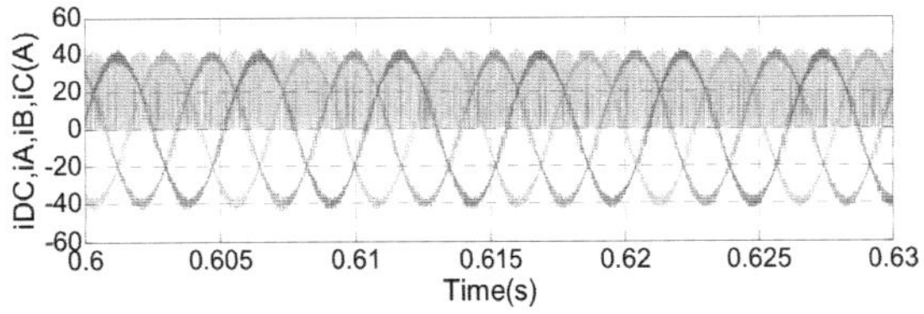

Figure 7. Phases and DC link currents of RSFSVPWM.

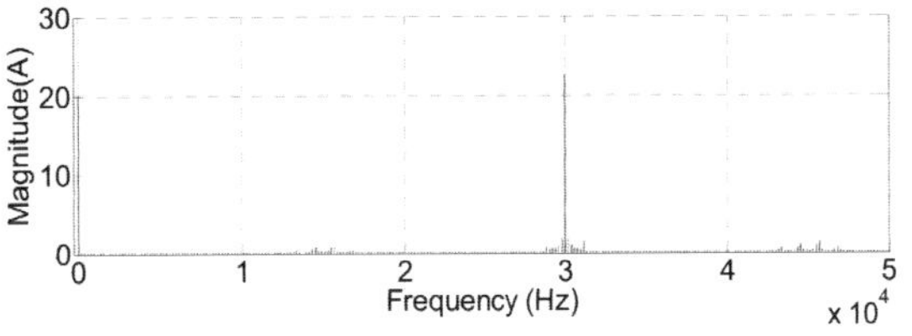

a) Deterministic SVPWM for 15000Hz.

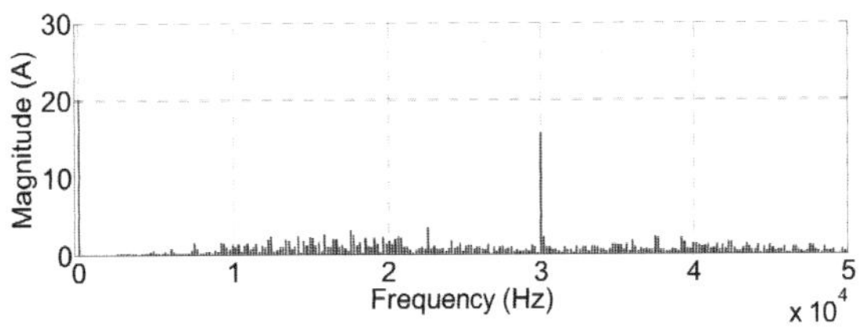

b) RZDSVPWM for 15000Hz.

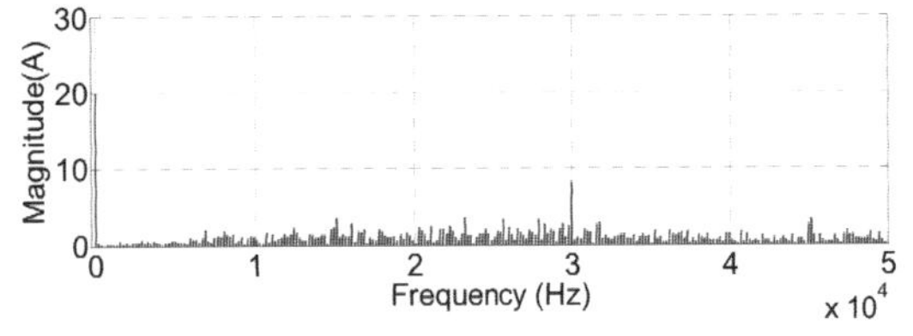

c) RPPSVPWM for 15000Hz.

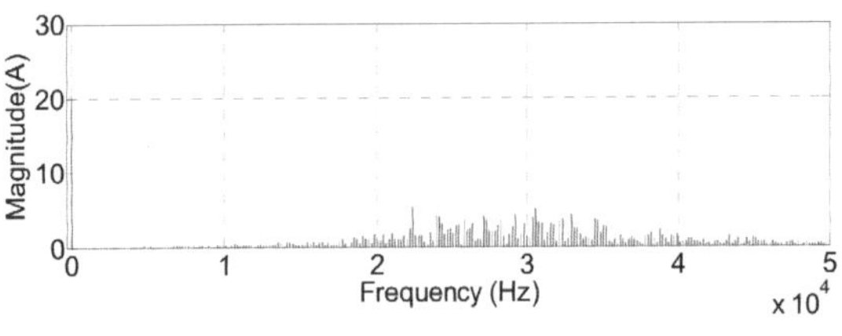

b) RSFSVPWM for [15000Hz, 20000Hz].

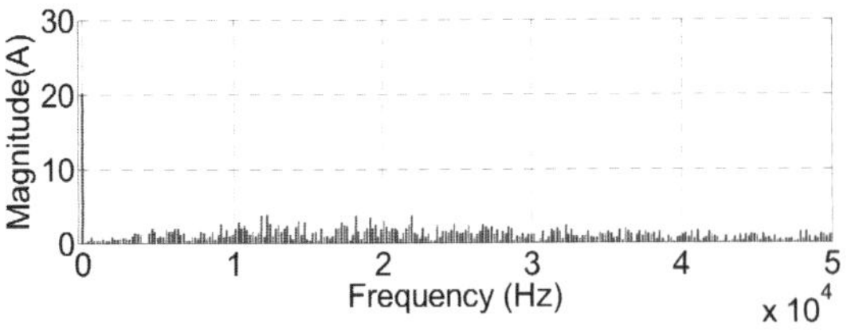

e) Hybrid random SVPWM for [10000Hz, 20000Hz].

Figure 8. Spectra of deterministic and random schemes.

The hybrid SVPWM scheme is a hybrid of not less than two of the three basic random schemes: RZDSVPWM, RPPSVPWM and RSFPWM schemes. The result shown in Figure 8(e) is corresponding to the hybrid of the above three random schemes.

The specified reference speed 1200rad/s is corresponding to the period $\pi/600$ s. In order to be compared intuitively and conveniently, the DC link currents of the random schemes are sampled in a whole period $\pi/600$ s. The discrete Fourier transform (DFT) function FFT in MATLAB is used to compute the spectra. It should be noticed that there is a little difference between the results in different sampling periods.

Figure 6 presents that the switching period is constantly changing with the time. The switching functions of the three phases are S_A, S_B and S_C that are shown in Figures 2 and 3, and the three phase currents are i_A, i_B and i_C. Figure 7 demonstrates that the DC-link current i_{DC} is the sum of three phase currents and expressed as $i_A S_A + i_B S_B + i_C S_C$ **[10]**. The randomization makes the pulse leading and trailing edges randomized to some extent, so the switching function and the three phase currents are also randomized to the same extent. Therefore the sum of the three phase currents is also randomized. The randomization can efficiently suppress the peak harmonics of the DC link current. From Figure 8, we can find that the magnitudes of the harmonics around 30000Hz are more than 20A for the deterministic SVPWM scheme, but they are about 15A for the RZDSVPWM scheme, and less than 10A for the RPPSVPWM, RSFSVPWM and hybrid SVPWM schemes. Especially for the hybrid SVPWM scheme, the magnitudes of the harmonics are less than 5A.

5 Conclusions

A simulation model for the DC link current analysis has been built in MATLAB and presented in this paper. The random SVPWM subsystem is the key to the simulation. The random SVPWM subsystem is built based on the Embedded Function block. In order to keep the random values those represent the variables constant in one switching period, the persistent variables are used. The results presented in the paper show that the random SVPWM scheme can efficiently suppress the harmonic peaks of the DC link current, especially the random switching frequency SVPWM scheme has most excellent performance. The simulation method is very flexible in analyzing the DC link current characteristic.

Acknowledgments

This work is supported by National Science Foundation of China (No. U1304525) and the Science and Technology Key Project of Education Department of Henan Province of China (No. 13B460027). The author would like to thank the anonymous reviewers for their valuable work.

References

1. F. Wu, J. Zhao, IEEE Trans. Power Electron. **31**, 1425 (2016)
2. N. Mutoh, M. Nakanishi, M. Kanesaki, J. Nakashima, IEEE Trans. Electromagn. Compat. **47**, 930(2005)
3. H. Shim, H. Kim, Y. Kwack, M. Moon, H. Lee, J. Song, J. Kim, B. Kim, E. Kim, Proc. IEEE Int. Symposium on Electromagn. Compat. (2015)
4. H. Saidi, R. Taleb, N. Mansour, A. Midoun, Proc. IEEE 6th Int. Renewable Energy Congress, 1(2015)
5. G. Zheng, S. Guo, S. Cai, S. Fan, Y. Ren, Proc. IEEE 7th Int. Power Electronics and Motion Control Conf. **4**, 2489(2012)
6. N. Rouhana, N. Patin, G. Friedrich, E. Negre, S. Loudot, 17th European Conf. on Power Electronics and Applications, 1(2015)
7. J.W. Kolar, S.D. Round, IEEE Proc.-Electr. Power Appl. **153**, 535 (2006)
8. T. D. Nguyen, N. Patin, G. Friedrich, Proc. IEEE Vehicle Power and Propulsion Conf., 1(2011)
9. J. Hobraiche, J. P.Vilain, C. Plasse, IEEE 35th Annual Power Electron. Specialists Conf. **5**, 3336 (2004)
10. X. Pei, W. Zhou, Y. Kang, IEEE Trans. Power Electron. **30**, 5401(2015)
11. X. Mao, J. A. Kumar, A. Rajapandian, IEEE Trans. Power Electron. **26**, 1954 (2011)
12. D. G. Holmes, T. A. Lipo, *Pulse Width Modulation for Power Converters: Principles and Practice.* (IEEE Press, USA, 2003)
13. K. Shahriyar, M. Javad, A. Ali, IEEE Trans. Ind. Electron. **54**, 2333(2007)
14. S. H. Na, Y. G. Jung, Y. C. Lim , S.H. Yang, IEEE Proc.-Electr. Power Appl.**149**, 195(2002)
15. H.W. Wang, J. Zhang, T. F. Yang, Electron. Measurement Technol. **34**, 18(2011)
16. L. Zheng, H. Q. Long, Chinese J. Automotive Eng. **4**, 319(2014)
17. M. L Du, Z. M. Xu, L. X. Ding, Automotive Eng. **36**,734(2014)
18. G. Q. Chen, Z.H. Wu, Y. Zhu, J. Tongji University **40**, 1111(2012).
19. G. Q. Chen, M. J. Zhang, J. W. Zhao, Adv. Inf. Sci. Serv. Sci. **4**, 242(2012)
20. A. Kiselev, A. Kuznietsov, R. Leidhold, Proc. 9th Int. Conf. on Compat. and Power Electron., 566 (2015)
21. K. Roumani, B. Schmuelling, Proc. IEEE Conf. on Energy Conversion, 201(2014)

Research on Stealthy Headphone Detector Based on Geomagnetic Sensor

Ya Liu[1], Si Chen[2]and Zhihui Luo[1,a], Liqing Pan[1]

[1]*Research Institute for Magneto-electronics & nT Magnetic Field Detection,China Three Gorges University, Yichang 443002,China*
[2]*College of Science, China Three Gorges University, Yichang 443002,China*

Abstract.A kind of stealth headphone detector based on geomagnetic sensor has been developed to deal with the stealth headphones which are small, extremely stealthy and hard to detect. The U.S. PNI geomagnetic sensor is chosen to obtain magnetic field considering the strong magnetic performance of stealth headphones. The earth's magnetic field at the geomagnetic sensor is eliminated by difference between two geomagnetic sensors, and then weak variations of magnetic field is detected. STM8S103K2 is chosen as the central controlling chip, which is connected to LED, buzzer and LCD 1602. As shown by the experimental results, the probe is not liable to damage by the magnetic field and the developed device has high sensitivity, low False Positive Rate (FAR) and satisfactory reliability.

1 Introduction

Stealthy headphone is used as a common tool for cheating and illegal monitoring at present and has brought great trouble to the inspection work. As shown in Figure 1, stealthy headphone is characterized by small size, light weight, easy to use and so on[1]. Stealth headphone is mainly divided into two types, namely rare earth type and electronic type[2]. The rare earth type is mainly composed of ndfeb magnets, and the vibrations coming from the audio frequency magnetic field enable the wearer to hear sounds. The electronic type mainly relies on coil of inductive electromotive force in the audio field, and the active ceramic drives trumpet to make sounds[3].

Figure 1. The Surface of Stealthy Headphone.

For each type of stealthy headphone, its internal contains magnetic material, whose near magnetic field strength is about 5~10Gs. But when the detect distance is 8cm, the magnetic field strength is attenuated to be below 10^{-3}Gs (the geomagnetic field strength is about 0.5Gs)[3]. The main methods for detecting stealth headset are as follows:

1) Visual inspection. The inspector observes whether there is obvious foreign body inside the candidate's ears through naked eye. Because the volume of the stealthy headset which can be stuffed into the external auditory canal directly is smaller and smaller, it is hard for the inspector to find the rice type stealthy headset through visual method.

2) Using electromagnetic wave detection device. Though traditional electromagnetic detection equipment can detect electromagnetic wave, the detection device fails to effectively detect when the carriers don't open the headphone[4, 5].

3) Using metal detectors. Metal detectors which are sensitive to metal materials with certain volume are easy to lead misinformation when influenced by irrelevant metal, such as belt buckle, key, hairpin and zipper, which are always on the object to be inspected. In addition, it is hard for the metal detector to detect the tiny devices, let alone the rice granular stealthy headphone[5].

In this paper, the U.S. PNI geomagnetic sensor is chosen to obtain magnetic field considering the strong magnetic performance of stealth headphones, the magnetic field is eliminated by the magnetic field of the geomagnetic sensor to obtain the weak magnetic field [6]. As shown in Table 1, PNI geomagnetic sensor has many characteristics, this is, small size, low power consumption, high resolution and low cost. The device can effectively detect changes of the magnetic field and eliminate the magnetic field interference. In general, it has good application prospect.

[a] Corresponding author: zhihui_luo@126.com

Table 1. Sen-XY-f Characteristic

Power Consumption	Package Size	Measuring Range	DIP	Temperature Range
3.3VDC, i<100μA	6.0×2.1×2.1mm (XY)	±1100μT(±11Gs)	13nT(0.13mGs)	-40to 85°C

2 Principle of geomagnetic sensor

The operating circuit of magnetic sensor PNI includes LR oscillation measured positively and reversely which consists of the bias resistors, Schmitt trigger and inductive sensors.

As shown in Figure2, when the circuit works, the Schmitt trigger outputs high voltage level and the inductance sensor and resistor connect the earth, then the positively measure begins to work[7]. The partial pressure on resistor keeps increasing, the trigger turns back when the voltage at point A is higher than the threshold voltage of Schmidt trigger.

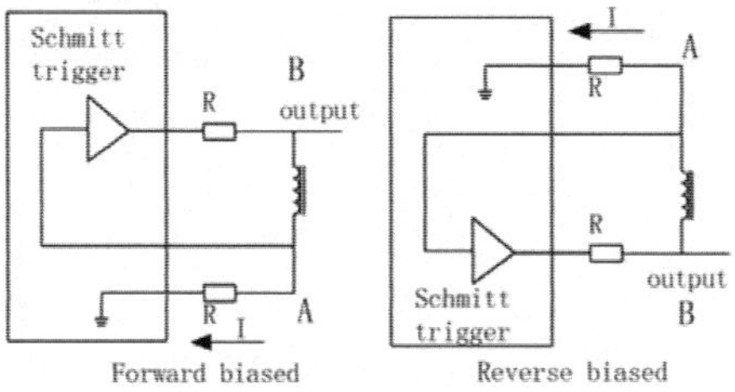

Figure 2. LR Oscillator Circuit Biasing Diagram

The energy storage of the inductance sensor begins to release, the voltage at point A decreases to low voltage level. According to the above steps repetitively, the oscillation circuit based on LR is formed. The period of oscillation directly depends on the size of the resistance and inductance. The process of the reversely measured is similar to the positively measured, in which the current flows through the sensors in the opposite direction. The period of output signal at Point B will be tested by dedicated IC in digital way[8].

When the inductance value of inductive sensors is affected by external magnetic field, the oscillation signal cycle at the output point will change. For a specific sensor, the inductance value is directly decided by the total magnetic field strength. The size of the total magnetic field is

$$H = k_o I + H_E \quad (1)$$

Here H_E is the external magnetic fields, H is magnetic field generated by electric current I, and k_o is a constant which is determined by the sensor material[9,10].

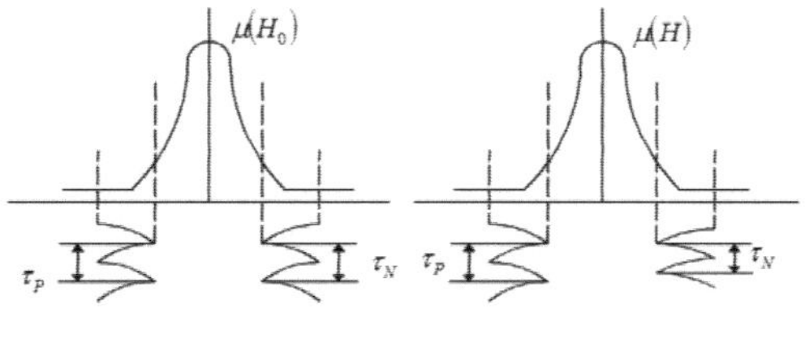

Figure 3. The Oscillation Cycle at Point A

When the external magnetic field is zero, the inductance value of the inductive sensor depends on the magnitude of the current. The oscillation period at point A for the positive and reverse measurements is the same. As shown in the Figure 3, when the sensor is influenced by the external magnetic field, positive total magnetic intensity increases. The oscillation period at point A τ_N increases when positively measured and decreases when reversely measured. By the differential treatment to the positive and negative oscillation period, the influence of the sensor's own magnetic field can be eliminated and the digital signal proportional to the external magnetic field H_E can be obtained.

3 Hardware design

Figure 4 shows the hardware structure of the detection device based on the magnetic sensor. STM8S103K2 is selected as the core control chip, of which the SPI interface, UART interface, multiple IO ports and other resources are used.

In the design, the power module consists of power whose control chip is 5V and DC is 3.3V. 5V supplies to control chip STM8S103K2 and 3.3V supplies to a magnetic sensor driver chip. The control chip STM8S103K2 and magnetic sensor driver chip MagI2C communicate through SPI interface. Interface RS232 is equipped with voltage conversion chip MAX232, STM8S103K2 downloads program through the serial[11]. Stealthy headphone detection device based on terrestrial magnetic sensor uses differential probe circuit to eliminate the influence of magnetic field, the interval of the two SEN-XY-f arranged on the PCB board is 1.5cm, 1cm is obtained by experiment. Geomagnetic sensor driver chip MagI2C converts analog magnetic field signal into digital pulse signal, which then passed to the chip STM8S103K2. The threshold of buzzer which controls the buzzer chirping is 500nT, as obtained through the experiment. Set different threshold of LED to control the corresponding LED light, as the measuring result increases through the control process, the number of lighted LED will also increase, different results of the two signal are showed on LCD1602 and presented accurately to inspectors in real time.

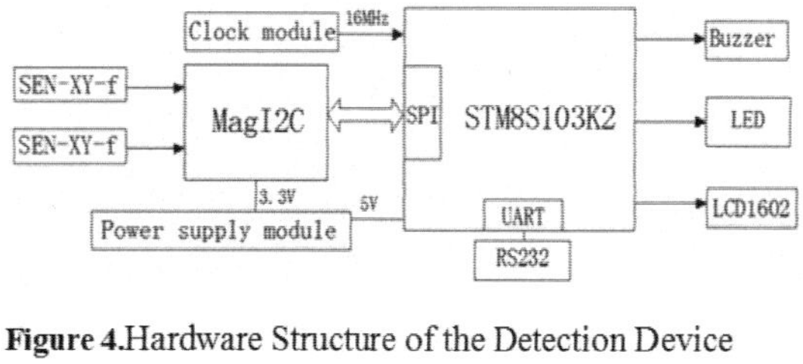

Figure 4. Hardware Structure of the Detection Device

A Sen-XY-f sensor coil serves as the inductive element in a simple LR relaxation oscillation circuit which is composed of an external bias resistor along with digital gates and a comparator internal to the MagI2C. The coil's inductance is a function of the magnetic field strength. As such, the frequency of the circuit oscillation changes with the strength of the total magnetic field parallel to the sensor. To make a measurement, one side of the coil is grounded while the other side is alternately driven with positive and negative current through the oscillator. The circuit is driven for a set number of circuit oscillations, and the time to complete the oscillations is measured using the MagI2C's internal high-speed clock[12]. The MagI2C next switches the bias connection to the coil and makes another measurement. The side previously grounded is now charged or discharged while the other is grounded.

4 Experiment and Discussion

Consider that the stealthy headphone will be placed on external auditory canal, the length of adult external auditory canal is about 2.0~2.5 cm. When inspectors hold stealth headphone detection device, the distance to the human ear is usually in the range of 2.0~ 4.0 cm. Magnet generates magnetic field in space which attenuates in accordance with the following equation[8].

$$H_0 = K\left(\frac{L+d}{L}\right)^3 H \qquad (2)$$

To test the magnetic field H_0 on the surface of the magnet, H is the magnetic field measured by probe. In this design, the measured value is observed by changing the interval of two SEN-XY-f, this is the change of the two SEN-XY-f differential component. Make a contrast experiment to the stealth headset and magnetic earrings.

As shown Figure 5.a. and 5.b, the X axis presents the distance between the first magnetic sensor and stealth headset, Y axis presents the magnetic field strength. Each of the three figures shows the cases when the distance between the two sensors is 1 cm, 1.5 cm and 2 cm, respectively. In the experiment, the measuring distance is 4.0~6.5cm, the magnetic field distribution is contrasted through changing the interval of two sensors. As shown in figure 5.a, two SEN-XY-f are presented. When the measuring distance is fixed, the measured values increases as the interval increases. With the decrease of measuring distance, the measured value of the magnetic field presents growth law of third power. When the value of interval is 2cm and the measuring distance is 4.0 ~ 6.5 cm, the measured results for the earrings with magnetic will be more than 500nT, this case leads to misjudgment. When the value of interval is 1cm and the measuring distance is 6.5 cm, the measured results for stealth headset will be less than 500nT, the headset fails to be detected in this case. When the value of interval is 1.5cm and the threshold is 500nT, the above problems can be effectively solved.

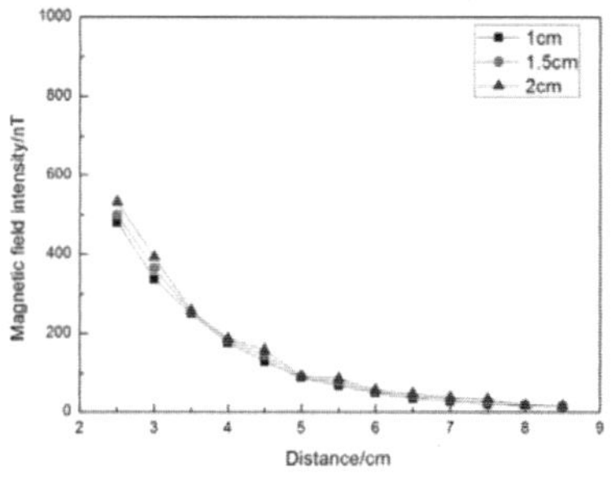

Figure 5.a. Measured Results for Earrings.

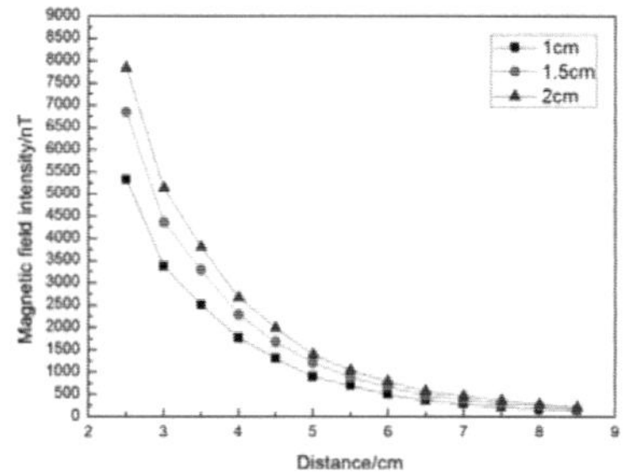

Figure 5.b. Measured Results for Stealthy headphone.

5 Conclusions

As shown by the experimental results, the developed device owns high sensitivity, low False Positive Rate (FAR) and satisfactory reliability, has a probe that's not liable to damage by the magnetic field, and is small in size, which has a very good application prospect.

Acknowledgment

This work is supported by the Graduate Student Research Innovation Foundation of China Three Gorges University (NO. 2015CX126) and Key Technology Support Program of Hubei Province, China (Grant No: 2014BEC060).

References

1. XuHuanhuan, SongShilei.Exploration on the mechanism of anti cheating in Higher Education.Hubei entrance examination, **36**(2012)
2. YangHanling.Talking about the examination of anti cheating equipment.Modern educational equipment in China, 04(2012)
3. WangYun.Stealth headset wits electronic test cheating.China Quality Miles, 10(2006)
4. ZhaoYantian. Modern anti cheating equipment in the examination room. China anti fake Report. 07(2009)
5. ZhangBing, JingYan. Development and application of invisible earphone detector. China Radio, 06(2011)

6. LiuYa, LuoZhihui. Three axis weak magnetic field detection system based on LabVIEW. Transducer and Microsystem Technologies, 07(2015)
7. ChenYongzhi, ZhangWei. Schmitt trigger and its sensor experiment for example,Middle school physics teaching reference, 11(2015)
8. ZhuBo, YuQi, NingNing. New type of electric reset circuiton low power consumption, Microelectronics, 01(2013)
9. Cirillo, Carla, García-Santiago, Antoni, Joan Manel Hernandez. Magnetic properties of double exchange biased diluted magnetic alloy ferromagnet antiferromagnet trilayers. Journal of Physics: Condensed Matter, 25, 17(2013)
10. V.R.Skal' skii,B.P.Klim,E.P.Pochapskii.Distribution of the induction of a quasi-stationary magnetic field created in a ferromagnet by an attachable electromagnet. Russian Journal of Nondestructive Testing, 48, 1(2012)
11. YanAijun, FangHaiming, ZhouYun. Touch button design based on single chip microcomputer STM8S103 and ST05A. Chemical Defence on Ships, 1(2013)
12. LuLingjie, QingHuibing. Design of intelligent home lighting system based on STM8S103 and WIFI. electric automatization, 03(2014)

MATEC Web of Conferences **61**, 02016 (2016) DOI: 10.1051/matecconf/20166102016
APOP2016

Review of Arc Models in Distribution Networks

Qi Yin, Rui Ding

State Grid Deyang Power Supply Co. Ltd., Sichuan, 61800, China

Abstract. The incipient fault in underground cable is recognized as arc fault. Then the arc model selection is very important for the incipient fault detection. The arc features and some typical models have been introduced in detail, including traditional thermal based models, arc models in low voltage and models of arc in long free air. At last, the Kizilcay's model is recommended to analyze the incipient fault in underground cable for its accuracy and widely utilized. Finally, some conclusions are summarized.

1 Introduction

Arc is defined as a discharge of electricity through a gas, normally characterized by a voltage drop in the immediate vicinity of the cathode approximately equal to the ionization potential of the gas. It's a continuous luminous discharge of electricity across an insulating medium which is changed into a conducting medium due to a huge number of free electrons and ions. The arc is a complex entity, often mobile, always intensely hot and capable of many forms[1-4]. The arc is affected by many factors. Even under steady-state conditions, arc behaviour depends to a large degree on such factors as electrode material, nature and state of the gas medium, range of current magnitudes, and whether the arc is relatively long or relatively short. There are large number of tests have been done under different conditions. The volt-ampere curves are always different under different test conditions.

Under dynamic conditions, arcs show important additional properties which depend to an essential degree on rates of current change. Often arc behavior is so interrelated with electrical properties of the connected circuit that it is just as responsive to circuit conditions as it is to the physical state of the arc's surroundings[5]. The different of free burning arcs with low currents and high currents are as follows[7]: At low currents, i.e. less than 30A in air, the controlling physical process is natural convection. It is assumed that the arc is vertical and that the input electrical energy produces an arc plasma which is carried upwards by natural convection. At very high currents, arc properties are also determined largely by convection, but the convection is determined by the self-magnetic field of the arc. The increase of pressure inside the arc is greatest at the cathode spot where the arc cross-section is least and it is this high pressure which induces the axial flow

2 The Development of Arc Analysis

Although the arc is harm to the system operation, it is also have been utilized in electric arc furnace, and electric arc welding for it can produce high temperature. In this report, the different types of arc models are introduced, especially the arc in the fault.

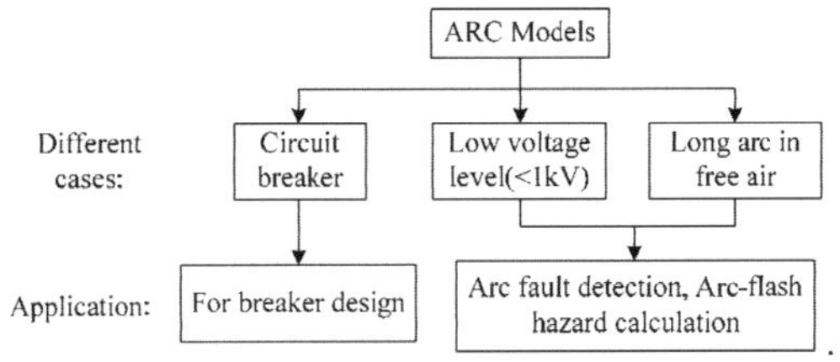

Figure 1. Different types of Arc model and its application.

The initial application of arc analysis is utilized in breaker design to extinguish the arc for it would damage the equipment. In recent three decades, the accurate arc models are developed for the arc fault detection. The accurate model is very important for the computer simulation to substitute large number of field tests which is high cost. And we can get the accurate features under different fault cases to verify the proposed methods.

The arc fault detection has been developed for many years. The arc in low voltage may cause fire hazard easily, so the arc detection is very important to prevent the fire disaster caused by arc. The transient faults in transmission system are always with arc in which the breaker should reclose as soon as possible to enhance the supply reliability. The incipient fault in power system is always the arc fault, then incipient fault detection became the arc fault detection. Also, the high impedance in medium-voltage distribution system always happen with arc, so the arc detection is very important to identify the high impedance fault.

3 The Arc Models

The arcs in power system have the similar nonlinear features which have been introduced in section 1, but the model is different with the different voltage level, medium, length and so on. Then the general characteristics of AC arc are introduced in 3.1. The instantaneous models in different conditions are introduced in 3.2. The arc models are different in different applications, including arc in breaker, low AC voltage system [8-22]. It is noted that the arc models for the energy calculation and secondary arc models are not introduced in this chapter.

3.1. Steady Properties

From the electrical properties of an arc under steady conditions (volt-ampere characteristic) point of view, a number of equations are derived from the experimental studies. The best known is that obtained by Ayrton.

$$U = A + BL + \frac{C + DL}{I} \tag{1}$$

where U is arc voltage, I is arc current and L is arc length, A is the anode/cathode voltage drop, B is the voltage gradient, C has the dimension of power, and D has the dimension of rate of power change over the arc length.
If in(1), L is made sufficiently large (the long arcs case), the terms involving parameters A and C may be neglected, and the characteristic equation becomes approximately

$$U = \left(B + \frac{D}{I} \right) L \tag{2}$$

If in (2) current is sufficiently large (the high current long arcs case), the arc voltage becomes a function only of the arc length, according to the following equation:

$$U = BL \tag{3}$$

Here, parameter B represents the voltage gradient in the arc column. It is almost independent of arc current, so the long high current arc voltages are essentially determined by the arc length L.

3.2. Dynamic equation

The basic assumption upon which mathematical "model" of the arc column are based on that the column conductance per unit length can be expressed as a prescribed function of the additional stored energy per unit length (often called heat content), Q, associated with the conducting state of the arc column:

$$g = \frac{1}{R} = F(Q) = F[\int (W - N)dt] \tag{4}$$

Where
g and R is conductance and resistance per unit length,
Q is excess energy content per unit length,
W is electric power input, watts per unit length,
N is rate of energy loss, watts per unit length, and
t is time.
In differential form, this equation is

$$\frac{dg}{dt} = \frac{d}{dt}\left(\frac{1}{R} \right) = (W - N)\frac{\partial F(Q)}{\partial Q} \tag{5}$$

Particular arc model equation depend upon the forms assumed for F(Q) and N.

3.2.1 Mayr's equation

The equation proposed and used by Mayr assumes:

$$F(Q) = K \exp(Q / Q_0) \tag{6}$$

Where
K is an unspecified constant,
Q_0 is a constant energy quantity,
N_0 is a constant rate of energy loss, all per unit length of arc column.

This quantity, Q_0, combined with the rate of energy loss, N0, determines the characteristic response frequency, f_0, of the arc, or inversely the arc's time constant, which may be given the symbol θ and conveniently defined by relation

$$\theta = \frac{Q_0}{N_0} \tag{7}$$

Substitute of these three relations in the general equation (5) yields Mayr's arc model differential equation

$$R\frac{d}{dt}\left(\frac{1}{R} \right) = \frac{1}{\theta}\left(\frac{W}{N_0} - 1 \right) \tag{8}$$

If W is expressed as I^2R, this becomes

$$\frac{d}{dt}\left(\frac{1}{R} \right) + \frac{1}{\theta}\left(\frac{1}{R} \right) = \frac{I^2}{\theta N_0} \tag{9}$$

Which is a linear first order equation if the current I=I(t) is determined solely by the connected circuit. If, on the other hand, W is written as E^2/R, where E is the column voltage gradient, equation (8) becomes

$$\frac{dR}{dt} - \frac{R}{\theta} = -\frac{E^2}{\theta N_0} \tag{10}$$

which is a linear first order equation in R if the arc column gradient E=E(t), is determined by the connected circuit alone.

3.2.2 Cassie's equation

A second arc model assume a column conductance simply proportional to Q, expressible by the relation

$$\frac{1}{R} = \frac{Q}{\theta E_0^2} \tag{11}$$

And a power loss per unit length also proportional to Q according to

$$N = \frac{Q}{\theta} \tag{12}$$

where E0 is a characteristic voltage gradient. Substituting these relations in the basic equation (5) yield Cassie's differential equation which can be written

$$R\frac{d}{dt}\left(\frac{1}{R} \right) = \frac{1}{\theta}\left[\left(\frac{E}{E_0} \right)^2 - 1 \right] \tag{13}$$

The linear first order equations may be obtained from (13) for cases where the circuit is an "infinitely stiff" source of either current or voltage. For impressed current, the equation can be written more conveniently

$$\frac{d}{dt}\left(\frac{1}{R^2}\right)+\frac{2}{\theta}\left(\frac{1}{R^2}\right)=\frac{2}{\theta}\left(\frac{I}{E_0}\right)^2 \tag{14}$$

which is linear in $1/R^2$. For impressed voltage, the equation may also be slightly simplified to

$$\frac{dR}{dt}=\frac{R}{\theta}\left(1-\frac{E^2}{E_0^{\,2}}\right) \tag{15}$$

which is linear in R.

4 Arc models in low-voltage AC system

Arcing faults have been recognized as a potential hazard in low-voltage systems as far back as the 1920s, but the number of arcing-fault events began to rise in the 1960s as electrical systems changed to meet greater load demands [8]. There are many models are used for incident energy calculation, but only several models could be utilized for instantaneous value prediction which have been summarized in [9].

4.1 Matthews model

An arcing fault in a complex circuit can be described by the following differential equation:

$$V_{\max}\sin(\omega t)=R\cdot i_{arc}+L\frac{di_{arc}}{dt}+V_{arc} \tag{16}$$

The general solution for the arc current is

$$i_{arc}(t)=\frac{V_{\max}}{|Z|}\cdot\left[\sin(\omega t-\phi)-\frac{m}{\cos\phi}+\left(\frac{m}{\cos\phi}-\sin(\omega t_a-\phi)\right)\varepsilon^{\frac{-\omega(t-t_a)}{\tan\phi}}\right] \tag{17}$$

The conduction angle, γ, can be determined with the FIND function in Mathcad from the following expression:

$$\varepsilon^{-\gamma/\tan\phi}=\frac{\dfrac{m}{\cos\phi}-\sin(\omega t_a+\gamma-\phi)}{\dfrac{m}{\cos\phi}-\sin(\omega t_a-\phi)} \tag{18}$$

In the first half-cycle, the arc current begins to flow at ta, when the supply voltage equals the restrike voltage. The current continues to flow until the conduction angle has been reached. At this point, the current becomes zero and the stored magnetic energy has been dissipated. The arc current in the second half cycle is a mirror image of the current in the first half-cycle.

4.2 Stokes and Oppenlander model

The models are expressed as differential equation, the arc currents are solved using a Runge-Kutta numerical method. It incorporates a current-dependent arc voltage based on an equation.

$$V_S=Ri_{arc}+L\frac{di_{arc}}{dt}+(20+534\cdot g)i_{arc}^{0.12} \tag{19}$$

The arc voltage in (19) was derived from the pseudo constant portion of an ac signal, i.e., near the peak, and was determined from over 200 digitally sampled arc tests with more than two million arc current and voltage data points. The model was verified with the conditions—arc gaps from 5 to 500mm, arc current from 0.1 to 20000A and arc powers from 50W to over 30MW.

4.3 Lawrence's model

The Lawrence's model incorporates an arc voltage based on the product of arc current and Fisher's equation for arc resistance.

$$V_S=Ri_{arc}+L\frac{di_{arc}}{dt}+(25\sqrt{39.37\cdot g})i_{arc}^{0.15} \tag{20}$$

The gap is expressed in meters in (20), the arc voltage term incorporated in (20) was used in Fisher's method to directly determine the rms arc current, without solving for the instantaneous current values.
It is believed that this equation is approximately correct for 120 to 600volt, 60Hz AC stable arcing burn down in a panel board or switchboard in which the arc length is approximately 2.5 inches.

4.4 Others

It is noted that there are many other models, such as the model proposed in [23] which new model derived by probability method according to the characteristics of three-phase fault arcs. Ref. [24] proposed a simplified model allows analyzing and determining basic parameters such as conduction angle, peak value, reduction factor of arc current, and power and energy supporting the understanding of the phenomenon of the arc flash.

5 Arc Models in Breaker

Breaker arc models can be classified into three categories [10]. Thermal and dielectric recovery models that describe the arc dynamical behaviour considering the impact of different interrupter parameters such as nozzle size and geometry, type of quenching medium and speed of flow, pressure, etc.. The second type of models is based on empirical forms. Both categories are efficient in determining the internal dimensions and quenching medium parameters of the interrupter. The third category of models is concerned with the arc external characteristics such as Cassie and Mary which have been introduced in 3.2. Although there are some modified method have been proposed for the analysis [11][12], they are not suit for the fault analysis. Then the arc models are not introduced in detail.

6 Summary

There are many types of arc models have been proposed for different destinations under different fault cases. It can be found that the model proposed by Kizilcay is the most widely used for arc fault analysis. The Kizilcay's model is time dependent dynamic resistance representation with emphasis of an empirical approach. It is flexible and is suit for different cases by changing the parameters in the model. However, the main problem of the model is the parameters selection: the unknown parameters must be estimated from test data or from experience.

References

[1] Stokes A D, Oppenlander W T. Electric arcs in open air[J]. Journal of Physics D: Applied Physics, 1991, 24(1): 26.
[2] Goda, Yutaka, et al. "Arc voltage characteristics of high current fault arcs in long gaps."Power Delivery, IEEE Transactions on 15.2 (2000): 791-795.
[3] Terzija, Vladimir, et al. "New Static "AirArc" EMTP Model of Long Arc in Free Air."Power Delivery, IEEE Transactions on26.3 (2011): 1344-1353.
[4] Elkalashy, Nagy I., et al. "Modeling and experimental verification of high impedance arcing fault in medium voltage networks."Dielectrics and Electrical Insulation, IEEE Transactions on14.2 (2007): 375-383.
[5] Browne T E. The electric arc as a circuit element[J]. Journal of the Electrochemical Society, 1955, 102(1): 27-37.
[6] Fisher, Lawrence E. "Resistance of low-voltage ac arcs."Industry and General Applications, IEEE Transactions on 6 (1970): 607-616.
[7] Lowke J J. Simple theory of free-burning arcs[J]. Journal of physics D: Applied physics, 1979, 12(11): 1873.
[8] Gammon, Tammy, and John Matthews. "The historical evolution of arcing-fault models for low-voltage systems." Industrial & Commercial Power Systems Technical Conference, 1999 IEEE.. IEEE, 1999.
[9] Gammon, Tammy, and John Matthews. "Instantaneous arcing-fault models developed for building system analysis."Industry Applications, IEEE Transactions on37.1 (2001): 197-203.
[10] Darwish, Hatem A., and Nagy I. Elkalashy. "Universal arc representation using EMTP."Power Delivery, IEEE Transactions on20.2 (2005): 772-779.
[11] Tseng, King-Jet, Yaoming Wang, and D. Mahinda Vilathgamuwa. "An experimentally verified hybrid Cassie-Mayr electric arc model for power electronics simulations."Power Electronics, IEEE Transactions on12.3 (1997): 429-436.
[12] Schavemaker P H, Van der Slui L. An improved Mayr-type arc model based on current-zero measurements [circuit breakers][J]. Power Delivery, IEEE Transactions on, 2000, 15(2): 580-584.
[13] Terzija, Vladimir V., and H-J. Koglin. "On the modeling of long arc in still air and arc resistance calculation."Power Delivery, IEEE Transactions on19.3 (2004): 1012-1017.
[14] Djuric, M. B., and V. V. Terzija. "A new approach to the arcing faults detection for fast autoreclosure in transmission systems."Power Delivery, IEEE Transactions on10.4 (1995): 1793-1798.
[15] Kizilcay, M., and T. Pniok. "Digital simulation of fault arcs in power systems."European Transactions on Electrical Power1.1 (1991): 55-60.
[16] Kizilcay, Mustafa, and Piergiovanni La Seta. "Digital simulation of fault arcs in medium-voltage distribution networks."15th PSCC, Liege(2005): 22-26.
[17] Kizilcay, M., and K - H. Koch. "Numerical fault arc simulation based on power arc tests."European Transactions on Electrical Power4.3 (1994): 177-185.
[18] Funabashi, Toshihisa, et al. "Influence of fault arc characteristics on the accuracy of digital fault locators."Power Delivery, IEEE Transactions on16.2 (2001): 195-199.
[19] Idarraga Ospina, G., D. Cubillos, and L. Ibanez. "Analysis of arcing fault models."Transmission and Distribution Conference and Exposition: Latin America, 2008 IEEE/PES. IEEE, 2008.
[20] Xu, Zhihan.Fault location and incipient fault detection in distribution cables. Diss. The University of Western Ontario, 2011
[21] Elkalashy, Nagy Ibrahim.Modeling and detection of high impedance arcing fault in medium voltage networks. Helsinki University of Technology, 2007.
[22] Johns, A. T., R. K. Aggarwal, and Y. H. Song. "Improved techniques for modelling fault arcs an faulted EHV transmission systems." IEE Proceedings-Generation, Transmission and Distribution141.2 (1994): 148-154.
[23] Wu, Huaren, et al. "Arc fault model for low-voltage AC systems."Power Delivery, IEEE Transactions on 20.2 (2005): 1204-1205.
[24] Parise, Giuseppe, Luigi Martirano, and Marco Laurini. "Simplified arc-fault model: the reduction factor of the arc current."Industry Applications, IEEE Transactions on49.4 (2013): 1703-1710.

MATEC Web of Conferences **61**, 02017 (2016) DOI: 10.1051/matecconf/20166102017
APOP2016

The Fault Causes of Overhead Lines in Distribution Network

Li Wang

State Grid Nanchong Power Supply Information & Telecommunication Co. Ltd., Sichuan, 637000, China

Abstract. The paper introduces the typical fault causes in distribution system, especially the fault in overhead line. The data in the paper are from the surveys by different agencies all around the world. It is noted that the statistics are different with each other for many reasons, and some of them are old which was finished in 1980s. The fault causes in different areas (Nordic, North American and China) are introduced separately. Some typical fault causes (trees, animals, lightning, and vehicles) are subdivided and analyzed in detail. Some waveforms of different faults are shown in the paper too. At last, the most frequent fault causes which should be focused on are proposed by the paper.

1 Introduction

There are lots of fault causes could lead to power outage, including the equipment failure, lightning, trees and so on. Some utilities and agencies have surveyed the power outage causes in power system 1-4. It is reported that most of the power outages are caused by the fault in distribution system (about 80% based on the statistics in Norway5) for the large number of distribution lines distributed in the complex environments. This paper focuses on the investigation about fault causes in distribution system, especially the fault on overhead line and underground cable.

The fault is inevitable in distribution system for many uncontrollable factors, such as animals, weather related factors. The different kinds of fault causes have the similar fault features. The fault causes can be identified by the fault feature analysis which is useful for the fault-finding and fault-clearing when a fault happened. It is noted that some statistics in the paper is about transmission line fault causes, although the fault causes are similar, the percentages of the fault causes are different.

A large EPRI study was done to characterize distribution faults in the 1980s at 13 utilities monitoring 50 feeders67. The fault causes are classified into 11 types: lightning, tree contact, equipment failure, animal, wind, dig-in, vehicle accident, ice/snow, vandalism, construction activity and other. In ref 1 and 2, the power outage was grouped into 8 possible causes by EATON, including animal, faulty equipment/human error, planned, unknown, vehicle accident, weather/trees, theft/vandalism, over demand. It is note that not all the power outage is caused by fault. ENTSOE(European network of transmission system operators for electricity) classified the fault causes into 7 types in 3, including lightning, other environmental cause, external influence, operation and maintenance, technical equipment, other and unknown. Alberta municipal affairs classified the damage on overhead line contacts in detail in 4. The reported faults in 4 are caused by mankind, including intentional and unintentional. The underground utility system fault causes are classified into excavating equipment, vehicles hitting transformers, pedestals, etc. There are 6 types of fault causes in urban distribution system classified by State Grid in China, including external factors, natural factors, improper maintenance and operation, improper installation, equipment failure and customer cause, as shown in Fig.1. The mentioned fault cause types can be subdivided into some specific fault causes.

2 Influence factors

The types of fault causes are similar, but the proportions are different in different countries, under different weather conditions and so on. Ref10 analyzed the weather related factors. Ref 11 introduced the influence of conductor type and environment. Ref 12 compared the fault causes data in different area of Duke power distribution systems.

All in all, the influence factors can be summarized as follows:

(1) The different grounding system: the system can be divided into effectively grounded system(including solidly grounded and low resistance grounded system which are used in north America) and non-effectively grounded system(including ungrounded, high resistance grounded and resonant grounded which are mainly utilized in Nordic and China)

(2) The conductor type: the fault causes in overhead line, covered conductor overhead line, underground cable and air cable are always different. The overhead line and the underground cable are the most commonly used in distribution system which are considered in this paper.

(3) The geographical environment: the fault causes in plain and forest are different; the tree caused fault and animal caused fault in forest area would be more.

(4) The weather/climate/season: the lightning fault is more in summer and the high wind would cause more tree related fault.

(5) Some other factors, such as the development of the area. There would be more construction caused faults in developing countries, such as China.

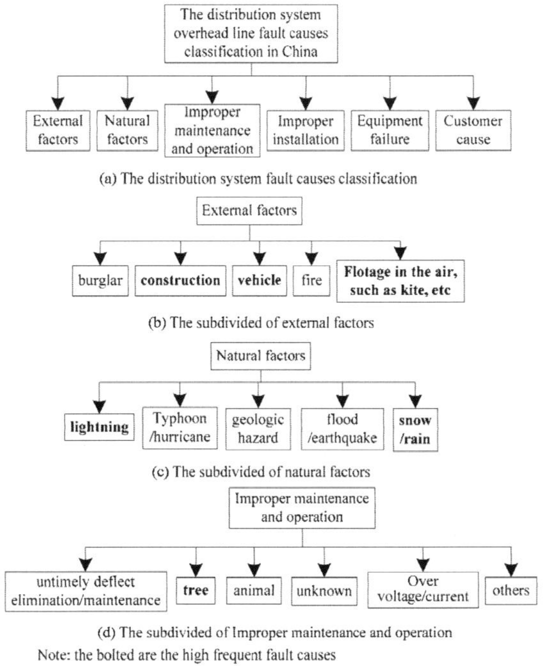

Figure 1. Fault causes classification in China.

3 Overhead line fault causes

3.1 The overhead line faults in North American

3.1.1 The statistics from EPRI

A large survey done by EPRI was finished in the 1980s at 13 utilities monitoring 50 feeders(in north America) 7. The fault causes reported in the EPRI is shown in Figure 2.1[67]. Approximately 40% of the faults occurred during periods of "adverse" weather, which include rain, snow, and ice. The top three fault causes are lightning, tree contact and equipment failure.

3.1.2 The statistics from Duke power system

Because of the factors mentioned last section, the statistics of the fault causes are different with each other. The fault causes and its distribution in Duke power system(north America, data between 1987 and 1990) are: trees(19%), animals(18%), equipment failure(14%), lightning(9%), public accidents(10%), other(13%) and unknown(17%).

3.2 The overhead line faults in Nordic

The fault in distribution system is different with the fault in transmission system. Some countries have analyzed the fault causes in distribution system, such as Norway5 and Finland13. The division of the fault causes in Finland was as follows: snow burden 35%, fallen trees 27%, boughs on pole transformers 9%, diggers 6%, lightning impulses 6%. The rest of the faults (17%) were probably caused by animals such as birds and squirrels, because the faults were cleared by an additional delayed reclosing. The causes of the disturbances cleared by auto-reclosing remained unknown. However, about one third of these events occurred during thunderstorms or windy weather[14].

3.3 The overhead line faults in China

The distribution system fault causes in China contain external factors, natural factors and some of the improper maintenance factors. The external factors caused fault is high all over the year, especially the construction, vehicle and flotage caused. It can be found that fault causes are affected by the seasons apparently. The fault number in summer is much more than other seasons for the terrible weather, such as lightning and high wind.

4 Typical Fault Analysis

In this section, some typical faults will be introduced in detail and some of the waveforms would be attached. The tree, animal and lightning caused faults have been analyzed by many research groups for the faults have the similar features. The partial discharge[1516] and the cable joint faults[8] identification in cable are also introduced in many references. But the man-made faults in overhead line (vehicle and construction related) and the external faults (construction related) in underground cable are hard to be identified.

4.1 Tree caused faults(overhead line)

Trees can cause faults in several ways: growth into conductors, failing trees or branches bridging gaps or pushing conductors together, and failing trees or branches causing mechanical damage17. Results of utility surveys show that growth is normally less than 15% of permanent interruptions. Dead trees or branches account for about 30 to 40% of tree faults, and trees with significant defects account for another significant portion. Faults caused by trees generally occur from a handful of conditions: 1) Falling trees or major limbs knock down poles or break pole hardware; 2) Tree branches blown by the wind push conductors together; 3) A branch falls across the wires and forms a bridge from conductor to conductor; 4) Natural tree growth causes a bridge across conductors.

The results from several utilities outlined below show that broken tree branches or falling trees account for the majority of interruptions. Growth only accounts for a small percentage of interruptions. Simpson reported on a survey of tree-caused faults at Eastern Utilities Associates (a small utility in Massachusetts that is now a part of National Grid). The main results were that tree-caused outages broke down as follows: 63% from broken branches, 11% from falling trees, 2% from tree growth. Document surveyed by BC Hydro of their tree-related interruptions that were as follows: 70% from tree failure, 18% from branch failure, 12% from growth.

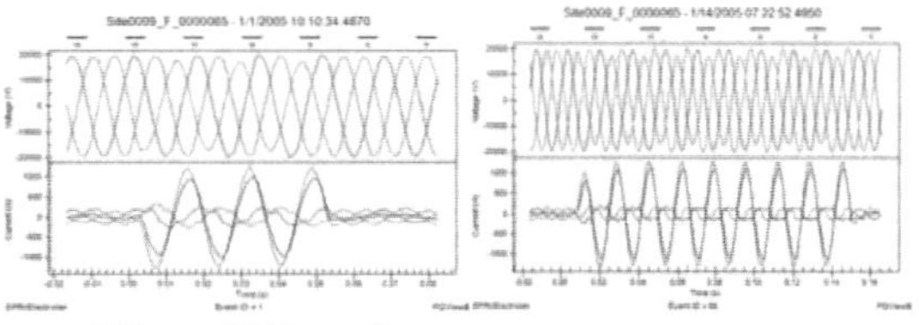

(a) Tree-caused fault (tree outside right of way, fall/lean on primary)

(b) Tree-caused fault (tree/limb growth)

Figure 2. Tree-caused fault waveforms (ref. 20)

4.2 Animal caused faults(overhead line)

4.2.1 Statistics of the animal-caused faults

There are many types of animals could cause fault in power system, including tree squirrel, birds, snake and so on. Also the animal-caused faults may occur on overhead distribution lines, underground distribution circuits, substations, and transmission lines. The EPRI work group has surveyed some utilities in North American about the animal-caused fault between 1993 and 1997. The statistics shows that thirty-five of 84 utilities(42%) estimate overhead distribution lines are associated with 79% of historic animal-caused outage events. In this paper, only the overhead line related animal-caused faults would be introduced.

Eighty utilities repot outage frequency is apportioned between mammals (64%) and birds (36%). The most problematic mammals are tree squirrels and raccoons, thirty-eight utilities indicate tree squirrels are responsible for the greatest frequency of all animal-caused overhead outages; raccoons are listed by another five utilities. Birds are responsible for the second greatest frequency of all distribution overhead outages. Twenty utilities(25%) report most bird outage are caused by unknown bird species.

In fact, some equipments are more susceptible to animal-caused damage and outages. Utilities were provided a list of common distribution items and asked to rank their top five animal-caused outage duration and frequency problem items: transformers, cutouts, surge arresters, jump wire contacts, phase to neutral contacts.

Table 1. Animals associated with the top six distribution overhead items associated with animal-caused outages(ref. 18)

Items Associated with Animal-Caused Outages	Most Common Animals and Number of Reporting Utilities
Power transformers	**Mammals – 46:** Tree Squirrels (21), Raccoons (14), Ground Squirrels (5), Other (6) **Birds – 22:** Unknown (8), Hawks (3), Blackbirds (2) , Other (9)
Cutouts	**Mammals – 18:** Tree Squirrels (10), Raccoons (3), Ground Squirrels (2), Cats (2), Other (1) **Birds – 16:** Unknown (4), Blackbirds (4), Hawks (3) , Other (5)
Surge arresters	**Mammals – 24:** Tree Squirrels (14), Raccoons (5), Ground Squirrels (2), Cats (2), Other (1) **Birds – 15:** Unknown (8), Blackbirds (2), Owls (2), Other (3)
Jumper wires contacts	**Mammals – 18:** Tree Squirrels (7), Raccoons (4), Cats (3), Other (4) **Birds – 14:** Unknown (5), Blackbirds (3), Hawks (3), Other (3)
Phase-to-neutral contacts	**Mammals – 18:** Tree Squirrels (6), Ground Squirrels (5), Cats (2), Raccoons (2), Other (3) **Birds – 19:** Unknown (7), Eagles (2), Hawks (2), Owls (2), Other (6)
Reclosers	**Mammals – 29:** Tree Squirrels (12), Raccoons (8), Ground Squirrels (4), Other (5) **Birds – 15:** Unknown (7), Hawks (3), Blackbirds (3), Other (2)

4.2.2 The waveforms of the animal-caused faults20

The waveforms of animal-caused faults are from DOE/EPRI National database20. Some of the typical fault are shown as follows, but the fault process and fault position are unknown. The different types of animal-caused fault are introduced in 19.

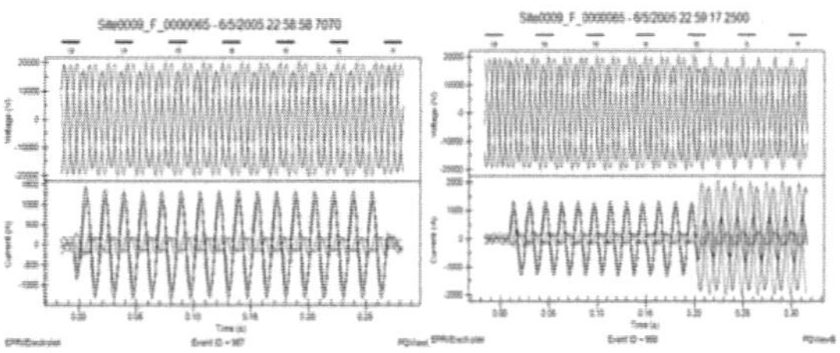

(a) The snake caused fault

(b) The bird caused fault

Figure 3. Two animal-caused fault cases(ref. 20)

4.3 Lightning caused faults(overhead line)

In the majority of cases, lightning causes temporary faults on distribution circuits; the lightning arcs externally across insulation, but neither the lightning nor the fault arc permanently damages any equipment. Normally, less than 20% of lightning strikes cause permanent damage. Lightning causes most damage by directly striking an overhead phase wire and injecting an enormous current surge that creates a very large voltage. The voltage impulse easily breaks down most distribution-class insulation unless it is protected with a surge arrester. Almost all direct lightning strokes cause flashovers. In addition, the lightning current may start a pole fire or burn through conductors. Also, nearby lightning strokes that don't hit the line may couple damaging voltages to the line. These induced voltages may fail equipment or cause flashovers.

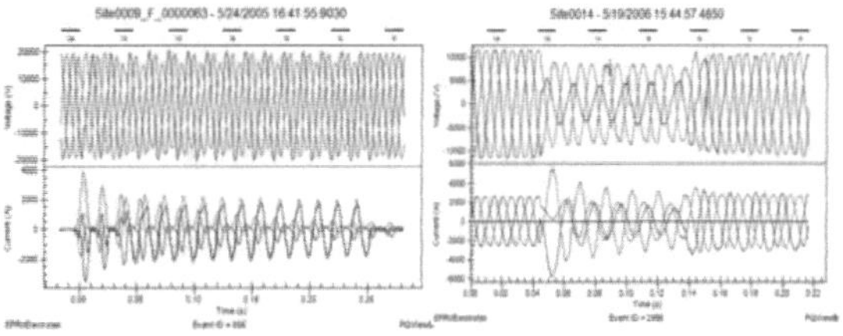

(a) Lightning struck transformer, burned primary down(overhead primary failure)

(b) Lightning caused fault(simultaneous lighting on lines during a lighting storm)

Figure 4. Lightning caused fault waveforms (ref. 20)

DOI: 10.1051/matecconf/20166102017

4.4 Vehicle caused faults (overhead line)

The feature of this type fault is irregular for the stochastic fault process. The faults always cause pole damage. Some of the fault waveforms are shown in Fig. 5.

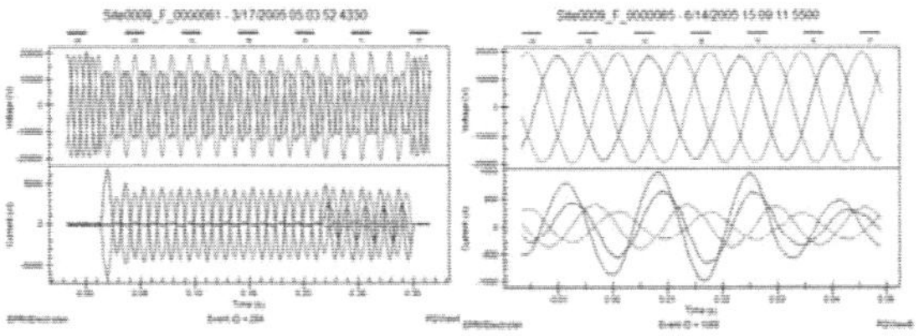
(a) Vehicle caused fault(car hit pole) (b) Vehicle caused fault(car hit guy wire)

Figure 5. Vehicle caused fault waveforms (ref. 20)

4.5 The relationship between fault type and fault causes

The different fault type happened in distribution and the relationship with fault causes have been surveyed by some references7917. It is reported that almost 80% of the faults measured involved only one phase either in contact or with ground. As another data point, measurements on 34.5kV feeders found that 75% of faults involved ground (also 54% were phase to ground, and 15% were phase to phase). Most faults are single phase because most of the overall length of distribution lines is single phase (in North American), so any fault on single-phase sections would only involve one phase. Equipment faults and animal faults tend to cause line-to-ground faults. Trees can also cause line-to-ground faults on three-phase structures, but line-to-line faults are more common. Lightning faults tend to be two or three phases to ground on three-phase structures7. The relationship between fault causes and fault type is also surveyed by 9, which is shown in Table 3.

Table 2. Number of phases involved in each fault measured in the EPRI fault study (ref. 7)

Fault	Percentage
One phase to neutral	63%
Phase to phase	11%
Two phases to neutral	2%
Three phase	2%
One phase on the ground	15%
Two phases one the ground	2%
Three phases on the ground	1%
Other	4%

Table 3. The relationship between some fault causes and fault types (ref. 9)

Event	Single phase faults	Multiphase faults	Total cases
Animal contact	53(86.88%)	8(13.12%)	61
Tree contact	66(73.33%)	24(26.67%)	90
Lightning induced	8(42.1%)	11(57.9%)	19

The fault cause is also related to the weather. For one northeastern US utility, "Tree Contact" leads all other outage categories during stormy weather; and it especially stands out during wind and ice/snow. This highlights the fact that tree caused faults increase rapidly during storms.

Faults are either temporary or permanent. A permanent fault is one where permanent damage is done to the system. This includes insulator failures, broken wires, or failed equipment such as transformers or capacitors. Virtually all faults on underground equipment are permanent. Most equipment fails to a short circuit. Permanent faults on distribution circuits usually cause sustained interruptions for some customers. A temporary fault does not permanently damage any system equipment—if the circuit is interrupted and then reclosed after a delay, the system operates normally. Temporary (non-damage) faults make up 50–90% of faults on overhead distribution systems. Some causes of faults are more likely than others to be permanent. The causes of temporary faults include lightning, conductors slapping together in the wind, animals that cause faults and fall off, and insulator flashovers caused by pollution7.

5 Summary

The fault causes in distribution system have been investigated in this paper. The fault in overhead line and underground are introduced separately. Some common faults(tree, lightning, animals and so on) are introduced in detail, and waveforms are shown in the paper. Some features of the different fault causes can be conclude as follows:

- The overhead line fault is high related to the weather condition: tree caused fault is more frequent under high wind condition, lightning fault is easy to identify by the weather. The information about weather is useful for the fault feature identify.
- The animal caused fault is frequent in north America which is always happen in spring, and fault resistance and fault type are regular.
- The man-made fault(construction and vehicle) in overhead line is hard to identify for its stochastic fault condition.
- The equipment failure related fault (including the flash over in overhead line and partial discharge in the underground cable) always caused by the insulation decaying, so there would be some disturbances before the fault which can be used as criterion.
- The external caused fault(construction) in underground cable is mainly because excavator digging which would damage the cable, the fault features are irregular, so it is hard to identify.

The fault causes have been introduced in this paper, but the fault process and some more details about the fault haven't been presented which would be investigated in the future. The fault process is dynamic and fickle, so the dynamic of the model should be considered for the accurate fault analysis and identification.

References

1. Power outage annual report. *Blackout Tracker Canada annual repot 2013* (Get from: pqlit.eaton.com/ll_download_bylitcode.asp?doc_id=24005)
2. Power outage annual report. *Blackout Tracker United States annual repot 2013* (Get from: pqlit.eaton.com/ll_download_bylitcode.asp?doc_id=24003)
3. European Network of Transmission System Operators for Electricity. *Nordic grid disturbance statistics 2012.* (Get from: http://www.fingrid.fi/fi/voimajarjestelma/voimajarjestelmaliitteet/S%C3%A4hk%C3%B6n%20toimitusvarmuus/Pohjoismainen_hairiotilasto_2012.pdf)
4. Alberta Municipal Affairs. *Statistics electrical incidents in Alberta*(January 1 to December 31, 2013)(Get from:http://www.municipalaffairs.alberta.ca/cp_electrical_forms_and_reports.cfm)
5. Gerd H.Kjolle. *What do fault statistics tell us regarding causes resulting in power outages?*(Get from: https://www.sintef.no/project/Vulnerability%20and%20security/Publications/Presentations/Workshop%2020110927_fault%20statistics.pdf)
6. Burke, James J., and D. J. Lawre. "Characteristics of fault currents on distribution systems." *Power Apparatus and Systems, IEEE Transactions on* 1 (1984): 1-6.
7. *Power Quality Implications of Distribution Construction,* EPRI, Palo Alto, CA: 2004. 1002188
8. Mehairjan, R. P. Y. *I. Application of Statistical Life Data Analysis for cable joints in MV Distribution Networks-An Asset Management Approach.* Diss. MSc Thesis Report, Delft University of Technology: Electrical Engineering, 2010.
9. Kulkarni, Saurabh, et al. *"Waveform characterization of animal contact, tree contact, and lightning induced faults."* Power and Energy Society General Meeting, 2010 IEEE. IEEE, 2010.
10. Campbell, Richard J. *"Weather-related power outages and electric system resiliency."* Congressional Research Service, Library of Congress, 2012.
11. Lehtonen, Matti. *"Fault rates of different types of medium voltage power lines in different environments."* Electric Power Quality and Supply Reliability Conference (PQ), *2010*. IEEE, 2010.
12. Chow, Mo-Yuen, and Leroy S. Taylor. *"A novel approach for distribution fault analysis."* Power Delivery, IEEE Transactions on 8.4 (1993): 1882-1889.
13. Kimmo Kivikko. *Assessment of electricity distribution reliability-interruption statistics, reliability worth, and applications in network planning and distribution business regulation.* Doctor thesis, 2010, Tampere university of technology
14. Hanninen, S., M. Lehtonen, and T. Hakola. *"Earth faults and related disturbances in distribution networks."* Power Engineering Society Summer Meeting, *2001*. Vol. 2. IEEE, 2001.
15. Zhang, Xiang, et al. *"Estimation of the lifetime of the electrical components in distribution networks."* Power Delivery, IEEE Transactions on 22.1 (2007): 515-522.
16. Densley, John. *"Ageing mechanisms and diagnostics for power cables-an overview."* Electrical Insulation Magazine, *IEEE* 17.1 (2001): 14-22.
17. *Power Quality Implications of Transmission and Distribution Construction: Tree Faults and Equipment Issues*, EPRI, Palo Alto, CA: 2005. 1008506.
18. *Mitigation of animal-caused outages for distribution lines and substations*, EPRI, Palo Alto, CA: 1999.Report TE-114915
19. *Distribution Wildlife and Pest Control,* EPRI, Palo Alto, CA: 2001. 1001883
20. *DOE/EPRI National Database Repository of power system events* (get from: http://expertmonitoring.com/doelibrary/)

Research and Application of Green Lighting in Power Plant

Zixia Cheng[1], Bingbing Du[1, 2, a], Yu Wang[3], Zhenzhen Niu[2], Xiu Zheng[1]

[1]*School of Electricity Engineering, Zhengzhou University, Zhengzhou 450001, China*
[2]*Henan Province Airport Group Co.,Ltd, Zhengzhou, 451161, China*
[3]*Bo Hui Fang Zhou Development Co.,Ltd, of Henan Province, Zhengzhou, 450052, China*

Abstract. To study the application of green lighting in power plant, this paper based on the analysis and comparison of the technical and economic of lighting source and energy-saving lamps and accessories, recommended the use of high luminous efficiency of lamps and lanterns, to achieve energy-saving effect. From the point of lighting design, through the calculation of the illumination value and the use of reasonable lighting control to achieve the effect of green lighting. By adopting the new energy saving lamps and light sources, the operation cost of the power plant and the lighting energy consumption can be reduced, the project investment and operation cost can be saved, and the social and economic benefits will be obtained.

1 Introduction

In the past few decades, with the rapid development of economy, energy shortage has become a serious problem; energy conservation has become a hot topic recently. "green lighting project" has been listed as one of the ten key energy-saving projects in our country, it will be used in China's new construction and renovation projects,such as: the efficient energy-saving lighting system, promotion of rare-earth trichromatic fluorescent lamps, high luminous efficiency, long service life of the lamp.

At present,domestic and foreign advocate green lighting, it can improve the efficiency of energy utilization, it is conducive to the improvement of the quality of life and urban ecological environment.It also can save energy, alleviate the pressure of environmental protection, in line with the national industrial policy, to achieve energy-saving and reduce the emissions of contaminants.

This paper mainly studies the application of green lighting in the power plant, and gives its application implementation plan.

2 Green lighting and energy-saving

2.1 The definition of green lighting

The concept of green lighting was proposed by the national environmental protection agency in early 1990s. Complete green lighting connotation contains 4 indicators: energy saving, environmental protection, safety, comfort, and so on. Energy efficient means to obtain adequate lighting and consume less electricity, resulting in a significant reduction in power plant air pollutant emissions to achieve the purpose of environmental protection; Safe, comfortable refers to the clear , soft light, and harmless light without light pollution, such as ultraviolet rays and glare.

To promote green lighting is the key work of energy saving in our country; it should be advanced technology, reasonable economy, the use of safe, convenient maintenance and management in power engineering design, actively the implementation of green lighting.

2.2 Lighting energy saving

At present, the development of Chinese power industry is very fast, but the shortage of power supply and low power efficiency is still severe, which will continue to exist in a considerable period of time in the future. Our lighting electricity largely accounts for the total generating capacity of 10-12%. At present, the power plant 2/3 for thermal power, the main energy for coal. Lighting energy-saving means to reduce emissions of harmful gases such as CO_2, SO_2, NOx, so wo should protect the atmospheric environment from a high degree of awareness and treatment of energy-saving lighting, promoting green lighting project.

In thermal power, for example, the production of 1 kWh of electricity, the pollutants generated to the air is shown in table 1.

The goals of energy saving are to build a high quality, high efficiency, economic and comfortable, safe and reliable, good environment and improve people's quality of life, improve work efficiency, and establish a lighting environment for the protection of physical and mental health. It should use advanced and reasonable lighting design methods, good-quality and efficient lighting equipments, to achieve a good effect in energy saving aspect.

[a] Bingbing Du: 2010dubing666@163.com

Table 1. Pollutants produced.

Pollutants	Coal	Fuel	Gas
SO_2	9 g	3.7 g	0 g
NOx	4.4 g	1.5 g	2.4 g
CO_2	1100 g	860 g	640 g

The lighting energy-saving evaluation index is the lighting power density (LPD), the power density of some rooms or places of the power plant should not be greater than the value in table 2. When the illumination value of the room or place is higher or lower than the corresponding value, the value of the lighting power density should be increased or reduced according to the proportion.

Table 2. Power plant lighting power density.

Room or place	LPD(W/ m^2)		Corresponding illumination value(lx)
	Current value	Target value	Gas
turbine room operation layer	7.0	6.0	200
turbine room bottom, pipe layer	4.0	3.5	100
bottom of boiler room	5.0	4.5	100
main control room, computer room	16.0	14.0	500
electronic equipment room	9.5	8.0	300
coal, water control room	9.5	8.0	300
high and low voltage power distribution room	7.0	6.0	200
cable interlayer	3.0	3.0	50
underground coal unloading, coal pier	3.0	3.0	50

The main ways of energy-saving lighting:

①The vigorous use of energy-efficient electric light source and lighting electrical accessories.

② The selection of lighting efficiency, reasonable light distribution.

③ The scientific and reasonable design of energy-saving lighting control and optimization.

In addition, a good job in daily maintenance management is important.

2.3 The problems existing in the domestic power plant lighting design

According to the field investigation and research in some power plants, there are the following problems in the power plant lighting:

(1) Lighting electrical products with low efficiency, short life and unstable performance.

Some power plants using low light efficiency, short life of high-pressure mercury lamp; T12 thick tube fluorescent lamp is used in a few plants, in the corridor, stair and ordinary places still use incandescent lamp as the light source. Although high efficient lamps and light source is applied in many plants, but the matching accessories of the lamp, especially advanced energy-saving ballast products still cannot get due attention. Many high strength gas discharge lamps and fluorescent lamps are still equipped with the old power inductance ballast.

(2) The unreasonable lighting design

There is no use of three-phase five wire system in the longer lighting line that leading to the line loss is too large, the voltage drop, power down, and then make the illumination flux decrease; plant road lighting without using partition power supply, and adopted centralized power supply; Part of the power plant of outdoor and stair lighting control adopts optically controlled method, while adopting light-operated switch can play a certain energy-saving effect during the day, but in the middle of the night, some unmanned place lamp Chang Ming will cause the waste of electricity; In selected lamps and lanterns of the plant is not suitable for light distribution curve and the exchanges, cause the cliff uneven illumination, cause the intensity of illumination is too centralized or decentralized; The glare phenomenon is serious in centralized control room of many power plants, lamps are wrongly designed in the interference area, that leads to rework after lighting construction, it is a great waste of resources.

3 Green lighting application in power plant

Lighting power is the most basic power demand; the quality of lighting has a direct relationshipthe for the power plant safety production, labor productivity, labor health and occupational health.

The power plant lighting electricity consumption (L) can be expressed by the following formula:

$$L=WT^{*}（EA/FK）=EAT/k\eta$$

Among:

W-electric power consumption per lamp， kW / each;

T-time of the light turn on， h;

E-average design illumination， lx;

A-floor area， m^2; ;

F-each lamp lumens， lm;

K-maintenance coefficient;

η-the comprehensive efficiency of the light bulb, F / W;

Therefore, in order to reduce lighting energy consumption, it must try to use efficient lighting, increase the rate of repair of the lamps; or reduce time to turn on the lights, maintain proper illumination and maximize the use of local lighting and so on. However, energy-saving lighting is the principle of saving energy in the premise of ensuring adequate lighting brightness and quality. Thus, choose a reasonable illumination (E), efficient lighting lumens (f) and improve comprehensive efficiency (η), strengthening the operation management, have very important significance for energy-saving lighting.

Lighting energy saving measures in power plant should start with the lighting design. Advanced technology, reasonable economy, the use of safe, convenient maintenance and management should be ensured to promote green lighting project in comprehensive engineering application and promotion.

3.1 High efficiency energy-saving electric light source and lighting accessories

Production and application of high quality and efficient lighting equipment is an important means for the implementation of green lighting, one of the most important is efficient light source. There are different types of efficient light source which have their own characteristics and applicable places; they should be selected by using conditions and requirements.

Compact fluorescent lamps (energy-saving lamps) compact size, easy to use, equipped with electronic ballast, which have good color, no flicker, light efficiency higher merit, can very convenient alternative to incandescent lighting light source. High power tube diameter straight tube fluorescent lamps and high-strength gas discharge lamp (HID) have the higher light efficiency, more advantages in price performance ratio, and is suitable for different use places.

In the comprehensive promotion and selection of various types of efficient light source at the same time, we should also pay attention to the application of efficient lighting and light source accessories, especially the advanced and energy-saving lamps and other lighting accessories.

3.1.1 The light source

Light source with high luminous efficiency, long life should be used, such as thin and straight tube fluorescent lamp, compact type fluorescent lamp and metal halide lamp, high pressure sodium lamp. It is better to use the general color rendering index (Ra) that greater than three fluorescent lamp 80 in adopting fluorescent lamp. The height of the lower room should adopt thin and straight tube fluorescent lamp, compact fluorescent lamp and low power metal halide lamp. High industrial plant and should be according to the requirement of production,or using metal halide lamp and high-pressure sodium lamp, using high power thin and straight tube fluorescent lamp.

Compact fluorescent lamp,also called single ended fluorescent lamp,or energy saving lamp. Energy-saving lamps more structural type, good color, the color is strong; high light efficiency is 5-7 times that of incandescent lamp; long service life is 5 times that of incandescent lamps, the average service life of 5000 hours, under the condition of low voltage can start lighting, and can achieve the normal voltage brightness, and no noise and flashes of light. At the power plant, corridors, staircases, toilet, coal handling system, coal layer, grey base places can be substantial use energy-saving lamps.

The light source is required three factors, namely: high luminous efficiency, long life, high color. The main parameters of the common light source refer to table 3.

Table 3. Luminous efficiency and technical parameters of common light source.

Type of light source	Rated power (W)	luminous efficiency (lm/W)	service life (h)
Ordinary incandescent lamp	15~1000	7~19	1000~1500
Tungsten halogen lamp	500~2000	19.5~21	1500
Fluorescent lamp (T8)	18~58	75~86	8000~10000
Fluorescent lamp (T5)	14~80	78.5~94	8000~12000
Compact fluorescent lamp	5~36	57~80.5	4000~15000
High pressure mercury lamp	80~400	47.5~55	2500~5000
Tube type hernia lamp	1500~100000	20~37	500~1000
High pressure sodium lamp	50~1000	68~128	2400~32000
Metal halide lamp	70~1000	75~110	9000~20000

3.1.2 Energy saving comparison

Energy saving compared with high efficiency and energy saving light source (the same illumination) is shown in table 4.

Table 4. Energy saving comparison of energy-saving light source.

Ordinary incandescent lamp(W)	Self ballast compact fluorescent lamp (lm/W)	Power saving effect	Power saving rate
100	25	75	75
60	16	44	73
40	10	30	73

It can be seen that the use of efficient energy-saving light source, has the following advantages: saving energy, prolong life, improve light efficiency, reduce material consumption, etc.. When the rated life is long, the energy saving effect is more obvious.

The use of energy-saving lamps can save electricity charges about 28 yuan than incandescent lamp per year, to the metal halide lamp and mercury lamp, the power savings is about 544 yuan a year.(Compared with mercury lamp, metal halide lamp can save electricity charges about 544 yuan per one per year) On the engineering, 2 units of 660 MW unit new factory use about 1000 fluorescent lamps, metal halide lights around 1100. If we use high efficiency and energy saving light source, it can save electricity of about 620000 yuan a year.

3.1.3 The LED light source

Compared with other light sources, LED is a new energy-saving light source which has the advantages such as: luminous efficiency highest, the longest life expectancy, color performance is good, energy saving and environmental protection, long service life, security, small volume, high flicker, convenient regulation and control. The disadvantage is that the stability of product quality is poor, service life and theory of life vary widely, the price is higher than other gas discharge light source several times. The technical indices of the common LED lamp are shown in table 5.

Table 5. The technical indices of common LED lamp.

Name	Mode	Working voltage (V)	Power (W)	Lamp holder model
4.8V red LED	L-ZE06R	4.8	0.6	E27
4.8V green LED	L-ZE06G	4.8	0.6	E27
12V yellow LED	L-ZF10Y	12	1.2	E27
12V blue LED	L-ZF24B	12	2.4	E27
Three colours lamp	D3F	12	0.6	E27
Candle lamp	ZGE	12	1.25	E27

3.1.4 Lighting accessories

Accessories for common lamps are: ballast, starter, trigger and capacitor. The energy consumption of the ballast is generally accounted for 20%~30% of the energy consumption of the whole lamp, so the selection of ballast is mainly studied. The commonly used ballast is divided into two categories: inductive ballast and electronic ballast. The inductive ballast is divided into two types: traditional type and energy saving type.

36W/40W performance comparison of various ballast, see table 6.

Table 6. Comparison of energy-saving inductance ballast and electronic ballast.

Model	Self power consumption (W)	Price ratio	Light efficiency ratio	Boot surge(times)
Ordinary inductance ballast	9	1	0.95~0.98	1.5
Energy-saving inductance ballast	4~5.5	1.6	1.02~1.05	1.5
Domestic standard electronic ballast	≦ 3.5	3~4	1.10	10~15
Imported electronic ballast	≦ 3.5	4~7	1.10	8~10
Domestic H type electronic ballast	≦ 3.5	1.3~1.8	1.10	15~20

3.1.5 Lighting lamps with high efficiency and light distribution

We should choose reasonable light distribution, high reflectivity and good durability of the reflective lamp. To improve the utilization rate of the light emitted by the light source, the light energy distribution of the light source to where it is needed, to prevent glare caused by light, and the light source is not influenced by external force, damp and harmful gases.

The efficiency of the high efficiency and high quality lighting should not be less than the value in table 7 and table 8.

Table 7. The efficiency of fluorescent lamps.

The lamp light outlet form	Open	Protective cover (glass or plastic)		grille
		Transparent	Frosted、prism	
Lamp efficiency	75%	65%	55%	65%

Table 8. The efficiency of high strength gas discharge lamp.

The lamp light outlet form	Open	Grille or transparent cover
Lamp efficiency	75%	60%

3.2 Scientific energy-saving lighting design

3.2.1 Setting reasonable illumination value

The different working place of the power plant requires different illumination. A point on the surface of the illumination is incident on the bin contains point of luminous flux, divided by the bin area. The average illumination can be calculated by using the coefficient method.

$$E_c = \frac{\Phi \times N \times CU \times K}{A}$$

Among:

E_c -average illumination of working face (lx);

Φ -light flux (lm);

N -number of light sources (sets);

K -lamp maintenance coefficient, depending on the room's pollution level and the cleaning cycle of the room and lighting;

A - Area of working faces (m^2);

CU -Utilization coefficient, generally provided by the lamps and lanterns manufacturer.

According to the requirements of visual work in each work place, the illumination standard value (average illumination value) of each place is determined. In the actual design, design allows the illuminance value deviation and standard value of 10. Of course, the selected illuminance values should take the feel of people working in this environment into consider, to ensure that there is a certain degree of visual satisfaction, 100~300lx range is the most continuous work of the indoor environment of the appropriate illumination.

3.2.2 Reasonable and integrated lighting layout

It should be based on the use of light sources, can not blindly pursue the uniform distribution of light.

On the one hand, the comprehensive utilization of general lighting and local lighting mixed lighting. General lighting just to satisfy the basic requirement of illumination, it should set local lighting in a special places (such as high and low pressure heater water level gauge of the main building and deaerator water level meter, etc.). On the other hand, make full use of natural light, the correct choice of natural lighting. In the workshop design of building, if possible, the window settings should be the most of the sun, from a number of points rather than from a injection workshop, and no harsh feeling, it is best to make the sunshine from both sides of the incident, so the incoming light is uniform.

At the same time, make full use of indoor smooth reflective, can effectively improve the utilization rate of light, such as the white walls of the reflection coefficient is up to 70% ~ 80%, it can also play a very good energy saving effect.

3.2.3 Reasonable lighting control mode

Reasonable lighting control is an important means to achieve energy saving. Using intelligent lighting controller, through the energy saving control module and energy optimization algorithm, so that the lamp work in the best condition. Using the combination of centralized control and local control mode of lighting switches, the position and number of the indoor switch should be appropriate to according to the needs of the light turn on or turn off the light, so as to achieve energy-saving.

Boiler body, roads and other places of outdoor lighting should adopt the method of synchronous reduce the luminous flux, it is in the line of lamps and lanterns control module, when outdoor lighting free will all light flux decrease to the same level, both to ensure the illumination function and achieve energy saving purpose. Corridors and stairwells can use acoustic, electric or compound control mode.

3.2.4 Reasonable power supply mode and line

It is appropriate to install the compensation capacitor, to ensure its power factor in the place for a large amount of gas discharge lamps. Consider adding capacitance compensation function in the electronic ballast, will enable the loss reduction, a reasonable choice of wire cross is selected based on the calculation.

Optimizing distribution mode. Replace single-phase power supply with three phase four wire, the line loss can be 75~80 lower than the original. We should use three-phase four-wire circuit in the long lighting, such as the lighting circuit of roads and coal pier; In addition, the voltage change of the power plant between day and night often form 5 to12, after midnight, the line voltage higher than rated value 3~5, this time running lights power often exceeds the rated value by 10%, high voltage is not only a waste of energy but also shorten the service life of the lights. The pressure limiting measures can be adopted, one is to use the limiting pressure of the reactor, and the other is to use the thyristor to control the voltage.

3.2.5 Safety of lighting system

The safety lighting system should include two aspects: on the one hand, the safety of the lighting system itself, on the other hand, it is the emergency lighting system which is relative to the normal lighting.

In order to ensure the safety of the lighting system itself, all the live parts of the lamp must be separated by insulating material, which is known as the protection against electric shock. According to the calculated current of each wire to verify the calculation of the current to select the conductor section, the boiler body and other high temperature sites should use heat conductor. Lighting distribution lines and branch lines shall use copper wire or cable, and the section of the branch shall be not less than 2.5mm^2. At the same time, all the socket circuit should be based on single-phase three wire wiring (with protection line), the branch circuit conductor section of the minimum is 2.5~4mm^2.

3.2.6 Good maintenance and management

The lighting value of the lighting equipment will fall after the use of a period of time, we should clean lighting lamps regularly, and replace the light source in accordance with the normal maintenance procedures. Regularly check the lamp shade, if its aging or discoloration, should be replaced in a timely manner to improve the light transmission efficiency of lamps and lanterns. At the same time, with the convenient maintenance and cleaning of the new type of lamp replacement of old or damaged lamps, can also be divided into areas for the measurement and assessment of electricity, effective energy saving.

4 Concluding remarks

Power plant electrical lighting is a comprehensive discipline, it should fully understand the structure of the building structure, process requirements under the premise of comprehensive consideration of economic, energy saving, rationality. The economy should not only focus on the number of initial investment, but also take into account the total annual lighting costs (equipment depreciation costs, management fees and the sum of the electricity tariff). Under the same illumination conditions, should give priority to select the lamps and lanterns of luminous efficiency highest lighting light source and the efficiency of the highest, at the same time from lighting design, lighting control, makes use of natural light and other aspects of strengthening maintenance and management, in order to obtain the double benefits of lighting and energy-saving.

There is no doubt that the promotion and use of green lighting has a very important significance to energy saving of lighting in power plant, it also in line with the green lighting energy policy. In power plant, reasonable selection of efficient energy-saving lamps and lanterns in lighting energy-saving design, electric light source and accessories, the improvement of lighting electricity utilization rate and effective control, will gain great social and economic benefits.

References

1. W. Hongmei, L. Ke, Technology Innovation and Application,**16**,26(2013)
2. Y. Sufang, Chinese Consulting Engineers,**3**,51(2013)
3. W. Chenghui,Scinence and Technology& Innovation, 7,66(2014)
4. R. Wenguang, L. Kun, Q. Haixian, Telecom Word, 15,93(2014)
5. L. Qishu, C. Ming, Technology Wind,125(2014)
6. C. Haihua, Zhejaing Electric Power,9,29(2013)
7. Z. Li, Technology Wind,133(2014)
8. T. Lianhua, Guangdong Science & Technology, 20, 93(2013)
9. L. Xiang, L. Lingru, Journal of Jiangxi Vocational and Techincal College of Electricity,3,16(2011)
10. S. Jingyu, Construction&Design for Engineering, 65(2008)
11. L. Guangxing, Inner Mongolia Water Resources, 1, 42(2013)
12. H.Rong,Sichuan Electric Power Technology, **12**, 55(2009)
13. S.Dianyang,R.Yan,Y.Di,Lamps and Lighting, **3**, 45(2014)
14. L.Song,Z.Kailin,Guangdong Science&Technology, **22**, 126(2013)
15. L. Hong, G. Fei, China Illuminating Engineering Journal, **2**,6(2006)
16. Z. Huangchagn, China Planning Press,DL/T 5390-2014

MATEC Web of Conferences **61**, 02019 (2016) DOI: 10.1051/matecconf/20166102019
APOP2016

Cyclic Wiener Filtering Algorithm in Discrete Cosine Transform Domain for Vibration Signal

Zhang Feng[1], Yang Shangjun[2], Hu Yantao[3] and Shi Xianfeng[4]
[1]Associate Professor, Xi'an Technological University, China
[2]Master Graduate student, Xi'an Technological University, China
[3]Master Graduate student, Xi'an Technological University, China
[4]Lecturer, Xi'an Technological University, China

Abstract. In order to solve the problem that the effect of using cyclic Wiener filter directly to remove the noise on the non-stationary vibration signal is poor, the paper applies discrete cosine transform to the cyclic Wiener filter, proposing the cyclic Wiener filtering algorithm in discrete cosine transform domain for the vibration signal. Using the energy concentration characteristic of discrete cosine transform and the linear phase characteristic of cyclic Wiener filtering, the paper adopts the method of combining both of them with segmented processing to give full play to the performance of the Wiener filter and achieves a better filtering effect with a lower filter order. Combining with industrial field turbine vibration signal, paper makes a simulation analysis for this algorithm. The result of simulation shows that the algorithm has a good noise filtering effect, and the filtered signal has no obvious phase distortion. So, the algorithm is suitable for removing noise on vibration signal and the filtering effect is better than just using cyclic Wiener filter or DCT filter only.

1.Introduction

It has the magnificent meaning to insure the safety and stable operation for mechanical equipment, which is based on the analysis of vibration signal with the advantage of online, true time, non-ablative, convenient and accurate diagnose for mechanical fault diagnosis [1].On the process of acquirement and transmit for the vibration signal, it is unavoidable to be disturb by the noise of environment and equipment itself. So that the operation status of rotating machinery cannot be truly reflected, which influence the analysis and dispose in the later stage, even maybe cause the erroneous judgment[2].So to require accurate character of vibration signal and using it to detect and diagnose the fault for the turbine running state, the priority is to denoise the vibration signal.

For the common filter technique, it is difficult to balance the quality between filtering effect and linear phase. Whereas, the linear phase of vibration signal filtering is strict, this is not allowed to produce the phase distortion clearly. Document[3] uses cyclic Wiener filtering to filter the self-adaption noise for the vibration signal, and it achieves the better effect, but it has a bad effect by using the cyclic Wiener filtering directly to filter non-stable random signal, meantime, the self-adaption filtering algorithm is complicated and the computation is heavy; The method of non-phase deviation filtering in DCT domain for vibration signal, presented by document[4], not only can guarantee the signal's phase character, but also achieves the reduction of data size, however, the filtering in DCT domain has a question that the filtering effect is bad while processing the long data. Soft threshold wavelet filtering for vibration signal, which is presented in document [5], achieves a good effect in filtering, but this algorithm needs to estimate the noise's square error, also it has difficulty in choosing the threshold. The paper on the basis of analyzing the vibration signal character and filter requirement, puts forward to cyclic Wiener filtering algorithm in Discrete Cosine Transform Domain for vibration signal, to guarantee the better filtering effect on the phase character, meanwhile achieves the data compression.

2. Vibration signal sampling and filtering requirement

2.1 Vibration signal sampling

In order to make this research and analysis more targeted, we sample the turbine vibration signal in industrial scene. The parameter of data sampling is:32 frequency multiplication sampling, rotating speed of 32 times, with the cooperation for the back of the FFT calculation, sampling points are set to 2 of the whole power, $N = 2^M$ .Also for the facilitate processing, each channel samples 128 dots continuously. The sampling signal waveform is shown as Figure 1. In order to research the denoising performance, adding the random noise to the signal, vibration signal after adding noise is shown as figure 2[6].

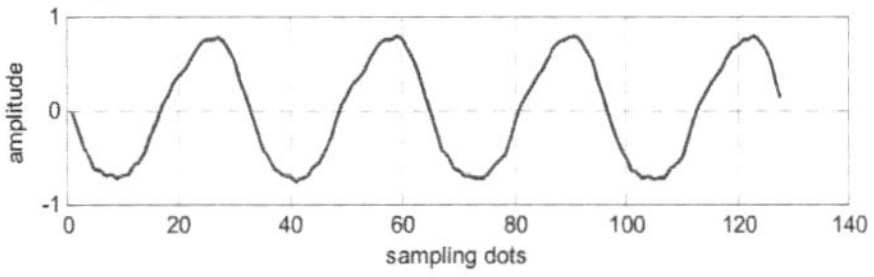

Figure1. Original sampling of vibration signal

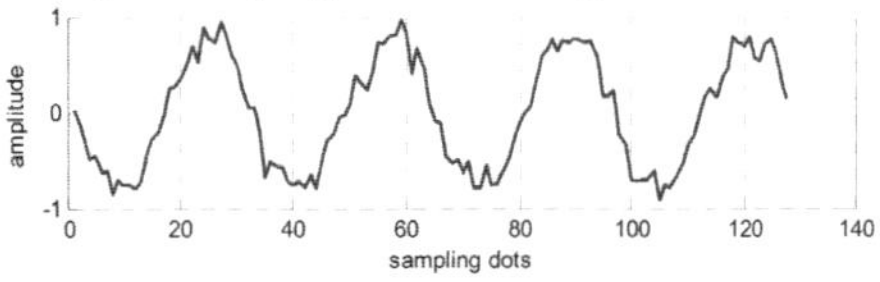

Figure2. Vibration signal after being added noise

2.2 Vibration signal filtering requirement

In order to acquire turbine's running status accurately, two basic requirements should be met for the turbine's vibration signal denoising as follow;

(1) The information of each harmonic phase for the vibration signal includes shaft vibration's position information. So after denoising, the signal should not produce phase distortion or less phase distortion. The nonlinear of phase frequency characteristic for filtering algorithm may cause the waveform distortion, and this kind of distortion is a manifestation of phase distortion. So this requires that filtering algorithm should have a good linear phase property.

(2) Under the premise of ensuring the phase characteristics, it is better to filter the noise to obtain the vibration signal that can truly reflect the turbine's operation. Especially in the following-up spectral estimation for vibration signal. The noise will have a great influence on variance performance for spectral estimation algorithm and the accuracy for the spectral estimation. So removing the noise is as much as possible.

3.Signal's filtering processing in DCT Domain

3.1 Discrete Cosine Transform

Discrete cosine transform is a kind of orthogonal transform in the real number field. For the sequence $x(n)$, the length of N. The discrete cosine transform is represented by $x_c(k)$, define:

$$X_c(k) = DCT[x(n)] = \sqrt{\frac{2}{N}} c(k) \sum_{n=0}^{N-1} x(n) \cos[\frac{(2n+1)k\pi}{2N}] \quad (1)$$

Where:
$$c(k) = \begin{cases} \frac{1}{\sqrt{2}}, & k = 0 \\ 1, & 1 \le k \le N-1 \end{cases} \quad (2)$$

The inverse transform is:

$$x(n) = IDCT[X_c(k)] = \sqrt{\frac{2}{N}} \sum_{n=0}^{N-1} c(k) X_c(k) \cos[\frac{(2n+1)k\pi}{2N}] \quad (3)$$

DCT has the property of solution correlation in large degree, after the transforming, the correlation of the signal becomes reduced. Most of energy for signal only concentrates on a few of transformation coefficients also these transformation coefficients represent signal's lower frequency energy. So abandoning the following data can still keep the signal's characteristics. Also the noise of the mixed coefficient is removed; meanwhile, the data volume is reduced and compressed. The main process of denoising by using DCT is: input the data then transform DCT by Formula (1); According to the characteristic of energy concentration, retain the previous data, abandon the behind data, then to fill in the data of 0 to the original data length; By formula (3) DCT inverse transform for the data, then obtain the new data after filtering.

3.2 DCT filtering effect analysis

Signal denoising based on DCT is a filtering method of non-phase deviation. But the filtering effect depends on the number of reserved coefficient which is in DCT domain. Using this algorithm to filter the noisy vibration signal then compared to the original signal, the result shows in Figure 3.

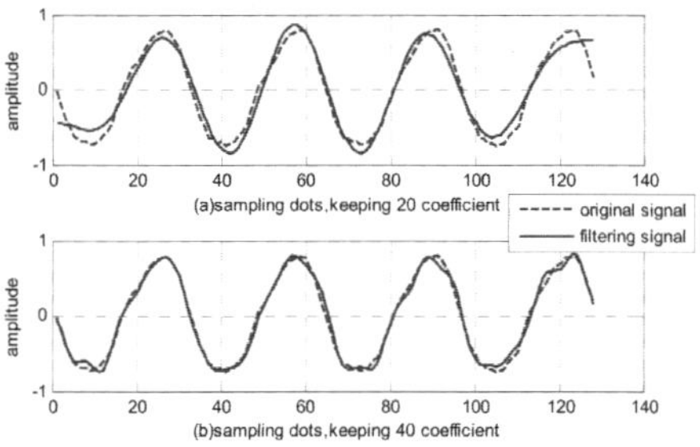

Figure3. Turbine Vibration signal after being filtered by DCT

The Figure3 (a) shows that when the coefficient of retention is low, the noise can be filtered in large degree, but after filtering the waveform is too smooth, which generates the obvious distortion. The reason is that while using DCT filtering to abandon the behind data, if the retained data is too less, then the lost energy is too much, after restoration, the original signal cannot be presented very well.

The Figure3 (b) shows that when the coefficient is more, the filtering effect becomes worse obviously. A lot of noise and disturbance can't be effectively removed. For DCT filtering, the latter coefficients which containing the adding noise is abandoned, whereas the former coefficients which the containing the noise is restrained completely.

Although applying the DCT to turbine vibration signal filtering, which can take effects in some degree, this algorithm is a kind of non-linear filtering, it needs to choose the proper length for keeping, if the abandoned data is too less, the noise cannot be removed effectively; if the abandoned data is too more, the signal's information for itself will loss too much; In addition, when the length of signal sampling is long, this choice of method will be much more difficult. So this algorithm directly using for vibration signal denoising still have some kind of problems, for the long data, it is better to deal with it by the method of periodic segment before dealing [7-8].

4.Cyclic Wiener Filtering Algorithm

Wiener filter is the linear filter which is based on the minimum mean square error. So it can be designed as finite duration impulse response or infinite duration impulse. Taking the FIR Wiener filter as an example, basic principle is shown in Figure 4.

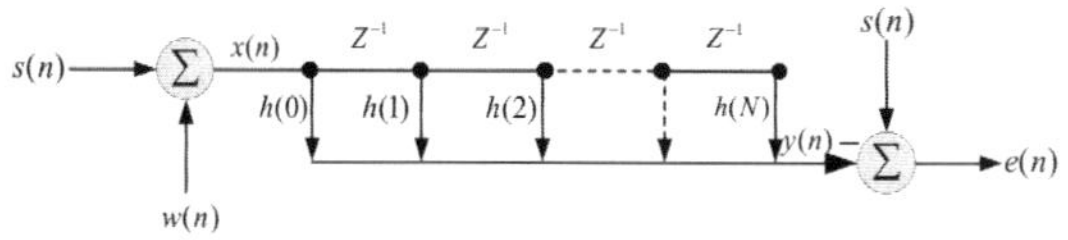

Figure4. Wiener filter schematic

In Figure 4, $s(n)$ is the real signal, $w(n)$ is the additive noise or measurement error, $x(n)$ is the observed signal, $h(n)$ is the Wiener filtering output signal, $e(n)$ is the processing error. By Figure 4, we can know:

$$x(n) = s(n) + w(n) \quad (4)$$

$$y(n) = \sum_{k=0}^{N} h(k) \cdot x(n-k) \quad (5)$$

$$e(n) = s(n) - y(n) \quad (6)$$

So the filtering algorithm's mean square error is:

$$E[e^2(n)] = E\{[s(n) - \sum_{k=0}^{N} h(k) \cdot x(n-k)]^2\} \quad (7)$$

By the principle of minimum mean square error, the value of equation above is the smallest, so Wiener filter's weighting coefficient $h(k)$ should meet the linear equation as follows:

$$\sum_{k=0}^{N} \varphi_{xx}(m-k) \cdot h(k) = \varphi_{sx}(m) \qquad 0 \le m \le N \quad (8)$$

In the equation, $\varphi_{xx}(m)$ is the auto-correlation sequence of the observed signal, N is the wiener filter's order, $\varphi_{sx}(m)$ is the cross-correlation sequence between desired signal and observed signal[9-10].

For Wiener filter, the signal should be stationary. Technically, the vibration signal which is sampled in the industrial scene is non-stationary signal, which cannot directly used by Wiener filtering. However, the vibration signal has a certain stable periodic characteristics. When the signal has a more obvious periodicity, we can use the cyclic Wiener filter to process the data by segmentation and circle. A whole segment of the original signal is divided into a small range by a certain rule, then Wiener filtering is used in each small range in order to form a circle, in later the segmentation data is linear integration. The cyclic Wiener filter is the improvement of Wiener filtering, which has a better filtering effect at the lower order. The simulation result of cyclic Wiener filter is shown as Figure 5:

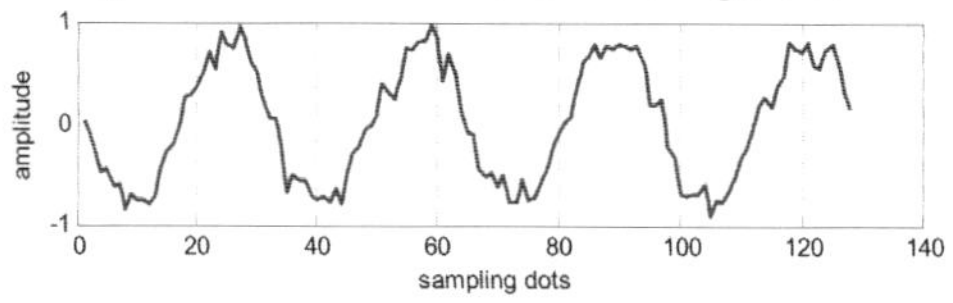

Figure5. Simulation result of cyclic Wiener filter

5.Cyclic Wiener filtering algorithm in Discrete Cosine Transform

5.1 The combination of DCT and cyclic wiener filtering

When using DCT filtering, the noise which is added in coefficient preserving data cannot be removed that can be filtered again by Wiener filter to improve the filtering effect. When the length of signal is longer, the energy concentration of DCT transform is reduced, the requirement of Wiener filter's order is also improved, so after segmentation, the Wiener filtering in DCT domain should be considered. Generally speaking, the vibration signal has the characteristic of stationary in cycle, so the vibration signal can be segmented by the whole period, and then each period is transformed into DCT domain then processing by Wiener filtering. It's called cyclic Wiener filtering in DCT domain. This combination not only makes up for the lack of DCT filtering, but also plays the better role of Wiener filtering. The algorithm process of cyclic Wiener filtering in DCT domain is as follow:

① The observation signal and ideal signal are segmented respectively according to the same role;

② Each segment is for Discrete Cosine Transform respectively

③ The first segment data by DCT transform is for Wiener filtering, the result is saved and retained.

④ The upper filtering result is as the next desired signal segment for filtering processing.

⑤ A number of segments of the signal which is processed for a simple combination.

⑥ Inverse Discrete Cosine Transform for recombination data then gets the filtering effect.

5.2Filtering result analysis

Analysis of algorithm principle, DCT filtering is a kind of filtering that has zero phase shift or none phase shift. Choosing the best reserved length can obtain a certainly of filtering effect. Also the characteristic of the waveform signal is not obviously lost. Meanwhile, cyclic Wiener filter is a kind of linear filtering, so this algorithm can achieve a better phase characteristics on the whole.DCT filtering removes the noise of coefficient of superposition, then the cyclic Wiener filtering can further enhance the filtering effect. The segmentation processing, that can greatly reduce the order of Wiener-Holf equation. So the problem mentioned above can be easily resolved.

Simulation result of cyclic Wiener filtering in DCT domain is shown as Figure 6.From the actual simulation results we can see: this algorithm not only achieve the better noise filtering effect, separates the noise and signal in maximum, also the features of the original signal are well preserved, meanwhile the good phase characteristics are obtained.

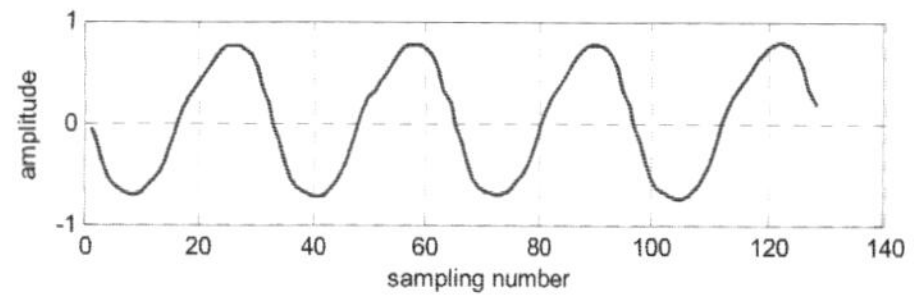

Figure6. Simulation result of cyclic wiener filtering in DCT domain

Practice shows the algorithm that cyclic Wiener filtering in Discrete Cosine Transform Domain can meet the filtering demands for this kind of vibration signal; also it achieves the effective compression for the data which reduces the amount of data.

6 Conclusion

This paper presents an algorithm for cyclic Wiener filtering algorithm in Discrete Cosine Transform Domain for vibration signal. This algorithm combines the DCT filtering and cyclic Wiener filtering effectively, which achieves the better filtering effect. Simulation result shows that this algorithm can achieve ideal effect for steam turbine's vibration signal denoise processing, and it can ensure the good linear phase, and the filtering processing for vibration signal does not produce obvious phase distortion. Also at the same time to ensure the linear phase characteristics this algorithm achieve the better filtering effect, which performance is better than Wiener filtering algorithm alone or DCT filtering algorithm alone. The algorithm which processes the data in segment, so the order of Wiener filtering is reduced, solves the problems that the energy concentration of long data in DCT domain is low. Using the characteristics of DCT energy concentration, it further reduces the order of computation, meanwhile it realizes the vibration signal data compression ,The basic principle of the algorithm can be also applied to other signal denoise processing which have the similar filtering requirements. So it has a very good value and application prospect.

References

1. ZHANG Jun, WANG Jin-ping, ZHU Bo. Signal Filtering and Feature Extraction Technology of Rotating Machinery Vibration Signal Process. Journal of Nanjing Institute of Technology :Natural Science Edition,**3**,15,(2009)
2. WANG Jing-fang. On DCT Domain Wiener Filtering for Speech Enhancement. Journal of Hunan International Economics University,**1**,83,(2014)
3. Ming Yang, CHEN Jin, DONG Guang-ming. Rolling Bearing Fault Diagnosis Based on Cyclic Wiener Filter and Envelop Spectrum. Journal of Vibration Engineering,**5**,537,(2010)
4. GUAN Bo, Hu Jin-song. Study of the DCT-Based Vibration Signal Zero Phase Distortion Filtering Method in Rotating Machinery. Turbine Technology,**4**,285,(2007)
5. G.K Singh, Saleh Al Kazzaz Sa'ad Ahmed. Vibration signal analysis using wavelet transform for isolation and identification of electrical faults in induction machine. Electric Power Systems Research, **68**,119-136,(2003)
6. WANG Jin-ping, DENG Ai-dong, DING Ke-ke. Research on and Design of DSP-Based Turbine Vibration Signal Acquisition System[J].Power System Engineering,**24**,49,(2008)
7. CHEN Jia-yuan, ZHANG Qiu-ju, YU Wen-kai. Research on Method for Vibration Signal Compression for a Compressor Based on DCT .Compressor Technology,**4**,5,(2009)
8. GUAN Bo, HU Jin-song. Research on Reduction Method for Vibration Signals of Rotating Machinery Based on DCT .Wind Turbine Technology,**3**,80(2007)
9. Wu Qing-ju, Tian Xiao-bo. Receiver Function Estimated by Wiener Filtering. Earthquake Research in China.**17**,387-388(2003)
10. W. Gardner. Cyclic Wiener Filtering: Theory and Method. IEEE Transaction on Communications.**41**,151-163(1993)

Visual Tracking Using L2 Minimization

Zhijun Pei[1,a] and Lei Han[1]

[1]*Department of Electronic Engineering, Tianjin University of Technology and Education, Tianjin, China*

Abstract. Visual tracking has been an active research topic in the computer vision applications. By modeling the target appearance with a sparse approximation over a template set, sparse representation has been applied to the visual tracker, which called L1 tracker. Due to the need to solve the L1 norm related minimization problem for many times, this L1 tracker is very computationally demanding. Although various fast numerical solver is developed to solve the resulting L1 norm related minimization problem, the framework is still a L1 norm related minimization model. Similar to the face recognition problem, sparse approximations may not deliver the desired robustness and a simple L2 approach to the visual tracking problem is not only robust, but also much faster. It may be possible to apply the L2 minimization, instead of L1 minimization, to the visual tracking problems, which has been verified by experiments on challenging sequences in the paper.

1 Introduction

Visual tracking is a critical task in many computer vision applications and has been widely applied in various fields such as the automated surveillance, vehicle navigation robotics. The challenges in designing a robust visual tracking algorithm are caused by the presence of noise, occlusion, varying viewpoints, background clutter, and illumination changes [1]. Although a variety of tracking algorithms have been proposed to overcome these difficulties in last two decades, many challenging problems still remain when designing a practical visual tracking system. An effective appearance model is of prime importance for the success of a tracking algorithm. Tracking algorithms can be generally categorized as either discriminative or generative based on their appearance models [2]. Tracking problem is formulated as a binary classification problem in discriminative tracking methods. Discriminative trackers locate the object region by finding the best way to separate object from background. In another way, generative tracking algorithms typically learn a model to represent the target object and then use it to search for the image region with minimal reconstruction error. And the appearance model is often dynamically updated during the tracking to adapt to pose and illumination changes of the object.

Recently, motivated by recent advances in sparse representation and its applications in computer vision such as robust face recognition, sparse representation have been applied to tracking problem, where a tracking candidate is sparsely represented by target templates and trivial templates [3]. In these approaches, the sparse representation is obtained via solving a L0-norm related minimization or L1-norm related minimization problem [4]. However, the computational complexity of this tracker is rather high, thereby limiting its applications in real time scenarios. It is well known that L0-norm related minimization is an NP-hard problem and the large scale L1-norm related minimization is also a challenging problem. There have been great progresses on fast numerical methods for solving large-scale L1-norm related minimization problems such as the so-called accelerated proximal gradient (APG) method [5]. Although the introduction of very fast numerical method to solve the resulting L1 norm minimization problems may lead to a real time L1 tracker, it still built on the same framework of the L1 tracker.

In the L1 tracker framework, hundreds of L1-norm related minimization problems need to be solved for each frame during the tracking process. An efficient solver for the L1 norm related problems has been the key in practice. As for the application of compressive sensing (CS) to the problem of face recognition, it has been shown that faster, more accurate, and more robust methods may be achieved by modelling outliers explicitly and using the L2-norm instead of L1-norm [6]. In many applications of compressive sensing, the sparsity is assumed, rather than proven or measured. However, the sparsity assumption may not supported by data of the practical problems. This implications are important for the application of compressive sensing to face recognition but also to other problems where sparsity is assumed rather than proven, such as visual tracking problems. So the possibility of the application the L2 minimization for the visual tracking problems is discussed, and which has been verified by the experiments in the paper.

2 L1 minimization tracking

A brief review on the L1 tracker within the particle filter framework is first given bellow.

[a] Corresponding author: peizj@tute.edu.cn

2.1 Particle filter

Tracking can be considered as an estimation of the state for a time series state space model. The problem is formulated in probabilistic terms. The particle filter, also known as the sequential Monte Carlo method, is one of the most popular approaches to provide solutions. It recursively constructs the posterior probability density function of the state space using Monte Carlo integration. It has been developed in the computer vision community and applied to tracking problems [7].

The particle filter is a Bayesian sequential importance sampling technique for estimating the posterior distribution of state variables characterizing a dynamic system. It provides a convenient framework for estimating and propagating the posterior probability density function of state variables regardless of the underlying distribution. It consists of prediction step and update step. Let x_t denote the state variable describing the affine motion parameters of an object at time t. Given all available observations $z_{1:t-1}=\{z_1,z_2,\cdots,z_{t-1}\}$ up to time t−1, the predicting distribution of x_t is recursively computed as

$$p(x_t|z_{1:t-1}) = \int p(x_t|x_{t-1})\,p(x_{t-1}|z_{1:t-1})\,dx_{t-1}$$

At time t, the observation z_t is available and the state vector is updated using the Bayes rule

$$p(x_t|z_{1:t}) = p(z_t|x_t)\,p(x_t|z_{1:t-1})\,/\,p(z_t|z_{1:t-1})$$

In the particle filter, the posterior $p(x_t|z_{1:t})$ is approximated by a finite set of N samples $\{x_{it}\}$, i=1,$\cdots$,N with importance weights w_{it}. The candidate samples x_{it} are drawn from an importance distribution $q(x_t|x_{1:t-1},z_{1:t})$ and the weights of the samples are updated. The samples are resampled to generate a set of equally weighted particles according to their importance weights to avoid degeneracy.

In the tracking framework, the object motion of two consecutive frames is model with the affine image warping. The state variable x_t is modeled by the parameters of the affine transformation parameters. The image region of interest z_t is cropped from the image as the target templates. A Gaussian distribution is employed to model the state transition distribution $p(x_t|x_{t-1})$. The observation model $p(z_t|x_t)$ reflects the similarity between a target candidate and the target templates, which is formulated from the error approximated by the target templates using L1 minimization.

2.2 L1 minimization tracking

The global appearance of one object under different illumination and viewpoint conditions is known to lie approximately in a low dimensional subspace. With template image columns are stacked to form a one dimension vector, the target template set can represented as

$$T=[t_1 .. t_n] \in R^{d\times n}, (d \gg n)$$

Then a tracking result $y \in R^d$ approximately lies in the linear span of T, that is

$$y \approx Ta$$

where a is the target coefficient vector.

In many visual tracking scenarios, target objects are often corrupted by noise or partially occluded, which may affect any part of the image and appear at any size on the image. The locations of corruption can differ for different tracking images and are unknown to the computer. The trivial templates I = [i_1, i_2, ...,i_d] can be used to capture the occlusion. To incorporate the effect of occlusion and noise, a tracking object can represent as

$$y=Ta+Ie=As \quad (1)$$

where the trivial template is a vector with only one nonzero entry, here I can be chosen as an identity matrix, and e is called trivial coefficient vector.

The system in equation (1) is underdetermined and does not have a unique solution for the vector s. The error caused by occlusion and noise typically corrupts a fraction of the image pixels. Therefore, for a good target candidate, there are only a limited number of nonzero coefficients in vector e that account for the noise and partial occlusion. In order to have a sparse solution, considering the problem as a L1-regularized least squares problem,

$$\min_s \|As-y\|_2^2 + \lambda\|s\|_1 \quad (2)$$

Where $\|.\|_1$ and $\|.\|_2$ denote the L1 norms and L2 norms respectively.

Then the tracking result can be found by finding the smallest residual after projecting on the target template subspace. Therefore, the tracking result is the sample of states that obtains the smallest error or the largest probability.

3 L2 minimization tracking

Just as discussed above, the visual L1 tracking frame caste the tracking problem as finding a sparse approximation in a template subspace, where handling occlusion using trivial templates such that each trivial template has only one nonzero element. During tracking, a target candidate is represented as a linear combination of the template set composed of both target templates obtained from previous frames and trivial templates. The number of chosen target templates are far fewer than the number of trivial templates. Intuitively, a good target candidate can be efficiently represented by the target templates. This leads to a sparse coefficient vector, since coefficients corresponding to trivial templates, named trivial coefficients, tend to be zeros. In the case of occlusion or other unpleasant issues such as noise corruption or background clutter, a limited number of trivial coefficients will be activated, but the whole coefficient vector remains sparse. A bad target candidate, on the contrary, often leads to a dense representation. The sparse representation is achieved through solving an L1-regularized least squares problem. Then the candidate with the smallest target template projection error is chosen as the tracking result. After that, tracking is led by the Bayesian state inference framework in which a particle filter is used for propagating sample distributions over time. The target template set is dynamically updated to keep the representative templates throughout the tracking procedure. This is done by adjusting template weights by using the coefficients in the sparse representation.

In the proposed L1 tracker, hundreds of L1-norm related minimization problems need to be solved for each frame during the tracking process. An efficient solver for the L1 norm related problems has been the key to use the L1 tracker in practice. Based on the interior point method, the solver for the L1 norm minimizations turns out to be too slow for tracking. In order to reduce the number of particles, which equal to the number of the 1 norm minimizations for solving, a minimal error bounding strategy can be introduced, but it is still far away from being real time. Although using accelerated proximal gradient approach can further improve the L1 Tracker, L1-norm related minimization problems still need to be solved for each frame [5].

Similar to the face recognition problem, one argument with the analysis of the L1 tracker might be that the L1 term is intended to achieve robustness, rather than indicating a belief in the sparsity of the coefficients. This is an important distinction. The L1-norm is used in compressive sensing as a tractable alternative to the L0-norm. Sparseness does not necessarily lead to robustness to the presence of outliers. To achieve robustness, one could use L1-regression as follows:

$$\min_{\alpha} \|y - A\alpha\|_{L1} \quad (3)$$

which avoids overly penalizing gross outliers.

The L1-norm in compressive sensing and L1-norm in L1-regression are, however, two unrelated uses of the same norm. The two applications differ in the quantities to which the L1 norm is applied. Robustness cannot therefore be used to justify applying the L1-norm to the coefficients without further explanation. No such explanation is given, however. The problem with (3) is that solving such a linear program can be computationally expensive as the data size grows. Rather than resort to approaches, however, a faster, more accurate, and more robust methods may be achieved by modelling outliers explicitly and using the L2-norm.

In contrast to the L1 case, it is possible to estimate using the L2 norm by solving

$$\arg\min \|y - A\alpha\|_{L2}^2$$

Even when the system is overdetermined, the optimal solution, in the sense of the smallest reconstruction error, can be recovered by

$$\alpha = (A^T A)^{-1} A^T y$$

To efficiently solve the related pseudo-inverse, by QR factorisation of A, then A = QR. where Q forms a orthonormal basis, and R is an upper triangle matrix. Therefore, we can have

$$\alpha = R^{-1} Q^T y$$

Here $R^{-1}Q^T$ remains the same for all y. So it just need to computed once and stored for each frame during the tracking process. Once the coefficients are estimated, one can find the state location of identified object via minimizing the residuals

$$c^*(y_t^i) = \arg\min_{x_t^i} \|y_t^i - T_t c_T^i\|_{L2}^2$$

where i is the particles index of t frame; c_T^i is the sub-vector consisting of components of a and T_t is a submatrix of A, both corresponding to the basis of particle i.

The particle filter provides an estimate of posterior distribution of random variables related to Markov chain. In visual tracking, it gives an important tool for estimating the target of next frame without knowing the concrete observation probability. At the frame t, denote x_t which describes the location and the shape of the target, $y_{1:t-1} = \{y_1, y_2, \cdots, y_{t-1}\}$ denotes the observation of the target from the first frame to the frame t − 1. The optimal state for the frame t is obtained according to the maximal approximate posterior probability. The observation model reflects the similarity between a target candidate and the target templates, and the observation likelihood of state x_t^i is given as

$$p(y_t \mid x_t^i) = \exp\{-a\|y_t^i - T_t c_T^i\|_{L2}^2\} / \Gamma$$

where α is a constant controlling the shape of the Gaussian kernel, Γ is a normal factor. Then, the optimal state x_t^* of frame t is obtained by

$$x_t^* = \arg\max_{x_t^i \in S_t} p(y_t \mid x_t^i)$$

In addition, a template update scheme is also adopted to overcome pose and illumination changes.

4 Experiments and discussions

In order to evaluate the performance of the proposed tracking framework with L2 minimization method, two videos are used in the experiments. The proposed tracker is implemented in MATLAB2013b on an Intel Core(TM) i3 CPU with 2.13 GHz.

The first test sequence is an infrared (IR) image sequence from VIVID benchmark dataset PkTest02 [8]. Some samples of the final tracking results are demonstrated in Figure 1, where six representative frames of the video sequences are shown. The frame indexes are 26, 46, 56, 66, 76 and 96. The target to background contrast is very low and the noise level is high for these IR frames. From Figure 1, it is shown that the tracker is capable of tracking the object all the time even with severe occlusions by the trees on the roadside. It keeps tracking it in the rest of the sequences. There is no case that fails to track the target in the some index frame. Similar to the proposed L1 tracker, our proposed method can also avoids fails problem and is effective under low contrast and noisy situation. The proposed tracker with L2 minimization achieves similar results to L1 method but more fast.

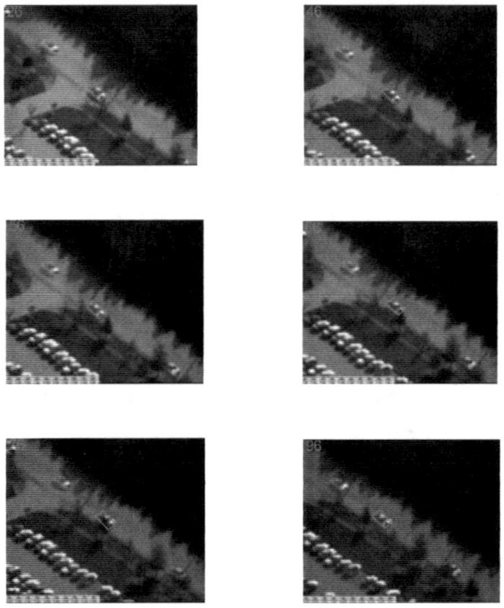

Figure 1. Some samples of the final tracking results

We also evaluate our tracking algorithm with the challenging sequences such as David indoor sequence, where there exits scale, pose and illumination change [2]. For the David indoor sequence shown in Figure 2, the illumination and pose of the object both change gradually. When using the feature templates proposed by [2], where the adopted features are similar to generalized Haar-like features which have been shown to be robust to scale and orientation change, the proposed L2 methods perform well on this sequence. Even if a shift happen, it can recover the failure and track the target properly again. The proposed tracker is robust to pose and illumination changes as the object appearance can be modelled well by random projections, based on compressed sampling, and templates online update is used. Our tracker achieves similar results to compressing tracking method.

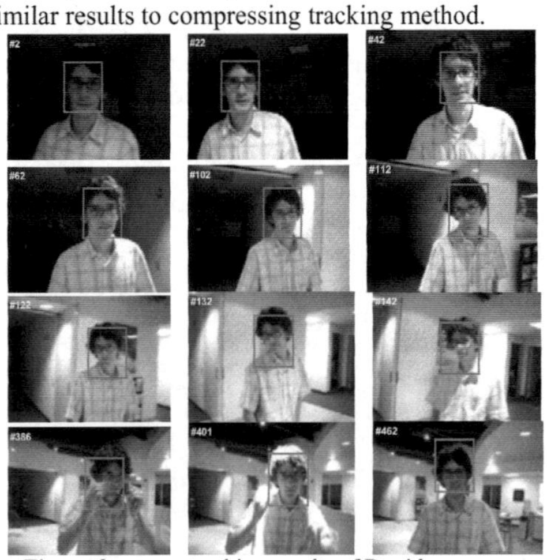

Figure 2. some tracking results of David sequence

5 Conclusion

Despite that numerous algorithms have been proposed, object tracking remains a challenging problem. As for L1 trackers, sparse representation has been applied to visual tracker by modeling the target appearance using a sparse approximation over a template set, which needs to solve an L1 norm related minimization problem for many times. While these L1 trackers showed impressive tracking accuracies, they are very computationally demanding and the speed bottleneck is the solver to L1-norm minimizations. Based on the accelerated proximal gradient approach, a fast numerical solver is developed to solve the resulting L1 norm related minimization problem, But it still build on a L1 norm related minimization model. Among the face recognition researches, it has been shown that sparse approximations may not deliver the robustness or performance desired and a simple L2 approach to the face recognition problem is not only more accurate, it is also more robust, and much faster. So the possibility of the application the L2 minimization for the visual tracking problems is investigated. The running time efficiency and tracking accuracy of the proposed tracker is validated with the evaluation involving two challenging sequences. The proposed tracking algorithm with L2 minimization have demonstrated good robustness in various tracking environments.

Acknowledgement

This work is supported by the Science and Technology Development Foundation of Tianjin Colleges and Universities under Grant No. 20140716. And this work is also partially supported by Research Foundation of Tianjin University of Technology and Education under Grant No. KJY1312.

References

1. A. Yilmaz, O. Javed, and M. Shah, ACM Computing Survey, Vol .38, No.4 (2006)
2. Kaihua Zhang, Lei Zhang, and Ming-Hsuan Yang, European Conference on Computer Vision (ECCV 2012), Florence, Italy, October (2012)
3. X. Mei and H. Ling, IEEE Trans. on Pattern Analysis and Machine Intelligence (PAMI), 33(11) (2011)
4. X. Mei and H. Ling, IEEE International Conference on Computer Vision (ICCV), Kyoto, Japan (2009)
5. C. Bao, Y. Wu, H. Ling and H. Ji, IEEE Conf. on Computer Vision and Pattern Recognition (CVPR), Rhode Island, 157(10) (2012)
6. Qinfeng Shi, Anders Eriksson, Anton van den Hengel, Chunhua Shen, IEEE Computer Society Conference on Computer Vision and Pattern Recognition (CVPR 11), Colorado Springs, USA, June 21-23, 42(7) (2011)
7. S. K. Zhou, R. Chellappa, and B. Moghaddam, IEEE Trans. Image Processing, 11 (2004)
8. R. Collins, X. Zhou, and S. K. Teh, IEEE Workshop on Performance Evaluation of Tracking and Surveillance, (PETS'05), Breckenridge CO,Jan (2005)

Efficient Adaptive Deinterlacing Algorithm Using Bilateral Filter

Jin Wang[1,a], Zhensen Wu[1], Jiaji Wu[2]

[1] *School of Physics and Optoelectronic Engineering, Xidian University, Xi'an Shaanxi, China*
[2] *School of Electronic Engineering, Xidian University, Xi'an Shaanxi, China*

Abstract. This paper introduces an efficient intra-field deinterlacing algorithm using bilateral filter while considering both closeness and similarity between the interpolated pixel and the neighbor pixels. Instead of estimating the edge orientations via limited candidate directions as previous intra-field deinterlacing methods, we propose an adaptive spatial locality- and similarity-based deinterlacing method. Many deinterlacing methods described in the literature only consider the similarity between neighbor pixels and interpolated pixels; however awareness of distance correlation and spatial locality has not yet been studied. Experimental results indicate that the proposed scheme outperforms a number of existing approaches, in terms of both the objective and subjective performance.

1 Introduction

Interlaced scanning has been widely employed as the standard of TV broadcast systems, such as NTSC and SECAM, and some of the high definition television (HDTV) formats, because this method allows for not only the reduction in bandwidth and memory but also the maintenance of the existing transmission channel. However, this interlaced scanning format introduces several problematic artifacts, known as line crawling, edge flickering, interline flickering, field aliasing, and so on1-3. Moreover, as the screen size increases, the artifacts become more troublesome. Deinterlacing methods have been developed to minimize these issues.

Deinterlacing methods can be roughly classified into two categories. One is the intra-field spatial domain method that only uses one field[3-8] and the other is an inter-field temporal deinterlacing method that uses multiple fields. Inter-filed temporal deinterlacing includes motion adaptive and motion compensation (MC) methods[].

Intra-field deinterlacing methods are widely used in real-time application because of their simplicity. Intra-field deinterlacing methods have been proposed with different degrees of complexity and performance. The line average (LA) method3 is a simple method that interpolates the pixel using an average value of the upper and lower pixels. LA has generally been used due to its clarity and simplicity in addition to the fact that it exhibits no motion artifacts. However, the vertical resolution of the input image is halved before the image is interpolated which causes a reduction in the amount of details in the progressive image.

In order to avoid this issue, Doyle proposed an edge-based line-average (ELA) [4] method, which uses edge directional correlation to interpolate a missing line between two adjacent lines in the interlaced images. ELA is a well-known edge-based intra-field deinterlacing method, which categorizes the edge direction as 45, 90, and 135 degrees and then linearly interpolates in the chosen direction. ELA performs well with very little complexity and can be used to eliminate the blurring effect of LA in the regions where the edge is estimated correctly. However, this method has poor visual performance in high spatial frequency regions and can inaccurately estimate the edge detections, which leads to image degradation.

In an effort to remedy these deficiencies, numerous solutions based on the ELA algorithm have been proposed. The efficient ELA (EELA) [5] introduces two useful measurements for estimating the directional spatial correlation more efficiently. In addition, deinterlacing methods depending on content analysis [6] or the horizontal edge pattern [7] have been proposed. Kim proposed a modified ELA (MELA) [6]. In addition to the direction measures of EELA, MELA includes another vertical measure. Therefore, MELA considers three directional spatial correlations that cover vertical, anti-diagonal. The low complexity interpolation method for deinterlacing (LCID) [7] uses four directional differences that are simpler than those in MELA, which are diagonal, anti-diagonal, vertical and horizontal. Specifically, the direction-oriented interpolation (DOI) [8] method uses a window-based measure and is more robust and efficient when compared to the previous ELA methods. DOI can increase the resolution of direction and shows good performance for an image with local, single strong edges or horizontal edges. However, DOI has a high complexity for searching edge patterns within a large search range and a poor performance for regions with similar and repetitive edge patterns.

The aforementioned deinterlacing methods can be roughly separated into two classes. One is simple and highly efficient but with no consideration, for edge direction information or incorrect edge direction estimation which degrades the image detail. The other is complex with high computation overhead but has good edge direction estimation.

[a] Corresponding author: jwang0608@xidian.edu.cn

 DOI: 10.1051/matecconf/20166102021

In this paper, we proposed an intra-field deinterlacing algorithm using both closeness and similarity that successfully estimates edge direction. The missing pixel is estimated using using a nonlinear spatial average without degrading the edges by means of a nonlinear combination of nearby image values. Moreover, we combine gray levels based on their geometric closeness and their similarity, and prefer near values to distant values in the context of both closeness and similarity. The deinterlacing method using closeness and similarity (DCS) is based on the assumption of closeness and similarity between the neighborhood values where two pixels can be close to one another.

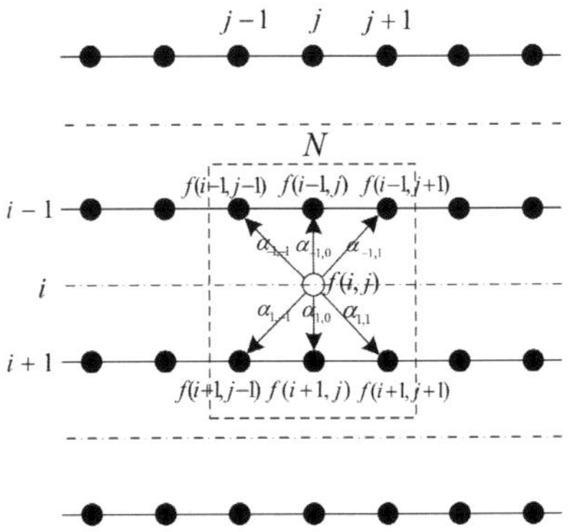

Fig. 1 Six neighbor pixels used for deinterlacing.

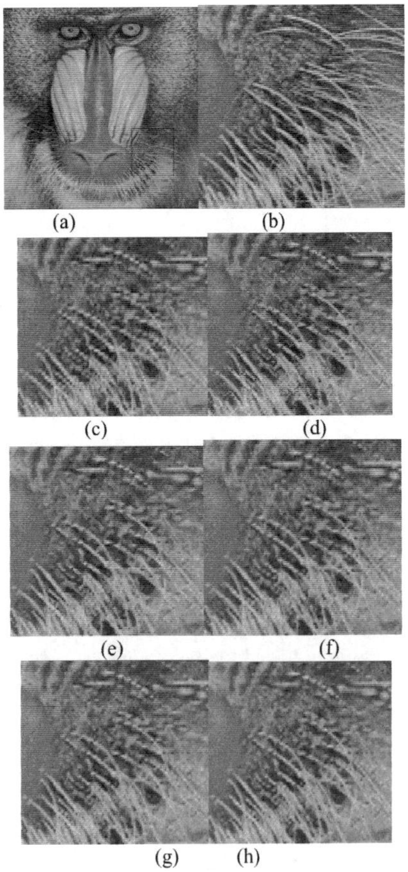

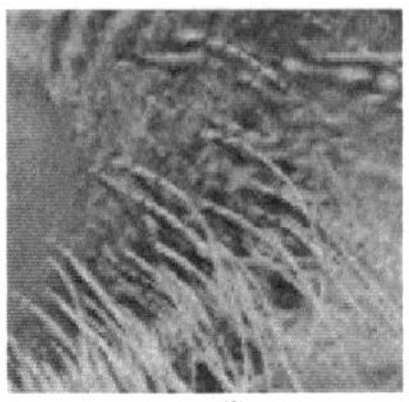

(i)

Figure 2. (a) The original image and (b) part of the original zoomed image. Perceived image quality comparison using various deinterlacing methods: (c) LA, (d) ELA, (e) EELA, (f) MELA, (g) LCID, (h) EMD, (i) proposed method.

2 The proposed algorithm

The pixel value at a given location is a function of the values in a small neighborhood at the same location. In the conventional deinterlacing method, the pixel values are averaged under the assumption that images typically vary gradually over space and neighbor pixels are likely to have similar values. The conventional deinterlacing method can be expressed as follows:

$$\hat{f}(i,j)=\frac{\sum_{(m,n)\in N}\alpha_{m,n}f(i+m,j+n)}{\sum_{(m,n)\in N}\alpha_{m,n}} \qquad (1)$$

where N is the set of the neighbor pixel locations, m and n are vertical and horizontal displacement parameters, respectively, , and $\alpha_{m,n}$ is the weight. Figure 1 shows an example of six neighbor pixels used in Equation (1) and LA is used as an example to explain Equation (1). For LA, we only use the pixels at the position (i-1, j) and (i+1, j), and the parameters $\alpha_{-1,0}=\alpha_{1,0}=1$.

There are two drawbacks to the existing deinterlacing methods, and they are listed below:

(1) The weight $\alpha_{m,n}$ is constant within some limited number of fixed candidate edge directions;

(2) These methods only consider the similarity between pixels, and awareness of the closeness or spatial locality is not taken into account.

The assumption is that slow spatial variations fail at the edges, which are consequently blurred. To address the first drawback, an adaptive nonlinear function is required that averages the image values with weights, which decay with dissimilarity and also depend upon image intensity. Generally speaking, the neighbor pixels that are markedly different from the current interpolated pixel should be assigned a small weight. In contrast, the neighbor pixels most similar to the current interpolated pixel should have

Table 2. PSNR and CPU time for different images for the proposed and existing methods

Method	Image	Lena	Boat	Pepper	Finger	Baboon	Zelda	Man	Avg.
LA	PSNR(dB)	37.668	35.316	33.712	31.862	23.487	41.809	31.897	33.670
	CPU time (sec)	0.015	0.021	0.015	0.021	0.015	0.015	0.021	0.0175
ELA	PSNR(dB)	35.854	32.326	34.030	28.862	23.056	38.968	30.814	31.987
	CPU time (sec)	0.218	0.234	0.218	0.221	0.249	0.234	0.234	0.2297
EELA	PSNR(dB)	37.021	33.627	34.083	29.504	23.103	40.038	31.841	32.716
	CPU time (sec)	0.249	0.259	0.249	0.312	0.335	0.249	0.249	0.2717
MELA	PSNR(dB)	37.915	35.338	34.165	31.466	23.602	41.727	32.185	33.771
	CPU time (sec)	0.265	0.296	0.265	0.256	0.280	0.280	0.280	0.2745
LCID	PSNR(dB)	37.383	33.843	34.112	30.848	23.511	40.91	31.867	33.210
	CPU time (sec)	0.254	0.296	0.254	0.280	0.280	0.254	0.254	0.2674
EMD	PSNR(dB)	36.848	32.26	33.731	31.220	23.547	33.568	31.054	31.746
	CPU time (sec)	0.665	0.698	0.665	0.715	0.698	0.698	0.698	0.691
DCS	PSNR(dB) (Ψ_{GFS})	37.986	35.598	34.393	31.929	24.048	42.006	32.183	34.021
	CPU time (sec)	0.078	0.089	0.079	0.091	0.091	0.078	0.078	0.0834

a large weight. Then, we can design an adaptive function that is a monotonic decreasing function of the absolute value of the difference in pixel value between the current interpolated pixel and the neighbor pixel, shown as Equation (2).

$$\alpha_{m,n} = g\left(\left|\hat{f}(i,j) - f(i+m, j+n)\right|\right) \tag{2}$$

where m and n are vertical and horizontal displacement parameters respectively, $m = 2k-1, n = l, (k,l) \in \mathbf{Z}$, $\hat{f}(i,j)$ is the prediction of the current interpolated high resolution pixels obtained based on the average of the two nearest pixels or two pixels in the vertical direction. We choose g(x) as

$$g_1(x) = \exp\left(-x^2\right) \tag{3}$$

Comparing the performance of several deinterlacing methods reveals that LA has a better quality peak signal to noise ration (PSNR) when compared to many of the edge-based methods with very low complexity. That is because the neighbor pixels used in LA are the shortest distance from the interpolated high resolution pixel, while, other methods just consider direction estimation, but ignore the spatial locality distance, or closeness, which results in poor performance. In other words, closeness is also an essential concept. Thus, we designed another weight, βm,n, which incorporates closeness by weighting pixel values with coefficients that fall off with distance. We can also design a monotonic decreasing function to assign a weight to the distance between the current interpolated pixel and the neighbor pixel.LA performs well when considering the nearest neighbor pixels. However, direction estimation is also an important issue due to edge regions and LA causes some artifacts because direction is not considered. In order to solve all of the above-mentioned problems, we propose a DCS method that consider closeness and similarity in both the smooth and edge regions.

$$\hat{f}(i,j) = \frac{\sum_{(m,n)\in N} \alpha_{m,n}\beta_{m,n} f(i+m, j+n)}{\sum_{(m,n)\in N} \alpha_{m,n}\beta_{m,n}} \tag{4}$$

$$\alpha_{m,n} = \exp\left(\frac{-\left(\hat{f}(i,j) - f(i+m, j+n)\right)^2}{2\sigma_s^2}\right) \tag{5}$$

$$\beta_{m,n} = \exp\left(\frac{-\left(m^2 + n^2\right)}{2\sigma_c^2}\right) \quad (6)$$

Here, σ_s and σ_c are regularization factors for $\alpha_{m,n}$ and $\beta_{m,n}$, respectively, used to control the performance of the algorithm and they are generally selected empirically.

3 Experiments

A simulation was performed using standard test 512×512 images Lena, Boat, Pepper, Finger, Baboon, Zelda, and Man in order to verify the superiority of the proposed method compared to the existing algorithms (LA, ELA, EELA, MELA, LCID, and EMD). We assessed the objective performance in terms of PSNR and CPU time in addition to the subjective performance, which is a widely adopted criterion in the literature. Our experiment was carried out on an Intel Core 2 Duo CPU E8500 @ 3.16 GHz with s set to 1.

our proposed algorithm can get good PSNR with fixed factor set. In terms of CPU time, our algorithm required only 30% of the complexity of other MELAs. Except for the simplest method (LA), our method required the least CPU time.For the subjective performance evaluation in terms of the visual effect of the resulting images, we also show part of the perceived image quality of Baboon in Fig. 2. It is apparent from Fig. 2 that the proposed method results in an improved visual quality in terms of the details of the fur compared to the other methods. These results indicate that the proposed method is superior to other intra-field deinterlacing methods in terms of objective and subjective qualities and has low complexity.

Acknowledgement

This work was partly supported by the National Natural Science Foundation of China (No. 61501359); this work was partly supported by the Fundamental Research Funds for the Central Universities (No. JB160504).

References

1. K. Jack, Video Demystified—A Handbook for the Digital Engineer,4th ed., Elsevier, Jordan Hill, Oxford (2005).
2. G. De Haan, "Television display processing: past & future," in Proc.IEEE ICCE'07, Las Vegas, pp. 1–2 (2007).
3. E. B. Bellars and G. De Haan, Deinterlacing: A Key Technology forScan Rate Conversion, Elsevier, Amsterdam (2000).
4. T. Doyle, "Interlaced to sequential conversion for EDTV applications," in Proc. 2nd Int. Workshop Signal Processing of HDTV, pp. 412–430(1998).
5. T. Chen, H. R. Wu, and Z. H. Yu, "Efficient deinterlacing algorithm using edge-based line average interpolation," Opt. Eng. 39(8), 2101–2105 (2000).
6. W. Kim, S. Jin, and J. Jeong, "Novel intra deinterlacing algorithm using content adaptive interpolation," IEEE Trans. Cons. Elect 53(3), 1036–1043 (2007).
7. P.-Y. Chen and Y.-H. Lai, "A low-complexity interpolation method for deinterlacing," IEICE Trans. Inf. Syst. E90-D(2), 606–608 (2007).
8. H. Yoo and J. Jeong, "Direction-oriented interpolation and its application to deinterlacing," IEEE Trans. Consum. Electron 48(4), 954–962 (2002).
9. D. Wang, A. Vincent, and P. Blanchfield, "Hybrid de-interlacing algorithm based on motion vector," IEEE Trans. Circuits Syst. Video Technol., 15(8), 1019–1025 (2005).
10. D. Gellies, M. Plantholt, and D. Westerkamp, "Motion adaptive field rate upconversion algorithms for 900 lines/100 Hz 2:1 displays," IEEE Trans. Consum. Electron., 36(5), 149–160 (1990).
11. Y. Wang and S. K. Mitra, "Motion/pattern adaptive interpolation of interlaced video sequences," IEEE Trans. Acoust., Speech, Signal Process., 2829–2832 (1991).
12. D. Hargreaves and J. Vaisey, "Bayesian motion estimation and interpolation in the interlaced video sequences," IEEE Trans. Image Process.,6(5), 764–769 (1997).
13. O. Kwon, K. Sohn, and C. Lee, "Deinterlacing using directional interpolation and motion compensation," in Proc. Image and Video Communications and Processing 2003, 604-611 (2003).
14. T. Jeong, Y. Kim, K. Sohn, and C. Lee, "Deinterlacing with selective motion compensation," SPIE Optical Engineering, 45(07), 077001 (2006).
15. Y. D. Kim, J. Chang, G. S. Shin, and M. G. Kang, "Interlaced-to-progressive conversion using adaptive projection-based global and representative local motion estimation," Journal of Electronic Imaging,17(2), 023008 (2008).

Research on Large Energy Power Enterprise Group total compensation conglomerates distribution systems Construction

Jia-xu CHENG [1,a], Hao-jie Liu[2], Dan Wang[1], Ying-hui HAN [3]

[1] *State Grid Energy Research Institute, Beijing, China*
[2] *State Grid Corporation of China, Beijing, China*
[3] *Beijing Institute of Technology, Beijing, China*

Abstract. In order to solve the problem of exist in W company salary distribution, such as the work guide is unknown, incentive enough, the target set properly and other issues, this paper especially focuses on the characteristics of enterprises, the company management and control model and other aspects to do in-depth study. On the basis of recognition control mode the pay distribution structure is determined, and then the appropriate indicators are sated. They are linked to further clarify the distribution of the guide, while the target value of properly designed and incentive mechanisms to ensure that the salary distribution model to follow on the basis of the relevant principles on the realization of economic promotion and to promote the strategic objectives.

1 Introduction

The Large Energy Power Enterprise Group total amount of remuneration in this article refers to enterprises or enterprise groups total paid remuneration to employees (including subsidiaries) directly in a given period, including enterprise (including subsidiaries) all wages, bonuses, overtime pay, job subsidies, labor insurance fees and expenses. On company-wide, it mainly refers to the sum of the total wages of all manner of employment programs. Reasonable salary distribution so that the total amount of remuneration into play to optimize the efficiency of labor costs, increase the role of incentive effect, and promote the healthy development of enterprises. W is large state-owned central enterprises, Large Energy Power Enterprise Group, its asset size reached 2 trillion Yuan, under a number of power enterprises, many industrial units, a number of financial institutions, more than the immediate professional support unit. The geographical scopes branched out into home and abroad, all kinds of branches and sub a total of more than 50 companies. And units' market positioning, stage of development, company size, positioning functions are varied. SASAC approved the annual W's total remuneration, and therefore need to complete a large total compensation distributed among the many molecules of the company[1].

For W company, total compensation control often involves two core issues. Firstly, under the supervision of the SASAC, how to develop a reasonable total salary to increase incentive effect on the basis of strengthening labor costs management. Secondly, how to pay the total amount of control and optimization of the balance between the subsidiary companies. Total salary distribution business is a core component of the total compensation budget management.

Since the W company was in the implementation of the total payroll budget management system, it constantly has explored scientific and rational plan of total wages, and accumulated some experience. With the intensification of human resource management continued to deepen, there are challenges to the total wage budget management. Firstly, headquarters labor costs for each unit in order to control the total amount of remuneration as the main means of control, only the overall level, the growth rate of labor cost and other considerations of human capital investment amount, and failed business performance, labor efficiency, and stable operation level human capital output capacity combine there is a big limitation considerations; different characteristics of each business sector companies, the lack of differentiation of total compensation allocation criteria and the plate , based on optimal balance between the various units, the total amount of remuneration has not yet formed the group balance of the system optimization mechanism and management system[2].

In this context, modern technology and human resource management concepts are used. Learning from the overall budget management theory, for critical issues, remuneration allocation model and optimization technology covering the total amount of enterprises are researched and developed including coordination grid enterprises, market-oriented industrial units, financial institutions. These will rich enterprise compensation management and human resource planning management theories, help companies improve the scientific, systematic and intensive lean level of human resource management[3].

[a] Corresponding author: shsrp@126.com

2 Total salary distribution system

2.1 Total compensation distribution system build path

Taking into account a complex industrial structure, salary distribution characteristics more realistic level, to build total remuneration allocation model should be in accordance with the "four steps" building strategies. It is hierarchical classification construction of the company's total salary distribution system. Specific process shown in Figure 1.

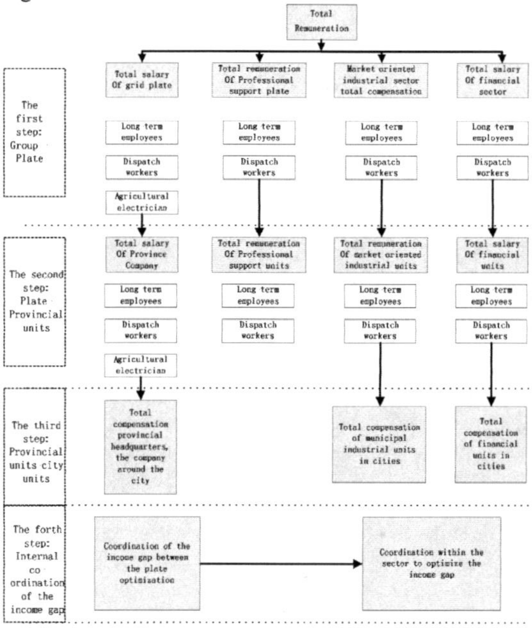

Figure 1. Overall configuration diagram of salary distribution model in company

2.1.1 Consider the heterogeneity of the business segments, the completion of Group - business segments salary distribution

The use of sections to divide the text of the paper is optional and left as a decision for the author. Where the author wishes to divide the paper into sections the formatting shown in Table 2 should be used.

The total remuneration of each segment constituted mainly by two content. One part as a standard part of the remuneration guarantee the income of workers, and the other part is the incentive for incentive compensation. Among them, the salary guarantee section includes the standard salary of long-term workers, the standard salary of dispatched workers, and agricultural electrician standard salary (grid plate only).

Total remuneration = the total salary of grid plate + the total salary of professional support units + the total salary of market-oriented industrial sector + the total salary of the financial sector

The total salary of grid plate=(the standard salary of long-term workers + the standard salary of dispatched workers + agricultural electrician standard salary)+ the incentive compensation of grid plate

The total salary of professional support units=(the standard salary of long-term workers + the standard salary of dispatched workers)+ the incentive compensation of professional support units

The total salary of market-oriented industrial sector=(the standard salary of long-term workers + the standard salary of dispatched workers)+ the incentive compensation of market-oriented industrial sector

The total salary of the financial sector=(the standard salary of long-term workers + the standard salary of dispatched workers)+ the incentive compensation of financial sector

Specific building process includes the following:

(1) To construct a standard pay fitting model. Selecting historical level of wage income based on different forms of the plate and employment key factors, the standard salary fitting model of each employment respective area lines is obtained by multivariate regression.

(2) To calculate the standard salary for each form of employment on each segment. Through the independent variable predicts, fitted value of the standard salary S_{0ij} for each plate in the each form of employment is gotten. which, $i=1,2,3,4$, representing the grid plate, the professional support plate, market-oriented industrial sector and the financial sector units. $j=1,2,3$, representing long-term workers, labor dispatch and agricultural electrician.

(3) To calculation incentive compensation. The total standard compensation of the company is gotten by each employment form of the standard salary on each plate plus. The all standard total compensation can be subtracted from the total amount of remuneration assigned to give incentive pay. Salary differences assigned specific role in the adjustment of $(S-S_0)$, known as incentive compensation for each of the plates.

(4) To complete the company incentive compensation allocation. Differentiate between the plate-adjusted per capita incentive compensation coefficient setting can be adjusted later. Adjustment coefficient of the plate includes a comprehensive and objective performance evaluation coefficient and the plate subjective strategic development incentive coefficient.

(5)To complete the plate salary distribution. The various portions of the plate in the form of labor remuneration are multiplied by forecast number of staff. the standard salary for each of the plate employment lines and incentive compensation are gotten. After the two parts have been added to general remuneration allocation amount for each of the plate employment forms.

2.1.2 Consider homogeneity within the sectors, the completion of the business segment - provincial unit salary distribution

For the grid plate, taking regression analysis, the per capita standard remuneration in the provincial companies is fitted by regression analysis. The provincial company's standard salary is obtained by multiplying per capita standard remuneration with the number of persons. Then, based on performance and human resources output efficiency, the total amount of incentive compensation on the grid plate was cut. Finally, the provinces per capita salary are adjusted by the objective region differences and per capita income differential[4].

For the professional support units, each unit based on the industries in which property and business service support features, is carded the immediate professional support plate. Immediate support plate professional salary management has many similarities with the grid plate. Therefore, in order to reflect the original management approach to succession and continuity, immediate professional support sector allocation model building process is similar to the grid plate. Immediate support professional sector allocation model in incentive compensation adjustment coefficient setting focuses more on support services for the company's strategic direction[5].

For market-oriented industrial units sector, considering its financial management control and strategic management control combining control mode, the Group Headquarters uses the "basic salary + incentive compensation" approach link the financial indicators and strategic indicators with basic pay and incentive pay. By enterprises self-reported economic target linked with base salary and incentive compensation, greatly stimulating endogenous power of the industry market segment unit enterprise. Its features of the market are more prominent. At the same time, the establishment of incentive factor in the market segment of industrial units decides mechanism, results from incentive compensation allocation linking the target completion. So, achieve economic efficiency has improved, on the other hand the realization of the company's strategic objectives has supported.

For the financial sector, considering its financial management control and strategic management control combining control mode, the Group Headquarters uses the "basic salary + incentive compensation" approach link the financial indicators and strategic indicators with basic pay and incentive pay. By enterprises self-reported economic target linked with base salary and incentive compensation, greatly stimulating endogenous power of the financial sector enterprise. Its features of the market are more prominent. At the same time, the establishment of incentive factor in the market segment of the financial sector decides mechanism, results from incentive compensation allocation linking the target completion. So, achieve economic efficiency has improved, on the other hand the realization of the company's strategic objectives has supported. Although the financial sector and the industry sector model idea is similar, in particular linked to indicators and incentive coefficient determining there are still significant differences in terms of mechanisms.

2.1.3 Consider the relationship distribution between different levels, the completion of provincial - salary distribution units cities

Power grid sector allocation model units around the city's mainly linked with job performance wage system. Basic security portion is related to salary point system. Incentive awards and other matters are related to the per capita average increase performance incentive compensation, the falling levels of responsible persons annual performance evaluation score, and investment and output efficiency coefficient in human resources. Immediate support professional sector, market-oriented industrial sector and the financial sector units in the municipal units salary distribution model substantially build on a reference model distribution.

2.1.4 Overall balance of the income gap among the plates, the completion of coordination and optimization of technology development

At the same income level evaluation standard premise, the plate (enterprise) per capita income equivalent theory value is calculated. This theory value is income levels in this environment. When the magnitude of the ratio of the different sectors (business), that its income gap ratio. Based on Moderate Income Gap of coordination and optimization techniques in the study, five-dimensional indicator system is constructed. Using factor analysis, theoretical income level equivalent value calculation model has been built.

2.2 The overall model structure of total compensation distribution system

According to the group company's business scope and form of employment structure sector, for each business segment, salary distribution models of the way the labor performed were constructed. Among them, the total amount of remuneration in the Group covers four sectors category, such as grid plates, the business professional support sector, market-oriented industrial sector and financial institutions sector. Based on the actual employment situation of each sector, the total amount of remuneration Distribution Model covered for various ways of long-term employment of workers, labor dispatch electricians and other sections were constructed. Research mainly includes five models and a coordinated allocation optimization techniques, as follows in Table 2 below.

Table 1. Model composition table.

No.	Achievement	Achievement category
1	The total compensation distribution model in Group	Distribution Model
2	The total compensation distribution model in grid plate	
3	The total compensation distribution model in business professional support sector	
4	The total compensation distribution model in market-oriented industrial sector	
5	The total compensation distribution model in financial institutions sector	
6	Coordination and Optimization Technology	Optimization techniques

On the relationship between the various models and the coordination and optimization technology, the distribution model from the company headquarters to the plate is the basis for salary distribution model. That determines the total amount allocated for the salary distribution model for each segment of each line of employment, but also for the salary distribution among sectors balance optimization. The each total labor remuneration allocation model for plates and lines of employment lays the foundation for the next step from the provincial units to municipal units in salary distribution model. Thus, the model building way from the plate, provincial to municipal level pay the total allocation layer by layer, on the one hand reflects the company's existing distribution model and method of succession and continuity, on the other hand deepens the model structure highlights the refinement and coordination of the remuneration distribution management optimization. At the same time, the coordination and optimization techniques solves fact lead to real limitations exist allocation model, and solves real problems for an orderly transition gap with the gap theory, achieves the income gap between the plate and the plate within the overall coordination.

3 Advantage analysis for the total amount of salary distribution system

3.1 The large and complex compensation distribution management process of Group is effectively decomposed, so that complex issues are broken down and the structure of the pattern are formed.

The original salary distribution in Group is only from headquarters to a provincial-level unit. The whole assignment process is single and general. And the way to build the total salary distribution model step by step increases the sector to provincial units, provincial to municipal units salary distribution model. This approach will be a complex process down to every level, step by step to show fully reflects the internal logic structure of salary distribution model, refine the salary distribution process and improve the transparency of salary distribution management.

3.2 The new system has expanded the range of management salary distribution management company, detailed the management level, fully implemented human resources "three-wide" management requirements.

Company existing salary distribution management system is to market-oriented as the classification criteria, all provincial units are divided into two categories as A, B. The amount of the total remuneration model above classification step to build is on the basis of the business characteristics, through the plate, provincial, municipal three levels, covering a variety of forms of employment. The new system has further expanded the longitudinal depth of salary distribution management, expanded the distribution management lateral coverage, detailed management level, deepen the human resource intensive management.

3.3 The new system has strengthen the company's control over distribution targeted, enhanced the salary distribution differentiated control the level of fine

To construct the overall model of Group total compensation by hierarchical classification reflects the Group assigned to different units differentiation control characteristics and control ideas in compensation distribution. This facilitates stratification of total Group salary distribution process are benefit to classification of control, prompt the company can be more direct and effective on all levels, in each category, salary allocation process various forms of employment improvement and optimization.

4 Conclusion

The contents of research are large, wide, deep level, efforts to solve the problem employee equity income of general interest. For this, first of all, top-level design is carried out, adding the plate level. The deconstruction of complex issues is carried out. At the same time, to carry out labor assign work hierarchically, in sub-lines, were constructed for each allocation model. In order to enhance the co-ordination of the distribution of work, especially the coordination and optimization techniques are developed. This innovative technology uses factor analysis method theoretical guidance value of large conglomerates income gap, as it is widely discussed and controversial issues of the income gap between the new idea. And the analysis is based on large amounts of data collected on the quantitative analysis of the revenue gap. However, due to the total amount of remuneration allocation model data collection is difficult, it brings greater confusion for the data analysis. So research group will continue to study further, to explore a more convenient, more efficient, more scientific allocation model.

References

1. Wang Jiaying. Research on total wage budget linkage mod el in WS Group subsidiary company[D]. Chang Chun : JiLin University, (2014).
2. Li Xin, Wu Fu-cheng, Yang Guang-ming. The Analysis on the Influence Factors of the Distribution of Gross Payroll in Power Supply Companies[J]. Hubei Electric Power, **9**(2013).
3. Grossman H. A General Model of Insurrenction[J]. American Economic Review, **81**(1991).
4. Agion , Pand Howitt P. *Endogenous Growth Theory*[M]. Cambridge: MIT Press, (1998).
5. Sala-ti-Martin.*The Distributing Rised of Global Income Inequality*[R]. NBER Working Paper WB. (2002).

Parametric Design and Rapid Prototyping of Installation Box for Vehicle Terminal PCB

Xingxing Wang[1,a], Hongjun Ni [1,b], Yu Zhu[1], Mingyu Huang[1] and Jiawen Deng[1]

[1] School of Mechanical Engineering, Nantong University, 226019, Nantong, China

Abstract. Installation box for vehicle terminal PCB (Printed Circuit Board) was took as research object, which is encountered in the process of project developing. Vehicle terminal PCB in actual development process was set as an example, point cloud data were acquired by three coordinate measuring method; Imageware software was used to reconstruct the vehicle terminal PCB model, basic size parameters of vehicle terminal PCB can be got and then design parameters of installation box for vehicle terminal PCB can be determined. Design of the installation box for vehicle terminal PCB was completed based on Solidworks software, then 3D modeling and 2D drawing of installation box for vehicle terminal PCB was gained. Up Plus 2 rapid prototype machine was used to manufacture installation box for vehicle terminal PCB rapidly based on 3D printing technology, then prototype of installation box for vehicle terminal PCB was obtained. It is of certain engineering significant for single (small amount) manufacturing of installation box for general PCB.

Key words: Parametric Design, 3D Modeling, Rapid Prototyping, 3D Printing, Installation Box

1 Introduction

Parametric design is an important thought of Revit Building software, and it is divided into two parts: parametric graph and parameterized modifying engine. Graphics primitive in Revit Building are all appeared in the form of component. The difference between these components is reflected by the adjustment of the parameters, and the parameters preserved all information of graphics primitive which is digitalization construction component. Compared with the parametric design software which is wildly used, there are four type of application software can be the mainstream and they are Pro/Engineer, UGNX, CATIA and Solidworks. These four softwares have their own characteristics and have occupied a certain market share in different areas respectively [1]. RPM (Rapid Prototyping Manufacturing) manufacture the digital description model which includes the combined information of prototype (or parts, components) such as the geometry, material, structure and so on[2], and input these information into the mechanical and electrical control system which is controlled by computer to complete the "add" of material, and then finish the manufacturing of products[3].

According to the instance of vehicle terminal PCB which is encountered in the process of project developing, completed the parametric design and rapid prototyping of vehicle terminal PCB's installation box, with the aid of modern measuring technology and rapid prototyping technology.

2 Obtain the point cloud date

Digital measurement is also called data collection and it is the basis of reverse engineering. There are two main methods to obtain the point cloud, such as contact measuring and non-contact measuring. Contact measuring device has high measuring precision and strong adaptability, but its efficiency is low and can only measure the hard surface. Non-contact measuring device has high efficiency and high precision. Through the digital measurement, get a lot of point cloud data. The quality of the point cloud data directly affects the progress of the subject and the model reconstruction. Improving the accuracy of digital measurement is the precondition of the follow-up work smoothly.

Coordinate measuring machine (CMM) is a kind of contact measuring and is widely used to measure the three-dimensional sizes, forms, and positions of workpieces. In this part CMM is used and its equipment type is MQ8106. This machine's frame moving bridge structure is all closed. Its workbench is made of marble and its guideway is air-suspending. The measuring head block used in the measuring system is RENISHAW PH10T which can automatically twin twist to divide degree. The machine is equipped with RENISHAW needle group and HEADER F-Scan Flexible laser scanning probe, and has a model measuring function based on three-dimensional model. Not only do contacting measure, but also can do the non-contact laser scanning.

Before measuring the vehicle terminal PCB, needed

[a] wangxx@ntu.edu.cn, [b] ni.hj@ntu.edu.cn (corresponding author)

to open the air pump first. Then, connected the data line of coordinate measuring machine with the computer and start the computer. Put the manual controller in a particular place of coordinate measuring machine and be convenient to do the measuring work.

Measurement process done as follows: first of all, put the parts on the work surface and let it be fixed, and realized the shape measurement by manual controller. Second, Debugged coordinate measuring machine to make it back to zero, and then start measuring. After the completion of measurement, save the file with the format of "STL"[4], this can be used directly when dealing with the point cloud. Figure 1 shows the picture of vehicle terminal PCB measurement.

The date can be measured by coordinate measuring machine should be saved to be dealing with, and the format of the file data is still "STL". Because the vehicle terminal PCB parts are too small and complex, there were still some size cannot be measured, such as some holes and the relative position of two parts. As a result, it was necessary to use vernier caliper to complete the measurement.

Figure 1. Measure of vehicle terminal PCB

3 The point cloud processing and key parameter extraction of vehicle terminal PCB

Imageware is a famous software in the field of the reverse engineering. Imageware software has strong processing capacity of point cloud, editing skills of surface and ability of surface construction. It is widely used in the area of automotive, aviation, aerospace, consumer appliances, mold design and manufacturing, computer parts, and so on [5].

Step 1: According to the three views in Imageware software, dealt with data. Put the point cloud data in the right place and corrected the sharp date. The reason for doing this was to make the point cloud data show more regulation in the software view, at the same time, it was convenient to correct the position and modify the data [6].

Step 2: Dealt with the point cloud data of vehicle terminal PCB (as shown in figure 2). Generated plane with the line existing, and made a mark on corresponding line and plane (as shown in figure 3). It was better to generate the corresponding entity in SolidWorks software [7].

Figure 2. Point cloud data of vehicle terminal PCB

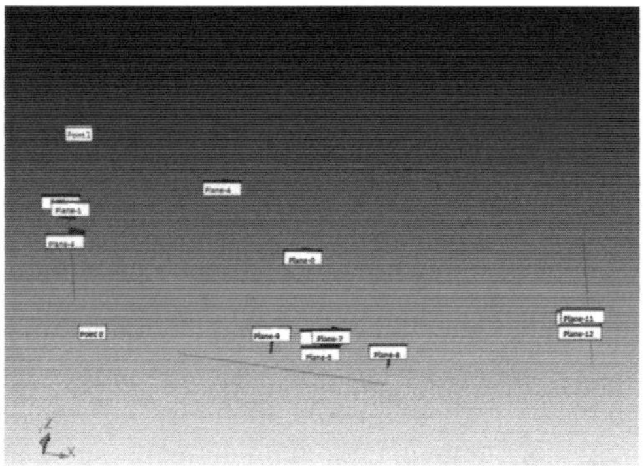

Figure 3. Relationship of line and plane of vehicle terminal PCB

It was necessary to get the information of key parameters of PCB, such as the size of the plane, the length of the line and the relationship of line and plane, in order to generate the corresponding 3D model better in the SolidWorks software. We sorted out the raw data and got the most direct parameters of vehicle terminal PCB. It was convenient to design the PCB install box based on the data. The date of line and plane of Vehicle terminal PCB is shown as table 1.

Table 1. Date of line and plane of vehicle terminal PCB

project	**parameter**
Length	125mm
Distance from Plane 1 to plane 0	15.5331mm
Distance from right Plane 2 to line 2	46.6746mm
Distance from Plane 5 to line 3	29.7657mm
Distance from Plane 7 to plane 0	10.7061mm
Width	89mm
Distance from left Plane 3 to line 0	46.2524mm
Distance from Plane 4 to plane 0	10.7450mm
Distance from Plane 6 to line 1	28.9435mm
Distance from Plane 8 to plane 0	9.6866mm

4 3D Modeling

4.1 Reconstruction of three-dimensional entity of vehicle terminal PCB.

The file including the information of line and plane was saved as "Iges" format in Imageware Software. Opened the file in Solidworks software and dealt with the line and plane to reconstruct the model we want. Opened the file in SolidWorks and made a series of operations, such as stretching the matrix convex, extruded cut and so on, combined with the line and plane generated in Imageware, completed reconstructing the 3D model of vehicle terminal PCB

4.2 Parametric design of vehicle terminal PCB's installation box.

We had generated the 3D model of vehicle terminal PCB. Then, need do the parametric design of the installation box of vehicle terminal PCB according to the data obtained [8]. Because the card slot of vehicle terminal PCB was opening up, it was necessary to open a skylight, above the box so that connecting to the outside world. There was an antenna connection port, so it needed an antenna interface slot in the side of box. The slot to insert a card (SIM) was connected with the outside world, so also need to open a card slot. Taken other characteristics of the board into consideration, and the appearance of installation box's upper cover was generally come out. Design a skylight lid which was cooperates with the Skylight and they were connected with screw. The bottom of the box was also connected with the box cover with screw thread, and the bottom of the box was set inside the box cover.

Completed the design of every component of the installation box, and assembled all the parts and then got the 3D model with the 3D design software. To finish assembling the bottom and the box cover, let the hole in the bottom of the lid be concentric with the hole in the box cover, and matching the plane in convex with the plane in threaded column hole. This step assembled the installation box, and assembled the model of vehicle terminal PCB with the box as well, realized the virtual assembly.

Figure 4 shows the explosive view of final assembly of vehicle terminal PCB and installation box.

5 Rapid Prototyping Manufacturing

Rapid prototyping manufacturing, is also 3D printing (Three-Dimentional printing). It's working principle is that the hot melt materials which is processed into the sharp of filamentous (such as ABS), will be sent to hot nozzle through the wire feeding mechanism[9]. Then, filamentous material is heated and melted within the nozzle. At the same time, the nozzle moves along the outline of layer parts and extrudes the molten material, making it deposit and solidify in the specified position. The molten material which is solidified will be bonding with the previous layer which materials have already formed, and stacking layer upon layer repeatedly and eventually formed the product. This subject completed the manufacture of vehicle terminal PCB installation box with rapid prototyping machine (3D printer), which type is UP Plus 2[10].

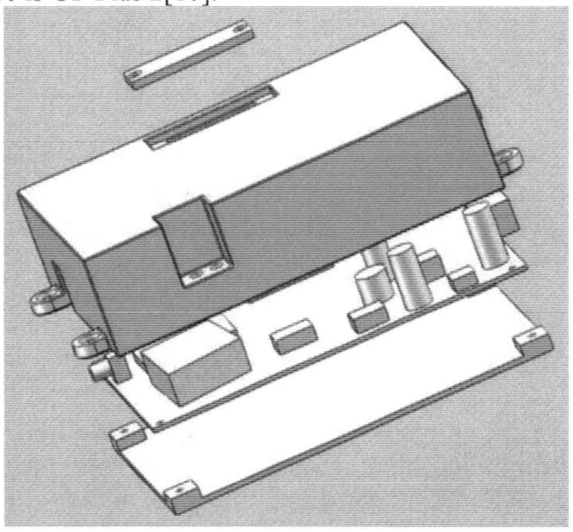

Figure 4. Explosive view of final assembly

Imported the STL file of vehicle terminal PCB's box cover into the computer. Because the size of installation box cover is larger than maximum working size of rapid prototyping machine, we divided the cover into two parts and then print it respectively. At the same time, the parameters were adjusted, such as the printing accuracy and the distance between nozzle and workbench. The printer can automatically print the product including support materials[11]. It took several hours during the printing process because the printer manufacture parts layer by layer.

There were four parts of the installation box: the left part of the box cover, the right part of the box cover, the bottom board, and the skylight lid, and they were all printed by the printer. Took the left part of the box cover as an instance. It took about 3 hours to complete the printing. Then got it out from the work plane of printer and done the post-processing work, such as removing the support material and polishing it slightly.

After printed all the components of the installation box, assembled them and got the box we want. Then, installed the Vehicle terminal PCB into the installation box. The result was fine, as shown in figure 5.

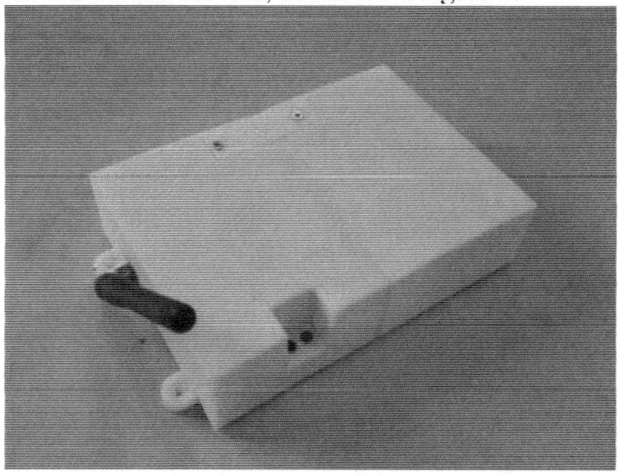

Figure 5. Vehicle terminal PCB installed

6 Summary

1) Obtained the point cloud data of vehicle terminal PCB, which was encountered in the process of project developing, by using coordinates measuring machine. Dealt with point cloud data of vehicle terminal PCB, extracted the basic size parameters, and determined the parameter requirements of designing vehicle terminal PCB installation box according to the Imageware software.

2) Completed the parametric design of vehicle terminal PCB based on Solidworks software, and finished the build of 3D model and 2D drawing of vehicle terminal PCB

3) With 3D printing technology, finished the manufacturing of vehicle terminal PCB installation box by 3D printer which type was UP Plus 2, and got the product prototype of the installation box.

4) During the process of project development, this method will have certain value in engineering application for the rapid and economic manufacturing of one piece (small batch) PCB installation box.

Acknowledgment

This work was financially supported by National Science and Technology Support Program of China (2011BAG02B10), A Priority Academic Program Development of Jiangsu Higher Education Institutions (PAPD), Frontier and Key Technical Innovation Program of Nantong (MS22015028), and Nantong City Applied Research Projects (BK2014053, GY12015028).

References

1. X.L. Li, K.C. Yu, X. Zhao. Manufacture Information Engineering of China. **39**(13): 67-70 (2010)
2. H.J. Ni, L. Chen, X.B. Feng, et al. Mechanical design and manufacture (Chinese). **01**: 137-139 (2014)
3. C. Weller, R. Kleer, F.T. Piller. Int. J. Prod. Econ. **164**: 43-56 (2015)
4. H. Zhu, Z.F. Yang, W. Zhuang. Machine Tool and Hydraulics (Chinese). **37**(6):186-189 (2009)
5. Y. Dan, B.F. Xie. *The base of Imageware and reverse modeling technology* (Tsinghua University Press, Beijing, 2006)
6. M. Vaezi, C. K. Chua. Int. J. Adv. Manuf. Tech. **52**: 275-284 (2011)
7. P. Feng, X.M. Meng, J.F. Chen, et al. Constr. Build. Mater. **93**: 486-497 (2015)
8. H.J. Ni, Y. Pei, S.S. Lv et al. Advanced Materials Research. **889-890**: 14-21 (2014)
9. Y.Q. Li, T.C Shi. Machine Tool and Hydraulics, (Chinese), **38**(23): 65-67 (2010)
10. B. Ezair, F. Massarwi, G. Elber. Comput. Graph. **51**: 117-124 (2015)
11. M. Asadi-Eydivand, M. Solati-Hashjin, A. Farzad, et al. Robot Comput. Integr. Manuf. **37**: 57-67 (2016)

An interval number-based multiple attribute bid-decision making model for substation equipments

Lili Zhu[a],

State Grid Material Co., Ltd, 100120 Beijing, China

Abstract. By analyzing the characteristics of public bidding for substation equipments and combining with the research methods of multiple attribute decision-making problems, a multiple attribute bid-decision making model is presented. Firstly, the weight of interval numbers is specified by using the interval numbers theory and entropy theory. Secondly, the deviation degree of decision-making scheme is proposed. Then the schemes are sorted. A typical case is analyzed based on the above-mentioned.

1 Introduction

Multiple attribute decision making is to select optimal alternatives or to order schemes in consideration of multiple attributes, widely existing in engineering[1-2], economic, social[3] and daily life. The complexity and the fuzziness of the objective things makes the multiple attribute decision problems encountered mostly uncertain and fuzzy. The uncertain multi-attribute decision making as a hot research topic has already caused extensive attention of scholars.

Equipment bidding is an effective means to improve the quality of equipment procurement and reduce the purchasing cost. With the rapid development of the power grid, especially the large-scale construction of UHV power grid, the investment to substation equipments has increased correspondingly. Meanwhile, the substation equipment investment has some characteristics, such as large amount and high risk. The effective bidding methods to achieve efficient investment have become a major problem which needs to consider.

The uncertainty is mostly discussed with triangular fuzzy numbers, trapezoidal fuzzy numbers and interval numbers. This paper aims to establish a substation equipment bidding model with multiple attribute decision making theory based on interval numbers. The model ultimately helps to ensure appropriate substation equipment purchase strategy and to ensure the power grid projects carried out smoothly.

2 Substation equipment bidding characteristics

According to the difference of the subject matter, the bidding can be divided into engineering bidding, equipment bidding, material bidding and supervision bidding, etc. Equipment bidding can be divided into two types as general equipments and large-scale industrial equipments according to the generality and the level of standardization. The substation primary equipment because of its special purpose and huge investment obviously does not belong to general equipment, then it as a large-scale equipment bidding has the following characteristics.

(1) The contract features: the subject matter is of large amount, and the rights and obligations of the contract is from the beginning of the purchasing, until the equipment production and operation to the expiration of the warranty period.

(2) The equipment characteristics: The risk of the equipment out of order is huge. Thus, the reliability requirement of the substation equipments may be higher and the agreed quality is more complex.

(3) The requirements of the bidder's ability: the bidders are of higher qualification and ability requirements.

According to the substation equipment bidding process, the bidders offer the technology response and the business documents simultaneously.

3 Traditional equipment bidding method

Next, we study the method of selecting the final winning bidder after the bidders submit the response documents.
In practice, two common methods are the lowest price method and the comprehensive evaluation method. The lowest price method applies to equipments the quality of which can be compared easily, and the price can be considered as the only factor to select the winning bidder, such as simple equipment, semi-finished products, general standard equipment and raw materials. And the bidder with the lowest price is the winning. Comprehensive evaluation method is to carry out the overall evaluation and comparison of the response documents considering the price standard and non price standard. With the comprehensive evaluation method, the non-price factors are folded into price standard and are given corresponding weighted in order to determine the

[a] Corresponding author: lilizhu1223@yeah.net

lowest bidding price. The winning bidder evaluated as the lowest bid or the best bid is recognized as the best comprehensive evaluation of the bid.

For the power grid project, in the case of the substation equipment bidding, the main consideration is not the price but the special technology, quality assurance and compliance capabilities, etc., so the comprehensive evaluation method is generally used. However, the following reasons will lead to unfair using this method. Firstly, it is difficult to achieve the accuracy of the evaluation items and their quantitative standards. In practice, many evaluation factors are difficult to quantify, and the quantitative given value is not able to reflect the actual situation of bidders. In addition, the overall level of knowledge and subjective tendency will seriously restrict the accuracy and effectiveness of evaluation from judges. So it is necessary to use some mathematical tools for the bid-evaluation, in order to weaken the subjective will and to strengthen the accuracy and scientific of the evaluation. Existing several mathematical evaluation methods are based on fuzzy theory, neural network and real options theory and interval number theory.

4 Bid-evaluation model of interval number-based multi-attribute decision making

According to the multiple attribute decision making theory and traditional entropy theory, the weight of interval numbers is specified. Then, the deviation degree of decision-making scheme is proposed. Finally, the schemes are sorted[4-6].

Suppose an uncertain multiple attribute decision making set as $A=\{A_1, A_2,..., A_n\}$, an attribute set as $G=\{G_1, G_2,..., G_m\}$. Measure the i-th scheme A_i according to the j-th attribute G_j as the attribute value $\tilde{a}_{ij}=[a_{ij}^L, a_{ij}^U]$. Supposeing I_t (t=1, 2) is the subscripts of benefit indexes and cost indexes respectively, and $M=\{1, 2, ..., m\}$, $N=\{1, 2, ..., n\}$.

4.1 The entropy-based attribute weights

The steps to caculate attribute weights based on entropy theory are as follows.

(1) Transform the decision matrix $A=(a_{ij})_{n\times m}$ into the deviation degree matrix $Q=(q_{ij})_{n\times m}$, where

$$q_{ij}=D(a_{ij}, a_j^*) \tag{1}$$

and $a_j^*=[a_j^-, a_j^+]$, where

$$a_j^-=\max_i a_{ij}^-,\ a_j^+=\max_i a_{ij}^+,\ \text{for } j\in I_1, \tag{2}$$

$$a_j^-=\min_i a_{ij}^-,\ a_j^+=\min_i a_{ij}^+,\ \text{for } j\in I_2. \tag{3}$$

(2) Normalize the deviation degree matrix $Q=(q_{ij})_{n\times m}$ as matrix $P=(p_{ij})_{n\times m}$, where

$$p_{ij}=\frac{q_{ij}}{\sum_{i=1}^{n} q_{ij}}. \tag{4}$$

Obviously,

$$\sum_{i=1}^{n} p_{ij}=1, j\in M. \tag{5}$$

(3) Caculate the entropy of attribute G_j as

$$S_j=-k\sum_{i=1}^{n} p_{ij}\ln p_{ij},\ j\in M, \tag{6}$$

where k is a constant and $k>0$, ln denotes the natural logarithm, and

$$S_j\geq 0,\ j\in M. \tag{7}$$

If $p_{ij}=1/n$, $i\in N$, S_j gets the maximum, i.e.

$$S_j=k\ln n, \text{ and } 0\leq S_j\leq 1. \tag{8}$$

(4) Caculate the deviation degree coefficient u_j of each decision-making scheme for attribute G_j, and

$$u_j=1-S_j,\ j\in M. \tag{9}$$

(5) Calculate the weight ω_j of each attribute, and

$$\omega_j=\frac{u_j}{\sum_{j=1}^{m} u_j},\ j\in M. \tag{10}$$

4.2 The deviation degree-based sorting method

The steps to sort the decision-making schemes based on deviation degree of the interval numbers are putforwarded as follows.

(1) Constructing matrix $C=(c_{ij})_{n\times m}$ and normalize it as matrix $R=(r_{ij})_{n\times m}$ according to the evaluation target type, that is

$$c_{ij}=\frac{a_{ij}}{\max_i a_{ij}}=[\frac{a_{ij}^-}{\max_i a_{ij}^+}, \frac{a_{ij}^+}{\max_i a_{ij}^-}]=[c_{ij}^-, c_{ij}^+]\quad i\in N, j\in I_1$$

$$c_{ij}=\frac{\min_i a_{ij}}{a_{ij}}=[\frac{\min_i a_{ij}^-}{a_{ij}^+}, \frac{\min_i a_{ij}^+}{a_{ij}^-}]=[c_{ij}^-, c_{ij}^+]\quad i\in N, j\in I_2$$

and

$$r_{ij}=\frac{c_{ij}}{\max_i c_{ij}^+}=[r_{ij}^-, r_{ij}^+]. \tag{11}$$

(2) Considering the weighted vector ω, construct bilaterally augmented matrix B, which is weighted and normalized, and

$$B=\begin{array}{c} \\ A_1 \\ \vdots \\ A_n \\ A^* \\ A^- \end{array}\begin{array}{c} G_1 \quad \cdots \quad G_m \\ \begin{bmatrix} \omega_1 r_{11} & \cdots & \omega_m r_{1m} \\ \vdots & & \vdots \\ \omega_1 r_{n1} & \cdots & \omega_m r_{nm} \\ \omega_1 r_1^* & \cdots & \omega_m r_m^* \\ \omega_1 r_1^- & \cdots & \omega_m r_m^- \end{bmatrix} \end{array}, \tag{12}$$

$r_j^* =[(r_j^*)^-, (r_j^*)^+]$ is the ideal attribute value of attribute G_j, and

$$(r_j^*)^- = \max_i r_{ij}^-,\ (r_j^*)^+ = 1. \tag{13}$$

The scheme constructed from the ideal attribute value is the ideal scheme, denoted as A^*.

$r_j^- =[(r_j^-)^-, (r_j^-)^+]$ is the negative ideal attribute value of attribute G_j, and

$$(r_j^-)^- = \min_i r_{ij}^-,\ (r_j^-)^+ = \min_i r_{ij}^+. \tag{14}$$

The scheme constructed from the negative ideal attribute value is the negative ideal scheme, denoted as A^-.

(3) Calculate the total deviation degree d_i^* of scheme A_i to the ideal scheme A^*, and the total deviation degree d_i^- of scheme A_i to the ideal scheme A^-, $i \in N$, and

$$d_i^* = \sum_{j=1}^m D(\omega_j r_{ij}, \omega_j r_j^*),\ i \in N, \tag{15}$$

$$d_i^- = \sum_{j=1}^m D(\omega_j r_{ij}, \omega_j r_j^-),\ i \in N. \tag{16}$$

(4) Calculate the relative closeness d_i of scheme A_i to the ideal scheme A^*, $i \in N$, and

$$d_i = \begin{cases} \dfrac{d_i^-}{d_i^*} & d_i^* \neq 0 \\ +\infty & d_i^* = 0 \end{cases}. \tag{17}$$

(5) Sort the corresponding scheme A_i according to the size of the relative closeness d_i.

And if $d_p > d_q$, $A_p \succ A_q$, *for* $p,q \in N$.

5 Application in bid-evaluation for substation equipment

Taking *A* project as an example, a new substation equipment is needed for the power grid project. A total of 5 bidders (A_1, A_2, A_3, A_4, A_5) participate in the bidding. The main factors considered in the bid-evaluation are economy, maintainability and reliability. Divided the factors into five attributes as fixed cost G_1, operating cost G_2, noise G_3, maintainability G_4 and reliability G_5. G_1, G_2 and G_3 are cost indexes, and G_4 and G_5 are benefit indexes. The original decision-making matrix based on interval number is listed in Table 1 .

Table 1. Interval number decision-making matrix *A*

	G_1	G_2	G_3	G_4	G_5
A_1	[3.7, 4]	[5.9, 6.8]	[30, 40]	[3, 5]	[90, 100]
A_2	[2.5, 3]	[4.7, 5.7]	[66, 75]	[3, 5]	[70, 80]
A_3	[3.5, 4.5]	[4.6, 5.4]	[35, 45]	[8, 10]	[75, 85]
A_4	[4.2, 4.8]	[4.8, 5.9]	[35, 48]	[8, 12]	[95, 98]
A_5	[3.6, 4.2]	[5.6, 6.9]	[68,75]	[4, 7]	[76, 86]

5.1 Calculation of entropy-based attribute weights

According to the steps of the model, the attribute weights are calculated firstly.

The deviation degree matrix *Q* is obtained in Table 2.

Table 2. The deviation degree matrix *Q*

	G_1	G_2	G_3	G_4	G_5
A_1	1.56	1.91	0	8.6	5
A_2	0	0.32	50.21	8.6	32.02
A_3	1.8	0	7.07	2	20.62
A_4	2.48	0.54	9.43	0	2
A_5	1.63	1.8	51.66	6.4	23.6

Normalize the deviation degree matrix, the matrix P is obtained in Table 3.

Table 3. The normalized deviation degree matrix *P*

	G_1	G_2	G_3	G_4	G_5
A_1	0.21	0.42	0	0.34	0.06
A_2	0	0.07	0.42	0.34	0.38
A_3	0.24	0	0.06	0.08	0.25
A_4	0.33	0.12	0.08	0	0.02
A_5	0.22	0.39	0.44	0.25	0.28

Caculate the entropy of attribute S_j as

S_1 =0.85, S_2 =0.73, S_3 =0.68,

S_4 =0.8, S_5 =0.83.

Caculate deviation degree coefficient u_j as

u_1 =0.15, u_2 =0.27, u_3 =0.32,

u_4 =0.2, u_5 =0.17.

Calculate the weight $\bar{\omega}_j$ as

ω_1 =0.14, ω_2 =0.24, ω_3 =0.29,

ω_4 =0.18, ω_5 =0.15.

5.2 Sort of the schemes deviation degree-based

The normalized matrix R can be obtained in Table 4 as follows.

Table 4. The normalized interval number matrix R

	G_1	G_2	G_3	G_4	G_5
A_1	[0.53, 0.68]	[0.58, 0.79]	[0.56, 1]	[0.17, 0.42]	[0.86, 1]
A_2	[0.69, 1]	[0.69, 0.98]	[0.3, 0.46]	[0.17, 0.42]	[0.67, 0.8]
A_3	[0.47, 0.72]	[0.73,1]	[0.5, 0.86]	[0.45, 0.83]	[0.71, 0.95]
A_4	[0.43, 0.59]	[0.67, 0.97]	[0.47, 0.86]	[0.45, 1]	[0.9, 0.98]
A_5	[0.5, 0.69]	[0.57, 0.82]	[0.3, 0.44]	[0.22, 0.59]	[0.72, 0.87]

The ideal scheme A^* and the negative ideal scheme A^- can be denoted as follows in Table 5.

Table 5. The ideal and negative ideal schemes

	G_1	G_2	G_3	G_4	G_5
A^*	[0.69, 1]	[0.73, 1]	[0.56, 1]	[0.45, 1]	[0.9, 1]
A^-	[0.43, 0.59]	[0.57, 0.79]	[0.3, 0.44]	[0.17, 0.42]	[0.67, 0.8]

Considering the weight ω_j, construct and normalize the bilaterally augmented matrix B, listed in Table 6 as follows.

Table 6. The normalized bilaterally augmented matrix B

	G_1	G_2	G_3	G_4	G_5
A_1	[0.07, 0.1]	[0.14, 0.19]	[0.16, 0.29]	[0.03, 0.08]	[0.13, 0.15]
A_2	[0.1, 0.14]	[0.17, 0.24]	[0.09, 0.13]	[0.03, 0.08]	[0.1, 0.12]
A_3	[0.07, 0.1]	[0.18,0.24]	[0.15, 0.25]	[0.08, 0.15]	[0.11, 0.14]
A_4	[0.06, 0.08]	[0.16, 0.23]	[0.14, 0.25]	[0.08, 0.18]	[0.14, 0.15]
A_5	[0.07, 0.1]	[0.14, 0.2]	[0.09, 0.13]	[0.04, 0.11]	[0.11, 0.13]
A^*	[0.1, 0.14]	[0.18, 0.24]	[0.16, 0.29]	[0.08, 0.18]	[0.14, 0.15]
A^-	[0.06, 0.08]	[0.14, 0.19]	[0.09, 0.13]	[0.03, 0.08]	[0.1, 0.12]

Calculate the total deviation degree d_i^*, and

d_1^*=0.23, d_2^*=0.34, d_3^*=0.15,

d_4^*=0.13, d_5^*=0.4.

Calculate the total deviation degree d_i^-, and

d_1^-=0.23, d_2^-=0.13, d_3^-=0.32,

d_4^-=0.33, d_5^-=0.07.

Calculate the relative closeness d_i, and

d_1=1, d_2=0.38, d_3=2.13,

d_4=2.53, d_5=0.18.

Then, the sorting results of the schemes is

$A_4 \succ A_3, \succ A_1 \succ, A_2 \succ A_5$.

That is the scheme of bidder A_4 is optimal, and bidder A_4 is the winner of the bidding.

6 Concluding remarks

The interval number theory and entropy theory are applied to the substation equipment bid-evaluation multi attribute decision making problems in this paper. As its objective uncertainty analysis to the attributes, its helpful to weaken the subjective will and strengthen the accuracy and scientific of the bid-evaluation, .it has been proved that these theories for the procurement of substation equipments, not only greatly improve the efficiency of equipment procurement, but also greatly improve the benefit of equipment procurement. The model proposed in this paper is innovative method and thought innvation as well. The optimization of the model for the application of power grid equipment becomes the further research orientation.

References

1. L. Crhz, J. Mesa. Eur. J. Oper. Res.**135**(2011).
2. Y. Song. Syst. Eng. Theory Pract., **22**(2002).
3. M. F. Shipley, A. Korvin, K. Omer. Eur. J. Oper. Res. **135**(2011).
4. Z. Xu, Z. Le. J. Manage. Sci China. **5**(2002).
5. Q Zhang. Syst. Eng. Theory Pract., **19**(1999).
6. Q. Zhang, Z. Fan, D. Pan. Eng. Theory Pract., **19**(2000).

Study of Emergency Power Based on Solar Battery Charging

Lei Wang, Mengfu Zhu [a], Ping Chen, Cheng Deng, Zhimeng Liu, Yanan Wang
Institute of Medical Equipment, Academy of Military Medical Sciences, Tianjin 300161, China

Abstract. To study an emergency power based on solar battery charging. Based on the electric-generation principle of solar panel, solar energy is changed into electrical energy. Through voltage conversion circuit and filter circuit, electrical energy is stored in the energy storage battery. The emergency power realizes the conversion from solar energy to electrical energy. The battery control unit has the function of PWM (Pulse-Width Modulation) charging, overcharging protection, over-discharging protection and over-current protection. It also realizes the fast and safe charging of energy storage battery. The emergency power could provide both 12V AC power for emergency equipment such as miniature PSA oxygen concentrator and 5V USB for electronic equipment (mobile phone, GPS device, rechargeable light, etc.).

Keyword. Solar; emergency power; MCU; energy storage battery

1 Introduction

When natural disaster and accident occur, scores of injured people need treatment and care. Taking in oxygen without delay can reduce the rate of injury and death[1, 2]. With the characteristics of light weight, small size and easy carrying, miniature PSA oxygen concentrator is an important and effective oxygen device for the sick and wounded[3]. However, if the power station equipment are damaged or the power supply goes bad, it is usually unable for the miniature PSA oxygen concentrator to work properly, which severely affects the efficiency of rescuing the sick and wounded. Besides, when hikers are unable to get a fixed power in the wild or in remote areas, electronic device, such as mobile phone, GPS device, and rechargeable light, can't be normal used for long time, which usually affects the quality of people's lives, even triggers survival crisis. As a new clean, green and renewable energy resource, solar energy, is gathering increasingly attention. Solar power system can charge and store electricity by absorbing the solar energy. It is chiefly discussed in this paper that based on the principle of solar battery charging, study an emergency power, providing 5V USB power for miniature PSA oxygen concentrator and 12V DC power for electronic device without external power support, to improve people's quality of life or survival[4-6].

2 Scheme Design

The emergency power based on solar battery charging (hereafter referred to as emergency power) is mainly composed of solar panel, battery control unit, and energy storage battery. The overall frame of the emergency power is shown in Figure 1. The solar panel converts the solar energy into electrical energy with suitable charging voltage after DC-DC conversion circuit and filter circuit, in the meantime, the battery control unit realizes the over-charging protection and over-current protection of power supply circuit by monitoring terminal voltage and load current of the energy storage battery, which makes it efficiently and safely to store electrical energy in the energy storage battery so as to provide power supply at any time[7].

[a] Corresponding author: zmf323@163.com

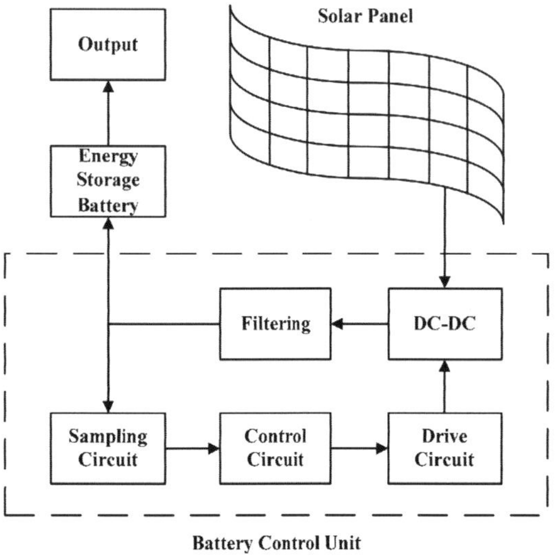

Figure 1. Overall frame of the emergency power

3 Hardware design

3 .1 Solar panel

According to the photovoltaic effect, in the emergency power solar panel is the most critical part that directly converts solar energy into electrical energy.

3.1.1 Working principle of solar panel

The monomer of solar panel is actually a PN junction which in light produces electromotive force, namely "photovoltaic effect". The working principle of converting solar energy into electrical energy of solar panel mainly includes three processes as following:

(1) While the solar panel absorbs photons, electron-hole pairs, namely photo-induced carriers, appear in PN junction. The electrons are negatively charged and the holes are positively charged;

(2) The photo-induced carriers with opposite polarity are separated in the electrostatic field produced by PN junction;

(3) If electrodes and lead wires are led out from both sides of PN junction and are connected with loads, in the effect of internal electric field, the electrons and holes of the photo-induced carriers are separately absorbed by positive electrode and negative electrode of the solar panel, which produces electrical current in the load circuit. Sequentially electric power is output and electric energy is gained.

The working principle of solar panel is shown in Figure 2.

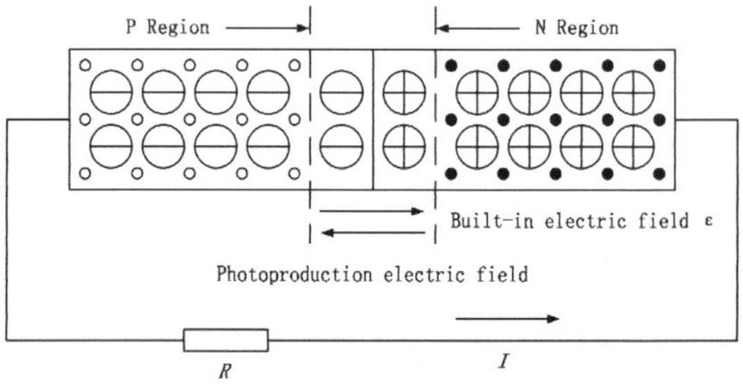

Figure 2. Working principle of solar panel

3.1.2 Equivalent circuit of solar panel

Figure 3 shows the equivalent circuit of solar panel. When sun light intensity is constant, the solar panel produce electric current called as I_{ph}, which doesn't change with the working status of solar panel, so the solar panel is equal to constant current source[8, 9]. The ends of the solar panel connect a diode and a bleeder resistance (R_{sh}) in parallel, and connect a load (R_L) with a series resistance (R_s). R_s mainly contains resistance of the panel itself, surface resistance and contact resistance of electrode connection materials. R_{sh} is a bleeder resistance that was caused by potential tiny cracks of the solar panel's edges and metal electrodes during the course of panel fabrication, which is the defects of silicon itself. R_s and R_{sh} are both self-resistances of solar panel and can be neglected in ideal model calculation. R_s is generally small, and the bigger resistance, the greater line loss, the lower efficiency of the solar panel. The value of R_{sh} is in thousands of the above[10, 11].

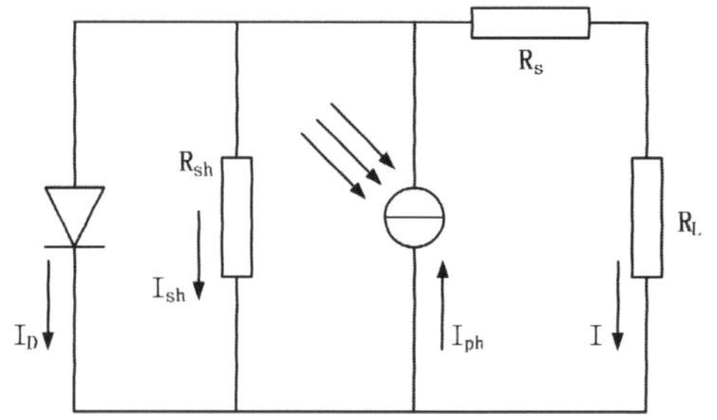

Figure 3. Equivalent circuit of solar panel

3.1.3 Selection of solar panel

Choosing energy storage battery with 12V/24Ah for guaranteeing continuous power supply of DC 12V was feasible. It was essential for the solar panel to determine the total capacity and amount of solar cell panels in series or in parallel according to the voltage, power and lighting, etc. In order to charge up energy storage battery, also considering factors such as high charging efficiency, not too large volume and easy carrying, it was necessary to choose several silicon solar cell panels with output of 18V/10W and size of 360×120mm, all of which were in parallel for increasing charging current and uprating power.

3.2 Battery control unit

Battery control unit is the main control module of the emergency power, and the controller technology plays an important role in solar system.

3.2.1 Selection of Microprogrammed Control Unit (MCU)

The system requires high demand of the speed of processor and built-in module. So ARM (Advanced RISC Machine) 7 or 8-bit MCU could be chosen. Although with fast processing speed, ARM owns a high price and many of its internal resources are useless and seriously wasted when it is applied to solar charging system. Given this, 8-bit MCU is a better choice. AVR series, C8051F series, and STC12 series are all common 8-bit MCUs. Their processing speeds all can be up to 1MIPS/MHz, but C8051F series and AVR series should not be used by considering utilization ratio of internal resources and clock frequency. Therefore it was better to adopt STC12C5412AD from STC12 series as main control chip, of which the instruction code is fully compatible with traditional 8051 series and the clock frequency can be up to 48MHz. There are abundant I/O interfaces and strong encryption function that could preferably protect the control system[12, 13].

3.2.2 Function design of battery control unit

Battery control unit includes voltage/current acquisition unit, battery temperature collector, LCD, alarm module, power module, and serial communication module. The function block diagram is shown as Figure 4.

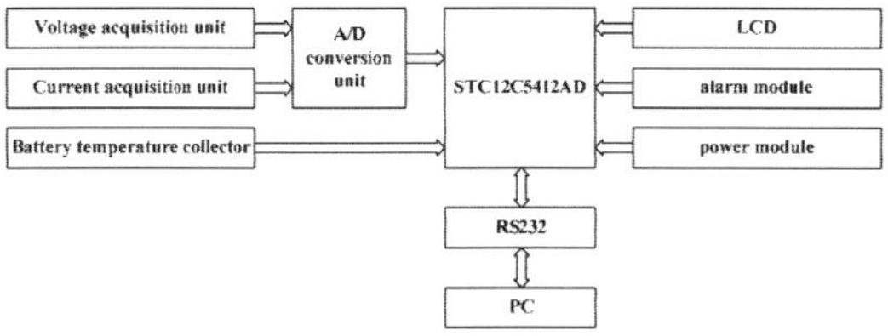

Figure 4. Function block diagram of battery control unit

Voltage acquisition unit identifies the strength of illumination and get the terminal voltage of energy storage battery. Current acquisition unit is used for over-current protection and load power detection. Battery temperature collector is used for energy storage battery's temperature compensation[14]. LCD is used to display voltage, current and load condition. Alarm module gives alarm signal if solar panel and energy storage battery are connected reversely, or energy storage battery is over-charging or over-discharging. Power module provides appropriate power supply for MCU and other components. Serial communication module plays a role of MCU in downloading programs, uploading system operation parameters and realizing remote control.

3.2.3 Signal acquisition circuit

Signal acquisition circuit is an important part of battery control unit. The protection of circuit is realized by means of that voltage acquisition unit monitors and measures terminal voltage, and current acquisition unit load current in real time, and then acquired signals are passed to main control chip through A/D conversion unit to calculate and process. When the acquisition of the battery voltage is too low, the power supply circuit is shut down and the power supply is stopped to prevent over-discharging. And when the voltage is too high, the charge circuit is shut down to prevent over-charging. If the current is too large, the power supply circuit is cut off to protect the critical load. Take a method of partial pressure of precise resistance (error is less than 1%) to regulate the signals measured by voltage acquisition circuit so as to make voltage reach the range the ADC port can accept[15]. Current acquisition circuit adopts a relatively safe and cost-effective scheme of low voltage resistance detection. Add an operational amplifier in circuit to magnify voltage into an ideal ADC inputting range, which helps to improve the precision of measurement and improve control performance. Figure 5 shows the signal acquisition circuit.

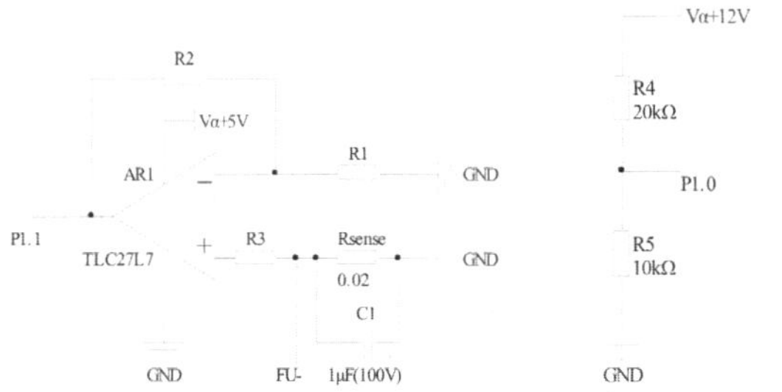

(a) Current acquisition circuit (b) Voltage acquisition circuit

Figure 5. Signal acquisition circuit

3.3 Energy storage battery

For the sake of suitable and longer power supply, it is better to choose batteries, single 2V, and combine them into 2 pieces of energy storage batteries with 12V/12Ah in parallel. In the design of charging, the chief charging circuit based on series type of PMW reduces voltage loss of the charge circuit by nearly half compared with using diodes, which could leave energy storage battery more sufficient response time, reduce gas evolution and improve charge efficiency. Compared to the parallel controller, it is more efficient in using solar energy and reducing calorific value. Pulse charging is a cyclic method that charging with pulse current and stopping charging for a while carries out alternately. Charging impulse makes a full charge. Interval eliminates concentration polarization and ohm polarization naturally so as to reduce the internal pressure of energy storage battery, which will make the next round of constant current charge smoother. The charging circuit is shown in Figure 6.

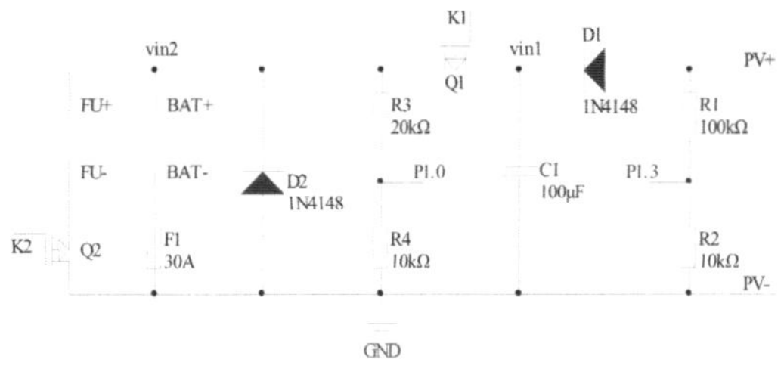

Figure 6. Charging circuit of the energy storage battery

PV+ and PV- connect the output of the solar panel to charge the battery. FU+ and FU- link electric equipment to power the load. BAT+ and BAT- connect both ends of the energy storage battery.

D1 is a backflow protection diode. D2 is a reverse polarity protection diode. If the battery polarities are reversed, battery would be short-circuit by the breakover of D2 and produce a great current to fuse the fuse protector (F1), which plays a role in reverse polarity protection. MOSFET Q1 is in series in the charging circuit. When the rechargeable voltage has achieved the average charge voltage, Q1 would get the state of PWM charging to prevent over-charging. MOSFET Q2 is the discharge switch of battery. When the terminal voltage is less than over-discharging voltage during battery discharging, Q2 cut off the circuit of battery and load to avoid empty battery and damaging battery, which is over-discharging protection. When solar panel reapplies power and end voltage ascends to float charging voltage, load circuit is reconnected by Q2.

Figure 7 shows the power supply circuit. The chip 78L05 can change the voltage from 12V to 5V for USB port. Capacitors are used to filter to make the power stable.

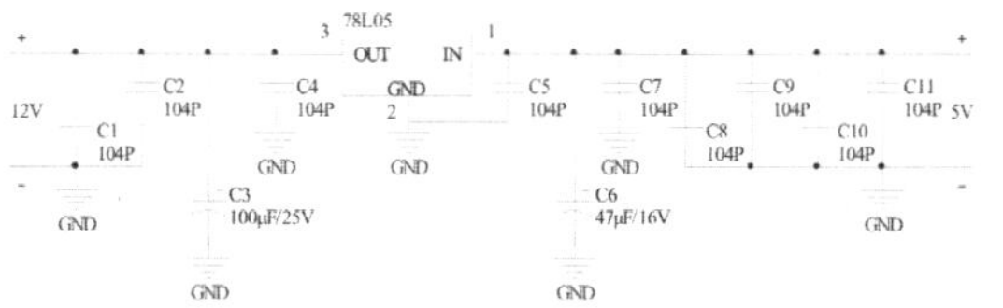

Figure 7. Schematic diagram of power supply circuit

4 Software design

The software design of the emergency power is as following: Real-time monitoring and protection of charging and power supply are carried out by that voltage/current acquisition unit monitors and measures the terminal voltage/load current of the battery in real time and main control chip processes the data gotten through A/D sampling port later. If the battery voltage collected is lower than 11.9V, the power supply circuit is shut down and the power supply is fast charged; if equal to 13.7V, PMW charging stage is entered; if higher than 14.4V, the charging circuit is shut down to avoid overcharge.

According to the requirements of the design, the software flow pattern is shown in Figure 8.

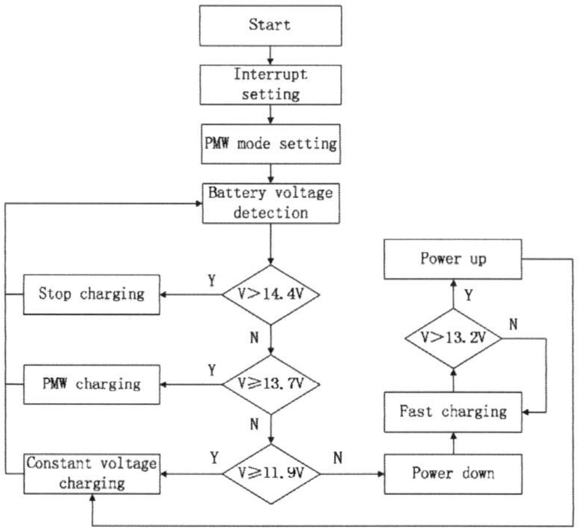

Figure 8. Software flow pattern

5 Conclusion

This paper adopts monocrystalline silicon solar panel in parallel as charging power supply, and takes advantage of the electric-generation principle of solar panel. Through conversion circuit and filter circuit, electrical energy converted from solar energy is stored in the energy storage battery. The emergency power realizes the safety of charging process through over-charging, over-discharging and over-current protection control. It can also provide a safe, effective and auxiliary 12V AC power for emergency equipment such as miniature PSA oxygen concentrator, and 5V USB power for electronic equipment. For example, mobile phone, GPS device, rechargeable light and so on. The emergency could add field service time of equipment effectively and improve people's quality of life or survival.

Acknowledgements

This work was supported by the army project (CWS12J109).

References

1. W.N. Liang, S.Y. Wang, J.Z. Tian. *Emergency Medical Response*. Beijing: People's Medical Publishing House (2012).
2. T. Li, G.D. Chen, G. Sun. *Emergency Rescue Handbook*. Beijing: Military Medical Science Press (2010).
3. P. Chen, M.F. Zhu, J. Yu, C. Deng, L. Zhu, H.B. Su, Y.H. Yuan. Development of Portable Oxygen Concentrator. Chinese Medical Equipment Journal, **34(10)**, 77-79 (2013).
4. Z.M. Liu, M.F. Zhu, P. Chen, C. Deng, Y.H. Yuan, H.B. Su. Design of intelligent battery management system for miniature PSA oxygen concentrator. Electronic Design Engineering, **22**, 63-65,68 (2014).
5. F. Gao, C.Q. Sun, Q.G. Liu. The Status and trends of solar energy utilization. World Sci-tech R & D, **4**, 35-39 (2001).
6. R.Q. Cui, C. Wang. Basic principle of solar cell testing system. Solar & Renewable Energy Sources, **3**, 36-40 (2008).
7. E. Trishan, L. Patrick. Chapman. Comparison of Photovoltaic Array Maximum Power Point Tracking Techniques. IEEE Transactions on Energy Conversion, **22(2)**, 439-449 (2007).
8. J. Chatzakis, K. Kalaitzakis, N. C. Voulgaris, S. N. Manias. Designing a new generalized battery management system. Industrial Electronics IEEE Transactions on, **50(5)**, 990-999 (2003).
9. P. Zhang, X.J. Wang, S.L. Wang. Design of miniwatt intelligent solar controller. Modern Electronics Technique, **18**, 177-180 (2007).
10. T. Kawamura, K. Harada, Y. Ishihara, T. Todaka, T. Oshiro, H. Nakamura, M. Imataki. Analysis of MPPT characteristics in photovoltaic power system. Solar Energy Materials & Solar Cells, **47(s 1–4)**, 155-165 (1997).
11. A. Zahedi, Smieee. Development of an electrical model for a PV/battery system for performance prediction. Renewable Energy, **15(s 1–4)**, 531-534 (1998).
12. T.X. Guo. *New concept C language tutorial of 51 singlechip: introduction, improvement, development and expansion*. Beijing: Publishing House of Electronics Industry (2009).
13. Y.G. Zhang. *New MCS - 51 single chip microcomputer application design*. Harbin: Harbin Institute of Technology Press (2003).
14. S. Rodrigues, N. Munichandraiah, A. K. Shukla. Review of state-of-charge indication of batteries by means of a.c. impedance measurements. Journal of Power Sources, **87(1-2)**, 12-20 (2000).

15. E. Phillip. Pascoe, H. Adnan. Anbuky. A VRLA battery simulation model. Energy Conversion & Management, **45(s 7–8)**, 1015-1041 (2004).

MATEC Web of Conferences **61**, 02026 (2016) DOI: 10.1051/matecconf/20166102026
APOP2016

Increasing luminous efficiency of led by joint dimming mode with adjustable peak current

Yongtao.Li, Jierong.Zhao, Youran.Li, Yuncui.Zhang, Nianyu.Zou and Jinpeng.Wang[a]

Research Institute of Photonics,Dalian Polytechnic University,116034 Dalian,China

Abstract. Luminous efficiency and color temperature of LED chip are affected by current and temperature. PWM dimming mode only works at two kinds of current levels-rated current and zero current. It has smaller effect on color temperature compared with linear dimming mode, but it will reduce the luminous efficiency of lamps. Linear dimming mode has higher luminous efficiency by comparison. The two dimming modes above cannot keep color temperature stable and Luminous efficiency high. Therefore, the paper designed a joint dimming mode by integrating advantages of the two modes. The dimming method not only adjusts peak current, but also changes the duty cycle of dimming signals. The injected average current is kept stable by reducing peak current and increasing the duty cycle. The decreased luminous efficiency by the raise of temperature was made up for. Then according to the mathematical model of total flux and current, the author calculates the mathematical mode of flux affected by temperature. It proves the fact that adjusting peak current slightly can compensates for the reduced luminous efficiency affected by temperature through theoretical calculation. In addition, the final peak current after calculation can contribute to the adjustment parameter of joint dimming modes.

1 Introduction

The technology of semiconductor develops constantly, and significant changes have taken place in illuminating lamps [1]. LED luminaire products are becoming more and more popular. As people have a higher demanding of illuminating environment, dimming technology emerges so as to provide a proper environment for people [2].Such as J. J. Zhang analysed radiant efficiency and thermal analysis of high-power LEDs at linear and PWM dimming modes and got that linear dimming mode's radiant efficiency is higher than PWM dimming mode when input power is same [3], X. S. Fu designed a novel isolating PWM dimming LED driver and the result showed that power exceed 0.9,the efficiency exceed 85% ,and the dimming range of 2%～100%[4].And X. B. Jiang designed the device of analog dimming mode on ZigBee technology, it realized the cooperation of the ZigBee internet gateway, monitoring and controlling remotely[5]. X. H. Qu proposed a led current sharing circuit with high luminous efficacy and individual dimming; it could adjust independently and improve the luminous efficiency [6]. Combining Standard for lighting Design of Urban Road and taking account of lighting comfort, it adjusts lighting parameters by changing output current, output power and duty cycle[7-9]. Theories of the technology include TRIAC dimming mode, partition dimming mode, linear dimming mode and PMW dimming mode [10-12]. However, there are some problems in the practical application of the above theories [13]. For example, there is the problem about that the output of linear dimming is not very dexterous, and the dimming-dynamic range is narrow [14-17].Traditional PMW dimming can caused electromagnetic interference frequency, and produced color cast as the temperature rises[18].

The paper combined the small effect PMW dimming mode had on color temperature with the advantage of high luminous efficiency of analog dimming, proposing a joint dimming mode based on these two dimming modes. The new mode not only kept color temperature of the lamps stable, but also made luminous efficiency remain at a high level. By looking up luminous flux and changing trend of current of a lamp in OSRAM Company, the author established a mathematical model related to them, and gave the compensation current through calculating. So it provided reference for adjusting current.

2 The traditional dimming mode

The TRIAC dimming mode achieved the function of dimming through silicon controlled rectifier dimmer working with the drive circuit. The dimming mode was commonly anywhere. It mainly was used on thermal radiation source, such as the incandescent lamp. It would redesign the driver circuit if the dimming mode was used on LED. There are some problems that LED lamp flickered when it was dimming, the illuminance was not balance, and it would produce the audio-frequency noise.

Analog dimming technology is called linear dimming; it changes the injected forward current by adjusting the parameters of potentiometer in peripheral circuit. Then the lamp's illuminance level is adjusted. The dimming mode had the following advantages, such as simple peripheral circuit, controllable, high luminous efficiency

[a] Corresponding author: wangjp@dlpu.edu.cn

and it could not introduce the electromagnetic interference in operation. But the color temperature drift was obvious because the junction temperature was raised when it adjusted the current in the wide range.

Partition dimming mode was using switch frequency in a certain time to adjust the illuminance of lamp. For example, a panel switch can be changed three sections. The first time of pressing down was full bright; the second was the percent of seventy-five of full light; the third was the percent of fifty of full light; the forth was the percent of twenty-five of full light. The dimming mode did not need the extra wiring, but it could not adjust illuminance continuously.

PWM is abbreviation of the Pulse Width Modulation.it kept the rated current in steady-state, changing the average current by adjusting the duty cycle in dimming operation. So it achieves the objective of dimming. The dimming mode is working in two types of current, such as the rate current and the zero current. So the color temperature drift is stable compared with linear dimming mode. Generally speaking, it used the physiological characteristics of the human eyes needing a short adjustment time to make them not detect to the light constantly blinking. And the frequency of dimming signal in operating is set to 200HZ～20KHZ. But it causes the ceramic capacitor to make noise in the circuit if the dimming signal is working in the high frequency. At the same time, the dimming circuit introduced the additional PWM signal generating module leading to be more complex. It was very bad that the luminous efficiency reduced compared with linear dimming mode.

Designing a new dimming mode is impending. The functions are shown following: high luminous efficiency, stable color temperature and dimming range from 0～100%. So the joint dimming mode was designed for realizing the functions above.

3 The design of joint dimming mode

The signal generating circuit's core chip is the PT4115. The chip integrated the EMI filtering function to ensure that the circuit is not affected by jamming of harmonic in the grid. And the chip's input voltage range from 8V to 30V.It integrated the DC/DC-BUCK highly to keep the output current constant. The circuit is integrated with the LED display module. It displayed the value of the duty cycle facilitating adjusting the effective-average current. The pins of chip packaging are shown in Figure 1.

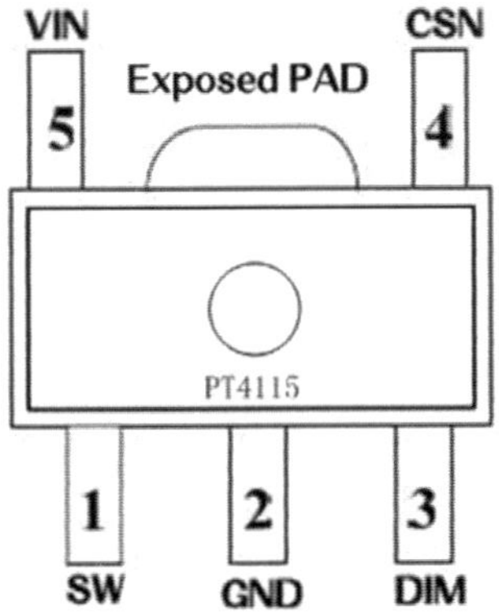

Figure 1. The package pin map of PT4115.

The function of all pins is as follows: 1.SW: Switch Output, SW is the drain of the internal N-CH MOSFET switch; 2.GND: Signal and power ground, connect directly to ground plane; 3.DIM: Logic level dimming input. Drive DIM low to turn off the current regulator. Drive DIM high to enable the current regulator; 4.CSN: Current sense input; 5.VIN: Input Supply Pin. Must be locally by passed;

The PT4115 is the core chip of the dimming circuit as same as the signal generating circuit. And the PWM dimming circuit is a constant-current driver. The PWM signal is loaded into the DIM pin of dimming module in operation, as shown in Figure 2 (a). It will turn off LED current, when the PIN voltage is less than 0.3V, and it will open the LED current, if the PIN voltage is more than 2.5V.

The joint dimming mode combined the advantages of the analog dimming and PWM dimming modes. It kept not only color temperature stable but also luminous efficiency remaining in a high level. The joint dimming mode increased luminous efficiency by reducing the peak current. The potentiometer was connected between the DIM and GND pins to adjust the peak current, as shown in Figure 2 (a). The dimming module was adjusted average current by changing duty cycle of the dimming signal, when it kept the potentiometer invariable.

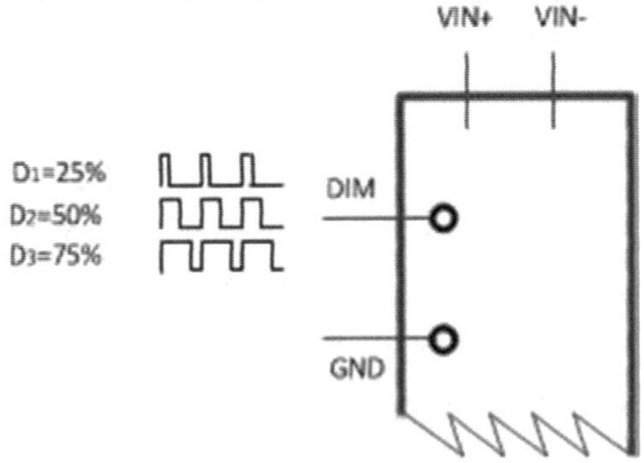

(a) PWM dimming signal access local.

DOI: 10.1051/mateconf/20166102026

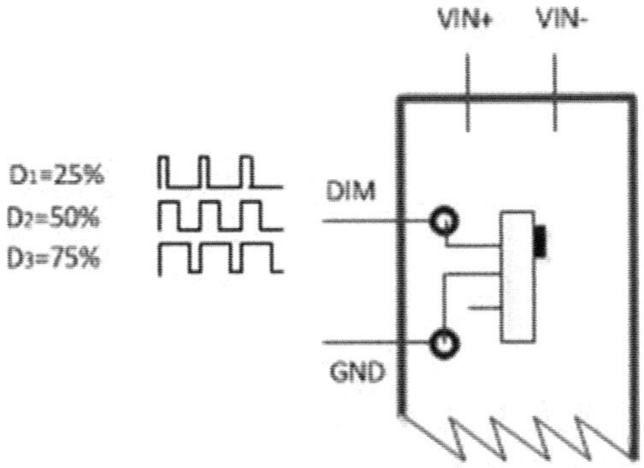

(b) Joint dimming mode's pin access local.

Figure 2. The local pin map of two dimming modes.

Generally speaking, a dimming mode only adjusted one parameter. But the joint dimming mode could adjust the duty cycle and peak current independently by improving the original PWM dimming circuit. So it kept the average input power constantly. The relation between duty cycle and peak current is expressed 1.

$$I_{avg} = I_{peak} * D \tag{1}$$

I_{avg} is the average-output current, I_{peak} is the peak current, D is the duty cycle of PWM signal. It could be known that there was the inverse relation between peak current and duty cycle when the average current was constant. The duty cycle would decrease when peak current increased and the peak current would decrease when the duty cycle increased.

4 Establishing Mathematical Model

The luminous efficiency of white LED would reduce when the current increased. Because, all parameters will be affected by the rose temperature, such as electron mobility reducing, the radiative recombination rate of electrons and holes in the well trap reducing. The luminous efficiency would be affected by the factors. But the effect of temperature on luminous efficiency was reversible. It says that the luminous efficiency has a tendency to recover, when the temperature decreases. The main factors of influencing luminous efficiency are current and temperature, so the paper just considers the effects of current and temperature. It got the relation between luminous efficiency and current by analysing the relationship between flux and current. Then it compensated the luminous efficiency to make it remain at a high level by the joint dimming mode.

It got the trendy of flux and forward current by analysing a lamp of the OSRAM, as shown in Figure 3(a). According to information, the relation between the flux and current was linear without other influencing factors. But the effect of temperature increased gradually with the increasing current. It led to reducing the growth rate of flux, decreasing the luminous efficiency. The front of curve was linear with the increased current. But the slope would reduce with the increased current when the current increased to a certain value. Therefore the relation for actual measured data between flux and current was consistent with the trend of theoretical analysis. The choice of lamp was random, so the analysis about the luminaire was universal. It can reflect the characteristics of the most LEDs.

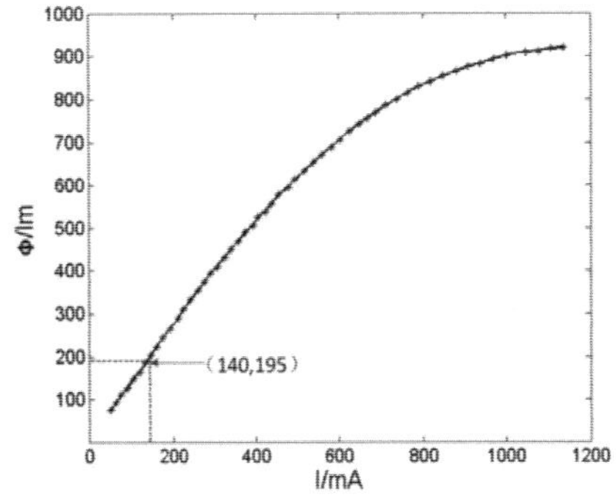

(a) The relation between and current.

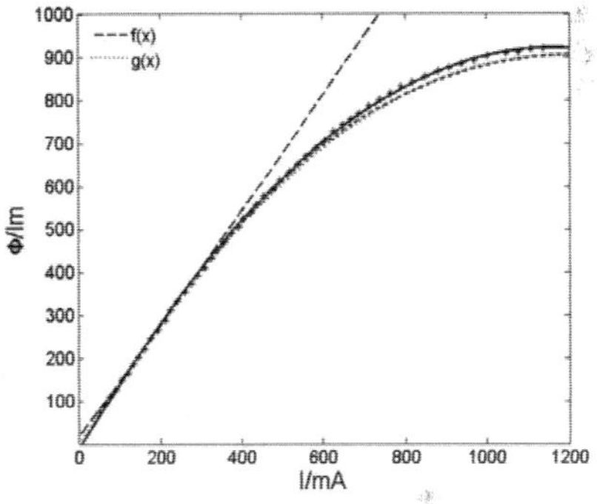

(b) Mathematical model of piecewise function.

Figure 3. The local pin map of two dimming modes.

It could be seen an obvious Cut-off point between the line and curve. The coordinates of the Cut-off was (140,195). So we could build a mathematical model of piecewise function, as shown in Figure 3 (b). It got a linear relationship by fitting the sampling points at the front of Curve, the relation was expressed 2. The relation between flux and current was nonlinear when the current was greater than the cut-off value. It got a nonlinear relation by fitting the sampling points, the relationship was expressed 3.

$$f(x) = b_1 * x + c_1 \tag{2}$$

$$g(x) = a_2 * x^2 + b_2 * x + c_2 \tag{3}$$

The expression of piecewise function was shown 4.

$$T(x) = \begin{cases} b_1 * x + c_1 & , x < 140 \\ a_2 * x^2 + b_2 * x + c_2, & x \geq 140 \end{cases} \tag{4}$$

The reasons for reducing luminous efficiency were caused by temperature. The relation between flux and temperature was got by adding equation 1, 2 together. It was expressed 5.

$$h(x) = a_3 * x^2 + b_3 * x + c_3 \tag{5}$$

DOI: 10.1051/matecconf/20166102026

The parameters of fitting equation were calculated by the sampling points. Taking the sampling points' coordinated into the mathematical model, it got the best degree of fitting as original data. So it could calculate the parameters. They were shown in table 1.

Table 1. the parameters of fitting equation.

equation 's subscripts	a	b	c
1	/	1.34	8.97
2	-0.66×10^{-3}	1.55	-5.41
3	0.66×10^{-3}	-0.21	14.38

It got the relationship between flux and current by analyzing the mathematical model above. And it got the relation between flux and temperature too. The compensation current of was calculated. It kept the range of flux same for increasing the luminous efficiency by reducing the peak current, as expressed 6.

$$\triangle f(x) = \triangle h(x) \tag{6}$$

Assuming the range of flux was (652,689), it could get the (537,583) of $g(x)$ and the (479,505) of $f(x)$. It knew that the results of calculation were consistent with theoretical data. It kept luminous efficiency remaining in a high level by reducing the peak current.

According to the calculated range of current, luminous efficiency could be increased by joint dimming mode. It calculated the value of duty cycle by equation 1, so as to increase the luminous efficiency.

5 Conclusions

The author focused on the relationship between luminous effect and current, and the relationship between luminous efficiency and temperature. A method of increasing luminous efficiency through reducing peak current was proposed. It was based on traditional dimming mode, such as linear dimming mode and PWM dimming mode. And a joint dimming mode was proposed by combining high luminous efficiency of analog dimming with the advantage of stable color temperature PWM dimming possesses. The new dimming mode could change the duty cycle of dimming circuit, and adjust peak current of circuit to achieve the function of adjusting two parameters-duty cycle and current independently. Then it established a related piecewise function by the relation between flux and current. The piecewise function was made up of the part of linear relation and the part of curve relation. There was an obvious Cut-off point. Then it got the mathematical model about luminous efficiency and temperature by calculating. The required current was calculated, and it was the reference of the joint dimming mode.

Acknowledgements

This research was supported by Project of the National Natural Science Foundation of China (61402069), 2012 key supported project for undergraduate education by Education Dept. of Liaoning province([2012]108), 2016 Project of the Natural Science Foundation of Liaoning province(2016020027), 2012 Education reform project for undergraduate by Education Dept. of Liaoning province ([2012]130), 2014 Dalian municipal science and technology plan project（2014A11GX050）, 2014 Dalian municipal science and technology plan project(2014A11GX052).

References

1. H.M. Chen, Semiconductor.Technol.**35**,101(2010)
2. Z. L. Liu, Z. N. Guo, Z. W. Hu, J. HuaQiao Uni, **34**, 14(2013)
3. J. J. Zhang, T. Zhang, Q. Zheng, J .OELE • Laser. **24**,50(2013)
4. X. S. Fu, T. Wang, M. Q. Wang TRA. CHI .ELETECHN.SOC,**30**,75(2013)
5. X. B. Jiang, K. Han, H. P. Chen, ZM.GONGCHEN G XB, **21**, 4 (2010.)
6. X. H. Qu, Y. P. Tang, S. C. WONG, Proc. CSEE, **34**, 1(2014)
7. R. Feng, C. Z. J. Ge, L. Zhu, JX S, **29**,424(2011)
8. P. Qian, S. Zheng, H. Wang, M. Chen, POW. Electron. **44**, 3,(2010)
9. J. Wang, X. Huang, L. Liu, CHI J LUMINE, **29**, 358(2008)
10. J. M. Alonso, D. G. Vaquero, IEEE. Trans Indu. Elec, **59**,1689(2012)
11. C. D. Bi, F. H. Zhang, B. A. Qi, J SHANXI. UNI & TEC **29**,56(2011)
12. P. C. Song, S. S. Wen, W. P. Guan, J Opt. elec. Laser, **26**,1895(2015)
13. Z.Vaitonis,A.Stonkus,A.Žukauskas, IET .OPT.ELEC, **6**, 52(2012)
14. Y.L. Lin, J. Huang, IND Elec. IEEE, **61**,4632(2014)
15. S. K. Jin, J. A. Yoon,Opt.Mat,**47**, 78(2015)
16. C. J. Tian, X. Y. Zhang, J. Zou, CHI. J LUMIN, **31**, 96(2010)
17. C.Terry, EC&M. Elec. Cons & Main , **114**, 10(2011)
18. Q. B. Feng, H. J. He, D. Han, Opt Laser Technol, **77**, 106(2015)

MATEC Web of Conferences **61**, 02027 (2016) DOI: 10.1051/matecconf/20166102027
APOP2016

Research and Development of Miniaturized Fuel Evaporation Closed Chamber for Motorcycle

Hong Bin[1,a], Zhu Di[1] and Wang Hongmei[2]

[1] Internal Combustion Engine Research Institute of Tianjin University, China
[2]Tianjin Tianbo Science & Technology Co,.Ltd., China

Abstract. With the increasingly serious environmental pollution, pollution emissions of motorcycles had gradually been concerned in all developed countries. But the fuel evaporative emissions had not attracted sufficient attention and its test equipment was far from perfect. Thus, this paper addressed the miniaturized problems of closed chamber. On the basis of analyzing fuel evaporative emission regulations and the features of closed room, technical solutions are determined and a prototype was successfully developed. The paper successfully resolved the seal problem and innovatively developed the pressure balance system and the temperature and humidity regulation system. After experimental verification, the prototype had reliable performance and stable operation. And it met the requirements of the relevant standards and it was also equivalent to imported equipments. Thus, this paper successfully resolved the miniaturized problems of closed chamber.

1 Introduction

With the increasingly serious environmental pollution, pollution emissions of motorcycles have gradually been concerned in all developed countries, and more stringent emissions regulations for motorcycles have been promulgated. There are three major sources of emissions: emissions, crankcase blow-by gas and fuel evaporation. Among them, the exhaust emissions and crankcase blow-by gas have caused concern of motorcycle manufacturers and the appropriate control measures have been taken. But the fuel evaporative emissions have not attracted sufficient attention. Currently, in the world, only a few countries and regions implement the fuel evaporative emission regulations for motorcycles, and China also too.

However, the associated test equipment is far from perfect. Currently, internationally accepted fuel evaporative emission measurement methods for motorcycles are collection method and airtight chamber method. The test equipment of collection method is relatively simple, but has the poor inaccurate, poor reproducibility and poor consistency. Closed chamber method, currently the mainstream testing method, has the accurate data, good repeatability, but sophisticated equipment and high cost. Existing closed chambers, mainly for automobile, have too large volume and too high price to Purchase for gasoline Engine Manufacturer (more than 200 million Yuan per set). Therefore, in accordance with national standards and indicators gasoline engine requirements, the paper studied the miniaturized problems of the fuel evaporation closed chamber, and developed the equipment in line with national standards, suitable for motorcycle and gasoline engine.

[a] Corresponding author: tianbohb@126.com

2 Hardware Structure

Hardware Structure of the closed chamber consisted of HC detector, pressure balance system, temperature/humidity regulation system, wind circulation system and sealing system. The Hardware Structure of the Closed Chamber is shown as Figure 1.

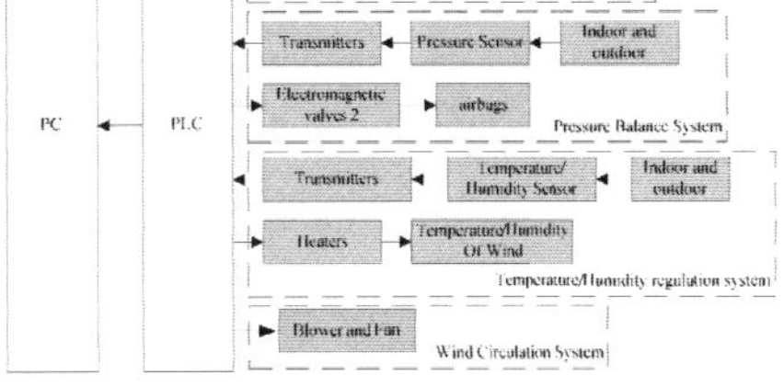

Figure 1. Hardware Structure of the Closed Chamber

3 System Components

3.1. HC Detector

Selecting the operating range of detector is so important that you can get the best resolution in measurement, calibration, leak detection and other processes. Hydrocarbon analyzer frequency of twice per minute automatic recording of data processing systems, and can be kept permanently. Analyzer automatically records the testing data with twice per minute, and permanently saves data.

3.2 Pressure Balance System

The closed chamber is completely sealed in the process of testing. When the ambient temperature changes indoor, the barometric pressure will change as the temperature varies. In order to simulate the real situation, the test must be done under constant pressure, and standard differential pressure within ± 500Pa between indoor pressure and atmospheric pressure. Therefore, we used a pressure balance system (shown as Figure 2) to control the balance of pressure caused by temperature changes.

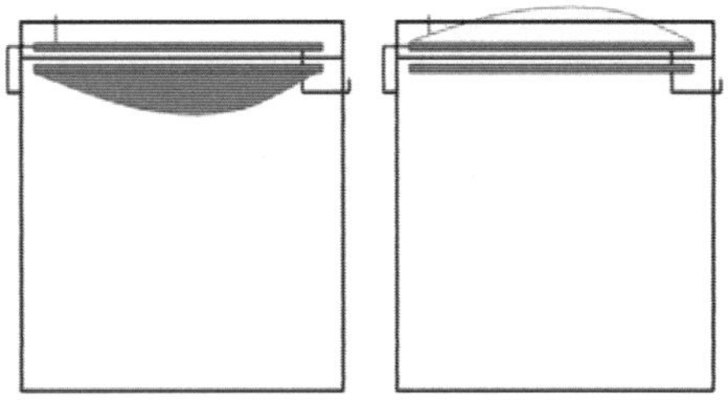

Temperature Decreased Temperature Rised

Figure 2. Changes of the Airbags

When the indoor temperature decreases in the process of testing, the outdoor air goes into the indoor airbags via a control valve, and the net volume of sealed chamber decreases. Thus, the indoor pressure is kept correspondence with the outdoor pressure. When the indoor temperature rises, the indoor air goes into the outdoor airbags via another control valve and the net volume of sealed chamber increases. Thus, the indoor pressure is also kept correspondence with the outdoor pressure. In the process of testing, the difference of indoor and outdoor pressure must be controlled within ±200Pa. Pressure recording system automatically records and shows the pressure data with twice per minute, and permanently saves data. And it automatically calculates and displays the pressure difference Δp, too. The accuracy of the pressure recording system is within ± 5Pa and its resolution is ± 1Pa.

3.3 Temperature/Humidity Regulation System

The Closed chamber is equipped with temperature and humidity regulation system in order to achieve control of the ambient temperature and humidity in the process of testing. The system uses PID intelligent control system. Its deviation is ± 3℃, better than the standard required by 5℃. The system automatically records and shows the temperature data and humidity data. The accuracy of the temperature/humidity regulation system is within ±1.0℃ and its resolution is ± 0.1℃.

The system uses intelligent PID control that is combined the intelligent control and conventional PID control. It is adaptive and parameters of its controller online change. The system does not rely on accurate mathematical model so that it has good adaptability, strong anti-interference ability and robustness. Its block diagram is shown as Figure 3.

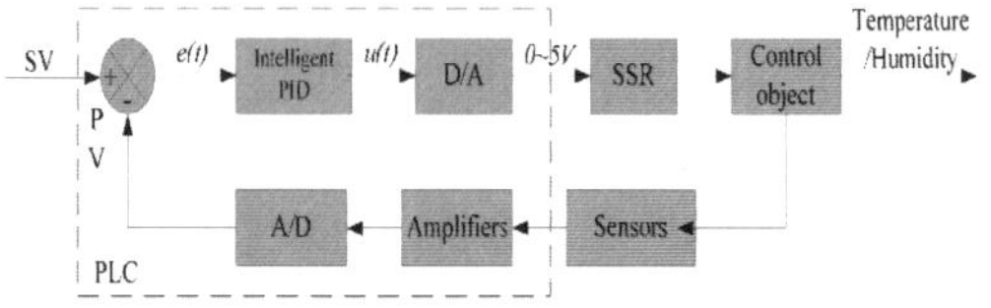

Figure 3. Schematic of PID Control

In practice, the control law adopts the sectional PID control scheme. When the deviation of the actual temperature and the set temperature is large, system will heat or cool at a large rate. When approaching a thermostat, it don´t cause oscillation because of too much inertia to reduce the transition time.

3.4 Wind Circulation System

To fully mixing the indoor atmosphere to ensure uniform temperature and HC concentration, wind circulation system is configured. The air flow generated is unable to face trial motorcycle. When you open the door, interior HC concentration can be reduced to the environment level. Large air fan is installed in the exhaust port. The indoor gas is replaced again every 16 seconds. Before and after the test, HC concentration can quickly reduce to the HC concentration level of the environment. Five mixing fans are equipped in the closed chamber and the range of their air volume is 10.8-14 m^3/min. The system is capable of sufficiently mixing the indoor gas to ensure indoor temperature and HC concentrations homogeneous.

3.5 Sealing System

Seal is an important parameter of the closed chamber, but it also is a technical difficulty. If the seal is poor, it will directly affect the measurement accuracy. Standard provides: the leakage rate must be less than 4% within four hours. In order to ensure its seal, 316 stainless steel was used as the inner wall, the interface was welded, and the punched holes were fully welded. Tray is installed under air vents. When the tray rises and close air vents, ring on the tray is pressed. Ring has a certain flexibility and generates a certain deformation to ensure the seal when the tray is closed.

Door of the closed chamber often open, so inflatable silicone ring was used as sealing structure. In the natural state, the ring hides in U groove structure. When inflated, the ring outward bulge because of internal pressure to impinge on the door frame so that the seal is ensured.

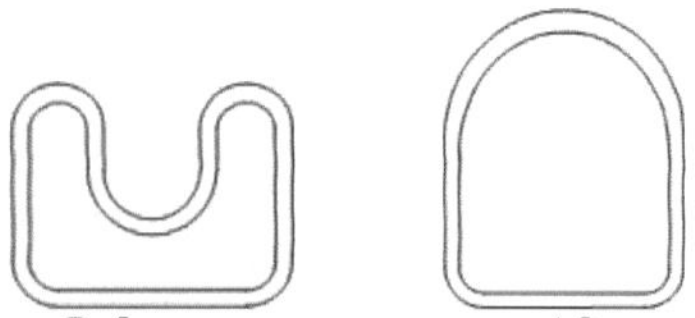

Before After

Figure 4. Cross Section of Sealing Ring

DOI: 10.1051/matecconf/20166102027

When installed, the sealing ring is placed in U grooves of door and its concave outward. Its inlet was connected with air compressor via fast access. When used, the control valve opened and compressed air gone into the cavity of the inflatable ring after the sealed chamber door was closed. Pressure of the cavity increased so that the concave of the inflatable ring expanded outwardly to form arc-shaped. Therefore, the arc and the door frame obtained an interference fitting and its seal had good effect.

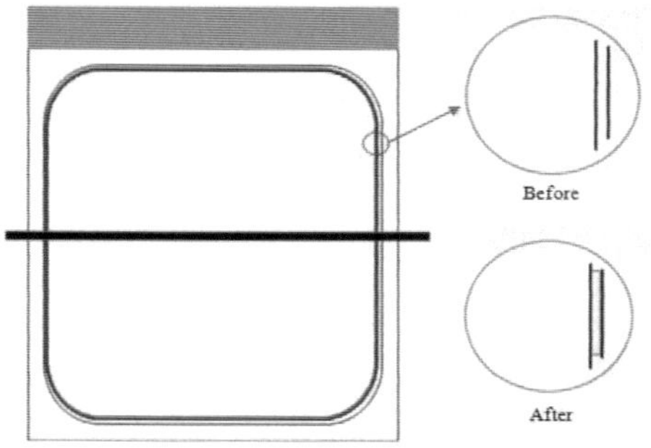

Figure 5. Schematic of Sealing Method

4 Software

The system uses WINDOWS operating systems, and uses industrial control computers and programmable control system as a control center. The system has good man-machine interface and digital control mode. Test interface can display real-time data, such as indoor and outdoor temperature, fuel temperature, steam temperature, indoor and outdoor pressure. The main interface is shown as Figure 6.

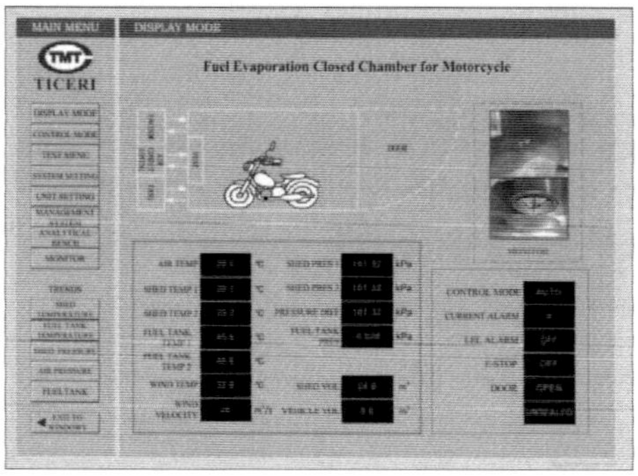

Figure 6. the Main Interface

5 Results and Analysis

In order to verify the seal of the closed chamber, uniformity and seal experiment were done in accordance with national standards GB/T 20998. The experimental curve is shown as Figure 7.

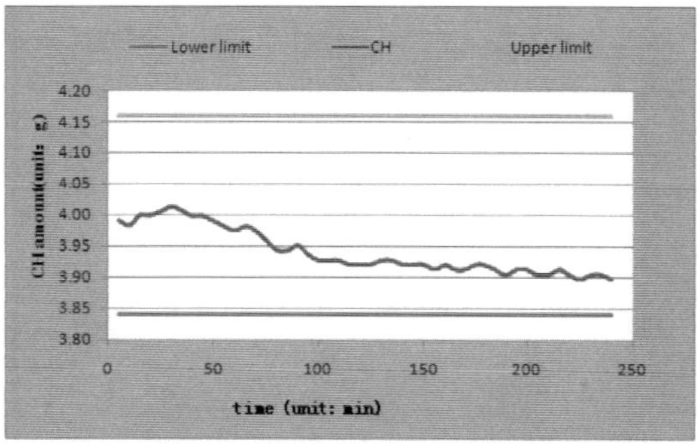

Figure 7. the Curve of CH Amount

From the above chart shows: before the experiment beginning, 4.0g hydrocarbons were injected into the closed chamber via sampling port and mixed for 5 minutes. Test Result shown: HC amount is 3.98g and its deviation is -0.5%. The deviation is in the range of national standards GB/T 20998 requirements (±2%) and it proofed that indoor HC concentration is uniform. After 4 hours (240min), HC amount is 3.90g and its deviation is -2.5% and less than ± 4% (required by national standards GB/T 20998). It also proofed that the seal of the closed chamber is very good.

In addition, this equipment was compared with similar imported equipments. The pressure deviation of the equipment is ±50Pa. The actual fuel heating curve fluctuated around the theoretical curve and its deviation was controlled in ±0.5 ℃ . The equipment met the requirements of the relevant standards and also equivalent to imported equipments. Comparative results were shown as Figure 8 to Figure 10.

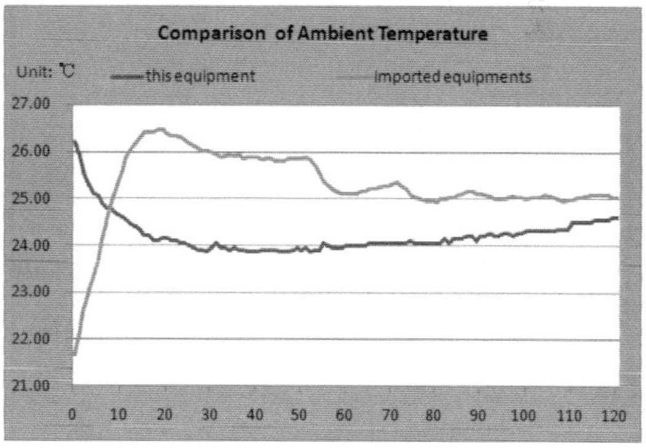

Figure 8. Comparison of Ambient Temperature

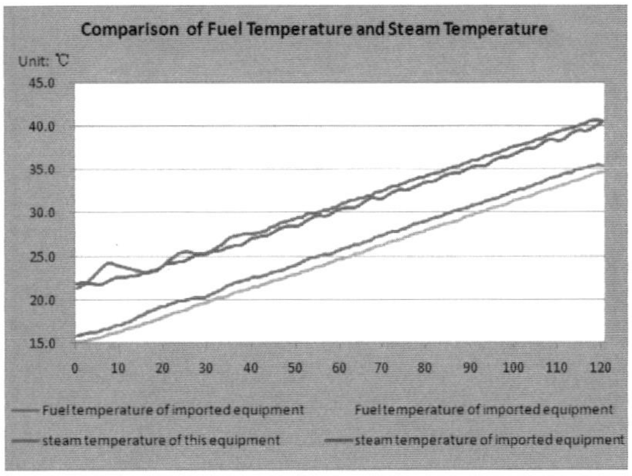

Figure 10. Comparison of Fuel Temperature and Steam Temperature

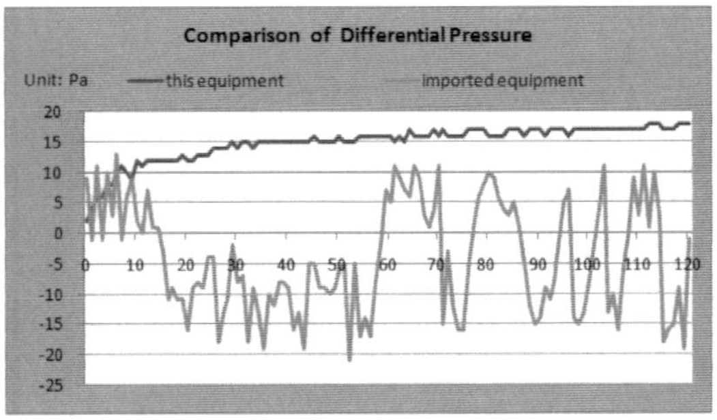

Figure 9. Comparison of Differential Pressure

6 Conclusion

This project addressed the miniaturized problems of closed chamber. On the basis of analyzing fuel evaporative emission regulations and the features of closed room, technical solutions are determined and a prototype was successfully developed. Experimental results shown: the miniaturization closed chamber had reliable performance and stable operation. Experimental comparative data shown: the equipment met the requirements of the relevant standards and it was also equivalent to imported equipments. Thus, this equipment greatly reduced the cost and resolved the miniaturized problems of closed chamber.

References

1. Miao Lei. Study on fuel evaporative emission property of motor vehicles [D]. Tianjin University, 2010.
2. Cheng Shili. Experimental research of the fuel evaporative control system for motorcycles[D]. Tsinghua University, 2012.
3. Jia Quanli. A brief introduction to the test method for fuel evaporative emission of motorcycle(II)[J]. Motorcycle Technology, 2009, 11:38-41.
4. Qiu Jie. Quality of fuel evaporative controlling system of motorcycle and study on related problem[J]. Motorcycle Technology, 2007, 5:16-19
5. GB20998-2007, Limits and measurement methods for evaporative pollutants from motorcycles and mopeds [S].
6. Small Off-Road Engine and Equipment Evaporative Emission Test Procedure, California Environmental Protection Agency, Air Resources Board, El Monte, CA, 2004.
7. Test Procedure for Determining Diurnal Evaporative Emissions from Small Off-Road Engines and Equipment, California Environmental Protection Agency, Air Resources Board, El Monte, CA, 2004.
8. Xie Jianguang. A Study of an Evaporation Emissions Control System for a Vehicle with Petrol Engine[J]. Small internal combustion engine, 2000, 29(4): 30-34.
9. Tan Linbo, Liu Shengji, Lu Cunhao, Wang Jian. CFD analysis and experiment research of non-road gasoline engine cooling system[J]. Journal of Chinese Agricultural Mechanization, 2014, 35(1): 189-192.
10. Delia D'Agostino, Paolo Maria Congedo, Rosella Cataldo. Computational fluid dynamics (CFD) modeling of microclimate for salts crystallization control and artworks conservation[J]. Journal of Cultural Heritage . 2013
11. G. Tong, D.M. Christopher, B. Li. Numerical modelling of temperature variations in a Chinese solar greenhouse[J]. Computers and Electronics in Agriculture . 2009 (1)

Establishment of Combustion Model for Isooctane HCCI Marine Diesel Engine and Research on the Combustion Characteristic

Biao Li[1, a] and Hongtao Gao[1]

[1]*The College of Marine Engineering, Dalian Maritime University, 116026, Dalian, China*

Abstract: The homogeneous charge compression ignition (HCCI) combustion mode applied in marine diesel engine is expected to be one of alternative technologies to decrease nitrogen oxide (NO_X) emission and improve energy utilization rate. Applying the chemical-looping combustion (CLC) mechanism inside the cylinder, a numerical study on the HCCI combustion process is performed taking a marine diesel engine as application object. The characteristic feature of combustion process is displayed. On this basis, the formation and emission of NO_X are analyzed and discussed. The results indicate that the HCCI combustion mode always exhibit two combustion releasing heats: low-temperature reaction and high-temperature reaction. The combustion phase is divided into low-temperature reaction zone, high-temperature reaction zone and negative temperature coefficient (NTC) zone. The operating conditions of the high compression ratio, high intake air temperature, low inlet pressure and small excess air coefficient would cause the high in-cylinder pressure which often leads engine detonation. The low compression ratio, low intake air temperature and big excess air coefficient would cause the low combustor temperature which is conducive to reduce NO_X emissions. These technological means and operating conditions are expected to meet the NO_X emissions limits in MARPOL73/78 Convention-Annex Ⅵ Amendment.

1 Introduction

Since the first industry revolution, diesel engine was used on ship and has contributed significantly to shipping industry. At the same time, environment pollution problem caused by harmful emissions especially the nitrogen oxide (NO_X) from marine diesel engine is more serious, which have attracted wide international attention [1]. Convention-Annex Ⅵ Amendment to MARPOL73/78 worked out by the International Maritime Organization (IMO) for the prevention of air pollution from ships has became effective on May 19, 2005. In addition, in October 2008, the Marine Environmental Protection Committee (MEPC) of IMO agreed upon progressively stricter limitations for NO_X emissions from vessels based on their date of engine installation with the standards of Tier Ⅰ, Tier Ⅱ and Tier Ⅲ. These standards have taken effect beginning on July 1, 2010.

As shown in Table 1, the Tier Ⅲ standard is the strictest mandatory requirement applied for the engine of which installation date is later than January 1, 2016 in designated Emission Control Areas (ECA). And the Tier Ⅱ requirement is still performed outside the ECA. Compared with the requirements of Tier Ⅱ and Tier Ⅰ, NO_X emissions must decrease by 74%-79% and 80% respectively to meet the Tier Ⅲ requirement.

Table 1. NO_X emissions limits in MARPOL73/78 Convention-Annex Ⅵ Amendment [1]

Standard	$n<130$ /g·(kWh)$^{-1}$	$130\le n\le 2000$ /g·(kWh)$^{-1}$	$n>2000$ /g·(kWh)$^{-1}$
Tier Ⅰ *	17.0	$45\times n^{-0.2}$	9.84
Tier Ⅱ	14.36	$44\times n^{-0.23}$	7.66
Tier Ⅲ**	3.40	$9\times n^{-0.23}$	1.97

"n" refers to rated engine speed (r·min^{-1}).
All the NO_X emissions limits exclude ships with marine diesel engines less than a power output of 130 kW or ships solely for emergency use such as emergency diesel engine, lifeboat engines etc.
*Annex VI entered into force in 2004, but it applies retroactively to new engines larger than a power output of 300kW installed on ships on or after January 1, 2000.
** Tier III applies only in emission control areas.

In the face of the new increasingly stringent NO_X emission reduction regulations and demand for the higher energy utilization rate. The development and application of effective energy-saving and emission reduction technology adapted in the marine diesel engine are has becoming one of the most basic way [2, 3], except the guidance of the government policies. Homogeneous charge compression ignition (HCCI) is one of the most attractive modes for automotive engines to achieve both high efficiency and low nitrogen oxide (NO_X) and particulate matters (PM) [4, 5].

The HCCI combustion technology has received increased attention for its potentials of ultra-low NO_X and PM emissions and high thermal efficiency. A significant

[a] Corresponding author: libiaoyx@126.com

feature of HCCI is that the compression ignitions are happened in multiple regions. Therefore, the temperature inside the combustion chamber is relatively uniform, the cyclic variations are small and there are no high-temperature zones in combustion process. These characteristics can improve the combustion roughness and fuel economy and the formations of NO_X and PM are controlled effectively.

Applying the chemical-looping combustion (CLC) mechanism inside the cylinder, a numerical study on the HCCI combustion process is performed taking the MAN-B&W6L23/30-A marine diesel engine as application object. The characteristic feature of combustion process is displayed. On this basis, the formation and emission of NO_X are analyzed and discussed.

2 Establishment of computational model

The elementary chemical reactions of diesel combustion in the cylinder are very complex and extremely large in quantity because of the multiple components and uncertainty component concentration in diesel fuel [6, 7]. Consequently, the other single component fuels are always selected as replacements for diesel fuel in combustion simulation. Isooctane is a kind of ideal alternative fuel whose cetane number is approximately of diesel fuel [8].

In this modeling, it is assumed that the thermodynamic pressure, thermodynamic temperature and composition concentration of inflammable mixture are completely uniform in cylinder combustion chamber. The convection heat transfer of cylinder wall is considered, but the radiation heat transfer is ignored. The mixed gas (working fluid) in cylinder, which is from system boundary, and residual exhaust gas are supposed to be thoroughly mixed instantaneously during the intake stroke. The change of cylinder volume caused by piston motion is simplified to a slider-crank mechanism.

3 Results and analysis

In the simulation calculation, MAN-B&W6L23/30-A marine medium-speed diesel engine in marine power plant laboratory of Dalian Maritime University is chosen as the simulation research object. Table 2 shows the structural specifications of MAN-B&W6L23/30-A diesel Engine [9].

The TDC (Top Dead Center) of the intake stroke is set as the zero point of crank angle (0°CA). The computational domain starts from the starting point of compression stroke, ends with terminal point of expansion stroke. So the range of crank angle is from 180°CA to 540°CA. The fuel injection way adopts the port fuel injection.

Table.2 Structural specifications of MAN-B&W6L23/30-A diesel Engine

Items	Specification parameter
Cylinder bore/mm	225
Length of stroke/mm	300
Number of cylinders	6
Geometry compression ratio	13:1
Connecting rod length/mm	600
Cylinder clearance volume /m^3	9.94×10^{-4}
Release valve open phase degree /°CA (BDC)	-37
Rated speed / $r\cdot min^{-1}$	900

Figure 1 is the Heat release rate curve of isooctane HCCI combustion. The curve exhibits two peak values which are the combustion releasing heats of low-temperature reaction (347°CA) and high-temperature reaction (354°CA). The combustion releasing heat of high-temperature reaction accounts for above 95% of total releasing heat.

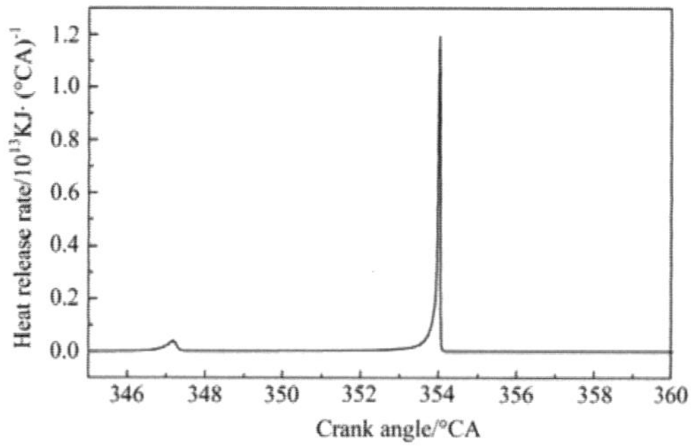

Figure 1. Heat release rate curve of isooctane HCCI combustion

Figure 2 shows the division of ignition time and combustion phase. The ignition points of low-temperature reaction zone and high-temperature reaction zone correspon to the two combustion releasing heats. The combustion phase is divided into low-temperature reaction zone, high-temperature reaction zone and negative temperature coefficient (NTC) zone.

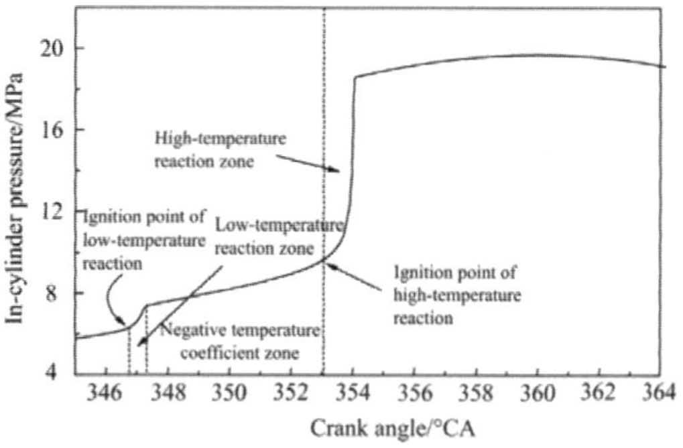

Figure 2. Division of ignition time and combustion phase

3.1 Impact analysis of compression ratio

Figure 3 and Figure 4 show the effects of the compression ratio on combustor temperature and in-cylinder pressure. According to the observation, the compression ratio affects the temperature and pressure greatly. With the compression ratio improves, both the temperature and pressure increase continuously and the rates are slowed down. On the whole, the peaks of combustor temperature are always low in HCCI combustion model compare with the conventional diesel engine with the same operation condition. When the compression ratio exceeds 16, the peak of combustor temperature has already over 2000K that reached the level of conventional diesel engine.

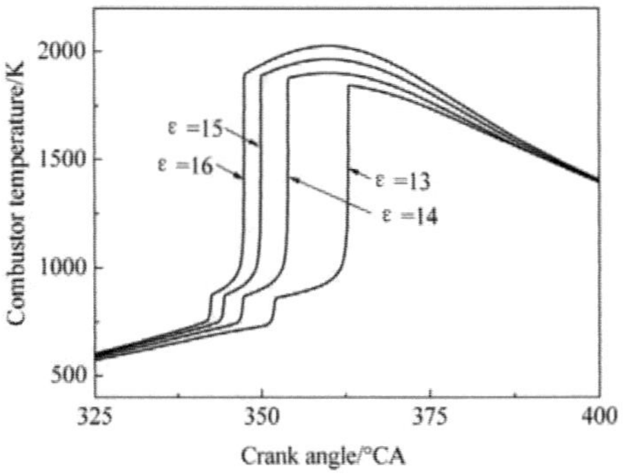

Figure 3. Effect of the compression ratio on combustor temperature
(Intake pressure p_0 =0.25MPa, intake air temperature T_0 =320K, excess air coefficient λ =2.8.)

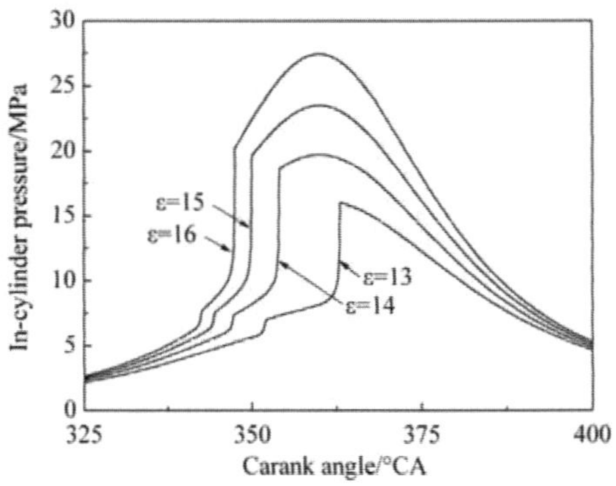

Figure 4. Effect of the compression ratio on in-cylinder pressure
(Intake pressure p_0 =0.25MPa, intake air temperature T_0 =320K, excess air coefficient λ =2.8.)

The modern marine diesel engine commonly adopts high compression ratio that favorable for the thermal efficiency and fuel economical efficiency significantly. Due to the lower combustor temperature, the NO_X emissions in HCCI combustion mode will be controlled effectively.

3.2 Impact analysis of intake air temperature

Figure 5 and Figure 6 show the effects of the intake air temperature on combustor temperature and in-cylinder pressure. With the rise of intake air temperature, the combustor temperature increases in the case of unchangeableness of the other initial conditions, but the max firing pressure slightly decreases. The ignition points of high-temperature reaction zone are brought forward with intake air temperature. These facts will promote the NO_X emissions for the combustor temperature play important role in the chemical reaction rate of NO_X formation [10].

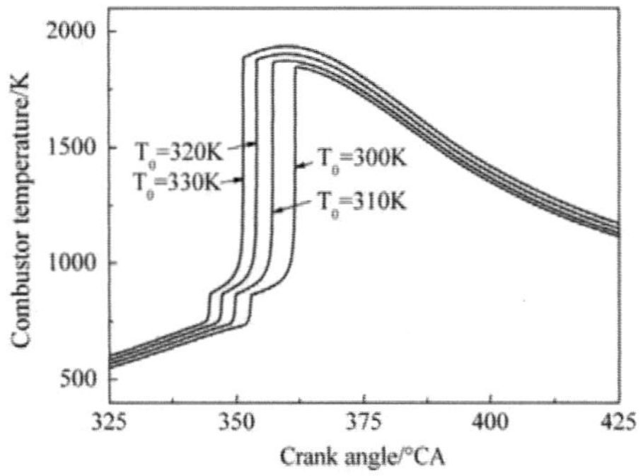

Figure 5. Effect of the intake air temperature on combustor temperature
(Intake pressure p_0 =0.25MPa, compression ratio ε=13, excess air coefficient λ =2.8.)

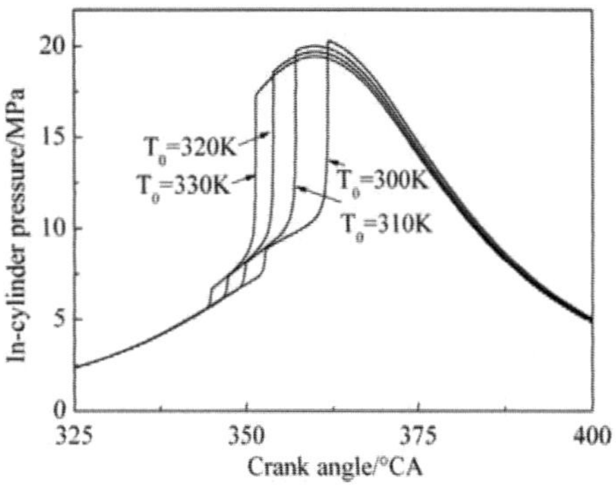

Figure 6. Effect of the intake air temperature on in-cylinder pressure
(Intake pressure p_0 =0.25MPa, compression ratio ε=13, excess air coefficient λ =2.8.)

3.3 Impact analysis of intake pressure

Figure 7 and Figure 8 show the effects of the intake pressure on combustor temperature and in-cylinder pressure. It is shown that the increase of intake pressure leads to rising of in-cylinder pressure, but the maximum combustor temperatures often remain about the same which have a relatively small impact on the NO_X formations and emissions. Also, the ignition points of high-temperature reaction zone are brought forward with intake pressure that would extend the chemical reaction time of oxygen (O_2) and nitrogen (N_2) in high temperature zone.

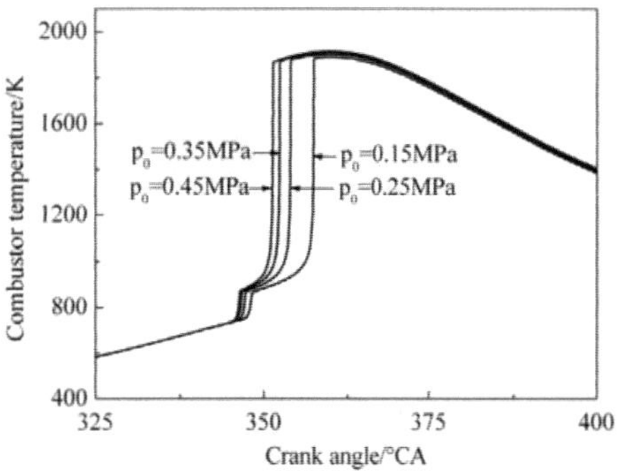

Figure 7. Effect of the initial pressure on combustor temperature
(Intake air temperature T_0 =320K, compression ratio ε=13, excess air coefficient λ =2.8.)

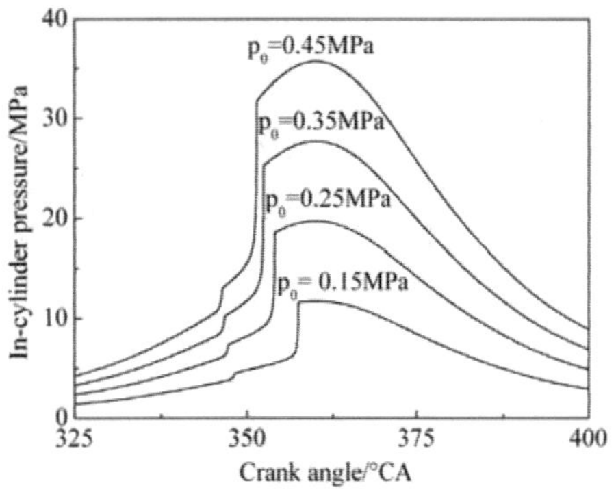

Figure 8. Effect of the initial pressure on in-cylinder pressure
(Intake air temperature T_0 =320K, compression ratio ε=13, excess air coefficient λ =2.8.)

3.4 Impact analysis of excess air coefficient

Figure 9 and Figure 10 show the effects of the excess air radio on combustor temperature and in-cylinder pressure. The simulation results confirm that, the impacts of excess air radio to the combustor temperature and in-cylinder pressure are remarkably. Overall, both the combustor temperature and in-cylinder pressure decrease with the increase of the excess air radio. The ignition points of high-temperature reaction zone are postponed that would decrease the chemical reaction time of oxygen (O_2) and nitrogen (N_2) in high temperature zone. These factors are not beneficial to the formations of NO_X. Related research also confirmed HCCI engines have extraordinarily lower NO_X emissions with the dilute homogeneous air and fuel mixture. The NO_X emissions of HCCI marine diesel engine can be controlled by adjusting the excess air radio.

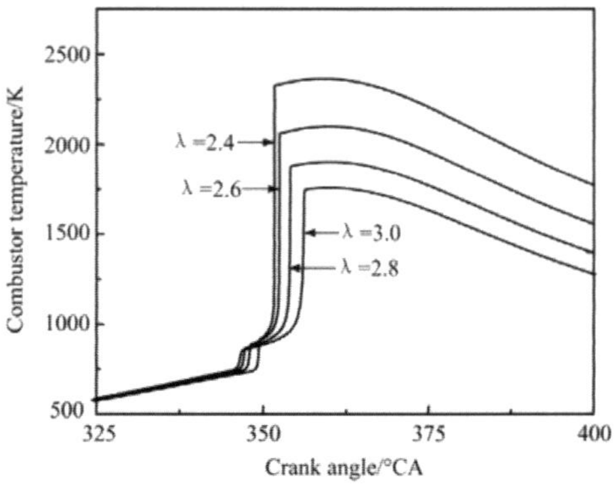

Figure 9. Effect of the excess air radio on combustor temperature
(Intake air temperature T_0 =320K, intake pressure p_0 =0.25MPa, compression ratio ε=13.)

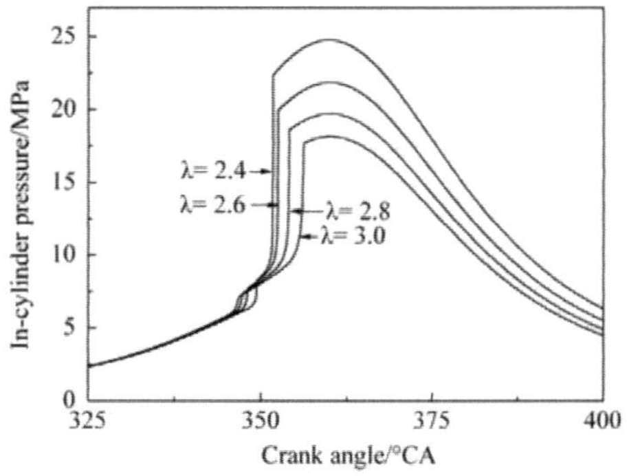

Figure 10. Effect of the excess air radio on in-cylinder pressure
(Intake air temperature T_0 =320K, intake pressure p_0 =0.25MPa, compression ratio ε=13.)

4 Conclusions

(1) Results from simulation confirm that HCCI combustion mode has the combustion releasing heats of low-temperature reaction and high-temperature reaction. The combustion phase is divided into low-temperature reaction zone, high-temperature reaction zone and negative temperature coefficient zone (NTC).

(2) The operating conditions of the high compression ratio, high intake air temperature, low inlet pressure and small excess air coefficient would cause the high in-cylinder pressure which often leads engine detonation.

(3) Applying HCCI combustion mode in marine diesel engine, the low compression ratio, low intake air temperature and big excess air coefficient would cause the low combustor temperature which is conducive to reduce NO_X emissions. These technological means and operating conditions are expected to meet the NO_X emissions limits in MARPOL73/78 Convention-Annex Ⅵ Amendment.

 DOI: 10.1051/matecconf/20166102028

Acknowledgements

This work was financially supported by the Research funds of the Maritime Safety Administration of the People's Republic of China (2012_27).

References

1. C. Yugma, J. Blue, S, Dauzère-Pérès, A, Obeid. Emission from international sea transportation and environmental impact, J. Geophys. Res. **108** (D17):129-144 (2003).
2. US EPA. *Regulatory impact analysis: Control of emissions of air pollution from category 3 marine diesel engines*, New York, (2009)
3. MAN Diesel & Turbo. *Emission project guide MAN B&W two-stroke marine engines*, (2013)
4. L. Xue, C.H. Zhang, J. Wang. Cyclical variation of an HCCI engine fuelled by n-heptane. J. Chang'an Univ., **34** (6): 157-161 (2014)
5. M. Rashed, P.F.A. Macconnell, A.F. Stronach, P. Acarnley. Sensorless indirect-rotor-field-orientation speed control of a permanent-magnet synchronous motor with stator-resistance estimation. IEEE Transactions on Industrial Electronics, **54**(3): 1664-1675 (2007)
6. S.K. Kim, Y.M. Kim. Prediction of detailed structure and NO_X formation characteristics in turbulent non-premixed hydrogen jet flames. Combust. Sci. Technol., **156(**1-6): 107-137 (2000)
7. Stephen R. Turns. *An introduction to combustion: Concepts and Applications,* McGRAW-HILL International Editions, (2000)
8. Z.L. Zheng, M.F. Yao. Numerical study on the chemical reaction kinetics of n-Heptane for HCCI combustion process. Trans. Csice, **22**(3): 227-234 (2004)
9. L.P. Xu. *The research of reducing the nitrogen oxide emission from marine diesel engine by wet inlet air with spraying water.* Dalian: Dalian Maritime University, (2003)
10. W. Palmer. Cost-benefit study of marine engine NO_X emissions control systems, A Case Study of the MV Cabot, 02, (2000)

Design of Speed and Current Controllers Based on Online Particle Swarm Optimization Method for IPMSM Drives

Zhengqiang SONG[1], Huiling YANG[1]

[1]Yangzhou Polytechnic College, China

Abstract. In this paper, a novel online particle swarm optimization method is proposed to design speed and current controllers of vector controlled interior permanent magnet synchronous motor drives considering stator resistance variation. In the proposed drive system, the space vector modulation technique is employed to generate the switching signals for a two-level voltage-source inverter. The nonlinearity of the inverter is also taken into account due to the dead-time, threshold and voltage drop of the switching devices in order to simulate the system in the practical condition. Speed and PI current controller gains are optimized with PSO online, that means single irritation optimization computation can be completed within single sample time. In addition, the fitness function is changed according to the system dynamic and steady states. The proposed optimization algorithm is compared with conventional PI control method in the condition of step speed change and stator resistance variation, showing that the proposed online optimization method has better robustness and dynamic characteristics compared with conventional PI controller design.
Key words: PMSM, Online PSO, Parameters Variation

1 Introduction

Proportional-integral (PI) control technique has been widely used in high performance field orientation controlled IPMSM drives. However, fixed PI gains have to be designed based on a precise mathematical model of the drive for guaranteeing certain control performance. It is unavoidable that the uncertainties are caused by parametric variations such as flux linkage or stator resistance and unstructured dynamics in a practical IPMSM drives [1, 2].

Thus, to obtain a high dynamic performance for an IPMSM drive, current controllers should be optimized together with speed controller at the same time, because the current controller influences directly the drive dynamics.

There are numerous researches on the applications of computational intelligence techniques to controller parameters design for PMSM [3-4]. Among these, Particle Swarm Optimization [4], first introduced by Kennedy and Eberhart in 1995, is one of the modern heuristic algorithms. Because of its simplicity and computational efficiency, PSO has been widely used to solve a broad range of optimization problems, such as adaptive tuning of controller gains and parameters identification. However, there still exist some problems/limitations with this method on the optimization of controller gains. Firstly, the PSO optimization applications in designing controller gains for PMSM rely on offline precise calculations of responses of PMSM using mathematical model [3]. It makes optimization effectiveness rely on the fixed PMSM model excessively. Secondly, the online PSO for PMSM controller optimization, however only applied to the speed controller gains without considering current controller optimization. It is difficult to achieve high dynamics and robustness for PMSM drive system. Thirdly, much effort has been made on real-time PSO application in parameters identification, and results show effectiveness because parameters of PMSM are changed slowly [3]. This should be incorporated with the optimization process online.

In this paper, a discrete simulation model of IPMSM is set up firstly considering dead-time of inverter to present a real-time simulation condition. Then by measuring speed and current, an online PSO method is applied to design speed and current PI controller parameters for PMSM drives. To achieve good dynamic and steady states performance, different fitness function has been adopted. Numerical simulation test has been carried out with a step speed reference, while the stator resistance is varying. Simulation results show that this online optimization method is model free and has better robustness and faster optimization ability.

2 Real-Time Simulation Model Of Ipmsm Drive

The IPMSM model in the rotor reference frame is

$$\begin{bmatrix} u_{d(t)} \\ u_{q(t)} \end{bmatrix} = \begin{bmatrix} R_s & 0 \\ 0 & R_s \end{bmatrix} \begin{bmatrix} i_{d(t)} \\ i_{q(t)} \end{bmatrix} + \begin{bmatrix} p & -\omega_{re(t)} \\ \omega_{re(t)} & p \end{bmatrix} \begin{bmatrix} \phi_{d(t)} \\ \phi_{q(t)} \end{bmatrix} \quad (1)$$

[a] Corresponding author: songyang@alumni.sjtu.edu.cn

$$\phi_{d(t)} = L_d i_{d(t)} + \phi_f \quad (2)$$

$$\phi_{q(t)} = L_q i_{q(t)} \quad (3)$$

$$T_{em(t)} = \frac{3}{2} P_n \cdot (\phi_{d(t)} \cdot i_{q(t)} - \phi_{q(t)} \cdot i_{d(t)}) \quad (4)$$

$$T_{em(t)} - T_{L(t)} = \frac{J}{P_n} \frac{d\omega_{re(t)}}{dt} \quad (5)$$

where $u_{d(t)}$, $u_{q(t)}$, $i_{d(t)}$, $i_{q(t)}$, L_d, L_q, $\phi_{d(t)}$ and $\phi_{q(t)}$ are the stator voltages, stator currents, inductances and flux linkages on *dq*-axis respectively; ϕ_f is permanent flux; $p = \frac{d}{dt}$ is the differentiation operator; P_n is the number of pole pair; $\omega_{re(t)}$ is rotor velocity electrical; $T_{em(t)}$ is generated torque; $T_{L(t)}$ is load torque.

The $d-$ and $q-$axis current controllers are PI whose transfer functions are $k_{id}(1+\tau_{id}s)/s$ and $k_{iq}(1+\tau_{iq}s)/s$, respectively. The reference of $q-$axis current i_q^{ref} is provided by the speed controller ($k_{s\_p} + k_{s\_i} \cdot \frac{1}{s}$). A low-pass filter is applied to the measured speed and a rate limitation is applied to the speed reference signal. They are used to reduce the overshoot and settling associated with step commands, while retaining a fast disturbance rejection. The online PSO scheme is implemented with a S-Function. The input variables of the S-Function are references and feedback values of q axis current and rotor speed respectively, the output variables are gains of the q axis current and speed PI controllers respectively.

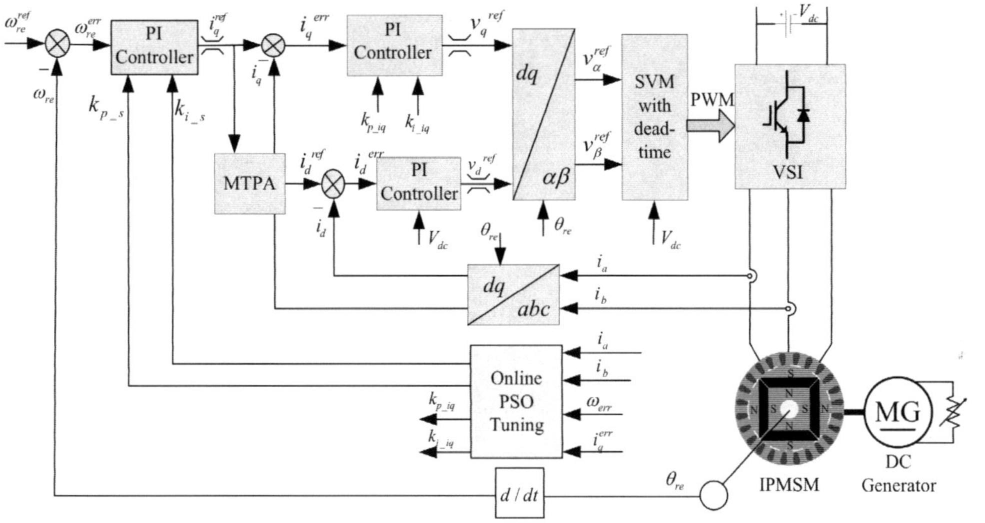

Fig.1. Block diagram of a vector-controlled PMSM drive based on PSO controller

3 Online PSO for Controller Gains Tuning

3.1. Basic principle of the PSO algorithm

Particle swarm optimization algorithm is an evolutionary computation technique developed by Kennedy and Eberhart in 1995. It finds global optimum solution in search space through the interactions of individuals in a swarm of particles. Similar to other evolutionary algorithms, the PSO algorithm firstly produces initial swarm of particles in search space. Each particle represents a candidate solution to the problem and it has its own position *X* and velocity *V*. Each row in the position matrix *X* shows each particle's position, through which we can acquire the evaluation value of the particle. At each iteration, each particle memorizes and follows the tracks of its personal best (*Pbest*) and the global best position (*Gbest*) vectors to update the velocity matrix *V*.

Known these two best positions, particles can change velocities and positions using the following rules:

$$v_{j,g}^{(t+1)} = w \cdot v_{j,g}^{(t)} + c_1 \cdot r_1 \cdot (pbest_{j,g}^{(t)} - x_{j,g}^{(t)}) + c_2 \cdot r_2 \cdot (gbest_g^{(t)} - x_{j,g}^{(t)}) \quad (6)$$

$$X^{(t+1)} = X^{(t)} + V^{(t)} \quad (7)$$

where *j*=1, 2, ... *m*; *g*=1, 2,... *n*. The superscripts *t* and *t+1* denote the time index of the current and the next iterations respectively. The parameters c_1 and c_2 are called acceleration constant which adjust the maximum step of the particle flight towards *Pbest* and *Gbest* position. Usually, parameters c_1 and c_2 are equal to 2, r_1 *and* r_2 are uniformly distributed random numbers in the interval (0, 1). Parameter w is inertia weight factor that usually decreases linearly from 0.9 to 0.4 in according to (8) over the course of the run. In this way, the algorithm can easily escape from local optimal solution in the early iteration stage as well as speed the convergence in the later iteration stage, and increase the reliability of finding the global optimal solution.

$$w^{(t)} = w_{\max} - t \cdot (w_{\max} - w_{\min}) / iter_{\max} \quad (8)$$

where $w_{\max}$ and $w_{\min}$ are the maximum and minimum values of w, and $iter_{\max}$ is the maximum iteration times.

In order to reduce the likelihood of particles leaving the search space, the value of each dimension of the velocity $v_{j,g}^{(t)}$ is clamped to the interval $[-v_g^{\max}, v_g^{\max}]$. The value of $v_g^{\max}$ is usually chosen to be

$$v_g^{\max} = k \cdot x_g^{\max}, \ 0.1 \le k \le 0.5 \quad (9)$$

where $x_g^{\max}$ is the upper bound of search region in the *g*-th dimension.

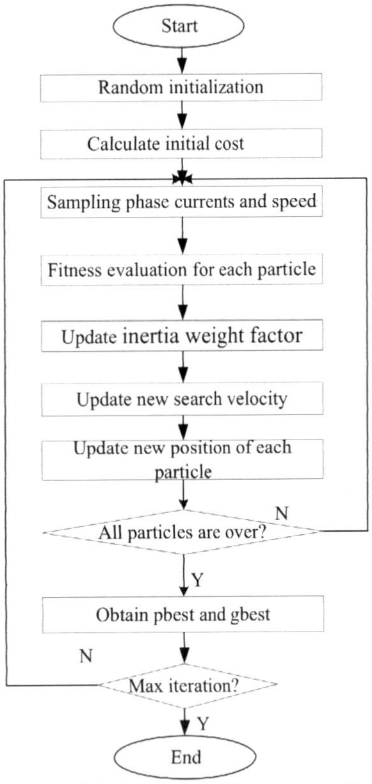

Fig.2 Online PSO flow frame within one sampling time

3.2 Definition of the Evaluation Function

To optimize the overall response of the motor drive, the fitness function is a weighted sum of several performance index based on measurements of the speed and currents outputs. It is shown in (10):

$$F(k_{p1}, k_{p2}, k_{i1}, k_{i2}) = \sum_{i=1}^{3} (a_i \cdot f_i) \quad (10)$$

$$f_1 = (\omega_{re}(k) - \omega_{re}(k-1))^2 \quad (11)$$

$$f_2 = (\omega_{re}(k) - \omega_{re}^{ref})^2 \quad (12)$$

$$f_3 = ((i_a(k) - i_a^{ref}(k))^2 + (i_b(k) - i_b^{ref}(k))^2 + ((i_c(k) - \hat{i}_c(k))^2] \quad (13)$$

and a_i represent positive weights, and f_i are three performance indices defined in the following:

1) f_1, speed transient response index

2) f_2, speed steady index

3) f_3, current reference oscillations for constant speed reference.

ω_{re}, i_a, i_b, i_c are speed and phase current measurement values, respectively, from IPMSM output; ω_{re}^{ref}, i_a^{ref}, i_b^{ref}, i_c^{ref} are speed and phase current references; when IPMSM is in transient state, $a_i = [0.6, 0.1, 0.3]$,when IPMSM nearly reach steady state, $a_i = [0.1, 0.6, 0.3]$.

4 Simulation Results

The block diagram of the drive system used for the simulation is shown in Fig.1 The flow chart of online PSO of controllers is shown in Fig. 2. The simulation model includes the encoder, sensor and control input nonlinearities. The controllers are in discrete time with the Euler method, to make the simulation to mimic the real experiment closely. The controller sampling time is chosen as 100 μsec, which is the same as the controller sampling time of real experiment.

IPMSM parameters are presented in TABLE I, and resistance can be changed from 5.8 to 7.8 for testing the robustness of proposed method. Initial swarm size of particles is 30, and each particle has four variables ($k_{p\_iq}, k_{i\_iq}, k_{p\_s}, k_{i\_s}$) representing its position vectors in search space. After 30 iterations, the results of *Gbest* would be updated as optimized outputs.

TABLE I: PARAMETERS of IPMSM

Phase resistance, R_s	5.8-7.8Ω
d-axis inductance L_d	0.0448H
q-axis inductance L_q	0.1027H
Rated voltage	230V
Moment of inertia	0.001329 $kg \cdot m^2$
Rated speed	4000 rpm
Permanent magnet flux	0.533 wb
Pole pair	2

Fig.3 shows the change of fitness value of PSO for IPMSM drives and its decline rate is very fast. In order to test the proposed method in this paper, two different controllers were applied on IPMSM drive, and comparison results between conventional PI controllers and online PSO-PI controllers for IPMSM drives are presented in Fig.4 and Fig.5. Where (a) is i_d current response curve; (b) is i_q current response curve; (c) is speed response curve; (d) is torque response curve; (e) are three phase current curves. These results clearly show the effectiveness of the PSO method proposed in this paper.

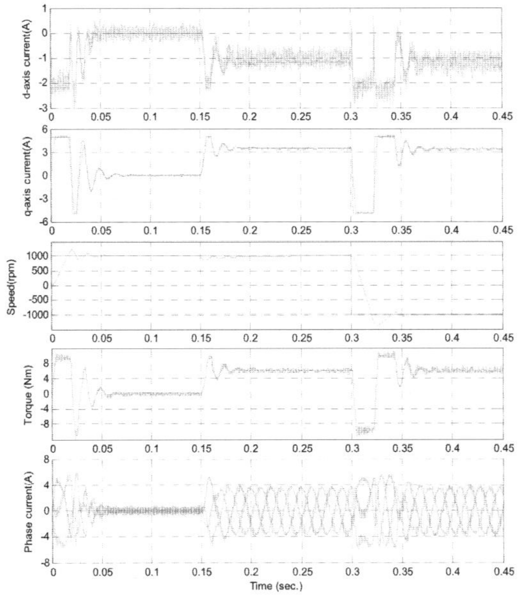

Fig.4 Results of IPMSM drive with conventional PI controllers

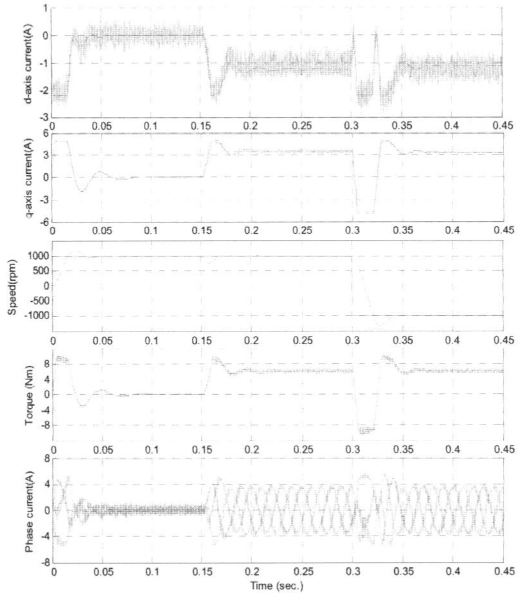

Fig.5 Results of IPMSM drive with online PSO-PI controllers

5 Conclusion

In this paper, a novel online particle swarm optimization method is proposed to design speed and current PI controllers of vector controlled interior permanent magnet synchronous motor drives, taking into account stator resistance variation. The speed and current controllers are optimized online as PSO update is carried out in each sampling cycle for the position and velocity of particle which represent the performance of the controllers. Furthermore, optimization process does not rely on the preciseness of IPMSM model excessively, because when stator resistance changes within 15%, IPMSM drive system still has a good dynamic and robustness. In addition, the dynamic and steady-state performance of the drive system has been improved by optimization of speed and current controller at the same time.

References

1. Shin-Hung Chang, Shin-Hung Chang, "Self-tuning gains of PI controllers for current control in PMSM," *IEEE 5th Conference on Industrial Electronics and Applications*, pp.1282-1286, (2010)
2. Muhammed Fazlur Rahman, Md. Enamul Haque, "Problems associated with the direct torque control of an Interior permanent-magnet synchronous motor drive and their remedies," *IEEE Trans. on industrial electronics,* vol.51, no. 4, pp. 799-808, 2004.
3. Wenxin Liu, Li Liu, Il-Yop Chung, David A. Cartes, "Real-time particle swarm optimization based parameter identification applied to permanent magnet synchronous machine," *Elsevier, Applied Soft Computing*, pp. 2556–2564, 2011.
4. Kennedy J, Eberhart RC. ''Particle swarm optimization,'' in *Proceedings of International Conference on Neural Network (ICNN)*, Vol. 4, pp. 1942–1948 Perth, Australia 1995.

Cross-site scripting attacks procedure and Prevention Strategies

Xijun Wang[1,a], Weigang Zhang[1]

[1]*Xuchang Vocational and Technical College, Xinxing Road Xuchang City No. 4336, China*

Abstract. Cross-site scripting attacks and defense has been the site of attack and defense is an important issue, this paper, the definition of cross-site scripting attacks, according to the current understanding of the chaos on the cross-site scripting, analyzes the causes and harm cross-site scripting attacks formation of attacks XXS complete process XSS attacks made a comprehensive analysis, and then for the web program includes Mobility there are cross-site scripting filter laxity given from ordinary users browse the web and web application developers two the defense cross-site scripting attacks effective strategy.

1 Introduction

Cross-site scripting attacks and defense has been an important topic of Defense website, along with attack technology advances, many theories about cross-site scripting attacks cannot meet the needs of today's attack and defense, but also because of the awareness of cross-site scripting confusion, resulting in a lot of programs now include dynamic network there are cross-site scripting filter lax problems.

1.1 Definition of cross-site scripting attacks

XSS (Cross-Site Script, abbreviated XSS) [1] refers to the network of human intruders insert some malicious script code to a dynamic web page, when a user opens these pages, the browser will automatically download malicious code embedded in Web pages inside the malicious script code will be interpreted, an attacker can bypass the document Object model (DOM) of security restrictions, to steal Cookies information, change account settings Web application, Web-mailing worm spread a series of malicious actions, so as to achieve illegal special purpose steal user information.

1.2 Harm cross-site scripting attacks

Cross-site scripting attack is a passive attack in many large portals (such as Yahoo, Microsoft, EBay) are on the existence of such loopholes, it is very popular in hacker circles, the danger should not be overlooked.

Intruder exploit XSS vulnerabilities, with the vulnerable Web sites to attack other visit the related website users, steal user browsing session such as user names and passwords (which may be included in the cookie) sensitive information, execute malicious code inserted by the user hanging horse attack. XSS vulnerabilities can also be used by attackers to tamper pages, but in most cases in order to maximize the economic benefits, the attacker does not directly tampering.

XSS attacks can collect user information [2], the intruder by inserting JavaScript in vulnerable Web applications,VBScript, ActiveX or Flash user information such as deception, once succeeded, you can easily obtain a user account illegally, Further modify user settings, false advertising, view the host information.

XSS attacks can also lead to a denial of service attack. An attacker could use a script to run on a recurring, so a simple message box is enough to cause a Dos attack, so that site administrators keep track of.

XSS attacks can also be combined with browser vulnerabilities, modify the system configuration, view system files, or install a backdoor Trojan and execute malicious code.

1.3 Cross-site scripting attack causes

Web server lack of information entered by the user verify the legitimacy or verification is not enough, and the information entered by the user will be returned to the client is the main cross-site scripting vulnerabilities generation [3], there are two points:

Web Server allows users to irrelevant pages in the table or in the edit box and enter the required characters. As edit box requires the user to enter the phone number, which is valid for a smart digital, any other form of symbols are illegal if the programmer did not verify the validity of this, there is a loophole.

Web server to the user's input is stored and returned to the end user on the page display, the designers did not echo the illegal character filtering or re-encoded. If an attacker entered a seemingly normal but hidden content XSS code, such as adding new users on the Web site, to enhance their permissions, end user browser will accept and execute the code, will be on the Web server Web applications and cause unpredictable hazards.

[a] Corresponding author: rwgwz@163.com

DOI: 10.1051/matecconf/20166103001

2 Cross-site scripting attacks

Belonging to the passive XSS attacks [4]. Or by e-mail when an attacker to construct a cross-site page, use script, <IMG>, <IFRAME> and other ways to lure users to browse this seemingly normal malicious links or visit the page, the trigger for the attack site is http request attacker returns to the user a web page that contains malicious code, engaged in malicious behavior can be achieved posing issued a document capture multiple attacks object permissions. If the embedded script with an additional capability to interact with other legitimate Web server, an attacker can use it to send an unauthorized request, using the data on the legitimate server. The procedure is shown in Fig.

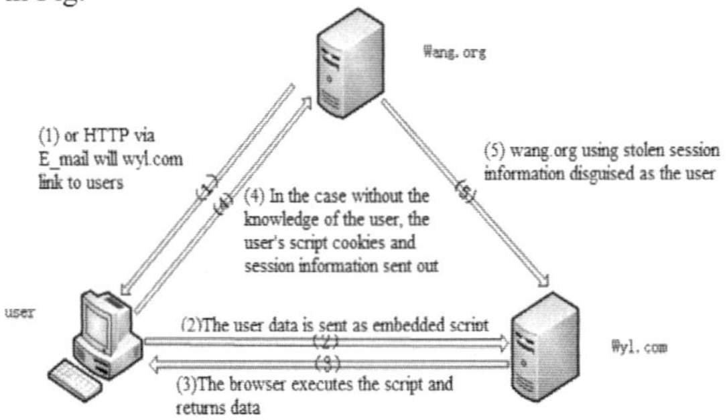

Figure 1. cross-site scripting attacks

In Figure 1, an attacker would first of a seemingly normal HTTP link or by E-mail wyl.com link to the user, the user clicks on the link, the browser will execute script in the user without the knowledge of, and the user cookie and session information is sent to the site wang.org attacker making, then, the attacker will be able to view the private information contained within the cookie and use the information to steal session came disguised as the user with real wyl.com conversation.

3 Cross-site scripting attacks step

A successful cross-site scripting attacks there are two necessary conditions: first, the need cross-site scripting vulnerability web applications; Second, the user needs to click on a link or visit a page [5]. Specific attack process is as follows.

3.1 Found XSS vulnerabilities

At present, people browse the pages are all based on HTML (HTML) created, XSS attack is through to the HTML script code artificially inject malicious scripts, the script is specified HTML tags: <script > </ script>, in the absence of filtering character case, just keep complete and error-free script tag can trigger XSS.

User to display an image in the page, you need to use <img>tag,such as<imgsrc= http://127.0.0.1/wyl_xss.gif">, this time through the browser's task is to explain the img tag to access the src attribute URL address assignment and display pictures, but browser src attribute value assigned does not validate correctly, which resulted in an opportunity for the attacker is known, javascript have a pseudo-URL protocol that allows users to use "javascript:" plus any form of javascript code when the browser loaded this URL, automatically execute the javascript code to form a classic XSS case:<Img src = "javascript: alert ('XSS');">

3.2 Injection of malicious code

Find pages containing XSS vulnerability, the attacker can start trying to write and inject malicious code. The purpose is to inject malicious code when the cheater is visited pages containing this malicious code, attackers can achieve the purpose.

Get cookie typical code:javascript:window.location=' http: //www.xcitc.edu.cn/cgi-bin/cookie.cgi?'+ document. cookie. In which, window.location role is to make the pag e automatically jump to another page; the role is to read d ocument.cookie cookie, users browse the page, the user's cookie will be read and passed as a parameter to http://w ww.xcitc.edu.cn/cgi-bin/cookie.cgi,after displaying the co ntents of the cookie. These malicious code, will be issued a visitor's information to a remote attacker Cookie hands, or to enhance the user's permission to upload arbitrary fil es and other malicious actions.

If a site for receiving user input pages of <,>, ', "special characters filtering, we need to use coded form into the browser's default .IE uses UNICODE coding, coding scripts can & # ASCII way to write the XSS transcoding support decimal and hexadecimal form, and is assigned value for the property, codes are as follows:javascript: window.location = & # x31http:? //www.xcitc.edu.cn/cgi-bin/cookie.cgi & # x31 + document.cookie

3.3 To deceive the user access

When the attacker malicious code into a web page, the next thing to do is to trick target users to access the malicious page, "indirectly" by the target user to complete the attacker's purpose.

4 Cross-site scripting attacks defense

XSS attack main goal is not the Web server itself, but the logged-on user websites for XSS attack defense, the need for defense from ordinary users browse the web and WEB application developers in two ways.

4.1 Ordinary person browsing the web

Click the link to pay special attention to the site, e-mail or instant messaging software, rice open suspicious links, especially when they appear to contain a HTML script code but cannot be easily opened.

For the XSS vulnerability, no kind of web browser has obvious security benefits, in order to get more security, you can install some browser plug-ins, users should try to access the regular large-scale portal, to avoid access may be a problem site.

There is no absolute safe and effective in the world, the average user should try to avoid visiting suspicious websites, such as the provision of information and hacking tools to crack software, indecent photographs of the site free of charge. These types of sites tend to take advantage of the current browser vulnerabilities to compromise the operating system.

4.2 Web application designer

Should concentrate on submission to all users a reliable input validation. These include submission URL, query keywords, post data, allowing only the use of legitimate characters only accept within a predetermined length range, the characters in the appropriate format, blocking, filtering, or ignore any other things.

Using the session tag (session tokens), the verification code technology to protect users of all confidential information, to avoid being robot automatically be executed or perform other illegal websites.

When developing a web application must support users to submit HTML script code, avoid application security by devastating decline, we must use the relevant technology protected when developing web sites, such as received HTML script code to confirm the contents are properly formatted, contains only minimized, secure tag (absolutely no JavaScript), to remove all references to remote content (especially CSS style sheets and JavaScript) and the like.

To make use of filtering and limiting the input method of XXS attack prevention [6], for all dynamic pages input and output should be encoded, so the maximum to avoid cross-site scripting attacks.

Has been completed for the site, you can filter all the special characters input from the web page by adding a dynamic IIS components. This component can usually intercept the Request goal from ASP pages, the table, cookie, request content strings and procedures for testing, which can effectively prevent cross-site scripting attacks.

5 Conclusion

Cross-site scripting attacks, while a passive attack, but because there are many sites on this loophole, likely to cause leaks and illegal data server to steal, the danger is very large, as long as ordinary users browse the web carefully, WEB applications developers tight design, this attack is difficult to achieve.

References

1. Wang Z.C.,Tonghua Teachers College, 08, P36-37(2012)
2. Cheng J.Q. Zhang Y.Q., Computer Engineering, 06, P49-51(2010)
3. Liu J., Zhang X.Z., Computer Development & Applications, 10, P28-30 (2011)
4. Sun W.Y., Zhang Y.R., Neijiang Science and Technology, 05, P33-34(2010)
5. Li J.M., Guo Y.L., network security and hacker attack and defense, P412-413(Electronic Industry Press,2010)
6. Deng W.W., information network security, 09, P23-25(2007)

MATEC Web of Conferences **61**, 03002 (2016) DOI: 10.1051/matecconf/20166103002
APOP2016

Keyword-based Ciphertext Search Algorithm under Cloud Storage

Ren Xunyi[1,a] ,Yan Shiyang[1]

[1] College of Computer Nanjing University of Posts and Telecommunications, 210003 Nan Jing, China

Abstract. With the development of network storage services,cloud storage have the advantage of high scalability,inexpensive,without access limit and easy to manage. These advantages make more and more small or medium enterprises choose to outsource large quantities of data to a third party. This way can make lots of small and medium enterprises get rid of costs of construction and maintenance,so it has broad market prospects. But now lots of cloud storage service providers can not protect data security.This result leakage of user data ,so many users have to use traditional storage method.This has become one of the important factors that hinder the development of cloud storage. In this article,establishing keyword index by extracting keywords from ciphertext data. After that,encrypted data and the encrypted index upload cloud server together.User get related ciphertext by searching encrypted index,so it can response data leakage problem.

1 Introduction

Cloud storage has the advantages of cost-effective,high scalability,inexpensive,without access limit and easy to manage.These advantages can enable small businesses spend less cost getting rid of the high cost of construction of the storage system, which greatly reduce their storage costs.Therefore, cloud storage has a high market prospect.The behavior of confidential data is saved to third parties caused the company's great worry,which limit the development of cloud storage.Many enterprises have to adopt the traditional storage model.This has become an important factor of restricting the development of cloud storage.

Under these conditions,encrypted storage produced.In the conditions of large amount of ciphertext data stored in server-side ,users need to put all of the ciphertext data downloaded to the local and retrieve after decrypting data. Although this strategy protects the privacy of data, but users need to waste a lot of resources, consume large amounts of network bandwidth, which also poses a serious obstacle to the development of cloud storage.At the same time, a huge cloud server computing capacity were idle, and became a simple data pool.In order to exert cloud server computing power and huge storage resources, keyword-based search technologies emerged.[1]After encrypting ,client upload documents and inverted index to the cloud server.When searching, users only need to enter a keyword.Cloud Server can search for relevant ciphertext file and then return to the client.Local users can use the key to decrypt the ciphertext file.This program not only protect the security of data,but also use computing power and large storage resources.

2 The Method of Ciphertext Search

First,The user selects a file and extracts the keywords in the file.Then, the keyword and the document identification combined to form an inverted index.Files use symmetric key encryption algorithm to encrypt and inverted indices also use a pseudo-random function to encrypt. Ciphertext and the encrypted indices upload to the server with storage.When the user queries, client hosts use the search algorithm to process the keyword to a query pointer and submit to the server.The server uses the pointer to call the user's encrypted search index.Server use the user's encrypted index to find the corresponding ciphertext and returns to the client, the client can use the key to decrypt the ciphertext.

With this idea, the cloud server will not get any useful information about files.Server just know the ciphertext, encrypted index, and search pointer and can not get any effective plaintext and keyword information.Therefore, using this search method will give the ability of transparently querying to cloud server .

3 Implementation Model of Ciphertext Search Method

This article discusses the method involving two entities.The one is the owner of the confidential data.They want the data to be stored in the cloud server, they need to protect the confidentiality of data, such

[a] Corresponding author: renxy@njupt.edu.cn

entities are called users.The other is a provider of cloud storage services.They offer storage interface for users, and perform some operations of retrieving on the data , such entities are called servers.

From the foregoing, in order to protect the confidentiality of data, the cloud server can not get any useful information throughout the search process.All data processing on locally, including file encryption, encryption of the index,and the process of generating search pointer by keywords.

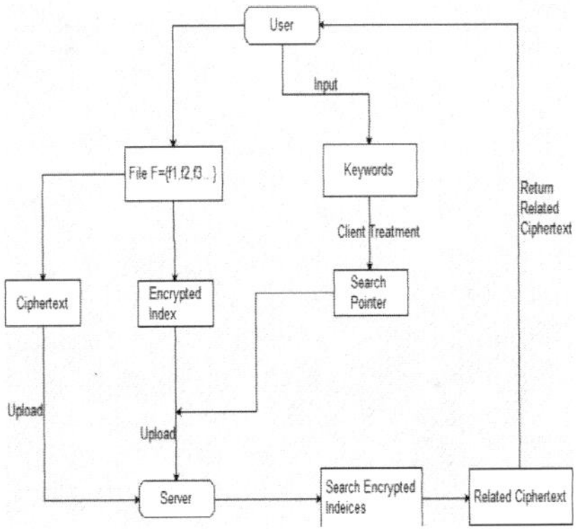

Figure 1. Search model under the cloud

From the figure above,the users select a file to be stored with the client and pretreat the file.Then,users upload this treated file.Pretreatment is divided into two portions.The one is to use a symmetric key encryption algorithm to encrypt the plaintext.The other is to encrypt indices composed by a set of keywords and get encrypted index. After that,users upload encrypted files and encrypted indices to server and store.It is not necessary for users to care store details. When the user queries, client hosts use the search algorithm to process the keyword to a query pointer and submit to the server.The server uses the pointer to call the user's encrypted search index.Server use the user's encrypted index to find the corresponding ciphertext and returns to the client, the client can use the key to decrypt the ciphertext.

The key to the searchable encryption algorithm is to encrypt file index.This programme use pretreatment method.First,we extract a few keywords from each file you want to upload, and then use the symmetric key encryption algorithm to encrypt the plaintext.Each keyword contain files generate list called keyword list.The node structure of linked list include file identification, the location of the previous node and the location of last node.All keyword lists consist of index called inverted index[3].Using pseudo-random function encrypt inverted index and get encrypted index.This method make server can not get effective information.Then,these encrypted indices are stored in random locations on the array(A_s).The first node for each keyword list is stored uniformly in the dictionary called(T_s).Encryption array and dictionary are stored on the server and the server can not get effective information.When the user queries, client hosts use the search algorithm to process the keyword to a query pointer and submit to the server.The server uses the pointer to call the user's encrypted search index.Server use the user's encrypted index to find the corresponding ciphertext and returns to the client.By these steps can complete the search process[4].

4 The Design of Algorithm

In this article,we use Random oracle to generate search index and use Pseudo-random function to generate encrypted index. We choose $\{0,1\}^m \rightarrow \{0,1\}^m$ as Random oracle,and use H_1、H_2 to express respectively.We choose $\{0,1\}^n \times \{0,1\}^s \rightarrow \{0,1\}^s$ as Pseudo-random function ,and use F、G、P to express.Using η to express search pointer.In this program,we design Symmetric Searchable Encryption(SSE).[2]Algorithm provides file encryption, key generation, encryption index,pointer search, file decryption and other functions.

This algorithm can be expressed as $SSE = \{Enc, Gen, IndEnc, IndGen, Search, Dec\}$.

Enc represent the algorithm of encrypting files.Client operate the algorithm to encrypt plaintext file. Gen represent the algorithm of generating key,and generate the key of symmetric key encryption algorithm and pseudo-random function. $IndEnc$ represent the algorithm of encrypting keyword and generate encrypted index.Client perform this algorithm to generate encrypted index and upload encrypted index and ciphertext to server. $IndGen$ represent the algorithm of generating search pointer.Client perform this algorithm to transform keyword to search pointer,and upload to server. $Search$ represent the algorithm of searching encrypted indices. This algorithm run in server,and server can find related ciphertext when client submit search pointer. Dec represent the algorithm of decryption.Client decrypt ciphertext return from server by using key.

4.1 The Generation of Key

The process of generating key: $IndEnc$:The algorithm of encrypting keyword,three randomly selected m-bit key length string ,they are k_1, k_2, k_3 as the key of pseudo-random function. $Gen(s)$:Given a system security parameter named s as input parameter and generating $K_4 = Gen(s)$ as key of symmetric encryption algorithm.Then,outputting $K = \{K_1, K_2, K_3, K_4\}$ .The key generating by this algorithm is stored in client,and encrypt files and keywords.

4.2 The Encryption of Files and Indices

Files encryption are to encrypt the plaintext document by using symmetric key encryption algorithm.[6]

The process of index encryption: $Enc(K, F, \lambda)$: K is key, $F = \{f_1, f_2, f_3.....\}$ is files collection , λ is inverted index, A_s is the array that store encrypted index, T_s is the dictionary that store first node of keyword list, V is the collection of keywords, $\$V$ is the number of keywords, L_f is the list consist of all keywords by file f , $\$L_f$ is the number of keywords in file f , L_v is The file list consist of keyword v , $\$f_{v_i}$ is the number of files including v .

(1) Initialize array A_s and dictionary T_s.

(2)Perform following operations for each $v_i \in V$:(a) Create linked list named L_{v_i} , the node of linked list are $(N_{i,1}, N_{i,2} .... N_{i,\$f_i})$,including $\$f_{v_i}$ nodes.These nodes are stored in random position of array A_s .The structure of node is $N_{i,j} := (< loc_s(N_{i-1,j}), loc_s(N_{i+1,j}), id_i > \oplus H_1(P_{K_1}(v)))$.
$loc_s(N_{i-1,j})$ represent the front location of $N_{i,j}$, $loc_s(N_{i+1,j})$ represent the back location of $N_{i,j}$, id_i represent file identification of node.(b) The first node of every linked list are stored in T_s.The structure of node is $T_s := < loc_s(N_{i,1}) > \oplus G_{k_2}(v)$, $loc_s(N_{i,1})$ represent location of first node of every linked list.

(3) These free nodes of A_s and T_s are filled with random strings.

(4) Each f_i is encrypted and generate $c_i = Enc(K, f_i)$.

(5) The algorithm output $\mu := (A_s, T_s)$ and $c = \{c_1, c_2, ... c_{\$F}\}$.

Encrypted files and encrypted indices are uploaded to cloud server.Specific storage operations are done by cloud server, the client is not involved in this work.[7]

4.3 Implement of Search

When user want to search files,they upload keyword and client transform keyword to η by using *IndEnc* algorithm .Server use search index to read encrypted index. Then,server use *Search*() algorithm to search encrypted index,and get related ciphertext collection.These ciphertext all return to client,and client decrypt these ciphertext to plaintext.[8-10]This completes the whole search process.Specific search process is as follows:

(1) $IndEnc(K, v)$: Input v and K ,output $\eta := \{F_{k_1}(v), G_{k_2}(v), P_{k_3}(v)\}$

(2) $Search(c, \eta, \mu)$:Input μ , η ,after following these steps:(a)Find the location of v_i in T_s,this location name α_i .(b) Using α_i find the coordinate of keyword in A_s and find N_i,then find id_i and find related ciphertext.(c)Using $I_{v_i} = \{id_1, id_2, ...id_m\}$ represent searched file mark collection.Then,finding every mark's corresponding ciphertext and outputting $\{c_i\}$.

(3) $Dec(K, c)$:Input K and $c = \{c_1, c_2, ...c_m\}$,using $f_i = SSE.Dec(K, c_i)$ to get plaintext.

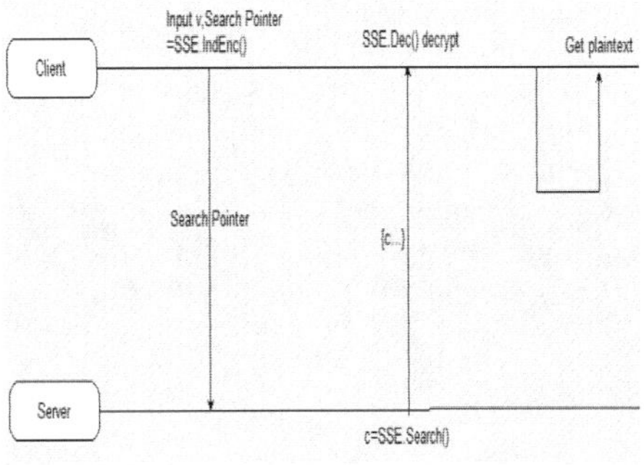

Figure 2. The whole search process of chart

5 Conclusion

Through the above analysis,programme extract keywords from user's files,and transform plaintext files and keywords to ciphertext and encrypted index respectively. These two steps can be done on the client.Ciphertext and encrypted indices are stored on cloud server.When the user queries, client hosts use the search algorithm to process the keyword to a query pointer and submit to the server.The server uses the pointer to call the user's encrypted search index.Server use the user's encrypted index to find the corresponding ciphertext and returns to the client, the client can use the key to decrypt the ciphertext.

References:

1. Fei X, Liu C Y, Fang B X, et al. Research on ciphertext search for the cloud environment[J]. Journal on Communications, 2013, 34(7):143-153.
2. KAMARA S,PAPAMANTHOU C,ROEDER T.Dynamic searchable symmetric encryption[A].Proceedings of the 2012 ACM conference on Computer and Communications Security(CCS 12)[C].New York,NY,USA,2012.965-976
3. Liu A F, An effective and Dynamic Ciphertext Search Method for Cloud Storage[D],2013
4. Neng-Hai Y U, Hao Z, Jia-Jia X U, et al. Review of Cloud Computing Security[J]. Tien Tzu Hsueh Pao/acta Electronica Sinica, 2013, 41(2):371-381.

5. Qi W U, Wan C X. Multi-user Conjunctive Keyword Search Scheme over Ciphertext[J]. Computer Science, 2013.
6. Yuan F,Zhou C,Zhang Q L, et al. Ciphertext search method based on word-for-word indexing: CN, CN 101937464 B[P]. 2012.
7. Qi W U, Wan C X. Multi-user Conjunctive Keyword Search Scheme over Ciphertext[J]. Computer Science, 2013.
8. Borisov N, Mitra S. Restricted queries over an encrypted index with applications to regulatory compliance[C]// Proceedings of the 6th international conference on Applied cryptography and network security. Springer-Verlag, 2008:373-391.
9. Peng L, Li R, Wang H, et al. An Encrypted Index Mechanism in Ciphertext Retrieval System[C]// Proceedings of the 2011 Eighth Web Information Systems and Applications Conference. IEEE Computer Society, 2011:131-136.
10. Kang H C, Mun D W, Kim Y H. Method for indexing encrypted column: US, US8281153[P]. 2012.

MATEC Web of Conferences **61**, 03003 (2016) DOI: 10.1051/matecconf/20166103003
APOP2016

A Review of Virtual Machine Attack Based on Xen

Ren xun-yi[1,a], Zhou yu-qi[1]
[1] College of Computer Nanjing University of Posts and Telecommunications, 210023 NanJing, China

Abstract: Virtualization technology as the foundation of cloud computing gets more and more attention because the cloud computing has been widely used. Analyzing the threat with the security of virtual machine and summarizing attack about virtual machine based on XEN to predict visible security hidden recently. Base on this paper can provide a reference for the further research on the security of virtual machine.

1 XEN virtual machine security character

Virtual machine infrastructure is consist of VMM, VM and Domian0, then discussing each of these in the following sections from function that undertaking in the XEN platform. The paragraph show that in terms of security to analyze excising problems

1.1 VMM

VMM belong to the software layer which locating between computer hardware and operating system and running in the privilege level[1]. It is in charge of managing and isolating VM which running at the higher layer to offer security and department running environment. Meanwhile, VMM are valid for a virtual environment that is irrelevant to the real hardware with virtualization, such as monitor, hard disk, CPU, memory, network hard etc[1,3]. In conclusion, VMM is not only a multitasking management tools but also a solution with high security and strong reliability. On the basis of the location of VMM in the computer can divide it into three kinds of modes: independent monitoring mode, host mode and mixed mode. VMM is also called XEN Hypervisor or XEN and VM is called Domain in the XEN[3].

1.2 Domain0

A special virtual machine exist in the XEN called Domain0 by reason of mixed mode used in the XEN.Domian0 is defined as privileged domain, it supervise other Domain(called DomainU) to provide virtual source service, particularly the other Domain access to I/O devices[3].

Domain0 is unique as assistant of XEN, it create with XEN running and become the first Domain which running in the XEN. Domain0. Dom0 possess Native Drive and have privilege to access hardware device. By using of control interface to control other Domain. Control Panel running in the Domain0 can manage creation, destroy, configuration and migration of other Domain. While Device Manager is responsible for initialization of device driver and visit of manage device. Furthermore, Domain0 can empower other Domain to access device, all of these Domain which was equipped with power to access real hardware device by using of Native Device. Thus it can be seen that VMM and Domain0 become the focus of attack object, it is easy to acquire information of DomainU once attacker access to the highest authority of virtual machine.

1.3 VM

VM possess independent ISA structure and veritable hardware device which simulated in the hardware platform. Different operating system can execute in each virtual hardware system, called Guest OS, each one of them access real hardware source via VMM[1].

2 VM attack

The previous section introduced XEN virtualization simply, and point out key character of VM. In this section will introduce all kind of VM attacks at present.

On account of the special of virtual system, it exist more attack point than traditional computer. Mode of VM attack could be divided into four types: gust to host, VM to VMM, VM to VM and outside-attack of VMM.

2.1 VM to Domain0

Domain0 possess the privilege of managing other VM, Attack of VM to Domain mainly attack host via VM to acquire administration of the host. The most famous of which is escape attack of VM, it utilize loophole of software running in the VM to conquer the field of attack or control VM system. VM via leakage of software to bypass VMM monitor, by this way hacker can access Dom0 directly and acquire privilege of

[a] Corresponding author:renxy@njupt.edu.cn

Domian0.Then, attacker can control all the VM in this system.

In 2015, XEN exposed a loophole named Dome Breaking and numbered XSA-148/CVE-2015-7835. This loophole stared to arise in version 3.4 and in version 4.6 have been impacted. It is easy for attacker to come true escape when they utilize a VM running in the mode of PV. Furthermore, attacker can take control of XEN Hypervisor, Domain0 and other VM running in the single physical machine. Dome Breaking exist in memory management facility. On security grounds, XEN set page table page as not be writable. So, PV DomU can not modify its page table directly, it only via Hypercalls to apply for altering page table. XEN Hypervisor will instead of PV DomU to modify page table while request checked off by it. However, XEN provide a quick selection that is to say request is able to skip strict process while the request be verified security by XEN Hypervisor. This strategy cause serious problem that quick update logic allow new Page Directory Table Entry(PDE)to carry the flag of _PAGE_PSE and _PAGE_RW when XEN Hypervisor update item of page table in Page Directory Table(PDT)[4]..

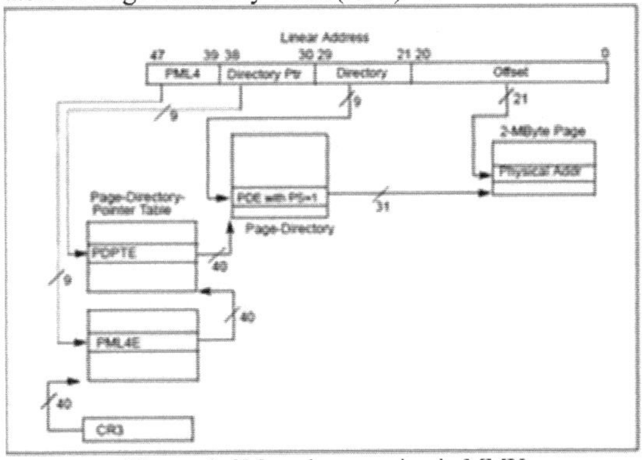

Figure 1. 2M paging mapping in MMU

Flag of _PAGE_PSE in PDE can lead MMU to allow process to use page in the size of 2M.In this case, whether allow to execute quick update operation applying from PDE request will decided by o12e12e_has_changed(o12e, n12e, _PAGE_PRESENT) routine. This function check that physical page frame number in o12e and n12e is consistent with _PAGE_PRESENT flag, if all of them is match well, updating PDE directly via n12e. At the moment, n12e set the flag of _PAGE_PSE which allow process to access physical memory in the size of 2M requested by n12e rather than 4KB stipulated by XEN, superfluous memory may not belong to this VM. Meanwhile, if o12e include the flag of _PAGE_RW at the same time, aforesaid memory can be written willfully. Analyzing form the point of loophole usage, by means of set the flag of _PAGE_PSE and _PAGE_RW in the PDE, users can push through limitation of 4KB and access memory in the size of 2M. In this condition, attacker can read Page Table maliciously which save in 2M memory aforementioned. Using this method to find breaking the limitation of reading Page Tables only for VM in XEN, attacker can read and write physical memory willfully. Finally, attackers can bypass all the security mechanism and execute malicious code in the context of XEN Hypervisor and Dom0[4].

2.2 Outside attack of VMM

Nowadays, there are two major types of outside attack to VMM, Rootkit attack based on VMM (VMBR) and malicious code.

1) VMBR

Attacker via Rootkit to hide and acquire authority of root, they will leave backdoor next. VMBR write program code into memory and running before VMM start. All of VM will be controlled once program running success. Fairly well known VMBR is Blue pill at present.

2) Malicious code

Hacker using long-distance attack to implant malicious code and remote management technology are mostly using HTTP/HTTPS in VM system. On account of these, attacker can exploit loophole of HTTP to implant malicious code. Such as XSS (cross-site scripting) loophole in API HTTP, attack VM by using of browser to execute malicious script[7,8].

2.3 Attack between two VM

VM to VM mode via access common source to attack, covert channel is famous in all of attack ways which implant code into physical machine because of error in process, memory and other information.

Attack between two VM is called cross-VM attack. In this mode, mainly research mode is coresident that means two VM share common source of physical machine. Attacker can obtain information from cache, CPU and shared memory.

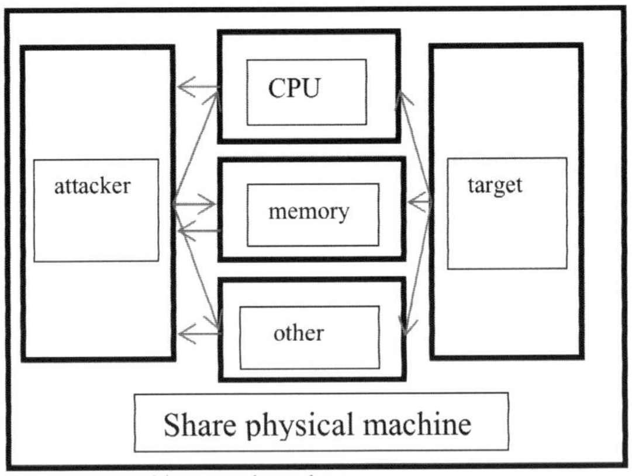

Figure 2. coresident attack mode

Theory of covert channel: Running program with changing in the time, energy and memory. Because the program is different, take time, memory size will be different too. With the number of program running increasing, the leakage of possibility of information increasing too, such as usage of shared memory, CPU utilization, IP address and other information. Different program need different running time, CPU occupancy and memory utilization, large program is able to make the shortage of CPU resource, CPU will always in the condition of high load operation. Attackers can collect

DOI: 10.1051/matecconf/20166103003

changing of utilization to analyse data, transfer regular data through covert channel to acquire valuable information[6].

Take XEN for example, Hypervisor save a table called M2P, it store a relation of machine addresses to pseudo physical addresses. Each Domain can read M2P and via HYPERVISOR_MMU_UPDATE to update table item in the scope of its address. This feather was proved to be used in creating convert channel between two VM. Communication on both of the VM share a structure data, one of the first domain as a flag (two side know each other), and the last domain contains hidden message of actual transmission. In the process of actual transmission, one side update shared data structures to list item in the range of address, the other side traverse M2P items to find opposite side flag, then, according to the excursion in the structure to find hidden message.

CPU load changes can also be used to construct hidden channels. Assume two VCPU in the two VM is mapped to the same physical CPU (or CPU kernel) and running same task, VM as a hidden message receiver execute this task always, as transfer via execute and not execute this task to acquire temporal variation in the receiver. Both of side can agree that the sender does not perform this task transfer 0 bit, and execute this task transfer 1bit. So the receiver analyse bit string form opposite side to grasp utilization of CPU[5].

3 Security of VM

Security of VM can be divided into four types:

1) Security of Hypervisor, ensure hypervisor security, prevent attacker take control of Hypervisor, isolate source between VM, assure security of data storage and confidentially and integrity of machine communication.

2) Trusted computing, establish a credible security domain or trusted computing platform to ensure upper security.

3) Access control, improve access control mechanism constantly.

4) Security application, via security of sand box, intrusion detection and honeypot system for protection.

4 Conclusion

This paper talk about security of XEN mainly and introduce attack mode of VM at present, guest to host, VM to VMM, VM to VM and outside attack to VM. On account of para-virtualization, XEN include two privilege level, VMM and Domain. It is easy for attacker to control other VM and access memory, once they acquire management of VMM and Domain. Attack VMM and Domian0 more frequent than other, second is covert channel or system loophole. Analyse usage of memory or load of CPU and collect running information of other VM. So, when consider to attack a VM, firstly attack host to acquire management of it, this way is easy and difficult. In XEN, possible loophole still threat security, what can we do is improve it in the use constantly.

Acknowledgement

Supported by National Natural Science Foundation of China (61073188), China Postdoctoral Science Foundation (20100471355), the Talent Project of Nanjing University of Posts and Telecommunications (NY208006)

References

1. Lei Shi, Deqing Zou,Hai Jin, M.XEN virtualzation,Wuhan: Huazhong University of Science and Technology Press (2009)
2. Zhongyuan Qin, Risheng Sheng, Qunfang Zhang, Yuxing Di, J. Summarize of security of VM pl. Application Research of Computers(2012)
3. Yingxu Lai, Shaolong Hu, Zheng Yang, J. Research for Security of VM pl. Journal of University of Science and Technology of China(2011)
4. http://xenbits.xen.org/xsa/advisory-148.html
5. Ke Ai, J. Discuss security of virtualization based on cloud pl. Information&Communication, 12(2015)
6. Hang Li, D. Research for security of VM pl. Xidian University(2014)
7. Olivier Heen, Christoph Neuamam, Luis Montalvo, Serge Defrance, D. Improving the Resistance to Side-channel Attacks On Cloud Storage Service pl. IEEE, 6(2012)
8. K. Suzaki, K. Iijima, T. Yagi etc, J. Memory deduplication as a threat to the guest pl. In Proceeding of the Fourth European Workshop on System Security, 1-6(2011)

The algebraic immunity and the optimal algebraic immunity functions of a class of correlation immune H Boolean functions

Huang Jinglian[a], and Wang Zhuo

School of Electrical Engineering, Northwest University for Nationalities, Lanzhou, 730030, China

Abstract. We put forward an efficient method to study the algebraic immunity of H Boolean functions with Hamming weight of $2^{n-1}+2^{n-2}$, getting the existence of the higher-order algebraic immunity functions with correlation immunity. We also prove the existing problem of the above 2-order algebraic immunity functions and the optimal algebraic immunity functions. Meanwhile, we solve the compatibility of algebraic immunity and correlation immunity. What is more, the main theoretical results are verified through the examples and are revealed to be correct. Such researches are important in cryptographic primitive designs, and have significance and role in the theory and application range of cryptosystems.

1 Introduction

The cipher security is the core of the cryptosystem, and only a cryptosystem with good security has an existing significance. Boolean functions with a variety of secure cipher properties are the key factors to design the cryptosystem with the ability to resist multiple cipher attacks and good safety performance. It is of great importance for a security cryptosystem to study some properties of Boolean functions, which make the cryptosystem resist various attacks [1], such as high algebraic degree, high nonlinearity, the strict avalanche criterion and propagation, higher-order correlation immunity and higher-order algebraic immunity. Therefore, there are some important research problems, such as the existence, the feature, the design, the construction and the count of Boolean functions with some kind of secure cryptographic property. Among them, the algebraic immunity of Boolean functions is current central issues [2~6].

The H Boolean functions have the propagation property. When we study the correlation immunity and algebraic immunity of the H Boolean functions, it is equivalent to study the compatibility of the propagation property, algebraic immunity and correlation immunity of Boolean functions. It is the basis of studying the correlation immunity and algebraic immunity of the H Boolean functions to study the correlation immunity and algebraic immunity of the H Boolean functions with the Hamming weight of $2^{n-1}+2^{n-2}$. In this paper, using the derivative and e-derivative [7~9] as research tools, we study the importance of study cryptographic properties of H Boolean function with Hamming weight of $2^{n-1}+2^{n-2}$.

2 Preliminaries

To study cryptographic properties of H Boolean functions, we proposed the concept of the e-derivative. The e-derivative and derivative are defined here as Definition 1&2.

Definition 1: The e-derivative (e-partial derivative) of n-dimensional Boolean functions $f(x)=f(x_1,x_2,\cdots,x_n)\in GF(2)^{GF(2)^n}$ for r variables $x_{i1},x_{i2},\cdots,x_{ir}$ is defined as

$$\begin{aligned}&ef(x)/e(x_{i1},x_{i2},\cdots,x_{ir})\\&=f(x_1,x_2,\cdots,x_{i1},x_{i2},\cdots,x_{ir},\cdots,x_n)\cdot\\&f(x_1,x_2,\cdots,1+x_{i1},1+x_{i2},\cdots,1+x_{ir},\cdots,x_n)\\&(1\le i\le n,1\le i_1\le i_2\le\cdots\le i_r\le n,1\le r\le n)\end{aligned}\quad(1)$$

If $r=1$, (1) turns into the e-derivative of $f(x)=f(x_1,x_2,\cdots,x_n)$ for a single variable, which is denoted by $ef(x)/ex_i$ $(i=1,2,\cdots,n)$. As a result, the simplified form below can be easily derived.

$$\begin{aligned}&ef(x)/ex_i\\&=f(x_1,x_2,\cdots,x_{i-1},1,x_{i+1},\cdots,x_n)\cdot\ (i=1,2,\cdots,n).\\&f(x_1,x_2,\cdots,x_{i-1},0,x_{i+1},\cdots,x_n)\end{aligned}$$

Definition 2: The derivative (partial derivative) of n-dimensional Boolean functions $f(x)=f(x_1,x_2,\cdots,x_n)\in GF(2)^{GF(2)^n}$ for r variables $x_{i1},x_{i2},\cdots,x_{ir}$ is defined as

$$\begin{aligned}&\partial f(x)/\partial(x_{i1},x_{i2},\cdots,x_{ir})\\&=f(x_1,x_2,\cdots,x_{i1},x_{i2},\cdots,x_{ir},\cdots,x_n)+\\&f(x_1,x_2,\cdots,1+x_{i1},1+x_{i2},\cdots,1+x_{ir},\cdots,x_n)\\&(1\le i\le n,1\le i_1\le i_2\le\cdots\le i_r\le n,1\le r\le n)\end{aligned}\quad(2)$$

If $r=1$, (2) turns into the derivative of $f(x)=f(x_1,x_2,\cdots,x_n)$ for a single variable, which is

[a] Huang Jinglian: huangjlstudy@163.com

denoted by $df(x)/dx_i$ $(i=1,2,\cdots,n)$. As a result, the simplified form below can be easily derived.

$$\begin{aligned}&df(x)/dx_i\\&=f(x_1,x_2,\cdots,x_{i-1},1,x_{i+1},\cdots,x_n)\\&+f(x_1,x_2,\cdots,x_{i-1},0,x_{i+1},\cdots,x_n)\end{aligned}\ (i=1,2,\cdots,n).$$

Definition 3: The cascade function $f(x)$ $(x=x_0x_1x_2\cdots x_n\in GF(2)^{n+1})$ cascade with two *n*-dimensional Boolean functions $f_1(x')$ and $f_2(x')$ $(x'=x_1x_2\cdots x_n\in GF(2)^n)$ is defined as

$$\begin{aligned}f(x)&=(1+x_0)f_1(x')+x_0f_2(x')\\&=f_1(x')+x_0(f_1(x')+f_2(x'))\end{aligned},$$

$f(x)$ is denoted by $f(x)=f_1(x)\|f_2(x)$.

According to Definition 1~3, we can get Lemmas 1 easily.

Lemma 1: For any arbitrary Boolean function $f(x)$, the following equations are true:

$$\begin{aligned}&f(x)\\&=f(x)df(x)/dx_i+ef(x)/ex_i\end{aligned}\quad(i=1,2,\cdots,n),\quad(3)$$

and

$$\begin{aligned}&w_t(f(x))\\&=w_t(f(x)df(x)/dx_i)+w_t(ef(x)/ex_i)\\&=2^{-1}w_t(df(x)/dx_i)+w_t(ef(x)/ex_i)\\&(i=1,2,\cdots,n)\end{aligned}.\quad(4)$$

3 The algebraic immunity and optimal algebraic immunity functions of a class of correlation immune H Boolean functions

We discuss the cryptographic properties of H Boolean functions $f(x)$ with the Hamming weight of $2^{n-1}+2^{n-2}$ (There is $w_t(f(x))=2^{n-2}$ correspondingly), such as algebraic immunity, annihilators, optimal algebraic immunity functions and correlation immunity.

In Theorem 1, we discuss the relationship of the algebraic degree of the lowest algebraic degree annihilators among $f_1(x)$, $f_2(x)$ and the cascade function $f(x)$ ($f(x)=f_1(x)\|f_2(x)$). Theorem 1 is a preliminary theorem preparing for Theorem 2.

Theorem 1: If $f(x)=f_1(x)\|f_2(x)$, $g_1(x)$ are annihilators of the lowest algebraic degree of $f_1(x)$ (or $1+f_1(x)$), $g_2(x)$ are annihilators of the lowest algebraic degree of $f_2(x)$ (or $1+f_2(x)$), and $\deg g_1(x)=m_1$, $\deg g_2(x)=m_2$.

1. If $g_1(x)=g_2(x)$ (or $g_1(x)=1+g_2(x)$), $AI(f(x))=m_2=m_1$.

2. If $g_1(x)\neq g_2(x)$ and $g_1(x)\neq 1+g_2(x)$,

1) If $m_1\geq m_2+1$ and $g_2(x)f_1(x)\neq 0$, $AI(f(x))=m_2+1$.

2) If $m_1=m_2=m$,and there doesn't exist $g_x(x)$ which makes $g_x(x)=g_2(x)$ or $g_x(x)=1+g_2(x)$, and g_x is an annihilator of the lowest algebraic degree of $f_1(x)$,

a. If *m*-degree terms of $g_1(x)$ and $g_2(x)$ are unequal, or at least one *m*-degree term is unequal in many *m*-degree terms, $AI(f(x))=m+1$.

b. If all the *m*-degree terms of $g_1(x)$ and $g_2(x)$ are equal, but the second high-degree terms are unequal, or at least one of many second high-degree terms is unequal, $AI(f(x))=m$.

Remark: The results of Theorem 1 are based on the Definition 3 of cascade in this paper.

For the proof of Theorem 1, attention should only be paid that we should take $(0\|g_2(x))f(x)=0$ (or $(0\|g_2(x))(1+f(x))=0$ correspondingly) in 1) of 2. Proofs of other clauses in Theorem 1 are easy, so we omit it here.

Theorem 2 will further discuss the existence of correlation immunity H Boolean functions whose algebraic immunity is not less than 2 (that $AI(f(x))\geq 2$), and the existence of the optimal algebraic immunity H Boolean functions with correlation immunity.

Theorem 2: $f(x)$ are H Boolean functions with $w_t(f(x))=2^{n-1}+2^{n-2}$. If $n\geq 5$,

$$w_t(\partial f(x)/\partial(x_{n-1}x_n))=2^{n-2},\quad(5)$$

and

$$w_t(ef(x)/e(x_{n-1}x_n))=2^{n-1}+2^{n-3}.\quad(6)$$

Taking

$$\begin{aligned}&g(x)\\&=\partial(f(x)df(x)/dx_n)/\partial(x_{n-1}x_n)\\&+e(ef(x)/ex_n)/e(x_{n-1}x_n)(x_{n-1}+x_n)\\&+e(f(x)df(x)/dx_n)/e(x_{n-1}x_n)\end{aligned},\quad(7)$$

then we can get

1) There are

$$\begin{aligned}&f(x)\\&=g(x)f(x)df(x)/dx_n\\&+d(g(x)ef(x)/ex_n)/dx_n\end{aligned},\quad(8)$$

and

$$\begin{cases}w_t(g(x))=2^{n-1}\\eg(x)/ex_n=0\end{cases}.\quad(9)$$

2) $g(x)$ is an annihilator of the lowest algebraic degree of $1+f(x)$.

3) If $w_t(\partial g(x)/\partial(x_{n-3}x_{n-2}x_{n-1}x_n))=2^{n-1}$, $AI(f)\geq 2$.

4) When $n\geq 5$, there exists the optimal algebraic immunity functions. If we can construct more than four(or three3) *n*-dimensional optimal algebraic immunity functions, and highest-degree terms of

annihilators g of any two optimal algebraic immunity functions are equal, the second high-order terms of $g_1(x)$, $g_3(x)$ and $g_4(x)$ are not absolutely equal, we can construct $n+2$-dimensional optimal algebraic immunity functions.

5) In $f(x)$ satisfy $AI(f)\geq 2$, there exist functions with $CI(f)=1$.

Proof: 1)

$$\begin{aligned}&dg(x)/dx_n\\&=d(\partial(f(x)df(x)/dx_n)/\partial(x_{n-1}x_n))/dx_n\\&+(x_{n-1}+x_n)d(e(ef(x)/ex_n)/e(x_{n-1}x_n))/dx_n\\&+e(ef(x)/ex_n)/e(x_{n-1}x_n)d(x_{n-1}+x_n)/dx_n\\&+d(e(f(x)df(x)/dx_n)/e(x_{n-1}x_n))/dx_n\end{aligned} \quad (10)$$

Since (5),(6), $w_t(f(x))=2^{n-1}+2^{n-2}$ and $w_t(df(x)/dx_i)=2^{n-1}$ $(i=1,2,\cdots,n)$,we can know, if we equally divide the range of value $0\sim 2^{n-1}$ of independent variable x into small ranges of value,which the numbers of 2^{n-2}:

$$0\sim 2^2-1,2^2\sim 2^3-1,\cdots\cdots,2^n-2^2\sim 2^n-1. \quad (11)$$

The 2-variable function which is divided by $f(x)$ in the small range of value of (11) would be

$$\begin{cases}g_{01}=x_{n-1}+x_n \ and \ g_{02}=1\\ g_{11}=1+x_n+x_{n-1}x_n \ and \ g_{12}=1+x_{n-1}+x_{n-1}x_n\end{cases}. \quad (12)$$

In (12), $g_{01}(x)$ and $g_{02}(x)$ must appear in pairs.

So we can know, in (10), there must be $d(e(ef(x)/ex_n)/e(x_{n-1}x_n))/dx_n=0$. Moreover there must be

$$\begin{aligned}&dg(x)/dx_n=\\&d(\partial(f(x)df(x)/dx_n)/\partial(x_{n-1}x_n))/dx_n\\&+e(ef(x)/ex_n)/e(x_{n-1}x_n)\\&+d(e(f(x)df(x)/dx_n)/e(x_{n-1}x_n))/dx_n\\&=1\end{aligned}, \quad (13)$$

and $dg(x)/dx_{n-1}=1$.

So

$$\begin{aligned}&d(g(x)ef(x)/ex_n)/dx_n\\&=ef(x)/ex_ndg(x)/dx_n\\&+g(x)d(ef(x)/ex_n)/dx_n\\&+dg(x)/dx_nd(ef(x)/ex_n)/dx_n\\&=ef(x)/ex_n\end{aligned}. \quad (14)$$

Because

$$\begin{aligned}&d(g(x)f(x)df(x)/dx_n)/dx_n=\\&f(x)df(x)/dx_ndg(x)/dx_n\\&+g(x)d(f(x)df(x)/dx_n)/dx_n\\&+dg(x)/dx_nd(f(x)df(x)/dx_n)/dx_n\\&=f(x)df(x)/dx_n+g(x)(df(x)/dx_n\\&+f(x)d(df(x)/dx_n)/dx_n\\&+df(x)/dx_nd(df(x)/dx_n)/dx_n)\\&+dg(x)/dx_nd(f(x)df(x)/dx_n)/dx_n\\&=f(x)df(x)/dx_n+g(x)df(x)/dx_n\\&+df(x)/dx_n\\&=(1+g(x)+f(x))df(x)/dx_n\end{aligned}, \quad (15)$$

we can get

$$\begin{aligned}&f(x)df(x)/dx_nd(g(x)f(x)df(x)/dx_n)/dx_n\\&=g(x)f(x)df(x)/dx_n\end{aligned}. \quad (16)$$

Due to $e(f(x)df(x)/dx_n)/ex_n=0$,

$$e(g(x)f(x)df(x)/dx_n)/ex_n=0. \quad (17)$$

Therefore, thanks to (15) and (17), there is

$$\begin{aligned}&g(x)f(x)df(x)/dx_n\\&=g(x)f(x)df(x)/dx_nd(g(x)f(x)df(x)/dx_n)/dx_n\\&+e(g(x)f(x)df(x)/dx_n)/ex_n\\&=g(x)f(x)df(x)/dx_nd(g(x)f(x)df(x)/dx_n)/dx_n\end{aligned}. \quad (18)$$

Known by (16) and (18),there is

$$\begin{aligned}&(f(x)df(x)/dx_n+g(x)f(x)df(x)/dx_n)\cdot\\&d(g(x)f(x)d(g(x)f(x)df(x)/dx_n)/dx_n\\&=0\end{aligned}. \quad (19)$$

By (7), there is

$$\begin{aligned}&g(x)f(x)df(x)/dx_n\\&=\partial(f(x)df(x)/dx_n)/\partial(x_{n-1}x_n)f(x)df(x)/dx_n\\&+e(ef(x)/ex_n)/e(x_{n-1}x_n)(x_{n-1}+x_n)f(x)df(x)/dx_n\\&+e(f(x)df(x)/dx_n)/e(x_{n-1}x_n)f(x)df(x)/dx_n\\&\neq 0\end{aligned} \quad (20)$$

By (20) and (18),we have

$$\begin{cases}g(x)f(x)df(x)/dx_nd(g(x)f(x)df(x)/dx_n)/dx_n\\ \neq 0\\ f(x)df(x)/dx_nd(g(x)f(x)df(x)/dx_n)/dx_n\neq 0\end{cases}.$$

Hence we can know that both $g(x)f(x)df(x)/dx_n$ and $f(x)df(x)/dx_n$ are not the annihilators of $d(g(x)f(x)df(x)/dx_n)/dx_n$. So from (18),(20) and (19),there must be

$$g(x)f(x)df(x)/dx_n=f(x)df(x)/dx_n. \quad (21)$$

By (14),(21) and Lemma 2,we have

$$\begin{aligned}f(x)&=f(x)df(x)/dx_n+ef(x)/ex_n\\&=g(x)f(x)df(x)/dx_n+d(g(x)ef(x)/ex_n)/dx_n\end{aligned},$$

that is,(8) is established.

Since $dg(x)/dx_n = 1$ in (13), $\begin{cases} w_t(g(x)) = 2^{n-1} \\ eg(x)/ex_n = 0 \end{cases}$. That is,(9) is established too.

2) By (12),we have

$$\begin{aligned}
&\partial g(x)/\partial(x_{n-1}x_n) \\
&= \partial(\partial(f(x)df(x)/dx_n)/\partial(x_{n-1}x_n))/\partial(x_{n-1}x_n) \\
&+\partial(x_{n-1}+x_n)/\partial(x_{n-1}x_n)e(ef(x)/ex_n)/e(x_{n-1}x_n) \\
&+(x_{n-1}+x_n)\partial(e(ef(x)/ex_n)/e(x_{n-1}x_n))/\partial(x_{n-1}x_n) \\
&+\partial(x_{n-1}+x_n)/\partial(x_{n-1}x_n)\partial(e(ef(x)/ex_n)/ \\
&e(x_{n-1}x_n))/\partial(x_{n-1}x_n) \\
&+\partial(e(f(x)df(x)/dx_n)/e(x_{n-1}x_n))/\partial(x_{n-1}x_n) \\
&= 0
\end{aligned} \quad (22)$$

Since (9) is established, which means $w_t(g(x)) = 2^{n-1}$ and $eg(x)/ex_n = 0$. By (22),we can know

$$\begin{aligned}
&eg(x)/e(x_{n-1}x_n) \\
&= g(x) + g(x)\partial g(x)/\partial(x_{n-1}x_n) = g(x)
\end{aligned}. \quad (23)$$

By (9),(22) and (23),we can get 2-variable functions of the numbers of 2^{n-2},which are gotten after dividing $g(x)$ in the small range of value of (11),there are only

$$h_1 = x_{n-1} + x_n \quad and \quad h_2 = 1 + x_{n-1} + x_n . \quad (24)$$

Moreover by (7),we can know, $g(x)$ should only take $x_{n-1} + x_n$ in $ef(x)/ex_n$ to guarantee the degree won't add owing to the difference of cascade functions in concatenation.

By (23) and (21),we have

$$d(g(x) + f(x)df(x)/dx_n)/dx_n = ef(x)/ex_n . \quad (25)$$

And

$$\begin{cases} d(f(x)df(x)/dx_n)/dx_n = df(x)/dx_n \\ ef(x)/ex_n + df(x)/dx_n = 1 \end{cases}, \quad (26)$$

by (7),(12) and (23)~(26),we can know, in each small range of (11), $g(x)$ is concatenated of the same or complementary 1st-degree functions, and there is $\deg(h_1(x)\|h_2(x)) = \deg(x_{n-2} + x_{n-1} + x_n) = 1$. But for $ef(x)/ex_n$, there is $0\|x_{n-1}$, thus there exists $\deg(0\|x_{n-1}) = \deg(x_{n-2}x_{n-1}) = 2 > \deg(h_1(x)\|h_2(x))$.

Known by (23),(25) and (26),

$$\deg ef(x)/ex_n = \deg df(x)/dx_n > \deg g(x) . \quad (27)$$

Since $g(x)$ is certain for each $f(x)$,and it's concatenated of completely the same 1st-degree functions. So known by (27), $g(x)$ is an annihilator of the lowest algebraic degree of $1 + f(x)$.

3) Known By (12) and the condition $w_t(\partial g(x)/\partial(x_{n-3}x_{n-2}x_{n-1}x_n)) = 2^{n-1}$, if dividing $f(x)$ into concatenations of 5-variable functions, $g_5(x)$ which is the annihilator of the lowest algebraic degree of 5-variable functions is concatenated of $g_{41}(x)$ and $g_{42}(x)$ which are annihilators of the lowest algebraic degree of two 4-variable functions,which means $g_5(x) = g_{41}(x)\|g_{42}(x)$, and $g_{41}(x) \neq g_{42}(x)$. So

$$\deg g_5(x) \geq 2 . \quad (28)$$

But there must be $AI(f) \leq \left\lceil \frac{n}{2} \right\rceil$. So there is $\deg g_5(x) = \left\lceil \frac{5}{2} \right\rceil = 2$,which means there exist the optimal algebraic immunity functions.

Since (28),we can know,the annihilators of $f(x)$ also have $\deg g(x) \geq 2$.

Therefore,we can know that $AI(f(x)) \geq 2$.

4) Suppose n-dimensional functions $f_1(x)$, $f_2(x)$, $f_3(x)$ and $f_4(x)$ are all optimal algebraic immunity functions, $g_1(x)$, $g_2(x)$, $g_3(x)$ and $g_4(x)$ which are gotten by (7) are annihilators of $f_1(x)$, $f_2(x)$, $f_3(x)$ and $f_4(x)$ respectively, and

$$\begin{cases} f_1(x)\|f_2(x) \neq f_3(x)\|f_4(x) \\ (\text{or taking } f_4(x) = f_1(x)) . \\ g_1(x)\|g_2(x) \neq g_3(x)\|g_4(x) \end{cases} \quad (29)$$

And the highest degree terms of $g_1(x)$, $g_2(x)$, $g_3(x)$ and $g_4(x)$ are equal, the second high-degree term of $g_2(x)$ and the second high-order terms of $g_1(x)$, $g_3(x)$ and $g_4(x)$ are not absolutely equal, and $g_3(x) = 1 + g_4(x)$.

For $n+2$ -dimensional functions $f(x)$, $f(x) = f_1(x)\|f_2(x)\|f_3(x)\|f_4(x)$, known by 2 and 3 of Theorem 2, $f(x)$ have annihilators of the lowest algebraic degree $g(x)$, $g(x) = g_1(x)\|g_2(x)\|g_3(x)\|g_4(x)$. Then

$$\begin{cases} \deg(g_1(x)\|g_2(x)) = \deg g_1(x) = \deg g_2(x) = \left\lceil \frac{n}{2} \right\rceil \\ \deg(g_3(x)\|g_4(x)) = \deg g_3(x) = \deg g_4(x) = \left\lceil \frac{n}{2} \right\rceil \end{cases} \quad (30)$$

So known by (29) and (30), the highest degree term of $g_1(x)\|g_2(x)$ and $g_3(x)\|g_4(x)$ are unequal. And known from Theorem 1,

$$\begin{aligned}
&\deg g(x) \\
&= \deg(g_1(x)\|g_2(x)\|g_3(x)\|g_4(x)) \\
&= \left\lceil \frac{n}{2} \right\rceil + 1 \\
&= \left\lceil \frac{n+2}{2} \right\rceil
\end{aligned}.$$

That is, $f(x)$ is a $n+2$ -dimensional optimal algebraic immunity function.

5) Since the 2-variable functions $g_{01}(x) = x_{n-1} + x_n$ and $g_{02}(x) = 1$ are matched (appearing in pairs of function $f_1(x)$ and $f_2(x)$, denoted by $f_1 \circ f_2$), there is

$$\begin{aligned} & w_t(g_{01} \circ g_{02} \mid x_i = 0) \\ & = w_t(g_{01} \circ g_{02} \mid x_i = 1)\ (i = n-1, n). \\ & = 2^{-1} w_t(g_{01} \circ g_{02}) \end{aligned} \quad (31)$$

For the $g_{11} \circ g_{12}$ which means $g_{11} = 1 + x_n + x_{n-1}x_n$ and $g_{12} = 1 + x_{n-1} + x_{n-1}x_n$ have been match, there also be

$$\begin{aligned} & w_t(g_{11} \circ g_{12} \mid x_i = 0) \\ & = w_t(g_{11} \circ g_{12} \mid x_i = 1)\ (i = n-1, n). \\ & = 2^{-1} w_t(g_{11} \circ g_{12}) \end{aligned} \quad (32)$$

In H Boolean functions with $w_t(f(x)) = 2^{n-1} + 2^{n-2}$, $g_{01} \circ g_{02}$ is necessary.Otherwise,it doesn't satisfy the weight condition of Boolean Functions and the request of $w_t(df(x)/dx_i) = 2^{n-1}$ $(i = 1, 2, \cdots, n)$. So we only need to require g_{11} and g_{12} in $f(x)$ to appear in pairs too.Thus,by (31) and (32), there must be

$$\begin{aligned} & w_t(f(x) \mid x_i = 0) \\ & = w_t(f(x) \mid x_i = 1) = 2^{-1} w_t(f(x)) \end{aligned} \quad (i = 1, 2, \cdots, n).$$

So we can know, there is existence of functions $f(x)$ which with $AI(f(x)) \geq 2$ and $CI(f(x)) = 1$.

The proof ends.

From the proof of 4) and 5) in Theorem 2,we can obtain the following Corollary 1.

Corollary 1: In H Boolean function $f(x)$ with $w_t(f(x)) = 2^{n-1} + 2^{n-2}$, there exists the optimal algebraic immunity functions with correlation immunity.

4 Conclusions

Using The cryptographic properties of Boolean functions are not only connected with the whole values of Boolean functions, but also connected with the values of two different properties and the structure of values(the amount and the distribution of the values of two different properties). In the paper, using the derivative and e-derivative defined by ourselves, we express the values of two different properties of Boolean functions and discuss the correlation immunity and algebraic immunity of the H Boolean functions with the Hamming weight of $2^{n-1} + 2^{n-2}$. The results show that we can better solve many problems of cryptographic properties of Boolean functions using the derivative and e-derivative.

In the paper, the results are very important. It is very useful to study cryptographic properties of Boolean functions using the derivative and e-derivative.

Acknowledgment

This work is supported by National Natural Science Foundation of China (Grant No. 61262085).

References

1. Q. Wen, X. Niu , Y. Yang. The Boolean Functions in modern cryptology. Beijing: Science Press (2000).
2. L. Qu, C. Li. On the 2^m -variable symmetric Boolean functions with maximum algebraic immunity. Science in China Series F: Information Sciences, 2008, 51, (2), pp. 120-127.
3. L. Qu, C. Li, K. Feng. A note on symmetric Boolean functions with maximum algebraic immunity in odd number of variables. IEEE Transactions on Information Theory, 2007, 53, (8), pp. 2908-2910.
4. Z. Tu, and Y. Deng. A conjecture about binary strings and its applications on constructing Boolean functions with optimal algebraic immunity. Designs, Codes and Cryptography, 2011, 60, (1), pp. 1-14.
5. Q. Wang, J. Peng, H. Kan, X. Xue. Constructions of cryptographically significant Boolean functions using primitive polynomials. IEEE Transactions on Information Theory, 2010, 56, (6), pp. 3048-3053.
6. J. Peng, Q. Wu, H. Kan. On symmetric boolean functions with high algebraic immunity on even number of variables. IEEE Transactions on Information Theory, 2011, 57, (10), pp. 7205-7220.
7. J. Huang, Z. Wang. The relationship between correlation immune and weight of H Boolean functions. Journal on Communications, Vol.33(2):110-118(2012). (In Chinese)
8. W. Li, Z. Wang, J. Huang. The e-derivative of boolean functions and its application in the fault detection and cryptographic system. Kybernetes, Vol.40(5-6):905-911(2011).
9. Y. Ding, Z. Wang, J. Ye. Initial-value problem of the Boolean function's primary function and its application in cryptographic system. Kybernetes, Vol.39(6): 900-906(2010).

The Nonlinearity of Sum and Product for Boolean Functions

Huang Jinglian[a], and Wang Zhuo

School of Electrical Engineering, Northwest University for Nationalities, Lanzhou, 730030, China

Abstract. In this paper, we study the relationship between the nonlinearity of Boolean function and the nonlinearity of the sum and product of Boolean function, while derivative and e-derivative are used to study the problem further . We obtain that the sum of two functions' nonlinearity is not less than the nonlinearity of the sum of two functions. The relationship between the nonlinearity of function and the nonlinearity of the sum and product of two functions are also obtained. Furthermore, we also get the relationship between the nonlinearity of the product of functions, and the derivative and e-derivative of function. Moreover, we also deduced some important applications on the basis of the above work.

1 Introduction

The cipher security is the core of the cryptosystem, and only a cryptosystem with good security has an existing significance. Boolean functions with a variety of secure cipher properties are the key factors to design the cryptosystem with the ability to resist multiple cipher attacks and good safety performance. It is of great importance for a security cryptosystem to study some properties of Boolean functions, which make the cryptosystem resist various attacks, such as high algebraic degree, high nonlinearity, the strict avalanche criterion and propagation, higher-order correlation immunity and higher-order algebraic immunity. Therefore, there are some important research problems, such as the existence, the feature, the design, the construction and the count of Boolean functions with some kind of secure cryptographic property [1~7].

The nonlinearity of Boolean functions is a measure of the ability to resist the affine approximation attack. The higher the nonlinearity of Boolean functions, the stronger the ability to resist the affine approximation attacks. It is an important task to study the nonlinearity of Boolean functions [8~9].

With the sum or product of two Boolean functions, the Boolean functions with good cryptographic properties can be obtained. As the sum of a 2 time homogeneous H Boolean function and a $n/2$ monomial can be a Bent function. Therefore, it is also necessary to study the cryptographic properties of the sum of the function or the product function of two Boolean functions, such as nonlinearity.

In this paper, we will discuss whether there is a triangle inequality between the nonlinearity of Boolean functions and the sum of two functions. At the same time, the derivative and e-derivative of the Boolean function are used to reveal the relationship between the nonlinearity of Boolean function and the sum or the product of the two functions.

[a] Huang Jinglian: huangjlstudy@163.com

2 Preliminaries

To study cryptographic properties of H Boolean functions, we proposed the concept of the e-derivative [10~12]. The e-derivative and derivative are defined here as Definition 1&2.

Definition 1: The e-derivative (e-partial derivative) of n-dimensional Boolean functions $f(x)=f(x_1,x_2,\cdots,x_n)\in GF(2)^{GF(2)^n}$ for r variables $x_{i1},x_{i2},\cdots,x_{ir}$ is defined as

$$\begin{aligned}&ef(x)/e(x_{i1},x_{i2},\cdots,x_{ir})\\&=f(x_1,x_2,\cdots,x_{i1},x_{i2},\cdots,x_{ir},\cdots,x_n)\cdot\\&f(x_1,x_2,\cdots,1+x_{i1},1+x_{i2},\cdots,1+x_{ir},\cdots,x_n)\\&(1\le i\le n,1\le i_1\le i_2\le\cdots\le i_r\le n,1\le r\le n)\end{aligned}\quad(1)$$

If $r=1$, (1) turns into the e-derivative of $f(x)=f(x_1,x_2,\cdots,x_n)$ for a single variable, which is denoted by $ef(x)/ex_i$ $(i=1,2,\cdots,n)$. As a result, the simplified form below can be easily derived.

$$\begin{aligned}&ef(x)/ex_i\\&=f(x_1,x_2,\cdots,x_{i-1},1,x_{i+1},\cdots,x_n)\cdot\ (i=1,2,\cdots,n).\\&f(x_1,x_2,\cdots,x_{i-1},0,x_{i+1},\cdots,x_n)\end{aligned}$$

Definition 2: The derivative (partial derivative) of n-dimensional Boolean functions $f(x)=f(x_1,x_2,\cdots,x_n)\in GF(2)^{GF(2)^n}$ for r variables $x_{i1},x_{i2},\cdots,x_{ir}$ is defined as

$$\begin{aligned}&\partial f(x)/\partial(x_{i1},x_{i2},\cdots,x_{ir})\\&=f(x_1,x_2,\cdots,x_{i1},x_{i2},\cdots,x_{ir},\cdots,x_n)+\\&f(x_1,x_2,\cdots,1+x_{i1},1+x_{i2},\cdots,1+x_{ir},\cdots,x_n)\\&(1\le i\le n,1\le i_1\le i_2\le\cdots\le i_r\le n,1\le r\le n)\end{aligned}\quad(2)$$

If $r=1$, (2) turns into the derivative of $f(x)=f(x_1,x_2,\cdots,x_n)$ for a single variable, which is denoted by $df(x)/dx_i$ $(i=1,2,\cdots,n)$. As a result, the simplified form below can be easily derived.

$$df(x)/dx_i = f(x_1,x_2,\cdots,x_{i-1},1,x_{i+1},\cdots,x_n) + f(x_1,x_2,\cdots,x_{i-1},0,x_{i+1},\cdots,x_n) \quad (i=1,2,\cdots,n).$$

Definition 3: The cascade function $f(x)$ $(x=x_0x_1x_2\cdots x_n\in GF(2)^{n+1})$ cascade with two n-dimensional Boolean functions $f_1(x')$ and $f_2(x')$ $(x'=x_1x_2\cdots x_n\in GF(2)^n)$ is defined as

$$f(x)=(1+x_0)f_1(x')+x_0f_2(x') = f_1(x')+x_0(f_1(x')+f_2(x')),$$

$f(x)$ is denoted by $f(x)=f_1(x)\|f_2(x)$.

According to Definition 1~3, we can get Lemmas 1 easily.

Lemma 1: For any arbitrary Boolean function $f(x)$, the following equations are true:

$$f(x) = f(x)df(x)/dx_i + ef(x)/ex_i \quad (i=1,2,\cdots,n),$$

and

$$w_t(f(x)) = w_t(f(x)df(x)/dx_i)+w_t(ef(x)/ex_i) = 2^{-1}w_t(df(x)/dx_i)+w_t(ef(x)/ex_i) \quad (i=1,2,\cdots,n).$$

3 The triangle inequality for the nonlinearity of the sum of two Boolean functions

We are always wondering if there exists a relationship among the nonlinearity of two Boolean functions and the sum function ,which can be represented by a triangle inequality. Theorem 1 just reveals the triangle inequality.

Theorem 1: Suppose Boolean functions $f_1(x),f_2(x)\in GF(2)^{GF(2)^n}$. The nonlinearity of $f_1(x)$ and $f_2(x)$ are N_{f_1} and N_{f_2}, the nonlinearity of the sum function $f_1(x)+f_2(x)$ of $f_1(x)$ and $f_2(x)$ are $N_{f_1+f_2}$. So N_{f_1}, N_{f_2} and $N_{f_1+f_2}$ are contacted with the following triangle inequality :

$$N_{f_1}+N_{f_2}\ge N_{f_1+f_2}.$$

Proof: Suppose there exists $l_1(x),l_2(x)\in L_n[x]$ which makes

$$N_{f_1}=\min_{l(x)\in L_n[x]} w_t(f_1(x)+l(x)) = w_t(f_1(x)+l_1(x)),$$

and

$$N_{f_2}=\min_{l(x)\in L_n[x]} w_t(f_2(x)+l(x)) = w_t(f_2(x)+l_2(x)).$$

And there exists $l_0(x)\in L_n[x]$ which makes

$$N_{f_1+f_2} = \min_{l(x)\in L_n[x]} w_t(f_1(x)+f_2(x)+l(x)) = w_t(f_1(x)+f_2(x)+l_0(x))$$

So

$$w_t(f_1(x)+f_2(x)+l_1(x)+l_2(x)) \ge w_t(f_1(x)+f_2(x)+l_0(x)).$$

There are

$$\begin{aligned} &w_t(f_1(x)+f_2(x)+l_1(x)+l_2(x)) \\ &= w_t(f_1(x)+l_1(x)+f_2(x)+l_2(x)) \\ &= w_t(f_1(x)+l_1(x))+w_t(f_2(x)+l_2(x)) \\ &-2w_t(f_1(x)f_2(x)+l_1(x)f_2(x) \\ &+l_2(x)f_1(x)+l_1(x)l_2(x)) \end{aligned}.$$

Therefore

$$\begin{aligned} &w_t(f_1(x)+l_1(x))+w_t(f_2(x)+l_2(x))-2w_t(f_1(x)\cdot \\ &f_2(x)+l_1(x)f_2(x)+l_2(x)f_1(x)+l_1(x)l_2(x)) \\ &\ge w_t(f_1(x)+f_2(x)+l_0(x)) \end{aligned}.$$

which is

$$\begin{aligned} &N_{f_1}+N_{f_2}-2w_t(f_1(x)f_2(x)+ \\ &l_1(x)f_2(x)+l_2(x)f_1(x)+l_1(x)l_2(x)) \\ &\ge N_{f_1+f_2} \end{aligned}.$$

Due to

$$w_t(f_1(x)f_2(x)+l_1(x)f_2(x)+l_2(x)f_1(x)+l_1(x)l_2(x)) \ge 0,$$

it must be

$$N_{f_1}+N_{f_2}\ge N_{f_1+f_2}.$$

The proof ends.

$l_0(x)\in L_n[x]$ of the formula $N_{f_1+f_2}=w_t(f_1(x)+f_2(x)+l_0(x))$ in Theorem 1, can help us calculate the nonlinearity $N_{f_1f_2}$ of the product functions $f_1(x)f_2(x)$ of $f_1(x)$ and $f_2(x)$. So we can get Theorem 2.

Theorem 2: For n-variable Boolean function $f_1(x)$ and $f_2(x)$ ($f_1(x),f_2(x)\in GF(2)^{GF(2)^n}$), if there exists $l_0(x)\in L_n[x]$ which makes the nonlinearity $N_{f_1+f_2}$ of $f_1(x)+f_2(x)$ are

$$N_{f_1+f_2} = \min_{l(x)\in L_n[x]} w_t(f_1(x)+f_2(x)+l(x)) = w_t(f_1(x)+f_2(x)+l_0(x))$$

then for $1+l_0(x)\in L_n[x]$, there must be

$$N_{f_1f_2} = \min_{l(x)\in L_n[x]} w_t(f_1(x)f_2(x)+l(x)) = w_t(f_1(x)f_2(x)+1+l_0(x))$$

Proof: Known by the condition, there exists $l_0(x)\in L_n[x]$ which makes

$$N_{f_1+f_2} = \min_{l(x)\in L_n[x]} w_t(f_1(x)+f_2(x)+l(x))$$
$$= \min_{l(x)\in L_n[x]} (w_t(f_1(x)+f_2(x))+w_t(l(x))$$
$$-2w_t(l(x)(f_1(x)+f_2(x)))$$
$$= w_t(f_1(x)+f_2(x))+w_t(l(x))$$
$$-2\max_{l(x)\in L_n[x]} w_t(l(x)(f_1(x)+f_2(x)))$$
$$= w_t(f_1(x)+f_2(x))+w_t(l(x))$$
$$-2\max_{l(x)\in L_n[x]} w_t(l(x)(f_1(x)+f_1(x)f_2(x)))$$
$$-2\max_{l(x)\in L_n[x]} w_t(l(x)(f_2(x)+f_1(x)f_2(x)))$$
$$= w_t(f_1(x)+f_2(x))+w_t(l_0(x))$$
$$-2w_t(l_0(x)(f_1(x)+f_1(x)f_2(x)))$$
$$-2w_t(l_0(x)(f_2(x)+f_1(x)f_2(x)))$$
$$= w_t(f_1(x)+f_2(x)+l_0(x))$$

So

$$\begin{cases} w_t(l_0(x)(f_1(x)+f_1(x)f_2(x))) \\ = \max_{l(x)\in L_n[x]} w_t(l(x)(f_1(x)+f_1(x)f_2(x))) \\ w_t(l_0(x)(f_2(x)+f_1(x)f_2(x))) \\ = \max_{l(x)\in L_n[x]} w_t(l(x)(f_2(x)+f_1(x)f_2(x))) \end{cases}.$$

There are

$$\begin{cases} w_t(l_0(x)f_1(x))-w_t(l_0(x)f_1(x)f_2(x))) \\ = \max_{l(x)\in L_n[x]} (w_t(l(x)f_1(x))-w_t(l(x)f_1(x)f_2(x))) \\ w_t(l_0(x)f_2(x))-w_t(l_0(x)f_1(x)f_2(x))) \\ = \max_{l(x)\in L_n[x]} (w_t(l(x)f_2(x))-w_t(l(x)f_1(x)f_2(x))) \end{cases}.$$

So

$$w_t(l_0(x)f_1(x)f_2(x)) = \min_{l(x)\in L_n[x]} w_t(l(x)f_1(x)f_2(x)).$$

So there have

$$C_{f_1f_2} = \max_{l(x)\in L_n[x]} w_t(f_1(x)f_2(x)+l(x))$$
$$= \max_{l(x)\in L_n[x]} (w_t(f_1(x)f_2(x))+w_t(l(x))$$
$$-2w_t(l(x)f_1(x)f_2(x)))$$
$$= w_t(f_1(x)f_2(x))+w_t(l(x))$$
$$-2\min_{l(x)\in L_n[x]} w_t(l(x)f_1(x)f_2(x))$$
$$= w_t(f_1(x)f_2(x))+w_t(l_0(x))$$
$$-2w_t(l_0(x)f_1(x)f_2(x))$$
$$= w_t(f_1(x)f_2(x)+l_0(x))$$

Therefore

$$w_t(f_1(x)f_2(x)+1+l_0(x))$$
$$= 2^n - w_t(f_1(x)f_2(x)+l_0(x))$$
$$= 2^n - C_{f_1f_2}$$
$$= N_{f_1f_2}$$

That is

$$N_{f_1f_2} = \min_{l(x)\in L_n[x]} w_t(f_1(x)f_2(x)+l(x))$$
$$= w_t(f_1(x)f_2(x)+1+l_0(x))$$

The proof ends.

Theorem 1&2 reveal that the nonlinearity of the sum of the two Boolean functions is not more than the sum of nonlinearity of two functions, and nonlinearity of the sum of the two functions and the nonlinearity of product of two functions. For the nonlinearity of the sum of two functions with two functions of the relationship between the nonlinearity of a single function, need to use e-derivative and derivative to discuss, back to do the work.

4 The nonlinearity of Boolean functions with the derivative and e-derivative

In Theorem 3, we will discuss the relationship between the nonlinearity of the sum of two Boolean functions, and the nonlinearity of a single function.

Theorem 3: Suppose Boolean functions $f_1(x), f_2(x) \in GF(2)^{GF(2)^n}$. There are $w_t(f_1(x)) > 2^{n-1}$, $w_t(f_2(x)) > 2^{n-1}$, and

$$\deg e(f_1(x)f_2(x))/ex_n = \deg ef_1(x)/ex_n$$
$$= \deg ef_2(x)/ex_n = 1.$$

1) If

$$w_t(f_1(x)df_1(x)/dx_n + f_2(x)df_2(x)/dx_n) < 2^{n-2},$$

then $\begin{cases} N_{f_1+f_2} > N_{f_1} \\ N_{f_1+f_2} > N_{f_2} \end{cases}$.

2) If

$$w_t(f_1(x)df_1(x)/dx_n) + w_t(f_1(x)df_1(x)/dx_n$$
$$+f_2(x)df_2(x)/dx_n) \geq 2^{n-1},$$

and $2^{n-2} \geq w_t(f_2(x)df_2(x)/dx_n) \geq w_t(f_1(x)df_1(x)/dx_n)$, then $N_{f_1+f_2} \leq N_{f_1} \leq N_{f_2}$ and

$$w_t(f_1(x)df_1(x)/dx_n + f_2(x)df_2(x)/dx_n) \geq 2^{n-2}.$$

3) There are

$$N_{f_1f_2} = w_t(f_1(x)df_1(x)/dx_n \cdot f_2(x)df_2(x)/dx_n).$$

4) There are $N_{f_1+f_2} = 2^n - N_{f_1} - N_{f_2} + 2N_{f_1f_2}$.

Proof: Because

$$\deg e(f_1(x)f_2(x))/ex_n = \deg ef_1(x)/ex_n$$
$$= \deg ef_2(x)/ex_n = 1,$$

so

$$e(f_1(x)f_2(x))/ex_n = ef_1(x)/ex_n \cdot ef_2(x)/ex_n$$
$$= ef_1(x)/ex_n = ef_2(x)/ex_n,$$

and $e(f_1(x)f_2(x))/ex_n \in L_n[x]$.

Known by Lemma and conditions $w_t(f_1(x)) > 2^{n-1}$ and $w_t(f_2(x)) > 2^{n-1}$,

$$\begin{cases} w_t(f_1(x)df_1(x)/dx_n) \leq 2^{n-2} \\ w_t(f_2(x)df_2(x)/dx_n) \leq 2^{n-2} \end{cases},$$

and

$$df_1(x)/dx_n ef_2(x)/ex_n = df_2(x)/dx_n ef_1(x)/ex_n$$
$$= df_1(x)/dx_n ef_1(x)/ex_n = df_2(x)/dx_n ef_2(x)/ex_n = 0.$$

So

$$\begin{aligned} & f_1(x)+f_2(x) \\ & =(f_1(x)df_1(x)/dx_n+ef_1(x)/ex_n) \\ & +(f_2(x)df_2(x)/dx_n+ef_2(x)/ex_n) \\ & =f_1(x)df_1(x)/dx_n+f_2(x)df_2(x)/dx_n \end{aligned},$$

and

$$\begin{aligned} & f_1(x)f_2(x) \\ & =f_1(x)df_1(x)/dx_nf_2(x)df_2(x)/dx_n \\ & +f_2(x)df_2(x)/dx_nef_1(x)/ex_n+ \\ & f_1(x)df_1(x)/dx_nef_2(x)/ex_n \\ & +ef_1(x)/ex_nef_2(x)/ex_n \\ & =f_1(x)df_1(x)/dx_nf_2(x)df_2(x)/dx_n \\ & +ef_1(x)f_2(x)/ex_n \end{aligned}.$$

Take $l_0(x)=ef_1(x)f_2(x)/ex_n\in L_n[x]$, there are

$$\begin{aligned} & N_{f_1+f_2} \\ & =\min_{l(x)\in L_n[x]} w_t(f_1(x)df_1(x)/dx_n \\ & +f_2(x)df_2(x)/dx_n+l(x)) \\ & =w_t(f_1(x)df_1(x)/dx_n+f_2(x)df_2(x)/dx_n, \\ & +1+l_0(x)) \\ & =2^n-w_t(f_1(x)df_1(x)/dx_n \\ & +f_2(x)df_2(x)/dx_n) \end{aligned} \quad (3)$$

$$\begin{aligned} & N_{f_1}=\min_{l(x)\in L_n[x]} w_t(f_1(x)+l(x)) \\ & =w_t(f_1(x)+l_0(x))=w_t(f_1(x)df_1(x)/dx_n) \end{aligned}, \quad (4)$$

$$\begin{aligned} & N_{f_2}=\min_{l(x)\in L_n[x]} w_t(f_2(x)+l(x)) \\ & =w_t(f_2(x)+l_0(x))=w_t(f_2(x)df_2(x)/dx_n) \end{aligned}, \quad (5)$$

and

$$\begin{aligned} & N_{f_1f_2} \\ & =\min_{l(x)\in L_n[x]} w_t(f_1(x)df_1(x)/dx_nf_2(x)df_2(x)/dx_n \\ & +ef_1(x)f_2(x)/ex_n+l(x)) \\ & =w_t(f_1(x)df_1(x)/dx_nf_2(x)df_2(x)/dx_n \\ & +ef_1(x)f_2(x)/ex_n+l_0(x)) \\ & =w_t(f_1(x)df_1(x)/dx_nf_2(x)df_2(x)/dx_n) \end{aligned}. \quad (6)$$

1) When $w_t(f_1(x)df_1(x)/dx_n+f_2(x)df_2(x)/dx_n)<2^{n-2}$, due to obtained permits $w_t(f_1(x)df_1(x)/dx_n)\le 2^{n-2}$, and $w_t(f_2(x)df_2(x)/dx_n)\le 2^{n-2}$, then known by (3), (4) and (5), there are

$$\begin{cases} N_{f_1+f_2}>N_{f_1} \\ N_{f_1+f_2}>N_{f_2} \end{cases}.$$

2) From the conditions, (3), (4) and (5), at a glance there are

$$N_{f_1+f_2}\le N_{f_1}\le N_{f_2}$$

and

$$w_t(f_1(x)df_1(x)/dx_n+f_2(x)df_2(x)/dx_n)\ge 2^{n-2}.$$

3) (6) already shows the results.

4) expanding (3), binding (4) and (5), there are

$$\begin{aligned} & N_{f_1+f_2} \\ & =2^n-w_t(f_1(x)df_1(x)/dx_n)-w_t(f_2(x)df_2(x)/dx_n) \\ & +2w_t(f_1(x)df_1(x)/dx_nf_2(x)df_2(x)/dx_n) \\ & =2^n-N_{f_1}-N_{f_2}+2N_{f_1f_2} \end{aligned}.$$

The proof ends.

Known from Theorem 3, when comparing the nonlinearity of the sum of two Boolean functions with each of the two functions under different conditions, it comes to different results . Under certain conditions, the nonlinearity of the sum and the product of two Boolean functions can establish contact.

In the following Theorem 4, using the derivative and e-derivative, we will discuss the nonlinearity of a class of separable variables Boolean function.

Theorem 4: If n is even, $\{y\}=\{x_1,x_2,\cdots,x_{\frac{n}{2}}\}$, $\{z\}=\{x_{\frac{n}{2}+1},x_{\frac{n}{2}+2},\cdots,x_n\}$, $\{x\}=\{x_1,x_2,\cdots,x_n\}$. There are $f(x)=R(y)S(z)$, $\deg dR(y)/dx_{\frac{n}{2}}=1$, $\deg dS(z)/dx_n=1$, $w_t(eS(z)/ex_n)<2^{n-2}$, and $w_t(eR(y)/ex_{\frac{n}{2}})<2^{n-2}$. Then there are

$$\begin{aligned} & N_f=w_t((1+S(z))R(y)dS(z)/dx_n) \\ & +w_t(R(y)eS(z)/ex_n) \\ & =w_t((1+R(y))S(z)dR(y)/dx_{\frac{n}{2}}) \\ & +w_t(S(z)eR(y)/ex_{\frac{n}{2}}) \end{aligned},$$

and

$$\begin{cases} 2^{n-2}<N_{S(z)}<2^{n-1} \\ 2^{n-2}<N_{R(y)}<2^{n-1} \end{cases}.$$

Proof: Known by the condition $\deg dS(z)/dx_n=1$, there are

$$\begin{aligned} & w_t(S(z)dS(z)/dx_n)=w_t((1+S(z))dS(z)/dx_n) \\ & =2^{-1}w_t(dS(z)/dx_n)=2^{n-2} \end{aligned}.$$

Also, because $w_t(eS(z)/ex_n)<2^{n-2}$, known by Lemma, there are $2^{n-2}<w_t(S(z))<2^{n-1}$.

Because the value of the structure of $eS(z)/ex_n$ and $S(z)dS(z)/dx_n$ is different, i.e. at x in $eS(z)/ex_n=1$, $dS(z)/dx_n=0$. And each linear function has only a single the value of the structure. Therefore,

$$\begin{aligned} & N_{S(z)}=\min_{l(x)\in L_n[x]} w_t(S(z)+l(x)) \\ & =w_t(S(z)+dS(z)/dx_n) \\ & =w_t(S(z))+w_t(dS(z)/dx_n)-2w_t(S(z)dS(z)/dx_n) \\ & =w_t(S(z)) \end{aligned}.$$

So $2^{n-2}<N_{S(z)}<2^{n-1}$, and

$$\max_{l(x)\in L_n[x]} w_t(l(x)S(z))=w_t(S(z)dS(z)/dx_n).$$

So

$$\max_{l(x)\in L_n[x]} w_t(R(y)S(z)l(x)) = w_t(R(y)S(z)dS(z)/dx_n).$$

Therefore

$$N_f = \min_{l(x)\in L_n[x]} w_t(R(y)S(z)+l(x))$$
$$= w_t(R(y)S(z)+dS(z)/dx_n)$$
$$= w_t((1+S(z))R(y)dS(z)/dx_n)$$
$$+w_t(R(y)eS(z)/ex_n)$$

Since $R(y)$ has the same conditions with $S(z)$ correspondingly, through the same deduction, we can get

$$2^{n-2} < N_{R(y)} < 2^{n-1}.$$

Also

$$N_f$$
$$= w_t((1+R(y))S(z)dR(y)/dx_{\frac{n}{2}}).$$
$$+w_t(eR(y)/ex_{\frac{n}{2}}S(z))$$

The proof ends.

Known from the proof of Theorem 4 , if $R(y)$ is changed while $S(z)$ remains the same , $2^{n-2} < N_{S(z)} = w_t(S(z)) < 2^{n-1}$ and $N_f = w_t((1+S(z))R(y)dS(z)/dx_n) + w_t(R(y)eS(z)/ex_n)$ still hold. So, if taking $R(y) = x_1x_2\cdots x_{\frac{n}{2}}$, we can get

$$N_f = w_t(dS(z)/dx_n) + w_t(R(y)S(z))$$
$$-2w_t(R(y)S(z)dS(z)/dx_n)$$
$$= 2^{n-1} + 2^{-1}\cdot 2^{\frac{n}{2}} - 2\cdot 2^{\frac{n}{2}-1}$$
$$= 2^{n-1} - 2^{\frac{n}{2}-1}$$

Then there are following Corollary.

Corollary: In Theorem 4, if taken $R(y) = x_1x_2\cdots x_{\frac{n}{2}}$, $S(z)$ unchanged, then for $f(x) = R(y)S(z)$, there have

$$N_f = 2^{n-1} - 2^{\frac{n}{2}-1}.$$

It found that there exists functions with high nonlinearity in the function $f(x) = R(y)S(z)$ which variables can be separable .

5 Conclusions

The nonlinearity of the sum of Boolean functions can be roughly estimated by the triangle inequality , so the triangle inequality has its practical significance. The relationship among triangle inequality, and the nonlinearity of the sum and the product of two Boolean functions in Theorem 2, plays a role on constructing Boolean functions with higher nonlinearity.

To estimate the lower bound of the nonlinearity of the sum of two Boolean functions,we can use derivatives and e-derivative to analysis as Theorem 3 . It shows that using the derivative and e- derivative to study nonlinearity of Boolean function is very useful.

Acknowledgment

This work is supported by National Natural Science Foundation of China (Grant No. 61262085).

References

1. Courtois, N., and Meier, W. Algebraic attacks on stream ciphers with linear feedback. Advances in Cryptology-EUROCRYPT 2003, Warsaw, Poland, 2003, LNCS, 2656:345-359.
2. Carlet C and Zeng X Y. Further properties of several classes of Boolean functions with optimum algebraic immunity.Designs, Codes and Cryptography, 2009, 52(3): 303-338.
3. Carlet C. A method of construction of balanced functions with optimum algebraic immunity. Proceedings of the First International Workshop on Coding and Cryptography, Fujian, 2007: 25-43.
4. Li Y, Yang M, and Kan H B. Constructing and counting Boolean functions on even variables with maximum algebraic immunity. IEICE Transactions on Fundamentals, 93-A(3): 640-643(2010).
5. Rizomiliotis P. On the resistance of Boolean functions against algebraic attacks using univariate polynomial representation[J]. IEEE Transactions on Information Theory, 56(8):4014-4024(2010).
6. Tu Z R and Deng Y P. A class of 1-resilient function with high nonlinearity and algebraic immunity. Ryptography ePrint Archive, Report 2010, 2010/179.
7. Wang Q, Peng J, Kan H, et al.. Constructions of cryptographically significant Boolean functions using primitive polynomials. IEEE Transactions on Information Theory, 56(6): 3048-3053(2010).
8. Li, C.,Zhang, H.,Zeng, X, et al. The lower bound on the second-order nonlinearity for a class of Bent functions. Chinese Journal of Computers, 35(8):1588-1593(2012).
9. Su S H, Tang X H. Construction of rotation symmetric Boolean functions with optimal algebraic immunity and high nonlinearity. Designs, Codes and Cryptography, 71(2): 183–199(2014).
10. J. Huang, Z. Wang. The relationship between correlation immune and weight of H Boolean functions. Journal on Communications, Vol.33(2):110-118(2012). (In Chinese)
11. W. Li, Z. Wang, J. Huang. The e-derivative of boolean functions and its application in the fault detection and cryptographic system. Kybernetes, Vol.40(5-6):905-911(2011).
12. Zhao, M.. Method of detecting special logic function based on Boolean e-derivative. Journal of Zhejiang University (Science Edition), 41(4):424-426(2014).

MATEC Web of Conferences **61**, 03006 (2016) DOI: 10.1051/matecconf/20166103006
APOP2016

Construction of the optimal algebraic immunity Boolean functions with correlation immunity from transformation RSBFs

Huang Jinglian[a]

School of Electrical Engineering, Northwest University for Nationalities, Lanzhou, 730030, China

Abstract. In this paper, we discuss the effect of the cryptographic properties of translation transformation rotation symmetric Boolean functions. On construction of rotation symmetric Boolean functions with the optimal algebraic immunity, we construct correlation immunity Boolean functions with the optimal algebraic immunity by translation transformation and concatenation transformation.

1 Introduction

The algebraic immunity, optimal algebraic immunity, nonlinearity, diffusion and correlation immunity are all important security properties of cryptographic functions and research contents of cipher security. The algebraic immunity and optimal algebraic immunity of algebraic attacks are the hot spots in the current research. The cipher security is the core of the cryptosystem, and only a cryptosystem with good security has an existing significance. Boolean functions with a variety of secure cipher properties are the key factors to design the cryptosystem with the ability to resist multiple cipher attacks and good safety performance. It is of great importance for a security cryptosystem to study some properties of Boolean functions, which make the cryptosystem resist various attacks, such as high algebraic degree, high nonlinearity, the strict avalanche criterion and propagation, higher-order correlation immunity and higher-order algebraic immunity. Therefore, there are some important research problems, such as the existence, the feature, the design, the construction and the count of Boolean functions with some kind of secure cryptographic property [1~7].

The Bent function and the rotational symmetric Boolean functions(RSBFs) are important functions in cryptography. If there is a Bent function in the RSBFs, and how to construct the optimal algebraic immune function with the Bent function are important problems to be studied [8~14]. In this paper, we discuss the existence and structure of rotational symmetric Bent function using derivative, translation and cascade, and the problem of the optimal algebraic immune function of the immune system, which is based on the Bent function of the rotational symmetry. It is more convenient to judge the diffusion by the derivative calculation than by the diffusion.

2 Preliminaries

To study cryptographic properties of Boolean functions, we proposed the concept of the e-derivative [15~17]. The e-derivative and derivative are defined here as Definition 1&2.

Definition 1: The e-derivative (e-partial derivative) of n-dimensional Boolean functions $f(x)=f(x_1,x_2,\cdots,x_n)\in GF(2)^{GF(2)^n}$ for r variables $x_{i1},x_{i2},\cdots,x_{ir}$ is defined as

$$\begin{aligned}&ef(x)/e(x_{i1},x_{i2},\cdots,x_{ir})\\&=f(x_1,x_2,\cdots,x_{i1},x_{i2},\cdots,x_{ir},\cdots,x_n)\cdot\\&f(x_1,x_2,\cdots,1+x_{i1},1+x_{i2},\cdots,1+x_{ir},\cdots,x_n)\\&(1\le i\le n,1\le i_1\le i_2\le\cdots\le i_r\le n,1\le r\le n)\end{aligned}\quad(1)$$

If $r=1$, (1) turns into the e-derivative of $f(x)=f(x_1,x_2,\cdots,x_n)$ for a single variable, which is denoted by $ef(x)/ex_i$ $(i=1,2,\cdots,n)$. As a result, the simplified form below can be easily derived.

$$\begin{aligned}&ef(x)/ex_i\\&=f(x_1,x_2,\cdots,x_{i-1},1,x_{i+1},\cdots,x_n)\cdot\\&f(x_1,x_2,\cdots,x_{i-1},0,x_{i+1},\cdots,x_n)\end{aligned}\quad(i=1,2,\cdots,n).$$

Definition 2: The derivative (partial derivative) of n-dimensional Boolean functions $f(x)=f(x_1,x_2,\cdots,x_n)\in GF(2)^{GF(2)^n}$ for r variables $x_{i1},x_{i2},\cdots,x_{ir}$ is defined as

$$\begin{aligned}&\partial f(x)/\partial(x_{i1},x_{i2},\cdots,x_{ir})\\&=f(x_1,x_2,\cdots,x_{i1},x_{i2},\cdots,x_{ir},\cdots,x_n)+\\&f(x_1,x_2,\cdots,1+x_{i1},1+x_{i2},\cdots,1+x_{ir},\cdots,x_n)\\&(1\le i\le n,1\le i_1\le i_2\le\cdots\le i_r\le n,1\le r\le n)\end{aligned}\quad(2)$$

If $r=1$, (2) turns into the derivative of $f(x)=f(x_1,x_2,\cdots,x_n)$ for a single variable, which is denoted by $df(x)/dx_i$ $(i=1,2,\cdots,n)$. As a result, the simplified form below can be easily derived.

[a] Huang Jinglian: huangjlstudy@163.com

$$df(x)/dx_i = f(x_1,x_2,\cdots,x_{i-1},1,x_{i+1},\cdots,x_n) + f(x_1,x_2,\cdots,x_{i-1},0,x_{i+1},\cdots,x_n) \quad (i=1,2,\cdots,n).$$

Definition 3: If $n \in I_+$, for random variable $x=(x_1,x_2,\cdots,x_n) \in GF(2)^n$, $k \in I_+$, and $0 \le k \le n-1$, give $\rho_n^k(x)=(\rho_n^k(x_1),\rho_n^k(x_2),\cdots,\rho_n^k(x_n))$, it defined that

$$\rho_n^k(x_i)=\begin{cases} x_{i+k}, i+k \le n \\ x_{i+k-n}, i+k > n \end{cases}.$$

For random variable $x=(x_1,x_2,\cdots,x_n) \in GF(2)^n$, there will be $f(\rho_n^k(x))=f(x)$, $0 \le k \le n-1$, then $f(x)$ is called Rotation Symmetric Boolean Functions (RSBFs).

Definition 4: For any arbitrary ω $(\omega=(\omega_1,\omega_2,\cdots,\omega_n) \in GF(2)^n,\ 1 \le w_t(\omega) \le m)$, $f(x)=f(x_1,x_2,\cdots,x_n) \in GF(2)^{GF(2)^n}$ holds: $w_t(f(x)+\omega x)=2^{n-1}$ ($\omega x = \omega_1 x_1 + \omega_2 x_2 + \cdots + \omega_n x_n$), then Boolean functions $f(x)$ was called a m-order correlation immune function, m was called order. The correlation immunity of order m and the correlation immunity order are both written as $CI(m)$.

Definition 5: For $f(x) \in GF(2)^{GF(2)^n}$, if $g_1(x) \in GF(2)^{GF(2)^n}$ make $g_1(x)f(x)=0$, it indicates $g_1(x)$ are annihilators of $f(x)$. If $g_2(x) \in GF(2)^{GF(2)^n}$ let $g_2(x)(1+f(x))=0$, it shows $g_2(x)$ are annihilators of $1+f(x)$. The algebraic degree of annihilators of the lowest algebraic degree in all nonzero annihilators of $f(x)$ and $1+f(x)$ are called algebraic immunity order which is written as $AI(f(x))$ or $AI(f)$. $f(x)$ are optimal algebraic immunity functions when $AI(f)=\left\lceil \frac{n}{2} \right\rceil$, and $\left\lceil \frac{n}{2} \right\rceil$ is called maximum algebraic immunity.

Definition 6: The cascade function $f(x)$ $(x = x_0x_1x_2\cdots x_n \in GF(2)^{n+1})$ cascade with two n-dimensional Boolean functions $f_1(x')$ and $f_2(x')$ $(x'=x_1x_2\cdots x_n \in GF(2)^n)$ is defined as

$$f(x)=(1+x_0)f_1(x')+x_0f_2(x') = f_1(x')+x_0(f_1(x')+f_2(x')),$$

$f(x)$ is denoted by $f(x)=f_1(x)\|f_2(x)$.

According to the above Definitions, we can get Lemmas easily.

Lemma: For any arbitrary Boolean function $f(x)$, the following equations are true:

$$f(x) = f(x)df(x)/dx_i + ef(x)/ex_i \quad (i=1,2,\cdots,n),$$

and

$$w_t(f(x)) = w_t(f(x)df(x)/dx_i)+w_t(ef(x)/ex_i) = 2^{-1}w_t(df(x)/dx_i)+w_t(ef(x)/ex_i) \quad (i=1,2,\cdots,n).$$

3 The impact of Translation transform on the properties of RSBFs

In this part, we first discusses the influence of translation transformation on the cryptographic properties of the symmetric Boolean functions. In this paper, we take $\overline{P}=(0,0,\cdots,0)$.

Theorem 1: If $f_0^r(x)$ $(1 \le r \le n)$ are completely homogeneous RSBFs with r degree, $f_0^r(x^{\overline{P}})$ are completely RSBFs, and $\deg f_0^r(x^{\overline{P}})=\deg f_0^r(x)$.

Proof: Since $f_0^r(\rho_n^k(x))=f_0^r(x)$ $(1 \le r \le n, 1 \le k \le n-1)$, and $\overline{P}=(0,0,\cdots,0)$, $x^{\overline{P}}=(\overline{x_1},\overline{x_2},\cdots,\overline{x_n})$, so it will still be $f_0^r(\rho_n^k(x^{\overline{P}}))=f_0^r(x^{\overline{P}})$. That is to say, $f_0^r(x^{\overline{P}})$ are completely RSBFs, it is still functions which constructing from the sum of all $\binom{n}{r}$ full terms of combinations of these new n variables $\overline{x_1},\overline{x_2},\cdots,\overline{x_n}$.

Calculating each item that only contains $\overline{x_{i1}},\overline{x_{i2}},\cdots,\overline{x_{ir}}$ of $f_0^r(x^{\overline{P}})$, the results own the same highest degree terms with $f_0^r(x)$. And among the newly generated item whose degree is less than $\deg f_0^r(x)$'s, the items with even number will be fully offset, and items with odd number will be retained. So $f_0^r(x^{\overline{P}}) \ne f_0^r(x)$, though $\deg f_0^r(x^{\overline{P}})=\deg f_0^r(x)$.

The proof ends.

Corollary 1:

$$f_0^2(x^{\overline{P}})=\begin{cases} 1+f_0^2(x), & (When\ n\ is\ an\ odd) \\ f_0^1(x)+f_0^2(x), & (When\ n\ is\ an\ even) \end{cases}.$$

Theorem 2: For any completely homogeneous RSBFs $f_0^r(x)$ with degree r ($0 \le r \le n$), there are

$$w_t(\partial f_0^r(x^{\overline{P}})/\partial(x_{i1},x_{i2},\cdots,x_{ir})) = w_t(\partial f_0^r(x)/\partial(x_{i1},x_{i2},\cdots,x_{ir}))$$
$$(1 \le i \le n, 1 \le r \le n, 1 \le i_1 \le i_2 \le \cdots \le i_r \le n)$$

That is $f_0^r(x^{\overline{P}})$ and $f_0^r(x)$ has the same diffusion degree.

Proof: Due to $\overline{P}=(0,0,\cdots,0)$, making $x+P=X$, then $x_1+P_1=X_1$, $x_2+P_2=X_2$, $\cdots$, $x_{i1}+P_{i1}=X_{i1}$, $x_{i2}+P_{i2}=X_{i2}$, $\cdots$, $x_{i_j}+P_{i_j}=X_{i_j}$, $\cdots$, $x_n+P_n=X_n$. And so

$$w_t(f_0^r(x^{\overline{P}}))=w_t(f_0^r(X))=w_t(f_0^r(x)).$$

Therefore

$w_t(\partial f_0^r(x^{\bar{P}})/\partial(x_{i1}^{\bar{P}_{i1}},x_{i2}^{\bar{P}_{i2}},\cdots,x_{ir}^{\bar{P}_{ir}}))$
$=w_t(\partial f_0^r(X)/\partial(X_{i1},X_{i2},\cdots,X_{ir}))$
$=w_t(\partial f_0^r(x)/\partial(x_{i1},x_{i2},\cdots,x_{ir}))$
$=w_t(\partial f_0^r(x^{\bar{P}})/\partial(x_{i1},x_{i2},\cdots,x_{ir}))$

The proof ends.

Theorem 3:

1) $N_{f_0^r(x^{\bar{P}})}=N_{f_0^r(x)}$;

2) $f_0^r(x^{\bar{P}})$ and $f_0^r(x)$ have the same order of correlation immunity;

3) $f_0^r(x^{\bar{P}})$ and $f_0^r(x)$ have the same order of algebraic immunity .

Proof:

1) Suppose there exists $l_0(x)\in L_n[x]$ which makes

$N_{f_0^r(x)}$
$=\min_{l(x)\in L_n[x]} d(f_0^r(x),l(x))$
$=\min_{l(x)\in L_n[x]} w_t(f_0^r(x)+l(x))$
$=d(f_0^r(x),l_0(x))$

Since $f_0^r(x^{\bar{P}})$ and $l_0(x^{\bar{P}})$ are respectively gated by $f_0^r(x)$ and $l_0(x)$ through translation $P=(1,1,\cdots,1)$, there must have $d(f_0^r(x^{\bar{P}}),l_0(x^{\bar{P}}))=d(f_0^r(x),l_0(x))$, and $N_{f_0^r(x^{\bar{P}})}=N_{f_0^r(x)}$.

2) The same proof with 1), we will also get $w_t(f_0^r(x^{\bar{P}})+(\omega x)^{\bar{P}})=w_t(f_0^r(x)+\omega x)$. That is $f_0^r(x^{\bar{P}})$ and $f_0^r(x)$ has the same order of correlation immunity, $CI(f_0^r(x^{\bar{P}}))=CI(f_0^r(x))$.

3) Since $f_0^r(x^{\bar{P}})$ are gated by $f_0^r(x)$ through translation $P=(1,1,\cdots,1)$, if $g(x)$ are annihilators of the lowest algebraic degree of $f_0^r(x)$, $g(X)$ must be annihilators of the lowest algebraic degree of $f_0^r(X)$, so $g(x^{\bar{P}})$ must be annihilators of the lowest algebraic degree of $f_0^r(x^{\bar{P}})$. Known by the proof of Theorem 1, we can get $\deg g(x^{\bar{P}})=\deg g(x)$ ($g(x^{\bar{P}})\neq g(x)$). So $AI(f_0^r(x^{\bar{P}}))=AI(f_0^r(x))$.

The proof ends.

4 Construction of higher-order algebraic immunity functions with correlation immunity

We know fully rotational symmetry Boolean function under the translational transform remained a lot of good cryptographic properties. Theorem 4 first looking for higher-order algebraic immune completely rotation symmetric Boolean function, on the basis of this again by translational transform to construct high order algebraic immunity related to immune function.

Theorem 4: Let x be a vector whose dimension is not less than $2n$, and n is an odd. Since $f_0^2(x)f_0^{n-2}(x)=f_0^{n-2}(x)$, and $\deg(f_0^{n-2}(x)f_0^n(x))>n$, $f_0^2(x)+f_0^n(x)$ are n-order algebraic immunity functions. And when the dimension of x is equal to $2n$, $f_0^2(x)+f_0^n(x)$ which meet the aforementioned conditions are optimal algebraic immunity functions.

Proof: In $f_0^{n-2}(x)f_0^n(x)$, each n-degree item is gated by the addition of items with a number of $\binom{n}{n-2}\binom{1}{1}$.

Since $\deg(f_0^{n-2}(x)f_0^n(x))>n$, $\binom{n}{n-2}$ must be an even number. Known n is an odd number, therefore $(n-1)/2$ must be an even number in $n(n-1)/2$. So, for $f_0^2(x)f_0^{n-1}(x)$, $\binom{n-1}{2}\binom{1}{1}=(n-1)(n-2)/2$ and $\binom{n}{2}\binom{2}{1}=n(n-1)$ must both be even numbers, $\binom{n+1}{2}\binom{1}{1}=(n+1)n/2$ must be an odd number (otherwise the product of $f_0^2(x)f_0^{n-1}(x)$ is 0). So $f_0^2(x)f_0^{n-1}(x)=f_0^{n+1}(x)$.

For $f_0^{n-1}(x)f_0^n(x)$, since $\binom{n}{n-1}\binom{1}{1}=n$ is an odd, we can get $f_0^{n-1}(x)f_0^n(x)=f_0^n(x)$ .So $f_0^{n-1}(x)$ is certainly not the annihilator of $f_0^2(x)+f_0^n(x)$.

And since $f_0^2(x)f_0^{n-2}(x)=f_0^{n-2}(x)$ and $\deg(f_0^{n-2}(x)f_0^n(x))>n$, $f_0^{n-2}(x)$ cannot be the annihilator of $f_0^2(x)+f_0^n(x)$.

If $h_1(x)$ are functions which the degree is less than $n-2$, we can get $\deg f_0^2(x)h_1(x)<n$. Since $\deg f_0^n(x)h_1(x)>n$, the annihilators of $f_0^2(x)+f_0^n(x)$ which the degree is less than $n-2$ cannot be exist. But if $h_2(x)$ are non-RSBFs whose degree is greater than or equal to $n-2$, less than n (i.e. $n-1$ degree), $h_2(x)$ only contains some of the items which possible $n-1$ items of all $\binom{2n}{n-1}$. If let the aforementioned items be $\sum x_{i_1}x_{i_2}\cdots x_{i_{n-1}}$, there must have n-degree terms which not contain $\sum x_{i_1}x_{i_2}\cdots x_{i_{n-1}}$ in $f_0^n(x)$, since there must have n-degree terms which not contain $x_{r1}x_{r2}$ in $\sum x_{i_1}x_{i_2}\cdots x_{i_{n-1}}$, we can get $\deg f_0^2(x)h_2(x)=n+1$. And $\deg f_0^n(x)h_2(x)=2n-1$, so $f_0^2(x)+f_0^n(x)$ doesn't contain the non-RSBFs annihilator whose degree is $n-2$ or $n-1$.

Therefore, the annihilators of the lowest algebraic degree of $f_0^2(x)+f_0^n(x)$ must be $1+f_0^2(x)+f_0^n(x)$. Also $\deg(1+f_0^2(x)+f_0^n(x))=n$, it is $AI(f_0^2(x)+f_0^n(x))=n$.

Clearly, when $f_0^2(x)+f_0^n(x)$ are $2n$ variables functions, which are the optimal algebraic immunity functions.

The proof ends.

Remark: Pay attention to the variables $2n$ in Theorem 4. If the variable $<2n$, Theorem 4 does not establish. For example, if $f_0^3(x)+f_0^5(x)$ is only a 5-variable function, it has a 2-degree annihilator $g(x)=x_3x_4+x_3x_5+x_2x_4+x_2x_5$. At the same time, $f_0^3(x)g(x)=0$ and $f_0^5(x)g(x)=0$ establish.

Corollary 2: $N_{f_0^2(x^{\bar{P}})+f_0^n(x^{\bar{P}})}=N_{f_0^2(x)+f_0^n(x)}$. If there exists $l_0(x)$ which makes $N_{f_0^2(x)+f_0^n(x)}=d(l_0(x),f_0^2(x)+f_0^n(x))$, $l_0(x^{\bar{P}})$ which meets $N_{f_0^2(x^{\bar{P}})+f_0^n(x^{\bar{P}})}=d(l_0(x^{\bar{P}}),f_0^2(x^{\bar{P}})+f_0^n(x^{\bar{P}}))$ also exists.

Example: The 98-variable function $f_0^2(x)+f_0^{49}(x)$ is an optimal algebraic immunity function.

In Theorem 4, $f_0^2(x)+f_0^n(x)$ are n-order algebraic immunity functions. Known from Theorem 3, $f_0^2(x^{\bar{P}})+f_0^n(x^{\bar{P}})$ are also *n*-order algebraic immunity functions. Using $f_0^2(x)+f_0^n(x)$ and its translation transformation functions, the *n*-order algebraic immunity functions with correlation immunity can be got. And Theorem 5 can be deduced.

Theorem 5: Let $f_0^2(x),f_0^n(x)\in GF(2)^{GF(2)^{2n}}$, and the annihilator of the lowest algebraic degree of $f_0^2(x)+f_0^n(x)$ is $1+f_0^2(x)+f_0^n(x)$. Then

1) $1+f_0^2(x^{\bar{P}})+f_0^n(x^{\bar{P}})$ are the annihilator of the lowest algebraic degree of $f_0^2(x^{\bar{P}})+f_0^n(x^{\bar{P}})$.

2) $$g(x')=1+f_0^2(x)+f_0^n(x)+x_0(f_0^2(x)+f_0^n(x)+f_0^2(x^{\bar{P}})+f_0^n(x^{\bar{P}}))$$

are the annihilator of the lowest algebraic degree of

$$f(x')=f_0^2(x)+f_0^n(x)+x_0(f_0^2(x)+f_0^n(x)+f_0^2(x^{\bar{P}})+f_0^n(x^{\bar{P}}))$$

(Among them $x'=x_0x_1x_2\cdots x_nx_{n+1}\cdots x_{2n}$). That is the $2n+1$-variable functions $f(x')$ are *n*-order algebraic immunity functions.

3) $f(x')$ are correlation immunity functions.

4) $N_{f(x')}=2N_{f_0^2(x)+f_0^n(x)}$.

Proof: 1) Known the conclusion is correct from the proof of Theorem 3.

2) Obviously $g(x')f(x')=0$ is hold. Known the conclusion holds by the known conditions and cascade conclusion.

3) Because $\bar{P}=(0,0,\cdots,0)$, then

$$f_0^2(x^{\bar{P}})+f_0^n(x^{\bar{P}})=f_0^2(x+P)+f_0^n(x+P).$$

There must be

$$\partial f(x')/\partial(x_0x_1x_2\cdots x_{2n})=0.$$

There are

$$w_t(f(x')|x'_i=0)=w_t(f(x')|x'_i=1)=2^{-1}w_t(f(x')).$$

So $f(x')$ are correlation immunity functions.

4) obviously there are

$$N_{f(x')}=d(f(x'),l_0(x)+x_0(l_0(x)+l_0(x^{\bar{P}}))).$$

Since $f(x')$ are $2n+1$-variable functions, $N_{f(x')}$ are the same as $N_{f_0^2(x)+f_0^n(x)}$. So the result is obviously established, and doesn't need to be proved in detail.

The proof ends.

Because of $\lceil\frac{2n+1}{2}\rceil=\lceil\frac{2n}{2}\rceil=n$, it has the following corollary 3.

Corollary 3: $f(x')$ of Theorem 5 are the optimal algebraic immunity functions with correlation immunity.

5 Conclusions

Translation transformation makes transformed function remain the original diffusion times, nonlinearity, correlation immunity order, algebraic immunity, and the original rotational symmetry. But after the cascade transformation, some properties of the function will remain, while other properties, such as the nature of diffusion, correlation immunity and rotational symmetry will lost. The characteristic of translation transformation makes it possible to construct optimal algebraic immunity function by cascade conversion and optimal algebraic immunity rotationally symmetric correlation immune Boolean function.

In this paper, the optimal Algebraic Immune Function of a rotational symmetric Boolean function is found. And the translational transformation and cascade is used to calculate the optimal algebra are not related immune rotation symmetric Boolean function transform as the optimum algebraic immunity related immune function, increase the ability to resist related attacks. Results and methods are very significant.

In this paper, we use the derivative to prove the correlation immunity and the diffusion property.

Acknowledgment

This work is supported by National Natural Science Foundation of China (Grant No. 61262085).

References

1. Courtois, N., and Meier, W. Algebraic attacks on stream ciphers with linear feedback. Advances in Cryptology-EUROCRYPT 2003, Warsaw, Poland, 2003, LNCS, 2656:345-359.
2. Carlet C and Zeng X Y. Further properties of several classes of Boolean functions with optimum algebraic immunity.Designs, Codes and Cryptography, 2009, 52(3): 303-338.
3. Carlet C. A method of construction of balanced functions with optimum algebraic immunity. Proceedings of the First International Workshop on Coding and Cryptography, Fujian, 2007: 25-43.
4. Li Y, Yang M, and Kan H B. Constructing and counting Boolean functions on even variables with maximum algebraic immunity. IEICE Transactions on Fundamentals, 93-A(3): 640-643(2010).
5. Rizomiliotis P. On the resistance of Boolean functions against algebraic attacks using univariate polynomial representation[J]. IEEE Transactions on Information Theory, 56(8):4014-4024(2010).
6. Tu Z R and Deng Y P. A class of 1-resilient function with high nonlinearity and algebraic immunity. Ryptography ePrint Archive, Report 2010, 2010/179.
7. Wang Q, Peng J, Kan H, et al.. Constructions of cryptographically significant Boolean functions using primitive polynomials. IEEE Transactions on Information Theory, 56(6): 3048-3053(2010).
8. Li, C.,Zhang, H.,Zeng, X, et al. The lower bound on the second-order nonlinearity for a class of Bent functions. Chinese Journal of Computers, 35(8):1588-1593(2012).
9. Su S H, Tang X H. Construction of rotation symmetric Boolean functions with optimal algebraic immunity and high nonlinearity. Designs, Codes and Cryptography, 71(2): 183–199(2014).
10. Sarkar S, Gangopadhyay S. On the second order nonlinearity of a cubic Maiorana-McFarland Bent Functions. International Journal of Foundations of Computer Science, 2010, 21(3):243-254.
11. Chen, Y., Zhang, Y., Tian, W.. Construction of Even-variable Rotation Symmetric Boolean Functions with Optimal Algebraic Immunity. Journal of Cryptologic Research,2014, 1(5): 437–448. (In Chinese)
12. Sarkar S, Maitra S. Construction of rotation symmetric Boolean functions with optimal algebraic immunity. Computation Systems, 2009, 12(3): 267–284.
13. Fu S, Qu L, Li C, et al. Balanced rotation symmetric Boolean functions with maximum algebraic immunity. IET Information Security, 2011, 5(2): 93–99.
14. Dong, D., Li, C., Qu, L., et al. Rotation symmetric Boolean functions in even-variable maximum algebraic immunity. Journal of National University of Defense Technology, 2012, 34(4): 85 - 89. (In Chinese)
15. J. Huang, Z. Wang. The relationship between correlation immune and weight of H Boolean functions. Journal on Communications, Vol.33(2):110-118(2012). (In Chinese)
16. W. Li, Z. Wang, J. Huang. The e-derivative of boolean functions and its application in the fault detection and cryptographic system. Kybernetes, Vol.40(5-6):905-911(2011).
17. Zhao, M.. Method of detecting special logic function based on Boolean e-derivative. Journal of Zhejiang University (Science Edition), 41(4):424-426(2014).

Research on the Teaching System of the University Computer Foundation

Xiaoyun Ji [1,a], Bin Hu [2]

[1]Department of College Fundation Education, Bohai University, Jinzhou, China
[2]College of information science and technology, Bohai University, Jinzhou, China

Abstract. Inonal students, the teaching contents, classification, hierarchical teaching methods with the combination of professional level training, as well as for top-notch students after class to promote comprehensive training methods for different students, establish online Q & A, test platform, to strengthen the integration professional education and computer education and training system of college computer basic course of study and exploration, and the popularization and application of the basic programming course, promote the cultivation of university students in the computer foundation, thinking methods and innovative practice ability, achieve the goal of individualized educ the College of computer basic course teaching, the specific circumstances of the need for students, professiation.

1 Introduction

May 23, 2015, the Ministry of education of Liberal Arts Computer Teaching and guidance of the committee, "computer based curriculum teaching requirements," the revised work conference held in Renmin University of China. The basic orientation of computer based course "application" is defined, and the basic principle of knowledge learning is to cultivate students "ability to solve problems. It is also the ability of each knowledge unit to solve the problem.

In recent years, the university computer based student training and classroom teaching results, since 2009, the process control management, the teaching effect has been improved, but the students' basic level difference is still relatively large, students are generally not very high acceptance of computer courses. This thesis mainly focuses on these differences, how to carry out in-depth research on the teaching system of College Students' computer basis.

2 The goal of the computer based teaching system

First, the construction of basic computer course in university students graded exemption, teaching content classification, hierarchical teaching, assessment and evaluation system;

Second, basic computer courses in University of students **grading** system and grading assessment methods, the contents of the evaluation and examination paper definition, reflecting students' ability in computer based classification of discrimination, to master the basic knowledge of the students implement exempt and exempt from computer basic course in the subsequent content learning and management;

Third, the study of different professional disciplines and the university computer based curriculum content, the formation of large and small professional computer based curriculum needs analysis, and for different disciplines to design teaching content and assessment methods, to achieve the combination and optimization of teaching content;

Fourth, according to the development of computer system, the new system, such as Office2010 and Windows7, to revise the teaching material, the subject needs as the core, to establish a new teaching material system;

Fifth, the establishment of the computer application of teaching content as the main body of the teaching program, to make up for the lack of classroom theory teaching;

Sixth, the establishment of scientific and technological innovation, competition platform, through the pre competition training and guidance, to strengthen the students after class learning. Greatly improve the quality of computer application, to encourage students to choose a higher degree of difficulty of the practice of topics, stimulate the students' potential, improve their analytical problems, solve the problem of ability and ability to do.

[a]Corresponding author: bhdxjxy@126.com
Project number: 2014-A101, project name: the research and practice of the teaching system of computer basic level in University

3 The specific content of the research on the computer based hierarchical teaching system

3.1 Research on the classification scheme of teaching object and its discrimination

For first-year undergraduate computer basic teaching object classification, the grading assessment methods, appraisal content and the appraisal of discrimination, through investigations, classification of good, common and poor foundation of students, distribution, research exemption system and exempt from students of computer basic teaching contents, methods, evaluation system; study different subjects of students grading assessment methods, and gradually form a perfect classification exemption system.

3.2 Research on the content level of computer basic teaching in University

According to the basic theory of software, the computer and the computer should be used to improve students' level of learning interest. According to the basic theory of computer science, such as the teaching content of computer, the research of the teaching method, the teaching method, the teaching method of computer science, the teaching method of the computer, the teaching method of the teaching, the teaching method of the computer, the teaching method, the teaching method of the computer, the teaching method of the teaching, the students' understanding of the teaching method. Knowledge of classroom teaching, autonomous learning, and after class reading.

3.3 Research on the teaching method of the integration of computer related technology and computer basic course in subject

The college or university of professional units, the scope of application of computer technology in the research of professional requirements, typical technology, for the students' basic knowledge of computer requirements, construction of new courses teaching case library, adjust and modify the courseware; for the professional course design work, Office show the computer technology in the application of professional in the competitions and other activities, improve the students' interest; division of teachers for professional, full-time teachers to set majors different, through continuous 2-3 years of teaching practice, deepen students' demand for professional categories and the subsequent development of tracking, and gradually improve the integration of students professional and related computer technology, training the classification system of computer basic course level form.

3.4 The competition of talent cultivation and innovation of science and technology "after class" plan

Long-term mechanism to establish the school program design contest or off campus program design competition, to create a once a year, the students are extremely attractive and brand school program design competition; establish mechanisms for training and competition in personnel, school science and technology innovation of training plan and training programs; organization of top-notch students to participate in some influential national and international competition and enhance students' innovation ability and practice ability, and strive to can get the first prize in the national competition within a few years.

4 Methods and measures for the study of the hierarchical teaching system

The basic computer proficiency test method, test for undergraduate computer basic level, the implementation of a, B, C classification and setting the exemption system, for a class of students can exemption experiment course or courses, some advanced experimental courses and theory courses elective curriculum.

Establish large specialty group, with arts, science, engineering category of coarse packet, different disciplines for basic computer courses teaching requirements, forming division of teaching knowledge points; according to the common features of the school curriculum, construction of basic courses module, namely software foundation, basic hardware, office based analysis, in different disciplines, organization optimization operating system, multimedia, computer network and advanced application experiments of different modules and the degree of difficulty, forming different disciplines teaching plan. See Table 1.

Table 1. Computer basic course requirements for each subject

Subject	Course requirements
Liberal arts	Common office softwar
science	Office software,
	drawing, the use of specialized software
engineering course	Office software, the use of a variety of tools

According to different professional organization of teachers establish the appropriate team continued the same professional teaching; teachers' organizational and professional teachers to communicate and study of computer basic courses and between different professional knowledge structure; organization teachers are professional relevant case base design, and carries on the analysis to the case, and gradually improve the courseware, forming large specialty knowledge structure graph.

Revision of teaching materials, according to the Windows7 and Office2010, design the content of experimental practice courses, according to the current network technology, mobile communications, multimedia and other technology to supplement textbooks.

Relying on the campus network platform, the establishment of students before class computer foundation for investigating the mechanism, design questionnaire, let each teacher master the class the students basic level, appropriate adjustments to the teaching schedule; the establishment of student feedback system, through the automatic evaluation platform of background data analysis students to grasp the weak link; open self-test questions, online answering system and knowledge evaluation system, the students through self assessment knowledge master and teacher's teaching situation, taking the opportunity of feedback to improve the teachers explain focus.

Reform of examination system, based on the study of computer operation and computer theory, combined with the formation of small papers, job development, advanced computer applications, such as a comprehensive assessment system.

Organize relevant teachers to participate in the National University computer foundation curriculum, training, and establish a school visit and inspection mechanism, and other college computer based courses teaching team to communicate, learn advanced teaching methods, exchange courseware material; through the system to discuss the typical teaching methods, for the current courseware.

Optimize the subject and content of the lecture plan, to create a continuous series of lectures for students interested in computer application, to add the students' understanding of the latest technology, improve learning interest.

The establishment of computer technology in the professional application of the small papers, continue to promote the Office job fair, try to use computer advanced application of the practice, improve students learning motivation.

Create a year of a brand, the school computer programming contest, to arouse students' interest in learning; organization of top-notch students to participate in some influential national and international competition and enhance students' innovation ability and practice ability; create competition platform, improve students' after-school training contents and related materials, according to the different organization of the competitions of the regular classroom training, promote the innovation ability of the students in the application of computer technology.

5 The teaching method of the hierarchical teaching system

In accordance with the requirements of the reform of the teaching model, the implementation of "rotation classroom mode to teaching target as guidance, combining the traditional teaching method and modern teaching means, use inspire, inquiry, discussion, case and competition so as to teaching methods, guide students to move the brain, hands, realize the teacher-student interaction, students interaction, stimulate students learning interest and enthusiasm in learning, improve the quality of teaching.

5.1 Class Preview

Teacher at least three days ahead of schedule to the students preview content. To establish the QQ group, used to assist the exchange of information between teachers and students, including the preparation of the task of publishing. At the same time, the establishment about study group (5-6), by small group leader is responsible for the preview, seminars, open research, homework collection work group member class, helping the teacher to complete the teaching of the classroom after class.

5.2 Discuss the test before class

Use before class 10-20 minutes time to test preview task completion, guide the students to master the basic knowledge of learning this lesson, and according to the test to determine the grades of students reading. Inspection methods can be flexibly used according to different contents.

5.3 Class teacher essence

Teachers according to the students to preview task completion, to explain the knowledge induction and focuses on the problem of presentation by 50 to 60 minutes at a time, in the middle can be interspersed combination of lecture and practice, or let the speaker student, the formation of the interaction between teachers and students.

5.4 Practical training after class

Use 10-20 minutes after class to focus on the course of the training of key knowledge, through the answer questions, solve the problem of the way to consolidate the focus of knowledge, cultivate the ability to solve practical problems.

5.5 Under class guidance

Teachers in the arrangement of the class of students to complete the task after the task, according to the requirements of students personalized counseling, discussion, answering questions to help students complete the task of learning, while the layout of the new curriculum preview task.

6 Summary

In college computer basic courses of University of hierarchical teaching, and Ministry of Education Computer Teaching Steering Committee to develop the college computer teaching requirements "coincide. The guiding ideology, the main problems and challenges of the computer foundation course, the teaching objectives and contents of the course, the emphasis and difficulty of the revision work, the task division of labor and the time schedule, etc.

In the process of writing, Thanks for the Department of leadership and colleagues, as well as the school's strong support!

References

1. Zhenting Li," Visual Basic Programing Course", Northern Jiaotong University press, January2003 (In Chinese)
2. Yongquan Lu , "Use of VB development environment of multimedia software", Chinese computer newspaper,Vol 22,June 1997 (In Chinese).
3. HaiZhen Wu,JiaFu Jiang , "VB language teaching reform of Oriented programming ability cultivation", J. Journal of Educational Science of Hunan University, vol 4(6), pp. 120-122, July 2005 (In Chinese)
4. K. Elissa, "Title of paper if known," unpublished.
5. Zhiwei Li, "The realization of Oracle Database's Backup Sync in VB Environment", Computer Engineering and Design, July 2009

MATEC Web of Conferences **61**, 03008 (2016) DOI: 10.1051/matecconf/20166103008
APOP2016

A Blind Source Separation Algorithm Based on Dynamic Niching Particle Swarm Optimization

Hongjie Li[1], Zhen Li[1,2] and Hongyi Li[1,2,*]

[1]*LMIB, School of Mathematics and Systems Science, Beihang University, Beijing 100191, China*
[2]*School of Software Engineering, Beihang University, Beijing 100191, China*
**Corresponding author (E-mail: Hongyili_buaa@163.com)*

Abstract. In this paper, the dynamic niching particle swarm optimization (DNPSO) is proposed to solve linear blind source separation problem. The key point is to use the DNPSO rather than particle swarm optimization (PSO) and fast-ICA as the optimization algorithm in Independent Component Analysis (ICA). By using DNPSO, which has global superiority, the performance of ICA will be improved in accuracy and convergence rate. The idea of sub-population in DNPSO leads to the greater efficiency compared with other methods when solving high dimensional cost functions in ICA. The performance of ICA based on DNPSO is investigated by numerical experiments.

1 Introduction

Blind source separation (BSS) is an advanced method in signal processing. In recent years, BSS has gained increasing interest in various fields of applications. The problem of linear BSS, in nature, is the problem of separating and estimating waveforms of the observed sources from their linear mixture, without knowing the characteristics of the transmission channels [1,8-10]. And the traditional method of linear BSS is Independent Component Analysis (ICA).

The main objective of ICA is to estimate an un-mixing matrix by using optimization algorithm to achieve the optimal solution. Since the traditional optimization algorithm, including gradient descent algorithm and natural gradient algorithm, are easily trapped into local optimum and have slow convergence rate, researchers proposed Particle Swarm Optimization (PSO) and the its improved algorithm as the optimization algorithm for ICA and have achieved remarkable experimental result [1]. However, the standard PSO algorithm also has the probability that being trapped into local optimum, and its convergence rate turns to be slow at the later time [6]. The Chaos-PSO algorithm proposed by Ding [7] improved the searching efficiency and accuracy of the standard PSO and shows high performance in optimizing low-dimensional functions, however, its accuracy still need to be improved in dealing with high-dimensional functions. [3] All that above demonstrates a requirement of an optimization algorithm with greater global superiority for BSS.

In this paper, we proposed ICA based on Dynamic Niching Particle Swarm Optimization (DNPSO), which has greater global superiority in optimization progress of ICA algorithm. Dynamic Niching Particle Swarm Optimization (DNPSO) maintains the diversity of its particle swarm by applying the idea of niche technique. The widely using of Niche Technique in Genetic Algorithm (GA) in recent years has interpreted its remarkable performance in global optimization of high-dimensions functions [5]. In DNPSO, the particle swarm is divided into several sub-swarms, which form independent searching spaces dynamically so as to ensure every extreme point being examined. Meanwhile, the free particles are always searching for new optimums, which can help the algorithm avoid being trapped into local optimum [6].

The remainder of this paper is organized as follows. Section 2 presents an overview of the BSS problem based on ICA. Section 3 describes ICA based on standard PSO. Section 4 presents the proposed ICA based on DNPSO. The simulation results will be presented in Section 5.

2 Blind Source Separation (BSS) Problem

2.1 Problem description of BSS

The problem of linear BSS can be modeled as follows:

$$s(t) = (s_1(t), s_2(t), \dots, s_n(t))^T$$

Assuming that there exists n source signals, which are mutually statistically independent and unknown at each time instantly. The BSS mixing model can be presented as

$$\mathbf{x}(t) = \mathbf{A}\mathbf{s}(t),$$

where A is an unknown $n*n$ mixing matrix with full rank, $x(t) = (x_1(t), x_2(t), \dots, x_n(t))^T$ are the observed signals. The task of BSS is to recover the source vector $s(t)$ by using the observed signals $x(t)$ and possible prior knowledge. The process of separation can be

**Corresponding author (E-mail: Hongyili_buaa@163.com)*

formulated that the $n*n$ separating matrix W has the following structure, which is called the un-mixing equation as follows

$$\mathbf{y}(t) = \mathbf{W}\mathbf{x}(t) = \mathbf{W}\mathbf{A}\mathbf{s}(t),$$

where $y(t) = (y_1(t), y_2(t), ..., y_n(t))^T$ is the estimation of the source signal $s(t)$, W is the separating matrix. Without considering the additive noise, the basic thought of BSS is to separate the independent sources from the mixtures by estimating the separating matrix so as to realize Blind Source Separation.

2.2 Independent Component Analysis (ICA) and its optimization algorithm

Independent Component Analysis (ICA) is currently the major method to solve BSS problem. The most widely used algorithm based on ICA is a kind of adaptive optimization algorithm, which find the separation matrix by using optimization algorithm to a determined cost functions after utilizing the measure of independence between separated signals. In the process of using ICA to solve BSS problem, the choosing of cost function, which estimate the similarity between source signals and separation signals, and the optimization algorithm for determining separation matrix will greatly influence the separation effect [1].

The core strategy of improving ICA method lies on the global optimization of the cost function. In traditional ICA, gradient descent algorithm and natural gradient algorithm are used as optimization algorithm, but these methods based on gradient exists the disadvantage of being easily trapped in local optimum[2] In 2007, Vrins, Pham and etc. proved that the cost function of non-gaussian maximum principle has a large amount of local optimums[4]. This demonstrates that when using the optimization algorithm with poor global superiority, the separation precision of ICA method may not be ideal enough as the algorithm based on gradient will be easily trapped into local optimums.

3. ICA based on particle swarm optimization

The Particle Swarm Optimization (PSO) is an optimization algorithm inspired by the process of searching food of bird flock and fish school, especially Swarm Theory. In PSO, each searching individual is regarded as a particle, which represents a potential solution in the D-dimension solution space. All the particles in the solution space combined are regarded as the Particle Swarm.

Assuming that there are m particles in the Particle Swarm, $p_i = (p_{i1}, p_{i2}, ..., p_{iD})$ is the D-dimensions position vector of the $i-th$ particle. The superiority of the current position in solution space can be estimated by calculating the fitness of p_i with the cost function determined. PSO disperses each particle in the whole solution space and adjust its velocity according to its previous velocity and independent cognition (the best solution position that this particle found) as well as the social cognition (the best solution position that the whole particle swarm found), so as to converge the particle swarm on the position of the optimal solution.

The PSO-based ICA algorithm can be implemented by the following iterative procedure [7]:

1) Set the cost function $f(x)$
2) Preprocess the source signals and randomly initialize particle swarm's positions and velocities.
3) The best previous position which giving the best fitness value of the ith particle is updated by the following expression, which also updates the best position among the whole particles:

$$p_i(t+1) = \begin{cases} p_i(t) \ if \ f(x_i(t+1)) < f(p_i(t)) \\ x_i(t+1) if \ f(x_i(t+1)) \geq f(p_i(t)) \end{cases}$$

4) Update the position and velocity vector of each particle:

$$\mathbf{v}_{is}(t+1) = \mathbf{v}_{is}(t) + c_1 r_1 (\mathbf{p}_{is}(t) - \mathbf{x}_{is}(t)) + c_2 r_2 (\mathbf{p}_{gs}(t) - \mathbf{x}_{is}(t))$$

$$\mathbf{x}_{is}(t+1) = \mathbf{x}_{is}(t) + \mathbf{v}_{is},$$

where learning factors c_1 and c_2 are nonnegative real numbers, r_1 and r_2 are independent random numbers subject to the uniform distribution between [0, 1].

5) Repeat 2) and 3) until convergence.
6) Use the particle w with the best fitness value as the separation matrix and compute the separated signals

$$\mathbf{y}(t) = \mathbf{W}\mathbf{x}(t)$$

In traditional PSO algorithm, each individual particle shares the best position of the whole particle swarm, which leads to the high possibility that the whole particle swarm converge on the optimum found early with slow convergence rate [6]. These limitations will require an efficient and accurate method.

4. ICA based on dynamic niching particle swarm optimization

Dynamic niching Particle Swarm Optimization (DNPSO), proposed by Nickabadi in 2008, was based on the niche technique. [6], and was mainly applied as the optimization algorithm for multi-extreme value problems.

DNPSO uses the concept of sub-populations. In this method, the particles are divided into some sub-swarms and a group of free particles. The sub-swarms are created and destroyed dynamically at each generation and the number of the particles in each swarm changes. The best fitted neighbor is determined as $lbest$ and is called the master of the particle. Any particle that is its own best neighbor can be a candidate solution or a niche in the objective function and is referred to as a self-master particle, the particle's master is replaced with the master of its master repeatedly until a self-master particle is reached. And if two masters are too close to each other, the particle with the better fitness value is set as the master of the other particle and form a sub-swarm. All of the particles that are not a member of any sub-swarm are set as free particles [6].

Owing to the using of GCPSO method in sub-swarm searching and the cognition-only model in free particles searching, the sub-swarms will form a niching search area near a promising solution and converge quickly to the optimum point while the free particles can find all of the extreme value points in searching space and form a new sub-swarm around it, which protects the algorithm from being trapped in local optimum The DNPSO improves the exploration capability of the standard PSO and ensures both the accuracy and efficiency of the results while being applied to high-dimensional functions with multi local optimums.

The most widely used cost function in ICA method such as Non-gaussian maximum principle and mutual information estimation, has been proved to have multiple local optimums in practice using.[4], Considering how the multiple local optimums affects the performance of ICA, which have been discussed above, we use DNPSO in the ICA method for BSS. The ICA based on dynamic niching particle swarm optimization (DNPSO) can be implemented as follows:

1) Set the cost function $f(\mathrm{x})$; processing pretreatment and whitening process to the blind source signals; make the signals zero mean and unit covariance:

$$\mathbf{X} = \mathbf{X} - E[\mathbf{X}], \mathbf{H} = \mathbf{D}^{-\frac{1}{2}}\mathbf{F}^T\mathbf{X} ,$$

2) Initialize the parameters and maximum iteration times of DNPSO; randomly generate the initial positions z_{id}^0 and velocity v_{id}^0 based on the dimensions of the signal and calculate the fitness value. Each position of the particles indicates a potential solution of separation matrix W needed in ICA.

3) Use the cognition-only model PSO algorithm for free particles group (At the beginning all the particles are free particles), the positions and velocities is updated according to the following equations(p_{id}^k is the best position of the id_{th} particle, w is inertia weight, c is the learning factor, r is a random parameter between [1,2])

$$v_{id}^{k+1} = w * v_{id}^k + c * r * \left(p_{id}^k - z_{id}^k\right)$$

$$z_{id}^{k+1} = z_{id}^k + v_{id}^{k+1}$$

4) For each particle, if it is not self-master, the particle's master is replaced with the master of its mater repeatedly until a self-master particle is reached.

5) If two self-masters particles' Euclidian distance is smaller than the threshold value $d_{\min}$, the particle with the better fitness value is set as the master of the other particle and form a sub-swarm.

6) Train the velocity and position of the particles of each sub-swarm according to GCPSO, specifically as the equation follows [6]: (p_{best}^k is the best position of this sub-swarm)

$$v_i^{k+1} = \omega v_{gbest}^k + c_1 * r_1 * \left(p_{id}^k - z_{id}^k\right) + c_2 * r_2 * \left(p_{gbest}^k - z_{id}^k\right)$$

$$z_i^{k+1} = z_{pbest,i}^k + v_i^{k+1}$$

7) Update the members of each sub-swarm: If the Euclidian distance between any particles in two sub-swarms is smaller than the threshold, combine these two sub-swarms.

8) Repeat 3) to 7) until converge or the maximum iteration times is reached or all the particles have converged.

9) Use the particle w with the best fitness value as the separation matrix and compute the separated signals

5. Simulation Results

ICA based on DNPSO was tested with three speech signals shows in Figure.2, where totally 100000 sampling pixels are set as abscissa and amplitude are set as ordinates.

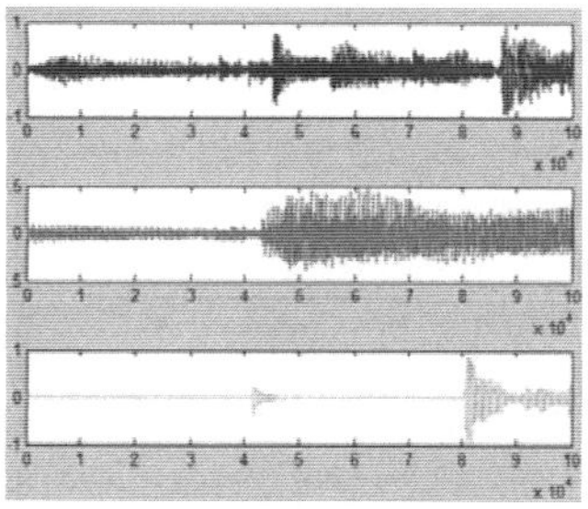

Figure 1. Source signals.

The mixture signals are generated by mixing the source signals with a random 3*3 matrix, which is as the following:

$$\mathbf{H} = \begin{pmatrix} 0.3912 & 0.4423 & 0.4265 \\ 0.2524 & 1.1225 & 0.3527 \\ 0.5223 & 0.4265 & 0.3127 \end{pmatrix}$$

The mixture signals are showed in Figure. 2:

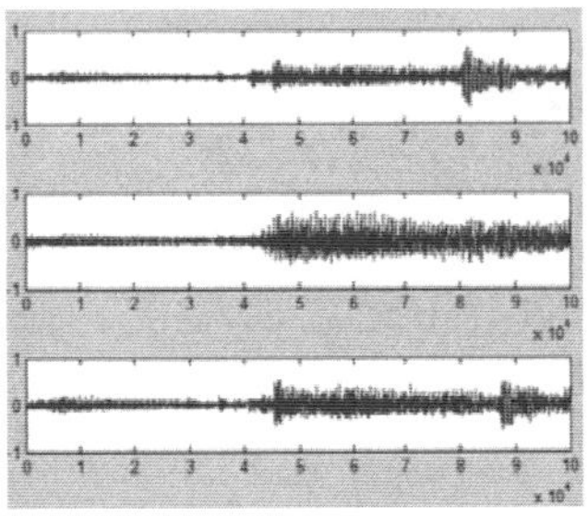

Figure 2. Mixture signals.

In this paper we use the widely-used negative entropy as the cost functions of ICA, its common approximation algorithm is as follows:

$$J(X) = \sum_{i=1}^{n} \left(E\left[f_i(X) \right] - E\left[f_i(G) \right] \right)^2$$

where G is the random Gaussian vectors with zero mean and unit covariance. And f is a none-quadratic function which can be chosen according to the requirement, $f(\mathrm{x}) = \frac{1}{a}\ln(\cosh(\mathrm{ax}))$ is used in our simulation where a = 1.5.

The parameters of DNPSO are set as follows. The number of particles n = 150. The maximum number of sub-swarms M = 30 , learning rate c = 1.5 , The maximum number of particles in a sub-swarm $MinPts = 5$, $w_{max} = 0.9$, $w_{min} = 0.1$, maximum iterate times $G_{max} = 700$.

To estimate the results of ICA based on DNPSO, we performed the blind source separation with the proposed algorithm along with two other algorithms, including the well-known FastICA and ICA based on standard PSO. Their separation results are showed in Figure. 3 to Figure. 5.

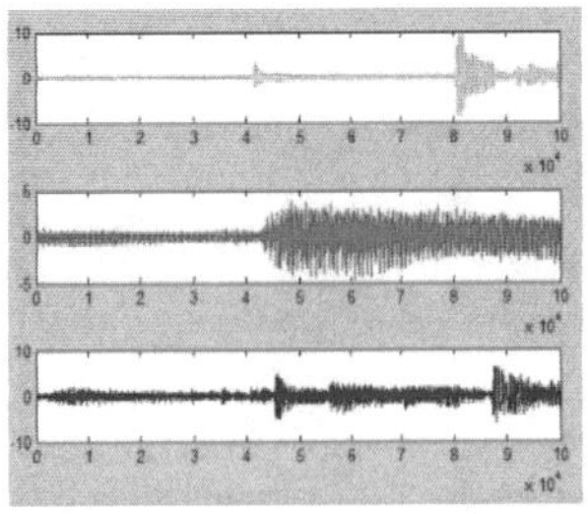

Figure 3. Separated signals using FastICA.

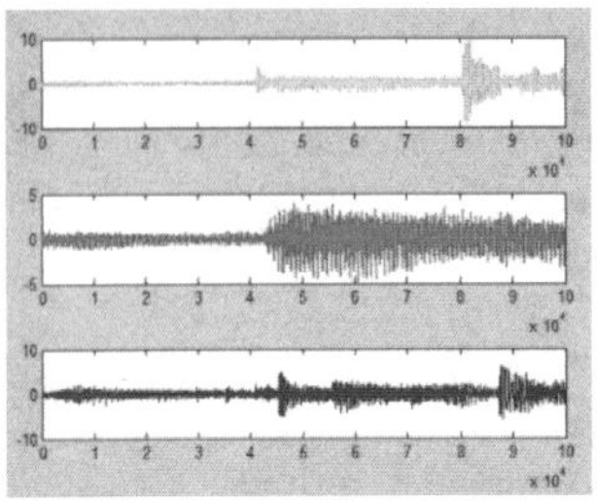

Figure 4. Separated signals using PSO

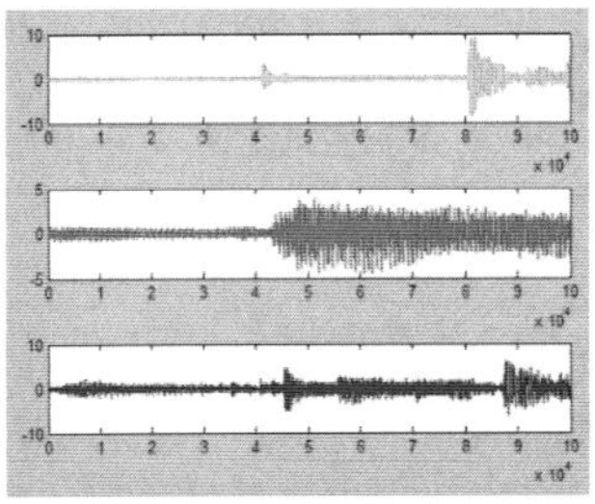

Figure 5. Separated signals using the proposed method.

We used measurement index of similarities between signals to estimate the results quantitatively (y indicates the separation signals while s indicates the source signals):

$$\varphi_{ij} = \varphi(y_i, s_i) = \frac{\left| \sum_{t=1}^{T} y_i(t) s_j(t) \right|}{\sqrt{\sum_{t=1}^{T} y_i^2(t) * \sum_{t=1}^{T} s_j^2(t)}}$$

The results of the three used algorithm is showed in the following tables:

Table 1. Performance of three algorithm in simulation

Methods	Proposed	PSO-ICA	Fast-ICA
Similarity of signal 1	0.9989	0.9670	0.9840
The similarity of signal 2	0.9892	0.9368	0.9897
The similarity of signal 3	0.9879	0.9644	0.9743
Time cost (in second)	4.321	9.917	0.553

The performance results show that ICA based on DNPSO outperforms both FastICA and ICA based on standard PSO in separation precision. As to the coverage rate, ICA based on DNPSO also shows remarkable superiority over ICA based on standard PSO.

Owing to its fast converge rate, ICA based on DNPSO requires fewer iteration times and has great superiority in stability. In all of the methods except the proposed algorithm, the exploration will probably focuses on the optimum found earlier and been trapped in the nearby domain. But in ICA based on DNPSO, the free particles are always search for better solutions. In general, ICA based on DNPSO solves the problem of low separation accuracy in blind source separation with an ideal processing rate.

DOI: 10.1051/matecconf/20166103008

6. Conclusion

In this paper, an ICA method based on Dynamic Niching Particle Swarm Optimization (DNPSO) is proposed for blind source separation in view of the local optimum and slow convergence rate difficulty. The DNPSO is for the first time in this paper to be used in ICA method. The division of main swarm in DNPSO not only help guarantee the global optimum to be found in searching domain, but also improves the separation precision of blind source separation remarkably. According to the simulation results which compare the proposed method with Fast-ICA and PSO-ICA, the proposed method significantly improves BSS's performance and gives better results.

As the parameter setting of ICA base on DNPSO still has much space to be optimized, and the additive noises is not concerned in this paper. Our future work will focus on the optimization of parameter as well as taking the additive noises into consideration, and to further deal with the nonlinear BSS problem.

Acknowledgement

This work was supported by the National Natural Science Foundation of China (No.61379001) and the Key Project of Beihang University (No. 201412) ("Exploration and Practice of Innovation Ability Cultivation in Mathematical Education").

References

1. Y. Gao, S. Xie. A blind source separation algorithm using particle swarm optimization. In Proceedings of the Circuits and Systems Symposium on Emerging Technologies: Frontiers of Mobile and Wireless Communication, **1**, 297-300 (2004)
2. T. Kurihara, K. Jinno. Analysis of convergence property of PSO and its application to nonlinear blind source separation. In Proceedings of IEEE Congress on Evolutionary Computation, 976-981 (2013)
3. C. L. Lin, S. T. Hsieh, T. Y. Sun. PSO-based learning rate adjustment for blind source separation. In Proceedings of International Symposium on Intelligent Signal Processing and Communication Systems, 181-184 (2005)
4. F. Vrins, D. T. Pham, M. Verleysen. Mixing and non-mixing local minima of the entropy contrast for blind source separation. IEEE Transactions on Information Theory, **3(3)**, 1030-1042 (2007)
5. R. Brits, A. P. Engelbrech, F. V. Bergh. A niching particle swarm optimizer. In Proceedings of Asia-Pacific Conference on Simulated Evolution and Learning, **2**, 692-696 (2002)
6. A. Nickabadi, M. M. Ebadzadeh, R. Safabakhsh. DNPSO: A Dynamic Niching Particle Swarm Optimizer for Multi-modal Optimization. In Proceedings of IEEE Congress on Evolutionary Computation, 26-32 (2008)
7. W. Ding. A New Method for Image Noise Removal using Chaos-PSO and Nonlinear ICA. Procedia Engineering, **4**, 111-115 (2011)
8. H. Li, C. Wang, D. Zhao. An Improved EMD and Its Applications to Find the Basis Functions of EMI Signals. Mathematical Problems in Engineering, (2015)
9. H. Li, Z. Song, D. Zhao, P. Wang, J. Chen. A Single Channel EMI Signal Separation Method Based on Directly-mean Empirical Mode Decomposition. Journal of Infromation & Computation Science, **12(17)**, 6333-6340 (2015)
10. D. Zhao, H. Li, On the computation of inverses and determinants of a kind of special matrices. Applied Mathematics and Computation, **250**, 721-726 (2015)

An Integration Algorithm for Bistatic Radar Weak Target Detection

Jiajun Chang[1]

[1]School of Electronic Engineering,University of Electronic Science and Technology of China,Chengdu,China

Abstract. The bistatic radar weak target detection problem is considered in this paper. An effective way to detect weak target is the long time integration. However, range migration (RM) will occur due to the high speed. Without knowing the target motion parameters, a long time integration algorithm for bistatic radar is proposed in this paper. Firstly, the algorithm utilizes second-order keystone transform (SKT) to remove range curvature. Then the quadratic phase term is compensated by the estimated acceleration. After that, SKT is used once more and the Doppler ambiguity phase term compensation is performed. At last, the target energy is integrated via FT. Simulations are provided to show the validity of the proposed algorithm in the end.

1 INTRODUCTION

Bistatic radars have most advantages due to the geometrical properties of the system. Apart from its relative immunity to some forms of attack, bistatic radars can utilize the echo signal, which does not travel over the same path as the transmitted signal, to detect weak targets [1-3].Therefore , this paper discuss the weak target detection methods for bistatic radar.

It is known that the long time integration can increase the signal-to-noise-ratio (SNR) and thus improve the radar detection ability [4-7]. It includes coherent integration and non-coherent integration. By compensating the phase fluctuation between different pulses, the performance of coherent integration is much better than the non-coherent integration. MTD is known as the classic coherent integration method to increase SNR of echo signals. However, for the high speed targets, RM will occur within the coherent integration interval. Keystone transform (KT) is studied to remove the RM in [8]. Paper [9] proposes an algorithm to integrate the echo signal energy, which is named RFT. However, KT and RFT can only detect the targets with a constant radial velocity. In order to address targets with radial acceleration, SKT is used to remove the range curvature in paper [10]. However, paper [10] hasn't considered the Doppler frequency ambiguity (DFA) problem, which is bring about by the acceleration.

This paper proposes a long time integration method for bistatic radars, which can solve the RM and DFA problem. Firstly, the algorithm utilizes SKT to remove the range curvature. Then the quadratic phase term is compensated by the estimated acceleration. After that, SKT is used once more and the Doppler ambiguity phase term compensation is performed. Finally, the target energy is integrated via FT.

The remainder of this paper is organized as follows. In Section II, the echo signal model is analyzed. In section III, the integration algorithm is proposed. Simulations are provided to demonstrate the effectiveness of the proposed method in Section IV. Finally, the conclusions are made in Section V.

Corresponding Author:electriciankevin@163.com

2 ECHO SIGNAL MODEL

2.1 Target Motion Model

The bistatic radar geometry is shown in Figure 1. The bistatic radar is located at the X coordinate. $(-L,0,0)$ denotes the location of the transmitter T_x and $(L,0,0)$ denotes the location of the receiver R_x. The origin of the coordinates system is located at the middle point of the two radars. The original location of the target is (x_0, y_0, z_0). $\vec{V}$ denotes the speed vector of the target.

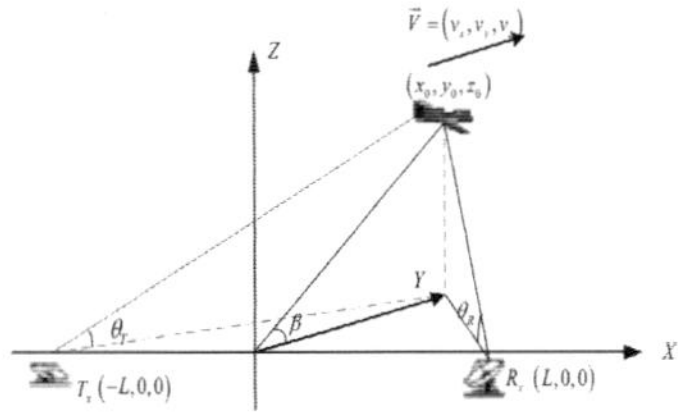

Fig.1. bistatic radar geometry

Suppose that v_x, v_y and v_z denote the coordinate values of target velocity along X coordinate direction, Y coordinate direction and Z coordinate direction, respectively. According to Figure 1, the sum of transmitter-object and object-receiver instantaneous distances can be calculated as follows

$$R_s(t)=\sqrt{(x_0+v_xt+L)^2+(y_0+v_yt)^2+(h_0+v_zt)^2}+\sqrt{(x_0+v_xt-L)^2+(y_0+v_yt)^2+(h_0+v_zt)^2} \quad (1)$$

According to the Taylor series theory, neglecting the high order components, we have

$$R_s(t)\approx R_0+vt+\frac{1}{2}at^2 \quad (2)$$

Where R_0 is the sum of transmitter-object and object-receiver initial distances, v, a denote the target velocity and acceleration respectively, which are relative to the direction of bistatic radar line of sight.

2.2 Echo Signal Model

Suppose that the radar transmits the linear frequency modulation (LFM) signal as follows

$$a(\tau)=rect\left(\frac{\tau}{T_p}\right)\exp\left(j\pi k\tau^2\right) \quad (3)$$

Where T_p is the pulse width, B is the band width, $k=B/T_p$ denotes the modulation rate.

The echo signal from a point target can be written as follows [11]

$$s_r(\tau,t_m)=A_0\cdot rect\left(\frac{\tau-R_s(t_m)}{T_p}\right)\times\exp\left(j\pi k\left(\tau-\frac{R_s(t_m)}{c}\right)^2\right)\times\exp\left(j2\pi f_c\left(t-\frac{R_s(t_m)}{c}\right)\right) \quad (4)$$

Where A_0 is the echo amplitude, $t_m=mT_r$ is the slow time, $m=0,1,\cdots,M$, M is the number of radar pulse, T_r is the pulse repetition interval, c denotes the speed of light, f_c is the carrier frequency, $R_s(t_m)$ can be given by (2).

After demodulation and range compression, the compressed signal in the range frequency-slow time (f_r-t_m) domain can be given as

$$\begin{aligned}S_r(f_r,t_m)&=A_1rect\left(\frac{f_r}{B}\right)\exp\left[-j\frac{2\pi(f_r+f_c)}{c}R_s(t_m)\right]\\&=A_1rect\left(\frac{f_r}{B}\right)\exp\left[-j\frac{2\pi(f_r+f_c)}{c}R_0\right]\times\\&\exp\left[-j\frac{2\pi(f_r+f_c)}{c}vt_m\right]\times\exp\left[-j\frac{\pi(f_r+f_c)}{c}at_m^2\right]\end{aligned} \quad (5)$$

Where A_1 is the compressed signal amplitude, the range frequency is f_r, λ denotes wavelength.

Due to the high speed of the target and the low radar pulse repetition frequency (PRF), under-sampling would occur. Therefore, the target speed can be expressed as

$$v=v_0+d\cdot v_{amb} \quad (6)$$

Where $v_{amb}=\lambda/T_r$ is the bistatic radar blind velocity, d is the Doppler center ambiguity number, $v_0=\mathrm{mod}(v,v_{amb})$ is the unambiguous velocity which satisfies

$$|v_0|<\frac{v_{amb}}{2}.$$

Substituting (6) into (5),we have

$$\begin{aligned}S_r(f_r,t_m)&=A_1rect\left(\frac{f_r}{B}\right)\exp\left[-j\frac{2\pi(f_r+f_c)}{c}R_0\right]\times\\&\exp\left[-j\frac{2\pi(f_r+f_c)}{c}v_0t_m\right]\times\exp\left[-j\frac{\pi(f_r+f_c)}{c}at_m^2\right]\\&\times\zeta(f_r,t_m)\end{aligned} \quad (7)$$

Where $\zeta(f_r,t_m)=\exp\left[-j\frac{2\pi(f_r+f_c)}{c}dv_{amb}t_m\right]=\exp\left[-j\frac{2\pi f_c}{c}dv_{amb}t_m\right]$.

Performing the inverse Fourier transform (IFT) of (7) in the range frequency domain, we have

$$s_r(\tau,t_m)=A_1B\mathrm{sinc}\left[B\left(\tau-\frac{R(t_m)}{c}\right)\right]\exp\left[\frac{-j2\pi f_cR_s(t_m)}{c}\right] \quad (8)$$

(8) shows that the compressed signal envelope moves along t_m, which is called RM, and the acceleration would result in the DFS. Both RM and DFS would result in

target energy being defocused.

3 THE INTEGRATION ALGORITHM

3.1 Performing SKT to Remove Range Curvature

We don't introduce the SKT in detail because it has been discussed in papers [10] before. The formula is

$$t_m = \left(\frac{f_c}{f_c + f_\tau}\right)^{\frac{1}{2}} t_\alpha \tag{9}$$

Where t_α is the new slow time after SKT.

Substituting (9) into (7), we have

$$\begin{aligned} S_r(f_\tau, t_\alpha) = A_1 rect\left(\frac{f_\tau}{B}\right) \exp\left[-j\frac{2\pi(f_\tau + f_c)}{c}R_0\right] \times \\ \exp\left[-j\frac{2\pi(f_\tau + f_c)}{c}v_0\left(\frac{f_c}{f_c + f_\tau}\right)^{\frac{1}{2}} t_\alpha\right] \times \exp\left[-j\frac{\pi f_c}{c}at_\alpha^2\right] \\ \times \zeta(f_\tau, t_\alpha) \end{aligned} \tag{10}$$

Where $\zeta(f_\tau, t_\alpha) = \exp\left[-j\frac{2\pi f_\tau}{c}dv_{amb}\left(\frac{f_c}{f_c + f_\tau}\right)^{\frac{1}{2}} t_\alpha\right]$

In (10), it can be seen that the range curvature is removed. However, the RM still exists and the second-order phase term have to be compensated.

3.2 Compensating the Quadratic Phase Term

The quadratic phase term should be compensated before correcting the residual RM. In (10), we can see that the phase in the cross range dimension is a chirp signal. Therefore, the method in paper [12] can be used to estimate the acceleration a .With the estimated value $\tilde{a}$, we can construct the phase compensation function

$$H(\tilde{a}) = \exp\left[j\frac{\pi f_c}{c}\tilde{a}t_\alpha^2\right] \tag{11}$$

Multiplying (11) with (10), if $\tilde{a} \approx a$, we have

$$\begin{aligned} S_r(f_\tau, t_\alpha) \approx A_1 rect\left(\frac{f_\tau}{B}\right) \exp\left[-j\frac{2\pi(f_\tau + f_c)}{c}R_0\right] \times \\ \exp\left[-j\frac{2\pi(f_\tau + f_c)}{c}v_0\left(\frac{f_c}{f_c + f_\tau}\right)^{\frac{1}{2}} t_\alpha\right] \times \zeta(f_\tau, t_\alpha) \end{aligned} \tag{12}$$

3.3 Correcting the Residual RM

In (12), the residual RM still exists. The SKT is used to correct the RM once more. Substituting the formula $t_\alpha = \left(\frac{f_c}{f_c + f_\tau}\right)^{\frac{1}{2}} t_\beta$ into (12), we get

$$\begin{aligned} S_r(f_\tau, t_\beta) = A_1 rect\left(\frac{f_\tau}{B}\right) \exp\left[-j\frac{2\pi(f_\tau + f_c)}{c}R_0\right] \times \\ \exp\left[-j\frac{2\pi f_c}{c}v_0 t_\beta\right] \times \zeta(f_\tau, t_\beta) \end{aligned} \tag{13}$$

Where $\zeta(f_\tau, t_\beta) = \exp\left[-j\frac{2\pi}{\lambda}dv_{amb}\frac{f_\tau}{f_c + f_\tau}t_\beta\right]$.

3.4 Compensating the Doppler ambiguity phase term

In (13), we can see that the range curvature and RM are removed. However, the Doppler ambiguity phase term still exists, which have to be compensated. We construct the compensation function

$$\xi(f_\tau, t_\beta, d') = \exp\left[j\frac{2\pi}{\lambda}d'v_{amb}\frac{f_\tau}{f_c + f_\tau}t_\beta\right] \tag{14}$$

Multiplying (14) with (13), we have

$$\begin{aligned} S_r(f_\tau, t_\beta) = A_1 rect\left(\frac{f_\tau}{B}\right) \exp\left[-j\frac{2\pi(f_\tau + f_c)}{c}R_0\right] \times \\ \exp\left[-j\frac{2\pi f_c}{c}v_0 t_\beta\right] \times \exp\left[j\frac{2\pi}{\lambda}(d' - d)v_{amb}\frac{f_\tau}{f_c + f_\tau}t_\beta\right] \end{aligned} \tag{15}$$

If $d' = d$, (15) can be expressed as

$$\begin{aligned} S_r(f_\tau, t_\beta) = A_1 rect\left(\frac{f_\tau}{B}\right) \times \exp\left[-j\frac{2\pi f_c}{c}v_0 t_\beta\right] \\ \times \exp\left[-j\frac{2\pi(f_\tau + f_c)}{c}R_0\right] \end{aligned} \tag{16}$$

Performing the IFT of (16) in the range frequency f_τ domain, we have

$$s_r(\tau, t_\beta) = A_2 \operatorname{sinc}\left[B(\tau - R_0/c)\right] \times \exp\left[-j\frac{2\pi f_c}{c}(R_0 + v_0 t_\beta)\right] \tag{17}$$

In (17), it is shown that the signal is a complex sinusoid signal over t_β . Therefore, the Doppler center ambiguity number can be calculated as follows

$$\tilde{d} = \underset{d}{\arg\max}\left|FT_{t_\beta}\left\{IFT_{f_\tau}\left[S_r(f_\tau, t_\beta)\xi(f_\tau, t_\beta, d')\right]\right\}\right| \tag{18}$$

Where IFT_{f_τ} denotes the IFT over f_τ dimension, FT_{t_β} denotes the FT over t_β .

With the estimated Doppler center ambiguity number $\tilde{d}$, the compensation function can be defined as $\xi(f_\tau, t_\beta, \tilde{d}) = \exp\left[j\frac{2\pi}{\lambda}\tilde{d}v_{amb}\frac{f_\tau}{f_c + f_\tau}t_\beta\right]$ to compensate the Doppler ambiguity phase term.

3.5 Integration by FT

In (17), we know that the RM and DFA are addressed, which is helpful to the integration of the echo signals energy. Applying FT to (17) over t_β, we have

DOI: 10.1051/matecconf/20166103009

$$s_r(\tau,t_\beta) = A_3 \sin c\left[B(\tau - R_0/c)\right] \times \sin c\left[T\left(f_{t_\beta} + \frac{v_0}{\lambda}\right)\right] \quad (19)$$

Where $T = NT_r$ is the integration time.

From (19), we know that the target energy is integrated. If the peak of (19) is larger than a

TABLE I. SIMULATION PARAMETERS

Pulsewidth T_p $10\mu s$	Transmitter location $(-8000,0,0)$
Bandwidth B $15MHz$	Receiver location $(8000,0,0)$
Carrier frequency f_c $10GHz$	Target initial location $(4000,6000,10000)$
Pulse repetition frequency f_r $300Hz$	Target initial velocity $(400,600,70)$
Number of integrated pulses M 150	SNR -8dB
Sampling rate f_s $30MHz$	

given threshold, the target will be detected.

4 SIMULATIONS

Simulations are carried out to evaluate the validity of the proposed method. The parameters used in the simulations are listed in Table I. Suppose there is a point target located in the scene.

Figure 2 shows the absolute error of the approximate formula (2). We see that the error is much less than the range sampling cell $\Delta r = \frac{c}{f_s} = 10$. Therefore, the approximate formula (2) is proper. Figure 3(a) shows the echo after demodulation. The result after range compression is shown in figure 3(b). In figure 3(b), serious range migration occurs due to the target's high speed. Figure (4) shows the result of the proposed method. Figure 4 (a) shows the 2-D image of the result after compensating the Doppler ambiguity phase term, which shows that the RM is removed. Figure 4(b) is the result of the proposed method, which shows the target energy is integrated. The target can be detected if the peak of figure 4(b) is larger than a given threshold. Therefore, the simulation result demonstrates the effectiveness of the proposed algorithm. Figure 5 shows the result of the other methods. Figure 5(a) and Figure 5(b) show that the two methods, i.e. MTD and KT, are invalid because the target radial velocity is changing during the integration interval. Figure 5(c) shows the result of the algorithm proposed in [10]. Due to DFA, the target energy hasn't been integrated in figure 5(c).

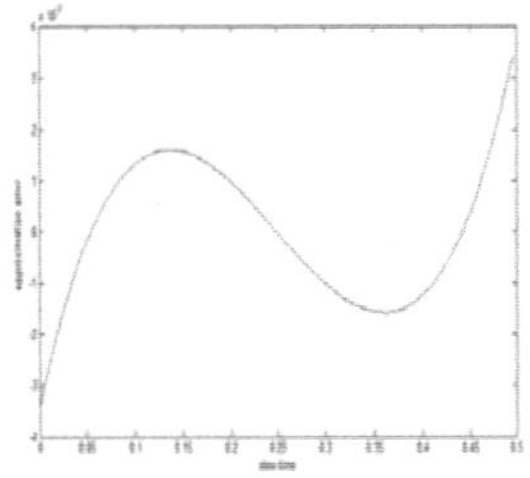

Fig.2 the absolute error of the approximate formula (2)

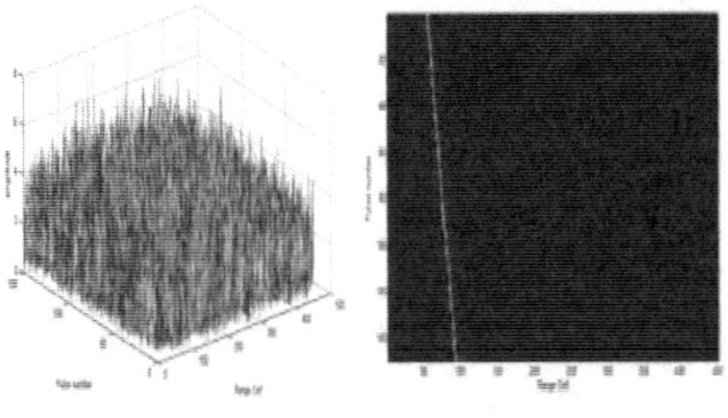

(a) (b)

Fig.3 target echo:(a) The echo after demodulation, (b) The result after range compression

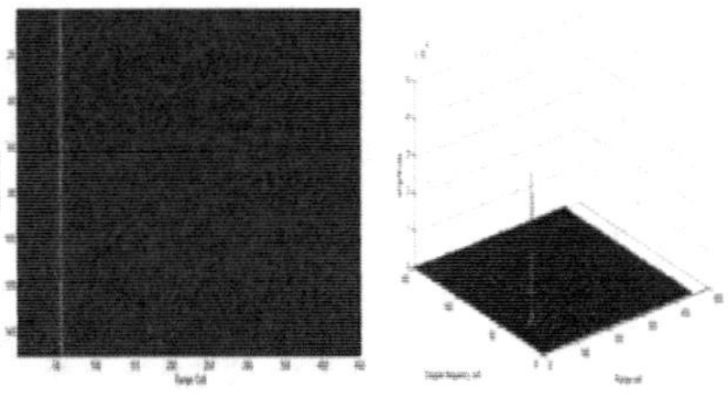

(a) (b)

Fig.4.the result after processing:(a) The 2-D image of the result after compensating the Doppler ambiguity phase term, (b) The result of the proposed method

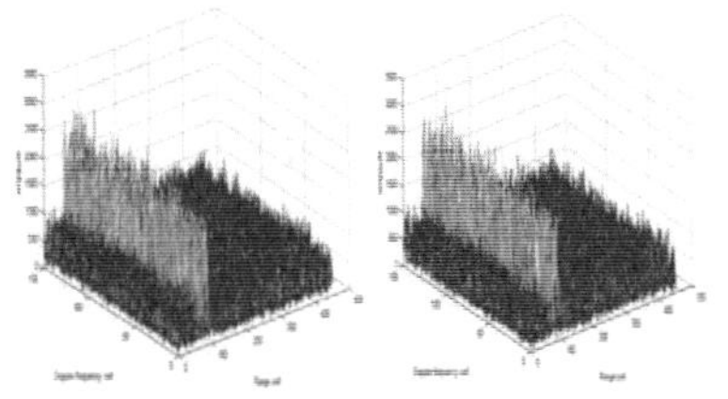

(a) (b)

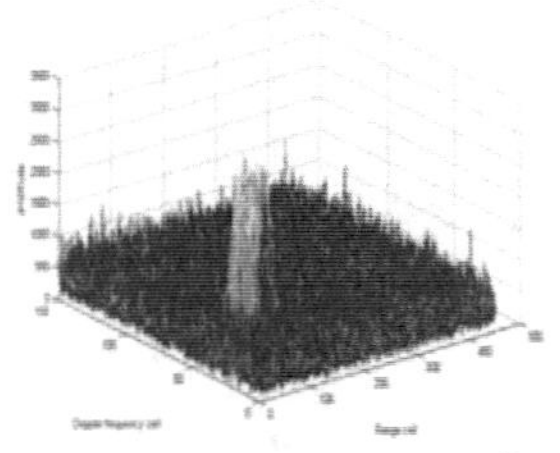

(c)

Fig.5.the results of other methods: (a) The result of MTD, (b) The result of KT, (c) The result of method in [10]

5 CONCLUSION

This paper proposes a long time integration method for bistatic radar weak target detection. Without knowing the target motion information, the algorithm firstly corrects the range curvature of the target echo using the SKT. Then the quadratic phase term is estimated and compensated. After compensating the Doppler ambiguity phase term, the target energy can be integrated by FT. Simulations have shown the validity of the proposed method.

References

1. Kreucher C. Exploiting narrowband bistatic radar measurements for dismount detection and trackin [J] .IEEE antennas and propagation magazine, 2011, 53(1) : 98-105.
2. Hussain M. Ambiguity functions for monostatic and bistatic radar systems using UWB throb signal [J] . IEEE trans on aerospace and electronic systems. 2011, 47 (3) : 1710-1722.
3. Christiansen J, Olsen K. Range and Doppler walk in DVB-T based passive bistatic radar [C] . IEEE international conference, 2010, 620-626.
4. Skolnik M . Role of radar in microwaves [J] . IEEE trans on microwave theory and techniques, 2002, 50 (3) :625-632.
5. Carretero-Moya J , Gismero-Menoyo J , Asensio-Lopez A, et al. Small-target detection in high-resolution heterogeneous sea-clutter: an empirical analysis [J] . IEEE transactions on aerospace and electronic systems , 2011 , 47(3) :1880-1898.
6. Carlson B D, Evans E D, Wilson S L. Search radar detection and track with the Hough transform, Part I: system concept[J]. IEEE transactions on aerospace and electronic systems. 1994, 30(1) : 102-108.
7. Mo L, Wu S, Li H. Radar detection of range migrated weak target through long-term integration [J]. Chinese Journal of Electronics. 2003, 12(4) : 539-544.
8. R. P. Perry, R. C. Dipietro, and R. L. Fante,"SAR imaging of moving targets,"IEEE Transactions on Aeropace and Electronic Systems, vol.35, no.1, pp.188-200, January 1999.
9. J. Xu, J. Yu, Y. N. Peng, and X. G. Xia,"Radon-Fourier transform(RFT) for radar target detection(I): Generalized Doppler filter bank processing,"IEEE Transaction on Aerospace and Electronic Systems, Vol. 47, no. 2, pp. 1186-1202, April 2011.
10. D.Kirkland,"Imaging moving targets using the second-order keystone transform," IET Radar Sonar and Navigation, vol. 5, no. 8, pp. 902-910, October 2011.
11. F. Comblet, A. Khenchaf, A. Baussard, and F. Pellen, "Bistatic synthetic aperture radar imaging: Theory, simulations and validations," IEEE Trans.Antennas Propag., vol. 54, no. 11, pp. 3529-3540, Nov. 2006.
12. Li Ya-chao, Su Jun-hai, and Xing Meng-dao, et al.. Research on ISAR imaging of the complexly moving target based on the time-chirp distribution [J]. Journal of Xidian University, 2008, 35(1): 1-7.

Design and Research of a New secure Authentication Protocol in GSM networks

Ai-qin QI[1,a], Yong-jun SHEN[2]

[1]Northwest University For Nationalities ,China
[2]Lanzhou University , China

Abstract. As the first line of defense in the security application system, Authentication is an important security service. Its typical scheme is challenge/response mechanism and this scheme which is simple-structured and easy to realize has been used worldwide. But these protocols have many following problems In the GSM networks such as the leakage of user indentity privacy, no security protection between home registers and foreign registers and the vicious intruders' information stealing and so on. This paper presents an authentication protocol in GSM networks based on maths operation and modular square root technique . The analysis of the security and performance has also been done. The results show that it is more robust and secure compared to the previous agreements.

1 Introduction

Global mobile communications network GSM originated in 1980s as the Pan-European digital cellular system standard and it is now accepted as the worldwide wireless communication. They convey all sorts of data and messages through wireless channel and can't require the cables. With the development of network applications, security problems arises . As the first defense line of secure application system, the authentication is the most important secure service. However, it is not feasible for GSM networks to use traditional security mechanisms because of its inherent limitations. A kind of simple and efficient authentication protocol needs to be taken into account.

In this paper, we propose a suitable GSM authentication protocol that use mathematical and modular square residual to complete the authentication process. The protocol security analysis has been done. Compared to the previous agreements, it is more robust and secure.

The rest of the paper is organized as follows: Section 2 discusses related work. Section 3 presents the proposed authentication protocol. Section 4 analyzes the security and the performance of the proposed protocol. Section 5 finally concludes the paper.

2 Related work

User authentication is an important topic for communication security and there are many schemes existed for the purpose in the literatures. In 1981, Lamport[1] proposed a user authentication scheme for communication in insecure channel. With the development and application of wireless networks and mobile technologies, people begin to focus on GSM authentication protocol . In 2004, M. S. Hwang and L. H. Li[2] presented an authentication scheme, but it was found that this solution had weakness and a modified version to overcome them has been presented by M. S. Hwang, I.E.Liao and C. C. Lee[3]. The study in the literature [4] has showed that the scheme in [3] also fails to provide anonymity. Two recent studies [5][6] has shown that all above schemes are incapable of providing the anonymity. Later security weaknesses of the approaches in [2-4] have been concluded and an enhanced authentication scheme[7] with anonymity named as EASA has been presented. Through the analysis of the protocol in the literature [7], we find that the scheme is still not to provode anonymity ,it is also vulnerable to insider attack, off-line password guessing attack and forgery attack etc.

In this paper, we propose a suitable GSM authentication protocol using mathematical and modular square residual[8] to complete the certification process. This paper also makes the analysis of security and performance. Compared to the previous agreement, it is more robust and secure.

3 The proposed protocol

Supposing that a mobile user Mu whose home agent is HA will visit a foreign network and the foreign agent is FA. HA Selects two large prime numbers p, q and calculates

[a] Corresponding author: qi-133@163.com

n=pq as the public key. HA selecsts a random number X_{HA} and a primitive g of the multiplicative group Zn* and calculates $Y_{HA} = g^{X_{HA}} \bmod n$. (X_{HA} , Y_{HA}) makes up the master key of HA and the key length is long enough such as 512 bits. For each foreign agent, H_A calculates the corresponding session key which meets $SK_{HF} = h(ID_{FA} \| X_{HA} \| ID_{HA})$ by using Diffie-Hellman key distribution protocol and hash function such as SHA-1 algorithm.

In this section, we present our authentication protocol which has five phases: initialization ,login, authentication , refreshment of session key and refreshment of password. Table 1shows the symbol that would be used in this paper.

Table 1. the symbol description

symbol	Description
Mu	mobile user
ID_x	The identity of mobile user x
pw_u	The passward of mobile user Mu
HA	Home register of mobile user
FA	Fforeign register
$\|\|$	Connection operation
$\oplus$	Exclusive-OR operation
$h(\cdot)$	Hash function

3.1 Initialization

Mobile user Mu sends his identity Id_U to his home register HA by a secure channel. HA computes $TH_u = h(ID_{mu} \| Y_{HA})$, $SK_{uH} = h(ID_{mu} \| X_{HA})$ and injects the information $\{TH_u, SK_{uH}, h(.), n\}$ into the smart card. Then HA sends the smart card to the mobile user Mu. After receiving the smart card, Mu selects a random password pw_u ,a random number b and calculates $SK_{uH}' = h(ID_{mu} \| pw_u) \oplus SK_{uH}$, $V_{uH} = TH_u \oplus h(ID_{mu} \| h(pw_u \| b))$, $H_{uH} = h(TH_u) \oplus h(ID_{mu} \oplus b)$. $\{V_{uH}, H_{uH}, SK_{uH}', h(.), n, b\}$ is the final information of the smart card.

3.2 Login

Mu inserts the smart card into the terminal and inputs ID_{mu} , pw_u to calculate $TH_u^* = V_{uH} \oplus h(ID_{mu} \| h(pw_u \| b))$, $H_{uH}^* = h(TH_u^*) \oplus h(ID_{mu} \oplus b)$. Mu checks if H_{uH} * equals H_{uH} , if equal the user is a legal user. Then Mu selects a random number n_{mu} ,computes $m_1 = (ID_{mu} \| ID_{FA} \| TH_u \| n_{mu})^2 \bmod n$ and sends m_1 to the foreign agent FA.

3.3 Authentication

The authentication phase includes five steps as follows:

Step 1: After receiving the information m_1, FA selects a random number n_{FA} and computes $req_{FA} = n_{FA} \oplus SK_{HF}$ and sends the information $m_2 = \{m_1, req_{FA}, h(m_1 \| req_{FA} \| SK_{HF})\}$ to HA.

Step 2: After receiving FA's message, HA decrypts m_2 by using the prime p,q to get $ID_{mu}, ID_{FA}, TH_u, n_{mu}$ .Then HA checkes the validity of ID_{mu} and judges if $h(ID_{mu} \| Y_{HA})$ equals TH_u . If correct, HA computes $h(ID_{FA} \| ID_{HA} \| X_{HA})$ and checks if $h(m_1 \| req_{FA} \| h(ID_{FA} \| ID_{HA} \| X_{HA}))$ calculated equals $h(m_1 \| req_{FA} \| SK_{HF})$ received from FA, if the above condition holds, FA is authenticated by HA. HA computes $n_{FA}^* = h(ID_{FA} \| ID_{HA} \| X_{HA}) \oplus req_{FA}$, $K_1 = h(ID_{mu} \| ID_{FA} \| n_{mu}) \oplus h(SK_{HF} \| n_{FA}^*)$, $K_2 = h(h(n_{mu}) \| h(n_{FA}^*) \| SK_{uH})$, $S_1 = h(K_1 \| K_2 \| n_{FA}^* \| SK_{HF})$ and sends $m_3 = \{K_1, K_2, S_1\}$ to FA.

Step3: When receiving the message m_3, FA uses the session key SK_{HF} to compute $h(K_1 \| K_2 \| n_{FA} \| SK_{HF})$ and judge if it equals S_1 . If it holds,HA is authenticated by FA. FA calculates $h(ID_{mu} \| ID_{FA} \| n_{mu}) = K_1 \oplus h(SK_{HF} \| n_{FA})$, $K_3 = h(ID_{mu} \| ID_{FA} \| n_{mu}) \oplus h(n_{FA})$ and sends $m_4 = \{K_2, K_3\}$ to the mobile user Mu.

Step4: Mu receives the message m_4 , computes $h(n_{FA}) = h(ID_{mu} \| ID_{FA} \| n_{mu}) \oplus K_3$ and judeges if $h(h(n_{mu}) \| h(n_{FA}) \| SK_{uH})$ equals K_2 , if it holds , FA is authenticated by Mu. Mu computes the session key $SK = h(h(n_{FA}) \| h(ID_{mu} \| ID_{FA} \| n_{mu}))$ which is used to communicate with FA. Mu computes $h(h(n_{FA}) - 1)$ amd sends $m_5 = h(h(n_{FA}) - 1)$ to FA.

Step5: FA computes $h(h(n_{FA}) - 1)$ with the number n_{FA} he selected and compares if $h(h(n_{FA}) - 1)$ equals m_5 . If equal, Mu is authenticated by FA. Then FA computes the session key $SK = h(h(n_{FA}) \| h(ID_{mu} \| ID_{FA} \| n_{mu}))$ to be used to communicate with Mu. The authentication proceess ends.

3.4 Refreshment of Session key

When Mu wants to communicate with FA frequently, the session key needs to update periodically. FA sends a temporary certificate $(TCert_u)_{SK}$ that includes life time and other informations to Mu. If Mu wants to visit FA in i-th time, Mu sends $\{TCert_u,(n_i \| TCert_u)_{SK_i}\}$ to FA. The i-th sessinon key $SK_i = h(n_{i-1} \| h(ID_{mu} \| ID_{FA} \| n_{mu})), i = 1,2,...,n$ ($n_0 = h(n_{FA})$). FA checks the validity of the certificate, decrypts $(n_i \| TCert_u)_{SK_i}$ with SK_i and compares if $TCert_u$ equals the certificate decrypted. If it holds, FA saves n_i as the next communication. The session key refreshment ends.

3.5 Refreshment of password

Mu inserts the smart card into the terminal to verify the legitimacy of the identity Mu. If it holds, Mu selects a new password pw' and a new random number b' and computes $V_{uH}* = TH_u \oplus h(ID_{mu}) \| h(pw' \| b'))$, $TH_u* = V_{uH}* \oplus h(ID_{mu} \| h(pw' \| b'))$, $H_{uH}* = h(TH_u^*) \oplus h(ID_{mu} \oplus b')$. The smart card information is $\{V_{uH}*, H_{uH}*, SK_{uH}*, h(.), n, b'\}$ and the refreshment of password ends.

4 Security and performance analysis

4.1 Security Analysis

In Our scheme, the anonymity of Mu is obtained by hash function and Exclusive-OR operation. Assume that an attacker has extracted the information stored in Mu's smart card and the used meaages tranmitted among Mu,FA and HA. However, the attacker can not derive the real identity of Mu without knowing the value of (p, q) and the random number n_{mu}.

In our scheme, the information of the smart card is is $\{V_{uH}, H_{uH}, SK_{uH}', h(.), n, b\}$. When the user's smart card is stolen or losses ,an attacker can guess the user's identity and password to compute SK_{uH}'' and TH_u^*. In order to verify the correctness of SK_{uH}'' and TH_u^*, the attacker must derive the information from $\{m_1, m_2, m_3, m_4, m_5\}$, but the malicious attacker does not know the value of (p, q). According to the residual knowledge of modular square, it is not possible to get the message m_1, n_{mu}, ID_{mu}, TH_u in a polynomial time. So SK_{uH}'' and TH_u^* can not be derived from the information $\{m_1, m_2, m_3, m_4, m_5\}$. Our scheme can resist offline password guessing attack

Our scheme also can resist a foreign agent attack. When A malicious foreign agent such as FA_{j+1} replays the used message m_1 to a legitimate foreign agent such as FA_j, in this case, FA_{j+1} will get the message $m_4 = \{K_2, K_3\}$, but he does not know the news $h(ID_{mu} \| ID_{FA} \| n_{mu})$ and can't get $h(n_{FA})$, so he can not send a response message $m_5 = h(h(n_{FA}) - 1)$. Malicious agents can not mimic legitimate agent identity to deceive mobile users because malicious agents do not know the key SK_{HF} and the random number n_{FA}.He can not generate messages m_4 to the user, the attack failed.

Our scheme also can resist forgery attack. Supposing that malicious attackers forgery into a mobile user or home agent to try to attack. They don't know $\{TH_u, ID_{mu}\}$, malicious attackers can not finish the legitimate user authentication through the home agent. They do not know the key (p, q), malicious attackers can not impersonate the home agent to decrypt the message m_1 and perform the authentication through mobile users and foreign agent.

The proposed protocol also can resist known-session key attack , internal attack and replay attack. This scheme also has the characteristics of forward secrecy and so on , which enhance the performance of the protocol.

4.2 Performance Analysis

Our proposed protocol contains only five news exchange in terms of communication load and does not need clock synchronization mechanism, which reduces the consumption of clock synchronization and the timestamp setting. In terms of computational complexity, our protocol uses hash function and Exclusive-OR operation because of the wireless environment restrictions on energy and computing power. The proposed protocol performs not more than 2 msec in OpenSSL running time. Compared with other relevant protocols , our protocol is more simple and efficient.

5 Conclusions

In this paper, we propose an efficient authentication protocol for GSM networks and analyze the security and performance of the protocol. The results show that the protocol is secure and robust.

Acknowledgement

The work is supported by Scientific and technological projects for Central colleges and universities. (No. 31920150079).

References

1. I.L .Leslie, Comm.ACM **24**, 770 (1981).
2. J.Zhu , J.Ma, IEEE.Tran.C.E **50**,230(2004).
3. C.C.Lee,M.S.Hwang ,I.E.Liao, IEEE. Tran.C.E **53**, 1683(2006).
4. C. C. Wu, W.B.Lee , W. J. Tsaur, IEEE. Com.L **12** , 22(2008).
5. J. S.Lee, D. H. Lee, IEEE.Com. L **13**, 292(2009).
6. P. Zeng,Z.Cao,K.K.R.Choo, S.Wang, IEEE.Com.L **13**,70(2009).
7. C.C.Chang,C.Y.Lee ,Y. C. Chiu,Com.Stan.I **32**,611(2009).
8. J.Y.Liu, *Applied cryptography*(2008).

ETC technology and its Application in the Campus Intelligent Vehicle Management System

Yanhua Ren, Ruijun Zhang, Simin Yin[a], Kui Fu

School of Management, Wuhan University of Science and Technology, Wuhan, China

Abstract: In order to solve the problem that the vehicles within the university campus are increasing steadily and a large number of outside vehicles drive through the campus arbitrarily and park disorderly. In this paper, a campus intelligent vehicle management system based on ETC is designed, and the ETC technology is introduced. On the basis of analyzing the system's hardware architecture and workflow, a design model of lane is given. Finally, it discusses the communication theory of Winsock, customs a format of data frame and proposes a Winsock-based intelligent data synchronization algorithm. The daily traffic volume has been decreased by 65% since the system is put into operation in a university.

Keywords: ETC; Intelligent Vehicle Management System; Model; Winsock; Data Synchronization

1 Introduction

As the living standard of people improve ceaselessly, the vehicles within the university campus are increasing steadily, and a large number of outside vehicles drive through the campus arbitrarily and park disorderly. In order to control the flows of the outside vehicles, ease the campus transport pressure, guarantee the traffic safety of the teachers and students and the traffic smooth and create a safe, green, harmonious campus environment, we urgently need a campus intelligent vehicle management system which can identify automatically a variety of types of vehicles and has a high degree of automation.

Some scholars made a lot of research in the vehicle management system, such as Hao Zhuo designed the structure of hardware and software for the vehicle management system based on RFID [1]; Wentao Guo etc, designed and researched intelligent parking management system based on RFID [2], but they didn't discuss the traffic safety issues when the vehicle access lanes and the data synchronization and consistency between the client and server.

In the ETC technology, Zhengdong Liu etc, introduced the design of hardware and software and workflow for the lane control system [3]; Fei Wang Designed and implemented the ETC lane toll software for highway [4], but they didn't apply the ETC technology to the area of the campus intelligent vehicle management system.

In the campus intelligent vehicle management project for a university, we put the ETC technology into it for the first time, gave a design model of lane and proposed a Winsock-based intelligent data synchronization algorithm. Finally, a campus intelligent vehicle management system based on ETC is designed and implemented.

[a] Corresponding author: Simin Yin@yinsimin0570@126.com.org

2 Introduction of ETC technology

Electronic Toll Collection System (ETC System) is developed on the RFID technology. It integrates computer technology, automatic control technology, detection technology and communication technology and other high-tech means. When the vehicles pass the toll lane, they are charged automatically by the system according to their type [3]. The system consists of the on board units (OBU), the road side unit (RSU) and loop sensor etc. When the vehicles pass the lane installing RSU, loop sensor perceive the vehicles, the RSU sends out enquiry signal and the OBU gives the response to complete the two-way communication and data exchange, and then complete the transaction between them according to various operation parameters, ultimately achieve the automatic collection of tolls.

3. System design

3.1 Hardware architecture

The hardware architecture of the system is shown in figure 1. The system consists of the IPC, the antenna (RSU), electronic label (OBU), controller, induction coil, vehicle detectors, alarm light, sound box, LED display, lane traffic lights, camera, automatic barrier, IC card reader, ballot box and other equipment. The microwave antenna installed in sentry box and the electronic label installed in the vehicle exchange information through a dedicated short range microwave communication (DSRC) in this system.

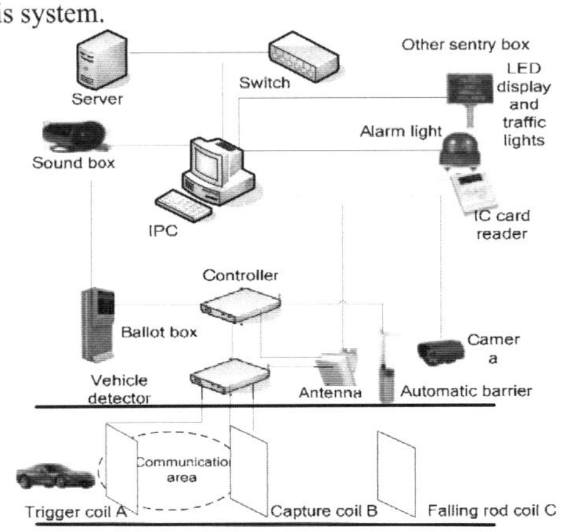

Figure 1. Hardware architecture diagram.

3.2 System workflow

According to the users' identity, the vehicle is divided into two categories which are staff vehicles and temporary user vehicles in this system. For the staff vehicles, we adopt remote identification technology, so it needn't stop when the vehicles pass the lane. For the temporary user vehicles, it must obtain a temporary card from the ballot box at the entrance and return the card at the exit. After paying by charging management practices, it is allowed to pass. The system workflow is shown in figure 2.

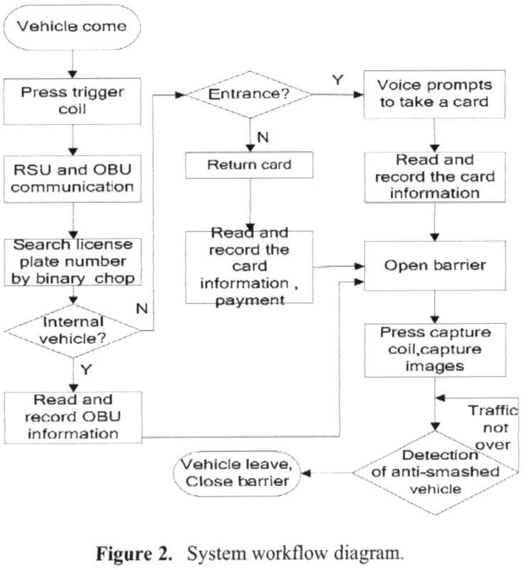

Figure 2. System workflow diagram.

4 Design model of ETC lane

The layout of ETC lane is shown in figure 3. When the vehicles enter the ETC lane of the sentry box, in order to ensure the traffic safety, realize smooth transaction between OBU and RSU and improve the efficiency of traffic, this paper puts forward with the design model of ETC lane. The model is confirmed by lane geometric parameter [5], which includes lane driving speed, the minimum safe distance between vehicles, the distance between the lane induction coils, lane length, lane width, etc. The specific design model as follows:

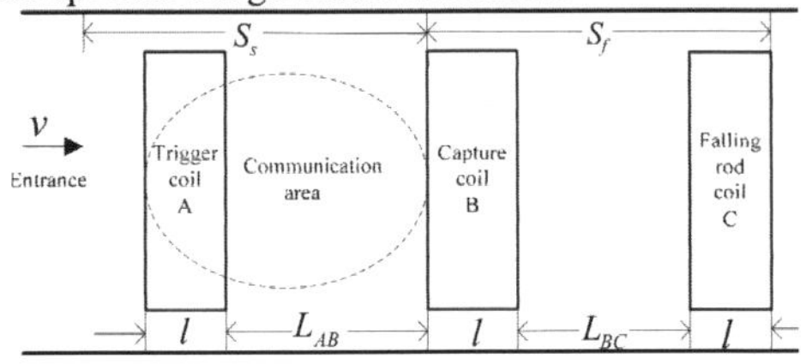

Figure 3. Layout diagram of ETC lane.

4.1 lane driving speed

As the vehicle passes the deceleration strip before it enters the ETC lane, so the lane driving speed is confirmed by initial value and the deceleration parameter of the deceleration strip.

$$v = v_0 - \lambda \ (0 \le \lambda \le v_0) \tag{1}$$

v_0 is the maximum speed limit for the vehicle on the campus, λ is set by the deceleration value of the deceleration strip, suppose v_0 = 20 km/h , λ = 10 km/h.

4.2 The minimum safety distance between vehicles

In general, two vehicles drive into the ETC lane one after the other, when the leading driver finds the traffic light being red and brakes immediately, the following driver can also brake immediately via the leading vehicle's braking information until stopping which needs two stages that is the driver's braking response time named t_r and continuous braking time. So we can work out the minimum safety distance between vehicles by working out the following vehicle's driving distance in those two stages which is named S_f and the leading vehicle's driving distance during the continuous braking time which is named S_l .

$$S_s = S_f + S - S_l = \frac{v_f t_r}{3.6} + \frac{v_f^2}{2g(f \pm i) \times 3.6^2} + S - \frac{v_l^2}{2g(f \pm i) \times 3.6^2} \tag{2}$$

S is the distance between the two vehicles after they are stopping, v_f and v_l are the initial speed of two vehicles when they are braking, g is the gravity accelerated speed, f is the resistance coefficient of road and i is the longitudinal grade of road, the uphill take mark "+" and the downhill take mark"-".Suppose v_f = v_l =10 km/h, S=1.5 m, g=9.8 m/ s^2 , f=0.6, i=0, t_r =1.2 s ,we can get $S_s \approx 4.83$ m.

4.3 The distance between trigger coil A and capture coil B

According to the manufacturer's product specifications, we know the projection of the microwave antenna in the lane is an ellipse that is communication area. The vehicle enters into the trigger coil A, and then it drives away the communication area to enter into the capture coil B. In the whole process, the OBU and the RSU communicate with each other through three stages that is the opened time of RSU, the waken time of OBU and the transaction time between the OBU and the RSU. They are named t_o, t_w and t_t. Supposing the length of the coil is l, we can figure out the distance between trigger coil A and capture coil B which is named L_{AB}.

$$L_{AB} = v(t_o + t_w + t_t)/3.6 - l \tag{3}$$

For example, according to the manufacturer's instruction, the opened time of RSU is generally between 60 ms and 300 ms, the waken time of OBU is less than 5 ms, the transaction time about 250 ms to 300 ms. Suppose t_o=300 ms, t_w=5 ms, t_t=300 ms, v=10 km/h, l=0.5 m, we can get L_{AB}=1.18 m. We set it as 1.5 m in this project.

4.4 The distance between capture coil B and falling rod coil C

When the Vehicle leaves communication area and enters into the capture coil B, if it is recognized as a temporary user vehicle, the alarm light is warning, the traffic lights is red and the automatic barrier is closed, then the driver begin braking in reaction time until stopping. According to S_f, we can get the distance between the capture coil B and the falling rod coil C which is named L_{BC}.

$$L_{BC} = S_f - 2l = \frac{v_f t_r}{3.6} + \frac{v_f^2}{2g(f \pm i) \times 3.6^2} - 2l \tag{4}$$

For example, according to B, v_f=10 km/h, t_r=1.2 s, g=9.8 m/s^2, f=0.6, i=0, l=0.5 m, we can get $L_{BC} \approx$ 3.00 m.

4.5 lane length

As shown in figure 3, the temporary user vehicle leaves the communication area until encountering the red light to stop. The entire route plus the minimum safety distance between vehicles is the lane length which is named L, we can get it according to B.

$$L = S_f + S_s = 2\left(\frac{v_f t_r}{3.6} + \frac{v_f^2}{2g(f \pm i) \times 3.6^2}\right) + S - \frac{v_l^2}{2g(f \pm i) \times 3.6^2} \tag{5}$$

For example, according to B and D, we can see that $L = S_s + S_f = L_{BC} + 2l + S_s$ = 3.00+0.50+4.83=8.83 m. We set it as 9 m in this project.

4.6 lane length

The lane width is related with the vehicle type, vehicle width, transverse safety distance and other factors. According to the City Road Design Standard Regulation, we set it as 3.50m in this project.

5 Intelligent data synchronization algorithm based on WinSock

In the implementation of the system, the SQL Server database of each sentry box and the Oracle database of management center server must exchange data with each other to realize the data synchronization between the server and each sentry box. In the field of distributed database, we usually adopt the technologies which are data replication, data pipeline, XML and Winsock to realize data synchronization between heterogeneous databases [6]. As for data replication technology, it is very complex for the configuration and maintenance of the database and prone to replication block, and what's more, there is some system delay. Data pipeline technology lacks of response mechanism, so it's hard to know whether the other has been received the data. Meanwhile, it also depends on high-performance network. Once the network breaks down, the transmission will come to naught. XML technology is a popular technique at present, but it requires setting up Web server and high-performance hardware. It also requires additional applications to complete data transmission, so it increases the system load. Therefore, this paper adopts Winsock communication mechanism to realize data synchronization. On the one hand it can greatly reduce the delay of data synchronization to realize complete intelligent. On the other hand, it has fault-tolerance mechanism and supports breakpoint continuingly. It can retransmit the data to adapt to more challenging environments.

5.1 the communication theory of Winsock

The TCP/IP protocol is the most important protocol in Internet network. Winsock conduct network communication on the basis of TCP/IP. It contains the UDP protocol and TCP protocol, in which the TCP protocol is connection-oriented and can provide high reliability service; the UDP protocol is connectionless and can provide efficient services. As the system requires relatively high for the reliability of data, so we use the TCP protocol. The data flow diagram of client/server based on Winsock is shown in figure 4.

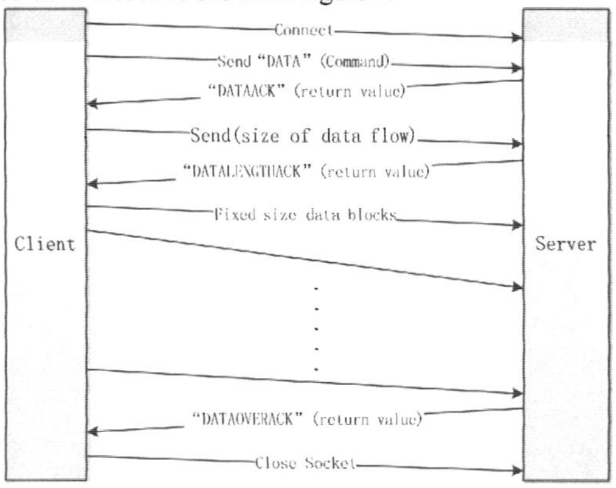

Figure 4. Data flow diagram of client/server.

5.2 Custom data frame

It is crucial to set a format of data frame for ensuring the synchronization and consistency of the sending data and receiving data. According to TCP protocol, when we set the data frame format, we must meet the following principles of communication protocol: (1) Handshake, when one starts to send data, it must notify the other. (2) Safety, it must ensure that only the two sides reached a handshake agreement, can they transmit data mutually. (3) Fault tolerance, when the data block is error during transmission, it must inform the other and make appropriate treatment immediately. Based on the above principles, we define the format of data frame as shown in Figure 5.

HEAD	CMD	IP	LEN	DATA	CRC	END

Figure 5. The format of data frame.

The symbols have the following meanings: "HEAD" is the head of the frame and its length is 2 bytes; "CMD" is the control instruction and its length is 4 bytes; "IP" is the receiver's address and its length is 4 bytes; "LEN" is the length of the sending data and its length is 2 bytes; "DATA" is the content of the data, and its length can not exceed 255 bytes; "CRC" is the check code and its length is 2 bytes; "END" is the end of the frame and its length is 2 bytes.

5.3 the implementation of intelligent data synchronization algorithm

The design of data synchronization algorithm is divided into two parts: client design and server design. In this paper, the system takes the database table as a unit to achieve data synchronization. Take the records of vehicle in or out table named t_IO for example, specific implementation as follows:

Step 1: Write a service program named Server and add a control array of Winsock named wskServer(0) in the server, then add a control of Winsock named wskClient in the client system which is running in each sentry box. Set IP address and port number, and transmission protocol for client and server. The transmission protocol is TCP protocol. Go to step 2.

Step 2: The Server monitors the connection request from each client by calling the listen method. It doesn't accept the client's request until the accept method is executed. Go to step 3.

Step 3: Each client calls the connect method to establish a connection with the Server. If the connection is failed, the system prompts error and handles it; if the connection is successful, the system check the temporary records of vehicle in or out table named t_tempIO if there are some records from local database. If there is record, remove the record and write it into the buffer first, then send it to the Server via the sendData method, delete this record from the t_tempIO at last, go to step 5; otherwise, go to step 4.

Step 4: When the vehicle passes through the lane, each client stores the vehicle access record to the t_IO of the local database, meanwhile, the system check if the connection is normal. If the connection is normal, write the record into the buffer via the sendData method, then remove the record from the buffer and send it to the Server; if the connection is failed, insert the record into the t_tempIO of local database.

Step 5: The Server receives the data from the local buffer by calling getData method, and then stores the record to the t_IO of the server database. Finally, the Server sends the message which whether processed successfully or not to the client. It goes to step 3 after the client makes appropriate treatment.

6 CONCLUSION

Applying ETC technology to this system, it had achieved obvious effect since it is put into operation at a university. On the one hand, the recognition rate of internal vehicles reaches over 99.95% which ensure the vehicle traffic of campus smooth; On the other hand, the daily traffic flow had decreased from 20000 to 7000 which alleviate the transport pressure of campus efficiently and create a harmonious campus environment.

Acknowledgement

This research is supported by the National Natural Science Foundation under Grant 90924026 from the China Government, and by Natural Science Foundation under Grant 2014CFB804 from Hubei Province, China.

References

1. Hao Zhuo. The vehicle management system based on RFID [J]. Computer and Digital Engineering, 2008, 36(1): 111-113.
2. Wentao Guo, Yigang He. The research and design of management system of intelligent parking based on RFID [J]. Techniques of Automation and Application, 2010, 29(6):60-64.
3. Zhengdong Liu, Lianguai Dai. Design and implementation of line control system for ETC [J]. Computer and Digital Engineering, 2009, 37(8):183-186.
4. Fei Wang. Design and implementation of the ETC lane toll software for highway [Master thesis]. Xi'an University of Electronic Science and Technology, 2010.
5. Yani Chen. Research of the combined ETC and optimization design of the lane [Master thesis]. Changsha University of science and technology, 2008.
6. Limiao Deng. Research and application of data synchronization in distributed database [Master thesis]. Huazhong University of science and technology, 2006.
7. An Gong. Based on Winsock Control of VB to Realize C/S Network Communication [J]. Science Technology and Engineering, 2006.5, 6(10): 1447-1450.
8. David Lxvinson, Elva Chang. A model for optimizing electronic toll collection systems. Transportation Research PartA, 2003(03):293-301.
9. Wei-Hsun Lee, Shian-Shyong Tseng, Ching-Hung Wang. Design and implementation of electronic toll collection system based on vehicle positioning system techniques [J]. Computer Communications, 2008, 31(12):2925-2933

Feature Fusion Algorithm for Multimodal Emotion Recognition from Speech and Facial Expression Signal

Han Zhiyan[1,a] and Wang Jian[1]

[1]*Bohai University, China*

Abstract. In order to overcome the limitation of single mode emotion recognition. This paper describes a novel multimodal emotion recognition algorithm, and takes speech signal and facial expression signal as the research subjects. First, fuse the speech signal feature and facial expression signal feature, get sample sets by putting back sampling, and then get classifiers by BP neural network (BPNN). Second, measure the difference between two classifiers by double error difference selection strategy. Finally, get the final recognition result by the majority voting rule. Experiments show the method improves the accuracy of emotion recognition by giving full play to the advantages of decision level fusion and feature level fusion, and makes the whole fusion process close to human emotion recognition more, with a recognition rate 90.4%.

1 Introduction

In recent years, emotion recognition has become a hot spot in human-computer interaction. There are two broad categories: single mode emotion recognition and multimodal emotion recognition. The single mode emotion recognition obtains emotion state from single information channel, as from speech signal, facial expression signal or physiological signal.

For speech emotion recognition, MIT media lab constructed an emotion editor in 1990[1]. Yan [2] used non-uniform subband filter to dig useful information for speech emotion. It increased the identification of all kinds of emotions, and improved the performance of emotion recognition. Mao [3] used parametric filter and fractal dimension to recognize emotion, and got better performance. Zou [4] proposed a emotion recognition method based on improved fuzzy vector quantization, and effectively improved recognition rate. Attabi [5] studied the effectiveness of anchor models applied to the multiclass problem of emotion recognition. Zheng [6] proposed a novel speech emotion recognition method based on least square regression model. This method achived better recognition performance. Mao [7] proposed to learn affect-salient-features using convolutional neural networks. This method led to stable and robust recognition performance in complex scenes.

Ekman [8] developed facial action coding system (FACS) to detect subtle changes in facial expression. Essa [9] proposed dynamic expression description method FACS+ based on video. It solved the problem that there was no time description information in FACS. Rahulamathavan [10] presented a system that addresses the challenge of performing facial expression recognition when the test image is in the encrypted domain. Zheng [11] proposed a novel group sparse reduced-rank regression (GSRRR) model to describe the relationship between the multi-view facial feature vectors and the corresponding expression class label vectors.

For physiological signal emotion recognition, petrantonakis [12] used higher order crossing analysis to extract feature from Electroencephalogram (EEG). Liu [13] used Ant Colony Optimization (ACO) to recognize the emotion from respiration signals. Zacharatos [14] studied the importance of body posture and movement for emotion recognition.

Although single mode emotion recognition has some achievements in scientific research, there are many limitations. Due to human beings express emotion information in many ways. So it has the complexity of expression and the relative property of culture [15]. When the signal is disturbed by noise, the multimodal emotion recognition method can generate complementary effect in some ways, and make up for the shortage to single mode emotion recognition. So the research for multimodal emotion recognition is very necessary. Kim [16] fused electromyogram, electrocardiogram, skin conductivity and respiration changes to recognize the emotion. Zhao [17] fused speech signal and electrocardio signal to recognize the emotion, and obtained a higher fusion recognition rate. But the above methods are all fused physiological signal, the measure of physiological signal must contact body. But the acquisition of signal has a certain difficulty. So the fusion of speech signal and facial expression signal has been widely studied. Busso [18] analysed the complementarity of speech emotion recognition and facial expression recognition. Hoch [19] fused speech and facial expression signal under vehicle environment, and realized the recognition for the three emotions, positive, negative and neural. Sayedelah [20] studied audio-visual feature decision level fusion for spontaneous emotion estimation in speech conversation. In a certain sense, the fusion of different channel information is bottleneck problem for multimodel

[a] Corresponding author: hanzyme@126.com

emotion recognition, and it is directly related to the accuracy of emotion recognition.

So, this paper takes speech signal and facial expression signal as the research subjects, takes the extracting of features, the fusion algorithm of features and recognition algorithm as the research contents, and finally accomplishes recognition for five kinds of human emotion (joy, anger, surprise, sadness, fear).

2 System structure frame

The system structure frame is shown in Figure 1. Firstly, emotion signal undergoes a series of preprocessing course. Secondly, we extract speech emotion feature and facial expression feature. Finally, make fusion recognition.

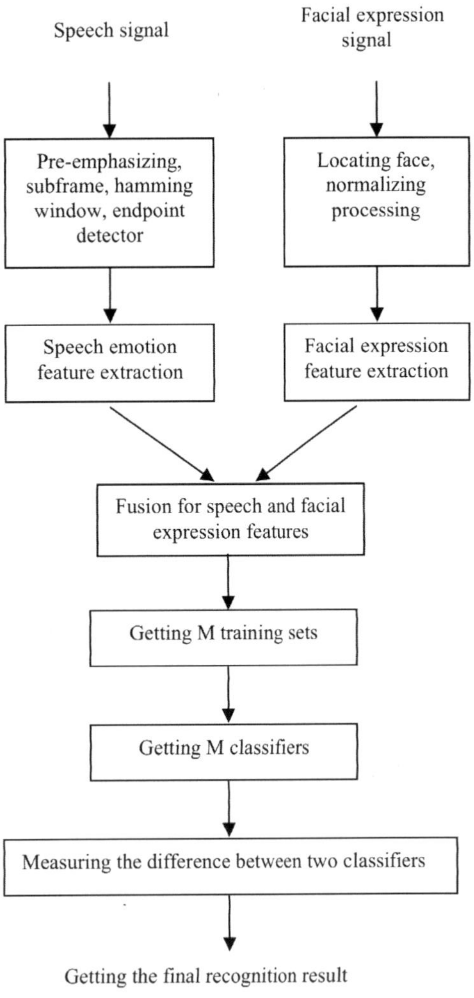

Figure 1. The system structure frame.

3 Feature extraction

3.1 Speech emotion feature extraction

In the past, the effective extraction of the emotion feature was mainly based on prosodic features. However, in recent years, through deep research, the combination of acoustic features and prosodic features could be more accurate to recognize emotion. Tao [21] found that acoustic feature had a good effect on the distinction between activate dimension relatively close to the emotion, and confirmed the correlation between acoustic feature and valence dimension.

In order to make full use of the emotion information contained in speech signal, we select the ratio of sentence pronunciation duration and corresponding calm statement duration, the average value of pitch frequency, the maximum pitch frequency, the difference of average value of pitch frequency and corresponding calm statement average value of pitch frequency, the difference of maximum pitch frequency and corresponding calm statement maximum pitch frequency, amplitude average energy, dynamic range of amplitude energy, the difference of amplitude average energy and corresponding calm statement amplitude average energy, the difference of dynamic range of amplitude energy and corresponding calm statement value, the average value of the first resonance peak frequency, the average value of the second resonance peak frequency, the average value of the third resonance peak frequency, the mean of harmonic noise ratio, the maximum harmonic noise ratio, the minimum harmonic noise ratio, the variance of harmonic noise ratio, as the emotion recognition features.

3.2 Facial expression feature extraction

Depending on the nature of the image, the extraction of facial expression feature is divided into static image feature extraction and sequence image feature extraction. The deformation features of facial expression are extracted from static images, and the motion features of facial expression are extracted from sequence image. In this paper, we take the static image as the research object, and use gabor wavelet transform to extract facial expression feature. The specific process is as follows:

(1) Put the preprocessed image divide into 25×25 pixels. Every image has 4 rows 3 columns.

(2) Make convolution with gabor wavelet and grid image. The formula is as follows:

$$r(x,y) = \iint I(\varepsilon,\eta) g(x-\varepsilon, y-\eta) d\varepsilon d\eta \qquad (1)$$

where $I(\varepsilon,\eta)$ is the pixel value of pixel coordinate (ε,η), $g(x,y) = \frac{k^2}{\sigma^2}\exp(-\frac{k^2(x^2+y^2)}{2\sigma^2})(\exp(ik \cdot \begin{pmatrix} x \\ y \end{pmatrix}) - \exp(-\frac{\sigma^2}{2}))$, $k = \begin{pmatrix} k_v \cos\varphi_u \\ k_v \sin\varphi_u \end{pmatrix}$, $k_v = 2^{-\frac{v+2}{2}}$, $\varphi_u = u\frac{\pi}{k}$, σ is a constant related to the frequency bandwidth of the wavelet, and the value is $\sqrt{2}\pi$. The wavelength of Gabor filter is determined by v, and the value is 0, 1, 2. u is the direction of the Gabor kernel function, and the value is 1, 2, 3, 4, 5, 6. k indicates the total number of directions, and the value is 6.

(3) Get the mean and variance of $\|r(x,y)\|$ as the facial expression feature.

(4) Make dimensionality reduction using Principal component analysis (PCA).

4 Algorithm description

Specific implementation steps are as follows:

Step 1: collect speech signal and facial expression signal through noise stimulation and watching television. For speech signal, undergoes a series of preprocessing course including pre-emphasizing, subframe, hamming window, and endpoint detector. For facial expression signal, firstly, locate face using skin color model, and then normalize processing for image geometry and optical properties.

Step 2: extract speech emotion feature according to section 3.1 and 3.2.

Step 3: combine speech emotion feature and facial expression feature in order, and get multimodal feature vector $u_1,u_2,\cdots,u_r,\cdots,u_W, r=1,2,\cdots,W$. W is sample number of original training sample set.

Step 4: Training sample set S_1 is obtained by putting back sampling N times for multimodal feature vector set, and then continue to obtain training sample set $S_2,\cdots,S_M$.

Step 5: Train sample set using BP neural network, and then the classifier is obtained on each training sample set.

Step 6: measure the difference between two classifiers by double error difference selection strategy, and then pick out the classifier that more than average difference as recognition classifier. The difference formula $Div(i,j)$ of classifier H_i and $H_j (i \neq j)$ is as follows:

$$Div(i,j) = \frac{num^{00}}{num^{00} + num^{01} + num^{10} + num^{11}} \quad (2)$$

where num^{ab} represents the sample number of correct/error for two classifiers. $a=1$ and $a=0$ respectively represent the correct and error of classifier H_i . $b=1$ and $b=0$ respectively represent the correct and error of classifier H_j .

Step 7: get the final recognition result by the majority voting rule.

5 Simulation experiments and results analysis

We did recognition experiment by using the method explained above. In our experiment, five discrete emotional states (joy, anger, surprise, sadness, fear) are classified throughout the work. 100 data per emotion have been used for the training, while a disjunctive set of 100 data per emotion were used testing. The data were recorded over a period of half a year to avoid anticipation effects of the actors.

Table 1 shows the results of the emotion evaluation only using speech emotion feature. Table 2 shows the results only using facial expression feature. Table 3 shows the results of the emotion evaluation by simple combination speech emotion feature and facial expression feature. Table 4 shows the results using proposed method. Columns represent the emotion elected in first choice belonging to the emotion of each row, where "J" stands for joy, "A" stands for anger, "P" stands for surprise, "S" stands for sadness, and "F" stands for fear.

Table 1. The results of the emotion evaluation only using speech emotion feature.

Emotion	J	A	P	S	F
Joy	86%	0	11%	2%	0
Anger	4%	81%	0	7%	8%
Surprise	20%	1%	77%	2%	0
Sadness	5%	4%	0	88%	3%
Fear	5%	10%	6%	4%	75%

Table 2. The results of the emotion evaluation only using facial expression feature.

Emotion	J	A	P	S	F
Joy	85%	2%	13%	0	0
Anger	0	79%	7%	10%	4%
Surprise	0	0	81%	9%	10%
Sadness	0	20%	4%	66%	10%
Fear	3%	8%	2%	9%	78%

Table 3. The results of the emotion evaluation by simple combination speech emotion feature and facial expression feature.

Emotion	J	A	P	S	F
Joy	92%	1%	6%	1%	0
Anger	1%	88%	2%	7%	2%
Surprise	4%	0	90%	4%	2%
Sadness	5%	8%	2%	85%	0
Fear	1%	3%	2%	6%	88%

Table 4. The results using proposed method.

Emotion	J	A	P	S	F
Joy	96%	2%	2%	0	0
Anger	2%	88%	2%	8%	2%
Surprise	0	3%	91%	6%	0
Sadness	3%	7%	4%	84%	2%
Fear	0	5%	1%	1%	93%

From the table 1, we could see that the average recognition correct rate for only using speech signal is 81.4%; From the table 2, we could see that the average recognition correct rate for only using facial expression signal is 77.8%. Due to human beings express emotion information in many ways. It has the complexity of expression and the relative property of culture. So there are many limitations for single mode emotion recognition. From table 3, we could see that the average recognition correct rate by simple combination speech emotion feature and facial expression feature has some increase, but the improvement is not obvious. So the fusion of different channel information is bottleneck problem for multi model emotion recognition, and it is directly related to the accuracy of emotion recognition. From table 4, we could see that the average recognition correct rate by using proposed method is 90.4%. So the method improves the accuracy of emotion recognition by giving full play to the advantages of decision level fusion and feature level fusion, and makes the whole fusion process close to human emotion recognition more.

6 Conclusion

This paper proposed a new multi model emotion recognition method, and improved the accuracy of emotion recognition. But this paper merely aimed at given text to recognize emotion, it has a certain distance with practical level. So the emotion recognition for non specific text will be our next research direction. Certainly, many shortages are lying the selecting of the features, so to find more efficient features and to do further analysis and experiment in a wide field will be our future work.

Acknowledgment

The authors wish to deeply thank graduate students who collaborated in the experiments and in the development of the system. The work is also supported by grant from the National Natural Science Foundation of China (No. 61503038, No. 61403042).

References

1. L.L. Yu, Z.X. Cai, M.Y. Chen, Study on emotion feature analysis and recognition in speech signal: an overview, J. Circuits and Systems, **12**, 4 (2007): 76–84.
2. Y.H. Yan, Y. Zhou, Y.Q. Sun, Feature Extraction Method for Speech Emotion Recognition. China: 2010102729713 (2010).
3. X. Mao, L.J. Chen, Speech emotion recognition based on parametric filter and fractal dimension, IEICE T INF SYST, **93**, 8 (2010): 2324–2326.
4. C.R. Zou, L. Zhao, Speech Emotion Recognition Method Based on Improved Fuzzy Vector Quantization. China: 2008101228062 (2008).
5. Y. Attabi, P. Dumouchel, Anchor models for emotion recognition from speech, TAC, **4**, 3 (2013): 280–290.
6. W.M. Zheng, M.H. Xin, X.L. Wang, A novel speech emotion recognition method via incomplete sparse least square regression, IEEE SIGNAL PROC LET, **21**, 5 (2014): 569–572.
7. Q. R. Mao, M. Dong, Z. W. Huang, Learning salient features for speech emotion recognition using convolutional neural networks, IEEE MULTIMEDIA, **16**, 8 (2014): 2203–2213.
8. P. Ekman, W. Friesen, Facial action coding system: a technique for the measurement of facial movement. Palo Alto: Consulting Psychologists Press, (1978).
9. L.H. Liang, H.Z. Ai, G.Y. Xu, A survey of human face detection, Chinese J. Computers, **25**, 5 (2002): 449–458.
10. Y. Rahulamathavan, R.C. W. Phan, J.A. Chambers, Facial expression recognition in the encrypted domain based on local fisherdiscriminant analysis, TAC, **4**, 1 (2013): 83–92.
11. W.M. Zheng, Multi-view facial expression recognition based on group sparse reduced-rank regression, TAC, **5**, 1 (2014): 71–85.
12. P.C. Petrantonakis, L.J. Hadjileontiadis, Emotion recognition from EEG using higher order crossings, IEEE T INF TECHNOL B, **14**, 2 (2010): 186–197.
13. S.L. Lin, G.Y. Liu, H.L. Zhang, Application of ACO algorithm to emotion recognition research based on RSP signal, IJCEA, **47**, 2 (2011): 169–172.
14. H. Zacharatos, H. Gatzoulis, Y.L. Chrysanthou, Automatic emotion recognition based on body movement analysis: a survey, IEEE COMPUT GRAPH, **34**, 6 (2014): 35–45.
15. Z. Zeng, M. Pantic, G.I. Roisman, A survey of affect recognition methods: audio, visual, and spontaneous expressions, IEEE T PATTERN ANAL, **31**, 1 (2009): 39–58.
16. J. Kim, E. Andre, Emotion recognition based on physiological changes in music listening, IEEE T PATTERN ANAL, **30**, 12 (2008): 2067–2083.
17. C.W. Huang, Y. Jin, Q.Y. Wang, Multimodal emotion recognition based on speech and ECG signals, JSEU (NSE), **40**, 5 (2010): 895–900.
18. C. Busso, Z. Deng, S. Yildirim, Analysis of emotion recognition using facial expressions, speech and multimodal information, ICMI 2004, (2004): 205–211.
19. S. Hoch, F. Althoff, G. Mcglaun, Bimodal fusion of emotional data in an automotive environment, ICASSP 2005, (2005): 1085–1088.
20. A. Sayedelahl, R. Araujo, M.S. Kamel, Audio-visual feature-decision level fusion for spontaneous emotion estimation in speech conversations, ICMEW 2013, (2013): 1–6.
21. R. Tato, R. Santos, R. Kompe, Emotion space improves emotion recognition, ICSLP 2002, (2002): 2029–2032.

MATEC Web of Conferences **61**, 04001 (2016) DOI: 10.1051/matecconf/20166104001
APOP2016

Strength and erosion mechanism of soft bedrock under high shear stresses

Li Zhijing[1, 2, a] and Jin Zhongwu[1, 2]
[1] *Changjiang River Scientific Research Institute, Wuhan 430010, China*
[2] *Collaborative Innovation Center for Geo-Hazards and Eco-Environment in Three Gorges Area, Hubei Province, Yichang 443002, China*

Abstract. Bedrock erosion plays a critical role in mountain river channel evolution, especially, where there are hydro-projects with high velocity flow conditions. In this study, laboratory experiments are carried out to investigate strength and erosion mechanism of soft bedrock under high shear stresses by high velocity open channel flows. Prototype bedrock from the upper reaches of Yangtze River is employed as testing samples. New observed dataset is generated concerning the flow, bedrock strength and evolution of bedrock erosion, which can be exploited to evaluate bedrock erosion rate and test mathematical bedrock erosion models.

1 Introduction

Bedrock erosion processes are believed to play important roles in the evolution of mountain river channels and their associated hydro-projects. During the last decade, there has been a rapid advancement in the development of models of mountain landscape evolution. These models have contributed to the understanding of the fundamental mechanisms underlying landscape evolution [1-6].

Generally, the main mechanisms responsible for bedrock erosion including hydraulic scour (plucking of rock fragments by fluid shear stress or differential fluid pressure) and abrasive scour (interaction between rock and moving sediment particles) [7]. It is possible that hydraulic scour may dominate on one reach, whereas abrasive scour may dominate on another. Alternatively, both mechanisms may operate concurrently to erode the bedrock on some rivers. A discussion of how to qualitatively identify the erosion mechanisms operating on a reach was presented by Whipple et al. (2000) [8]. A review was also provided by Lai and Greimann (2009) [9]. However, it is cautioned that the mechanisms are interrelated. Hydraulic scour can occur without the assistance of sediment particles but is influenced by sediments through particle-rock collision and sediment deposition. Similarly, abrasive scour is also strongly linked to hydraulic flow characteristics such as shear stress and turbulence.

In this study, only the hydraulic processes are considered, and particular attention is paid to the bedrock erosion in the upper reaches of the Yangtze River. Clear water alone can cause bedrock scour by instantaneous or progressive failure of closed-end rock joints or by dynamic ejection of single rock blocks [10]. Hydraulic scour may be an important mechanism for weak or highly fractured rock, as reviewed by Sklar and Dietrich (2004) [4]. Additional review was also provided by Tomkin et al. (2003) and Sklar and Dietrich (2006) [11, 12]. In the upper reaches of Yangtze River, many alluvial rivers are characterized by the presence and exposure of soft bedrock that consists of mainly mudstone and sandstone with relatively lower compressive strengths. During flood season, the floods have become more and more frequent. These floods significantly erode the exposed soft bedrock and lead to damage to bridge pier foundations, diversion structures, and other infrastructure. Evaluation of bedrock erosion rate requires quantification of bedrock characteristics, together with hydraulic processes. Key geomorphic controls, such as rates of river incision or hillslope erosion, are often unknown, as valid data describing the relationships between rates of bedrock erosion and hydraulic processes are not sufficient.

Here we present 18 runs of experiments on soft bedrock erosion. A flume is purposely designed to generate high velocity open channel flow, and prototype bedrock including mud-rock and sand-rock from the upper reaches of Yangtze River are employed as testing samples. For each run, detailed measurements of the flow and bedrock erosion processes are conducted.

2 Experimental Set-Up

The experiments were conducted in a tilting bed flume at Changjiang River Scientific Research Institute (Figure 1). The tilting flume is made by Plexiglas, has a width of 0.4 m, a length of 20 m long and a bed slope fixed at 0.035. In order to relate experimental studies to field conditions, as we propose to do, the upstream of this flume is equipped with a high head water tank, by which high velocity open channel flows with a maximum value of 6.0 m/s can be generated. In the test section of the flume, there is a notch designed to fix bedrock samples. Before each run, a bedrock sample is inserted into the notch, flush with the surface. The flat sample area exposed to

[a] Corresponding author: lzjketty@126.com

the flow is round with a diameter of 7.0 cm. The magnitude of erosion was determined by weighing the erodible sample block before and after the experiment and obtaining the mass loss.

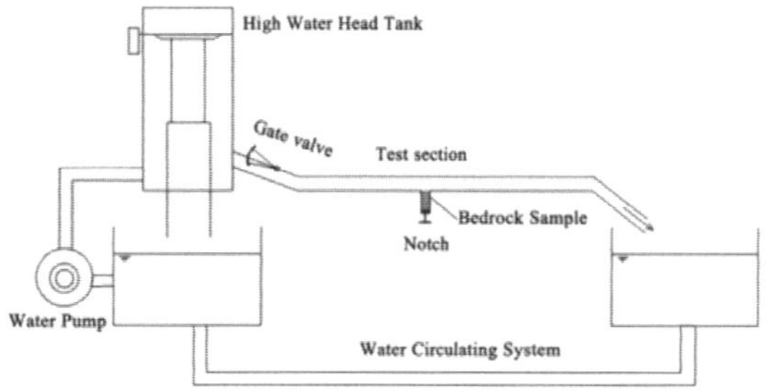

Figure 1. Illustration of the flume used to conduct bed erosion experiments.

3 Field Survey

The upper reaches of the Yangtze River cover a region from the Three Gorges Dam to the River's origin. The total area is about 1005 thousand km^2 with a total population of 160 million, which accounts for 58.9% of the area and 35% of the population in the valley, respectively [13, 14]. Within this area, 90% or more of the land is mountains and plateaus, in which the ecosystem is very fragile and subject to soil and bedrock erosion due to inappropriate use of land. On the other hand, the Yangtze River Basin is of great economic significance, with its economic productivity amounted to almost half of China's Gross National Product (GNP) during the 1990s. In addition, within the Yangtze River Basin, bedrock erosion problems associated with energy projects such as the Three Gorges Dam (TGD) and the South-to-North Water Diversion Project [15, 16].

Soft bedrock samples including mud-rock and sand-rock were collected from the main river channel within the area of Chongqing Province .These relatively weaker sample bedrocks eroded more readily than other hard rocks, and also more susceptible to hydraulic erosion, thus shortening the experimental duration needed to produce measurable wear. Samples of bedrock are taken at possible foundation locations, boring method was used, and the erodible samples were cast into cylinder blocks with diameter of 7.0 cm and later cut into flume-ready samples. All samples were kept in a sealed box after sampling until laboratory analysis (Table 1).

Table 1. Physical and mechanical characteristics of bedrock samples

Sample	Density (kg/m3)	Porosity	Tensile Strength (Mpa)	Remarks
M-A	2670	7.53	0.85	Mud-rock
M-B	2678	7.34	1.09	
M-C	2685	7.12	1.21	
S-A	2713	6.34	1.58	Sand-rock
S-B	2720	6.02	1.74	
S-C	2725	5.87	1.96	

4 Results and Discussion

We conducted 18 runs of erosion experiments in which we varied hydraulic condition and bedrock strength (Table 2). The procedure for each experiment was as follows: An erodible sample was weighed and placed in the flume, its top surface level with the flume bed, and the flume was filled with still water until sample was completely water-saturated (after 48 hour this work). During the run, we documented the depth, velocity and slope of the flow. The duration of the experiment was chosen to be long enough to obtain significant wear, but not so long that the erodible sample degraded significantly lower than the flume bed. At the end of each run, the sample was removed, examined for wear marks, and weighed on subsequent days until they equilibrated to their dry weight. Typically only one run was obtained from the same sample before it was replaced due to extensive wear, but when possible, we obtained multiple runs from the same sample to test repeatability.

Table 2. Experimental Runs, Measured Variables, and Results

Run	Sample	Mean flow velocity (m/s)	Flow depth (cm)	Water slope	Eroded mass (g)	Duration (s)
1	M-A	3.0	6.7±1.0	0.035	27.60	7200
2	M-A	3.5	14.0±1.5	0.035	25.92	5200
3	M-A	4.0	22.9±2.0	0.035	25.52	3400
4	M-B	3.0	6.7±1.0	0.035	26.48	8000
5	M-B	3.5	14.0±1.5	0.035	27.36	6000
6	M-B	4.0	22.9±2.0	0.035	26.24	4200
7	M-C	3.0	6.7±1.0	0.035	26.88	8800
8	M-C	3.5	14.0±1.5	0.035	27.36	6800
9	M-C	4.0	22.9±2.0	0.035	27.04	5000
10	S-A	3.0	6.7±1.0	0.035	27.28	9200
11	S-A	3.5	14.0±1.5	0.035	26.16	7400
12	S-A	4.0	22.9±2.0	0.035	26.40	5600
13	S-B	3.0	6.7±1.0	0.035	28.16	10000

14	S-B	3.5	14.0±1.5	0.035	26.08	8200
15	S-B	4.0	22.9±2.0	0.035	26.24	6400
16	S-C	3.0	6.7±1.0	0.035	27.60	10800
17	S-C	3.5	14.0±1.5	0.035	26.72	9000
18	S-C	4.0	22.9±2.0	0.035	26.32	7200

To analyze the erosion process with different conditions, we define the erosion rate as the ratio of the eroded mass and the erosion duration, and calculate the bed shear stress by $\tau = \rho g h J$, where ρ is the density of water, g is the gravity acceleration, h is the flow depth, and J is the water slope. Figure 2 shows the plot of erosion rate versus bed shear stress for different bedrock samples. It can be seen that erosion rate generally increases with the increase of shear stress for all samples, and the increasing trend decreases with the increase of tensile strength.

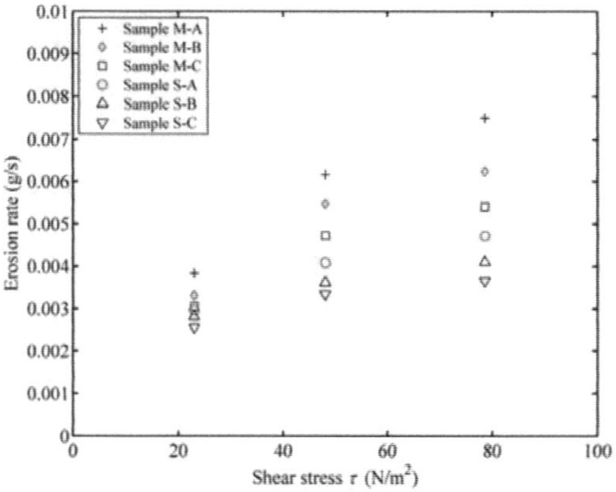

Figure 2. Plot of erosion rate versus bed shear stress for different bedrock samples.

Figure 3 presents the plot of erosion rate versus tensile strength for different bed shear stresses. It shows that if keep the shear stress constant, erosion rate decreases approximately with the square of the tensile strength. Hsu et al. (2008) conducted experimental study of bedrock erosion by granular flows, and explored the relationship between erosion of a synthetic bedrock sample and variables such as grain size, shear rate, water content, and bed strength [17]. It is interesting to see that their work also suggested that the bedrock erosion rate increases with the increase of shear rate and decreases with the increase of tensile strength. Particularly, our experiments employed prototype bedrock samples and high velocity open channel flow condition, which is closer to natural scale. It is acknowledged that to make the analysis of bedrock erosion more mechanistic, more field measurements are needed.

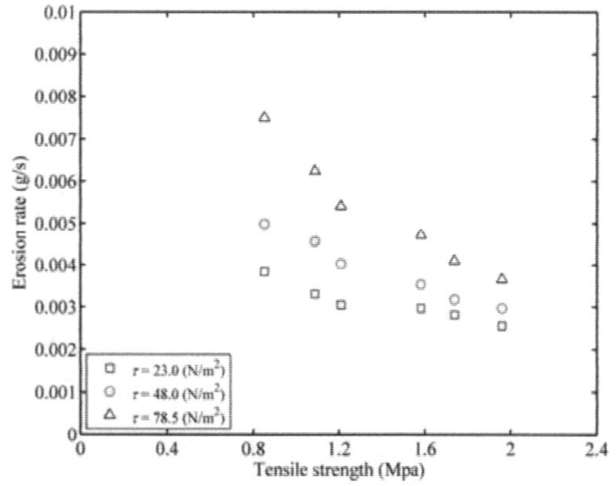

Figure 3. Plot of erosion rate versus tensile strength for different bed shear stresses.

5 Summary

Experiments on soft bedrock erosion have been conducted. A flume is purposely designed to generate high velocity open channel flow, and prototype bedrock including mud-rock and sand-rock from the upper reaches of Yangtze River are employed as testing samples. New observed dataset is generated concerning the flow, bedrock strength and evolution of bedrock erosion, which can be exploited to evaluate bedrock erosion rate and test mathematical bedrock erosion models.

Acknowledgements

The research is funded by Natural Science Foundation of China (Grants No. 51479009 and 51339001), and the special fund for basic scientific research business of central public research institutes (CKSF2016011/HL).

References

1. Howard, A. D. (1994). "A detachment-limited model of drainage basin evolution." Water Resour. Res., 30, 2261-2285.
2. Tucker, G. E., and Slingerland, R. L. (1994). "Erosional dynamics, flexural isostasy, and long-lived escarpments: A numerical modeling study." J. Geophys. Res., 99, 12,229-12,243.
3. Whipple, K. X., and Tucker, G. E. (2002). "Implications of sediment-flux dependent river incision models for landscape evolution." J. Geophys. Res., 107(B2), 2039, doi: 10.1029/2000JB000044.

4. Sklar, L. S., and Dietrich, W. E. (2004). "A mechanistic model for river incision into bedrock by saltating bed load." Water Resour. Res., 40, W06301, doi: 10.1029/2003WR002496.
5. Gasparini, N. M., Bras, R. L., and Whipple, K. X. (2006). Numerical modeling of non-steady-state river profile evolution using a sedimentflux-dependent incision model, in Tectonics, Climate and Landscape Evolution, edited by S. Willett et al., GSA Special Paper 398, Penrose Conference Series, Geol. Soc. of Am., 127-141.
6. Lamb, M. P., Dietrich, W. E. and Sklar, L. S. (2008). "A model for fluvial bedrock incision by impacting suspended and bedload sediment." J. Geophys. Res., 113, F03025, doi: 10.1029/2007JF000915.
7. Lai, Y. G., Greimann, B. P., and Wu, K. (2010). "Soft bedrock erosion modeling with a two-dimensional depth-averaged model." J. Hydraul. Eng., 137(8), 804-814.
8. Whipple, K. X., Hancock, G. S., and Anderson, R. S. (2000). "River incision into bedrock: Mechanics and relative efficacy of plucking, abrasion, and cavitation." Geol. Soc. Am. Bull., 112(3), 490-503.
9. Lai, Y. G., and Greimann, B. P. (2009). "Rock erosion modeling for rivers in Taiwan and progress report of bank erosion and turbidity current modeling." Technical Rep. SRH-2009-40, Bureau of Reclamation, U.S. Dept. of the Interior, Denver.
10. Bollaert, E. F. R., and Schleiss, A. J. (2005). "Physically based model for evaluation of rock scour due to high-velocity jet impact." J. Hydraul. Eng., 131(3), 153-165.
11. Sklar, L. S., and Dietrich,W. E. (2006). "The role of sediment in controlling steady-state bed channel slope: Implications of the saltation-abrasion incision model." Geomorphology, 82(1-2), 58-83.
12. Tomkin, J. H., Brandon, M. T., Pazzaglia, F. J., Barbour, J. R., and Willet, S. D. (2003). "Quantitative testing of bedrock incision models for the Clearwater River, NW Washington State." J. Geophys. Res., 108(B6), 1-19.
13. Li, C. H, Gu, S. Z. (1992). "A study on the characteristics of soil and water loss in the upper reaches of the Yangtze River and corresponding countermeasures." Bulletin of Soil and Water Conservation, 12, 1-6 (in Chinese with English abstract).
14. Wang, X. T. (1998). "The control of soil and water loss in the upper reaches of the Yangtze River and agricultural and rural sustainable development." Issues in Agricultural Economy, 12, 40-44 (in Chinese with English abstract).
15. Zhang, J., Zhang, Z.F., Liu, S.M.,Wu, Y., Xiong, H., Chen, H.T. (1999). "Human impacts on the large world rivers: Would the Changjiang (Yangtze River) be an illustration?" Global Biogeochemistry Cycles, 13, 1099-1105.
16. Chen, J.Y., Chen, X.Q., Chen, S.L. (2003). Impacts and countermeasures of the South-To-North Water Diversion Projects on the ecological environment of Changjiang Estuary. In: Chen, J.Y. (Ed.), Impacts and Countermeasures of the South-To-North Water Diversion Projects on the Ecological Environment of Changjiang Estuary. Publication of ECNU, Shanghai, pp. 1-12.
17. Hsu, L., Dietrich, W. E., and Sklar, L. S. (2008). "Experimental study of bedrock erosion by granular flows." J. Geophys. Res., Earth Surface (2003–2012), 113(F2).

Measurement Uncertainty Evaluation of Digital Modulation Quality Parameters: Magnitude Error and Phase Error

Zhiqiang Zhan[1a], Junwen Xia[1], Yunling Wang[1], and Lei Yu[1]

[1] Division of Electricity and Optics Measurement Technology, Shanghai Institute of Measurement and Testing Technology, China

Abstract. In digital modulation quality parameters traceability, the Error Vector Magnitude, Magnitude Error and Phase Error must be traced, and the measurement uncertainty of the above parameters needs to be assessed. Although the calibration specification *JJF1128-2004 Calibration Specification for Vector Signal Analyzers* is published domestically, the measurement uncertainty evaluation is unreasonable, the parameters selected is incorrect, and not all error terms are selected in measurement uncertainty evaluation. This article lists formula about magnitude error and phase error, than presents the measurement uncertainty evaluation processes for magnitude error and phase errors.

1 Preface

Shanghai Institute of Measurement and Testing Technology (SIMT), which is also the National Center of Measurement and Testing for East China, undertakes the task of measurement traceability in East China. When tracing the measurement of digital modulation quality parameters, the most important traceability parameters are Error Vector Magnitude (EVM), magnitude error and phase error. Relevant calibration specifications have been already published, such as *JJF1128-2004 Calibration Specification for Vector Signal Analyzers*, which specifies the traceability method of EVM, magnitude error and phase error. But in *JJF1128-2004 Calibration Specification for Vector Signal Analyzers*, the uncertainty assessment methods are not so reasonable, not all the uncertainty components are included, and the assessment method and procedure are also questionable. This article discusses and assesses the source of measurement uncertainty of magnitude error and phase Error, and shows the result of National Comparison of Digital Modulation Quality Parameter organized by China Radio Metrology Technology Committee ([1]). The value of this assessment result is basically same as the result assessed by National Institute of Metrology (NIM).

2 Measuring method

According to *JJF1128-2004 Calibration Specification for Vector Signal Analyzers*, digital modulation quality parameters are calibrated by direct comparison method. Schematic diagram is shown in Figure 1.

Connect calibrated Digital Signal Generator (DSG) to a Vector Signal Analyzer (VSA) or a Spectrum Analyzer with the analyzing function of vector signal. The time base output of Vector Signal Generator (VSG) should be connected to the time base input of VSA, and the time base of VSA should be set as external time base. Set the output frequency of the calibrated DSG, and generally the output level is set to -10 dBm. Select the required standard format or the customized modulation parameters, such as modulation, filter type, filter factor α and symbol rate. Set the input frequency of the VSA equal to the one of calibrated DSG, and input level is -10 dBm or select **Level Auto Range** button to regulate the input level of VSA automatically. Select the required modulation standard format or set the modulation system, filter type, filter factor α and symbol rate which is consistent with the VSG. Read the digital modulation quality parameters from VSA.

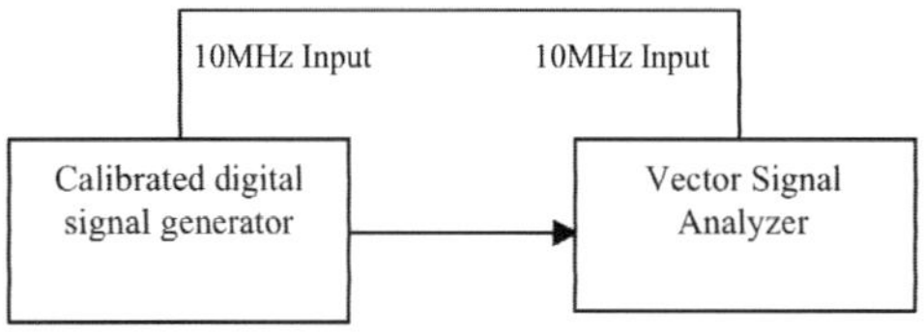

Figure 1. Schematic diagram of digital modulation quality parameter calibration

3 Definition of digital modulation quality parameter

For various vector modulation, the signal magnitude and phase can both be represented by the points on the constellation diagram. However, due to the effects such as non-ideal performance of vector modem hardware and the precision of vector modulation algorithm, there are errors occurring between signal vector in real and ideal situation. Vector modulation error defines the error between real constellation points and corresponding

[a] Corresponding author: zhanzq@simt.com.cn

reference constellation points in every symbol period. It is also called Digital Modulation Quality Parameter. The major parameters are magnitude error (MagErr), phase error (PhaseErr) and EVM, shown in Figure 2. Those parameters are usually represented by root mean square (RMS) ([2-3]).

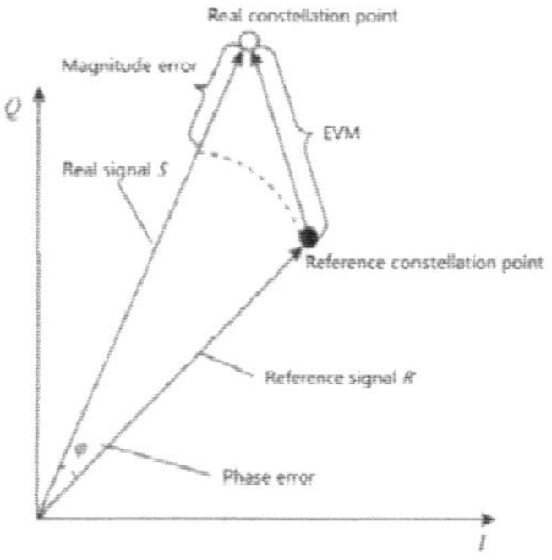

Figure 2. Definition of vector modulation error

As Eq. (1)~(3), the subscript rms is expressed as the root mean square value.

$MagErr_{\text{rms}}$ is expressed as:

$$MagErr_{\text{rms}} = \sqrt{\frac{\sum_{i=0}^{i=N-1}\left(|S_i|-|R_i|\right)^2}{\sum_{i=0}^{i=N-1}|R_i|^2}} \times 100\% \qquad (1)$$

$PhaseErr_{\text{rms}}$ is expressed as:

$$PhaseErr_{\text{rms}} = \sqrt{\frac{1}{N}\sum_{i=0}^{i=N-1}\left(\arg S_i - \arg R_i\right)} \qquad (2)$$

EVM_{rms} is expressed as:

$$EVM_{\text{rms}} = \sqrt{\frac{\sum_{i=0}^{i=N-1}|S_i - R_i|^2}{\sum_{i=0}^{i=N-1}|R_i|^2}} \times 100\% \qquad (3)$$

In Eq. (1)~(3), S is real signal vector, and R is reference signal vector.

4 Evaluation of measurement uncertainty of magnitude error

EVM will be first introduced before the evaluation of measurement uncertainty of digital modulation quality parameters are introduced.

An ideal vector modulation signal expression is shown in the Eq. (4).

$$V(t) = I(t)\cos(2\pi f_c t) - Q(t)\sin(2\pi f_c t) \qquad (4)$$

Assume I/Q gain imbalance degree g (g=1 means complete balance), phase error φ, and interference signal $D\sin(2\pi f_i t)$ are exist. Phase noise $\theta(t)$ is a random variable which variance is σ_P. $n(t)$ is Gauss white noise with a variance of σ_G.

Then Eq. (4) turns to Eq. (5):

$$V(t) = I(t)\cos\left[2\pi f_c t + \frac{\phi}{2} + \theta(t)\right] - gQ(t)\sin\left[2\pi f_c t - \frac{\phi}{2} + \theta(t)\right] + n(t) + D\sin(2\pi f_i t) \qquad (5)$$

It is concluded that EVM and the error factor can be obtained by the derivation ([3]).

$$EVM_{\text{rms}} = \left[2 - (1+g)\cos\left(\frac{\varphi}{2}\right)\sqrt{\frac{1}{\frac{(g^2 + g^2\sigma_p^2 + \sigma_p^2 + 1)}{2}}} + ISR\right]^{\frac{1}{2}} \qquad (6)$$

$$ISR = \frac{\sigma_G^2 + \frac{D^2}{2}}{I^2(t) + Q^2(t)} \qquad (7)$$

In Eq. (7):

I/Q gain imbalance degree g is a linear value, not a logarithm value.

Unit of phase error φ is radian.

ISR is the power ratio of interfering signals and useful signals, and is also a linear value.

If φ=0, EVM in Eq. (6) equals $MagErr_{\text{rms}}$.

$$MagErr_{\text{rms}} = \left[2 - (1+g)\sqrt{\frac{1}{\frac{(g^2 + g^2\sigma_p^2 + \sigma_p^2 + 1)}{2}}} + ISR\right]^{\frac{1}{2}} \qquad (8)$$

Eq. (8) is the expression of measurement uncertainty of magnitude error in digital modulation quality parameters traceability.

4.1 Uncertainty source

In *JJF1128-2004 Calibration Specification for Vector Signal Analyzers*, an example of measurement uncertainty evaluation is also given. For magnitude error and phase error, its uncertainty source is the error of vector signal generated by the VSG, resolution of VSA and measurement repeatability. When the VSG is to be calibrated, the VSA is the standard. The connection graph is the same as Figure 2, which means calibration of digital modulation quality parameters of VSG and digital modulation quality parameters of VSA is mutual cycle.

From Eq. (8) it can be concluded, when using the DSG to calibrate the digital modulation magnitude error of the VSA, the magnitude error is related to the following factors: I/Q gain imbalance (including inband flatness) g, signal-noise ratio ISR, phase noise, I/Q origin offset, resolution and measurement repeatability ([2, 4, 5, 6]).

4.2 Uncertainty analysis

The evaluation process of the measurement uncertainty component is introduced below. For the Eq. (8) is complex, in the calculation of the uncertainty of the component, consider the impact of individual components, and finally combine the components.

4.2.1 Standard measurement uncertainty components produced by I/Q gain imbalance g

I/Q gain imbalance g includes inband flatness, which means g is related to signal bandwidth. Bandwidth of the signal becomes wider, then I/Q gain imbalance increaser, and magnitude error also increaser.

Only consider the effect of *I/Q* gain imbalance on magnitude error, it can be obtained from Eq. (8).

$$MagErr_{\text{rms}} = \sqrt{2-(1+g)\sqrt{\frac{2}{1+g^2}}} \qquad (9)$$

According to the technical specification, *I/Q* gain imbalance is 1.01. It can be obtained from Eq. (9).

a_1=0.5%,

Assuming it's a uniform distribution, the coverage factor is $\sqrt{3}$.

u_1=0.29%

4.2.2 Standard measurement uncertainty components produced by signal-noise ratio

Signal-noise ratio is the ratio between signal power and noise power. The greater the SNR is, the stronger the receiver's receptivity.

Only consider the effect of *SNR* on magnitude error, it can be obtained from Eq. (7).

$$ISR=1/SNR \qquad (10)$$

Put Eq. (10) into Eq. (8),it can be obtained.

$$MagErr_{rms} = \sqrt{2-2\sqrt{\frac{1}{1+ISR}}} \qquad (11)$$

In fact, SNR can reach more than 55dB in VSA. It can be obtained from Eq. (11).

a_2=0.21%

Assuming it's a uniform distribution, the coverage factor is $\sqrt{3}$.

u_2=0.13%

4.2.3 Standard measurement uncertainty components produced by phase noise

Phase noise is related to the power spectral density of the single side. With reference to the technical specification, the phase noise parameter of the VSA is roughly as follows:

-80dBc/10Hz, -96dBc/100Hz, -105dBc/1kHz, -127dBc/10kHz,-126dBc/100kHz, -144dBc/1MHz, -146dBc/10MHz.

Then:

N/(10Hz~100Hz)=(-80-96)/2+10lg(100-10)=-68dB,

Similarly,

N/(100Hz~1kHz)=-71dB,
N/(1Hz~10kHz)=-76dB,
N/(10kHz~100kHz)=-77dB,
N/(100kHz~1MHz)=-75dB,
N/(1MHz~10MHz)=-75dB,

Formula of phase jitter is

$$phase = \sqrt{10^{\frac{N}{10}} \times 2} \qquad (12)$$

Then:

$phase/(10\text{Hz-}100\text{Hz})=5.6\times10^{-4}$
$phase/(100\text{Hz-}1\text{kHz}) =4.0\times10^{-4}$
$phase/(1\text{kHz-}10\text{kHz}) =2.2\times10^{-4}$
$phase/(10\text{kHz-}100\text{kHz})=2.0\times10^{-4}$
$phase/(100\text{kHz-}1\text{MHz})=2.5\times10^{-4}$
$phase/(1\text{MHz-}10\text{MHz})=2.5\times10^{-4}$

Then:

$$\sigma_P = phase/(10Hz \sim 10MHz) = 8.3\times10^{-4} \text{ rad}$$

Ignoring other factors, it can be obtained by the effect of phase noise on EVM.

$$MagErr_{rms} = \sqrt{2-2\sqrt{\frac{1}{1+\sigma_p^2}}} \qquad (13)$$

Put the calculation result of σ_P into Eq. (13), then a_3=0.083%. Assuming it's a uniform distribution, the coverage factor is $\sqrt{3}$.

u_3=0.05%

4.2.4 Standard measurement uncertainty components produced by I/Q origin offset

Demodulator imbalance will cause the carrier leakage, which generates an interference signal. It interferences signal vector and deviates it from the correct position, then origin drift. Therefore, origin offset is carrier leakage signal components of demodulated signal.

Ignoring other factors, it can be obtained by the effect of interference signal on magnitude error. It can be obtained from Eq. (8).

$$MagErr_{rms} = \sqrt{2-2\sqrt{\frac{1}{1+ISR}}} \qquad (14)$$

ISR is *I/Q* origin offset. According to the technical specification of VSA, *I/Q* origin offset is -60dB. It can be obtained from Eq. (14).

a_4= 0.1%

Assuming it's a uniform distribution, the coverage factor is $\sqrt{3}$.

u_4=0.06%

4.2.5 Standard measurement uncertainty components produced by resolution

When the vector signal is demodulated by the VSA, the minimum resolution is 0.01.

a_5=0.005%

Assuming it's a uniform distribution, the coverage factor is $\sqrt{3}$.

u_5=0.003%

4.2.6 Standard measurement uncertainty components produced by measurement repeatability

Measure the magnitude error for 10 times when standard value is 4.5%.

Times	Measurement Value
1	4.51%
2	4.48%
3	4.51%
4	4.49%
5	4.48%
6	4.49%
7	4.52%

8	4.52%
9	4.52%
10	4.49%

According to the Bessel formula:

$u_6 = s = 0.013\%$

4.3 Calculation of the combined standard uncertainty

4.3.1 List of uncertainty

No.	The uncertainty component					
	The sources of the uncertainty	Type	Symbol and value	Distribution	The coverage factor	The symbol and value of standard uncertainty
1	*I/Q* gain imbalance	B	a_1=0.5%	Uniform	$\sqrt{3}$	u_1=0.29%
2	signal-noise ratio	B	a_2=0.22 %	Uniform	$\sqrt{3}$	u_2=0.13%
3	phase noise	B	a_3=0.08 %	Uniform	$\sqrt{3}$	u_3 =0.05%
4	*I/Q* origin offset	B	a_4=0.1%	Uniform	$\sqrt{3}$	u_4=0.06%
5	resolution	B	a_4=0.005 %	Uniform	$\sqrt{3}$	u_5=0.003%
6	measurement repeatability	A	/	/	/	u_6=0.013%

4.3.2 The synthesis of the standard uncertainty

Each Standard uncertainty component is independent from each other with no correlation, then consider the larger value of resolution and measurement repeatability involved in the synthesis of the standard uncertainty.

$$u = \sqrt{u_1^2 + u_2^2 + u_3^2 + u_4^2 + u_6^2} = 0.328\%$$

4.4 The expanded uncertainty

The coverage factor is 2.

U=0.66%(k=2)

5 Evaluation of measurement uncertainty of phase error

5.1 Uncertainty source

Phase error is related to the following factors: *I/Q* gain imbalance, *I/Q* origin offset, resolution and measurement repeatability.

5.2 Uncertainty analysis

5.2.1 Standard measurement uncertainty components produced by I/Q gain imbalance

The formula of the effect of *I/Q* gain imbalance on phase error is:

$$PhaseErr = 45 - \frac{180}{\pi} arctg\left(\frac{1}{g}\right) \tag{15}$$

According to the previous hypothesis, g=1.01. Error is 1%.

a_1=0.29°

Assuming it's a uniform distribution, the coverage factor is $\sqrt{3}$.

u_1=0.17°

5.2.2 Standard measurement uncertainty components produced by I/Q origin offset

Figure 3 as an example, the effect of *I/Q* origin offset on the measurement results is introduced. According to the analysis, the magnitude of the error vector is perpendicular to X axis, and the *I/Q* origin is shifted in X axis, which has the greatest influence on the phase error.

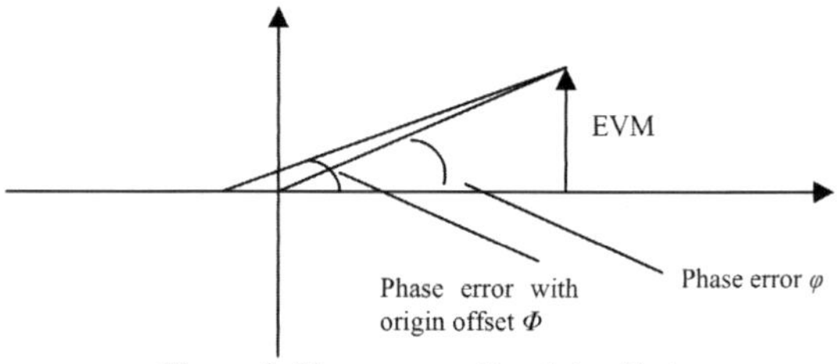

Figure 3. Phase error with origin offset

Assuming the origin offset is -60dB, which is 0.1%. According to Figure 3, the influence of the phase error is approximately equal to the origin offset value, which is in radian. It needs to multiply $\frac{180^\circ}{\pi}$,then turns to degree.

a_2=0.06°

Assuming it's a uniform distribution, the coverage factor is $\sqrt{3}$.

u_2=0.04°

5.2.3 Standard measurement uncertainty components produced by resolution

Resolution is 0.01°.

a_3=0.005°,

Assuming it's a uniform distribution, the coverage factor is $\sqrt{3}$.

u_3=0.003°

5.2.4 Standard measurement uncertainty components produced by measurement repeatability

Measure 10 times.

Times	Measurement Value
1	3.49°
2	3.50°
3	3.48°
4	3.49°
5	3.50°
6	3.49°

7	3.50°
8	3.51°
9	3.51°
10	3.51°

Then

$$u_4 = s = 0.010°$$

5.3 Calculation of the combined standard uncertainty

5.3.1 List of uncertainty

No.	The uncertainty component					
	The sources of the uncertainty	Type	Symbol and value	Distribution	The coverage factor	The symbol and value of standard uncertainty
1	*I/Q* gain imbalance	B	a_1=0.29°	Uniform	$\sqrt{3}$	u_1=0.17°
2	*I/Q* origin offset	B	a_2=0.06°	Uniform	$\sqrt{3}$	u_2=0.033°
3	resolution	B	a_3=0.005	Uniform	$\sqrt{3}$	u_3 =0.003°
4	measurement repeatability	A	/	/	/	u_4=0.010°

5.3.2 The synthesis of the standard uncertainty

Each Standard uncertainty component is independent from each other with no correlation, then consider the larger value of resolution and measurement repeatability involved in the synthesis of the standard uncertainty.

$$u = \sqrt{u_1^2 + u_2^2 + u_4^2} = 0.174°$$

5.4 The expanded uncertainty

The coverage factor is 2.

$$U = 0.35°(k=2)$$

6 Conclusion

In *JJF1128-2004 Calibration Specification for Vector Signal Analyzers*, the measurement uncertainty evaluation of magnitude error is unreasonable and the measurement uncertainty evaluation formula of phase error is not perfect. In this article, the uncertainty source of the magnitude error and phase error are comprehensively analyzed. In line with the principle of non repetition and no missing, measurement uncertainty of magnitude error and phase error are assessed. This article is uncertainty evaluation content when participating in the National Comparison of Digital Modulation Quality Parameters organized by China Radio Metrology Technology Committee. Table 1 and Table 2 are part of the result of National Comparison of Digital Modulation Quality Parameters.

Table 1. Magnitude error comparison result (1800MHz, 16QAM modulation, symbol rate 4Mbit/s, filter RRC, filter coefficient 0.22)

Company	Measurement value	Uncertainty (k=2)
National Institute of Metrology,China(Comparison leading laboratory)	4.49%	0.61%
Ministry of Industry and Information Technology Communication Measurement Center	4.5%	1.1%
China Aerospace Science&Industry Corporation 203	4.51%	0.45%
Measurement and testing center of research laboratory equipment of General Armament Department	4.5%	1.0%
Keysight Technologies,Inc.	4.52%	0.82%
Rodhe&Schwarz(China)Technology Co.,Ltd	4.5%	1.0%
Beijing Oriental Institute of Measurement and Test	4.5%	1.0%
National Instruments Corporation	4.55%	0.70%
Shanghai Institute of Measurement and Testing Technology	4.51%	0.66%
Jiangsu Institute of Metrology	4.5%	1.0%
Guangdong Provincial Institute of Metrology	4.54%	0.62%
The Fifth Electronics Research Institute of Ministry of Industry and Information Technology	4.47%	0.08%

Table 2. Phase error comparison result (1800MHz, 16QAM modulation, symbol rate 4Mbit/s, filter RRC, filter coefficient 0.22)

Company	Measurement value	Uncertainty (k=2)
National Institute of Metrology,China(Comparison leading laboratory)	3.50°	0.28°
Ministry of Industry and Information Technology Communication Measurement Center	3.51°	0.61°
China Aerospace Science&Industry Corporation 203	3.51°	0.28°
Measurement and testing center of research laboratory equipment of General Armament Department	3.50°	0.62°
Keysight Technologies,Inc.	3.52°	0.57°
Rodhe&Schwarz(China)Technology Co.,Ltd	3.50°	0.62°
Beijing Oriental Institute of Measurement and Test	3.5°	0.5°
National Instruments Corporation	3.52°	0.20°
Shanghai Institute of Measurement and Testing Technology	3.50°	0.35°
Jiangsu Institute of Metrology	3.5°	0.6°
Guangdong Provincial Institute of Metrology	3.53°	0.72°
The Fifth Electronics Research Institute of Ministry of Industry and Information Technology	3.48°	0.28°

From table 1 and table 2, the evaluation of measurement uncertainty is basically same as the result assessed by Comparison leading laboratory (NIM).

Acknowledgements

The author wishes to thank Key Laboratory of Shanghai On-line measurement and control technology and all colleagues who provided technical support.
This work was supported by the National Science & Technology Pillar Program (Grant No. 2014BAK02B04), the Project of General Administration of Quality Supervision, Inspection and Quarantine of the People's Republic of China (Grant No.2014QK138).

References

1. Fang Hong, Bian Xin ,He Zhao,Zhou Xin, Research on Measurement Method of Vector Modulation Error based on Constellation Diagram, Chinese Journal of Scientific Instrument, Vol.34 No.1 Jan. (2013) pp128~132
2. Fang Hong, Bian Xin ,Zhou Xin, Liu Ke, The Numerical Simulation and Experiment Research for Measurement of Error Vector Magnitude(EVM), Applied Mechanics and Materials, Vol.103(2012) pp199-204
3. Zhang Rui, Zhou Feng,Guo Long Qing, Wireless Communication Instrument and Test Application, Posts&Telecom Press, pp135~138.2011
4. Cheng Yi Xin, Measurement and Testing Technology for the Uncertainty Analysis of Error Vector Magnitude, Vol 37 No.12(2010) pp62~68
5. Wu You Zhang, Zhao Hai Ning, Yu Hui Dong, Calibration and Measurement of Digital Modulation Quality Parameters, Acta Metrologica Sinica, 2005, 26(3) : pp271~274
6. Khald M. Gharaibeh, Kevin G. and Michael B. Steer. Accurate Estimation of Digital Communication System Metria-SNR, EVM and p in a Nonlinear Amplifier Environment. ARFTG Microwave Measurements Conference, Fall 2004. 64th, 2~3 Dec. 2004:41~44.
7. The people's Republic of China National Metrology Technical specification, *JJF 1128-2004 Calibration Specification for Vector Signal Analyzers*, China Metrology Publishing House, 2004
8. The people's Republic of China National Metrology Technical specification, *JJF 1174-2007 Calibration Specification for Digital Signal Generator*, China Metrology Publishing House, 2007

MATEC Web of Conferences **61**, 04003 (2016) DOI: 10.1051/matecconf/20166104003
APOP2016

A Reliability Accelerated Test of High-speed Punch Based on Failure Analysis

Lan Chen[2], Xinzhou Zhang[1,a], Shuyuan Gan[1] and Yuan Wang[1]

[1]School of Mechanical Engineering, Jiangsu University, Zhenjiang, 212013, China
[2] Jingjiang College, Jiangsu University, Zhenjiang, 212013, China

Abstract. By analyzing the maintainability data of a certain high-speed punch, its main fault modes, such as oil/gas parts damage, parts damage and leakages, were identified. According to the fault signal measurability and the accelerated failure mechanism, the content and scheme of a reliability accelerated test (RAT) were planned specifically, which was partly verified by some tests on a high-speed punch. This paper provides a basis for the RAT of high-speed punch.

1 Introduction

High-speed punch is an efficient, high-precision and high-automation punching machine. The typical use is in manufacturing of huge quantities products and it has been widely used in aeronautics and astronautics, automobile, traffic and transportation, metallurgical chemical industries and so on.

Reliability accelerated test (RAT) technology, which gets more reliability information with less work, is widely used to stimulate the potential defects and improve the reliabilities of products by failure analysis [1]. However, the domestic application of the RAT technology is not so widely as it used aboard because of the limit of test technology and equipment [2]. Nelson [3] described a model and methods for analyzing age-degradation data on product performance and provided estimates of the distribution of time for product performance to degrade to a specified value while under a specified stress. Oliveira [4] analyzed some estimation methods, such as analytical, the numerical or the approximated, which can be used to obtain the time-to-failure distribution and presented an example of the application and analysis of degradation tests. Chinnam [5] demonstrated the feasibility of such an approach using 'general' polynomial regression models for degradation signal modeling. Wang [6] gave some suggestions to reduce early failures for machining center. Zhang [7] provided an optimum distribution model which can be determined as Weibull distribution, providing a theoretical basis for the failure analysis and prediction, and reliability enhancement of the numerical control machine tools. Adamyan [8] represented a methodology that can be used for identifying the failure sequences and assessing the probability of their occurrence in a manufacturing system. Tseng [9] made an optimal step-stress accelerated degradation test plan for gamma degradation processes. Wang [10] discussed the reliability of the vibration control system with uncertain parameters. Zhang [11] proposed reliability enhancement testing plan based on the analysis of the failure mechanism of NC rotary table.

As stated previously, there is few study on application of the RAT technology in high-speed punch. In this paper, the failure analysis of high-speed punch is analyzed, and then the content and scheme of RAT is proposed and carried out.

2 Failure analysis of high-speed punch

2.1 Failure analysis

High-speed punch is a complex and high automatic system, which is composed by many parts and accessories, and the malfunction of the parts is multifarious [12]. In order to make scientific accurate analysis and calculation, find out the weak link and provide the safeguard, it is divided into nine subsystems, such as Numerical Control System(NC), Main Transmission System(T), Electric System(E), Lubrication system(L), Balance System(B), Detecting System(D), Feed System(F), Pneumatic System(P) and Other System(R). Based on the failure data of high-speed punch, the failure positions statistics is shown in Table 1 and the histogram is shown in Fig. 1.

From the Table 1 and Fig. 1 in can see, in the malfunctions of high-speed punch, the Main Transmission System failures (including more failures of the clutch) occur mostly, about 23.5% of the total ones. The Electric System failures, including the main failure such as wiring fault, components damage and so on, account for 22.9% of the total malfunction.

[a] Corresponding author: zhangxinzhou1984@126.com

Table 1. Frequency of failure positions.

Code	Failure positions	Times	Frequency
T	Main Transmission System	113	0.235
E	Electric System	110	0.229
P	Pneumatic System	87	0.181
L	Lubrication system	73	0.152
D	Detecting System	46	0.096
R	Other System	29	0.060
B	Balance System	17	0.035
F	Feed System	5	0.010
NC	Numerical Control System	1	0.002

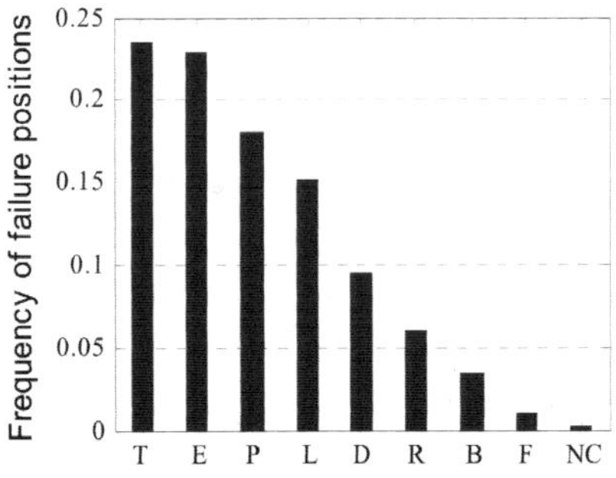

Figure 1. Histogram of failure positions.

Taken together, the Main Transmission System, the Electric System, the Pneumatic System and the Lubrication system are more positions of occurring failures in the high-speed punch and the failure of the four subsystems is about 79.7% of the total malfunction.

2.2 Failure mechanisms analysis

The main failure reason of the oil/gas leakage is because of the failure of airproof system. The common airproof system is usually composed by the bolt-flange-gasket connection system or the airproof ring. The main failure form of the bolt-flange-gasket connection system are the intensity failure caused by bolt failure and the gasket failure, and the airproof failure caused by leakage. The failure of airproof ring is mainly due to the deformation and abrasion of the airproof ring which causes the leakage.

The electrical line malfunction is mainly caused by the disconnection of the line and the open circuit, short circuit and undesirable contact of the cable. The main reasons of the malfunctions are that the circuit layout is not so optimization and the circuit installment quality is not so high.

2.3 Failure measurability analysis

Based on all kinds of failures of high-speed punch, analysis of failure measurability is shown in Table 2. According to the measurability analysis one can see that, some failure modes of the high-speed punch can be detected directly or indirectly, while some others cannot. But it is still possible to predict the occurrences of those non-detectable failures by monitoring some relative parameters. For depth analysis of the failure and performance degradation mechanism of the high-speed punch, reliability testing system need to be designed based on the measurability analysis above and other related theories. Through the analysis of experimental data, its failure and performance degradation mechanisms can be obtained.

Table 2. Analysis of failure measurability.

Failure Mode	Failure phenomenon	Analysis of measurability
Oil/gas components damage	Oil/gas tube break	Unmeasured directly, can be monitored indirectly by oil/gas pressure, flow and so on.
	Oil/gas pump break	
	Oil/gas valve break	
Components damage	Fastener break	Hard to measure
	Locking gasket break	Measured by press load
	Belting fracture	Measured by transmission torque
Oil/gas leakage	Seal break	Measured indirectly by pressure
Electrical components damage	Transformer break	Measured indirectly by voltage, electric current and so on.
	Control switch break	Hard to measure
Oil/gas blocking	Oil/gas blocking	Measured by cleanliness
Big/abnormal noise of machine	Big noise	Measured by noise
Inadequate lubrication	Inadequate lubrication	Measured by lubrication pressure

3 Reliability accelerated degradation test of high-speed punch

3.1 Accelerated analysis of failure mechanisms

The accelerated test can be used to stimulate some specific features within a relatively short time by an appropriate higher stress, and usually the same features will come out spontaneously over a long time in normal operation. It is necessary to analyze whether the accelerated test can be used to research the failure process of high-speed punch. The judgment principles are as follow [13]: 1) The consistency of failure mechanisms: The failure mechanisms remain unchanged under different stress levels and it can be guaranteed by experimental design. 2) The regularity of failure process: There is a certain relationship between the product life and stress, namely the existence of accelerated model. 3) The identity of failure distribution: The product life under different stress levels follows the same distribution.

These conductions are essential for experimental design and analysis in accelerated test. According to the analysis of the bottom dead center (BDC) data of the high-speed punch, one can know that the repetitive dynamic accuracy of the BDC reduces with the increase of stamping speed, which is shown in Fig. 2. Therefore, the acceleration is feasible during the accuracy degradation of the bottom dead center. As the acceleration will not change the failure mechanisms, the performance degradation of the BDC and other parts can be accelerated by increasing the punch speed. Consequently, the purpose of RAT for high-speed punch can be achieved by scientific experimental design without change of failure mechanisms.

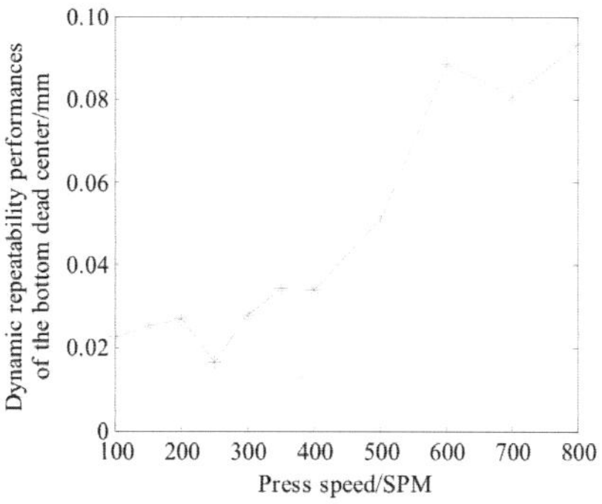

Figure 2. The change curve of bottom dead center precision.

3.2 Accelerated accelerating stress

Accelerated stress should be able to stimulate failure or accelerate performance degradation of parts, and cannot change the failure mechanisms in normal use. Increasing the punch load and press speed of the punch can check the reliability of the transmission system consists of motor, flywheel, clutch, crankshaft, connecting rod, and the slide block and other components. As the working load and speed increased, the friction torque of the crankshaft gets larger and brings out a large amount of heat, which will result in thermal deformation and transmission precision reduction. In addition, as the punch load and press speed increased, the inertial load and the vibration stress are also increased. This will stimulate failure or accelerate performance degradation of parts, such as mechanical fatigue, mechanical wear and fracture.

Through the failure analysis of high-speed punch, the punch load and press speed can be selected as accelerated stresses. Considering the fact that the punch load is harder to control, press speed is finally selected as accelerated stress.

3.3 Reliability accelerated test

All the technical parameters of operation need to be monitored in reliability test. According to the failure mechanism analysis of the high-speed punch, the test content is made as shown in Fig. 3, based on which the performance degradation and failure condition during operation can be analyzed. The test object, which is customized and fewer in number, is high in cost. Comprehensive considering all the circumstances, the number of test object is preliminarily determined to be 1~3, but the specific number is determined by the actual circumstances. The test is on the premise of not destroying high-speed punch with censored time as its termination conditions.

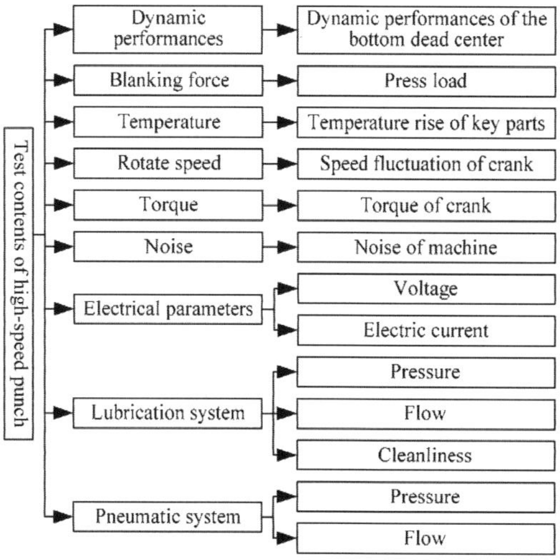

Figure 3. The test content of high-speed punch.

4 Preliminary test verification

A dynamic accuracy test of slider block was carried out on a certain high-speed punch whose nominal force is 800KN. Three displacement sensors are located at a block fixed on the workbench to measure the dynamic accuracy of the slider block along X-, Y- and Z-direction, respectively, see in Fig. 4.

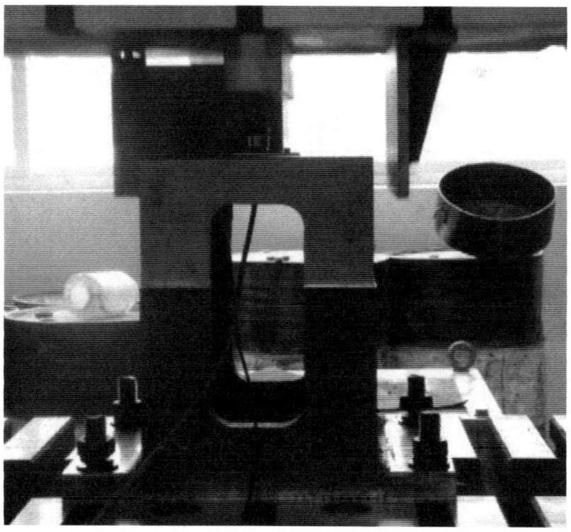

Figure 4. Dynamic precision test of slide block.

Measurement scheme is as follows: In idle load cases, the dynamic accuracies of the slider block along three directions were measured simultaneously with the punching speed increasing from 100SPM to 800SPM gradually. After each speed-up, the signals were acquired while the speed was stabilized. The averaged positions and averaged repetitive positioning accuracies calculated from the testing results are shown in Fig. 5.

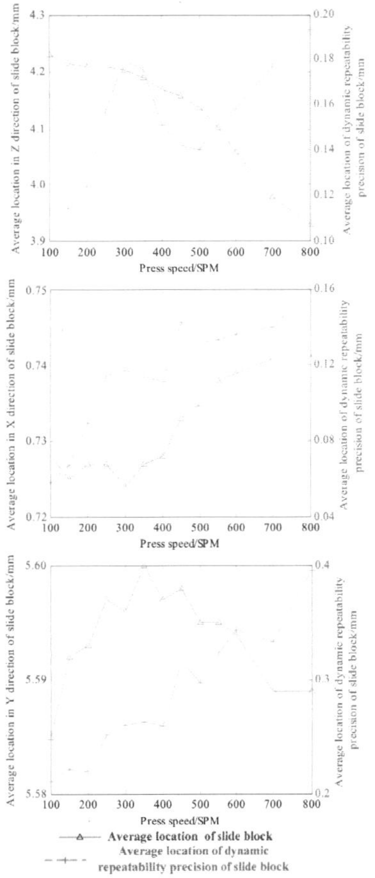

Figure 5. Dynamic precision of slide block.

From the Fig. 5 one can see that, as the punching speed increasing, the averaged position of the BDC of slide block falls down subsequently, and the mean value of X direction position of slide block gets bigger, this error is the primitive error, which is in proportion to punch speed. Simultaneously, the mean value of Y direction of slide block first increases and then decreases, the difference value between the maximum and minimum value is 0.015mm which is not very big, and there is no regularity. Therefore, slide block Y direction position mean is approximate seen the same in the speed change process. From the overall trend analysis, the dynamic repetitive positioning accuracy of slide block increases with the intensity of punch speed. With the increase of punch speed, the vibration and thermal deformation of the punch increase, therefore the slide block dynamic accuracy is declined.

According to the survey data analysis, the dynamic accuracy of slide block: Dynamic accuracy of X direction is higher, dynamic accuracy of Z direction takes second place, and dynamic accuracy of Y direction is worst. The test results provide some important information for the reliability growth of the high-speed punch.

5 Summary

The failure analysis of high-speed punch was analyzed based on the failure and maintenance data and the weakest links of the high-speed punch were pointed out. The reliability accelerated test system of high-speed punch was built after the analysis of failure measurability and failure mechanisms. The pretest, which is demonstrated reasonable and effectively by the analysis of the reason of the dynamic repeatability precision decline of the bottom dead center, laid the foundation for the comprehensive test of the high-speed punch.

Acknowledgments

This work was financially supported by Fundamental Research Project of the Natural Science in Colleges of Jiangsu Province of China (Grant No. 15KJB460005), Jiangsu Planned Projects for Postdoctoral Research Funds (Grant No. 1501102C), Advanced Talent Foundation of Jiangsu University of China (Grant No. 14JDG138).

References

1. D.Aimin, C. Xun, Z. Chunhua, W. Yashun. Acta Armamentarii **28**, 8 (2007)
2. C. Xun, Z. Chunhua. Journal of Mechanical Engineering **45**, 8 (2009)
3. W. B. Nelson. IEEE Transactions on Reliability **30**, 2 (1981)
4. V. R. B. Oliveira, E. A. Colosimo. Quality and Reliability Engineering International 20 (2004)
5. R. B. Chinnam. Quality and Reliability Engineering International 18 (2002)
6. W. Yiqiang, J. Yazhou, J. Weiwei. Reliability Engineering & System Safety **72**, 1 (2001)
7. Z. Yingzhi, J. Yazhou, Z. Xuewen, D. Qinglin. Journal of Jilin University (Engineering and Technology Edition) **34**, 2 (2004)
8. A. Adamyan, H. David. Reliability Engineering & System Safety **76**, 3 (2002)
9. S. T. Tseng, N. Balakrishnan, C. C. Tsai. IEEE Transactions on Reliability **58**, 4 (2009)
10. W. Shipeng, X. Yancai. International Journal of Advancements in Computing Technology **3**, 8 (2011)
11. Z. Genbao, X. Zhi, H. Wenhui, T. Lin. China Mechanical Engineering **22**, 8 (2011)
12. H. Ye. Changchun: Jilin University (2008)
13. S. Hiroshi. *Reliability engineering*(Beijing: Science Press, 1982)

Research on Safety Monitoring of Oxygen Cylinder Filling

Fangcheng WANG[1, a], Lijing ZHANG[1], Gang TAO[1]

[1]Jiangsu Key Laboratory of Hazardous Chemicals Safety and Control,College of safety science and Engineering,Nanjing Tech University Nanjing 210009

Abstract. In recent years, the process of filling oxygen explosion accidents have occurred, causing great casualties and property losses. Great pressure on the safety production of enterprises, resulting in adverse effects on the society.This paper analyzes and summarizes the oxygen bottle filling process types and causes of the accident, found the shortcomings of traditional monitoring methods, and propose a new monitoring methodology, which is based on infrared detection technology oxygen cylinder filling process safety monitoring system.The method can accurately monitor the temperature changes of oxygen cylinder filling process , timely detection of anomalies in the filling process, provide some security for the oxygen cylinder filling operations..

1 Introduction

Oxygen as an ideal combustion gas is widely used in industrial production, with the rapid development of China's economy, the demand for oxygen is increasing, and oxygen-related accidents are increasing.At 15:50 on December 22, 2013, Shanxi Province, a company in cylinder explosion occurred, the place of the cylinder explosion is located in oxygen filling station, two people were killed and two people were injured[1]. For the Oxygen accident-prone situation, the party and the country attaches great importance.It has been issued many regulations, technical conditions and other documents about compression, gas production and cylinder filling areas[2].Enterprises and related researchers also carried out a detailed accident investigation and analysis, and gives many countermeasures in the aspect of management and technical.Based on previous research, this paper discusses the current lack of oxygen filling process management and monitoring technology, through practice and research to design a oxygen filling process safety monitoring system.

[a]Corresponding author: 1006133376@qq.com

2 infrared temperature measurement technology and principle

2.1 Infrared temperature measurement technology.

Infrared temperature measurement technology was originally used in the military field, the current application in industrial production is also very extensive, especially in metallurgy, coal, electric power and other industries.Infrared temperature measurement technology has the following advantages:One is that its measurement does not interfere with the measured temperature field, which has a measurement with high accuracy; Second, the wide temperature range, no measurement limit in theory, can be measured relatively high temperatures; Third, the detector response time is short, fast response, easy to fast and dynamic measurement;Fourth, you do not have contact with the measured object, which is easy to operate; Fifth, tiny target temperature can be determined[3].

Currently, the device application of infrared temperature measurement technology is more, infrared thermometer, infrared television, infrared camera and so on. Infrared thermometer is researched and developed at home and abroad earlier for its simple principle and structure,which is the most common form of non-contact type single-point thermometer. In practical application, its price is lower and the infrared thermometer for single point temperature measurement accuracy. It have been widely used in basic level for its simple operation and easy to master[4].

2.2 Infrared temperature measurement principle.[5]

In nature, an object that its temperature is above absolute zero (-273℃), will radiate electromagnetic waves due to thermal motion of molecules, wherein the wavelength of 0.75μm ~ 1000μm band called infrared. At a given wavelength and temperature, radiation of the object having a maximum energy , which is called a blackbody, and its reflection coefficient is set to 1, the other materials reflection coefficient is less than 1, called gray body.

Blackbody spectral radiant power and temperature to meet the Planck radiation Theorem:

$$p_b(\lambda T) = \frac{c_1 \lambda^{-5}}{e^{\frac{c_2}{\lambda T}} - 1} \tag{1}$$

In Eq.1, $p_b(\lambda T)$ is the exit of the blackbody radiation, W/(m^2·μm); λ is wavelength, μm; T is the absolute temperature, K; c1, c2 radiation constant.

From the Eq.1,we know that: (1) As the temperature increases, the stronger the radiation energy of the object; (2) at absolute temperature, the wavelength of the radiation emitted per unit area of a blackbody is blackbody radiation power.

Can be obtained the black-body radiation spectrum curve according to (2). From the graph, the radiation energy of the object is stronger with the temperature rising, the radiation peak shifts to shorter wavelength (to left), and meet the Wien displacement Theorem.

$$T \cdot \lambda_m = 2897.8 \mu m \cdot K \tag{2}$$

Infrared temperature measurement must be based on wavelength division temperature range, high temperature in shortwave, low temperature in longwave, its sensitivity was high (steep curve), stronger anti-interference.

we get the Stephen - Boltzmann Theorem that blackbody radiation is proportional to the fourth power of the temperature,when the Planck theorem integral to wavelength from 0 to ∞.

$$p_b(\lambda T) = \sigma T^4 \tag{3}$$

In Eq.3, $P_b(T)$ is the total blackbody radiation energy radiated from the blackbody per unit area and unit time at temperature T, called the total irradiance,W/m^2; σ is Stephen - Boltzmann constant; T is the temperature of the object.

Blackbody radiation law is the theoretical basis of infrared temperature measurement technology. In general, gray body radiation power is always less than the blackbody radiation power that monochromatic radiation of the object is smaller than the total monochromatic radiant emittance of a blackbody. Their ratio $\varepsilon(\lambda)$ is called monochrome blackness of the object, namely the degree of actual objects close to the black body.

$$\varepsilon(\lambda) = P(T) / P_b(\lambda T) \tag{4}$$

Monochrome blackness of the object is a constant that not varies with wavelength. This object is called a gray body. The constant varies with the material. Only blackbody is 1, and the general body of gray is between 0 and 1.So:

$$P(T) = \varepsilon P_b(\lambda T) \tag{5}$$

$$P(T) = \varepsilon \sigma T^4 \tag{6}$$

Then the measured target temperature is :

$$T = \left(\frac{P(T)}{\varepsilon \sigma} \right)^{\frac{1}{4}} \tag{7}$$

Eq.7 is the mathematical description of thermal radiation thermometry of an object.

3 Infrared temperature measurement system

3.1 System Design.

Infrared temperature measurement system is based on infrared radiation characteristics of the object, using an infrared thermometer to receive infrared radiation *P(T)* from oxygen cylinder surface. Get the electric signal corresponding to the cylinder surface temperature T with corresponding signal processing according to equation (7). We obtained a corresponding digital signal of the temperature T by the A / D conversion and then displayed on the LCD.Set an alarm temperature, when the temperature reaches the alarm, the alarm action alert. Notify the relevant personnel to evacuate or take emergency measures, safety interlock of filling system operation, automatic stop filling. System design block diagram shown in Figure 1.

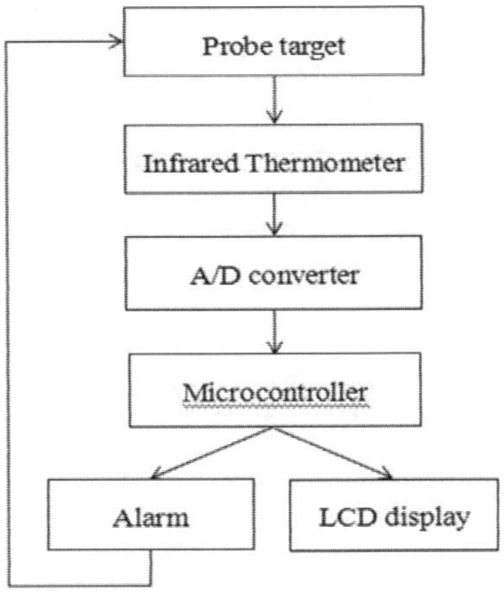

Figure 1 Block diagram of the overall infrared temperature measurement system design

3.2 Each part of the system design.

According to the total system design ideas, the following is the various parts of the system detailed design and presentation.

(1) infrared thermometer: Infrared thermometer consist of the optical system, photoelectric detectors, signal amplifiers and signal processors and other components.

The optical system receives goals infrared radiation in its field. Optical parts and its locations determine the size of the field of view. Infrared radiant energy focused on the optical detector and converted to electrical signals, amplified through the amplifier, and the signal would be corrected.

(2) A / D converter: The signal that Infrared thermometer get is an signal (analog signal), while the microcontroller can only handle digital signals. Therefore, the temperature of the electric signal first convert to digital signal by the A / D conversion,then the digital signal supplied to the microcontroller for data analysis and processing.

(3) Microcontroller : Digital signal got by the A / D converter is treated in the microcontrollert. SCM is an integrated circuit chip.It is a micro-computer system, which is small but perfect piece of silicon, consist of the CPU has a data processing capability, random access memory RAM, a read only memory ROM, a variety of I/O port and interrupt system, timers/counters, etc. It complete the process by the use of VLSI technology[6]. MCU of this system needs to be done the following functions: First, accurate measured target temperature, and sent to LCD display; Second, can set the temperature by ourselves, send alarm signal to alarm when the real temperature is over; Third, can set a alarm temperature rise speed, when the actual temperature rises too fast, alarm alarm; Fourth,fuction of data storage; Fifth, input and output functions.You can print and copy a variety of data inside memory.

(4) LCD display: Liquid crystal display (LCD) displays real-time monitoring of temperature. Suitable display size and suitable resolution to meet the basic requirements of a clear and accurate display.

(5) Alarm: Alarm receives the alarm signal from the microcontroller and alerts. Audible alarm is one of the alarm to alert personnel via visual and auditory stimuli double.

4 Example applications

In order to verify the feasibility of infrared temperature measurement system,I choose a oxygen cylinder filling Enterprise as a pilot. The company has an oxygen filling line.

4.1 Selection of major equipment

(1) infrared thermometer

Parameters of the infrared thermometer selected are shown in Table 1.

(2) Display control device

Parameters of display control instrument selected in are shown Table 2.

Table 1 The main parameters of infrared thermometer

Measuring range	-50～600℃（-158～1112℉）
Temperature measurement accuracy	±1 ℃
Measure the distance ratio	50:1
Response time	500ms & (8～14) um
Operating temperature	0～50℃(32～122℉)

Table 2 Display control device parameters

input signal	0～375Ω、0～75mV、0～30mV、0～5V、1～5V、0～10V、0～10mA、 0～20mA、4～20mA
Display Range	-1999～9999
measurement accuracy	Non-linear signal： 0.2%，Linear signal：0.1%
Transmitter output signal	0～20mA、4～20mA

4.2 Molding equipment assembly.

According to the system design, the assembled device is shown in Figure 2.

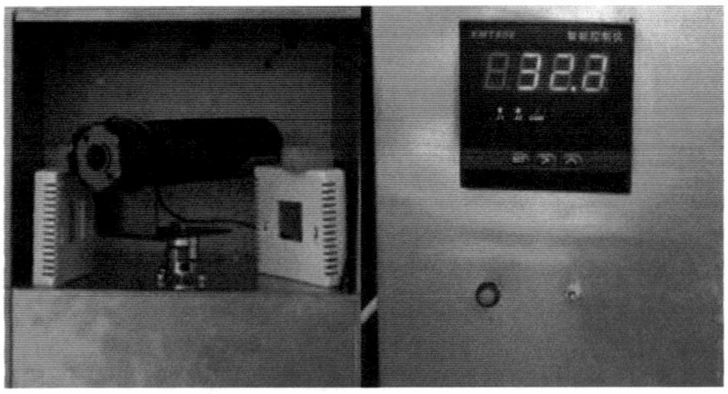

Figure 2 Front and side view of the device

4.3 System operation.

After the system is assembled, one oxygen cylinder for being in the process of filling is in for real-time monitoring by the system. The system runs smoothly and normal. Part of the monitoring data is as follows:

(1)Alarm temperature is set to 60 centigrade, room temperature is 20 centigrade. See Table 3.

(2)Alarm temperature is set to 45 centigrade, room temperature is 20 centigrade. See Table 4.

(3)Alarm temperature is set to 30 centigrade, room temperature is 20 centigrade. See Table 5.

The above analysis of the data shows that: First, With the filling pressure increases, the surface temperature of the oxygen cylinder increases;Second, Pre-filled cylinder temperature rises faster, the late temperature rise slowly;Third, Normal filling, oxygen cylinder surface temperature will not exceed 60 centigrade; Fourth, the average temperature of the three measurements was 325.4K (52.4 centigrade) in the maximum filling pressure.

Table 3 The alarm monitoring data of the temperature of 60 centigrade

Filling Pressure[P/MPa]	3	5	7	9	11	13	14.7
Temperature display[T/K]	294.8	309.3	316.1	319.9	322.5	324.3	325.4
Alarm [Yes / No]	No	No	No	No	No	No	No

Table 4 The alarm monitoring data of the temperature of 45 centigrade

Filling Pressure[P/*M*Pa]	3	5	7	9	11	13	14.7
Temperature display[*T*/K]	295.1	309.5	316.0	320.1	322.4	324.6	325.2
Alarm [Yes / No]	No	No	No	Yes	Yes	Yes	Yes

Table 5 The alarm monitoring data of the temperature of 30 centigrade

Filling Pressure[*P*/MPa]	3	5	7	9	11	13	14.7
Temperature display[*T*/K]	295.3	309.7	316.3	319.8	322.2	324.4	325.5
Alarm [Yes / No]	No	Yes	Yes	Yes	Yes	Yes	Yes

5 Conclusion

From the oxygen cylinder explosion, we analyze the cause and type of the explosion and pointed out shortcomings of current oxygen cylinder filling process safety monitoring. Designed a oxygen cylinder filling process monitoring system based on infrared detection technology. Its practical and scientific was verified by examples. The results show that the use of infrared temperature measurement system for oxygen cylinder filling process temperature monitoring is scientific and feasible. The system can accurately detect and display the oxygen cylinder filling process temperature, and alarm when the temperature exceeds the set alarm. Can provide some security to the oxygen cylinder filling process.It have greater economic benefits to the enterprise production safety.

References

1. Yang Xu. Reason analysis of oxygen cylinder explosion accident. China Special Equipment Safety, **30**(009) (2014) 64-68
2. Ye yi. On safety issues of filling oxygen cylinders. Cryogenic technology.**6**(2005)46-48
3. Li Jun, Liu Meidong, Zeng Yike.Research on Non-contact infrared temperature measurement.PIEZOELECTRICS & ACOUSTOOPTICS.**23**(3) (2001) 202-205
4. Qian Wei.Design of real-time on-line monitoring system for switchgear temperature based on infrared sensor.Nanjing University of Science Master Thesis.(2009)
5. Yan Ming,Peng Chiwu,Yan Yonghong.Principle and Error Analysis of Infra-Red Temperature Measurement.Journal of Hunan University(Natural Sciences).**31**(5) (2004) 110-112
6. Zhang Xuezhong,Wang Xiujun.Closed-loop control of stepper motor based microcontroller At89S52.AGRICULTURAL EQUIPMENT & TECHNOLOGY. **3**(2011) 31-32.

DOI: 10.1051/matecconf/20166104005

Comparison between Measurements and Models For Daily Global Solar Radiation of Urumuqi

Hao Yang [a], Yingni Jiang [b], Jingbo Liu [c] Zhengguo Gong [d]
Architecture Engineering Department, Logistics University of PAP, Tian'jin, china
[a] E-mail address:849253712@qq.com, [b] E-mail address: jyingni@sina.com
[c] E-mail address:568818102@qq.com, [d] E-mail address:782101251@qq.com

Abstract. In this study, daily global solar radiation data at Urumuqi in 1995-2004 were investigated. Sunshine based models were used to estimate the global radiation. Estimated values were compared with measured values in terms of statistical error tests such as mean percentage error (MPE), mean bias error (MBE), root mean square error (RMSE). The model performing best was selected. The global solar radiation estimated from the best model was compared with measured values. It was determined that the predicted values have good agreement with the measured values at high daily global solar radiation.

1. Introduction

Daily solar radiation is an important variable to any vegetative, biophysical and hydrological model that simulates mass, water and energy fluxes. An accurate knowledge of the global solar radiation data at a particular geographical location is of vital importance. However, there are many meteorological stations that do not measure solar radiation but do register other variables such as precipitation, pressure, sunshine hours and temperature. Thus, using models to estimate the global radiation has an excellent foreground. In order to achieve this, several empirical models have been developed to predict the solar radiation all over the world using various parameters. These parameters include extraterrestrial radiation, sunshine hours, mean temperature, maximum temperature, soil temperature, relative humidity, number of rainy days, altitude, latitude and cloudiness [1–10].The most commonly used parameter for estimating global solar radiation in the publications is sunshine duration. Sunshine duration can be easily and reliably measured and data are widely available. The most widely used method is that of Angström [1], who proposed a linear relationship between the ratio of average daily global radiation to the corresponding value on a completely clear day and the ratio of average daily sunshine duration to the maximum possible sunshine duration. The main objective of this paper is to estimate the daily global solar radiation with measured daily sunshine duration data from Urumuqi, and to find the most suitable model for global radiation estimation at Urumuqi, China.

2. Data Used And Methods Of Computation

2.1. Database

Urumuqi meteorological station is chosen for the case study. Urumuqi (43.8°N, 87.7°E) is located in northwest China, at an altitude of 935 m. A database containing daily measured global solar radiation and sunshine duration of

Urumuqi station was obtained from China Meteorological administration. In this paper ten years' recorded daily solar radiation from 1995 to 2004 are used.

2.2. Methods of computation

In the present work, the following correlations were used to express the dependence of global radiation on various parameters:

$$\frac{H_g}{H_0} = a_1 + a_2 \frac{S}{S_0} \quad (1)$$

$$\frac{H_g}{H_0} = a_3 + a_4 \frac{S}{S_0} + a_5 \left(\frac{S}{S_0}\right)^2 \quad (2)$$

$$\frac{H_g}{H_0} = a_6 + a_7 \frac{S}{S_0} + a_8 \left(\frac{S}{S_0}\right)^2 + a_9 \left(\frac{S}{S_0}\right)^3 \quad (3)$$

where H_g is daily global solar radiation on a horizontal surface, H_0 is the daily value of the extraterrestrial radiation, S is sunshine duration, S_0 is the maximum possible sunshine duration, a are empirical coefficients.

The daily value of the extraterrestrial radiation on a horizontal surface was defined as [11]:

$$H_0 = \frac{24\times3600}{\pi} I_0 f(\cos\lambda\cos\delta\sin\omega_s + \frac{\pi}{180}\omega_s \sin\lambda\sin\delta) \quad (4)$$

where I_0 is the new solar constant (=1367 W/m2 [12]), f is the eccentricity correction factor, λ is the latitude of the site, δ the solar declination and ω_s the sunrise hour angle. The eccentricity correction factor, solar declination and sunrise hour angle can be respectively calculated as [11]:

$$f = 1 + 0.33(\cos\frac{360n}{365}) \quad (5)$$

$$\delta = 23.45\sin\left[\frac{360(284+n)}{365}\right] \quad (6)$$

$$\omega_s = \arccos(-\tan\lambda\tan\delta) \quad (7)$$

where n is the number of day of year starting from first of January.

The maximum possible sunshine duration (S_0) can be obtained as [11]:

$$S_0 = \frac{2}{15}\omega_s \quad (8)$$

Fig.1 presents the plot of H_g / H_0 against S / S_0 at Urumuqi (1995-2004). It can be seen that the relationship between daily global solar radiation and daily sunshine duration is rather strong.

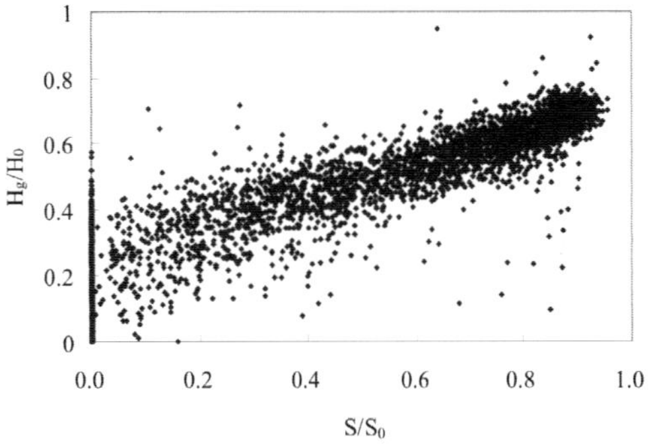

Fig.1. Relationship between H_g / H_0 and S / S_0 of Urumuqi (1995-2004) obtained from the observed data

2.3. Methods of comparison

In this study, three statistical tests, mean percentage error (MPE), mean bias error (MBE) and root mean square error (RMSE) were used to estimate the accuracy of the correlations described above.

The mean percentage error is defined as:

$$MPE = \frac{1}{N}\sum_{i=1}^{N}\left(\frac{D_{ie} - D_{im}}{D_{im}}\right)\times 100 \quad (9)$$

The mean bias error is defined as:

$$MBE = \sum_{i=1}^{N} (D_{ie} - D_{im}) / N \quad (10)$$

The root mean square error is defined as:

$$RMSE = \sqrt{\sum_{i=1}^{N} (D_{ie} - D_{im})^2 / N} \quad (11)$$

where D_{ie} is the ith estimate value, D_{im} is the ith measured value and N is the total number of observations.

The mean bias error test provides information on the long term performance. A low MBE is desired. A positive value gives the average amount of over estimation of an individual observation, which will cancel an under-estimation in a separate observation. The root mean square error RMSE gives information on the short term performance of the correlations by allowing a term by term comparison of the actual deviation between the calculated and measured values. The smaller the value, the better the model's performance is.

3. Results and discussion

The obtained empirical coefficients of the Eqs.(1)-(3) are expressed as follows:

Model 1

$$\frac{H_g}{H_0} = 0.206 + 0.535\frac{S}{S_0} \quad (12)$$

Model 2

$$\frac{H_g}{H_0} = 0.197 + 0.615\frac{S}{S_0} - 0.087\left(\frac{S}{S_0}\right)^2 \quad (13)$$

Model 3

$$\frac{H_g}{H_0} = 0.197 + 0.615\frac{S}{S_0} - 0.087\left(\frac{S}{S_0}\right)^2 + 0.803\left(\frac{S}{S_0}\right)^3 \quad (14)$$

Table 1 shows MPE, MBE and RMSE errors for the three daily global radiation models (models (1-3) corresponding to Eqs.(12)-(14) respectively) of Urumuqi station during the period 1995-2004. From this table, the MBE values obtained from models (1, 2, 3) are positive, which shows that the models over estimate global radiation. The RMSE value, which is a measure of the accuracy of estimation, has been found to be smallest for model 3 (2.182 MJ/m2), see Table 1. Also, MPE, MBE and RMSE are shown in Fig.1-3, respectively. As a result, model 3 performs better than the other models. The MPE, MBE and RMSE of model 3 are 46.1%, 0.0666 MJ/m2 and 2.182 MJ/m2.

Table 1 Prediction errors for daily global solar radiation of three models

	MPE (%)	MBE (MJ/m^2)	RMSE (MJ/m^2)
Model 1	50.9	0.0653	2.186
Model 2	48.7	0.066	2.182
Model 3	46.1	0.0666	2.182

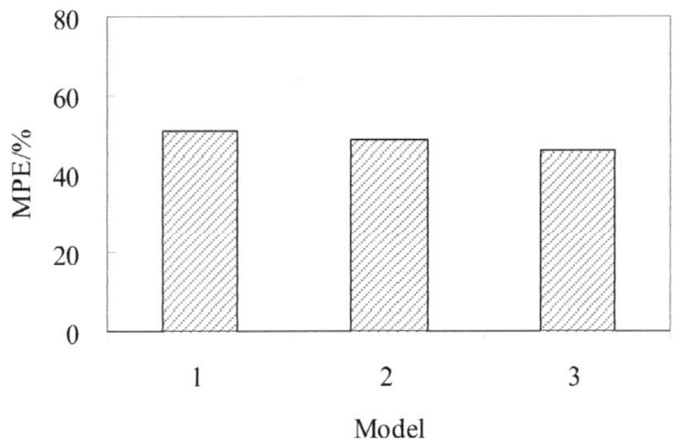

Fig.2. MPE of the three models

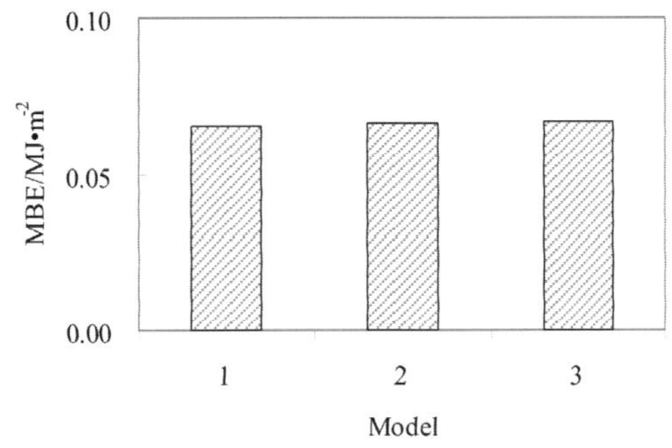

Fig.3. MBE of the three models

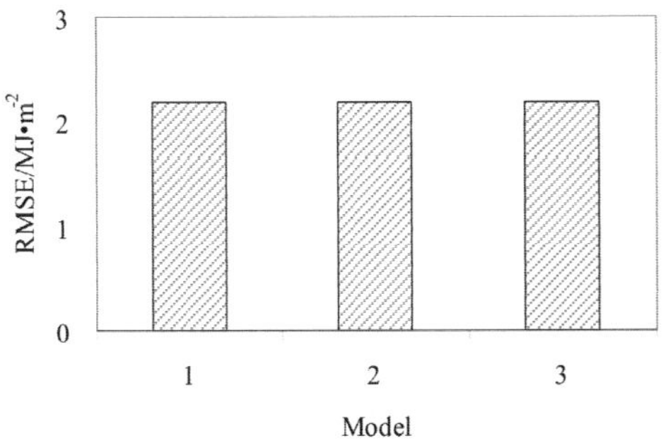

Fig.4. RMSE of the three models

To compare observed and estimated global radiation values of model 3 for Urumuqi (1995–2004), Fig.5 is provided. Model 3 has large errors in estimating observed solar radiation at low daily global solar radiation. But it is of little consequence for solar energy operated systems, because low global radiation constitutes only a small part of the whole radiation. Fig.5 suggests the good fitness between the observed and the estimated values when daily global solar radiation values are large.

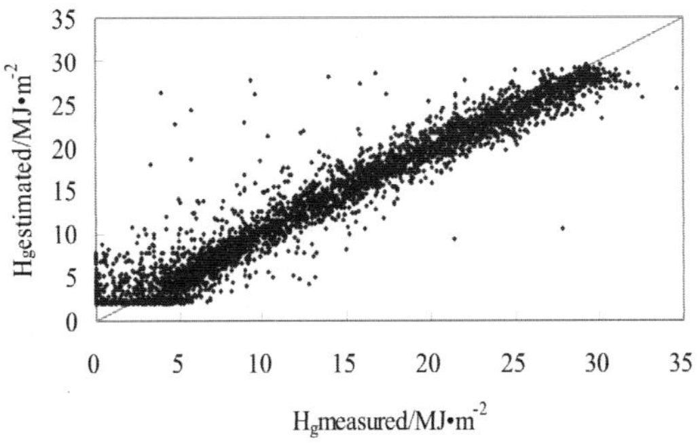

Fig.5. The measured and predicted (by model 3) values of daily global solar radiation for Urumuqi

4. Conclusions

This work demonstrated the efficient computation of daily global solar radiation from an estimation of H_g vs. S/S_0 using data from Urumuqi, China. The data set covers a period of 10 years (1995-2004). Linear, quadratic and cubic models were tried to estimate H_g from S/S_0. It was found that H_g can easily be estimated using the cubic model when the value of H_g is large.

Acknowledgement

The project was financially supported by Natural Science Basic Research Plan in Shaanxi Province of China (Program No. 2006E142). The author would also like to thank China Meteorological administration for its data support.

References

1. Angström A. Solar and terrestrial radiation. Quarterly Journal of Royal Meteorological Society 1924; 50: 121-126.
2. Tiba C. Analysis of monthly time series of solar radiation and sunshine hours in tropical climates. Renewable Energy 2004; 29(7): 1147-1160.
3. Ehnberg JSG., Bollen MHJ. Simulation of global solar radiation based on cloud observations. Solar Energy 2005; 78(2):157–162.
4. Skeiker K. Correlation of global solar radiation with common geographical and meteorological parameters for Damascus province, Syria. Energy Conversion and Management 2006; 47(4): 331-345.
5. Forero NL, Caicedo LM, Gordillo G. Correlation of global solar radiation values estimated and measured on an inclined surface for clear days in Bogota. Renewable Energy 2007; 32(15): 2590–2602.
6. Sabziparvar AA, Shetaee H. Estimation of global solar radiation in arid and semi-arid climates of East and West Iran. Energy 2007, 32(5):649–655.
7. Bulut H, Büyükalaca O. Simple model for the generation of daily global solar-radiation data in Turkey. Applied Energy 2007; 84(5): 477–491.
8. Sabziparvar AA. A simple formula for estimating global solar radiation in central arid deserts of Iran. Renewable Energy 2008; 33(5): 1002-1010.
9. Abraha MG, Savage MJ. Comparison of estimates of daily solar radiation from air temperature range for application in crop simulations. Agricultural and Forest Meteorology 2008; 148(3): 401–416.
10. Chineke TC. Equations for estimating global solar radiation in data sparse regions. Renewable Energy 2008; 33(4): 827–831.
11. Duffie JA,Beckman WA, Solar engineering of thermal processes [M]. New York:Wiley, 1991.
12. Frohlich C, London J, editors, "Revised instruction manual on radiation instruments and measurements," World climate research program, Publication Series No.7, WMO/TD-No.149. World Meteorological Organisation:Geneva, 1986, 140 pp.

CFD Simulation Evaluation for Flow Distribution of the Closed Chamber

Hong Bin[1,a], Wang Hongmei[2]

[1] *Internal Combustion Engine Research Institute of Tianjin University, China*
[2]*Tianjin Tianbo Science & Technology Co,.Ltd., China*

Abstract. Taking a closed chamber flow field as object, this paper establishes the 3D model of the Closed Chamber. With the conclusion of flow distribution and gasoline vapor concentration distribution of the Closed Chamber reached via the CFD software of Fluent and experiment research, this paper analyzes the flow distribution and concentration distribution of gasoline vapor. Several improvement projects are proposed, such as improving the wind circulation system to optimize the indoor gas concentration. By improving the structure of wind circulation system, it makes flow distribution more reasonable and significantly improves the uniformity of indoor gas concentration. Through implementation of this improvement, the reliability and performance of the containment chamber obviously program to improve.

1 Submitting the manuscript

With the development of computer technology, the use of computers to solve numerical simulation of indoor air control equations has also been greatly developed. This method is a combination product of computer technology and hydrodynamics, namely computational fluid dynamics CFD. Today, CFD simulation has been more and more into each field and gradually become one indispensable tool of the research projects.

With the rapid development of motorcycles manufacturing industry, pollution emissions of motorcycles have gradually been concerned in all developed countries and more stringent emissions regulations for motorcycles have been promulgated. THE CLOSED CHAMBER is a important test equipment for measuring fuel evaporative emissions of motorcycles and it should also be more sophisticated with strict regulations. Taking the Closed Chamber as object, this paper establishes the 3D model of the Closed Chamber and uses CFD simulation calculate flow distribution and gasoline vapor concentration distribution of the Closed Chamber and it provides a basis for solving practical engineering needs.

2 SIMULATION ANALYSIS

When applying CFD method to simulate the Flow distribution of the Closed Chamber, the basic equations and Theoretical equations are firstly established closed chamber. Starting from the basic principles (such as fluid mechanics, thermodynamics, heat and mass transfer and other conservation law), Conservation equations of turbulence characteristics method are built. These equations constitute a nonlinear partial differential equations, we cannot use the classical analytical method can only be solved by numerical methods. To solving the equations, the geometry and dimensions of the model must first be given, the calculated area and the proper import and export are determined. At the same time, the boundary conditions of the free surface are given. CFD simulation generally follows the following steps:

(1)Physical model of research subjects is first established closed chamber, and it becomes abstract to mathematical or mechanical model. Then its spatial influence area of model is determined.

(2) Establishing a CAD model of calculating area and gridding the outer surface of the grid geometry and the entire calculation area. Sparse grid and grid cell shape will have a huge impact on the future of computing. Different algorithms format to ensure the stability and efficiency calculations, the general requirements of the grid are not the same.

(3) Add Solution required initial conditions, such as boundary conditions at the inlet and the outlet is generally velocity pressure.

(4) Select the appropriate algorithm, some of the conditions set specific control procedures and the accuracy of solving the required analysis to solve problems and save the data file was found.

(5) Select the appropriate post-processor to read the results file, analysis and display.

3 CAD MODEL

Figures and tables, as originals of good quality and well contrasted, are to be in their final form, ready for reproduction, pasted in the appropriate place in the text. Try to ensure that the size of the text in your figures is approximately the same size as the main text (10 point). Try to ensure that lines are no thinner than 0.25 point.

[a] Corresponding author: tianbohb@126.com

DOI: 10.1051/matecconf/20166104006

3.1. Geometry

Establish geometric model based on design drawings. Its size: 3000 (L) × 1800 (W) × 2000 (H). Hexahedral and tetrahedral meshes mixed model mesh, the total number is about 850,000 grids.

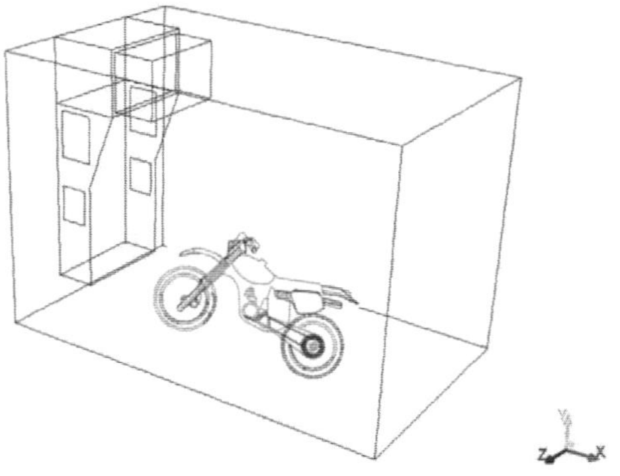

Figure 1. Geometry

3.2 Mathematical Model

According to the characteristics of the project, Selection the K-epsilon turbulence model:

$$\frac{\partial}{\partial t}(\rho k)+\frac{\partial}{\partial x_i}(\rho k u_i)=\frac{\partial}{\partial x_j}\left[\left(\mu+\frac{\mu_t}{\sigma_k}\right)\frac{\partial k}{\partial x_j}\right]+G_k+G_b-\rho\varepsilon-Y_M+S_k$$

$$\frac{\partial}{\partial t}(\rho \varepsilon)+\frac{\partial}{\partial x_i}(\rho \varepsilon u_i)=\frac{\partial}{\partial x_j}\left[\left(\mu+\frac{\mu_t}{\sigma_\varepsilon}\right)\frac{\partial \varepsilon}{\partial x_j}\right]+G_{1\varepsilon}\frac{\varepsilon}{k}(G_k+C_{3\varepsilon}G_b)-C_{2\varepsilon}\rho\frac{\varepsilon^2}{k}+S_\varepsilon$$

Where:

Gk: the turbulent kinetic energy represented by the laminar flow velocity gradient produced;

Gb: turbulent kinetic energy generated by buoyancy;

YM: compressible turbulence due to the fluctuation of the transition diffusion;

C1, C2 and C3: the constant;

σk, σe: turbulent Prandtl number of k equation and ε equation;

Sk, Se: are for the user-defined.

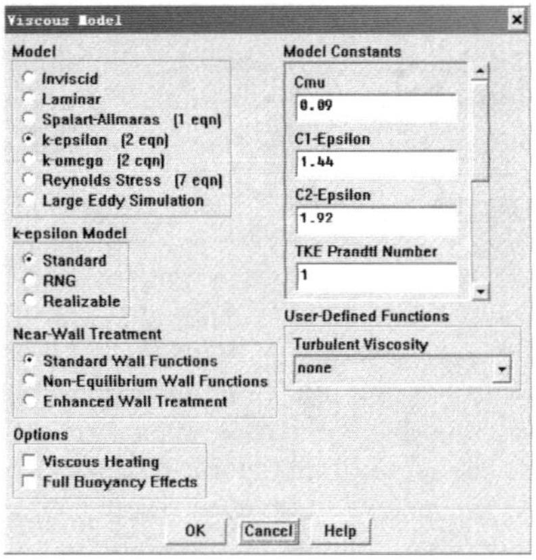

Figure 2. Mass transport model

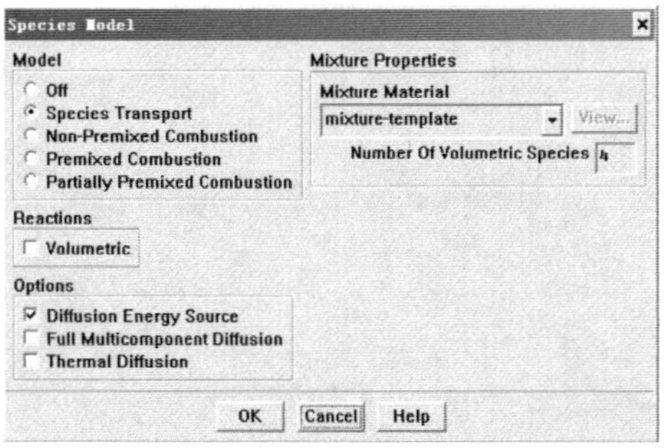

Figure 3. Gasoline vapor diffusion in air

3.3 Boundary conditions

Boundary conditions of the system are as follows:

i) Volatile border of gasoline vapor: Volatile gasoline vapor to the mass flow boundary conditions. The evaporation rate of gasoline is 2g / h.

ii) Boundary fan: fan capacity: 2280m3 / h, wind speed: 2.87m / s, wind pressure: 55Pa.

iii) Air Conditioning border: conditioned air volume: 2280m3 / h, wind speed: 2.3m / s, temperature: 15.5 ~ 35.5 degrees.

4 RESULTS AND ANALYSIS

To verify the reliability of the results of this simulation, the author conducted an actual experiment checked. Using CH analyzers to measuring concentration of CH, and using handheld anemometer measuring wind speed. The boundary condition in the preamble is given by the measurement. All measuring points are taken as the height of the transverse plane Y = 1.6m ~ 1.8m, which the section according to X and Z directions equidistant take several measuring points respectively. During the experiment, the indoor flow distribution of HC is shown in Figure 4 and the concentration distribution of gasoline vapor is shown in Figure 5.

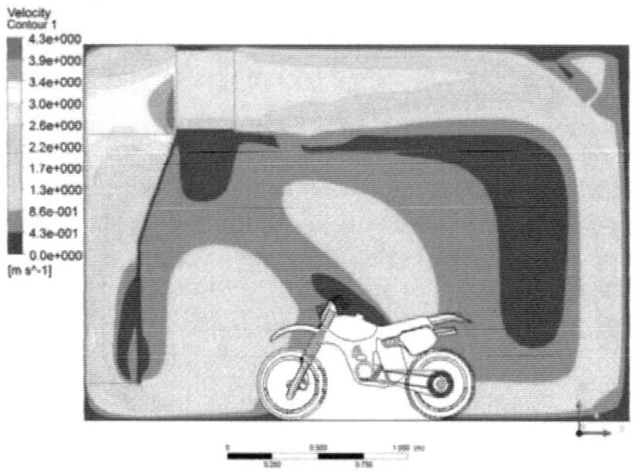

After 5 minutes

 DOI: 10.1051/matecconf/20166104006

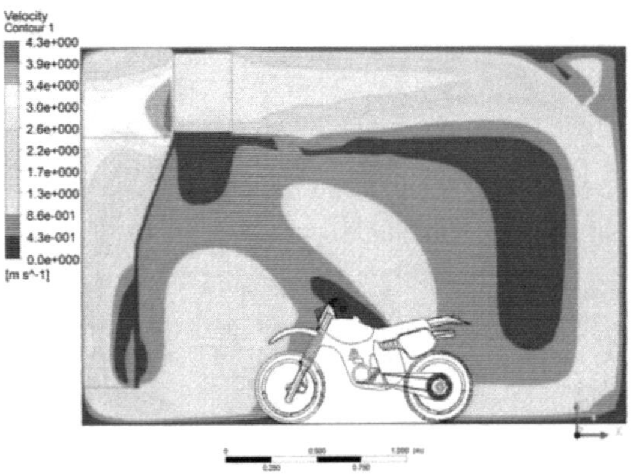

After 60 minutes

Figure 4. Flow distribution

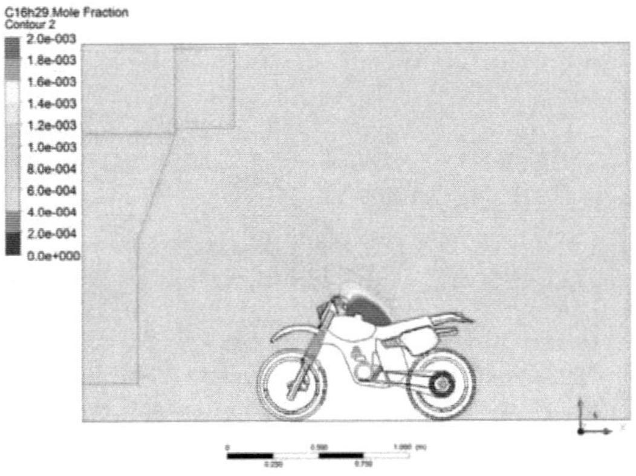

After 5 minutes

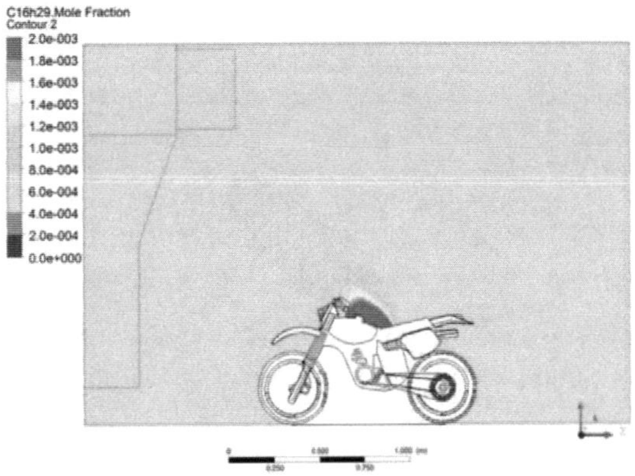

After 60 minutes

Figure 5. Gasoline Vapor Concentration Distribution

The above chart shows that the error of the edge indoor is larger. In addition to itself error of the testing equipment, the main reason is that the actual experimental control conditions is different from the simulation model. The simulation assumes that the flow is steady, that is, the speed and temperature values of each measuring point does not change with time. The interior is in fact due to the turbulent flow, velocity changes over time in the vicinity of the average random fluctuations. In addition, there is the time factor, as compared to temperature fluctuations over time smaller field, the error also small. The result is a simplified simulation model simulation and measured values. Another reason for the error, the problem will improve as well as improve the capacity of computer models of turbulence and continue to improve.

5 CONCLUSION

Taking the flow distribution as object, structural of the Closed Chamber improved to the containment chamber through the flow field analysis and it makes flow distribution more reasonable and significantly improves the uniformity of indoor gas concentration. The conclusions are summarized as follows: 1) With the conclusion of flow distribution and gasoline vapor concentration distribution of the Closed Chamber reached via the CFD software of Fluent and experiment research, this paper analyzes the flow distribution and concentration distribution of gasoline vapor. 2) CFD simulation showed that: the rational design of the Closed Chamber is the key factor to ensure a uniform concentration of indoor gasoline. Reasonable flow rate can significantly improve the indoor gas concentration uniformity. The practical application of these simulation methods improved the performance and reliability of the Closed Chamber.

References

1. SUN Hailong. The CFD Analysis Computation and Optimization Research of Locomotive Electrical Cabinet Ventilation System[D]. Dalian Jiaotong University, 2015.
2. Tan Linbo, Liu Shengji, Lu Cunhao, Wang Jian. CFD analysis and experiment research of non-road gasoline engine cooling system[J]. Journal of Chinese Agricultural Mechanization, 2014, 35(1): 189-192.
3. Miltner M, Jordan C, Harasek M. CFD Simulation of straight and slightly swirling turbulent free jets using different RANS-Turbulence models[J]. Applied Thermal Engineering, 2015, 89:1117-1126.
4. Vyskocil L, Schmid J, Macek J. CFD simulation of air–steam flow with condensation[J]. Nuclear Engineering & Design, 2014, 279:147–157.
5. Nesiadis Athanasios, Nikolopoulos Nikolaos, Margaritis Nikolaos, et al. Optimization of a log wood boiler through CFD simulation methods[J]. Fuel Processing Technology, 2015, 137:75-92.
6. Jia Quanli. A brief introduction to the test method for fuel evaporative emission of motorcycle(II)[J]. Motorcycle Technology, 2009, 11:38-41.
7. Qiu Jie. Quality of fuel evaporative controlling system of motorcycle and study on related problem[J]. Motorcycle Technology, 2007, 5:16-19.
8. Terashima M, Goel R, Komatsu K, et al. CFD simulation of mixing in anaerobic digesters[J]. Bioresource Technology, 2009, 100(7):2228-2233.

Study on Coupled Medium effects of Complex Slotted Charge Blasting

Xie Hua-gang, Wu Ling-li
Tongling University, Anhui Tongling 244000

Abstract: Based on the theory of complex slotted charge water, air coupling medium directional fracture controlled blasting analysis and simulation test of similar, draw the complex slotted charge water coupling charge directional fracture controlled blasting effect is better than that of the air coupling blasting; peak pressure than the air coupled charge under the conditions of quicker acting time, effect of higher value, longer duration; water coupling blasting conditions, unit speed peak value than air coupled charge under the conditions of higher.

1 Introduction

Due to the rapid development of human society, the development of space has been unable to meet the needs of social development, so in accordance with the west, nineteenth Century is a railway bridge of the century, twentieth Century is the century of high-rise buildings, and in twenty-first Century will be the development and utilization of underground space century. According to the situation of our country, urbanization process is accelerating, the western big development is ongoing, China may road and bridge, high-rise buildings and underground space development and utilization of the three go hand in hand with the situation [1]. However, with the implementation of the western development, highway, railway, mining, water conservancy and hydropower, buildings, airports and ports and other infrastructure construction projects will be more and more, the rock soil engineering construction method and means is facing unprecedented challenges, and geotechnical engineering as important Shi construction method of drilling and blasting method is facing unprecedented challenges, the growing demand for its higher and higher, especially decorative stone mining [2], tunnel excavation [3-6], civil air defense engineering [7-10], cutting slope [11-13], coal mine gas drainage [14-16] and other engineering. The amount of application of cutting seam cartridge blasting method, so we need to study it further. Analysis of the blasting research workers of blasting the application and mechanism of the composite type cutting seam cartridge of provides reliable basis. The complex slotted charge instantly in the explosion due to the interaction of PVC pipe before and after, and on different coupling medium blasting characteristics of different [17,18], making explosion energy flow distribution is getting more and more complex. Therefore, it is necessary to study it further.

2 Composite slit charge explosion cracking mechanism

The complex slotted cartridge blasting crack mechanism of see the compound cutting seam cartridge cracking mechanism and numerical simulation experiment. The theory in the role of specific blasting parameters derived compound cutting seam cartridge than conventional cutting seam cartridge explosion in cutting seam direction of shear force more than twice as high as the [19]. That compound cutting seam cartridge in the slit direction of the cumulative effect of more high. In this paper, the numerical simulation method of further research in the role of different coupling mediums, composite type cutting seam cartridge blasting effect, to better use of composite type cutting seam cartridge of the shaped charge effect, and improve the utilization rate of explosion energy.

3 Numerical simulation test research

Blasting damage process only in a few milliseconds even tens of microseconds completed within the corresponding many mechanical properties is difficult to get from the test; also in order to reduce the test cost and time constraints, most researchers choose numerical simulation method to the corresponding research [20-22], they in the constitutive model of rock material established a large number of studies [23], which provides a theoretical basis for the damage and fracture of rock material.

In this paper, the finite element software ANSYS10.0/LS-DYNA is used for numerical calculation. For the convenience of calculation and the calculation model is simplified as an "associate" state plane, using a quarter model, dimensions are 200 mm ×150 mm×5mm, with respect to x, Y axis symmetry, the blasting parameters see Table 1. In the model, BURN *EOS_JWL is used to simulate the change of pressure and volume in the process of explosive explosion, and the relationship between the pressure P and the relative volume V is in the process of detonation:

$$p = A\left(1-\frac{\omega}{R_1V}\right)e^{-R_1V} + B\left(1-\frac{\omega}{R_2V}\right)e^{-R_2V} + \frac{\omega E_0}{V}$$

In the formula, A、B、R_1、R_2、ω is the material constant, E_0 is the initial specific energy. Explosive density by RDX, 1.63g/cm3, D is 6500m/s C-J, detonation velocity, pressure is 18.5GPa. The density, elastic modulus and Poisson's ratio of PVC tube were 1.3 3.1MPa, g/cm3 and 0.38, respectively. The density of water is 1.02 g/cm3. Air density is 1.29 x 10-3g/cm3. KINEMATIC *MAT_PLASTIC_, which is used to model the rock, is a mixture model with the strain rate and the failure of isotropic, kinematic hardening or isotropic and kinematic hardening, and the parameters are shown in table 1. The stress-strain relationship is as follows:

$$\sigma_Y = \left[1+\left(\frac{\dot{\varepsilon}}{C}\right)^{1/p}\right]\left(\sigma_0 + \beta E_P \varepsilon_P^{eff}\right)$$

Table 1 Parameter of rock

Coupled medium	explosive	db/mm	ρ/103kg·m-3	σ/MPa	E/MPa
Water	RDX	16	1.10	1.60	169.45

In the formula, σ_0 is the initial yield stress; $\dot{\varepsilon}$ is strain rate; C, P is symonds Cowper strain rate parameter; ε_P^{eff} is equivalent plastic strain; $E_P = E_{\tan}E/(E-E_{\tan})$ plastic hardening modulus; β for hardening parameter (β = 0 only with kinematic hardening, β = 1 isotropic hardening); stress strain characteristics can only be given at a temperature.

3.1 Blasting effect

Figure 1 shows the complex slotted cartridge in the air under the condition of the coupling, the initial explosion t = 2.7us explosion effect chart, from the chart can be seen, PVC inner layer tube in under the action of the detonation products to the outer PVC pipe direction. When the internal and external PVC pipe contact, in PVC inner layer pipe crack tip and crack opening took place. Finally in the outer PVC pipe cutting seam crack occurred at the. So a composite type of cutting seam charge air coupled directional fracture effect than water coupled directional fracture effect. It also can by comparing Figure 1 and Figure 2 has been proved.

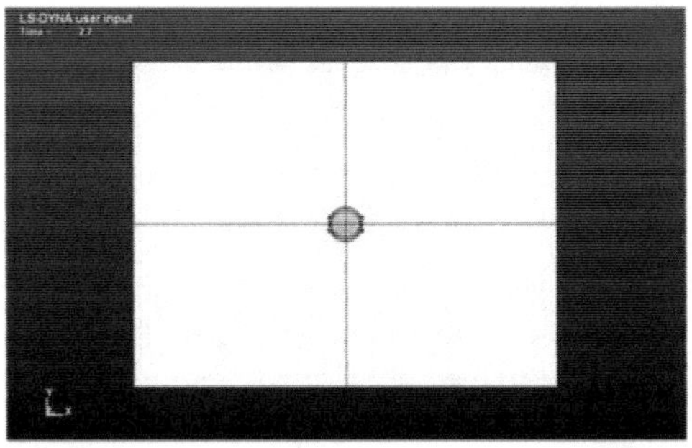

Fig.1 Blasting effect

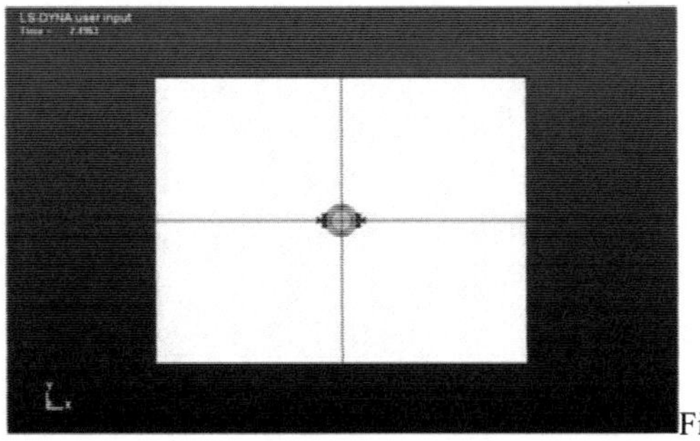

Fig.2 Blasting effect

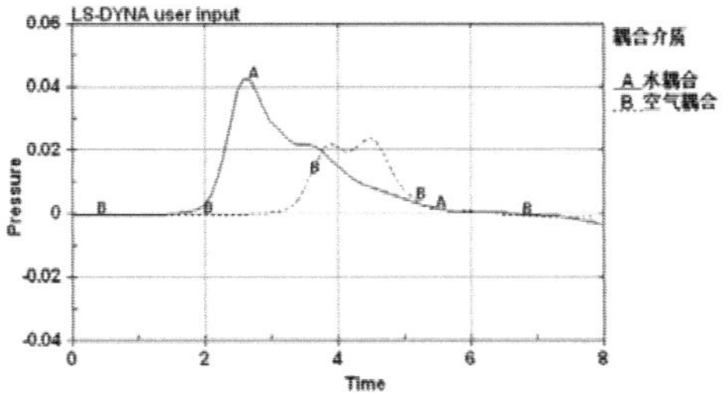

Fig.3 Two kind of coupling medium unit pressure contrast

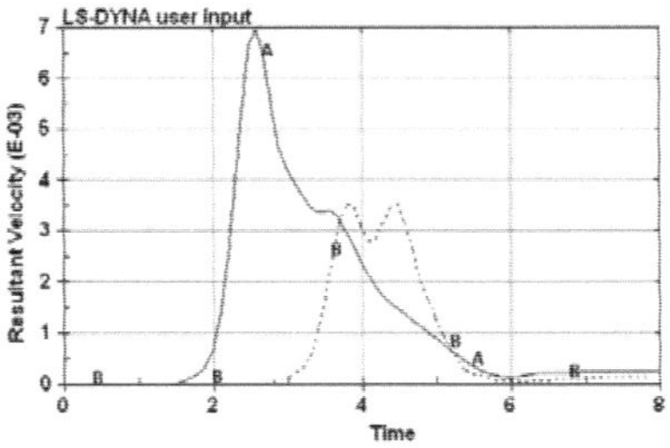

Fig.4 Two kind of coupling medium unit resultant velocity contrast

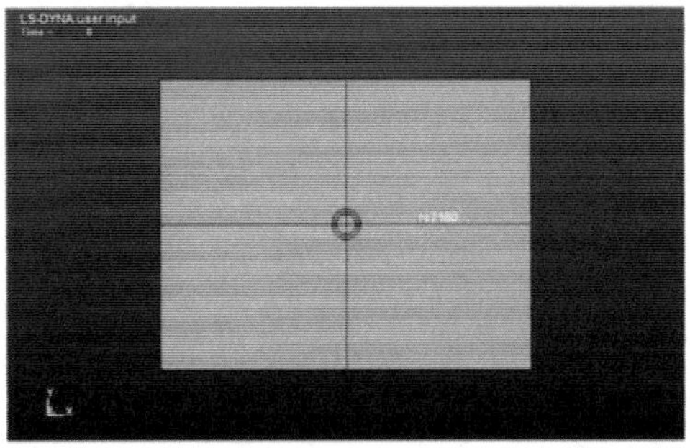

Fig.5 Unit position

3.2 Comparison of mechanical properties of two kinds of coupling medium

Figure 3 and Figure 4 were given compound cut under the condition of water and air coupling charge suture and cutting seam direction unit pressure and velocity contrast curve. Location of the unit is shown in figure 5. Compared with the curve A, the curve B can get the following conclusion:

①under the condition of water coupling, the peak value of unit pressure is 1.92 times of the air coupled condition. The explosive power of the same kind of explosive is different in the two kinds of coupling medium. Due to the water and air with different density (water density ρ=1024.6 kg / m3, and air density ρ =1.226 kg / m3, the density of water about to 835 times the air) and different compression of air is a compressible of, and the water compressibility is usually only air 30 000 to 20 000. It is generally believed that is incompressible). When explosive explosion, the compressibility of the coupling medium water is very small, the ability to save energy is very low, it can be used as a good conductor, so that the blast shock wave can be transmitted to the hole wall, so the explosion pressure in water is higher than in the air.

②the initial pressure peak time is earlier than the air coupling. The compressibility of water is small, and the air is compressed. Therefore, when the explosion occurs, the potential energy of the coupling medium water is very small, the explosion shock wave directly acts on the wall of the blasting hole, and in the air of the coupling medium, the explosion product first compressed air medium, so that the potential energy of the air can reach a certain extent.

③the duration of the pressure is 2 times of the air coupling. In the coupled medium water explosion, the pressure acting time is close to 4us; and the explosion in the air of the coupling medium, the pressure acting time is 2us.

④ under the condition of water coupling, the peak value of unit velocity is 2 times of the air coupled condition.

4 conclusions

Based on the composite cutting simulation test comparison analysis and numerical charge theory, draws the following conclusion:

①The complex slotted cartridge water coupling blasting effect is better than that of air coupling blasting;

② water coupling blasting conditions, the peak pressure than the air coupled loading fast under the condition of drug action time, action value, lasted for a long time.

③water coupling blasting under the condition of unit peak velocity ratio of air coupled charge under the conditions of high.

Fund Project: the natural science research project of Anhui Education Department (KJ2013B299), the talent research foundation of Tongling University (2012tlxyrc09), the teaching research project of Tongling University (JY13035)

References

1. Gao Dazhao. Review and Prospect of geotechnical engineering [M]. Beijing: China Communications Press, 2001
2. Xie Wen-bin, Zhou Xiang. Application Research of Incising the White Marble Material with Grooved Hole-well Blasting[J]. Blasting, 2007, 24(2): 45-48.(In Chinese)
3. The Yan army. Inclined joint development under the condition of tufa formations tunnel vault shaped control blasting technology [D]. Changsha: Central South University, 2012.
4. Mei Huihao. Underground bauxite mine advanced detection and Study on tunnel smooth blasting technology [D]. Changsha: Central South University, 2013
5. Huang Qiang. Study on [D]. Wuhan stress of tunnel surrounding rock based on viscoelastic theory: Wuhan University of Technology, 2012
6. Yan Chun. After the earthquake are Wen Road special heights should force phenomenon and the tunnel rock explosion prevention [D]. Chengdu: Chengdu University of science and technology, 2012.
7. Zhong Guang. Research on mechanical behavior and environmental impact of construction of city tunnel by mining method [D]. Beijing: China Academy of Railway Science, 2013
8. Yang Kewen. Study on Key Technologies of urban tunnel through air defense hole [D]. Beijing: Beijing Jiaotong University, 2013
9. Wang Yafei. Experimental study and control of vibration reduction for tunnel drilling and blasting under complex environment [D]. Beijing: Beijing Jiaotong University, 2013
10. Zhao Lu-xue. Analysis on selection of TBM city track traffic engineering[J]. Modern Tunneling Technology, 2013(02): 7-13+33.(In Chinese)
11. Zhao Ke. The stability analysis of high and steep slope of the double tunnel through the landslide section [D]. Chengdu: Southwest Jiao Tong University, 2014
12. Wang Zhiqiang. Gansu Yintao project of major engineering geological problems research [D]. Lanzhou: Lanzhou University, 2006
13. Huangdan. Shallow buried, large span tunnel of controlled blasting and the scheme of the controlled blasting of [D].[place of publication unknown]: Chongqing Jiaotong University, 2012.
14. Wang Shuang-kai. Low permeability coal seam gas extraction technology[J]. China Science and Technology Information, 2014(02): 45-46.(In Chinese)
15. Hu Jin-hong. Application of high ground stress of outburst coal seam directional pressure relief and permeability increasing technology[J]. Zhongzhou Coal, 2014(02): 58-59+62.(In Chinese)
16. Guo You-hui, Feng Zeng-chao, Liu Jian-guang, et al. Application of deep hole high pressure hydraulic cutting seam technology to increase permeability in

coal seam with outburst danger[J]. Shanxi Coal, 2014(03): 73-75.(In Chinese)

17. Zhong Ming-shou, Long Yuan, Xie Quan-min, et al. Analysis of different borehole coupling medium on explosion seismic signal of carbonate rocks influence[J]. Chinese Journal of Rock Mechanics and Engineering, 2011(04): 702-708.(In Chinese)
18. [18] Zhong Ming-shou, Long Yuan, Li Xing-hua, et al. The wall of the hole hole of different coupling medium explosive loading and specific energy function analysis based on time[J]. Journal of Vibration and Shock, 2011(07): 116-119.(In Chinese)
19. Xie Hua-gang, Ruan Huai-ning, Wu Ling-li. Complex Ligamented Charge Holders Slitting Mechanism and Numerical Simulation Experiment[J]. Chinese Journal of High Pressure Physics, 2012, V26(02): 205-210.(In Chinese)
20. Wang T, Yu W-l, Dong S-q, et al. Influence of the Cell Wall Thickness of Core on the Dynamic Response of Square Honeycomb Sandwich Plate Subjected to Blast Loading[J]. Advanced Materials Research, 2011, 160: 1732-1737.
21. Shao X-n, Xu Y, Wang H-b. Influence of Blasting Vibration on Roadway Loose Range[J]. Advanced Materials Research, 2011, 243: 2518-2522.
22. Tu S, Wang F, Lu Y, et al. Practice and Prospect of Fully Mechanised Mining Technology for Thin Coal Seams under Complex Conditions in China[J]. 2011
23. Lu Y, Xu K. Modelling of dynamic behaviour of concrete materials under blast loading[J]. International Journal of Solids and Structures, 2004, 41(1): 131-143.

Heat-balance Thermal Protection with Heat Pipes for Hypersonic Vehicle

Rong Yisheng[1, 2, a], Wei Yuechuan[1, 2], Duan Dongli[1] and Zhan Renjun[1]

[1]*Engineering University of the Armed Police Force, Xi'an 710086, PR China*
[2]*National University of Defense Technology, Changsha 410073, PR China*

Keywords: Heating Heat-balance Thermal Protection, Heat-pipe, Hypersonic, Aerodynamic Heating

Abstract. Heat-balance thermal protection is non-ablating thermal protection for leading edge of hypersonic vehicle. Heat will be quickly transferred from high aerodynamic heating area to low aerodynamic heating area, where the energy will be released by radiation. The temperature of high aerodynamic heating area could be reduced to protect the designed structure from being burned down. Heat-balance thermal protection is summarized. The research on heat-pipe for heat-balance thermal protection is introduced.

1 Introduction

There is serious aerodynamic heating during hypersonic flight. The heat flux on leading nose whose radius is 20mm can reach 2~3MW/m2 and the wall temperature can reach 1400K when the vehicle Mach number is 7 and flight altitude is 24km. When traditional ablating thermal protection is used, the ablating material is burned out. During the ablation, aerodynamic heating is absorbed and released. However, the designed aerodynamic shape may be burned down, especially the sharp leading edge. The vehicle will be unable to fly in high Mach number if its structure is seriously ablated.

Heat-balance thermal protection is different from traditional thermal protection. Heat will be quickly transferred from high aerodynamic heating area to low aerodynamic heating area, where the energy will be released by radiation. The temperature of high aerodynamic heating area could be reduced to protect the designed structure from being burned down. It is non-ablating thermal protection.

2 Heat-balance thermal protections

According to thermal protection mechanism, heat-balance thermal protection can be divided into single heat-balance thermal protection and combinational heat-balance thermal protection. In practice, single heat-balance thermal protection is hardly used. There are usually more two methods used together to achieve better thermal protect effect, which is shown in Table 1.

As widely used methods, the research on heat-pipe for heat-balance thermal protection is introduced.

[a] Corresponding author: rong83117@163.com

Table 1. Different heat-balance thermal protections

Type	Thermal Protect Way	Materials	Thermal Protect Mechanism
Single	Using high thermal conductivity material	Metal, high thermal conductivity carbon material	Heat is conducted quickly with high thermal conductivity.
	Using fluid	Air, water, liquid Na, liquid Li	Heat is conducted with fluid flowing
Combinational	Nano-fluid	Nano-sized particles with fluid	Heat is better conducted with fluid flowing and high thermal conductivity of Nano-sized particles.
	Phase-changeable material in special matrix	Matrix is made of high thermal conductivity material filled with phase-changeable material.	Phase-changeable material will absorb heat while phase is changing. After phase change, material becomes fluid. Matrix made of high thermal conductivity for heat transfer will also keep phase-changeable material from leaking out.
	Heat-pipe	Container is made of high heat capacity material and filled with full wetting working fluid.	Working fluid evaporates in evaporator first. Vapor will be transported to condenser, where vapor condenses. Then condensate will return to evaporator through a wick of suitable capillary structure.
	Embedded heat-pipe	Heat-pipe is embedded with high thermal conductivity material	Heat-pipe and high thermal conductivity material can both transfer heat isothermally. And embedded structure is more stable.

3 Heat-balance thermal protections with high thermal conductivity material

Heat-pipe mainly uses fluid to transfer heat. Heat pipes transfer heat nearly isothermally by the evaporation and condensation of a working fluid, as illustrated in Fig. 1. The heat is absorbed within the heat pipe by evaporation of the working fluid. The evaporation results in a slight internal pressure differential that causes the vapor to flow from the evaporator region to the condenser region, where it condenses and gives up heat. The cycle is completed with the return flow of the liquid condensate to the evaporator region by the capillary action of a wick. When heat-pipe is used to cool hypersonic leading edge, it can be design as Fig. 2.

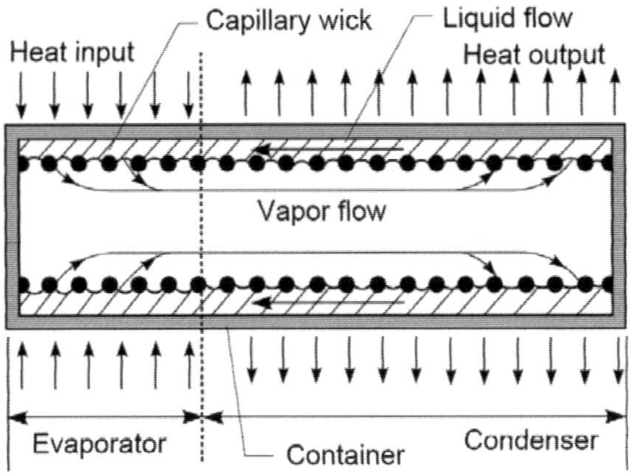

Figure 1. Schematic diagram of the operation of a heat pipe showing the heat-pipe container, working fluid, and wick.

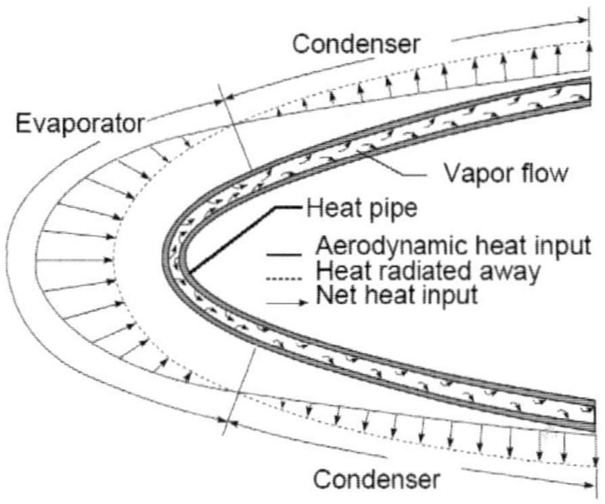

Figure 2. Schematic diagram of a heat-pipe-cooled leading edge showing regions of net heat input (evaporator) and net heat output (condenser)

Working fluid is an important factor in heat-pipe working effect. The choice of working fluid is related with working conditions and will affect heat transfer. There are appropriate working fluids at different temperatures for heat-pipe in Fig. 3 [1]. From Fig. 3, it is known that metals such as K, Na and Cs could be used as working fluid when temperature is high. In the same temperature region, like 550K-740K, Cs is a better choice for working fluid than K. But K will be better when temperature is higher than 740K [2].

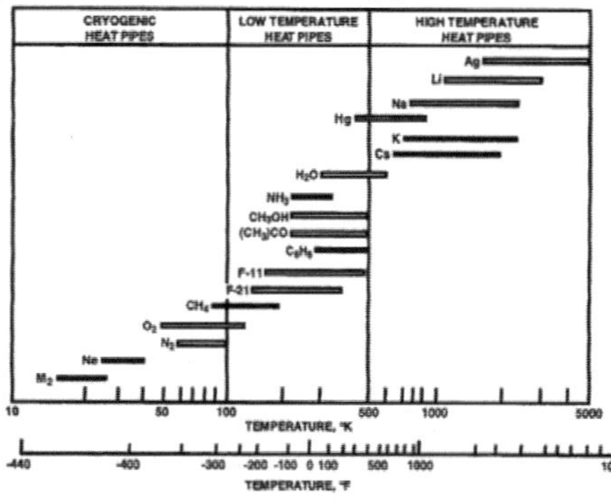

Figure 3. Appropriate working fluids at different temperatures for heat-pipe [1]

Heat-pipe does not need special energy supply to work, and it has high thermal conductivity. The hard container can be used as support structure. These advantages enable heat-pipe to be widely used in heat control and heat-balance thermal protection. In early 1966, a first W-Li super high temperature heat-pipe was developed [3]. In last century 70's, heat-pipe was used for heat control on space vehicles [4-6]. Donovan [7] used heat-pipe as heat-balance thermal protection for missile fin. Buffone [8-9] used liquid metal (Na and Li) heat-pipe to cool rocket nozzle wall.

The research on heat-pipe cooled leading edge for hypersonic vehicle began in last century 70's. NASA Langley Research Center (LaRC), through a contractual study, analytically verified the viability of heat pipes for cooling stagnation regions of hypersonic vehicles [10]. Niblock [11] proposed four space shuttle wing leading edge concepts, in one of which is a liquid-metal/superalloy heat-pipe-cooled concept. The heat-pipe-cooled concept was determined to be a feasible and durable design concept, but was slightly heavier than the other candidate concepts. In 1973, Camarda [12] fabricated a half-scale shuttle-type heat-pipe-cooled leading edge to verify feasibility of the concept (see Fig. 4). This model was tested by a series of radiant heat and aerothermal tests at NASA LaRC from 1977 to 1978 to verify heat-pipe transient, startup, and steady-state performance [12-14]. In 1979, to optimize a heat-pipe-cooled wing leading edge for a single-stage-to-orbit vehicle, Peeples's [15] study indicated that the mass of a shuttle-type heat-pipe-cooled leading edge could be reduced by over 40 % by use of a more efficient structural design. In 1986, Boman [16] designed and fabricated a sodium/superalloy heat-pipe- cooled leading edge component for an advanced shuttle-type vehicle. This advanced shuttle-type heat pipe was 1.83m long and was tested by radiant heating and by induction heating [17-18].

In 1988, a Haynes 25/sodium leading-edge-shaped heat pipe with a "D-shaped" cross section was designed and tested by Clark [19]. The heat pipe had a length of 30cm, and a wall thickness of 0.076cm. Thermal performance, vibration sensitivity, and critical heat flux tests were conducted. Straight niobium/lithium heat pipes with a "D-shaped" cross section and a 0.06cm thick wall have also been fabricated and tested by Wojcik[20-21]. These heat pipes had an oxidation resistance coating on the outer surface.

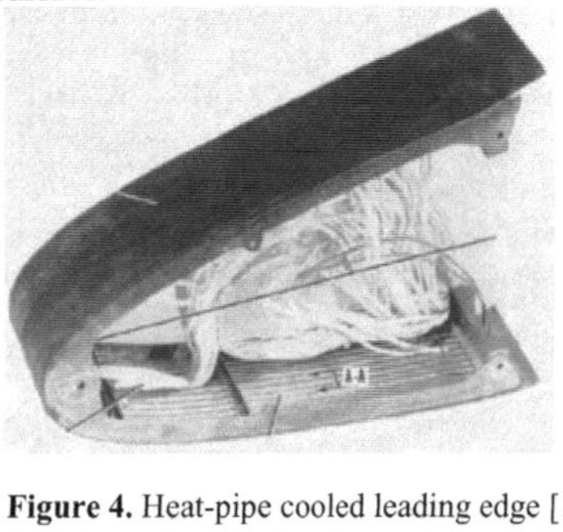

Figure 4. Heat-pipe cooled leading edge [12]

In 1992, Glass designed a heat-pipe-cooled leading edge which could reduce the leading-edge mass by over 50% compared to an actively cooled leading edge and could completely eliminate the need for active cooling, and has the potential to provide failsafe and redundant features [22]. It was a Mo-Re heat-pipe, used lithium as the working fluid and embedded in C/C (see Fig. 5). From 1992 to 2006, Glass [22-29] did a lot of research on heat-pipe-cooled leading edge by theory analysis, numerical simulation and experiments. During their tests, the heat-pipe-cooled leading edge worked nearly isothermally.

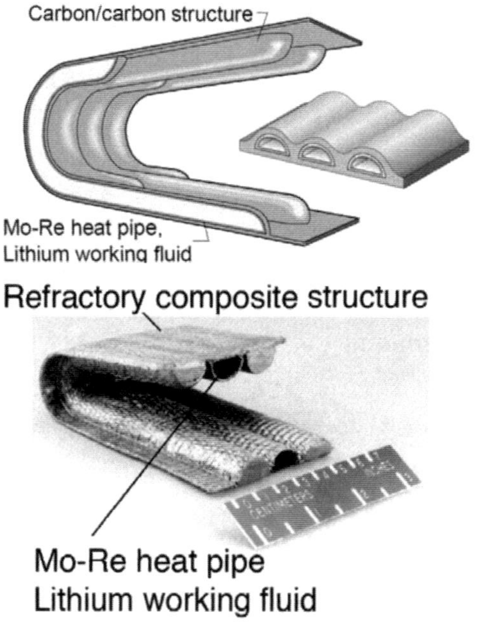

Figure 5. Shape and materials for heat-pipe-cooled leading edge

In 2007, Steeves [30] examined the feasibility of metallic leading edge heat pipes, made of Cb-752 and used Na as the working fluid, for hypersonic vehicles in Mach 7 flight. The heat pipes integrated within metallic leading edges provided very high effectively thermal conductivity which allowed heat to be transferred from the hot leading edge to large cool surfaces for radiation into space. Different materials were analyzed by Steeves [31]. Compared with Inconel 625 heat pipes used Na as working fluid, Cb-752 heat pipes used Li as working fluid have better thermal protection effect.

Chen [32-34] conducted experiments on heat pipes, used alkali metals as working fluid (see Fig. 6), for thermal protection system for hypersonic vehicle within the atmosphere. Compared with composite material, the stagnation point temperature was reduced by 9.5% and temperature at the contact position of the sphere and cylinder was reduced by 10.4%. And wall temperature at the position 130mm away from the stagnation point was increased by 14.6%. The results of experiments showed that heat pipes provided very high effectively thermal conductivity which allowed heat to be transferred from the hot stagnation region to cool surfaces. The feasibility of heat pipe used as heat-balance thermal protection for stagnation region of hypersonic vehicle was validated.

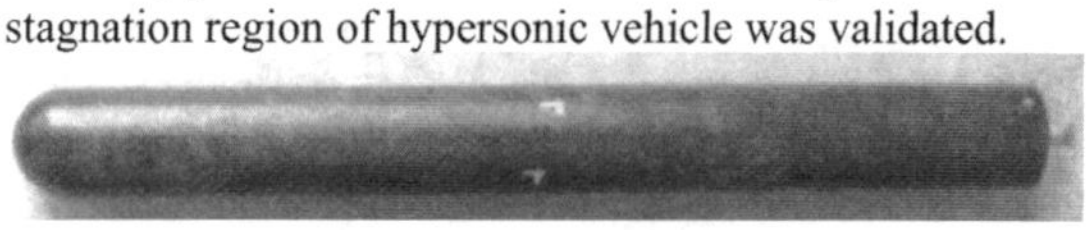

Figure 6. Photograph of high temperature heat pipe

From 2009 to 2012, Liu [35-39] conducted numerical and experimental research on GH600 high temperature heat pipes, used Na as working fluid and embedded in C/C coating, for vehicle leading edge structure. Discontinuous Galerkin finite element method was used for thermomechanical coupling calculation to simulate heat transfer of high temperature heat pipe cooled thermal protection structure. Two way heat pipes distribution was designed to reduce the stagnation point temperature efficiently and to increase structure reliability greatly. When reserving assembly gap was used, contact stress on the structure interface would decrease observably while thermal contact resistance would increase which raised the stagnation point temperature. Experiments on high temperature thermal contact resistance between 3D braided C/C composite and high temperature alloy GH600 were conducted. The changing of thermal contact resistance with interface stress, interface roughness and interface temperature was obtained, which offered reference for design and safety evaluation on new high temperature heat pipe cooled thermal protection structure.

Due to the reported time, Table 2 is summarized.

Table 2. Heat-pipe-cooled leading edge for hypersonic vehicle

Time	Author	Working fluid	Container	Coating	Ref.
1966	Romano	Li	W	No coating	[3]
1978	Camarda	Na	Hastelloy-X	No coating	[12-14]
1988	Clark	Na	Haynes 25	Oxidation resistance coating	[19]
1990	Boman	Na	Hastelloy-X	No coating	[16-17]
1991	Wojcik	Li	Niobium Alloy	Oxidation resistance coating	[20-21]
1992	Busse	Li	W-Re	No coating	[40]
1994	Rovang	Li	-	C/C coating	[41]
1996	Merrigan	Na	Hastelloy-X	No coating	[18]
1992	Glass	Li	Mo-Re	C/C coating	[22-29]
2007	Steeves	Na	Cb-752	No coating	[30]
2009	Steeves	Na	Inconel 625	No coating	[31]
2009	Steeves	Li	Cb-752	No coating	[31]
2009	Chen	Alkali metals	-	No coating	[32-34]
2010	Liu	Na	GH600	C/C coating	[35-39]
2011	Qu	Na & Li	Niobium alloy	No coating	[42]

These research works by theory analysis, experiments and numerical simulation validated the feasibility and effect of heat-balance thermal protection by using heat-pipes. With these works, heat-balance thermal protection could be known qualitatively and quantitatively.

4 Working Condition of Heat-balance Thermal Protections

During hypersonic flight, there is serious aerodynamic heating, especially on leading edge. The heat flux on leading nose whose radius is 20mm can reach 2~3MW/m^2 and the wall temperature can reach 1400K when the vehicle Mach number is 7 and flight altitude is 24km. And the heat flux on leading nose whose radius is 1~2mm can reach 4~5MW/m^2 and the wall temperature can reach 1600~1900K when the vehicle Mach number is 6 and flight altitude is 25km. When the vehicle Mach number is 10~20 and flight altitude is about 100km And the heat flux on leading nose whose radius is 10~30mm can reach 10MW/m^2 and the wall temperature can reach 2000K.

When wall temperature is larger than critical temperature such as melting point, ultimate strength, and the aerodynamic structure has to be kept from burning out, heat-balance thermal protection could be a good choice for thermal protection system. However, wall temperature should be lower than container melting point to prevent working fluid from leaking. Working fluid should be chosen according to working conditions to make sure part of working fluid remains liquid to keep working cycle going on.

Heat-pipe-cooled leading edge is used as heat-balance thermal protection for hypersonic vehicle [29]. It started to work when wall temperature is 2255K and the leading edge was nearly isothermal at 1870K. The heat pipes provided very high effectively thermal conductivity which allowed heat to be transferred from the hot leading edge to large cool surfaces for radiation into space.

5 Conclusions

Heat-balance thermal protection is non-ablating thermal protection, which can prevent the aerodynamic structure of hypersonic vehicle from burning down with the serious aerodynamic heating. Heat-balance thermal protection provides very high effectively thermal conductivity which allows heat to be transferred from the hot leading edge to large cool surfaces for radiation into space. It is validated by research with theory analysis, numerical simulations and experiments. In China, we are still at the primary stage. Learning abroad is a good way to make fewer mistakes while developing quickly. In order to develop heat-balance thermal protection, we could consider:

- Develop or choose new materials for container and working fluid;
- Develop or choose new materials for heat pipe coating;
- Design new structure for heat pipes;
- Design new heat-balance thermal protection.

Heat-balance thermal protection is a different thought changing from heat proof into heat control, which keep the whole vehicle as isothermal as possible. The research on heat-balance thermal protection will offer more efficient practical solutions to solve the seriously aerodynamic heating on hypersonic vehicle.

DOI: 10.1051/matecconf/20166104008

Acknowledgement

This research is supported by Natural Science Foundation Project of Shaanxi Province (No. 2014JQ7275) and China Postdoctoral Science Foundation Grant (No. 2015M582911). The authors would like to express their appreciations for the above fund supports.

References

1. S. W. Chi: *Heat Pipe Theory and Practice* (Hemisphere Publishing Corporation, Washington, DC., 1976).
2. P. M. Dussinger, W. G. Anderson and E. T. Sunada: *3rd International Energy Conversion Engineering Conference*, San Francisco, California, 2005.
3. A. Romano, A. Fleitman and C. Klamut: *Proceedings of the Symposium on Alkali Metal Coolants*, Vienna, 1966.
4. J. Roukis, J. Rogovin and B. Swerdling: *AIAA 6th Thermophysics Conference*, 1971, AIAA-71-0410.
5. T. R. Jr. Scollon: *AIAA 6th Thermophysics Conference*, 1971, AIAA-71-0412.
6. M. Tawil, J. Alario, R. Prager and R. Bullock: *AIAA 6th Thermophysics Conference*, 1971, AIAA-72-0272.
7. B. D. Donovan, W. S. Chang and J. M. Gottschlich: AD 358663, 1998.
8. C. Bruno and C. Buffone: *23rd International Symposium on Space Technology and Science*, Matsue, Japan, 2002.
9. C. Buffone, C. Bruno and K. Sefiane: *39th AIAA/ASME/SAE/ASEE Joint Propulsion Conference and Exhibit*, 2003, AIAA-2003-4452.
10. C. C. Silverstein: NASA CR-1857, 1971.
11. G. A. Niblock, J. C. Reeder and F. Huneidi: Journal of Spacecraft and Rockets Vol. 11 (1974), p. 314.
12. C. J. Camarda: NASA TN D-8468, 1977.
13. C. J. Camarda: NASA TP-1320, 1978.
14. C.J. Camarda and R.V. Masek: Journal of Spacecraft and Rockets Vol. 18 (1981), pp. 71-78.
15. M. E. Peeples, J. C. Reeder and K. E. Sontag: NASA CR 159096, 1979.
16. B. L. Boman, E. C. Citrin, E. C. Garner and J. E. Stone: NASA CR 181922, 1990.
17. B. L. Boman and T. Elias: AIAA Paper 90-1759, 1990.
18. M. A. Merrigan, J. T. Sena and D. E. Glass: *Proceedings of the ASME Heat Transfer Conference*, Houston, TX, 1996.
19. L. T. Clark and G. S. Glenn: AIAA Paper 88-2679, 1988.
20. C. C. Wojcik: *8th Symposium on Space Nuclear Power Systems*, Albuquerque, NM, 1991.
21. C. C. Wojcik: AIAA Paper 91-1400.
22. D. E. Glass and C. J. Camarda: *Progress in Astronautics and Aeronautics*, AIAA, Washington, DC, 1992.
23. D. E. Glass, C. J. Camarda, J. T. Sena and M. A. Merrigan: AIAA 97-3876, 1997.
24. D. E. Glass, M. A. Merrigan and J. T. Sena: NASA/CR-1998-207642, 1998.
25. D. E. Glass: NASA CR-1998-208962, 1998.
26. D. E. Glass, C. J. Camarda, M. A. Merrigan, J. T. Sena and R. S. Reid: AIAA-99-4866.
27. D. E. Glass, C. J. Camarda, M. A. Merrigan and J.T. Sena: Journal of Spacecraft and Rockets Vol. 36 (1999), p. 79-86.
28. D. E. Glass, N. R. Merski, and C. E. Glass: NASA/TM-2002, 2002.
29. D. E. Glass: NASA/CR-2006, 2006.
30. Steeves, C. A., He, M. Y., Valdevit, L. et al. Metallic Structural Heat Pipes as Sharp Leading Edges for Mach 7 Vehicles[C], Proceedings of IMECE 2007, November 11-15, 2007, Seattle, USA.
31. C. A. Steeves, M. Y. He, L. Valdevit et al.: Journal of Applied Mechanics Vol. 76 (2009), p. 031014-1-9.
32. L. Chen, D. Ou and D. Liu: Frontier Science Vol. 3 (2009), p. 41-45.
33. L. Chen and D. Ou: Journal of Experiments in Fluid Mechanics Vol. 24 (2010), p. 51-54.
34. D. Ou, L. Chen, H. Chen, J. Yu and S. Chen: Aerospace Materials Technology No. 2 (2014), p. 54-57.
35. D. Liu, X. Zheng and Y. Liu: Computer Modeling in Engineering and Sciences, Vol. 39 (2009), p. 263-299.
36. D. Liu, X. Zheng and Y. Liu: Tsinghua Science and Technology Vol. 50 (2010), p. 1094-1098.
37. D. Liu, X. Zheng and Y. Liu: Acta Materiae Compositae Sinica Vol. 27 (2010), p. 43-49.
38. D. Liu, X. Zheng and Q. Huang: Acta Aeronautica et Astronautica Sinica, Vol. 31 (2010), p. 2189-2194.
39. D. Liu and X. Shang: Acta Aeronautica et Astronautica Sinica, Vol. 33 (2012), p. 1834-1841.
40. C. A. Busse: *Heat Pipe Science, Advances in Heat Pipe Science and Technology*.(Int. Academic Publishers, 1992).
41. R. D. Rovang and A. J. Juhasz: AIAA-94-4065, 1994.
42. W. Qu and H. Wang: Journal of Chemical Engineering, Vol. 62 (2011), p. 77-81.

Research on the assessment indicators for crime prevention lighting in residential areas based on AHP and Entropy Weight

Youran Li, Ailin Chen, Yangling Zhou, Nianyu Zou, Xiaoyang He and Jinpeng Wang[a]

Research Institute of Photonics, Dalian Polytechnic University, 116034 Dalian, China

Abstract. Lighting can affect the probability of crime. In order to establish safe and secure residential areas' lighting environment, the elements for crime prevention are researched. Originally propose 14 assessment indicators of lighting environment which can be recognized subjectively and may influence crime rate. They are horizontal illuminance、illuminance uniformity、surround ratio、vertical illuminance、three-dimensional、color rendering、glare、lamp pole height、light pole distance、lamp aesthetic、lamp conciseness、color temperature、lamp distribution、light source. The data came from residents in China. Through screening and giving weights by Analytic Hierarchy Process, there are 7 key assessment indicators left. Then give weights to the ultimate 7 key assessment indicators by Entropy Weight to verify their rank. The results show that 7 key assessment indicators have the same rank in contrast of the two methods. According to the crime prevention influence of the lighting environment, the sort is: vertical illuminance, horizontal illuminance, three-dimensional, color temperature, glare, uniformity of illuminance, color rendering.

1 Introduction

Darkness has a certain relationship with the occurrence of crime [1-2]. The crime prevention lighting aims at reducing crime occurrence through lighting design [3-4]. The main purpose using it in residential area is to provide security and comfort environment for people [5-6].

In someplace of America, the level of street lighting was improved in order to prevent crime, since then, the crime rate was reduced by 37% [7-8]. In Essex, turn off the street lights between midnight and 5.30am of one year, and the night-time crime has almost halved in Saffron Walden and reduced by over a third in Dunmow [9]. In Japan, the crime rate showed a downward trend when using anti-encroachment lamps, and they had proved that blue light contained in the lamp have deterrent effect on the criminal [10]. In China, Tongji University studied the impact of spectral power distribution of LED on crime prevention lighting. The results showed that yellow-rich LED was better for facial recognition compared to blue-rich LED [11-12]. Chongqing University researched the indicators of crime prevention lighting, and verified that the semi-cylindrical illumination and three-dimensional elements are the most appropriate indicators in the evaluation of crime lighting environment [13].

The study of crime prevention lighting is mainly focus on facial recognition. But for residential areas, lighting should create a security environment from psychological and physiological. In the paper through investigating residents in China, analyze the data by Analytic Hierarchy Process and Entropy Weight. To reduce crime rate, improve the sense of security, research the lighting assessment indicators and their rank for crime prevention.

2 Investigation methods

2.1 Investigation process

Faced with the current status of lighting, and existing problems of residential lighting environment, our school undergraduates investigated crime prevention lighting environment in residential areas. Through professional training, students understood the purpose and significance of investigation deeply. After the written test and interview, trainers exercised investigation to meet the fieldwork. Investigators used their holiday to visit the residential areas in hometown. 450 questionnaires were issued, which cover 24 provinces and cities. And 425 of them were returned, the effective rate was 86.82%.

2.2 Investigation content

Contrasting national standards of lighting environment in residential areas and related research, extract the factors that can prevent crime. Combine with lighting experts' recommendations, 14 preliminary elements obtained, as shown in the Table 1.

Questionnaire is scoring form. Scores are from -10 to 0 to 10. Negative scores mean this element promotes crime. Positive scores mean this element prevents crime.

[a] Corresponding author: wangjp@dlpu.edu.cn

Table 1. The original crime prevention elements.

No.	Element	Presentation
1	Horizontal illuminance	Bright degree of the road
2	Illuminance uniformity	Light uniformity degree on road
3	Surround ratio	Brightness difference of roads and surroundings
4	Vertical illuminance	Deciding action and face whether can to be identified
5	Three-dimensional	Determine deviation degree of facial expressions
6	Color rendering	The color deviation degree compared original object with irradiated by light
7	Glare	The light produce an uncomfortable feeling
8	Lamp pole height	The height of the light pole
9	Lamp pole distance	The distance between the poles on one side
10	Lamp aesthetic	Lamp shape is beautiful or not
11	Lamp conciseness	Lamp shape is simple or not
12	Color temperature	Light color
13	Lamp distribution	Mutual position between the poles
14	Light source	Incandescent lamp、CFL、LED and so on

0 point indicates that the factors have no effect on crime prevention lighting environment. Low absolute values indicate little influence; High absolute values represent great influence. With the increasing of scores in the questionnaire, the degree of impact is increasing at the same time.

3 Investigation results

Firstly, analyze the reliability of survey data. Cronbach's alpha coefficient can measure the consistency of the project. When between 0.7 and 0.98, it can be considered a high confidence value. After calculation, the coefficient is 0.837. So the questionnaire is reasonable and data is consistency.

Negative scores can be seen as the option of "negative affect". Positive scores can be seen as the option of "positive affect". And 0 point can be seen as the option of "no affect". Figure 1 and 2 show the results of every assessment indicator. In Figure 2, each element has three styles. Low、midterm and high color temperature represent the color of warm、warm white and cold. Lamp distribution has three styles which are symmetrical、stagger and unilateral distribution. And select three kinds of typical light sources. They are incandescent lamp、CFL and LED. As we can see, for residents, each element will give them a different sense of security. For most elements, the proportion of "positive affect" is higher than "negative affect" except of glare and lamp pole distance. And some elements have the highest proportion of "no affect". Remove the elements of which the "no

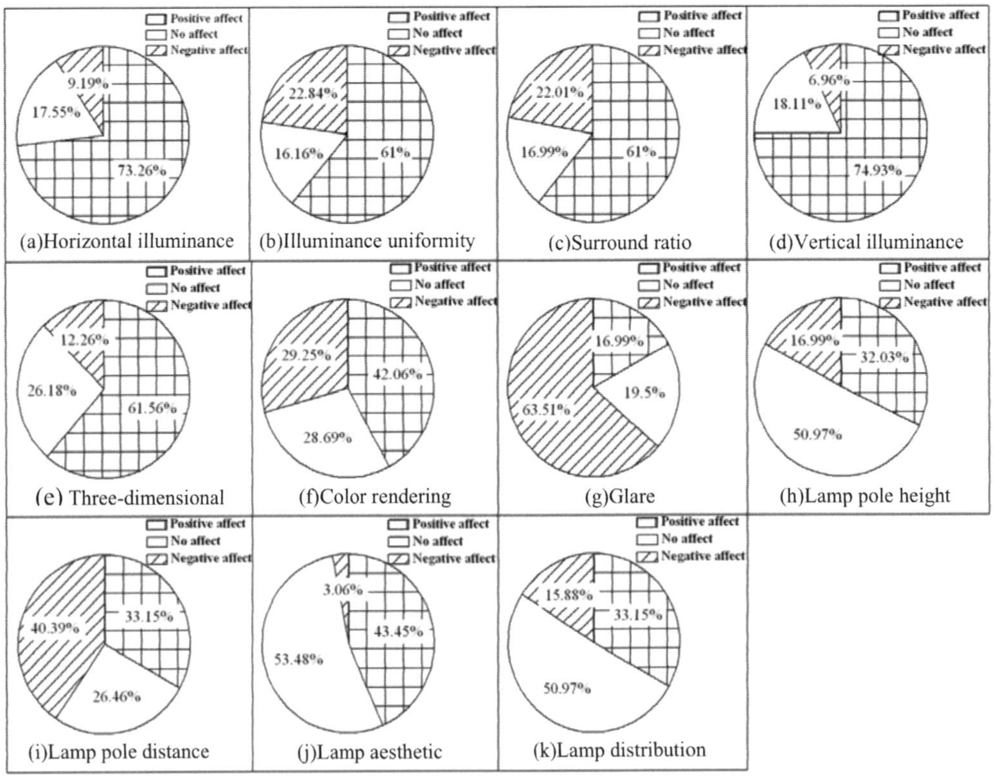

Figure 1. The results of every assessment indicator.

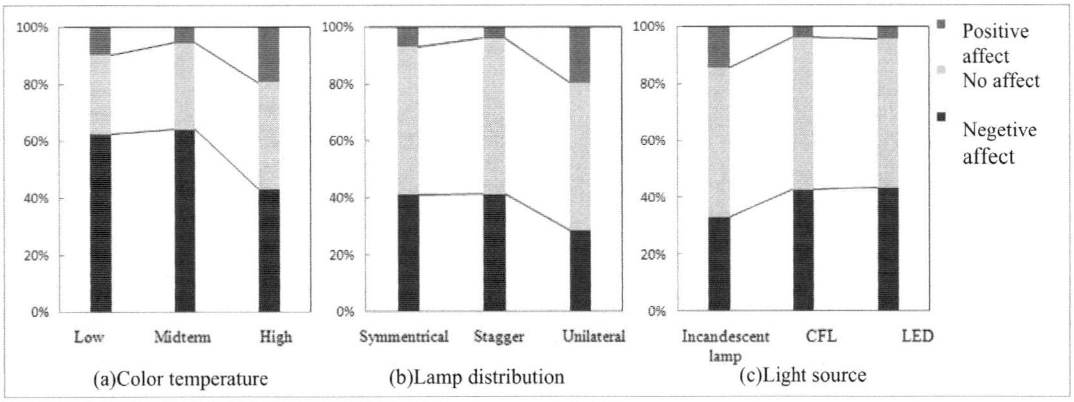

Figure 2. The results of color temperature、lamp distribution and light source.

affect" percentage is more than 50%. So the elements of lamp pole height、lamp aesthetic、lamp conciseness、lamp distribution、light source are removed. And the rest elements are : horizontal illuminance 、 illuminance uniformity、surround ratio、vertical illuminance、three-dimensional 、 color rendering 、 glare 、 light pole distance、color temperature.

4 Analysis

4.1 AHP

The Analytic Hierarchy Process (AHP) is a systematic approach that can solve complex multi-program decision-making problems. Calculate the weight of levels and total sorting by the method of qualitative indicators fuzzy quantification. It is now widely used in the environmental assessment and selection decision programs [14-15].

Use AHP evaluation method to screening deep for nine factors after initial screening. This method main utilize eigenvector method (EM) [16]. The importance ratio of every elements form a comparison matrix. The weight vector is the right feature vectors of the comparison matrix.

According to these elements, B1-B9 are: horizontal illuminance、illuminance uniformity、surround ratio、vertical illuminance 、 three-dimensional 、 color rendering 、 glare 、 light pole distance 、 color temperature. After comparing the importance of each element, the comparison matrix is:

$$A=\begin{bmatrix} 1 & 2 & 5 & 1 & 1 & 2 & 1 & 8 & 1 \\ 1/2 & 2 & 1 & 1/2 & 1/2 & 1 & 1/2 & 4 & 2 \\ 1/5 & 1/2 & 1 & 1/6 & 1/4 & 1/2 & 1/4 & 2 & 1/4 \\ 1 & 2 & 6 & 1 & 2 & 3 & 2 & 9 & 2 \\ 1 & 2 & 4 & 1/2 & 1 & 2 & 1 & 6 & 1 \\ 1/2 & 1 & 2 & 1/3 & 1/2 & 1 & 1/2 & 3 & 1/2 \\ 1 & 1/2 & 4 & 1/2 & 1 & 2 & 1 & 6 & 1 \\ 1/8 & 1/4 & 1/2 & 1/9 & 1/6 & 1/3 & 1/6 & 1 & 6 \\ 1 & 1/2 & 4 & 1/2 & 1 & 2 & 1 & 6 & 1 \end{bmatrix}$$

After calculation, the weight vector is $W=(w_1, w_2, w_3, w_4, w_5, w_6, w_7, w_8, w_9)^T$=(0.1616, 0.0993, 0.0365, 0.2188, 0.1408, 0.0724, 0.1244, 0.0218, 0.1244)^T.

Define index trade-off weight ξ =0.05. As the results, B3's relative weights w_3=0.0365, B8's relative weights w_8=0.0218, less than ξ .They are weak weight indicators, so get rid of surround ratio and light pole distance. The remaining 7 indicators were normalized. The weight vectors are W= (0.1716, 0.1054, 0.2324, 0.1495, 0.0769, 0.1321, 0.1321)^T.

4.2 Entropy Weight

Entropy Weight is an objective weighting method which can determine the index weight based on the amount of information contained in each index [17]. According to basic principles of information theory, information is a measure of ordering degree in the system, but the entropy is a measure of disorder degree in the system [18-19]. Information augment means the entropy reduction. In the questionnaire with the increasing of the information amount for a factor's data, its entropy is decreasing to the contrary. It should be given a greater weight.

Validate the rank of the final 7 elements by Entropy Weight. Through calculation, the weights are W'=(0.1933, 0.1904, 0.1669, 0.1601, 0.1118, 0.0912, 0.0863)^T.

4.3 Comparison

In contrast of the two methods' results, the weights of 7 elements are different, but they have the same rank. According to the crime prevention influence of the lighting environment, the sort is: vertical illuminance、horizontal illuminance 、 three-dimensional 、 color temperature、glare、 illuminance uniformity、color rendering. The impact of each key element is shown in Table 2. The only negative factor is glare, and the others can prevent crime.

Table 2. The impact of each key element.

No.	Element	AHP	Entropy Weight	Effect
1	Vertical illuminance	0.2324	0.1933	Allow pedestrians to see the people's face expression, posture around them, to find danger in an emergency.
2	Horizontal illuminance	0.1716	0.1904	Allow pedestrians to see the road ahead, avoid obstacles timely, and protect their own safety.
3	Three-dimensional	0.1495	0.1669	Make people identify objects' true original better, thereby reduce the fear of unkown objects.
4	Color temperature	0.1321	0.1601	Different color temperatures effect different psychological. Color temperature can warn criminal mentality when use reasonably.Thereby reducing the crime rate.
5	Glare	0.1321	0.1118	Extreme glare will cause a loss of visibility. If someone commit a crime, it would threaten residents' personal and property safety.
6	Illuminance uniformity	0.1054	0.0912	Low level illumination uniformity make pavement emerge staggered stripes of bright and dark. Strong chiaroscuro effect pedestrians' vision. However, if there isn't extreme contrast, the human eye can distinguish sudden dangerous due to the slower moving.
7	Color rendering	0.0769	0.0863	The object whether restore the true color. High color rendering helps pedestrians distinguish the faces and bodies come across. But at night, the human eye has llittle sensitive to color. So it rank at the bottom.

5 Conclusions

Originally propose 14 assessment indicators of light environment which can be recognized subjectively and may influence crime rate. They are horizontal illuminance、illuminance uniformity、surround ratio、vertical illuminance 、 three-dimensional 、 color rendering、glare、lamp pole height、light pole distance、lamp aesthetic、lamp conciseness、color temperature、lamp distribution、light source. Remove the elements of which the "no affect" percentage is more than 50%. So the elements of lamp pole height、lamp aesthetic、lamp conciseness 、 lamp distribution 、 light source are removed . Through screening further by AHP, get rid of surround ratio and light pole distance. In the removed elements, most of them are related to lamp shape but not light. Thus it can be seen that these elements are not be noticed in the night, nor impact the psychological of resident. But in the daytime, design these elements reasonably will show the residential area better, improve residents' quality of life.

Give weights to the ultimate 7 key assessment indicators by AHP and Entropy Weight. The results show that 7 key assessment indicators have the same rank in contrast of the two methods. According to the crime prevention influence of the lighting environment, the sort is: vertical illuminance, horizontal illuminance, three-dimensional, color temperature, glare, uniformity of illuminance, color rendering. When design the lighting environment, emphasizing these assessment indicators will reduce crime rate, improve the sense of security.

Acknowledgements

This research was supported by Project of the National Natural Science Foundation of China (61402069), 2012 key supported project for undergraduate education by Education Dept. of Liaoning province([2012]108), 2016 Project of the Natural Science Foundation of Liaoning province(2016020027), 2012 Education reform project for undergraduate by Education Dept. of Liaoning province ([2012]130), 2014 Dalian municipal science and technology plan project（2014A11GX050）, 2014 Dalian municipal science and technology plan project （2014A11GX052）.

References

1. C. Knight, Lighting Res. Technol. **42**, 313(2010)
2. S. Fotios, P. Raynham, Lighting Res. Technol. **43**,129(2011)
3. Tarriconep, Lighting Des. App. **44**,32（2014）
4. J. Mohammad , D. Reza, K. Ahad, Util. Pol. **32**,19 (2015)
5. Y. Li, W. Shi, D. Li, Optik，**126**,4887(2015)
6. M. Dong, S. Fotios, Y. Lin, Lighting Res. Technol.**47**,693(2015)
7. P. Kate, Infl. Str. Lig. Imp. Cri. fear Ped. Str. Landscape and Urban Planning, **35**, 193 (1996)
8. S. Fotios, J. Unwin, S. Farrall, Lighting Res. Technol.**47**,449(2015)
9. Christine, S. Linda, J Env. Psy. **39**,22(2014)

10. T. Matsui, Con. Urb. Des. Cri. Pre. J Ill. Eng. Ins. **89**,25 (2005)
11. L. Hao, X. Yang, Y. Lin, Chin Ill. Eng. Ins. **24,**55 (2013)
12. X. Yang, L. Hao, Y. Lin, Chin Ill. Eng. Ins,**25**,37 (2014)
13. Y. Ma, Z. Chen, Y. Hu. ,Light & Lighting, **32**, 5(2008)
14. D. Rakesh, C. Vaishali, International J. Log. Sys. Man. **20**,260 (2015)
15. R. Lekurwale, M. Akarte, D. Raut, Int. J. Adv. Man. Technol. **75**,565 (2015)
16. H. Xu, C. Luo, Z. Gang, Chin Off. Oil Gas, **19**,415 (2007)
17. Q. Wang, C. Wu, Y. Sun, J. Air Trans. Man,**42**,55(2015)
18. T. Wang, J. Chen, T. Wang, Nat. Haz. **76**,747(2015)
19. J. Casquilho, NS,**6**,545(2012)

MATEC Web of Conferences 61, 04010 (2016) DOI: 10.1051/matecconf/20166104010
APOP2016

Research of System Building Basing on the Low Carbon Economy About Carbon Accounting for the Enterprise

Yao Liqiong[1,2,a]

[1]School of Management, Wuhan University of Technology, Wuhan, P.R.China, 430070
[2]School of Economics and Management , Ningbo University of Technology, Ningbo, P.R.China, 315211

Abstract. As global warming has become truth, is developing as a new economic model,The new economic development model has given rise to an important branch of environmental accounting, namely carbon accounting. At first, this paper discusses the carbon accounting theoretical foundation comprehensively, and then analyzes the environment of the construction of the carbon accounting system. The focus of the article is to build enterprise carbon accounting system, it covers the confirmation and measurement ,record and information disclosure of the enterprise carbon accounting on the way of low carbon economy , its core is the processing of carbon emission rights, information disclosure mode and content, etc.; The purpose of this paper is to build enterprise carbon accounting system which is suitable for China's national conditions, in order to provide certain reference and theoretical support for the low carbon economy development of our country.

Key words. Low Carbon Economy; Carbon accounting confirmation; Carbon Accounting System; Carbon accounting

1 The Development Background and Theory of the low Carbon Economy

1.1 The development background of the low carbon economy

Low-carbon economy was published by the British for the first time in energy white paper" in 2003 , the concept of low carbon economy was caused the wide attention and recognition by the international community, low carbon economy is recognized as the most effective and critical way to solve the problem of global climate . In the midst of the heavy industrialization, China promotes low carbon economic development model in all levels actively for seeking a new road to industrialization , at the same time, the domestic accounting profession also began to discuss the related problems of the enterprise carbon emissions and trading accounting under the background of the low carbon economy for China's national conditions quietly , Combining with the economic situation of home and abroad, this requires that we must change the traditional development model of the dependence on resource consumption and environmental costs excessively for economic growth.

1.2 Concepts and theories of Low carbon economy

Low carbon economy is an economic development form which is in order to build a win-win between the social ecological development and environment protection. It reduces the high carbon energy consumption coal, oil, and greenhouse gas emissions as much as possible by the technical innovation, system innovation, industrial transformation, new energy development and other means under the guidance of the concept of sustainable development. The characteristics of low-carbon economy is to reduce emissions of greenhouse gas and construct economic development system of low energy consumption, low pollution , including low carbon energy system, low carbon technology and low carbon industry system. Low carbon energy system is to reduce carbon dioxide emissions by the developing of clean energy, including wind, solar, nuclear, geothermal energy and biomass energy instead of coal, oil and other fossil energy to(Ratnatunga 2012). Low-carbon technologies include clean coal technology (IGCC) and carbon dioxide capture and storage technology (CCS), and so on. Low carbon industry system include thermal power reduction, new energy vehicles, energy saving building, industry energy saving and emission reduction, circular economy, recycling, environmental protection equipment, energy-saving materials and so on.

[a] E-mail: yaoliqiong@163.com

2 The Concepts and Theories of Carbon Accounting

2.1 The concepts of carbon accounting

The development of low carbon economy has brought the development of carbon accounting; carbon accounting is the accounting system to promote deep integration and coordination between environment and economy development by reflecting and controlling the social environment resources on the basis of the traditional accounting. The latest research progress of carbon accounting is defined as a foreign currency as the main unit of measurement, mainly to the enterprise performance of the low carbon responsibility of saving energy and reducing pollutant emissions, its purpose is to help enterprises to realize energy conservation and pollution reduction. Carbon accounting mainly related to carbon emissions associated with carbon accounting information disclosure, risk or uncertainty management and strategic development, and so on.

2.2 The theories of carbon accounting

2.2.1 External theory

For a long time, the space of carbon dioxide emission belongs to the category of public goods, externalization is obvious. Specification of carbon dioxide emissions is the main content of the carbon emissions trading(David 2008). Ultimately, the market capitalization, quantification approaches of emissions of carbon dioxide should be clear; it should become public goods and the resources which are gotten from paying a price in production and business operation process. It has rep arability and exclusiveness in the process of using. The externalization of carbon dioxide is not only related to its amount but also because of not reasonable using by the human, the irrational economic structure and technology is not enough and so on.

2.2.2 Property rights theory

Property rights is very wide in the modern economics implication, it can be understood as a right which can chose the multiple uses of some economic goods through social force. It can be divided into two categories: private property and public property. Property rights theory is a kind of non-government interventionist scheme which solve the problem of externalizations by introducing the market mechanism, can be made through the trading policy of environment of economic management of emissions generally, the seller can sell pollution index from the reduction of pollution emission for the compensate of the external economy, the buyer buy the shortage of pollution index because of not to reduce pollution by the regulation, in order to realize the enterprise private cost close to its social cost, can achieve the optimal allocation of environmental resources. It is not hard to imagine that the result is the destruction of the resource, waste and excessive development if you can't definite or guarantee property rights clearly. If carbon resources rights are clear and can be transferred, carbon resource cost and value can be detailed evaluation measured; also, resources can be allocated effectively.

2.2.3 Sustainable development theory

Thought of Sustainable development is the long-term development of human society; it reflects a reflection of the relationship between the development of human beings and natural environment resources. Sustainable development is proposed by the world commission of environment and development in the report of "our common future" in 1987, the definition of "sustainable development" is to satisfy the need of modern people without hurting the ability of future generations to meet development needs in generally(Gregg 2002). Enterprises must take the environment effects from the production and business operation activities into account in the process of sustainable development, and measure, reflect and control the environmental resources, in order to provide services for improving the environment, using resources rationally and promoting the sustainable development of the enterprise. Implementation of Carbon accounting will affect the enterprise decision-making and economic benefits, the relevant transaction will also affect the enterprise's financial position, operating results, in order to meet the requirements of information users to understand the carbon accounting information, it requires companies have to provide the information system of carbon accounting information .So, the sustainable development theory is important for deepening the understanding of the "low carbon", as well as the formation and development of carbon accounting.

3 The Relationship Between Low Carbon Economy and Carbon Accounting

3.1 The relationship of the unity and opposites

The opposite of low carbon economy and carbon accounting is that low carbon economy is a description of development concept, but carbon accounting is a new form of economic development for management activities, which is material; The unity of them is mainly manifested in the consistency of the target , the main goal of low carbon economic model's is to respond to climate change and energy security, so as to promote the sustainable development of society and economy, carbon accounting is based on how to compensate the cost of natural resources and the accounting , its main purpose is to use accounting to reflect the problem of environment and resources, and make the environment and resources information to influence people's behavior. So target of carbon accounting is consistent with the low carbon economy. Visible, low carbon economy and carbon accounting attaches great importance to the ecological environment and material circulation rule, reasonable development and utilization of resources, efforts to improve social benefit and environmental benefit, so as to achieve sustainable development. to achieve coordination between man and nature system.

3.2 low carbon economy is the theoretical basis to realize carbon accounting

Low carbon economy is a concept and mode of sustainable development which correlates energy, environment and economy. On theory, low carbon economy is a kind of development, and accounting is a kind of management activities to reflect and control the enterprise economic activity .According to the development of economy, it can improve and enrich the traditional theories and methods, in order to reflect and control the consume energy and its external environment damage from enterprise production activities, as to make the enterprise accounting can reflect the economic activities, and assess their effectiveness fully and truly. In fact, Environment and resource factors must accounted in the economic activity of enterprise, which requires the accounting system to reflect. The establishment of low carbon economic development model provides the necessary theory preparation and practice basis for carrying the environmental resources and carbon emissions into the accounting system.

3.3 carbon accounting is the necessary means to realize low carbon economy

Low carbon economy model is asked to reduce dependence on natural resources; we can not only achieve the efficient use of energy but should pay attention to the protection of the ecological environment in the process of social and economic development. As to realize the sustainable development of economy. As the leading force of the social production, enterprises are the body of the quantity, and the development of low carbon economy. To build and implement the feasible carbon accounting system is to apply quantitative carbon accounting information to fulfill its obligations to the environment to the enterprise, the evaluation of environmental responsibility, To evaluate whether the enterprise can reduce energy consumption, reduce pollution emission and improve efficiency is the objective requirement of low carbon economy development, also is the essential means to realize low carbon economy.

4 Construction of Carbon Accounting System

4.1 Environmental analysis of building carbon accounting system

Carbon accounting is from the environmental accounting. Its development can not leave from a certain macroscopic environment and microcosmic environment, macro environment is primarily policy factors, the micro environment factors are from the market, before the constructing of carbon accounting system, we have to analysis policy environment and market environment to prepare for it well.

4.1.1 Macroscopic environment

The clean development mechanism is our country to participate in most project mechanism, the main content is: in the Kyoto protocol, developed countries carry out the project level with developing countries by providing funding and technology, they make project investment which not only conform to the requirements of the sustainable development policy, but have greenhouse gas emission reduction effect in developing countries. In return to get some or all of the emissions reductions for investment project ,as a part of its emissions reduction obligations, the lines is defined as a certified carbon emission reductions in the clean development mechanism (CDM).

Carbon emissions trading mechanisms refers to that countries transferred its carbon emissions reduction obligations which is over fulfilled to another country directly by trading. It is from annex 1 in the "Kyoto protocol", including the "emission reduction unit", "emission reduction certified", "distribution of the number of units", "removal of units" such as carbon reduction unit certified(Larry 2013).

Joint performance of the mechanism, is refers to the developed countries fulfill project based in return for carbon emission through trading in the transitional developing countries. Carbon emissions produced in the joint performance of the project is called reduction unit.

4.1.2 Micro environment

Carbon markets are the soil of accounting for growing, the development of accounting can not leave from the market environment, carbon emissions trading market is important in the development of carbon accounting market environment. In order to lay a foundation for the later construction enterprise carbon accounting system. We must understand the relevant market environment. For the moment, in carbon emissions trading market, compliance of trading market is given priority; voluntary compliance trading market is minority. According to the report by world bank in 2011 , From carbon emissions trading volume and the trading amount in the global main large carbon emissions trading market , the carbon emissions trading market dominant is based on the quota of carbon emissions trading, the trading of several large carbon emissions trading system is in large number. Carbon emissions trading on the basis of the project is mainly based on CDM project of certified carbon emissions, it makes amount of carbon emissions trading on the basis of the project, most of the joint project implementation mechanism of emissions trading unit and the number of carbon emissions trading market is little voluntarily(Janet 2008). Based on the above analysis, you can see that global carbon emissions trading market has begun to take shape, it is feasible to obtain the fair value of the carbon emissions.

4.2 Confirmation of carbon accounting

4.2.1 Confirmation of measurement principle

To distinguish the type of market. By above knowable, at present, carbon emission rights have two different types of trading markets mainly, one is a mandatory compliance, the other is a voluntary compliance trading.

To distinguish the type of transaction. Because there may be a variety of transaction type, so we cannot treat as the same even in the same transaction market. even if there is a mandatory compliance, but type can be based on the quota, may also be based on the project (Jonatan 2011).

To distinguish the purpose of use. Carbon emission rights obtained by enterprises are for different purposes, for different purposes, the choice of enterprise must be different. Even if the market and the transaction type is same, but because of different uses, the accounting measurement still have differences, so we also need to distinguish purposes.

4.2.2 To confirm the measurement content

Because of the different nature of trading, the measurement of carbon emission rights is also different, as mandatory compliance market transactions for example, make division according to the different combination of the transaction type and uses.

① Transaction type: on the basis of quota, use: emissions plan for private use. Confirm the measuring specific content as shown in table 5-1.

Table 1 The confirmation of measurement based on quota trading plan for private use

Project	specific content
Initial confirm ation	Carbon emissions that are acquired from the government's free accounting treatment; From a government auction for carbon emissions or purchase from the secondary market on the basis of quota of processing of carbon emissions
Subseq uent measur ement	Emissions of greenhouse gases yes accounting: carbon emission rights value changes in accounting: amortized during the period of accounting: final carryover and cases submitted to the accounting treatment: enterprise accounting change hold carbon emissions purposes

② the transaction type: on the basis of quota, application: permits plans to sell. Confirm the measuring specific content such as table 5-2

Table 2 The confirmation of measurement based on quota trading plans to sell

Project	specific content
Initial confirmation	Carbon emissions that are acquired from the government's free accounting process, from the government by auction of carbon emissions accounting treatment, from the secondary market to buy based on the quota of the accounting treatment of carbon emissions
Subsequent measurement	Carbon emissions accounting value changes, to sell carbon permits to accounting, the accounting treatment of the final presentation, enterprise change hold carbon emissions accounting purposes

③Transaction type: on the basis of the project, use: emissions plan for private use. Confirm the measuring specific content as shown in table 5-3.

Table 3 The confirmation of measurement plan to use based on project

Project	specific content
Initial confirmation	Through project cooperation in carbon emissions accounting, on the secondary market to buy based on the project of carbon accounting
Subsequent measurement	The same as ①

④Transaction type: on the basis of the project, use: permits plans to sell. Confirm the measuring specific content as shown in table 5-4.

Table 4 The confirmation of measurement plan to sell based on project

Project	specific content
Initial confirmation	Through the project cooperation in carbon emissions accounting confirmation, on the secondary market to buy based on the project of accounting confirmation.
Subsequent measurement	The same as ②

4.3 Accounting records of carbon

4.3.1 Account Settings

Based on the above analysis, when the accounting account Settings to make it satisfy the requirement of conventional financial accounting work, and meet the needs of the carbon accounting work. When in the specific design, can from the following aspects:

①Add general ledger subjects "carbon emissions", in view of the carbon in the new situation, new business accounting and management of the new requirements set, it can reflect the carbon emissions of the specific subject, and can judge the implementation of its environmental responsibility, in a timely manner so as to take measures in a timely manner to improve governance.

②Add detail course, set under the "carbon emissions" subject "personal use" and "sell" two subsidiary subjects, according to the different USES of company carbon emission rights classification collection summary.

4.3.2 Accounting treatment

As mandatory compliance market transactions for example

①Transaction type：Based on the quota, application: emissions plan for private use

Carbon emissions way free from the government. In accordance with the fair value as obtained confirmed that involves accounting account for "carbon emissions - for

DOI: 10.1051/matecconf/20166104010

personal use" and "deferred revenue", processing is as follows:

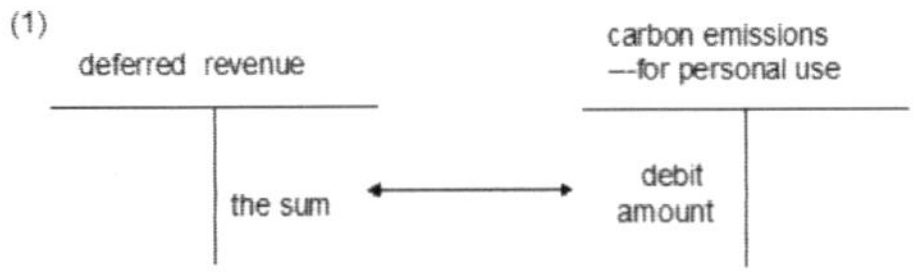

Carbon emission right can be got from government by auction. When the sale cost is greater than the fair value, it is confirmed by auction cost. Processing is as follows:

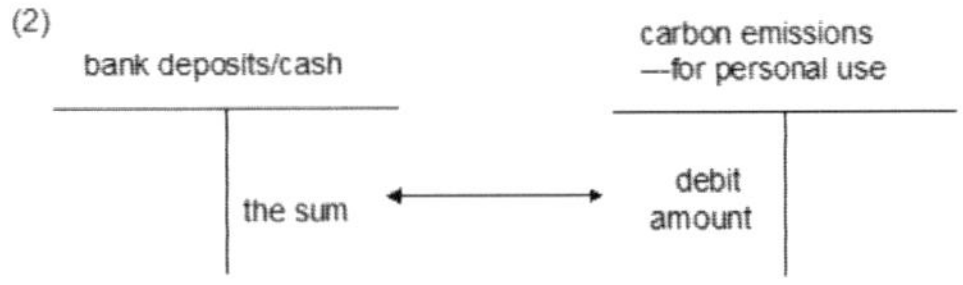

When auction cost < fair value, at fair value, involving accounting account again, borrowers' difference "deferred revenue", and a process is as follows:

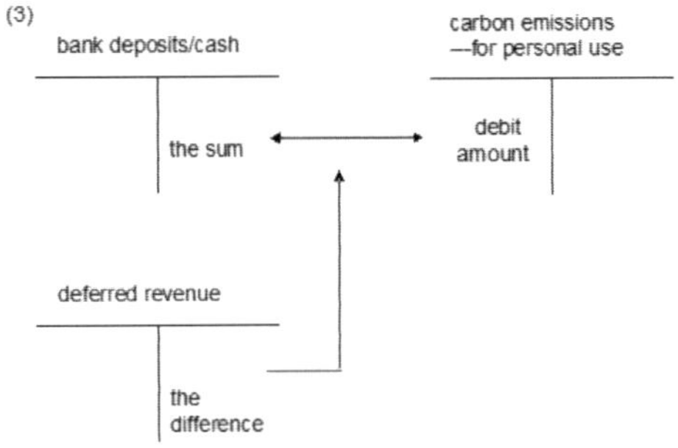

Carbon emissions from the secondary market to buy. Carbon emission rights obtained by this way, the purchase cost is equal to the fair value, this is in the market of fair trade both parties of a voluntary deal price, according to the purchase cost can be confirmed, involving accounting account for "carbon emissions - for personal use" and "bank/cash on hand", processing as follows.

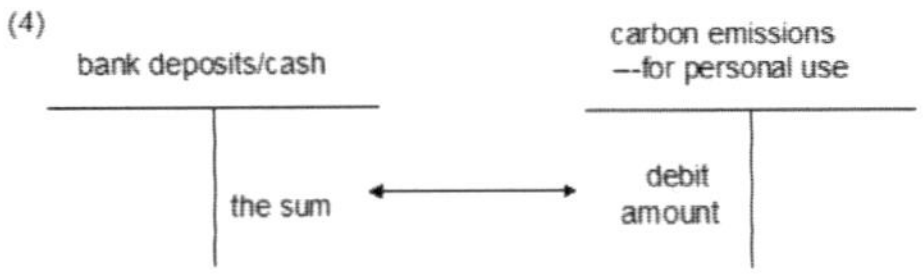

Deferred revenue. Deferred revenue according to adjusted the book balance of the processing, the non-operating income at the current, the premise according to the use of carbon emission rights plan, and adjust the early spread of deferred earnings. When the deferred revenue is a credit balance, processing as "deferred revenue" debit, "non-operating income" lenders, when the deferred revenue for the debit balance, "non-operating income" debit, "deferred revenue" lenders. Greenhouse gas emissions. In accordance with the fair value of the liabilities, confirm the enterprise cost, fair value for emissions, and specific treatment for "manufacturing cost/management fee/cost of sales/production costs, such as "debit," estimated debts lenders. Adjustment is expected liabilities, according to the fair value of the carbon emissions(Kenneth 2004). The fair value increases, the estimated debts to adjust its book value, capital reserves also adjust accordingly; the specific process is as follows:

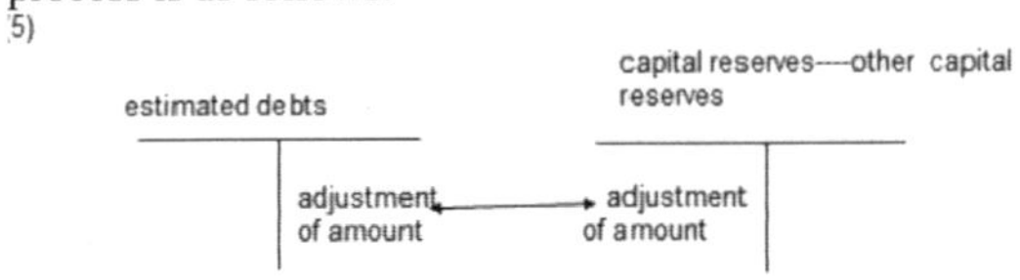

Fair value decrease, the capital reserves to offset (if the preexisting for appreciation and confirmation), capital reserves enough write-downs, would need to be accounted for as non-operating income account, the specific process is as follows:

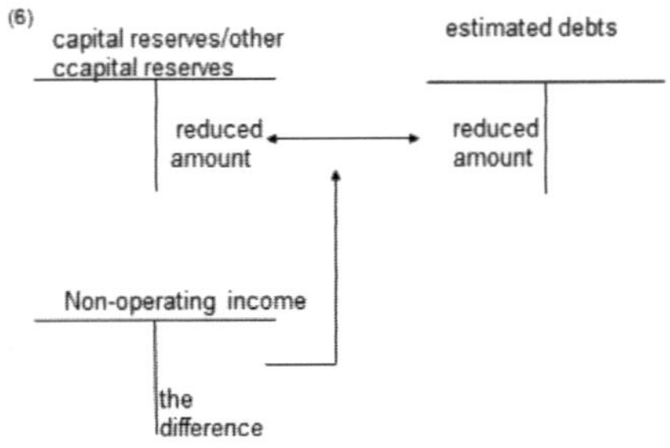

The final balance sheet date accounting. The final accounting, follow the principle of first in first out, return to the government carbon emissions, and the actual emissions equivalent to the final at the same time carry forward capital reserves and carbon emission rights impairment provision, should correspond to actual emissions.

②Transaction type: on the basis of the project, use: emissions plan for private use

Emissions through project cooperation, in accordance with the fair value can be confirmed as obtained (usually can choose acquisition cost), involving accounting accounts for "carbon emissions - for personal use" and "bank deposit cash/library", processing is as same as (2)

Carbon emissions in the secondary market to buy, and initial recognition through project cooperation in confirm the identity of the practice of carbon emissions. Described in the subsequent measurement and (1) accounting, no longer here.

③ Transaction type: on the basis of quota, use: permits plans to sell. Company plans to sell carbon permits to accounting, method is similar to in stock, but with inventory and not identical, to establish accounting account for "carbon emissions - for sale". "Is the explanation of China's accounting standards for stock inventory is refers to the enterprise in everyday activities held for the sale of finished goods or commodities, in the production process in the product, in the production process or provide labor services process consumption of materials, materials, etc." The explanation shows that companies hold for plans to sell carbon permits, the basic characteristics and the definition of inventory, but on the other hand, from the definition of inventory, inventory is a physical form, in contrast, carbon emissions, is no physical form. In conclusion, if you plan to sell carbon permits to directly recognized as the inventory, it is not appropriate, you should add "carbon emissions - sell" course, confirm measurement of carbon emissions. Carbon emissions way free from the government. In accordance with the fair value as obtained confirmed that involves accounting account for "carbon emissions - sell" and "deferred revenue", processing is as same as (1)

Carbon emissions from the government by auction. When the auction cost or fair value, confirmed by sale cost, involving accounting account for "carbon emissions - for personal use" and "bank/cash on hand", processing is as same as (2)

When auction cost < fair value, at fair value, involving accounting account again, borrowers difference "deferred revenue", a process is as same as （3）

Carbon emissions from the secondary market to buy. Carbon emission rights obtained by this way, the purchase cost is equal to the fair value, this is in the market of fair trade both parties of a voluntary deal price, according to the purchase cost can be confirmed, involving accounting account for "carbon emissions - for personal use" and "bank/cash on hand", processing is as same as (2)

Subsequent accounting treatment is mainly related to the carbon emissions of self-use value change, can be divided into value-added and impairment of two cases. Carbon emissions rise in value, if the initial measurement involves a deferred income, carbon emissions and the book value of the deferred income increase, processing for the increase in "carbon emissions - for" debit, "deferred revenue" credit, if the initial measurement without involving deferred income, processing for the "carbon emissions - for" debit, "capital reserves, other capital reserve" lenders, in addition, still need to write down impairment provision if there is a provision for impairment loss of the original; Carbon emission rights impairment, if initial measurement involves the deferred income, when processing to reduce the forehead, "deferred revenue" debit, "carbon emissions - for private use" credit, if the initial measurement without involving deferred income, processing to reduce the amount of the asset impairment loss "debit," carbon emission rights impairment provision "lenders, value-added have previously included in the capital reserves, need to offset capital reserves.

Deferred revenue. Deferred income according to adjusted the book balance of the processing, the non-operating income at the current, the premise according to the use of carbon emission rights plan, and adjust the previous share of deferred income, treatment as "deferred revenue" debit, "non-operating income" lenders. Impairment test. Plans to sell carbon emissions if made free from the government or auction, when cost is < the fair value, to compare costs and net realizable value of carbon emissions, according to both the smaller adjust the book value. That there are two kinds of circumstances, when the entry value > net realizable value, the difference between the two confirmed as carbon emission rights impairment provision, involving accounting account for "asset impairment loss" and "carbon emission rights impairment provision", the process is as follows:

(7)

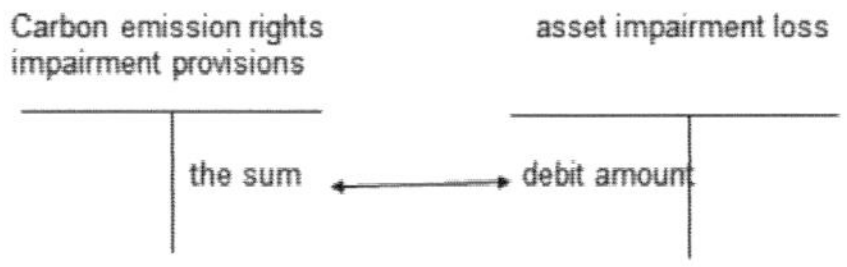

When the entry value < net realizable value, the two cases, one if the original impairment provision, should be in the provision of the amount of the impairment of converted back and there is no provision for impairment of a second if the original is made of, does not make the accounting treatment. Plans to sell if carbon emissions from the secondary market to buy or obtained from the auction, when purchase cost or fair value, are "lower of cost and net realizable value" for impairment test. , same as above is divided into two cases. The sale of carbon emissions. Plans to sell carbon emissions in accordance with the sale income amount is recognized as the other business income, namely debit "bank/cash", "other business income" is credited to account the corresponding cost, shall follow the principle of first in first out, and the specific processing is: borrow other businesses "cost", credit "carbon emissions - for sale", if there is provision for impairment loss, debit "other business cost", "carbon emission rights impairment provision", credit "carbon emissions - for sale".

5 Conclusion and Prospect

With the development of low-carbon economy, the importance of carbon accounting has been self-evident, theory and practice in our country are making unremitting efforts. In this chapter, from the perspective of the enterprise accounting practice case carbon, carbon accounting practice processing has carried on the concrete discussion, and the future application of carbon accounting are further analyzed. Believe that through the joint efforts of theory and practice, the theoretical framework and method of carbon accounting system will be constantly to improve and develop, thus for enterprise development provide effective guidance of carbon accounting practice.

References

1. Jane k T.D.Ratnatunga, Kashi R. Balachandran. Carbon Business Accounting: The Impact of Global Warming on the Cost and Management Accounting Profession[J]. Journal of Accounting Auditing and Finance. 2012 (2): 333-355

2. AnsKolk, David Levy, Jonatan Pinkse. Corporate Responses in an Emerging Climate Regime: The Institutionalization and Commensurable of Carbon Disclosure[J]. European Accounting Review, 2008(6): 175-183
3. Tristram O.West, Gregg Marland. A synthesis of carbon sequestration, carbon emissions, and net carbon flux in agriculture: comparing tillage practices in the United States[J]. Agriculture, Ecosystems and Environment, 2002(7): 85-92
4. Ans Kolk, David Levy, Jonatan Pinkse. Corporate Responses in an Emerging Climate Regime:The Institutionalization and Commemoration of Carbon Disclosure[J]. European Accounting Review, 2011 (4): 719-745
5. Janet Ratnatunga, Stewart Jones. An Inconvenient Truth about Accounting: The Paradigm Shift Required in Carbon Emissions Reporting and Assurance[C]. American Accounting Association Annual Meeting, Anaheim CA, 2008(10):89-93
6. Larry Lohmann. Toward a different debate in environmental accounting: The cases of carbon and cost-benefit[J]. Accounting, Organizations and Society, 2013(34): 499-534
7. Kenneth R. RIchards, Carrie Stokes. A Review of Forest Carbon Squestration Cost Studies: A Dozen Yeas of Research[J]. Climatic Change, 2004(63): 41-48

MATEC Web of Conferences **61**, 04011 (2016)
APOP2016
DOI: 10.1051/matecconf/20166104011

A Clustering Method For Electromagnetic Interference Signals Based On Particle Swarm Optimization

Hongyi Li [1,2], Shengyu Chen [1,2] and Di Zhao [1,a]

[1]*LMIB, School of Mathematics and Systems Science, Beihang University, Beijing 100191, China*
[2]*School of Software Engineering, Beihang University, Beijing 100191, China*

Abstract. This paper proposes a method which clusters EMI signals based on particle swarm optimization. And the advantages of this method compared to other clustering algorithms are present in this paper. From the clustering experiments on 158 sets of data and comparison with other classical algorithms, the clustering algorithm proposed has certain accuracy and feasibility.

1 Introduction

With the rapidly increase of electronic devices such as mobile phones, the electromagnetic interference (EMI) has become more and more urgent, considering its huge interference to normal electronic communication and signal analysis. It is therefore for of great significance to analyze unknown EMI source signals.

The current state-of-the-art technologies to address this problem include the fast independent component analysis (Fast ICA) [1-2], wavelet transform [2-3], the fast Fourier transform (FFT) [4-5]. An indispensable procedure of these techniques is the clustering step.

Existing clustering method are K-means [12], fuzzy clustering [13], wavelet analysis [3, 14], Fourier analysis [5]. K-means is an intuitive method, and easy to implemented, so that it is one of the most widely used clustering approach. However, it also has obvious defects, that it can only handle numeric attributes and is sensitive to initialization. It also require the number of cluster centroids beforehand, and could only obtain a local optimal clustering results [6]. Similar to K-means, Fuzzy c-means could not overcome the local minimum, and the sensitivity to initialization. Moreover, it is hard to select an appropriate weighting parameter M, which has a significant influence on the clustering performance. Wavelet analysis and Fourier analysis are limited to dimension reduction clustering processing of signal in the time domain.

Due to the high dimension that EMI signals have and its non-linear characteristic [15-17], it's necessary to reduce the dimension of the data in a proper way. In addition, to avoid problems like local minimum discussed above, we decided to use PSO based clustering algorithm which can handle this task well. Therefore, this paper proposes a new algorithm procedure for extracting electromagnetic interference characteristics to reduce the dimension of the data, and uses PSO based clustering algorithm to cluster the low dimensional data. Through the experiments, this algorithm has a good performance on accuracy and efficiency in signal processing and clustering.

The rest of this paper is organized as follows: in the second section, we a feature extraction method for electromagnetic inference signals. Subsequently, we propose the clustering algorithm based on PSO. Section 4 shows comparison experimental results. Finally, we draw a conclusion in the last section.

2 A low dimensional representation of EMI signals

In this section, we propose a powerful representation of EMI signals with low dimensionality.

EMI signals usually contain thousands of discrete point whether in time domain or frequency domain. Original EMI signal data usually has high dimensions that is from hundreds to thousands. EMI signals we handled in this paper have more than 6000 dimensions. In addition, EMI signal data value changes dramatically. The intensity of EMI signal data is also a significant feature we interested in.

High dimensional data will make cluster algorithm hard to converge and result in deadly problems like dimension crisis. On the other hand, high dimensional data will cost algorithm much more calculation and spend more time.

Because EMI signals have their special characteristics, we need special features to describe them. In this paper, we present a new method to reduce data dimension. EMI signal data are extracted into five aspects of feature: exception, variance, bandwidth, number of thorns and the number of peaks and troughs. Exception value is a macroscopic feature describing the data value on average. And variance describes degree of dispersion on data value. We use bandwidth to show the volatile of the signal. Signal has larger bandwidth may contain more energy. The number of thorns reflects the instability of

[a] *Corresponding author(E-mail:zdhyl2010@163.com)*

the signal and represent the number of the extreme values. Values in extreme prominent will be recognized as thorns. The number of peaks and troughs shows the overall trend of the signal. This describe the shape of the signal which contains trend information. By doing feature extraction, all information we interested is extracted. And through these features we can discriminate different EMI signals in appropriate time.

3 EMI signal cluster

In this section, we will introduce the PSO clustering algorithm, and describe our proposed method.

3.1. PSO cluster algorithm

In 2002, Omran et al. [10] proposed a swarm optimization based unsupervised image classification algorithm. This is the earliest PSO based clustering algorithm in literature. Most of the PSO clustering algorithms follow the basic framework of this algorithm, where the number of clustering centroids is user defined. Each particle in the swarm contains a vector representing a candidate of cluster centres. The whole particle swarm contains variety of partition of data set. Firstly, a set of random cluster centre is assigned to each particle. For each particle, data are clustered according to the nearest candidate centre. Subsequently, the algorithm utilizes the PSO algorithm to find the optimal particle, of which the positon is an approximate optimal partition on a data set.

Based on Omran's work, Merwe et al. proposed basic PSO clustering algorithm for the clustering to general data set [7]. Fitness function of this PSO clustering algorithm is J_e.

$$J_e = \frac{\sum_{j=1}^{N_c}\left[\sum_{\forall z_p \in C_j} d(z_p, m_j)\right] / |C_j|}{N_c} \tag{1}$$

$$m_j = \frac{1}{n_j}\sum_{\forall z_p \in C_j} z_p \tag{2}$$

$$\mathrm{d}(\boldsymbol{z_p}, m_j) = \sqrt{\sum_{k=1}^{N_b}(z_{pk} - m_{jk})^2} \tag{3}$$

N_b——dimension of the data; N_c ——number of cluster; z_p——data vectors; n_j——number of samples in set C_j; m_j——mean(centre) of set C_j;

The outline of the basic PSO clustering algorithm is as follows [7]:

Algorithm 1 PSO Clustering Algorithm

Input: normalized data vectors z_p

Output: the optimal particle z_{pb} that contains best cluster centre

1: Randomly chosen cluster centre and assigned to each particle, randomly generated particle velocity.
2: Divide each particle on minimum distance principle, and calculate each particle's fitness value by (1) and update individual extreme value.
3: According to each particle's extreme value, find the global extreme value and its position.
4: Based on the PSO velocity formula, update each particle's velocity, and limit it in VMAX.
5: Based on the PSO position formula, update each particle's position.
6: Jump to step 2 until satisfy some condition
7: Output the position of the optimum particle z_{pb}, which contains the optimal N_c cluster centre.

The basic PSO clustering algorithm provides an effective way to solve the problems of many traditional clustering algorithms. Traditional hierarchical methods need to search and estimate large amount of objects or clusters. This results in bad scalability. In addition, these methods may be hard to deal with overlapping clusters. On the other hand, partition methods like K-means can only divide spherical clusters. And it's easy to get in local optimum. However, PSO clustering algorithm can solve these problems.

3.2 The proposed algorithm

Here, we will present out method. Firstly, we extract a low dimensional feature from source EMI signals, and standardize the extracted features as follows:

$$\boldsymbol{x_{ij}} = \frac{x_{ij} - \min(x_{kj})}{\max(x_{kj}) - \min(x_{kj})} \tag{4}$$

where x_{ij} is the j-th component of the i-th vector, $\max(x_{kj})$, $\min(x_{kj})$ are the maximum and minimum values for the j-th component of all vectors, respectively.

After extraction, all the EMI signal data is presented as vectors as follows:

$$\boldsymbol{z_{pi}} = (x_{i1}, x_{i2}, \dots, x_{in}), \tag{5}$$

Where i is the serial number of the data; n is the dimension of data after feature extraction;

For each particle p_i, it is an $n \times N_c$ matrices, representing the positions of N_c clustering centres, where

$$p_i = (m_{i11}, m_{i12}, \cdots, m_{i1n}, \cdots, m_{iN_c1}, \cdots, m_{iN_cn}) \tag{6}$$

At the beginning, the positions and velocities of particles are randomly initialized. As described in subsection 3.1, we then divide every particle with the minimum distance principle, calculate the fitness and update the individual optimal value as follows

$$J_e = \frac{\sum_{j=1}^{N_c}\left[\sum_{\forall z_p \in C_j} d(z_p, m_j)/n_j\right]}{N_c} \tag{7}$$

$$m_j = \frac{1}{n_j}\sum_{\forall z_p \in C_j} z_p \tag{8}$$

$$\mathrm{d}(z_p, m_j) = \sqrt{\sum_{k=1}^{N_b}(z_{pk} - m_{jk})^2} \tag{9}$$

WhereN_b is the dimension of data; N_c is the number of clusters; z_p is the data vector; n_j denotes the number of samples in C_j; m_j stands for centres of samples in C_j.

 DOI: 10.1051/matecconf/20166104011

Based on the individual optimal position, we search for a global optimal particle together with its position. According to the velocity formula [11] as below, we update the velocity of particles with a limited maximum value v_{max}.

$$v_{id}(t+1) = \omega v_{id}(t) + c_1 r_{1d}(t)\big(p_{id}(t) - x_{id}(t)\big) + c_2 r_{2d}(t)\left(p_{gd}(t) - x_{id}(t)\right) \quad (10)$$

$$x_{id}(t+1) = x_{id}(t) + v_{id}(t+1) \quad (11)$$

Where $v_{id}(t)$ is the velocity of the particle i in the t-th iteration; $x_{id}(t)$ is the position of the particle i in the t t-th iteration; ω is the inertia weight; and non-negative c1 and c2 are learning factors; r1 and r2 are independent random functions obeying uniform distribution in the range of [0, 1].

The termination condition can be set as a maximum iteration number, or the residual error.

At last, the optimal position of each particle is output as a cluster centre.

The overall outline of the proposed algorithm is shown in Figure 1.

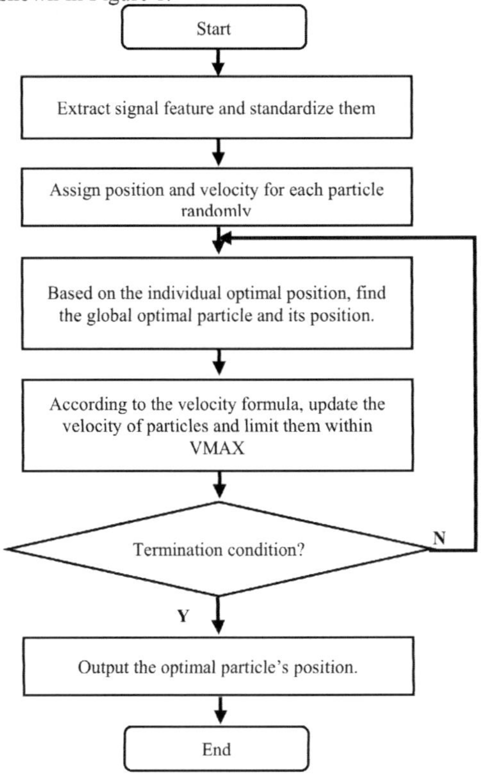

Figure 1. Caption of the Figure 1. Below the figure.

4 Experimental results and analysis

4.1. Data format and other clustering algorithm

The original electromagnetic signal data used in this paper is the frequency domain information in two dimension——frequency (in ascending order) and voltage. We have totally 158 groups of categorized data and each group contains 2431 to 3069 two-dimensional vectors.

For other different mixed electromagnetic signals, such as time domain signals or mixed signals, they can also be used as the input to our method. When the signals are mixed, this method can cluster closely related signal together, which will be helpful for further separation and analysis.

For some commonly used clustering algorithm, PSO algorithm has obvious advantages. Compare to methods based on neural networks, PSO clustering algorithm has a more simple structure, which means it is easy to converge. In addition, PSO clustering algorithm is not sensitive on initialization, and it has a better scalability.

4.2. Parameter settings

In all our experiments, the number of particles N is set as 85. c1 and c2 are set to 1.2, while the inertia weight range WMAX is chosen as 0.9, and WMIN is set to 0.4. the maximum iteration M is 200. Finally, we set the number of clusters K as 7.

4.3. Experimental results

4.3.1 Groups of samples clustering

Here, we first demonstrate the low dimensional feature extracted by using the method described in section 2, as shown in Table 1.

Table 1. Data after feature extraction.

Exception	Variance	Bandwidth	Thorns	Peaks & trough
38.1077	119.2337	6.672511	12	13
37.3756	94.9141	6.594406	11	11
46.6688	19.2154	3.672786	3	1
65.3996	101.838	3.144071	17	2
37.8156	131.037	9.222	0	2
36.5505	129.9987	9.270943	0	0
-1.8437	78.2903	6.438411	13	7
-4.6797	46.608	3.317348	13	10
23.5268	102.1685	3.465579	0	1
39.9496	176.6618	4.889012	0	6
38.4838	139.4067	4.538699	2	9
37.9304	145.5721	4.744465	1	5
15.2725	180.7833	5.200211	0	2
31.7017	396.7352	6.818715	12	8
40.0647	157.7335	6.711977	1	6
40.8305	117.545	6.203311	0	5
9.2589	232.1797	9.173041	0	2

We then normalize the feature, and conduct clustering by suing PSO clustering algorithm. The result is shown in Table 2.

Table 2. Cluster result by using the low dimensional feature and PSO clustering algorithm.

Class	Data serial number
1	17
2	5,6
3	1,2
4	9,10,11,12,13,15,16
5	14
6	7,8
7	3,4

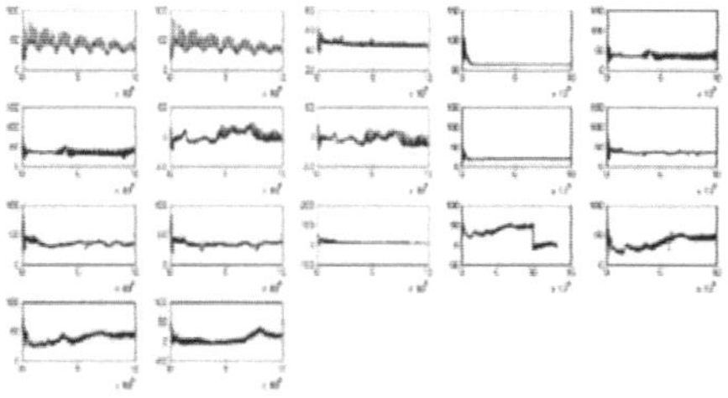

Figure 2. Curves of the original EMI signal data.

The final output clustering centres can be seen in Table 3.

Table 3. Output centres using the proposed method.

Class	Clustering centre
1	(0.1989 , 0.5641 , 0.9840 , 0 , 0.1538)
2	(0.5974 , 0.2948 , 0.9960 , 0 , 0.0769)
3	(0.6053 , 0.2327 , 0.5695 , 0.6765 , 0.9231)
4	(0.5480 , 0.3350 , 0.3205 , 0.0336 , 0.3736)
5	(0.5191 , 1.0000 , 0.5998 , 0.7059 , 0.6154)
6	(0.0202 , 0.1145 , 0.2830 , 0.7647 , 0.6538)
7	(0.8664 , 0.1094 , 0.0431 , 0.5882 , 0.1154)

According to the comparison with original data images (see Figure 2), we can draw the conclusion that the algorithm has good effect on clustering.

Acquiring the clustering centre, we can compare each feature's difference between two classifications based on their clustering centre. For example, the first class and the second class are similar in bandwidth and the number of thorns, but are different in exception and variance. Furthermore, we can also calculate the difference accurately.

For the same data, this paper also carried on comparison experimental by using the SOM. The results are as follows:

The final clustering result is as shown in Table 4:

Table 4. Clustering result by using SOM.

Class	Data serial number
1	11,13,15,16
2	17
3	5,6
4	1,7
5	2
6	14
7	8
8	3,4,9,10,12

As can be seen from the result, by using SOM, data 1 and data 7 are divided into the same class. Data 2 and 8 are divided into different classes. According to the classification label, PSO clustering algorithm has a better solution that data 1 and 2 are divided together and data 7 and 8 are in the same class (see Figure 2).

Besides, the result of SOM clustering is influenced by subjective factors. On the contrary, PSO clustering can draw a clear clustering results and output each classes' clustering center so as to quantitatively reflect the difference between each category.

4.3.2 Comparison of 158 groups of data clustering algorithms

Here we use 158 groups of categorized EMI signal data as test data. To present the advantages of our method, we use 4 other common clustering algorithms as comparison. To show the significance of dimension reduction, we also use original data as input vector as a comparison. During experiment, it spends hours of time to calculate. After doing the same number of recursion, results are disappointing. Clusters are hard to converge, which means almost all the signal are divided into a same class.

The comparison clustering results by using different approaches are shown in Table 5.

Table 5. Comparison results by using different approaches.

	Accuracy	Computation Time	Whether need clustering number
PSO+ feature extraction	93.04%	257ms	Yes
PSO+ original signals	10.17%	>1h	Yes
K-means	70.89%	985ms	Yes
SOM	91.77%	102ms	No
FCM	86.08%	364ms	Yes

Through repeated experiments for clustering, as can be seen, the algorithm proposed in this paper has high accuracy and efficiency in electromagnetic signal clustering. However, compared to SOM network, this algorithm requires the number of clustering, which is one disadvantage of the algorithm.

5 Conclusion

In order to separate interference factor from the system signal and use clustering to find out typical electromagnetic interference factor, this paper presents a method for feature extraction and clustering of electromagnetic signal data by PSO clustering algorithm, and describe the advantages of this method compared to traditional clustering algorithms such as K-means, SOM, FCM, which show that our method outperforms the comparison approaches, and require less running time.

Acknowledgement

This work was supported by the National Natural Science Foundation of China (No.61379001) and the Key Project of Beihang University (No. 201412) ("Exploration and Practice of Innovation Ability Cultivation in Mathematical Education").

References

1. Y. Ye, Z. Zhang, J. Zeng, L. Peng. A fast and adaptive ICA algorithm with its application to fetal electrocardiogram extraction. Appl. Maths. Comput., **205(2)**, 799-806 (2008)
2. M.T. Akhtar, W. Mitsuhashi, C.J. James. Employing spatially constrained ICA and wavelet denoising, for automatic removal of artifacts from multichannel EEG data. Signal Processing, **92(2)**, 401-416 (2012)
3. D. Zhao, S. Wu. Method of electromagnetic signal leakage detecting based on wavelet transform. Elec. test, **8(1)**, 1-3 (2008)
4. C. Zhang, S. Liu, Q. Yang, L. Jin, L. Feng. Research on signal processing of electromagnetically induced acoustic emission based on the FFT and wavelet packet transform. CTATEE'09 , 283-287 (2009)
5. S. Yang, Q. Zhong. Low-tension arc fault detection based on FFT and electromagnetic radiation. Modern elec. Tech, **35(18),** 86-88 (2012).
6. Y. Jiang, Z. Zhang, P. Qiu, D. Zhou. Clustering algorithm used in data mining. Journal of elec. & info. Tech., **27(4)**, 655-662 (2005)
7. D.W. Merwe, A.P. Engelbrecht. Data clustering using particle swarm optimization. Proc of IEEE CEC, 215-220, (2003)
8. L. Yang, H. Su, Y. Zhang, J. Chu. A method of data mining based on SOM clustering and its application. Computer eng. & sci., **29(8)**, 133-136, (2007)
9. T. Kohonen. Self-Organized Formation of Topologically Correct Feature Maps. Bio. Cybernetics, **43(1),** 59–69, (1982)
10. M. Omran, A. Salman, A.P. Engelbrecht. Image classification using particle swarm optimization. Proc of the 4th APCSEL, 370-374 (2002)
11. N. Sun, Q. Zhang. Research on PSO algorithm. Journal of Hengshui Univ., **12(1)**, 25-28 (2012)
12. L. Chen, E. Zhang, R. Shen. The sorting of frequency hopping signals based on K-means algorithm with optimal initial clustering centers. Journal of national Univ. of defense tech., **2,** 70-75 (2009)
13. Z. Ding, Z. Hu, J. Yang, Z. Tang, Y. Wu. A fuzzy-clustering-based approach to image segmentation. Computer research & devel. , **7,** 536-541 (1997)
14. Z. Chen, J. Deng, H. Chen, P. Ni. De-noising research of magneto telluric signal based on wavelet transform. Chinese journal of eng. Geophys., **6**, 732-737 (2012)
15. H. Li, C. Wang, D. Zhao. An Improved EMD and Its Applications to Find the Basis Functions of EMI Signals. Mathematical Problems in Engineering, (2015)
16. H. Li, Z. Song, D. Zhao, P. Wang, J. Chen. A Single Channel EMI Signal Separation Method Based on Directly-mean Empirical Mode Decomposition. Journal of Infromation & Computation Science, **12(17)**, 6333-6340 (2015)
17. D. Zhao, H. Li, On the computation of inverses and determinants of a kind of special matrices. Applied Mathematics and Computation, **250**, 721-726 (2015)

MATEC Web of Conferences **61**, 04012 (2016) DOI: 10.1051/matecconf/20166104012
APOP2016

Risk Assessment for Bridges Safety Management during Operation Based on Fuzzy Clustering Algorithm

Hanyu Xia [1,a], Lijing Zhang [1], Gang Tao [1], Bing Tong[1] and Haiou Zhang [2]

[1]Jiangsu Key Laboratory of Urban and Industrial Safety, College of Urban Construction and Safety Engineering, Nanjing Tech University, Nanjing, 210009, China;

[2]Nanjing Architectural Design & Research Institute CO., LTD,Nanjing 210005,China

Abstract. In recent years, large span and large sea-crossing bridges are built, bridges accidents caused by improper operational management occur frequently. In order to explore the better methods for risk assessment of the bridges operation departments, the method based on fuzzy clustering algorithm is selected. Then, the implementation steps of fuzzy clustering algorithm are described, the risk evaluation system is built, and Taizhou Bridge is selected as an example, the quantitation of risk factors is described. After that, the clustering algorithm based on fuzzy equivalence is calculated on MATLAB 2010a. In the last, Taizhou Bridge operation management departments are classified and sorted according to the degree of risk, and the safety situation of operation departments is analyzed.

1 Introduction

Bridges operation departments are important for bridges safety, because they conduct the regular inspections, testing, maintenance and debugging for the main body of the bridge and the auxiliary road facilities during operation. As the number of large bridges increased, many bridges accidents occur due to the inadequate management, management risk of bridge operation departments has become an important part of phase risk factors which cannot be ignored during operation. Identifying the risk factors of the bridges operation departments and building a risk assessment system and analyzing the risk distribution are benefit of bridges operation departments to adopt appropriate risk control measures. And it has an important significance for reducing the risk of bridge operation and improving the overall safety of the bridge during operation.

2 Fuzzy clustering algorithm steps

2.1 Original data obtaining

The objects been studied are supposed as $X=(x_1,x_2,\cdots,x_n)$, the i-th object consists of m indicators to characterize its properties, namely$x_i=(x_{i1},x_{i2},\cdots,x_{im})(i=1,2,\cdots,n)$, then the original data are represented as a matrix:

$$X=\begin{bmatrix} x_{11} & x_{12} & \cdots & x_{1m} \\ x_{21} & x_{22} & \cdots & x_{2m} \\ \vdots & \vdots & & \vdots \\ x_{n1} & x_{n2} & \cdots & x_{nm} \end{bmatrix}$$

2.2 Standardization of data

When dealing with practical problems, because the dimension and magnitude of the m index data may be different, the larger magnitude indexes may have a more significant impact on the resulting classification in the calculation, while the impact of the smaller magnitude

[a] Corresponding author: 810562002@qq.com

indexes may be too small to be looked. In order to eliminate the different impacts of data dimension and magnitude in comparison process, data should be converted appropriately. Then each index value is in the range of united values, the data comparison will be more meaningful. The mean and variance of j-th column in the matrix X are calculated, formula is shown in Eq. 1 and Eq. 2 and Eq. 3. [1]

$$x'_{ij}=\frac{x_{ij}-\overline{x_j}}{s_j},(i=1,2,\cdots,n;j=1,2,\cdots,m;) \tag{1}$$

$$\overline{x_j}=\frac{1}{n}\sum_{i=1}^{n}x_{ij}, \tag{2}$$

$$s_j=\left[\frac{1}{n}\sum_{i=1}^{n}(x_{ij}-\overline{x_j})^2\right]^{\frac{1}{2}} \tag{3}$$

2.3 Fuzzy similar matrix structuring

Process of structuring a fuzzy similar matrix is also known as calibration, which is calculating statistic $r_{ij}(i,j=1,2,\cdots,n)$ to measure the degree of similarity between the analyzed data, thereby the fuzzy similar matrix R which expresses similar relationship is established.

$$R=(r_{ij})_{n\times n}=\begin{bmatrix} r_{11} & r_{12} & \cdots & r_{1n} \\ r_{21} & r_{22} & \cdots & r_{2n} \\ \vdots & \vdots & & \vdots \\ r_{n1} & r_{n2} & \cdots & r_{nn} \end{bmatrix}$$

r_{ij} is calculated as Eq. 4 and Eq. 5:

$$r_{ij}=1-cd(x_i,x_j),(i,j=1,2,\cdots,n). \tag{4}$$

$$d(x_i,x_j)=\left[\sum_{k=1}^{m}(x_{ik}-x_{jk})^2\right]^{\frac{1}{2}}. \tag{5}$$

Wherein, c is parameters which is chosen to satisfy $0\le r_{ij}\le 1$, $d(x_i,x_j)$ represents the distance from x_i to x_j [2].

2.4 Fuzzy equivalent matrix structuring

Fuzzy clustering analysis requires that matrix to be analyzed satisfies the requirements: reflexive, symmetric, and transitive [3]. To make the matrix R transitive, it is needed to make the following changes: Starting from the matrix R, square in turn $R\to R^2\to R^4\to\cdots$, when $R^k\cdot R^k=R^k$ first appears, it is considered fuzzy matrix R^k transitive, then fuzzy equivalent matrix $R^*=R^k$ [4].

2.5 Dynamic Clustering

Dynamic clustering is conducted by using transfer closure method which is based on fuzzy equivalence relation. That is, when the fuzzy equivalent matrix element values $\lambda(\lambda\in R^*)$ declines from 1 to 0, the category finally is classified as a class, which is a dynamic clustering process.

3 Evaluation index system establishing

3.1 Risk indicators description and quantification of operation departments

The following indicators are mainly considered in risk quantification of the operation departments:

3.1.1 Description and quantification of security managers

Bridges operation companies should set up safety management departments to manage the bridges overall safety, other operation departments should select full-time in charge of operation safety [5]. The number of safety managers in each departments has a direct impact on the effect of safety management; the quantification method is counting "the number of" the full-time safety managers in operation management departments.

3.1.2 Description and quantification of safety training

Safety training can increase safety knowledge of employees and improve safety management level of managers. There are many safety training methods, such as propagandizing road transportation safety laws or regulations and accident cases, etc. [6]. The quantification method for safety training is calculating the ratio of personnel who participates in safety training and the total number of each department personnel, namely "security training rate", units: "%".

3.1.3 Description and quantization of fire alarm detection system

Most of bridges operation management departments locate in different office locations and fire alarm detection system is an effective way to prevent and control office fires, so it has a direct impact on the safety of bridges operation management departments. The quantization method for fire alarm detection system is: assigning a value based on the effectiveness of each department's fire alarm system. Ineffective is "1"and effective is "2".

3.1.4 Description and quantization of fire-fighting equipment and facilities

The effectiveness of fire-fighting equipment and facilities is an important factor to measure the operation safety risk of each department. The quantization method for fire-fighting equipment and facilities is calculating the ratio of each department firefighting equipment and facilities coverage and the total area of the department, namely "firefighting equipment and facilities coverage rate", units: "%".

3.1.5 Description and quantization of hazard sources

Hazard sources are important part of the bridges operation risk; each management department should identify and assess regular hazard sources and develop control measures according to the assessment results. The quantification method of hazard sources is identifying "the total number of hazard sources" of each department by using the same hazard identification methods.

3.1.6 Description and quantization of the high-grade hazard sources

High-grade hazard sources are the main threat to each bridge management department. The quantification method of high-grade hazard sources is identifying "the total number of high-grade hazard sources" of each department by using the same hazard identification methods.

3.1.7 Description and quantization of safety activities

Bridges operation departments can effectively improve the safety awareness of employees and increase the level of safety management by holding periodic safety activities and publicizing safety knowledge. The quantification method of safety activities is counting the annual average number of holding safety activities times. Units: "times / year".

3.2 Construction of evaluation system

Based on the above analysis, "A-safety management risk of operation departments" is set as evaluation target layer for safety management risk assessment. Factors layer are "A1-Number of safety managers"; "A2-Safety training rate"; "A3- Effect of fire alarm detection system"; "A4-Firefighting equipment and facilities coverage rate"; "A5-Total number of hazard sources"; "A6-Total number of high-grade hazard sources"; "A7-Safety activities". Evaluation System is detailed in Figure 1.

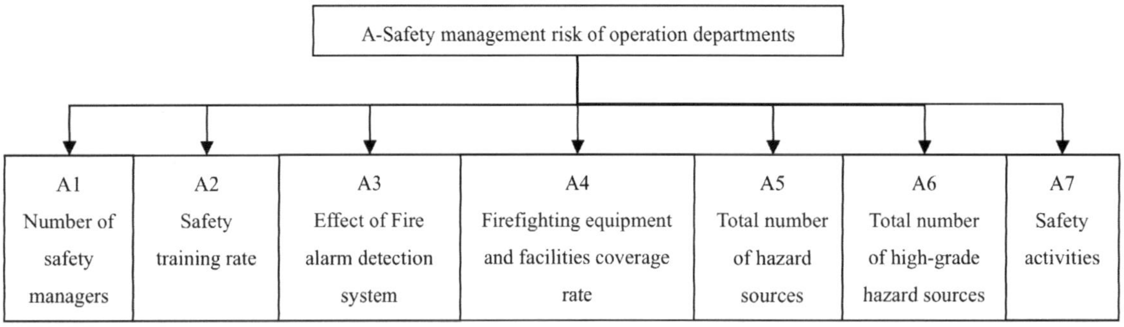

Figure 1. The whole safety risk assessment system of bridges during operation

4 Operation safety risk assessment of Taizhou Bridge

4.1 Evaluation target selecting

Frontline operation management departments are chosen to assess safety risk of Taizhou Bridge. The assessment includes each toll station; service area and production scheduling center along Taizhou Bridge; repairing department; troubleshooting departments; equipment maintenance center and departments which directly relate to the bridge operation management. Evaluation targets are shown in Figure 2.

4.2 Evaluation data obtaining

According to the indicators (A1~A7) which relate to the risk of Taizhou Bridge management departments, research about 13 departments (K1~K13) of Taizhou Bridge is conducted and internal data are referred. Safety situation of operation management in 13 departments is quantized and the original data are obtained as Table 1 shows.

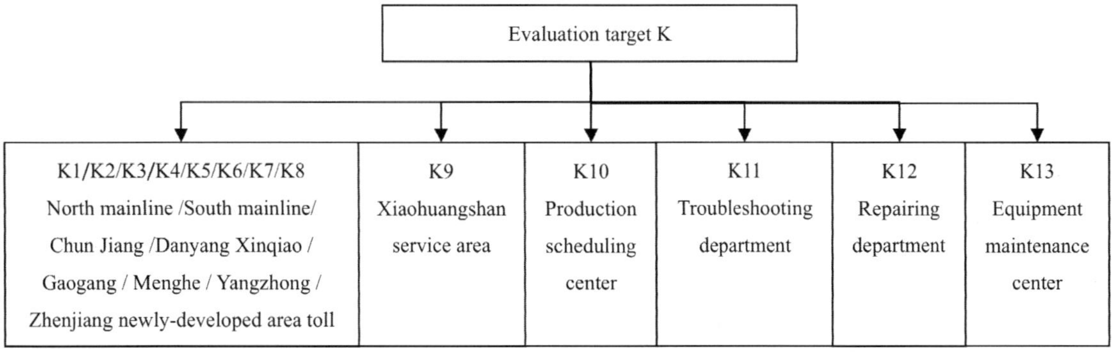

Figure 2. Internal management risk assessment object of Taizhou Bridge

Table 1. Original data collection of safety situation for 13 departments of Taizhou Bridge

Risk indicators / Research objectives	A1	A2	A3	A4	A5	A6	A7
	Number of safety managers	Personnel safety training rate[%]	Effect of fire alarm detection system	Firefighting equipment and facilities coverage rate[%]	Total number of hazard sources	Total number of high-grade hazard sources	Safety activities /[times/year]
K1	3	95	2	90	72	2	24
K2	3	95	2	85	93	1	28
K3	1	70	1	70	79	3	10
K4	1	75	1	72	92	4	12
K5	2	82	2	75	85	3	18
K6	1	85	2	83	86	4	20
K7	1	75	2	84	62	6	14
K8	1	78	2	70	83	3	12
K9	2	78	2	73	174	3	20
K10	2	83	1	85	15	1	8
K11	1	85	2	65	28	1	10
K12	1	76	2	70	42	2	8
K13	1	78	1	68	16	1	12

Note: "A5-Total number of hazard sources"and"A6- Total number of high-grade hazard sources" are derived from the internal report " Hazard sources identification and assessment report in operational phase of Taizhou Bridge " [7].

4.3 Assessment analysis

The data in Table 1 are converted into the form of a matrix and the above-mentioned fuzzy clustering algorithm steps are used to assess safety risk of Taizhou Bridge operation departments. For a more clear analysis of clustering results, Figure 3 is plotted to do further analysis on the dynamic clustering process.

It can be seen from Figure3: When λ = 0.93, K1 and K2 are first classified as a category; When λ = 0.89, K5 and K9 are classified as a category; When λ = 0.84, K6 and {K5, K9} are classified as a category; When λ =0.8307, K8 and K12 are classified as a category; When λ =0.8284, K7 and {K5, K9,K6 } are classified as a category; When λ =0.77, K3 and K4 are classified as a category; When λ =0.74, K10 is added to the aforementioned category; When λ =0.44, K13 is also added to the aforementioned category, thereby all the departments form a category.

From dynamic cluster analysis process and analysis, it can be seen that the risk level of the Taizhou Bridge operations departments (K1~K13) is {K1,K2}<{K5,K9} < {K6} < {K8,K12} < {K7} < {K3,K4} < {K11} < {K10}<{K13}.

It can be seen from Figure 3: When λ =0.8307, K8 and K12 are classified as a category; When λ =0.8284, K7 and {K5, K9, K6} are classified as a category; strictly speaking, the operational risks {K8, K12} < {K7}. Since the difference in value 0.8307 and 0.8284 is small, that in the actual management the risk difference can be ignored.

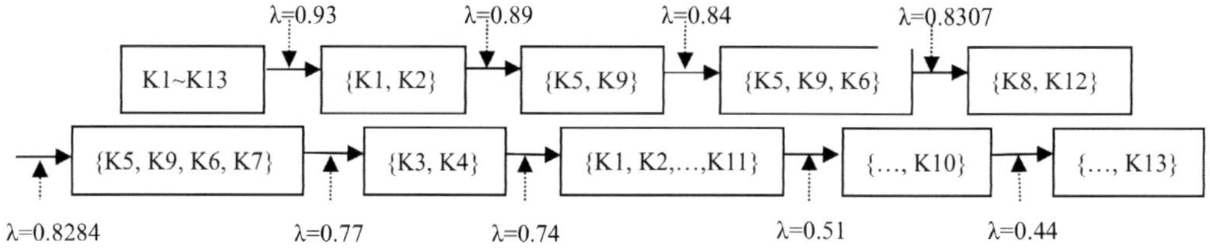

Figure 3. Further analysis of dynamic clustering of Taizhou Bridge (K1~K13 departments)

5 Conclusions

From the above analysis it can be seen: in Taizhou Bridge operation departments K1~K13, high-risk departments are troubleshooting department, production scheduling center and equipment maintenance center. As these departments are mainly responsible for road-block removing, inspection scheduling and equipment maintenance and repairing across the board in Taizhou Bridge, there are more persons, vehicles and other electrical equipment Etc., the risk level is higher. Accordingly, the assessment results match the actual operation situation.

References

1. Xie Jiping, Liu Chengping, Fuzzy mathematics and its applications, Wuhan, 2000.
2. Gao Xinbo, Fuzzy cluster analysis and its application, Xi'an, 2004.
3. Jiang Zejun. Fuzzy Mathematics tutorials, Beijing, 2004.
4. Wu Fubao, Li Qi. Delivery Methods of Fuzzy Clustering Analysis, J. Journal of Southeast University, Natural Science. **29**(2) (1999) 105-110.
5. Yan Weiwen, Zhang Wenhai WANG Lei, Chen Guohua. Present Status of Quality of Safety Personnel in Enterprises and Suggestions on Their Training, J. China Safety Science Journal, **16**(10) (2007) 51-55.
6. Dun Xuhua, Li Shouzhao. Function of Case Teaching and Discuss Teaching in the Safety Training Job, J. Journal of Safety Science and Technology, **2**(1)(2006)95-97.
7. Hazard sources identification and assessment report in operational phase of Taizhou Bridge, R. Nanjing, Jiangsu Academy of Safety Science and Technology, 2014.

DOI: 10.1051/matecconf/20166104013

Analysis of Daily Global Solar Radiation Data for Wuhan

Zhengguo Gong, Yingni Jiang, Hao Yang and Jingbo Liu

The Logistic University Of PAPF, Tianjin 300300, China

Abstract. In order to improve knowledge on solar resources, this paper analyses the models correlating daily global solar radiation with sunshine duration for Wuhan, China. Estimated values were compared with measured values in terms of statistical error tests such as mean percentage error (MPE), mean bias error (MBE), root mean square error (RMSE). The model performing best was selected. The global solar radiation estimated from the best model was compared with measured values. It was determined that the predicted values have good agreement with the measured values at high daily global solar radiation.

1. INTRODUCTION

Solar radiation data are essential for both energy efficient building design and solar systems developments. Unfortunately, for many cases, solar radiation measurements are not easily available due to the cost and maintenance and calibration requirements of the measuring equipment. In order to achieve this, several empirical models have been developed to predict the solar radiation all over the world using various parameters. These parameters include extraterrestrial radiation, sunshine hours, mean temperature, maximum temperature, soil temperature, relative humidity, number of rainy days, altitude, latitude and cloudiness [1–10].

The most commonly used parameter for estimating global solar radiation in the publications is sunshine duration. Sunshine duration can be easily and reliably measured and data are widely available. The most widely used method is that of Angström [1], who proposed a linear relationship between the ratio of average daily global radiation to the corresponding value on a completely clear day and the ratio of average daily sunshine duration to the maximum possible sunshine duration.

The main objective of this paper is to estimate the daily global solar radiation with measured daily sunshine duration data from Wuhan, and to find the most suitable model for global radiation estimation at Wuhan, China.

2. DATA USED AND METHODS OF COMPUTATION

2.1. Database

Wuhan meteorological station is chosen for the case study. Wuhan (30.6°N, 114.1°E) is located in the interior of China, at an altitude of 23.1m. A database containing daily measured global solar radiation and sunshine duration of Wuhan station was obtained from China Meteorological administration. In this paper ten years' recorded daily solar radiation from 1995 to 2004 are used.

2.2. Methods of computation

In the present work, the following correlations were used to express the dependence of global radiation on various parameters:

$$\frac{H_g}{H_0} = a_1 + a_2 \frac{S}{S_0} \tag{1}$$

$$\frac{H_g}{H_0} = a_3 + a_4 \frac{S}{S_0} + a_5 \left(\frac{S}{S_0} \right)^2 \quad (2)$$

$$\frac{H_g}{H_0} = a_6 + a_7 \frac{S}{S_0} + a_8 \left(\frac{S}{S_0} \right)^2 + a_9 \left(\frac{S}{S_0} \right)^3 \quad (3)$$

where H_g is daily global solar radiation on a horizontal surface, H_0 is the daily value of the extraterrestrial radiation, S is sunshine duration, S_0 is the maximum possible sunshine duration, a are empirical coefficients.

The daily value of the extraterrestrial radiation on a horizontal surface was defined as [11]:

$$H_0 = \frac{24 \times 3600}{\pi} I_0 f (\cos\lambda \cos\delta \sin\omega_s + \frac{\pi}{180} \omega_s \sin\lambda \sin\delta) \quad (4)$$

where I_0 is the new solar constant (=1367 W/m2 [12]), f is the eccentricity correction factor, λ is the latitude of the site, δ the solar declination and ω_s the sunrise hour angle. The eccentricity correction factor, solar declination and sunrise hour angle can be respectively calculated as [11]:

$$f = 1 + 0.33 (\cos \frac{360n}{365}) \quad (5)$$

$$\delta = 23.45 \sin \left[\frac{360(284+n)}{365} \right] \quad (6)$$

$$\omega_s = \arccos(-\tan\lambda \tan\delta) \quad (7)$$

where n is the number of day of year starting from first of January.

The maximum possible sunshine duration (S_0) can be obtained as [11]:

$$S_0 = \frac{2}{15} \omega_s \quad (8)$$

Fig.1 presents the plot of H_g / H_0 against S / S_0 at Wuhan (1995-2004). It can be seen that the relationship between daily global solar radiation and daily sunshine duration is rather strong.

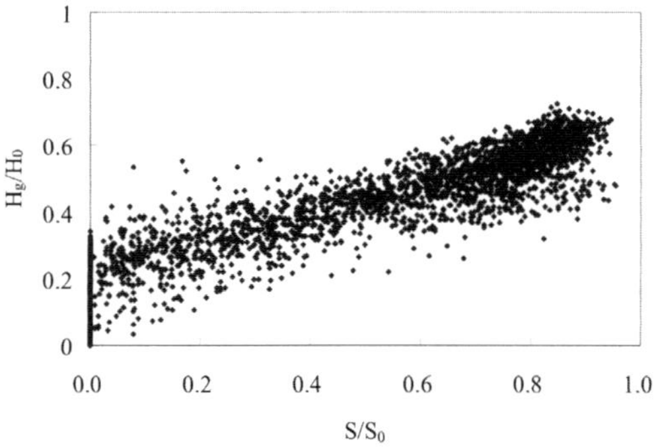

Fig.1. Relationship between H_g / H_0 and S / S_0 of Wuhan (1995-2004) obtained from the observed data

2.3. Methods of comparison

In this study, three statistical tests, mean percentage error (MPE), mean bias error (MBE) and root mean square error (RMSE) were used to estimate the accuracy of the correlations described above.

The mean percentage error is defined as:

$$MPE = \frac{1}{N} \sum_{i=1}^{N} \left(\frac{D_{ie} - D_{im}}{D_{im}} \right) \times 100 \quad (9)$$

The mean bias error is defined as:

$$MBE = \sum_{i=1}^{N} (D_{ie} - D_{im}) / N \quad (10)$$

The root mean square error is defined as:

$$RMSE = \sqrt{\sum_{i=1}^{N} (D_{ie} - D_{im})^2 / N} \quad (11)$$

DOI: 10.1051/matecconf/20166104013

where D_{ie} is the ith estimated value, D_{im} is the ith measured value and N is the total number of observations.

The mean bias error test provides information on the long term performance. A low MBE is desired. A positive value gives the average amount of over estimation of an individual observation, which will cancel an under-estimation in a separate observation. The root mean square error RMSE gives information on the short term performance of the correlations by allowing a term by term comparison of the actual deviation between the calculated and measured values. The smaller the value, the better the model's performance is.

3. Results and Discussion

The obtained empirical coefficients of the Eqs.(1)-(3) are expressed as follows:

Model 1

$$\frac{H_g}{H_0} = 0.134 + 0.547\frac{S}{S_0} \quad (12)$$

Model 2

$$\frac{H_g}{H_0} = 0.120 + 0.786\frac{S}{S_0} - 0.292\left(\frac{S}{S_0}\right)^2 \quad (13)$$

Model 3

$$\frac{H_g}{H_0} = 0.112 + 1.240\frac{S}{S_0} - 1.741\left(\frac{S}{S_0}\right)^2 + 1.106\left(\frac{S}{S_0}\right)^3 \quad (14)$$

Table 1 shows MPE, MBE and RMSE errors for the three daily global radiation models (models (1-3) corresponding to Eqs.(12)-(14) respectively) of Wuhan station during the period 1995-2004. From this table, the MBE values obtained from models (1, 2, 3) are all negative, which shows that these models under estimate global radiation. The RMSE value, which is a measure of the accuracy of estimation, has been found to be smallest for model 3 (2.362 MJ/m2), see Table 1. Also, MPE, MBE and RMSE are shown in Fig.1-3, respectively. As a result, model 3 performs better than the other models. The MPE, MBE and RMSE of model 3 are 131.4%, -0.0041 MJ/m2 and 2.362 MJ/m2.

Table 1: Prediction errors for daily global solar radiation of three models

	MPE (%)	MBE (MJ/m^2)	RMSE (MJ/m^2)
Model 1	159.7	-0.0219	2.511
Model 2	141.6	-0.0116	2.417
Model 3	131.4	-0.0041	2.362

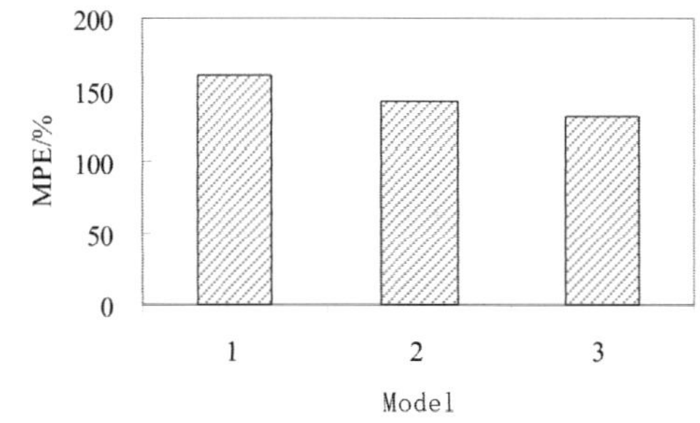

Fig.2. MPE of the three models

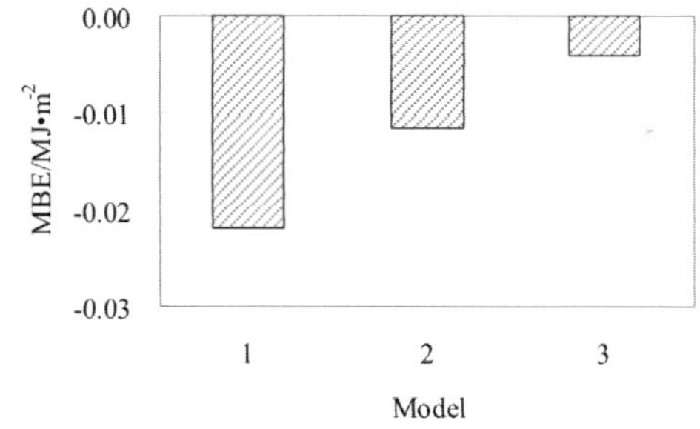

Fig.3. MBE of the three models

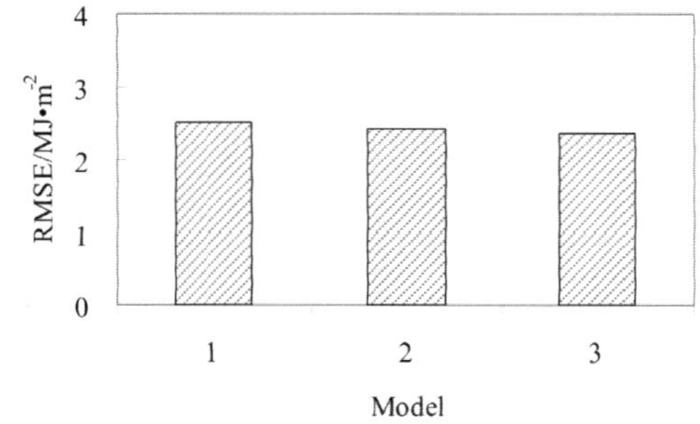

Fig.4. RMSE of the three models

To compare observed and estimated global radiation values of model 3 for Wuhan (1995–2004), Fig.5 is provided. Model 3 has large errors in estimating observed solar radiation at low daily global solar radiation. But it is of little consequence for solar energy operated systems, because low global radiation constitutes only a small part of the whole radiation. Fig.5 suggests the good fitness between the observed and the estimated values when daily global solar radiation values are high.

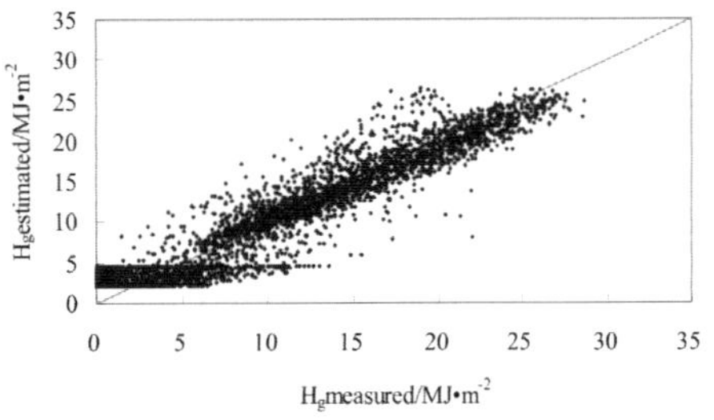

Fig.5. The measured and predicted (by model 3) values of daily global solar radiation for Wuhan 1

4. Conclusions

This work demonstrated the efficient computation of daily global solar radiation from an estimation of H_g vs. S/S_0 using data from Wuhan, China. The data set covers a period of 10 years (1995-2004). Linear, quadratic and cubic models were tried to estimate H_g from S/S_0. It was found that H_g can easily be estimated using the cubic model when the value of H_g is large.

References

[1] Angström A. Solar and terrestrial radiation. Quarterly Journal of Royal Meteorological Society 1924; 50: 121-126.

[2] Tiba C. Analysis of monthly time series of solar radiation and sunshine hours in tropical climates. Renewable Energy 2004; 29(7): 1147-1160.

[3] Ehnberg JSG., Bollen MHJ. Simulation of global solar radiation based on cloud observations. Solar Energy 2005; 78(2):157–162.

[4] Skeiker K. Correlation of global solar radiation with common geographical and meteorological parameters for Damascus province, Syria. Energy Conversion and Management 2006; 47(4): 331-345.

[5] Forero NL, Caicedo LM, Gordillo G. Correlation of global solar radiation values estimated and measured on an inclined surface for clear days in Bogota. Renewable Energy 2007; 32(15): 2590–2602.

[6] Sabziparvar AA, Shetaee H. Estimation of global solar radiation in arid and semi-arid climates of East and West Iran. Energy 2007, 32(5):649–655.

[7] Bulut H, Büyükalaca O. Simple model for the generation of daily global solar-radiation data in Turkey. Applied Energy 2007; 84(5): 477–491.

[8] Sabziparvar AA. A simple formula for estimating global solar radiation in central arid deserts of Iran. Renewable Energy 2008; 33(5): 1002-1010.

[9] Abraha MG, Savage MJ. Comparison of estimates of daily solar radiation from air temperature range for application in crop simulations. Agricultural and Forest Meteorology 2008; 148(3): 401–416.

[10] Chineke TC. Equations for estimating global solar radiation in data sparse regions. Renewable Energy 2008; 33(4): 827–831.

[11] Duffie JA,Beckman WA, Solar engineering of thermal processes [M]. New York:Wiley, 1991.

[12] Frohlich C, London J, editors, "Revised instruction manual on radiation instruments and measurements," World climate research program, Publication Series No.7, WMO/TD-No.149. World Meteorological Organisation:Geneva, 1986, 140 pp.

Experimental Study on Radon motion during Rock Compression

Jiankun Zhao[1,2], Liangquan Ge[1,2,a], Yaoyao Luo[1] and Yi Gu[1,2] Qingxian Zhang[1,2]

[1] *Chengdu University of Technology, Chengdu 610059, China*

[2] *Key Laboratory of Applied Nuclear Techniques in Geosciences, Chengdu 610059, China*

Abstract: A mathematical model to describe the radon motion outside the surface of rocks during the sustained stress on the rocks has been established and an experimental device for monitoring changes of radon concentration has been designed. The theoretic and experimental results show that there is a negative correlation between radon concentration emitted from granite and sandstone rock specimens and the axial pressure loaded on the rock specimens. The maximum error of the radon concentration between theoretic and experimental results is about 10.74%.

Keywords: radon motion, rock compression, mathematical model, experiments

1 Introduction

As an effective natural tracer, radon has been widely used in many applied physics fields such as exploration, engineering investigation and other applied geophysics. It is also considered a sensitive physical method to predict the earthquake and has been ordinarily monitored in a number of Earthquake Monitoring Stations[1-3]. There are many experts that, at home and abroad have investigated the relevance between the earthquake and the anomalies of radon concentration in rocks for a long time. As early as 1944, on the eve of the 8.3 magnitude earthquake taking place in southwest of Japan, the mutation of Radon concentration has aroused the great attention of Hatuda[4]. In 1991, Houlub put forward his research result about the condition of radon exhalation under uniaxial pressure [5]. The physical mechanism of radon motion is qualitatively explained that the tectonic stress impacts on the primary fracture of geologic body, which influences the migration channels of radon[6-8]. This paper attempts to provide a mathematical model to describe the radon motion outside the surface of rocks and designs an experimental device to prove it.

2 The Relationship between Stress and Radon Concentration on the Surface of Rocks

During the diagenesis process, there are a large number of finer textures, such as fractures and pores, existing insiderocks upon the effects of natural conditions including cooling, dilation and stress. These finer textures are regarded as the natural systems and storage spaces for free radon. If the macroscopic stress effected on the rock continually, the finer textures will change with it. A power function can be utilized to express the relationship between porosity and pressure [9,10]:

$$P = P_0 - A\sigma^n \qquad (1)$$

Where, P and P_0 is the porosity of rocks at the stress pressure with σ and at normal pressures (%), respectively; A and n are both constants (n=0.73); σ is effective stress (MPa/cm^2).

The radon concentration on the surface of rocks can be defined as:

$$N_K = kN_P \qquad (2)$$

Where, N_K is the radon concentration on the surface of

[a] Corresponding author: glq@cdut.edu.cn

rocks (Ci/cm^{-3}); N_P is the radon concentration in fractures of rocks (Ci/cm^{-3}); k is called the escaping coefficient of radon from a rock which is defined as the ratio of the radon concentration on the surface of a rock to the radon concentration in fractures of the rock.

If the decay of radium and radon is in radioactivity decay equilibrium in which the decay rate of radium is equal to that of radon in rocks, the radon concentration in fractures of rock can be given by [2]

$$N_P = P\rho g\eta \qquad (3)$$

Where, ρ is the density of the rock (g/cm^3); g is the specific activity of the natural ^{226}Ra in the rock (g/g) ; η is the emanation coefficient of radon of the rock, which is defined as the ratio of the radon concentration in fractures of a rock (N_P) to the total radon concentration of the rock.

Eq. (2) and Eq. (3), we have:

$$N_K = kP\rho g\eta \qquad (4)$$

We denote an effective escaping coefficient K which is equals to $k\eta$. From Eq. (1) and Eq. (4), we have:

$$N_K = K\rho g(P_0 - A\sigma^n) \qquad (5)$$

Based on above equations, the radon concentration on the surface of rocks is not only direct proportional to the specific activity of ^{226}Ra in the rock, but also to the effective stress (σ) loaded on the rock. When the compressive stress loaded on a rock increases, the radon concentration on the rock will decreases in the way of n power function of σ.

Through the way of the free diffusion, the radon in air migrates along the distance that can be described as:

$$N_x = N_K e^{-\sqrt{\frac{\lambda}{D}}\cdot x} \qquad (6)$$

Where, x is the migration distance in air(cm); λ is the decay constant of ^{222}Rn ($2.1\times10^{-6}\cdot s^{-1}$); D is the diffusion coefficient at the normal temperature and pressure (cm^2/s) .

$$N_x = K\rho g(P_0 - A\sigma^n)\ e^{-\sqrt{\frac{\lambda}{D}}\cdot x} \qquad (7)$$

Taking differential form of Eq. (7), then we get the Eq. (8):

$$\frac{dN_x}{d\sigma} = -K\rho gnAe^{-\sqrt{\frac{\lambda}{D}}\cdot x}\sigma^{n-1} \qquad (8)$$

Where, the ratio of dN_x to $d\sigma$ is defined as the sensitivity of the radon concentration effected by stress (Ci/cm^{-3}Pa); minus indicates the negative correlation; if the density and specific activity of rocks are given, the sensitivity is only related to the stress loaded on the rock.

3 Experiment and Measurement of Radon

3.1 Specimen Preparation and Experimental Facility

Both granite and red sandstone rocks are specimend for the experiments of radon. The main mineral compositions of granite, specimend from Longmen Mountains Fault Zone where Wenchuan Ms 8.0 earthquake took place in 2008, are mica, quartz, feldspars and pyroxene; the red sandstone is composed of calcium-iron cementation and the grain sizes of minerals is from 0.25mm to 0.5mm. The main physical parameters of granite and red sandstone are listed in **Table 1**. The specific activity of the granite and red sandstone are determined by the Key Laboratory of Applied Nuclear Techniques in Geosciences, Sichuan, China. The design and cutting size of experimental specimens refer to the national standard 'The standard of the engineering rock mass test method' (GBT50266- 1999), and then specimens do the drying treatment for 24h.

Table 1 The main physical parameters of granite and sandstone specimens

specimens	Size/cm	Specific activity of Ra-226/Bq/kg	Specific activity of Th-232 Bq/kg	Compressive strength/ MPa
Granite	ϕ5×10	10.27	13.31	100
Sandstone	ϕ5×10	18.74	33.55	19

The Uniaxial stress experiment is accomplished in the State Key Laboratory of Geohazard Prevention and Geoenvironment Protection with triaxial creep testing machine. The pressure loaded on the rock specimen strictly maintains during the experiment. For the purpose, a double-deck sealing device without air convection was designed as shown in **Fig. 1**. The device is made up with the rock specimen, radon accumulating detectors and sealing cover which consists of an oblong-shaped rack and double thin plastic coating. The rock specimen is located in center and three radon detectors are parallel to each other 10cm away from the specimen. Meanwhile, three detectors are placed into another device without specimens to measure the background count [11].

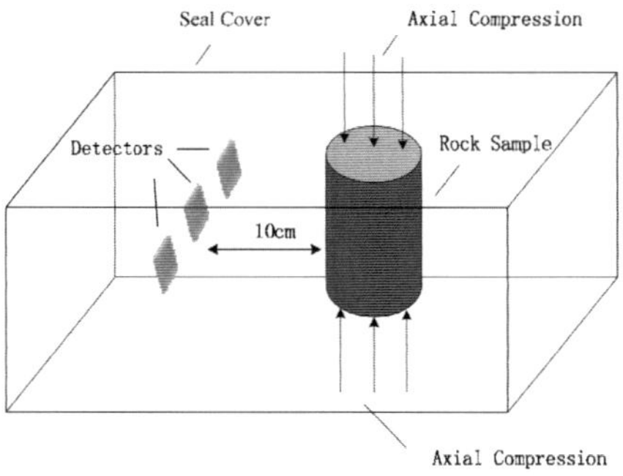

Fig.1 Schematic sketch of Radon measurement device

At first, a test of breaking, in which the pressure loaded on specimens is high enough to break the rock, has been done to determine the maximum uniaxial pressure. The results show that the maximum uniaxial pressures of granite and sandstone are about 100MPa and 19MPa, respectively. So, the proper gradient is set as: 30MPa、60MPa、90MPa for granite and 5MPa、10MPa、15MPa for red sandstone. Under the given uniaxial stress, the radon is accumulated for 10 days. According to the law of radon decay and accumulation, 10 days is about three times of half-life for ^{222}Rn and the accumulating amount of ^{222}Rn in 10 days is about 84% of total accumulation amount.

3.2 Detection and Calibration

CR-39 solid-state track detector is chosen as the radon accumulation detector. When a alpha particle emitted from radon and its daughters strike the CR-39 detector, a damaged latent track may be formed along alpha passed path. The latent track can expand to be observed and counted with optical microscope by means of chemical etching method. The solution composition of Chemical etching is: the amount-of-substance concentration of KOH solution of 6.5mol/L and 300 ml is need for every 10 detectors. Keep the detectors in the etching solution 10 hours at 70 degrees Celsius by water-bath heating. Then count the amount of the latent tracks on the materials through the 'α Image Acquisition System' developed by CDUT.

The relationship between track population on the detector and radon concentration can be expressed as:

$$N_x = \frac{L}{T \cdot F_R \cdot S} \quad (9)$$

Where, Nx is the air radon concentration on the point where the radon detector located (Bq/m^3); L is the number of the tracks; F_R is the radiation coefficient (cm^{-2}/Bq · h/m^3); T is the irradiation time (h); S is the effective-area of the detector (cm^2). As the irradiation time is given, the radon concentration is proportional to the track population. Thus, Eq. (9) becomes:

$$N_x = aI \qquad (10)$$

Where, a=1/($T{\cdot}F_R$) is defined as the calibration factor; $I=L/S$ is the track population. The calibration for CR-39 solid-state nuclear track detector is accomplished in the Radon Chamber in CDUT. The calibration factor is 0.0356 Bq/m^3 and the reproducibility is 2.9%.

4 Results and Discussion

Radon concentrations in the closed device under the different pressure are listed in **Table 2**. The first column in **Table 2** is the axial pressure loaded on the rock specimens for 10 days. The means of concentration on three detectors are listed in the second column. The radon concentrations, according to Eq. (9), are shown in the second column. **Fig. 2** represents the radon concentrations for granite and red sandstone. The results at 100MPa pressure for granite and at 19MPa pressure for red sandstone are obtained under the conditions of broken rock specimens.

The concentration decrease as the uniaxial stress increases. The concentration just 9.6 Bq · m^{-3} at 90MPa pressure loaded on granite specimens that is only 32% of the natural specimen at ambient pressure conditions. Obviously, it can be explained as radon is much more difficult to migrate outside from the pores and fractures of granite specimens that is narrowed or closed when uniaxial stress is loaded on the granite specimens.

The uniaxial stress experiments show a complex result (**Fig. 2** and **Table 2**). Firstly, when 5MPa pressure is loaded on red sandstone specimens, the radon concentration from the red sandstone is higher than that of the natural sample at ambient pressure conditions. It is because that sandstone consists of hard mineral grain and weak cement, so, it is easy to expand the pores and fractures when primary pressure loaded. Secondly, with the uniaxial stress increasing the concentration are decreasing. The concentration at 15MPa are just 60% of those at 5MPa.

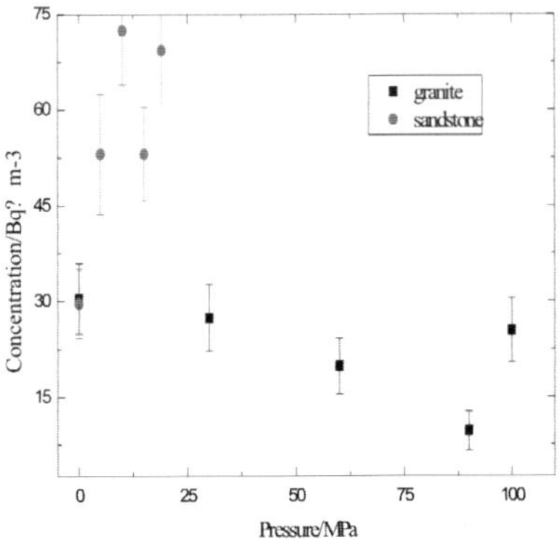

Fig.2 relationship between number of tracks and axial pressure

Table 2 Data showing the relationship between radon concentration and axial pressure

	Pressure /MPa	Concentration /Bq · m^{-3}
granite	0	30.4±5.5
	30	27.4±5.2
	60	19.8±4.4
	90	9.6±3.1
	100	25.4±5.0
sandstone	0	29.6±5.4
	5	88.6±9.4
	10	72.4±8.5
	15	53.1±7.3
	19	69.3±8.3

When the rock specimens are broken, such as, at 100MPa and at 19MPa in **Fig. 2**, the concentration of the radon increase. That is to say, the Eq. (3) will be no long applicable. Evidence to date indicates that there is a positive correlation between rupture degree and radon exhalation rate [12].

Table 3 is the normalized results of theoretical calculation and experiment. The theoretical data of radon concentration is based on Eq. (7). **Fig.3** is the normalized theoretical curves and the experimental scatter diagrams of the relationship between the radon concentration from rock specimens and uniaxial stress loaded on rock specimens. It shows a good agreement between theoretical and experimental results. The relative errors between them are 10.74% and -8.68% under the condition of 60MPa and 90MPa for granite while 6.7% and 10.37% under the condition of 10MPa and 15MPa for red sandstone. From **Fig.3** and **Table 3**, we also see that the concentration of the radon emanated from red sandstone is more sensitive to uniaxial stress than that from granite to the changes of axial pressure. Also, the radon concentration from red sandstone is 3 to 5 times than that from granite.

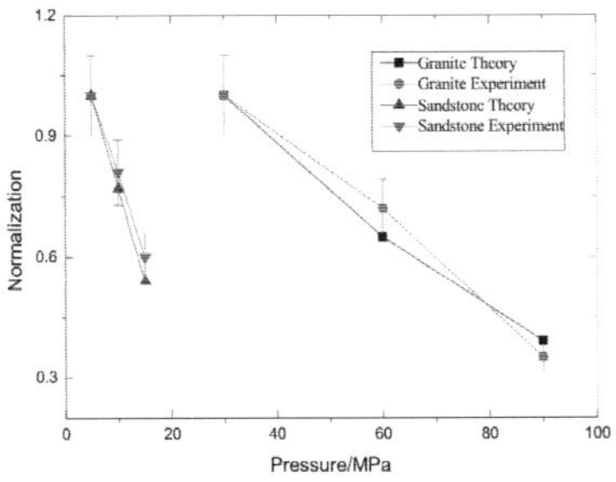

Fig.3 The relationship between radon concentration and axial pressure after normalization

Table 3 The contrast analysis of theoretical calculation and experimental results

Pressure /MPa	**Normalization**		**Relative error/%**
	Theory	Experiment	
30	1.00	1.00	/
60	0.65	0.72	10.74
90	0.39	0.35	- 8.68
5	1.00	1.00	/
10	0.77	0.81	6.70
15	0.54	0.60	10.37

5. Conclusions

Both the theoretic and experimental results indicate that there is a negative correlation between the concentration of radon emitted from rock and the uniaxial pressure loaded on the rock. The relative error of radon concentrations obtained from theoretic calculation and experimental results is less than 10.74%.

Acknowledgements

The authors gratefully acknowledge the financial support of: (1) National Natural Science Foundation of China through Grant No. 41074093 entitled ''Research on mechanism of earthquake radon anomaly'', (2) The National High - tech Research and Development Program of China through Grant 2012AA061803 entitled ''Development of high precision nuclear radiation spectrometry''.

References

1. Y Zhang, R Z Hua, B S Shi. *Radioactive Prospecting Methods* (BJ Atom Ener Pres, 1985)
2. Chengdu Geological Institute. *Radioactive Exploration Methods.* (BJ Atom Ener Pres, 1978)
3. Y Gu, L Q Ge, G X Wang, et al.Jour of Engi Geol. J. **17**, 296(2009)
4. A H Sultanhuojiaye. *The application of radioactive gas in the study of geological process* (BJ Seis Pres, 1981)
5. M Singh, R C Ramola, B R Singh, et al. Nucl Trac and Radi Meas. J.**19**,417(1991)
6. X H Han, G X Wang. Jour of Taiyuan Univ of Tech.J. **31**, 101 (2000)
7. L Chen, J L Xie, L Huang. Rad Pro Bul. J. **18**, 28(1998),
8. J Yan, L Q Ge, G J Zou, et al. Nucl Tech. J. **36**, 27 (2013)
9. J Geertsma. AIME. R. 331 (1957)
10. K W Vander. AIME. R. 179 (1959)
11. J K Zhao, L Q Ge, G J Zou, et al. Nucl Tech. J. **36**, 110404 (2012)
12. G W Luo, X Z Shi. Acta Seis Sini. J. **2**, 198(1980)

A Correlation Model of Mixed Service Qoe and Qos in Integration Network Environment

Shanji Chen[1], Weidong Fang[2], Guoqing Wu[1], Guoqing Jia[1]

[1]College of Physics and Electronic Information Engineering, Qinghai University for Nationalities, Xining, China
[2]School of Communication and Information Engineering, Shanghai University， Shanghai, China
My602dx@163.com, 173935311@qq.com, wgq_28@163.com, 13709759761@139.com

Abstract: With the rapid development of Internet and gradual enriched multimedia services, diversified needs of internet services are put forward. At the same time, people's requirements on quality of service are more stringent, traditional QoE only considers the evaluation method of network quality has not fully response service quality. At present, many research study association model of QoE and QoS only for single network and service. In this paper, a correlation model of mixed service integration network based on the environment is proposed.

Keywords- QoE;QoS; Path selection; mixed service; correlation model

1 Introduction

With the gradually mature of wireless communication technology and rapid development of digital multimedia technology such as network television (IPTV), video on demand (VOD), and the growing demand of dynamic services and the services based on IP, such as telemedicine, video surveillance and data center protection. But all these services need high quality, however, the current wireless communication technology and network is clearly unable to meet the growing demand of user for service and the rate, the importance of high speed broadband access technology is more and more outstanding. On the other hand, for service providers, is more concerned about the capacity of new system service. How to evaluate the service quality? All the time, it tends to use the QoS parameter to measure network traffic carrying capacity, such as single index of bandwidth, delay, packet loss. However, from the existing network can be found that high bandwidth does not represent the system providing high quality service, and for fusion network, many network damaging factors such as delay and jitter, can affect service performance, thereby affect service performance and quality of the whole network. However, each service sensitive index is different, For example, for voice service, sensitive index is time delay,

response service capacity accurately and comprehensively, On the other hand, from the user's perspective, the user cares not the number of packet loss, delay but weather the call is clear, image is a Mosaic and video playback is smooth. In this case, the evaluation method for user experience quality QoE emerges as the times require.

The user experience is not only universal impact of service, but also the operators in the market competition ability; It is an important factor of the successful transformation of theory research to practice. In addition, to explore QoE evaluation method of researched integration network can monitoring network real-time and accurately for the operator, improve the quality of user experience, apply the theory research to practice, realize the industrialization of new technology.

2 Technical analysis and the method

A. Background

With the rapid development of Internet, and gradual enriched multimedia services, diversified needs of Internet services are put forward. At the same time, people's requirements on quality of service are more stringent, traditional QoE only considers the evaluation method of network quality has not fully response service. The traditional QoS generally respond to the quality of the network through discrete indexes and parameters, such as packet loss, delay, jitter, bandwidth, etc. But these parameters are not separate, unrelated, but interaction. For example, when the loss exceeds a certain threshold, jitter will increase exponentially [1]. This QoS single indexes and parameter are not sufficient to evaluate network performance quality accurately, for some parameters of QoS, such as packet loss, if the lost packet is not critical, it will not affect the transmission of video, from the user's perspective, the user is not concerned with packet loss but has or not influent on video quality, So, service providers gradually shifted the interest to the user experience quality. In the face of this trend, the concept of QoE emerges as the times require.

Various organizations and literature has not a uniform definition of QoE. For example, in ITU-T P P.10/G.100, QoE is defined as terminal user's subjective overall satisfaction for used service [2]. It is closely related and end-users feeling experience, including the overall network end-to-end performance of the system. In literature [3], QoE is described as the user experience on the application layer, among them, application layer experience is similar to the overall results of the individual quality of service. The literature [4] indicated that QoE is used for total quality assurance Service mechanism to guarantee sound and smooth video transmission in IP network. In literature [5], QoE refers to the "performance" evaluation of the whole system from the user's point of view, QoE is from end to end performance levels to measure the user's experience, it implies the user satisfaction with the system. But the narrow sense QoS usually refers to the underlying data transmission performance indicators, including the packet loss rate, delay, delay jitter, bandwidth, error etc. QoS stands in the view of the network, and QoE is standing in the user angle to experience QoS. The all definition of QoE showed that QoE is standing on a higher level than the narrow QoS concept, if you want to get better QoE, you must provide a good QoS. Therefore, in a sense, QoE can be regarded as an extension and expansion of QoS narrow definition on terminal user perspective [6].

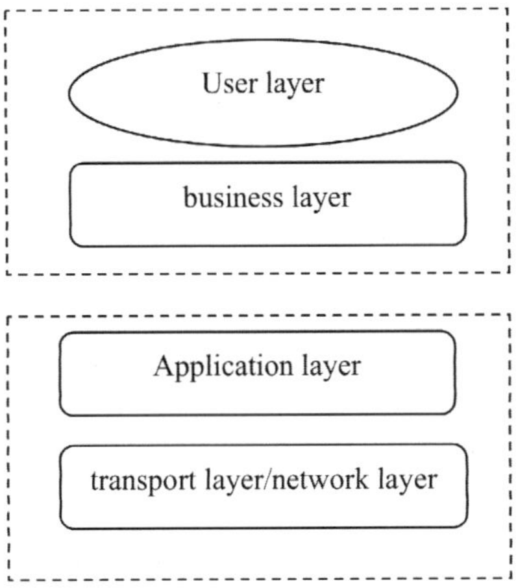

Figure 1. Hierarchical model of QoS and QoE

The relationship between QoE and QoS can be depicted as Figure 1, QoE is defined at the user level, the user is actually feel QoS, and QoS is closely related with application layer and the transport/network layer, from the user and the service level, QoS is a kind of demand, in order to meet this demand, the network layer adopts corresponding mechanism.

B. Research on related technology

In recent years, the international organizations and research institutions have an evaluation method on the network performance of QoE. ITU-T P.861 puts forward the method of PSQM (Perceptual Speech Quality Measure), British Telecom proposes PAMS (Perceptual Analysis Measurement System), PESQ and E-Model methods.

ITU-T initially definite QoE on IPTV service, ITU-T TP.10/G.100 defines QoE as "terminal user's overall subjective perception and satisfaction to their using service ". Among them, 12th study group set up a sub topic of "multimedia QoS/QoE performance evaluation method". ITU-T is currently working on a parametric model of video quality assessment, and has released the ITU G.1070 model; the model indicates the Five-point scale MoS value quality of experience. The QoE header field is added before some extended field of real-time stream media protocol In 3GPP R6 specification to ensure end-to-end negotiation of QoE parameters.

The influence factor of IPTV QoE is classified by ATIS-0800004 (Alliance for Telecommunications Industry Solutions), the QoE influence factors are divided into the objective evaluation and subjective evaluation, the objective QoE corresponds to the related factor of service provider, and it can be understood as IPTV QoS; and the subjective QoE corresponds to user related factors. The IPTV QoS parameter is divided into two layers: QoS of application layer and network layer. The application layer QoS represents the quality and performance of the application layer; network layer QoS represents network transmission quality. According to the characteristics of the application layer QoS it is further divided into three sub layers: the transmission quality, content quality, and multimedia stream quality.

Digital subscriber line forum (DSL Forum) divided IPTV QoE directly into three layers: service layer, application layer and the transport layer.

Index of service layer includes availability, reliability, and robustness.

The application layer includes the control level and the data level, the control level index includes channel change rate, scalability with certain load and responding time of user application interface. The data level includes bit rate, video codec quality of application layer, pretreatment, rate matching index.

The transport layer also comprises a control level and data level, the control level includes IGMP (The Internet Group Management Protocol) processing time, cross flow time, the data level indicators include delay, packet loss, jitter.

C. A correlation model of mixed service QoE and QoS in integration network environment

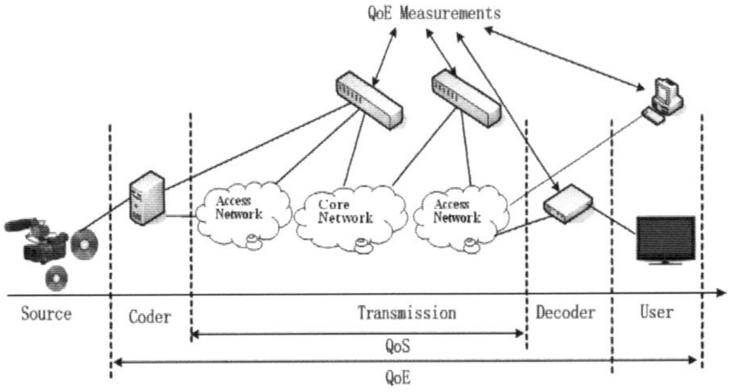

Figure 2. Application of association of QoE with QoS

At present, many research on association model of QoE and QoS only consider a single service, study a single network, the paper proposes a correlation model of mixed service based on the integration network environment, the specific method and described as follow(mixed service with IPTV and VoIP two kinds of service (representing the interaction class and conversational services)).

1. Gained the weights of each service

First of all, according to the network specified service priorities determine the weight of each service. Secondly, establish the hierarchical model of QoE index determine the transmission service in the network, as well as the sensitive different index for each service. Then establish the priority relation matrix X, according to the formula (1) for the line and get r_i, then use the formula (2), converted r_i into r_{ij}, and get fuzzy consistent matrix.

$$X = \begin{bmatrix} r_{11} & r_{12} & L & r_{1p} \\ r_{21} & r_{22} & L & r_{2p} \\ M & M & O & M \\ r_{n1} & r_{n2} & L & r_{np} \end{bmatrix}_{n \times p}$$

$$r_i = \sum_{k=1}^{n} r_k, i = 1, 2, ..., n \quad (1)$$

$$r_{ij} = \frac{r_i - r_j}{2n} + 0.5 \quad (2)$$

Plug r_{ij} into (3) get ω_i, then get the weight corresponding to each service sensitive index. Where $a = (n-1)/2$.

$$\omega_i = \frac{1}{n} - \frac{1}{2a} + \frac{\sum_{j=1}^{n} r_{ij}}{na}, i = 1, 2, ...n \quad (3)$$

For the convenience of calculation, the influence factor of network QoE only consider the IPTV, VoIP PLR, delay and jitter the three QoS parameters.

2. To obtain the correlation of single service from The IQX (exponential interdependency of quality of experience and quality of service) hypothesis.

$$QoE = a \cdot e^{-\beta \cdot QoS} + \gamma \quad (4)$$

The QoS index in the IQX hypothesis represents the packet loss rate, delay, and jitter. These single indexes represent the QoS damage, the damage is bigger, and the user satisfaction is lower. QoS and QoE is Negative exponential relationship.

3. The mathematical relationship between IQX and QoE

To the IPTV service, the mathematical model of relationship between PLR, delay, jitter and QoE is shown as:

$$QoE_{IPTV}(PLR) = a_1 \cdot e^{-\beta_1 \cdot PLR} + \gamma_1$$

$$QoE_{IPTV}(delay) = a_2 \cdot e^{-\beta_2 \cdot delay} + \gamma_2$$

$$QoE_{IPTV}(jitter) = a_3 \cdot e^{-\beta_3 \cdot jitter} + \gamma_3$$

To the VOIP service, the mathematical model of relationship between PLR, delay, jitter and QoE is shown as:

$$QoE_{VOIP}(PLR) = a_4 \cdot e^{-\beta_4 \cdot PLR} + \gamma_4$$

$$QoE_{VOIP}(delay) = a_5 \cdot e^{-\beta_5 \cdot PLR} + \gamma_5$$

$$QoE_{VOIP}(jitter) = a_6 \cdot e^{-\beta_6 \cdot jitter} + \gamma_6$$

4. By (3) to obtain the weight ω_i of each service sensitive index, calculated for each service QoE_i:

$$QoE_{IPTV} = \omega_1 QoE_{IPTV}(PLR) + \omega_2 QoE_{IPTV}(delay) + \omega_3 QoE_{IPTV}(jitter) \quad (5)$$

$$QoE_{VOIP} = \omega_1 QoE_{VOIP}(PLR) + \omega_2 QoE_{VOIP}(delay) + \omega_3 QoE_{VOIP}(jitter) \quad (6)$$

5. To obtain Service weight by network service priority (normalized according to the standard) and gain ω_{IPTV} and ω_{VOIP}, then the QoE of the whole system.

Plug (5), (6) into (7),

$$QoE = \sum_{i=1}^{n} QoE_i \cdot \omega_i \quad (7)$$

For the fusion network, QoE is shown as:

$$QoE = \omega_{IPTV} * QoE_{IPTV} + \omega_{VoIP} * QoE_{VoIP} \quad (7)$$

3 Conclusion

This paper for hybrid service transmission in IMT-A heterogeneous network fusion environment proposes a association model of QoE and QoS, the method is in the application of the original single service and single network, introduce scenarios of mixed service and heterogeneous network, and for each parameters in the related model, explains the method of obtaining.

Acknowledgment

This work was supported by the project of Qinghai sc ience and Technology Department(2015-ZJ-721),Educa tion Ministry Chunhui Plan (Z2012106, Z2011024, Z20

11017, Z2012107, Z2012105, Z2012112, Z2012104), International Science & Technology Cooperation Project of Qinghai (2013-H-811), Tianjin University - Qinghai University of Nationalities Cooperation Innovation Fund.

References

1. O. Issa, W. Li, H. Liu. Performance Evaluation of TV Over Broadband Wireless Access Networks[J]. IEEE Transactions on Broadcasting,2010,56(2):202-210..
2. ITU-T P.10/G.100.Vocabulary for performance and quality of service[S], July 2006.
3. M.Siller and J. Woods. Improving quality experience for multimedia services by QoS arbitration on a QoE framework[C]. Proceedings of the 13th Packed Video Workshop 2003,France, 2003.
4. T. M. O' Neil. Quality of Experience and Quality of Service for IP Video Conferencing. Polycom, Whitepaper, 2002.
5. J. Goodchild. Integrating data, voice and video Implementation and planning guide[R]. United States Telecom Association, 2005.
6. D. Lopez, F. Gonzalez, L. Bellido, A. Alonso. Adaptive multimedia streaming over IP based on customer oriented metrics[C]. Proceedings of the Seventh IEEE International Symposium on Computer Networks,2006: 185-191.

A hybrid approach to calculate the Shielding Failure-Caused Trip-out Rate

Liang Zhou[1], Chengli Liu[1], Ying Wang[2,a], Jinghua Yang[3]

[1]*Sichuan Electric Power Design & Consulting Co. LTD, 610041 Chengdu, China*
[2]*College of Electrical Engineering and Information Technology, Sichuan University, 610065, Chengdu, China*
[3]*Kunming Power supply Bureau, Yunnan Power Grid Co. LTD, 650011, Kunming, China*

Abstract. Lightning has become a big threat to the safe operation of the main transmission line. Reasonable and accurate calculation of shielding failure rate plays important role in transmission line and tower design. This paper proposes a hybrid approach to calculate the shielding failure-caused trip-out rate, based on the typical electro-geometric model and the regulation method. The case study prove the validity and correctness of this approach, by comparing with the actual operation shielding failure rate.

1 Introduction

The transmission line goes through the different landscapes and is exposed to nature in a long-term, it is therefore influenced and damaged by the natural phenomena easily. Lightning has become a big threat to the safe operation of the main transmission line.

Lightning shielding failure means lightning strikes on the transmission line, bypassing the lightning conductor. It is the main factor to threat the safety and reliability of EHV transmission line. The typical method to calculate the shielding failure-caused trip-out rate (SFTR) includes: regular method [1-2], electrogeometric model (EGM) [3-4], leader progression model (LPM) [5-6], and so on. Regular method is sample, but only considering the protection angle and height of tower. EMG provides a rational basis for taking account of the influence of structure height and is essentially independent of any assumptions regarding attractive distances to ground. The shortcomings of EGM is that the arc over rate and some other related the electrical characteristics are regardless. LPM is a model of lightning channel progression towards the earth is given based on the physics of discharge on long air gaps. The physical assessment of the electric field is done with the lightning lead progression, however, it does not consider the randomness of the lightning lead progression, the modelling results doesn't actually describe the actual lightning channel.

This paper proposes a new approach to calculate SFTR, based EGM and considering the electrical characteristics. Using the proposed approach to calculate a 500kV transmission line SFTR, compared with SFTR by EGM and the actual trip rate in operation, prove the effectiveness of the proposed approach.

2 SFTR Calculation

2.1 SFTR

Based on the Chinese regulation, SFTR is the number of trips which caused by shielding in per hundred kilometers, Nsf is SFTR in Eq. 1.

$$N_{sf} = N_s \eta P_\alpha P_2 \tag{1}$$

where N_s is the number of lightning per year per hundred kilometer, η is arc over rate, P_α is the probability of shielding strike flashover, P_2 is the probability of shielding lightning withstand level.

2.2 Calculation of N_s

Based on N_g is the number of lightning per year per square kilometer, and it is determined by thunderstorm day T_d, according to IEEE Std 1243-1997 [1].

$$N_g = 0.04T_d^{1.25} \tag{2}$$

N_s is determined by Eq. 3 as follows.

$$N_s = N_g\left((28h^{0.6} + b)/10\right) \tag{3}$$

where h is the height of tower, and b is the distance between the two ground wire.

2.3 Calculation of η

Trip is determined by arc over rate, after the insulator flashover. The probability from the impulse flashover into power frequency arc is arc over rate η, which is described in Eq.4.

$$\eta = [4.5(\frac{U_n}{\sqrt{3}L_d})^{0.75} - 14] \times 10^{-2} \tag{4}$$

U_n is the rated voltage, kV; L_d is the insulation distance of the insulator string, m.

[a] Corresponding author: 769429505@qq.com

2.4 Calculation of P_2

The different insulator flashover voltage $U_{50\%}$ is determined by the different type and number of insulator. For the different project, $U_{50\%}$ is different. The strike lightning withstand level is in Eq.5.

$$I_2 = U_{50\%} / 100 \quad (5)$$

From the strike lightning withstand level, the probability of shielding lightning withstand level is P_2 as follows.

$$p_2 = 10^{-\frac{I_2}{88}} \quad (6)$$

2.5 Calculation of P_a

EGM proposed in [4] used here to calculate the probability of shielding strike flashover P_a. The diagram of electrical geometry model is shown in Fig.1. D is the hanging point of ground line, E is hanging point of conductor, A is the point of intersection of the circle, whose center is D, and the vertical line from the tower center to horizontal. B is the intersection of the two circles, center is D and center is E, respectively. C is the intersection of the circle, whose center is E, and the straight line of the ground attractive distance.

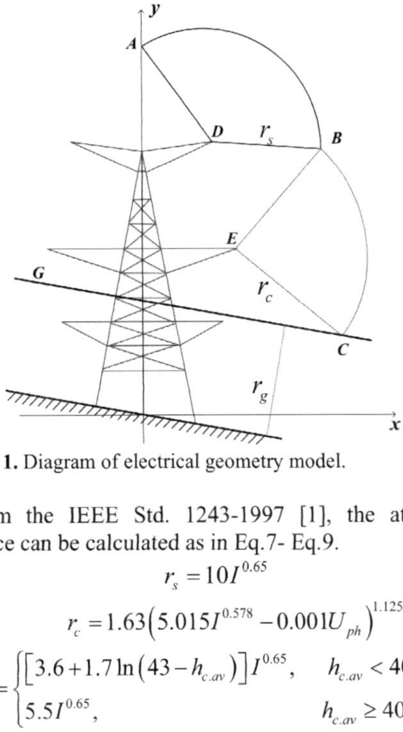

Figure 1. Diagram of electrical geometry model.

From the IEEE Std. 1243-1997 [1], the attractive distance can be calculated as in Eq.7- Eq.9.

$$r_s = 10 I^{0.65} \quad (7)$$

$$r_c = 1.63\left(5.015 I^{0.578} - 0.001 U_{ph}\right)^{1.125} \quad (8)$$

$$r_g = \begin{cases} \left[3.6 + 1.7\ln\left(43 - h_{c.av}\right)\right] I^{0.65}, & h_{c.av} < 40m \\ 5.5 I^{0.65}, & h_{c.av} \geq 40m \end{cases} \quad (9)$$

where I is lightning current, r_s is the attractive distance from lightning to ground line; r_c is the attractive distance from lightning to conductor; r_g is the attractive distance from lightning to ground; U_{ph} is the instantaneous voltage when lightning striking; $h_{c.av}$ is the average height of the phase conductor, which can be obtain in Eq. 10.

$$h_{c.av} = h - \frac{2}{3} S_d \quad (10)$$

where S_d is sag of conductor.

In Figure.1, the striking point is on the ground line, if the lightning leader progression reaches Arc AB first. The striking point is on the conductor line, if the lightning leader progression reaches Arc BC first. The thundercloud discharge to earth, if the lightning leader progression reaches Straight CG first. Arc AB is protection arc, and Arc BC is exposure arc.

When angle of lightning progression is 0°, lightning current perpendicular to x axis. In Fig. 1, $y_B < y_D$, and the projection of protection surface on x axis is longer than $(x_B - x_A)$, as shown in Eq. 11 and Figure.2.

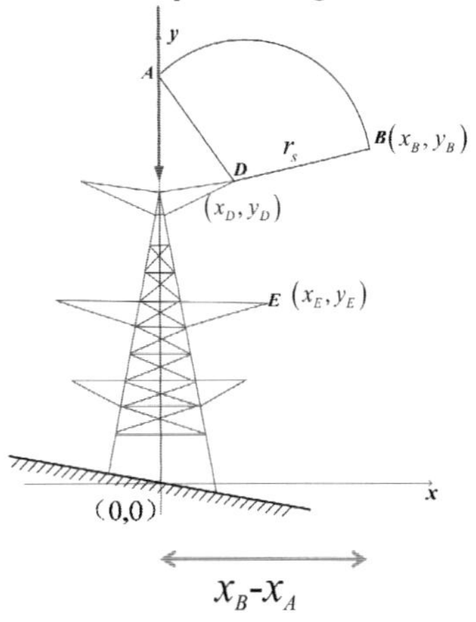

(a)The protection arc when $y_B \geq y_D$.

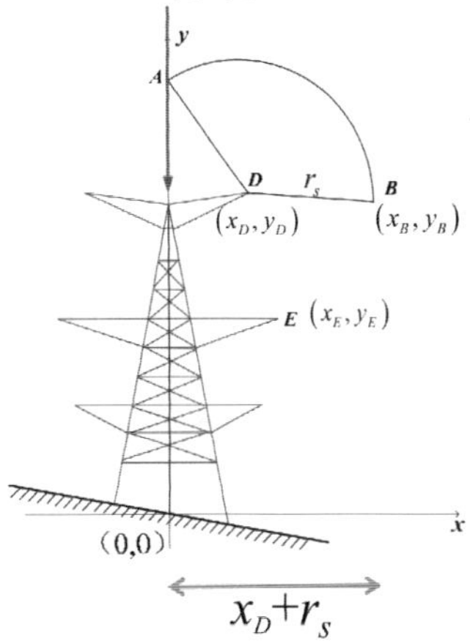

(b)The protection arc when $y_B < y_D$.

Figure 2. Diagram of the protection arc.

$$l_{AB} = \begin{cases} x_B - x_A, & y_B \geq y_D \\ x_D + r_s, & y_B < y_D \end{cases} \quad (11)$$

Coordinate of point C can be obtained by solving equations of straight CG and the circle of center E.

$$y = kx + b \quad (12)$$

where $k = \tan\theta_g, b = r_g / \cos\theta_g, \theta_g$ is ground inclination.

The circle of center E can be described in Eq. 13:

$$(x - x_E)^2 + (y - y_E)^2 = r_c^2 \quad (13)$$

Simultaneous circular and linear equation, get the coordinate of C.

$$\begin{cases} x_C = (-b_1 + \sqrt{b_1^2 - 4ac}) / (2a) \\ y_C = kx_C + b \end{cases} \quad (14)$$

a, b1 and c can be calculated by the Eq. 15-17.

$$a = 1 + k^2 \quad (15)$$

$$b_1 = -2x_E + 2k(b - y_E) \quad (16)$$

$$c = x_E^2 + (b - y_E)^2 - r_c^2 \quad (17)$$

From Figure.1, because of $y_C < y_E$, the projection of Arc *BC* on *x* axis is longer than $(x_C - x_B)$, l_{BC}, the projection of Arc *BC* on *x* axis, is calculated as in (18). The diagram of exposure arc is shown as in Figure.3.

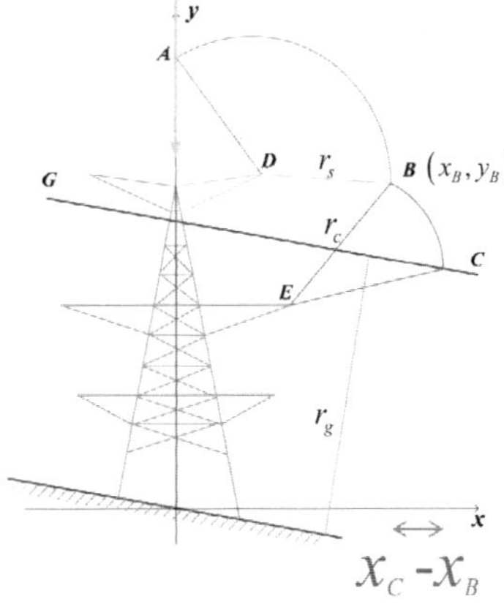

(a)The protection arc when $y_C \geq y_E$.

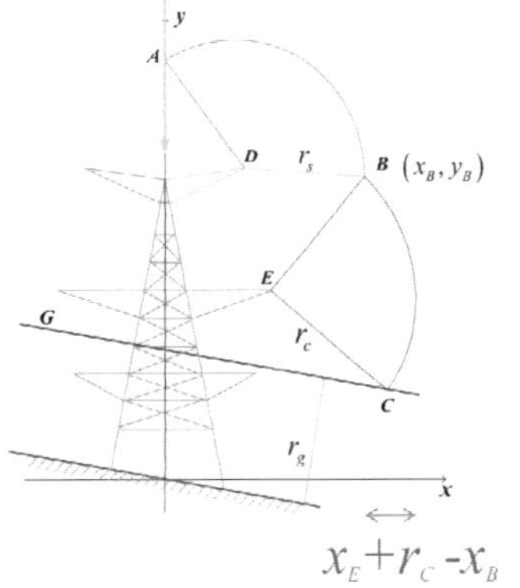

(b)The protection arc when $y_C < y_E$.

Figure 3. Diagram of the exposure arc.

$$l_{BC} = \begin{cases} x_C - x_B, & y_C \geq y_E \\ x_E + r_c - x_B, & y_C < y_E \end{cases} \quad (18)$$

P_α is the probability of shielding strike flashover. When angle of lightning progression is 0°, P_α is as follows.

$$p_\alpha(I) = l_{BC} / (l_{AB} + l_{BC}) \quad (19)$$

From the calculation result of N_s, η, P_α and P_2, N_{sf} in Eq. 1 can be obtained.

3 Case Study

A 500kV transmission line in operation in the Southwest China is as an example here, there are a total of 340 towers in the transmission line. The example here is a selection of the tower number is #N227 to #N340 , SFTR of these tower is calculated by the proposed calculation method. 50% of these towers are located in the plateau, which is higher than 3000 meters. The average thunderstorm days T_d of the transmission line are up to 80 days. The mountain slopes of each tower is obtained from the GoogleEarth software, and assumed the positive slope is at the right side of the tower, the negative slope is at the left side of the tower. Insulator flashover voltage $U_{50\%}$ is 2520kV.

The calculation result of SFTR is shown in Figure.4 and Figure.5. The Figure.4 is the SFTR of the right side of towers, which is mentioned above that assumed the positive slope is at the right side of the tower. The Figure.5 is the SFTR of the left side of towers. The maximum SFTR of two sides of one tower is the SFTR for this tower, which is shown in Figure. 6. The largest SFTR for the towers in the example is 0.1778 times/100km per year, the calculation result is close to the statistics operating SFTR in last two years, which proves the validity and correctness of this approach.

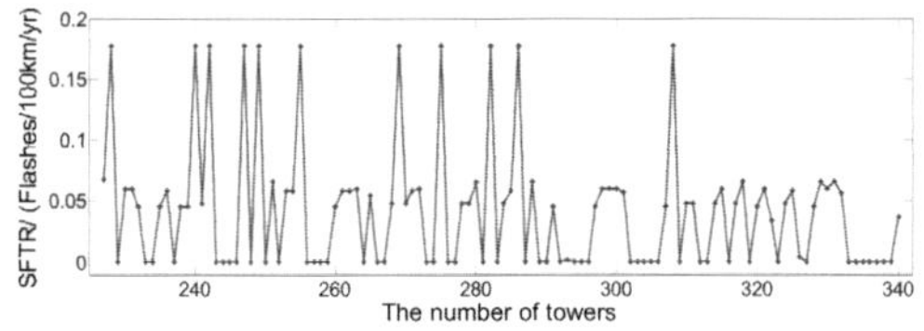

Figure 4. SFTR of the right side of towers.

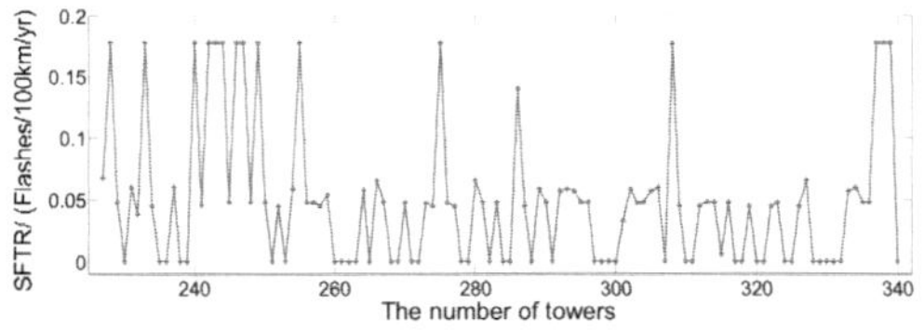

Figure 5. SFTR of the left side of towers.

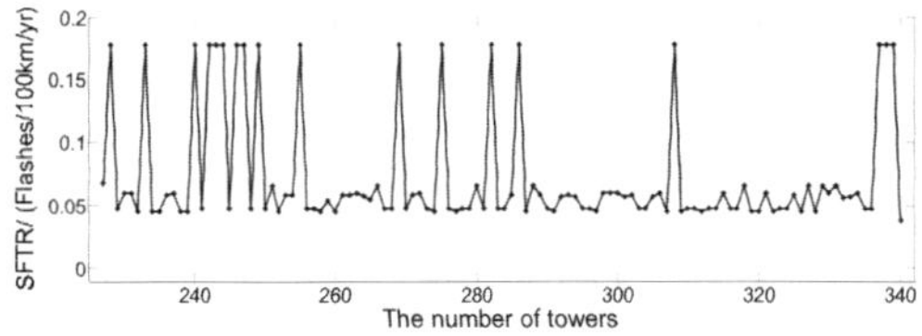

Figure 6. SFTR of the towers.

4 Conclusion

Transmission line fails by lightning has been a big threat to safe operation in power system. Shielding failure is most lightning strike type in all of the world. Reasonable and accurate calculation of shielding failure rate plays important role in transmission line and tower design. This paper proposes a hybrid approach to calculate the shielding failure-caused trip-out rate, based on the typical electro-geometric model and the regulation method. The case study prove the validity and correctness of this approach, by comparing with the actual operation shielding failure rate.

References

1. IEEE international standard, IEEE Std 1243-1997, IEEE guide for improving the lightning performance of transmission lines.
2. DL/T 620-1997, Overvoltage protection and insulation coordination for AC electrical installations.
3. Li Ruifang, Wu Guangning, Cao Xiaobin, et al. : Improvement of electro-geometric model with considering lightning strike angle. Insulators and Surge Arresters. (2009), p. 23-26.
4. Peng Qian, Li Jun, Bian Peng, Et al. : Application of Improved Electrical Geometry Model in Calculation of Lightning Shielding Failure-Caused Trip-out Rate of 1000 kV Power Transmission line. Power System Technology. Vol. 34 (2010), p. 155-159.
5. Dellear.L, Garbagnati.E : Lightning stroke simulation by means of the leader progression model. I. Description of the model and evaluation of exposure of free-standing structures. IEEE Transactions on Power Delivery. Vol. 5 (1990), p. 2009-2022.
6. Rizk, Farouk A.M. : Modelling of transmission line exposure to direct lightning strokes. IEEE Transactions on Power Delivery. Vol. 5 (1990), p. 1983-1997.

A Novel Crawling In-pipe Robot Design

Xu Kejie[1,a], Xu Hong[2], Yu Qikuan[1] , Wang Zhiqiang[2], Xu Wenjiao[1]

[1]College of Engineering, Zhejiang Normal University, Jinhua, Zhejiang 321004, China,

[2]Xingzhi College, Zhejiang Normal University, Jinhua, Zhejiang 321004, China

[a]m18395969159_1@163.com

Abstract: A novel crawling in-pipe robot was designed to improve the adaptive ability with self-locking principle applied. In the moving, the robot was adaptable to horizontal, vertical and bending pipes with different diameters and sections without exerting additional pressure to the shell. The robot used a telescopic umbrella stand as a basic structure. The force of telescopic mechanism, self-locking principle of supporting mechani sm and shell were deduced from design requirements. The design of supporting and coupling mechanism was completed. The relationship between screw rod thrust and parallelogram driving force was established. A prototype was made and tested according to the suggested design methods. The results show that the robot with powerful traction, good self-locking performance and smooth passage through curve can move forward in the pipe with 90~150 mm inner diameter.

Keywords: robot; pipe; crawling; self-locking; screw-driven

1 Introduction

In recent years, with the development of pipeline transportation of petroleum, natural gas and other fluids, it was urgent and important to conduct regular inspection and maintenance[1]. Especially on August 1st, 2014, the explosion of gas pipeline in Kaohsiung, Taiwan reminds people to give sufficient weight to installation and maintenance of pipelines. The limitations of pipe restrict man's active operability. As an effective testing tool carrier, a micro in-pipe robot can go deep into confined space while many common testing tools can't do so. Nowadays, various in-pipe robots as the main testing carriers have emerged in response to the solution to the industrial problem of pipe inspection[2-5]. Due to complicated environment yet limited space, in-pipe robot should adopt simple unit structure and self-adaptability[6].

For successful march, a robot should have: 1.the ability to adaptable changes in the pipeline; 2. sufficient traction;3. miniaturized dynamical system and transmission mechanism. In terms of ways of moving in the pipeline, there are different types of in-pipe robots such as wheeled robot, crawler-type robot, hydraulic robot, vibrating robot and bionic crawling robot, etc. The vibrating robot was used for pipe with rigid shell; the wheeled drive was a frequently-used mobile carrier by in-pipe robot. the wheeled robot has advantages of simple structure, fast speed, and powerful traction; while the crawling robot with compact structure can be miniaturized, it was widely applied in minor-caliber pipe testing[7]. Based on force condition, the maximum tractive effort from crawling robot was equal to the maximum static friction force of moving mechanism and shell. To increase system traction, it was necessary to raise pressure between system and shell of pipe; On the other hand, with increasing pressure, there was a growing demand for dynamical system and moving speed will be restricted. So the crucial part of design was to deal with the contradiction between system traction and positive pressure[8-9]. The study attaches importance to take advantage of friction against pipe to drive robot.

In the passage, a telescopic in-pipe robot based on self-locking mechanism was adaptable to horizontal, vertical and bending pipes with different diameters and sections. According to mechanics principles, the force of telescopic mechanism, self-locking principle of supporting mechanism and shell are deduced to develop telescopic in-pipe robot prototype. Furthermore, related tests are completed successfully.

2 In-pipe robot structures

Figure 1 shows the basic structure for in-pipe robot.

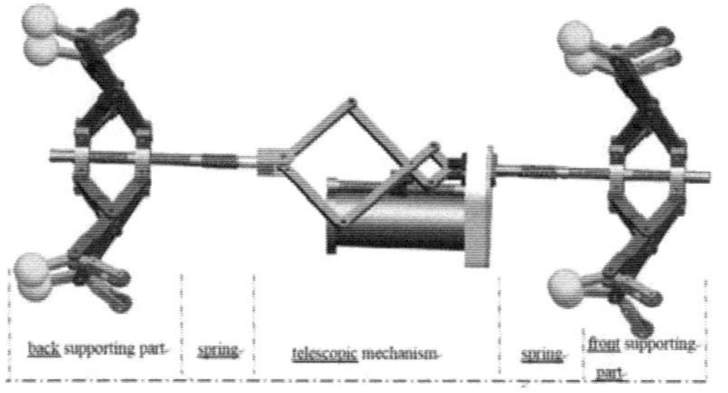

Fig.1 the in-pipe robot overall structure

It can be seen from Fig.1 that the robot was composed with three parts, including front, back supporting structures and central telescopic structure. The front and back supporting structures use umbrella stand to support pipc shcll and movcmcnt for robot; central telescopic structure adopts parallelogram structure to stretch out and draw back driven by motor and screw rod to move ahead. The soft connection by spring among three parts makes robot to meet the need of turning and restoring to its original state after turning for better movement.

Cycle of robot moving ahead was shown in Figure 2.

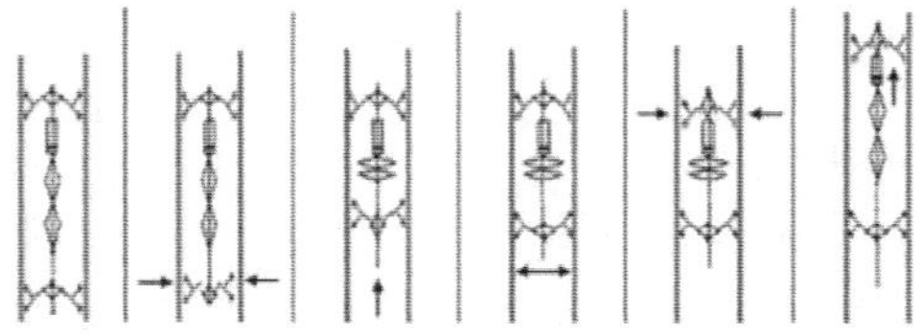

Fig.2 schematic drawing principle of in-pipe robot movement

It can be seen from Fig.2 that after central parallelogram starting to stretch out, the robot need finish the following 6 actions in one cycle: contracting back supporting, parallelogram, stretching out back supporting, front supporting, stretching out parallelogram and front supporting. It should be noted that the needed movement range of front, back stretching out and drawing back was different from that in Fig.2. The purpose of contraction was to reduce friction between support and pipe shell.

3 Parameter design of in-pipe robot

3.1 Supporting structure design.

The main requirements of supporting structure design were: ① with enough friction between supporting structure and pipe shell to offer driving force to move forward when parallel mechanism draws back;②with manual and automatic adjustment for pipes with different sizes and sections; ③with guide mechanism to pass corner led by pipe shell[10]. To meet the above requirements, the supporting structure was designed as shown in Figure 3.

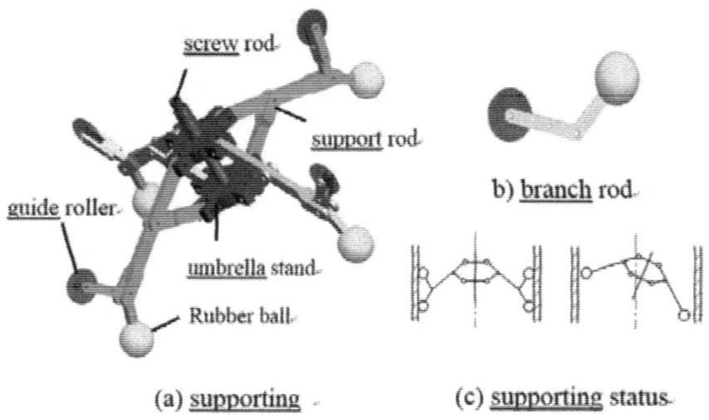

Fig.3 the in-pipe robot supporting structure

The fundamental supporting form in Fig.3(a): it adopts umbrella stand as overall structure and two umbrella stand bearings up and down are connected with

DOI: 10.1051/matecconf/20166104017

screw rod. At the top of umbrella rod there was a branch rod that can be turned as shown in Fig.3(b); each supporting rod has two supporting points to keep the supported surface perpendicular to pipe axes in case of falling down as shown in Fig.3(c). The aluminium truckle at the front serves as guide wheel to keep pure rolling with pipe shell; there was a rubber ball at the back to increase frication with pipe shell.

Supporting mechanism was provided with self-locking function from structural design. Its self-locking principle was shown in figure 4. The direction shown in Fig.4(a) indicates that mechanism can move upward easily while downward movement was not allowed. AC and intermediate shaft connected with slider can be moved up and down; EF and intermediate shaft are connected with helical pair. The relative location of EF and intermediate shaft can be changed manually. Because of spring, when diameter was changed, EF rod was fixed and B,D points bear pressure outward or inward. And AB,CD separately revolve about central points with relevance in order to shorten or extend the distance between AC and EF. The force from spring may support umbrella stand to stretch out or draw back until supporting ball and supporting roller come into contact with inwall of pipe. The self-adaptable diameter of section ranges from 90 mm to 150 mm as shown in Fig.5. The principle of downward movement can be seen from Fig.4(b).

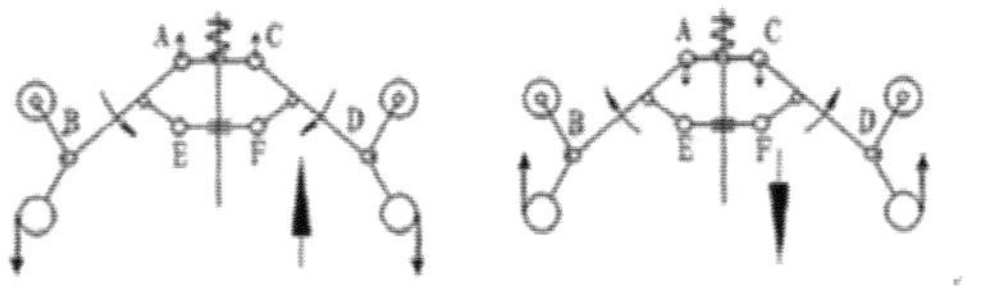

(a) supporting structure moves up (b) supporting structure moves down

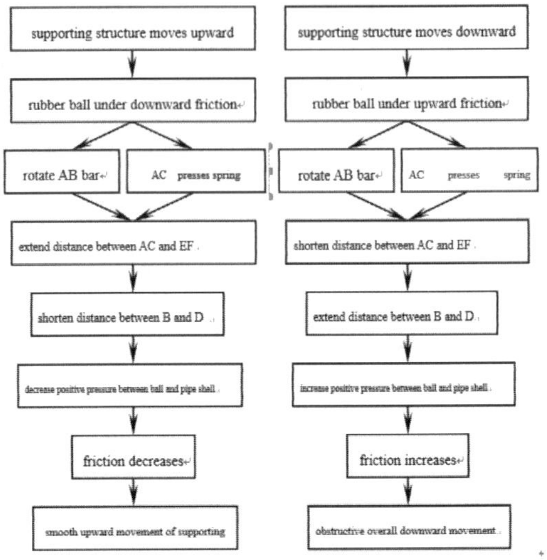

(c) motion principle of supporting structure moves up (d) motion principle of supporting structure moves down

Fig.4 Self-locking principle of supporting mechanism

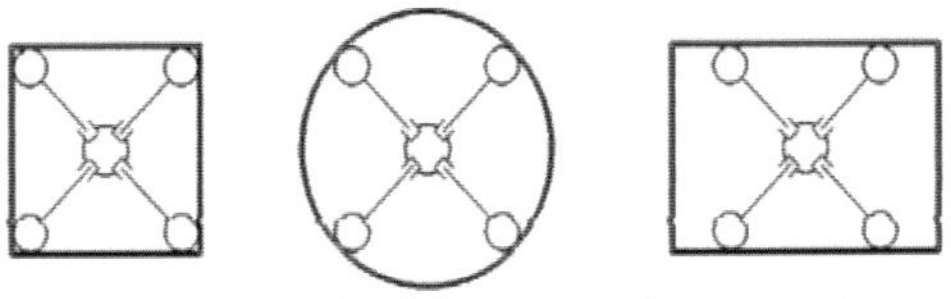

(a) square (b) round (c) rectangle

Fig.5 Self-adaptability of supporting mechanism to section

3.2 Coupling mechanism design.

The basic mode of robot was set up in the above-mentioned structural design. The crucial part were composed with connection between back, front supporting and telescopic structures. There are two couplings between supporting structure and telescopic structure as supporting bands. Taken easy turning at corner into consideration, turnning center should be placed near the supporting structure, so large degree of freedom was set at front, back supporting structures, while telescopic structure adopts common hinge to couple.

Due to turning towards uncertain direction in the movement, coupling methods should be provided with large degree of freedom. Thus, front, back supporting structures and telescopic structure are connected with spring. When the robot goes through the bend, it can return to intermediate state to make better movement.

Figure 6 show the coupling between telescopic structure and back supporting.

 DOI: 10.1051/matecconf/20166104017

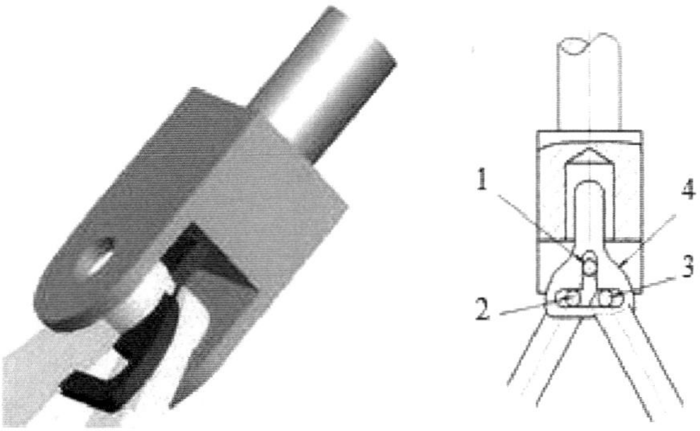

(a)coupling between back supporting and telescopic mechanism (b)T-shaped sliding block mechanism

Fig. 6 coupling methods of back supporting and telescopic mechanism

Figure 6 shows the coupling requires free rotation of relative parallelogram pin. On the other hand, the connected cylinder in Fig.6(a), the better use of restoration function, cylinder axes and diagonal of parallelogram based on the back supporting structure by spring should be coincided. So the degree of freedom should be limited to ensure free movement of parallelogram and that axes of spring and diagonal of parallelogram always coincide.

To implement the control of degrees of freedom, T-shaped sliding block structure was chosen as shown in Fig.6(b). Axis 1 works as rotary shaft for bar 2 and pin 2, 3 are separately installed in 2 bars of telescopic structure. If the bar turns around, hinge pin 1,2,3 can only move in the "T-shaped" slot of sliding block 4. The slot shape makes axis of sliding block 4 and diagonal of parallelogram always coincided.

4 Design calculation

4.1 Relationship between screw rod thrust and parallelogram impetus.

Figure 7 show the inequilateral parallelogram mechanism to shorten mechanism length effectively. The structure not only ensures magnifying power of movement but also improves transmission efficiency.

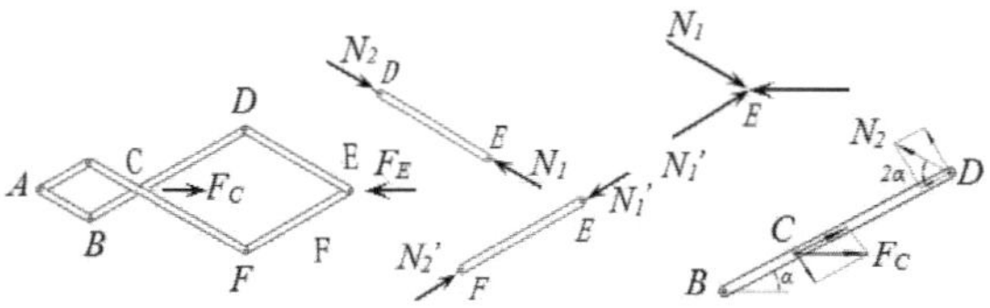

Fig.7 force analysis of telescopic mechanism

1) DE,EF both are two force bars. The force status was shown in Fig. 7, $N_1 = N_2$, $N_1' = N_2'$.

2) For E point, due to all symmetrical bars so here $N_1 = N_1'$. For any angle α, here

$$F_E = 2N_1 \cos\alpha \tag{1}$$

2) For BD bar, based on the banlance of moment of force, there

$$F_C L_{BC} \sin\alpha = N_2 L_{BD} \sin 2\alpha \tag{2}$$

$$F_C L_{BC} = 2N_2 L_{BD} \cos\alpha \tag{3}$$

$$N_2 = F_C \frac{L_{BC}}{2L_{BD}\cos\alpha} \tag{4}$$

substitute formula (2) into formula (1) to get:

$$F_E = F_C \frac{L_{BC}}{L_{BD}} \tag{5}$$

From formula (5), it could be found that the application of force from quadrangle agsint supporting mechanism, proportional to scew rod thrust, not depend on BD bar or the parallelogram angle.

4.2 Relationship between motor power and robot impetus FE.

Figure 8 show the transmission from motor to screw rod according to the transmission mechanism that has been worked out.

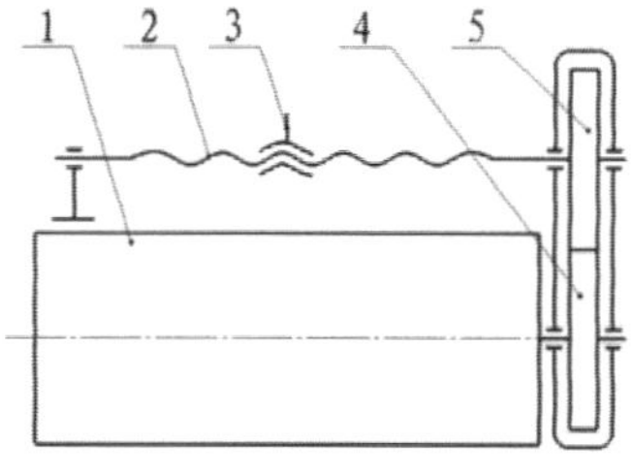

1-motor 2-screw rod 3-nut 4,5 gear bank

Fig.8 motor transmission diagram

In Figure 8, motor 1 transmits driving force to gear 4 by built-inspeed reducer. And force was transmitted to screw rod 2 by gear set 4-5. Screw rod 2 drives nut 3 to return rotary motion to rectilinear motion. By pin, nut 3 puts parallelogram structure into motion to drive robot.

If output power of motor was P, while rotative speed was n.

1)Output power of screw rod P_0 was

$$P_0 = P \cdot \eta = 0.9P \tag{6}$$

2) M6 screw rod and triangle screw thread are adopted and equivalent frictional coefficient fv was

$$f_v = f / \sin(90° - \beta) = f / \cos\beta \tag{7}$$

In the formula, β represents half of thread angle of common triangle screw thread.

Equivalent friction angle was

$$\varphi_v = \arctan f_v \tag{8}$$

3) Pressured by external load FC, the required maximum torque of screw rod rotation was

$$T = F_C \cdot d_2 \cdot \tan(\alpha + \varphi_v) / 2 \tag{9}$$

and in the formula, α represents helix angle of screw thread.

4)The relationshop between torque and input power of screw rod was

$$T = 9.55 \times 10^6 P_0 / \cdot n \tag{10}$$

According to formulas (5)-(6), (9), the relatonship between motor power and suffered thrust by robot can be deduced as

$$F_E = \frac{L_{BC}}{L_{BD}} \cdot \frac{2 \times 9550 \times 10^3 \cdot \eta \cdot P}{d_2 \tan(\alpha + \varphi_v) \cdot n} \tag{11}$$

The effeiency of spur cylindrical gear 4, 5 was 0.9; pitch diameter of screw rod d2=5.355 mm; helix angle $\alpha = 3°24'17''$; half of thread angle $\beta = 30°$; assign f=0.1; in parallelogram mechanism, LBD=15 mm,LBC=63 mm; output power of motor was P=2.4 W, rotative speed n=420 r·min-1, substitute them into above formula to get

$$F_E = \frac{L_{BC}}{L_{BD}} \cdot \frac{2 \times 9550 \times 10^3 \cdot \eta \cdot P}{d_2 \tan(\alpha + \varphi_v) \cdot n} = 51.97(N)$$

5 Experimental verification

The relative in-pipe robot prototype was made according to the above parameters from theory and caculation. It weighs only about 0.4 kg with the maximum load of 17. 4 N and it was adaptable to 90~150 mm inner diameter. Figure 9 shows test of in-pipe robot.

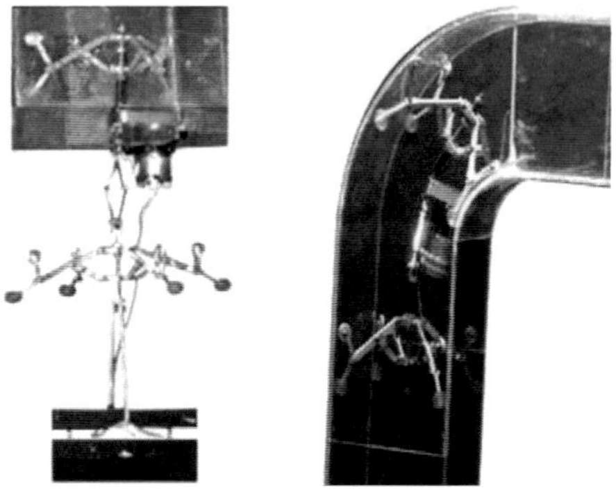

a)load self-locking test b)turning test

Fig.9 in-pipe robot test

The load test of Figure 9(a) with an iron block of 1.5 kg to show good self-locking performance of prototype. Figure 9(b) shows ability test to pass throught pipe. Taking control of screw rod, trial prototype moves back and forth four times between two ends of 2-meter straight pipe. The mesured average speed in the straight pipe reaches 2.7 m·min^{-1}.

To test motion characteristics of in-pipe robot, it requires to test electric current (I) and voltage(U) under different load and operating conditions. The results are shown in Table 1.

Table 1 power test of in-pipe robot under different load conditions

Load G[kg]	Voltage U[V]	Horizontal linear motion		Curve-way motion		Vertical upward	
		I[A]	P[w]	I[A]	P[w]	I[A]	P[w]
0	12	0.07	0.84	0.12	1.44	0.11	1.32
1.0	12	0.09	1.08	0.13	1.56	0.16	1.76
1.5	12	0.09	1.08	0.132	1.58	0.21	2.52

From the above analysis result, output power P in idle load was mainly to overcome the friction between robot and pipe shell. The resistance of friction was about 18 N, increasing with raising load. The robot need maximum power in upward vertical motion. The maximum external load was 17.4 N which was 4 times larger than its own weight.

6 Conclusion

1) The crawling in-pipe robot was adaptable to horizontal, vertical and bending pipe with good self-locking performance without exerting additional pressure to the shell.

2) The robot was adaptable to pipes with 90~150 mm in square, round and rectangle sections with the automatic adjustment.

3) Proper coupling methods and restraint function of pipe shell made robot to pass through a nearly 90°right-angled bending pipe successfully.

4) The telescopic mechanism uses inequilateral parallelogram to shorten transmission screw rod effectively which ensures magnifying power in the movement.

References

1. Li Peng,Ma Shugen,Li Bin,et al.Design and motion analysis of an in-pipe robot with adaptability to pipe diameters,J.Journal of Mechanical Engineering,2009,45(1):154-161.(in Chinese)
2. Liu Qingyou,Chen Yonghua,Ren Tao,et al.Optimized inchworm motion planning for a novel in-pipe robot[J].Journal of Mechanical Engineering Science,2014,228(7):1248-1258.
3. Schempf H, Mutschler E,Gavaert A,et al.Visual and nondestructive evaluation inspection of live gas mains using the ExplorerTM family of pipe robots[J].Journal of Field Robotics,2010,27(3):217-249.
4. Xiong Juntao,Zou Xiangjun,Chen Lijuan,et al.Fruit recognition and positioning technology of litchi picking manipulator[J].Journal of Jiangsu University: Natural Science Edition,2012,33(1):1-5.(in Chinese)
5. Li Peng,Ma Shugen,Li Bin,et al.Self-rescue mechanism for screw drive in-pipe robots[C]// Proceedings of the 2010 IEEE/RSJ International Conference on Intelligent Robots and Systems.Taipei: IEEE Computer Society,2010:2843-2849.
6. Qiao Jinwei,Shang Jianzhong,Goldenberg Andrew.Development of inchworm in-pipe robot based on self-locking mechanism[J].IEEE/ASME Transactions on Mechatronics,2013,18(2):799-806.
7. Zhang Yanheng,Pang Zenghui.Design of a novel crawling pipe robot[J].Machinery Design & Manufacture,2010(4):13-15.(in Chinese)
8. Nakazato Y,Sonobe Y,Toyama S.Development of an In-pipe micro mobile robot using peristalsis motion[J].Journal of Mechanical Science and Technology,2010,24(1):51-54.
9. Qiao Jinwei,Shang Jianzhong,Chen Xun,et al.Development of an inchworm in-pipe robot based on the cam self-locked principle[J].Journal of Mechanical Engineering,2010,46(11):83-88.(in Chinese)
10. Jeon W,Kim I,Park J,et al.Design and control method for a high-mobility in-pipe robot with flexible links[J].Industrial Robot:An International Journal,2013,40(3):261-274.

Research on Identification of Dust Particles on COF

Jiayi Zhang [1], Qiang Gao[2,a]

[1]*School of Mechanical and Electric Engineering, Soochow University, 215021 Su Zhou, China*
[2]*School of Mechanical and Electric Engineering, Soochow University, 215021 Su Zhou, China*

Abstract. Chip On Film(COF) is the key component of electronic products, and is different from Printed Circuit Board(PCB). The properties of high flexibility, thin thickness, lightweight and high wiring density make it difficult to inspect COF, especially dust particles interference. Dust particles are similar to defects, and it is hard to identify dust particles from defects, so dust particles interference of quality test is the difficulty of automatic surface defect detection. In this paper, a new method to identify dust particles is discussed from abnormal area called junction points detection and machine learning method Support Vector Machine(SVM) according to the characteristics of dust particles. As a result, a 94.8% correct rate of dust particles images identification has been achieved with the method.
Key Words. COF, Dust, Identification, Junction Point, SVM

1 Introduction

COF is a product using IC chip placing and flexible substrate carrying technologies[1], as shown in Figure 1. COF with $25\mu m$ pitch was achieved by using a new etching technique[2], therefore, COF meets light, thin, short and small design requirements of electrical products, and is a typical product of micro-nano manufacturing.

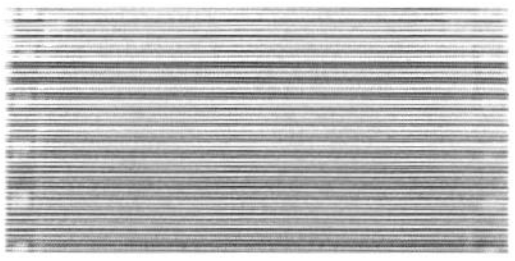

1) Front side 2) Back side

Figure 1. Chip On Film.

With the rapid development of Information Times, the properties of high flexibility, thin thickness, lightweight and high wiring density make the demand for COF enhance greatly, at the same time, the demand for automatic test of COF is higher and higher. Currently, quality test of COF is most completed by artificial, and it has high labor intensity and strong subjective interference[3], especially for dust particles identification. While there are AOI(Automatic optical Inspection) equipments used in the quality detection of COF, but dust particles are often as a defect detection error, which greatly reduces the production efficiency.

For dust particles identification, some researchers conducted some corresponding exploration. YoungJun Roh etc. put forward using the fast Fourier transform and fast Fourier inverse transformation technologies for reference images, operating subtraction on original image and reference image, then making segmentation on result image from subtraction operation, extracting data from segmentation image with analysis of margin and intensity, and setting up Bayesian model for identification. However, the method is only applicable to linear circuit, cannot be applied to curved lines[4]. Rambabu K etc. proposed a method using template matching and analysis of many directions[5], but contour of dust particles is complex, and dust particles have different gray levels, which makes determining a template difficult, analyzing many directions complex, and realizing the method hard. Chia-Te Liao etc. prenented a statistical learning method on the basis of SVM, but reference images are needed. In addition to this, the method involves image registration and Hausdorff distance calculation, requirement of accuracy is high, and amount of calculation is large[6]. Jin-soo Kim proposed a method based on reference images and parameters extraction from local binary pattern with location correction technology, which requires high accuracy[7]. Some of methods mentioned above are only suitable for linear circuit, and the other methods refer to reference image matching, position correction and similarity calculation technologies, which are realized intricately and whose calculation amount is large. According to these limitations and that papers about surface defect detection of COF have little been published, a dust particles identification method without reference images is put forward on the basis of junction point detection and SVM in this paper, which provides a new method for solving dust particles interference in the quality inspection process.

2 Dust Particles Identification Algorithm

Irregular contour, different gray levels and uncertain position of dust particles lead to that the traditional image segmentation technique is unable to achieve segmentation

[a] Corresponding author: gaoqiang@suda.edu.cn

of dust particles. The image contrast technology requires reference images and high accuracy, so it is demanding. According to the characteristic that boundary of dust particles is irregular in this paper, a method without reference images combining abnormal region detection through geometrical junction point and identification by using SVM algorithm is proposed.

The information of gray images is complex, and gray levels of different images are different, especially Cu conductors because of light diffraction, mechanical vibration and illumination variation during the image acquisition process. Therefore, the adaptive threshold method is adopted to make original images segmentation to achieve binary images that are simple and do not change the state of dust particles' boundary, the original dust particles images are show in Figure 2, and the binary images are shown in Figure 3.

Figure 2. Original Dust Particles Images.

Figure 3. Binary Images.

Skeleton as a thinning structure, is the important topological description of target image, and keeps original shape of the binary image containing important shape information. The essence of thinning: objects in the image are simplified to single pixel width line, which shows its topological properties[8]. Skeleton extraction is based on image thinning algorithm[9], and look-up table thinning algorithm and layer by layer thinning algorithm are used commonly[10]. In this paper, the look-up table thinning algorithm based on the eight neighborhood is applied to extract the skeleton[11].

A junction point is defined as the point at the position, where two skeleton lines cross, and the angular difference is greater than zero[12]. The flow chart of the junction point detection is shown in Figure 4. The result of skeleton extraction is shown in Figure 5, and the white part is skeleton. The result of the junction point detection is shown in Figure 6, and the white points are junction points in skeleton images. The result of the junction point detection in dust region is shown in Figure 7, and the red points are junction points, which indicates that junction points exist in dust particles.

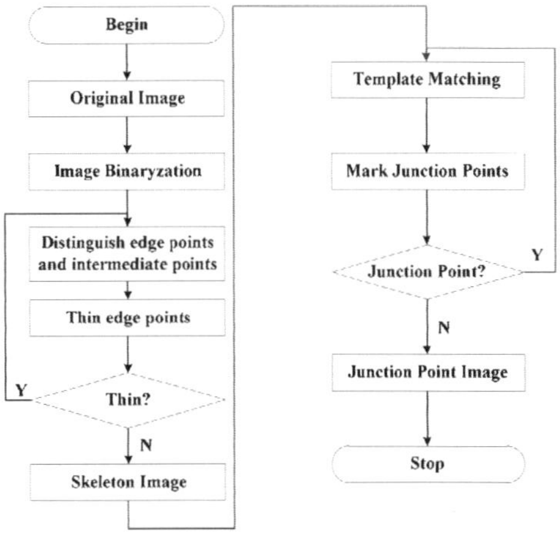

Figure 4. Junction Points Detection Flow Chart.

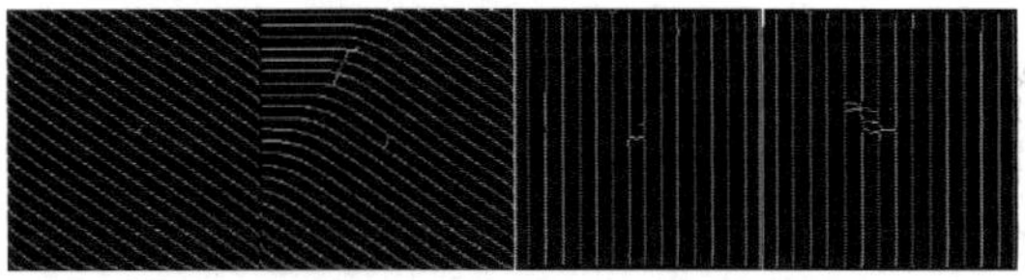

Figure 5. Skeleton Images.

Figure 6. Junction Point Images.

Figure 7. Junction Points in Dust Region.

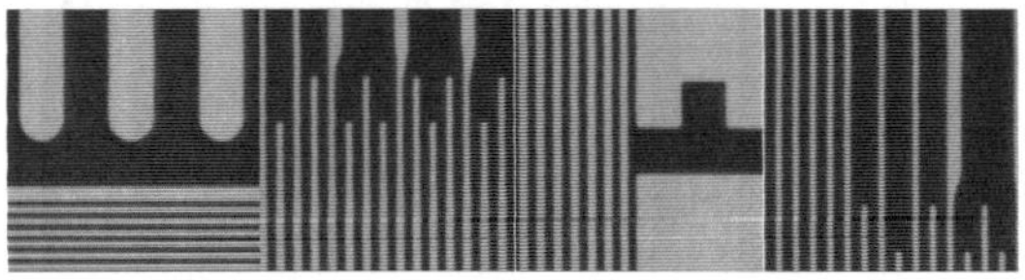

Figure 8. Normal Images with Junction Points.

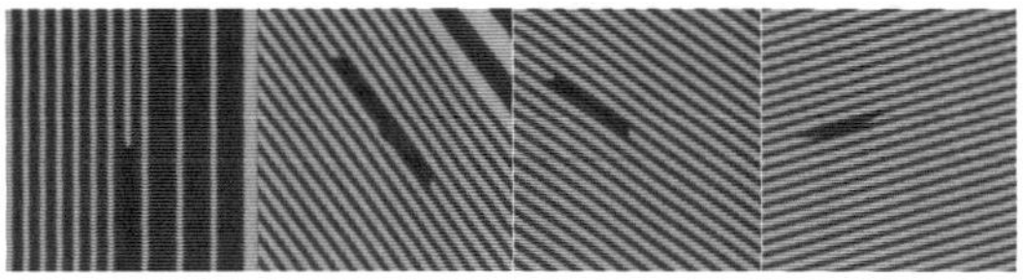

Figure 9. Defect Images with Junction Points.

Figure 7 shows that junction points exist in dust particles, but some normal images and defect images also have the same feature. Some normal images with junction points are shown in Figure 8, and some defect images with junction points are shown in Figure 9.

The result shows that dust particles identification can

not be solved completely by detecting junction points, so a further determination is needed for images with junction points. After analyzing regions near junction points, a discovery can be made that gray level of dust particles changes significantly, and the gray level value is great in that most dust particles exist on insulating substrate. According to great gray level value and obvious gray level change of dust region, mean value, range, variance and weber coefficient are applied to present characteristics of dust particles. The template to extract features is shown in Figure 10, and the red square presents junction point detected using the method mentioned before. The definition of mean value, range, variance and weber coefficient is list specifically below.

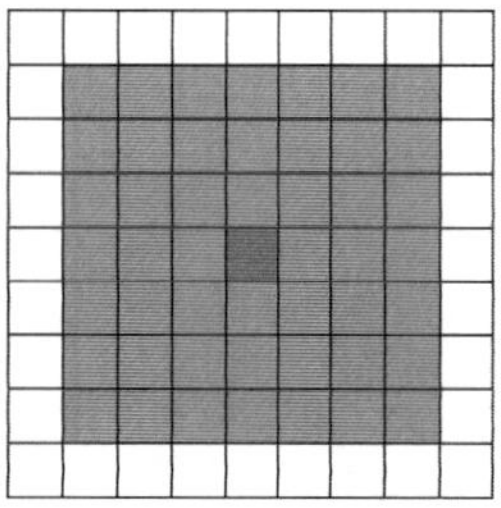

Figure 10. Template.

The mean value of the region for extracting features is defined as

$$\overline{P} = \frac{1}{n}\sum_{i=1}^{n} P_i \tag{1}$$

P_i presents gray level value of pixel point in region, n presents the number of pixel points, and $n = 81$ here.

The range of the region for extracting features is defined as

$$R = P_{\max} - P_{\min} \tag{2}$$

$P_{\max}$ presents the maximum gray level value of pixel point in region, $P_{\min}$ presents the minimum gray level value of pixel point in region.

The variance of the region for extracting features is defined as

$$D = \frac{1}{n}\sum_{i=1}^{n} (P_i - \overline{P})^2 \tag{3}$$

P_i presents gray level value of pixel point in region, $\overline{P}$ presents the mean value of the region, n presents the number of pixel points, and $n = 81$ here.

The weber coefficient[13] of the region for extracting features is defined as

$$\overline{W} = \frac{1}{n}\sum_{i=1}^{n} W_i \tag{4}$$

$W_i = \Delta(P_i) / P_i$, $\Delta(P_i) = \sum_{k=1}^{m} | P_{i,k} - P_i |$, W_i presents the weber coefficient of pixel point, n presents the number of pixel points, and $n = 49$ here, which presents the 7×7 blue region in the template. P_i presents gray level value of pixel point in region, $\Delta(P_i)$ presents the absolute value of pixel gray difference with its neighborhood pixels, $P_{i,k}$ is the k neighbor pixel point of P_i, k is the number of neighbor pixel points, and $k = 8$ here, because of using eight neighborhood.

SVM is a data mining approach on the basis of statistiacal learning theory, which can solve regression problems(time series analysis) and pattern recognition problems(classification and discrimination)[14]. In other words, SVM is a process of machine learning, which looks for a classification hyperplane in high dimensional space to separate date points of different categories, and maximize the largest distance between points of different categories[15].

According to the theory of SVM, the dust particles identification is classified as a binary classification problem in this paper, one kind is dust particles image class, and the other kind is non-dust particles image class which includs normal and defect images. The mean value, range, variance and weber coefficient of an image are defined as the average of eigenvalues of regions for extracting features. Given the training sample set $(\mathbf{x}_i, y_i)$, $i = 1,2,...,I$, $\mathbf{x} \in R^n$, $y \in \{\pm 1\}$, hyperplane is $(\mathbf{w} \cdot \mathbf{x}) + b = 0$. In order to make the classification plane classify samples correctly, and have classification interval, the constraint $y_i[(\mathbf{w} \cdot \mathbf{x}_i) + b] \geq 1$, $i = 1,2,...,I$ requires to be met. The classification interval is $2/\|\mathbf{w}\|^2$. To seek the optimal hyperplane problem is converted to minimize $\|\mathbf{w}\|^2$ under constraint.

Dust particles image training sample set is $(\mathbf{x}_m, 1)$, $m = 1,2,...,M$, non-dust particles image training sample set is $(\mathbf{x}_n, -1)$, $n = 1,2,...,N$, x is an eigenvector including mean value, range, variance and weber coefficient. For forecast sample set $(\mathbf{x}_k, y_k)$, $k = 1,2,...,K$, $\mathbf{x} \in R^n$, $y \in \{\pm 1\}$, the classification is predicted by using the SVM classifier to determine that the sample belongs to dust particles($y = 1$) or non-dust particles($y = -1$).

3 Results

Using images from the production line, 1807 dust particles images and14652 non-dust particles images of nine image sets that are needed to be verified are achieved. Automatic feature extraction program is written to collect sample sets of two kinds. According to the optimum principle that positive and negative samples are $1:1$, positive and negative samples are selected randomly to establish the training sample set, cross validation method is adopted to determine the parameters for the training process ending up with a better classifier model, which is used to predict and classify test images.

Automatic detection program is written with junction point detection and SVM prediction to classify images from image set IPC00009 which is to be verified, test results are shown in Table 1. The junction point detection method can detect 97.7% of dust particles images, and the SVM can identify 80.9% of dust particles images on the basic of it. According to it, the method including junction point detection and SVM achieve a 94.8% correct rate of dust particles images identification.

Table 1. Test result.

IPC00009		Number /Picture	Dust Number/ Picture	Missing Rate/%	Failure Rate/%	Correct Rate/%
No-Junction Point		2710	26	3.8%	1.4%	94.8%
Junction Point	No-Dust	1961	190			
	Dust	995	913			

4 Conclusion

According to the result above, for dust particles identification of COF, the method proposed in this paper with combination of junction point detection and SVM has certain feasibility, and the method is easy to realize without complex calculation, reference images and image registration. Therefore, it has certain reference significance to solve dust particles interference in COF quality test, and provides a reference for the actual production on the basis of the 94.8% correct rate.

Acknowledgements

This research was supported by KGK Precision Electronics Technology Development(Suzhou) Co.Ltd.

References

1. K. Toyosawa, N. Nakamura, K. Fukuta, Y. Chikawa, COF(Chip-On-Film) Technology for LCD Driver ICs Using Reel-to-Reel System, *Sharp Corporation Technical Report* **8**, 56 (2001)
2. J.Q. Cai, The current status and perspective of COF tape for large scale LCD, *Short Comment & Introduct* **6**, 10(2012)
3. X.F. Dong, Z.Y. Han, S.Y. Liao, X.X. Yi, Study on Semiconductor Surface Defect Detection Based on Machine Vision, *Metrology & Measurement Technology* **5**, 22(2014)
4. Y.J. Roh, C.W. Kim, C.O. Jung, D. Jeong, Defect Classification Using Bayesian Approach for Tape Substrate Inspection System, *Proceedings of the 17th World Congress on The International Federation of Automatic Control*, Seoul, Korea, July 6-11, 8177(2008)
5. R. K, D.K. M, V.K. K, S.P. Nath, Optical Pattern Inspection for Flex PCB – Challenges & Solution, *Proceedings of the 17th World Congress on The International Federation of Automatic Control*, Seoul, Korea, July 6-11, 8196(2008)
6. C.T. Liao, W.H. Lee, S.H. Lai, A Flexible PCB Inspection System Based on Statistical Learning, *J Sign Process Syst* **67**, 279(2012)
7. J.S. Kim, COF Defects Detection and Classification System Based on Reference Image, *Journal of the Korea Institute of Information and Communication Engineering* **17**, 1899(2013)
8. P. Jiang, X. B. Xu, M.Fang, Study and Implementation of Skeleton Extraction Algorithm Based on Morphology, *Computer Application* **23**, 136(2003)
9. L. Lam, S. W. Lee, C. Y. Suen, Thinning Methodologies-A Comprehensive Survey, *IEEE TRANSACTIONS ON PATTERN ANALYSIS AND MACHINE INTELLIGENCE* **14**, 869(1992)
10. L. Yu, G.W. Wang, Fast Thinning Algorithm of Binary Images, *Science and Technology Vision* **5**, 100(2012)
11. W. Yang, K. Guo, Y.K. Wei, An Efficient Index Thinning Algorithm of Fingerprint Image Based on Eight Neighbourhood Points, *JOURNAL OF SICHUAN UNIVERSITY OF SCIENCE & ENGINEERING (NATURAL SICENCE EDITION)* **21**, 61(2008)
12. A.B. Naeem, H. Allan, Granulometry based Detection of Junction and End Points in Patent Drawings, *7th International Symposium on Image and Signal Processing and Analysis*, Dubrovnik, Croatia, September 4-6, 307(2011)
13. X.L. Zou, G.C. Feng, Weber's Law-based Transition Region Extraction and Thresholding, *Science Technology and Engineering* **13**, 4217(2013)
14. S.F. Ding, B.J. Qi, H.Y. Tan, An Overview on Theory and Algorithm of Support Vector Machines, *Journal of University of Electronic Science and Technology of China* **40**, 2(2011)
15. J.B. Xie, Visual Machine Learning 20, *Tsinghua University Press*, 88(2015)

MATEC Web of Conferences **61**, 04019 (2016) DOI: 10.1051/matecconf/20166104019
APOP2016

Light Environment Simulation of Xinyang Traditional Residence in Haicang District, Xiamen

He Miao[1,a]

[1] *College of Civil Engineering and Architecture, Xiamen University of Technology, Xiamen 361024, China*

Abstract. Applying the analysis software ECOTECT, which is related to the constructing environment, the paper attempts to make a simulated experiment on the daylight environment factors in traditional residence in Xinyang Village. The conclusion is that the lighting coefficient of local dwellings cannot reach the regulation design standard of daylight, and daylight uniformity of most dwellings is poor. Towards these problems, some influencing factors are put forward through the field research, and by using ECOTECT, we get the distribution diagram and distribution percentage of lighting coefficient, and then proceed the quantization and the merits and demerits of each variable, which can offer a reference for the dwelling design.

1 Instruction

Daylight environment is closely linked to people's life, which not only affects people's physical health, but psychological health as well. It is an important indicator of creating the comfortable indoor environment. In recent years, with the gradual execution of the new national rural policy and the improvement of residents' quality of life, people pay much attention to the indoor daylight environment of traditional residence. On account of the living habit of residents in early stage, high-density construction of towns and houses, and the layout of construction itself, the current situation of natural light of traditional residence in Southern Fujian needs urgen improving. The paper selects the traditional residence in Xinyang Village, Haicang District, Xiamen as the object of study, and applies the analysis software ECOTECT to proceed the simulated analysis on the indoor daylight, and studies the influence of the daylight on people's life, and based on this, the paper puts forward the daylight design strategies in the construction of residence in this region.

2 The current situation of traditional residence in Xinyang Village

Haicang District is located in the southwest of Xiamen, Fujian province, which belongs to Jindong Plain. Its sunshine duration is between 1500-2400 hours throughout the year, which belongs to the IV in light climate division, and its sunshine duration can exceed 2100h per year,which means light resource is quite abundant.

In the village, although there exist many undamaged traditional residence, the indoor light has a throng of problems. A large portion of the reason why many aborigines move away from the old houses, leaving them empty is that the indoor light is poor.Hence,it accelerates the aging of the buildings. The aborigines said that though artificial lighting can solve this problem, people would feel more comfortable and pleasant in the natural

a Corresponding author: 88407567@qq.com

light.

Through the field research, we tentatively find the influencing factors that lead to the poor light:

-First, due to the local geographical characteristics and folk customs, the residences have bright lobby and dim bedroom (the lobby functions as sacrifice and living room, so the lobby which orients the side of the inner garth does not have walls, but open instead; however, in order to reduce the solar radiant heat, especially reduce the window area in the room and kitchen, the rooms have to be dim.).

-On the structure, the proportion between window and land, the location of the window, sunshade and indoor reflectance ratio are also vital factors.

-On its layout, the orientation and the density of the building are also included in the influence factors.

The research will apply the analysis software ECOTECT to verify these influence factors and draw conclusions.

3 Indoor daylight simulation of traditional residence in Xinyang Village

3.1 Simulating object

We select Chen's house, one of the typical traditional residence in Xinyang Village, as the object of this study. This building orients north by east 10° , which is made up of red brick, with two-entrances, three standard width and one layer height. Its plane layout presents as a quadrangle courtyard (see Figure 1 and Figure 2). Chen's house has the most typical features, that is, the subject is a composite structure with masonry and wood. Except the lobby, the exterior is entire open. The windows are few and the apertures are small. The windows in the bedroom do not exceed 0.2. The density between the buildings is large. Compared with the other residences, it belongs to such kind of the residence with steep slopes and long eave in order to be shaded and insulated.

Figure 1. Exterior of traditional residence Chen's

3.2 Experiment conditions and methods

(1) Basic conditions: its northern latitude is 24.27, and the east longitude is 118.06. It is 12:00 in the high noon, July 21st. The coefficient K of light climate is 1.10.

(2) The indoor parameter: The ceilings in Chen's house are made up of brown wood, and the floors are red brick. And the walls are painted as cement mortar, with reflection coefficient 0.3, 0.33, 0.32 respectively. From the investigation, the cleanliness degree of the window is quite low, and some of them are fencing windows, so the cleanliness is set as 0.75.

(3) The model of sky illumination: in the process of design, the model of sky illumination is the whole-cloud, and its critical illumination in the outside is 4500lx. In this calculative model, the natural light in the outside is only the diffusion light, which cannot form the shade, so the orientation of the window does not have any effect on the indoor daylight.

(4) The selection of simulating area: in Southern Fujian, in the house with two-entrance, three standard width, the large room is used as the main bedroom, and Jutou (Pinyin, a room used as kitchen) is used as kitchen, and the main hall is used as living room. Because the orientation of the window could not affect the experiment, three rooms in Chen's house can be respectively simplified as three simulating zones A, B, C (Figure2) .

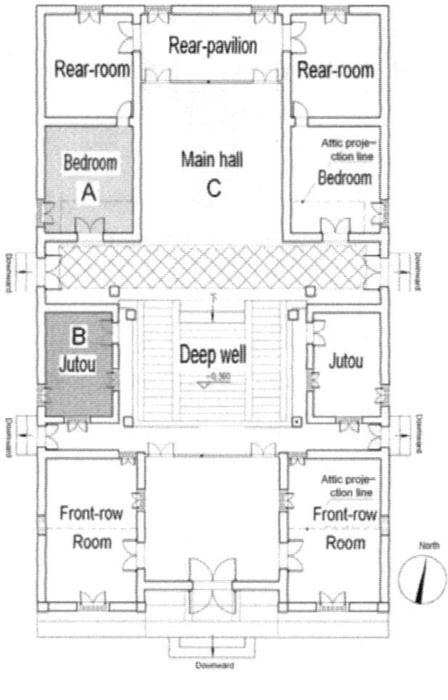

Figure 2. Ichnography of the traditional residence Chen's House

(5) Experiment methods: This simulation uses grid method, and chooses the plane height 0.8m, and then analyzes the sunshade ,window and the indoor reflectance ratio.

Table 1.Daylight standard value for residential buildings

Lighting Level	Rooms	Sidelighting			
		Old minimum value of d lighting coefficient (%)		New minimum value of lighting coefficient (%)	
		Standard value	Corrected value	Standard value	Corrected value
Ⅳ	Living room Bedroom, Study, Kitchen	1.0	0.91	2.0	1.82
Ⅴ	Toilet, Aisle, Dining room, Staircase	0.5	0.45	1.0	0.91

3.3 Simulating results

Because the outward illumination is changing, the indoor illumination is also changing; thus, the result here is not fixed. And there are two indicators to evaluate the light environment: the indoor lighting coefficient and the uniformity of illumination. So the simulating experiment will apply the indoor lighting coefficient and the uniformity of illumination to summarize the changing rules and the lighting conditions of each zone. ECOTECT presents the results as follows:

3.3.1 Minimum value of lighting coefficient

In architectural lighting design standards GB50033-2013, as for minimum value of lighting coefficient (Cmin), that in bedroom and living room should not be lower than 1.82%, and that toilet, passage, staircase and dining room should not be lower than 0.91% (See Table 1).

From the simulating analysis, we get that Cmin of the bedroom A is 0.07% , and Cmin of the left Jutou B is 0.09%, and Cmin of the main hall C is 0.81%. The former two, A and B, cannot meet the daylight design standard, and though Cmin of the main hall C is higher than that of A and B, it is still lower than the national rules for the daylight. It is 12:00, which is the best time for the daylight throughout the day, so at other time, zone A, B and C still cannot meet the daylight requirement.

3.3.2 Uniformity of daylight

According to Figure3, 4, 5 and Table 2, the variation range of daylight coefficient can be divided into ten classes. Table 2 presents the daylight condition of room A, whose variation range of daylight coefficient is between 0% and 20%, and the amplitude of variation is 20%. The range 0~2% occupies 80.13%, but this indoor space still cannot satisfy the requirement. Although the zone near the window is brighter and changes obviously, the illumination of the whole room remains low and it is not distributed enough, either.

The daylight condition of Jutou B is reflected in Figure 3, whose variation range of daylight coefficient is also between 0% and 20%,and the amplitude of variation is 20%. The range 0~2% occupies 80.00%, but this indoor space still cannot satisfy the requirement. The illumination of the whole room is low, but the illumination of the daylight opening is high. The daylight opening is far away, so the daylight uniformity of the whole room is still poor.

Figure 4 shows the daylight condition of main hall C. Because it is lighting indirectly through Zisun alley, the variation range of daylight coefficient centers on the

Table 2. Percentage of the original lighting coefficient of each test point

Test point / Lighting coefficient (%)	0~2	2~4	4~6	6~8	8~10	10~12	12~14	14~16	16~18	18~20
Region A(Bedroom)	80.13%	10.27%	3.20%	1.47%	1.07%	0.80%	0.67%	0.53%	0.40%	0.53%
Region B(Kitchen)	82.00%	8.53%	2.67%	2.00%	0.93%	1.20%	0.40%	0.93%	0.53%	0.27%
Region C(Jutou)	10.13%	38.13%	43.47%	8.27%	0	0	0	0	0	0

former four classes, that is, the variation range of daylight coefficient is between 0% and 8%, and the amplitude of variation is 8%. The lighting coefficient between

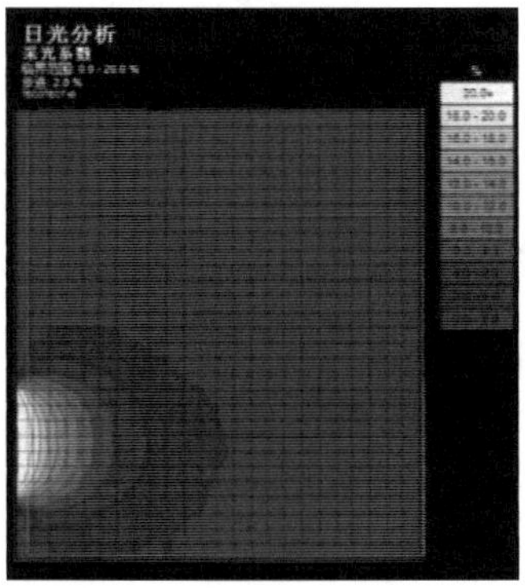

Figure 3. Region A's lighting coefficient changing curve chart

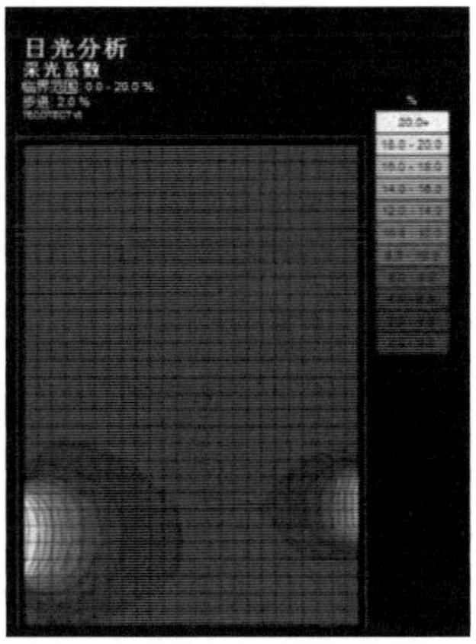

Figure 4. Region B's lighting coefficient changing curve chart

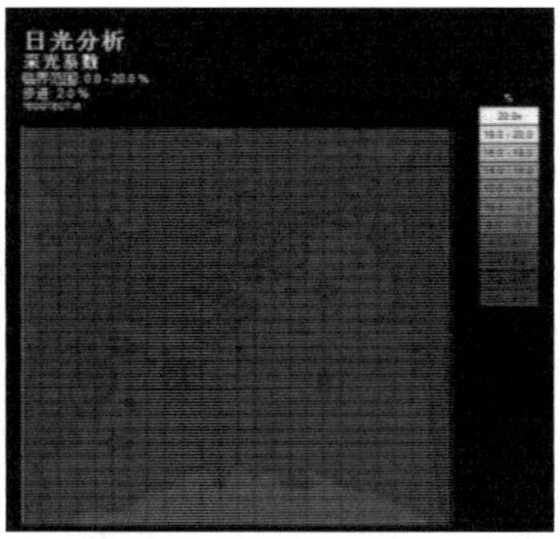

Figure 5. Region C's lighting coefficient changing curve chart

0%-2%only occupies 0.86%, and except for this, the other spaces all meet the daylight requirement, and the uniformity of daylight along the room is high. The simulating results reflect the feature "bright lobby and dim bedroom" of residence in Southern Fujian.

4 Optimization simulation of light environment

The side of the main hall is open and its window is high, so the optimizable space is not that much; while, as minor space, Jutou is primarily used as kitchen and storeroom. The left room is used as main space, whose fuction is like main bedroom in nowadays; thus, we select zone A, the main bedroom, as our object to study the optimization simulation of light environment, and then observe the influence each variable has on the indoor daylight. Taking the difficulty of reforming the indoor daylight environment of traditional residence into consideration, and following the principles that styles and features should be protected during the reforming of traditional residence, we can select the variable which is harmonious with the local styles and features.

4.1 Opening the door

Since the climate of Southern Fujian is hot in summer and warm in winter, it has longer sunlight time in summer. Local traditional residences are primarily designed to shield the sun and summer heat for the sake of lessened solar radiant heat, especially minimizing the windows in the room and kitchen, which results in dim indoor environment. Meanwhile, leave the side entrance that is beneath the eave and orients courtyard open. By doing so, it is effective to avoid direct sunlight and good for ventilation. The simulation results are shown in Figure 6 and Table 3: as no sunshade measures on windows and high illumination about, the lighting coefficient near the window appears its variability, mainly between 2% and 20%. Whereas, for the reason that the outward side entrance which is not only under the eave but also has Jutou shade from the sun, the lighting coefficient near the

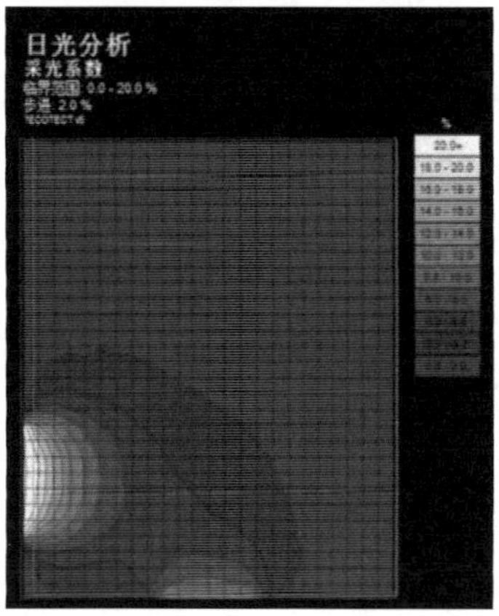

Figure 6. Region A's lighting coefficient changing curve chart of opening the door

door varies from 2% to 12%.

On the whole, Table 3 shows that the lighting coefficient distribution from classes of 2% ~ 4%, 4% ~ 6%, 6% ~ 8%, 8% ~ 10% and 10% ~12% have increased to some certain extent, at the same time, the value from 0 to 2% is reduced by 11.06%, but still accounts for 69.07% of daylight area. The daylight area that is below national minimum standard is yet the majority.

Therefore, opening the door only improves partial illumination of the room, but contributes little to the overall illumination and lighting uniformity of the room.

Table 3. Distribution percentage comparison for lighting coefficient of the bedroom (Close and open door as variable)

Test point / **Lighting coefficient (%)**	**0~2**	**2~4**	**4~6**	**6~8**	**8~10**	**10~12**	**12~14**	**14~16**	**16~18**	**18~20**
Region A(Close the door)	80.13%	10.27%	3.20%	1.47%	1.07%	0.80%	0.67%	0.53%	0.40%	0.53%
Region A(Open the door)	69.07%	12.67%	8.40%	3.20%	1.73%	1.73%	0.80%	0.53%	0.40%	0.40%
D-value(Open-Close the door)	-11.06%	2.40%	5.20%	1.73%	0.66%	0.97%	0.13%	0	0	-0.07%

4.2 Open the door and remove the cornice

As most sunshade is situated towards south(with door, facing courtyard and overhang eave is relatively huge, while there is little sunshade on the side of eastward window. So opening the door is to illuminate what effects no-sunshade measure has on the daylight. We found that compared with mere door-opening measure, daylight uniformity has a slight increase (see Table 4). But if compare Figure 6 with Figure 7, we cannot find any evident change on the eastern side, but the side of entrance. That is because the cornice above the eastern window is small and distant from the window, however, the one on the side of entrance is deep and lower than the window. Therefore, illumination on the entrance changes greatly.

Hence, the measure of opening the door and removing the sunshade has little influence on indoor light uniformity, but has significant impact on illumination.

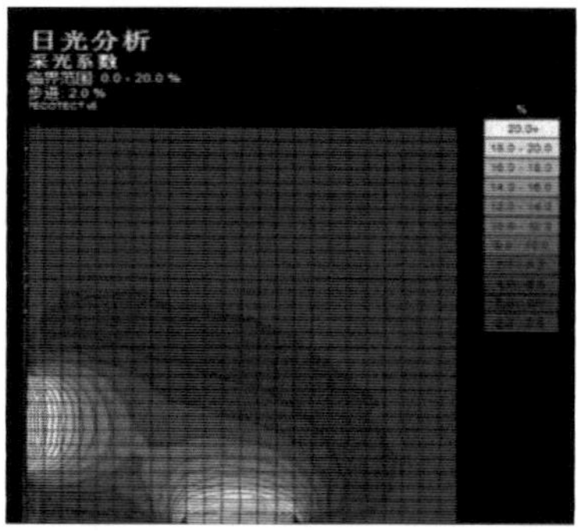

Figure 7. Region A's lighting coefficient changing curve chart of opening the door and removing the cornice

4.3 Increase glass tiles and demolish attic

Table 4. Distribution percentage comparison for lighting coefficient of the bedroom(Open door and remove cornice as variable)

Test point / **Lighting coefficient (%)**	**0~2**	**2~4**	**4~6**	**6~8**	**8~10**	**10~12**	**12~14**	**14~16**	**16~18**	**18~20**
Region A(Open the door)	80.13%	10.27%	3.20%	1.47%	1.07%	0.80%	0.67%	0.53%	0.40%	0.53%
Region A(Open door and remove the cornice)	69.07%	12.67%	8.40%	3.20%	1.73%	1.73%	0.80%	0.53%	0.40%	0.40%
D-value(Open the door-Open door and remove cornice)	-11.06%	2.40%	5.20%	1.73%	0.66%	0.97%	0.13%	0	0	-0.07%

 DOI: 10.1051/matecconf/20166104019

Taking into account the fact that the skylight efficiency is higher than that of side-window lighting, and the roof of Chen's house is wooden-framed with red tiles on, so the red brick tiles can be replaced by glass ones. The simulation result after increasing the amount of tiles is shown in Figure 8:because of the wooden attic's shade, when tiles increased, partial illumination has somewhat improved, but beneath the projected area in the attic the illumination is still very low, nor is the indoor illuminate uniformity. In the meantime, considering the attic is now available as storeroom full of sundries and dust, inconvenient for cleaning but easy to remove, it is suggested to increase tiles simultaneously demolish the attic for better lighting effects. As shown in Figure 9 the simulation result after tearing down the attic, daylight uniformity and illumination have notably improved. The distribution coefficient in the class of 0~2% is 0,2%~4% is only 1.87%, and 4%~6%, 6%~8%, 8%~10% are respectively increased by 19.20%, 38.06%, 22.53%. The area ratio of lighting coefficient has improved much and the lighting coefficient is mainly from 6%~8% (See Table 5).

It proves that the measure of increasing tiles and demolishing attic can significantly improve the indoor illumination and lighting uniformity.

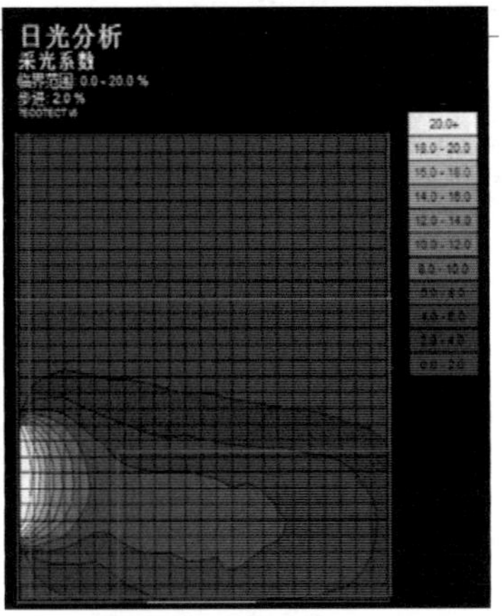

Figure 8. Region A's lighting coefficient changing curve chart of increasing glass tiles

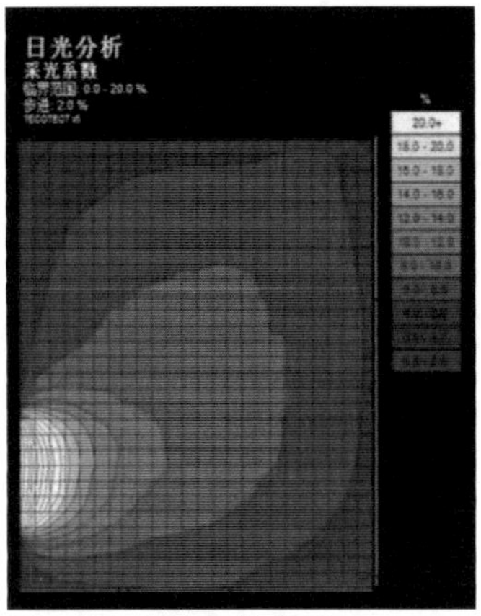

Figure 9. Region A's lighting coefficient changing curve chart of removal attic and increased glass tiles

Table 5. Distribution percentage comparison for lighting coefficient of the bedroom
(Removal attic and increased glass tiles as variable)

Test point **Lighting coefficient (%)**	**0~2**	**2~4**	**4~6**	**6~8**	**8~10**	**10~12**	**12~14**	**14~16**	**16~18**	**18~20**
Region A(Open the door)	80.13%	10.27%	3.20%	1.47%	1.07%	0.80%	0.67%	0.53%	0.40%	0.53%
Region A(Removal attic and increased glass tiles)	69.07%	12.67%	8.40%	3.20%	1.73%	1.73%	0.80%	0.53%	0.40%	0.40%
D-value(Open the door- removal attic and increased glass	-11.06%	2.40%	5.20%	1.73%	0.66%	0.97%	0.13%	0	0	-0.07%

4.4 Improve the indoor reflectance ratio

Interior decoration materials is dated and gloom, with brown wooden-structured ceiling and attic, red-bricked floor and cement-mortared wall. Their reflection coefficients are respectively set as 0.3, 0.33 and 0.32. From the research, we explore that the cleanliness of windows is quite low, so it should be set to0.75. Because of the dim room found in the field research, interior material's reflectance ratio is heightened to do comparative research in the simulation. Assuming they are all whitewashed walls, whitewashed attic's bottom

 DOI: 10.1051/matecconf/20166104019

side and the ceiling, thus the reflection coefficient of wall and attic plate is changed to 0.75, besides, window cleanliness is altered to 0.9 (clean). As is shown in Table 6 and Figure 10 the simulation results,daylight area that is less than the national standard is reduced by 63.46%, lighting area between 2% and 4% is increased by 56.40%, which means the lighting effect has obviously improved.

From Figure 10 we can analyze that except illumination in the daylight opening remains unchanged, others are even, most of which are

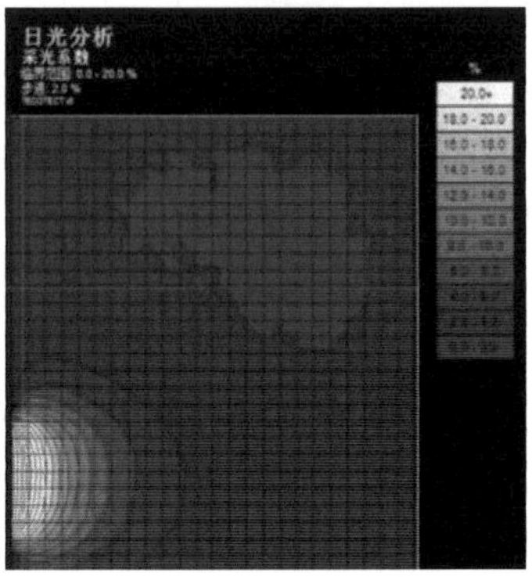

Figure 10. Region A's Lighting coefficient change curve of improving the indoor reflection ratio

Table 6. Distribution percentage comparison for lighting coefficient of the bedroom (Raise reflectance ratio as variable)

Test point lighting coefficient (%)	**0~2**	**2~4**	**4~6**	**6~8**	**8~10**	**10~12**	**12~14**	**14~16**	**16~18**	**18~20**
Region A(Open the door)	80.13%	10.27%	3.20%	1.47%	1.07%	0.80%	0.67%	0.53%	0.40%	0.53%
Region A(Raise reflectance ratio)	69.07%	12.67%	8.40%	3.20%	1.73%	1.73%	0.80%	0.53%	0.40%	0.40%
D-value(Open the door-Raise	-11.06%	2.40%	5.20%	1.73%	0.66%	0.97%	0.13%	0	0	-0.07%

between 2% and 4%. There is only one side window in the largest room and near the southeast corner of the east wall, so the dimmest place in the room (i.e. the lowest lighting level 0-2%) are mainly distributed diagonally closed to the window, accounting for only 16.67% daylight area. Therefore, to improve the reflection ratio of the wall space can significantly improve the indoor lighting uniformity, furthermore, substantially reduce the lighting coefficient space that is not up to the standard.

4.5 Variable comparison

According to analysis of the simulation results and Figure 11, we know that opening the door and removing the sunshade can improve the illumination, but have poor effects on better daylight uniformity.

The measure that mere increase the amount of glass tiles can raise the illumination and daylight uniformity for places other than the shaded area of attic. While the measure of enhancing interior reflectance ratio and demolishing the attic can much improve the overall indoor lighting uniformity, meanwhile reduce the space whose lighting coefficient cannot satisfy the standard, especially the latter measure.

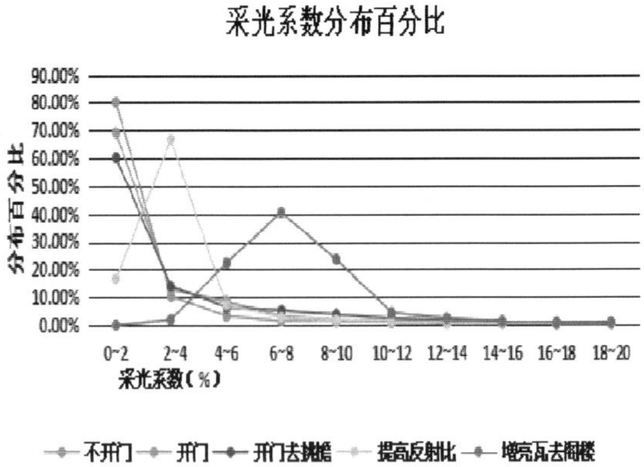

Figure 11. Lighting coefficient changing curve of improving the indoor reflection ratio of Region A

5. Conclusions

This research is done by ECOTECT, a built environment

analysis software, to do simulated experiment on factors that affect indoor luminous environment of traditional residence in Southern Fujian. It finds that generally most of them cannot meet the national design standard of daylight, either poor is the most indoor lighting uniformity. And we also use the indoor lighting coefficient's distribution diagram and per centum that analyzed by ECOTECT to do quantitative analysis and comparison on elements that affect Southern Fujian traditional residence, which is thought to provide a reference for lighting design in newly constructed residence

References

1. He Miao. The analysis on the thermal environment and energy Conservation of Masonry-wood residence in southern Fujian. The Journal of Xiamen University of Technology, 2015,6(23):56-62.
2. Ministry of Construction of People's Republic of China. Architectural Lighting Design Standards GB50033-2013. Beijing: China Building Industry Press, 2013:13.
3. National Standard GB5700—1985 The measuring Methods of indoor illumination.
4. Architectural Physics. Guangzhou: The Press of South China University of Technology, 2002:202—244.
5. Lin Zurui, Li Hengyan, Simulation and study on the indoor natural lighting of traditional residences in Yingtan Village, Xingtai, Hebei Province. Huazhong Architecture, 2013(10):36-40.

MATEC Web of Conferences **61**, 04020 (2016) DOI: 10.1051/matecconf/20166104020
APOP2016

Estimated Daily Global Radiation for Zhengzhou and Its Comparison With Observations

Jingbo Liu [a], Yingni Jiang , Zhengguo Gong , Hao Yang

Architecture Engineering Department, Logistics University of PAP. Tianjin, China

Abstract. Three models relating daily global solar radiation to sunshine duration has been investigated in order to estimate daily global radiation for Zhengzhou, China. The data from January 1995 to December 2004 have been used in the models. Comparison of the estimated and measured values showed that the cubic model was able to estimate global radiation at high daily global solar radiation.

1 Introduction

The amount of solar energy reaching the surface of the earth is a primary source of renewable energy. A pre-requisite for the estimation of the daily global solar radiation on horizontal surface is the inclusion of the sky conditions in the computations at a locality for design purposes. Unfortunately, for many cases, solar radiation measurements are not easily available due to the cost and maintenance and calibration requirements of the measuring equipment. In order to achieve this, several empirical models have been developed to predict the solar radiation all over the world using various parameters. These parameters include extraterrestrial radiation, sunshine hours, mean temperature, maximum temperature, soil temperature, relative humidity, number of rainy days, altitude, latitude and cloudiness [1–10].

The most commonly used parameter for estimating global solar radiation in the publications is sunshine duration. Sunshine duration can be easily and reliably measured and data are widely available. The most widely used method is that of Angström [1], who proposed a linear relationship between the ratio of average daily global radiation to the corresponding value on a completely clear day and the ratio of average daily sunshine duration to the maximum possible sunshine duration.

The main objective of this paper is to estimate the daily global solar radiation with measured daily sunshine duration data from Zhengzhou, and to find the most suitable model for global radiation estimation at Zhengzhou, China.

2 Data used and methods of computation

2.1 Database

Zhengzhou meteorological station is chosen for the case study. Zhengzhou (34.7°N, 113.7°E) is located in the interior of China, at an altitude of 110.4 m. A database containing daily measured global solar radiation and sunshine duration of Zhengzhou station was obtained from China Meteorological administration. In this paper ten years' recorded daily solar radiation from 1995 to 2004 are used.

[a] Corresponding author: Jingbo Liu, 568818102@qq.com

2.2 Methods of computation

In the present work, the following correlations were used to express the dependence of global radiation on various parameters:

$$\frac{H_g}{H_0} = a_1 + a_2 \frac{S}{S_0} \quad (1)$$

$$\frac{H_g}{H_0} = a_3 + a_4 \frac{S}{S_0} + a_5 \left(\frac{S}{S_0}\right)^2 \quad (2)$$

$$\frac{H_g}{H_0} = a_6 + a_7 \frac{S}{S_0} + a_8 \left(\frac{S}{S_0}\right)^2 + a_9 \left(\frac{S}{S_0}\right)^3 \quad (3)$$

where H_g is daily global solar radiation on a horizontal surface, H_0 is the daily value of the extraterrestrial radiation, S is sunshine duration, S_0 is the maximum possible sunshine duration, a are empirical coefficients.

The daily value of the extraterrestrial radiation on a horizontal surface was defined as [11]:

$$H_0 = \frac{24 \times 3600}{\pi} I_0 f(\cos\lambda\cos\delta\sin\omega_s + \frac{\pi}{180}\omega_s \sin\lambda\sin\delta)$$

(4)

where I_0 is the new solar constant (=1367 W/m2 [12]), f is the eccentricity correction factor, λ is the latitude of the site, δ the solar declination and ω_s the sunrise hour angle. The eccentricity correction factor, solar declination and sunrise hour angle can be respectively calculated as [11]:

$$f = 1 + 0.33(\cos\frac{360n}{365}) \quad (5)$$

$$\delta = 23.45\sin\left[\frac{360(284+n)}{365}\right] \quad (6)$$

$$\omega_s = \arccos(-\tan\lambda\tan\delta) \quad (7)$$

where n is the number of day of year starting from first of January.

The maximum possible sunshine duration (S_0) can be obtained as [11]:

$$S_0 = \frac{2}{15}\omega_s \quad (8)$$

Fig.1 presents the plot of H_g / H_0 against S / S_0 at Zhengzhou (1995-2004). It can be seen that the relationship between daily global solar radiation and daily sunshine duration is rather strong.

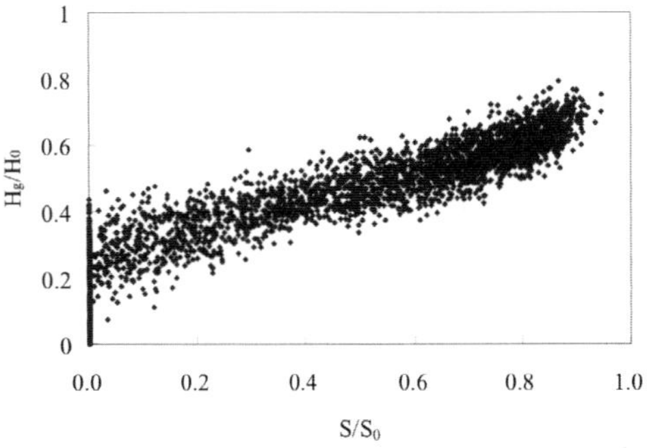

Figure 1. Relationship between H_g / H_0 and S / S_0 of Zhengzhou (1995-2004) obtained from the observed data.

2.3 Methods of comparison

In this study, three statistical tests, mean percentage error (MPE), mean bias error (MBE) and root mean square error (RMSE) were used to estimate the accuracy of the correlations described above.

The mean percentage error is defined as:

$$MPE = \frac{1}{N}\sum_{i=1}^{N}\left(\frac{D_{ie} - D_{im}}{D_{im}}\right) \times 100 \quad (9)$$

The mean bias error is defined as:

$$MBE = \sum_{i=1}^{N}(D_{ie} - D_{im}) / N \quad (10)$$

The root mean square error is defined as:

$$RMSE = \sqrt{\sum_{i=1}^{N}(D_{ie} - D_{im})^2 / N} \quad (11)$$

where D_{ie} is the ith estimated value, D_{im} is the ith measured value and N is the total number of observations.

The mean bias error test provides information on the long term performance. A low MBE is desired. A positive value gives the average amount of over estimation of an individual observation, which will cancel an under-estimation in a separate observation. The root mean square error RMSE gives information on the short term performance of the correlations by allowing a term by term comparison of the actual deviation between the calculated and measured values. The smaller the value, the better the model' s performance is.

3 Results and discussion

The obtained empirical coefficients of the Eqs.(1)-(3) are expressed as follows:

Model 1

$$\frac{H_g}{H_0} = 0.187 + 0.538\frac{S}{S_0} \quad (12)$$

Model 2

$$\frac{H_g}{H_0} = 0.172 + 0.749\frac{S}{S_0} - 0.259\left(\frac{S}{S_0}\right)^2 \quad (13)$$

Model 3

$$\frac{H_g}{H_0} = 0.158 + 1.344\frac{S}{S_0} - 2.159\left(\frac{S}{S_0}\right)^2 + 1.470\left(\frac{S}{S_0}\right)^3 \quad (14)$$

Table 1 shows MPE, MBE and RMSE errors for the three daily global radiation models (models (1-3) corresponding to Eqs.(12)-(14) respectively) of Zhengzhou station during the period 1995-2004. From this table, the MBE values obtained from models (1, 2, 3) are negative, which shows that these models under estimate global radiation. The RMSE value, which is a measure of the accuracy of estimation, has been found to be smallest for model 3 (1.939 MJ/m2), see Table 1. Also, MPE, MBE and RMSE are shown in Fig.1-3, respectively. As a result, model 3 performs better than the other models. The MPE, MBE and RMSE of model 3 are 84.7%, -0.051 MJ/m2 and 1.939 MJ/m2.

Table 1. Prediction errors for daily global solar radiation of three models.

	MPE (%)	MBE (MJ/m^2)	RMSE (MJ/m^2)
Model 1	101.9	-0.0784	2.149
Model 2	93.2	-0.0233	2.045
Model 3	84.7	-0.051	1.939

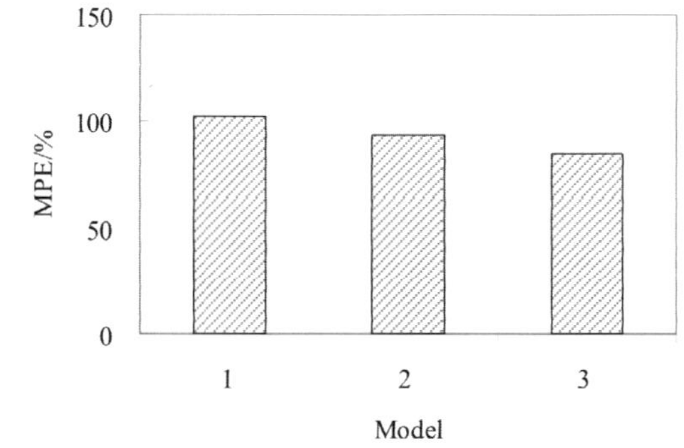

Figure 2. MPE of the three models

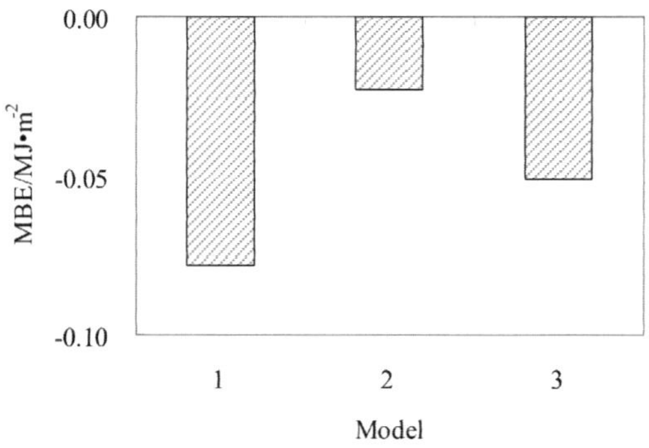

Figure 3. MBE of the three models

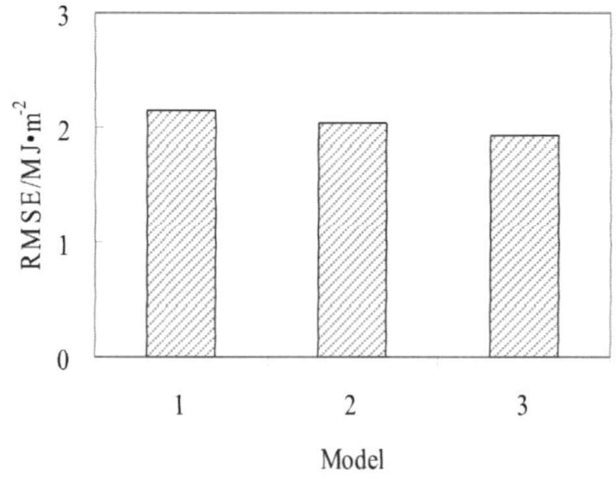

Figure 4. RMSE of the three models

To compare observed and estimated global radiation values of model 3 for Zhengzhou (1995 - 2004), Fig.5 is provided. Model 3 has large errors in estimating observed solar radiation at low daily global solar radiation. But it is of little consequence for solar energy operated systems, because low global radiation constitutes only a small part of the whole radiation. Fig.5 suggests the good fitness between the observed and the estimated values when daily global solar radiation values are high.

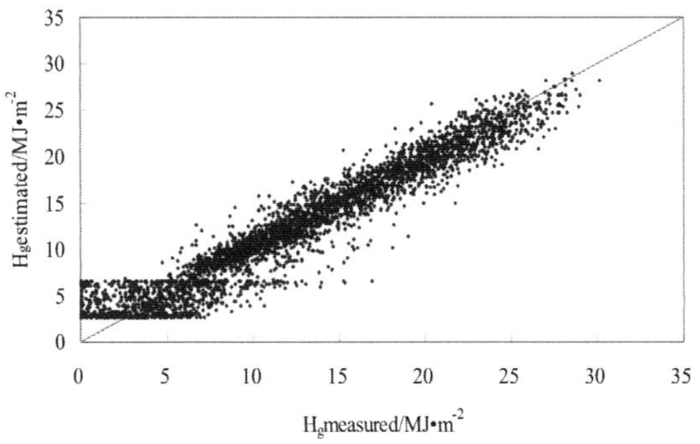

Figure 5. The measured and predicted (by model 3) values of daily global solar radiation for Zhengzhou

4 Conclusions

Three sunshine-based models for estimating daily global solar radiation on horizontal surfaces were compared with the measured data at Zhengzhou, China. Model 3 performs better than the other models. However, all these models have large MPE in estimating observed solar radiation at low daily global solar radiation. It is of little consequence for solar energy operated systems, because when the global radiation is very small, it constitutes only a small part of the whole radiation. It was found that H_g can easily be estimated using model 3 when the value of H_g is large.

Acknowledgement

The project was financially supported by Natural Science Basic Research Plan in Shaanxi Province of China (Program No. 2006E142). The author would also like to thank China Meteorological administration for its data support.

References

1. Angström A. Solar and terrestrial radiation. Quarterly Journal of Royal Meteorological Society: **50**,121-126(1924)
2. Tiba C. Analysis of monthly time series of solar radiation and sunshine hours in tropical climates. Renewable Energy: **29**, 1147-1160(2004).
3. Ehnberg JSG., Bollen MHJ. Simulation of global solar radiation based on cloud observations. Solar Energy: **78**, 157–162 (2005).
4. Skeiker K. Correlation of global solar radiation with common geographical and meteorological parameters for Damascus province, Syria. Energy Conversion and Management: **47**,331-345(2006).
5. Forero NL, Caicedo LM, Gordillo G. Correlation of global solar radiation values estimated and measured on an inclined surface for clear days in Bogota. Renewable Energy: **32**, 2590–2602(2007).
6. Sabziparvar AA, Shetaee H. Estimation of global solar radiation in arid and semi-arid climates of East and West Iran. Energy: **32**, 649–655(2007).
7. Bulut H, Büyükalaca O. Simple model for the generation of daily global solar-radiation data in Turkey. Applied Energy: **84**, 477–491(2007).
8. Bulent Yaniktepe, Yasin Alperen Genc.Establishing new model for predicting the global solar radiation on horizontal surface. International Journal of Hydrogen Energy: **40**, 15278-15283(2015).
9. Seyyed Mohammad Mousavi, Elham S. Mostafavi, Alireza Jaafari, Arefeh Jaafari, Fariba Hosseinpour.Using measured daily meteorological parameters to predict daily solar radiation. Measurement:**76**, 148 - 155(2015).
10. Milan Despotovic,Vladimir Nedic, Danijela Despotovicb, Slobodan Cvetanovic.Review and statistical analysis of different global solar radiation sunshine models. Renewable and Sustainable Energy Reviews: **52**, 1869 - 1880(2015).
11. Duffie JA,Beckman WA, Solar engineering of thermal processes [M]. (New York:Wiley, 1991).
12. Frohlich C, London J, editors, “Revised instruction manual on radiation instruments and measurements,” World climate research program, Publication Series No.7, WMO/TD-No.149. World Meteorological Organisation:Geneva, 140(1986).

Pressure Transient Analysis of Multi-Fractured Horizontal Well in Tight Gas Reservoirs

Zhao Ermeng[1, 2, a], Yin Hongjun[1, 2], Zhong Huiying[1, 2], Xu Zhitao[1, 2] and Liu Wei[1, 2]

[1]Key Laboratory of Ministry of Education PRC, Northeast Petroleum University, Daqing, Heilongjiang 163318, China
[2]Accumulation and Development of Unconventional Oil and Gas, State Key Laboratory Cultivation Base Jointly-constructed by Heilongjiang Province and the Ministry of Science and Technology, Daqing, Heilongjiang 163318, China

Abstract. Multi-fractured horizontal well is applied in tight gas reservoirs due to the low permeability. A new pressure transient model of multi-fractured horizontal well based on discrete-fracture model in which the hydraulic fractures are discretized as 2D entities is built in this paper, The model is divided into hydraulic fracture region and formation region. The model can be solved using the Galerkin finite element method, then the pressure transient type curves are plotted by computer programming. The results show that there are five different flow regimes observed in type curves including early linear flow, early radial flow, elliptical flow, later pseudo-radial flow and boundary response regime. A sensitivity analysis is conducted to study impacts of hydraulic fracture number, hydraulic fracture half-length, hydraulic fracture spacing, and hydraulic fracture conductivity on pressure transient type curves. The new model and obtained results in this paper not only enrich the well testing models, but also play a guiding role in analyzing pressure transient response of multi-fractured horizontal well in tight gas reservoirs.

1 Introduction

The research on tight gas reservoirs becomes the hot spot in petroleum development. Tight gas reservoirs have low matrix permeability, horizontal wells with multiple hydraulic fracturing technology have been an effect way to produce from tight gas reservoirs.

The pressure transient behavior of multi-fractured horizontal well has been investigated a lot, and several flow models have been developed over the past decades. Gringarten firstly solved the flow problems of two and three dimensions with Green's functions with Green's functions[1]. Ozkan developed a new procedure to generate the solutions of flow equations in porous media combining Laplace transformation with source function[2]. Based on Gringarten 's approach, Raghavan presented a semi-analytical model to study the pressure transient response characteristic of multi-fractured horizontal wells[3]. Zerzar combined the Laplace transformation and source function to present an analytical model for fractured horizontal well, the reservoir can be anisotropic or homogenous or naturally fractured, boundary can be closed or semi-infinite[4]. Al-Kobaisi presented a numerical-analytical flow model of multi-fractured horizontal wells intercepted by a vertical fracture[5]. Amini developed a distributed volumetric sources method to solve the flow problems of horizontal wells with multi-stage fractures in a box-shaped reservoir[6]. Zhao used source function to analyze the transient pressure response of the multi-fractured horizontal well in shale gas reservoir[7]. Yao and Liu obtained the analytical solution of pressure in a horizontal well with multiple hydraulic fractures by using mirror-injection and superposition theorem[8].

Ozkan and Brown developed a tri-linear flow model to study multi-fractured horizontal wells performance[9-10]. Stalgorova and Mattar improved the tri-linear flow model to simulate fractures that is surrounded by a simulated region of limited extendibility[11]. Later, Stalgorova and Mattar extended the tri-linear flow model to a five regions flow model[12]. Then some researchers (Samandarli et al, 2014; Clarkson et al, 2013, 2014) successfully used these models to study the complex well behavior exhibited in the field[13-15]. Recently, Sureshjani and Clarkson developed a new enhanced fracture-region model based on the five regions flow model, and the new model is more general and rigorous than previous models[16]. These models are simple and flexible, however, the models assume that all fractures have the same properties (fracture half-length, conductivity, and fracture aperture, etc.), as a result, some flow regimes cannot be observed.

In this paper, we present a new flow model based on discrete-fracture method of multi-fractured horizontal well in tight gas reservoirs, and the Galerkin finite element method is applied to solve the model. This model can accurately and effectively describe the effects of transient pressure behavior of multi-fractured horizontal wells, which can provide theoretical guidance for the development of multi-fractured horizontal well in tight gas reservoirs.

[a] Corresponding author: Zhao Ermeng18745953495@163.com

2 Theoretical model

2.1. Physical model

Fig. 1 illustrates the physical model of multi-fractured horizontal well used as the basis for the mathematical model developed in this work. The length of horizontal well is L, and the net pay thickness is h. The gas reservoir is stimulated with n_f hydraulic fractures, hydraulic fracture half length is x_f, and the space between fractures is d_f.

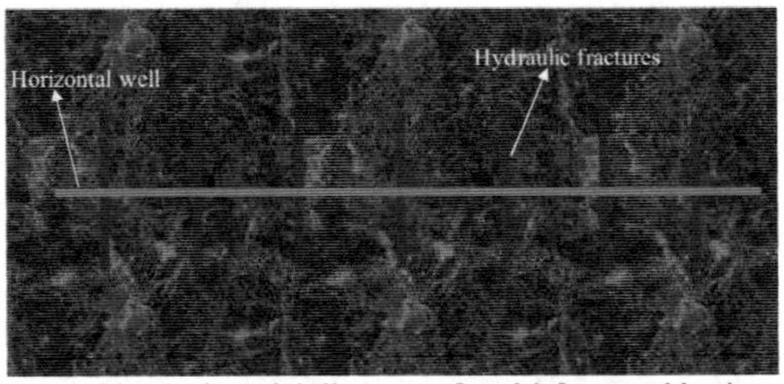

Figure 1. Physical model diagram of multi-fractured horizontal well in tight gas reservoir

The following assumptions are made to simplify the derivation of mathematical model: (1) The reservoir is homogeneous, isotropic, and bounded by an upper and lower closed boundary;(2) The reservoir is composite with the hydraulic fracture region and formation region; (3)The reservoir is produced through a multi-fractured horizontal well, and the wellbore is intersected by several fully penetrating vertical fractures; (4) The gas flow follows Darcy's law and the gravitational and frictional effects is neglected.

2.2. Mathematical model

The mass conservation equation is given by

$$-\nabla\cdot(\rho v)-\rho q\delta(M-M')=\frac{\partial(\rho\phi)}{\partial t} \tag{1}$$

where, $\nabla\cdot$ is the divergence operator, ρq is a source term, ρ is the gas density, q is the flow rate per unit volume, δ is the Dirac delta function, if M equal M', $\delta(M\text{-}M')$ is 1, otherwise $\delta(M\text{-}M')$ is 0.

The Darcy's Law in cartesian coordinate is given by

$$v=-\frac{K}{\mu}\nabla p \tag{2}$$

The rock compressibility is defined by

$$\phi=\frac{1}{\phi}\frac{\mathrm{d}\phi}{\mathrm{d}p} \tag{3}$$

After integration, it is given by

$$\phi=\phi_i e^{C_f(p-p_i)} \tag{4}$$

The real gas density can be written as

$$\rho=\frac{pM}{ZRT} \tag{5}$$

Where M is the molecular weight of gas, R is the gas constant, Z is the gas deviation factor, T is the temperature.

The isothermal gas compressibility can be expressed as follows

$$C_g=\frac{1}{\rho}\frac{\mathrm{d}\rho}{\mathrm{d}p}=\frac{1}{p}-\frac{1}{Z}\left(\frac{\mathrm{d}Z}{\mathrm{d}p}\right)_T \tag{6}$$

Substituting (2) , (4) , (5) and (2) into (1), we can see that

$$\nabla\cdot\left(K\frac{p}{\mu Z}\nabla\cdot p\right)-\frac{p}{Z}q\delta(M-M')=\frac{p}{Z}\phi C_t\frac{\partial p}{\partial t} \tag{7}$$

Where, C_t is the total compressibility, $C_t=C_g+C_f$.

To simply the equation, the real gas pseudo-pressure is introduced, which can be defined as

$$\psi(p)=2\int_{p_0}^{p}\frac{p}{\mu Z}\mathrm{d}p \tag{8}$$

The diffusivity equation becomes

$$K\Delta\psi-\frac{2p}{Z}q\delta(M-M')=\phi\mu C_t\frac{\partial\psi}{\partial t} \tag{9}$$

Where, Δ is the Laplacian operator, $\Delta=\frac{\partial^2}{\partial x^2}+\frac{\partial^2}{\partial y^2}+\frac{\partial^2}{\partial z^2}$.

For hydraulic fractures, the diffusivity equation of pseudo-pressure can be written as

$$K_F\Delta\psi_F-\frac{2p}{Z}q\delta(M-M')=\phi_F\mu C_{tF}\frac{\partial\psi_F}{\partial t} \tag{10}$$

The initial condition for this problem is

$$\psi_F\big|_{t=0}=\psi_i \tag{11}$$

Where, ψ_i is the original reservoir pseudo-pressure.

It is assumed that the hydraulic fractures and reservoir region are perfectly connected, so that pseudo-pressure is continuous at the interfaces

$$\psi_F\big|_{\partial\Omega_R\,I\,\partial\Omega_F}=\psi_R\big|_{\partial\Omega_R\,I\,\partial\Omega_F} \tag{12}$$

The diffusivity equation for reservoir region is given by

$$K_R\Delta\psi_R=\phi_R\mu C_{tR}\frac{\partial\psi_R}{\partial t} \tag{13}$$

The initial condition is given by

$$\psi_f\big|_{t=0}=\psi_m\big|_{t=0}=\psi_i \tag{14}$$

Pseudo-pressure continuous condition between the natural fractures and hydraulic fractures is given by

$$\psi_{\mathrm{f}}\big|_{\partial\Omega_{\mathrm{f}}\cap\partial\Omega_{\mathrm{F}}}=\psi_{\mathrm{F}}\big|_{\partial\Omega_{\mathrm{f}}\cap\partial\Omega_{\mathrm{F}}} \tag{15}$$

No-flow condition at the outer reservoir boundary is given by

$$\frac{\partial\psi_{\mathrm{R}}}{\partial x}\Big|_{x=x_e}=\frac{\partial\psi_{\mathrm{R}}}{\partial y}\Big|_{y=y_e}=\frac{\partial\psi_{\mathrm{R}}}{\partial z}\Big|_{z=z_e}=0 \tag{16}$$

Dimensionless variables are defined by

$$\psi_{j\mathrm{D}}=\frac{\pi K_{\mathrm{f}}hT_{\mathrm{sc}}}{q_{\mathrm{sc}}p_{\mathrm{sc}}T}(\psi_{\mathrm{i}}-\psi_j)\qquad (j=\mathrm{F,R}),$$

$$t_{\mathrm{D}}=\frac{K_{\mathrm{R}}t}{\phi_{\mathrm{R}}\mu C_{\mathrm{tR}}L^2},\ x_{\mathrm{D}}=\frac{x}{L},$$

$$y_{\mathrm{D}}=\frac{y}{L},\ z_{\mathrm{D}}=\frac{z}{L},\ q_{\mathrm{D}}=\frac{qZ_{\mathrm{sc}}T_{\mathrm{sc}}p_{\mathrm{wf}}}{q_{\mathrm{sc}}ZTp_{\mathrm{sc}}}hL^2.$$

Where, F, R are hydraulic fractures and reservoir region respectively.

So the dimensionless mathematical model of hydraulic fractures can be written as

$$\begin{cases}\dfrac{K_{\mathrm{F}}}{K_{\mathrm{R}}}\Delta\psi_{\mathrm{FD}}+2\pi q_{\mathrm{D}}\delta(M_{\mathrm{D}}-M'_{\mathrm{D}})=\dfrac{\phi_{\mathrm{F}}C_{\mathrm{tF}}}{\phi_{\mathrm{R}}C_{\mathrm{tR}}}\dfrac{\partial\psi_{\mathrm{FD}}}{\partial t_{\mathrm{D}}}\\ \psi_{\mathrm{FD}}\big|_{t_{\mathrm{D}}=0}=0\\ \psi_{\mathrm{FD}}\big|_{\partial\Omega_{\mathrm{R}}\cap\partial\Omega_{\mathrm{F}}}=\psi_{\mathrm{RD}}\big|_{\partial\Omega_{\mathrm{R}}\cap\partial\Omega_{\mathrm{F}}}\end{cases} \tag{17}$$

Dimensionless mathematical model of reservoir region can be written as

$$\begin{cases}\Delta\psi_{\mathrm{RD}}=\dfrac{\partial\psi_{\mathrm{RD}}}{\partial t_{\mathrm{D}}}\\ \psi_{\mathrm{RD}}\big|_{t_{\mathrm{D}}=0}=0\\ \psi_{\mathrm{RD}}\big|_{\partial\Omega_{\mathrm{R}}\cap\partial\Omega_{\mathrm{F}}}=\psi_{\mathrm{FD}}\big|_{\partial\Omega_{\mathrm{R}}\cap\partial\Omega_{\mathrm{F}}}\\ \dfrac{\partial\psi_{\mathrm{RD}}}{\partial x_{\mathrm{D}}}\Big|_{x_{\mathrm{D}}=x_{\mathrm{eD}}}=\dfrac{\partial\psi_{\mathrm{RD}}}{\partial y_{\mathrm{D}}}\Big|_{y_{\mathrm{D}}=y_{\mathrm{eD}}}=\dfrac{\partial\psi_{\mathrm{RD}}}{\partial z_{\mathrm{D}}}\Big|_{z_{\mathrm{D}}=z_{\mathrm{eD}}}=0\end{cases} \tag{18}$$

3 Solution of the mathematical model

In this paper, hydraulic fractures are described by discrete fracture model (DFM). The main idea of the DFM is a 2D representation of the fractures[17].

Galerkin weighted residual method and finite-element discretization are used in this work to solve the equations. The reservoir geometry is discretized using tetrahedron elements for the reservoir region and triangular elements for the hydraulic fractures.

The hydraulic fracture element pseudo-pressure is approximated as

$$\psi_{\mathrm{FD}}^{e}=N_{\mathrm{F1}}\psi_{\mathrm{FD1}}^{e}+N_{\mathrm{F2}}\psi_{\mathrm{FD2}}^{e}+N_{\mathrm{F3}}\psi_{\mathrm{FD3}}^{e} \tag{19}$$

Where ψ_{FD1}^{e}, ψ_{FD2}^{e} ,and ψ_{FD3}^{e} are pseudo-pressure of the three nodes, N_{F1}, N_{F2}, and N_{F3} are piecewise-linear interpolation functions defined on triangles.

Then, the discrete form of the Eq. 17 can be written as

$$\begin{aligned}&w\frac{K_{\mathrm{F}}}{K_{\mathrm{f}}}\iint_{\bar\Omega_{F,e}}\nabla\boldsymbol{N}_{e,F}^{\mathrm{T}}\nabla\boldsymbol{N}_{e,F}\mathrm{d}\bar\Omega_{F,e}\boldsymbol{\psi}_{\mathrm{FD}}^{e}\\&+w\frac{\phi_{\mathrm{F}}C_{\mathrm{tF}}}{\phi_{\mathrm{m}}C_{\mathrm{tm}}+\phi_{\mathrm{f}}C_{\mathrm{tf}}}\iint_{\bar\Omega_{F,e}}\boldsymbol{N}_{e,F}^{\mathrm{T}}\boldsymbol{N}_{e,F}\mathrm{d}\bar\Omega_{F,e}\frac{\partial\boldsymbol{\psi}_{\mathrm{FD}}^{e}}{\partial t_{\mathrm{D}}}\\&-2\pi w\iint_{\bar\Omega_{F,e}}\boldsymbol{N}_{e,F}^{\mathrm{T}}q_{\mathrm{FD}}\delta(M_{\mathrm{D}}-M'_{\mathrm{D}})\mathrm{d}\bar\Omega_{F,e}=0\end{aligned} \tag{20}$$

Where $\boldsymbol{N}_{e,F}=\left[N_{\mathrm{F1}}\ N_{\mathrm{F2}}\ N_{\mathrm{F3}}\right]$, $\boldsymbol{\psi}_{\mathrm{FD}}^{e}=\left[\psi_{\mathrm{FD1}}^{e}\ \psi_{\mathrm{FD2}}^{e}\ \psi_{\mathrm{FD3}}^{e}\right]^{\mathrm{T}}$,

The same operation is implemented to tetrahedron elements for reservoir region. The pseudo-pressure of formation region is approximated as

$$\psi_{\mathrm{RD}}^{e}=N_{\mathrm{R1}}\psi_{\mathrm{RD1}}^{e}+N_{\mathrm{R2}}\psi_{\mathrm{RD2}}^{e}+N_{\mathrm{R3}}\psi_{\mathrm{RD3}}^{e}+N_{\mathrm{R4}}\psi_{\mathrm{RD4}}^{e} \tag{21}$$

Where ψ_{RD1}^{e}, ψ_{RD2}^{e}, ψ_{RD3}^{e} ,and ψ_{RD4}^{e} are four pseudo-pressure of reservoir region, N_{R1}, N_{R2}, N_{R3} ,and N_{R4} are piecewise-linear interpolation functions defined on tetrahedrons.

Then, the discrete form of Eq.18 are as follows

$$\iiint_{\Omega_{R,e}}\nabla\boldsymbol{N}_{e,R}^{\mathrm{T}}\nabla\boldsymbol{N}_{e,R}\mathrm{d}\Omega_{R,e}\boldsymbol{\psi}_{\mathrm{RD}}^{e}+\iiint_{\Omega_{R,e}}\boldsymbol{N}_{e,R}^{\mathrm{T}}\boldsymbol{N}_{e,R}\mathrm{d}\Omega_{R,e}\frac{\partial\boldsymbol{\psi}_{e,\mathrm{RD}}}{\partial t_{\mathrm{D}}}=0 \tag{22}$$

Where, $\boldsymbol{N}_{e,R}=\left[N_{\mathrm{R1}}\ N_{\mathrm{R2}}\ N_{\mathrm{R3}}\ N_{\mathrm{R4}}\right]$,

$\boldsymbol{\psi}_{\mathrm{RD}}^{e}=\left[\psi_{\mathrm{RD1}}^{e}\ \psi_{\mathrm{RD2}}^{e}\ \psi_{\mathrm{RD3}}^{e}\ \psi_{\mathrm{RD4}}^{e}\right]^{\mathrm{T}}$,

Combining the equations of reservoir region and hydraulic fractures by the relationship of nodes, the only thing we should notice is that hydraulic fractures and reservoir region have the same node, as below

$$\boldsymbol{A}^{R}\boldsymbol{\psi}_{\mathrm{RD}}+\boldsymbol{B}^{R}\frac{\partial\boldsymbol{\psi}_{\mathrm{RD}}}{\partial t_{\mathrm{D}}}=\boldsymbol{Q}^{f} \tag{23}$$

Where,

$$\boldsymbol{A}^{R}=\sum_{e=1}^{N_R}\iiint_{\Omega_{R,e}}\nabla\boldsymbol{N}_{e,R}^{\mathrm{T}}\nabla\boldsymbol{N}_{e,R}\mathrm{d}\Omega_{R,e}+w\frac{K_{\mathrm{F}}}{K_{\mathrm{R}}}\sum_{e=1}^{N_F}\iint_{\bar\Omega_{F,e}}\nabla\boldsymbol{N}_{e,F}^{\mathrm{T}}\nabla\boldsymbol{N}_{e,F}\mathrm{d}\bar\Omega_{F,e},$$

$$\boldsymbol{B}^{R}=\sum_{e=1}^{N_R}\iiint_{\Omega_{R,e}}\boldsymbol{N}_{e,R}^{\mathrm{T}}\boldsymbol{N}_{e,R}\mathrm{d}\Omega_{R,e}+\frac{w\phi_{\mathrm{F}}C_{\mathrm{tF}}}{\phi_{\mathrm{R}}C_{\mathrm{tR}}}\sum_{e=1}^{N_F}\iint_{\bar\Omega_{F,e}}\boldsymbol{N}_{e,F}^{\mathrm{T}}\boldsymbol{N}_{e,F}\mathrm{d}\bar\Omega_{F,e},$$

$$\boldsymbol{Q}^{R}=2\pi w\sum_{e=1}^{N_F}\iint_{\bar\Omega_{F,e}}\boldsymbol{N}_{e,F}^{\mathrm{T}}q_{\mathrm{FD}}\delta(M_{\mathrm{D}}-M'_{\mathrm{D}})\mathrm{d}\bar\Omega_{F,e}.$$

To get the solutions, the time derivative terms are discretized by forward difference. We can obtain the (n+1)th pseudo-pressure of reservoir region by nth pseudo-pressure, as below

$$(\boldsymbol{A}^{R}+\frac{\boldsymbol{B}^{R}}{\Delta t_n})\boldsymbol{\psi}_{\mathrm{RD}}^{n+1}=\boldsymbol{Q}^{R}+\frac{\boldsymbol{B}^{R}}{\Delta t_n}\boldsymbol{\psi}_{\mathrm{RD}}^{n} \tag{24}$$

Where Δt_n is the timestep.

DOI: 10.1051/matecconf/20166104021

4 Discussion of the results

As shown in Fig 2, the pseudo-pressure and its derivative curves can be divided into five flow regimes. (1) Early linear flow regime, the pseudo-pressure and its derivative curve have a slope of 1/2;(2) Early radial flow regime, the pseudo-pressure derivative curve is seems like a horizontal straight line;(3) Elliptical flow regime, which is characterized by a slope of 0.36 on the pseudo-pressure derivative curve in this regime;(4) Later pseudo-radial flow regime, the pseudo-pressure derivative curve is almost a horizontal straight line;(5) Boundary response regime, the pressure has spread to the outer boundary for closed outer boundary, the pseudo-pressure and its derivative curve tend to an asymptote with a slope of 1.

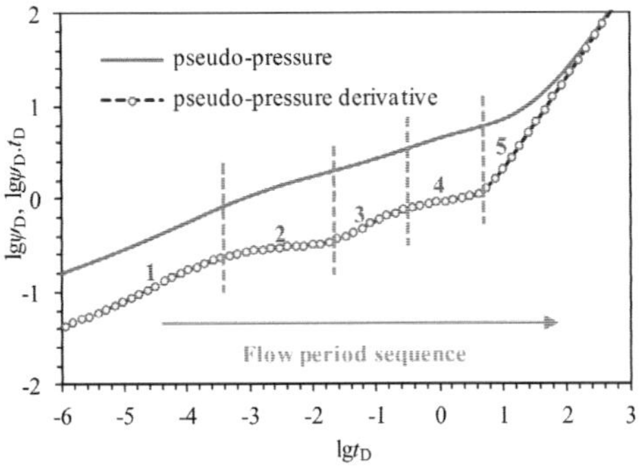

Figure 2. Pseudo-pressure and its derivative curves

Fig.3 shows the effect of hydraulic fracture number (n_f) on pseudo–pressure and its derivative curves. As shown in Fig.3, the n_f has great effect on early and intermediate regimes of the curves, with the decrease of n_f, the occurrence time of elliptical flow shows later, even disappears when there is only one hydraulic fracture. Meanwhile, pressure drop will be smaller for more hydraulic fractures because the increased formation permeability around wellbore caused by hydraulic fractures will decrease the pressure drop around wellbore.

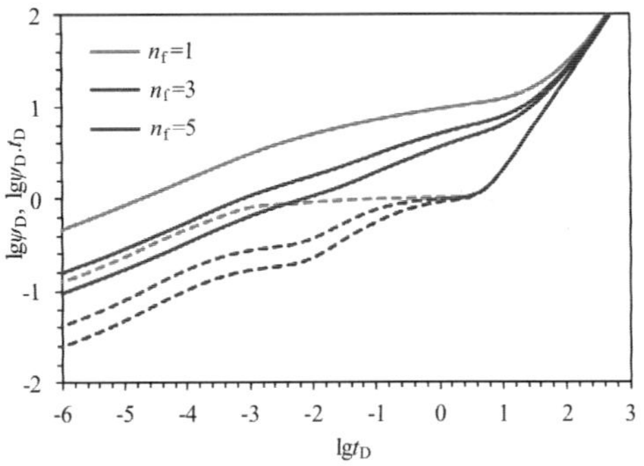

Figure 3. Effect of n_f on the type curves

Fig.4 shows the effect of hydraulic fracture half-length (x_f) on pseudo–pressure and its derivative curves. The effect of x_f on the curves is mainly on early linear flow and early radial flow regime, with the decrease of x_f, the appearance time of early radial flow will be earlier, and the longer the duration will last. In addition, the pressure drop will become larger for a bigger x_f, because the formation permeability will be bigger around wellbore caused by hydraulic fractures.

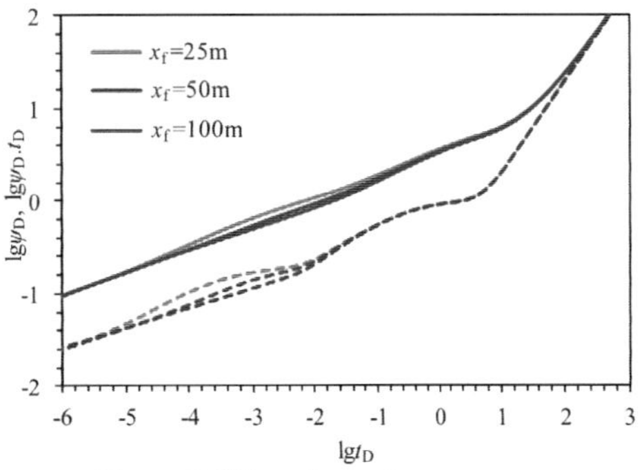

Figure 4. Effect of x_f on the type curves

Fig.5 shows the effect of hydraulic fracture spacing (d_f) on pseudo–pressure and its derivative curves. With the increase of d_f, the early radial flow becomes longer and then the occurrence of elliptical flow regime postpones. In addition, the smaller d_f shows, the larger the corresponding pressure drop becomes, this is mainly because the interference between the hydraulic fractures will be more serious for a smaller d_f.

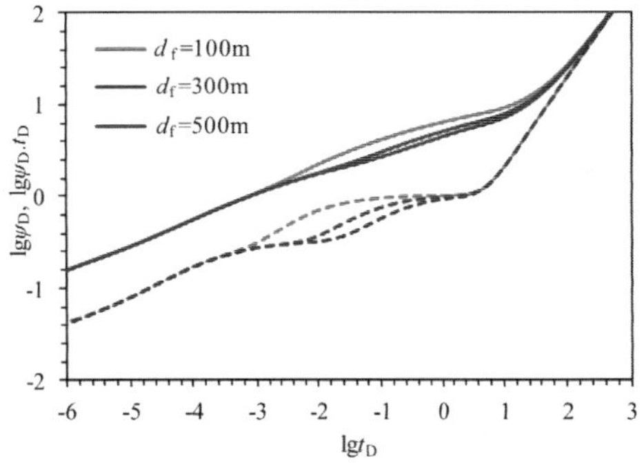

Figure 5. Effect of d_f on the type curves

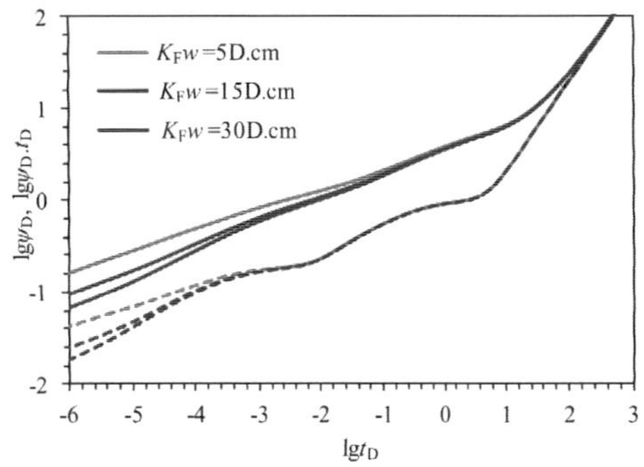

Figure 6. Effect of K_{Fw} on the type curves

Fig.6 shows the effect of hydraulic fracture conductivity (K_{Fw}) on pseudo–pressure and its derivative curves, hydraulic fracture conductivity is the product of hydraulic fracture permeability and its width. It can be seen that the K_{Fw} mainly affects the early linear flow regime, and the bigger K_{Fw} is, the the smaller the pressure drop will be.

5 Conclusions

(1) In this paper, a new pressure transient model of multi-fractured horizontal well based on discrete-fracture model has been built and solved by finite element method.

(2) In this new pressure transient model, five main flow regimes can be observed in transient pressure and its derivative curves.

(3) Hydraulic fracture parameters have great effect on the pressure transient curves. With the decrease of hydraulic fractures, the occurrence time of elliptical flow shows later, even disappears when there is only one hydraulic fracture. The smaller the hydraulic fracture half-length is, the earlier the early radial flow will appear, and the longer the duration will last. The smaller the hydraulic fracture spacing shows, the larger the corresponding pressure drop becomes.

Acknowledgment

The authors would like to acknowledge the financial support of the Science and Technology Project of China Petroleum and Chemical Industry Association (2014-01-02).

References

1. Gringarten, A. C., Ramey, H. J., 1973. The use of source and green's function in solving unsteady-flow problem in reservoir. SPE Journal, 285-296, SPE 3818.
2. Ozkan, E., 1988: "Performance of horizontal wells," Ph.D. Dissertation, The University of Tulsa, May (1988).
3. Raghavan, R., Chin-Cheng Chen, Bijin Agarwal, 1997. An analysis of horizontal wells intercepted by multiple fractures. SPE Journal, SPE-27652.
4. Zerzar, A., Bettam, Y., Tiab, D., 2003. Interpretation of multiple hydraulically fractured horizontal wells in closed systems. In: SPE 84888 presented at SPE International Improved Oil Recovery Conference in Asia Pacific, Kuala Lumpur, Malaysia.
5. Al-Kobaisi M., Ozkan E., 2004. A hybrid numerical analytical model of finite conductivity vertical fractures intercepted by a horizontal well. In: SPE 92040 presented at SPE International Petroleum Conference in Puebla, Mexico.
6. Valko, P.P. and Amini, S., 2007. The method of distributed volumetric sources for calculating the transient and pseudosteady-state productivity of complex well-fracture configuration. In: SPE Paper 106279 Presented at the SPE Hydraulic Fracturing Technology Conference, College Station, TX.
7. Zhao, Y. L., Zhang, L. H., Zhao, J. Z, et al., 2013. "Triple porosity"modeling of transient well test rate decline analysis for multi-fractured horizontal well in shale gas reservoirs. J. Petrol. Sci. Eng. 110, 253e262.
8. Yao, J., Liu, P. Y., Wu, M. L., Well test analysis of fractured horizontal well in fractured reservoir. Journal of China University of Petroleum, 2013, 37(5):107-113+119.
9. Ozkan, E., Brown, M., Raghavan, R., et al., 2009. Comparison of fractured horizontal well performance in conventional and unconventional reservoirs. In: SPE Paper 121290-MS Presented at SPE Western Regional Meeting, San Jose, California, USA.
10. Ozkan, E., Brown, M., Raghavan, R., et al., 2011. Comparison of fractured-horizontal well performance in tight sand and shale reservoirs. SPE Reser. Eval. Eng. 14 (2), 248–259.
11. Stalgorova, E., Mattar, L., 2012a. Practical Analytical model to simulate production of horizontal wells with branch fractures. In: SPE Paper 162515 Presented at SPE Canadian Unconventional Resource Conference, Calgary, Alberta, Canada.
12. Stalgorova, E., Mattar, L., 2012b. Analytical model for history matching and forecasting production in multifrac composite systems. In: SPE Paper 162516-MS Presented at SPE Canadian Unconventional Resource Conference, Calgary, Alberta, Canada.
13. Samandarli, O., McDonald, B., Barzola, G., et al., 2014. Understanding shale performance: performance analysis workflow with analytical models in eagle ford shale play. In: SPE Paper 169004-MS Presented at the SPE Unconventional Resources Conference, Woodlands, Texas, USA.
14. Clarkson, C. R. 2013. Production data analysis of unconventional gas wells: workflow. Int. J. Coal Geol. 109–110 (1 April): 147–157.
15. Clarkson, C. R., Williams-Kovacs, J. D., Qanbari, F., et al., 2014. History matching and forecasting tight/shale gas condensate wells using combined analytical, semi- analytical, and empirical methods. In: SPE Paper 171593-MS Presented at the SPE/CSUR Unconventional Resources Conference–Canada, Calgary, Alberta, Canada.
16. Sureshjani M. H., Clarkson, C. R.. An analytical model for analyzing and forecasting production from multifractured horizontal wells with complex branched-fracture geometry. SPE Reservoir Evaluation & Engineering, 2015, 18(3). SPE-176025-PA.
17. Karimi-Fard M. and Firoozabadi A., Numerical simulation of water injection in fractured media using the discrete-fracture model and the galerkin method. SPE Reservoir Evaluation & Engineering, 2003, 6(2), 117–126. SPE-83633-PA.

The Application of Smart Textiles in the Brand Fashion Design

Hong-Ying Deng[1,a] and Yu-Mei Cui [2,b]

[1]College of Fashion and Design, Donghua University, West Yan'an Road No.1882, Shanghai, China, 200051
[2]College of Fashion and Design,Donghua University, West Yan'an Road No.1882,Shanghai, China, 200051

[a]ypnypj@126.com, [b]cymyhyez@126.com

Abstract. With the economic and social development, material life in the era of abundance is not only in meeting the basic needs of life, but also modern consumers become forced to pursue the spiritual and cultural needs. On the other side, clothing will not just fulfill the basic functions of beauty and suitability, thus, more consumers begin to pay closer attentions to apparel textile's individuation expression and technological elements, or to some other deeper emotional requirements etc. Smart textiles originally belongs to the cutting-edge scientific field of fashion industry, however, with the booming development of internet industry and smart phone devices, acute apparel manufacturers must have to take a ride on advanced tech-trends and launch a wide expansion of smart textile fibres's applications into the clothing industry. This thesis would present a basic introduction on the concept and classifications of the smart textile fibres, and then like to deploy a profound analysis of smart textiles applied in the brand fashion design.

Keywords: Smart Textiles, Brand, Fashion Design, Technology and Internet.

1 Introduction

The wearable smart technology has always been concerned among the computer academia and industry. Because its high cost and technical sophistication, most of all for its comparably low consumption demands, wearable smart mobile devices has just remained at the conceptualization phase. Along with the development of mobile Internet, advanced technology and the rising concerns of the fashion industry on the smart textiles, this newly-type textile materials then begin to come upon on a commercial mode. For example, international technology giants (like Google or Apple) have increased wearable technology investment into the fashion design field, which in turn, the whole fashion industry would then correspondingly gain several inspirations. Manufactures of different levels of clothing markets, no matter the mass market or Paris's haute couture, all have to anticipate in a more intense market competition. Textile materials, as the most basic component of fashion design, play a key role, therefore, some famous designer brands almost ready to launch a full swing, boldly apply the smart textile fibres, which being easy to human body, for their target to improve the odds on increasing the sales amounts in the fierce commercial battle.

2 A basic introduction to smart textiles

2.1 A Basic Concept of Smart Textiles

The so-called intelligence refers that clothing obtains the function of reading human body's language, or by saying, clothing functions would change with the changing human body and environment. It is an indication that intelligent apparel forms the functions of saving energy and information but also conveying or transferring. [1]Smart textiles means the simulation of living bio-system, and meantime both combines with perceiving and driving actions. More specifically, this type of textiles cannot just perceive a subtle alteration at the inner state or the outer atmosphere, but also through itself feedback mechanism responses simultaneously changes of the properties as expected. Perception, feedback and response, the main three elements, altogether make up smart materials. Smart textile materials defines as an immediate response stimulated by environmental conditions or other factors, but temporarily still retains the inherent nature and clothing functions. [2]

2.2 The Main Classifications of Smart Textiles in the Apparel Application

The production of clothing by smart textiles has possessed additional functions in the daily life or work routine; as a consequence, smart textile material would

Corresponding author: Yu-Mei Cui, cymyhyez@126.com

become more focused and popular. Smart textile fibres and fabrics could stimulate the outside environment(such as heat, chemistry, light, humidity, machinery, electromagnetism etc.), good self-adaptive, self-judging, and self-healing abilities actually cannot also lack either.[3] In general, smart textile fibres has main classifications as followed: memory fibres, cameleon fibres, thermo-regulated fibers, smart gel fibres and electronic intelligent fibers, etc.

2.3 The Development Tendency of Smart Textiles

Nowadays, smart textiles have been at a state of rapid advance along with the development of science and technology, especially the wide population of internet technology and smart mobile devices has forced intelligent textiles to come into the mass marketization. From early period's single function to multi-purpose directions, it again combines technology of smart mobile devices and Bulk data handling techniques, then develops multi-functions: health monitoring, media entertainment system, GPS position(Global Positioning System) etc, which formerly needs to be configured separately by certain electronic equipment.

Green and Environmental Protection still stays a constant theme in the near future, naturally, the development of smart textiles should coherently run towards an ecological green direction. Apparel manufacturing process must be strictly under the scientific monitor and detection without causing any harm to human beings and environment, thus, how to convert sustainable, recycled, reclaimed materials to smart textiles would be a must future tendency and important research direction.

3 The Application Analysis of Smart Textiles in the Brand Fashion Design

3.1 The Importance of Smart Textiles' Application in the Brand Fashion Design

Under the background of economic globalization, information and techniques and technology would no longer be the barriers between competitors among the clothing enterprise. And nowadays as for a more heated white-hot competition in the international fashion market, customers behave more particular about products influenced by individual, scientific frontier philosophy. Therefore, fashion brands cannot only rely on the marketing team to make up a fantastic and touching thematic story, or just struggle to find a catchy phrase, or say a single core item cannot meet complete needs of target customers, which has not attached any extravagant hope to it. At present, consumers become more concerned about the visual effects and comfort index when dressing clothes, to more excitements, it could not be better when affording fancy high-tech products experiences.

In terms of the applications of smart textiles, the most commonly known one is the space suit wearing by space astronauts. It is undeniable that a space suit is a mixture of the most advanced sci-tech levels which smart textiles can keep, but in the daily life smart textiles become most widely applied by some sports clothing brands. For example, at the World Cup Brazil in 2014, PUMA put *ACTV* (a type of sport elastic tape) into the clothes, which could provide a massage function with a slight effect of strength. It even helps working muscles deliver energy more quickly and efficiently, then sequentially boosting players' match performance. In 2012, Hot pants launched by Adidas were developed especially for British cycling team, this statement racing suit can be conductive to keeping warm temperature of athletes, then obtain more sustained motivation. [4] Sports brands become the foregoers of applying smart textiles, which also proves that the application of smart textiles plays a more important role in the brand fashion design.

3.2 Smart Textiles' Brand Clothing to be a Consumers' Demand

Today in the 21st century, modern people praise highly lifestyles of personality, high-tech and intelligence; intellectualization gradually develops into an inevitable tendency. Daily trips or maintenance can be wholly implemented by smart mobile devices with just a swipe of your fingers. On the other hand, particular intellectualization of clothing to a great extent would stride to be more popularized or routinized. For instance, Madura from India has recently released *Icetouch* series shirts, which can react to lower body surface temperature by five degree and play a function of intelligence temperature control. In addition, famous lingerie brand Victoria's secret has also launched a functional bra of *Incredible* combined with Heart Rate Monitoring band, and it could check for heart rate concurrently, which in turn escort for body health. [4]From above clothing brands cases, they show willingly to adopt smart textiles to meet consumers' higher needs, therefore, without any exaggeration, it proves to be a role of boosting the application of smart textiles among fashion industry.

4. Features of Smart Textiles as a Perfect Interpretation in the Brand Fashion Design

4.1 The Recreational function of Smart Textiles in the Brand Fashion Design

The integration of music and apparel makes the application of smart textiles more fresh and amusing; there is no doubt that music has been an essential way of entertainment and leisure for most in the daily life. Though music player can be made to be a small and exquisite looking, yet try to imagine, if what we dress itself can furnish like a music player, it appeals promising market attractiveness factors among young communities. Now in Europe, famous jeans brand Levi's has just popularized such a music jacket. Even more wonderful, it can be provided as FM radio including saving and playing functions. For the purpose of such mentioned functions, it needs additional energy to absorb solar energy, wind energy, thermal energy and other physical energy sources, which altogether perfectly interprets smart textiles' wide applications in the brand fashion design. Moreover, Alyce Santoro personally invented an innovative Sonic fabric, the sonic is a durable and versatile audible fabric made from 50% recycled polyester and 50% reclaimed audio-cassette tape, which is formed by the hues of the polyester warp combined with the cassette tape weft. It is extremely durable, but also remarkably soft even with a sparkly sheen. The strands of tape could form a collage of sounds, such as, Beethoven, the Beatles, and sounds collected in the Peruvian rain forest etc.

4.2 The Health Care Function of Smart Textiles in the Brand Fashion Design

Scientists make an attempt to insert embedded photoconduction fibres into clothing fabrics, and then this sort of smart textile fibres can anytime inspect human beings' heart rate and respiratory rate. It certifies a concentrated embodiment and main symbol about smart textile fibres's medical care values. For example, American Textronics has already introduced the technology to produce several items of underwear or sport suit, equipped with functions of checking heart rate, respiratory rate, temperature and blood pressure or other life cycle information. Once this type of smart clothes was widely used for prevention from SIDS(Sudden Infant Death Syndrome)[5]. In all, it passes for a perfect annotation on the promising application of smart textiles in the brand fashion design.

4.3 The High-tech Features of Smart Textiles in the Brand Fashion Design

Smart Textiles incorporates with many relevant fields of high-tech, such as, Bulk data analysis technology, sensor technology, wireless network technique and etc. After adding special performance of materials, smart textiles can then turn into a data receiver formed by numberless sensors, which by virtue of combining wireless network technique and smart phones or computers finally achieve real time GPS positioning. However, previously, it only can be implemented functionally by expensive instruments and apparatus for analyzing motion data. For instance, Nike company together hand to cooperate with Google or Apple launch gym shoes, which could track real-time footprints by use of Google Earth, wearing this pair of shoes cannot check movement routes at any time, but control different massive motion data. [5] High-tech features can be greatly combined with smart clothing, which again proves a perfect presentation in the brand fashion design.

5 Conclusions

Smart textiles, considered as a new classification of apparel fabrics, simultaneously obtain entertainment, medical care, and other high-tech properties in the integrated; therefore, smart textiles can be served as ideal materials for fashion designers. However, smart textiles have just been applied at an initial stage. Internet technology's soaring development, Bulk Data widening application, and more concentrations by modern people, smart textiles still has a vast market prospect. For the sake of brand clothing becoming received by more market favors, how to develop smart textile fibres tends to be a step of necessity.

References

1. Zhu Songwen, The New Development of Clothing Fabrics, Cotton Textile Technology, 1999.08.
2. Cao Lihui, A Research on the Development and Application of Smart Clothing Fabrics, Journal of Zhejiang Textile & Fashion Vocational College, 2010.09:No.3.
3. Yao Lianzhen, Yang Wenfang, Qiao Yanli, The Research Progress of Smart Textile Fibres and Fabrics, Textile Printing: Textile College of Tianjin Polytechnic University, 2012.No.12.
4. Information on http://qtnews.zjol.com.cn/culture/text.asp?id=517130.
5. Information on http://gold.hexun.com/2008-10-20/110131964.html.

Analysis on Real time Perception Technology of Wireless Sensor Network in CPS

Zhou Benhai[1,a]

[1]Teaching Department of Computer Science, Shenyang Institute of Engineering, Shenyang City, Liaoning Province, 110136, China

Abstract. Cyber Physical Systems (CPS) combines physical and computing systems tightly. Node operating systems (OS) are fundamental units in CPS. There are still many problems unsolved when designing CPS especially CPS node OS in aspects of predictability, reliability, robustness, etc. Aiming at the problem, this paper proposes the effective shortest time priority algorithm and the adaptive shortest time priority algorithm. Experimental results show that, compared with the traditional FIFO (advanced first in first out) and LSF (least slack first) algorithm, the algorithm proposed in this paper effectively reduce the deadline miss ratio, as a result, the real-time performance of the CPS are effectively improved.

1 Introduction

Cyber and physical system (CPS) is a complex system, which contains the network and the physical environment. CPS is closely integrated with many technologies, such as computing, communication and control. The system must be aware of the constraints caused by environmental changes in real time. At present, the CPS application in the medical, military and other related countries and people's livelihood in important areas, has a very important role, which is a higher requirement for real-time performance of the system. Therefore, the research of CPS real-time scheduling algorithm can meet the requirements of high real-time performance of CPS system, and it is also conducive to accelerate the construction of information in various fields[1-7].

Since CPS have diverse computational and resource characteristics, they are complicated and there are still many research challenges. One example is how to describe architecture of cyber physical systems. In this paper, we propose a 4-layer architecture for CPS.

This layer is at the bottom of CPS architecture including a large number of CPS nodes on which node operating systems are running. This layer is directly responsible for the interaction with physical systems. Nodes collect physical data. They not only transmit data up to the upper layer, but also accept feedback data. They control physical operations of reactors. Node operating system controls interaction devices and other hardware facilities of nodes to complete the feedback loop.

2)Data Internet Layer

This layer should receive sensor data and send control data seamlessly to ensure information two-way exchange within horizontal and vertical systems transparently.

3) Information Integrating Layer

The data collected from lower layers are different in types and formats. Even some data are uncertain. So CPS must do the corresponding work (data cleaning, data reorganizing, data fusion, etc) to pick out useful information from raw data. Meanwhile, this layer also transforms the reverse interaction information to corresponding control data for distribution to the lower layer.

4)Application Service Layer

This layer is the interface layer between CPS and upper applications. It could encapsulate all functions from lower layers as various services to provide to applications as interface specifications or standards.

Real time scheduling is an important part of real-time system[6-8] and it is also a critical part of CPS system, which is the key to guarantee the real-time task. It is also a widely studied problem. In the current research, a lot of real-time scheduling methods are proposed for various task types. But in the CPS, important parameters of real-time tasks with the release time and deadline will be more dependent on nodes in the CPS location and migration time and other factors. Therefore, traditional real-time scheduling algorithms of the physical factors influence considered insufficient, easy to cause the system overload operation, which makes the real-time task deadline missed. The system's real-time perceiving performance is seriously affected, causing major production safety hazards as well as economic losses. In view of these problems, this paper proposes a CPS real-time scheduling algorithm considering the physical environment factors.

[a] Corresponding author: 13734610r@qq.com.

This paper is supported by Research foundation of Liaoning Provincial Education Department (L2014532, L2014520, L2014523), National Social Science Fund(15CGL050).

2 The real time scheduling model of CPS

In this paper, we propose a real-time scheduling algorithm for the mobile real-time service node (CPS), which has many nodes in the network. This paper proposes a CPS real-time scheduling algorithm, which enable the system to achieve the maximum deadline satisfaction rate. Therefore, based on the real-time scheduling algorithm under CPS network framework of the mobile real-time service node, the model is describe as follow:

l_i : slack time of the task T_i (exponential distribution average $1/\lambda$).

e_i : execution time of task T_i (evenly distributed on [0, E]).

m_{Ti} : migration time (Uniform distribution on [0, M]).

DMR : Rate of meeting deadline(DMR=tasks of meeting deadline/all of tasks)

In the real time CPS of the mobile node, in the absence of task conflict, the real-time task DMR is the relaxation time of Ti, which is more than that of the real-time service node. Real time task DMR Ti is the probability that the relaxation time of Ti is greater than that of real-time service nodes. Given As distribution of l_{Ti} is $\lambda e^{-\lambda t}$, the DMR of T_i ($DMR_{Ti}(\lambda, m)$) is computed as follows:

$$DMR_{Ti}(\lambda,\mathrm{m}) = \int_m^{\infty} \lambda e^{-\lambda t}dt = e^{-\lambda m} \tag{1}$$

As m is assumed to evenly distributed [0,M], an average DMR is:

$$Mean(DMR_{Ti}(\lambda,\mathrm{m})) = \frac{1}{M}\int_0^M e^{-\lambda m}dm = \frac{1}{\lambda M}(1-e^{-\lambda M}) \tag{2}$$

In case of conflict between the two real time tasks, the paper will give the FIFO (first in first service), LST (shortest time priority), and ELST (CPS effective shortest time first) algorithm to calculate the average DMR:

3 Improved effective LSF algorithm

Effective LSF algorithm is an optimal algorithm in real-time scheduling algorithm. However, when the real-time tasks are scheduled in CPS system, DMR is needed to consider factors of the physical environment. In the real-time service CPS, the real-time service node migration time is one of the physical factors that need to be considered. Therefore, the effective shortest relaxation time of the real-time service node is considered. Therefore, when the real-time task is scheduled, the task with the most short time of slack time is first scheduled. We can use $l_{elsf,Ti} = l_{Ti} - m_{Ti}$ denoting the effective slack time of T_i .

Calculation methods are as follows: As the distribution function of l_{Ti} be $\lambda e^{-\lambda t}$, the distribution function of $l_{elsf,\mathrm{Ti}}$ >0 is below.

$$\frac{1}{M}\int_0^M \lambda e^{-\lambda(t+m)}dm = \frac{e^{-\lambda t}}{M}(1-e^{-\lambda M}) \tag{3}$$

When the distribution of $l_{elsf,\mathrm{Ti}}$ in (-M,0), the function is as follows:

$$\frac{1}{M}\int_{-t}^M \lambda e^{-\lambda(t+m)}dm = \frac{1}{M}(1-e^{-\lambda(M+t)}) \tag{4}$$

When T1 and T2 has not conflict, the DMR of T1 should satisfy the probability of $l_{elsf,\mathrm{T_1}}$ >0 which is described as follows.

$$\begin{aligned} Mean(DMR_{T_1}(\lambda,\mathrm{m})) &= \frac{1}{M}\int_0^{\infty} \frac{e^{-\lambda t}}{M}(1-e^{-\lambda M})dt \\ &= \frac{1}{\lambda M}(1-e^{-\lambda M}) \end{aligned} \tag{5}$$

The deadline meet ratio of task B following task A is:

$$\begin{aligned} & Mean(DMR_{T2}(\lambda,\mathrm{m_1},\mathrm{m_2},\mathrm{e_1})) \\ &= \int_{m1+m2}^{\infty} \frac{e^{-\lambda t}}{M}(1-\mathrm{e}^{-\lambda M})\,\mathrm{dt} \\ &= \frac{(1-\mathrm{e}^{-\lambda\mathrm{M}})}{\lambda M}e^{-\lambda(\mathrm{m_1}+\mathrm{e_1})} \end{aligned} \tag{6}$$

As we assume m1 and e1 are evenly distributed on [0,M] and [0, E], respectively, mean DMR of m2 is computed as:

$$\begin{aligned} & Mean(DMR_{T2}(\lambda,\mathrm{m_1},\mathrm{e_1},\mathrm{m_2})) \\ &= \frac{1}{ME}\int_{-t}^{\mathrm{E}}\int_{\mathrm{m_1}+\mathrm{e_1}} \frac{e^{-\lambda t}}{M}(1-e^{-\lambda M})?\, e^{-\lambda(\mathrm{m_1}+\mathrm{e_1})}d_{m_1}d_{e_1} \\ &= \frac{(1-e^{-\lambda M})^2(1-e^{-\lambda E})}{\lambda^3 M^2 E} \end{aligned} \tag{7}$$

It can be concluded from the results of the ELSF, average deadline meet ratio($Mean(DMR_{T1}(\lambda,\mathrm{m}))$) of T1 equals average deadline meet ratio ($Mean(DMR_{T2}(\lambda,\mathrm{m_1},\mathrm{e_1},\mathrm{m_2}))$) of T2. In addition to ELSF, the other parameters are the same as the FIFO algorithm and C algorithm except $l_{elsf,\mathrm{T}i} = l_{Ti} - m_{Ti}$. Also, FIFO algorithm applies $l_i > m_i$, and ELSF algorithm is applied by $l_{elsf,\mathrm{T}i} = l_{Ti} - m_{Ti}$. Both of DMR are the same.

Therefore, FIFO and ELSF algorithm to meet the deadline rate is the same. It can be concluded that the ELSF algorithm cannot effectively improve the real-time performance, so the optimization of ELSF algorithm is studied in this paper,

Let p be the probability of DMR of firstly scheduled T1.And then, let q be the probability of DMR of the secondly scheduled T2.

$$\begin{aligned} p &= Mean(DMR_{T1}(\lambda,\mathrm{m})) = \int_0 \frac{(e^{-\lambda t}1-e^{-\lambda M})}{M}dt \\ &= \frac{1-e^{-\lambda M}}{\lambda M} \end{aligned} \tag{8}$$

$$q = Mean(DMR_{T2}(\lambda, m_1, e_1, m_2))$$
$$= \frac{1}{ME}\int_{-t}^{E}{}_{1+e_1} \frac{e^{-\lambda t}}{M}(1 - e^{-\lambda M}) \times e^{-\lambda(m_1+e_1)} d_{m_1} d_{e_1}$$
$$= \frac{(1 - e^{-\lambda M})^2 (1 - e^{-\lambda E})}{\lambda^3 M^2 E} \quad (9)$$

Optimal ELSF algorithm is designed in this paper, considering the real-time service node in the mobile time and relaxation time, so as to improve the CPS nodes task deadline satisfaction rate. When the task T1 and T2 conflict, ELSF algorithm will maximize the system to meet the deadline rate. In order to $T_1 \rightarrow T_2$ task sequence, it will be divided into the following four scheduling methods.

4 The Description of Scheduling Method

Now , we describes the main fuction of optimal ELSF algorithm.

Initialization:

```
#include <stdio.h>
#include <iostream>
#include <stdlib.h>
#include <ctype.h>
#include <string.h>
#include <math.h>
#define queuesize 10
#define movetime M
#define MAXTIME 150 /
#define MINLLF 9999
#define PRO_LRU_T 0
```

Function:

//find the least slack time process

```
process *elsf(sqqueue *dui, int currenttime)
{
sqqueue *q1 = dui;
process *currentpro, *pro;
int minllf = MINLLF , llf;
int pro_LRU_t = PRO_LRU_T;
int front = q1->front, rear=q1->rear;
currentpro = q1->data[front];
if( currenttime <= MAXTIME )
{
//compute currentpro， if there was only a process
if( front==rear )
{ return currentpro;
printf("%dms%s%d%d\n",currenttime,currentpro-
        >pname,currentpro->cycle,llf);
} //more than a process
else
{
do
{
if(front!=rear )
{
pro=q1->data[front];
if(pro->arivetime <= currenttime && currenttime<=
        MAXTIME)
{
//Slack time equals to period is multiplied by execution times
llf=(pro->deadtime)*(pro->cycle)-     pro->lefttime     -
        currenttime;
if(minllf>=llf) //get the least slack time task
{ if( pro->LRU_t >= pro_LRU_t )
{
pro_LRU_t = pro->LRU_t ;
minllf=llf;
currentpro=pro;
} } }
front=(front+1)%queuesize; }
else
break;
}while(front!=rear ); //test queue} }
return currentpro; }
```

// get the preemption process

```
process *leastlaxityfirst(sqqueue *dui,int currenttime)
{
sqqueue *q1=dui;
process *pro=NULL, *nextpro;
int front= q1->front, rear= q1->rear ;
/*
Find the process which the slack time is zero.
*/
while(front!=rear)
{
nextpro=q1->data[front];
/*
pro->latestarttime
(pro->deadtime)*(pro->cycle) - pro->servetime;
pro->latestarttime - currenttime
pro->latestarttime - currenttime <= 0 ,
*/
if( nextpro->latestarttime <= currenttime )
break;
else
front=(front+1)%queuesize;
}
//if the queue is empty, return pro
if(front==rear)
return pro;
// if the queue is not empty, nextpro is the running process which can preempte the others.
else
return nextpro;
}
```

5. Experiments and analysis

In this paper, we test the performance of the real-time scheduling algorithm through simulation experiments. To verify the performance of the algorithm in the CPS system, the performance of the ELSF algorithm is verified in the case of multi task conflict. In order to verify the performance of real-time scheduling algorithm, the following parameters are configured in this paper:

Ei is denoted by execution time at node i with exponential distribution of average 1/ λ . In the experiments, we set the λ =0.02 and λ =0.2. Di is deadline at the node i with variable range. S is represented by the moving distance between computing node c and serviced node i. let s be $s = \sqrt{(x_i - x_c)^2 - (y_i - y_c)^2}$. The node C and service node I are distributed in a square of 100 by 100. Ri is release time, the value of Ri is 0. N is denoted by the number of simulations, which is 1000 times.

The paper uses DMR to test performance of optimal ELSF algorithm, LSF algorithm and LSF algorithm in task conflict situations. Experimental results are shown in fig 2.

The paper uses the DMR to test performance of optimal ELSF algorithm, LSF algorithm and LSF algorithm in task conflict situations. Experimental results are shown in below.

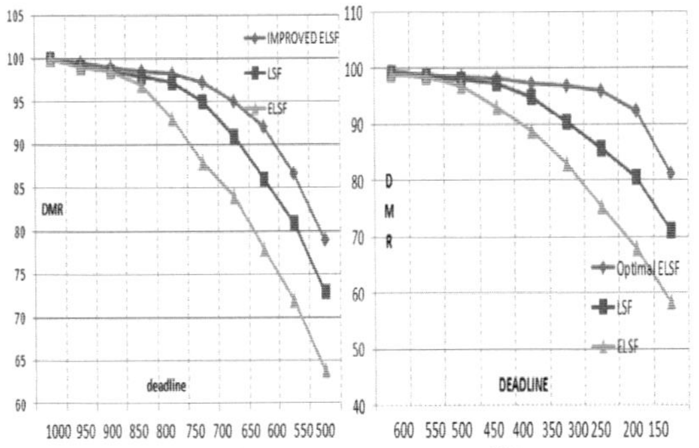

(a) (b)

Fig 1. (a)When cyber factor is quite big in comparison to physical factor.(b)When cyber factor is quite small in comparison to physical factor.

Experimental results show that the ELSF algorithm and LSF algorithm have different performance in the case of both information and physical factors (ei and mi), regardless of the specific gravity of the information factor, and the traditional LSF algorithm is better than the ELSF algorithm in the special condition. From the simulation results of Figure 3, the ELSF scheduling algorithm can show better performance when the information factor of CPS is relatively large. On the other hand, the traditional LSF algorithm has better performance than the ELSF algorithm when the physical factors are more than the information factor. However, in the above two cases,the optimization of the algorithm ELSF are able to exhibit better performance. The optimal ELSF scheduling algorithm can crease the DMR by up to 20% compared to other ways.

References

1. J. Stankovic, I. Lee, A. Mok, R. Rajkumar, Opportunities and obligations for physical computing systems, IEEE Computer, Volume 38, Issue 11, 23-31, 2005.
2. John A. Stankovic, When Sensor and Actuator Networks Cover the World, ETRI Journal, Vol. 30, No. 5, 627-633, 2008.
3. Fang-Jing Wua, Yu-Fen Kaob and Yu-Chee Tseng, From wireless sensor network towards cyber physical systems, Pervasive and Mobile Computing, Vol. 7, Issue 4, 397-413, 2011.
4. Edward AL and Sanjit AS, Introduction to Embedded Systems, A Cyber-Physical Systems Approach. NY: Lee & Seshia, 2011:119-158.
5. Hussain D, Illon T.Event. Cyber-Physical Systems Method. In proceedings of Object Component Service Oriented Real-Time Distributed Computing, Beijing, IEEE press,2011: 355-349.
6. Smita P, Prabha K. A Qos Based Mac Protocol for Wireless Multimedia Sensor Network[J]. Journal of Electronics and Communication Engineering, 2012,1(5): 30-35.
7. Yong F, Qiang G. Design of a Wireless Sensor Network Platform for Real-Time Multimedia Communication [C]. In Proceedings of International Conference on Communication, Electronics and Automation Engineering, NY, IEEE press, 2013:1305-1311.
8. Gann H. A Distributed Wireless Sensor Networks Mobile Communication Technology Research[C]. In Proceedings of Multimedia Information Networking and Security (MINES), UK, 2012, 203-207.
9. Shah, A. Cross-Layer Framework for QoS Support in Wireless Multimedia Sensor Networks[J], IEEE Transactions on Multimedia, 2012, 14(5):1442-1455.
10. Hong W, Research on Visual Minor Platform Based on ARM and DSP [D], Da Lian: Dalian University of Technology, 2009.
11. Eswaran A, Rajkumar R. Energy-aware memory firewalling for QoS-sensitive applications[C]. In Proceedings of the Euromicro Conference on Real-Time Systems, Palma de Mallorca, Spain:IEEE press, 2005:11–20.
12. Maria MG, Pedro M and David S. Model Checking Dynamic Memory Allocation in Operating Systems [J], Journal of automated reasoning, 2009,42(2):229-264.

Accurate Assessment of RSET for Building Fire Based on Engineering Calculation and Numerical Simulation

Zhenzhen Yan[1], Xuefeng Han[1,a] , Mengjing Li[2]

[1]*Jiangsu Key Laboratory of Urban and Industrial Safety, College of Urban Construction and Safety Engineering, Nanjing Tech University, Nanjing, 210009, China*
[2]*Sunning Real Estate Group Co. Ltd, Nanjing Jiangsu, 210042, China , China*

Abstract. In order to obtain the Required Safety Egress Time (RSET) accurately, traditional engineering calculation method of evacuation time has been optimized in this paper. Several principles and fact situations were used to optimize the method, such as detecting principle of the fire detecting system, reaction characteristics of staff being in urgent situation, evacuating queuing theory, building structure and the plugging at the porthole. Taking a three-storey KTV as an example, two methods are used to illustrate the reliability and scientific reasonability of the calculation result. The result is deduced by comparing the error (less than 2%) at an allowable range between two results. One result is calculated by a modified method of engineering calculation method, and the other one is given based on a Steering model of Pathfinder evacuation simulation software. The optimized RSET has a good feasibility and Accuracy.

1 Introduction

With the development of people's life quality greatly in recent years, places of public entertainment has expanded rapidly. As the typical public places of entertainment, KTV has achieved great success. However the complex of internal spatial structure, high personnel density and there are a large amount of flammable and combustible decoration materials indoors. Obvious fire load density support and sustain the fire growth, and go against fire detection, fire fighting and evacuation. So it's an important subject to promise that everyone in the fire must evacuate to safety places quickly, Required Safety Egress Time is the key factor of fire protection design of buildings.

RSET is the time that needed when the people evacuate to safety places, including detection time, preparation time and evacuation movement time[1]. It's an important parameter in the design of fire safety evacuation escape. At present, there are three ways to determine RSET, experimental method[2-4], empirical formula method[5,6] and computer simulation method[7-9]. experimental method is used for scientific research generally, but it is costly. The evacuation design that based on computer software simulation has become an important tool in the calculation of RSET, but empirical formula method has its unique advantages , such as simple, quick and easy to operate. Traditional empirical formula method is not considering the detection time of the fire, and think the evacuation movement time is the time of person through the horizontal direction and the queuing the exit time, without considering other structure channel such as evacuation of stairs in the building and blockage of people during the evacuation. In this paper, traditional engineering calculation method of evacuation time has been optimized, based on several principles and fact situations, such as detecting principle of the fire detecting system, reaction characteristics of staff being in urgent situation, evacuating queuing theory, building structure and the plugging at the porthole. Taking a three-storey KTV as an example, using the optimized engineering calculation method of RSET, detection time, preparation time and evacuation movement time are calculated respectively. Meanwhile using Pathfinder evacuation simulation software simulate evacuation process to illustrate the reliability and scientific reasonability of the calculation result. Then put forward to reduce the RSET and take some preventive measures for the retention and blockage status at the exit and the corridor when people evacuate. Besides, it has good practical guiding significance to the performance-based design on building fire safety evacuation .

2 Engineering calculation of RSET

RSET is mainly composed of detection time T_d , preparation time T_{pre} , and evacuation movement time T_t, that is $T_{RSET}=T_d+T_{pre}+T_t$.

2.1 The prediction of detection time T_d

Traditional empirical formula method is not considering the detection time of the fire, however, the fact is that the process from outbreak of fire to fire alarm is also an important part of the evacuation time. detection time Td is determined by detector, and this process is divided into two stages: (1) The fire outbreak and develop to a certain heat release rate, the flue gas and the detector began to heat exchange ,then the detector start to work ,and this stage is the start working time t_w; (2) From the heat exchange of flue gas and the detector to sending out an alarm signal when reaching the response temperature, this stage is named t_r. Therefore, there is $T_d=t_w+t_r$.

As the optimal installation position of the detector is the place of maximum gas velocity or of maximum smoke temperature , the prediction of the detector's start working time t_w can be based on the model of t^2 [10].

$$Q=\alpha.t^2 \quad (1)$$

[a] Corresponding author: safety@njtech.edu.cn

In which, Q is the heat release rate of the fire, kW; α is development coefficient of fire, kW/s^2;t is Fire development time,s.

Fire growth rate α characterizes the speed of fire spread. According to the classification of the American Fire Protection Association, fire growth rate of different fires are shown in Table 1.

Table 1. Fire growth rate of different fires.

combustible material	Fire type	Fire growth rate $\alpha(kW/s^2)$
unnoted	Slow speed fire	0.0029
non-cotton product Polyester mattress	Medium speed fire	0.0117
Plastic foam Stacked wood Mail bags full of mail	High speed fire	0.0469
methanol The rapid combustion upholstered seats	Over speed fire	0.1876

The flue gas and the detector begin to heat up to the response time of the alarm signal. Based on a review of the relevant research[11-13], and combined with statistical analysis of KTV fire accidents over the years, the Calculation formula of response Time t_r,(the time from the heat exchange of flue gas and the detector to sending out an alarm signal when reaching the response temperature) is as followings:

$$\begin{cases} t_r=-\frac{RTIr^{5/12}}{0.444Q^{1/6}H^{1/4}}\ln[1+\frac{H(T-T_0)}{5.38(Q/r)^{2/3}}] & r>0.18H \\ t_r=-\frac{RTI}{0.973(Q/H)^{1/6}}\ln[1+\frac{(T_0-T)H^{5/3}}{16.9Q^{2/3}}] & r\leq 0.15H \end{cases} \quad (2)$$

In which, RTI ----- the characteristic index of response time of the detector, $m^{1/2}.s^{1/2}$; r----- the radial position of the detector under the roof, m;Q-----the heat release rate of the fire, kW; H ----- height of ceiling，m.T_0 is----- the ambient temperature ℃ ;T-----response time of detector, ℃.

2.2 The prediction of T_{pre}

In paper[14] , pre-evacuation time has something to do with some factors such as structure, function of buildings, the characteristics of people's response to emergency, fire alarm system of the buildings and property manage system. In order to balance the research results of theory and practice, this article use the calculation empirical formula to determine the preparation time Tpre, which was recommended by the book named Planning pointer of Construction disaster

$$T_{pre}=120+A_0^{1/2}+0.4H \quad (3)$$

In which, A_0-----the area of building，m2; H-----height of the building,m.

2.3 The prediction of evacuation movement time T_t

Taking into account security considerations, most of the high-rise buildings are not allowed to take elevators during a fire, but along the stair evacuation. But traditional empirical formula method consider the evacuation movement time is the sum of time through the horizontal direction and queuing time through the exit, without considering the evacuation of stairs in non-horizontal direction. According to architectural structure, this article divided the evacuation structure into horizontal evacuating passage, exit and stairway. In the process of evacuating, the calculation formula of evacuation movement time as followings:

(1) horizontal passageway

$$t_c=\frac{l_{max}}{v} \quad (4)$$

In which, l_{max}-----the greatest distance of horizontal channel, m ; v-----the average speed when people in emergencies, m/s.

According to the model of Fruin[15], when the flow density meets 0.2 person/m^2<ρ<4.0 person/m^2 on the horizontal channel, the relationship between movement speed and the flow density is as followings:

$$v=1.427-0.3549\rho \quad (5)$$

(2) Exit (entrance)

$$t_e=\frac{P_e}{f_e w_e} \quad (6)$$

In which , P_e-----the number of people gathered at the exit, person; f_e-----the flow rate when people queen through the exit, person/(m.s);w_e -----the width of the exit, m.

Traditional empirical formula method is just the simple superposition of evacuation time at each stage without considering the congestion status during the evacuation, when $t_c<t_e$, namely when people evacuate from the greatest distance to the exit, others are not all through the exit, but some stranded at the exit. At this time, the evacuation time from the greatest distance to move through the export is t_e . When $t_c>t_e$, namely when people evacuate from the greatest distance to the exit, others have all passed the exit, and nobody is delayed. At which the time from the greatest distance to move through the export is t_c .That means, the evacuation movement time (people evacuate from the greatest distance to the exit in the region) t=Max（t_c,t_e） .

(3) Stairs:

Pauls[16] puts forward that the time named t_{stair} when people through stairs have something to do with the effective width of the stairs and the number of people using the stairs, as shown in the formula (7):

$$t_{stair}=4.579(\frac{P_{stair}}{w_{stair}})^{0.73} \quad (7)$$

DOI: 10.1051/matecconf/20166104024

In which ,w_{stair}----- effective width of the stairs ,m; P_{stair} ----- number of people who use the stairs, person.

3 Assessment of RSET and calculation case

This paper takes a three-store KTV as an example, calculating the required safety egress time of fire in the KTV. There are three floors in the KTV building and it covers an area of 441 m^2. The average density of people in the single room is about 0.3 person /m^2,and the ceiling height is 3.2 meters. Heat fire detection system has been installed in the KTV room according to the requirement of the standard. The distance between the fire source and the nearest fire detector is about 2m. Exponents of the response time about the fire detector is $98m^{1/2}.s^{1/2}$ and responsive temperature is 58 ℃ . Assuming ambient temperature named T_0 is 25℃. In the paper, the fire source is set at the most unfavorable part where is on the sofa at the end of the corridor in the second floor in the KTV, as shown in figure 2.

When the fire happens, the evacuation route in the KTV room is: 1. The room to the exit of the room; 2. Through the exit of the room; 3. Through the corridor passage to the entrance of the staircase; 4.Through the entrance of stairs; 5. The stairs to the first floor corridor; 6. The first floor corridor to safe exit. The specific evacuation route is shown in figure 1:

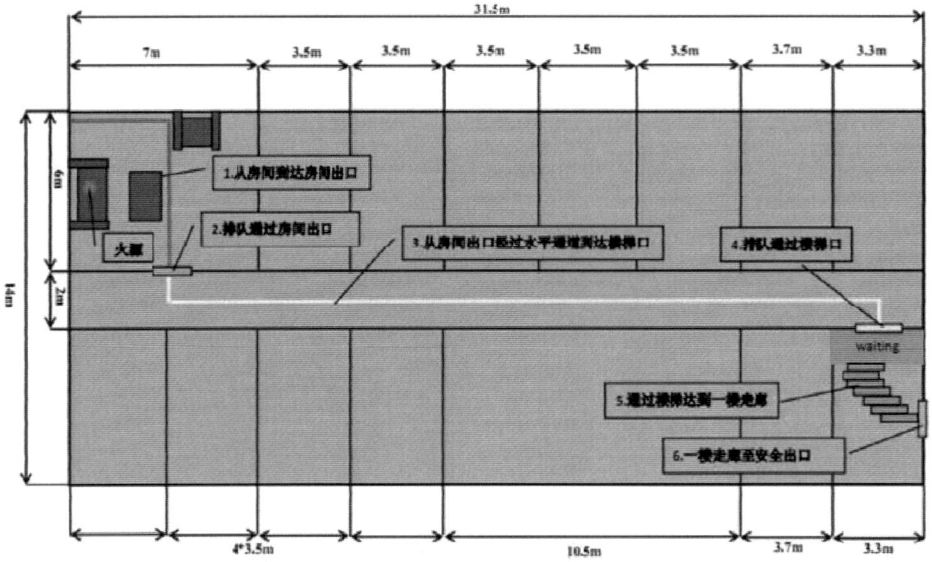

Figure 1. Evacuation route in KTV room fire

(1)There are a lot of flammable and combustible materials such as sofas, carpets, etc in the KTV. Once fire, the room will be ignited and become over speed fire quickly. Therefore, this paper choose α=0.1876kW/s^2 as fire growth rate according to table 1. When HRR reaches 300kW, the heat fire detector begins to exchange heat as a response and then start the alarm device. According to the formula (1) ,we get t_w=40s.

From the model ,we can get r=0.63H. According to formula (2), we can gain the reaction time t_r=101s when smoke and the detector began to exchange heat, and then the detector achieves a response temperature to alarm. The detection time $T_d=t_w+t_r$=141s.

(2)According to formula (3) ,we can obtain the preparation time T_{pre}=145s when people are ready to evacuate after hearing the alarm .

According to the statistical data of personnel flow coefficient[17], parameters of KTV model and formula (5) , we can fundamental parameters of different evacuation routes are shown in table 2.

Table 2. Fundamental parameters of different evacuation structure

No.	Evacuation route	Evacuated people *P* (person)	Average movement speed *v* （m/s）	Flow rate *f* (person/(m.s))
1	room-room exit	6	1.32	
2	Room exit	12		0.93
3	Corridor passage	107	0.82	
4	Corridor exit	107		1.3
5	stair	214	0.35	
6	First floor Corridor –safe exit	214		1.3

Note: the above-mentioned rooms show the general rooms of KTV.

According to the fundamental parameters of the different evacuation structure in table 2, the evacuation movement time of different evacuation route of figure 2 are calculated by the formula (4), (6), (7) as shown in table 3. Considering the interior space of the KTV room which comes about fire and has furniture such as sofa, tables and chairs is very big, people need to bypass the obstacles to get to the room export during evacuation, so the route of evacuating is the shape of L[18]. That is, the maximum horizental distance between room and room export meets $l_{\max}=\frac{7}{2}+6=9.5$m.The longest distance across corridor passage meets $l_{\max}=\frac{7}{2}+3.5\times5+3.7+\frac{3.3}{2}=26.35$m。

Table 3. Evacuation movement time of different evacuation route

No.	Evacuation route	maximum horizontal distance $l_{\max}$（/m）	average movement speed v（m/s）	Width *w*（/m）	Evacuated people *P*(person)	Flow rate f(person/(m.s))	Evacuation time t（/s）
1	room-room exit	9.5	1.32				7.2
2	Room exit			1.2	12	0.93	10.7
3	Corridor passage	26.35	0.82				32.1
4	Corridor exit			1.6	107	1.3	51.4
5	stair			1.6	214		163.3
6	First floor Corridor –safe exit			1.8	214	1.3	91.4

Therefore, the evacuation movement time in KTV is T_t=Max(t_1,t_2)+Max(t_3,t_4)+Max(t_5,t_6)=225.4s, and T_{RSET}=T_d+T_{pre}+T_t=511.4s. At the same time, from the evacuation time of each evacuation route in table 3 ,we know that people are stranded and blocked at the exit of room，corridor and stairs in the process of evacuating.

4 Numerical simulation of RSET

Evacuation movement time can be simulated and calculated through the Pathfinder software. Pathfinder software developed by American Thunderhead engineering company is a simulator based on evacuation movement, which consists of Steering model and SFPE model. Unlike SFPE model of linear escape route, the escape route of Steering mode is curve. When people are in disorder and congestion situation, Steering model can choose to different directions, which is more in line with the actual escape behavior in an emergency. Therefore, this paper uses Steering model of Pathfinder software to establish the Pathfinder evacuation model and design the solving method about the emergency evacuation problem of some three-store KTV building in fire emergency.

The parameters setting of the model: the average density of persons is 0.3 person/m^2, and the movement speed meets the theory of performance-based fire prevention and design in building, and the average movement speed is 0.9 m/s, and the maximum is 1.32m/s, and the minimum is 0.65m/s. The setting speed obeys the normal distribution characteristics. Through Steering model of Pathfinder software, the evacuation simulation diagram of each moment and the personnel evacuation time of all layers and people flowrate at the safe exit are gotten as shown in figure 2 to 4 respectively.

From figure 3, the evacuation time of the second-layer is shortest and the third-layer is longer. The total evacuation of the first-layer is longest. This is because people of the second-layer and third-layer personnel are both evacuated to the first layer, resulting in more and more people to the first-layer to evacuate the longest time. From figure 4, the flow of the emergency exit changes basically in line with the linear growth trend as time changes in 0-25s. After 25s, the flow of the emergency exit is in 1.28 person/s and 1.74 person/s between ups and downs, which is in line with the average flow of the unit width of the engineering calculation. This is because with the increasing of evacuees, evacuation passage is gradually saturated and the flow shows gradually stability and trend of fluctuation. From the evacuation simulation dynamic figure, in 25s people in the corridor exports strand and congestion occurs on the stairs. The evacuation movement time of some three-store KTV is obtained through simulation analysis, that means, T_t=216.3s, and T_{RSET}=502.3s . We can see that the results of two methods to deal with the RSET comes to a great degree by comparing the error (less than 2%) at an allowable range between two results. One result is calculated by comparing a modified method of engineering calculation method with the other one based on a Steering model of Pathfinder evacuation simulation software. It confirmed that the optimized RSET and the result of RSET of three-store KTV fire have a good feasibility and accuracy.

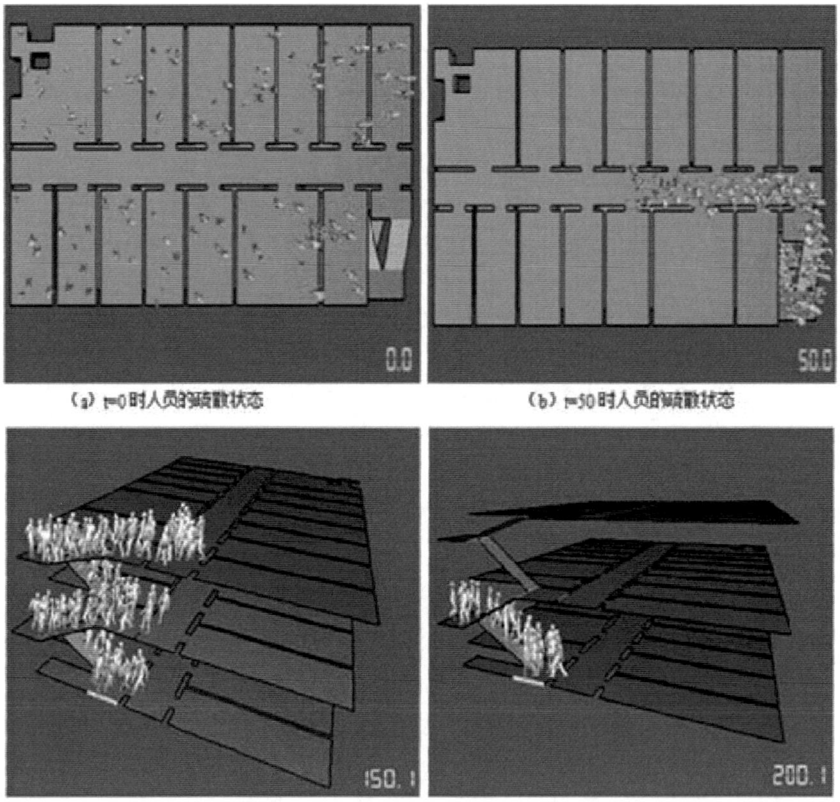

（c）t=150时人员的疏散状态　　（d）t=200s时人员的疏散状态

Fig.2 Dynamic diagram on personnel evacuation

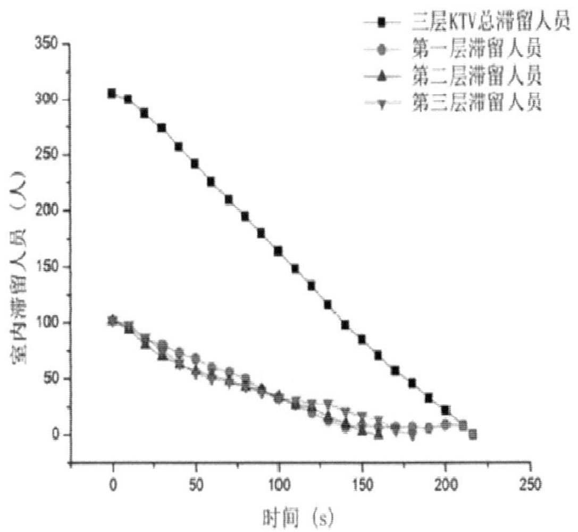

Fig. 3 Stayers in fire vs. time

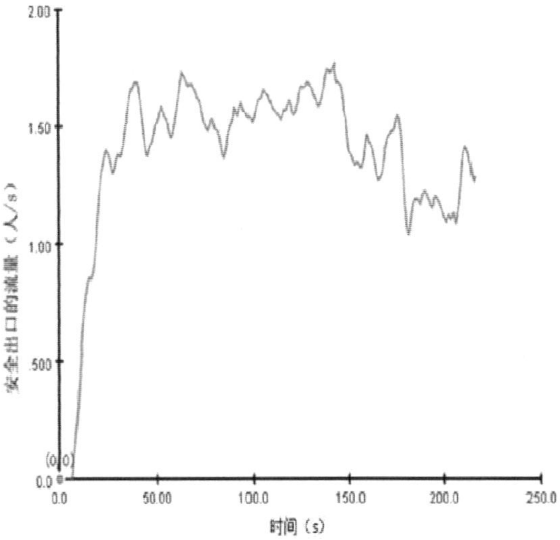

Fig. 4 People flowrate at the safe exit

5 Conclusion

Through engineering calculation and the dynamic simulation diagram of Pathfinder software, evacuees on the stairs exports, the stairs, the first-layer corridor strand and jam phenomenon occurs which affect the evacuation time. Therefore, to reduce the personnel required safety evacuation time, with the optimization of building evacuation design, some improvement measures can be appropriately increased.

(1) The response temperature of the chosen fire detector by the author is 58 ℃ in the calculation example. The detector begins to conduct heat exchange to start alarm device when the fire source power is 300KW. In the actual situation, to reduce the detection time of the fire detection system, the high sensitivity of fire alarm device can be used such as the smoke detector used in the current projects which can detect the 100KW (or less) fire source and the detection time can be down from 141s to 98s.

(2) The exits and stairs are the bottleneck of personnel evacuation to escape. So the width of the exits and stairs can be increase appropriately. In addition, people during emergency evacuation in case model can only be evacuated to downstairs after running from the room to the end of the corridor, which not only extends the evacuation time, but also causes the personnel crowded and jam. If evacuation stairs respectively are built at both ends of the corridor, the evacuation movement time in the KTV fire can be down from 225.4s to 141.4s and people don't need to spend much time in the long corridor and crowded on the stairs and rooms.

References

1. R. HUO, Y. Hu, Y.Z. Li, Introduction to building fire safety engineering, Hefei: Press of University of Science and Technology of China (1999)
2. J.H. Wang, Z.Y. Jin, Z.G. Tian et al, Pre-evacuation Time Experiment in Public Assembly Occupancy of School Under the Fire. J. Science & Technology Review **30**, 11 (2012)
3. X.Y. Duan, L.Y. Dong, G.Y. Wang, Experimental Investigation on Evacuation from Classroom Considering Internal Layout and Exit width. J. Journal of Shanghai University(Natural Science) **19**, 6 (2013)
4. Z.G. Jiang, F.Z. Huo, X.D. Liu et al, Experimental research of pedestrian evacuation from an underground market in emergency. J. Fire Safety Science **22**, 3 (2013)
5. H.X. Wang, X.Z. Liu, X.J. Ma, Research on Safe Evacuation In Long Road Tunnel during Fire. J. Journal of Highway and Transportation Research and Development **27**, 11 (2010)
6. Construction Ministry, Integrated fire protection design of buildings, Tianjin: Tianjin science and technology translation and publishing company (1994)
7. R. Huo, H.Y. Yuan, Performance-based Fire Prevention and Design in Building, Hefei: Anhui Science and Technology Press (2003)
8. Y.M. Tian, Application of Building EXODUS in predicting crowd evacuation time. J. Fire Science and Technology **34**, 4 (2015)
9. J. Zhang, X.J. Liu, Numerical Simulation of Evacuation in a Supermarket Fire Based on FDS. J. Industrial Safety and Environmental Protection **37**, 5 (2011)
10. ISO/CD 13388—Fire Safety Engineering. Design Fire Scenarios and Design Fires [DB/CD], May 1997.
11. L.M. Yuan, W.C. Fan, The Prediction of Safety Evacuation Building Fires. J. Journal of Natural Disasters 6, 2 (1997)
12. R. L. Alpert. Calculation of Response Time of Ceiling- Mounted Fire Detectors. J. Fire technology, 8 (1972)
13. T.J. Shields, K.E. Boyce, A study of evacuation from large retail stores. J. Fire Safety Journal 35, 1 (2000)

14. H. Ma, Z.C. Shi, Z.H. Yuan, Performance-based evaluation of people evacuation from a large exhibition center. J. Fire Science and Technology 32, 2 (2013)
15. J. Fruin, Metropolitan Association of Urban Designers and Environmental Planners[M]. New York: ASCE,1971
16. P. Jake. Movement of People. SFPE Handbook of Fire Protection Engineering (1994)
17. L.C. Xiong, L.B. Yang, J.H. Chen et al, An revised optimized model for urgent human evacuation based on graph theory and queuing theory. J. Journal of Safety and Environment, 3（2014）
18. T. Yu, Y.F. Zhang, Z.Z. Hou et al, Comparative study of evacuation time predict ion methods. J. Fire Science and Technology 28, 3 (2009)

MATEC Web of Conferences **61**, 04025 (2016) DOI: 10.1051/matecconf/20166104025
APOP2016

The Impact of Subway on Urban Traffic Network Based on Complex Network Theory

Cao Hongmei [a], Zhao Fang, Liu Hao and Yu Tong

Inner Mongolia University, China

Abstract. In order to explore changes in accessibility and connectivity of urban traffic network before and after the construction of subway, traffic networks within the Second Ring road, Hohhot, was analyzed based on complex network theory. Firstly, the urban road traffic network and the urban compound traffic network with the adjunction of the subway to former were mapped. Secondly, the node degree, clustering coefficient, average path length and node strength of the two networks were calculated based on SuperMap. Finally, the eigenvalues obtained above were compared and analyzed. Results show that both of the two traffic networks have some characteristics of small clustering coefficient reflected in random network and scale-free network. The latter, compared to the former, network diameter and average shortest path reduce, the average node degree and average node intensity increase. That is to say, with the adjunction of the subway, the connectivity and the accessibility of the entire network are significantly improved, and its carrying capacity increases remarkably.

1 Introduction

Subway plays an important role in improving traffic conditions as a high capacity and long distance and fast transportation mode, and with the adjunction of subway, the structure and evolution of urban traffic network will be impacted. Because topological structure of urban traffic system are similar to that of complex networks, many scholars make traffic networks researches based on complex network theory. Some studies on urban compound traffic network which is a combination of subway network and road network are conducted respectively in Statistical topological Characteristics, carrying capacity and evolution [1-3]. Based on complex network theory, in this paper, the impact of subway on character, structure and carrying capacity of the urban traffic network were concluded by comparing and analysing its statistics eigenvalues before and after the construction of subway. It will help to improve the performance of urban traffic network in management practice, and provide guidelines for transportation planning in future.

2 Urban traffic network constructions

In order to explore changes in accessibility and connectivity of urban traffic network before and after the construction of subway, Urban Road Traffic Network and Urban Compound Traffic Network were constructed. For research convenience, the former only considered the expressways, primary arterials and secondary arterials within the Second Ring road, Hohhot. The latter contained subway based on the former. In addition, it's assumed that both of them were undirected networks.

2.1 Urban road traffic network

With links and nodes representing road segments and junctions, the urban road network is built. Geert man and van Eck claims that this option is attractive as many existing modelling programmes can be used without much change [4]. By use of GIS integrated graphics and attribute data management capabilities, edit and update road network information, combined with the characteristics of urban road traffic road network of Hohhot weighted network dataset is added as a GIS layer. The network consists of 169 nodes, 290 edges, as shown in Figure 1.

Figure 1. Hohhot road traffic network

2.2 Urban compound traffic network

With the subway stations as nodes, the lines connecting sites as sides, the urban compound traffic network with

[a] Corresponding author: Cao Hongmei, xghcao@163.com

the adjunction of urban subway was constructed. The urban compound traffic network consists of 172 nodes, 294 edges, as shown in Figure 2.

Figure 2. Hohhot compound traffic network

3 Calculation and comparative analysis of eigenvalues

3.1. Comparison and analysis of the degree and degree distribution function

Degree is defined as the degree of node i, it represents the number of edges connected to node i, the greater is, the greater the influence node i has on the entire network; represents the degree distribution function, which refers to the probability that a node has a number of k edges connected in network, and it reflects the macro-statistical characteristics of the network system. The distribution is in line with Poisson distribution indicating that the network has the characteristics of small-world network, in line with the distribution of power indicating that the network has a scale-free property.

Table 1. Node degree Characteristics of complex network models of different topologies

Degree	Road traffic network			Compound traffic network		
	Number of nodes	Percentage	Average	Number of nodes	Percentage	Average
1	2	0.012	3.592	2	0.012	3.890
2	5	0.030		5	0.029	
3	53	0.314		50	0.291	
4	109	0.645		91	0.529	
5	0	0		3	0.017	
6	0	0		20	0.116	
7	0	0		0	0	
8	0	0		1	0.006	

Table 1 showed us that both networks haven't presented strictly Poisson or exponential probability distribution.

3.2 Comparison and analysis of diameter and average path lengths

Distance between two nodes i and j in network is defined as the number of edges of the shortest path connecting the two nodes. The average path length A refers to the mean shortest path length between any two nodes, which can be formulated as:

$$A = \frac{1}{\frac{N(N+1)}{2}} \sum_{i\geq j} d_{ij} \quad (1)$$

where: N= the number of nodes in network

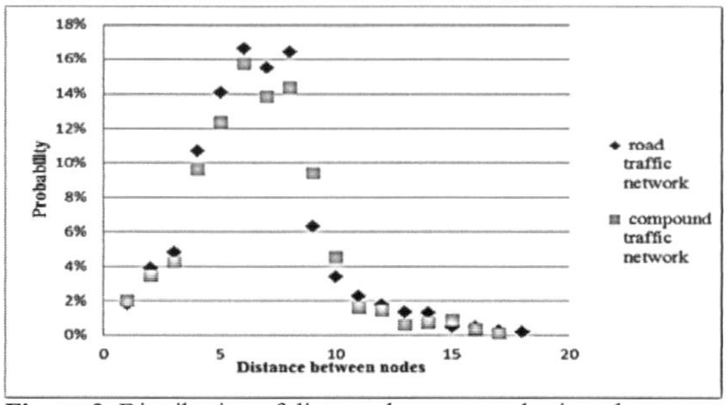

Figure 3. Distribution of distance between nodes in unban traffic network

The diameter and average path lengths of the network can be calculated based network analysis tools on Supermap Deskpro. The results show that Diameter of the urban compound traffic network is 17, the average shortest path length is 6.2731 instead of 18 and 6.9036 in the urban road traffic network respectively. These decreases reflect that fewer numbers of sections are needed for nodes to reach each other. The distributions of distance between nodes in two networks are shown at Figure 3. It shows that the dispersion degree about distribution of shortest path between nodes in compound traffic network reduced, the probability in the central region [5,10] become larger, and the tightness of the network nodes increase.

3.3 Comparison and analysis of clustering coefficient

Clustering coefficient of a node refers to the proportion of the number of edges between its neighboring nodes and the largest possible number of connected edges, which can be formulated as:

$$C_i = \frac{2E_i}{k_i(k_i-1)} \quad (2)$$

where: E_i= the number of edges between node i and its neighboring nodes.

The clustering coefficient C of entire network is average of all clustering coefficient C_i of node i, where, $0 \leq C \leq 1$. C=0 means that all nodes are isolated nodes; C =1 means the network is a global coupled, in other words, any two nodes in network are directly connected. Because clustering coefficient the network may show a characteristic of random nature or small world [6].

Table 2. Clustering coefficient characteristics of complex networks

Clustering coefficient	Road traffic network		Compound traffic network	
	Percentage	Average	Percentage	Average
0	91.12%	0.01874	93.02%	0.01395
1/15	0		0.58%	
1/6	6.51%		4.65%	
1/3	2.37%		1.75%	

The average clustering coefficient of road traffic network and compound traffic network are 0.01874 and 0.01395 respectively. They are less than 1 to a large extent which shows that they all don't have small world characteristic. Table 2 presents a comparison of clustering coefficient characteristics between road traffic network and compound traffic network. The proportion of nodes that have 0 clustering coefficient reaches to more than 90%, showing the lack of overall network partial connectivity and less intensive

3.4 Comparison of the intensity of node

Edge intensity is a measure of the importance of edges in the network. Generally, different types of edge have different intensity value as shown Table 3. Intensity of a node is the sum of all connected edge intensity. Edge intensity and node intensity limit the capacity of the network.

Table 3. Edge intensity

Types of edge	Subway	Expressway	Main road	Secondary road
Traffic capacity(Million passengers / hour)	4.400	1.344	1.176	0.672
Edge intensity	1	0.31	0.27	0.15

Table 4. Node intensity

Node intensity	Maximum	Minimum	Mean
Road traffic network	1.24	0.15	0.9251
Compound traffic network	5.08	0.15	1.2440

Maximum, minimum and mean of node intensity are showed in Table 4. The intensity of node in network that connects to subway lane is obviously higher than others. Statistics obtained that 56% of the subway stations are connected to the intersection of two trunks to achieve superimposed.

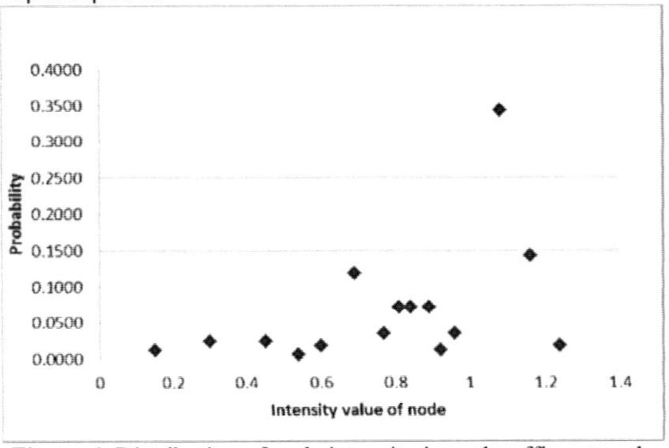

Figure 4. Distribution of node intensity inroad traffic network

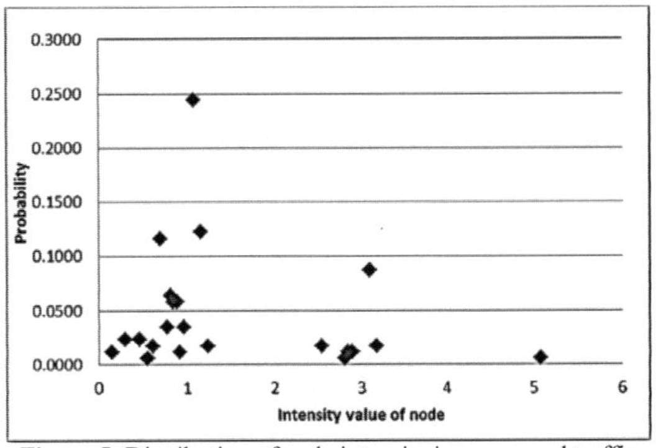

Figure 5. Distribution of node intensity in compound traffic network

4 Summary

Compare and analysis the overall network performance as is shown in Table 5, diameter and average path lengths in compound traffic network reduce by 5.6% and 0.9% respectively, compared to road traffic network, reflecting the connectivity of network has significantly improved, the amount of traffic carried by network increases, residents can choose more routes than before, the distance and time required in a travel become shorter, the accessibility of the entire network is improved remarkably. In addition, as it presented in Table 4, statistical eigenvalues of key nodes in network change a lot, which makes the traffic distribution becomes more balanced, more stable the network traffic is, especially the junction of two subway lines, Xinhua Square Station, the degree and intensity of it increase by 100% and 370% respectively, indicating its influence in the whole network has been greatly improved, to some extent, it shares the traffic pressure of Hailiang Plaza, however, strengthen of guidance and supervision also are necessary.

Table 5. Statistical indicators

Statistical indicators	Road traffic network	Compound traffic network
Average degree	3.592	3.890
Average path lengths	6.9036	6.2713
Network diameter	18	17
Average clustering coefficient	0.01874	0.01395
Average node intensity	0.9251	1.2440

References

1. ZL. Wei, YJ. Gan and P. Zhao. Characteristic Research of Urban Complex Traffic Network ［ J ］ .Journal of Transportation Systems Engineering and Information Technology. **15** (2015)
2. P. Zhao. Collaborative Study on Multi-Modal Urban Traffic Network Based on Complex Network Theory—Urban Road Traffic Network and Urban Rail Transit Network ［ D ］ .Beijing :Beijing jiaotong university(2014)

3. W. Liu. Research on the effect of mass transit on complex network theory[D]. Harbin Institute of Technology(2009)
4. Geertman and Ritsema van Eck. GIS and models of accessibility potential: an application in planning. Int. J. Geographical Information Systems, **9**, 67-80(1995)
5. XF. Wang, X. Li and GR. Chen. *complex networks and applications Edited* [M] Beijing: Tsinghua University Press(2006)
6. Watts D J, Strogatz S H. Collective dynamics of 'small-world' networks [J]. Nature, 393: 440-442. (1998)
7. Albert-László, Barabási, Erzsébet Ravasz, et al.Deterministic Scale-Free Networks[J]. Physica A: Statistical Mechanics and its Applications, 29963(3-4): 559-564(2001)
8. Albert R, Barabasi A L. Statistical mechanics of complex networks [J]. Reviews of ModernPhysics. **74**: 47-97(2002)
network on the urban transportation network based
9. H. Corley and D. Y. Sha, "Most vital links and nodes in weighted networks," Operations Research Letters, **1**, 157-160(1982)
10. Albert-L´aszl´o Barab´asia, Erzs´ebet Ravasza, Tam´as Vicsekb Deterministic Scale-Free Networks(2008)
11. KM. Chen and BJ. Yan. Analyze of the road capacity, Beijing: People's Communication Press(2003)
12. Dalvi M. Q. and Martin K. M. The measurement of accessibility: Some preliminary results. Transportation, **5**, 17-42 (1976)
13. M Barthélemy. Crossover from Scale-free to Spatial Networks [J]. Europhys. Lett., **63**(6),915-921(2003)
14. YM. Lu and JM. Tang. Fractal Dimension of a Transportation Network and its Relationship with Urban Growth: a study of the Dallas — Fort Wortharea [J]. Environment and Planning B: Planning and Design, **31**,895-911(2004)

MATEC Web of Conferences **61**, 04026 (2016) DOI: 10.1051/matecconf/20166104026
APOP2016

The Evaluation of Node Importance in Urban Road Network Based on Complex Network Theory

Hongmei Cao[1,a], Hao Liu[1], Fang Zhao[1], Yuanyuan Li[1] and Mingkun Du[1]

[1]Inner Mongolia university, China

Abstract. Identifying and protecting traffic hub is vital to keep the stability and reliability of urban road traffic network. In this paper, a comprehensive evaluation method of node importance in urban road traffic network was introduced based on identification theory of key nodes in complex network. Firstly, taking into account the impact of road network topology and traffic flow characteristics on node's capacity, node importance in topological network and betweenness, urban road grade and node traffic flow are selected as the evaluating indexes of node importance, and AHP is used to determine the weight of each index. Secondly, the associated value of the indexes are calculated and obtained based on SuperMap Deskpro and practical investigation, and then, combined with TOPSIS, a sequence of node importance is determined. Finally, we applied the method to part district of Hohhot road traffic network, it can be seen that the key nodes obtained with the method stay in the important position in the the traffic network and have greater traffic flow, the results showed it is effective and practical feasible.

1 Introduction

Complex networks are highly abstract of complex systems which often have a large number of nodes[1]. The studies of network stability have shown that as long as 5% of the key nodes being attacked the network will be basically paralyze[2]. The applications of the research on node importance in real life are very meaningful. For example, we can identify the key character and focus fire against him in the terrorist organization network[3], and the congestion problem can be solved by finding the nodes with the most flow in a communication network[4].

The research on the evaluation of node importance in urban road traffic network which is a kind of complex and integrated network is also very pregnant. Wang Wei proposed that key node is refer to the node that is crucial to overall characteristic of network, and its damage will cause the loss of the function of network, simultaneously make the network connectivity, reliability and the road network performance drop rapidly[5]. We can infer that the protection of the vital nodes in the urban traffic network has great significance to ease the congestion situation, improve the invulnerability ability and enhance the reliability of urban road traffic network.

In order to explore the method of identifying the vital nodes in urban road traffic network, this paper presented a kind of method of evaluating comprehensively the node importance of urban road traffic network. And this method will be introduced in the following.

2 Selection of indexes

2.1 Selection of network topological characteristic indexes

Domestic methods of evaluation of node importance in complex network have been relatively mature. Chen Yong[6] and Tan[7] evaluated node importance by means of node deletion and node constriction. Since the two methods have no local analysis on the nodes, the final results are not accurate or convincing. Then many scholars further studied the methods of evaluating node importance in complex network. Zhou put the node contribution and node efficiency as the considering factors[1]. A method which is based on node connectivity and node criticality is presented[8]. Qin and Yang listed six indexes, and seven of them are finally selected for the evaluation by principal component analysis[9] (PCA).

When complex network theory is applied to the urban traffic network, it must be concerned that the degree of the nodes in urban road traffic network is much too single because of the existence of intersections and t-junctions. So some above methods are not suitable for the research on road network, for example it is not suitable for Qin and Yang's method to apply to the evaluation of node importance in urban road network, because degree of the nodes in urban road network is mostly 3 or 4. In addition, considering

[a] Cao Hongmei: xghcao@163.com

the lengths of the roads between the intersections make a great difference, the inter-node length can be concerned in the process of selecting indexes.

The paper chose node importance in topological network and betweenness as network topological characteristic indexes. This paper will recommend them as follows.

Node importance in topological network (C_i) is the product of the node efficiency (I_i) and the sum of the importance contribution(H_i) of its adjacent nodes, as given by Eq. (1), Eq. (2) and Eq. (3).

$$C_i = I_i \times H_i \tag{1}$$

$$H_i = \sum_{j=1, j \neq i}^{n} \delta_{ij} D_j I_j \Big/ <k>^2 \tag{2}$$

$$I_j = \frac{1}{n} \sum_{m=1, m \neq j}^{n} \frac{1}{d_{jm}} \tag{3}$$

Where δ_{ij} is the adjacency coefficient, D_j is the degree of node v_j , d_{jm} is the distance between node v_j and node v_j , and k is the average degree of the selected network.

Node betweenness(B_i) refers to the number of shortest paths which pass node v_j between all the pairs of nodes in the network, was calculated as Eq. (4).

$$I_j = \frac{1}{n} \sum_{m=1, m \neq j}^{n} \frac{1}{d_{jm}} \tag{4}$$

Where n_{jk} is the number of the shortest paths between node v_j and node v_k , and $n_{jk}(i)$ is the number of the shortest paths which pass node v_i between node v_j and node v_k .

2.2 Selection of traffic flow characteristic indexes

It has just started in recent years that complex network theory was applied to the research on node importance of urban road traffic network, the literature of this aspect is not a lot. Wang screened out six indexes including delay, queue length and degree of saturation from eleven indexes by AHP[4], although the results obtained by this method is accurate, it is difficulty to get the indexes and it would cost huge manpower resources to evaluate large networks. Zhong proposed a kind of weighted method of the nodes base on urban road grade, and calculated the node importance according to the node traffic flow and topology value which is got by connection degree and weights of nodes[10].

This paper chose node traffic flow and urban road grade as the traffic flow characteristic indexes. Traffic flow of the intersections can intuitively reflect the significance of the intersections in urban road network. And the significance of the intersections consist of the roads in different grade differs from each other. Stated below .

In the light of the status, traffic function and serve function of urban roads, the roads can be divided into four categories: expressway (k), major arterial (z), secondary arterial (c) and collector (g). The quantitative analysis of urban road grade can be conducted on the basis of Table 1. For example, the weight the intersection consist of a expressway and a mayor arterial is: G_i=2×10+2×8=36. The weight is the value used in conducting quantitative analysis of urban road grade in this paper.

Table 1. Calculation method of node weight[10]

Road	Expressway	Major arterial	Secondary arterial	Collector
Weight	10	8	5	3

We define the node traffic flow as the sum of traffic flow on the roads which connect the node, as given by Eq. (5).

$$Qi = \sum_{n=1}^{m} f(i,n) \tag{5}$$

Where m is the number of the nodes that are connected to node i and $f\ (i,n)$ is the traffic flow which is in the direction of entering node i on the road which connect node n and i .

3 Evaluation procedure

The evaluation is conducted around data collection and data processing. The specific steps of the evaluation as follows:

(1) Load urban road network diagram into the software SuperMap Deskpro, then get the urban road network topological diagram by using the function of topological data processing of the software.

(2) Calculate the node importance in topological network by the functions of adjacency point analysis and measuring the distance of SuperMap Deskpro, then analysis and calculate node betweenness by the function of best path analysis of the software. Finally get the data about node traffic flow and urban road grade by investigation, searching the Internet and inquiring the traffic administration.

(3) Determine indexes weights by analytic hierarchy process(AHP). Firstly, determine the target layer, criterion layer and index layer, then set up hierarchical analysis model, as shown in Figure 1. Secondly, establish judgment matrix by 1~9 method in Table 2. Finally calculate the weights of indexes according to the judgment matrix and verify conformance.

(4) Calculate and sort the node importance by using TOPSIS. Firstly, establish a date matrix which has been normalized. Secondly, find the most ideal solution Z^+ and the least ideal solution Z^- , then calculate the distance respectively from each node to Z^+ and Z^- , and record the result as D^+ and D^- . It needs to be stressed that this calculation has to take

into account the weights obtained above. Thirdly, figure up the node importance by Eq. (6). Finally, Sort node importance.

$$E_i = \frac{D^+}{D^+ + D^-} \quad (6)$$

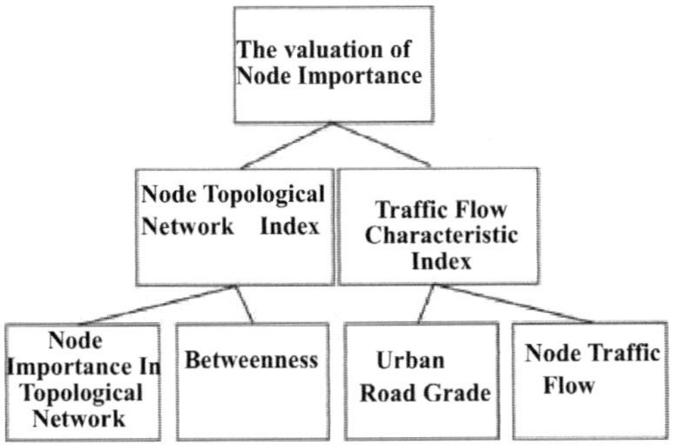

Figure 1. Hierarchical analysis mode

Table 2. 1~9 scale method

Scale	Meaning
1	Two factors are equally important
3	The former is more important than the latter
5	The former is more important than the latter obviously
7	The former is more important than the latter strongly
9	The former is more important than the latter extremely
2, 4, 6, 8	The intermediate value between adjacent judgments
Reciprocal	The importance of scale of the latter than the former

4 The case study on Hohhot urban road traffic network

4.1 Selection of the study area and the nodes

The planning and construction of urban road traffic network get increasingly the attention of people with the speeding up of urbanization process, and the traffic management puts a high value on the revamping of traffic trunk network. This paper selected the study area which encircled by West Second Ring Road, South Second Ring Road, West XinHua Street, South TongDao Road, DaBei Street, DaNan Street and ZhaoJun Road which are showed in Figure 2. And the topological network diagram can be obtained by SuperMap Deskpro. It needs to be added that the paper abstracted the roads into non directional edges, and the branches with small traffic flow, narrow alleys and streets, as well as the small intersections without traffic lights are neglected in the process of topology operation. In the topological network, this paper selected sixteen nodes as the research objects, and labeled them as Figure 3. Their specific composition is as shown in Table 3.

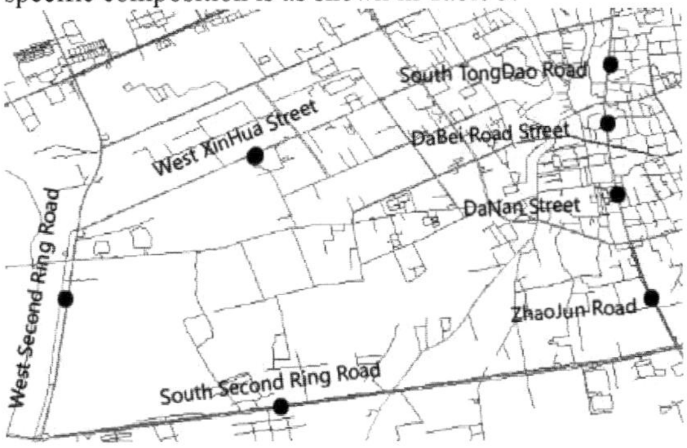

Figure 2. Network diagram of part district of Hohhot road traffic network

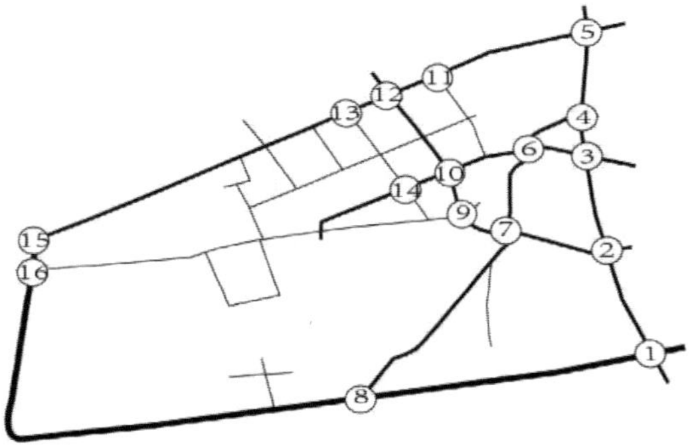

Figure 3. Serial number of the selected nodes

Table 3. List of nodes

Node	Node composition
1	ZhaoJun Road-South Second Ring Road
2	DaNan Street-Ordos Street
3	DaBei Street-DaNan Street-DaDong Street-DaXi Street
4	South TongDao Road-DaBei Street-XiShun City Street-West ZhongShan Road
5	South TongDao Road-North TongDao Road-XinHua Street-West XinHua Road
6	XiShun City Street-XianFu Street-DaXi Street
7	XiShun City Street-YunZhong Road-Ordos Street
8	South Second Ring Road-YunZhong Road
9	South Bayinnaoer Road-Ordos Street-West Ordos Street
10	South Bayinnaoer Road-XianFu Street
11	West XinHua Road-South AJiLaQin Road
12	South Bayinnaoer road-North Bayinnaoer road-West XinHua Street
13	North ALaShan Road-West XinHua Street
14	South ALaShan Road-XianFu Street
15	GuangMing Street-West Second Ring Road
16	West Ordos Street-West XinHua Road

4.2 Data preparation

The steps of getting the date are as follows, and the date is listed in Table 4.

Table 4. The list of date

Node	Topological network importance of node ($\times 10^{-4}$)	Betweenness	Node traffic flow	Weighted value of the node
1	2.7	0	6072	32
2	12.5	2	5124	32
3	29.4	6	5016	32
4	25.5	9	6144	32
5	9.2	1	7692	32
6	37.2	25	5280	32
7	11.3	26	4280	32
8	3.3	0	4586	32
9	29.9	27	3912	26
10	41.6	29	3564	32
11	23.8	4	5904	24
12	29.6	14	3340	32
13	20.5	8	2928	24
14	23.6	11	3264	26
15	14.5	4	4376	24
16	17.7	9	3664	26

(1) Calculate node importance in topological network. The node importance can be got by above step 2 and the definition of node topological importance.
(2) Calculate the betweenness. The betweenness would be got by using the analysis function of optimal path in SuperMap Deskpro.
(3) Count node traffic flow. This paper recorded the peak hour flow of the nodes as the node traffic flow by on-the-spot investigation.
(4) Conduct quantitative analysis of the nodes considering the urban road grade. The Quantitative analysis can be conducted by the method in 2.2.

4.3 Determination of indexes weights

The weights of the indexes can be determined by using AHP, and the weights obtained are as given in Table 5.

Table 5. Weights of the indexes

Index	Node importance in Topological network	Betweenness	Node traffic flow	Urban road grade
Weight	0.375	0.125	0.100	0.400

4.4 Sorting of nodes importance

This paper calculated the node importance by TOPSIS, the orders are as Table 6.
It can be seen that node 6, 10 and 9 are the first three in the ranking of node importance from Table 6 and node 6 is the first. The first three nodes are ranked respectively in second, first and third place in node topological network importance and forth, first and second place in betweenness, they are the hinges for the vehicles which connected to the north and south, east and west. On the other hand, there is more traffic flow in node 6 than node 10 and 9, that is why node 6 is most important. If node 6 is removed, many of the vehicles in the network will have to bypass and it will affect traffic efficiency of the vehicles. After the analysis, the nodes obtained by the method can be consistent well with the actual situation.

Table 6. Sorting of nodes importance

Ranking	Node	Node importance
1	6	0.247
2	10	0.317
3	9	0.388
4	4	0.478
5	3	0.502
6	12	0.503
7	11	0.551
8	7	0.564
9	14	0.573
10	5	0.624
11	2	0.660
12	13	0.674
13	16	0.687
14	15	0.743
15	1	0.748
16	8	0.847

5 Summary

An evaluation method of node importance in urban road traffic network was introduced based on identification theory of key nodes in complex network in the paper. This paper tried to avoid lengthy evaluation process under the premise of making the results comprehensive and accurate. The nodes obtained by this method are in good agreement with the reality compared with the actual urban road network. In the future research on the evaluation, scholars will try to make the evaluation process more concise and the evaluation results more accurate. Through the research is getting harder, there is no doubt that the research is bound to promote the development of urban road traffic better.

References

1. X. Zhou, Finding vital node by node importance evaluation matrix in Complex networks[J]. Acta Physica Sinica, **61**(5): 201-208. (2012)
2. X. Wang, Y. Ling, YL. Fei, Personalization recommendation system based on web log&cache data mining[J]. Journal of the China Society for Scientific And Technical Information, **24**(3): 324-328. (2005)
3. S. Haykin, Cognitive radio: Brain-empowered wireless communi-Cations[J]. IEEE Journal on Selected Areas in Communications, **23**(2): 201-220. (2005)
4. FX. B, J. Budazik, K.J. Hammond. Mining navigation history for recommendation[C]. Proc of the International Conf on Intelligent User Interfaces, ACM. New Orleans. LA, ACM, 106-112. (2000)
5. W. Wang, Research on the method of key node and section distinguish of urban rode traffic network. Jilin University. (2015)
6. Y. Chen, AQ. Hu, X. Hu. China Institute Commun, **25**, 129 (2004)

7. YJ. Tan , J. Wu, HZ. Deng, Syst. Eng. Theory & Practice. **26**, 79 (2006)
8. J. Chen, LF. Sun. Evahiation of node importance in complex networks[J]. Journal of Souzhwest Jiaotong University, **44** (3): 426-429. (2009)
9. L. Qin, ZL. Yang, Synthesis evaluation method for node importance in complex networks. Computer Science, **42** (2): 60-63. (2015)
10. R. Zhong, Research of key nodes and important sections of the road network[D]. BUPT. (2013)

MATEC Web of Conferences **61**, 04027 (2016) DOI: 10.1051/matecconf/20166104027
APOP2016

College Teaching Quality Evaluation Based on System Dynamics Model

Zheng kang-ning[1,2,a], Sun shi-min[1] ,Yu hai[1]

[1]*Economic management department, Beijing City University , Beijing 100083*
[2] *Economic management institute, Nanjing University of aeronautics and astronautics, Nanjing 210016*

Abstract. This paper analyzes the main factors that influence the teaching quality and the cause and effect relationship between them, using system dynamics to establish the evaluation model of teaching quality in Colleges and universities. Taking a college A for example, this model is induced in the simulation of teaching quality, then the change process of teaching quality and feedback mechanism between these effective factors is proposed, as well as the teaching quality change under different policy parameters. And with the purpose of improving teaching quality, some measures are put forward, including staff quality, management quality, organization quality and environment quality.

1 Introduction

With the function of teaching quality in colleges and universities is teaching at the mouth of the regularity of metrics, shall be formulated by the educate and the social comprehensive requirements[1]. Enrollment expansion of colleges and universities to promote the development of higher education in our country, but also brings the new challenges to college teaching and education work. Teaching quality is the core of education activities in colleges and universities, teaching quality evaluation is to improve the teaching activities in colleges and universities, the key to improve the level of education in colleges and universities. Teaching quality evaluation is to follow the law of education teaching within the premise, according to the related policy, regulations and school personnel training, using the theory of education evaluation and scientific technology in teaching process and its result was analyzed and the quantitative judgment technical management style[2]. At present, one more is the reform of the teaching quality and teaching quality evaluation of qualitative research [3-5], the other is from the Angle of quantitative teaching quality evaluation [6-8], but these can't reveal the evaluation method of teaching interaction effects between various factors in the system. From 1956 since the founding of system dynamics, the social, economic, ecological, and other fields has been widely applied, has been hailed as a social economic system of laboratory. Using system dynamics modeling, can consider a dynamic and complex system of all kinds of main influence factors, the system can be more fully analysis, the conclusion can also conforms to the actual situation, system dynamics has been widely used in the system evaluation. This paper constructs the system dynamics model of university teaching quality evaluation, the model considering the factors of affecting the quality of teaching and the relationship between various factors and form an overall structure and quantitatively describe the complexity of the relationship between the factors, can be used to analyze the development trend of the teaching quality, find out the regulation of variables, the parameters and structure in different policy under the condition of the teaching quality, in order to put forward effective measures for the reform of the teaching quality.

[a] Corresponding author: zhengkangning@163.com

2 System analysis of the teaching quality

2.1 Limit analysis of teaching system

Analysis of the teaching quality system should first preliminary delineate the boundaries of the teaching quality system, in order to determine the endogenous variables and exogenous variables and input, determine the behavior of the system reference model. Draw system line should be refined to simplify, what interests will consist of the behavior of the system must be variable. Teaching activity is “teaching” and “learning” bilateral activity, including teacher's “teaching” and students “learning”, “tube” of teaching management, supplemented by a variety of teaching media. From the perspective of the nature of teaching system, teaching system by the “teacher”, “students”, “teaching environment”, “teaching management ”of four elements (figure 1). In figure 1 Scientific management refers to the system through certain measures to the specific requirements of the teacher how to perform the duties of a teaching, culture management is through certain incentives to pass higher education idea to teachers, to fully arouse the enthusiasm of teachers and potential to work creatively carry out personnel training.

In this paper, considering the four essential factors of the teaching system and the factors affecting the quality of teaching, teaching environment, teaching management quality, quality of teachers' quality, teaching organization as the main variables which influence the teaching quality, according to the four variables, divide the system boundaries. Teaching environment and teaching quality management quality with the above analysis, including teachers teaching echelon, the knowledge structure of teachers' quality, the concept of teacher education. Teaching echelon including teacher education titles, etc.; Teachers' knowledge structure including teacher's practical knowledge and implicit knowledge; Teachers education concept is based on the higher education idea formed by the understanding of the concept. Teaching organization quality mainly includes the teaching contents, teaching methods and teaching quality. Any links and stages in the process of teaching, all have important influence on the quality of teaching, the factors affecting the quality of teaching is various, multi-level and variety.

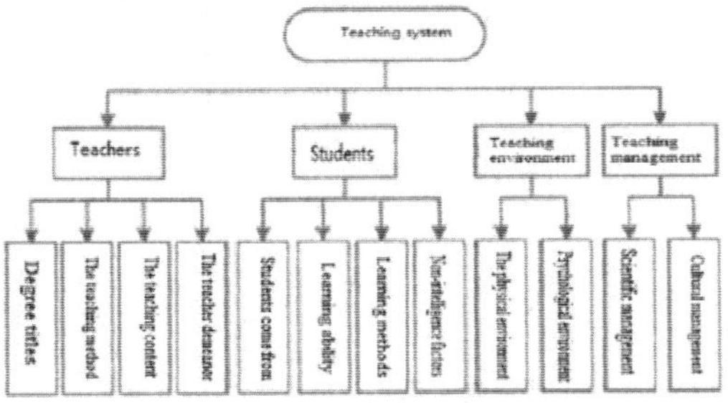

Figure 1. teaching system

2.2 Causality analysis of college teaching quality system

Able to clearly express the causal feedback loop in the system set victory of the relationship between the elements, is the key to the system dynamics research. Based on the system boundary and the relationship between the various factors, can be concluded that teaching quality evaluation system dynamics model of causality diagram (figure 2), figure in the causal chain with a plus or minus sign indicates that the effect of nature, plus said arrow pointing in the direction of variables will increase as the arrow source variable, or reduce the; Minus sign says the opposite relationship between variables.

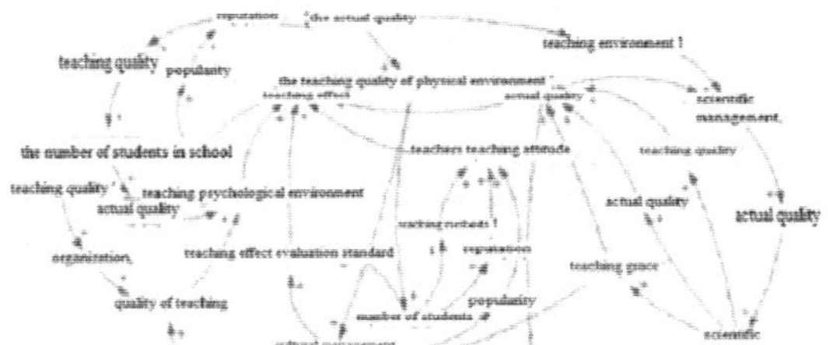

Figure 2. Causality diagram in teaching quality evaluation system

Figure 2, the main positive feedback loop ①-⑥ and negative feedback loop ⑦all landowners

①the actual quality↑, teaching quality↑, reputation and popularity↑, the number of students in school↑, teaching fund input↑, the teaching quality of physical environment↑, teaching environment↑, the actual quality↑

②the actual quality↑, teaching quality↑, reputation and popularity↑, the number of students in school↑, expect quality↑, teaching effect evaluation standard↑, teaching management quality↑, the actual quality↑

③the actual quality↑, teaching quality↑, reputation and popularity↑, the number of students in school↑, expect quality↑, teaching effect evaluation standard↑, scientific management↑, quality of teaching organization↑, the actual quality↑

④the actual quality↑, teaching quality↑, reputation and popularity↑, the number of students in school↑, expect quality↑, teaching effect evaluation standard↑, cultural management↑, the teaching methods↑, quality of teaching organization↑, the actual quality↑

⑤the actual quality↑, teaching quality↑, reputation and popularity↑, the number of students in school↑, expect quality↑, teaching effect evaluation standard↑, cultural management↑, teachers teaching attitude↑, teaching grace↑, teaching psychological environment↑, quality of teaching organization↑, the actual quality↑

⑥the actual quality↑, teaching quality↑, reputation and popularity↑, the number of students in school↑, expect quality↑, teacher quality standards↑, improving teachers' cognitive degree↑, structure of teachers↑, teachers' quality↑, the actual quality↑

⑦teachers' quality↑,to improve teachers' cognitive degree↑, introducing high level teachers↑, teachers structure↑, the quality of teaching staff↑

3 Collges and universities teaching quality evalution system dynamics model

3.1 Model building

Causality diagram describes the system structure of the basic party, reflects the variable trend of increase or decrease, is qualitative analysis. Flow graph can clearly describe the structure, the relationship between state and the rate of change regularity and gives a quantitative relation between variables, describe feedback system dynamic performance of the cumulative effect. To grasp objectively and quantitative evaluation of teaching quality, according to the causality diagram analysis of college teaching quality evaluation, established the dynamic model of university teaching quality evaluation system of flow diagram (figure 3).

Figure 3 Teaching quality evaluation system flow chart

The model uses the number of students in school, school famous for level variables, in order to get as table function, other variables as auxiliary variables or exogenous variables. Determined by expert scoring, the quality of teaching quality of teaching environment, teaching management quality of teachers, teaching quality, the weight, namely, teaching quality = 0. 3* teaching environment quality +0. 2 *teaching management quality + 0. 4 * teaching organization quality + 0. 1 * the teachers quality.

3.2 The results of simulation analysis model

Using the model to evaluate the teaching quality of A university. A simulation, system parameters of the initial value is derived from the experts on the actual data of the school and the school's scoring record. In 2007 33,14 students, will all the data into the equation, school popularity comprehensive score of six points, vensim software, running on the school from 2007 to 2016, the teaching quality of simulation, the simulation results intuitively shows the trend of the development of teaching quality.

Gradually increased, according to the results of the school teaching quality and growth fast after a slow first, the quality of teaching from 6 in 2007 increased to 7. 77 in 2016. The main reason is that as the school students enrollment increased by 2007, 33114 to 2016, 34443 people, teaching fund input by 18212. 7￥ up to 18943. 9 ￥, the teaching environment quality improvement, to improve teaching quality, to improve the school's reputation and popularity, and public expectations of school quality improvement, improve the quality of our expectations will require schools teaching reform actively, such as improve the teaching management and teaching organization, improve the quality of teachers, improve the quality of teaching as a whole.

Through software toolbar, can intuitively, specific analysis of the change of the main factors of affecting the quality of teaching.

By the figure can visually see the quality of teaching and the quality of teaching environment, teaching management, teaching quality organization quality and strength of the influence factors and changes in the quality of teachers. Teaching management quality, teaching quality, teachers' quality present relatively steady growth, this is because in the model, assuming that under the expected level of quality of teaching, the school in the three aspects of the reform of investment is carried out step by step, so the three steady growth. The teaching quality of the environment changes, for example, analysis the influence of different policy parameters on the quality of teaching.

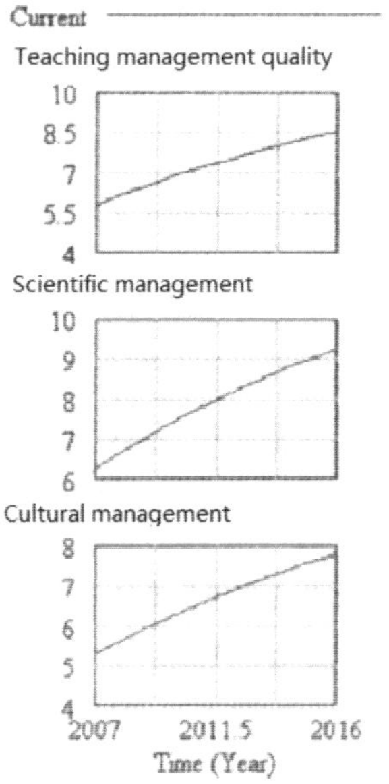

Figure 4 teaching management quality impact factor changes

Teaching environment quality is not steady growth, can from figure5, the not steady growth is mainly due to the instability of teaching psychological environment. Analysis of the factors influencing teaching psychological environment, teachers teaching steady growth, students study style slightly decreased. Students study the influence of students' quality and students' quality is influenced by the gross enrolments ratio and school acceptance rate. As the growth of the gross enrolments ratio, our country gradually from the elite higher education popularization education level, the student's overall quality has fallen. According to the ministry of education puts forward the mouth of the standard in our country, the gross enrolments ratio will reach 36% by 2015, to 40% in 2020. If the school to maintain the original acceptance rate, so as the growth of the gross enrolments ratio, the quality of students will also be presented as shown in figure 9. The quality of the students reduce can improve through strengthening the construction of study style. If the school on the school tradition construction effect is not ideal, can also be admitted through lower proportion, improve admission to improve students' quality.

To strengthen the construction of study style and lower enrolment two kinds of measures to the simulation of model, analysis the influence of different policy parameters on the quality of teaching, the simulation results as shown in figure.

By the same token, the quality of teaching management, teaching quality and teaching quality can also be simulated analysis by changing the parameters of policy, seek a more reasonable strategies to improve the teaching quality.

4 Improving the quality of the teaching strategy

Teaching quality evaluation system dynamics model not only can evaluate the quality of teaching, can also predict the development trend of the quality of teaching, simulation teaching quality under different policy trends, to provide decision support for improving the teaching quality. From model analysis shows that the quality of teaching reform from teaching environment, teaching management quality, organization of teaching quality and teaching quality of four parties and, according to the specific situation of the school, take the appropriate measures.

(1) the teaching reform of environmental quality. Environment quality can improve teaching from teaching physical environment and psychological environment for both sides. Teaching and physical environment, increase the teaching facilities, equipment, site, such as fixed asset investment, introduce advanced technology at the same time to improve teaching methods. With the advent of the era of mass education, college students' quality on the whole, at this point, the school should pay attention to the cultivation of the students study more, at the same time in order to strengthen the management of teachers' culture, improve teachers' education concept and teaching attitude, promote the formation of good teaching.

(2) the quality of teaching management reform. Schools must clear their own talent training goal and the education idea, and make it into the teaching management. There can improve the quality of the teaching management from the scientific management and culture management for both sides. Scientific management, reasonable teaching plan, improve the system of teacher evaluation, supervise and urge teachers to complete the teaching work; Culture management, the innovation incentive constraint mechanisms, deepen understanding teachers for higher education target, to standardize the behaviour of teachers and arouse teachers' potential, improve teaching level.

(3) the quality of teaching organization reform. First, select teaching materials and teaching reference books to form a complete knowledge system; Secondly through the teaching management, formulate reasonable, conform to the talented person education target teaching content; Through scientific management, supervise and urge the teacher fully preparation, teaching system and to build their own style, using new teaching means, to form the unique teaching methods.

(4) reform of teachers' quality From the point of teaching team, pay attention to introduce high levels of teachers, improve the quality of teachers from the number. At the same time should pay attention to all teachers and professional development, pay attention to teachers' personal development orientation, education idea, training teachers to attach importance to practical experience and professional development for teachers, improve teachers' professional ability, and improve the quality of teachers from the quality.

Acknowledgments

1. Beijing City University professional comprehensive reform project

2. Private education development promotion project in 2015, service area economic and social development of education teaching reform project, Beijing City University bases and experimental teaching demonstration centre construction project

References

1. Y. D. Wang, Views on the current high school undergraduate teaching quality, China University Teaching, No.3, 4-14(2008)
2. Sh. Liu, Sh. Li, Theory of the science of teaching quality evaluation system of institutions of higher learning, Modern Education Science, No.7, 155-157(2011)
3. B. Z. Gong, L. B. Li, The popularization of higher education under the condition of talent training quality evaluation, Journal of Xiangtan Normal University(Social Science Edition), No.5, 179-181(2009)
4. Y. Liu, Based on the self update effective English classroom teaching and research, Journal of Hunan University of Science & Technology(Social Science Edition), No.4,137-140(2010)
5. C. X. Zhu, Give full play to the teaching enthusiasm to improve the quality of undergraduate teaching, China University Teaching, No.3,50-51(2014)
6. Y. H. Gao, W. L. Huang, J. Liu, University teachers' talent competence quality factor model empirical research, Journal of Hunan University of Science & Technology(Social Science Edition), No.5,79-83(2011)
7. X. H. Wang, F. H. Huang, J. Liu, The teaching quality evaluation model based on BP neural network and its application, Research in Higher Education of Engineering,, No.5,78-81(2007)
8. Q. D. Wang, H. J. Hou, J. Liu, Based on interval number variable weight principle of teaching quality evaluation model, Systems Engineering-Theory & Practice, Vol.28, No.3,151(2013)

Experimental Study on the Porosity Creep Properties of Broken Limestone

Shun-cai Li[1, 2,a], Qiang Li[2], Jing-na Guo[2]

[1]School of Mechanical and Electrical Engineering, Jiangsu Normal University, Xuzhou, Jiangsu, 221116, China
[2]State Key Laboratory for Geo-mechanics and Deep Underground Engineering, China University of Mining and Technology, Xuzhou Jiangsu 221008, China

Abstract. In the underground engineering, the long-term stability of the surrounding rocks (especially the broken rocks containing water) and the ground settlement resulted from the seepage-creep coupling above goaf have been the important research subjects concerning the deep mining. For the broken rock, its porosity is an important structural parameter determining its creep properties, and the porosity change rate is more superior to describe the creep characteristics compared with the strain change rate at a certain direction. In this paper, MTS815.02 Rock Mechanics Test System is used to carry out the creep experiments on water-saturated broken limestone, and then the time curves of porosity and of the porosity change rate are obtained. By regression, we have got the relation equation between the porosity change rate with the instant porosity and the stress level during the creep. The study indicates that when the stress retains a constant level, the relation between the porosity change rate and the instant porosity can be fitted with a cubical polynomial. The obtained creep relation equation between the porosity change rate and the instant porosity and the instant stress provides a necessary state equation for studying the coupling between the seepage and the creep of the broken rock. Furthermore, the seepage in the broken rock has been verified to satisfy the Forchheimer's non-Darcy flow according to our previous studies, and its seepage properties, k, β and c_a can all be expressed respectively as the polynomial of the porosity, so, by combining with these three state equations we have obtained the four essential state equations for solving the coupling problems of the seepage and the creep for the broken rocks.

1 Introduction

As we know, the rock creep has great effect on the long time stability of the surrounding rock and surface subsidence. For broken rock, the porosity is the most basic factor determining its creep properties, but in the existing literature the stress and strain rate was mainly used to describe the creep properties of broken rock. For example, Wang [1] carried out the numerical simulation on creep damage process of the rock surrounding the roadway under high stress in deep engineering of different depth; Zhu and Ye [2] discussed the creep law affected by water content through the comparison of the rock creep test results under the dry and saturated state. Zhang and Luo[3]studied on the rock creep characteristics under different stress level . Liu and Yang [4] carried out tri-axial creep tests on the coal and rock through step loading method. Li [5] and Zhang [6] studied the creep test of marble and soft rock separately; A. k. Parkin[7]used compression meter to study the rheological property of granular materials. Shen [8] proposed three parameters model of rock-fill creep [9] through the rheological experiments on limestone; Cheng [10] proposed rock-fill creep model of nine parameters and obtained its parameter indexes through the tests. Liang[11] pointed out that the creep law of rock-fill decays exponentially with time according to the tests; Guo[12] proposed modified three-parameter rheological model of coarse granular material; Wang[13] summarized the rheological status of coarse-grained materials and pointed out that the experiment studies on granular materials were not sufficient. They proposed that it is necessary to study the micro deformation mechanism of coarse-grained materials in consideration of the scale effect of indoor test.

For the granular material, porosity is a physical quantity which can reflect the creep characteristics comprehensively. Usually, in the previous literature we can get the creep curve of the axial strain changing with time under each stress level, but the changing rules of porosity in the creep process have not been given directly.

This paper argues that the creep characteristics of rock are mainly determined by the current level of porosity and stress. The data of porosity changing can be achieved within a short time under the high porosity and high stress levels. As a result, the test time can be shortened and the cost can be reduced. In this paper, we carried out the creep tests on granular limestone saturated by water with MTS815.02 Rock Mechanics Test System [14][15], then we got the relations among the change rate of porosity , the instant porosity and the stress level in the creep process.

2 Test object and the scheme

We carried out the experiments on the broken rock by the test system [14], as shown in Figure 1.

[a] Corresponding author: zsclsc@263.net

DOI: 10.1051/matecconf/20166105001

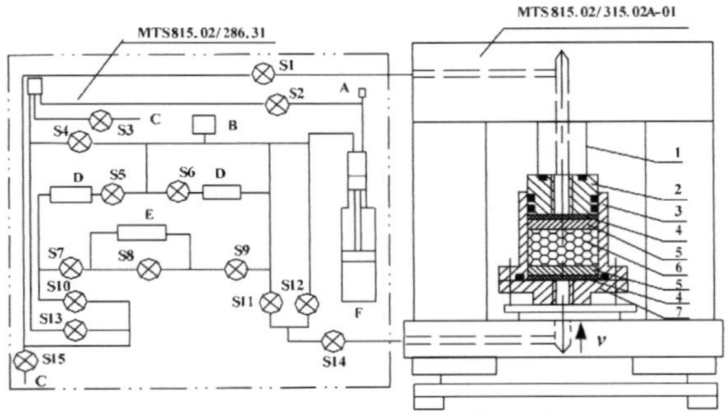

note: A–pressure transducer; B–pressure relief valve; C–drainage; D–stabilizer; E–differential pressure transducer; F–pressure intensifier (booster); S1-S15 are switches.
1–spherically seated upper plate for axial loading; 2–piston; 3–cylinder tank; 4–felt; 5–permeable plate with many holes; 6–rock sample; 7–base plate

Figure 1. MTS815.02 rock mechanics test system[14]

Test object is broken limestone with particle size 10 ~ 15 mm, and its mass density is 2700kg/m^3, the axial compressive strength is $\sigma_c = 62.3\text{MPa}$. Test scheme is as follows:

According to the increasing of the axial load, we set five stress levels. The corresponding axial load is 20 kN, 40 kN, 80 kN, 160 kN and 320 kN respectively, and the creep time of five stress level is 15 minutes, 30 minutes ,60 minutes, 120 minutes and 240 minutes respectively, as shown in Table 1

Table 1. each stress level and corresponding creep time

level	1	2	3	4	5
Force / kN	20	40	80	160	320
Loading time/s	40	120	120	120	120
Creep time/min.	15	30	60	120	240
Stress σ /MPa	1.58	3.16	6.32	12.63	25.26

The relationship between the porosity change rate and the instant stress and porosity can be regressed by two steps. Firstly, fit the relationship between the rate of change of porosity and the porosity (rather than the initial porosity) under the fixed stress level, which gets a_0, a_1, a_2 and a_3. Secondly, fit the relationship between a_0, a_1, a_2, a_3 and the current stress (rather than the initial stress). As a result, we can save test time, and reduce the cost of test.

Test flowchart are as follows:

(1) Determine the initial porosity of the sample.

Weigh the sample quality m, and by the density ρ of core we can calculate the volume V_0 before the rock sample was broken for $V_0 = \dfrac{m}{\rho}$. After the sample was placed into the cylinder, we can obtain the initial volume V_0' of the sample and the initial porosity n_0 by the height h_0 of sample in the cylinder and the inner diameter $2a$ of the cylinder, we have

$$V_0' = \pi a^2 \cdot h_0 ,\quad n_0 = \frac{V_0' - V_0}{V_0'} \tag{1}$$

(2) Saturate the rock sample;

(3) Apply the axial force gradually until its value reach the defined value , and keep the rock sample creep as the pre-designed time in Table 1. By record the height change Δh of rock sample during the loading and creeping process, we can calculate the porosity

$$n = \frac{\pi a^2 (h_0 - \Delta h) - \dfrac{m}{\rho}}{\pi a^2 (h_0 - \Delta h)} \tag{2}$$

(4) Calculate the porosity change rate $\dot{n}$ at each sampling time by the finite difference method

$$\dot{n}\big|_{i+1} = \frac{n_{i+1} - n_i}{\Delta t} \tag{3}$$

By regression, the relationship between $\dot{n}$ and $\dot{n}$ at each stress level is obtained by $n - \dot{n}$ scatter diagram, further more we can get therelationship between a_0, a_1, a_2, a_3 and the stress level.

$$\dot{n} = a_0 + a_1 n + a_2 n^2 + a_3 n^3 \tag{4}$$

3 The test results

3.1 $n-t$ and $\dot{n}-t$ curves under various stress levels

According to the collected axial displacement and the Equation (2), we can calculate the porosity of each time under each stress level. Thus we can draw time history curve of the porosity and porosity change rate under various stress levels, and draw relation curve of the porosity and the porosity change rate, as shown in Figure 1 ~ 10 respectively. As a result, at the same stress level, the larger the porosity n, the greater the absolute of change rate $\dot{n}$. By regression analysis, we can obtain $\dot{n} - n$ relation curve of each stress level, which can be fitted by cubic polynomial.

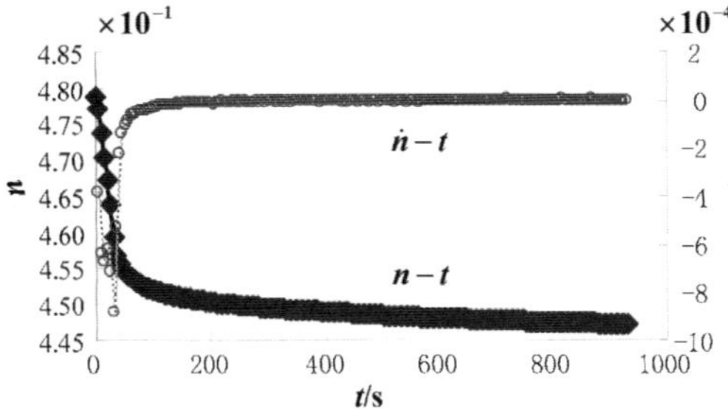

(a) $\dot{n} - t$ curve and $n - t$ curve

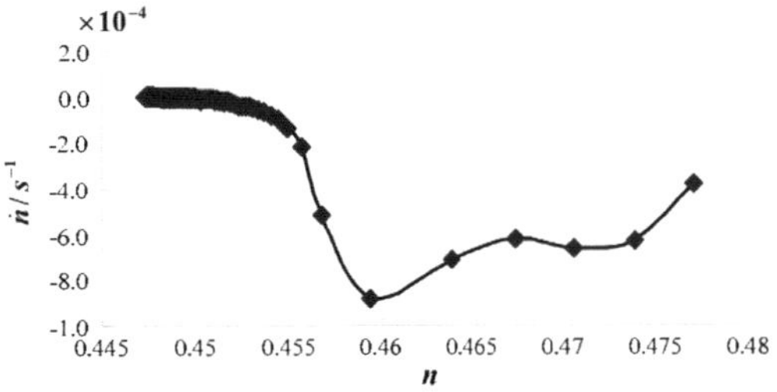

(b) $\dot{n}-n$ curve

Figure 2. curves of porosity and its change rate at σ_a = 1.58 MPa

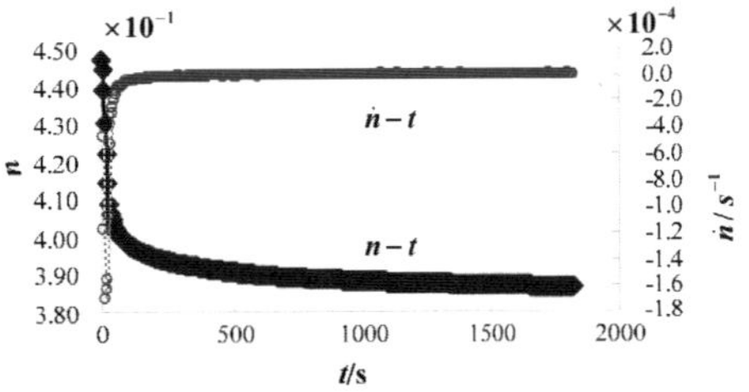

(a) $\dot{n}-t$ curve and $n-t$ curve

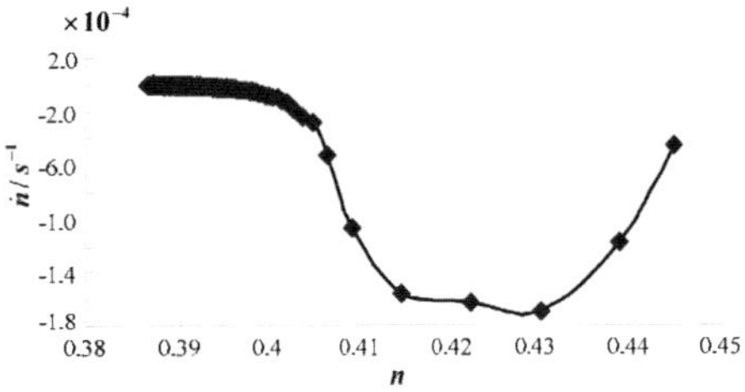

(b) $\dot{n}-n$ curve

Figure 3. curves of porosity and its change rate at σ_a = 3.16 MPa

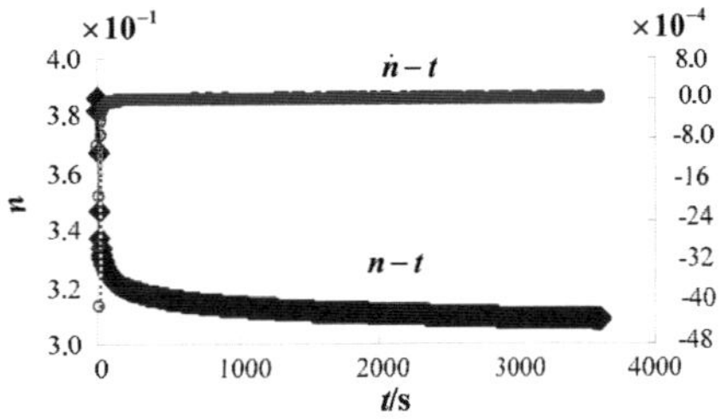

(a) $\dot{n}-t$ curve and $n-t$ curve

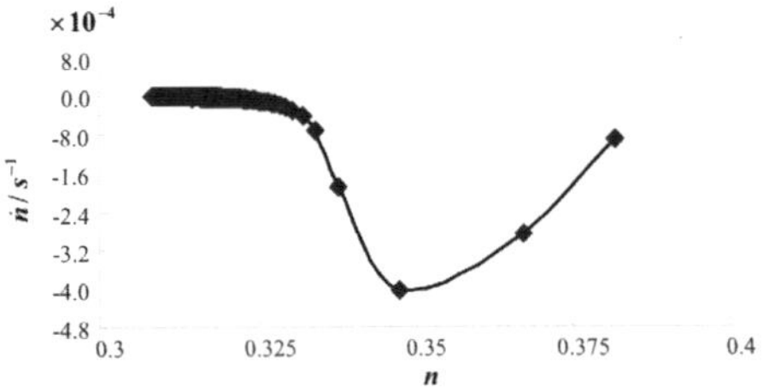

(b) $\dot{n}-n$ curve

Figure 4. curves of porosity and its change rate at σ_a = 6.32 MPa

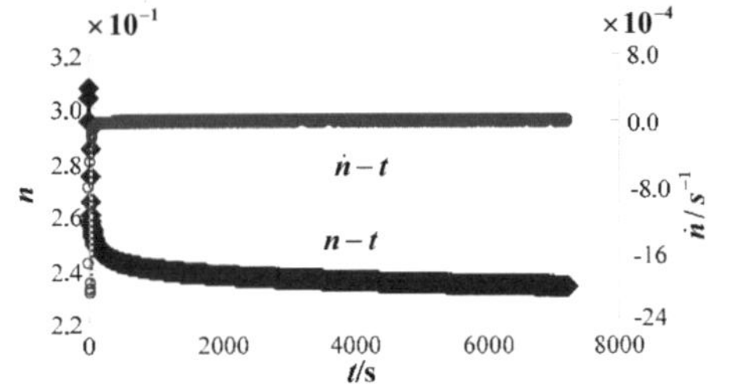

(a) $\dot{n}-t$ curve and $n-t$ curve

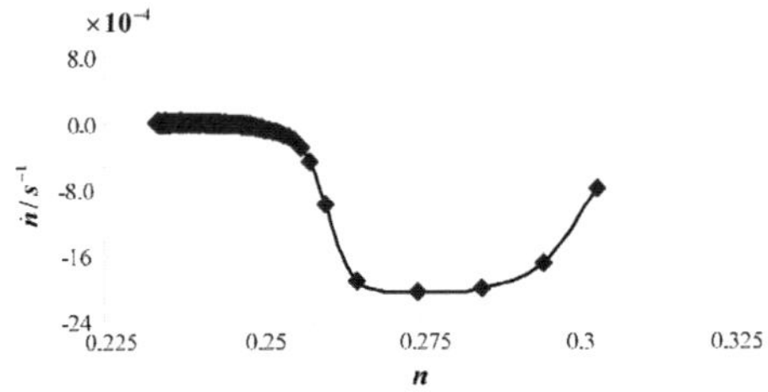

(b) $\dot{n}-n$ curve

Figure 5. curves of porosity and its change rate at σ_a = 12.63 MPa

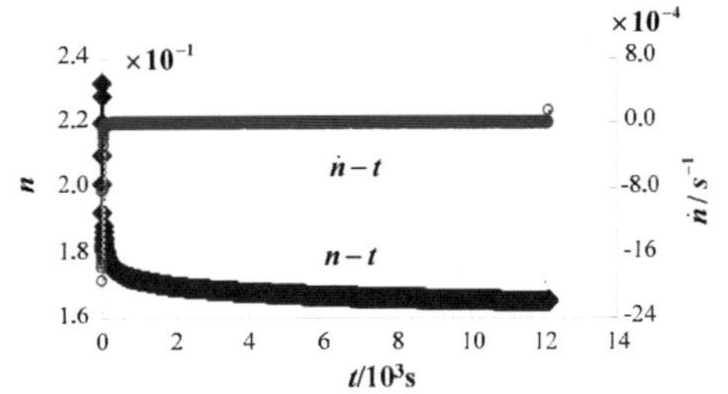

(a) $\dot{n}-t$ curve and $n-t$ curve

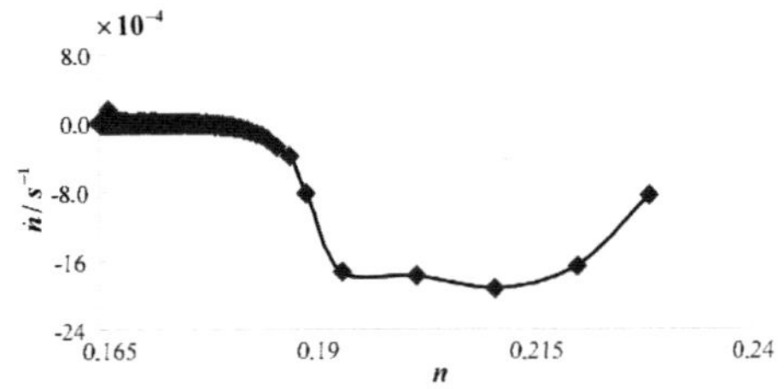

(b) $\dot{n}-n$ curve

Figure 6. curves of porosity and its change rate at σ_a = 25.26 MPa

Under various stress levels, the polynomial coefficients obtained by the curve fit are as shown in Table 2, where σ_a is the axial stress and σ_c is the uniaxial compressive strength of limestone core.

Table 2. coefficients of the cubic polynomials

σ_a(MPa)	σ_a/σ_c	a_0	a_1	a_2	a_3
1.58	0.0254	-18.6	122	-265	192
3.16	0.0507	-4.41	32.4	-9.3	64.5
6.32	0.101	-2.07	18.6	-55.3	54.6
12.6	0.202	-0.566	6.57	-25.2	32.0
25.3	0.406	-0.244	3.90	-20.6	25.8

3.2 The relation equation between the porosity change rate and the present porosity and stress

In Table 1, polynomial coefficient a_0, a_1, a_2, a_3 is related to the ratio $\frac{\sigma_a}{\sigma_c}$ at various levels. By further regress we can get

$$a_0 = -0.0543\left(\frac{\sigma_a}{\sigma_c}\right)^{-1.5479}$$

$$a_1 = 1.09\left(\frac{\sigma_a}{\sigma_c}\right)^{-1.2245}$$

$$a_2 = -7.22\left(\frac{\sigma_a}{\sigma_c}\right)^{-0.9029}$$

$$a_3 = 11.8\left(\frac{\sigma_a}{\sigma_c}\right)^{-0.6806}$$

Namely, the coefficients a_0、a_1、a_2、a_3 can be expressed by the power function of ratio $\frac{\sigma_a}{\sigma_c}$ respectively.

4 The application of test results in the theoretical analysis and calculation

The creep characteristics of rock (including broken rock) are determined by the rock pore and the fracture structure instead of the stress or strain of the rock in a certain direction. Therefore, the description of creep characteristics by porosity is superior to that by the stress or strain. When we study the seepage-creep coupling of the broken rock we can find that the mass conservation equation (continuity equation) and the momentum equation (equation of motion) can not constitute a set of complete control equation. In order to obtain a set of close control equation of seepage-creep system, we need to supplement the state equation. Equation (4) is actually a state equation on the porosity; it provides the basis for the analysis of coupled seepage-creep problems.

A large number of experiments prove that the rock seepage accord with the Forchheimer relationship [16]. There are three parameters to describe the seepage parameters: the permeability k, the non-Darcy flow factor β and acceleration coefficient c_a. Although the logarithmic function, the power function and the exponential function can be used to fit the relationship between the seepage parameters and the porosity, there are some distinct errors. Only the polynomial fitting can get the best correlation. Certainly, the high order we fit the polynomial, the better effect we can get. However, this will complicate the analysis and the operation. Cubic polynomial can be chosen to give the following relationship.

$$\begin{cases} k = k_0(a_0 + a_1 n + a_2 n^2 + a_3 n^3) \\ \beta = \beta_0(b_0 + b_1 n + b_2 n^2 + b_3 n^3) \\ c_a = c_a^0(c_0 + c_1 n + c_2 n^2 + c_3 n^3) \end{cases} \quad (5)$$

When we carry out numerical calculation, if the stress/strain state at the initial moments is known, we can use the strain state of a point to obtain porosity and determine the seepage parameters according to the Eq.(5), then calculate the porosity at the next moment according to the porosity creep equation (4), then calculate the stress/strain, pore pressure and seepage at the next moment according to the control equations. Cycle like this, we can get the dynamic stress/strain field for solid skeleton and dynamic seepage field

5 Conclusions

By the MTS815.02 rock mechanics testing system, we carried out the creep experiments on the water-saturated limestone at five levels of stress. As a result, we obtained the time curve and the relation curve of the porosity and its change rate. Then, we analyzed the test results and got the following conclusions:

(1) For broken rock the porosity change rate is more suitable to describe its creep properties than strain rate in a certain direction.

(2)The creep characteristic of broken rock is mainly determined by the porosity and the stress level, the greater the porosity and the higher the stress level, the greater the porosity change rate is.

(3)At the same stress level, we can use cubic polynomial to fit the porosity change rate and the current porosity; Under different stress levels, the polynomial coefficient depends on the ratio between the axial stress and the uniaxial compressive strength of the cores, and the coefficient can be fitted by the power function of this ratio.

(4) The obtained creep equation of the porosity change rate relating with the current porosity and stress provides an essential state equation for the study of creep-seepage coupling effect.

Acknowledgement

The authors gratefully acknowledge the financial support from the National Natural Science Foundation of China (51574228, 51227003), Natural Science Foundation of Jiangsu Province University (14KJB440001).

References

1. Wang Y, Qi J, Yang C, Wei J. A study of nonlinear creep law in deep rocks. Rock and Soil Mechanics, 2005, 26(1): 117-121
2. Zhu H, Ye B. Experimental study on properties of rock creep in saturation. Chinese Journal of Rock Mechanics and Engineering, 2002, 21(12), 1791-1796

3. Zhang Z, Luo J. Study on creep properties of rock under step load. Chinese Journal of Rock Mechanics and Engineering, 2004, 23(2), 218-222(in Chinese)
4. Liu J,Yang C, Li X. Jiang D. Testing study on creep of coal rocks in the tunnel of Wankai Speedway. Chinese Journal of Rock Mechanics and Engineering, 2004, 23(22), 3794-3798(in Chinese)
5. Li H, Li Z, Su C. Testing study on Creep characteristics of marble, Chinese Journal of Rock Mechanics and Engineering, 2004,23(22), 3745-3749(in Chinese)
6. Zhang X, Li Y, Zhang S, Huo B. Creep theory of soft rock and its engineering application, Chinese Journal of Rock Mechanics and Engineering, 2004,23(10), 1635-1639(in Chinese)
7. Parkin AK. Creep of rockfill (Part A). Maranhadas Neves E. Advances in rockfill structure. London: Kluwer Academic Publishers. 1992:221-239.
8. Shen Z, Zuo Y. Test research on the rheology character of rockfill [A]. The Sixth Proceedings of Geo-mechanics and Basic Engineering [C]. Shanghai: Press of Shanghai Tongji University, 1991, 443-446(in Chinese)
9. Shen Z, Zhao K. Back analysis of creep deformation of rockfill dams. Journal of Hydraulic Engineering, 1998, (6):1-6
10. Cheng Z, Ding H. Creep test for rockfill. Chinese Journal of Geotechnical Engineering, 2004, 26(4): 473-476
11. Liang J, Liu H. Creep test for rockfill of CFRD. Chinese Journal of Geotechnical Engineering. 2002, 24(2):257-259(in Chinese)
12. Guo X, Wang D, Cai X, Dong L. Rheological analysis of concrete faced rock-fill dam. Journal of Hydraulic Engineering, 1999, (11): 42-46
13. Wang M, He X, Cheng Z. Current situation and prospect of studies on rheology property of course stuff. Rock and Soil Mechanics, 2003(supp), 451-454
14. Liu Y, Li Y, Sun M. A new method of permeability test for loose rock. Ground pressure and strata control, 2002, 19(4):108-110
15. Liu W, Miao X, Chen Z. A Testing Method for Determining the Permeability of Over broken Rock. Journal of Experimental Mechanics, 2003, 18(1):57-61
16. Chen Z. Bifurcation of Dynamic System of Non-Darcy Flow in Post-Failure Rock [Ph. D. Thesis][D]. Xuzhou: China University of Mining and Technology, 2003

Analysis of Characteristics of Ore about Iron Deposit of Da Hong Mountain in Yun Nan Province

Zhang Yuefeng[1a], LI Jun[1], Tian Yimin[1], Yao Yanling[1], Zhang Yaguang[1], Ma Mengling[2]
[1] *Faculty of Land and Resource Engineering, Kunming University of science and technology, Kunming, Yunnan*
[2] *Regional Geology and Mineral Resources Survey Institute of Hebei Province Lang fang*

Abstract: This thesis aims to analyse the deposit characteristics about Da Hong Mountains Iron ore in Yunnan province. The texture and structure, especially the chemical composition, is different in every section of deposit after comparing. Moreover, the content of SiO_2 is much higher than general iron ore. However, the content of other noble metals cannot reach the lowest industrial grade. Da Hong Mountains Iron ore has unique features because of metallogenic periods.

1 Introduction

The great red iron ore in Xinping County Yunnan province .It is a large Iron-copperdeposit, a typical submarine volcanic eruption and depositional deposit. There are unique form between different section of mineral deposit with the multiply metallogenic period

2 Distribution of Ore Bodies in the Mining Area

According to the distribution of ore bodies、output position、buried depth、tectonic boundaries and other factors. Mining area Can be divided into five ore blocks, that is shallow iron ore、Deep iron ore、No. 1 East of iron-copper、the north shore of Mangang river and Hamubai group iron.

Shallow iron or: Shallow iron mine is located in the south of Mangang river, the north of F_2 fault, which exposed surface buried shallow orebody. The main ore body located in more than 700 meters in elevation, it included industrial orebody 30.

Deep iron ore is located in between F1 and F2 fault location, buried deep blind ore bodies. Most main orebody reserves distributed in below 600 meters in elevation.

No. 1 East of iron-copper is located in between the eastern F3 and the east A49, which contain iron and copper ore bodies.

Mangang River is located the north of the river surface of the exposed surface of the iron ore. The ore body is the same as that in the southern bank of the south bank. The ore body is buried in shallow depth, the depth is not large, the scale is small, and there are 4 industrial ore bodies [1].

Hamubai group iron is the ore deposit that is between the easts of A39 line to the A45 Line East Hamubai group stratum. The orebody and the south shore shallow strata are the same horizon, the surface of the scattered, small size, a total of 3 industrial ore body.[2] （photo1）

The main of metal mineral is magnetite, siderite, followed by chalcopyrite. The main of gangue minerals is feldspar, quartz, garnet, biotite, and dolomite. According to the main of metal mineral and gangue mineral is divided into five types: Quartz with brass magnet ore, felsic brass magnet ore, garnet biotite contain brass magnet ore, felsic dolomite with brass magnet siderite mixed ore, felsic dolomite with brass siderite ore.

3 The Texture and Structure

Metal mineral is mainly composed of magnetite, followed by hematite, with a small amount of martite (magnetic hematite), ilmenite, pyrite and chalcopyrite, occasionally it can see bornite. Gangue mineral is mainly composed of sodium feldspar, quartz, followed by white mica, carbonate and iron silicate minerals (mainly chlorite). Although the content of apatite and tourmaline is less, but its distribution is broad, and occasionally it can see rutile, zircon, epidote and hornblende, etc.

The main structure of mineral is granular structure, followed by slab structure and lamellar structure. The part of the structure is porphyritic structure, commonly it can see metasomatic texture, and the local of structure is granular clastic texture. The structure characteristic of

[a]Corresponding author: 18687210822@163.com

 DOI: 10.1051/matecconf/20166105002

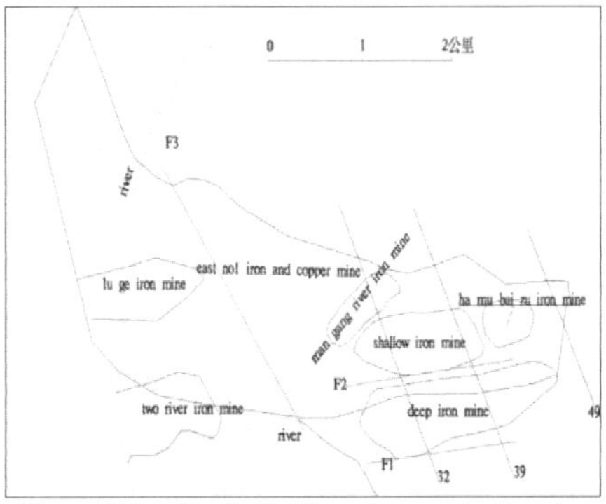

Figure 1 Ore block dividing plane projection draw

Magnetite is granular structure (**Figure** 2). The structure characteristics of hematite is allotriomorphic granular texture (**Figure** 3), the structure of hematite also has slab structure and lamellar structure. Magnetite is phenocryst of the mixed ore and the mixed ore has porphyritic structure.Hematite replaces the magnetite, which results the replacement texture,normally it can see that the hematite fills and replaces magnetite particles along its edge, cleavage and fissure , the shape of hematite replaces magnetite is needle, lattice, grid or net vein, and even when magnetite is replaced by hematite, the shape of the residual magnetite seems like an island or all the magnetite is replaced by hematite and appears the illusion of hematite(magnetic hematite);Granular blasted texture is only can be seen in Striped and banded magnet.

The ore structures have various styles. Such as disseminated structure ,banding and barred structure, piebald structure, breccia structure, mottling structure, patch structure, massive structure and dense massive structure[3](**Figure** 4、5、6)

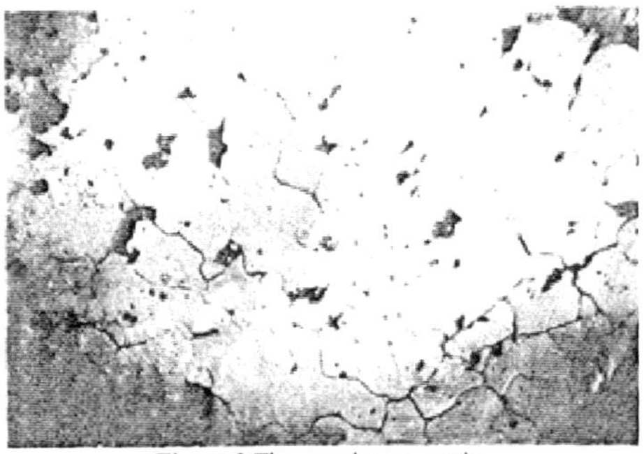
Figure 2 The massive magnetite
Equigranular and subhedral mosaicstructure

Structure characteristics of the mineral composition and structure of the iron ore bodies and mineralization with common characteristics，But different ore belt iron ore body also has obvious difference. Especially in the deep iron ore more features，that the main metal mineral and gangue mineral, structure, ore types and mineralization, and so on have its own way, making it different from other Iron Orebody

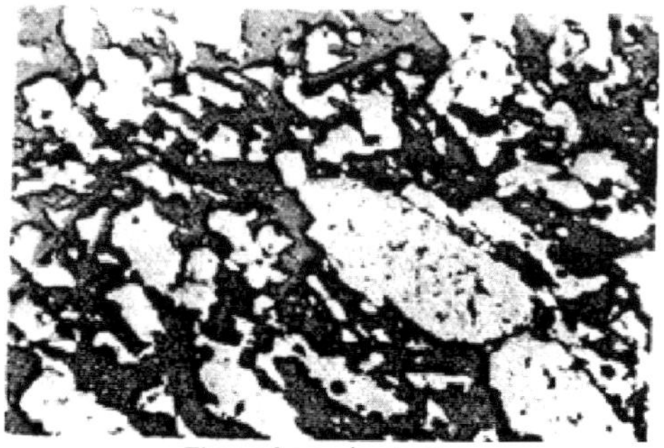
Fi gure 3 granular structure
the white object is hematite，subhedral and anhedral form hematite

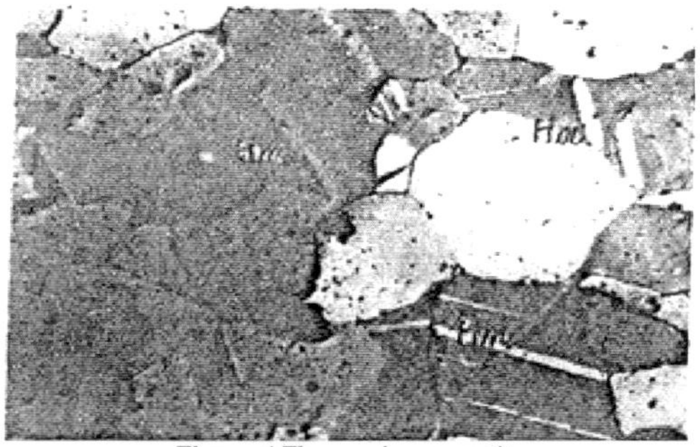
Fi gure 4 The massive magnetite
Equigranular and subhedral mosaicstructure

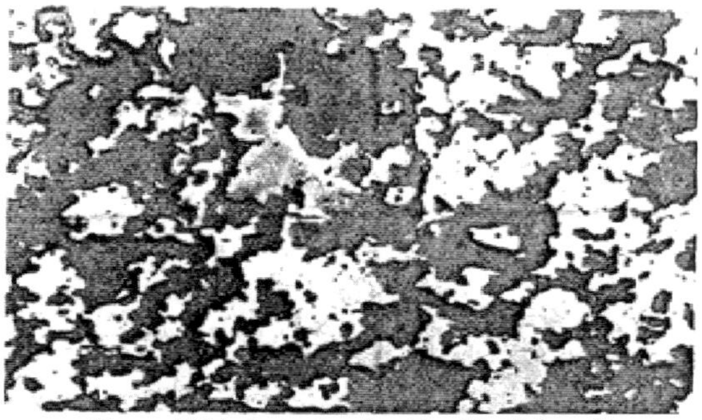
Fi gure 5 Veinule structure thewhite object is hematite

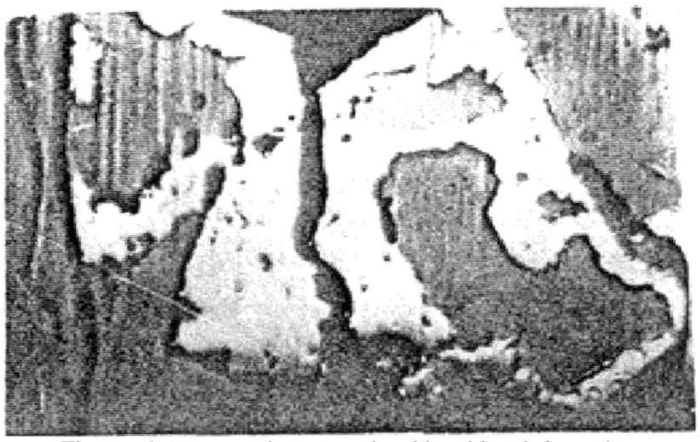
Fi gure 6 metasomatic texture thewhite object is hematite

4 Ore chemical composition analysis

Main element content of different ore block of Iron orebody percentage shown in the following table (1)
Rich iron ore. The scope of TFe is51.39～59. 08%the average content of TFe is 54.71% .poor iron ore, the scope of TFe is30.36～34.90% the average content of TFe is 52.74% .TFe in the second deep orebody is higher than shallow orebody. The content of the Sio2 in iron ore is usually higher, In Rich iron ore it is12.49～24.19% and the average is about 18.91%.in Poor iron ore the data is 37～44.55%, the average is 40.60%.

Table 1 The main elements content of ore

Orebody numbering	Ore type	Ore level	The number	The main elements content									
				TFe	SFe	FeO	S	P	SiO_2	Al_2O_3	CaO	MgO	K_2O
V3	Magnetite	Poor	5	31.33	30	14. o6	0.02	0.24	40.45	4.26	2.25	1.44	0. O5
		Rich	1	59. O8	58.52	26.89	0.OI	0.14	12.49	0.76	1.76	0.65	0.05
V2	Magnetite	Poor	15	30.36	29.03	16. 0I	0.03	0.45	37	8. O3	3.52	2.12	0.08
		Rich	5	56.40	56.35	26.22	0.00	0.12	15.22	1.51	1.64	0.36	
IV1-1	Magnetite and red-magnetite	20～25	1	24.15	23. O5	13.25	0.10	0.11	41.49	11.80	1. O2	3.40	
		25～30	3	28.82	28.15	7.94	0.O3	0.15	43.59	8.12	1.34	0.59	0.50
		Poor	3	32.60	32.17		0.O4	0.15	44.55	3: 56	0.82	1. O5	0.35
		Rich	7	51.95	50.87	23.19	0.01	0. O7	24.19	0.73	1.O5	0.33	0.06
		Average		38.75	38.13	17.21	0. O3	0.13	37.90	3.90	1.18	0.81	0.32
III1	Magnetite	20～25	4	22.24	21.63	11.12	0.08	0.12	40.86	11.75	2.51	1.79	
		25～30	2		28.44	10.53	0. OO	0.14	40.38	7.63	1.93	0.98	1.69
		Poor	1	34.90	32.37	11.31	0.06	0.08	41.98	3.70	2.64	0.88	
		Everage		25.27	24.50	1.76	0.06	0.12	40.77	8.64	2.35	1.32	1.69_
II1	Magnetite and red-magnetite	20～25	44	21.99	21.65	5.83	0.05	0.16	45.75	10.24	1.87	1.13	1.04
		25～30	50	27.41	27.16	5.92	0.03	0.17	45. 03	5.83	1.74	0.63	0.61
		Poor	51	34.53	34.24	8.96	0.02	0.17	39.01	4.34	1.77	0.61	0.48
		Rich	174	51.39	49.96	13.77	0.02	0.13	23.73	1.26	1.19	0.44	0.09
		Everage		39.91	39.62	10.38	0.02	0.15	30.08	3.65	1.49	0.59	0.33

There is an inverse correlation between the content of SiO2 and TFe, especially in the rich ore usually, the value of TFe+SiO2 is a constant (Table 2)[4], which indicates that mineral composition is simple and consist of iron minerals and quartz. For example, enriched iron ore of II lmetallogenic belt, which has a high grade of TFe(more than 55%), the value of TFe+SiO^2 is 73.50%. Magnetite with 60% grade of TFe is consisting of 84.25% magnetic mineral and 13.50% SiO^2, that means the value of magnetic mineral and quartz is 97.75%. Because lean ore has other minerals like albeit and other silicate, the law is not as regular as rich ore.

Table 2 relationship between different grade ore TFe and SiO_2

Grade (%)	TFe (%)	Si02 (%)	TFe+Si02 (%)	For note
65 以上	66.71	5.73	72.44	As TFe is 55～65%
60～65	62.46	10.90	73.36	TFe60.06%、Si0213.94%
55～60	57.73	16.97	74.70	TFe+Si02=74%
Average	62.30	11.20	73.50	

MyO, CaO:The content of CaO and MgOis low in the Iron orebody, The content of CaO usually is 1.53～2.51%, The highest content of the Cao is 7.61%, The lowest levels of the Cao is 0.65% and the average content is 2.39%; The content of MyO usually is 0.59～2.50%and the average content is 1.34%。The content of CaO in Late Palaeozoic basin of the first and third ore-group and the average content is 4.04%, The content of MyOis relatively high Iron Orebody, The content of MyO usually is 1.5 ～ 4.5 % and the average content is 3.38 % .The content of Different grades and different types Iron Orebody changed little, irrelatively stable.

Acid-base Ratio: Thanks to the content of CaO and MyO are low however, Sio2 is high, the Acid-base Ratio of Iron Orebody is small. The ratio of major Iron Orebody is 0.04～0.15%. They belong to the acidic ore.[5]The ratio of Different grades, different types Iron Orebody changed little and there are the trends of the ratio decreases change low with The decline of ore grade. By the magnetite and hematite ratio from large to small (table3).

Table 3 II_1Acid-base Ratio

Ore type	TFe (%)	slaggingcomponents (%)				$\frac{CaO+MgO}{SiO_2+Al_2O_3}$
		CaO	MgO	Si02	Al_2O_3	
Rich magnetite	55.01	1.18	0.56	19.46	1.19	0.084
Red-magnetite	45.14	1.48	0.45	31.76	1.88	0.057
rich Red-magnetite	40.60	1.13	0.38	32.47	4.12	0.041
rich Red-magnetite	50.62	1.02	0.33	24.84	0.83	0.052
Poor magnetite	34.10	· 1.80	0.83	36.33	5.88	0.062
Poor Red-magnetite	37.75	2.17	1.14	29.88	5.65	0.093
Poor magnetite	27.78	1.13	0.43	46.09	6.14	0.029
Poor Red-magnetite	30.95	0.96	0.20	46.35	3.53	0.023

5 Beneficial Associated component content and Occurrence state

Through the combination of analysis, artificial placer, optional test, in a single iron ore body was not found with comprehensive utilization value of accompanying useful components, but in the I 、III ore Au 、Ag、Co、Pt、Pd can be integrated recovery, Its content is shown in table 1-4. From table（4）ore grade is low, no separate recovery value, but it can get in sulfide concentrates after beneficiation different degree of enrichment, that can consider to comprehensive recovery.

Gold, silver: IN ore belt iron copper (I 3、 I 2) Au 0.08 and 0.11 g/t, Ag0.53 and 0.63 g/due to the run of mine ore content with little, Au, Ag are not found in mineral alone. But we can get in sulfide concentrates after beneficiation good enrichment. This description Au, Ag occurrence in sulphide minerals easy for comprehensive recovery. The distribution of its distribution is obviously related to the Cu, namely when the Cu-rich, Au, Ag content is also relatively increased.

According to the III copper ore analysis the mineral sample, Au 0.33 g/t, Ag, 2.6 g/t. According to the selected 18 sample test results show that the copper concentrate in Au enrichment cannot get enrichment, but it can get enriched in iron concentrate.

Table 4 Copper associated components

Orebody numbering	head grade (g / ton)					concentrate grade(g / ton)				
	Au	Ag	Co(%)	Pt	Pd	Au	Ag	Co(%)	Pt	Pd
I 3	0.08	0.53	0.012	0.00	0.01	1.62	11	0.027	0.03～0.09	0.14
I 2	0.11	0.63	0.012	0.00	0.02	1.69	13	0.0594	0.03～0.09	0.17
III	0.33	2.60	0.006	0.00	0.012	0.28	12.84	0.045	0.00	0.02

Co: No.1ore containing Co generally not less than 0.01%, average 0.012%. Sulfide concentrate containing Co 0.027 ~ 0.0594%, compared with about five times ore enrichment, different types of ore concentrate containing Co are quite different taste, marble-type ore is higher, up to 0. 514%; Biotite schist type ore is low. Cobalt in class in the sulfide mass with like and adsorption state, mainly because some of the iron cobalt replace when entering the pyrite lattice, so the value of the lattice parameters of pyrite increases

6 Conclusions

1.From rock slice, original rock of ore district can be deduced as mafic-intermediate volcanic rocks, which was obviously affected at hydrothermal alteration from early time, some rocks got regional metamorphism later.

2.Big red mountain iron ore is a kind of magmatic eruption - sedimentary hydrothermal deposit, which suffered metamorphism during the ore-forming process. Especially, the mineralization of ore is obviously at the ore enriched zone.

3.There is an inverse correlation between the content of SiO_2 and TFe, especially in the rich ore. Usually, the value of TFe+SiO_2 is a constant

4. The content of CaO and MgO is poor and stable, and it doesn't change a lot at different grade and types of iron ore.

5.Except the content of Au and Ag for parts of No.1 or3 mine can reach the grade of pay utilization; other ore can't reach the grade. Moreover, though Au can't enrich at the copper concentrate, Au can enrich at the iron concentrate. It's a new question waiting for further research.

References

1. Li Shengrong. Crystallography and Mineralogy. Geological Publishing House.2008.in China
2. McSween H. Y, Murchie S. L, Crisp J. A, et al. Chemical, multispectral, and texturconstraints on the composition and origin of rocks at the Mars Pathfinder landing site[J].Journalof Geophysical Research Atmospheres, 1999, 104(E4):8679–8715.
3. WangHunxun, HeChunshan. About the test procedure and method of grinding. nonferrousmetals, 1980, (6)in China
4. Zhaole. Identifying correlation of iron grade and silicon dioxidecontent in iron- mineral concentrate using regression analysis[J]Metallurgical Standardization & Quality2006,(44)17-21in China
5. Ying-chieh Lee, Wei-HuaLu, et al. Influence of SiO_2 Addition on the Dielectric Properties and Microstructureof Ceramics[J]. International Journal of Applied Ceramic Technology, 2009,(6):692-701.

 DOI: 10.1051/matecconf/20166105003

An Investigation on Corrosion Behavior of a Multi-layer Modified Aluminum Brazing Sheet

Wei LIU[1,a], Zhenming YANG[1,b], Long LI[1,c], Dejing ZHOU[1,d]

[1]Yin Bang Clad Material Co., Ltd., Wuxi Jiangsu, China

[a]wei.liu@cn-yinbang.com, [b]zhenming.Yang@cn-yinbang.com

[c]long.li@cn-yinbang.com, [d]dejing.zhou@cn-yinbang.com

Abstract. The corrosion behavior of a multi-layer modified aluminum brazing sheet (AA4045/3003Mod./AA7072/AA4045) was investigated. The results shows that, the existence of BDP, which forms at the interface between clad and core layer during brazing, changes the corrosion form of the air side of the material from inter-granular corrosion to local exfoliation corrosion. The addition of anti-corrosion layer makes the corrosion form of the water side from inter-granular corrosion into uniform exfoliation corrosion. Compared to the normal triple-layer brazing sheet at the same thickness, the time to perforation of the modified four-layer brazing sheet is increased by more than 200%.

Key words: aluminum alloy; brazing sheet; corrosion behavior

1 Introduction

Aluminum alloys are of interest for various applications, particularly due to their high strength/weight ratio, good formability and good corrosion resistance in vehicles, household items, infrastructures, constructions, aerospace, etc. It has been calculated that for each kilogram of weight saved in a vehicle, a saving of 20 Kg of CO_2 emissions can be achieved. Therefore the use of aluminum in the automotive industry in order to produce more fuel-efficient vehicles and to reduce the energy consumption and air pollutions has increased greatly over the last few decades, from 20 Kg in 1960 to a predicted level of more than 160 Kg per vehicle in 2010 and250 to 340 Kg by 2015[1]. One of the increasing applications of Al alloys in vehicles is in heat exchangers (with tube and fin components) such as radiators, evaporators, engine cooling and air conditioning systems. Corrosion behavior of the brazed sheet is critically important since this is the major failure mode for automotive heat exchangers, especially under the trend of lightweight of automotive parts. By using the test methods of copper accelerated salt spray (CASS), acetic acid salt spray (AASS), sea water acid accelerated test (SWAAT) and the analysis techniques of optical microscopy (OM), scanning electronic microscopy (SEM), scanning Kelvin probe force microscopy (SKPFM), many studies have been done to characterize the corrosion behavior of the brazed AA4xxx/AA3xxx alloys. Initially, Scott et al.[2] and Isobe et al.[3] considered that noble Si needles acted as active cathodes thus make the surrounding aluminum dissolved. While a subsequent study[4,5] shown that Si particles do not have a detrimental effect on the corrosion behavior of the brazing sheets. And it is coincident with the study of Tohma et al.[6,7] that the oxygen reduction current density on silicon is too low to anodically polarize

DOI: 10.1051/matecconf/20166105003

aluminum. The study of Tierce et al.[8] considered that the micro-battery effect between α-Al(Fe,Mn)Si phase, which generates during the brazing process[9], and the Al matrix are responsible for the corrosion in acidic or neutral solutions. In order to protect the core material, a type of modified AA3xxx aluminum alloy, which a certain amount of Cu is added, is widely used in place of the normal AA3xxx. And a investigation of the modified AA3xxx aluminum brazed sheet shown that the diffusion of Cu element to the re-solidified clad material during brazing will reduce the cathodic protection towards the core material[10]. Besides, a band of dense precipitates (BDP)[11,12]is formed in the interface between the clad and core materials, as a consequence of the diffusion of Si and Cu elements during brazing. And it is said that the existence of BDP will affect the corrosion behavior of the brazed material[13], however, there is no definitive mechanism about the effect above.

However, the studies above are mainly focused on the triple-layer brazed aluminum alloys. The study on the four-layer brazing sheet which an anti-corrosion layer was added is rare. With respect to the 7xxx series aluminum alloy, exfoliation corrosion is the typical form in the acidic solutions. Therefore, it is an effective way to increase the corrosion life by the means of adding a 7072 layer between the clad and core materials. This study aims to investigate the corrosion behavior of a self-developed four-layer modified Aluminum brazing sheet, and to be the theories storage for the new material development.

2 Experiment

Materials. A4 size sheet (600μm thick) was cut from a rolled modified AA3xxx (3003 Mod.) aluminum alloy clad on one side with an AA4045 (60μm thickness) and another side with an AA7072 (180μm thickness) and AA4045(60μm thickness). The bulk chemical analysis of the clad, core and the anti-corrosion layer material is presented in Table 1.In order to distinguish easily during the test, two names are given to the each side of the material, namely, the air side (AA4045) and water side (AA4045+AA7072). All sheets used in these experiments before brazing were O temper.

The sheet was properly cleaned in ultrasonic bath with ethanol, dried and placed in a controlled atmosphere brazing (CAB) furnace. CAB is brazing in a dry, inert nitrogen gas atmosphere. Brazing cycle that was applied is shown in Fig.1. The material is heated up to 605℃ and is kept at that temperature for 3min. Then it is cooled down with a specified ramp (30℃/min) in the furnace. And samples will be taken out and cooled in the air when the temperature is down to 390 ℃. The holding temperature was selected to be close to the liquidus temperature of the clad material, reported as 612℃ by Lacaze et al.[14],while still being far below the solidus temperature of the core material which is about 640℃.

Table1 Chemical composition (wt%) of the core, anti-corrosion layer and clad material

	Si	Fe	Cu	Mn	Mg	Zn
Clad(AA4045)	10.0-11.0	0.45	0.1	/	0.02	/
Core(3003Mod)	0.6	0.7	0.4-0.7	1.0-1.5	0.03	/
Anti-corrosion layer(AA7072)	Si+Fe≤0.7		0.1	0.1	0.1	0.8-1.3

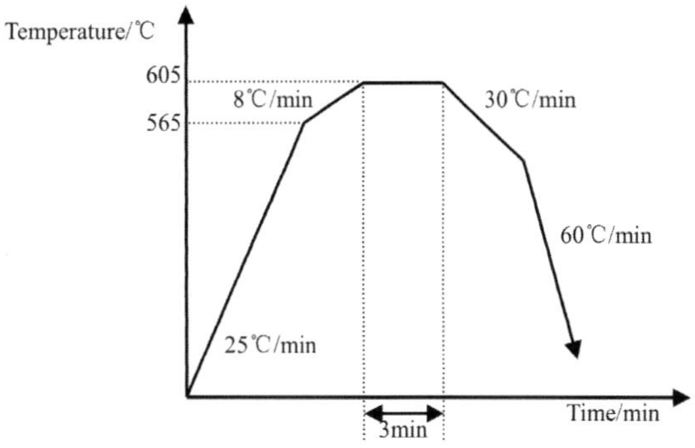

Fig.1 Standard controlled atmosphere brazing thermal cycle

SWAAT procedure. Samples of 110mm×60mm were used. The back side was covered with 3 M Scotch brand tape and the edges were protected with bee wax. In addition, another set of samples that cut from a normal triple-layer (AA4045/AA3003/AA4045) brazing sheet were prepared with the same brazing progress for SWAAT lifetime comparison. The samples were placed in the cabinet, tape side down, at an angle of 20°and assessed every 7 days for corrosion attack mechanism and morphology of the cross section. The test comprises a repetitive two hours cycle consisting of 30 min spray followed by a 90 min soak at above 98% humidity[15]. The spray is a solution of 42 g/L of sodium chloride and is acidified with 10 ml/L of glacial acetic acid. The pH is then adjusted with 10 wt% NaOH solution to a value between 2.6 and 2.8. The temperature during spray and soak was kept constant at 49℃.

Material characterization. Using ZEISS Imager.A2m Optical Microscopy (OM), Zeiss Sigma Scanning Electron Microscope (SEM) and Energy-Dispersive X-ray Spectroscopy (EDS), the microstructure of the material was examined both before and after brazing. For chemical characterization (elemental analysis) spectral imaging data sets (generated X-rays) were collected at 12 kV. Image-Pro Plus was used to measure the size, number distribution density of particles in cross section. In order to make the concentration of the elements comparable, the same color scale has been applied for each element in different maps.

3 Results

Material characterization. Optical microstructures of the material under investigation both before and after brazing are shown in Fig.2. The image of the material before brazing shows the clad layer with a higher contrast. And the equiaxed or elongated coarse particles with a dark contrast in the clad layer are Si particles. The core material has a lower contrast, and some small particles exist with different chemical composition. For the anti-corrosion layer, it has the lowest contrast, and its approximate interface between the other layers is shown in Fig.2(a).

After brazing, as it is shown in Fig.2(b), Al-Si eutectic phase is accumulated at the surface of the residual clad and in between is the pre-eutectic aluminum grains. And a 30~40μm width diffusion zone is formed during the interaction between the liquid clad and the solid core material, which is called BDP (the band of dense precipitates), as it is shown in Fig.2(c).

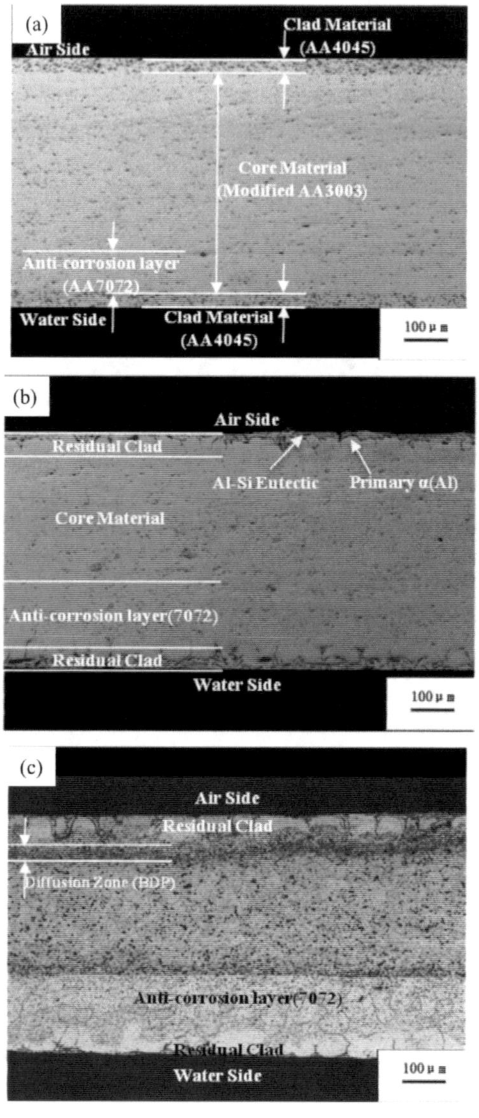

Fig.2 Optical microstructures of the brazing sheet: (a) before brazing; (b) after brazing; (c) after brazing (etched with keller reagent)

EDS map analysis for the major elements of the material under investigation both before and after brazing are shown in Fig.3 and 4, respectively. Comparison of the element maps in Fig.4 and Fgi.5 shows that it is the diffusion of Si and Mn elements during the brazing process that lead to the formation of BDP.

A further EDS pitting analysis for the size and composition of the precipitates after brazing are shown in Fig.5 and table 2, respectively. As a result of resolution and diffusion, the residual clad material is comprised of three types of phases: α(Al), eutectic Si and α-Al(Fe,Mn)Si[11]. For the diffusion zone, namely BDP, two types of particles are distinguished, regarding the size. One is the particles which are larger than 1~2μm(Spectrum5), and its size can reach up to 5~10μm. The other type contains the particles (Spectrum4) with the size below 1μm. And the result of Table 2 shows that the two types of particles are manly Al(Fe,Mn)Si phases.

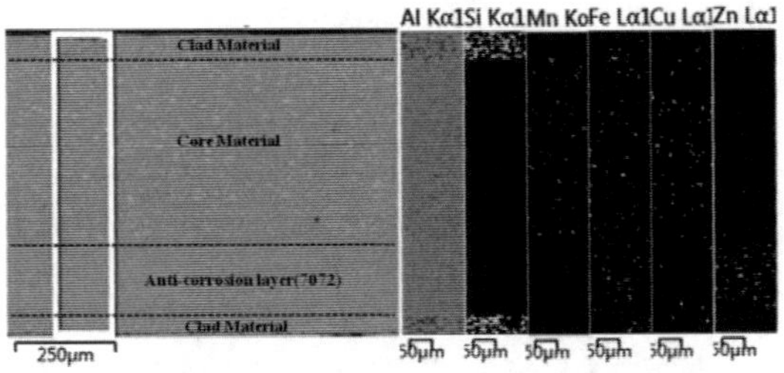

Fig.3 EDS element maps of the modified material before brazing

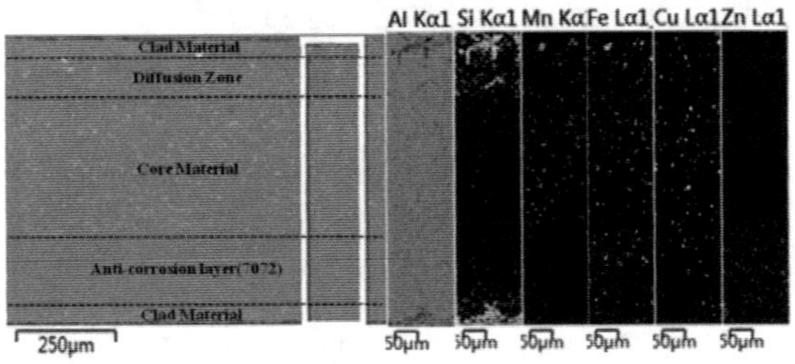

Fig.4 EDS element maps of the modified material after brazing

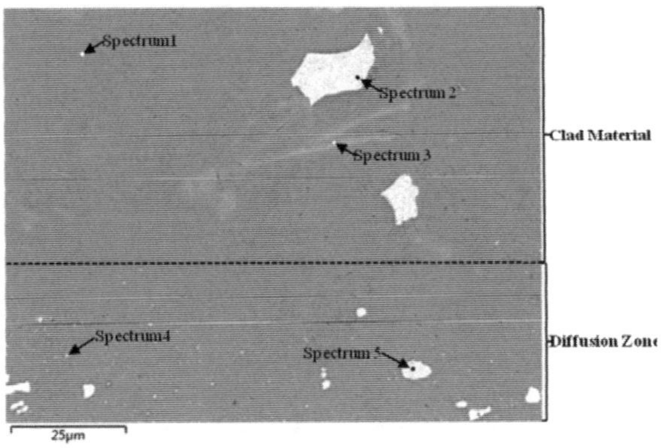

Fig.5 EDS pitting analysis of the modified material after brazing

Table 2 The composition of the precipitates after brazing (at%)

Label	Al	Si	Ti	Mn	Fe	Cu
Spectrum1	98.67	1.07	0.01	0.07	0.00	0.18
Spectrum2	70.75	10.65	0.05	6.80	11.48	0.27
Spectrum3	9.78	90.20	0.02	0.00	0.00	0.00
Spectrum4	91.34	4.24	0.06	3.47	0.80	0.09
Spectrum5	69.54	12.43	0.03	10.02	7.80	0.18

The images managed by Image-Pro Plus shows the distribution of the precipitates in different areas away from the interface of modified the clad(AA4045) and core(AA3003+Cu) materials are shown in Fig.6. And the statistical results are listed in Table 3. As can be seen in table 3, the number density of areas that 30~50μm away from the interface is much higher than areas that were not affected by element diffusion during brazing progress, which is good agreement with the BDP position identified in Fig.2(c).

Corrosion morphology characterization. The air side corrosion morphologies of the modified brazed material in different periods of SWAAT are shown in Fig.7. The corrosion begins in the eutectic Al at the surface in the form of pitting. Then it develops along the grain boundaries of the primary α(Al) to the diffusion zone, and the residual clad is partly exfoliation as a result of the corrosion of the diffusion zone, as it is shown in Fig.7(a) and (b). Finally, the material was failure in the form of IGC at the corrosion period of 28 days.

The water side corrosion morphologies of the brazed modified material in different corrosion periods are shown in Fig.8. As with the brazing side, the corrosion begins in the eutectic Al at the surface in the form of pitting. Then it develops along the grain boundaries of the primary α(Al) to the anti-corrosion layer (7072 layer), and the residual clad is uniformly exfoliation as a result of the corrosion of the 7072 layer, as it is shown in Fig.8 (a) and (b). Finally, the material was failure in the form of IGC at the corrosion period of 28 days. The result of the SWAAT shows that the time to perforation of the modified four-layer brazed sheet in the sea water solution is less than 28 days.

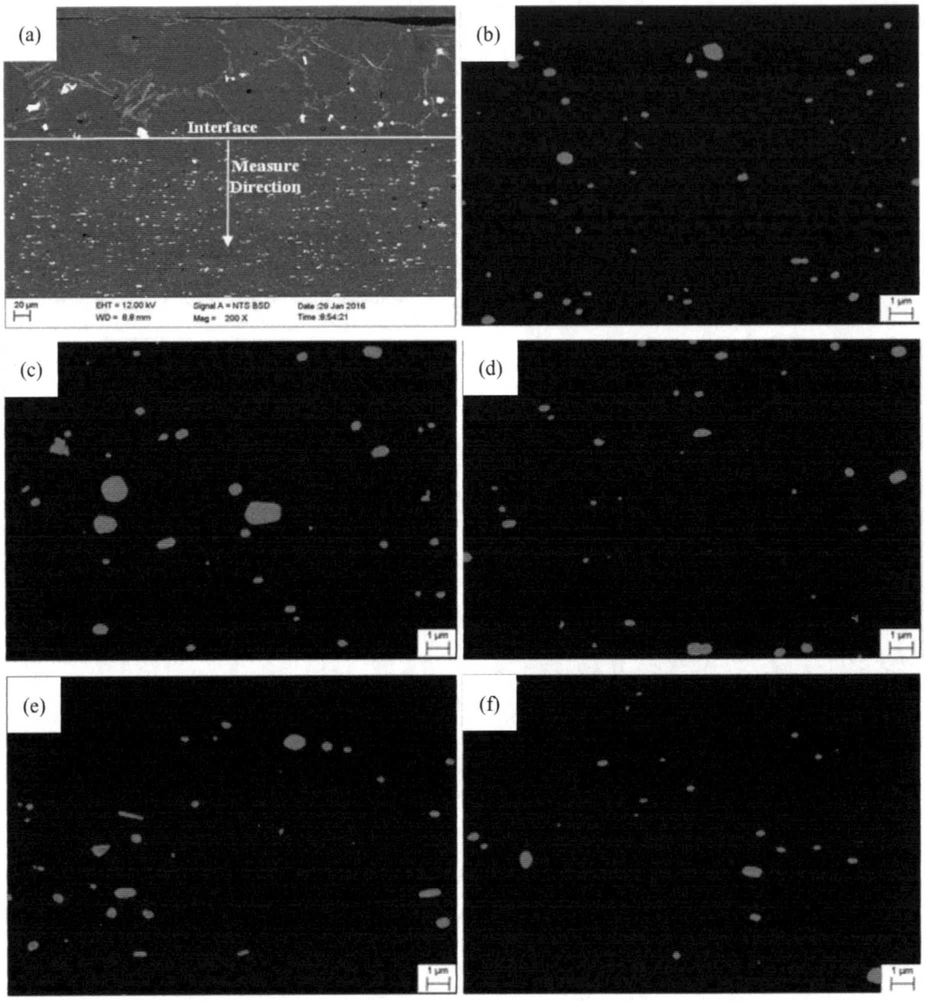

Fig.6 The distribution images of the precipitates in different areas away from the interface between the clad and core materials:

(a) the diagram of the observation; (b)30um; (c)50um; (d)70um; (e)90um; (f)110um

TABLE 3 STATISTICAL NUMBERS OF THE PRECIPITATES IN DIFFERENT AREAS AWAY FROM THE INTERFACE BETWEEN THE CLAD AND CORE MATERIALS

	Distance to Interface (um)	Precipitate Number	Number Density (per um^2)
Diffusion Areas(BDP)	30	53	0.604087
	50	48	0.547097
	70	45	0.512904
Diffusion-free Areas	90	36	0.410323
	110	28	0.319140

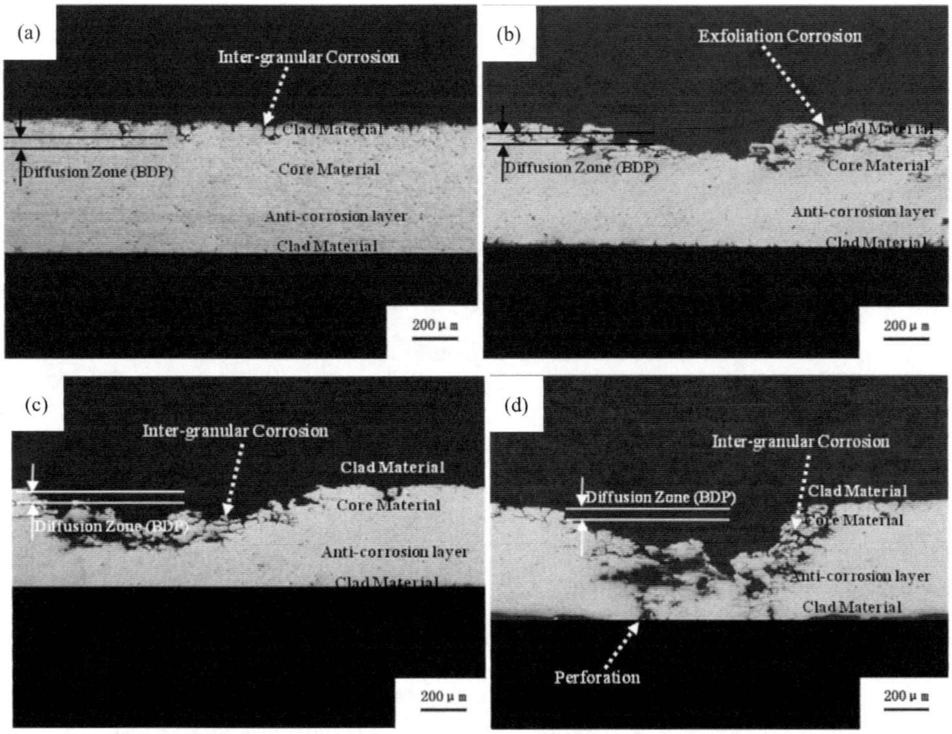

Fig.7 The air side corrosion morphologies of the modified material in different corrosion periods: (a)7d; (b)14d; (c) 21d; (d)28d

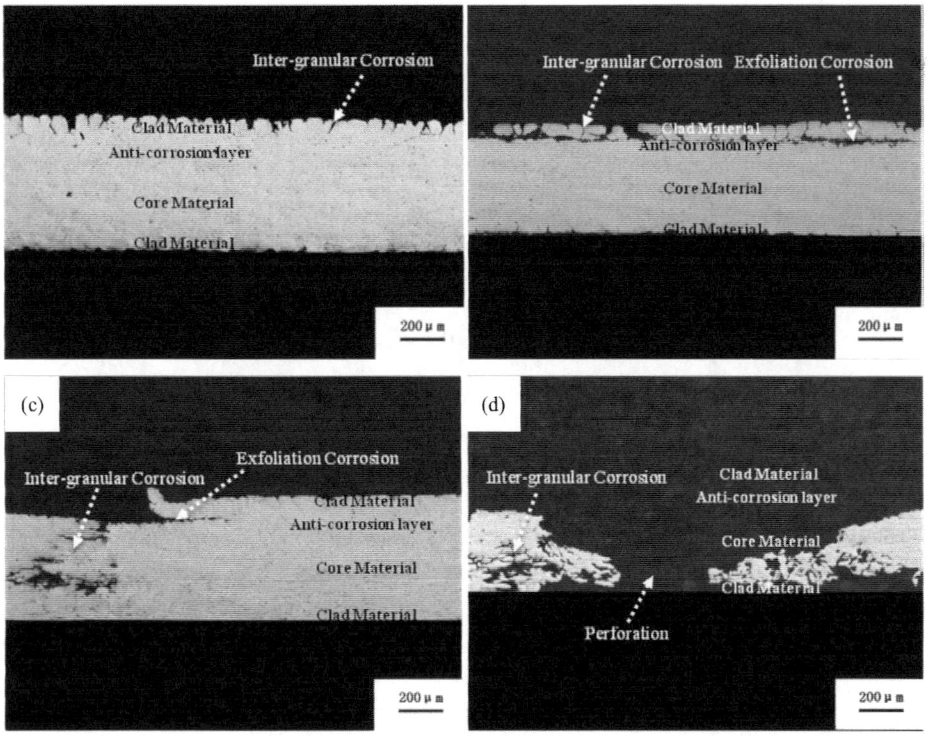

Fig.8 The water side corrosion morphologies of the brazed modified material in different corrosion periods: (a)7d; (b)14d; (c)21d; (d)28d

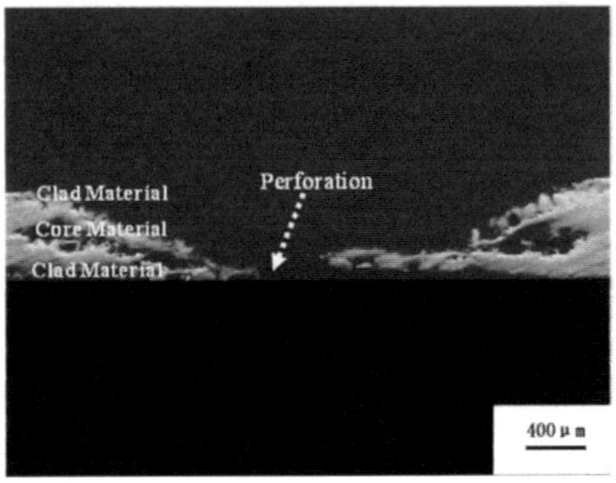

Fig.9 Corrosion morphologies of a normal triple-layer brazed material at the corrosion period of 7 days

The corrosion morphologies of a normal triple-layer brazed material at the corrosion period of 7 days is shown in Fig.9. In contrast to the normal triple-layer brazed material at the same thickness, the time to perforation of the modified four-layer brazing sheet is increased by more than 200%.

4 Discussion

The varies corrosion forms of the material during the SWAAT test is the result of the mcriostructure evolution duiring the brazing process. According to the Al-Si binary phase diagram, duiring the barzing process, the primary α(Al) precipitates at the interface of liquid clad and solid core materials prior to the Al-Si eutectic, then Al-Si eutectic accumulates at the surface and in between the primary α(Al) grains at a lower temperature (Fig.2c). Besides, the α-Al(Fe,Mn)Si phases, which are transformed from β-AlFeSi due to the diffusion of Mn element during brazing, is in favor of accumulating at boundaries of the α(Al) and Al-Si eutectic. As it is shown in Fig.1(b), the 30~40μm width areas with a darker contrast is the so-called BDP, which is the consequence of the diffusion of Si and elements during brazing[8]. The major phases of BDP has been reported as Al_6(Fe,Mn) and α-Al(Fe,Mn)Si phases[11](Fig.5 and Table 2).

The eutectic (Al) of residual brazing layer has a high sensitivity of pitting in solutions containing Chloride. During the sea water acid accelerated test, the corrosion form of the air side changed from pitting to the form of inter-granular corrosion (IGC) due to the accumulation of the α-Al(Fe,Mn)Si phases, as shown in Fig.7(a). When corrosion spread to the band of dense precipitates (BDP), as a consequence of the lower corrosion potential of the dense precipitates (Al_6(Mn,Fe) and α-Al(Fe,Mn)Si phases[17]), the corrosion form is replaced by the partly exfoliation. Therefore, the existence of BDP is beneficial to the prolonging of material's perforation time. The corrosion behavior changed to typical pitting and IGC, while it comes to the core (Fig.5c).

As with the air side, the corrosion of the residual clad began in the typical form of pitting and IGC. When it comes to the 7072 layer, the corrosion form becomes the typical uniform exfoliation. It was found that the 7072 layer that acted as an active cathode to protect not only the core but also the residual clad from corrosion will greatly improves the corrosion life of the material. Finally, the material was failure in the form of IGC and local pitting when corrosion propagated to the core.

5 Conclusion

1. The existence of BDP(the band of dense precipitates), which forms at the interface between clad (AA4045) and core (AA3003+Cu) layer during brazing, changes the corrosion form of the air side of the material from inter-granular corrosion to local exfoliation corrosion.
2. The addition of anti-corrosion layer makes the corrosion form of the water side from inter-granular corrosion into uniform exfoliation corrosion.
3. Compared to the normal O temper triple-layer (AA4045/AA3003/AA4045) brazing sheet at the same thickness, the time to perforation of the modified four-layer brazing sheet is increased by more than 200%.

Acknowledgments

Authors would like to thank the support from Jiangsu Key Laboratory for Clad Materials (No:BM2014006), International Science & Technology Cooperation Program of Wuxi city (No:CZE02H1504), Program on Jiangsu Science and technology achievements transformation (No:BA2015079).

References

1. Ali Davoodi, Mechanistic studies of localized corrosion of Al alloys by high resolution in-situ and ex-situ probing techniques, Ph.D Thesis, Royal Institute of Technology, Stockholm, Sweden, 2007: p.1.
2. A.C. Scott, R.A. Woods, J.F. Harris, Accelerated corrosion test methods for evaluating external corrosion resistance of vacuumed brazed aluminum heat exchangers, in: Conference Proceedings of SAE International Congress, SAE, Warrendale, USA, 1991.
3. Y. Isobe, K. Takeuchi, M. Tanaka, M. Mori, S. Yamauchi, K. Namba, Development of corrosion resistant brazing sheet for drawn cup type evaporators part 2:Application to evaporator, in: SAE: Conference Proceedings of SAE International Congress, SAE, Warrendale, USA, 1993.
4. F.M. Al-Kharafi, W.A. Badawy, Corrosion and passivation of al and al–si alloys in nitric acid solutions: 2 – effect of chloride ions, Electrochimica Acta 40 (1995)1811.
5. S.S. Abdel Rehim, H.H. Hassan, M.A. Amin, Chronoamperometric studies of pitting corrosion of Al and (Al–Si) alloys by halide ions in neutral sulphate solutions, Corrosion Science 46 (2004) 1921.
6. S. Kuroda, K. Tohma, Aluminum alloys, in: Conference Proceedings of ICAA6, Japan Institute of Light Metals, Toyohashi, Japan, 1998, p.1543.
7. S. Meijers, Corrosion of aluminum brazing sheet, Ph.D. Thesis, TU Delft, 2002, ISBN: 90-805661-3-6.
8. S. Tierce, N. Pebere, C. Blanc, C. Casenave, G. Mankowski, H. Robidou, Corrosion behavior of brazing material AA4343, Electrochim. Acta 52 (2006) 1092–1100.
9. Y.J. Li, L. Arnberg, Quantitative study on the precipitation behavior of dispersoids in DC-cast AA3003 alloy during heating and homogenization, Acta Materialia 51(2003): 3415-3428.
10. F. Norouzi Afshar, J.H.W. de Wit, H. Terryn, J.M.C. Mol, Scanning Kelvin probe force microscopy as a means of predicting the electrochemical characteristics of the surface of a modified AA4xxx/AA3xxx(Al alloys) brazing sheet, Electrochimica Acta 88 (2013): 330– 339.
11. S.Tierce, N. Pe´be`re, et.al., Corrosion behavior of brazed multilayer material AA4343/AA3003/AA4343: Influence of coolant parameters, Corrosion Science 49(2007) 4581–4593.
12. F. Norouzi Afshar, J.H.W. de Wit, H. Terryn, J.M.C. Mol, The effect of brazing process on microstructure evolution and corrosion performance of a modified AA4XXX/AA3XXX brazing sheet, Corrosion Science, 58 (2012) 242–250.
13. G.J. MARSHALL, et.al. Microstructural Control in an Aluminum Core Alloy for Brazing Sheet Applications, METALLURGICAL TRANSACTIONS A, 24A(1993): 1935-1942.
14. J. Lacaze, S. Tierce, M.C. Lafont, Y. Thebault, N. Pebere, G. Mankowski, C. Blanc, H. Robidou, D. Vaumousse, D. Daloz, Study of the microstructure resulting from brazed aluminum materials used in heat exchangers, Mater. Sci. Eng., A 414 (2005): 317–321.
15. ASTM Standard, Designation G85, Standard Practice for Modified Salt Spray(Fog) Testing, 2006.
16. D.T.L. Alexandre, A.L. Greer, Acta Mater. 50 (2002) :2571.
17. N. Birbilis, R. G. Buchheit, Electrochemical Characteristics of Intermetallic Phases in Aluminum Alloys: An Experimental Survey and Discussion, Journal of The Electrochemical Society, 2005,152 (4): B140-B151.

Solvothermal synthesis and characterization of hierarchical spindle-like $BiPO_4$ aggregates

Ming Ou[1], Lin Ma[1,a], Limei Xu[1], Haizhen Li[1], Zhuomei Yang[1] and Zhifeng Lan[1]
[1]*School of Chemistry and Chemical Engineering, Institute of Physical Chemistry, Lingnan Normal University, Zhanjiang 524048, P. R. China*

Abstract. Monodispersed spindle-like $BiPO_4$ aggregates with uniform morphology and size were prepared by a solvothermal route with ethylene glycol (EG)/water as mixed solvent, bismuth nitrate and sodium tripolyphosphate (STPP) as raw materials. The products were characterized by X-ray powder diffraction (XRD), scanning electron microscopy (SEM) and UV-vis absorption spectroscopy. The results indicate that obtained BiPO4 products are hexagonal phase and have good crystallinity and high purity. Additionally, it is found that the both STPP and mixed solvent play a crucial role in the formation of the spindle-like $BiPO_4$ aggregates, which are constructed through oriented aggregation of nanoparticles. The possible formation process was initially presented. Moreover, the bandgap energy of these novel spindle-like $BiPO_4$ aggregates is also estimated as 3.46eV from the UV-vis absorption spectrum.

1 Introduction

Inorganic nanomaterials with controllable size and shape have been one of the hottest research fields based on the supposition that there is a strong correlation between these parameters and their properties and potential applications [1]. Specially, hierarchical micro/nanostructures, which are assembled from nanoparticles, nanorods or nanowires, and nanosheets or nanoplates as building blocks, have recently attracted significant attention. It is believed that the hierarchical structures can inherit the unique advantages of their pristine building blocks and obtain intriguing properties from their superior secondary architecture [2]. To date, various compouds with hierarchical micro/nanostructures have been successfully prepared and exhibited novel photo/electrochemical properties and a wide spectrum of potential applications [3-6] .

As one kind of important bismuth-containing oxysalts, bismuth phosphate ($BiPO_4$) is a typical compound that crystallizes in two polymorphs such as hexagonal phase and monoclinic phase, which have found potential applications in catalysis, sensors, and many other functional materials for separating radioactive elements [7-9]. In addition, $BiPO_4$ was reported to be an excellent host for rare-earth ions that can yield luminescence and a novel photocatalyst for degrading organic dyes [7, 10]. Up to now, $BiPO_4$ micro/nanomaterials with different morphologies have been synthesized such as nanoparticles [11, 12], nanorods [13-18] and hierarchical architectures [19, 20]. To obtain such micro/nanomaterials, a variety of methods including chemical vapor deposition21, sonochemical method [13, 22], microwave irradiation [19], electrochemical deposition [15], hydrothermal/solvothermal treatment [14, 18, 23, 24] have been employed. In spite of a few approaches developed to fabricate $BiPO_4$ hierarchical micro/nanostructures, which followed complicated and tedious experimental procedure and might cause pollution, a facile synthesis of $BiPO_4$ hierarchical architectures is still an ongoing challenge.

In this paper, we report a facile solvothermal method to prepare spindle-like $BiPO_4$ hierarchical aggregates with sodium tripolyphosphate (STPP) and bismuth nitrate as reactants and ethylene glycol (EG)/water as solvent. It is found that the as-prepared $BiPO_4$ microspindles are assembled by numerous nanoparticles in an oriented way and both STPP and the volume ratio of EG to water have great influences on the morphology of the final product.

2 Experimental procedure

2.1 STPP-assisted synthesis of BiPO4 spindle-like aggregates

The experimental procedure was designed as follows. 1mmol of $Bi(NO_3)_3{\cdot}5H_2O$ was dissolved in 15 ml ethylene glycol to form a solution A and 1mmol of STPP ($Na_5P_3O_{10}$) was dissolved in 45 ml deionized water to form a solution B. Then the solution B was slowly dropped into the solution A under vigorous stirring to form a mixture. After stirring about 30 min, the mixture was finally transferred into a 100 ml Teflon-lined stainless steel autoclave. It was sealed tightly and maintained at 150 °C for 12 h. After that, the autoclave was allowed to cool down naturally. The products were collected, and washed three times using deionized water and absolute ethanol, then dried for 12 h under vacuum at 60 °C.

[a] Corresponding author: ml_buct@126.com

2.2 Characterization of $BiPO_4$ samples

XRD analysis was performed on a D/Max-2550 X-ray diffractometer with monochromatized CuK α radiation (λ =0.1540562 nm). SEM images were taken with FEI SIRION-100 field-emission scanning electron microscope. Absorption spectrum was measured on a UV-vis spectrophotometer (UV-2550) in the wavelength range of 200-700 nm.

3 Results and discussions

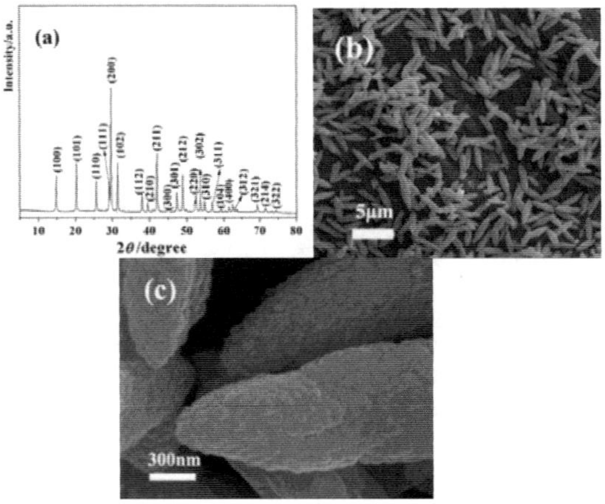

Figure 1. XRD patterns and SEM images of $BiPO_4$ spindle-like aggregates prepared by a STPP-assisted solvothermal route.

The crystal structure and phase purity of the products are determined by XRD. A typical XRD pattern of the $BiPO_4$ microspindles is presented in Fig. 1a. The major detectable diffraction peaks can be readily indexed to the hexagonal phase of $BiPO_4$ consistent with the standard powder diffraction file of $BiPO_4$ (JCPDS 15-0766). No impurity phase can be found. The high and sharp peaks are indicative of good crystallinity in spite of slightly broadened diffraction peaks. The XRD result confirms that the well-crystallized hexagonal $BiPO_4$ products can be easily prepared in our synthetic strategy. The morphology and microstructure details of the $BiPO_4$ products were investigated with SEM. Fig. 1b shows a typical SEM image to exhibit the overall view of the $BiPO_4$ products, clearly demonstrating a well-dispersed distribution of spindle-like rods with uniform morphology and size. It is estimated that the microspindles of a mean length about 3.15 μm and a middle width about 790 nm have an aspect ratio approaching 4:1, and that a high yield close to 100%. Fig. 1c clearly reveals surface structures of these microspindles, from which it can be seen that the surfaces are rather rough and numerous flat nanoparticles with size of 70 nm orientedly arrange on the surface. Therefore, it is reasonable to speculate that these novel microspindles with hierarchical structures should be assembled by many nanoparticles through an oriented aggregation.

Figure 2. SEM images of $BiPO_4$ products prepared (a) with Na_2HPO_4 as reactant; (b) with water as solvent and (c) with EG as solvent.

It can be seen from Fig. 2a that the obtained $BiPO_4$ products are scattered irregular particles when STPP is replaced by equivalent amount of Na_2HPO_4. This result explicitly indicates that STPP plays a crucial role in the formation $BiPO_4$ spindle-like aggregates. In addition, in order to investigate the effect of solvent, the $BiPO_4$ products were also prepared by a similar process with pure water and pure EG as solvent, respectively. Fig. 2b manifests that the $BiPO_4$ products obtained with water as solvent are microspindles with ununiform morphology and size. Moreover, these spindles are interconnected side by side. With EG as solvent, the products are uniform spheres assemble by nanoparticles. The above control experiment results demonstrate that an appropriate ratio of EG to water also acts as a key factor to obtain spindle-like $BiPO_4$ aggregates.

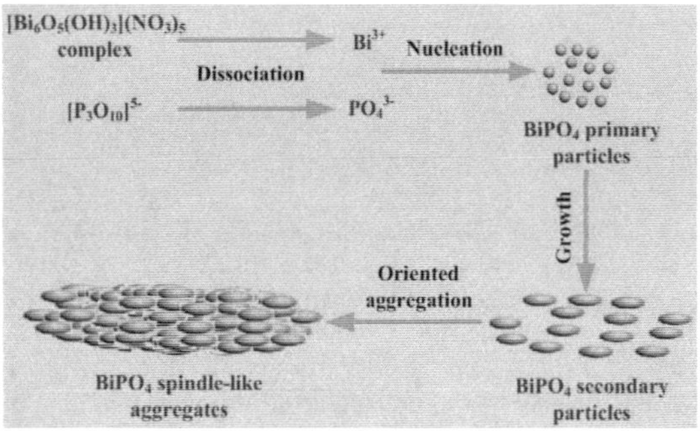

Figure 3. Illustration of the formation process of the spindle-like $BiPO_4$ aggregates.

Based on the experimental results, the probable formation mechanism of the spindle-like $BiPO_4$ aggregates may be ascribed to an oriented aggregation process in our current experiment. First, $Bi(NO_3)_3$ could alcoholyze and hydrolyze in EG and water to form a bismuth oxide hydroxide nitrate complex which might be in the form of $[Bi_6O_5(OH)_3](NO_3)_5$ and then Bi^{3+} was generated at a release speed by the dissociation of the complex [14, 19]. Meanwhile, PO_4^{3-} ions were also provided by gradual dissociation of STPP in EG/water system. Subsequently, PO_4^{3-} combined with Bi^{3+} to form $BiPO_4$ primary nanoparticles and would further grow into flat secondary particles due to the capping effect of EG. EG-water system with proper volume ratio could not only guarantee an opportune nucleation rate but also have appropriate viscosity that allowed the $BiPO_4$ secondary particles to find suitable low-energy configuration interfaces and thus aggregated in an oriented way to form the final microspindles. The detailed formation mechanism deserves further investigations. The schematic illustration for the formation of the spindle-like aggregates is shown in Fig. 3

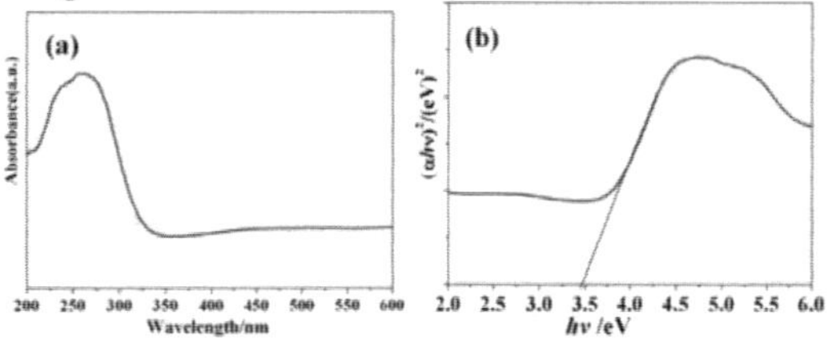

Figure 4. UV-vis absorption spectrum of the spindle-like $BiPO_4$ aggregates.

Fig. 4 shows the UV-vis absorption spectrum of the spindle-like $BiPO_4$ aggregates, which displays the photoabsorption property from the UV light region. The steep shape of the spectrum indicates that the UV-light absorption was due to the band gap transition. As a crystalline semiconductor, the optical aborption near the band edge is based on the equation $\alpha h\nu=A(h\nu-Eg)^{1/2}$(where α is the absorption coefficient, A is a constant, hv is the photo energy, and Eg is the bandgap energy). From the curve of the $(\alpha h\nu)^2$ vs. photo energy (hv), the bandgap energy of the spindle-like $BiPO_4$ aggregates is estimated to be 3.46 eV, slightly lower than or similar to those of $BiPO_4$ materials reported previously [13-20].

4 Conclusions

In summary, well-dispersed hexagonal spindle-like $BiPO_4$ aggregates with uniform morphology and size were fabricated by a facile solvothermal process in EG/water solvent. The STPP and mixed solvent play a key role in the formation of spindle-like $BiPO_4$ aggregates, which are assembled by oriented aggregation of nanoparticles. The strategy presented in this work is easily controllable and well reproducible and may be feasible to develop into the scale-up production.

Acknowledgement

This work is financially supported by the Guangdong Natural Science Foundation for Joint Training Innovative Talents in East-West-North Guangdong (2014A030307030, 2016A030307654), the Science and Technology Planning Project of Guangdong Province (2014A010106032), the Stong Innovation School of Engineering Program of Department of Education of Guangdong Province (Distinctive Innovation Project No. 2014KTSCX157), Natural Science Foundation of Lingnan Normal University (LZL1402, LZL1502).

References

1. Y. Wang, Y. Xia, Adv. Mater. **25**, 5336 (2013).
2. H.B. Yao, H.Y. Fang, X. H. Wang, S. H.Yu. Chem. Soc. Rev. **40**, 3764 (2011).
3. H. Zhou, H. Zhang, Y. Wang, Y. Miao, L. Gu, Z. Jiao. J. Colloid. Interface. Sci. **448**, 367 (2015).
4. X. Y. Liu, W. G. Bian, C. Y. Tian. Mater. Lett. **112**, 1 (2013).
5. M.Y. Li, F.P. Wang, Z. Y. Wang, M. Z. Iqbal, Q. U. A. Javed, Y. Z. Lu, M. Xu, Q. S. Li. Mater. Lett. **112**, 81 (2013).
6. S. Liu, H. Liu, Q. Q. Shi, Z. Wang, Y. Xu. Nanosci. Nanotech. Lett. **6**, 268 (2014).
7. J. Xu, L. Li, C. S. Guo, Y. Zhang, W. Meng. Appl. Catal. B-environ. **130**, 285 (2013).
8. M. Sheng, L, Gu, R. Kontic, Y. Zhou, K. Zheng, G. Chen, X. Mo, G. R. Patzke. Sensor Actuat. B-chem. **166**, 642 (2012).
9. S. Kagaya, Y. Araki, N. Hirai, K. Hasegawa. Talanta **67**, 90 (2005)
10. B. S. Naidu, B. Vishwanadh, V. Sudarsan, R. K. Vatsa. Dalton. Trans. **41**, 3194 (2012).
11. C. Pan, Y. F. Zhu. J. Mater. Chem. **21**, 4235 (2011).
12. S.L. Lee, S. W. Yong, K. O. Hartini. Asian J. Chem. **25**, 7205 (2013).
13. J. Geng, W. H. Hou, Y. N. Lv, J. J. Zhu, H. Y. Chen. Inorg. Chem. **44**, 8503 (2005).
14. F. Xue, H. Li, Y. Zhu, S. Xiong, X. Zhang, T. Wang, X. Liang, Y. T. Qian. J. Solid State Chem. **182**, 1396 (2009).
15. M. Yang, N. K. Shrestha, R. Hahn, P. Schmuki. Electrochem. Solid State Lett. 13, C5 (2010).
16. Y. F. Liu, X. G. Ma, X. Yi, Y. F. Zhu. Acta Phys. Chim. Sin. **28**, 654 (2012).
17. Y. Y. Zhu, Y. F. Liu, Y. H. Lu, H. Wang, Q. Ling, Y. F. Zhu. Acta Phys-Chim. Sin. **29**, 576 (2013).
18. L. Wang, J. Tang, H. Yin. Asian J. Chem. **25**, 2146 (2013).
19. G. Li, Y. Ding, Y. Zhang, Z. Lu, H. Sun, R. Chen. J. Colloid. Interface Sci. **363**, 497 (2011).
20. L. Li, J. Xu, C. Guo, Y. Zhang. Front Env. Sci. Eng. **7**, 382 (2013).
21. Y. F. Lin, H. W. Chang, S. Y. Lu, C. W. Liu. J. Phys. Chem. C **111**, 18538 (2007).
22. M. H. Fulekar, A. Singh, D. P. Dutta, M. Roy, A. Ballal, A. K. Tyagi. Rsc Adv. **4**, 10097 (2014).
23. Y. Wang, X. Guan, L. Li, G. Li. Crystengcomm **14**, 7907 (2012).
24. Y. He, J. Wang, L. Guo, X. An, Y. Lv, Y. Liu, W. Fa. J. Syn. Crystals **41**, 1410 (2012).

MATEC Web of Conferences **61**, 05005 (2016) DOI: 10.1051/matecconf/20166105005
APOP2016

Numerical simulation of premixed Hydrogen/air combustion pressure in a spherical vessel

GUO Han-yu[1], TAO Gang[1,a], and ZHANG Li-jing[1]

[1]Jiangsu Key Laboratory of Urban and Industrial Safety, College of safety science and engineering, Nanjing Tech University, Nanjing, 210009, China

Abstract. In order to study the development process of hydrogen combustion in a closed vessel, an on-line chemical equilibrium calculator and a numerical simulation method would be used to analysis the combustion pressure and flame front of mixed gas, which based on 20L H_2/air explosion experiments in spherical vessel(Crowl and Jo,2009). The results showed that, the turbulent model could reflect the process of combustion, and the error of combustion pressure by simulation is smaller than the Chemical Equilibrium Calculation. The heat loss and incomplete combustion are the main reason to cause the error.

1 Introduction

Gas explosion in the closed vessel is a typical accident in the process of industrial production. A small ignition can cause explosions once combustible gases mixed with air when it leakage and diffusion in the process of production, transport, storage and use, and the gas mixture components are located on the combustible gas explosion limit of range. According to the statistics, there are more than 300 combustible gas explosion accidents which death toll more than 3 in our country in 2010~2015 [1], in particular of Dong Huang Qingdao Sinopec oil pipeline leakage and explosion accidents, which left 62 dead, 136 wounded and a direct economic loss of up to 751.72 million yuan, had extremely serious consequences [2]. The similar accident would cause a large number of casualties and economic loss and affect the stability of social development directly.

With the fossil fuel exploitation and use and the earth's environmental problems challenging, new energy will be used more and more widely. The fossil fuel does not meet the long-term needs of modern industrial development because of their inherent scarcity. And it also has a large number of CO, CO_2 and SO_2 are produced in the use of fossil fuel, which can cause global warming and a lot of secondary disasters. As a new energy, hydrogen have a high calorific value, low pollution, renewable and advantages of rich, which makes it can be widely applied to chemical production, automobile manufacturing, aerospace and food industries. But hydrogen also may cause heavy casualties and losses because of the flammable and explosive properties, which is a problem that people use the hydrogen in the future.

There have lots of research on combustion process of hydrogen we can achieve. Crowl and Young-do Jo measurement of different concentrations of flammable gases such as hydrogen, methane explosion limits in air by experiment, comparison of traditional fuels and hydrogen in industrial production process, and calculated explosion limits of combustible gases by second derivation method [3,4]. Hansen and Crowl estimate the explosive limit by thermodynamic calculations and empirical formula method, and the results showed that explosion limits of gases and oxygen concentration exists linear relationship at fixed initial conditions [5]. Liu studied the ignition temperature and ignition delay time by changed the initial pressure, and explained the reasons for the formation of the third explosive limit form the perspective of chemical kinetic [6]. Young-do Jo predict the hydrogen-air premixed combustion characteristics of explosion parameters in different concentration of hydrogen, and the results showed that the maximum explosion pressure and K_G does not appear in stoichiometric concentration [7]. Tao calculated the fixed nitrogen concentration of hydrogen/oxygen/nitrogen explosion characteristics of premixed gas by Chemical Equilibrium Calculation, and get the general equations of the maximum explosion pressure and KG in the fixed nitrogen concentration [8, 9]. Molkov have simulated the Hydrogen-air combustion in open space by the Large Eddy Simulation, and the results show that the turbulence is the main cause of flame instability [10]. Xiao study on the combustion of hydrogen-air premixed gases in the pipe by experiment, and make a simulation by using laminar flow thickness of flame model to analyze the causes of formation the Tulip flame [11]. Bao study on the characteristics of the maximum explosion pressure in

[a] Corresponding author: taogang@njtech.edu.cn

different initial conditions by using constant volume combustion bomb experiment [12].

This paper focused on the hydrogen-air premixed gas explosion pressure variations by using Chemical Equilibrium Calculation and numerical simulation for a 20L spherical vessel, to study of premixed gases explosion in the vessel by comparing the calculated data and experimental data.

2 Chemical Equilibrium Calculation

The full combustion of premixed gas combustion parameters could be calculated by using the initial status parameter based on the thermodynamic theory. According to a known set of initial conditions, an online calculation of chemical equilibrium, which based on the iterative computation of STANJAN, can be used to calculate the explosion pressure. According to initial condition in the experiment of Crowl, which the initial pressure is 1atm and the initial temperature of 300K, the parameters can be calculated. Considered the formation of NO_x, the products include H, H_2、H_2O(g), OH, N, N_2、N_2O、NH_3, NH_2, NH, O, O_2, NO_2, NO and H_2O(l) [8]. The hydrogen-air premixed gases explosion pressure in different concentration by calculation as shown in Fig.1 [13]. From chart show, there are a small differences in middle of hydrogen content, but the calculated results with experimental chemical equilibrium value larger than the difference in the hydrogen concentration near the explosion limits. The actual burning condition and heat dissipation is the main cause of the difference. In the premixed gas combustion process, content of active substances is so less that could not enough to support a smooth chain reaction when the oxygen or hydrogen content at low levels. But it would be calculated in accordance with the combustible gases react at all in the Chemical Equilibrium Calculation, which is the main reason to cause the calculation results bigger than the experimental results. Meanwhile, a small amount of heat loss in the process of experiment, which due to slow chemical reaction near the explosion limits, also is a reason lead to lower combustion temperature and pressure at the terminal state.

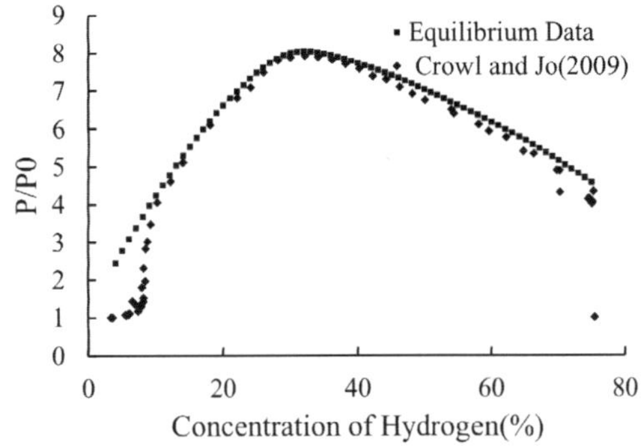

Figure 1. Maximum Pressure in equilibrium and experimental Data vs. Concentration of Hydrogen

2 Numerical methods

The governing equations. Because of the complexity of the combustion process, laminar flame propagation model only to explain the simplified characterization of flame propagation. In fact, flame propagation folds due to the instability of the flame in the actual combustion process, therefore turbulent propagation model can explain the flame propagation process better than laminar model. Numerical simulation select spherical vessel as the physical model based on the experiment of Crowl and Jo, meanwhile, for ease of calculation, the model is reduced to 2D. Mass equation, momentum equation and energy are as follows:

$$\frac{\partial \rho}{\partial t} + \frac{\partial (\rho u_j)}{\partial x_j} = 0 \tag{1}$$

$$\frac{\partial (\rho u_i)}{\partial t} + \frac{\partial}{\partial x_j}(\rho u_i u_j) = -\frac{\partial p}{\partial x_i} + \frac{\partial \tau_{ji}}{\partial x_j} \tag{2}$$

$$\frac{\partial (\rho E)}{\partial t} + \frac{\partial}{\partial x_i}(u_i(\rho E + p)) = \frac{\partial}{\partial x_i}\left(k_{eff}\frac{\partial T}{\partial x_i} - \sum_j h_j J_j + \tau_{ij,eff} u_j\right) + S_h \tag{3}$$

Where t is the time, ρ is the density, E is the total energy of micro-turbulence, p is the pressure, k_{eff} is the heat conductivity, J_j is the diffusion flux of component J, S_h is the chemical reaction heat source and other volume heat source.

k-ε equations would be selected as the turbulent model, and the equations are as follows:

$$\frac{\partial (\rho k)}{\partial t} + \frac{\partial (\rho k u_i)}{\partial x_i} = \frac{\partial}{\partial x_j}\left(\left(\mu + \frac{\mu_t}{\sigma_k}\right)\frac{\partial k}{\partial x_j}\right) + G - \rho\varepsilon \tag{4}$$

$$\frac{\partial (\rho \varepsilon)}{\partial t} + \frac{\partial (\rho \varepsilon u_i)}{\partial x_i} = \frac{\partial}{\partial x_j}\left(\left(\mu + \frac{\mu_t}{\sigma_k}\right)\frac{\partial \varepsilon}{\partial x_j}\right) + C_1 G\frac{\varepsilon}{k} - C_2 \rho\frac{\varepsilon^2}{k} \tag{5}$$

Where G is the turbulent kinetic energy caused by velocity gradient, C_1、C_2 are the empirical constant.

Composition PDF transport, which based on the chemical reactions, would be seledted as the combustion model, calculation method is SIMPLE. And the 19-step reaction mechanism of hydrogen would be used in the simulation. The initial pressure is 1atm, and the initial temperature of 300K. The ignition location at the central of vessel, and ignition temperature for 2000K. monitor point would be selected on the wall, to detect the change of pressure in the combustion process.

3 Results and discussion

According with the model based on the above numerical methods, the simulation results of CFD are showns in the Table 1. And the relative error δ, which is the result of combustion pressure between the Chemical Equilibrium Calculation, simulation and experimental at the selected concentration of hydrogen, are also shown inthe Table 1.

DOI: 10.1051/matecconf/20166105005

Near the stoichiometric concentration, the relative error of Chemical Equilibrium Calculation are less than 4%, and the relative error of Numerical simulation are less than 3%. But both errors are higher in the areas nera the explosion limits. In general, the result of simulation is better than Chemical Equilibrium Calculation. And the reason of error as mentioned above.

Table 1. The Combustion Pressure in different Concentration of Hydrogen by Calculated and Numerical

Concentration of Hydrogen	Calculated Data		Numerical Data	
	P_{max}(P /P0)	δ	P_{max}(P /P0)	δ
17.75%	6.19	1.48%	6.16	0.98%
22.73%	7.13	4.85%	7.16	5.29%
29.59%	7.95	1.92%	7.82	0.26%
35.33%	7.96	2.05%	7.89	1.15%
41.67%	7.63	3.11%	7.58	2.43%
45.67%	7.36	3.66%	7.27	2.39%
51.23%	6.91	3.13%	6.79	1.34%

Different from the Chemical Equilibrium Calculation, numerical simulation of premixed gases could simulate the combustion process, and chemical equilibrium calculation can only get the final result. So the premixed combustion process can be analyzed based on simulation results. The pressure curve of stoichiometric concentration is shown in the Fig.2. Compared with the experimental data, the time of maximum explosion pressure in experiment is longer than thesimulation required. In the simulation process, the pressure would rise at 2.5ms, and reached the maximum value at 6.8ms, but in the experiment process,the pressure rise is more gentle,it would be 5.5ms to reached the maximum totally. In the actual combustion process, the shape of the flame will change because of different flame burning speed in the different directions, which due to the influence of gravity, would delay the combustion process. However, the effects of buoyancy on flame kernel would not be considered in the simulation, which lead to flame propagation speeds in different directions almost identical in flame. After the acceleration of turbulence, the time of the combustion process would be reduced.

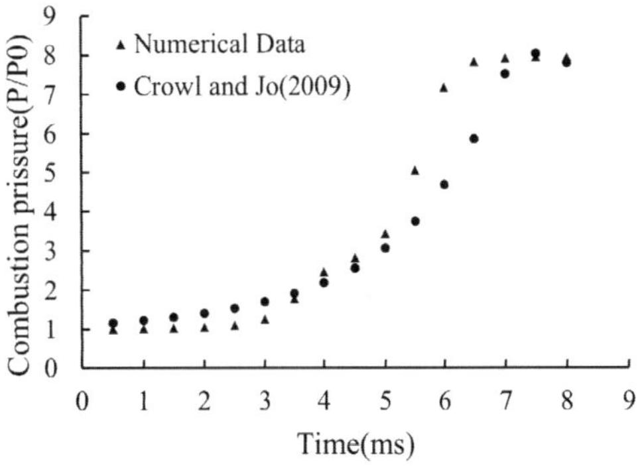

Figure 2. Comparison of combustion-explosion pressure between experiment and Simulation

With the change of hydrogen concentration, combustion pressure and combustion process would vary, the curve of combustion pressure at different hydrogen concentration as shown in Fig.3. The fastest burning is not at stoichiometric concentration. from the figue, we can found the rate of pressure rise at equivalence ratio Φ=1.3 was higher than stoichiometric concentration, and this conclusion is consistent with experimental results [7], As the Φ increases (or decreases), combustion pressure gradually reduces, and the time to the maximum combustion pressure are gradually increases.

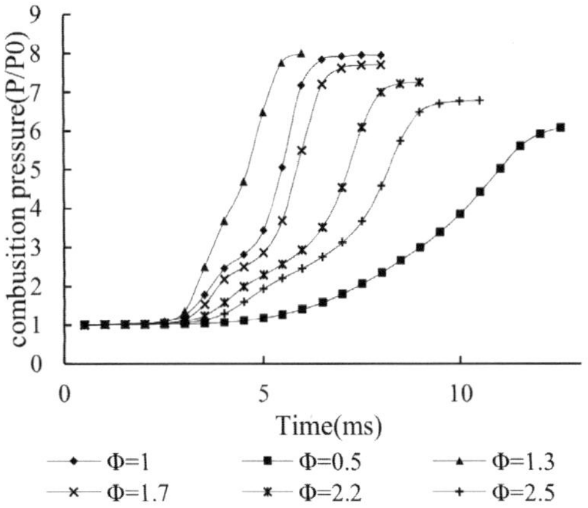

Figure 3. Combustion pressure vs. Time in different equivalence ratio

4 Conclusions

An online Chemical Equilibrium Calculation was used to calculated the combustion pressure of hydrogen-air premixed gas, and the calculation resultsagrees well with experimental results in the area of near stoichiometric concentration, but the error was big at explosion limits. And the main reason causing the error is the heat loss in the actual explosion and incomplete combustion.

The composition PDF transport model can be used to simulated the combustion process of hydrogen-air premixed gas in the spherical vessel. Compared the simulation and experiment, the results show that the turbulent combustion model can well simulate the combustion process, but the area near the explosion limts is still need to study.

By comparing different stoichiometric ratio of combustion process, the results show that the maximum explosion pressure and the rate of pressure rise is not occur at stoichiometric concentration and it is consistent with experimental results. it indicates that the most dangerous concentrations of premixed gases are not the stoichiometric concentration in the actual explosion.

Acknowledgments

this study is financially supported by the Jiangsu Government scholarships (No. KYLX_0783) and Michigan Technological University; Prof. Daniel A. Crowl also contributed to this work.

References

1. Information on http://media.chinasafety.gov.cn:8090/iSystem/shigumain.jsp (2015)
2. Information on http://www.chinasafety.gov.cn/newpage/Channel_21679.htm (2015)
3. Crowl D A, Jo Y D. The hazards and risks of hydrogen. J. Journal of Loss Prevention in the Process Industries, **20**, 2 (2007)
4. Crowl D A, Jo Y. A method for determining the flammable limits of gases in a spherical vessel. J. Process Safety Progress, **28** (2009)
5. Hansen T J, Crowl D A. Estimation of the flammability zone boundaries for flammable gases. J. Process Safety Progress, **29** (2010)
6. Liu Y, Pei P. Asymptotic analysis on autoignition and explosion limits of hydrogen–oxygen mixtures in homogeneous systems. J. International Journal of Hydrogen Energy, **31** (2006)
7. Jo Y D, Crowl D A. Explosion characteristics of hydrogen-air mixtures in a spherical vessel [J]. Process Safety Progress, **29** (2009)
8. Tao G, Crowl D A. Comparison of the Maximum Gas Combustion Pressure of Hydrogen/Oxygen/Nitrogen-between Chemical Equilibrium Calculations and Experimental Data [J]. Procedia Engineering, **62** (2013)
9. Tao G, Crowl D A, Guo H Y. Equations for Flammability Parameters, Pmax and KG, AIChE/GCPS, New Orleans, 2014.
10. Molkov V V, Makarov D V, Schneider H. Hydrogen-air deflagrations in open atmosphere: large eddy simulation analysis of experimental data. J. International Journal of Hydrogen Energy, **32** (2007)
11. Xiao H, Sun J, Chen P. Experimental and numerical study of premixed hydrogen/air flame propagating in a combustion chamber. J. Journal of hazardous materials, **268** (2014)
12. Bao X, Liu F, Chen C. Experimental study on hydrogen constant volume combustion [J]. Explosion and Shock Waves, **34** 5 (2014) Information on http://navier.engr.colostate.edu/~dandy/code/code-4/index.html (2016)

The Reform and Practice of Architectural Teaching System in Local Universities under the Background of Energy Conservation

Bai Ling ZHOU[1,a], Jun YE[2] and En Tian QIE[3]

[123] College of Urban Construction, Wuhan University of Science and Technology, Wuhan, Hubei 430065, P.R. China

Abstract. In order to realize the diversification and differentiation development of architecture professional teaching in local universities under the background of energy conservation, we must scientifically determine the training objectives, arrange reasonable relevant courses, adjust to enrich teaching contents, and continuously enrich the teaching methods. Thus we will form professional framework adapting to the development of local, which reflects the different characteristics of architectural specialty.

1 Introduction

On the whole, Chinese modern architectural education began in 1927, which founded.in National Central University (Southeast University now). Since then, there were only a dozen universities that set up the department of architecture in the whole country during the course of more than half a century. It seemed that architecture had become a veritable "noble" discipline from then on. Since the reform and opening up of China in the 1980's, the architectural speciality of our country's colleges and universities has made a rapid progress. There are over 220 universities which have set up architecture department by the end of the year 2012. And 75 of them have put on record in department of personnel and education in Ministry of construction, 35 of them have passed the evaluation of authorizing bachelors degree of architectural education.

On one other hand, the academism of Europe and the United States which is represented by Germany Bauhaus and Russian Bxytemac has a profound influence in contemporary Chinese architectural education system. On the other hand, China is facing a historical period of transition from developing country to developed country at present. With the rapid advance of the urbanization process, some natural villages on the edge of city have gradually developed into towns and towns have gradually developed into the city, which causes a great demand for talents of urban planning and architecture. Due to the social economic system in China has been transferred from planned economy to market economy, the "uniform" professional teaching mode of the past has significantly lagged behind the current situation of economic and social development.

All though in recent years, various colleges and universities have tried to adjust the educational system which is relatively closed and conservative, but there is no change in the overall breakthrough. For example, how to adapt the architecture departments of local colleges and universities to the development of society? How to research and practice in line with its own characteristics of the architectural education model according to local conditions with a clear mind? How to deal with variety of negative issues such as their relatively blurry positioning and the indistinct objectives of their talent cultivation.

Different levels of colleges and universities should have different training objectives. At the beginning of the founding of new China there were eight schools in architectural education which were first to offer urban planning or architecture departments in college, including Tsinghua University, Southeast University, Tianjin University, Tongji University, Harbin University of Architecture (now Harbin Institute of Technology), South China University, Chongqing Architecture University (now Chongqing University) and Xi'an Metallurgical Construction Institute (now Xi'an University of Architecture and Technology). The latterly four schools were to balance the layout of architecture education in various regions of the country in the wave of adjusting disciplines in PRC in1952. The formerly institutions possess powerful teachers and long academic lineage (mainly inherited in the United States, Europe and Japan).

These eight schools constitute the first echelon of the professional education of architecture in our country, which have trained a large number of professionals in the architecture of high-end talent since the beginning of the establishment, making a great contribution to the

[a] Corresponding author: cps.zhou@gmail.com

economic development of our country. Up to now, International professionals with innovative spirit and professional leaders are still the orientation of their talent training objectives.

Since China's reform and opening up in 1979, China's higher education has been greatly changed. So far we have formed the structure of the three modes coexist: the national "211 Project" universities, universities co-established by the province and the ministry, local colleges and universities. Most of the local colleges and universities (provincial universities) rely on local financial support, which are allocated funds from local administrative departments. They aim to serve the economic and social development of the region and strive to cultivate high quality talents. Since local college students and graduates employment whereabouts are with obvious regional color, the Architectural Education in these colleges should take the main goal of training local talents as the main goal of running a school.

2 Current Situation

The architecture departments in local universities have much in common with Chinese famous universities from all aspects, such as talent cultivation target approximation, syllabus convergence, curriculum setting, and the teaching method and so on. The Architectural Education in local colleges and universities has a few disadvantages, such as obsolescence of teaching system, lack of engineering application and innovation ability, which cannot meet the needs of local social and economic development. In addition, they are lack of policy and law guarantee, special funds to support. That the insufficient internal demand of design enterprise in college-enterprise joint cultivation led to that the connection between school and enterprise is not close enough. Some problems are particularly prominent, such as the gap between higher engineering education and the demand of talent market. Therefore, we must establish the architectural profession education system which attaches importance to the training of engineering application and innovation ability, strengthens the training of employment and Entrepreneurship, and then we can develop applied excellent engineers to meet the needs of the community.

At present, the architecture specialty course in Engineering University in China is mainly composed of 4 modules: the design course module whit architectural design, planning and design, landscape design and so on as the core; technical course modules represented by building structures, building equipment, construction technology, materials and construction courses; the architectural theory curriculum module represented by architectural history; the performance of the class module based on art training and computer software teaching.

This kind of course generally uses the teaching method of classroom teaching, adding the drawings or exercises to practice. The duck-feeding mode makes the students are no sense of developing innovative ideas. They are lack of proactive creative thinking, only accepting the knowledge passively. For example, the teaching thought of the course of building structure is still a continuation of the teaching thinking of the students of civil engineering, and focuses too much on the derivation of the formula for the calculation of mechanics and structure, which ignores the demand for technology integrated design innovation capability in the course of the architect's practice. The courses of construction technology and materials are obviously lack of comprehensive training content.

"The first rule is academic history" is a sentence we often speak of, which shows the importance of professional history in various disciplines. The level of mastering architectural history (Theory) determines the extent of the students' vision, which deeply influences students' professional development in the future.

But the relative lack of the books in the history of architecture, especially in the architectural criticism, has made many students is inadequacy in the accumulation of architectural theory. The theoretical study or discussion of undergraduate students in professional learning is almost a blank in the existing architecture teaching system in our country, whether it is in science and engineering or Art College. A considerable part of the students' understanding of the building are often equated with the copy of surface form or function. Although their conversations buzzed with talk of "high tech" and "minimalism", but they do not know what it is, where it is produced from, what is the meaning of existence?

Generally speaking, most colleges and universities have chosen to arrange the design course and practice according to the building type. First grade arrangements for basic and architectural art courses; thereafter the second grade to the fifth grade, each semester has a plurality of different types of architectural design project from the shallow to the deep, through simple function to complex functions. The first is a reading room, small restaurants, independent residential, kindergarten and then to the club, museums, office buildings, hotels, theaters, and so on. It is very difficult to understand all aspects of building design only through a few limited building types. These conditions cause the serious gap between the professional education and the engineering practice become bigger and bigger.

18 weeks of practice Design Institute in the current architecture teaching syllabus of our country is shorter than any other country. At the same time the architects of Design Institute have no defects in the guidance of the students of architecture department because of the pressure of their own business. This caused most of the students' training project is just going through the motions, which didn't play a real role at all. The section headings are in boldface capital and lowercase letters. Second level headings are typed as part of the succeeding paragraph (like the subsection heading of this paragraph).

3 Reform Ideas

From the professional teaching goals, the architecture school's professional education must help the students to establish the correct concept of architecture by understanding the relationship between cultural property and engineering property of architecture, so that they will deeply understand the social, economic, technical and other factors which the architectural design constraints. Let them understand the experience of a large number of complex and detailed work procedures of a building from conceptual design to the final completion. So as to improve the comprehensive ability of students.

From the point of view of curriculum, we must adhere to the base of enhancing the comprehensive ability and improving the overall quality. We integrate the traditional basic course module, the design module, the technical course module, the teaching arrangement of the theoretical course module together by updating the content of technical courses, strengthening the horizontal connection of technical courses and design courses, exploring new curriculum outline. So as to form a new teaching link and mode. We will enrich the new knowledge, theory and technology to the teaching content, and provide students with the teaching content in line with the needs of the times.

We set up a number of main courses to break through the single course model which takes the architectural design as the main. These series of courses are relying on each other, from easy to difficult, step by step, which highlight the building energy saving, environmental protection and other emerging technologies.

In terms of specific methods, we put forward the teaching method of architecture specialty based on the integration principle of art and technology. The reform and practice of curriculum system construction consists of three aspects in order to achieve the ideal teaching purpose: the integrity of the teaching organization, the continuity of the teaching process and the dynamic evaluation of the teaching process.

We carry out architectural teaching activity on the integration principle of art and technology by taking the architectural design courses as a link and taking the conceptual design as the center. In the course, we combine architecture design with structure design, power design, water supply and drainage design, HVAC design, paying particular attention to fire protection design and environmental landscape design. In the process of integrated design, we take the apprenticeship model (teacher - student): Each tutor and student carries out two-way choice according to their interests and study direction. Determine the respective design issues and optimize the teaching process on the basis of the full discussion. We do will all flowers bloom together. We should also pay attention to the communication between teachers and students, especially among the instructors.

We broke the parallel state of each technologic curriculum in the traditional teaching mode. We closely integrated multi courses such as building design principle, building structure selection, architectural physics (sound, light and heat), and building tectonic, building security by taking the student's design blueprint in design course module as the core. We open up the original system of the lack of coordination of different knowledge finally.

We highlight the core status of the architectural design curriculum module by integrating the teaching process of program, technology and art. The teachers of the relevant courses arrangement unit training session surrounding the drawings of architectural design course module of last term or this term on the basis of the theory. We will eventually build the architectural design module drawings as an important indicator of the expiration of the assessment, which have completely abandoned the traditional evaluation system (Usually work adds final examination). That prevents students being great at rote memorization, but ill-prepared for creating progress and change in the world.

4 Conclusions

With the rapid economic development in China, more and more energy is consumed for the improvement of people's living conditions. The building sector accounts for 33.32% of Chinese energy requirements currently, which was only 15% in 1992 and 27.45% in 2004. In particular, residential buildings represent nearly 61% of the total energy consumption in the building sector. Furthermore, CO_2 emissions from residential buildings are the fifth largest key source of greenhouse gas emissions in China. They accounted for 12% of total greenhouse gas emissions in 2010. Significant potential exists to reduce the rate of future gas emissions in the building sector by promoting energy efficiency in buildings.

The opening words of the citation when Zaha Hadid was named as the first woman to win the prestigious Pritzker Prize for architecture in 2004 were: "Her architectural career has not been traditional or easy." You could call her work baroque modernism. Baroque classicists like Borromini shattered Renaissance ideas of a single viewpoint perspective in favour of dizzying spaces designed to lift the eyes and the heart to God.

Modern citizens need aesthetic appreciation from architecture so that they could go through their invariable life. The famous foreign architects, who designed these large-scale public buildings for Beijing 2008 Olympic Games, are considered to declare the coming of the age of globalization in Chinese architecture design. The dreaming spaces will give people novel aesthetic experience with much of debate.

In the practice courses, we should also promote the architecture professional practice teaching reform by orienting on the needs of the enterprises, combining with engineering background, taking engineering technology as the main line. We should summarize the achievements of the practice teaching of architecture specialty and draw lessons from the successful experience of foreign countries. We strive to enhance the students' engineering accomplishment through close cooperation between universities and industry enterprises, which will cultivate students' engineering practice ability, engineering design ability and engineering innovation ability.

Acknowledgement

This research was financially supported by the Teaching And Research Project of Wuhan University Of Science And Technology (2013X051) and Year 2011 the "Twelfth Five Year Plan" of Educational Science in Hubei Province (2011B446) and Year 2011 Higher Education Teaching Research Project in Hubei Province (2011204)

References

[1] Ma Cheng- Song, Zhu Shi-Tan, Status quo architecture educational abroad and its inspiration to us, Journal of Architectural Education in Institutions of Higher Learning. vol. 31 (2004) no. 2, p. 30-32.

[2] Ni Qi, Wang Yi, Undergraduate Architecture Education for Kyushu University in Japan, Urbanism and Architecture, vol. 11 (2006) no. 5, p. 90-92

Enhancing the Precision of Timing Gear Used in Automobile by Making Use of Azotizing Deformation

Yangang WEI [1], Huifang LIU [1&2] and Jing ZHAO[3]

[1] School of Mechanical Engineering, Dalian Jiaotong University, Dalian 116028, China

[2]Dalian Chuangxin Gear Case Manufacture Company Limited, Dalian 116314, China

[3] School of Physical and Electronic Engineering, Liaoning Normal University, Dalian 116026 China

Corresponding author: weiyg@djtu.edu.cn

Abstract. Statistic analysis about the effect of heat treatment deformation on tooth profile deviation distribution of the timing gear used in automobile is done with mathematical statistics. Based on the research on the effect of the heat treatment deformation on tooth profile deviation, helical deviation and common normal length etc, the processing measure, tolerance and correcting quantity before and after azotizing treatment can be adjusted to optimize the gear process and to enhance processing precision of the gear.

1 Introduction

The timing gear used in automobile is mostly involute gear. The process of gear made of middle carbon steel or middle carbon alloy steel, hardening and tempering, is generally hobbing or shaping, shaving or extruding, heat treatment, honing[1]..

The high quality and capability of gears not only depend on advanced design method, but also high manufacture technical level. Process of gear modification is developed along fining orientation as the demand for gear accuracy become high. Gear modification is one of key technology among gear design and manufacture. A lot of errors are produced during the process of gear manufacture. The tooth profile is in the range of modification K form chart, but middle dent of tooth profile and addendum thick or dedendum thick of tooth profile appears sometimes[2~3]. A large number of trial and gauging data are demanded to analize the causes of the errors in order to get idea gear modification. The errors are possibly caused by cutters, machine tool, processing errors and heat treatment deformation etc[4]. So the whole process of gear manufacture should be known and the procedures effecting gear modification be studied to get optimized gear modifying effect. In this way, different measures are able to be used aiming at the different causes and optimized process be decided according to theory calculation and adequate error analysis to raise the modifying effect.

Taking a timing gear mounted in crane shaft used in an automobile for example, the effect of heat treatment on the gear process is researched in the present paper.

2 Brief Introduction of Process of the Timing Gear Mounted in the Crane Shaft

The gear made of 40Cr with tempering and azotizing treatment. Its hardness is 26~32 HRC and its surface hardness is 280~740 HV1 with depth 0.2~0.4mm since tooth surface and shaft surface mounting the gear are ion carburized.

The process flow of the gear is blanking, forging blank, rough turning, tempering, ball blasting, fine turning the ends of the blank, hobbing of teeth, removing rag, marking, drilling of wheel disk holes, grinding of inner diameter hole and front edge surfaces, grinding of tooth profile arris, drilling and boring of pin hole, shaving of teeth, azotizing treatment, sanding of inner diameter hole, cleaning of the gear and inspecting.

The azotizing treatment deformation plays a key role on the effect on last precision of the gear according to the process of the gear. Therefore, a large number of the azotizing treatment deformation of the gear are analized in order to find the basic law of the azotizing treatment deformation and the processing measure, tolerance and correcting quantity before and after azotizing treatment to be adjusted to enhance processing precision of the gear.

3 Mathematical Statistic Analysis of Gauging Data of the Timing Gear Mounted in the Crane Shaft

To limit the paper length, only one of tooth profile gauging diagrams of ten years before and after azotizing are showed in Fig.1 and only probability density function curves of gauging data of the left profiles of teeth are showed in Fig.2 and Fig.3.

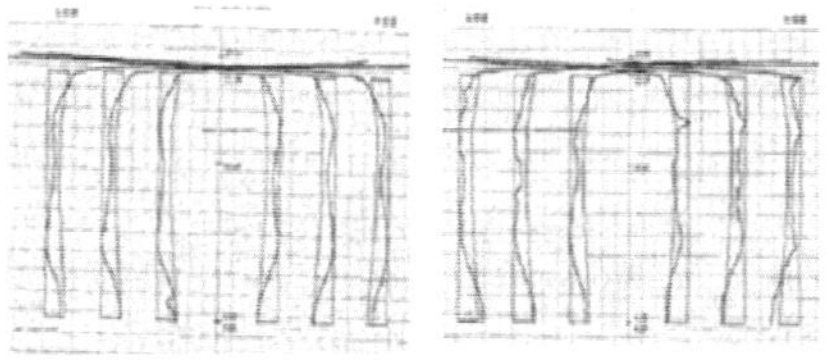

(a) Before (b) after

Fig.1 Gauging tooth profile map before and after azotizing

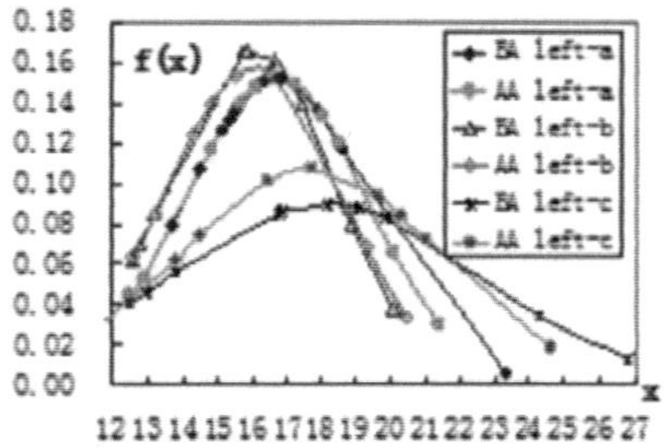

Fig.2. Probability density curves of profile total deviation distribution

Ten gears before and after azotizing are gauged and the gauged data are analyzed with Mathematical Statistics. The data are measured from six points at same diameter circle across right and left profiles of three teeth uniformly distributing and selected randomly. "BA" means before azotizing and "AA" means after azotizing in the table. The points a, b, c before azotizing are corresponding to the points a, b, c after azotizing.

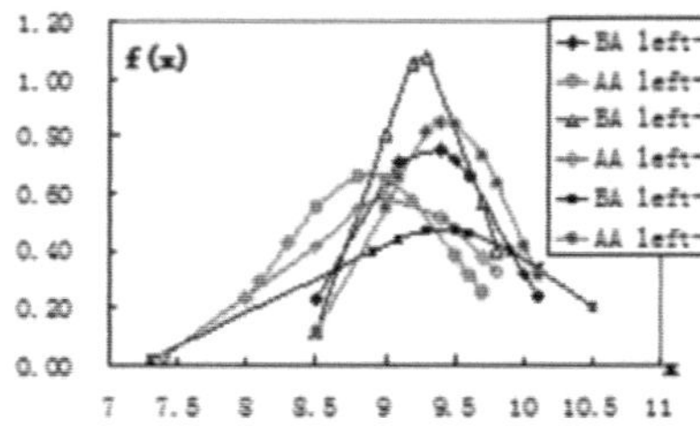

Fig.3 Probability density curves of tooth helical deviation distribution

4 Comparison and Mathematical Statistic Analysis of Tooth Profile Gauging Data before and after Azotizing Treatment.

The gauging data of tooth profile total deviations of ten gears are listed in Tab.1 and Fig.2 is its probability density function curves. The expectation and standard deviation of the tooth profile total deviations are given in the table also.

Tab.1 Gauging Data Of Tooth Profile Total Deviations （Unit：Mm）

Gear number	Left profile						Right Profile						Average
	Point a		Point b		Point c		Point a		Point b		Point c		
	BA	AA	BA	AA	BA	AA	BA	AA	BA	AA	BA	AA	AA-BA
1	23.3	17.2	24.3	14.8	15.8	20.3	16.3	21.5	16.9	19.4	23.3	19.6	-1.18
2	18.6	18.5	13.8	20.5	15.7	17.7	13.9	20.8	13.7	17.7	15	18.9	3.9
3	16.3	12.9	13	19.3	16.6	19.7	13.5	25.8	15.9	15.6	16.1	15.3	2.87
4	15.3	15.6	18.2	12.6	18.9	13.8	12.4	14.7	17.5	21.2	15.5	15.3	-0.77
5	14.5	14.8	16.7	16.5	12.8	24.6	15.2	22	12.7	18.3	14.3	18.3	4.72

6	17.9	20.1	19.9	18.8	20.1	17.6	19.3	17.1	17.3	22.4	16.6	14.4	-0.12
7	15.1	12.5	26.8	14.3	16.7	14.5	15.6	14.2	28.1	14.8	21.6	14.3	-6.55
8	15.5	18	12.5	14.8	12.6	16.4	14.1	15.1	17.8	14.6	14.3	15	1.18
9	16.8	21.4	16.8	13.2	13.2	11.3	14.3	16	15.3	18.2	12.9	17.6	1.4
10	13.7	16.1	19.1	15.5	17.4	21	20	15.2	18.8	16.2	15.3	21.8	0.25
expectation	16.7	16.71	18.11	16.03	15.98	17.69	15.46	18.24	17.4	17.84	16.49	17.05	0.57
standard deviation	2.62	2.76	4.45	2.54	2.40	3.71	2.34	3.78	3.99	2.50	3.16	2.43	3.01

It is found through researching Fig.2 and Tab.1 that before azotizing, the average value of the expectation of the tooth profile total deviations is 16.69μm and the average value of the standard deviation is 3.16μm and after azotizing, the average value of the expectation is 17.26μm and the average value of the standard deviation is 2.95μm. The average value of the expectation after azotizing is 0. 57μm more than that before azotizing and the average value of the standard deviation after azotizing is 0. 21μm less than that before azotizing

5 Comparison and Mathematical Statistic Analysis of Tooth Helical Deviation Gauging Data before and after Azotizing Treatment.

The gauging data of tooth helical deviations of ten gears are listed in Tab.2 and Fig.3 is its probability density function curves. It is found through researching Fig.3 and Tab.2 that before azotizing, the average value of the expectation of the tooth helical deviations is 9.24μm and the average value of the standard deviation is 0.51μm and after azotizing, the average value of the expectation is 9.03μm and the average value of the standard deviation is 0.54μm. The average value of the expectation after azotizing is 0. 21μm less than that before azotizing and the average value of the standard deviation after azotizing is 0. 03μm more than that before azotizing.

Tab.2 Gauging data of tooth helical deviations （unit：μm）

Gear number	Left profile						right profile						average
	Point a		Point b		Point c		Point a		Point b		Point c		
	BA	AA	BA	AA	BA	AA	BA	AA	BA	AA	BA	AA	AA-BA
1	8.5	8.3	9	7.4	7.3	9.4	8.1	7.9	9	8.7	9.2	9	-0.07
2	9.1	8.5	8.5	9	8.9	8.5	9.5	9.1	9.5	9	9.8	9.5	-0.28
3	8.5	8	9.3	8.8	10.1	10	8.4	8.2	8.4	8.2	10	9.5	-0.33
4	10	9.5	9.2	9.4	9.5	9.5	8.6	8.1	8.9	8.5	9.4	9.1	-0.25
5	10.1	9.7	9.3	8.5	9.1	9.8	8	7.8	9.6	9.5	9.4	9.7	-0.08
6	9.4	8.1	9.3	9.8	9.9	9.7	9.6	9.7	9.5	9.6	9.7	9.5	-0.17
7	9	8.8	9.7	9.5	10.5	10.1	8.5	8.3	9.1	8.8	9	8.9	-0.23
8	9.5	9.2	9.8	9.7	9.6	9.1	9.9	9.7	9.6	9.5	9.6	9.3	-0.25
9	9.6	9	9.7	9.5	9.3	9	8	7.8	9.5	9.3	9.1	9.2	-0.23
10	9.4	9.6	9	9	9.9	9.3	8.5	7.6	9.5	9.5	9.6	9.8	-0.18

expectation	9.31	8.87	9.28	9.06	9.41	9.44	8.71	8.42	9.26	9.06	9.48	9.35	-0.21
standard deviation	0.52	0.60	0.37	0.68	0.84	0.47	0.66	0.75	0.38	0.47	0.30	0.28	0.08

6 Comparison and Mathematical Statistic Analysis of Common Normal Length Gauging Data before and after Azotizing Treatment.

The probability density function curves of gauging data of the common normal length are showed in Fig.4. The expectation of the common normal lengths is 42.3297mm and the standard deviation is 4.5μm before azotizing. The expectation is 42.3406 mm and the standard deviation is 4.9μm after azotizing,. The expectation after azotizing is 10. 9μm more than that before azotizing and the standard deviation after azotizing is 0. 7μm more than that before azotizing.

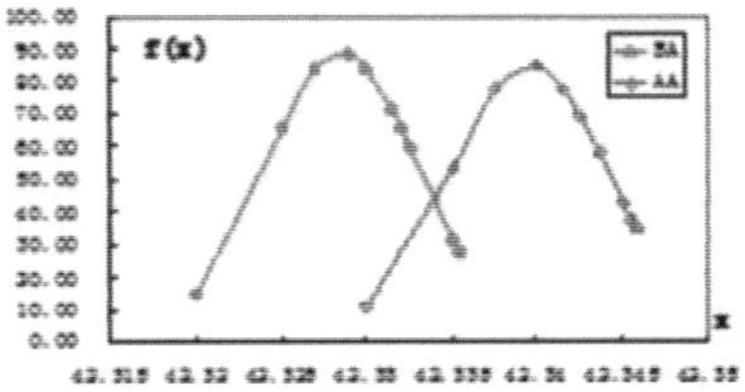

Fig.4 Probability density curves of common normal length distribution

7 Summary

The average value of the standard deviation of the tooth profile deviations is biggest, the average value of standard deviation of tooth helical deviations is smaller than that of the tooth profile deviations and the average value of standard deviation of common normal lengths is smallest. This fact proves that the distribution of tooth profile deviations is the least approaching to standard normal distribution and the distribution of common normal lengths is the most approaching to standard normal distribution.

According to the research of the statistic analysis, modification principle of reverse deformation can be adopted before heat treatment, that is, the tooth profile deviation is reduced by about 0.6μm, the tooth helical deviation is increased by 0.21μm and the common normal length is reduced by 11μm to advance precision of the gear manufacture and to realize design tooth profile during shaving the gears.

References

1. WEI Yangang, Quan Kueide. Research on the Longitudinal Modification of Involute Spur Cylindrical Gear Pair with Axial angle Errors Used in Timing Transmission System of Automobile. ICMET 2011 Proceedings of International Conference on Mechanical and Electrical Technology Vol.3,P:607-612
2. Lizheng Huang and Sonh Qi etc. Prof ile Modif ication Method for K2teeth while Gear Shaper Cutting [J]. Tool Technology, 2004, 38(5):33～35
3. Qiang Wang and Jinjin Zhou. Development and technology for trimming gear teeth [J]. Machine Engineer,2002, 2(2):5～8
4. Feng Li. Computer Estimating for the Errors of Tooth Prof ile and Helix of Transmission Gears [J]. Machine Tool and Hydraulic Pressure, 2004, 8(8):161～162

Axisymmetric Thermo-elastic Deformation of the Cylinder with Two-dimensional Inhomogeneity of Material

V.I. Andreev

Moscow State University of Civil Engineering, Yaroslavskoye sh., 26, 129337 Moscow, Russia

asv@mgsu.ru

Abstract. The article discusses the use of a numerical method the calculation of finite cylinders into account the dependence of physical and mechanical properties of the material on temperature. If we have two-dimensional temperature field characteristics of the material depends on two coordinates. - r and z from which follows that the problem of thermoelasticity is also a two-dimensional. Using the numerical method allows to solve the problem for any state of the cylinder (plane stress or plane strain) and consider arbitrary boundary conditions at its ends.

Keywords: cylindrical shell, reinforced concrete, heat conductivity, thermoelasticity, inhomogeneity.

1 Introduction

In [1-6] considered the problem of calculating thick-walled cylinders under the action of force and temperature loads at one-dimensional (radial) inhomogeneity of the material. Temperature inhomogeneity was due to axisymmetric not varying along the axis of the cylinder temperature field. In many practical problems, a heat source inside the cylinder can be regarded as a point. In this case, the temperature field remaining axisymmetric varies along the cylinder axis, resulting in a two dimensional an inhomogeneity due to which you cannot apply a numerical-analytical method [1, 7].

2 State of the Problem

As an example, below we consider the calculation of stresses in the radiation-heat shield of the reactor plant. This design, also known as "dry protection," designed to reduce the effects of radiation and heat generated when operation of the reactor. In constructional sense thermal protection is fixed at the base a thick-walled cylindrical shell made of heat-resistant reinforced concrete [8]. Fig. 1 shows the design scheme of the problem. In constructional sense thermal protection is a fixed at the base thick-walled cylindrical shell made of heat-resistant reinforced concrete. Fig. 1 shows the design scheme of the problem.

To reduce the thermal load there is envisage prorolling air cooling in the clearances between the shell of the reactor, the design of thermal protection and biological protection of the reactor. To solve the problem is used the variational-difference method [9-11], while this method is used for solving the problem of heat conduction subject to the availability of internal sources of heat and coefficient of thermal conductivity depending on the temperature, and to solve the problem of determining the stress-strain state of the cylinder.

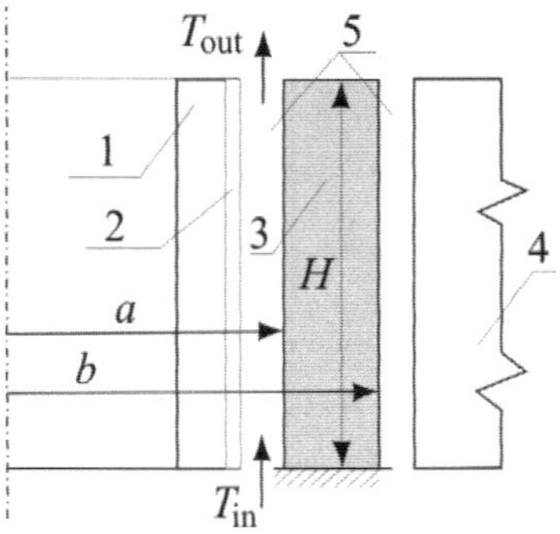

Fig. 1. Design scheme of heat shield:
1 - reactor vessel; 2 - heat insulation;
3 - radiation-heat shield;
4 - biological shield;
5 - cooling channels

3 Solution of the Problem of Heat Conduction

In the stationary temperature field heat equation

$$\lambda\nabla^2 T + W(r,z) = 0 \quad (1)$$

for the axisymmetric problem in cylindrical coordinates, taking into account the dependence F (h) is written as

$$\frac{1}{r}\frac{\partial}{\partial r}\left[r\lambda_T(T)\frac{\partial T}{\partial r}\right] + \frac{\partial}{\partial z}\left[\lambda_T(T)\frac{\partial T}{\partial z}\right] = -W(r,z) \quad (2)$$

,

where λ_T – coefficient of thermal conductivity, W - the intensity of heat sources per unit volume.

For the function $W(r,z)$, describing the density of the internal heat sources caused by ionizing radiation, use dependence [12]:

$$W(r,z) = W_0 + W_1 \exp[-\delta(r-a)]\sin\frac{\pi z}{H} \quad (3)$$

,

in which W_0, W_1 and δ – empirical parameters.

The boundary conditions on the side and end surfaces of the cylinder taken as:

$$r = a; \quad -\lambda_T\frac{\partial T}{\partial r} = \alpha_a(T - T_{cp}^a);$$
$$r = b; \quad -\lambda_T\frac{\partial T}{\partial r} = \alpha_b(T - T_{cp}^b); \quad (4)$$
$$z = H; \quad -\lambda_T\frac{\partial T}{\partial z} = \alpha_H(T - T_{cp}^H);$$
$$z = 0; \quad T = T_0(r).$$

Here T_{amb}^a, T_{amb}^b, T_{amb}^H – ambient temperature (cooling air) onto the side surfaces and upper (end) surface; $\alpha_a, \alpha_b, \alpha_H$ – heat transfer coefficients for these surfaces are listed in [13]. The last boundary condition of the first kind can be accepted, given that the mass base is significantly greater than the mass the structure under consideration, and the temperature at the contact of two bodies – a known function $T_0(r)$.

Dependence $\lambda_T(T)$ approximated by a polynomial of the type

$$\lambda_T(T) = \lambda_0 \cdot \sum_{n=0}^{N} \beta_n T^n$$

on the basis of the data presented in [14].

Fig. 2 shows the graphs of the internal heat source constructed in accordance with (3). Fig.3 shows temperature dependences taken into account that the temperature of the cooling air in channel height varies linearly, at values $\lambda_{T0} = 1.5W/(m \cdot {}^{o}C)$; $T_{in} = 30\,{}^{o}C$; and $T_{out} = 80^{o}C$. The dimensions of the cylinder are: $a = 2\,m$, $b = H = 3\,m$.

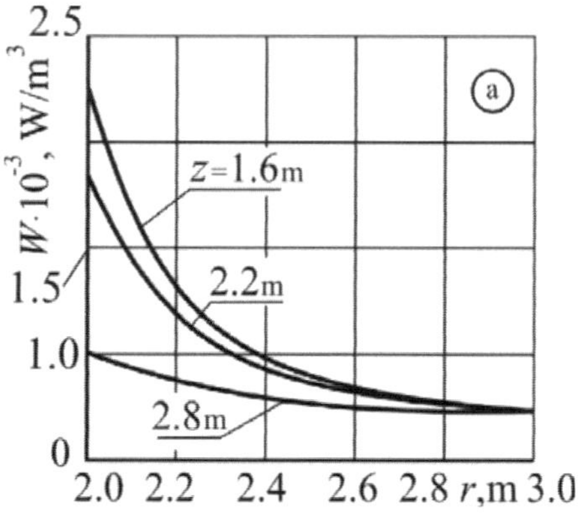

Fig. 2. Graph of the internal heat sources at different levels of the cylinder:

On the basis of performed calculations it can be concluded that in this temperature range the dependence $\lambda_T = \lambda_T(T)$ leads to a slight increase of temperature (not more 10%).

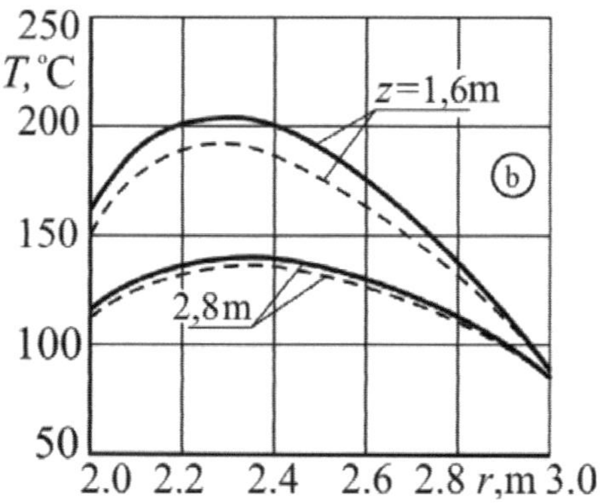

Fig. 3. Temperature dependences at different levels of the cylinder.
- - - - Homogeneous material – $\lambda_T = \lambda_{T0}$;
—— Inhomogeneous material – $\lambda_T = \lambda_T(T)$

The solution of the temperature problem by a variation-difference method is convenient from the point of view that the resulting calculation of net function $T(r_i, z_j)$ and its gradient is used in the future to solve the problem of thermoelasticity in the same manner and on the same grid.

4 Calculation of Stress State

Based on the obtained temperature distribution at half section of the cylinder $T(r,z)(a \le r \le b; 0 \le z \le H)$ in accordance with the [14] are set dependences $\lambda(r,z)$, $\mu(r,z)$ and $K(r,z)$, that allows us to solve the boundary value problem for an inhomogeneous body, described by equations [1, 7]

$$\mu\nabla^2 u + 3(\lambda+\mu)\frac{\partial\varepsilon_{cp}}{\partial r} - \frac{\mu}{r^2}u + 3\frac{\partial\lambda}{\partial r}\varepsilon_{av} + 2\frac{\partial\mu}{\partial r}\frac{\partial u}{\partial r} + \frac{\partial\mu}{\partial z}\left(\frac{\partial w}{\partial r}+\frac{\partial u}{\partial z}\right) - 3\frac{\partial}{\partial r}\left(K\varepsilon_f\right) = 0; \quad (5)$$

$$\mu\nabla^2 w + 3(\lambda+\mu)\frac{\partial\varepsilon_{cp}}{\partial z} + 3\frac{\partial\lambda}{\partial z}\varepsilon_{av} + 2\frac{\partial\mu}{\partial z}\frac{\partial w}{\partial z} + \frac{\partial\mu}{\partial r}\left(\frac{\partial u}{\partial z}+\frac{\partial w}{\partial r}\right) - 3\frac{\partial}{\partial z}\left(K\varepsilon_f\right) = 0 \quad (6)$$

with mixed boundary conditions (see. Fig. 1):

$$z = 0, \quad u = w = 0;$$
$$z = H, \quad \sigma_z = \tau_{rz} = 0;$$
$$r = (a,b), \quad \sigma_r = \tau_{rz} = 0.$$

In (5, 6) $\varepsilon_{av} = (\varepsilon_r + \varepsilon_\theta + \varepsilon_z)/3$ – average strain, ε_f – forced (temperature) deformations.

Some results of calculations made by the variational-difference method shown in Fig. 4. Mechanical characteristics correspond to concrete and equal to $E = 2\cdot10^4$ MPa; $\nu = 0,2$; $\alpha_T = 1\cdot10^{-5} 1/{}^\circ C$. Provided graphs correspond to the homogeneous ($E = E_0,\ \lambda_T = \lambda_{T0}$) and inhomogeneous ($E = E(T),\ \lambda_T = \lambda_{T0}$) materials. The results of calculations obtained at $E = E(T),\ \lambda_T = \lambda_T(T)$, taking into account a increase the temperature in the last case, will correspond to curves lying between shown.

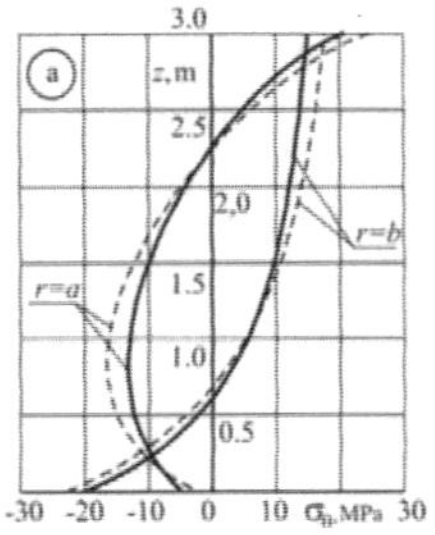

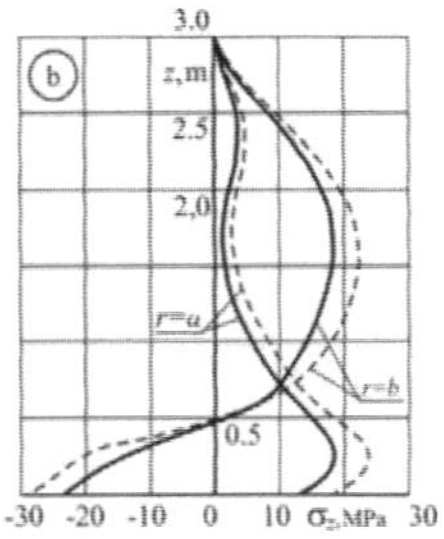

Fig. 4. Diagrams of thermal stresses σ_θ (a) and σ_z (b)
- - - - Homogeneous material; —— Inhomogeneous material

From these results it follows that both (compressive and tensile) stresses reach a very significant absolute values comparable and higher than the design strength of the concrete. This should be reflected in the appointment of the scheme of reinforcement structure. Accounting the temperature inhomogeneity of the elastic modulus leads to a significant reduction (up to 20%) in stresses that agrees qualitatively with the results of numerous calculations obtained earlier [6].

In [15] there was considered a similar problem of calculation the orthotropic cylinder with discrete changes in temperature on its inner surface. It was an analytical solution heat conduction problem in the Fourier series and the calculation of the stress state of the cylinder carried out variational-difference method. In the above-mentioned work was investigated the influence of two-dimensional temperature inhomogeneity of deformation properties of the material and orthotropy on the stress state related to the reinforcement of concrete cylinder.

5 Conclusions

From the presented results it is clear that the inhomogeneity of concrete at high temperatures leading to a decrease in the stiffness leads to reduction in stresses. This allows in design to reduce the thickness or decrease the reinforcement shell. Calculation of the shell located in a two-dimensional temperature field allows you to define areas tensile and compressive stresses. Averaging the temperature field provides only a rough estimate of the stress state.

Acknowledgement

Scientific work is realize under support of Ministry education and science Russian Federation (grant of President Russian Federation No14.Z57.14.6545-SS)

References

1. V.I. Andreev, Some problems and methods mechanics of nonhomogeneous bodies, Publ. house ASV, Moscow, 2002.
2. V.I. Andreev, Optimization of thick-walled shells based on solutions of inverse problems of the elastic theory for inhomogeneous bodies, Computer Aided Optimum Design in Engineering XII (OPTI XII). WIT Press. (2012) 189-201
3. V.I. Andreev, A.S. Chepurnenko, B.M.Yazyev, Model of Equal-stressed Cylinder based on the Mohr Failure Criterion. Advanced materials Research. 887-888 (2014) 869-872
4. V.I. Andreev, Stress-strain State of Weightless Inhomogeneous Array with Cylindrical Hole, Advanced Materials Research. 919-921 (2014) 740-743

5. V.I. Andreev, Inverse problems of the inhomogeneous theory of elasticity for thick-walled shells, Journal "Computational Methods and Experimental Measurements", 2 No. 2 (2014) 202-216
6. V.I. Andreev, D.A. Kapliy, Stress state of a thick-walled cylindrical shell under the combined action of radiation and temperature field, Advanced Materials Research. 1006-1007 (2014) 177-180
7. V.I. Andreev, Numerical-analytical solution of two-dimensional problem for elastic radially inhomogeneous thick-walled cylinder, Applied Mechanics and Materials. 752-753 (2015) 642-647
8. G.I. Zholdak, V.I. Ivanov, Construction dry protection nuclear reactor nuclear power plant // Problems of Atomic Science and Technology. Design and construction. 2 (1984) 35 - 40
9. N.P. Abovskiy, I.P. Andreev, A.P. Deruga, Variational principles of the theory of elasticity and the theory of shells, Nauka, Moscow,1978.
10. A.A. Samarskiy, V.B. Andreev, Difference methods for elliptic equations, Nauka, Moscow, 1976.
11. K. Vasidzu, Variational methods in the theory of elasticity and plasticity, Mir, Moscow, 1987.
12. B.K. Pergamenshchik, G.A. Lavdanskaya, Temperature field in a thick-walled cylinder with an internal heat source, Proceedings MISI. 56 (1968) 18-24
13. V.B. Dubrovsky, Z. Oblevich, Building materials and constructions protection against ionizing radiation, Stroyizdat, Moscow, 1983
14. SNiP 2.03.04-84 Concrete and reinforced concrete structures, designed to work in conditions of elevated and high temperatures. CITP Gosstroy USSR, Moscow, 1985
15. V.I. Andreev, M.V. Belyj, S.P. Kirillov, Calculation of axisymmetric thermostressed state radially inhomogeneous cylinder, Building mechanics engineering designs, 6 (1996) 27-30

MATEC Web of Conferences **61**, 05009 (2016) DOI: 10.1051/matecconf/20166105009
APOP2016

Influential factors of 2-chlorobiphenyl reductive dechlorination by highly dispersed bimetallic nanoparticles

Junrong Jiang[1,2], Chong Fu[2] and Deming Zhao[2a]
[1]*School of Biological & Chemical Engineering, Taizhou Vocational & Technical College, Taizhou 318000, China*
[2]*College of Chemical Engineering, Zhejiang University of Technology Hangzhou 310014, China*

Abstract. Highly dispersed Pd-Fe^0 bimetallic nanoparticles were prepared in the presence of 40 kHz ultrasonic irradiation in order to enhance disparity and reactivity, and simultaneously avoid agglomeration. Influential factors of 2-chlorobiphenyl (2-Cl BP) reductive dechlorination by highly dispersed Pd-Fe^0 nanoparticles were investigated. Experimental results showed that highly dispersed Pd-Fe^0 nanoparticles prepared in the in the presence of ultrasound could further improve the dechlorination efficiency of 2-Cl BP, meanwhile the biphenyl (BP) formation rates increased obviously and increased from 47.4% (in the absence of ultrasound) to 95.3% (in the presence of ultrasound) within 300 min. The catalytic reductive dechlorination effciency of 2-Cl BP was dependent on Pd-Fe^0 nanoparticles prepared methods, Pd-Fe^0 nanoparticles dosage, Pd loading percentage over Fe^0 and initial pH values

1 Introduction

Polychlorinated biphenyls (PCBs) are a sort of organic compounds of which there are 209 distinct chemical species known as congeners. PCBs as a class share a variety of properties (e.g., chemical and thermal stability) that made them popular for industrial applications in the latter part of the 20th century [1]. The high stability, low aqueous solubility, and high organic affinity of PCBs make them difficult to treat [2,3]. This raises an urgent need for efficient reductive dechlorination methods to eliminate chloroaromatics from both concentrated industrial effluents and diluted polluted groundwater.

Many remediation technologies,such as physical, chemical, and biological methods have been proposed, including dredging, landfilling, incineration, biodecomposition, in *situ* capping, and electrochemical degradation [2-4]. Among these technologies, the use of reactive metals such as nanoscale zero-valent iron (nZVI), as a mild and cost-effective reducing agent, has been documented to work efficiently for the reductive chlorinated organic compounds [5]. A rapid and complete catalytic reductive dechlorination method of chloroaromatics involving the use of bimetallic Pd-Fe^0 nanoparticles that led to the formation of non-chlorinated hydrocarbons have also been reported [6]. As one of the noble metal, Pd can utilize the produced H_2 from Fe^0 corrosion and accelerate the rates of dechlorination reaction [7].However, the presence of Pd not only reduces the accumulation of toxic byproducts, but also inhibits particle oxidation in air . Pd-Fe^0 bimetallic nanoparticles, when compared to the conventional large particles have some advantages with possessing highly dispersion, large specific surface area and high surface reactivity [6]. In order to obtain the highly dispersed, stabilized and high reactive Pd-Fe^0 bimetallic nanoparticles and decrease their agglomeration and accumulation in the effluent, ultrasound is applied to the preparation of Pd-Fe^0 bimetallic nanoparticles. Sonochemistry arises from acoustic cavitation, the formation, growth, and implosive collapse of bubbles in a liquid. Acoustic cavitation can increase the surface area of the reactive solids by causing particles to rupture [8].

In the present work, Highly dispersed Pd-Fe^0 bimetallic nanoparticles were prepared in the presence of 40 kHz ultrasound and influential factors of 2-chlorobiphenyl (2-Cl BP) reductive dechlorination by highly dispersed Pd-Fe^0 nanoparticles, such as Pd-Fe^0 nanoparticles availability, Pd content over Fe^0, initial 2-Cl BP concentration, reaction temperature and the initial pH values, were investigated.

2 Experimental

2.1 Chemicals

2-chlorobiphenyl (AP grade), biphenyl (CP grade) and potassium hexachloropalladate (AP grade) were purchased from the J&K Chemical Reagent Co., Ltd., China, $FeSO_4 \cdot 7H_2O$ (AR grade), sodium borohydride (AR grade) purchased from Tianjin Chemical Reagent Research Institute. 2-Cl BP is dissolved in a methanol/deoxygenated deionized water solution (50:50, *V*/*V*) and stored at 4 ℃. Pd-Fe^0 nanoparticles were synthesized immediately before use.

2.2 Experimental procedures

In a 500 mL three-necked flask, Pd-Fe^0 bimetallic nanoparticles were prepared in the presence of 40 kHz and 150 W ultrasound under nitrogen gas. Nanoscale zero-valent iron (nZVI) particles were synthesized by

[a] Corresponding author: dmzhao@zjut.edu.cn

drop wise addition of stoichiometric amounts of $NaBH_4$ aqueous solution into a flask containing $FeSO_4{\cdot}7H_2O$ aqueous solution simultaneously with mechanical stirring at 25 ℃. nZVI particles were then rinsed several times with deoxygenated deionized water. nZVI particles were then rinsed several times with deoxygenated deionized water. Subsequently, $Pd\text{-}Fe^0$ nanoparticles were prepared by reacting with the wet nZVI particles in an aqueous solution of potassium hexachloropalladate under mechanical stirring. Batch experiments of 2-Cl BP catalytic reductive dechlorination were performed in the same flask into which nanoscale bimetallic $Pd\text{-}Fe^0$ particles were added. 2-Cl BP stock solutions and a certain amount of deoxygenated deionized water and methanol solution (50:50,*V*/*V*) were added into the flask containing fresh prepared nanoscale $Pd\text{-}Fe^0$ particles into 500 mL of total reaction volume. The reaction solution was stirred under nitrogen flow to simulate anaerobic environment at 25 ℃ . Aliquots of samples were periodically collected with glass syringes and the reaction was quenched by passing through 0.22 μm polyether sulphone (PES) membrane filters.

2.3 Methods of analysis

All fresh prepared synthesized bimetallic $Pd\text{-}Fe^0$ nanoparticles (with Pd content 0.3 wt. %) were immersed in absolute ethyl alcohol and dispersed by an ultrasonator. Brunauer–Emmett–Teller (BET) specific surface area of all synthesized bimetallic $Pd\text{-}Fe^0$ nanoparticles were measured using nitrogen adsorption method with a surface analyzer (ASAR2020M+, Micromeritics Instrument Corp., US). Before the analysis, the particles were dried in vacuum at 25 ℃ for 24 h and then hydrogen flow at 260 ℃ for further 4 h. Scanning electron microscope (SEM) images were obtained through a microscope (HITACHI S-4800 HITACHI Instruments Corp., JP). X-ray diffraction (XRD) analysis was performed by using X'Pert Pro advanced X-ray diffractometer (λ = 1.5418 A°). Organic compounds such as 2-chlorobiphenyl and BP were analyzed by Thermofish Trace 1310 Gas Chromatography. ECD and FID detector, Thermofish TR-5 Column (30 m×0.32 mm, 1.0 μm), inlet temperature 300 ℃, detector temperature 250 ℃, no split injection, injection volume 1 μL, carrier gas nitrogen (purity ≥99.99%), column flow 1.5 mL/min (constant current), program temperature: column temperature 100 ℃, holding 0.5min, 25 ℃/min was raised to 300 ℃, maintained 2 min.

3 Results and Discussions

3.1 Characterization of Pd-FeO Nanoparticles

The BET specific surface area of $Pd\text{-}Fe^0$ nanoparticles prepared in the presence and absence of 40 kHz ultrasound were 41.68 $m^2\ g^{-1}$ and 22.39 $m^2\ g^{-1}$, respectively. The SEM images of the fresh synthesized $Pd\text{-}Fe^0$ nanoparticles in the presence of 40 kHz ultrasound, and the aged (after 300 min) $Pd\text{-}Fe^0$ nanoparticles prepared in presence of 40 kHz ultrasound are shown in Fig. 1 (a, b). Fig. 1 (a) shows many $Pd\text{-}Fe^0$ nanoparticles aggregate to form flocks and dendrites. After 300 min of reaction, the white platelet-shaped crystals appeared on the surface of $Pd\text{-}Fe^0$ nanoparticles and fraction of $Pd\text{-}Fe^0$ nanoparticle's diameter getting larger (Fig. 1(b)), suggests the formation of iron oxides resulting from iron corrosion. These minerals were likely composed of FeOOH, or Fe_3O_4 [9]. This is generally agreement with our observation in XRD patterns. Fig. 1(e, f) shows the XRD patterns of the fresh and the 300 min reacted $Pd\text{-}Fe^0$ nanoparticles prepared in the presence of ultrasound. The XRD pattern for the fresh sample presents a strong peak 44.66° which corresponds to the body-centered cubic $N\text{–}Fe^0$ at the (110) plane. The peak in the XRD pattern of the reacted sample shows evidence of iron oxides, possibly Fe_3O_4 or Fe_2O_3, or their mixture. This agrees with the previous work [10].

(a)

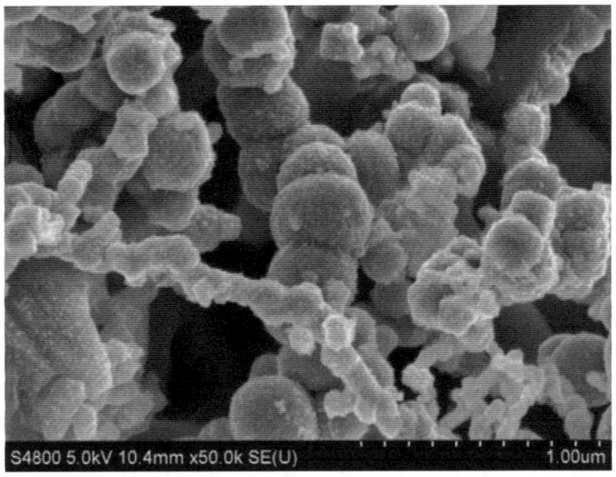

(b)

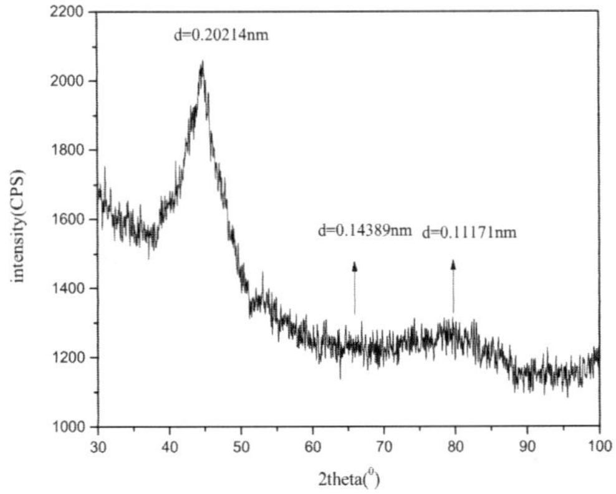

(c)

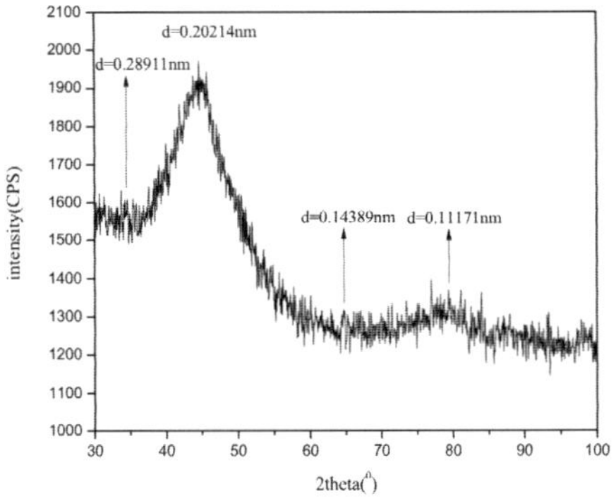

(d)

Figure 1. (a) SEM image of fresh synthesized Pd-Fe^0 nanoparticles in the presence of 40 kHz ultrasound, (b) SEM image of reacted Pd-Fe^0 nanoparticles synthesized in the presence of 40 kHz ultrasound (300 min), (c) XRD patterns of fresh synthesized Pd-Fe^0 nanoparticles in the presence of 40 kHz ultrasound and (d) XRD patterns of reacted Pd-Fe^0 nanoparticles synthesized in the presence of 40 kHz ultrasound.

3.2 Comparisons on 2-Cl BP catalytic reductive dechlorinated efficiency by Pd-Fe^0 nanoparticles synthesized in different methods

Effects of different Pd-Fe^0 nanoparticles prepared methods on 2-Cl BP catalytic reductive dechlorination were investigated at initial 2-Cl BP concentration of 10 mg L^{-1}, Pd-Fe^0 nanoparticles dosage of 3 g L^{-1}, Pd content of 0.8 wt. %, initial pH value of 3.0, reaction temperature of 25 °C and mechanical stirring speed of 600 rpm. Fig. 2 shows reductive dechlorinations of 2-Cl BP by Pd-Fe^0 nanoparticles prepared in different methods .

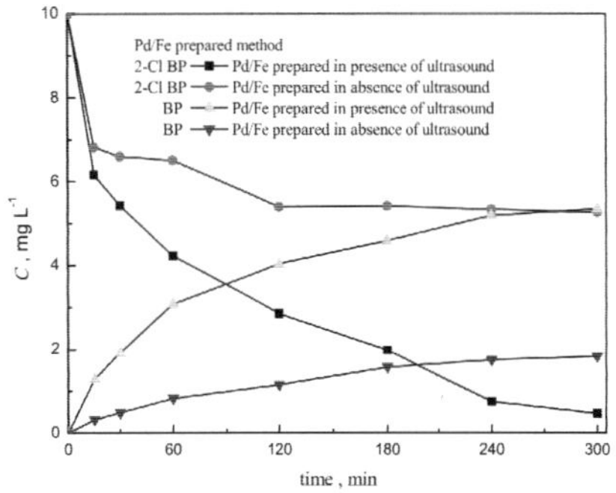

Figure 2. Effects of Pd-Fe^0 nanoparticles prepared methods on 2-Cl BP dechlorination

2-Cl BP was first adsorbed by Pd-Fe^0 nanoparticles, then reduced to *BP*. No other chlorinated intermediates or final organic products were found. The concentration of 2-Cl BP decreased rapidly and the removal rate reached 80.2% in 180 min, then further improved to nearly 95.3% in 300 min for Pd-Fe^0 nanoparticles prepared in the presence of 40 kHz ultrasonic irridation. However, only about 45.8% and 47.4% of the removal rate were obtained for Pd-Fe^0 nanoparticles prepared in the absence of ultrasonic irridation at the same stages, respectively. Obviously, the catalytic reductive chloronation process quickly proceeded with bimetallic Pd-Fe^0 nanoparticles synthesized in the presence of ultrasonic irridation than with bimetallic Pd-Fe^0 nanoparticles prepared in the absence of ultrasonic irridation. So, highly dispersed bimetallic Pd-Fe^0 nanoparticles were synthesized in the presence of 40 kHz ultrasonic irridation during the following experiments.

3.3 Effects of Pd content over Fe0 on 2-Cl BP ruductive dechlorination

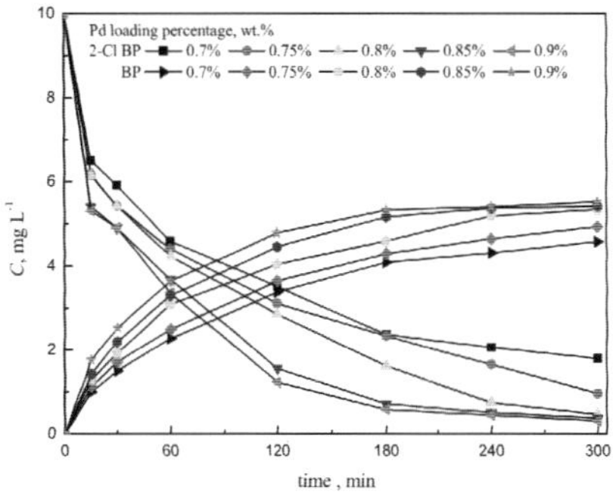

Figure 3. Effects of Pd loading content over Fe^0 in Pd-Fe^0 nanoparticles on 2-Cl BP dechlorination

Fe^0 can enhance the hydrogenolysis reaction in which a chlorine atom in the organochlorides is replaced by a hydrogen atom. Palladium is an excellent catalyst for the hydrogenolysis reaction. Co-existence of Pd and Fe^0 in the nanoparticles has been confirmed to be very effective to quicken the catalytic reductive dechlorination reaction. Then, the Pd content over Fe^0 on Pd-Fe^0 nanoparticles may be one of the important influential factors on the reductive dechlorinated ratio. As Fig. 3 shows the efficiencies of reductive dechlorination and BP formations were obivously promoted as Pd content over Fe^0 increased from 0.70 to 0.90 wt. %. The removal ratio of 2-Cl BP reached from 82 to 97% within 300 min. The increase in the Pd loading content over Fe^0 from 0.75 to 0.85 wt. % only caused slight improvement to the 2-Cl BP removal percentage. The most likely cause for this is that the maximum Pd coverage is less than one layer, and in this way, the loss of available catalytic reactive sites due to the overlapping between Pd atoms can be excluded. Nonetheless, the amount of hydrogen gas absorbed by Pd atoms promoted with the increasing Pd loaded content over Fe^0 [11], and so make possible to accelerate the ratios of dechlorination and BP formations. The slight improvement of the removal ratio at Pd content over Fe^0 >0.80 wt. % is that the accumulation of excessive hydrogen gas hinders the contact between 2-Cl BP and Pd-Fe^0 nanoparticles, and reduces the surface area available for 2-Cl BP dechlorination. This phenomenon is in agreement with previously reported studies [6,10]. Pd loading content over Fe^0 in bimetallic Pd-Fe^0 nanoprticles was selected at about 0.80 wt. % for optimized and efficient reductive dechlorination and yet minimal palladium content over Fe^0.

3.4 Effects of Pd-FeO nanoparticles dose on 2-Cl BP reductive dechlorination

Because the reductive dechlorination by Pd-Fe^0 nanoparticles happens on the surface of bimetallic Pd-Fe^0 nanoparticles, therefore bimetallic catalyst nanoparticles to 2-Cl BP ratio (g bimetallic Pd-Fe^0 nanoparticles /mg 2-Cl BP) is also an important influential factor. The quantity of available surface area is among the most significant experimental variables affecting 2-Cl BP reductive dechlorination efficiency. With increasing the dose of Pd-Fe^0 nanoparticles in the solution, it is insignificant to the final removal ratio in excess of Pd-Fe^0 nanoparticles dose. However, it will accelerate the initial reacted velocity and provide more active sites of Pd-Fe^0 nanoparticles for collision with 2-Cl BP during the reductive reaction. Effects of different Pd-Fe^0 nanoparticles dose on 2-Cl BP dechlorination were investigated as shown in Fig. 4. With the addition of the Pd-Fe^0 nanoparticles dose from 5 to 9 g L^{-1}, obviously differences were detected, the removal ratio of 2-Cl BP increased from 63 to 91% after 180 min of the reaction. Increasing the Pd/Fe nanoparticles dose means the larger calalysts surface area. The higher the catalysts surface area concentration is, the faster the reaction velocity. The removal ratio of 2-Cl BP is similar for a dose of Pd-Fe^0 nanoparticles of 7 or 8 g L^{-1} after 300 min of the reaction. Hence, the appropriate dose of Pd-Fe^0 nanoparticles in the level of 7 g L^{-1} is chosen for 2-Cl BP reductive dechlorination.

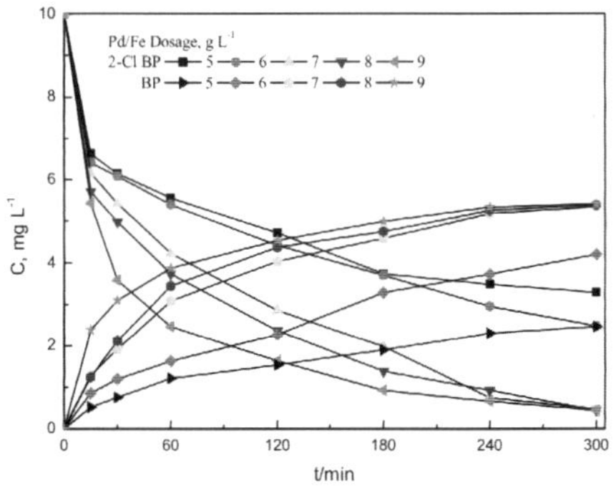

Figure 4. Effects of Pd-Fe^0nanoparticles dose on 2-Cl BP dechlorination

3.5 Effects of the initial pH value on 2-Cl BP reductive dechlorination

Initial pH value in aqueous solution is an important influential parameter in catalytic reductive dechlorination of organic chlorinated compounds using Fe^0. Low pH prefers more Fe^0 surface available for hydrogenolysis reaction with the chlorinated molecules or at least increase the corrosion velocity, leading to generate of chloride ions. Fig. 5 shows the effect of different initial pH values on the 2-Cl BP reductive dechlorination by bimetallic Pd-Fe^0 nanoparticles. Prior to initializing reaction, all reactant solutions were adjusted to different pH values by dilution with H_2SO_4 and NaOH, and during the reaction pH values were not adjusted. When the initial pH values changed from 3, 5, 7 to 9, the removal ratios of 2-Cl BP dropped from nearly 95.3, 70.6 to 54.8%, respectively in 300 min. The increase of initial pH values of the reactant solutions from 7 to 9, leads to the reactant solution change from acidic to alkaline condition, consequently the removal ratios of 2-Cl BP dropped obviously. It shows that the presence of H^+ largely promotes the 2-Cl BP reductive dechlorination rate. The feasible reasons may be that (1) the surface of Pd-Fe^0 nanoparticles would be oxidized inevitably during the synthesis and storage, which had been confirmed by the earlier published literatures [12]. However, at lower pH values, the oxides on the catalyst particles surface were dissolved, and the active sites of the catalyst particle surface were exposed and it was benefical to the reductive reaction; (2) at lower pH values, Fe^0 corrosion could be accelerated, producing enough hydrogen (or hydrogen atoms), which were benefical to hydrogenation reaction [13]; (3) Iron corrosion in solution of pH higher than 7 tends to form a passive film of iron oxides and hydroxide on Fe^0 surface, which inhibits further reductive reaction. In the previous study [10,14], detection of the solution pH during the entire period of the hydrogenation reaction, and ferrous ions/total iron ions produced in the reaction could further support our assumptions.

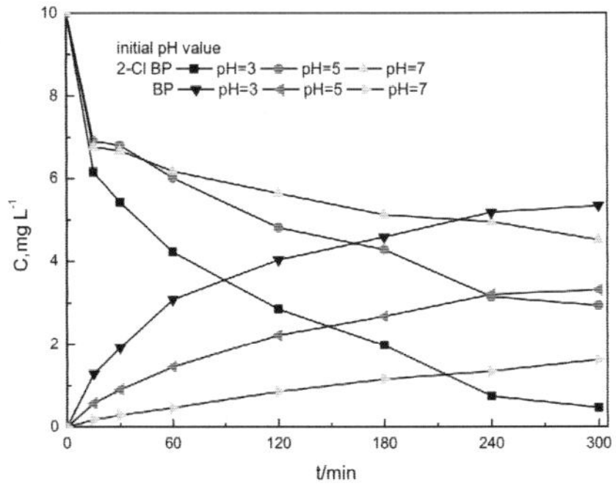

Figure 5. Effects of solution initial pH on 2-Cl BP dechlorination

4 Conclusion

The experimental results indicate that the prepared bimetallic Pd-Fe^0 nanoparticles in the presence of ultrasound was a better technique to enhance dispersion and the surface special properties. 2-Cl BP reductive dechlorination efficiency was dependent on a series of influential factors including highly dispersed bimetallic Pd-Fe^0 nanoparticles prepared method, Pd-Fe^0 nanoparticles dose, Pd loading content over Fe^0 and initial pH values. Up to 95.0% of 2-Cl BP was removed after 300 min reaction with these experimental conditions: initial 2-Cl BP concentration 10 mg L^{-1}, Pd loading content over $Fe^0$0.8 wt. %, Pd-Fe^0nanoparticles prepared in the presence of 40 kHz ultrasonic irradiation available dose 7 g L^{-1}, initial pH value 3.0, and reaction temperature 25 ℃ . The removal ratio of 2-Cl BP degradation increased with the increasing Pd-Fe^0 nanoparticles dosage, and with the decrease of pH values.

Acknowledgements

The authors are grateful for the financial support provided by Zhejiang Provincial Natural Science Foundation of China (LY15B070005) and by the Science and Technology Department of Zhejiang Province（No. 2015C33233）.

References

1. C.P. O'brien, M.C. Thies, D. A. Bruce, Environ. Sci. Technol. **39**,6839-6844 (2005)
2. H. Choi, S. Agarwal, S.R. Al-abed, Environ. Sci. Technol. **43**, 488–493 (2009)
3. J. Borja, D.M. Taleon, J. Auresenia, S. Gallardo, Process Biochem. **49**, 1999–2013 (2005)
4. P. Murphy, A. Marquette, D.D. Reible, G.V. Lowry, J. Environ. Eng. **132**, 787–794 (2006)
5. Y. Fang, S. R. Al-Abed, Environ. Sci. Technol. **41**, 6253–6258 (2007)
6. Z. Zhang, Q.H. Shen, N. Cissoko, J.J. Wo, X.H. Xu, J. Hazard. Mater. **182**, 252–258 (2010)
7. G.V. Lowry, M. Reinhard, Environ. Sci. Technol. **35** , 696-702 (2001)
8. K.S. Suslick, M. Fang, T. Hyeon, J. Am. Chem. Soc., **118**, 11960-11961 (1996)
9. D.H. Phillips, B. Gu, D.B. Watson, Y. Roh, L. Liang, S.Y. Lee, Environ. Sci. Technol.,**34**, 4169-4176 (2000)
10. D.M. Zhao, M. Li, D.X. Zhang, S.A.Baig, X.H. Xu, Ultrason Sono. 20, 864-871 (2013)
11. C. Grittini, M. Malcomson, Q. Farnando, N. Korte, Environ. Sci. Technol. **29**, 2898-2900 (1995)
12. J.T. Nurmi, P.G. Tratnyek, V. Sarathy, D.R. Baer, J.E. Amonette, K. Pecher, C.M. Wang, J.C. Linehan, D.W. Matson, R.L. Penn, M.D. Driessen, Environ. Sci. Technol. 39, 1221-1230 (2005)
13. Y.H. Kim, E.R. Carraway, Environ. Sci. Technol. **34**, 2014-2017 (2000)
14. D.M. Zhao, Y.Y. Zheng, M. Li, S.A.Baig, D.P.Wu, X.H. Xu, Ultrason Sono. **21**, 1714-1721 (2014)

MATEC Web of Conferences **61**, 05010 (2016) DOI: 10.1051/matecconf/20166105010
APOP2016

Fatigue Crack Growth Under Random Overloads with Retardation

XiaoLi Zou

School of Civil Engineering and Architecture, Wuhan Polytechnic University, Wuhan 430023, P. R. China

Abstract. A Monte Carlo simulation scheme is presented to study the behavior of fatigue crack propagation under random overloads. Overloads of Poisson flow with uniform distribution on base-line constant-amplitude cyclic loads are considered. The retardation effect of overload on fatigue crack growth is taken into account by using Wheeler retardation model. The fatigue crack growth curve from initial crack size till fatigue fracture is simulated step by step. With the large number of the simulated samples, the average fatigue crack propagation life is obtained and the influences of overload intensity and magnitude on fatigue lifetime are studied.

1 Introduction

For reliability design and maintenance plan of structures subjected in service to spectrum loading, statistical prediction of fatigue crack lifetime under random loading is one of the most important problems. However, the main difficulty in evaluating fatigue crack growth in metallic structural components under spectrum loading is associated with the loading interaction effects induced by extreme overload peaks. For engineering structural components, many loading histories can be represented, at first approximation, as a superposition of a constant-amplitude cyclic loading and random sequences of overload peaks. As is known, sharp overloads would delay succedent crack growth significantly. To take into account the retardation effect of overload, some stochastic models of fatigue crack growth under random loading have been proposed [1-4]. However, these theoretical models are based on lots of approximate assumptions. On the other hand, it is obviously unrealistic to do a great deal of experiments on fatigue crack growth under random loading. In this paper, the residual stress model proposed by Wheeler [5] is applied to account for retardation effect of overload and random process of crack growth is simulated by Monte Carlo method. Consequently, the statistical characteristics of fatigue crack lifetime under random loading are determined.

2 Computational model

Consider a body with a growing fatigue crack, subjected to constant-amplitude cyclic loading with superimposed rare random overloads. It is assumed that the probability of occurrence of overloads is history-independent and that the flow of overloads is stationary. Under these assumptions the number of overloads can be described by a Poisson process characterized by an exponential distribution of inter-overload times with parameter λ. The parameter λ represents the average number of overloads occurring within one unit time. In addition, the magnitude of the overload stress is also assumed as a random value. Therefore the sequence of overloads is a set of mutually independent random variables with identical probability distribution function.

The residual stress model assumes that due to a high stress concentration at the crack tip a plastic zone is created. The greater the overload peak is, the larger the plastic zone size is. Residual compressive stresses formed in this zone will reduce the magnitude of the tensile stresses during the next load cycle and tend to delay the crack growth.

On the basis of the residual stress concept, Wheeler proposed that the crack propagation rate after the overload is given by

$$(\frac{da}{dn})_R = C_R \frac{da}{dn} \tag{1}$$

where C_R is the retardation coefficient and da/dn the crack propagation rate under the baseline load. According to Paris-Erdogan equation

$$\frac{da}{dn} = c(\Delta K)^m \tag{2}$$

where c and m are material constants, ΔK the stress intensity factor range.

$$\Delta K = B\sqrt{\pi a}(\sigma_{\max} - \sigma_{\min}) \tag{3}$$

where B is geometric shape parameter, $\sigma_{\max}$ and $\sigma_{\min}$ the maximum stress and minimum stress of the base-line constant-amplitude cyclic loads.

In Eq. 1, the retardation coefficient C_R is described by Wheeler as

Corresponding author: xiaolizou@eyou.com

$$C_R = \begin{cases} \left(\dfrac{r_n}{a_p - a_n} \right)^k & a_n + r_n \le a_p \\ 1 & a_n + r_n > a_p \end{cases} \quad (4)$$

where k is an empirical constant (retardation exponent). As shown in Fig. 1, a_0 is the crack length at the onset of the overload, a_n is the current crack length after the overload ($a_n \ge a_0$) corresponding to the nth cycle of the baseline load, ($a_p - a_0$) is the radius of the plastic zone produced by the overload, r_n is the radius of the current plastic zone corresponding to the nth cycle. The dependence of the size of the plastic zone on the stress intensity factor leads to the following expression

$$r_n = \frac{a_n}{A}\left(\frac{A\sigma_{\max}}{\sigma_y}\right)^2 \quad (5)$$

where σ_y is the yield strength of the material, parameter A is equal to 1 for plane stress state and 3 for plane strain state.

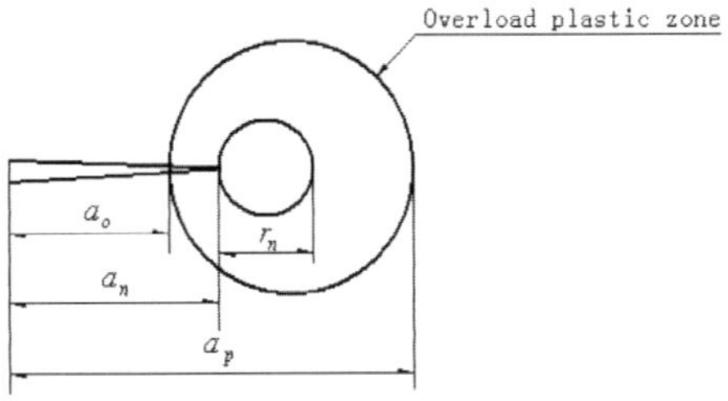

Figure 1. Definition sketch concerning Wheeler equation.

To begin the simulation of crack growth, a series of occurring moment and magnitude of overloads are generated by Monte Carlo method. Then according to above equations, the fatigue crack propagations from initial crack size till final fracture are computed cycle by cycle. In each step of computation, the radius of plastic zone corresponding to the current maximum stress of the base-line cyclic loads or overload stress is calculated from Eq. 5. If the size of the new plastic zone exceeds the old plastic zone, the new plastic zone will replace the old in the next step of computation. As the fatigue crack grows, final fracture will take place once the stress intensity factor K is greater than the fracture toughness of the material K_{IC}. There are two possibilities of final fracture: one is induced by overload, another by stress peak of base-line cyclic loads. Consequently, the fatigue crack lifetime is obtained.

3 Numerical example and discussion

Consider propagation of an edge in a semi-infinite body under a random sequence of overloads superimposed on base-line cyclic loading characteristic of minimum stress $\sigma_{\min} = 0$ and maximum stress $\sigma_{\max} = 300\text{MPa}$. The random sequence of overloads are characterized by uniform distribution of stress peaks in the range from σ_o^L to σ_o^U. In this case the stress intensity factor is $K = 1.12\sigma\sqrt{\pi a}$ and parameter $A = 1$ in Eq. 5. The material constants included in calculation are as follow: $\sigma_y = 1400\text{MPa}$, $K_{\text{IC}} = 95\text{MPam}^{1/2}$, $m = 3$, $c = 9.4 \times 10^{-7}$, $k = 2$.

Assuming initial crack size $a_i = 2\text{mm}$, the fatigue crack propagations from initial crack size till final fracture are computed cycle by cycle for parameters of overload as $\lambda = 2 \times 10^{-3}$, $\sigma_o^L = 450\text{MPa}$ and $\sigma_o^U = 470\text{MPa}$. The simulated samples of fatigue crack growth curve are shown in Fig. 2. The dot line in Fig. 2 is crack growth curve without overload and the other three curves with overloads. In the three samples, two final fractures are resulted from peak stress of base-line cyclic loads and the other one from overload. It can be seen from Fig. 2 that the retardation effect of overload is significant.

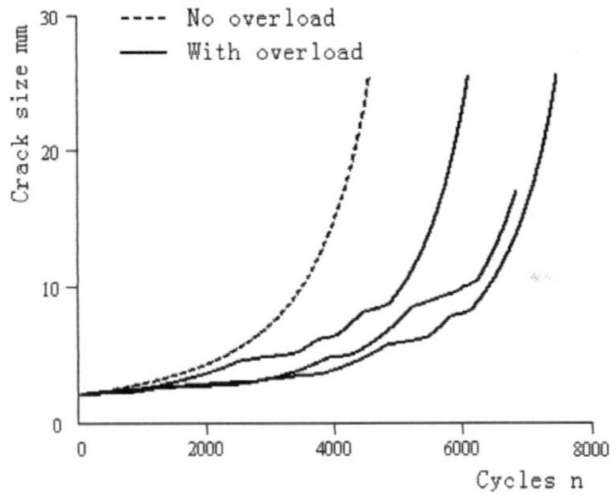

Figure 2. Simulated fatigue crack growth curves.

For studying the influence of overload intensity and magnitude on the fatigue lifetime, 200 samples of fatigue crack growth are computed for every different overload intensity and magnitude. Subsequently, the mean and standard deviation of fatigue lifetimes for various overload intensity and magnitude are obtained, as shown in Table 1. It can be seen from Table 1 that fatigue lifetime increases with overload magnitude. This is because the greater overload is, the more it delays crack growth. Besides, with increase of overload intensity, fatigue lifetime increases at the beginning but decreases afterward. The reason is that the contribution of overload retardation effect to fatigue lifetime is dominant for very small overload intensity but with the increase of overload intensity the possibility of early fracture due to overload increases, which results in decrease of average fatigue lifetime.

Table 1. The mean and standard deviation of fatigue lifetime for various overload intensity and magnitude.

(σ_o^L , σ_o^U)	(400,420)		(400,450)		(450,470)		(450,500)	
λ	Mean	Stad Dev.	Mean	Stad Dev.	Mean	Stad Dev.	Mean	Stad Dev.
9×10^{-4}	4757	298	4860	383	5275	645	5590	862
10^{-3}	4800	315	4889	391	5406	729	5576	897
2×10^{-3}	5086	418	5305	563	6439	911	7181	1372
3×10^{-3}	5458	437	6001	584	7594	1102	8555	1562
4×10^{-3}	5755	474	6429	621	8762	1171	10226	1683
5×10^{-3}	6140	511	6896	738	9618	1407	11498	1865
6×10^{-3}	6385	576	7366	812	10466	1481	12132	2043
7×10^{-3}	6698	656	7790	978	10738	1778	12456	2671
8×10^{-3}	6977	694	8139	1117	11300	2021	12706	3010
9×10^{-3}	7200	798	8461	1285	11186	2425	12673	3064
10^{-2}	7289	938	8562	1418	11041	2773	12142	3476
2×10^{-2}	7242	1686	7881	2245	8128	2556	8697	2894

References

1. J. Schijve, ASTM STP. **700**, 3 (1980)
2. R. Arone, Engng. Fract. Mech. **39**, 895 (1991)
3. R. Arone, Int. J. Fatigue **12**, 275 (1990)
4. J. Dominguez, J. Zapatero, B. Moreno, Engng. Fract. Mech. **62**, 351 (1999)
5. O. E. Wheeler, Basic Engng. ASME **94**, 181 (1972)

 DOI: 10.1051/matecconf/20166105011

Composites Similarity Analysis Method Based on Knowledge Set in Composites Quality Control

Haifeng LI [a]

Dalian Jiaotong University, DJTU, Dalian, China

Abstract. Composites similarity analysis is an important link of composites review, it can not only to declare composites review rechecking, still help composites applicants promptly have the research content relevant progress and avoid duplication. This paper mainly studies the composites similarity model in composites review. With the actual experience of composites management, based on the author's knowledge set theory, paper analyzes deeply knowledge set representation of composites knowledge, improves the similarity calculation of composites knowledge, and builds the composites similarity level model.

1 Introduction

In the actual composites quality control, we comprehensive understanding of the whole process of composites audit and then start thinking in practice about the existing problems about current composites audit. Which, in composites auditing, mostly is simply to declare application composites, and no form to analyze application composites, therefore, appeared on a particular research by the repetitive content, not only waste the money, also waste a great deal of resources.

Therefore, how to explore the research content repetitive composites can improve the effectiveness of the research. In the composites audit need to declare in similarity calculation between application composites and the previous composites, sure the similar degree between application composites and previous composites to judge whether the research content of application composites already improved, this is the deep analysis to the composites audit, need the support of the corresponding algorithms.

2 Knowledge Set Theory

Knowledge set theory is brought out by the author based on the extension theory and the analysis of the characteristics of knowledge, which is a kind of effective knowledge representation method.

2.1 Knowledge Set Related Concepts

Knowledge set (KS) is a collection of knowledge on knowledge carrier, said as the research object of knowledge construction unit. [2]

Various types of knowledge (scientific knowledge, experience knowledge, process knowledge) exist in different types of knowledge carrier (material, person, process). For a project management department, its knowledge system is the sum of this knowledge contained in these knowledge carriers. If take these knowledge in each carrier knowledge sets for a look, then, a knowledge system in project management={KS_1, KS_2, … , KS_n}. So the whole knowledge system can be described by a knowledge structure constituted by the relationship between knowledge sets.

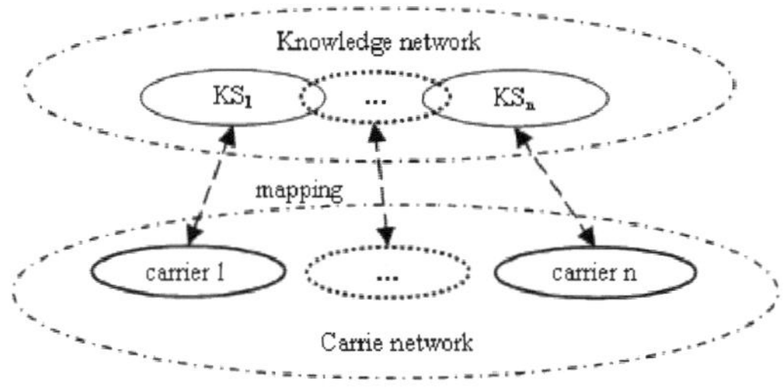

Fig. 1 The Mapped between Knowledge Set and Knowledge Carriers

3 Extension Representation of Knowledge.

3.1 Representation of Knowledge Itself.

With orderly triad KS = (N, C, V) to represent knowledge set, N, C, V three called three elements of knowledge set, including KS, N is knowledge carrier, C is carrier feature set {c_1, c_2,… , c_n}, V is quantum set of N about C {v_1, v_2,…, v_n}. C and V constitute

---

[a] Corresponding author: lihaifeng@djtu.edu.cn

binary group (C, V) called carrier N characteristics yuan, scope of about features c_i written for V (c_i), known as the amount of c_i domain.

Therefore, can define knowledge set as

$$KS = (\mathrm{N,C,V}) = \begin{bmatrix} \mathrm{N}, & c_1, & v_1 \\ & c_2, & v_2 \\ & c_3, & v_3 \\ & \cdots & \cdots \\ & c_n, & v_n \end{bmatrix}. \quad (1)$$

3.2 Representation of Knowledge Process

With orderly triad T = (D, H, U) to represent knowledge process, D is a verb, which means the way of knowledge transformation; H represents the characteristics of verbs, such as the object, object, time, place, degree, method, tool, etc. U is the value of D. So, the process of knowledge transformation can be expressed as:

$$T = (\mathrm{D,H,U}) = \begin{bmatrix} D, & h_1, & u_1 \\ & h_2, & u_2 \\ & h_3, & u_3 \\ & \cdots & \cdots \\ & h_n, & u_n \end{bmatrix}. \quad (2)$$

3.3 Representation of Knowledge Relations

With orderly triad R = (S, A, Z) to represent knowledge relations, where S represents the type of relationship, and A represents the relationship characteristics, Z represents the value of S on A. Well, the relationship between the knowledge sets can be expressed as:

$$R = (\mathrm{S,A,Z}) = \begin{bmatrix} S, & a_1, & z_1 \\ & a_2, & z_2 \\ & a_3, & z_3 \\ & \cdots & \cdots \\ & a_n, & z_n \end{bmatrix}. \quad (3)$$

3.4 Knowledge Set Correlation Analysis Method.

Use symbols ~ to represent knowledge set correlation. If KS_1 and KS_2 related, can remember for: $KS_1 \sim KS_2$. Concrete research, $KS_1=(N_1, c_1, c_1(N_1))$, $KS_2=(N_2, c_2, c_2(N_2))$, if $c_1(N_1) = f(c_2(N_2))$, including f(x) for correlation functions, says KS_1 and KS_2 about features (c_1, c_2) is related , remember $c_1KS_1 \sim f(c_2)KS_2$.

If KS_1 and KS_2 about certain characteristics with $(C_1, C_2)=\{(c_{11}, c_{21}), (c_{12}, c_{22}), \ldots, (c_{1n}, c_{2n})\}$ related, namely as $c_{1i}(N_1) = f_i(c_{2i}(N_2))$ (i=1, 2, …, n), including $F(x)=\{f_1(x), f_2(x), \ldots, f_n(x)\}$ for function sets, then the relationship of KS_1 and KS_2 can be expressed as

$$C_1KS_1 \sim F(C_2)KS_2 = \begin{bmatrix} c_{11} \\ c_{12} \\ \cdots \\ c_{1n} \end{bmatrix} KS_1 \sim \begin{bmatrix} f_1(c_{21}) \\ f_2(c_{22}) \\ \cdots \\ f_n(c_{2n}) \end{bmatrix} KS_2 . \quad (4)$$

Set $W=\{\omega_1, \omega_2, \ldots, \omega_n\}$ to the research weight of the corresponding feature $\{(c_{11}, c_{21}), (c_{12}, c_{22}), \ldots, (c_{1n}, c_{2n})\}$, then the correlation between KS_1 and KS_2, namely knowledge set degree can determine for

$$KsR(KS_1, KS_2) = \sum_{i=1}^{n} f_i(\omega_i) . \quad (5)$$

4 Composites Similarity Analysis Method

4.1 Knowledge Set Representation of Composites

The composites as a knowledge carrier, it contains the knowledge is called composites knowledge, according to knowledge set theory, composites knowledge can be represented as

$$KS_P = (N_P, C, V) = \begin{bmatrix} N_P & c_1 & v_1 \\ & c_2 & v_2 \\ & c_3 & v_3 \\ & \cdots & \cdots \\ & c_n & v_n \end{bmatrix}, \quad (6)$$

where N_P is composites, C is the composites' feature set $\{c_1, c_2, \ldots, c_n\}$, V is quantum set of N about C $\{v_1, v_2, \ldots, v_n\}$.

4.2 Similarity Calculation of Composites Knowledge

There are two composites $KS_{P1}=(N_1, c, c(N_1))$ and $KS_{P2}=(N_2, c, c(N_2))$, if $c(N_1) = f(c(N_2))$, including f(x) for correlation functions, says KS_{P1} and KS_{P2} is related, remember $KS_{P1} \sim f(c)KS_{P2}$; If KS_{P1} and KS_{P2} about certain characteristics with $C=\{c_1, c_2, \ldots, c_k\}$ is related, namely $c_i(N_1) = f_i(c_i(N_2))$ (i=1, 2, …, k), including $F(x) = \{f_1(x), f_2(x), \ldots, f_k(x)\}$ for function sets, hen the relationship of KS_{P1} and KS_{P2} can be expressed as

$$KS_{P1} \sim F(C)KS_{P2} = KS_{P1} \sim \begin{bmatrix} f_1(c_1) \\ f_2(c_2) \\ \cdots \\ f_n(c_n) \end{bmatrix} KS_{P2} . \quad (7)$$

Set $W=\{\omega_1, \omega_2, \ldots, \omega_n\}$ to the research weight of the corresponding feature $C=\{c_1, c_2, \ldots, c_k\}$, then the knowledge set degree between KS_{P1} and KS_{P2} can determine for

$$KsR(KS_{P1}, KS_{P2}) = \sum_{i=1}^{n} f_i(\omega_i) . \quad (8)$$

4.3 Composites Similarity Level Model.

Define composites similarity level matter-element [3] as

$$(R_{PP},G,X)=\begin{bmatrix} R_{PP} & g_1 & X_1 \\ & g_2 & X_2 \\ & \cdots & \cdots \\ & g_n & X_n \end{bmatrix}=\begin{bmatrix} R_{PP} & g_1 & <a_1,b_1> \\ & g_2 & <a_2,b_2> \\ & \cdots & \cdots \\ & g_n & <a_n,b_n> \end{bmatrix} \tag{9}$$

where R_{PP} is the composites similarity degree, g_i(i=1, 2, …, n) is the ith evaluation grade, Interval X_i $=<a_i,b_i>$ is the energy range of the R_{PP} ith evaluation grade g_i.

If the similarity with KS_{P1} and KS_{P2} is $R_{PP}(K_{P1}, K_{P2}) \in X_i$, then i is the similarity level between KS_{P1} and KS_{P2}, and according to the correlation degree function of matter-element theory can calculate the membership degree $K_i(R_{PP}(KS_{P1}, KS_{P2}))$ of the similarity level, and can understand more details of KS_{P1} and KS_{P2} similar degree.

5 Summary

Composites similarity analysis plays an important role in composites review. Composites similarity analysis not only is the necessary means in composites review, but also can help composites applicants know the situation about declare content in the related research, avoid duplication. Composites similarity analysis also provides methods for future composites integration.

Acknowledgements

This research was financially supported by the National Natural Science Foundation of China (Grant NO. 71301016), the Social Science Foundation of Liaoning Province (Grant NO. L15BGL002) and the Outstanding Talent Support Program of Liaoning Province (Grant NO. WJQ2015006).

References

1. X. Xuan and Y. Jiang, "The Thinking of Improving Local Science Operation Mode", Science , vol. 11 (suppl), pp. 66-68, 1997.
2. H. Li, Y. Dang. "Research on Knowledge Connotation and Representation in Science and Technology Composites Management", Composites management technology, vol. 8 (2) , pp. 29-34, 2010.
3. W. Cai, "Extension theory and its application", Chinese Science Bulletin, vol. 44(17), pp. 1538-1548, 1999.

The Research on Materials and Technique of "The lion rolling ball" parcel cloth of Qing Dynasty

Xueling Ren[a], Shaobo Wang

Qingdao University Fine Art College, Qingdao, Shandong, China

Abstract. By studying on the little cotton textures from relics excavation, there is not a clear answer to when the China cotton planting started. Although Qiru area was the north weaving center, there were not so much unearthed fabrics. "The lion rolling ball" parcel cloth of Qing Dynasty is one of the seldom printed cottons. It contains rich culture, life custom and the information of weaving technology with cotton in the Qilu region. From the view of history, this paper analyzes the pattern model, fiber material, weaving technology of "The lion rolling ball" parcel cloth of Qing Dynasty. By comparing it with modern Lu brocade using empirical analysis method, we found that the yarns in fabric is cotton, one of ancient Asia cottons, which showed"The lion rolling ball" parcel cloth of Qing Dynasty played an important role in the history of cotton clothes.

1 Introduction

In history, there is no clear answer to when the cotton was originate and introduced. But it is confirmed that the cotton products were considered treasure at the central plain area of China from 6th century to 12th century, and they were used only by the ruling class[1]. The cottons planted in Qiru area were Asia cotton[3] introduced from China south at the end of Song Dynasty and the earlier of Yuan dynasty[2].

Qiru area has been an important weaving center since ancient times. Zhoucun, as the "Land wharf" of ancient silk, was ever the biggest place of silk and textile, dyeing and sales in north China; Changyi of Weifang was the famous hometown of silk, well-known city in history. The silk produced inLiutuan of Changyi was exported regularly, then Liutuan's silk craft was certified as the non-material cultural heritage. Zhoucun and Changyi, as the important silk cities, laid a strong foundation for the growth of textile printing and dyeing in Qiru area[4].

Because cotton clothes were used by the poor in old time, and the weather in Shandong was not suitable to keep the cotton fabrics, cotton texture was seldom remained. So "The lion rolling ball" parcel cloth of Qing Dynasty is also very rare to us.

By analyzing the fiber material, pattern model of "The lion rolling ball" parcel cloth of Qing Dynasty, this paper studies its historical value, material characteristics and the relationship with the modern cotton texture.

2 The profile of "The lion rolling ball" parcel cloth of Qing Dynasty

2.1 Background

After looking up lots of documents and spot investigations, we learnt that the ancient cotton parcels were kept in some folk weaving artisans. Between 2006~2009 years, the author went to Changyi to do field investigation to search for the ancient cotton textures, and study the weaving skeleton and advanced dyeing culture.

During the investigation at Changyi in Feb. 2002, we succeeded in finding the aged women, Che xiuyun, who was born 8th years of the Republic of China. She is fromLiutuan county of Changyi City, and comes from one famous weaving family. Che xiuyun gave "The lion rolling ball" parcel cloth of Qing Dynasty, which she inherited from her mother, to us for further studying.

2.2 The brief of "The lion rolling ball" parcel cloth of Qing Dynasty

The pattern structure of "The lion rolling ball" parcel cloth of Qing Dynasty is axisymmetric, printed finely, and weaved delicately. It is the precious material for studying cotton fabrics of Qing Dynasty, and is also the typical specimens for studying cotton printing and dyeing technolgy.

According to Che xiuyun's memory, we know that this parcel is past to her through generations, and it is at least 300 years old up to now. It date bact about at the earler of 18 centry(Fig.1).

[a] Corresponding author :rxlz999@163.com

Fig. 1 cotton textile dye printing parcel pattern(collection of the author)

Only half of this parcel is left, and 36cm width, 106cm length. So the complete parcel should be 71cm X 106cm(1cm seam). The parcel was weaved manually, single-color printed. Although it is a little soiled and broken, the condition is good. It is white cloth, purple printed(faded blue), and the pattern is clear.

3. Analysis of "The Lion Rolling Ball" Parcel Of Qing Dynasty

3.1 Analysis of the parcel pattern

(1) "The lion rolling ball": Parcel pronounces "Baofu" in Chinese , which means good fortune,and people living in Qiru area think that the bigger the parcel is, the more fortune will get. This also implicits good wishes to daughter from her family. So the selected patterns always pose auspicious messages. For examples, "phoenix passing through peony" implicts beauty, brightness, happiness and richness(Fig. 2), " Carps Jumping over the Dragon Gate" implicts social climbing and good luck, " bursting Pomegranate" implicts more descendants, and more blessedness(Fig. 3), " lion rolling silk ball" means that lions send auspiciousness, and " two lions rolling silk ball" symbolizes the human reproductive ceremony[5].

Fig. 2 phoenix passing through peony

Fig. 3 bursting Pomegranate

(Original author: Zhang mingjian, folk artisan, famous inheritance man of China folk culture. Collected by the author of this paper)

(2) Characteristics of pattern: The pattern of "The lion rolling ball" parcel cloth of Qing Dynasty is full of luck message(Fig. 1). Based on the central axis, pattern is symmetrical on two sides. Two lions with different shapes and a pair of rolled silk balls compose of the main dermatoglyphic pattern. From inner to outside are the two-continual ribbon flower made up of inner edges, auspicious cloud, fret pattern and outer edges in turn. "The lion rolling ball" is one symbol of human reproductive ceremony in folk. The decorated auspicious cloud on lions symbolizes auspicious wishes, the repeated copper cash patterns inside the rolled silk ball means profits pouring in from all sides, and the surrounding lucky cloud pattern expresses tranquil and happy. The fret patterns are geometrical veins which are derived from thunder veins on pottery and bronze and their continuos distribution means successive rich. The complete pattern symbolizes auspiciousness and affluence[6].

In the selection of topic, "The lion rolling ball" parcel of Qing Dynasty is in accordance with the modern "The lion rolling ball" parcel. But they are a little difference in technique of expression. The main veins of "The lion rolling ball" Qing Dynasty are representational, the line style is elegant and smooth, and the atmosphere of picture is vivid; In other side, the main veins of modern JiNing color printed "The lion rolling ball" parcel are more patterning, the line style is standardized. The atomospher of picture is passinated with some solemnity.

3.2 material analysis of the cotton fiber

According to Che xiuyun's description, "The lion rolling ball" parcel cloth of Qing Dynasty is pure cotton with flat grain, printed in blue pattern. We tested the yarn diameter and size under DXS-19A microscope. The pictures under digital 3D video fiber measure system-HZ SCOPE are following. (Fig. 4, Fig. 5)

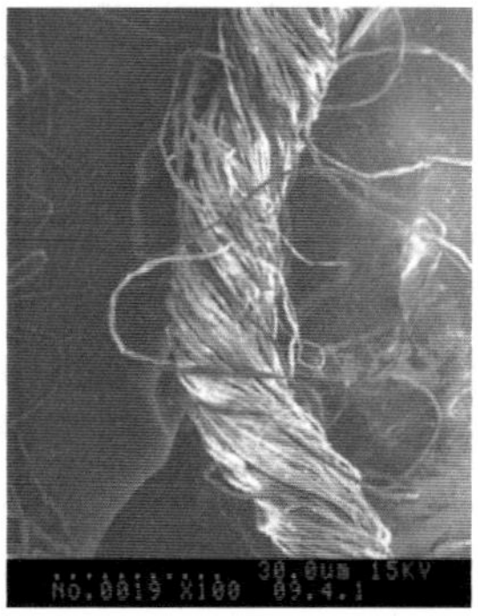

Fig. 4 100X cotton parcel yarn of Qing Daynasty

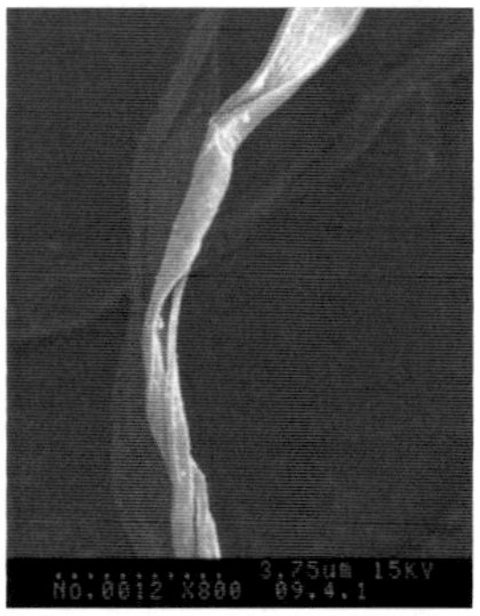

Fig. 5 800X cotton parcel yarn of Qing Daynasty(by DXS-10A electronic microscope)

"The lion rolling ball" parcel cloth of Qing Dynasty was always weaved using one-ply yarn with weak S-like twists. Under microscope, fiber is thick and short, and more matured fibers exist. The fibers have kinds of shape such as natural twist, helix with corner; Some fibers are judged as pure cotton texture without twists. Before 19 century, cottons planted in Shandong were all short fiber Asia cotton. From this we conclude "The lion rolling ball" parcel cloth of Qing Dynasty was weaved with Asia cotton[7]. (Fig. 6, Fig. 7)

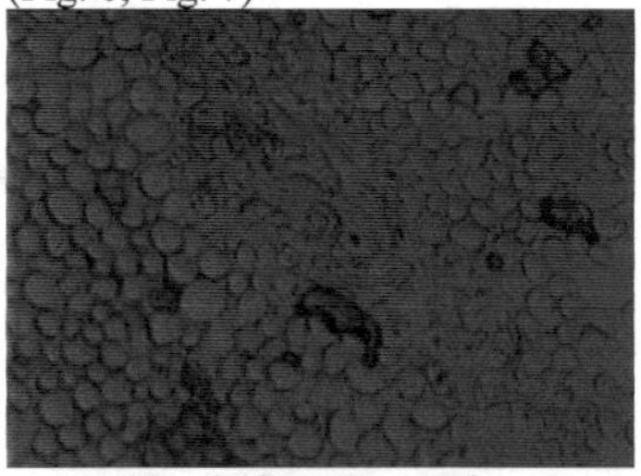

Fig. 6 100x ross section

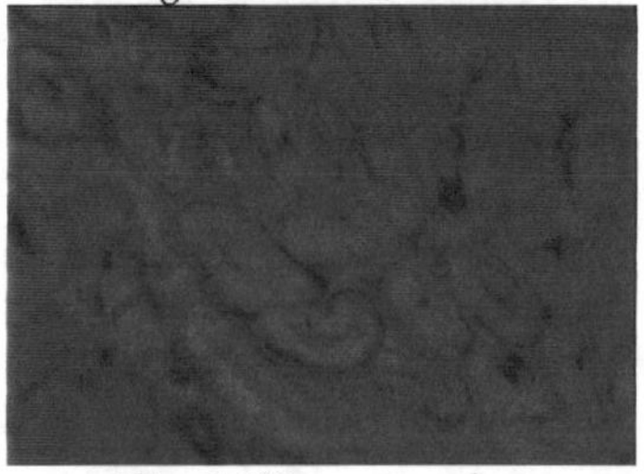

Fig. 7 400x ross section

With the help of digital 3D video fiber measure system – HZ SCOPE, the yarn trunk of "The lion rolling ball" parcel of Qing Dynasty has nearly no difference with modern Lu brocade in yarn diameter and uniformity. So we can come to a conclusion that their weaving tecnology is simlar(Fig. 8, Fig.9, Fig. 10). Detail specification is shown in Table 1. From these values we can predicate that the weaving technology is so advanced in Qing Dynasty.

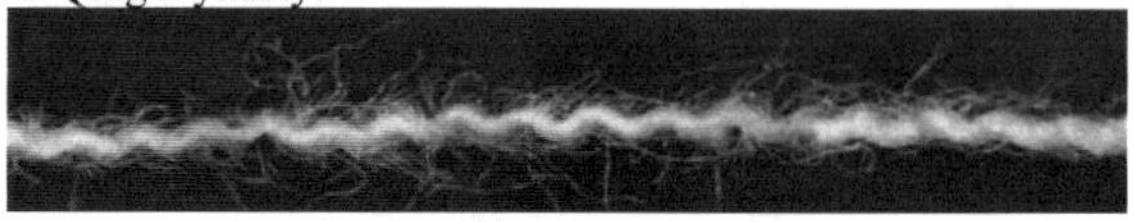

Fig. 8 the cotton of "The lion rolling ball" parcel of Qing Dynasty

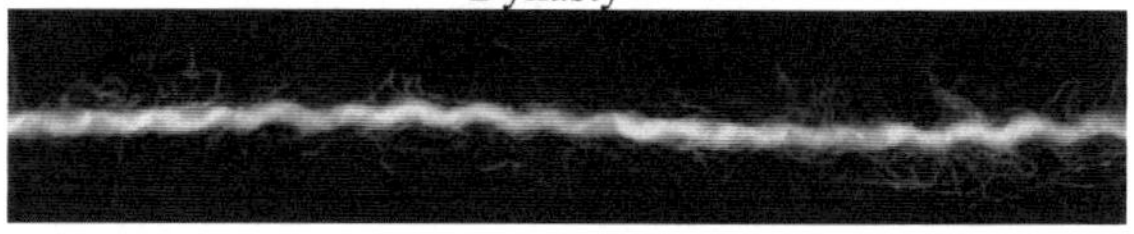

Fig. 9 modern Lu brocade

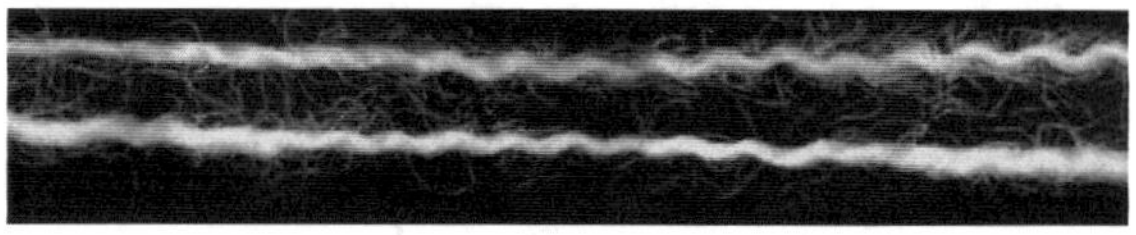

Fig. 10 the yarn comparison of "The lion rolling ball" parcel of Qing Dynasty with modern Lu brocade (by digital 3D video fiber measure system. Above is the cotton from "The lion rolling ball" parcel of Qing Dynasty, and the under is modern Lu brocade)

Table 1. the diameter and text value of "The lion rolling ball" parcel of Qing Dynasty and modern Lu brocade

specimens	*Yarn daimeter(μm)*	tex
"The lion rolling ball" parcel of Qing Dynasty	230	42.37
Modern Lu brocade	220	41.89

3.3 analysis of the weaving technology

Observed under microscope, "The lion rolling ball" parcel cloth of Qing Dynasty was plain woven fabrics, and the yarn thickness is not so even to indicate that it is made by spinning with manual twirling.

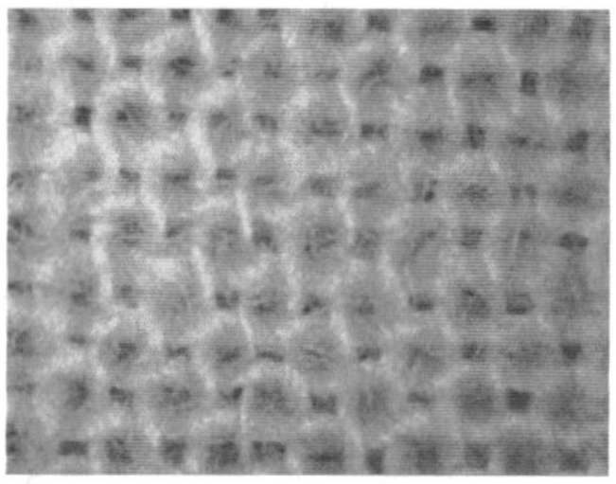

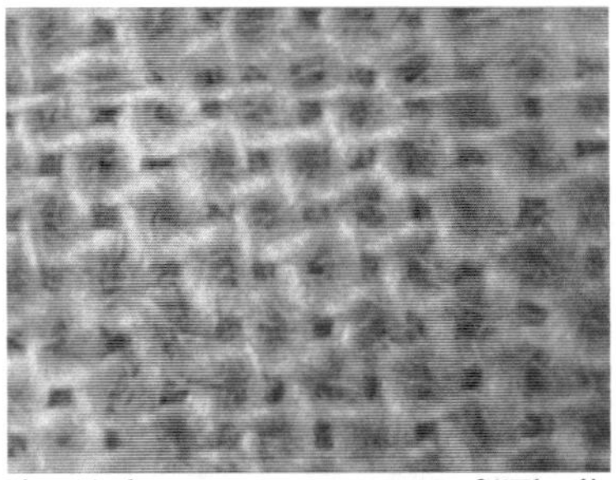

Fig. 11, Fig. 12 the yarn arrangement of "The lion rolling ball" parcel cloth of Qing Dynasty

The warp density and weft density of "The lion rolling ball" parcel cloth of Qing Dynasty are very even. Watched on the surface, Both the warp yarn and filling yarn are very straight, the warp density and weft density is very big, which warp density is 225strings/10cm and weft density is 220strings/10cm, and the size of parcel is 36cm, which is smaller than that of modern Lu brocade which is about 50cm wide. So we conclude that the manual weaving technology in Shandong at the end of Qing Dynasty is not lower than the modern manual weaving technology [8].

4. Summary

By analyzing the material of "The lion rolling ball" parcel cloth of Qing Dynasty, the author concludes that the parcel cloth is made of ancient Asia cotton; After studying the parcel yarn by instruments, we can say that at the beginning of using cotton as weaving material, cotton weaving technology is already very mature and advanced. By analyzing the parcel pattern, the author consider that "The lion rolling ball" parcel cloth of Qing Dynasty includes rich cultural, historical and technological information, and it is one strong evidence to study ancient cotton textile.

References

1. Wenbing LI. Shandong Cotton[M] Shanghai: Shanghai Science and Technology Press, 2011,pp.1,pp.2.
2. Jie SONG, Yinnan WU, etc,. Cotton Logging of LiaoCheng, Beijing: China Science and Technology Press, 1992, pp.1.
3. Zhen WANG. Nong Shu, Beijng: the Commercial Press, 09.28 of the Republic of China(1939.9).
4. Chronicle log of Shandong 18, agriculture part. Compilation committee of Shandong province. Jinan: ShangDong People's Publishing House. 2000.
5. Gu HUI.History of China silk fabric, Harbin: Heilongjiang Press, 1990, pp.148.
6. Feng ZHAO. History of silk art. Zhejiang art college Press, 1992,pp.157.
7. Weiji CHEN.History of China textile technology, Beijng: Science Press, 1984,pp.313-315.
8. Xueling REN. The Study on Culture, Art and Textile Technology of Lu Brocade .Donghua University Press,2012,pp.135.

Research on Microscopic Characteristics of Carbonate Reservoir Fracture fillings Based on Microscopic Infrared Spectroscopy

LI Jing[1,a] , ZHOU Han-guo[2, 3] , Yang Feng[4] , LIU Si-meng[5] ,PENG Cheng-le[1] , GUO Tian-tian[1]

[1]*College of Pipeline and Civil Engineering in China University of Petroleum, Qingdao 266580, China;*
[2]*Investment Development Department of SINOPEC Shengli Oilfield Company, Dongying, Shandong 257000, China;*
[3]*College of oil and natural gas engineering in Southwest Petroleum University, Chengdu Sichuan 610500, China;*
[4]*Research Institute of Petroleum Engineering of SINOPEC Shengli Oil Field. Dongying, Shandong 257000, China;*
[5]*China Petroleum Pipeline Bureau , Langfang, Hebei 065000, China*

Abstract. The characteristics of hole and seepage of carbonate rocks depend on the structural components of carbonate rocks and the combination relationship with these structural components under multi-scale. The quality of filling in reservoir fracture is an important part of fracture description. In this paper, the method of microscopic infrared spectral imaging was used, and the spectra cubic database was established by means of spectral imaging experiment. Each single band image which characterizes the related substances was extracted and the microcosmic characters of the filling in carbonate reservoirs fracture from Wumishan formation in Renqiu oilfield were studied. The spatial distribution maps of different chemical composition and groups in the core sample were obtained. The distribution of different hydrocarbon in the fracture of the reservoir and the connectivity of the pores etc. were better revealed
Keywords. Carbonate reservoirs; fracture; microscopic infrared spectral image.

1 Introduction

In the process of oil and gas exploration and development, fracture-cavity filling and connectivity in fracture-cavity carbonate reservoir is an important research topic [1-3]. The classic techniques commonly used are optical microscopy and X-ray diffraction, and often supplemented by scanning or transmission electron microscope. Although these detection methods have some achievements in the research of micro morphology, the common deficiency is that they can not distinguish the homogeneous or determination of partially ordered degree and other phase information, mainly because they stress on the spatial morphology analysis and lack of energy spectrum (mainly spectrum) analysis [3]. In the research of fracture fillings, the experimental method named imaging spectroscopy technology is urgently needed, which has the ability of spatial resolution and spectral resolution. The carbonate reservoir fractures of buried hill in Renqiu and microscopic characteristics of its fracture fillings were researched by microscopic infrared spectral imaging method, which provides a new way to fully grasp the physical properties of reservoir.

2 Imaging spectroscopy

In recent years, newly developed imaging spectroscopy system has combined spectroscopy and imaging technology organically [3] .The basic principle is to make spectrum (λ) for each pixel space site (X, Y) of the target, to form a so-called spectral cube (X, Y, λ), as shown in Fig.1.

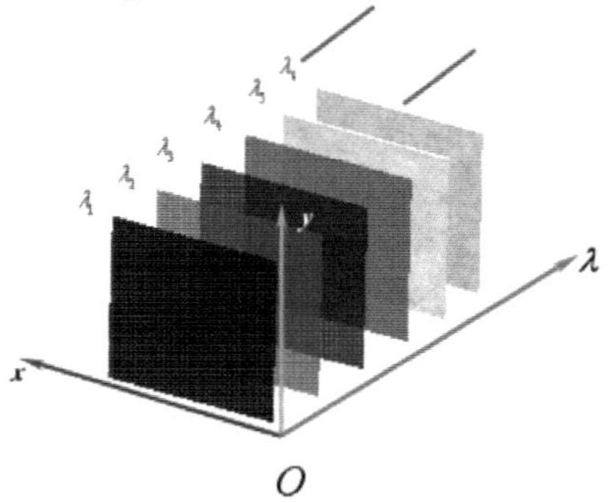

Fig.1 The sketch map of spectrum cube in imaging spectrum

It is can be seen from Figure 1 that the core of the imaging method is the integration of spatial information and spectral information of the target, which can analyze and identify the target in two aspects of the spectrum and space [3].

Fluid inclusion is one of the most important research objects in the study of fracture fillings[5,6]. Because of the mobility of fluid itself, it is only has high requirements of spectral resolution to keep the uniformity of the space in the range of entire fluid inclusions (especially fluid inclusions device of artificial simulation).

[a] Corresponding author: lijingupc@163.com

3 The microscopic characteristics of carbonate reservoir fracture fillings in Renqiu Oilfield

3.1 General geology of the study area.

Renqiu oilfield is located in the transitional area between northern Central Hebei Depression Raoyang Sag in the west of Bohai Bay Basin and Baxian Depression area. It's a buried hill oilfields with Lower Palaeozoic and Middle-Upper Proterozoic Erathem carbonate as reservoir, the second and third segments of lower tertiary of Shahejie formation as hydrocarbon source rock, and the first segment of Shahejie formation and the Dongying formation for cover. Renqiu oilfield is one of the large and high producing ancient buried hill oilfields in the world. Renqiu ancient buried hill is 30km from north to south, 7km from east to west, and the area is about 183km2. Jixian system Wumishan formation is the most important oil producing layer in Renqiu buried hill. The core samples taking from Ren 28 well of Wumishan formation in Renqiu oilfield were analyzed in this work.

3.2 Infrared imaging spectrum experiment.

The sample used in the experiment came from Ren 28 well of Renqiu oilfield in Central Hebei Depression, which belong to the Upper Proterozoic-Jixian system-Wumishan formation layer. The well depth is 3227.9m and the mainly sedimentary facies is a shallow marine. The lithology is grey brown dolomite. The instrument used in this infrared spectral imaging experiment is Spotlight400/400N Fourier transform infrared / near-infrared imaging system which is produced by American perken-Aylmer company. The wavelength range of imaging mode is 7800-720cm-1, and the wavelength range of single point mode is 7800-600cm-1. In order to locate a sample area of infrared spectral analysis image in the infrared spectral imager, it is need to select and delineate on the black and white image displayed on a computer monitor visible with a red rectangular box. The results were shown in Figure 2. IR imaging experiment results of all sites in the sample area encircled by the red box in Figure 2 were stored in the computer, which forming the spectral cube database were shown in Figure 1. According to the need, it is easy to extract different information from the cube database obtained previously.

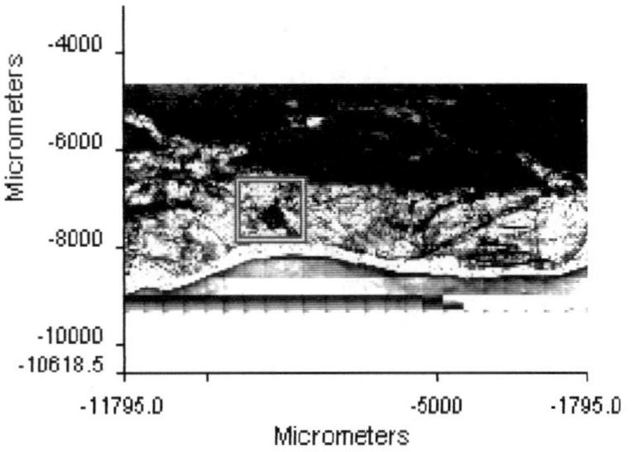

Fig.2 Visible black-white images of sample B008 of Ren28 (Red casing express the region of infrared imaging spectrum)

3.3 Experimental results and analysis

The infrared imaging spectral characteristics of the sample B008 of Ren 28 were studied as follows.

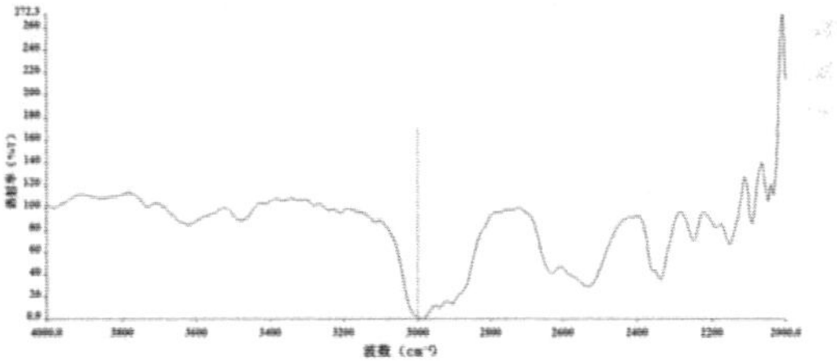

Fig.3 The infrared imaging spectrum of one space site of sample B008 of Ren 28

Figure 3 express the infrared spectrum of a certain C in the red box in Figure 2, i.e. the spectral intensity I (X_0, Y_0, λ) of a certain site (X_0, Y_0) in the spectral cube in Figure 1 changes along the wavelengthλ axis of the spectral cube. So the chemical composition of this site can be known. But it is only the spectral distribution of a certain site in the sample, which is not representative of the spectral distribution of various sites in the sample.

It is can be seen from Figure 3 that the infrared absorption spectrum (expressed by optical transmission spectra) of site C (shown in Figure 4) mainly includes symmetric and asymmetric stretching vibration absorption band of CH3–(methyl) and CH2–(methylene). It is indicates that the absorption of organic saturated hydrocarbons is in the dominant position of site C.

On the other hand, Fig. 4 indicates the absorbance spatial distribution of all the sites in the red region in Fig. 2. The absorbance intensity is expressed in a pseudo color, and on the right side of the diagram is a intensity scale. However, the absorbance of each site is the absorbance after spectral integration in the range of the entire measured spectrum, and the absorbance is the sum of all the substances in the sample. Since no spectral resolution, it is not known exactly what the chemical composition of the absorption. Therefore, it is necessary to seek the spectral images with spectral resolution according to the spectroscopy knowledge of the related substances.

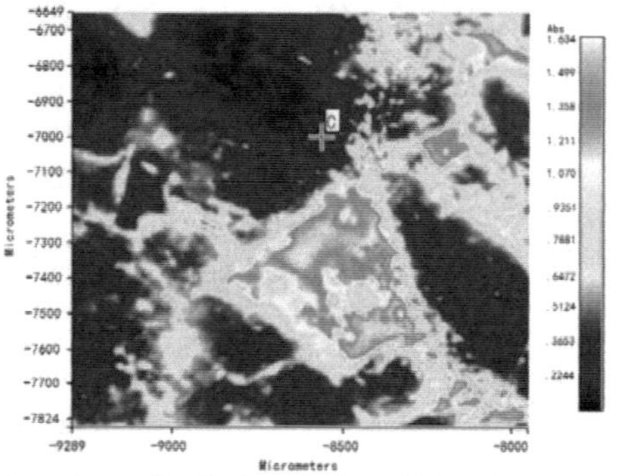

Fig.4 Space distribution imaging of infrared spectrum integral absorbance of sample B008 of Ren 28

As stated above, Fig. 4 is a pseudo color image using the integral value of the absorbance of the spectrum at the whole band. It is not only refers to the absorption of hydrocarbons, more are not the absorption of a specific chemical group in the hydrocarbons. It can not truly reflect the spatial distribution of oil and gas.

According to the spectral cube (as shown in Fig. 1) database obtained from the experimental measurements of infrared imaging spectra, each site (or pixel) of the sample select a specific chemical group in hydrocarbon, whose peak intensity of the absorption spectra or area integral value of absorption bands is performed imaging (in pseudo color). Fig. 5 and Fig. 6 express the spatial distribution of the sample at a single wavelength (or wave number).

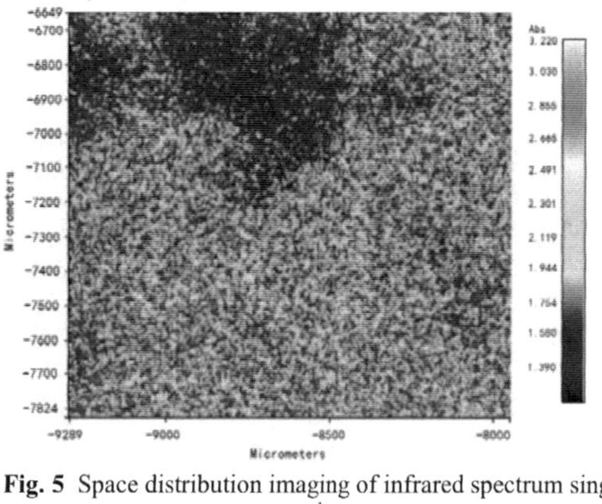

Fig. 5 Space distribution imaging of infrared spectrum single wave number（$3000cm^{-1}$）absorbance of sample

Fig. 5 is a single wave number spectrum imaging map, which characterizes CH_3–(methyl) and CH_2–(methylene) wave number of $3000cm^{-1}$. It shows that there are only a few small area in the sample have the group.

It should be pointed out that the absorbance of the sample is proportional to the peak of the spectral absorbance or the absorption band area. Therefore the absorbance scale (in pseudo color) of all imaging spectrum (whether single wave number, or spectral integral) has only relative significance before standard sample calibration. But in the characterization of the absorbance (chemical composition) on the relative distribution of space it is meaningful.

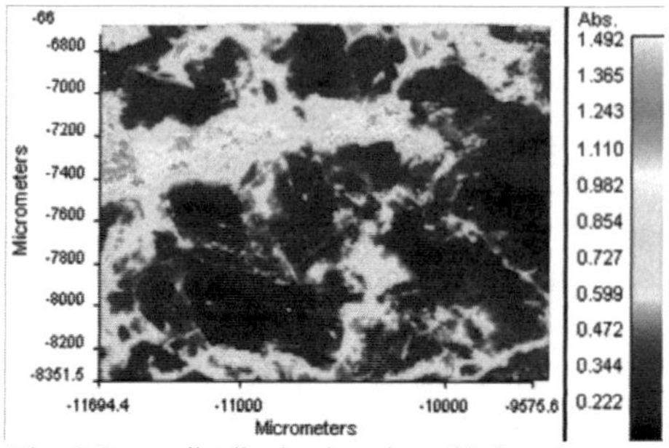

Fig. 6 Space distribution imaging of infrared spectrum single wave number（3500cm-1）absorbance of samples

Fig. 6 is a single wave number spectrum imaging map, which characterizes containing intermolecular or intermolecular association hydroxy wave number of $3500cm^{-1}$. It has strong absorption, indicating the group is obviously existence in the crack area of the sample [4]. This indicates that there is hydrocarbon organic matter in the crack area of the sample. It can be speculated that there has oil gas migration in the crack.

4 Summary

Analyzing the spatial distribution of chemical composition of the fracture fillings can not only be judged by the spectral characteristics of several sites of the samples, or only depends on the absorbance spatial distribution of the spectral integral of the samples. It can be achieved with the help of the spectral cube database established by spectral imaging experiment, continuously extracting the single spectrum band images of characterizations of related substances. A new method for reservoir characterization is provided by the study of infrared spectral imaging characteristics of carbonate reservoirs fracture fillings.

Acknowledgements

This work was financially supported by National Natural Science Fund of China(41272141) and National Great Science Fund of China(2016ZX05002-002).

References

1. Graham Wall B R, Girbacea R, Mesonjesi A, et al. . *Evolution of fracture and fault-controlled fluid pathways in carbonates of the Albanides fold-thrust belt.*(AAPG Bulletin) **90**, 8(2006)
2. Popp elreiter M, Balzarini M A, De Sousa P, et al. *Structural control on sweet-spot distribution in a carbonate reservoir: Concepts and 3-D models*(Cogollo Group, Lower Cretaceous, Venezuela, AAPG Bulletin) , **89**,12(2005)
3. LI Jing. *Research on Microcosmic Analysis and Fracture Prediction of Carbonate Reservoirs in Renqiu Buried Hills*(Paper of doctor's degree, China University of Petroleum, Qingdao,China), (2010).

The Assessment of Comprehensive Vulnerability of Chemical Industrial Park Based on Entropy Method and Matter-element Extension Model

Yan Jingyi[1, a], Zhang Lijing[1, b], Tao Gang[1, c], Jiang Shu[2, d] and Guo Hanyu[1, e]

[1]College of safety science and engineering, Nanjing Tech University, Nanjing
[2]Nanjing boiler and pressure vessel inspection institute, Nanjing

[a]330720125@qq.com, [b]zhanglj@njtech.edu.cn, [c]taogangs@163.com, [d] shujiang@njtech.edu.cn, [e]781994362@qq.com

Abstract. The paper focuses on studying connotative meaning, evaluation methods and models for chemical industry park based on in-depth analysis of relevant research results in China and abroad, it summarizes and states the feature of menacing vulnerability and structural vulnerability and submits detailed influence factors such as personnel vulnerability, infrastructural vulnerability, environmental vulnerability and the vulnerability of safety managerial defeat. Using vulnerability scoping diagram establishes 21 evaluation indexes and an index system for the vulnerability evaluation of chemical industrial park. The comprehensive weights are calculated with entropy method, combining matter-element extension model to make the quantitative evaluation, then apply to evaluate some chemical industrial park successfully. This method provides a new ideas and ways for enhancing overall safety of the chemical industrial park.

Keywords: Comprehensive vulnerability evaluation; Chemical industrial park; Entropy method; Matter-element extension model

1. Introduction

Chemical industrial park, as the most significant embodiment of chemical concentration in recent years, brings more obvious problems. The most dangerous is high risk-taking, so it is benefit for chemical industry park to measure overall safety situation by comprehensive vulnerability evaluation based on entropy method and matter-element extension model. Through effective method, it can figure out the level of vulnerability for providing compensation measures to avoid potential risk.

The earliest studies of vulnerability can be traced back to natural disasters 0. Now, it is becoming a hot point in evaluation of ecosystem and environment. P. Timmerman first proposed the concept of vulnerability which means the extent of a system may produce adverse effect after disasters 2. Li He etc. systematically analyzed the concept of vulnerability 3. Li Fengying etc. researched the vulnerability of Nanjing chemical industry based on GIS 4. Shi Peijun proposed broad and narrow disaster vulnerability assessment model 5. Li Qiujin analyzed vulnerability's concept and principle of chemical industrial park 6. Tan Zhaoyang applied fuzzy comprehensive method and PCA method to build area vulnerability by SVM (Support Vector Machine) 7. Chemical industrial park as target of regional vulnerability, it is necessary to introduce the comprehensive vulnerability assessment. At present, evaluation are most based on risk and it seldom considered the recovery of facilities, personnel density and other factors, however, the assessment of comprehensive vulnerability of chemical industrial park will benefit to monitor the dynamic management of chemical industrial park. Therefore, in this paper, it uses improved analytic hierarchy process (IAHP) combined with entropy method to determine the weight of each index layer, and then applies to Shizhuang chemical industrial park to get each level's correlation and the final vulnerability level of this chemical industrial park.

2 Matter-element Extension Model and Method

Matter element analysis method is used to solve the study of incompatibile problems, but this is often a difficult point in study, because it must consider things, characteristics and the corresponding quantity, and make a formal process to resolve conflicts. Currently, the matter element analysis method has gradually been applied in engineering fields including chemical industrial parks [8,9]. But the concept of vulnerability in chemical industrial park is fuzzy, so many comprehensive vulnerability assessment design index, and often the incompatibility between single evaluation index, so this article will be based on the fuzzy matter-element extension model to build the suitable

*Corresponding author: zhanglj@njtech.edu.cn

comprehensive vulnerability evaluation indexes for chemical industrial park.

3 Establish the Fuzzy Matter-element Matrix

We will take the vulnerability of chemical industrial park as the evaluation object, namely the context in the matter-element analysis, represented by *M*. The set of evaluation index is represented by *C*, its value is *u*. For *M*, which has n index, then the set *C* is expressed by *C*, $C = (C_1, C_2, \cdots, C_n)$, assuming that the *n*-th value of evaluation index C_n is u_n, so $u = (u_1, u_2, \cdots, u_n)$. The evaluation object *M*, the evaluation index set *C*, the corresponding evaluation indexes values *u*, these three components form an orderly triple $R = (M, C, u)$, which is classical matter-element matrix and used to describe the basic elements to be evaluated.

4 Determine the Classical Domain and Segment Field Matter-element Matrix

Chemical industrial park comprehensive vulnerability classic domain matrix can be $R_j = (M_j, C_n, u_{jn})$. In the equation, the evaluation object *M* will be divided into *j* evaluation levels, namely $M = (M_1, M_2, \cdots, M_j)$. So, each corresponding evaluation index C_n also has *j* evaluation levels, then $u_{jn} = (a_{jn}, b_{jn})$ is the standard interval for the evaluation index (evaluation parameters) of M_j, and satisfies $(a_{j1}, b_{j1}) \cap (a_{jn}, b_{jn}) = \emptyset$, (i≠j), said any two standard intervals are incompatible. Classical field matter element matrices can be expressed as:

$$R_j = \begin{bmatrix} M_j & C_1 & u_{j1} \\ & C_2 & u_{j2} \\ & C_3 & u_{j3} \\ & \vdots & \vdots \\ & C_n & u_{jn} \end{bmatrix} = \begin{bmatrix} M_j & C_1 & (a_{j1} b_{j1}) \\ & C_2 & (a_{j2} b_{j2}) \\ & C_3 & (a_{j3} b_{j3}) \\ & \vdots & \vdots \\ & C_n & (a_{jn} b_{jn}) \end{bmatrix} \tag{1}$$

M_j indicates the scope of all the evaluation level. In order to give a section domain matter-element matrix, we need to do the extension analysis on the standard interval of each evaluation level for the index C_k, to obtain the some development interval $u_{pk} = [a_{pk}, b_{pk}]$, become the segment field *k*. The expanding range$[a_{pk}, b_{pk}]$ should at least contain all the standard intervals of evaluation index C_k, means it meets the condition $[a_{pk}, b_{pk}] \supseteq \cup_{1 \le i \le j}[a_{jn}, b_{jn}]$. To give u_0 a specific value, then call u_0 the specific value of C_n for R_0, that the actual value of the evaluation index. The matter-element matrix of the vulnerability in chemical industrial park is expressed as$R_0 = (M_0, C_n, a_{0n} b_{0n})$.

5 Determine Correlation Function and Correlation Degree

In order to characterize 'relevant degree' between an element and a specific element interval, introduces the concept of 'function distance' in functional analysis to quantify the 'distance' from point to interval. Thus, the distance from the point x_i (value of the evaluation indexC_n) to the corresponding feature vector interval $x_{0ji} = [a_{0ji}, b_{0ji}]$ is represented by $\rho(x_i, x_{0ji})$.

$$\rho(x_i, x_{\rho i}) = \left| x_i - \frac{(a_{0ji} + b_{0ji})}{2} \right| - \frac{(b_{0ji} - a_{0ji})}{2} \tag{2}$$

With $|x_{0ji}| = |b_{0ji} - a_{0ji}| (i = 1,2, \cdots, n; j = 1,2, \cdots, m)$; x_i, x_{0ji}, $x_{\rho i}$ respectively with the quantity of chemical industrial park vulnerability, the scope of the classical field matter-element and the range of section domain matter-element's value.

We will expand the interval $[x_i, x_{oji}]$ to $[y_i, y_{oji}]$, and make it meet $[x_i, x_{oji}] \subset [y_i, y_{oji}]$, the interval expanded $[y_i, y_{oji}]$ is called *Y*, for further membership characterization between x_i and $[x_i, x_{oji}]$ interval, combined with the distance function, we can calculate the correlation $K_i(x_i)$ between x_i and $[x_i, x_{oji}]$, and determine the associated degree according to its value, shown as Eq.3.

$$k_i(x_i) = \begin{cases} -\dfrac{\rho(x_i - x_{0ji})}{|x_{0ji}|}, x_i \in X_{0ji} \\ \dfrac{\rho(x_i, x_{0ji})}{\rho(x_i, X_{\rho i}) - \rho(x_i, x_{0ji})}, x_i \notin X_{0ji} \end{cases} \quad and\ X \in Y \tag{3}$$

6 The Correlation Function of Each Level for Evaluation

If $K_j^i(P_0)$ indicates the correlation between the actual value of the j-th evaluation index C_n and the i-th measurement standard interval for the evaluation object, a_i indicates the weight of each evaluation index, then meet the equation:

$$K_j(P_0) = \sum_{i=1}^{n} a_i K_j(x_j) \tag{4}$$

If $K_j(x_j) = \max K_j(x_j), j = 1,2, \cdots, m,$ then the chemical industrial park vulnerability belong to level *j*. When $K_j(P_0) > 0$, said chemical industrial park comprehensive vulnerability P_0 is in the some standard level range, conforms to the characteristics of this level, the size of its value is proportionate to the conformity degree; when $-1 \le K_j(P_0) \le 0$, said chemical industrial park comprehensive vulnerability P_0 is beyond the stone village standard level range, deviate the characteristics of this level, but have the conversion conditions, under certain conditions can conform to the standard level, and the greater its value is, the more easy to convert; when $K_j(P_0) < -1$, said chemical industrial park comprehensive vulnerability P_0 is beyond the stone village standard level range, and can't convert to the standard level 10.

7 Determine Relevant Weight

This paper adopts Improved Analytic Hierarchy Process (IAHP) to quantify the incidence for each index layer and strengthen the reliability of results by entropy method to modify weight. Entropy is an ideal scale in different decision-making process. The less uncertainty, the lower entropy 11.

For positive indexes:

$$P'_{i,j} = \frac{P_{i,j} - \min P_{i,j}}{\max P_{i,j} - \min P_{i,j}} \quad (5)$$

For negative indexes:

$$P'_{i,j} = \frac{\max P_{i,j} - P_{i,j}}{\max P_{i,j} - \min P_{i,j}} \quad (6)$$

Calculate the entropy of j-th, the equation is expressed as:

$$e_j = -k \sum_{i=1}^{n} \frac{p_{ij}}{a_j} ln\left(\frac{p_{ij}}{a_j}\right) \quad (7)$$

With $a_j = \sum_{i=1}^{n} p_{ij}; k > 0, k = \frac{1}{ln(n)}, e_j \geq 0; j = 1,2 \cdots, n$

Calculate the difference coefficient of j-th, the equation is expressed as:

$$g_j = \frac{1 - e_1}{n - E_e} \quad (8)$$

With $E_e = \sum_{i=1}^{n} e_i, 0 \leq g_i \leq 1$

Using the weight to correct W_j to get the final weight, the equation is expressed as:

$$\theta_j = \frac{W_j w_j}{\sum_{j=1}^{n} W_j w_j} \quad (9)$$

8 Build the Comprehensive Vulnerability Assessment System of Chemical Industrial Park

8.1 Set Vulnerability Evaluation Indexes and Level of Chemical Industrial Park

There are many influence factors for comprehensive vulnerability evaluation of chemical industrial park and they are divided to three dimensions by Vulnerability Scoping diagram (VSD). It can be concluded from analysis that dominancing and causing vulnerability's factors can be summarized as two categories for chemical industrial park, structural and threaten vulnerability indexes6. The 21 index layers are established shown as Fig.1.

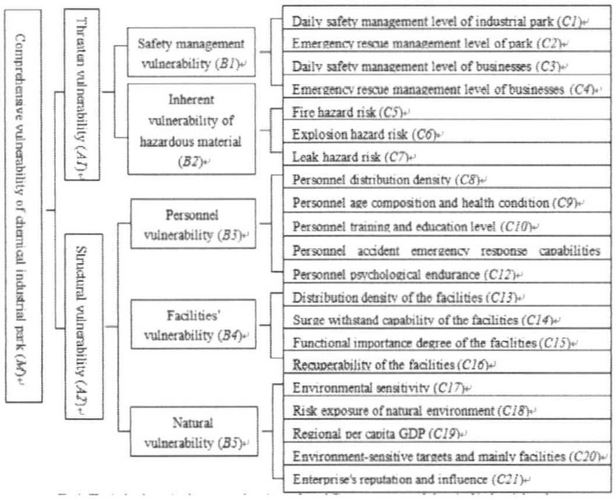

Fig.1 The index layer in the comprehensive vulnerability assessment of chemical industrial park

These 21 indexes cover all most evaluation indexes of chemical industrial park and it provides an important basis for measuring the chemical industrial park comprehensive vulnerability level accurately. This vulnerability evaluation will be divide into the following five levels: *M1*, *M2*, *M3*, *M4* and *M5* which means very vulnerable, vulnerable, medium, safe and very safe.

8.2 Determine Classical and Segment Field Matter-element Matrix of the Chemical Industrial Park

Considering the 21 indexes are difficult to be measured in a unified way, therefore, the indexes distribution was scored within the closed interval [0,100] in the form of expert scoring. The higher the score gets, the safer the index is. At the same time, to divide the closed interval into five equal parts and get standard interval set$C_k(1 \leq k \leq 21)$: $\{[0,20],(20,40],(40,60],(60,80],(80,100]\}$. The segment field matter-element matrix is established as bellow by matter-element extension model.

$$R_1 = \begin{bmatrix} M_1 & C_1 & [0,20] \\ & C_2 & [0,20] \\ & C_3 & [0,20] \\ & C_4 & [0,20] \\ & C_5 & [0,20] \\ & C_6 & [0,20] \\ & C_7 & [0,20] \\ & C_8 & [0,20] \\ & C_9 & [0,20] \\ & C_{10} & [0,20] \\ & C_{11} & [0,20] \\ & C_{12} & [0,20] \\ & C_{13} & [0,20] \\ & C_{14} & [0,20] \\ & C_{15} & [0,20] \\ & C_{16} & [0,20] \\ & C_{17} & [0,20] \\ & C_{18} & [0,20] \\ & C_{19} & [0,20] \\ & C_{20} & [0,20] \\ & C_{21} & [0,20] \end{bmatrix} R_2 = \begin{bmatrix} M_2 & C_1 & (20,40] \\ & C_2 & (20,40] \\ & C_3 & (20,40] \\ & C_4 & (20,40] \\ & C_5 & (20,40] \\ & C_6 & (20,40] \\ & C_7 & (20,40] \\ & C_8 & (20,40] \\ & C_9 & (20,40] \\ & C_{10} & (20,40] \\ & C_{11} & (20,40] \\ & C_{12} & (20,40] \\ & C_{13} & (20,40] \\ & C_{14} & (20,40] \\ & C_{15} & (20,40] \\ & C_{16} & (20,40] \\ & C_{17} & (20,40] \\ & C_{18} & (20,40] \\ & C_{19} & (20,40] \\ & C_{20} & (20,40] \\ & C_{21} & (20,40] \end{bmatrix} R_3 = \begin{bmatrix} M_3 & C_1 & (40,60] \\ & C_2 & (40,60] \\ & C_3 & (40,60] \\ & C_4 & (40,60] \\ & C_5 & (40,60] \\ & C_6 & (40,60] \\ & C_7 & (40,60] \\ & C_8 & (40,60] \\ & C_9 & (40,60] \\ & C_{10} & (40,60] \\ & C_{11} & (40,60] \\ & C_{12} & (40,60] \\ & C_{13} & (40,60] \\ & C_{14} & (40,60] \\ & C_{15} & (40,60] \\ & C_{16} & (40,60] \\ & C_{17} & (40,60] \\ & C_{18} & (40,60] \\ & C_{19} & (40,60] \\ & C_{20} & (40,60] \\ & C_{21} & (40,60] \end{bmatrix} R_4 = \begin{bmatrix} M_4 & C_1 & (60,80] \\ & C_2 & (60,80] \\ & C_3 & (60,80] \\ & C_4 & (60,80] \\ & C_5 & (60,80] \\ & C_6 & (60,80] \\ & C_7 & (60,80] \\ & C_8 & (60,80] \\ & C_9 & (60,80] \\ & C_{10} & (60,80] \\ & C_{11} & (60,80] \\ & C_{12} & (60,80] \\ & C_{13} & (60,80] \\ & C_{14} & (60,80] \\ & C_{15} & (60,80] \\ & C_{16} & (60,80] \\ & C_{17} & (60,80] \\ & C_{18} & (60,80] \\ & C_{19} & (60,80] \\ & C_{20} & (60,80] \\ & C_{21} & (60,80] \end{bmatrix} R_5 = \begin{bmatrix} M_5 & C_1 & (80,100] \\ & C_2 & (80,100] \\ & C_3 & (80,100] \\ & C_4 & (80,100] \\ & C_5 & (80,100] \\ & C_6 & (80,100] \\ & C_7 & (80,100] \\ & C_8 & (80,100] \\ & C_9 & (80,100] \\ & C_{10} & (80,100] \\ & C_{11} & (80,100] \\ & C_{12} & (80,100] \\ & C_{13} & (80,100] \\ & C_{14} & (80,100] \\ & C_{15} & (80,100] \\ & C_{16} & (80,100] \\ & C_{17} & (80,100] \\ & C_{18} & (80,100] \\ & C_{19} & (80,100] \\ & C_{20} & (80,100] \\ & C_{21} & (80,100] \end{bmatrix}$$

8.3 Determine the Weights for Each Layer Index

The weights are scored by experts for each various decision-making who marked the indexes in a range from 1 to 5 as shown in Table.1. There are m experts scoring for n indexes that a m×n matrix is composed of the scores represented by $P_{i,j}$ (with $i \in 1, 2, \ldots, m$; $j \in 1, 2, \ldots, n$). Standardizing the data in the matrix with Eq.5 and Eq.6.

Table.1 The Five-Level Scale

Level	1	2	3	4	5
Importance	General	In between	Important	In between	Very important

8.4 Case Study Applied to Shizhuang Chemical Industrial Park

There are 25 enterprises (including storage wharfs) in Shizhuang chemical industrial park and 20 of them are chemical enterprises. There are 11 major hazard installations enterprises, 6 high-risk process enterprises, and 17 have key regulatory dangerous chemical enterprises. Shizhuang chemical industrial park has some risk, so using the matter-element extension model based on entropy weight to evaluate chemical industrial park comprehensive vulnerability is necessary.

8.5 Determine the Weights for Each Layer Index of Chemical Industrial Park

According to the rating value by experts for chemical industrial park, the subjective relative weight of 21 comprehensive vulnerability indexes of chemical industrial park are calculated through Eq.7 to Eq.9. The results are shown as Table.2.

Table.2 Weight Of Each Index Layer In The Comprehensive Vulnerability Assessment

Index layer	*C1*	*C2*	*C3*	*C4*	*C5*	*C6*	*C7*	*C8*	*C9*	*C10*	*C11*
Weight	0.1818	0.0455	0.4545	0.3182	0.333	0.5833	0.0833	0.3714	0.2	0.1143	0.2857
Index layer	*C12*	*C13*	*C14*	*C15*	*C16*	*C17*	*C18*	*C19*	*C20*	*C21*	
Weight	0.0286	0.4545	0.0445	0.3182	0.1818	0.0286	0.2	0.2837	0.3714	0.1143	

8.6 Determine the Matter Element Matrix of Chemical Industrial Park

Each index layer can be measured by examining meteorological, geographical, daily safety management level and emergency rescue management level of industrial park. It carries on the analysis appraisal separately from safety management vulnerability, personnel vulnerability, inherent vulnerability of hazardous material, facilities' vulnerability and environmental vulnerability, then established the matter element matrix R_0. The collection of each evaluation index actual score value is $u_0 = [86.6, 76.6, 51.6, 71.6, 30, 75, 31, 6.5, 86.6, 75, 53.3, 56.6, 8.3, 68.3, 31.6, 21.6, 70, 26.6]$.

8.7 Correlation Coefficients of Different Indicators for Each Evaluation Level

The correlation coefficients of different indicators were drawn for all indexes combined with classical field, segment field, matter-element matrix and equations .The result is shown in Table.3.

Table.3 The Results Of Correlation Coefficients Of Different Indicators For Each Evaluation Level

$K_j(x_i)$	very vulnerable	vulnerable	medium	safe	very safe	Max$K_j(x_i)$	Results
$K_1(x_1)$	-0.833	-0.776	-0.655	-0.330	0.971	0.971	very safe
$K_2(x_2)$	-0.708	-0.610	-0.415	0.170	-0.127	0.170	safe
$K_3(x_3)$	-0.395	-0.193	0.210	-0.148	-0.369	0.210	medium
$K_4(x_4)$	-0.645	-0.527	-0.290	0.420	-0.228	0.420	safe
$K_5(x_5)$	-0.250	0.500	-0.250	-0.500	-0.625	0.500	vulnerable
$K_6(x_6)$	-0.687	-0.583	-0.375	0.250	-0.166	0.250	safe
$K_7(x_7)$	-0.268	0.362	-0.210	-0.473	-0.605	0.365	vulnerable
$K_8(x_8)$	0.250	-0.750	-0.875	-0.916	-0.937	0.250	very vulnerable
$K_9(x_9)$	-0.791	-0.722	-0.582	-0.165	0.246	0.246	very safe
$K_{10}(x_{10})$	-0.666	-0.555	-0.332	0.335	-0.200	0.335	safe
$K_{11}(x_{11})$	-0.832	-0.776	-0.665	-0.330	0.970	0.970	very safe
$K_{12}(x_{12})$	-0.687	-0.583	-0.375	0.250	-0.166	0.250	safe
$K_{13}(x_{13})$	-0.418	-0.225	0.162	-0.122	-0.363	0.162	medium
$K_{14}(x_{14})$	-0.457	-0.276	0.085	-0.072	-0.350	0.085	medium
$K_{15}(x_{15})$	-0.791	-0.721	-0.582	-0.165	0.246	0.246	very safe
$K_{16}(x_{16})$	0.415	-0.585	-0.792	-0.861	-0.896	0.415	very vulnerable
$K_{17}(x_{17})$	-0.603	-0.471	-0.207	0.354	-0.269	0.354	safe
$K_{18}(x_{18})$	-0.268	0.362	-0.210	-0.473	-0.605	0.362	vulnerable
$K_{19}(x_{19})$	-0.068	0.080	-0.460	-0.640	-0.730	0.080	vulnerable
$K_{20}(x_{20})$	-0.625	-0.500	-0.250	0.500	-0.250	0.500	safe
$K_{21}(x_{21})$	-0.198	0.330	-0.335	-0.556	-0.667	0.330	vulnerable

According multiplication rule to calculate comprehensive correlation calculations $K_j(P_0)$ of layers and unvalued matter-element by weighted summation. The results are shown as Table.4.

Table.4 The Results Of Comprehensive Correlation Calculations Of Chemical Industrial Park

Objective level	very vulnerable	vulnerable	medium	safe	very safe	Max$K_j(x_1)$	Results
B1	-0.577	-0.436	-0.155	0.037	-0.078	0.037	safe
B2	-0.503	-0.137	-0.318	-0.064	-0.358	-0.064	safe
B3	-0.406	-0.723	-0.676	-0.415	-0.034	-0.034	very safe
B4	-0.376	-0.451	-0.257	-0.274	-0.274	-0.257	medium
B5	-0.350	-0.074	-0.309	-0.132	-0.501	-0.074	vulnerable
A1	-0.530	-0.247	-0.258	-0.027	-0.255	-0.027	safe
A2	-0.384	-0.474	-0.512	-0.304	-0.217	-0.217	very safe
M	-0.477	-0.330	-0.351	-0.128	-0.242	-0.128	safe

9 Analysis of vulnerability assessment's results

Through the calculation of comprehensive correlation degree, we can draw the conclusion that Shizhuang industrial park's comprehensive vulnerability level is safe.

The main influence factors in Shizhuang chemical industrial park are Explosion hazard risk(*C6*), Daily safety management level of businesses(*C3*), Distribution density of the facilities(*C13*), Personnel distribution density(*C8*), Environment-sensitive targets and mainly facilities(*C20*) according to the weight (in Table.2) and their correlation calculations' result are safe, medium, medium, very vulnerable and safe (in Table.4). It represents that personnel distribution density need to focus on. On the one hand, this park is really high concentrations that causing mass causalities is a strong possibility. Actually, Hongyuan Metal Product Co., Ltd and Kingboard Chemical Holdings Ltd are labor-intensive enterprises who need more safety supervision or move out of park when conditions permit. On the other hand, Fire hazard risk (*C5*) is more important than other vulnerable factors (*C8*, *C16*, *C5*, *C7*, *C19*, *C21*), so its key points are to strengthen the management of hazardous material because fire especially pool fire is main type of accidents in this park.

Besides that, further analysis found that although *C5*,

DOI: 10.1051/matecconf/20166105014

C7, *C18*, *C19* and *C21* are vulnerable, the level of these five indexes have some different. To be specific, for index *C7*, its *M1*<*M3* which means the level of Leak hazard risk clearly tend to be medium. And for index *C19*, its *M1*>*M3* which means the level of Regional per capita GDP clearly tend to be very vulnerable. In consequence, this method not only can get the vulnerability level of each index, but also can determine the tendencies according to correlation calculations for finding very vulnerable indexes in order to provide focused measures.

10 Conclusion and Discuss

In traditional evaluation system, the risks of explosion, fire and leakage are more focused on ,but take less attention to structural vulnerability such as personnel, facilities' and natural vulnerability. This paper puts forward quantitative methods and scoring criteria, and drawn a conclusion with Safe Level for Shizhuang chemical industrial park. According to statistics from Safety Department, there was no fatal accident in the past 5 years. The safety situation of Shizhuang industrial park is good which is identical with the result of assessment, therefore, this method is objective and practical.

Besides, this method is able to sort the indexes in a same level which is helpful for searching the vulnerable links in industrial park. As you see, establishing assessment model of comprehensive vulnerability for chemical industrial park is an important channel to make sure it run in an efficient, safe way and reduce the risk. At present, research remains to be further deepened in the recoverability of vulnerability for chemical industry park. This research establishes the foundation for further study.

References

1. Huang Jianyi, Liu Yi, Review on the Theoretical Model and Assessment Framework of Foreign Vulnerability Research. Areal Research and Development, 31(5) (2012) 1-5.
2. Peter Timmerman. Vulnerability resilience and collapse of society: A Review of Models and Possible Climatic Applications. Journal of Climatology, 1981, 1(4):396-396.
3. Li He, Zhang Pingyu, Cheng Yeqing, Concept and Assessment of Vulnerability. Progress in Geography, 27(2) (2008) 18-25.
4. Li FengYing, Bi Jun, Qu Changsheng. Whole-process environmental risk assessment and management and its application. China Environmental Science, 6 (2010) 858-864.
5. Shi Peijun. Theory on disaster science and disaster dynamics. Journal of natural disasters, 11(3) (2002) 1-9.
6. Li Qiujin, Overview of research on Chemical industrial park vulnerability. COSHA academic conference proceedings, (2013).
7. Tan Zhaoyang. Study of Comprehensive Assessment Methods for Chemical Industry Park Vulnerability. Nankai University, (2012).
8. Zongqin G. Construction and Development of Chemical Industrial Parks in China. International Petroleum Economics, (2004).
9. Bohle H, Downing T, Watts M. Climate change and social vulnerability: toward a sociology and geography of food in security. Global Environmental Change, 4(1) (1994) 37-48.
10. Wu Huajun, Liu Nianfeng, He Jun. Comprehensive evaluation for ecological environment based on matter-element analysis of. Journal of Huazhong University of Science and Technology, 23(1) (2006) 52-55.
11. Zhang xianqi, Liang Chuan. Application of fuzzy matter-element model based on coefficients of entropy in comprehensive evaluation of water quality. Journal of Hydraulic Engineering, 36(9) (2005) 1057-1061.
12. Polsky C, Neff R, Yarnal B. Building comparable global change vulnerability assessments: The vulnerability scoping diagram. Global Environmental Change, 17 (2007) 472-485.

MATEC Web of Conferences **61**, 05015 (2016) DOI: 10.1051/matecconf/20166105015
APOP2016

Research on the Dynamics of Crankshaft Under the Condition of Operating Mode Based on LS-DYNA

Shaojun HAN[1, a], Ting WU[1,b], Shiquan Liu[2,c], Zihan Zhou[1,d], Chuan Liu[1,e], Qiang HE[1,f]

[1] School of Mechanical and Electronic Engineering, Wuhan University of Technology, Wuhan, China
[2]Shanghai Automotive Gear Works, Shanghai, China

[a]hanshjun@163.com, [b]hbwuting@163.com, [c]702057684@qq.com, [d]864426302@qq.com, [e]772837913@qq.com, [f]qianghe1992@163.com

Abstract: A model of crankshaft under working condition is established based on LS-DYNA. The stress distributions of crankshaft and its regularities of the stress along surface of crankshaft and layer depth at the fillet are studied. The results show that the obvious stress concentration exist in the upper end of fillet on main journal and the lower end of fillet on rod journal when the piston arrives at top dead center, and the maximum equivalent stress occurs at the fillet on rod journal, so the fracture of crankshaft is most likely to occur in the lower end of fillet on rod journal. With the increase of depth, the stress on the surface of fillet reduces correspondingly, and the equivalent stress begins to stabilize when the depth reaches up to 2.8 mm. The study of maximum working stress provides the theory basis for the selection of residual stress.

Keywords: crankshaft; explicit dynamics; distribution regularities; residual stress

1 Introduction

Crankshaft is one of the most important parts of the engine. Due to the impact of cyclic stress, the permanent deformation will occur at the crankshaft [1].How to ensure the crankshaft working life and prevent the occurring of fracture has been the research topic over the years. In order to study the stress distribution of crankshaft under the working condition, this paper builds and analyzes the model of crankshaft based on LS-DYNA, and gets the results of the stress distribution and the regularities of the stress along the fillet surface and layer depth in fillet of connecting rod journal.

The crankshaft fillet rolling strengthening is an ideal surface strengthening method. Compared with the traditional methods, rolling strengthening has many advantages like low cost, short processing time, obvious strengthening effect[2]etc. After rolling strengthening, the crankshaft fatigue strength can increase by 30%-200%[3]. However, how much the residual stress produced in the process of rolling strengthening is uncertain, and it's very hard to confirm the value of residual stress. This paper finds that the maximum working stress can provide the theory basis for the selection of residual stress.

2 Research of kinematics of crankshaft connecting rod system and load-bearing performance

2.1 Kinematics analysis of crankshaft connecting rod system

Figure 1 shows the structural diagram of crankshaft connecting rod. $\alpha = 0^0$, the piston moves to the top dead center position. $\alpha = 180^0$, the piston moves to the lower dead center position.

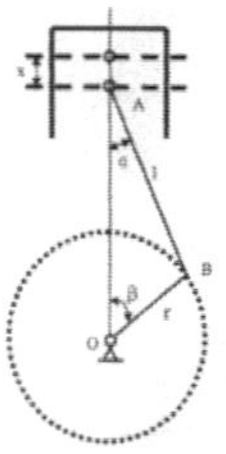

Fig.1 Structural diagram of crankshaft connecting rod

If $\lambda = l/r$, $\sin\alpha = \lambda\sin\beta$, the displacement formula is shown below

$$x = r[(1 + 1/\lambda) - cos\beta - [1 - \lambda^2 sin^2\beta]/\lambda]. \quad (1)$$

According to the piston displacement formula, it deduces the velocity and acceleration:

$$v = wr\left[sin\beta + \lambda/2\, sin2\beta\, [1 - \lambda^2 sin^2\beta]^{-\frac{1}{2}}\right]. \quad (2)$$

$$a = w^2 r\left[cos\,\beta + \lambda(1 - \lambda^2 sin\,\beta)^{-3/2}\left[cos\,2\beta(1 - \lambda^2 sin\,\beta)^2 + \frac{\lambda^2}{4} sin^2 2\beta\right]\right]. \quad (3)$$

2.2 Gas force and inertia force of connecting rod mechanism

The force of the cylinder is not only related to the size of the cylinder, but also to the pressure between inside and outside of the cylinder. The gas force formula is shown as follows:

$$F_g = (P_1 - P_0)\pi D^2/4\, cos\,\alpha. \quad (4)$$

α :swing angle of connecting rod; D :cylinder diameter; P_1 :absolute pressure of gas in cylinder; P_0:absolute pressure of gas inside the box.

In the theoretical analysis, the whole weight of connecting rod is translated into large mass m_1 and small mass m_2 by the static equivalent principle, m_1 is heavier than m_2. Reciprocating inertia force is generated by the piston and small mass. It is considered that the weight of piston assembly concentrates on the center of the pin. $M_p = \sum m_i$. The reciprocating inertia force is shown as follows:

$$F_1 = (M_p + m_2)rw^2\left[cos\,\beta + \lambda(1 - \lambda^2 sin\,\beta)^{-3/2}\left[cos\,2\beta(1 - \lambda^2 sin\,\beta)^2 + \frac{\lambda^2}{4} sin^2 2\beta\right]\right]. \quad (5)$$

Centrifugal inertia force is generated by the crankshaft and the large mass of connecting rod. The crankshaft comprises the main crankshaft journal, crank arm, connecting rod journal etc. In order to facilitate the theoretical calculation, the crankshaft's weight is converted to the connecting rod's weight, so the centrifugal inertia force after converting weight is equal to the centrifugal inertia force before converting. The single crankshaft's weight conversion formula is shown as follows:

$$m_k = m_a + 2m_b p/x. \quad (6)$$

m_a: weight of connecting rod journal; m_b: weight of crank arm; p: distance between the center of crank arm and rotating center of crankshaft; x: piston stroke. The centrifugal inertia force is shown as follows:

$$F_2 = (m_k + m_1)rw^2. \quad (7)$$

2.3 Solution of load

There're some parameters of crankshaft showed in table 1.

Table 1 Parameters of the crankshaft

Parameters	Value	Parameters	Value
crank radius ***r***	*39mm*	Bulk mass quality m_1	*0.342kg*
Connecting rod center distance ***l***	*130mm*	head mass quality m_2	*0.180kg*
main journal diameter D_1	*50.33mm*	single crank quality m_k	*0.281kg*
connecting rod diameter ***d***	*38.5mm*	piston quality m_p	*0.315kg*
cylinder diameter ***D***	*75mm*	angular velocity ***w***	*255.2rad/s*

This paper finds out the similar cylinder size and the size of the cylinder pressure, the maximum pressure in the 465Q cylinder is about 8.12MPa, the inner pressure of the box is about 0.12MPa.

$$F_g = (P_1 - P_n)\pi D^2/4\, cos\,\alpha = 35325N. \quad (8)$$

$$F_1 = (M_p + m_2)a = 1634N. \quad (9)$$

$$F_2 = (m_k + m_1)rw^2 = 1582N. \quad (10)$$

During the power stroke, the maximum pressure of crankshaft is shown as follows:

$$F = F_g - F_1 - F_2 = 32109N. \quad (11)$$

3 Dynamics analysis of the crankshaft under working condition

3.1 Parameter selection

The diameter of the main journal is 50.33mm, the diameter of the connecting rod journal is 38.5mm, the diameter of the cylinder is 75mm. Material of crankshaft is made of QT800 ductile iron, its density is $7.16 \times 10^3\, kg/m^3$, modulus of elasticity is $7.12 \times 10^5\, Mpa/mm^3$, the poisson's ratio is 0.27. The actual stress-strain curve of ductile cast iron is shown in figure 2. Connecting rod is made of 42CrMo, its density is $7.85 \times 10^3\ kg/m^3$, modulus of elasticity is $2.09 \times 10^5\ MPa/mm^3$, the poisson's ratio is 0.28. Piston pin is usually made of T10, its density is $7.86 \times 10^3\ kg/m^3$, modulus of elasticity is $2.08 \times 10^5\ MPa/mm^3$, the poisson's ratio is 0.28.

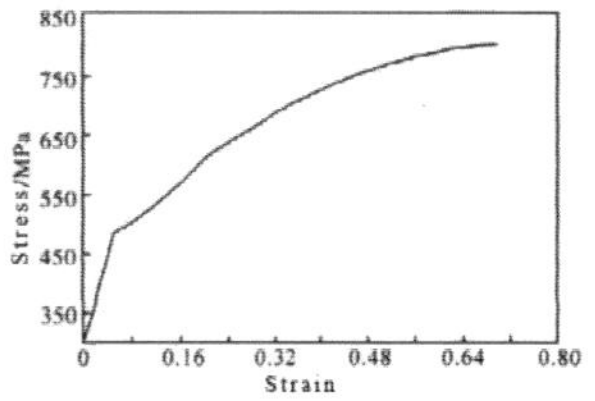

Figure.2 Stress-strain curve of ductile cast iron

3.2 Simplified simulation model

This paper makes an appropriate simplification of the crankshaft structure and takes a part of the crankshaft as the calculating object. The crankshaft fatigue point occurs in the third crank throw of connecting rod journal, so this paper chooses the third crank throw as the analysis model. The simplified model is composed of half single crank throw, connecting rod and piston pin. The simplified model is shown in figure 3a, and its element type is SOLID164 unit, it is shown in figure 3b.

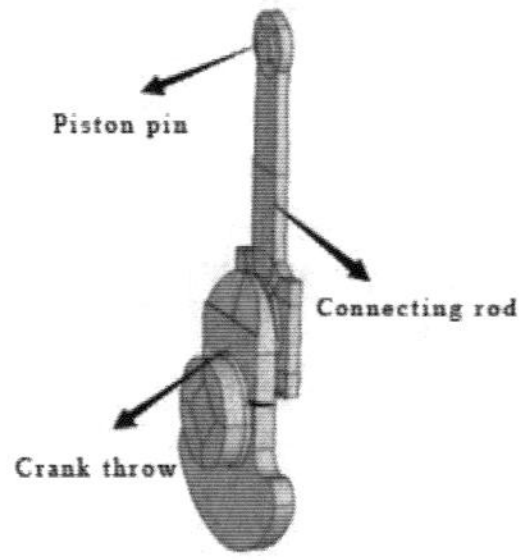

Figure.3a Solid model

Figure.3b SOLID164 unit

3.3 Apply Load

During the process of working, the impact load of the crankshaft is maximum when it's in the stage of power stroke which is the key point of leading to the fatigue of the crankshaft. This paper mainly analyzes the crankshaft when it's in the stage of power stroke.

The normal speed of the crankshaft is around 1500~3000r/min, this paper takes the constant speed 2400r/min for analysis speed. The load value is $F_{max}/2$. The loading force is not constant, when the gas bursts, the pressure in the cylinder increases rapidly to the maximum. As the piston moves down, the gas volume increases and the pressure begins to fall.

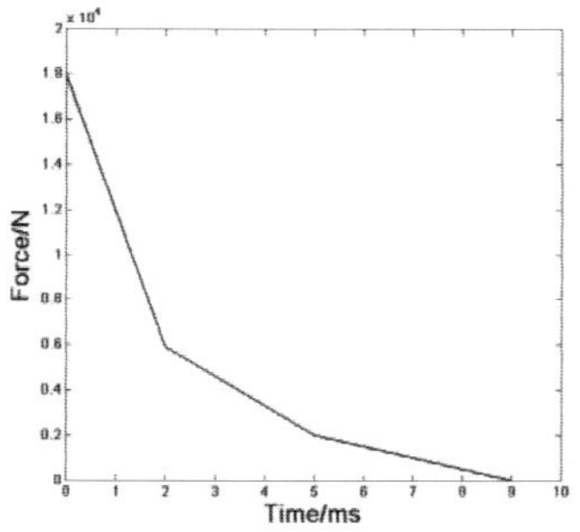

Figure.4 Force-time curve

In order to make the load and the actual gas pressure closer, the force-time curve is established in figure 4. With the increase of rotation angle of crankshaft, the load decreases rapidly. When the time is 0.009s, the load decreases to 0. The terminal time in this paper is 0.009s.

3.4 Solution and result processing

Figure 5a shows the equivalent stress of the main journal and figure 5b shows the equivalent stress of the connecting rod journal in the maximum load. From the simulation results, it can be seen that the upper end of the main journal and the lower end of the connecting rod journal exist obvious stress concentration phenomenon, and the maximum equivalent stress, 901MPa, occurs at the fillet of the connecting rod journal.

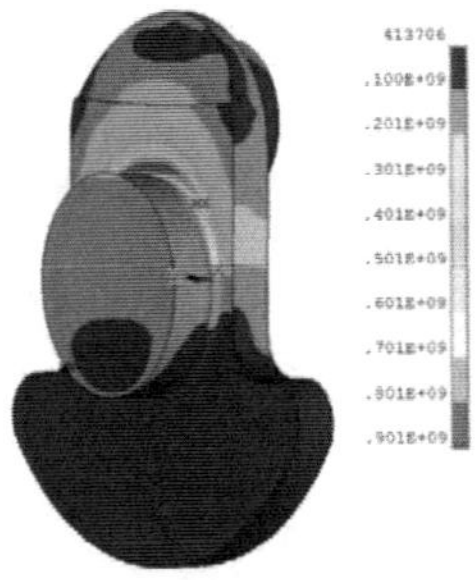

Figure.5a Equivalent stress of main journal

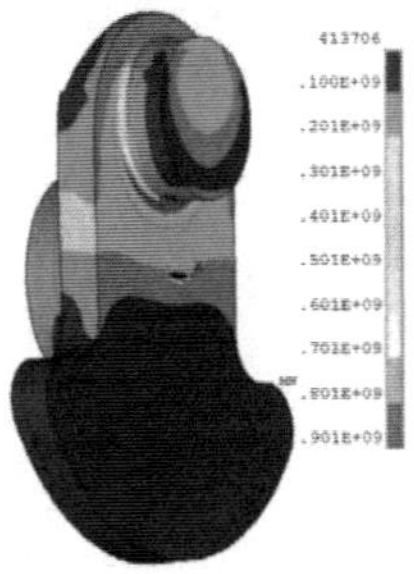

Figure.5b Equivalent stress of connecting rod journal

Analysis result shows that the upper end of the main journal withstands the pressure stress and the lower end of the connecting rod journal withstands the tensile stress. Rigid material can withstand the more pressure stress than tensile stress. Therefore, tensile stress in the fillet of the connecting rod journal is the key factor of leading to fatigue failure. This paper analyzes the tensile stress in the fillet of the connecting rod journal. In order to analyze it intuitively, the two paths in the circumferential direction and the rolling direction of the fillet of the connecting rod journal are selected. The paths in two directions are shown in Figure 6.

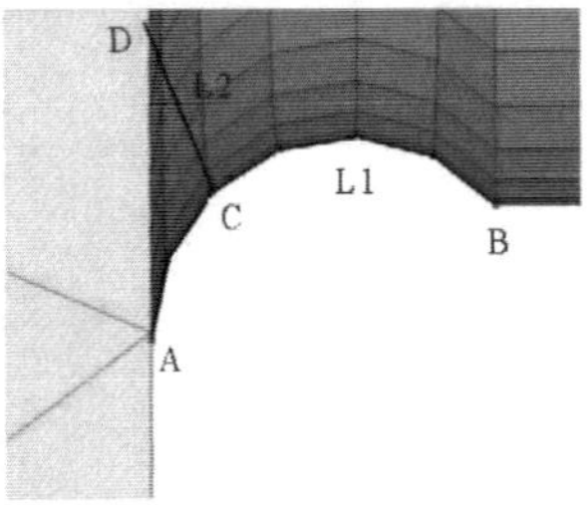

Figure.6 Paths in two directions

The path L1 is from A to B in the connecting rod journal along the circumferential direction, the path L2 is from C to D in the connecting rod journal along the rolling direction. The equivalent stress values and the stress distributions along path L1 and path L2 are obtained from the simulation which is shown in Figure 7a and 7b. In Figure 7a, it can be seen a parabolic form of the fillet surface equivalent stress curve, the maximum value of equivalent stress in the fillet of connecting rod journal is 888MPa and the point of maximum equivalent stress locates in between A and C. In Figure 7b, it can be seen that the stress in the fillet of the connecting rod journal gradually reduces with the increase of the depth, when the depth approaches 2.8mm, the equivalent stress is beginning to stabilize, and the maximum equivalent stress on the surface is 860MPa.

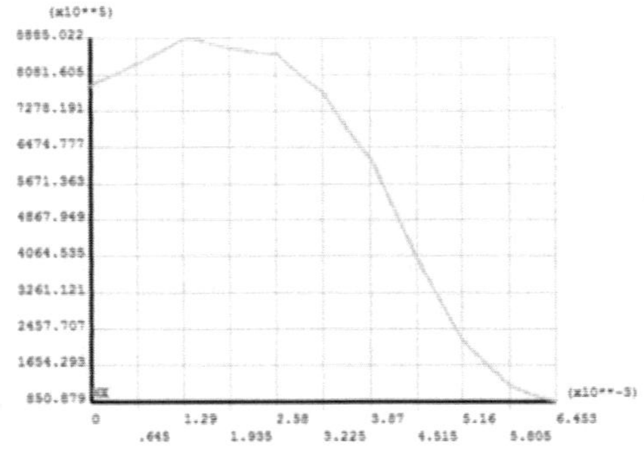

Figure.7a Simulation result of path L1

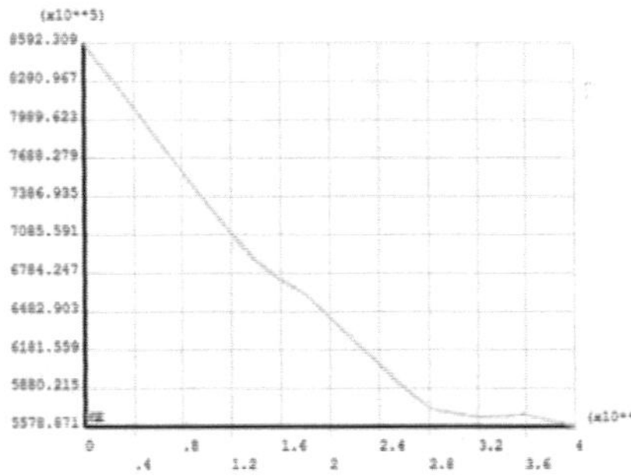

Figure.7b Simulation result of path L2

4 Summary

1. This paper researches the maximum equivalent stress, and gets the stress concentration phenomenon in the lower end of the connecting rod journal and the upper end of the main journal. The maximum equivalent stress occurs at the fillet of the connecting rod. Therefore, the key factor of leading to fatigue failure is the stress in the fillet of the main journal and the connecting rod.

2. The stress distributions along the path L1 and the path L2 are obtained by simulation. They can provide the theory basis for the selection of the depth of the carburizing layer.

3. The maximum working stress, 901MPa, is obtained by simulation. It can provide the theory basis for the selection of the residual stress in the process of crankshaft fillet rolling.

References

1. Chen Lidong, Ma Shuying, Liu Rongchang, Li Shuzhen, Shi Lei, Li Guofang, Zheng Lixin. Design of hydraulic system of crankshaft rolling strengthen machine, Machine Tool and Hydraulics, 2009,7(37): 66-68
2. Xue Longquan, Wang Yuqiu, Liu Rongchang, Zhang Hongjun, Wang Huiwu. Finite element analysis of residual deforming of crankshaft rolling process, Machine Tool and Hydraulics, 2005,10: 40-41
3. Wang Huiwu, Xue Longquan, Liu Hongzhao, Zhou Chunguo, Liu Rongchang. The design of eccentric fixture of crank for fillet rolling, Machine Tool and Hydraulics,2004,8:164-165
4. Liu Rongchang, Sun Huadong, Jiao Honglei, Ma Shuying, Xue Longquan. Explicit dynamic analysis of crankshaft rolling process, Transactions of Csice,2009,5(27):463-468

The Effect of Triangular Dimple Orientation on Load-Carrying Capacity

Yi Xie[1,a], Xiaohan Pei[1] and Songbo Wei [1]

[1]PetroChina Research Institute of Petroleum Exploration & Development, Beijing 100083, China

Abstract. In this study, the effect of triangular dimple orientation on load-carrying capacity is further investigated. The effect of geometric and operating parameters including the area ratio, the depth, the angular speed and the film thickness on load-carrying capacity is discussed under different orientation angles. The load-carrying mechanisms of the triangular dimples is analyzed.

1 Introduction

Surface modified technology is known to improve tribological performances of mechanical components, such as enhancing load-carrying capacity, reducing friction and wear etc. Several surfaces with dimples fabricated by surface modified technology have been applied in mechanical seals[1-5], thrust bearing[6-9], and piston rings[10-11], etc. The area density, the depth and feature distribution of the dimple are important geometric parameters which influence the tribological performance. The tribological performance of surfaces with dimples can be improved by optimizing the above three parameters [1-2,7-8].

In recent years, the effect of the dimple geometric shape on the tribological performance has been the focus of researcher interest, because various dimple shapes can be fabricated with the development of the surface processing methods, such as laser surface texturing, reactive ion etching and abrasive jet machining, etc. There have been many research reports [12-14] about the effect of the dimple geometric shape. At the same time, the triangular dimple with its special nature has aroused the interest of researchers. Siripuram et al.[15] carried out to numerically studied to find that the effect of the triangular dimple orientation on the leakage. Yu et al. [16] carried out theoretical studies to investigate the effect of the triangular dimple orientation on the hydrodynamic pressure. Xie et al. [17] have investigated tribological performance of triangular dimples in water lubrication by experimental methods. Guo et al. [18] have numerically studied the effect of various micro-dimple shapes on the pumping rate and the friction torque of radial lip seals and found that it was an efficient way to increase pumping rate and decrease friction torque of radial lip seals by changing the triangular dimple orientation. Meng et al. [19] also studied the effect of the triangular dimple orientation on the load-carrying capacity and leakage rate of the mechanical seal. However, it is still necessary for further research about the triangular dimple orientation. It is helpful for understanding the best choice of the triangular dimple orientation and its special properties.

In this paper, the effect of triangular dimple orientation on load-carrying capacity is further investigated. The effect of triangular dimple orientation angles on load-carrying capacity under different geometric and operating parameters including the area ratio, the angular speed, the depth and the film thickness is discussed. The load-carrying mechanisms of the triangular dimples is analyzed.

2 Theoretical model

2.1 Geometric model

To investigate the effect of triangular dimple orientation on load-carrying capacity, a pair of friction components composed of one upper ring and one lower ring are chosen as shown in Figure 1. The upper ring is the rotating ring and rotates along the anticlockwise direction at an angular speed ω, while the lower ring is the stationary ring. There is a fluid film between the rotating ring and the stationary ring, and the thickness of the film in non-dimple area are is expressed by h_0. The sizes of the rotating and stationary ring are the same. The outer diameter, r_o, is 25.4 mm and the inner diameter, r_i, is 20.75 mm. The surface roughness of the rotating ring is 0.18 μm and the surface roughness of the stationary ring is 0.26 μm. The triangular dimples are radioactively distributed on the stationary ring. The dimple numbers in circumferential and radial direction are 80 and 4, respectively. The dimple bottom is flat, and the dimple depth is expressed by h_g .

As is well known[1-2], the area density of dimples is an important parameter which affects the performance of dimples. So in this study the area density for triangular dimple can be defined as

[a] Corresponding author: xieyi0117@163.com

$$S_p = \frac{A_p}{A} = \frac{n_r n_c ab}{2\pi\left(r_o^2 - r_i^2\right)} \tag{1}$$

where A_p is the total dimple area, A is the total surface area, n_r is the dimple number in circumferential direction, n_c is the dimple number in radial direction, a is the bottom of the triangular dimple, b is the height of the triangular dimple.

The inclination of the triangular dimple is described by an orientation angle, α, which is shown in Figure 1. The shape dimension parameter, γ, can be described by the ratio of the height to the bottom length in the following

$$\gamma = \frac{b}{a} \tag{2}$$

The nominal film thickness equation can be expressed in the following:

$$h(x,y) = \begin{cases} h_0 + h_g & (x,y) \in \text{dimple area} \\ h_0 & \text{else} \end{cases} \tag{3}$$

where x and y are the coordinates in the surface with triangular dimple.

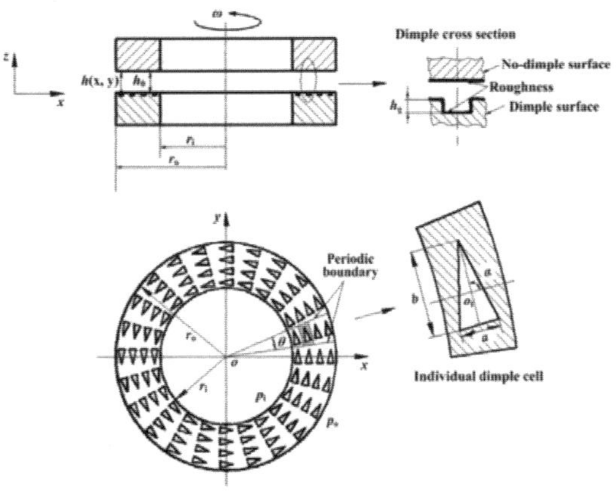

Figure 1. Structure schematic of a pair of friction with triangular dimples.

2.2 Mathematical model

In this study, the fluid between the rotating ring and the stationary ring is assumed to be laminar, incompressible Newtonian fluid. The hydrodynamic pressure can be solved by the Reynolds equation. According to the average flow model proposed by Patir and Cheng [20-21], the Reynolds equation which satisfies the mass conservation can be written in the Cartesian coordinate in the following:

$$\frac{\partial}{\partial x}\left(\phi_x \frac{\rho h^3}{12\mu}\frac{\partial p}{\partial x}\right) + \frac{\partial}{\partial y}\left(\phi_y \frac{\rho h^3}{12\mu}\frac{\partial p}{\partial y}\right) = \frac{\partial}{\partial x}\left[\frac{U}{2}\rho\left(h_T + \sigma\phi_s\right)\right] + \frac{\partial}{\partial y}\left[\frac{V}{2}\rho\left(h_T + \sigma\phi_s\right)\right] + \frac{\partial\left(\rho h_T\right)}{\partial t} \tag{4}$$

where p is the hydrodynamic pressure; h_T is the average local film thickness; U and V are respectively the speeds of the rotating ring in the x and y directions; ϕ_x and ϕ_y are respectively the pressure flow factors in the x and y directions; ϕ_s is the shear flow factor. ρ and μ are the viscosity and density of the fluid film. In the process of numerical simulation, the effect of the squeeze is not taken into account, so the squeeze term, $\partial h_T / \partial t$, is neglected.

2.3 Boundary condition

The periodic boundary condition and the boundary condition at the inner and outer diameters in computational domain are given by

$$p_{\theta-} = p_{\theta+} \tag{5}$$

where $\theta-$ is the initial boundary of triangular dimple column, and $\theta+$ is the final boundary of triangular dimple column.

The fluid pressure at the inner and outer diameters is assumed to be constant, so the boundary condition at the inner and outer diameters can be given by

$$\begin{cases} p(\sqrt{x^2+y^2} = r_o) = p_o \\ p(\sqrt{x^2+y^2} = r_i) = p_i \end{cases} \tag{6}$$

where p_o is the outside pressure and p_i is the inside pressure.

It is known that when the fluid local pressure falls below the cavitation pressure in operation, the fluid film will rupture and the cavitation occurs. It is very necessary to use an appropriate boundary condition for the cavitation problem. In the present, there are three boundary conditions: the half-Sommerfeld boundary condition, the Reynolds boundary condition and the JFO boundary condition. The JFO boundary condition which satisfies the mass conservative law is considered as an accurate method for handling the cavitation problem [22-23]. Therefore, the method in this study is based on the FEM proposed by Kumar and Booker [24]. The basic assumptions for the fluid film are: 1) the fluid film consists of the bi-phase lubricant mixture, including liquid and vapor or gas; 2) the liquid in full film zones is assumed to be incompressible, while the gas/vapor in the cavitation zones is assumed to be compressible with zero bulk modulus; 3) the density and the viscosity of gas/vapor in the cavitation zones are assumed to be zero. According to the basic assumptions, the relationship of the local viscosity and density of the bi-phase mixture can be shown in the following

$$\mu / \mu_{\text{liq}} = \rho / \rho_{\text{liq}} \tag{7}$$

where $0 \le \rho \le \rho_{\text{liq}}$, $0 \le \mu \le \mu_{\text{liq}}$. ρ and μ are the viscosity and density of the bi-phase mixture. ρ_{liq} and μ_{liq} are the density and the viscosity of the liquid in full film zones.

2.4 Dimensionless form

Under the above basic assumptions, a dimensionless form equation is expressed in the following

$$\frac{\partial}{\partial x^*}\left(\phi_x h^{*3}\frac{\partial p^*}{\partial x^*}\right)+\frac{\partial}{\partial y^*}\left(\phi_y h^{*3}\frac{\partial p^*}{\partial y^*}\right)=$$

$$\Lambda\left\{\begin{aligned}&\frac{\partial}{\partial x^*}\left[U^*\rho^*\left(h_T^*+\phi_s\right)\right]+\frac{\partial}{\partial y^*}\left[V^*\rho^*\left(h_T^*+\phi_s\right)\right]\\&+2h_T^*\frac{\partial\rho^*}{\partial t}\end{aligned}\right\} \tag{8}$$

The dimensionless terms in Eq. 8 are defined as follows

$$x^*=\frac{x}{r_i},\ y^*=\frac{y}{r_i},\ U^*=\frac{U}{r_i},\ V^*=\frac{V}{r_i},\ h^*=\frac{h}{\sigma},\ p^*=\frac{p}{p_i},$$

$$\rho^*=\frac{\rho}{\rho_{liq}},\ h_T^*=\frac{h_T}{\sigma}=\frac{h^*}{2}\left[1+erf\left(\frac{h^*}{\sqrt{2}}\right)\right]+\frac{1}{\sqrt{2\pi}}e^{-\frac{h^{*2}}{2}},$$

$$\Lambda=\frac{6\mu_{liq}r_i^2}{p_i\sigma^2}$$

The dimensionless hydrodynamic pressure p^* can be obtained by solving Eq. 8. The validity of the present method has been verified in ref.[25].

The dimensionless load-carrying capacity F can be expressed in the following

$$F=\iint_A p^* dA \tag{9}$$

3 Results and discussion

The following operating parameters and geometrical parameters are used in this study: outside pressure p_o=0.1 MPa, inside pressure p_i=0.1 MPa, cavitation pressure p_c=0.09 MPa, film thickness h_0=5 μm, angular speed ω=1000 rpm, viscosity of the liquid μ_{liq}=0.001 Pa.s, dimple depth h_g=10 μm, area density S_p=20%, ratio of the height to the bottom length γ=0.866. The parameters will be constant unless they are discussed.

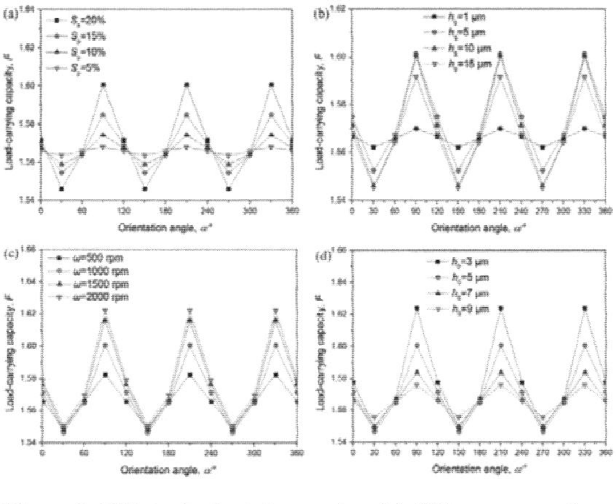

Figure 2. Effect of orientation angles: (a) different area ratios; (b) different dimple depths; (c) different angular speeds; (d) different film thicknesses.

3.1 Effect of orientation angles under different area ratios

Figure 2(a) shows the effect of orientation angles under different area ratios. It can be seen that *F* is dependent on the orientation angles. The *F* value shows periodic variation with alteration of *α*. So the variation of *F* value with *α* can be divided into three periods: 0°~120°; 120~240°; 240~360°. The maximum values of *F* appear in 90°, 210°, 330°, and the minimum values of *F* appear in 30°, 150°, 270° in three periods. So a conclusion can be found that the *F* could be improved by changing the orientation angles of regular triangular dimple in a period. Furthermore, the value of *F* is different under the same orientation angle when the area density is different. The larger the value of area density is, the larger the value of *F* is. In Figure 2(a), when the orientation angles are 90°, 210°, 330° and the area density is 20%, the values of *F* are maximum. So the result shows that the effect of orientation angles on the *F* is obvious when the area ratio is large.

3.2 Effect of orientation angles under different depths

Figure 2(b) shows the effect of orientation angles under different dimple depths. It can be also seen that *F* is dependent on the orientation angles under different dimple depths. The effect of orientation angles on the *F* is obvious at a dimple depth and *F* could be improved by changing the orientation angles of regular triangular dimple. It is indicated that the maximum values of *F* appear in the dimple depth h_g=5 μm and orientation angle α=90°, 210°, 330°.

3.3 Effect of orientation angles under different angular speeds

Figure 2(c) shows the effect of orientation angles under different angular speeds. It can be also seen that *F* is dependent on the orientation angles under different angular speeds. The effect of orientation angles on the *F* is obvious at an angular speed and *F* could be improved by changing the orientation angles of regular triangular dimple. But when the angular speed is increased gradually, the incensement rate of *F* value become small. It is indicated that the effect of the angular speed variation on *F* is little at an orientation angle when the angular speed is up to 2000 rpm.

3.4 Effect of orientation angles under different film thicknesses

Figure 2(d) shows the effect of orientation angles under different film thicknesses. It can be also seen that *F* is dependent on the orientation angles under different film thicknesses. The effect of orientation angles on the *F* is obvious at a film thickness and *F* could be improved by changing the orientation angles of regular triangular dimple. Furthermore, the value of *F* is different under the same orientation angle when the film thickness is different. The smaller the value of film thickness is, the larger the value of *F* is.

4 Conclusions

The effect of triangular dimple orientation on the load-carrying capacity is further investigated by numerical simulation, and the conclusions can be drawn.

(1) The load-carrying capacity is dependent on the orientation angles of the triangular dimple. The load-carrying capacity can be improved by changing the orientation angles of regular triangular dimple.

(2) The value of the load-carrying capacity shows periodic variation with the alteration of the orientation angle for the regular triangular dimple. So the variation can be divided into three periods: 0°~120°; 120~240°; 240~360°.

(3) The variation of the area ratio, the depth, the angular speed and the film thickness can change the effect of the orientation angles on the load-carrying capacity. Therefore, it is important to consider the variation of the geometric and operating parameters of the triangular dimple when the effect of the orientation angle is investigated.

Acknowledgements

This project was supported by China Postdoctoral Science Foundation (No. 2015M571063). The authors would also like to thank supports from PetroChina and China National Petroleum Corporation.

References

1. I. Etsion, L. Burstein, STLE Tribol. Trans. **39** 677-683 (1996)
2. I. Etsion, Y. Kligerman, STLE Tribol. Trans. **42** 511-516 (1999)
3. X.Q. Yu, S. He, R.L. Cai, J. Mater. Process. Technol. **129** 463-466 (2002)
4. Y. Wan, D.S. Xiong, J. Mater. Process. Technol. **197** 96-100 (2008)
5. S.X. Bai, X.D. Peng, J.Y. Li and S.G. Sheng, STLE Tribol. Lett. **38** 187-194 (2010)
6. I. Etsion, G. Halperin, V. Brizmer and Y. Kligerman, STLE Tribol. Lett. **17** 295-300 (2004)
7. X.L. Wang, K. Kato, K. Adachi and K. Aizawa, Tribol. Int. **36** 189-197 (2003)
8. V. Brizmer, Y. Kligerman, I. Etsion, STLE Tribol. Trans. **46** 397-403 (2003)
9. V.G. Marian, M. Kilian, W. Scholz, Proc. Inst. Mech. Eng. Part J: J. Eng. Tribol. **221** 771-778 (2007)
10. G. Ryk, I. Etsion, Wear **261** 792-796 (2006)
11. I. Etsion, E. Sher, Tribol. Int. **42** 542-547 (2009)
12. T. Nanbu, N. Ren, Y. Yasuda, D. Zhu and Q.J. Wang, STLE Tribol. Lett. **29** 241-252 (2008)
13. H. Yu, H. Deng, W. Huang and X. Wang, Proc. Inst. Mech. Eng. Part J: J. Eng. Tribol. **225** 693-703 (2011)
14. S.X. Bai, X.D. Peng, J.Y. Li and X.K. Meng, Sci. China Technol. Sci. **54** 659-662 (2011)
15. R. Siripuram, L.S. Stephens, ASME J. Tribol. **126** 527-534 (2004)
16. H.W. Yu, X.L. Wang, F. Zhou, STLE Tribol. Lett. **37** 123-130 (2010)
17. Y. Xie, Y.J. Li, Y.M. Wang, S.F. Suo and X.F. Liu, Sci. China phys. mech. **57** 273-279 (2014)
18. F. Guo, X.H. Jia, Z. Gao and Y.M. Wang, Sci. China phys. mech. **57** 1343-1351 (2014)
19. X.K. Meng, S.X. Bai, X.D. Peng, Tribol. Int. **77** 132-141 (2014)
20. N. Patir, H.S. Cheng, ASME J. Lubr. Technol. **100** 12-17 (1978)
21. N. Patir, H.S. Cheng, ASME J. Lubr. Technol. **101** 220-229 (1979)
22. B. Jakobsson, L. Floberg, Chalmers Tek Hoegsk Handl, **190** 1-116 (1957)
23. K.O. Olsson, Trans. Chalmers Univ. Technol. **308** 1-60 (1965)
24. A. Kumar, J.F. Booker, ASME J. Tribol. **113** 276-286 (1991)
25. Y. Xie, Y.J. Li, S.F. Suo, X.F. Liu, J.H. Li and Y.M. Wang, Sci. China phys. mech. **56** 1909-1919 (2013)

The Concrete Testing by Ultrasonic Pulse Velocity (UPV)

Weerapol and Namboonruang[1,a]

[1]Department of Civil Technology and Architecture, Faculty of Science and Technology, Muban Chombueng Rajabhat University, MCRU 70150, Thailand

Abstract. The study of an infrastructure development is continuously one of the targets of the national economic and social development plan. The purpose is to develop the local area sustainably and increase the ability of local authorities who have a direct role to promote and plan the other factors effectively. To success this plan the government agencies have to focus on the development dimension and taking into account other factors that may affect the way of life of the community.

Keywords: Rural infrastructure assessment, Half-cell potential, UPV test, Information engineering

1 Introduction

Recently, coastal and marine concrete structures were constructed from GGBS concrete because high volume of GGBS can contribute to reduction of temperature gradient [1,2]. Therefore, cement replacement materials such as fly ash, silica fume, blast-furnace slag, etc. may reduce greatly the probability of steel corrosion as well as the permeability of concrete [3,4,5,6,7]. In a study of the assessment of infrastructure development previous explained, the approach may not be sufficient to meet the in-depth local development in all dimensions. An integrating technology assessment research with public policy and It was proposed that the research process should include a means of engaging with the values of the political environment [8,9]. Therefore, the direction and knowledge determination need to be analyzed changes of circumstances which are suitable to the modern conditions occurred from the collaboration of multiple parties which requires the integration of various knowledge's for conducting the direction of local development as appropriate.

Fig 1. show the western rural organizations, Thailand (area of the study)

2 Methodology

Assessment of engineering infrastructure is starting role and one of the factors that is important to the development of local government agencies in Ratchaburi province in western region of Thailand in figure 1. The aim of this study is to focus on the integration of the physical dimension factor. This is to say the assessment of the local infrastructure conditions that area effected by various environmental conditions in each local area authority including society which need to assess the infrastructure needed by the people in the area. These two dimensions are integrated to understand the economic indicators, and policies for the planning application to compose a policy to meet the needs of the community and local development appropriately.

3 Results and Discussions

3.1 Laboratories and in-situ tests

As mentioned above, it was found that the concrete structure of local area authority's building is cracking. Disintegration of concrete and water permeability are the cause of corrosion of reinforcing steel, and some may not be able to use it in a secure environment. Considered by the strength components of 42 infrastructure examples in the western area of local organizations of Thailand, it was found that using non-destructive test by Rebound Hammer Test the strength value were approximately 84.19 percent compared to the coring test. According to this test it shows that the compressive strength of concrete appears likely to remain moderate and likely slightly lower than that of design.

[a]Corresponding author: bannork_civil@hotmail.com

Fig 2. show the laboratory test of reinforced concrete specimens

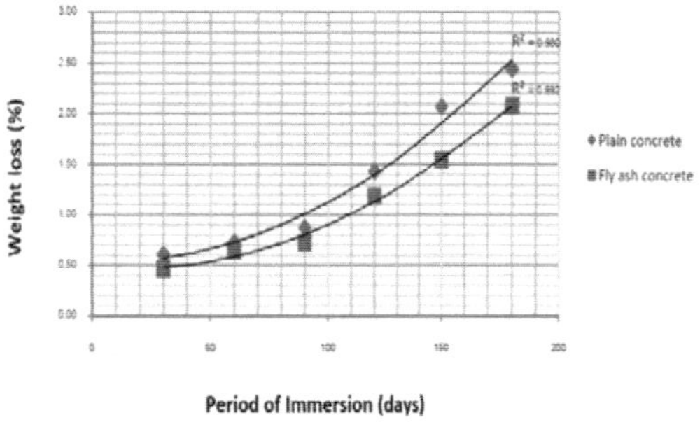

(a)

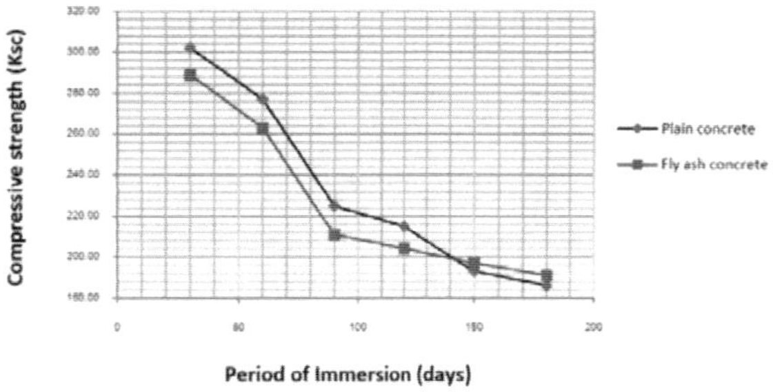

(b)

Fig 3. show relationship of weight loss (a) and compressive strength (b) of the reinforced concrete specimens

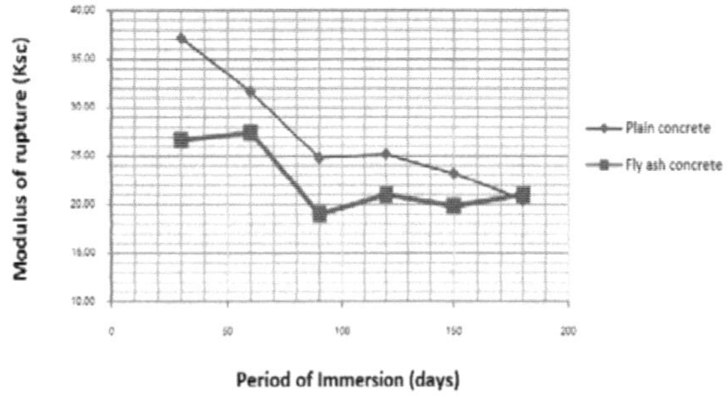

(c)

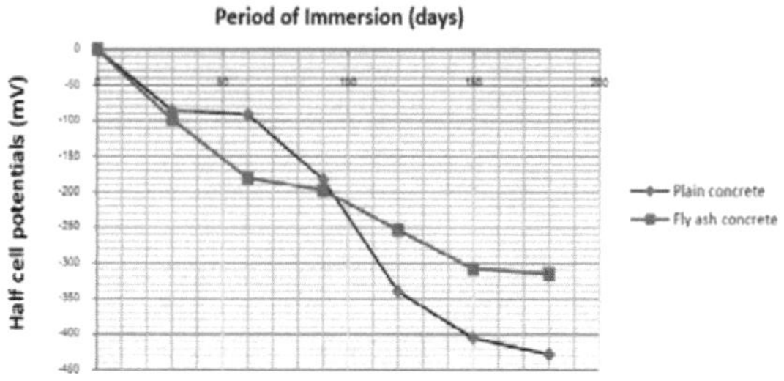

(d)

Fig 4. show relationship of weight loss (a), compressive strength (b), MOR (c) and half-cell potential (d) of the reinforced concrete specimens

Of the field tests considered by the weight loss of steel reinforcement in the concrete test of two example types compared to the test laboratory, results show that the electricity potential is approximately 350 mV (410 mV) higher than that of the overall test. So it could be estimated that the weight of steel reinforcing concrete tends to corrosion more than one time as is consistent with the laboratory tests for the age of 90-180 days, the rate of reduction of steel weight are 1.980 and 1.603 % of -428 mV and -315 mV respectively shown in figure 3-4. This result confirms that the potential of the field tests can adequately explain the tendency to corrosion and the reduction of the weight of steel. As well as the concrete testing by Ultrasonic Pulse Velocity (UPV) test showed that the concrete condition is fair which tends to correspond with the voltage of electricity and the value of the strength with significantly, that is, the PV reduced by increasing of amount of the electrical potential difference and decreasing of the strength. The average value is of 3,180 m/s for the average electrical potential difference at -410 mV of overall tests shown in figure 6.

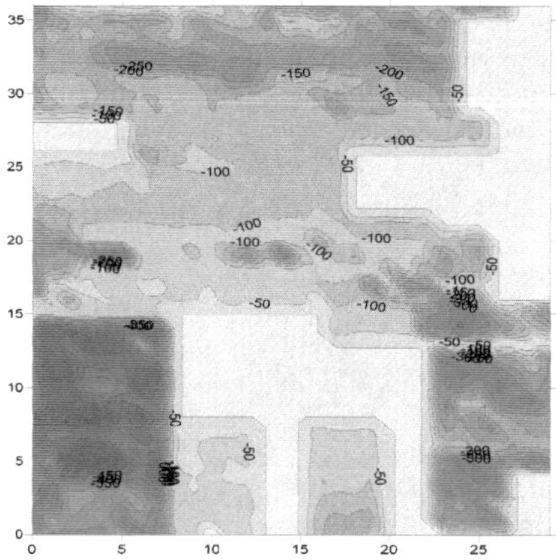

 DOI: 10.1051/matecconf/20166105017

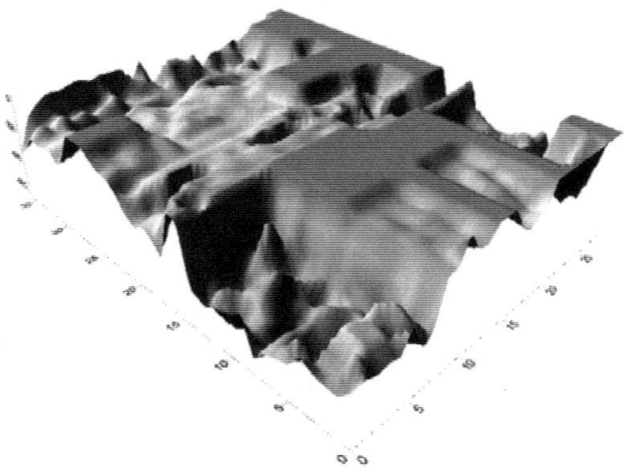

Fig 5. show potential profile of reinforced concrete slab by half-cell measurement

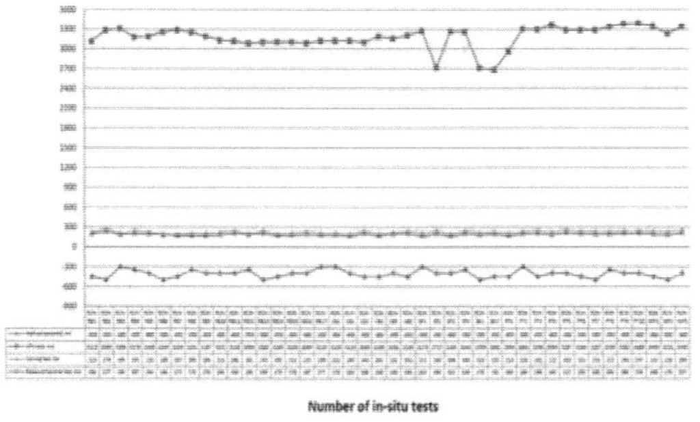

Fig 6. show relationship of half-cell potential, UPV, coring, and rebound hammer tests of infrastructure examples in the western area of local organizations of Thailand

3.2 An economic

For economic analysis and policy evaluation for assessment of infrastructure found that they should be repaired and maintained which will prolong the service life being more than 10 years compared to overall infrastructure. Of the above factors, when analyzed in terms of economics, the government benefits from the use of the building longer than that of design and operation (Service life$_{50}$) at percentage of 20.00 and 28.57 (Service life$_{35}$). If consider on an economic value of the maintenance of overall local infrastructures, government will get benefit about 94.57 million baths.

3.3 An evaluation of the information engineering database management system

For the participation of the community in view of the local infrastructure, it was found that the participants had the most satisfied for four factors; such as an overview of service infrastructure components ($\bar{x}$ = 4.33), benefits of the infrastructure policy ($\bar{x}$ =4.53), the appropriateness of the infrastructure ($\bar{x}$ = 4.26), and the knowledge that was available to meet the local development ($\bar{x}$ = 4.70) as well as an evaluation of Information Engineering system of local infrastructure. Therefore, the documents and information systems show an overall at a good level ($\bar{x}$ =3.64). The solved - Improved document system is at a high level ($\bar{x}$ =3.61), and satisfaction in the operating system($\bar{x}$ =3.69). Finally, the satisfaction of website is at the moderate level ($\bar{x}$ =3.43) shows in (e), (f), (g) and (h) of figure 7.

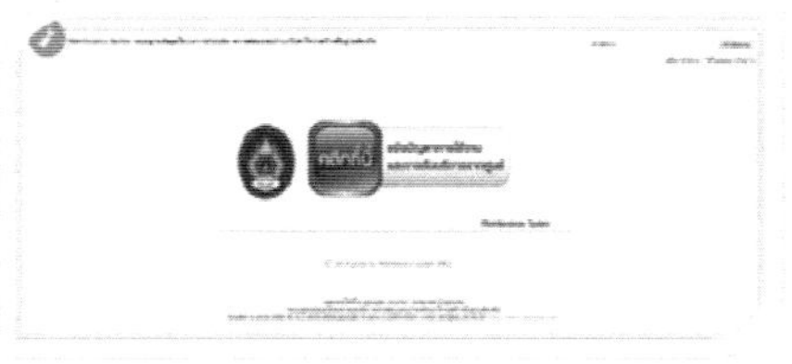

(e)

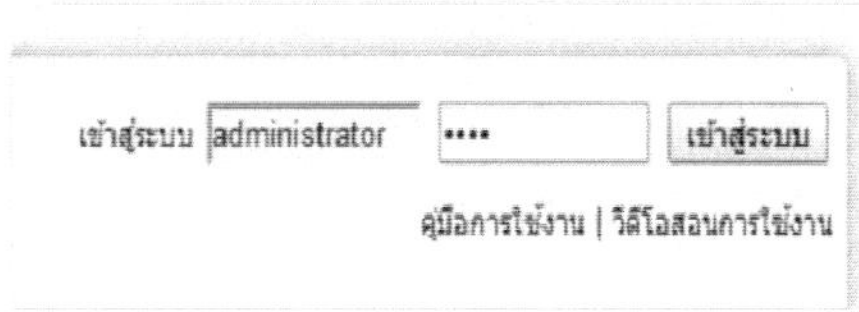

(f)

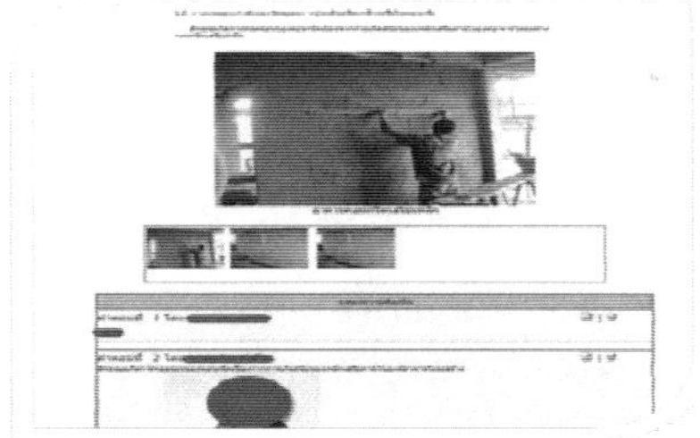

(g)

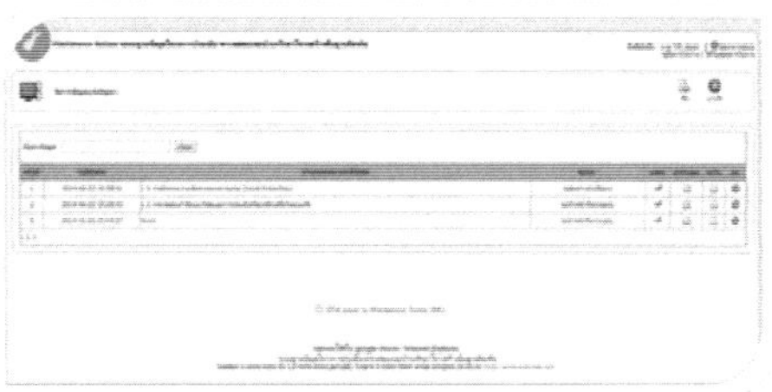

(h)

Fig 7. show the website pattern of the information engineering database management system

4 Conclusions

In RC floor slab of the rural infrastructure public building where its thickness is thin compared to its effective span length, the crack pattern generates due to the corrosion caused by the repetition of moisture and chloride condition. Generally, in the early stage of deterioration, cracks generate in perpendicular direction to the axis of the structure where the distribution of steel bars are less and weight of steel reinforcing concrete tends to corrosion more than one time as is consistent with the laboratory tests for the age of 90-180 days

The concrete testing by Ultrasonic Pulse Velocity (UPV) test showed that the concrete condition is fair which tends to correspond with the voltage of electricity and the compressive strength of concrete appears likely to remain moderate and likely slightly lower than that of design on in-situ tests.

The government benefits from the use of the infrastructure longer than that of design and operation of 20 % at Service $life_{50}$. For the evaluation of the use of information technology infrastructure by local overview of the four factors found that user satisfied in the use of information technology systems engineering for managing local engineering infrastructure at high level.

Acknowledgements

The author would like to express gratitude to researchers, namely Dr. Rattanakorn Rawangkul, STRR Engineering CO.,LTD, Thailand, Assoc. Prof. Dr. Wanchai Yodsudjai, Head of Department of Civil Engineering, Kasetsart University, Assoc. Prof. Dr.Trakool Aramraks, Department of Civil Engineering, Kasetsart University, Thailand for their invaluable help and constant encouragement throughout this research. Sincere thanks to the Department of Civil and Architecture for instrument test supporting and Muban Chombueng Rajabhat University, MCRU for partial financial support.

References

1. D.M. Roy, A. Kumar and J. Rhodes, ACI SP-91, V.M. Malhotra, ed., Vol. 1423, 1986.
2. Z.G. Matta, Concrete Intemat., 14(5) (1992) 47-48.
3. A.A. Ramezanianpour, Effect of curing on the compressive strength, resistance to chloride-ion penetration and porosity of concretes incorporation slag, fly ash or silica fume, Cem. Concr. Compos. 17(1995) 125– 133.
4. G.J. Osborne, Durability of portland blast-furnace slag cement concrete, Cem. Concr. Compos. 21 (1999) 11 – 21.
5. E.F. Irassar, M. Gonzalez, V. Rahhal, Sulphate resistance of type Vcements with limestone filler and natural pozzolana, Cem. Concr.Compos. 22 (2000) 361– 368.
6. K.M.A. Hossain, M. Lachemi, Corrosion resistance and chloride diffusivity of volcanic ash blended cement mortar, Cem. Concr. Res. 34 (4) (2004) 695–702.
7. O.E. GjArv, Effect of condensed silica fume on steel corrosion in concrete, ACI Mater. J. 92-M60 (1995) 591– 598.
8. OCce of Technology Assessment. Development of Medical Technology: opportunities for Assessment, U.S.Government Printing OCce, Washington, DC, 1976.
9. B. Bozeman, F. Rossini, Technology assessment and political decision-making, N.Z. Med. J. 104 (1991) 199–202.
10. ASTM C 42. Standard Method for Obtaining and Testing Drilled Cores and Sawed Beams of Concrete, West Conshohocken, USA, (2008).
11. ASTM C 597 Standard Test Method for Ultrasonic Pulse Velocity Through Concrete, West Conshohocken, USA, (2008).
12. ASTM C 805 Standard Test Method for Rebound Number of Hardened Concrete, West Conshohocken, USA, (2008).
13. ASTM C876 Standard Test Method for Half-Cell Potential of Uncoated Reinforcing Steel in Concrete, West Conshohocken, USA, (2008).

Study on energy transmission through metal wall based on piezoelectric material

Yun-zhe Sun[1], Ding-xin Yang[1] , Zheng Hu[1], Hai-feng Hu[1] and Bao-jian Hou[1]

[1] Laboratory of science and technology on integrated logistics support, National University of Defense Technology, Hunan Changsha, China

Abstract. The efficiency of electric energy transmission through metal wall by ultrasonic wave mainly depends on the relevant parameter of the sandwich structure system, which is consist of a transit piezoelectric transducer, a receive piezoelectric transducer and a metal wall. This paper discussed the characteristics measurement of piezoelectric energy transmission channel through metal wall based on a kind of piezoelectric transducer. First, the intrinsic impedance characteristic of piezoelectric transducer is measured using a high precision impedance analyzer. Then ultrasonic energy transmission channel through a metal wall testing system is analyzed. The voltage ratio under different frequencies of the channel is measured and the maximum energy transmission frequency is obtained. Mason model is verified and compared with the experimental results, which provides a beneficial reference for further researches on ultrasonic through metal wall energy transmission.

1 Introduction

Many sensing applications are used within metallic vessels to monitor extreme environmental conditions. With the disadvantages of feed-through wires, it proposed a great challenge to the energy supply system and the technology of transmitting energy through metal wall has become a hotspot. Under the influence of the Faraday shielding [1], traditional wireless electromagnetic energy could not transmit through metal barriers with no decay [2]. An alternative approach to transmit electric energy is ultrasonic which has perfect performance when through solid wall even it's a metal wall [3]. Because of its perfect performance in transmission capacity and power transmission efficiency, ultrasonic energy soon got the attention of the public [4]. Using signal generator can generate electric signal, which can be converted into ultrasonic vibrations by the transducer because of the inverse piezoelectric effect and then the ultrasonic waves penetrate the solid metal wall [5]. Simultaneously, the acoustic mechanical vibration energy of the metal wall can be received and converted into electric energy by the piezoelectric transducers [6] because of the piezoelectric effect, which provides a means for wireless energy transformation.

In order to efficiently transfer energy through a metal wall, it is important to modeling the acoustic-electric energy transmission channel. Numerous equivalent circuit theories have been developed. Most of the models so far described in the literature [7-9] have been used for only the simulation of the piezoelectric transducer. First, the channel of energy transmission of the theoretical analysis is introduced. Then the experimental of characteristics of through metal wall based on ultrasonic energy transmission channel are test. The experimental results and simulation results are discussed in last.

2 Piezoelectric Transducer Impedance and Acoustic-Electric Energy Transmis sion Channels

Piezoelectric ceramic can be made into piezoelectric devices for various purposes because of its piezoelectric effect. Using acoustic waves to transfer piezoelectric energy by transmit and receive piezoelectric transducer with the structure by coaxially aligning a pair of piezoelectric transducers on opposite sides of a metal barrier, just like a 'Sandwich', the structure described as shown in Fig. 1.Continuous-wave excite the outside transducer to create an acoustic beam which is captured by the inside transducer, forming an acoustic-electric link for power harvesting circuitry.

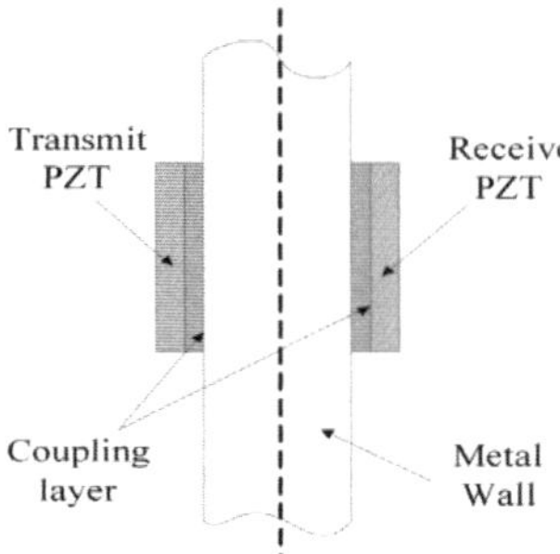

Figure 1. Structure of wireless energy transfer through metal wall

In the work of Hu [10] the piezoelectric and coupling layer was solved using the wave equation and linear equations of piezoelectricity. In order to eliminate the elements circuit between the top transformer and the node of the acoustic transmission and the first lumped equivalent circuit was proposed by Mason. The modeling of other theories based on equivalent circuits develop by Sherri [11] that can easily be simulated when connect to other circuits or networks. This modeling analysis conveniently simulates the dynamic behavior of the mechanical vibration and electricity conversion systems. The schematic of acoustic and electric equivalent circuit of Mason model is shown in Fig. 2.

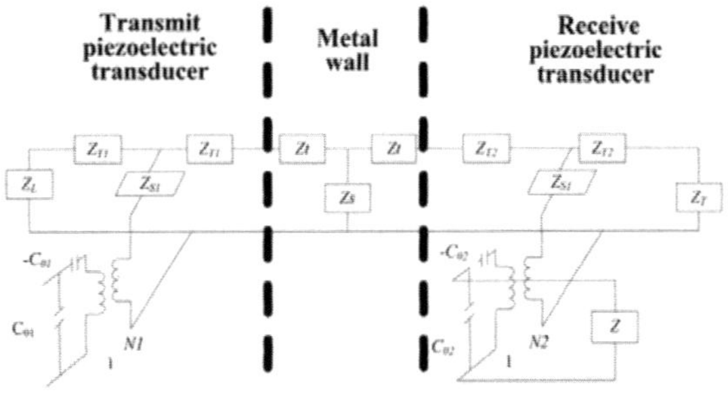

Figure 2. the schematic of acoustic and electric equivalent circuit of Mason model

The piezoelectric material parameters of modeling required are clamped complex permittivity ε 33S, complex electromechanical coupling kt, and open circuit complex elastic stiffness c33D. Besides ρ =density, t= thickness, A=area.

$$k_t^2 = e_{33}^2 / c_{33}^D \varepsilon_{33}^s = h_{33}^2 \varepsilon_{33}^s / c_{33}^D \tag{1}$$

$$h_{33} = k_t \sqrt{c_{33}^D / \varepsilon_{33}^s} \tag{2}$$

Mason Model of transmit piezoelectric transducer

$$C_{01} = \frac{\varepsilon_{33_1}^s A_1}{t_1} \tag{3}$$

$$N_1 = C_{01} h_{33_1} \tag{4}$$

$$Z_{01} = \rho_1 A_1 V_1^D = A_1 \sqrt{\rho_1 \varepsilon_{33_1}^S} \tag{5}$$

$$\Gamma_1 = \frac{\omega}{V_1^D} = \omega \sqrt{\frac{\rho_1}{c_{33_1}^D}} \tag{6}$$

$$Z_{T1} = iZ_{01} \tan\left(\Gamma_1 t_1 / 2\right) \tag{7}$$

$$Z_{S1} = -iZ_{01} \csc\left(\Gamma_1 t_1\right) \tag{8}$$

The configuration shown as Fig. 2 can also apply to voltage transformers and the model proceeds as with any network solution. In this approach the voltage on mechanical side of a transformer corresponds with the velocity of the surface. Voltage is multiplied by a factor of N when in the process of conversion of electrical energy to mechanical energy. Accordingly, Reverse process have to be divided by a ratio N. Starting from the model of equivalent circuit loads to the left, the impedance is as the following

$$Zx1 = Z \frac{Z02}{Z + Z02} \tag{9}$$

$$Zx2 = Zx1 - Z02 \tag{10}$$

$$Zx3 = N_2^2 Zx2 \tag{11}$$

$$Zx4 = Zs2 + Zx3 \tag{12}$$

$$Zx5 = Zx4 \frac{(Zt2 + ZtR)}{(Zt2 + ZtR + Zx4)} \tag{13}$$

$$Zx6 = Zx5 + Zt2 + Ztw \tag{14}$$

$$Zx7 = Zx6 \frac{Zsw}{Zx6 + Zsw} \tag{15}$$

$$Zx8 = Zx7 + Ztw + Zt1 \tag{16}$$

$$Zx9 = Zx8 \frac{(Zt1 + ZtL)}{(Zt8 + Zt1 + ZtL)} \tag{17}$$

$$Zx10 = Zx9 + Zs1 \tag{18}$$

$$Zx11 = \frac{Zx10}{N_1^2} \tag{19}$$

$$Zx12 = Zx11 - Z01 \tag{20}$$

$$Zx13 = Zx12 \frac{Z01}{(Zx12 + Zx01)} \quad (21)$$

Where the Zx13 is electrical impedance of the transmit transducer. According to the above formulas a variety of equivalent impedance can be calculated. Note that is the impedance value is dynamic changes because the magnitude and the phase of sinusoidal input changes.

3 Experiments and Simulation

The system of ultrasonic energy transmission through metal wall is consists of a signal generator of TEKRONIX-AFG302, the channel of metal wall coupling with piezoelectric transducers and oscilloscope of TEKTRONIX-TDS2012B. The detail of the experimental process as following:

Running the software of Agilent Connection Expert which will auto search the net port connecting to the PC and indication it. Then, we refresh the tmtool toolbox software of MATLAB and click the connect button to get handler functions. This protocol realizes effective and reliable communication between debugger and monitor and simplifies their design. In this system, the signal can be controlled by programming on PC, which gives full play to the advantage of the computer, by this way by using MATLAB software to control signal generator to generate sweeping frequency excitations. Meanwhile, we get the graph of frequency response received by oscilloscope. Operation schematic diagram of experimental process is shown in Fig.3.

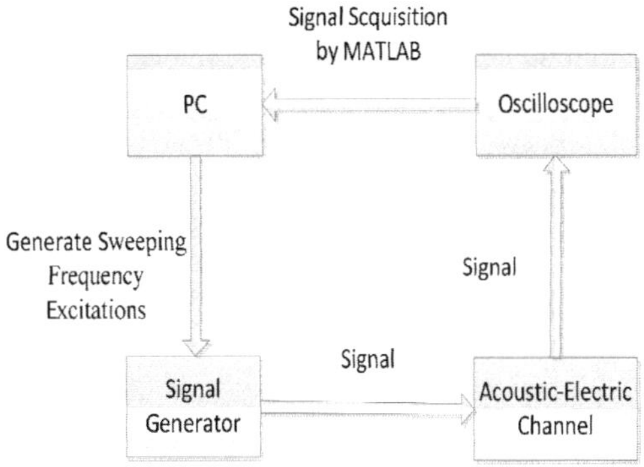

Figure 3. Operation schematic diagram of experimental process

The material properties used in the experimental are shown in Table 1.

Table 1. The material properties used in the experiment

Property	Value
Density of piezoelectric[g/cm^3]	7800
Area of piezoelectric transducer on wall[m^2]	0.01
Wall thickness[m]	0.022
Piezoelectric transducer thickness[m]	0.002
Piezoelectric transducer coefficient e_{33}[C/m^2]	23.2
Elastic stiffness at constant displacement c^D_{33} [N/m^2]	15.87
Mechanical quality factor Q	80

Mason equivalent circuit model of the receiving transducer parameters and transmit transducers take the same values because they use the same type of transmit transducer.

The experimental result is shown in Fig. 4, in order to test the theory of equivalent circuit model, we compared the result of simulation and experimental, from which we can see the theoretical simulation results show good agreement with experiment results.

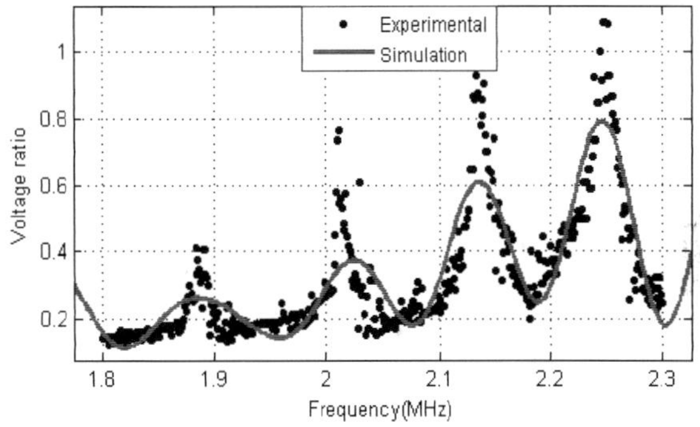

Figure 4. Simulation result compared with experimental

Measured result of signal piezoelectric transducer using a precision impedance analyzer is shown in Fig. 5, from which we can see that the minimum impedance of this kind of piezoelectric transducer is about 2.1MHz.

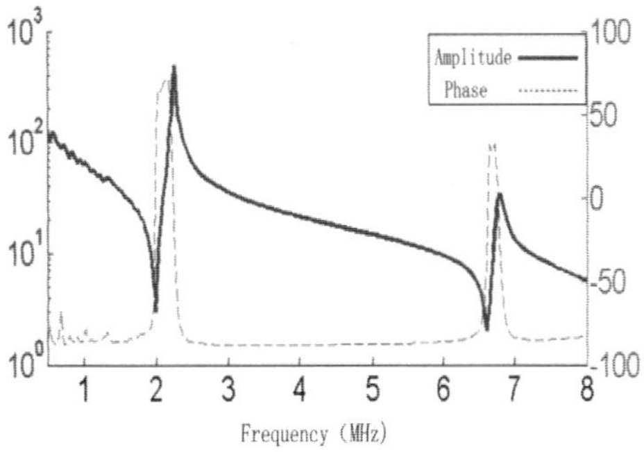

Figure 5. A signal piezoelectric transducer impedance character

Point to be noted that the physical parameters used in these models vary depending on piezoelectric transducer's statements: hanging freely in air or attached to the metal wall with epoxy or a cyanoacrylate adhesive. Piezoelectric transducer coupled with different metal wall and the result of is shown in table2.

Table 2. Results of coupled forms with different metal wall

	Intrinsic frequency	Aluminum (5mm)	Steel(10mm)
Impendence[Ω]	3.017	21.4	6.03
Frequency[MHz]	2.101	2.135	2.138

The result show effectiveness of the simulation modeling algorithm and be fit for application in the engineering. Meanwhile, wireless energy transmission using ultrasonic is feasible.

4 Summary

Although some important milestones were reached in the technology of wireless energy transmission using ultrasonic through metal wall with the effort of researches during recent decades, much work remains to be done in practical engineering application. The theory model must be developed further to allow for functionality capable for use in a practical design.

According to the engineering needs of wireless energy transmission through metal wall, we analyze the acoustic and electric equivalent circuit of Mason model. Based on the preliminary analysis theory of the model, we test the characteristics of through metal wall based on ultrasonic energy transmission with the experimental methods. Voltage is a measure of the energy carrier by the charge, strictly: voltage is the "energy per unit charge", thus, using voltage representing energy is reasonable.

ACKNOWLEDGEMENTS

The authors would greatly appreciate the support provided by National Nature Science Foundation of China No.51375485 and No.51305449 for this work.

References

1. Yang, Ding-Xin, et al, Through-Metal-Wall Power Delivery and Data Transmission for Enclosed Sensors: A Review. Sensors 15.**12** (2015) 31581-31605.
2. Lawry, Tristan J., et al, A high-performance ultrasonic system for the simultaneous transmission of data and power through solid metal barriers, Ultrasonics, Ferroelectrics, and Frequency Control, IEEE Transactions on 60.**1** (2013) 194-203.
3. Wilt, K. R., Lawry, T. J., Scarton, H. A., Roa-Prada, S., Saulnier, G. J., Ashdown, J. D., Das, P. K. and Pinezich, J. D. Mechanical Design Implications on Power Transfer Through Thick Metallic Barriers Using Piezoelectric Transducers, *ASME 2010 International Mechanical Engineering Congress and Exposition*, (2010).
4. Zhang, Q., Shi, S. and Chen, W. Research on Effective Electric-Mechanical Coupling Coefficient of Sandwich Type Piezoelectric Ultrasonic Transducer Using Bending Vibration Mode, *Advances in Mechanical Engineering*, **7** (1), 204370-204370, (2014).
5. Chang, Zensheu, et al, Power loss consideration in wireless piezoelectric acoustic-electric power feedthru, The 14th International Symposium on: Smart Structures and Materials & Nondestructive Evaluation and Health Monitoring. International Society for Optics and Photonics(2007)
6. Lawry, T. J., Wilt, K. R., Roa-Prada, S., Ashdown, J. D., Saulnier, G. J., Scarton, H. A., Das, P. K. and Pinezich, J. D. Electrical optimization of power delivery through thick steel barriers using piezoelectric transducers, *Proceedings of SPIE - The International Society for Optical Engineering*, **7683** (6), 768314-768314-12, (2010).
7. Krimholtz, Richard, David A. Leedom, and George L. Matthaei, New equivalent circuits for elementary piezoelectric transducers, Electronics Letters 6.13 (1970): 398-399.
8. Xiao, S. and Li, Y. Modeling and High Dynamic Compensating the Rate-Dependent Hysteresis of Piezoelectric Actuators via a Novel Modified Inverse Preisach Model, *IEEE Transactions on Control Systems Technology*, **21** (5), 1549-1557, (2013).
9. Paradies, R. and Schläpfer, B. Finite element modeling of piezoelectric elements with complex electrode configuration, *Smart Materials & Structures*, **18** (2), 1282-1294, (2009).
10. Hu, Yuantai, et al, Transmitting electric energy through a metal wall by acoustic waves using piezoelectric transducers, Ultrasonics, Ferroelectrics, and Frequency Control, IEEE Transactions on 50.7 (2003) 773-781.
11. Wilt, K. R., Scarton, H. A., Roa-Prada, S., Saulnier, G. J., Ashdown, J. D., Lawry, T. J., Das, P. K., Gavens, A. J., Wilt, K. R. and Scarton, H. A. Finite Element Modeling and Simulation of a Two-Transducer Through-Wall Ultrasonic Communication System, *ASME 2009 International Mechanical Engineering Congress and Exposition*, (2009).

MATEC Web of Conferences **61**, 05019 (2016) DOI: 10.1051/matecconf/20166105019
APOP2016

Scarf Repair of Composite Laminates

Zonghong Xie[1], Xiang Li[1,a] and Qun Yan[2]

[1]Northwestern Polytechnical University, Department of Aerospace Design Engineering , 710072 Xi'an, China
[2]Shenyang Aircraft Design & Research Institute, 110000 Shenyang, China

Abstract. The use of composite materials, such as carbon-fiber reinforced plastic (CFRP) composites, aero-structures has led to an increased need of advanced assembly joining and repair technologies. Adhesive bonded repairs as an alternative to recover full or part of initial strength were investigated. Tests were conducted with the objective of evaluating the effectiveness of techniques used for repairing damage fiber reinforced laminated composites. Failure loads and failure modes were generated and compared with the following parameters: scarf angles, roughness of grind tool and number of external plies. Results showed that scarf angle was the critical parameter and the largest tensile strength was observed with the smallest scarf angle. Besides, the use of external plies at the outer surface could not increase the repairs efficiency for large scarf angle. Preparing the repair surfaces by sanding them with a sander ranging from 60 to 100 grit number had significant effect on the failure load. These results allowed the proposal of design principles for repairing CFRP structures.

1 Introduction

Adhesively bonded scarf repairs provide a significant recovery of residual strength in damaged composite structure and are compared with external patch repairs, and scarf repairs have relatively high level to minimize aerodynamic disturbance[1]. For highly loaded advanced composite structures, scarf angles ranging from 20:1 to 60:1 are often required to restore a damaged structure to its as-designed ultimate strength.

The restore efficiency are affected by many parameters, typically including surface preparing, scarf angles and external plies. Harman and Wang [2] investigated the use of a low stiffness patch to repair carbon-fiber epoxy composites for scarf joint repairs. They found that a low stiffness patch improves the uniformity of adhesive stresses along the joint and consequently enhances the joint strength. Liu and Wang [3] used a 3D progressive damage model to investigate the effects of several repair parameters on the failure initiation strength, ultimate strength and failure mechanism of repaired open-hole composite plates in tensile tests. Their research showed that overlay thick patches will deteriorate the strength of the repaired structures. Campilho [4] investigated the tensile behavior of adhesive-bonded carbon/epoxy scarf repair using scarf angles ranging from 2° to 45°. Two distinct failure modes were observed experimentally: type A and type B failures. Type A was observed for the repairs with 15°, 25° and 45° scarf angles, consisting of an entirely cohesive failure of the adhesive layer, which occurred abruptly without visible crack initiation. Type B failure occurred for the repairs with 2°, 3°, 6° and 9° scarf angles, consisting of cohesive and interlaminar/intralaminar failure. Kumar [5] investigated adhesively bonded aerospace composite scarf joints, with scarf angles ranging from 0° to 5°, in uni-axial tension both experimentally and numerically. Results showed that fibre fracture and pull-out failure occurred for scarf angles less than about 2° and cohesive shear failure of the adhesive occurred for scarf angles more than 2°. The knockdown in tensile strength was most prominent for scarf angles less than about 1°. Charalambides [6,7] and Pinto [8] investigated the use of over-laminating plies covering the repaired region at the outer or both repair surfaces as an attempt to increase the repairs efficiency. Sung[9] found that preparing the repair surface by sanding them with a diamond sander ranging from 60 to 400 grit number does not significantly affect failure load.

The current experiments for scarf repairs are designed to study these parameters on the repair performance. The tests were conducted at room temperature. The failure load and failure strength were used to evaluate the repair efficiency.

2 Tensile experiment

Plain weave carbon fiber fabric prepreg plies(with T700 fiber and CYCOM970 resin system) were used in the construction of the parent laminate and the repair patch. The mechanical properties were shown in Table 1.

The plates are manually scarfed to the required scarf angle using a die grinder having a detachable abrasive pad of #60 grade and #100 grade. The scarf surfaces prior the repair are cleaned.

[a] Corresponding author: 895208876@qq.com

Two separate plates were joined together by the scarf techniques. Figure 1 shows the procedure sequence in the repair process of a damaged laminate.Initially a back plate was bonded at the back of the laminate in order to position the two parts. Then prepreg plies with tailored dimensions are placed on top of the plate. The patch materials are bonded to the laminate using adhesive film. Oneexternal plywas placed on the top and its boundaries are sealed with some adhesive. The electrically heated blanket is maintained until the reinforcement(patch) adherends to the parent component, awaiting solidification.

The manufacture process consisted of hand lay-up followed by curing in a hot-plates with 0.8 bar pressure. The heat cycle initiated warming up the plate at 2°C/min, followed by two hours curing at 180°C. Cooling was performed at 3°C/min up to room temperature (Figure 2).

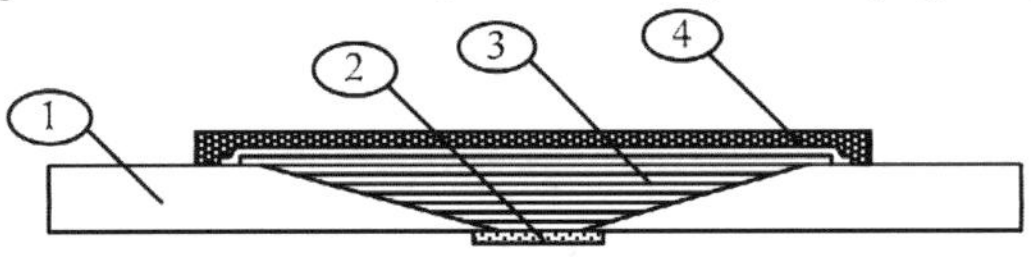

1. Parent Component 3. Repair Ply
2. Back Patch 4. Vacuum Bag

Figure 1. Schematic of the vacuum patch repair process.

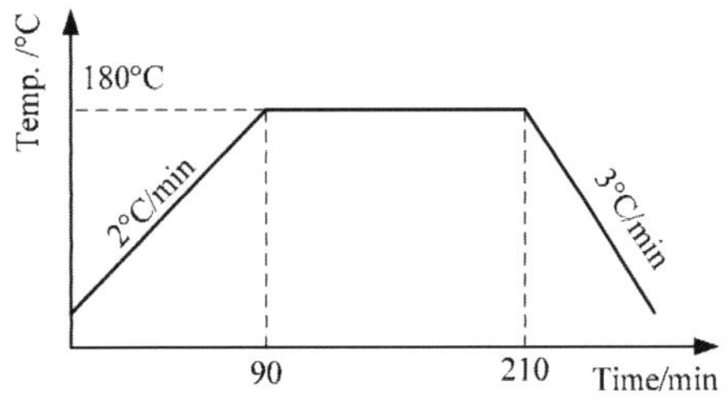

Figure 2. The temperature cycle used in making the repair.

The coupons are prepared by cutting the panel into the standard coupon sizes (ASTM D3039). The coupons dimensions were shown in Figure 3. Three different scarf angles were conducted, and α=2°, 4° and 6°. In order to investigate the external plies effect on the repair efficiency, two different laminates were designed with N=0 and 1. Besides, undamaged specimens, noted as scarf angle α=0°, were tested to verify the repairs recovery efficiency. A total of five specimens were tested in tension for each case described above. The complete test matrix was shown in Table 2.

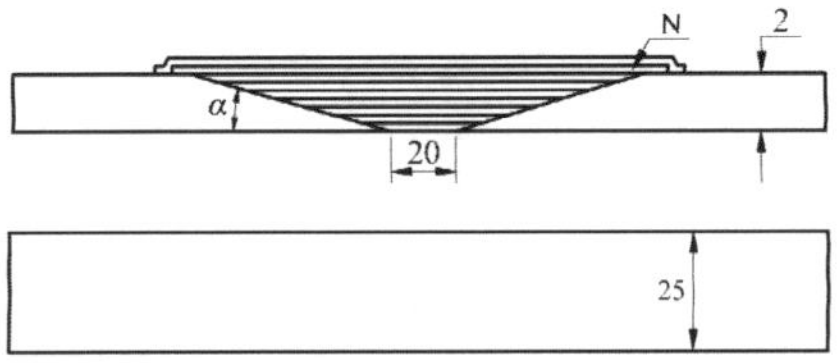

Figure3. Dimensions of the scarf repairs coupon.

The experiments were performed in a displacement control mode. All the specimens were loaded in tension until failure. Based on the failure load, the failure strength is calculated using Eq.(1).

$$\sigma_{max}=F_{max}/(Wt) \quad (1)$$

Strain gages were placed at several locations on the specimen as shown in Figure 4. The strain gage readings were used to analyze the damage initiation and damage growth path.

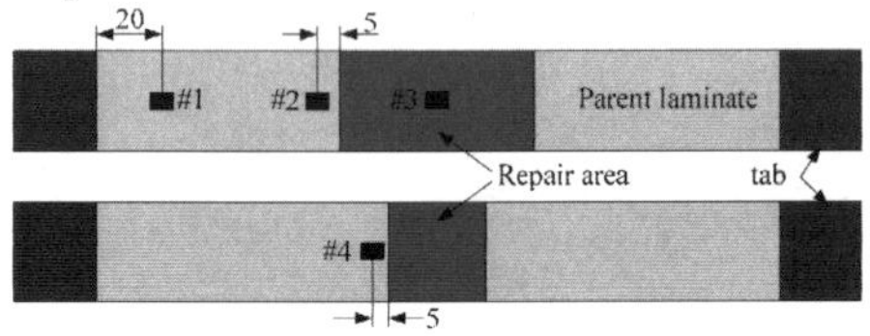

Figure 4. Strain gauge positions of the scarf joint coupon

Table 1. Mechanical properties of carbon-fiber/epoxy.

E1(GPa)	E2(GPa)	G12(GPa)	v12	
54	54	13	0.044	
Xt(MPa)	Xc(MPa)	Yt (MPa)	Yc(MPa)	S(MPa)
537	549	604	724	261

3 Test results of scarf repair specimens

The load-displacement curves for the repaired specimens and undamaged specimens are shown in Figure 5. The repair specimens' stiffness reduces as the adhesive modulus is much lower than the parent laminate. Besides, stiffness reduction experienced by the repairs prior to failure was caused by the softening of the adhesive layer at the bond edges.

The strains measured at critical locations were shown in Figure 6. For the 6° repair structure, the strain increase rate at location #4 decreases with the load whilst it increases at location #3 which indicates that the damage initiates from the bottom of the adhesive and grows along the bondline(see in Figure 7).

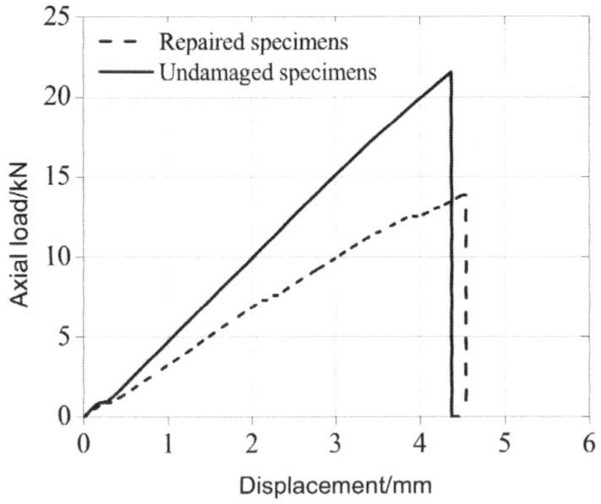

Figure 5. Typical load-displacement curvesof the specimens.

Figure 8 shows the typical damage modes observed in the tests conducted. For the case of 4°~6° repair structure, it was seen that the primary damage mode is adhesive debond. However, for the case of 2°repair structure, intra-laminar failure mode was also found.

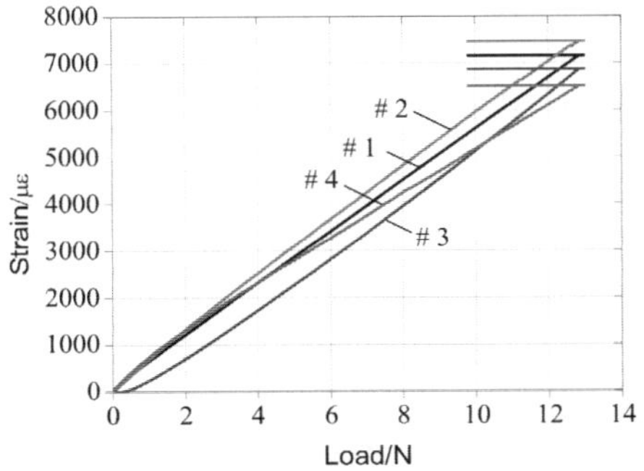

(a) α=2°

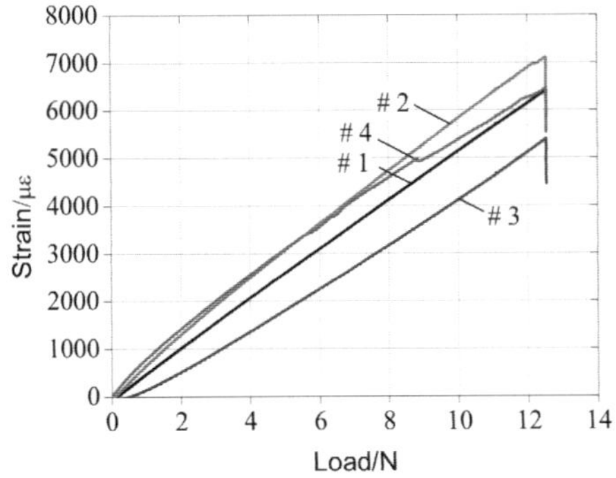

(b) α=4°

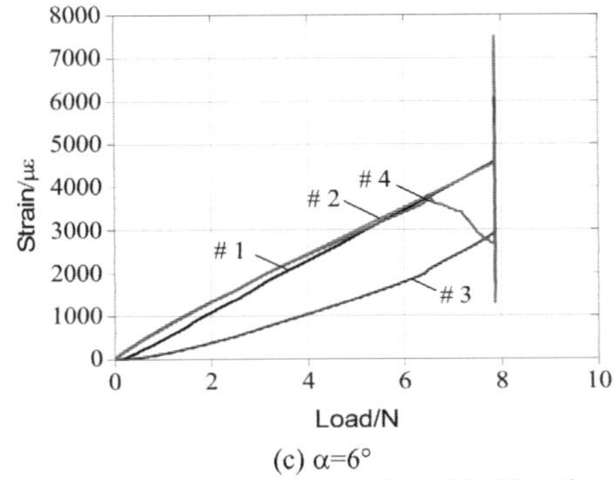

(c) α=6°

Figure 6. Strains measured at critical locations.

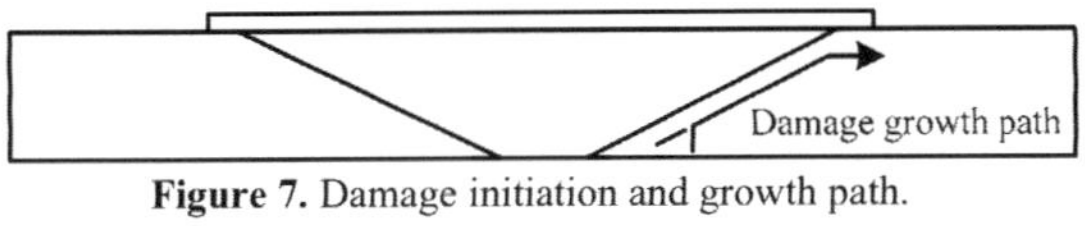

Figure 7. Damage initiation and growth path.

Figure 9 and Table 2 show the failure strength obtained from the scarf repair approach with different repair parameters. The failure strength of the parent material was approximately 455MPa.

Testing of the scarf joints/repair recovered 29%~62% of the parent laminate strength. The failure strength increases as the scarf angle decreases. A slight decreasing trend on failure strength was observed increasing the scarf angle, even though with some variations.

(a) α=2° (b) α=4° (c) α=6°

Figure 8. Failure modes in repaired specimens.

Customarily, prior to repair the repair surface of the base laminate is sanded. To evaluate the influence of sanding on the failure load of repaired specimens, the repair surfaces were sanded with different grit diamond sanders. The failure loads of specimens repaired using different grit sanders are shown in Figure 9. The data indicate that the grit number of the sander have a significant effect on the failure load. For the material and type of repair used in this study, a sander with greater grit number could increase the failure strength for lager scarf angle and inverse trend was observed for small scarf angle.

The addition of external plies on the surface of the specimen resulted in decreased loads. The repair patch applied on one side contributes to out-of-plane bending by affecting the symmetry which resulted in secondary bending effects that begin to add the adhesive peel stress. However, reverse effect was observed in Sung's work [9]. This was because that in Sung's work, a smaller scarf angle was used. For the smaller values of α, normal stresses, σ_n, diminish in relative magnitude to τ_{avg}. Under these conditions, shear stresses, τ_{tn}, govern the repairs behavior [10].

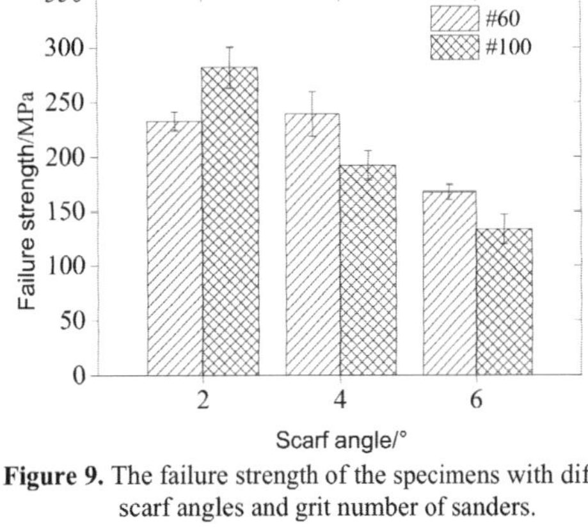

Figure 9. The failure strength of the specimens with different scarf angles and grit number of sanders.

Table 2. Overview of the scarf joint specimens used in the tests.

Scarf angle	Grit number	External plies	σ_{max}/MPa
0°	--	--	455.1
2°	#60	1	232.6
	#100	1	281.8
4°	#60	1	239.2
	#100	1	192.4
6°	#60	1	168.0
	#100	1	133.6
	#100	0	155.8

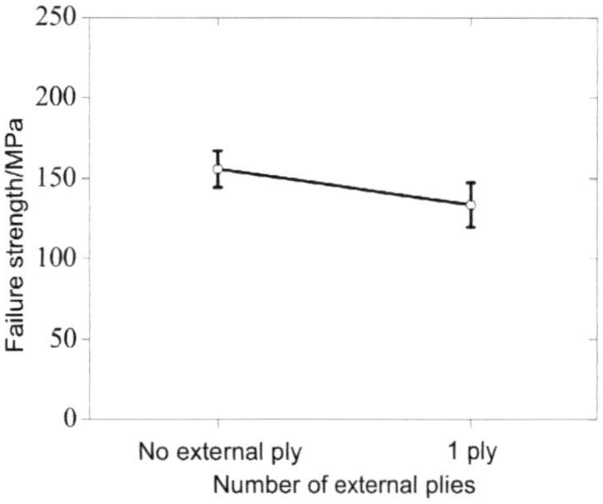

Figure 10. The effect of the number of external plies on the failure strength in scarf repairs.

4 Conclusions

The strength of scarf joints in composite structures has been thoroughly investigated. A series of scarf repair tests were performed to obtain their strength. The mechanical response and failure behavior of the specimens with different scarf angles and surface treatments method were evaluated via the comparison of experimental data. The experiment data shows that the optimal results are obtained from 2 degree using 1 layer for the external plies. The 2-degree repair results in intralaminar failure mode which is a higher strength failure mode. Repairs restored the strength to within 62% of the pristine panels. However, it should be noted that externalplies would reduce failure load for large scarf angle repairs as a result of secondary bending effect.

Acknowledgements

Project supported by National Natural Science Foundation of China (U1233202).

References

1. B. Whittingham, A.A. Baker, A. Harman, D. Bitton. Compos. Part A **40**, 1419 (2009).
2. A.B. Harman, C.H. Wang. Compos. Struct. **75**, 132, (2006).
3. X. Liu, G. Wang. Compos. Struct. **81**, 331, (2007).
4. R.D.S.G. Campilho, M.F.S.F. de Moura, A.M.G. Pinto, J.J.L. Morais, J.J.M.S. Domingues. Compos. Part B. **40**, 149 (2009).
5. S.B. Kumar, I. Sridhar, S. Sivashanker, S.O. Osiyemi, A. Bag. Mater. Sci. Eng. 132, 113 (2006).
6. M.N. Charalambides, R. Hardouin, A.J. Kinloch, F.L. Matthews. Compos. Part A. **29**, 1371 (1998a)
7. M.N. Charalambides, R. Hardouin, A.J. Kinloch, F.L. Matthews. Compos. Part A. **29**, 1383 (1998b)
8. A.M.G. Pinto, R.D.S.G. Campilho, M.F.S.F. de Moura, I.R. Mendes. Int. J. Adhes. Adhes. **30**, 329, (2010).
9. H.A. Sung, G.S. Springer. J. Compos. Mater. **32**, 11 (1998).
10. R.D.S.G. Campilho, M.F.S.F. de Moura, J.J.M.S. Domingues. J. Adhes. Sci. Technol. **21**, 855 (2007).

Breakout Prediction Based on BP Neural Network in Continuous Casting Process

ZHANG Ben-guo[1,a], ZHANG Xinjiang [1,b] and FAN Lifeng[2,c]

[1]Yancheng institute technology, Yancheng 264005, China
[2]Inner Mongolia University, Hohhot, 010070, China
[a]zhangbenguo@163.com, [b]zhangxinjiang1983@163.com, [c]251155328@qq.com

Abstract. An improved BP neural network model was presented by modifying the learning algorithm of the traditional BP neural network, based on the Levenberg-Marquardt algorithm, and was applied to the breakout prediction system in the continuous casting process. The results showed that the accuracy rate of the model for the temperature pattern of sticking breakout was 96.43%, and the quote rate was 100%, that verified the feasibility of the model.

Keywords: continuous casting; breakout prediction; BP neural network

1 Introduction

Continuous Casting is the process whereby molten steel is solidified into a semi finished slab, or some such elementary form, for subsequent rolling in the finishing mills[1]. Breakouts are of major concern in the continuous steel-casting process, because they can lead to severe damage to equipment, significant process downtime, and potential safety consequences. The loss caused by a typical breakout accident is close to 200,000 dollars[2]. Among many forms of breakouts, the sticking-type breakout has the highest incidence and the greatest harm, thus it is very critical to prevent the sticking type breakout occurrence for reducing incidence of breakout accident.

In this paper, based on the indepth study of a variety of breakout prediction systems, a breakout prediction system has been constructed based on a BP (Back Propagation) neural network model which was optimized with LM (Levenberg Marquardt) algorithm. And the system was trained and tested with the data collected at the continuous casting production site.

2 Principle of Sticking-type Breakout Prediction

The prediction of the sticking type breakout is to make the right judgments of the temperature patterns which may cause leakage. The principal of the means is to identify the abnormal temperature waves from normal ones measured by thermocouples embedded in the inner walls of the casting mould. Actually, it is a dynamic wave pattern recognition process, as shown in Fig. 1. Under normal continuous casting conditions, the temperature reading of thermocouples decreases gradually with the growth of the solidified shell' thickness, and the temperature reading measured by the thermocouples in the first row is higher than the temperature reading measured by the second row thermocouples in the same vertical plane, as shown in Fig. 1(a). The temperatures measured by thermocouples imbedded in the mould at each measuring point are relatively stable and can only fluctuate narrowly, as shown in Fig. 1(b). At the continuous casting progresses, the sticker occurs in the vicinity of the meniscus and remains during the negative strip time. A tear in the shell can then occur at the sticking point during the positive strip time when the forces acting on the shell exceed the shear strength of the shell. When the sticker reaches the upper thermocouple mounted close to the meniscus, the temperature reading rises, as shown in Fig. 1(c). At the point of the sticker, the strand sticks to the side of the mould more strongly than elsewhere under the action of the stresses in the strand shell, with the result that the speed of the shell is also reduced at sticking point. The temperatures measured by the upper thermocouples decrease with the gradual healing of tear, and the temperatures measured by the lower thermocouples rise with the crack propagation, as shown in Fig. 1(d). Once the sticker goes past the upper thermocouples, the temperature readings of them begin decreasing, such that it will be smaller than the lower thermocouple reading at some point. Then the breakout alarm is sent out, as shown in Fig. 1(e). As the sticker goes past the lower thermocouple its temperature reading begins decreasing too, as shown in Fig. 1(f).

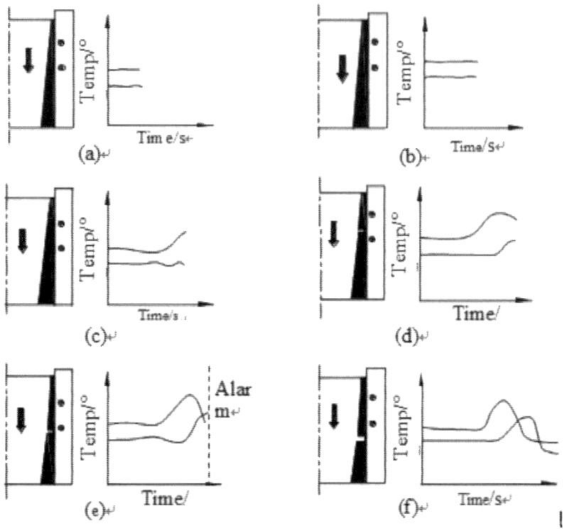

Fig. 1 The principle of sticking type breakout prediction in continuous casting process

3 Breakout Prediction Based on BP Neural Network of LM Algorithm

Currently, BP neural network is the most widely applied neural network, and 80%~90% artificial neural network models in practice are BP neural networks or its changes in various forms. Therefore, BP neural network is the core part of feed-forward neural networks, and is the most essential artificial neural network[3].That anyone of continuous functions in a closed interval can be approximated by a BP network with one hidden layer has been proved by Robert Hecht-Nielson in 1989, thus any n-dimension to m-dimensional mapping can be completed with a three-layer BP network[4]. In this paper, the structure of the BP neural network with the LM algorithm as the learning algorithm is the same as the standard BP network (see Fig.2).

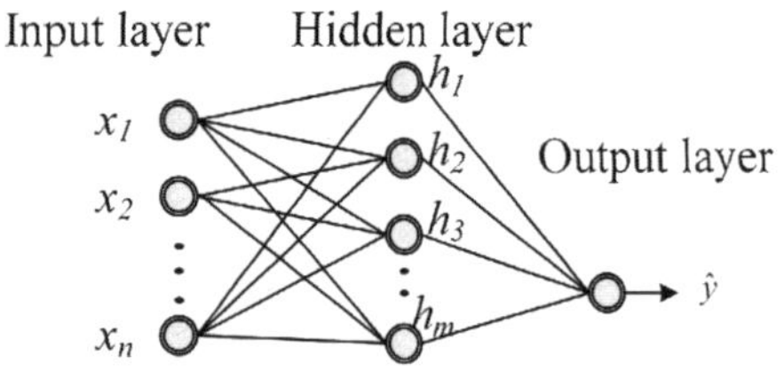

Fig.2 Three layers BP net

4 BP Neural Network of LM Algorithm

The algorithm of the standard BP is based on the gradient descent algorithm. In order for the network to learn, the value of each weight and threshold has to be incrementally adjusted proportionally to the contribution of each unit compared with the total error. The learning process is composed by forward and backward information disseminations: input sample data forward transmission and error information feedback. The mean squared errors of the expectations and the output of neural network tend to be minimized by adjusting the weights and thresholds[5]. It is actually to make a minimum mean square error of an approximate gradient descent algorithm. There are some shortcomings in the convergence process, e.g. slow rate of convergence, existing local minimum.

At present, there are two main methods for increasing the convergence rate of the BP network: (1) using heuristic information technology, e.g. adding momentum term to the learning algorithm; (2) using numerical optimization methods, e.g. conjugate gradient method, LM algorithm, etc. Although the BP algorithm using heuristic technology is simple and intuitive and to some extent can improve the convergence speed, its accuracy achievable is limited. Among the numerical optimization methods, LM algorithm has the fastest convergence rate and the smallest training error. Therefore, LM algorithm is chosen as an optimization algorithm for the BP neural network in the breakout prediction system. The iterative formula for the BP neural network is shown as follows:

$$w(n+1) = w(n)-[J^T J + uI]^{-1} \cdot J^T e(n) \tag{1}$$

Where J is the Jacobian matrix, as follows:

$$J = \left| \frac{\partial e(n)}{\partial w_1} \frac{\partial e(n)}{\partial w_2} \cdots \frac{\partial e(n)}{\partial w_i} \right| \tag{2}$$

The calculation steps are as follows:

1) Give the initialize weights and the value of training error allowed: ε, β, μ, set $n = 0, u = u_0$.

2) Calculate the output of the network $y(n)$ and the error $e(n)$.

3) Calculate the Jacobian matrix J.

4) Calculate the rectification $\Delta w(n)$ of the weights, and change the weights.

5) If $e(n) < \varepsilon$, turn to step (7), otherwise, set $w(n+1)$ as the weight and calculate the error $e(n+1)$.

6) If $e(n+1) < e(n)$, set $n = n+1$, $\mu = \mu/\beta$, and turn to step (2); otherwise, don't change the weights this time, set $w(n+1) = w(n)$, $u = u\beta$, turn to step (4).

7) Stop.

5 Data Preprocessing

In order to eliminate the effects of the order of magnitude on the output of the neural network and highlight the date with Breakout features, the data are normalized as follows:

$$X^*(i)=\begin{cases}\dfrac{X(i)}{\sqrt{\sum_{i=1}^{15}X^2(i)}} & (X_{\max}-X_{\min})\le\lambda \\ \dfrac{X(i)-X_{\min}}{X_{\max}-X_{\min}} & (X_{\max}-X_{\min})>\lambda\end{cases}$$

(3)

Where, $X=\{X_1,X_2,...X_n\}$ is the raw data, n is the dimension of the input column vector, and λ is the stability threshold.The temperature sequence is considered to be stable, if the temperature change is not greater than λ. Otherwise, the temperature sequence is considered to be volatile. Usually, the change of the temperature is greater than 35℃, when the breakout occurs. this paper takes λ= 25℃ through the analysis of historical data collected in a steel plant.

Using the general data normalization method would undermine this stability, and it is not conducive to the identification of the network model. So, the equation (3) is used for the data normalization in this paper, and that problem is well resolved.

6 The Determining of the Network Parameters

1) The nodes in the input layer and hidden layer: It is found that the time for the temperature fluctuation measured by a single thermocouple is about 30s during the breakout process. Considering the size and the sensitivity of the neural network, the number of the nodes in the input layer and the output layer is set to 15 and 1. The output of the neural network indicated the likelihood of the breakout.

2) The determination of the optimal number of hidden layer neurons: According to Kolmogorov's theorem, the approximate relationship between the number of the input layer neurons and the number of the hidden layer neurons in three-layer neural network is:

$$K\le 2m+1$$

(4)

Where K is the number of the hidden layer neurons and m is the number of the input layer neurons. So the number of the neurons in the hidden layer is less than or equal to 31 according to Eq.(4). In this paper, through repeated experiments, results analysis and comparison, the hidden layer nodes is determined as 30.

7 Training of the Neural Network Model

1) First of all, 100 groups of typical temperature trace containing 30 groups of normal and 70 groups with breakout pattern have been chosen from the historical data of the temperature trace collected from a continuous casting plant as the training samples for the neural network model.

2) The maximum training times is set as 1000, the maximum allowable error is set as10^{-6}, and the learning rate is set as 0.1.

3) The Neural Network Training Result is shown as Fig. 3:

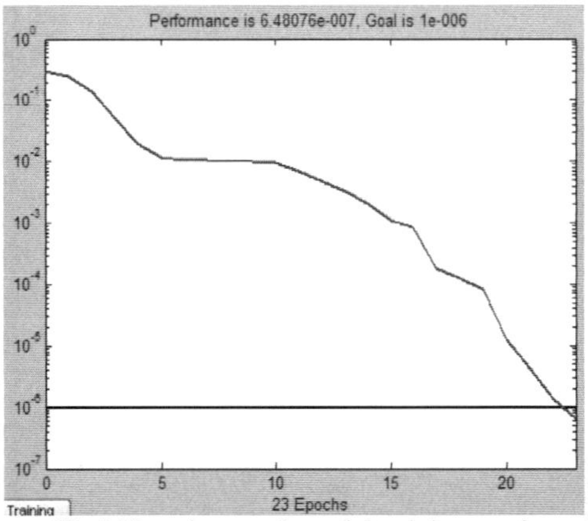

Fig 3 Neural network model training result

8 Testing of the Neural Network Model

The testing samples for the neural network model were 56 groups of typical temperature trace collected from the continuous casting plant, containing 27 groups of sticking type breakout samples, 9 groups of false alarm samples, and 20 groups of normal samples. In the period of testing, all the 27 true breakout alarms, and 1case of undesirable alarm were sent out by the model. So the accuracy rate of the model for the temperature pattern of sticking breakout was 96.43%, and the quote rate was 100%.

9 Conclusions

In this paper, the improved BP neural network model based on the LM algorithm was presented by modifying the learning algorithm of the traditional BP neural network. It has stronger real-time efficiency and faster convergence speed, compared with the traditional BP neural network model. The model was applied to the breakout prediction system in the continuous casting process; the identification capabilities and the forecast accuracy were greatly improved. The result of the test was that the accuracy rate of the model for the temperature pattern of sticking breakout was 96.43%, and the quote rate was 100%. The feasibility of the model for the breakout prediction system was well verified.

Acknowledgment

The authors would like to acknowledge the support of Basic Research Program of Jiangsu province (No. BK20150429).

DOI: 10.1051/matecconf/20166105020

References

1. Zhang B, Zhang R, Wang G, et al. Breakout prediction for continuous casting using genetic algorithm-based back propagation neural network model. Int. J. Modeling, Identification and Control, 2012,16(3):199-205.
2. Irani R, Nasimi R. Evolving neural network using real coded genetic algorithm for permeability estimation of the reservoir[J]. Expert Systems with Applications, 2011,38(8): 9862-9866.
3. Li Zhang, Jianhua Luo, Suying Yang. Forecasting box office revenue of movies with BP neural network J. Expert Systems with Applications, vol. 36, 2009, pp. 6580-6587.
4. Q. Ju, Z. Yu, Z. Hao, et al. Hydrologic simulations with artificial neural networks, in: Third International Conference on Natural Computation, IEEE Computer Society Publications, vol. 2, 2007, pp. 22–27.
5. S.H. Hsiang and Y.W. Lin. Application of fuzzy theory to predict deformation behaviors of magnesium alloy sheets under hot extrusion, Journal of Materials Processing Technology 201 (2008), pp. 138–144.

Kinetics of 2-chlorobiphenyl Reductive Dechlorination by Pd-fe^0 Nanoparticles

Junrong Jiang[1] Chong Fu[2] and Deming Zhao[2,a]
[1]School of Biological & Chemical Engineering, Taizhou Vocational & Technical College, Taizhou 318000, China
[2]College of Chemical Engineering, Zhejiang University of Technology Hangzhou 310014, China

Abstract. Kinetics of 2-chlorobiphenyl (2-Cl BP) catalytic reductive dechlorination by Pd-Fe^0 nanoparticles were investigated. Experimental results showed that ultrafine bimetallic Pd-Fe^{0}e nanoparticles were synthesized in the presence of 40 kHz ultrasound in order to enhance disparity and avoid agglomeration. The application of ultrasonic irradiation during the synthesis of Pd-Fe^0 nanoparticles further accelerated the dechlorinated removal ratio of 2-Cl BP. Up to 95.0% of 2-Cl BP was removed after 300 min reaction with the following experimental conditions: initial 2-Cl BP concentration 10 mg L^{-1}, Pd content 0.8 wt. %, bimetallic Pd-Fe^0 nanoparticles prepared in the presence of ultrasound available dosage 7 g L^{-1}, initial pH value in aqueous solution 3.0, and reaction temperature 25℃. The catalytic reductive dechlorination of 2-Cl BP followed pseudo-first-order kinetics and the apparent pseudo-first-order kinetics constant was 0.0143 min^{-1}.

1 Introduction

Historical and continued releases of polychlorinated biphenyls (PCBs) to the environment have resulted in polluted water, soil, and sediment [1]. These contaminated soils, sludges and groundwater need to be effectively remediated at ambient temperatures using either ex *situ* or in *situ* methods. Methods for PCBs dechlorination include chemical reduction, catalyzed hydrodechlorination, and electrolytic dechlorination. Most of these techniques use highly active reducing hydrogen from different sources to facilitate the dechlorination through the use of noble metal catalysts such as palladium (Pd) and rhodium (Rh), etc [2-4].

Pd-Fe^0 bimetallic nanoparticles have been improved to completely dechlorinate chloroaromatics and thoroughly purify a wide range of chlorinated organic compounds without leaving chlorinated side products [2-6]. As one of the noble metal, Pd can utilize the produced H_2 from zero-valent iron (ZVI) corrosion and improve the rates and efficiencies of dechlorination reaction [1,4]. In addition, the presence of Pd not only reduces the accumulation of toxic byproducts, but also inhibits particle oxidation in air [5]. Pd-Fe^0 bimetallic nanoparticles, when compared to the conventional large particles have some advantages with possessing large specific surface area and high surface reactivity [6]. In order to obtain the stabilized and high reactive Pd-Fe^0 bimetallic nanoparticles, ultrasound is applied to the preparation of Pd-Fe^0 nanoparticles because acoustic cavitation can enhance the surface area and surface properties of the reactive solids by causing particles to rupture [7].

In the present work, Pd-Fe^0 bimetallic nanoparticles were prepared in the presence of 20 kHz ultrasound and 2-chlorobiphenyl (2-Cl BP) dechlorination kinetic by Pd-Fe^0 bimetallic nanoparticles in aqueous solution was investigated. Meanwhile, other influential factors contributing to 2-Cl BP reductive dechlorination, such as Pd-Fe^0 nanoparticles availability, Pd content over Fe^0, initial 2-Cl BP concentration, reaction temperature and the initial pH values, were also studied

2 Experimental

2.1 Chemicals

2-chlorobiphenyl (AP grade), biphenyl (CP grade) and potassium hexachloropalladate (AP grade) were purchased from the J&K Chemical Reagent Co., Ltd., China, $FeSO_4·7H_2O$ (AR grade), sodium borohydride (AR grade) purchased from Tianjin Chemical Reagent Research Institute. 2-Cl BP is dissolved in a methanol/deoxygenated deionized water solution (50:50, *V*/*V*) and stored at 4 ℃. Pd-Fe^0 nanoparticles were synthesized immediately before use.

2.2 Experimental procedures

In a 500 mL three-necked flask, Pd-Fe^0 bimetallic nanoparticles were prepared in the presence of 40 kHz and 150 W ultrasound under nitrogen gas. Firstly, nanoscale zero-valent iron (nZVI) particles were synthesized by drop wise addition of stoichiometric amounts of $NaBH_4$ aqueous solution into a flask containing $FeSO_4·7H_2O$ aqueous solution

[a] Corresponding author: dmzhao@zjut.edu.cn

DOI: 10.1051/matecconf/20166105021

simultaneously with mechanical stirring at 25℃. nZVI particles were then rinsed several times with deoxygenated deionized water. nZVI particles were then rinsed several times with deoxygenated deionized water. Subsequently, Pd-Fe^0 nanoparticles were prepared by reacting with the wet nZVI particles in an aqueous solution of potassium hexachloropalladate under mechanical stirring. Batch experiments of 2-Cl BP catalytic reductive dechlorination were performed in the same flask into which nanoscale bimetallic Pd-Fe^0 particles were added. 2-Cl BP stock solutions and a certain amount of deoxygenated deionized water and methanol solution (50:50,*V*/*V*) were added into the flask containing fresh prepared nanoscale Pd-Fe^0 particles into 500 mL of total reaction volume. The reaction solution was stirred under nitrogen flow to simulate anaerobic environment at 25 ℃ . Aliquots of samples were periodically collected with glass syringes and the reaction was quenched by passing through 0.22 μm polyether sulphone (PES) membrane filters.

2.3 Methods of analysis

All fresh prepared synthesized bimetallic Pd-Fe^0 nanoparticles (with Pd content 0.3 wt. %) were immersed in absolute ethyl alcohol and dispersed by an ultrasonator. Transmission electron microscope (TEM) images were obtained through a microscope (JEOL JEM 200CX JEOL Electronics Co., JP). X-ray diffraction (XRD) analysis was performed by using X'Pert Pro advanced X-ray diffractometer (λ = 1.5418 A°). Brunauer–Emmett–Teller (BET) specific surface area of all synthesized bimetallic Pd-Fe^0 nanoparticles were measured using nitrogen adsorption method with a surface analyzer (ASAR2020M+, Micromeritics Instrument Corp., US). Before the analysis, the particles were dried in vacuum at 25 ℃ for 24 h and then hydrogen flow at 260 ℃ for further 4 h. Organic compounds such as 2-chlorobiphenyl and BP were analyzed by Thermofish Trace 1310 Gas Chromatography. ECD and FID detector, Thermofish TR-5 Column (30 m×0.32 mm, 1.0 μm), inlet temperature 300℃, detector temperature 250 ℃, no split injection, injection volume 1 μL, carrier gas nitrogen (purity ≥99.99%), column flow 1.5 mL/min (constant current), program temperature: column temperature 100 ℃, holding 0.5min, 25 ℃/min was raised to 300 ℃, maintained 2 min.

3 Results and Discussions

3.1 Pd-Fe^0 nanoparticles Characterization

Fig. 1 (a, b) shows TEM images of the freshly synthesized bimetallic Pd-Fe^0 nanoparticles in the presence and absence of ultrasound. The freshly prepared Pd-Fe^0 nanoparticles in the absence of ultrasound, were spherical in shape with particle size ranging from 20 to 100 nm, and appeared to aggregate together (Fig. 1(a)). However, Pd/Fe nanoparticles synthesized in the presence of ultrasound were also spherical, but the particle size ranges from 10 to 100 nm, mainly foused on 30 to 75 nm, and appeared to be smaller particles diameter and better dispersion (Fig. 1(b)). The sizes and specific surface area of most particles synthesized in the presence of ultrasound were obviously modified. The ultrafine and highly dispersion nanoparticles were obtained.

Fig. 2(a, b) shows the XRD patterns of the fresh and the 300 min reacted bimetallic Pd-Fe^0 nanoparticles synthesized in the presence of ultrasound. The XRD pattern for the fresh sample presents a strong peak 44.66° which corresponds to the body-centered cubic N–Fe^0 at the (110) plane. The peak in the XRD pattern of the reacted sample shows evidence of iron oxides, possibly Fe_3O_4 or Fe_2O_3, or their mixture. This is consistent with the previous work [8].

In addition, BET specific surface area of the synthesized Pd-Fe^0 nanoparticles in the presence and absence of 40 kHz ultrasonic irradiation were 41.68 m^2 g^{-1} and 22.39 m^2 g^{-1}, respectively.Therefore, bimetallic Pd-Fe^0 nanoparticles were synthesized by using ultrasound strengthened liquid phase reductive method in the following experiments.

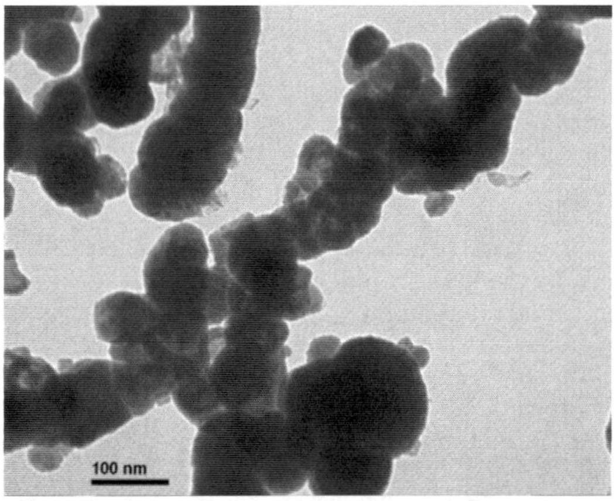

(a)

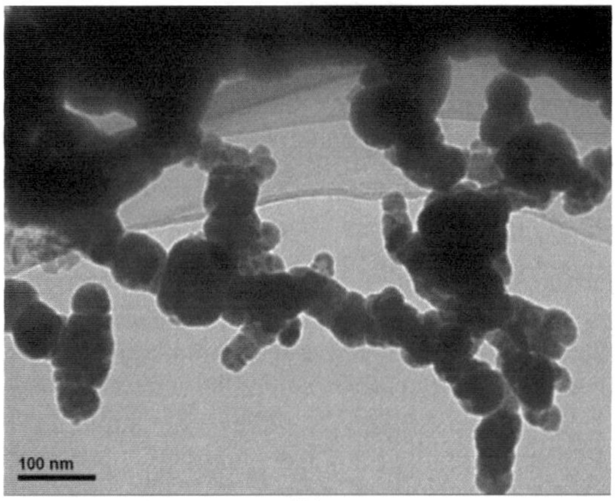

(b)

Figure. 1. (a) TEM image of fresh Pd/Fe nanoparticles prepared in the absence of ultrasound, (b) TEM image of fresh Pd/Fe nanoparticles prepared in the presence of 40 kHz ultrasound

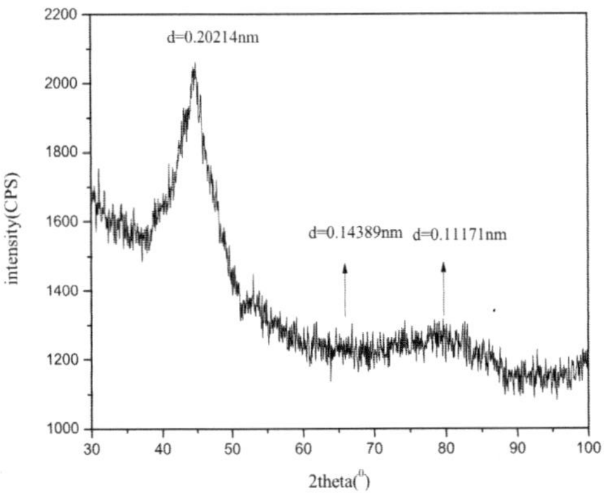

(a)

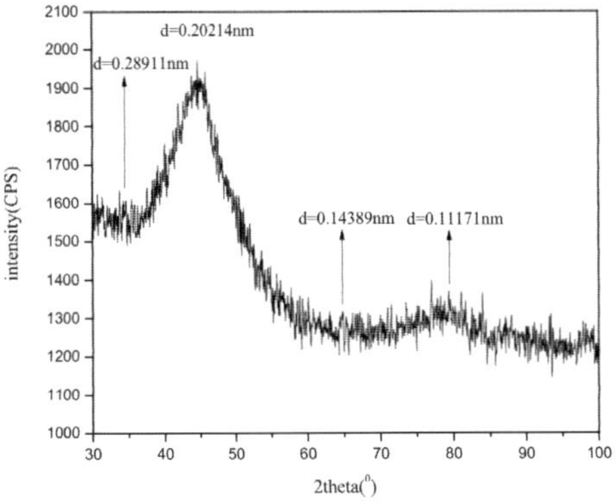

(b)

Figure. 2. (a) XRD patterns of fresh Pd/Fe nanoparticles prepared in the presence of 40 kHz ultrasound and (b) XRD patterns of reacted Pd/Fe nanoparticles prepared in the presence of 40 kHz ultrasound.

3.2 Kinetic modeling of 2-Cl BP catalytic reductive dechlorination by bimetallic Pd-Fe^0 nanoparticles.

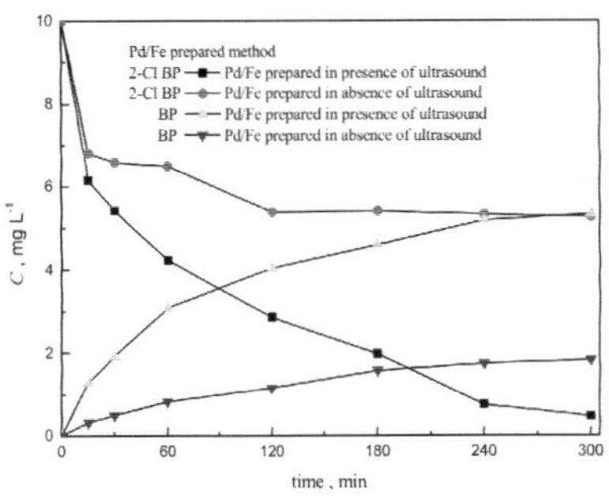

Figure. 3. Reductive dechlorinations of 2-Cl BP by Pd-Fe^0 nanoparticles prepared in different methods .

Figure 3 shows the hypothetical reductive dechlorated pathways of 2-chlorobiphenyl (2-Cl BP) by Pd-Fe^0 nanoparticles prepared in different methods. Biphenyl (BP) is the sole final organic product. In addition, ased on the previous studies, the catalytic reductive dechlorination of chloroaromatics by bimetallic Pd-Fe^0 nanoparticles involves three main steps [2,4,9-13]: (1) chloroaromatics in solution diffuse to Pd-Fe^0 nanoparticles surface and adsorb in the surface of catalyst; (2) the adsorbed chloroaromatics takes reduction reaction under the effect of Pd-Fe^0 nanoparticles; (3) the products of reaction desorbs from Pd-Fe^0 nanoparticles and diffuses to the bulk solution. It is well established that the pseudo-first-order kinetics could be applied in the reductive dechlorination of chloroaromatics if Pd-Fe^0 nanoparticles availability is excessive in the reaction. Therefore, the pseudo-first order reaction kinetics was adopted to model the 2-Cl BP dechlorination reaction by Pd-Fe^0 nanoparticles. Therefore, it is assumed that 2-Cl BP was hydrodechlorinated according to the following sequence of steps.

$$2-\mathrm{Cl\ BP} \xrightarrow{k_1} \mathrm{BP} + \mathrm{Cl}^- \quad (1)$$

The corresponding reaction rate equations for the disappearance of 2-Cl BP and the accumulation of BP in the batch system are shown as follows:

$$\frac{dC_{2-\mathrm{Cl\ BP}}}{dt} = -kC_{2-\mathrm{Cl\ BP}} \quad (2)$$

$$\frac{dC_{BP}}{dt} = kC_{2-\mathrm{Cl\ BP}} \quad (3)$$

The above simultaneous rate equations are integrated, leading to the following molar fractions:

$$\alpha_{2-\mathrm{Cl\ BP}} = e^{-kt} \quad (4)$$

$$\alpha_{BP} = 1 - \alpha_{2-\mathrm{Cl\ BP}} \quad (5)$$

Where α represents the molar fraction of the subscript organic compound to the initial concentration of 2-chlorobiphenyl. Since a fraction of 2-Cl BP and BP were absorbed on the larger surface of Pd-Fe^0 nanoparticles, the actual concentration of the 2-Cl BP and BP in aqueous phase has to be amended. As the production of BP was step wise, equilibrium time for organic compounds absorbed onto Pd-Fe^0 nanoparticles was little, and the total molar fraction of 2-Cl BP and BP did not change, the ratio of BP in the aqueous phase can be seen as invariable, as a result, Eq. (5) can be revised as follows:

$$\alpha_{Bp}{}' = \alpha_{BP}(1-a) \quad (6)$$

Where α' represents the molar fraction of the organic compounds which were adsorbed onto bimetallic Pd-Fe^0 nanoparticles to the initial concentration of 2-chlorobiphenyl. Then k values were derived from fitting the experimental data into Eq. (6) according to the non-linear least-square regression. The influential factors, such as the preparation method of bimetallic Pd-Fe^0 nanoparticles, Pd content over Fe^0, bimetallic Pd-Fe^0 nanoparticles dosage, reaction temperature, initial pH values and initial 2-Cl BP concentration on the 2-Cl BP dechlorination efficiency were investigated in the following sequences. Kinetic constant k values in different experimental conditions were listed in Table 1.

The experimental results showed that k values obviously increased from 0.0018 to 0.0143 min^{-1} under bimetallic Pd/Fe nanoparticles prepared in the absence and presence of ultrasound, respectively. They increased from 0.0119 min^{-1} to 0.0203 min^{-1} as the Pd content over Fe^0 was increased from 0.70 wt. % to 0.90 wt. %, and they increased from 0.0098 min^{-1} to 0.0309 min^{-1} as the bimetallic Pd-Fe^0 nanoparticles available dosage was increased from 5 g L^{-1} to 9 g L^{-1}. In short, k values increased with the increasing Pd content over Fe^0, Pd/Fe nanoparticles available dosage and reaction temperature, with the decrease of initial pH values.

4 Conclusion

The experimental results indicate that the prepared Pd-Fe^0 nanoparticles method in the presence of ultrasound was a better technique to enhance the surface special properties. 2-Cl BP dechlorination efficiency was dependent on a number of influential factors including bimetallic Pd-Fe^0 nanoparticles dosage, Pd content over Fe^0, reaction temperature, mechanical stirring speed and initial pH values. The degradation of 2-Cl BP followed pseudo-first-order kinetics reaction and k values increased with the increasing Pd/Fe nanoparticles dosage, stirring speed and reaction temperature, with the decrease of initial pH values.

Acknowledgements

The authors are grateful for the financial support provided by Zhejiang Provincial Natural Science Foundation of China (LY15B070005) and by the Science and Technology Department of Zhejiang Province（No. 2015C33233）.

Table 1 k values in different experimental conditions

Reaction conditions		a	k/ min^{-1}	R
1. Pd-Fe^0 nanoparticles prepared method	in the absence of ultrasound	0.3563	0.0018	0.91
	in the presence of ultrasound	0.3566	0.0143	0.99
2. Pd content over Fe^0	0.70 wt. %	0.4326	0.0119	0.99
	0.75 wt. %	0.3977	0.0125	0.99
	0.80 wt. %	0.3566	0.0143	0.99
	0.85 wt. %	0.3339	0.0164	0.99
	0.90 wt. %	0.3322	0.0203	0.99
3. Pd-Fe^0 nanoparticles dosage g L^{-1}	5.00	0.6944	0.0098	0.97
	6.00	0.4258	0.0069	0.97
	7.00	0.3566	0.0143	0.99
	8.00	0.3542	0.0168	0.99
	9.00	0.3755	0.0309	0.96
4. Reaction temperature	15 ℃	0.8811	0.0035	0.97
	25 ℃	0.3566	0.0143	0.99
	35 ℃	0.8133	0.0092	0.99
5. Initial pH values	3	0.3566	0.0143	0.99
	5	0.5622	0.0087	0.99
	7	0.8090	0.0038	0.99

Note: Exp.1.T=25 ℃, pH_{in}= 3.0, $C_{2\text{-}Cl\ BP}$=10 mg L^{-1}, $C_{Pd/Fe}$ =7 g L^{-1}, mechanical stirring speed at 600 rpm, Pd content was 0.8 wt. %; Exp.2.T=25 ℃, pH_{in}= 3.0, $C_{2\text{-}Cl\ BP}$=10 mg L^{-1}, $C_{Pd/Fe}$ =7 g L^{-1}, mechanical stirring speed at 600 rpm; Exp.3.T=25 ℃, pH_{in}= 3.0, $C_{2\text{-}Cl\ BP}$=10 mg L^{-1}, mechanical stirring speed at 600 rpm, Pd content was 0.8 wt. %; Exp.4. pH_{in}= 3.0, $C_{2\text{-}Cl\ BP}$=10 mg L^{-1}, $C_{Pd/Fe}$ =7 g L^{-1}, mechanical stirring speed at 600 rpm, Pd content was 0.8 wt. %; Exp.5.T=25 ℃, $C_{2\text{-}Cl\ BP}$=10 mg L^{-1}, $C_{Pd/Fe}$ =7 g L^{-1}, mechanical stirring speed at 600 rpm, Pd content was 0.8 wt. %. Bimetallic Pd-Fe^0 nanoparticles were synthesized by using ultrasound strengthened liquid phase reductive method in Exp.2 to 5.

References

1. T. Luigi, D Luca, Propulsion physics (EDP Sciences, Les Ulis, 2009
2. G.V. Lowry, K.M Johnson, Environ. Sci. Technol. **38**,5208-5216 (2004)
3. Z. Zhang, Q.H. Shen, N. Cissoko, J.J.Wo, X.H. Xu, J. Hazard. Mater. **182**,252-258 (2010)
4. T. Zhou, Y.Z. Li, T.T. Lim, Separation and Purification Technology.**76**, 206-214 (2010)
5. J.J. Wei, X.H. Xu, Y. Liu, D.H. Wang, Water Research.**40**,348-354 (2006)
6. H.L. Lien, W.X. Zhang, Colloid Surface. A. **233**,103-112 (2001)
7. W.H. Zhang, X. Quan, J.X. Wang, Z.Y. Zhang, S. Chen, Chemosphere. **65**,58-64 (2006)
8. K.S. Suslick, M. Fang, T.Hyeon, Journal of American Chemical Society.118,11960-11961 (1996)
9. D.M. Zhao, M. Li, D.X. Zhang, S.A. Baig, X.H. Xu, Ultrason Sono. **20**,64–871 (2013)
10. G.N. Jovanovic, P.Z. Plazl, P. Sakrittichai, K. Al-Khaldi, Ind. Eng. Chem. Res. **44**,5099-5106 (2005)
11. G.Yuan, M.A.Keane, Catal. Commun. **4**,195-201 (2003)

12. F. Cheng, Q. Fernando, N. Korte, Environ. Sci. Technol. **31**,1074-1078 (1997)
13. J.L. Chen, S.R. Al-Abed, J.A. Ryan, Z.B. Li, J. Hazard. Mater. B.**83**,243-254 (2001)
14. Z. Zhang, N. Cissoko, J.J. Wo, X.H. Xu, J. Hazard. Mater..**165**,78-86 (2009)

MATEC Web of Conferences **61**, 05022 (2016) DOI: 10.1051/matecconf/20166105022
APOP2016

Analysis of Several Key Parameters in the Design of Infrared Stealth Coating

HU Rui[1a]and WANG Xuanyu[1]

[1]Laboratory of Pyrotechnical Technology, Institute of NBC Defense, Beijing 102205, P.R.China

[a]guoguo_lily2010@sina.com

Abstract. In order to optimize the design and further determine the related parameters of the infrared stealth coating, this thesis derives the calculating formula of coating thickness d and surface emissivity ε based on optical thin film theory and conducts relevant designing targeted at multi-layer spraying existing in the stealth coating operations. The research results show that the proposed method extends the application of optical thin film theory in stealth coating designing and has theoretical significance for the future research of stealth coating.

1 Introduction

With the development of modern military infrared reconnaissance and detection technologies, especially the extensive use of thermal infrared homing guided weapon system, the possibilities that various types of military targets and weapons are spotted in the battle field and thus were destroyed turn out to be higher and higher. Under such circumstances, our military equipment including aircraft, tanks and vehicles are required not only with sound accuracy and maneuverability but also with outstanding concealment. Facing the big challenge of how to effectively minimize and avoid the possibility our military targets being spotted, the infrared stealth coating technology is a simple and economical approach which is easy to be carried out [5~8].

The core evaluation parameter of infrared stealth coating is generally surface emissivity ε [9~11], which is obtained from surface reflectivity R concluded through radiation transfer equation based on Kirchhoff's law. As the solution of the radiation equation is difficult in practical calculation, researchers tend to adopt multi-flow theory [12] instead which still requires a huge workload. To make it easy for calculation, the adoption of optical thin film design theory can facilitate the calculation of several key parameters in infrared stealth coating design and offer a fresh-new method for determining ε.

2 Determination of the Relationship between Filler and Base Material of Infrared Stealth Coating

The infrared stealth coating is generally composed of fillers, pigments and binders [13]. Without regard to the compatibility with visible light, the biggest influence on the infrared surface emissivity is the filler and binder, among which the emissivity of the binder is the definite value. Thus, the determination of the relationship between fillers' surface emissivity and matrix carries great significance. According to the refractive index matching conditions, when the infrared radiation is incident at a certain angle of inclination, the effective refractive index of component $\boldsymbol{P}$ and component $\boldsymbol{S}$ on different refractive index material interface is:

$$\eta_p = N/\cos\theta$$
$$\eta_s = N\cos\theta \tag{1}$$

Where:

η refers to the effective refractive index;

N refers to the refractive index of the incident material;

θ refers to the angle of incidence; degree.

Due to the different effective refractive index and reflection coefficient, the separation of the spectral reflectance (or transmittance) curve of component $\boldsymbol{P}$ and component $\boldsymbol{S}$ tends to occur and thus polarization effect may happen. On the interface of two kinds of refractive indexes n_H and n_L, when the incident angle satisfies the Brewster angle, the reflection of $\boldsymbol{P}$ polarized light is zero and there is partial reflection in $\boldsymbol{S}$ polarized light. In order to increase the reflectivity of $\boldsymbol{S}$ polarized light and to make the transmittance of $\boldsymbol{P}$ polarized light close to 1, two kinds of materials may be made into a multi-layer coating system to achieve the greatest degree of infrared stealth effect.

From Fresnel Principle, the relationship of the reflection coefficient of component $\boldsymbol{P}$ between the two materials with the refractive index of n_H and n_L is as follows:

$$r_{HLP} = \frac{n_H / \cos\theta_H - n_L / \cos\theta_L}{n_H / \cos\theta_H + n_L / \cos\theta_L} \tag{2}$$

Where:

θ_H and θ_L are the refraction angles where the light is at high and low refractive index coating of n_H and n_L.

If

$$n_H / \cos\theta_H = n_L / \cos\theta_L \tag{3}$$

Then

$$r_{HLP} = 0 \tag{4}$$

Assume that the angle of incidence of infrared radiation at the coatings is θ_0, and the refractive index of the base material is n_G.

$$\cos\theta_H = \left[1-(n_G / n_H)^2 \sin^2\theta_0\right]^{1/2}$$
$$\cos\theta_L = \left[1-(n_G / n_L)^2 \sin^2\theta_0\right]^{1/2} \tag{5}$$

Introduce(5) into(3):

$$n_G \sin\theta_0 = \frac{n_H n_L}{(n_H^{\ 2} + n_L^{\ 2})^{1/2}} \tag{6}$$

It can be concluded from (6) that there is relationship between base material and the conditions that the high and low refractive index (filler) shall meet and there is relationship between the refractive index n_G of matrix material and the determination of angle of incidence θ_0.

$$\begin{cases} n_H d_H = \dfrac{\frac{\lambda_{\text{Incident}}}{4}}{\sqrt{1-n_0^{\ 2}\sin^2\theta_0 / n_H}} \\ n_L d_L = \dfrac{\frac{\lambda_{\text{Incident}}}{4}}{\sqrt{1-n_0^{\ 2}\sin^2\theta_0 / n_L}} \end{cases} \Rightarrow \begin{cases} d_H = \dfrac{\frac{\lambda_{\text{Incident}}}{4}}{\sqrt{n_H^2 - n_0^{\ 2}\sin^2\theta_0}} \\ d_L = \dfrac{\frac{\lambda_{\text{Incident}}}{4}}{\sqrt{n_L^2 - n_0^{\ 2}\sin^2\theta_0}} \end{cases} \tag{9}$$

where:

d_H and d_L refer to high and low geometrical thickness; μm

λ refers to the wavelength of incident light; μm.

According to the computational formula of the thickness d of the coating deduced from refractive index matching condition(6) and phase matching

When the angle θ_0 is equal to $90°$, the value of n_G is maximum.

Then

$$n_G = \frac{n_H n_L}{(n_H^{\ 2} + n_L^{\ 2})^{1/2}} \tag{7}$$

Metal materials such as aluminum, aluminum alloy plate and iron plate are the common matrix materials. So the value of n_G is known which provides theoretical basis for the selection of the fillers at the early stage of the test and which provides a convenient method for the selection of some organic materials.

3 Determination of Thickness of Infrared Stealth Coating

Thickness value of infrared stealth coating ranges between 0~100 μm. Based on the vector calculation results of optical thin film, the coating is designed as $\lambda/4$, that is, the thickness in each layer is $\lambda/4$ and the selected coating thickness is ***S*** (The reflectivity increase with the increase of the number of layers)to make the reflected light interfere. With appropriate number of layers, ***S*** component of the incident beam reflects and the thickness is supposed to satisfy the condition:

$$n_H d_H \cos\theta_H = n_L d_L \cos\theta_L = \frac{\lambda}{4} \tag{8}$$

Hence, the whole coating is retained at an odd number of layers, but the equivalent thickness of the whole coating is the integer multiples of $2/\lambda$. Thus, the infrared stealth coating is achieved where the component ***S*** is reflected and component ***P*** is all through.

The thickness of each layer is:

conditions (8), as the atmospheric infrared window is composed of three bands, the corresponding thickness is also quite different.

4 Determination of Reflectivity of Infrared Coating

Assume that the number of layers of multiple coatings is 2n+1 (n=0,1,2,3,......)and the thickness is $\lambda/4$, According to EM theory, the characteristic matrix of coating is:

$$\begin{bmatrix} B \\ C \end{bmatrix} = \begin{bmatrix} 0 & i/\eta_H \\ i\eta_H & 0 \end{bmatrix} \begin{bmatrix} 0 & i/\eta_L \\ i\eta_L & 0 \end{bmatrix} \begin{bmatrix} 0 & i/\eta_H \\ i\eta_H & 0 \end{bmatrix} \text{L L} \begin{bmatrix} 0 & i/\eta_H \\ i\eta_H & 0 \end{bmatrix} \begin{bmatrix} 1 \\ \eta_G \end{bmatrix} \tag{10}$$

where:

η_H is the effective refractive index of high refractive index coating;

η_L is the effective refractive index of low refractive index coating;

η_G is the effective refractive index of base material.

$$\begin{bmatrix} B \\ C \end{bmatrix} = \begin{bmatrix} 0 & i^{2n+1}\eta_L^n/\eta_H^n \\ i^{2n+1}\eta_L^n/\eta_H^n & 0 \end{bmatrix} \begin{bmatrix} 1 \\ \eta_G \end{bmatrix} = \begin{bmatrix} i^{2n+1}\eta_L^n \eta_G/\eta_H^n \\ i^{2n+1}\eta_L^n/\eta_H^n \end{bmatrix} \quad (11)$$

The equivalent refractive index of the coating is:

$$Y = \frac{C}{B} = \frac{\eta_H^{2n+2}}{\eta_L^{2n}\eta_G} \quad (12)$$

The reflectivity of the coating is:

$$R = \left(\frac{\eta_G - Y}{\eta_G + Y}\right)^2 = \left(\frac{\eta_G - \eta_H^{2n+2}/\eta_L^{2n}\eta_G}{\eta_G + \eta_H^{2n+2}/\eta_L^{2n}\eta_G}\right)^2 \quad (13)$$

$$\eta_G = \eta_H = \eta_L = n_G / \cos\theta_G \left(n_0 = n_G\right) \quad (14)$$

When the light is tilted, the effective refractive index of each layer of the S polarization component is:

$$\begin{cases} \eta_G = n_G \cos\theta_G \\ \eta_H = n_H \cos\theta_H \left(n_0 = n_G\right) \\ \eta_L = n_L \cos\theta_L \end{cases} \quad (15)$$

As for P polarization component, introduce (14) into (12):

$$R_p = \left(\frac{\eta_G - \eta_H^2/\eta_G}{\eta_G + \eta_H^2/\eta_G}\right)^2 = \left[\left(\frac{n_G^2\cos\theta_H}{n_H^2\cos^2\theta_G} - 1\right) \Big/ \left(\frac{n_G^2\cos\theta_H}{n_H^2\cos^2\theta_G} + 1\right)\right]^2 \quad (16)$$

As for S polarization component:

$$R_s = \left[\frac{n_G^2\cos^2\theta_G - (n_H\cos\theta_H)^{2n+2}/n_G\cos\theta(n_L\cos\theta_L)^{2n}}{n_G^2\cos^2\theta_G + (n_H\cos\theta_H)^{2n+2}/n_G\cos\theta(n_L\cos\theta_L)^{2n}}\right]^2 \quad (17)$$

It can be obtained from (3):

$$\frac{n_H\cos\theta_H}{n_L\cos\theta_L} = \frac{n_H^2}{n_L^2} \quad (18)$$

Introduce into (16), then:

$$R_S = \left[\frac{n_G^2\cos^2\theta_G - (n_H\cos\theta_H)^2 n_H^{4n}/n_G n_L^{4n}}{n_G^2\cos^2\theta_G + (n_H\cos\theta_H)^2 n_H^{4n}/n_G n_L^{4n}}\right]^2 \quad (19)$$

It can be simplified into:

$$R_S = \left[1 - \frac{2n_G^2\cos^2\theta_G}{n_G^2\cos^2\theta_G + \frac{n_H^2\cos^2\theta_H}{n_G}\left(\frac{n_H}{n_L}\right)^{4n}}\right]^2 \quad (20)$$

It can be concluded from (15) that as long as the Brewster angle conditions are met, P polarized light reflectivity has nothing to do with the number of layers but depends on refractive index of coating and base material; It can be concluded from (19) that the reflectivity of infrared stealth coating depends on the layer $2n+1$, among which the value of n bigger, the R_s bigger.

Thus, the reflectivity of infrared coating is:

$$R = \frac{1}{2}(R_s + R_p) \quad (21)$$

5 Determination of Surface Emissivity ε of Coating

According to Kirchhoff's law:

$$\varepsilon = 1 - R = 1 - \frac{1}{2}(R_s + R_p) \quad (22)$$

$$\varepsilon = 1 - \frac{1}{2}\left[\frac{\left(\frac{n_G^2\cos\theta_H}{n_H^2\cos^2\theta_G} - 1\right)}{\left(\frac{n_G^2\cos\theta_H}{n_H^2\cos^2\theta_G} + 1\right)}\right]^2 - \frac{1}{2}\left[1 - \frac{2n_G^2\cos^2\theta_G}{n_G^2\cos^2\theta_G + \frac{n_H^2\cos^2\theta_H}{n_G}\left(\frac{n_H}{n_L}\right)^{4n}}\right]^2 \quad (23)$$

The ε of infrared stealth coating depends on n , for the bigger the n , the bigger the valve of ε and the calculation of ε can be achieved through ***Matlab*** procedure. Through the combination of theory and experiment, the target surface emissivity ε can be determined which is accompanied by the optimal number of layers.

6 Conclusion

Infrared radiation, like X rays, radio waves and light waves is also one form of electromagnetic waves but with different frequencies. So the theory of physical optics is also applicable to infrared radiation, that is, the interference, diffraction and polarization of light wave in the transmission process also exist in infrared radiation.

In general, when the organic compound is used as fillers, the molecular vibration and rotation of its functional groups are the main factors to be considered . According to the relationship between the filler and the base of the infrared camouflage coating derived from refractive index matching conditions, when the angle of incidence is 90 degree and the material value of the base material n_G is known, it can provide a convenient method for the selection of organic materials.

The layers n of infrared stealth coating are mostly determined by the experimental results and in the calculation formula that deduces the target surface emissivity ε, the optimal layers can be theoretically determined from the above formula.

In conclusion, it is theoretically feasible for this thesis to conduct the calculation of several key parameters in the design of infrared stealth coating by adopting the optical thin film theory. The components mentioned hereinabove constitute the design basis and offers a new method for determining ε, posing theoretical guidance for the future research of stealth coating.

References

[1] Zhou L.W. *Target Detection and Recognition*.(BITP, 2004.)

[2] Liang W. *Research on Infrared and Millimeter Wave Fusion Detection* [D]. Xi'an: Northwestern Polytechnical University. 2007.

[3] Chen D.Q., Xu H.J, Liu Z.Q. Computer Measurement and Con..11,18,2015.

[4] Cui D.P., Su J.P. Journal of Sichuan Ordnance.5,5,2015.

[5] Cheng C.L. *Research on 8~14μm Low-emissivity Infrared Stealth Coatings*[D].Nanjing: Nanjing Tech University, 2005.

[6] Huang Z. *Study on Thermal Performance and Artificial Accelerated Aging of 8~14μm Low-emissivity Infrared Stealth Coatings*[D]. Nanjing University of Aeronautics and Astronautics, 2010.

[7] Wang K.Z. Fine Chem. In. 8, 30,2013.

[8] Ren X.H. M. Fun.M. 5, 21,2014.

[9] Shao C.M. *Composition Structure and Properties of Low Infrared Emissivity Coatings*[D]. Nanjing University of Aeronautics and Astronautics, 2010.

[10] Qu S.Y. *Theoretical and Experimental Study of Low Emissivity Infrared Stealth Materials*[D].Xidian University, 2012.

[11] Tao Q.Y. *Preparation of An Infrared Laser Compatible Stealth Coating*[D]. Anhui University Of Sci. & Tec., 2013.

[12] Liu L.Y. S*tudy on the Mechanism of Infrared and Radar Wave Compatible Stealth Composite Coatings* [D]. Wuhan: Huazhong University of Sci. &Tec., 2006.

[13] Li Jianfang. *Deposition Techniques for Optical Thin Films*.(CEPP. 2013)

Study on the Dispersion of 1.5mm & 4mm Chopped Carbon Fiber in Triethyl Phosphate

Hu Rui[1,a], Wang Xuanyu[1] and Liu Zhilong[1]

[1]Laboratory of Pyrotechnical Technology, Institute of NBC Defense, Beijing 102205, P.R.China

[a]guoguo_lily2010@sina.com

Abstract. For the preparation of chopped CF (carbon fiber, CF) composite material, it must be scattered at first. Based on the theory of "direct observation" and "crowding factor", 1.5 mm and 4 mm short cut dispersion of CF which is done by acetone firstly in triethyl phosphate was studied in this paper, using triethyl phosphate as dispersant. Controlling the dosage of the dispersant and ultrasonic time to prepare for the dispersion of suspension, the macroscopic and microscopic structure of CF were tested before and after treatment. Test results show that to achieve good dispersibility for 4mm and 1.5mm CF with mass of 0.1g, the ultrasonic time should be controlled 2h and the amount of triethyl phosphate, the quality is 0.1 g of 4 mm and 1.5 mm of CF good dispersibility, respectively is 200 ml and 300 ml of dispersion conditions.

1 Introduction

Carbon fiber, as the representative of carbon series materials, is usually prepared into composite material for its features of low weight and price, good mechanical property etc. At present, it is widely used in the fields of building, chemistry, industry and aerospace etc. In the process of preparing composite materials, the problem of CF (carbon fiber, CF) dispersion in liquid medium has always been a difficulty. Many scholars at home and abroad have carried out extensive research on it and made great achievements[1~4]. With the target to form hydrophilic, oxygen-containing functional group on the surface of CF to improve its hydrophily[6], silane coupling agent, strong oxidant and chemical reagent, are usually used as three kinds of dispersing agent[5] for it. On the surface of CF ,then improve its hydrophily[6]; traditional evaluation methodology on CF dispersion are: the turbidity value method[4], the surface charge method[4], the resistivity test method[2] and the dispersion ratio method[1] , but the technological process is redundant and experimentation is complex. Based on such a situation, we have realized the CF dispersion by adopting triethyl phosphate as the dispersing agent, which has simple process and is easy to operate. We used methods of combing direct observation with 'crowding factor' to evaluate the dispersing performance of CF and made an analysis to its dispersing mechanism.

2 Experiment

2.1 Experiment Materials and Instruments

The carbon fiber in the experiment is produced by Yancheng Xiangsheng Carbon Fiber Products Co., Ltd., whose specific parameters are as follows: 1.5mm,4mm CF (diameter is about 7μm) , the tensile strength of 3800MPa, the tensile modulus of 228GPa,the density of 1.75g/cm^3 and the electrical resistivity of (1.0~1.6)×10^{-3} Ω·cm. We choose acetone(AR, Tianjin Guangfu Reagent Co.,500ml) to conduct pretreatment on CF. Triethyl phosphate is selected as the dispersing agent (CP, Tianjin Guangfu Reagent Co., 500ml). AE200 analytical balance (That accuracy is 0.0001g). KQ-250DB type numerical control supersonic cleaner(with temperature control range at 20~80℃, the ultrasonic electric power of 100~250W , ultrasonic cleaning time of 1~480min , Kunshan Ultrasonic Instruments Co., Ltd.).

3 Test Method

3.1 CF Pretreatment

Pretreatment is aimed at removing impurities on the fiber surface. First, we weigh appropriate 1.5mm and 4mm CF with the AE200 analytical balance; put them in the acetone solvent for soaking 2h in normal temperature; then, use ionized water to clean several times until the liquor is clear, finally we get the pretreatment products after natural drying.

3.2 The CF Dispersion in Triethyl Phosphate

Put the 1.5mm and 4mm CF that experienced pretreatment in 500ml- beakers respectively, and use the ultrasonic vibration to achieve dispersion then finally determine the optimal disperse conditions by adjusting the amount of CF and triethyl phosphate and controlling ultrasound time, through direct observation. Figure.1 and 2 are respectively the photo of 1.5mm and 4mm CF that before and after pretreatment.

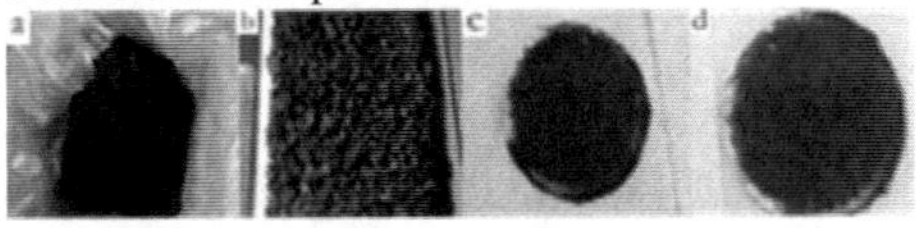

Figure 1. 1.5mmCF before and after pretreatment

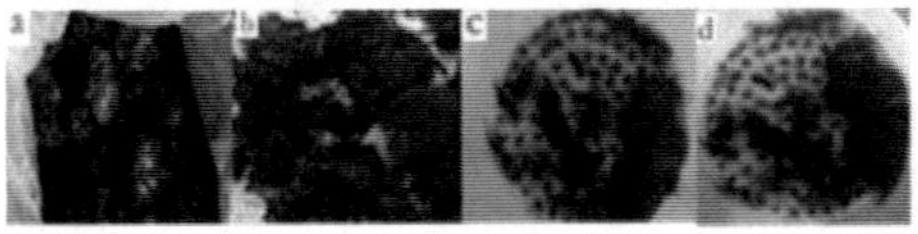

Figure 2. 4mmCF before and after pretreatment

3.3 Evaluation on Dispersion

The commonly used evaluation method is mainly dispersion ratio method[1], which is easy to calculate but complex for the experiment process to operate. In order to realize CF dispersion through simple process, we hereby shall adopt characterization method of dispersing performance for aramid fiber in references[7], using methods of combining direct observation and crowding factor calculation[8] to give a comprehensive assessment on the homodisperse state of CF's suspendability in triethyl phosphate. This method is easy to operate and hasn't been reported in CF dispersion. The method of direct observation is to observe suspension disperse condition in a visual inspection way. 'Crowding factor' is defined as the number of fiber in spherical medium (whose diameter equals the length of singer fiber) and represented by *N*. The definition is shown in formula (1) .When $1 < N < 60$[7], the dispersion of fiber in dispersing agent is good; therefore, 'crowding factor' can be used in quantitative description to fiber's dispersion situation.

$$N = \frac{2}{3} \cdot C_v \cdot \left(\frac{L}{d}\right)^2 \qquad (1)$$

Where,

N represents crowding factor, non-dimensional
C_v is the volume concentration of fiber, %;
L is the length of single CF, mm;
d is the diameter of single CF, mm.

4 Experiment Results and Analysis

4.1 The Dispersion of CF in Triethyl Phosphate

The dispersion condition for the "crowding factor" of CF in 100ml, 200 ml and 300ml was shown from Table 1.

Length	Mass	Triethyl phosphate	N
1.5mm	0.05g	100ml(fig. g)	9
		200ml(fig. h)	3
		300ml(fig. i)	2
1.5mm	0.1g	100ml(fig. j)	18
		200ml(fig. k)	9
		300ml(fig. l)	3
4mm	0.05g	100ml(fig. a)	65
		200ml(fig. b)	22
		300ml(fig. c)	21
4mm	0.1g	100ml(fig. d)	131
		200ml(fig. e)	65
		300ml(fig. f)	22

Table 1. The Dispersion Condition For The 'Crowding Factor' Of 1.5 Mm& 4mm Cf

From pictures of Figure.3 (fig.***a*** to fig.***l***), we could discover that: even though under the circumstance that being with same N, the dispersion condition is inconsistent cause of different ultrasound time. According to the definition formula of crowding factor, we can get to know, in the same situation, the length of CF is the main factor that influencing dispersion for CF with equal quality. That is, the bigger for the draw ratio, the more beneficial to the CF dispersion. This is because, the collision times for single CF increased with the function of ultrasound, and the shorter for single CF, the better for its dispersity as mutual repulsion of surface charge. When the single CF achieved balance in dispersion liquid, the surface charge would gather again and appear the phenomenon of second aggregation. Finally, we determined the optimal experiment condition based on methods of direct observation and 'crowding factor 'calculation. Test results show that the ultrasonic time 2h and the amount of triethyl phosphate, respectively is 200 ml and 300 ml of dispersion conditions, the quality is 0.1 g of 4mm and 1.5mm of CF good dispersibility.

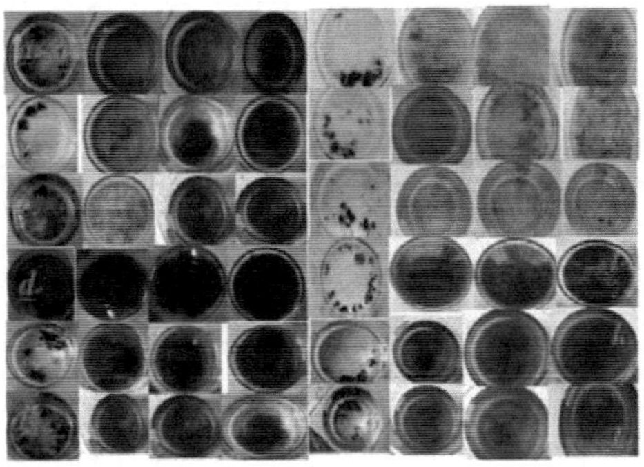

Figure 3. 1.5mm&4mm CF before and after pretreatment

4.2 The SEM Test Analysis Before and After CF Dispersion

The surface of new purchased CF takes on regular striation with obvious crack and groove. The kind of structure makes CF be difficult to spit into single ones

[6].Impurities on the surface of 1.5mm and 4mm CF were removed after pretreating by acetone, and original regular strips were destroyed with the appearance shown in Fig.4-a and Fig.4-b).After treating with triethyl phosphate on the basis, the streak and crack on the CF surface were wrapped and filled （see Fig.4-c and Fig.4-d），now, the surface of CF became bright and clean ,more hydrated than before, the dispersity of CF in the solvent was improved consequently.

Fig. 4-a

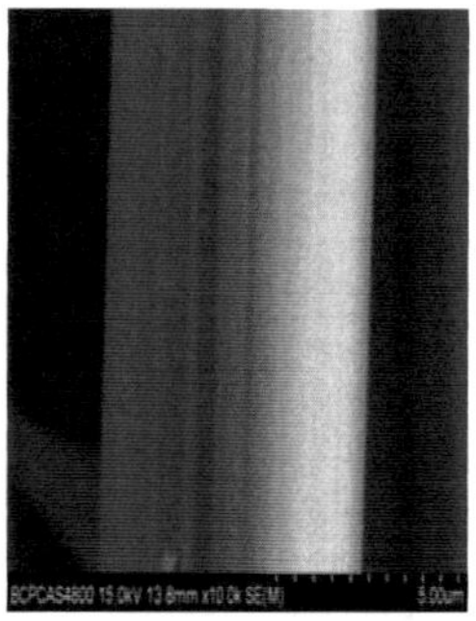

Fig. 4-b

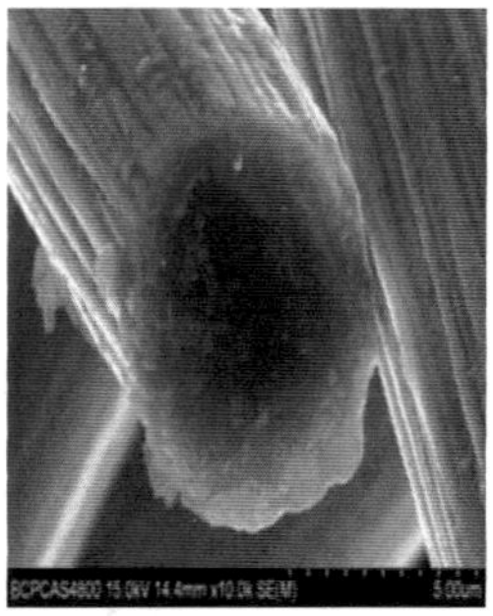

Fig. 4-c

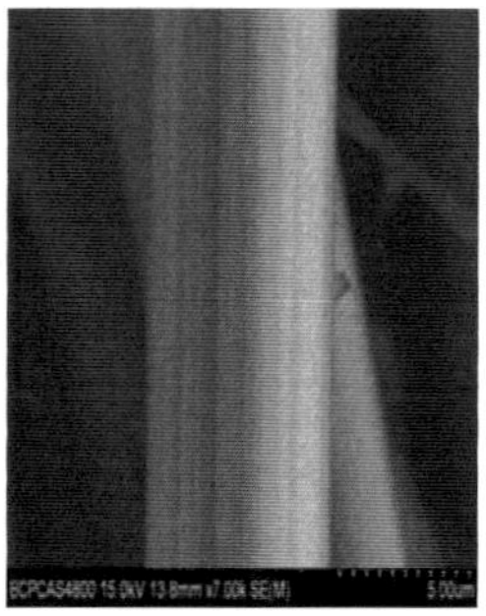

Fig. 4-d

4.3 Mechanism Analysis on How Triethyl Phosphate Improved the Dispersion of Short CF

According to the study of Heyongjia et al, PAN-based carbon fiber could form similar colloidal dispersion system by adsorbing polar bond, such as hydroxyl or carbonyl polar group, to attain the dispersion. In the triethyl phosphate solution, a lot of carboxide formed a skim on the surface of the CF, after a time of ultrasonic function, the repulsive force between each CF surface strengthened, and the dispersion of CF was consequently improved. With the increase of ultrasound time, optimal dispersion effect realized when the repulsive force of single CF surface reached balance in triethyl phosphate solution.

5 Conclusion

Experiments results show that the dispersity of 4mm and 1.5mm CF (with mass of 0.1g)is good under the disperse conditions of 2h of ultrasound time and triethyl phosphate addition are of 200 and 300ml respectively. We can theoretically determine the dispersity of short cut CF in organic solvent through “crowding factor” method. As the actual operation is influenced by ultrasound time [5], a lot of charge gathered on short CF surface will lead to second aggregation phenomenon, so it still needs combining with the observation method to finally determine the optimal situation of dispersion.

References

1. Wangchuang,Likezhi,Lihejun. Dispersity of Short Carbon Fiber in Different Dispersing Agents [J]. Fine Chemicals, 1, (2007)1-4.
2. Heyongjia，Jinshun，Lvlinnv.The Effect of HPMC on the Carbon Fiber Dispersity[J].Function Materials, 6, (2010)1034-1037.
3. Wangbaomn,Mahainan,Zhangyuan et al. The Surface Finish of Carbon Nanofiber and its Dispersion in Water Solution[J]. Materials Science and Technology, 4(2014) 41-48.
4. Cuiying, Sunyu,Wuwo et al. The Effect of Coating Treatment of Dispersing Agent on Dispersing Performance of CF in Oily Body [J]. Surface Technology,1(2015)， 112-116.
5. Lizhao. Ultrasonic Testing and Evaluation Technique Study on the Porosity of Carbon Fiber Composite [D]. Hangzhou：Zhejiang University. 2014
6. Zhaojun,Hujian, Liangyun et al. Surface Properties of Carbon Fiber and its Dispersity in Water[J]. China Papermaking,5(2008) 15-18.
7. Liujunhua. The Dispersity of Domestic Para-aramid Fiber Suspension System & its Mechanism Study [D]. Xi’an：Shaanxi University of Science and Technology.2014
8. Zhangzeng,Hujiang. Evaluation on the Disperstiy of Fiber [J]. Paper Science & Technology. 5(2001) 30-31.
9. Heyongjia，Jinshun，Lvlinnv.The Effect of HPMC on the Carbon Fiber Dispersity[J].Function Materials, 6(2010)1034-1037.

Impact of Standard Deviation and Reflectance of the Measured Surface on Laser Diode- Position Sensitive Detector System

Dianyou Song, Dan Xia[a] and Zhiyong Wang

School of Electronic Information Engineering, Tianjin Key Laboratory of Film Electronic and Communication Device, Tianjin University of Technology, Tianjin 300384, China

Abstract. For improving the accuracy and applicability of laser diode- position sensitive detector (LD-PSD) system, a model of roughness and reflectance of the measured surface is established. Combining this model with the light intensity response model of PSD and the space radiation model, we derived the relationship between the standard deviation (σ), reflectance (ρ) and the center-of-gravity of light intensity (X_0). This relationship shows that X_0 is disturbed by σ or ρ only if σ or ρ changes within the spot. We conducted a simulation and experiment to test this relationship. The results showed a good consistency with the theory, which proved that the variation of σ and ρ can greatly impact the accuracy and robustness of LD-PSD system, and a smaller light spot can improve this situation.

1 Introduction

Due to its wider spectral range, higher detection precision, higher response rate and simpler processing circuit, position sensitive detector (PSD) has received extensive attention in recent years and is widely used in a variety of measurement systems [1-5]. Typically the PSD is combined with the laser diode (LD) to detect micro-displacement, which is the so-called LD-PSD triangulation system. Previous researchers have conducted many studies on the properties of LD-PSD system. Rainer G.Dorsch et al. studied the uncertainty introduced by laser speckle in laser triangulation measurement [6]. C.Narayannan et al. analyzed the response of PSD under the modulated laser pulse [7]. Song Cui et al. proposed an algorithm to improve the linearity of the PSD response to the incident spot location [8]. And they analyzed the impact of the Gaussian spot size on the measurement [9]. But there is no deep research on the impact of the standard deviation and reflectance of the measured surface. In the research of high precision LD-PSD system, we found that large errors existed in measuring planes with different surface properties. So we estimated an optical model to derive the relationship between these properties of measured surface and the response of PSD, which was proved in our simulation and experiments.

The structure of this paper is as follows. The second section will introduce the measurement principle of the LD-PSD system. In the third section we will propose the optical model of system. And the relationship between the standard deviation, reflectance and the focus of light intensity will be derived by this model. Simulations and experiments are carried out in section 4 and section 5 respectively. Finally, we will summarize our study in section 6.

2 LD-PSD Systems

The schematic of LD-PSD system is shown in Figure 1. Monochromatic light issued by LD is transported through the focusing lens, and converges to the measured surface, which is reflected by the plane P_1. Diffuse reflection light collected by the imaging lens converges to the photosensitive surface of PSD to form a spot Q_1. The PSD determines the location of P_1 by detecting the position of Q_1. When the measured surface has a vertical displacement Δh, Q_1 will move to the position of Q_2. And the PSD determines the location of P_2 by detecting the position of Q_2. By comparing the photocurrent of the two spots, the system can calculate the displacement Δh. In order to ensure a constant object-image relationship with the movement of the measured plane, the position of the optical devices is required to meet the Scheimpflug principle. That is, in the observation plane as shown in Figure 1, the optical axis of LD, the principle plane of the imaging lens and the direction of the photosensitive surface of PSD should intersect at point O.

The key of LD-PSD system is the detection of spot position with PSD. So we need to introduce the principle of PSD. PSD is a kind of detectors based on the lateral photoeffect. When the light projects on the photosensitive surface of PSD, the two electrodes generate photocurrent i_{left} and i_{right} separately. Then we can derive the position of spot from the following equation:

$$X_0 = \frac{L}{2}\left(1 - \frac{i_{left} - i_{right}}{i_{left} + i_{right}}\right) \tag{1}$$

[a] Corresponding author: xiadan@tjut.edu.cn

where L is the length of the photosensitive surface of PSD. Typically, the spot size on the PSD cannot be ignored compared to the photosensitive surface size of PSD. In this paper, we use the model proposed by Lingchen Kong et al to divide the real spot into a series of small spots [10]. The size of each small spot can be ignored compared to L. So the generated photocurrent of these points is a linear superposition of each photocurrent generated by small points. In this model, the photocurrent of the two electrodes is as follows:

$$i_{left} = D\int_{X_l}^{X_r} \Phi(X)(1-\frac{X}{L})dX \quad (2)$$

$$i_{right} = D\int_{X_l}^{X_r} \Phi(X)\frac{X}{L}dX \quad (3)$$

where D is the generation rate of the electron-hole pairs induced by per unit light intensity, $\Phi(X)$ is the incident radiant flux at the point X, X_l and X_r are the left and right margins of the spot position. Take i_{left} and i_{right} into (1), X_0 can be written as:

$$X_0 = \frac{\int_{X_l}^{X_r} \Phi(X)XdX}{\int_{X_l}^{X_r} \Phi(X)dX} \quad (4)$$

The equation (4) shows that the PSD detection position is center-of-gravity of light intensity.

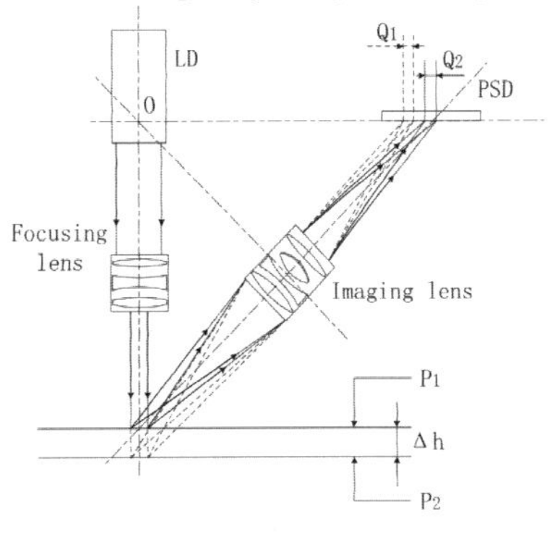

Figure 1. Schematic of LD-PSD system.

3 Theoretical analyses

There are three radiation models when light is irradiated to the interface. They are:

(1) When the interface is a smooth surface, the light produces only two types of behavior-absorption and specular reflection, as shown in Figure 2(a). Here, smooth surface is a plane whose standard deviation of height is far less than the wavelength of incident light.

(2) When the interface is a rough surface, the light also produces two types of behavior-absorption and diffuse reflection [11], as shown in Figure 2(b). This radiation model is commonly used in rough dielectric surfaces such as walls and ceramics.

(3) In most cases, the light at the interface produces four types of behavior-absorption, specular reflection, specular diffuse lobe and diffuse reflection [12], as shown in Figure 2(c). Generally, the object surface of LD-PSD system belongs to this model.

For the system as shown in Fig.1, Monochromatic light is vertically focused onto the measured plane. So the angles of incident and reflection are 0°. Contrasting Figure 1 with Figure 2(c), we can obtain that specular reflection light and specular diffuse lobe is close to the normal. So the imaging lens whose angle of axis and the normal is 45°can only receive the diffuse reflection light. Based on the micro-surface model proposed by Oren and Nayar [13], the region receiving light of the measured surface can be divided into many micro-surfaces. The diffuse reflection light of each micro-surface follows Lambert Law:

$$L_r = \frac{1}{\pi} L_i \rho \cos\theta_i d\omega \quad (5)$$

where L is the radiance of receiving direction, θ is the direction angle of receiving, ω is the solid angle of receiver and ρ is the reflectance of micro-surface.

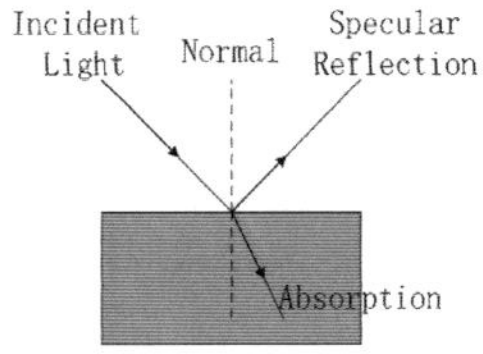

(a) The behaviour of light when it irradiates to an ideal smooth interface.

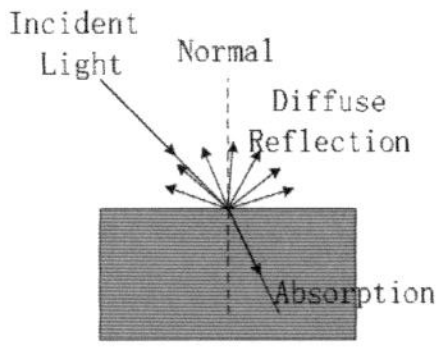

(b) The behaviour of light when it irradiates to an ideal rough interface.

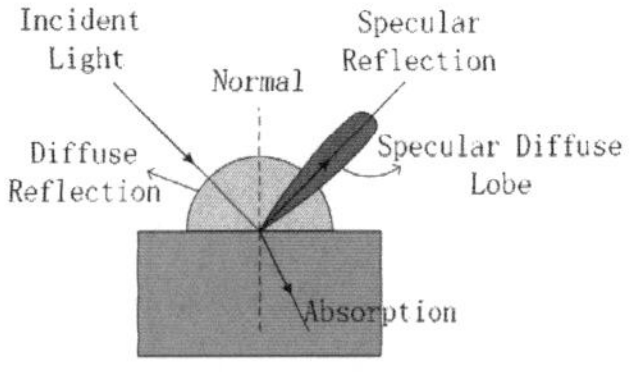

(c) The behaviour of light when it irradiates to the real interface.

Figure 2. Absorption-reflection model of interface.

Lawrence B. Wolff et al. have studied the diffuse reflected radiance of Gaussian distribution microfacet orientations, which is superposition of masking, shadowing and inter-reflection as follows[14]:

$$L_r(\theta_r,\theta_i,\phi_r-\phi_i;\sigma) = L_r^1(\theta_r,\theta_i,\phi_r-\phi_i;\sigma) + L_r^2(\theta_r,\theta_i,\phi_r-\phi_i;\sigma) \quad (6)$$

$$L_r^1(\theta_r,\theta_i,\phi_r-\phi_i;\sigma)=\frac{\rho}{\pi}L_i\cos\theta_i[C_1(\sigma)$$
$$+\cos(\phi_r-\phi_i)C_2(\alpha;\beta;\phi_r-\phi_i;\sigma)\tan\beta$$
$$+(1-|\cos(\phi_r-\phi_i)|)C_3(\alpha;\beta;\sigma)\tan(\frac{\alpha+\beta}{2})] \quad (7)$$

$$L_r^2(\theta_r,\theta_i,\phi_r-\phi_i;\sigma)=0.17\frac{\rho^2}{\pi}L_i\cos\theta_i\frac{\sigma^2}{\sigma^2+0.13}$$
$$\times[1-\cos(\phi_r-\phi_i)(\frac{2\beta}{\pi})^2] \quad (8)$$

$$C_1=1-0.5\frac{\sigma^2}{\sigma^2+0.33} \quad (9)$$

$$C_2=\begin{cases}0.45\dfrac{\sigma^2}{\sigma^2+0.09}\sin\alpha & if\cos(\phi_r-\phi_i)\geq 0\\ 0.45\dfrac{\sigma^2}{\sigma^2+0.09}(\sin\alpha-(\dfrac{2\beta}{\pi})^3) & otherwise\end{cases} \quad (10)$$

$$C_3=0.125(\frac{\sigma^2}{\sigma^2+0.09})(\frac{4\alpha\beta}{\pi^2})^2 \quad (11)$$

$$\alpha=\max[\theta_r,\theta_i] \quad (12)$$

$$\beta=\min[\theta_r,\theta_i] \quad (13)$$

where the definitions of θ_i, θ_r, φ_i and φ_r are shown in Figure 3, σ is the standard deviation of Gaussian distribution micro-surfaces, which is used to characterize the surface roughness [15-19]. In Ld-PSD system shown in Figure1, it can be derived that:

$$\theta_i=0,\ \theta_r=\pi/4,\ \varphi_i=\varphi_r=\pi/2$$

Take these angles into (6)~(13), the radiance of diffuse reflection can be written as follows:

$$L_r=\frac{\rho}{\pi}L_i(1-0.5\frac{\sigma^2}{\sigma^2+0.33})+0.17\frac{\rho^2}{\pi}L_i\frac{\sigma^2}{\sigma^2+0.13} \quad (14)$$

The relationship between radiance and radiant flux is:

$$L_r=\frac{d^2\Phi_e}{d\Omega dA\cos\theta} \quad (15)$$

where Φ_e is radiant flux, dA is the area of receiving surface, $d\Omega$ is solid angle at the given direction to light source and θ is the angle between the given direction and the normal of receiving surface. In LD-PSD system shown in Figure 1, the imaging lens collect diffuse reflection light and project to PSD. Therefore, radiant flux received by PSD is equal to radiant flux collected by imaging lens. Here, we assume that the receiving area of imaging lens is A. Divide A into the combination of micro-area A_n. The direction angle and the solid angle corresponding to A_n are θ_n and Ω_n. Then radiant flux received by PSD can be derived from equation (15) as follows:

$$\Phi_e=L_r\sum_n\Omega_n A_n\cos\theta_n \quad (16)$$

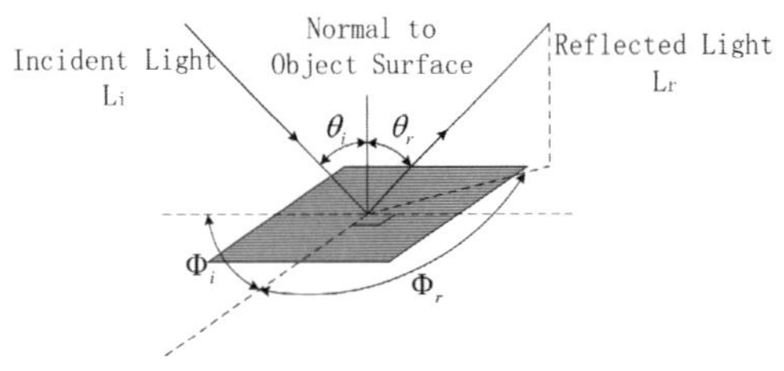

Figure 3. Definition of diffuse reflection parameters

The above analysis is based on the condition that the light spot on measured surface is an ideal point. However, as described in section 2, the light spot of LD has a size that cannot be ignored. For this reason, we divide the light spot into a series of micro-spots. The diffuse reflection radiant flux of each micro-spot obeys equation (16). So we can take Φ_e as a function of X into equation (4), where X is the position of micro-spot at the photosensitive surface of PSD.

$$X_0=\{\int_{X_l}^{X_r}[\rho(X)(1-0.5\frac{\sigma(X)^2}{\sigma(X)^2+0.33})$$
$$+0.17\rho(X)^2\frac{\sigma(X)^2}{\sigma(X)^2+0.13}]XdX\}$$
$$/\{\int_{X_l}^{X_r}[\rho(X)(1-0.5\frac{\sigma(X)^2}{\sigma(X)^2+0.33})$$
$$+0.17\rho(X)^2\frac{\sigma(X)^2}{\sigma(X)^2+0.13}]dX\} \quad (17)$$

where $\rho(X)$ and $\sigma(X)$ are reflectance and standard deviation of the measured surface corresponding to X. Equation (17) implies that center-of-gravity of light spot detected by LD-PSD system is related to the reflectance and standard deviation of measured surface. When LD-PSD system scans the measured surface, the light spot will move, while ρ and σ are variables. This will introduce uncertainty of X_0 to the measured result, which decreases the stability and robustness of LD-PSD system. However, if we can ensure that ρ and σ are constants between X_l and X_r at each moment while scanning, X_0 will no longer be related to ρ and σ. Then the system will get high stability and robustness.

4 Simulations

The reflectance and standard deviation of measured surface always change simultaneously. So we assume that t ($X_l < t < X_r$) is the critical point of the changing of ρ and σ. That is, in equation (17), $\rho(X)$ and $\sigma(X)$ are ρ_1 and σ_1 when $X_l < X < t$, while $\rho(X)$ and $\sigma(X)$ are ρ_2 and σ_2 when $t < X < X_r$. Under this assumption, equation (17) can be written as:

$$X_0=\frac{(a-b)t^2-aX_l^2+bX_r^2}{2(a-b)t-2aX_l+2bX_r} \quad (18)$$

where

$$a = \rho_1(1-0.5\frac{\sigma_1}{\sigma_1+0.33})+0.17\rho_1\frac{\sigma_1^2}{\sigma_1^2+0.13} \quad (19)$$

$$b = \rho_2(1-0.5\frac{\sigma_2}{\sigma_2+0.33})+0.17\rho_2\frac{\sigma_2^2}{\sigma_2^2+0.13} \quad (20)$$

Then we can derive the drift error of center-of-gravity of light spot as:

$$\Delta X = X_0 - \frac{X_l+X_r}{2} = \frac{(a-b)(t-X_l)(t-X_r)}{2(a-b)t-2aX_l+2bX_r} \quad (21)$$

From equation (21) we can derive that $a = b$ and $\Delta X = 0$ when $\rho_1 = \rho_2$ and $\sigma_1 = \sigma_2$, which is corresponding to the condition that ρ and σ are constants within the light spot. And when $\rho_1 \neq \rho_2$ or $\sigma_1 \neq \sigma_2$, a is no longer equal to b and ΔX changes with the change of t. The relationship between t and ΔX is depicted in Figure 4, where the horizontal axis is the position of critical point t while the vertical axis is the drift ΔX. The blue curve indicates that the light spot diameter is D, while the red curve indicates that the light spot diameter is $2D$. For the blue curve, when t is located out of the spot ($1 < t/D < 2$), ΔX is 0. While ΔX is no longer 0 when t is located within the light spot ($0 < t/D < 1$) and the variation trend is similar to the quadratic curve. Contrasted with the blue curve, the red curve has a larger range of error because of the larger spot diameter.

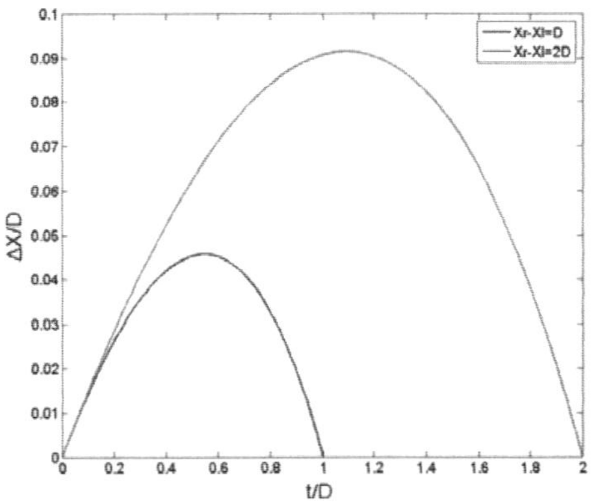

Figure 4. Simulation result of critical point position and errors.

5 Experiments

Before the experiment, we need to exclude the influence of other factors on the measurement results. Take (16) into (2) and (3):

$$i_{left} = D\int_{X_l}^{X_r}[\frac{\rho(X)}{\pi}L_i(1-0.5\frac{\sigma(X)^2}{\sigma(X)^2+0.33}) +0.17\frac{\rho(X)^2}{\pi}L_i\frac{\sigma(X)^2}{\sigma(X)^2+0.13}\sum_n \Omega_n A_n\cos\theta_n](1-\frac{X}{L})dX \quad (22)$$

$$i_{rignt} = D\int_{X_l}^{X_r}[\frac{\rho(X)}{\pi}L_i(1-0.5\frac{\sigma(X)^2}{\sigma(X)^2+0.33}) +0.17\frac{\rho(X)^2}{\pi}L_i\frac{\sigma(X)^2}{\sigma(X)^2+0.13}\sum_n \Omega_n A_n\cos\theta_n]\frac{X}{L}dX \quad (23)$$

The above two equations shows that photocurrent of PSD is also related to reflectance and standard deviation of measured surface. The A/D converter of LD-PSD signal processing circuit has a linear conversion range. When photocurrent changes the signal may exceed the linear conversion range, which will lead to a big circuit noise and the measured result is not the real position of X_0. So we use the signal band-pass filtering techniques to optimize the circuit system, which can modulate the intensity of photocurrent adaptively. The frequency response of band-pass filter is as follows:

$$A_u = \left|\frac{1}{1+jQ(\frac{f}{f_0}-\frac{f_0}{f})}\right| \quad (24)$$

where f_0 is the natural frequency and f is the modulate frequency. When the circuit finds the change of photocurrent, f will change adaptively through negative feedback to modulate the photocurrent. Finally, photocurrent will rapidly return to the linear region of the A/D converter. This allows the circuit system and the optical system to maintain a good consistency, thus excluding the impact of the circuit factors.

In the experiment, we use a standard roughness gauge block as the measured object. The block length is L and the flatness error is less than 1 μm. The centre line divides the roughness of block into $\mu 1$ and $\mu 2$. So the reflectance and standard deviation are ρ_1, σ_1 and ρ_2, σ_2 separately. We set the real distance of block to 50mm (d = 50mm) and use LD-PSD system to scan the block from left boundary to right. The resolution of LD-PSD system is 1 μm and the diameter of light spot is D and $2D$ ($D \ll L$). If the distance from spot right margin to the left boundary of block is x, the distance of the block measured by LD-PSD is depicted in Figure 5. For the first scanning, the diameter of light spot is D. When x is less than D, the measured results have large errors because the light spot has not entirely projected to the block surface. When $x = D$, the measurement starts to be normal and the measured results are equal to the real value. With the moving of the spot, when $x = L/2$, that is, ρ and σ begin to change within the spot, the measured results begin to be error. When $x = L/2+D$, the measured results return to the real value because the light spot have entirely accessed to the surface that the reflectance is ρ_2 and standard deviation is σ_2. For the second scanning with the $2D$ diameter of light spot, the trend of the curve is similar to that of the first scanning. From Figure 5, it can be seen that the trend of curves is agree with the simulation in Figure 4 when the spot goes through the mutative σ and ρ. And a bigger light spot will have a larger probability of error because the bigger spot is apt to reach the area that contains mutative reflectance and standard deviation, meanwhile it is easier to exceed the measured surface.

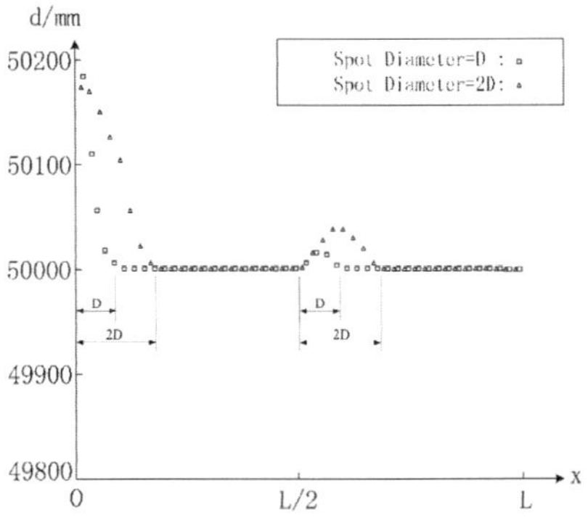

Figure 5. The distance measured by the scanning LD-PSD.

6 Conclusions

Starting from the reflection model of the general interface and combining the Lambert law and the metering center-of-gravity of light spot principle of PSD, we derive the relationship between the spot location and the reflectance and standard deviation of measured surface, which allows us to analyze the impact of the measured surface properties on LD-PSD system. By simulations and experiments, we demonstrate that error occurs when reflectance and standard deviation changing within the spot area, while the measured result is stability when reflectance and standard deviation are constant at every moment. In addition, a smaller laser spot can get smaller error and smaller error area. In other words, a smaller laser spot can improve the accuracy and robustness of LD-PSD system.

Acknowledgements

The research was supported in part by the Yumiao Foundation of Tianjin University of Technology (LGYM201106).

References

1. J. Contreras, L. Gomes, S. Filonovich, N. Correia, E. Fortunato, R. Martins, I. Ferreira, Opt. Express, **20**, 4583-4602 (2012)
2. I. Matsuya, R. Tomishi, M. Sato, K. Kanekawa, Y. Nitta, M. Takahashi, S. Miura, Y. Suzuki, T. Hatada, R. Katamura, T. Tanii, S. Shoji, A. Nishitani, I. Ohdomari, IEEJ T. Electr. Electr., **6**, 266-272 (2011)
3. V. Esmaeile Sani, A. Moussavi-Zarandi, M. Kafaee, Nucl. Instrum. Methods Phys. Res., Sect. A, **654**, 464-470 (2011)
4. Y.C. Sung, T. Miyasaka, T.K. Lin, C.Y. Wang, C.Y. Wang, Struct. Control Health Monit., **19**, 295-308 (2012)
5. H. Nakajima, K. Sumi, H. Inujima, IEEE Trans. Instrum. Meas., **59**, 3041-3048 (2010)
6. R.G. Dorsch, G. Hausle, J.M. Herrmann, Appl. Opt., **33**, 1306-1314 (1994)
7. C. Narayanan, A.B. Buckman, I. Busch-Vishniac, W. Wang, IEEE T. Electron Dev., **40**, 1688-1694 (1993)
8. S. Cui, Y.C. Soh, IEEE T. Electron Dev., **57**, 2310-2316 (2010)
9. S. Cui, Y.C. Soh, Opt. Quant. Electron., **42**, 721-728 (2011)
10. L. Kong, E. Li, D. Yang, H. Xing, IEEE T. Electron Dev., **59**, 835-840 (2012)
11. K.E. Torrance, E.M. Sparrow, J. Opt. Soc. Am., **57**, 1105-1114 (1967)
12. T. Kiuchi, K Ikeuchi, IAPR MAV' 92, 175-178 (1992)
13. M. Oren, S.K. Nayar, Int. J. Comput. Vision, **14**, 227-251 (1995)
14. L.B.Wolff, S.K. Nayar, M. Oren, Int. J. Comput. Vision, **30**, 55-71 (1998)
15. R. Reizer, Wear, **271**, 539-543 (2011)
16. G. Zhang, Z.S. Wu, Opt. Express, **19**, 7007-7019 (2011)
17. G. Zhang, Z.S. Wu, Opt. Express, **20**, 1491-1502 (2012)
18. J.F. Lingevitch, K.D. Lepage, IEEE J. Oceanic Eng., **35**, 199-208 (2010)
19. C.F. Cheng, D.P. Qi, D.L. Liu, S.Y. Teng, Acta Phys. Sinica, **48**, 1635-1643 (1999)

 DOI: 10.1051/matecconf/20166106002

The Secondary Optical Design and Fabrication for the Uniform Illuminating LED Spotlight Using TIR Lens

Yingchao Xu [a,1,2], Guanyu Chen [1,2], Yongxiao Chang [1,2], Hongyi Lin [1,2], Jiamei Zheng [1,2], Liangqin Gan [1,2]

[1]School of Opto-electronics and Communication Engineering, Xiamen University of Technology, Xiamen 361024, China
[2] Fujian Provincial Key Laboratory of optoelectronic information materials and devices, Xiamen University of Technology, Xiamen 361024, China

Abstract. Different with the fact that the traditional imaging optics emphasizes its ability to get clear image on the image plane, the purpose of LED illumination design is to meet the requirement of high availability of optical energy and homogeneity of illumination. As the study object, the LED track lighting has been designed with secondary optical design by the Lighttools design software. After using TIR lens which has two ways of concentrating: reflectors and lenses, the exit light of the spotlight will be adjusted and the illuminated surface will be with high homogeneity. Therefore, it will reduce the harm of the light to human's eyes. In this study, the LED spotlights have been produced and tested with the corresponding test results: Lighting power is 7W, luminous efficiency is82.57lm/W, color rendering index is 83.7, the main wavelength is 488.2nm, the ratio of red, green, blue is 13.1: 83.0: 3.9.

1 Introduction

The implementation of the "Ten cities, ten thousand lights" project in China and the LED lighting demonstration project for Beijing Olympics and Shanghai World Expo greatly enhanced the development speed of the LED lighting. Known as the fourth generation lighting source, LED's development is very rapid. And it is a trend to replace the traditional energy-saving lamp light sources gradually[1-2]. As the luminous efficiency of LED increases, its advantages such as the full color, environmental protection, energy saving become more and more significant. Although LED has already entered the field of lighting, the problem of secondary light distribution restricts its further development. LED's application will shine in the field of lighting, when the bottlenecks of secondary light distribution[3-4] and heat-release are solved. The development of advanced lighting design software and the appearance of free surface optical components, making the research works of improving secondary light distribution and the efficiency of LED lighting are more and more convenient.

In recent years, the research works have done on the secondary light distribution design of the LED sources at home and abroad. For example, LED array illumination light source and surface light source[3-4], the research on LED lamp[5], the secondary light distribution design on LED spotlights are also reported in the literature[6] , by adjusting each individual Fresnel lens ring with the inclination to optimize the distribution of the projected light. So far greater advantage of the Fresnel lens is applied in terms of the focus of light in the secondary light distribution field, but it is not the best means.

In this paper, we use the LED track spotlight as the example. Using the TIR lens as the light-uniform element has advantages that it integrates the reflector and the converging lens. Ultimately uniform illumination is achieved at the irradiated surface. We manufactured the LED lamp and the photoelectric parameter test results were given in the paper.

2 Introduction of TIR Lens And Its

Performance Evaluation

TIR lens is the total internal reflection lens, using the principle of total reflection optics, to collecte and process light. Fig.1 Shows the sketch of the TIR lens. TIR lens can complete the collection of light, and we can achieve a beam angle control by adjusting the radius of curvature of the TIR lens.

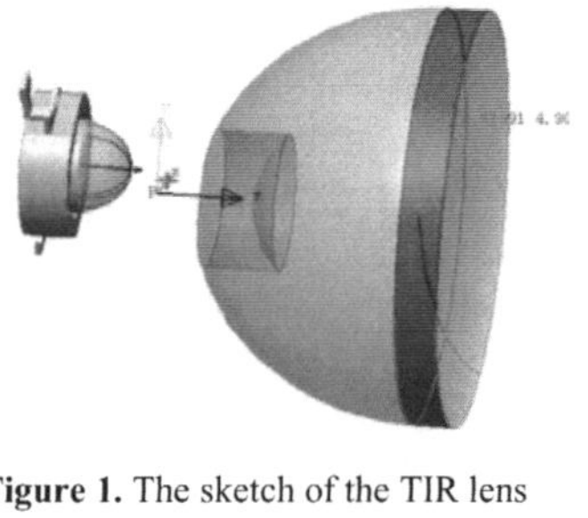

Figure 1. The sketch of the TIR lens

[a] Corresponding author: ycxu@xmut.edu.cn

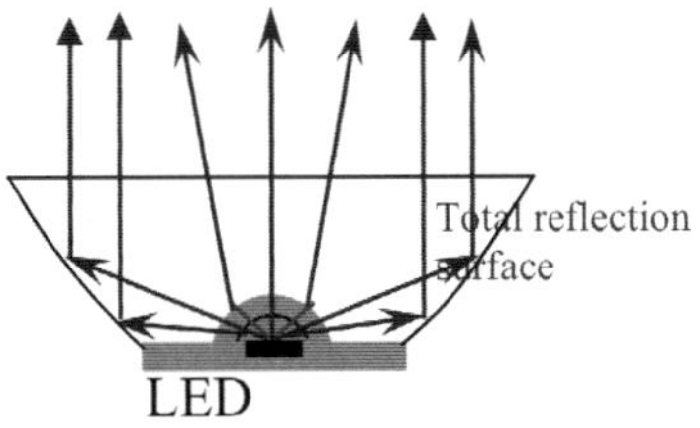

Figure 2. The diagram of the light distribution effect for TIR lens

Lens has the excellent light-gathering ability for the rays at a small angle. While the reflector has the advantage to converging the rays at a large angle. So both the lens and the reflector are combined together, as shown in Figure 2. Reflector and lens group combination into TIR lens, can effectively improve the luminous efficiency of LED lighting devices, avoid the waste of LED luminescence. In summary, Using the combination of the reflector and the lens, addition to the use of imitation compound eye micro lens array, the light concentrating ability of the LED lamp will be greatly improved. The efficiency and the uniformity will be improved greatly [7]. By adding 3D texture, we can also eliminate the glare of the lamp and the eye discomfort.

In beam shaping process of the LED light source, although tapered light pipe (TLP) [8] and compound parabolic concentrator (CPC)[9] both have very high optical collectting efficiency, but in order to improve the intensity of illumination uniformity of the system, usually makes the system very long. Obviously, in improving the illumination uniformity and reducing the size of the system to make the system more compact simultaneously, TIR lens has more advantages. So we adopt the TIR lens as secondary light distribution element for LED track lamp.

For the performance evaluation of TIR lens used in the illumination system, we mainly use the light energy utilization ratio and intensity of illumination uniformity [10].

[1]. Light energy utilization ratio

By calculating the ratios of luminous flux that reaches the target screen receiver and the total light flux emitted by LED light source, we can get the energy utilization of the system. The computation formula is as follows:

$$\eta' = \phi_{screen} / \phi_{LED} \quad (1)$$

[2]. Intensity of illumination uniformity

By dividing the target screen into m×n grid, and through the calculation of intensity of illumination B of each box ,we can work out the error between each region and the mean value S:

$$S = \sqrt{\frac{1}{m \times n - 1} \sum_{i=1}^{m \times n} (B_i - \overline{B})^2} \quad (2)$$

Calculated error S, and based on the average $\overline{B}$ has been calculated, seeking both ratios, you can get illuminance uniformity:

$$U = \frac{S}{\overline{B}} \quad (3)$$

In the fomula above, the smaller of the error S, the smaller of the intensity of illumination uniformity. That the intensity of illumination uniformity is the better.

3 Lighting System Design

In order to eliminate the harm to the human's eyes caused by the glare of the LED lamp, lighting system design must be executed. And the TIR lens is used to obtain better illumination uniformity

The light source used is the patch type white LED Lambertian light source, with a divergence angle of 120 °. The radiation intensity has nothing to do with the viewing angle, the radial brightness on the various angles are the same. Therefore, the light scattering characteristics as follows: the scattering surface is arbitrary; the probability of each scattering direction is same; Scattering follow normal direction rather than on the surface of the incidence angle; Each of the scattering light have the same energy. Light intensity emitted from the light source into a cosine distribution, formation of the spot center is very bright, along the radial attenuation [11].

Through research, we found that the arrangement have less effect on the intensity of illumination uniformity and light energy utilization rate, but the distance between LEDs on both greatly influenced. For rectangular arrangement of six LED module systems, the utilization of light energy decreases gradually with the increasing of the distance between the LED lights, and illumination uniformity decreases with the increase of distance, and then decreases with the increasing of the distance. For the annular arrangement,light energy utilization ratio decreases with increasing distance, intensity of illumination uniformity decreases with increasing distance change trend [12].

In this design, in accordance with the lamp bead emission angle with the scattering angle and uniformity of lens and light energy utilization and the relationship between the distance, the arrangement of seven LED lamp bead every three forms an equilateral triangle, the angle is 60° and distance is 2.2 r (r is the radius of the lens). The arrangement is shown in Figure 3, each dot represents a single LED lamp bead.

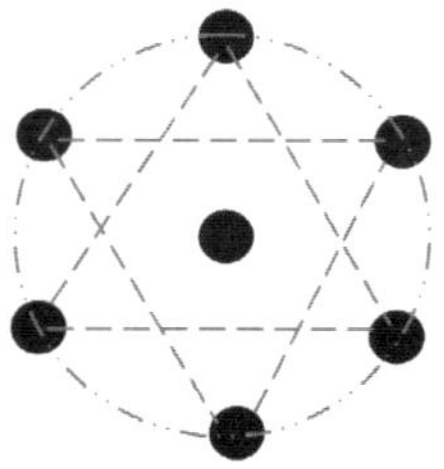

Figure 3. The arrangement of the seven LED lamps

According to the theory of nonimaging optics, the purpose of secondary light distribution design is to change the LED light Angle and optical intensity distribution, in order to satisfy the demands of the corresponding lighting.

Light distribution curve according to the beam angle is usually divided into:

$$\begin{array}{c}\text{Narrow light distribution} \ (<20^{o}) \\ \text{Medium light distribution}(20^{o}\text{- }40^{o}) \\ \text{Wide light distribution} \ (>40^{o})\end{array} \quad (4)$$

Because TIR lens is essentially a lens at center area, mainly for the light at small angles to carry out light distribution design. The surrounding area of the TIR lens is total reflection, mainly for the light at the large angles to carry out the secondary light distribution design, so TIR lens deals with the most perfect illumination uniformity.

In the design we use Lighttools lighting design software.Total reflection of both the middle section of lens, and peripheral reflection section, are used with quadric surfaces (it could be a parabolic, ellipsoid or hyperboloid). In fact which we used are hyperboloid surface shape. So that each surface is more than the traditional spherical surface of a quadric coefficient variation. In actual optimal design process, we set the two quadric coefficients of the surfaces for the lens in the center and peripheral total reflector as two optimization variables. And then add collimating optimization function, run ray tracing. The collimation embedded in the Lighttools software design optimization function, its principle is to use Monte Carlo ray tracing method will all fall in the cosine of the light receiving surface consistent.Final design layout and illumination intensity distribution is shown in figure 4. According to the following design results and formula (2), (3), we can get error S = 74.81, U = 33.85%.

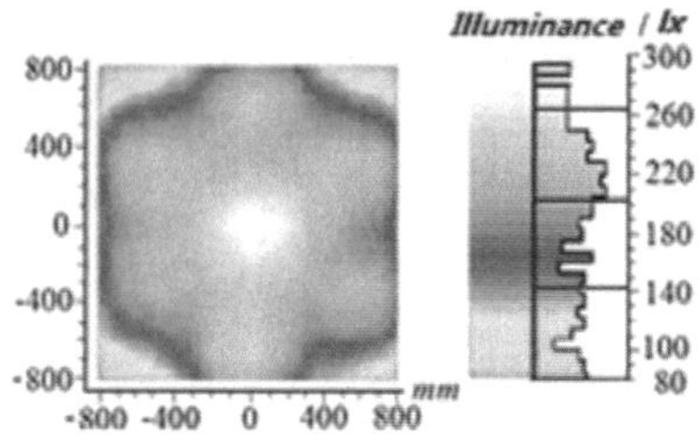

(b)

101	192	195	109
274	299	292	279
281	298	296	280
118	198	189	135

(c)

Figure 4. The illumination intensity effect of the lamp after optimized

(a) The diagram of the ray tracing for the lamp (b) The illumination intensity (c)The value of the illumination intensity divided by the grid of 16×16

The micro lens array stage imitating compound eyes is also used in lighting system,Its aim straight design for lighting uniformity has a good effect[13]。Therefore we have to add the above design a good set of TIR lens design of micro lens array（As shown in figure 5）. According to the design results and formula shown below (2), (3), we are able to get the error S = 40.43, U = 19.18%. Design result of light intensity of illumination figure shows that the illumination uniformity is improved significantly. But the disadvantage is that due to the addition of a micro lens array, the scattering light to the periphery has a small part of the increase. The average illumination of the lamp is reduced to 210.75lx by 221lx, reduce the rate of 4.64%. In the main pursuit of the effect of the illumination uniformity, reducing the harm of the lamp glare on the human eye, the light energy utilization rate of the range is acceptable.

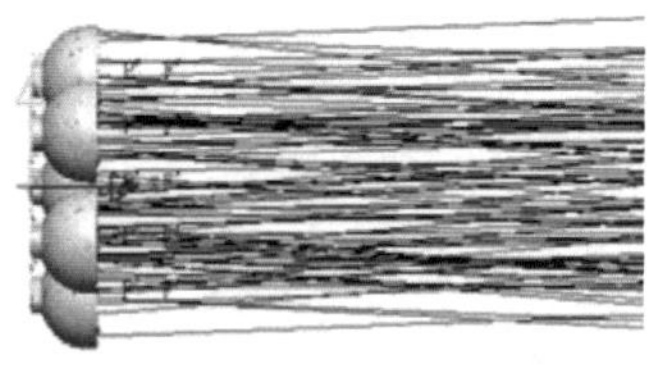

(a)

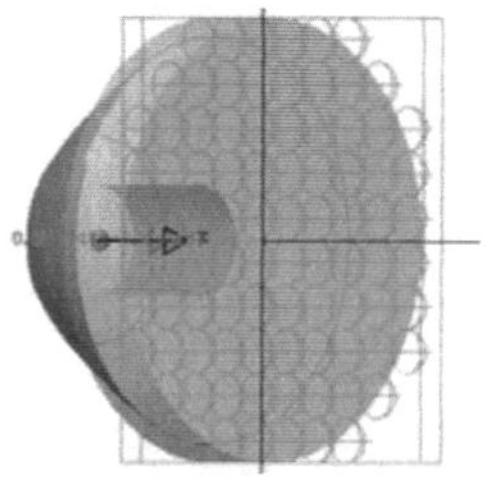

(a)

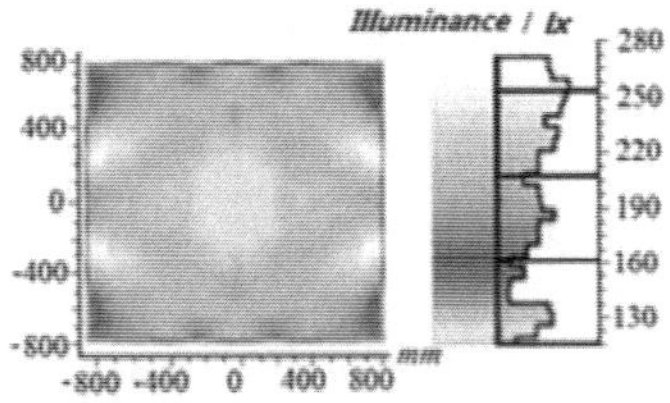

(b)

155	208	211	159
236	264	248	228
234	266	247	231
150	190	197	148

(c)

Figure 5. The micro-lens array for the TIR lens and the improvement of the illumination intensity uniformity (a) The micro-lens array added for the TIR lens (b) The illumination intensity (c) The value of the illumination intensity divided by the grid of 16×16

4 The Manufacture Of Lamps And Testing

4.1 The manufacture of the lamp

According to the driving characteristics of LED lamp, we designed the corresponding driver circuit, as shown in figure 6. Including voltage transformer, rectifier, filter, voltage regulator and other circuits.etc.

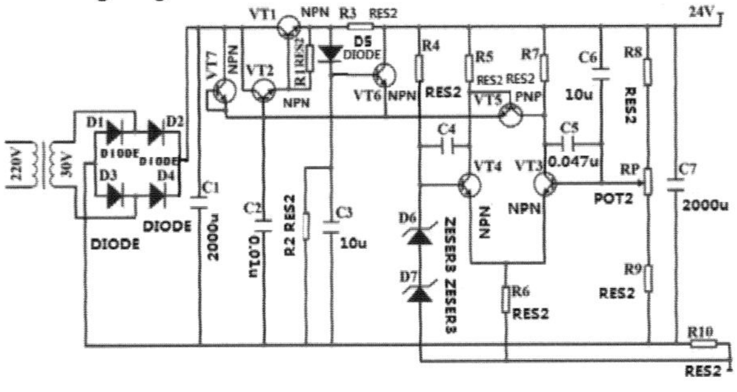

Figure 6. Driving circuit for LED light

We used Solidworks software to design the lamp housing, cooling systems, metal ring groove that fixed TIR lens. Ultimately,we obtained the the corresponding design and the manufactured LED track lamp is shown in Figure 7.

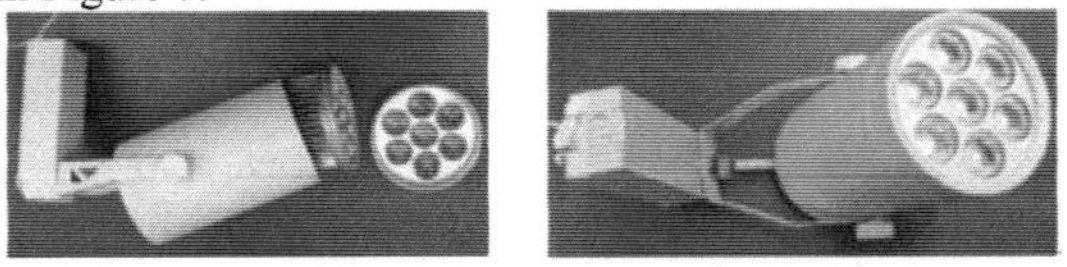

(a) (b)

Figure 7. The picture of the LED track spotlight

4.2 Lamp Test

The light intensity distribution curve of the lamp was tested by GO2000A which produced by Hangzhou Everfine Ltd.. The light intensity distribution curve is shown in Figure 8 and the basic parameters of the light source as shown in Table 1.

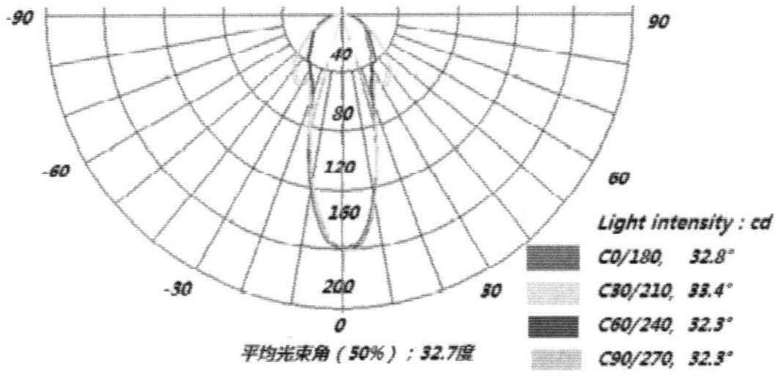

Figure 8. The curve of the intensity distribution

From Figure 8, we can see that the lamp light distribution is within ±15 degrees, the average beam angle is 32.7degrees, belonging to medium light distribution. The photoelectric parameters of the lamps were measured by integrating sphere, as shown in Figure 9 and figure 10. The results of the measurements are shown in the following table 1.

Table 1. Test results fixtures

<table>
<tr><td colspan="2">Total power</td><td colspan="2">7 W</td><td colspan="2">Luminous efficiency</td><td colspan="2">82.57 lm/W</td></tr>
<tr><td colspan="2">Peak intensity</td><td colspan="2">16306 cd</td><td colspan="2">The total flux</td><td colspan="2">577.99lm</td></tr>
<tr><td colspan="2">Average beam angle</td><td colspan="2">32.7 °</td><td colspan="2">Color Temperature</td><td colspan="2">6270 K</td></tr>
<tr><td colspan="2">Dominant wavelength</td><td colspan="2">488.2 nm</td><td colspan="2">Peak wavelength</td><td colspan="2">450.0 nm</td></tr>
<tr><td>Red ratio (%)</td><td>13.1</td><td>Green ratio (%)</td><td>83.0</td><td>Blue ratio (%)</td><td>3.9</td><td>Ra</td><td>83.7</td></tr>
</table>

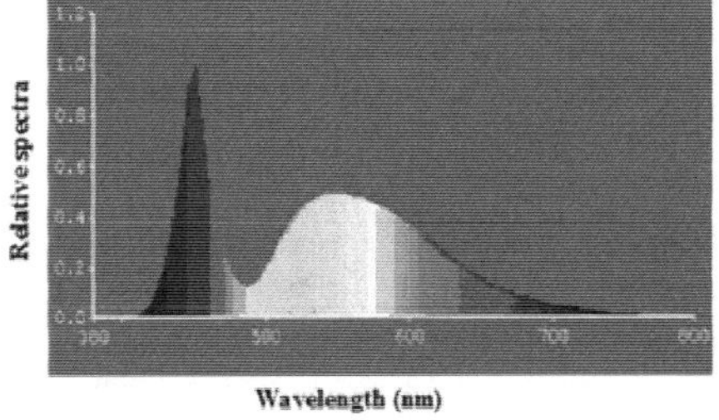

Figure 9. The spectral curve of the lamp

DOI: 10.1051/matecconf/20166106002

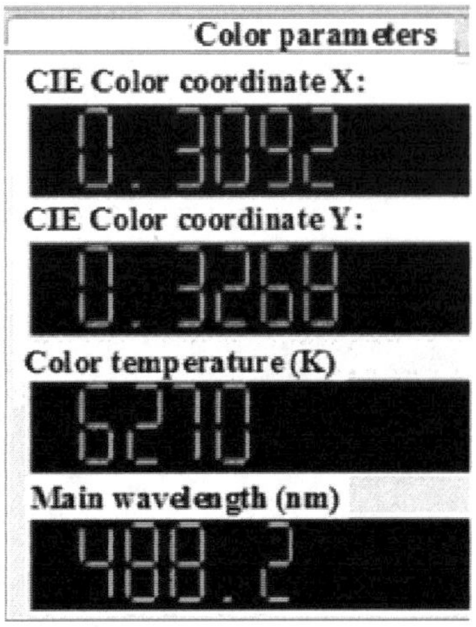

Figure 10. The photometric test reslut of the LED lamp

After testing, we obtain that the power of the LED lamp is 7W, the luminous efficiency is 82.57lm/W, the color rendering index is 83.7. From Figure 10 we can see the main wavelength is 488.2nm, and the ratio of red light, green light, and blue light is 13.1: 83.0: 3.9.

5 Conclusion

For LED lamps which just has been carry out the first light distribution, the light is not all projected to the expected range, but also after the package is not uniform distribution of light intensity, the illumination area always appears to diffuse. The secondary light distribution can make up for the waste of energy and distribution of unequal defects, and it can effectively eliminate glare, improve LED lighting illumination uniformity.

From the results of the design, manufacture and testing, the technical parameters of the LED track lamp in this paper has reached a good level in all aspects. In many indicators, especially for the process of illumination uniformity achieved superior market LED track lamp levels, it can be widely used clothing stores or large stage lighting and other aspects of the configuration, and highly practical.The development of the secondary light distribution in this paper for the new generation of LED lighting has a certain reference value for the same industry .

Acknowledgment

This work is supported by the National Natural Science Foundation of China (11304259) and the S&T plan projects of Fujian Provincial Education Department (JA12254)

References

1. R.lenk,C.Lenk,*Practical lighting design with LEDs*(IEEE Press,Wiley,2011)
2. R. Winston, J.C. Miflano, P. Benitez. *Nonimaging Optics*(Elserier, Amsterdam, 2005)
3. C.Sommer,E.Hartmann,P.Pachler.Optical Materials,**9** (2008)
4. Chang,Y.-C.,Ou,C.-J.,Juang,F.-S.Opt. Rev.**16** (2009)
5. Lee,H.,et al.Opt.Express,,**21** (2013)
6. Yi Ding , Peifu Gu ,Wei Lu.Engineering Science **41**,9 (2007)
7. F.Chen,K.Wang,Z.Qin,D,Wu,X.B.Luo,S.Liu,Opt Express **18**,20 (2010)
8. Bol'shukhin, V.A., Ilyasov, V.S., Soshchin, N.P., Ulasyuk, V.N.,J. Opt. Technol **78**,7 (2011)
9. C.P.Krijn,A.B.Salters,H.O.Willemsen.SPIE,**6196** (2006)
10. Jiang,J.,To,S.,Lee,W.B,Cheung,B.,Optik,**121** (2010)
11. Zhang Kang, Chengyue Su,Qian Fu,Tao Xue,Chunhua Zhang.Journal of Guang Dong University of Technology **30**,1 (2013)
12. Wang Bao , Rongbin Yu , Feng-guo Li.China Illuminating Engineering Journal,**5** (2013)
13. K.H.Jeong,J.Kim,L.P.Lee.Science,**312**(2006)

Construction of the Application Systems for Universal Education in Architectural Heritage – Multimedia Touch Screen Desk System

SATO. Reika[1,2,a], GUO Weimin[2]

[1] *Department of Design, Jiang Nan University, 214122, China*

[2] *Department of Digital Games, Osaka Electro-Communication University, 575-0063, Japan*

Abstract. The importance of popular education in architectural heritage is well known. It is an effective method for public education to develop applications that utilize digital copy of architectural heritage. We provide a case study focusing on the historical buildings of Toji Temple in Kyoto, Japan. We digitize the historical buildings in both overall construction and detailed structures. We build a demonstration and operating system using multimedia technologies such as multimedia touch screen desk system. The system will produce 3D digital models of architectural heritage with high resolution and use advanced technology to develop a variety of demonstration applications that can enhance your presentation and ease operation. The users can also have more opportunities and ways to improve their interests in learning and understanding about the architectural heritage knowledge, hence finding new ways to further improve the methodology of architectural heritage education with the application of multimedia interactive technologies and digital copy of historical buildings.

1 Introduction

Many countries attach great importance to conservation and education of cultural heritages in a city, a key part of which is architectural heritage. Architecture represents historical legacy for the modern society, telling stories of a nation's cultural development and transmitting various national development information. However, over time, many architectural heritages have disappeared or are disappearing, rendering conservation of these heritages so challenging a task in the modern development process. Popular education of architectural heritages becomes imminent, so that their history and charm can be learned and felt. Hard efforts must be made to discover the historical and cultural messages in the architectures so as to build awareness and energy for heritage conservation.

Digital copy of architectural heritage matters first in that it can visualize and present architectural research assumptions and results of vanished buildings. The digital copy is preserved and accessed as cultural heritage, and the non-visual (lost or non-public, unsplit, inaccessible, etc.) parts get visualized at the same time. In other words, the heritage is made more accessible and digestible, arousing greater interest to learn in the public.

Conservation of existing architectural heritages in itself contradicts with their exposure to public. Restoration and preservation efforts can be offset by long-time exposure and demonstration of the buildings. In contrast, demonstration of their digital copy is vivid, and free from constraints of location, time and space, thus more interesting, practicable and digestible. From the perspective of digital copy production, it's easier to produce, process and refine the contents to be demonstrated; transmission means is not constrained by location or time; and online promotion and education can reach more people.

In recent years, the touch screen technology is growing increasingly popular. Intuitive and fast in operation, it has been applied to home appliances, game consoles and mechanical equipment. Yet, large-scale application of this technology is no easy. As its alternative, the projector-based multi-media touch screen system, with a large display screen and touch-enabled, comes into being. The large screen reacts to touch based on the detecting function of the infrared radar.

Virtual reality technology can digitize architectural heritages. And with interactivity, optional viewpoints, optional dimensions, audio guide and other functions, it can better play its role in the popularization of architectural heritage. This technology comes down to computer graphics processing, human-computer interaction technology, sensor technology, artificial intelligence, etc. With this technology, viewers can be guided into virtual space for detailed observation and leaning, thus leading to better results in the popular education.

Kyoto is historically and culturally icon citie in Japan's development. This paper studies Toji Temple in Kyoto, so as to explore new approaches to copy and popularize architectural heritage via multimedia interactive technologies.

Toji Temple was built in the Heian period, with a history of over 1,200 years. It is a representative cluster of Buddhist architecture, and became a UNESCO world

[a] lihua@mwd.biglobe.ne.jp

heritage site in 1994. This historical building was burnt, and rebuilt and restored. Now they represent a key part in Japan's popular education of architectural heritage.

2 Construction Of The Multimedia Touch Screen Desk System For Toji Temple In Kyoto

Toji Temple was originally built in Heiankyo as Kyō-ō-gokoku-ji which indicates that it previously functioned as a temple providing protection for the nation, and then was given as a present by Emperor Saga to priest Kobo. There the Shingon sect of Buddhism introduced from China was lectured. The architecture comprises the Kondo Hall, Kodo Hall (existing lecture hall as is shown in Figure 1), shokudo Hall, Toji Pagoda, Mieido House and other buildings. Based on the 3DCG modeling technology, both the entire appearance of the architecture and its detailed structure are presented as in Figure 2 and Figure 3 respectively.

Figure 1. Kodo Hall in Toji Temple

Figure2. Modeling of the Kodo Hall

The importance of popular education in architectural heritage is well known. It is an effective method for public education to develop applications that utilize digital copy of architectural heritage. We provide a case study focusing on the historical buildings of Toji Temple in Kyoto, Japan. We digitize the historical buildings in both overall construction and detailed structures. We build a demonstration and operating system using multimedia technologies such as multimedia touch screen desk system. The system will produce 3D digital models of architectural heritage with high resolution and use advanced technology to develop a variety of demonstration applications that can enhance your presentation and ease operation. The users can also have more opportunities and ways to improve their interests in learning and understanding about the architectural heritage knowledge, hence finding new ways to further improve the methodology of architectural heritage education with the application of multimedia interactive technologies and digital copy of historical buildings.

Figure3. Detailed Structure of the Architecture

To enable interactive operation on a large touch screen with the copied 3D architecture, the multimedia touch screen desk system is designed and developed. As is shown in Figure 4, this system comprises an 80-inch frosted glass desktop, a projector with ultra-short focus and near-infrared radars.

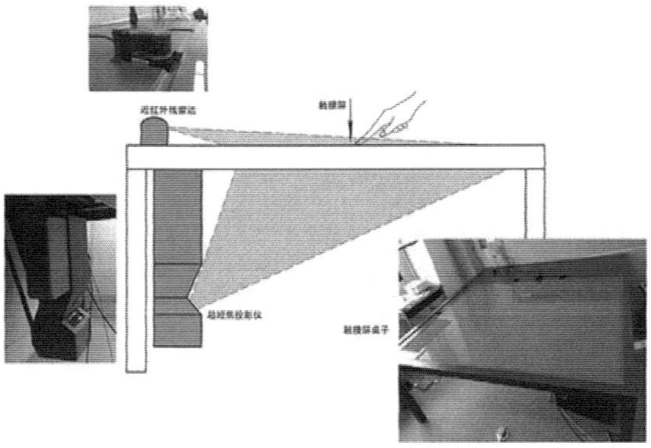

Figure 4. Components of the Multimedia Touch Screen Desk System

2.1 Desk-type touch screen

In the newly developed desk-type multimedia touch screen system, there is an 80-inch frosted glass desktop (a square with side length of 186cm); beneath the desktop an ultra-short focused projector (Sanyo LP-XL51) is installed to project images onto the glass desktop; and at the sides of the desktop, near-infrared radars are built in to scan and detect touch spots on the desktop. To prevent vibration from touch, the desk is made of heavy stainless steel. For the convenience of operation, the display screen desk and the projector are integrated. And in case that the

80-inch large touch screen would bend because of the dead weight, the desktop is made of 10mm-thick frosted glass. Four desk legs can be removable, and the height of the desk is 75cm, rendering it convenient to operate the system in a standing position, as is shown in Figure 5. Grooves are built along the four outer sides of the desktop to store near-infrared radars, enabling the installation of four radars at most for touch spot location. Because of those multiple radars, several layers of touch surface are combined, so that no blind angles would be missed and the detection accuracy would increase.

2.2 Near-infrared radar

Currently, many projector-based touch screen applications feature infrared camera detection of touch spots, and different infrared-relied systems have different strengths[1]. When an infrared radar is used, anything movement, not limited to finger touch, can be detected. So it also can project images to the floor or the wall. On top of the desktop use, adjustment of the projector's focal length enables floor or wall use. Then radars are responsible for tracing footsteps or other mobile objects in contact with the ground or wall. Radars can also detect multiple touch spots simultaneously, which means the system can be operated by several people at the same moment. In addition, radars are capable of identifying the shape of the target object, and based on this function, additional signals can be sent by touch. But the infrared radar has its disadvantage. It cannot detect the range behind itself. So in the development of the touch screen desk system, the operation range must be taken fully into consideration.

With high-speed beam scanning, near-infrared radars are able to identify touch spots on the desktop. To capture touch spots, the radar-based range measuring device incorporates UBG-04LX-F01 range sensors produced by Hokuyo. The radar has a view angle of 240 degrees and can finish one beam scanning in 28 microseconds, capable of detecting touch spots in various shapes. Assume that one touches the desktop 1m away from the radar, the beam scanning distance would be 6mm, so any touch spot of more than 6mm radius can be located.

Touch signals received by sensors as is shown in Figure 5 are in various shapes: ① for one finger, ② for two fingers, ③ for a palm and ④ for a fist. After the touch stays at the screen for over 28 microseconds, the touch spot will be detected[2].

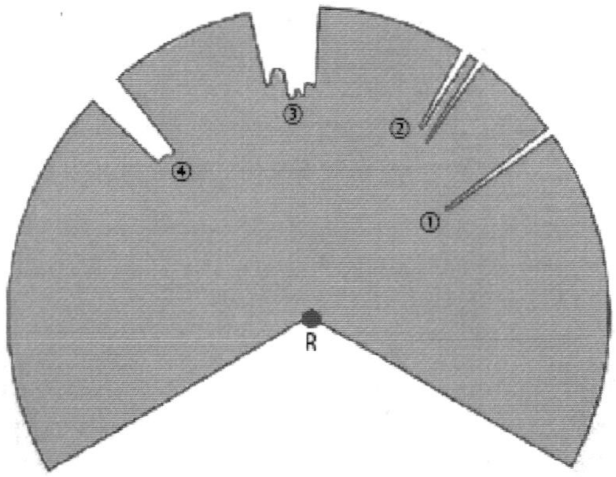

Figure5. Data Mapping of the Touch Spots Detected by Near-infrared Radars

Radars can detect touch made by objects substantially isotropic with radar beams, which can be a finger, a pointing stick, a chessman or other objects. Glass objects that can be pierced by radar beams or objects with mirrors that can reflect beams are not applicable. Any item used for touching is equivalent to a mouse delivering a detection point, so multiple items can be used simultaneously for giving various detection points. The scanning plane of radars must be accurately above the touch screen. The distance between the two surfaces is usually controlled within 5-10mm.

2.3 Application of digitized Toji Temple buildings

Use of the desk-type touch screen system in demonstrating Toji Temple buildings enriches the approaches to education of historical architectural heritage. In the demonstration to college students, this novel approach inspires their strong interest. And during college open days, the system catches the eyes of many children and elderly people, and is highly recognized.

3 COMMENTS ON THE APPLICATION SYSTEMS

It is a new try to apply the touch screen desk system in the education of architectural heritage. More applications need to be developed, and more technologies improved.

The touch screen desk system applies widely. It can demonstrate various documents, graphics and 3D representations. Participants can be one person or two to four people. The currently developed system realizes the demonstration of the copied buildings in Toji Temple. This system allows for both a bird's-eye view and a walking view of the architecture. Each part of the building can be zoomed in or out for detailed observation. Users can choose to view the heritage from their preferred height and angle, and within the preferred visual range. And when they use the near-infrared detection device, they will not experience inaccuracy or delayed response in detection.

Usually infrared detection-based touch screen systems only respond to warm human hands. In comparison, the near-infrared radar can detect touch spots but without the constraint of temperature. So projection and imaging can be done both on translucent glass or plastic plates, and on the ground or the wall, where touch spots can also be detected. In this sense, more projection screens are available.

When evaluated as a demonstrating screen, the touch screen desk system gets the highest score. In evaluation of its projection methods, the desktop screen and ground projection are both considered. In ground projection, a projector with a general focus length is used to project images onto the 5*5m floor, and the near-infrared radars to detect user locations on the ground; and the user controls the architecture representation by walking. Demonstration through ground projection requires cooperative footsteps. This means more attention to the architecture itself will be diverted to walking control. In comparison, wall projection is more applicable to detailed introduction of architectures. The narrator touches the wall with a pointing stick, and similarly the near-infrared radars detect the touch spots to control the image-viewing process. This method is similar to general PC demonstrations via a mouse, where the specific experience of operating is absent.

The digital copy of architectural heritage is visualized representation, just as drawings and photos. The difference is that it allows users to walk in virtual realistic space, free from the constraint of angle, scale and time. Based on the survey results, users have different observation focuses when operating the system from looking at photos shown in Table 1. This may lead to many new findings.

Table 1. Survey Results and Comparison

Observati on Focus	Observer Interpretation	Compared with Looking at Photos
Location and scale	Observer 1: it's like you're in real buildings. When you look down at them, you'll get a clear idea of how they're arranged. Observer 2: it's really like being there if you follow those moving lines. Observer 3: so large buildings! But not far from each other.	More detailed; (Observer) imagining himself/hersel f in the real building, and wishing to see more clearly
Appearan ce	Observer 1: people built so delicate buildings even hundreds of years ago! Observer 2: tiles on the roof are very beautiful. And Eaves that large must be very effective in blocking lights. Observer 3: those small sculptures on top of the roof have any special meaning? Doors are really high!	

Judging from the above comments on the demonstration method, virtual reality works are apparently more effective in inspiring users to see more and see into details. Stronger interest in architecture matters greatly in education. This is where the new operation method -- touch screen desk system stands out with its obviously stronger appeal for users.

In the production of a virtualized Toji Temple, the virtual reality technology is applied for delivering a more detailed copy of one of the buildings, the Toji Pagoda. The copy can represent every plane and section of the building, with introductory texts incorporated for each part and component. But the detailed production leads to a rather huge data file, making it slightly difficult to use touch screen for demonstration. A more appropriate demonstration method needs to be developed.

4 Summary and outlook

Thanks to the concerted efforts made by local administrative authorities, enterprises, research institutes and colleges and universities, many cultural heritage education and research centers have been established as part of the three-in-one programs combining business, education and research. These centers are where citizens learn about cultural heritages. There exhibits real heritage pieces and reference material with new demonstration devices and methods. Routine activities are open to citizens and sometimes organized specially for primary and secondary school students, which prove very successful. At many of those activities, the two systems built by this study also play their part. In the future, the study will focus on following the rapid progress of computer technologies and developing various new application systems to provide meaningful backup for educational activities. Meanwhile, the existing applications should be further improved, and more relevant data collected for more efficient education. This study analyzes the popular education of Japan's two representative historical buildings, and the methods employed and results achieved can be reference for other countries and regions.

References

1. J.Han. Low-cost multi-touch sensing through frustrated total internal refection, ACM UIST, 2005, 115-118.
2. T.Takami,etc. Design of the Multi-Touch Table with Near-Infrared Radars and its Application of Floor/Wall Display DiGRA JAPAN Game Studies, 2009,(1)26~30.

Intelligent Technique for Signal Processing to Identify the Brain Disorder for Epilepsy Captures Using Fuzzy Systems

Sasikumar Gurumurthy [a], B.K.Tripathy

School of Computing Science and Engineering, VIT University, India

Abstract. The new direction of understand the signal that is created from the brain organization is one of the main chores in the brain signal processing. Amid all the neurological disorders the human brain epilepsy is measured as one of the extreme prevalent and then programmed artificial intelligence detection technique is an essential due to the crooked and unpredictable nature of happening of epileptic seizures. We proposed an Improved Fuzzy firefly algorithm, which would enhance the classification of the brain signal efficiently with minimum iteration. An important bunching technique created on fuzzy logic is the Fuzzy C means. Together in the feature domain with the spatial domain the features gained after multichannel EEG signals remained combined by means of fuzzy algorithms. And for better precision segmentation process the firefly algorithm is applied to optimize the Fuzzy C-means membership function. Simultaneously for the efficient clustering method the convergence criteria are set. On the whole the proposed technique yields more accurate results and that gives an edge over other techniques. This proposed algorithm result compared with other algorithms like fuzzy c means algorithm and PSO algorithm.

1 Introduction

Epilepsy captures is considered as the maximum collective neurological disorder that disturbs 1–3% of the world's inhabitants. This one is considered through the amount of two or supplementary meaningless epileptic captures that remain irregular rhythmic exoneration of electrical movement of the brain. This spectrum of brain disorders ranges from severe, life-threatening and disabling, to ones that are much more benign. In epilepsy, the normal pattern of neuronal activity becomes disturbed, causing strange sensations, emotions and behaviour or sometimes convulsions, muscle spasms and loss of consciousness. [1] There are many possible causes of epilepsy with several types of seizures. Anything that disrupts the normal pattern of neuron activity (from illness to brain damage to abnormal brain development) can lead to seizures. Epilepsy may develop because of an anomaly in brain wiring, an unevenness of nerve signalling chemicals called neurotransmitters, changes in the vital features of brain cells called channels or a combination of these factors and other possible factors. Having a single seizure as the result of a high fever (called febrile seizure) Common diagnostic tests for epilepsy are measurement of electrical activity in the brain and brain scans such as magnetic resonance imaging (MRI) or computed tomography (CT).

Paroxysmal alteration of single or extra neurological occupations such as motor, behaviour and autonomic occupations is definite as a seizure. Epileptic seizures remain episodic, quickly developing fleeting actions, normally enduring for less than a minute. There are numerous studies to understand the machine after epileptic capture and however it is not entirely known yet, a seizure event can be developed as the enlarged network excitation of the neural networks with synchronous discharge as well as variable propagation in the brain. In crucial epilepsy, a specific brain region may candidate for the epilepsy event, but in comprehensive epilepsy the complete brain might remain candidate aimed at seizure events. The utmost broadly castoff quantity aimed at analysis of neurological syndromes such as epilepsy in medical backgrounds is electroencephalogram (EEG). Enduring intensive care of EEG is one of the best skilled conducts aimed at analysis of epilepsy. This one delivers evidence almost forms of brain electrical movement, category then occurrence of captures and seizure emphasis laterality. In abiding monitoring, ictal EEG footage is typically connected through the medical appearance of removal. One of the greatest defining features in seizure recognition before primary recognition is the assignment of electrodes [8]. Unknown footage location is where the capture attention is placed; the variations in EEG can happen earlier the medical appearances. But if the electrodes are located in remote place after the seizure beginning spot, the medical appearances could transpire earlier any graphic variations in EEG [4]. Professionals observing abiding EEG footages typically look on behalf of initial visually specious variations in EEG to classify ictal arrival.

[a] sasichief@gmail.com

2 Related Work

Epilepsy is one of the neurological disorders under extensive research and over the years there have been several methods of seizure detection. One of method was based on detecting strong rhythmic movements of the patient. The drawback in these methods is that, seizures do not always present strong movements. This setback led the detection problem to apply methods based on EEG signal analysis, for example, J.R. Ives & Woods (1974) described detection of large seizures discharges in several EEG channels by amplitude discrimination; T.L. Babb & Crandall (1974) designed an electronic circuit for seizures detection from intracranial electrodes [9].

2.1 Feature Extraction

Finding a set of measurements or a block of information with the aim of expounding in a well-defined way the data or an event present in a signal is called feature extraction. For detection, a classification or regression task in biomedical signal processing these measurements or features is the fundamental basis and is one of the key steps in the data analysis process [7]. These features constitute a new form of expressing the data, and can be binary, categorical or continuous, and also represent attributes or direct measurements of the signal. For example, features may be age, health status of the patient, family history, electrode position or EEG signal descriptors (amplitude, voltage, phase, frequency, etc.) [8]. More formally, feature extraction assumes we have for N samples and D features, a matrix N × D, where D represents the dimension of the feature matrix. That means, at the sample n from the feature matrix, we could obtain an unidimensional vector x = [x1, x2, ... , xD] called as "pattern vector." Guyon et al. (2006) consists several methods in EEG feature extraction [10].

The Time domain, frequency domain features, and entropy-based features were extracted from EEG segments. The four features used in this study were average amplitude, rhythmicity (coefficient of variation of amplitude), dominant frequency, and entropy. These features are known to contain the most discriminant information for detecting seizure events. Features extraction methods are expounded briefly in the following sections.

2.2 Average Amplitude

For time-based development of limited seizures, average amplitude (AVA) is a decent amount. EEG signals display rhythmic movement with a recurrence frequency among 3 and 30 Hz throughout limited seizures. Consequently, to calculate regular amplitude, EEG divisions were initially high-pass cleaned overhead 3 Hz to eliminate low-frequency sound. Then to identify peaks, a peak recognition algorithm created on the zero-crossings of the first derivative of EEG signals was used. Then, by taking the average of the amplitudes of their half waves the amplitudes of the peaks were computed. Lastly, the average amplitude () was calculated by taking the average of the amplitudes of the identified peaks [1].

2.3 Rhythmicity

Coefficient of variation of amplitude (CVA) is a measure of rhythmicity or regularities of ictal activities. During seizure evolution, the regularity of the amplitude of EEG tends to increase slowly; this increase is characterized by the CVA. In case of partial seizures, the signals exhibit strong rhythmic characteristics that are likely to have regularity in amplitude. The coefficient of variation (CVA) quantifies the increased regularity observed during partial seizures. The CVA is defined as the ratio of the standard deviation of absolute amplitude to the mean absolute amplitude as where is the standard deviation and is the mean of each EEG segment [1].

2.4 Entropy

Entropy was initially introduced by Shannon in 1948 and is briefly defined as a measure of "irregularity" or "uncertainty". The Shannon entropy () is computed as where are the probabilities of a datum in bin. Pincus and Goldberger introduced approximate entropy (ApEn) are considered more appropriate to compute the entropy for short and noisy time series data. A low value of the entropy indicates that the time series is deterministic, whereas a high value indicates randomness. Therefore, a high value of entropy indicates the irregularities in the EEG data. To compute ApEn, it is required to determine a run length and a tolerance window to measure the likelihood between runs of patterns [3]. The tolerance window and embedding dimension are the two important parameters in computation of ApEn. In this study, Sample Entropy (SampEn), which is a variant of approximate entropy to quantify entropy of EEG, was used considering its robustness over ApEn. Sample Entropy is the negative natural logarithm of an estimate of the conditional probability that segments of length that match point wise within a tolerance also match at the next point. Therefore, for investigating dynamics of biomedical signal and other time series this measure is a useful tool [1].

2.5 Domain Frequency

The peak with the maximum spectral power in the power spectrum of a signal is defined as dominant frequency O. This feature is particularly important in distinguishing ictal activities from interictal activities by quantizing the frequency signature information mostly found in partial seizures. This is characterized by a high-frequency activity at seizure onset and a low-frequency activity at the end of the seizures [9]. In this study, parametric spectrum estimation method, autoregressive-modeling (AR) approach, was used to estimate the spectral frequency band of the short EEG segments. The AR model order was chosen according to Akaike information criterion. The Burg method was used for computing the

AR coefficients for short EEG segments. Then, the spectral power of a given segment is estimated using these AR coefficients. For every spectral peak, the spectral frequency band was defined as the frequencies at rising and falling slopes of the peak with half the amplitude of the peak.

3 Fuzzy C-means clustering

Fuzzy c-means (FCM) is a method of clustering which allows a [13] datum to belong to more than one cluster by varying degree of membership described by a fuzzy matrix with n rows and c columns in which n is the number of data and c is the number of clusters: Fuzzy c-means algorithm is popularly used clustering algorithm and introduced by Bezek in 1974, FCM algorithm result totally depend on the initialization of center, the random selection of the center causes the local optima problem. To solve the shortcoming of the FCM algorithm, this proposal integrate the optimization algorithm with FCM [11]

Consider the set of vectors (X=(x1, x2,….xn) 2<=c <=n) clustering into c groups. The set of fuzzy partition matrices is denoted as

$M_{fc}=\{W \in R^{cn} \mid W_{ik} \in [0,1], \forall i,k; \sum_{i=1}^{c} W_{ik} = 1, \forall k;$

$0<\sum_{k=1}^{n} w_{ik} < n, \forall i$ (1)

where 1<= i <=c, 1 <= k <=n

FCM objective function computed by using following equation and Euclidian distance [11]

$J_m(W, P)=\sum_{\substack{1\le k\le n \\ 0\le i\le c}} (w_{ik})^{m} (d_{ik})^2$

$(d_{ik}) = \|x_k - p_i\|$ (2)

4 Firefly Algorithm

First proposed by Yang, a Firefly Algorithm (FA) is an optimization algorithm, which simulates the flash pattern and features of fireflies. The Firefly Algorithm is a population-based algorithm to discover the global optima of objective functions based on swarm intelligence. Individually firefly is enticed by the brighter luminosity of additional adjacent fireflies. The attractiveness decreases when far between the couples of fireflies [14].

Three idealized rules in firefly algorithm distinct by Yang:

1) All fireflies are unisex so that one firefly will be attracted to other firefly irrespective of their sex;
2) Attractiveness is proportional to their brightness. Thus, for any two fireflies, the less bright one will move towards the brighter one. If there is no brighter one than a particular firefly, it will move randomly;
3) The brightness of a firefly is from the objective function. For a maximization problem, the brightness can simply be proportional to the value of the objective function.

X-S Yang made use of firefly algorithm like an optimization algorithm for different purposes .In the design of real-world engineering problems, Azad and Azad and Gandomi showed that firefly algorithm shall be a good remedy for multimodal optimization problems. Banati and Bajaj showed that FA should be a potential future prospective for efficient performance in terms of time and optimality for feature selection when compared with other algorithms [11]. Horng qualified that firefly-based method (firefly-based algorithm) illustrated better high speed and good reconstructed images while compressing the digital images.

The attractiveness function β calculate from the distance r(i,j) of the firefly is determined by:

$\beta(r(i,j)) = \beta 0 e^{-\gamma r(i,j)^2}$ (3)

Where, is the attractiveness at and is the light absorption coefficient at the source. It should be noted that the which is the euclidence distance between any two fireflies and at and , where and are the spatial coordinate of the fireflies and , respectively.

The movement of a firefly *i*,which is attracted to another more attractive firefly *j*, is decided by:

$\beta(r(i,j))x_i = x_i + \beta 0 e^{-\gamma r(i,j)^2}(x_i - x_j) + \alpha\left(rand - \frac{1}{2}\right)$ (4)

α is the randomization parameter, in most of cases α between [0,1] , $\beta 0$=1 and $\gamma \in [0.01,100]$.

In various signal processing and pattern recognition applications fuzzy logic has been extensively used. Fuzzy rules can be defined using experts' knowledge for decision making which are simpler to implement and modular as well. Increasing the number of rules one can increase the accuracy of the model. With less complex mathematical analysis and modelling, processing speed can also be significantly improved. More over fuzzy logic is a useful method for non- linear input-output mapping which is effective in seizure detection or early detection applications. Artificial neural networks and support vector machines are other prominent detection methods which demands training, complex mathematical analysis, and modelling. All these major setbacks can be avoided in fuzzy logic. Therefore, in this study, we applied adaptive version of fuzzy logic system with a unique approach of merging information in feature as well as spatial and temporal domain.

5 Fuzzy Firefly Algorithm

Fuzzy-based Firefly algorithm (FFA) for data clustering is proposed. Each solution of the cluster is defined fuzzily and then the solution is updated using firefly algorithm

DOI: 10.1051/matecconf/20166106004

the same way it does with non-fuzzy solution. However, minimization of the objective function is calculated in the same way as FCM algorithm.in the proposed method cluster centers are the decision parameters to minimize the membership function. A single firefly represents the cluster centers vectors. Eq. 5 represents each firefly

$$X_i=(p_{i1},.....p_{ij},...., p_{ic}),\ 2 \leq j \leq c \quad (5)$$

Where f_{ij} represents the jth center vector. n number of fireflies represents n candidates solutions. This is a minimize problem the intensity of each firefly is equal to the values of the objective function of FCM.FCM will terminate when there is no scope for further improvement in the member function. FFA will terminate when the no changes in the 5 consecutive iterations. Improved FFA method has 50 is the maximum iteration or no changes in current best in 3 consecutive iterations.

Pseudo code of the algorithm:

Initialize the population of n fireflies with C random cluster centers of d-dimensions
Initialize the algorithm parameters
Repeat:

For i=1:n all n fireflies
For j=1:n all n fireflies
Calculate light intensity (membership function value) of each firefly
If ($I_i > I_j$)
Move firefly I towards j to update the position of fireflies

Rank the fireflies and find current best one
Until: maximum iteration or minimum changes in the member function

6 Results and Discussion

Our experimental study is to compare the proposed algorithm with other popular optimization algorithms like PSO, FFA, and FCM. The dataset collected from the "ictal" information records from Freiburg project, which must captures with at least 50 minutes of preictal data and postictal data with no specified duration. Consequently, the false recognition rate per hour is slight complex associating to additional approaches in the works but sensible seeing the assessment dataset.

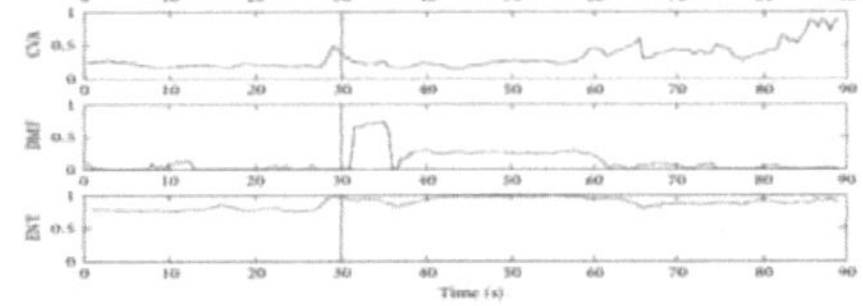

Figure 1. Seizure is marked by red line. Corresponding changes in features also shown. (a) Top subplot: an example of a seizure evolution in iEEG. (b) Bottom four subplots: corresponding changes in characteristics features: Average amplitude (AVA), coefficient of variation of amplitude (CVA), dominant frequency (DMF), and entropy (ENY). Seizure onset is marked by red vertical line. Early electrographic changes are visual in three of the four features.

For each of the patients, there are datasets called "ictal" and "interictal", the former containing files with epileptic seizures and at least 50 min pre-ictal data [10].

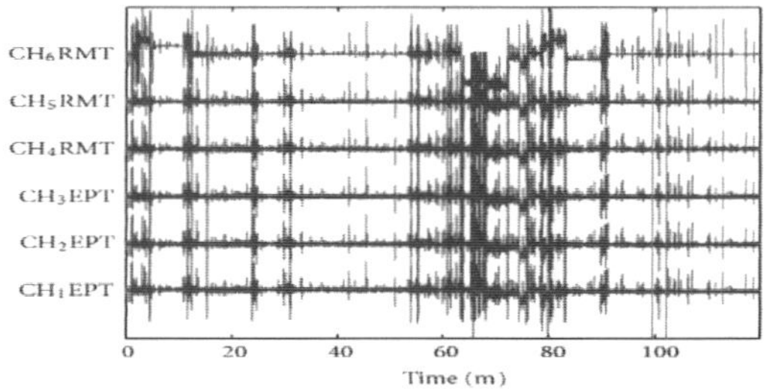

Figure 2. Seizure evolution profile in iEEG obtained from patient 10. Seizure onset and offset times are marked by red vertical lines, respectively. Acronyms: CH1EPT: Epileptic channel 1, CH4RMT: Remote channel 4.

Therefore, the false detection rate per hour is little higher comparing to other methods in the literature but reasonable considering the evaluation dataset. Till date, many algorithms for epilepsy and seizure detection have been developed with different degrees of success [2–15]. Here, we have discussed briefly some of these methods providing a scope of comparison with our method. In a recent study, Zhang et al. Proposed an automatic patient-specific method for seizure onset detection using a novel incremental learning scheme based on nonlinear dimensionality reduction. Feature sets were extracted using continuous wavelet transform (CWT). Considering computation time and resources, the choice of discrete wavelet transform might have been better. Their method was evaluated on iEEG recordings from 21 patients obtained from Freiburg project with duration of 193.8 hour and 82 seizures.

In this paper, different thresholds were used for different patients and a postprocessor was utilized to reduce the false detections in two steps. First short length detections (less than 5 s) and artifacts were rejected. Secondly, two consecutive detections were unified given that they are less than a predefined minimum time interval (set to 30 s). Chan et al. presented a novel patient-specific algorithm for seizure onset detection and accurate onset time determination. The algorithm extracts spectral and temporal features in five frequency bands within a sliding window and the feature windows were classified as containing or not containing a seizure onset using support vector machines SVMs). Support vector machine is a popular classification paradigm for epileptic seizure detection and prediction being used by many researchers in this area. In order to accurately localize the seizure onsets in time, the method makes use of clustering and regression analysis. Therefore, their algorithm yielded precise detection in time as reported in five of the six patients, at least 90% of the latencies are less than 3 s resulting median detection latency less than 100 ms with standard deviation less than 3 s. However, the method utilizing user-adjustable parameters allow tuning to achieve high detection sensitivity, low false positive rate, and low detection latencies. Standard cross-validation performance measures resulted sensitivities in the range of 80% to 98% and false positive rates from 0.12 to 2.8/h [6]. Gardner et al. [5] presented a detection latency which is negative in time (−7.58 s) however with a higher false detection rate of 1.56 false detections per hour.

6.1 Sensitivity

Sensitivity is significant quantity for event based system performance assessment. It measures the capability of a system to identify seizure suitably.

It is the amount of true positive rate and distinct as the ratio of the number of correctly detected seizure to the total number of seizures.

Sensitivity= (TP/ (TP+FN)) x 100 (6)

True Positive (TP): The system detects the actual seizures

False Negative (FN): The System misses the actual seizures

Table 1. Show the comparison sensitivity and detection latency for various algorithms.

Algorithm	Sensitivity	Average detection latency (s)	FDR /h
FCM	94.5	29	0.35
Fuzzy+ PSO	91.3	35	0.32
FFA	94.7	28	0.28
Improved FFA	96.8	13.5	0.23

Table 1: comparison of objective function value and no of iteration for the dataset for various algorithms. When compare to the other algorithm Improved version of FFA detect the seizure with minimum iterations.

Our method achieved average sensitivity of 96.8% with 0.23/h false detection rates. The average detection latency achieved was 13.5 seconds. Firefly algorithm minimizes the detection latency of the classification process.

7 Conclusions and Future Work

After extensive research, in this paper we existing a strong technique of classifying seizure using fuzzy firefly algorithm classification. Firefly algorithm helps the fuzzy system to obtain the better classification in minimum iteration. This proposed algorithm converge the result faster than the other algorithm with minimum iterations. Considering the challenge and important development in the capacity of seizure recognition, we primarily absorbed in designing effective seizure recognition classification with low complexity. In future our work will compare with other various classification algorithm like SVM and optimization algorithm like Genetic algorithm.

References

1. Alessandro, D.M, R. Esteller, G. Vachtsevanos, A. Hinson, J. Echauz, and B. Litt (2003). Epileptic Seizure Prediction Using Hybrid Feature Selection over Multiple Intracranial EEG Electrode Contacts: A Report of Four Patients. *IEEE Transaction on Biomedical Engineering*. Vol.50, No.5. pp. 603-615.
2. Carlos, G. M, A.M. Trigueros, A. N. Vazquez (2012). EEG Signal Processing for Epilepsy. *EEG Signal Processing for Epilepsy*. Histological, Electroencephalographic and Psychological Aspects, Dr. Dejan Stevanovic (Ed.). pp. 1-27.
3. EEG database: seizure Prediction project Freilburg http://epilepsy.uni-freiburg.de/freiburg-seizure-prediction-project/eeg-database
4. Izakian, .H and A. Abraham (2011). Fuzzy C-means and Fuzzy swarm clustering Problem. *Expert System with Application*. Vol.38, No.3. pp. 1835- 38.
5. Jitpakdee, P., and P. Aimmanee (2014). Image Cluster using fuzzy based firefly algorithm. Parisut ICICTES2014, Thailand. Vol.1, No.2. pp. 1-8.
6. Lotte, F., M. Congedo, A. Lécuyer, F. Lamarche and B. Arnaldi (2007). A review of classification algorithm for EEG-based brain-computer interfaces. *Journal of Neural Engineering*. Vol.4, No.2, pp.1-23.
7. Nandy, S., P. P. Sarkar, A. Das (2012). Analysis of a Nature Inspired Firefly-Algorithm Based Back Propagation Neural Network Training. *International Journal of Computer Applications*. Vol.43 (22). pp. 8-16.
8. Rabbi, .F and R.F. Rezai (2012). Fuzzy Logic System for Seizure Onset Detection in Intracranial EEG, *Computational Intelligence and NeuroScience. Hindawi Publication Corporation*. Vol. 2012, Article.ID.705140. pp. 1-12.
9. Runkler, T.A., and C. Katz (2006). Fuzzy clustering by Particle Swarm Optimization. *Proceedings of 2006, IEEE international Conference on Fuzzy Systems*, Canada. pp. 601-608.
10. Senthilnath, J., S.N. Omkar, V. Mani (2011). Clustering using Firefly algorithm: Performance Study. Swarm evolutionary computation. Vol.1, No.3. pp. 164-171.
11. Saibal K.P, C.S.Rai, A.P. Singh (2012). Comparative Study of Firefly Algorithm and Particle Swarm Optimization for Noisy Non Linear Optimization Problems. *International Journal of Intelligent Systems and Applications*. Vol.4, No.10. pp. 50-57.
12. Yang, X. S. (2010). Firefly algorithm Stochastic Test Functions and Design Optimization. *International Journal of Bio-Inspired Computation*. Vol.2, No.4. pp. 78-84.
13. Yang, X. S. (2008). Nature-inspired metaheuristic Algorithm. Luniver Press. pp. 741-751.

MATEC Web of Conferences **61**, 06005 (2016) DOI: 10.1051/matecconf/20166106005
APOP2016

A Supplementary Lighting System for Plant Growth with Lighting-Emitting Diode Based on DT TS&IC

Jierong.Zhao, Yongtao.Li, Yuncui.Zhang, Nianyu.Zou and Jinpeng.Wang[a]

Research Institute of Photonics,Dalian Polytechnic University,116034 Dalian,China

Abstract. In order to achieve quantitative supplementary lighting at different stages of plant growth, design a lighting system with LED that can provide different light quality. The system monitor illumination of light quality with specific wavelength in real-time,ambient temperature, the amount of light that plant really need, and control the brightness of red, blue LED lights by pulse width modulation (PWM) signal. The system can be used for different plants at different growth stages under different environments with different demand in light quantity and quality , with accurate, intelligent, low-power features. The system prevents plants from suffering light deficiency or excess, improve energy efficiency.

1 Introduction

Plant growth and blossom are inseparable from light regulation and participation.With the improvement of luminous efficiency of LED,good monochromaticity, small size, long life, low energy consumption, using direct current, and low heat and the feature of flexible,LED in the area of plant cultivation will gradually replace traditional lamps[1].In agricultural fields,fluorescent lamps or high pressure sodium lamps and other artificial lights were commonly used.Compared with these artificial lights, the light quality of LED is adjustable, the heat that LED produced is less, the quality of cultivation[2] is enhanced. For agricultural production with environmental system in closed spaces, such as greenhouses, plant factory, LED is a kind of suitable artificial light sources [3]. But R/B ratio of light intensity in most existing research and development plan for plant lighting and products is constant,which is not match with the selective plant absorption spectrum.And continuous irradiation ways are often used ,which cannot adjust light density, photoperiod and irradiation way in a quantitative modulation way, without considering the difference of light quality and light quantity that plants need at different stage ,resulting in supplementary lighting insufficient or excessive[4].The problem of high energy consumption and the waste of light are still not solved[5].

Research has shown that the light involved in photosynthesis is primarily red and blue light .The wavelength coverage of them are respectively 400-500nm and 600-700nm[6].The coverage of wavelength 640-660nm red light and 430-450nm blue light are the most important for plant growth[7].Red and blue light have different influence on growth and development of plant.Different blue light is beneficial to the growth of plant leaves, red light is beneficial to plant stem elongation[8]. Strong Blu-ray made the shape of plants lower, strong red light made plants stems too long and caused the phenomenon of yellowing[9]. Therefore, the appropriate red and blue luminous flux ratio of R/B is necessary to cultivate plants that own sound form [10].

In order to improve light utilization, and reduce energy consumption, we analysis light quality and quantity that plants really need, propose a precise method to adjust light quality and quantity,design a LED supplementary lighting system which is based on DT (detection technology), and TS(temperature sensing technology)[17,19] and IC(intelligent control technology)[12-16], all of which are modern electronic information technology[18].The system used STC12C5A60S2 single chip as core processor, PT4115 for LED drive module[20].According to the measurement result of temperature and light intensity,use PWM (pulse width modulation) signal adjusting LED light on plant aiming at meeting the needs of plants in different stages.

2 Supplementary lighting pattern

Light not only as a source of energy control the photosynthesis of plants, but also as a signal source affect plant growth and development in all its aspects. Blu-ray affect cryptochrome in plants, thereby regulating the phototropism of plants; Red light affects phytochrome of plant ,thereby regulating leaf differentiation of plant , so the photomorphogenesis of plants can be adjusted by changing R/B, thus breeding plants with specific form requirements. Red light of 610-720nm can reduce the content of gibberellins in plant, thus reducing section between plant height and length; R/B ratio has become an

[a] Corresponding author: Jinpeng .Wang.wangjp@dlpu.edu.cn

important assessment parameters on controlling plant growth. In the nature, as the season and weather change, the light intensity is different. Light intensity in winter and spring is weaker than that in summer and early autumn. And it is weaker in sunny days than that in rainy days. The demands of different plants on light intensity are different. The coverage of light intensity that Shade plants needs is 500~2500lx.The coverage of light intensity that neutral plants need is 2500~30000lx. The coverage of light intensity that positive plants needs is greater than 30000lx.

Considering that different photosynthetic pigment has different peaks of absorption spectrum, absorption peak of photosynthetic pigment of red light is at 625~640nm, the absorption peak of blue light is at 420~470nm, so the peak wavelength of 660nm LED as a red light, the peak wavelength of 450nm LED as a blue light [6]. In order to monitor light environment of plant growth in real-time, the spectrum of 400~500nm and 600~700nm were respectively detected by light sensor; Join the temperature detection module in this system at the same time; according to the real-time temperature, it is determined whether to supplement light. This method can improve the utilization rate of light source and prevent plants suffering harm that caused by supplementary lighting under the condition of excessive temperature stress.

Different spectral range has different effect on plant growth process. The ratio of light and the light quantity that plant growth need are different at different growth stage. Fixed light quality and light intensity of the supplementary lighting pattern is difficult to meet the demand of plant growth at different growth stage. And the light compensation point and light saturation point are different. If light intensity was lower than light compensation point, or more than the light saturation point, it will cause supplementary lighting insufficient or excessive, and cause invalid energy consumption and photoinhibition[7]. In this paper, a method that light quality is adjustable and the quantity of supplementary lighting is on-demand is presented. Users can set different value of light quality and light density according to light saturation point and light compensation point at different stages of plant growth. According to the differentials between current value and the value set before, light quantity and light quality that supplementary lighting needed in next cycle is obtained by periodic monitoring, then complete the feedback adjustment and realize the work mode that output of light intensity is on demand.

3 Prototype design of the system

As is shown in Figure 1, the modular design of this system consists of power module, detection module, control module, lighting module and user interface module. The power supply module uses the battery power supply mode. The red light detection module and the blue light detection module use 3V power supply; Temperature detection, buzzer, indicator light, display module use 5V power supply; Control module uses micro-controller as the core. According to the data collected by the system, set the threshold. The calculation of the duty cycle of the corresponding PWM control signal is realized and two PWM control signals output; detection module detects red, Blue light intensity and real time temperature, and the detection signal is filtered, amplified and transmitted into the MCU, thus realize the detection of relevant environmental information; In order to achieve accurate quantitative light ,the supplementary light module uses PWM current control function with two constant current drive circuit, respectively control the red, blue LED lighting lamp array ; Display module use the LCD screen to show the test results ; Keyboard is used to achieve the required threshold modification and other functions. Completing threshold modification and setting can effectively improve the convenience and expansibility of this system.

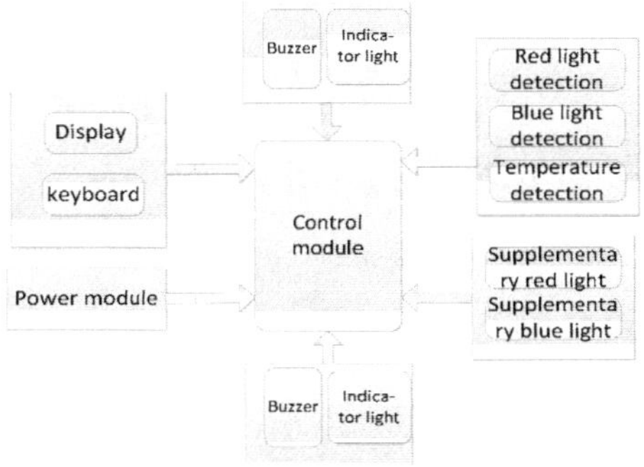

Figure 1. System principle block diagram.

4 Hardware design

4.1 Control module

The core CPU is STC12C5A60S2, it integrates the MAX810 dedicated reset circuit. And the CPU equips with two PWM outputs, four 16-bits timers, eight 10-bits-precision A/D conversion, and flash storage space of 56k and RAM of 1280 byte. They realize acquiring data, adjusting threshold and controlling intelligently. Provide foundation and guarantee for the functions of system. The circuit is shown in Figure 2. P0 is connected to the eight data ports of LCD;P1 is connected with sampling signal; the signal of double solid state reactance is exported by P1.0. the warning is produced by P1; P1.5 is accessed temperature the signal detection,P1.6,P1.7 are connected to the signals of red and blue light detection. So as to collecting data detected by sensor. P2 is connected to 4*4 Matrix Keyboard. P3.0, P3.1 are responsible for reading and writing data between the micro-controller and port, so as to realize programming the program.P3.2-P3.7 are connected to the terminal of controlling to LCD. P4.2 and P4.3 are the ports to export signal. Adjust brightness of light precisely through controlling the brightness of the light by the PWM signal of different duty cycle.

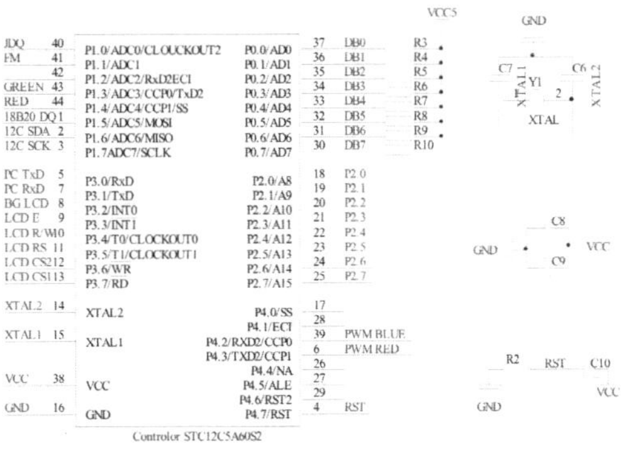

Figure 2. Principle diagram of the control module.

4.2 Light detection module

The module of illumination detection measured light intensity and temperature by relate sensors. And it fed back data to the micro-controller for processing; the schematic is shown by figure 3.The module of detecting temperature was composed by 18B20 and the standard adjustable circuit. It is acquired by the data-line accessing to P1.5. The detection of light included light intensity detection about red and blue. Light is filtered through different filters. The wavelength range is 400-500nm or 600-700nm.Then it detected the processed red or blue ray by light sensor of ISL29010. The detected signal was sent to SCM. The light sensor was connected to P1.6, P1.7 of SCM, so as to completing the light detection of two-band.

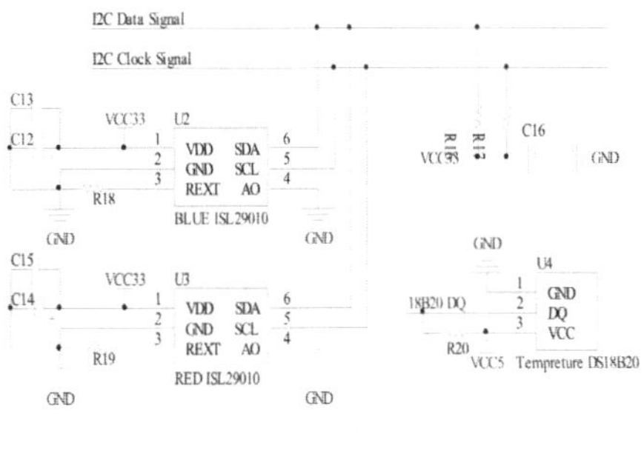

Figure 3. Light detection module.

4.3 Driver circuit module

LED array using step-down constant current drive mode in driver circuit module, constant color output of LED can be obtained. The module selected PT4115 constant current drive chip, driver circuit is shown in figure 4. PT4115 is a step-down constant current source with continuous inductance current conduction mode. It has a wide range of input voltage 6 V dc to 30 V , output current is adjustable, up to 1.2 A, it can drive series of small power LED. PT4115 built-in power switch, the use of high-end current sampling set the average current of LED, and DIM pin can accept the signal of analog dimming and a wide range of PWM dimming. The largest average current of LED is determined by RS which is connected with both ends of VIN and CSN resistance. Joining variable duty cycle of PWM signal by the DIM pin can adjust output current to achieve dimming. Output current can be calculated by formula I_{OUT} = 0.1 * D/RS, type D is in the PWM duty cycle.

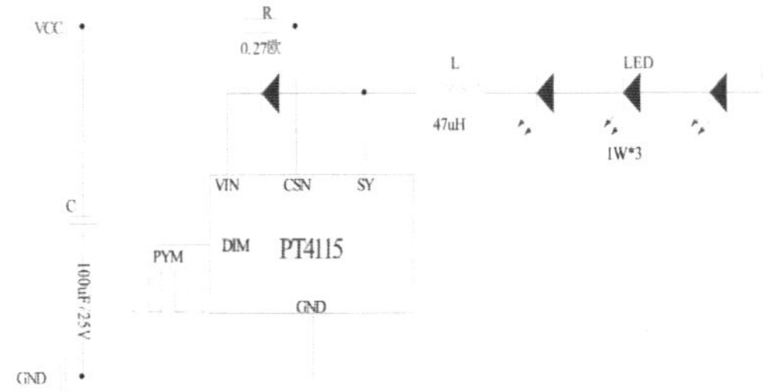

Figure 4. Principle of drive module.

4.4 Alarm light and keyboard module

Design detection alarm indicator module in this system. By computing amplitude changes of two adjacent light intensity in the system, and compared PWM output changes, the system automatically judge whether LED headlamp unit work properly or not. If the result of judgment is abnormal, the MCU P1.1 port will output warning signal, control buzzer and warning lights, achieve early warning function. User interaction module mainly include LCD screen and keyboard implementation system. Use 4 * 4 matrix keyboard system to set parameters.

4.5 System software

System is initialized at first, users can set temperature range at different growth stage according to photosynthesis effective temperature range. Choose value between the light compensation point and light saturation point as red blue light target parameters. Collect cycle temperature to determine whether meet the requirements of the temperature of the photosynthesis, otherwise turn off the supplementary light module when it is beyond the range; If the temperature is appropriate, then the red light blue light sensors detect the current light quality, according that the filter can get specific red and blue light intensity of illumination, and then judge whether the illuminance values is within the set threshold; If the value of detected light is greater than the target value set before, turn off supplement lighting module; Instead, calculate the red and blue light by formula, the amount of supplementary equals default values minus the actual values, then according to the red and blue light quality that need to supplementary, respectively adjust PWM duty ratio signal, then realize accurate supplementary light . After a period, detect red blue light quality again, if two test results do not conform to the requirements, and then start the buzzer alarm.

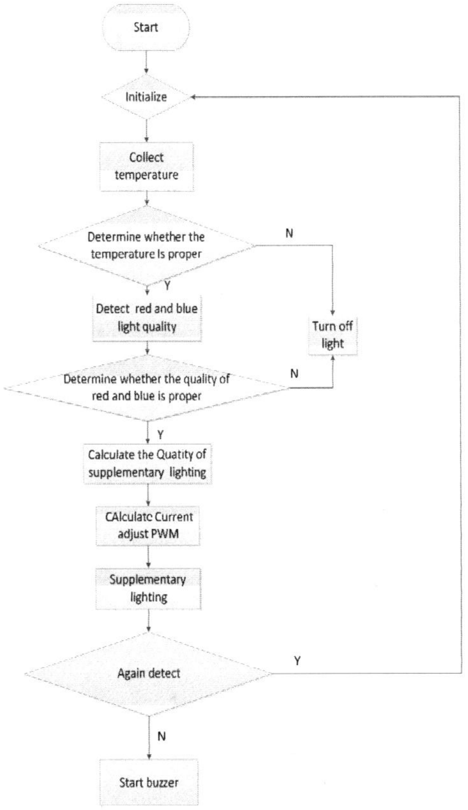

Figure 5. System software diagram.

5 Conclusions

Through analyzing the mechanism of supplementary lighting for plant growth during different periods, setting temperature and light intensity, light quality that plant required, developed a supplementary lighting system based on single-chip STC12C5A60S2. This system monitor real-time temperature, light intensity and other factors, and through PWM dimming, control LED, thereby control the current, realize precise supplementary lighting on demand, avoid supplementary lighting insufficient or excessive. The system can be used to cultivate plants in plant factory and greenhouse with the advantage of goal-oriented and high-efficiency supplementary lighting. The experimental results show that the system is stable. It can meet the demand for supplementary lighting for different plants at different stages with replicability.

Acknowledgements

This research was supported by Project of the National N atural Science Foundation of China (61402069), 2012 ke y supported project for undergraduate education by Educ ation Dept. of Liaoning province([2012]108), 2016 Proje ct of the Natural Science Foundation of Liaoning provinc e(2016020027), 2012 Education reform project for u- ndergraduate by Education Dept. of Liaoning province ([2012]130), 2014 Dalian municipal science and technolog y plan project(2014A11GX050,2014 Dalian municipal science and technology plan project(2014A11GX052).

References

1. Cq.Xue,Jw.Wang,Lj.Yang,Act.Tab.Sin.J. 20,2 (2014)
2. Wk.Liu,Qc.Yang.Lam . Light.J.17,3(2014)
3. Sh.Sun,Lam . Light.J.17,3(2014)
4. J.S.Navas,M.M.Risueno,C.Manzano,Plan.J.84,1(201 5)
5. Ga.Jorge,Ma.Núñez,PloS .One,9,1(2014)
6. Hl.Gao,Dc.Zhu,Lamp. Illum.J.29.4,(2014)
7. Wk.Liu,Qc.Yang.Light. Eng.J.25.4,(2014)
8. Wk.Liu,Qc.Yang.Sci.Tech.J.32.10,(2014)
9. Wp.Li,Zh.Shen,Wd.Li,Lamps. Light.J.2,(2015)
10. N.Li,Jr.Wang,Light. Eng.J.24.2,(2013)
11. Nn.Su,J.Cui,Chin.Veget.J.24,(2012)
12. Zp.Zhang,Da.Zhao,Yq.Zou,J.Agri .J.5,(2013)
13. Ol.Gon,Al.Lum,Com.Elecs .J.92,(2013)
14. Hy.Sun,Opti. Sci.Har.D.(2014)
15. Q.Yao,Da.Chen,Lig. Eng.J.26.4,(2015)
16. Sf.Wang,Opt.Des.D.(2013)
17. Yj.Wu,Zm.Chen, J.Anh.J.41.1,(2013)
18. J.Liu,Xian.Det.Aut.D.(2013)
19. Ns.Liu,J.Ni,Jf.Dong,J.Agri. J.21.30,(2014)
20. X.Yuan,K.Song,Ch.Cha,Elec.Des.Eng.J.**23**.6,(2015)

A New Remote Monitoring System Application in Laser Power Based on LabVIEW

Gaoqiang Liu, Weilin Wu, Mingyu Liu, Kun Zhang and Guannan Chen [a]
Key Laboratory of Optoelectronic Science and Technology for Medicine, Ministry of Education
Fujian Normal University
Fuzhou, 350007, China

Abstract: In this paper, a new remote monitoring system based on LabVIEW was proposed to measure laser power automatically and remotely. This system consists of four basic components: an DH-JG2 optical power meter, a NI-USB 6008 data acquisition card, a personal computer (PC), and HP laserJet 1020 Plus printer. Since power output of laser is generally so unstable that abnormal work situation could not retroaction to inspectors right away, new system was designed to solve this problem. The detection system realized function of remote control by TCP protocol and mobile phone. Laser power curve that is measured by detection system demonstrated that the design has a good performance in real-time detection and operability.

1 Introduction

Laser is the important component of technologies in the field of communication, manufacturing, as well as optical experiment because of its characteristics including high brightness, good monocolor, good coherence as well as good direction. However, as far application, one of the main problems is that its output power is not stable and the laser is easy to be destroyed because of transient circuit and voltage peak[1]. Lots of power meters were invented for measurement in the process of laser manufacture last decades. With the rapid development of the computer technology, manual operation desktop machines have the disadvantages of low efficiency, inconvenient data analysis and storage in the quality of laser testing[2]. Therefore, software-based automated testing system design is imperative. Measurement system of laser power in VC++[3], the software of optical power measurement and calibration and instrument control system based on LabVIEW have overcame the lack of the manual operation desktop machines, but they remains to be more automatical, and errors in the testing process can't be managed when failed in working in time[4].

With the rapid development of the computer technology and the software technology, there are more and more kinds of software development platforms and instrument functions which can be achieved by computers and related software NI-LabVIEW[5]. LabVIEW produced by National Instruments is the most widely used as virtual instrumentation system. It has the development and debugging tools of traditional programming language. The greatest feature of LabVIEW is the graphical programming language, which is easy to grasp, understand develop a friendly, realistic user interface. Furthermore, it has the powerful signal processing ability, and integrates almost all of the features of hardware communications such as data acquisition card. Virtual instrument is an individual computer testing equipment that has a virtual instrument panel. It brings a revolutionary change in the field of automatic testing system. To a testing platform, LabVIEW includes the storage functions of data collection, processing and

[a] Corresponding author:edado@fjnu.edu.cn

analysis. To meet the needs of various practical applications toolbox, it can use dynamic link (DLL) technology to users of the virtual instrument software to link to them. use links embedded object (OLE) technology will be embedded in other virtual instruments. LabVIEW offers a numbers of source-level instrument drivers, including DAQ, GPIB, PXI, VXI, RS232 as well as USB, users can choose their own hardware drivers according their own needs[6].

In this paper, a remote monitoring system in laser power based on Labview was designed and implemented. The system used optical power meter DH-JG2 as monitoring laser power component produced by DaHeng. NI-USB 6008 data acquisition card was utilized to collect the power data from power meter. A electrical energy supply system HSPY_36_03 produced by HSPY was employed for turning on or off the laser, Also HP laserJet 1020 Plus as the output device of the test report. Besides, using the real time software LabVIEW displayed and analyzed the power data, and a mobile phone was used to remotely monitor.

2 System composition

System design framework is shown in fig. 1, from which we can see the system consisting of Laser, power meter, data acquisition card, printer, PC machine, electrical energy supply system as well as mobile phone. In which, DH-JG2 is the power meter, responsible for collecting the actual laser power data, and outputting voltage signals. And the USB 6008 is a data acquisition card produced by NI(National Instrument), can collect voltage signals from DH-JG2, convert the signals into digital signals and send the digital signals to the PC machine by USB bus[7]. On the one hand, PC machine displays and analyses the power data, and generates a test report and sends it to the HP laserJet 1020 plus and then the report would be printed by the printer. In addition, the system can easily to turn off laser through sending the codes to HSPY_36_03 and then laser will stop working until the electrical energy supply down to zero. A mobile phone is needed in remotely monitor system.

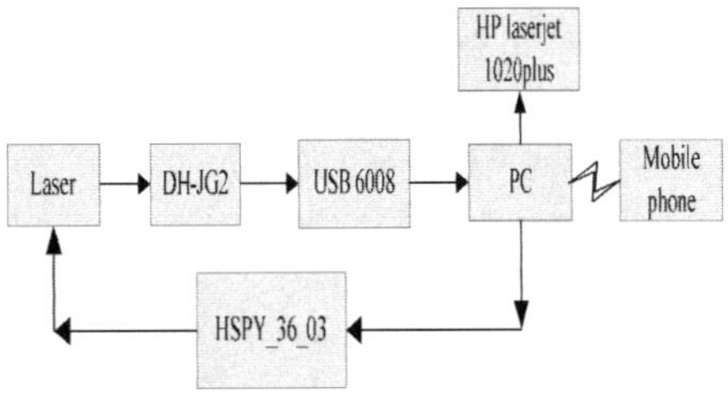

Figure 1. Framework map of system design

3 LabVIEW Program

The LabVIEW has been used to acquire the laser power data obtained from power meter, and display the raw data. After data of laser's power waveform acquisition, the data are sent to Express Virtual Instruments for signal processing. To evaluate the quality of the laser, the program will find the maximum, minimum as well as average value from the processed signal in order to calculate the stability of laser power. After that, the system needs to generate the test report and output it by printer. Virtual Instrument (VI) provides Report Generation tools and thus the test report is produced.

3.1 Data Acquisition

In this system, setting the measurement time interval and the total measurement time are required. After that we can easily read out the measurement results of fixed interval. Comapared to the traditional measurement of power stability that is measured by inspectors and need recorded measurements within 24 hours continuously, this system

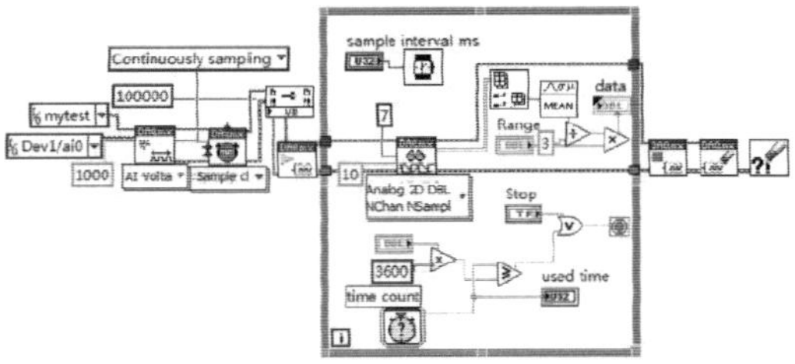

Figure 2. The program of data acquisition

can save labor force. The block program of data acquisition is shown as Fig. 2, which includes DAQmx acquisition functions, Mean filter and measurement count. When the power meter and the software connect to the DAQ card[8], we can measure the output the laser power automatically. And the data real-time displayed on the wave diagram, we can realize the trend of power. Fig. 3 shows the front panel structure of the data acquisition, in which the results of Measurement 4.28 is just the measured output power[9].

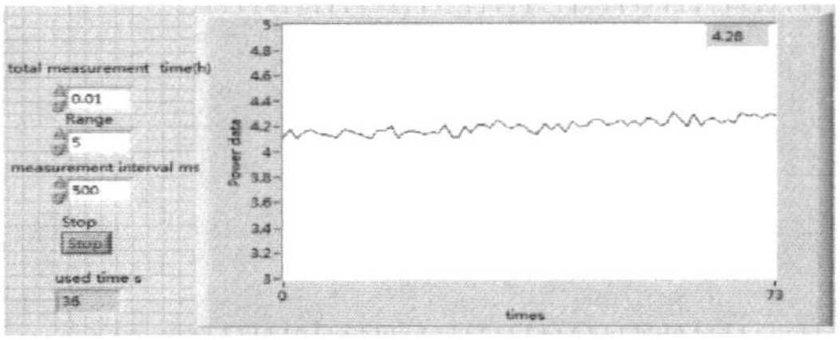

Figure 3. The front panel structure of the data acquisition

3.2 Data Analysis

The data analysis includes displaying real-time laser power data and extracting the data of the average, maximum, minimum, peak-peak and the stability value. The stability value was calculated in Equation 1.

$$S_w = \frac{(P_{\max} - P_{\min})}{P_a} \times 100\% \qquad (1)$$

Where, S_w is the stability of the laser power, and the $P_{\max}$, $P_{\min}$, and P_a represent maximum, minimum and average value of the laser power, respectively. The Statistics function was used to get the average, maximum, minimum, peak-peak value of the power data[10]. Fig. 4 and Fig. 5 show the front panel procedure and the block diagram procedure of LabVIEW of the laser's power data analysis.

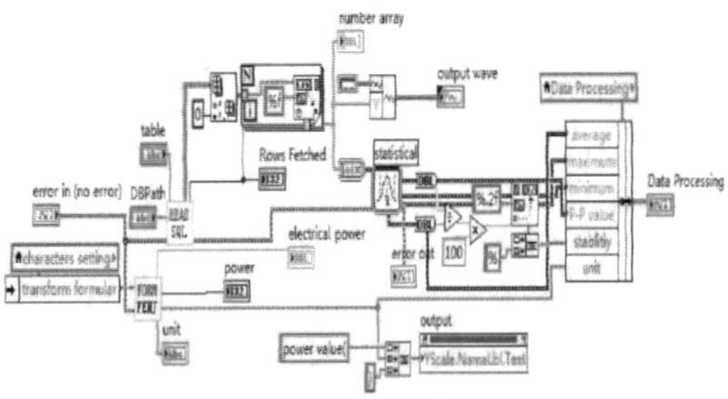

Figure 4. The program of data analysis

3.3 Application and testing

Fig. 6 shows the working model of the system. Before start testing, the Parameters Setting part that is the details about the laser and the test settings including testing time, collect interval, the type of power meter etc. should be

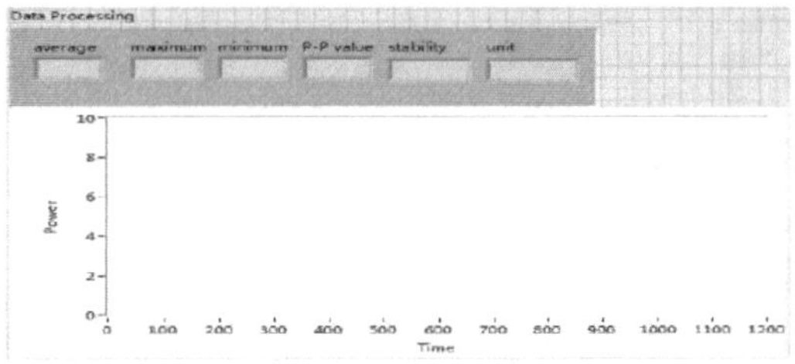

Figure 5. The block diagram procedure of LabVIEW

filled in. When we start collecting, the power data will be real-timely displayed in the Measurement result block like the blue swash. After collecting, the data analysis results block like the blue swash. After collecting, the data analysis results will be shown in the Data Processing block. In order to output the report, we use Report Generation tools. And the report is shown in Fig. 7.

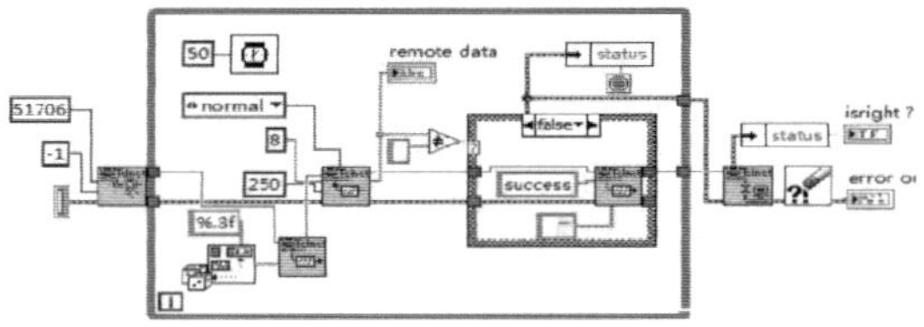

Figure 6. The working model of the system

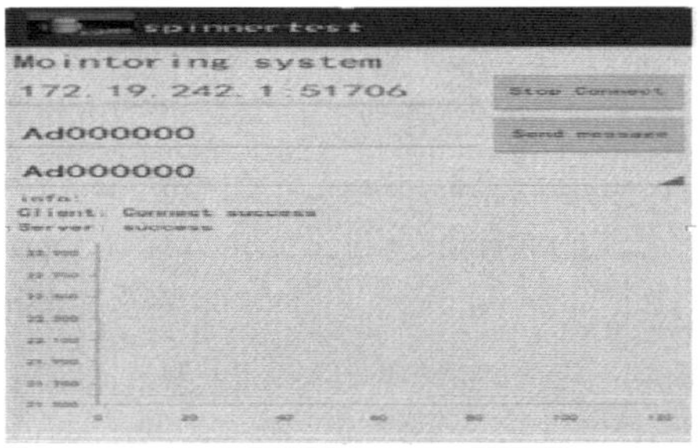

Figure 7. The remote client interface

4 Remote monitor

Remote monitoring of laser power is implemented by TCP communications. LabVIEW can achieve TCP communications by using the published TCP module. In this paper, we designed server mode by LabVIEW in PC machine and client mode by android that used in a mobile phone. The first part of the program is the server mode including data acquisition, data analysis, and the report generation laser controlling worked in the server model. Besides, the second part is the client model consisting of the management, display and remote monitor implemented by the mobile phone.

In server, using Telnet Create Listen receives the

 DOI: 10.1051/matecconf/20166106006

connection request of the client. The Telnet tools is based on TCP protocol .The establishment of TCP connection, using Telnet Read function nodes to receive the data which sent by the client and the analyze and process, using Telnet Write function nodes send over data and 'success' to the client. To send double-type data, the procedures first change the double type data into type of characters by the use of Format value node function, and then send to the network. The process diagram was shown in Fig. 8.

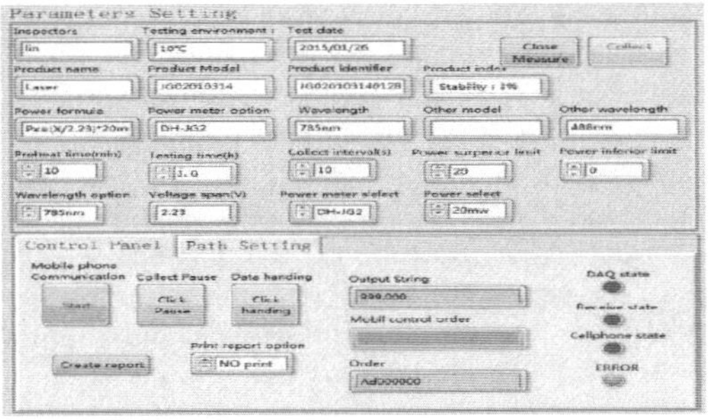

Figure 8. The parameter setting and control interface

In client, the procedure was mainly built by using the Java language in development platform of Eclipse. We can directly realize trend of the laser power data. If the laser works disorderedly, we can send the code that is 'Ss000000' to server model and turn off the laser. In addition, there are many codes with different functions, like as 'Rs000000' means restarting the laser; 'Sc000000' means startting the system to collect the laser power; 'Ar000000' means analyzing the laser power data and printing data etc. By this system, inspectors can conveniently get the working state of laser every second. And the Fig. 9 shows the remote monitoring in mobile phone.

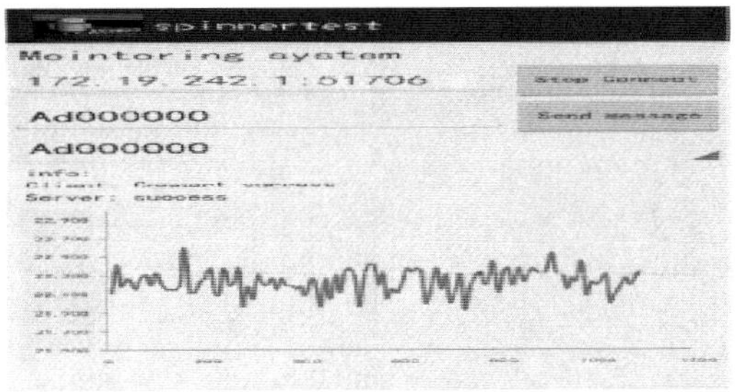

Figure 9. The client synchronization interface

5 Summary

In this paper, a new remote monitoring system based on LabVIEW was proposed to measure laser power automatically and remotely. LabVIEW software development platform, combined with data acquisition cards and TCP/IP protocol was used to achieve the power data acquisition, the process of remote monitoring, the data analysis, and report generation. Compared with the traditional data acquisition and processing system, it has the advantages of a cost-effective, high applicability and easy operation. Furthermore, it can make the users using a mobile phone as a remote monitor more convenient. LabVIEW is a powerful front-end design tool for testing system and the monitor system could be used potentially in laser power testing.

Acknowledgement

This work is supported by the National Natural Science Foundation of China (Nos. 81101110), the Science and Technology Project of Fujian Province (Nos.2015J01300), and Fujian Normal University foundation for excellent young teachers (No. fjsdjk2012058).

References

1. Swain N K, Anderson J, Singh A, et al. *Proceedings. IEEE.* 2003, pp. 1-6.
2. Stankovic, L, *Signal Processing*, vol. **44**(1996).
3. Rapuano, S, Zoino, F, IEEE T INSTRUM MEAS, vol. **55**, Oct. 2006 , pp. 1757 -1766.
4. Vargas, H. Sanchez-Moreno, J. Dormido, S. Salzmann, C. Gillet, D. Esquembre, F, COMPUT SCI ENG, vol. **11**, May.2009 , pp.36,doi: 10.1109/MCSE.2009.61
5. Pan Yu, Wu Guoan, Xu Qinfen, *Huazhong Univ. of Sci. & Tech. (Nature Science Edition)*, vol. **32**, Dec. 2004.
6. Ballesteros, J. Fernandez Palop, J. I. Hernandez, M. A, REV SCI INSTRUM, vol. **75**, Jan.2004 , pp.90 – 93.
7. X-Series Programmer's Guide. *USA Agilent Technologies*, Inc. October 2009.
8. N1911A/1912A P-Series Power Meters Service Guide, *USA Agilent, Inc.* December 2010.
9. HP 53131A/132A 225 MHz Universal Counter. *USA: Hewlett- Packard Company*. 1999.
10. GY/T 229.2-2008, *Academy of Broadcasting Planning, SARFT*. February 2008.

Determination of Total Volatile Basic Nitrogen (TVB-N) Content in Beef by Hyperspectral Imaging Technique

Shanmei Liu[1],Lu Gan[2,a]

[1] College of Informatics, Huazhong Agricultural University, Wuhan 430070,China;
[2] School of Electric and Electronic Engineering, Wuhan Polytechnic University, Wuhan 430023,china;

Abstract. Non-destructive determination of TVB-N content in beef using hyperspectral imaging (HSI) technique was evaluated. In order to create a robust model to predict the TVB-N content in beef, partition of sample set, spectral pretreatment, and the optimum wavelength selection were discussed. After the beef sample set was parted by concentration gradient (CG) algortithm, and the spectra of beef samples were preprocessed by standard normalized variate (SNV) combined with auto scale(AS), the partial least square regression (PLSR) model was established using the full spectral range, which had the best prediction abilities with R_{cv}^2 of 0.9124, R_p^2 of 0.8816, RMSECV of 1.5889, and RMSEP of 1.7719, respectively. After the optimum wavelengths which is closely related to the TVB-N content of beef samples was obtained using the competitive adaptive re-weighted (CARS) algorithm, a new PLSR model was established using the optimum wavelengths, which had outstanding prediction abilities with R_{cv}^2 of 0.9235, R_p^2 of 0.9241, RMSECV of 1.4881, and RMSEP of 1.4882, respectively.The study showed that HSI is a powerful technique to predict the TVB-N content in beef by a nondestructive way.

Keywords: beef, TVB-N content, hyperspectral imaging, data processing methods

1 Introduction

Due to the role of enzymes, microorganisms and other functions, beef is prone to spoilage during storage, transportation and processing. spoiled beef not only can not meet the needs of people's taste and nutrition, but also endanger people's health, cause disease. Therefore, the detection of beef freshness has important significance in public health.

Total volatile basic-nitrogen (TVB-N) is a substance produced in the process of beef spoilage, and its content in beef was proportional to the extent of beef spoilage. So it is an important index to evaluate the degree of beef freshness[1]. At present, the detection of beef freshness mainly includes sensory detection, physical and chemical detection, microbial detection, and so on[2-6]. However, these traditional detection methods are not conducive to the rapid detection of beef freshness in the product circulation.

With the development of sicence and technology, hyperspectral imaging (HSI) has been a very powerful tool for the nondestructive detection of agricultural and livestock products. There are also some applications of HSI in the measurement of meat food characteristics, such as tenderness, water-holding, marbling level and so on[7-9]. However, the studies on the use of HSI to evaluated the TVB-N content in beef were only a very few. In this study, nondestructive determination of TVB-N content in beef using HSI technique will be evaluated. Furthermore some data processing methods, such as partition of sample set, spectral pretreatment, and the optimum wavelength selection, will be discussed to achieve the optimal predicting results.

[a] Corresponding author:lugan81@163.com

2 Experiments and Methods

2.1 Experimental samples

The beef samples was collected from Wuhan Guoqiang poultry wholesale center in Hubei, China. After 24h post-mortem, the muscles of longissimus dorsi were dissected from Huangpi cattle and Enshi mountain buffalo carcasses. Then the muscles were packaged and transported to the laboratory to be cut into several samples. A total of 168 beef samples were prepared, including 88 Huangpi cattle beef samples and 84 Enshi mountain buffalo beef samples numbered from 1 to 80, respectively. Each sample was 8cm*8cm*4cm*3.5cm (length*width*thickness) chops, with sealed bag packaging and placed in the biochemical incubator at 0℃ ~4 ℃ .All the samples were measured in numerical sequence within 12 days, and 7 Huangpi cattle beef samples and 7 Enshi mountain buffalo beef samples were measured every 24 hours.

2.2 Experimental equipment

In this study, the HSI system used for acquiring the hyperspectral images of the beef samples was the Headwall Hyperspec hyperspectral imaging system developed by Headwall Photonics, Inc. The HSI system is mainly composed of a hyperspectral imager (VNIR Concentric, Headwall photonics, USA), an illumination source, a CCD camera with an objective lens (Schneider Optics, USA), a sample mobile station (Andor Technology Inc, USA), and a computer with a SDKs control platform which was developed based on LabVIEW software.

To measure the reference values of TVB-N content of beef samples, standard diffusion dish and microburette were required and reagents such as saturated solution of potassium carbonate, water soluble gum, absorption liquid, mixed indicator solution and standard titration solution of hydrochloric acid were needed.

2.3 Acquisition of hyperspectral data

Before the HSI system was used to acquire the hyperspectral data of the beef samples,it was turned on for preheating thirty minutes to ensure a stable performance. Then the interface of the SDKs control platform was opened and the parameters of the spectrometer were set as follows: wavelength range of 400~1000nm, resolution of 0.8nm, exposure time of 5ms. And the parameters of the sample mobile station were set as follows: step distance of 110mm, object distance of 650mm, and line scan length of 90mm.

In order to eliminate the effect of white and black background on the hyperspectral images, the correction of black and white board was needed. Firstly, The blackboard image was obtained by the way of closing the lens cap and the whiteboard image was obtained by collecting standard white board image. Then the calibration image was calculated according to formula 1.

$$I = \frac{I_0 - I_b}{I_w - I_b} \quad (1)$$

where I is the calibrated image, I_0 is the original image, I_w is the whiteboard image and I_b is the blackboard image. After the black and white board calibration, the hyperspectral data of all the beef samples were acquired with the hyperspectral imaging system and saved in the computer.

2.4 Measurement of the TVB-N content

After hyperspectral data acquisition, the reference value of TVB-N content in beef sample was measured by the microdiffusion method according to the Chinese standard GB/T 5009.44-2003[10].The statistical results of TVB-N content of the 168 beef samples are as follows: the minimum of 9.80mg/100g, the maximum of 32.44 mg/100g, the mean of 18.0093mg/100g, and the standard deviation of 5.6132mg/100g.

2.5 Data processing methods

In this study, some data processing methods such as abnormal sample detection, spectral extraction, partition of sample set, spectral pretreatment, model establishment, optimum wavelength selection were discussed. The software ENVI4.7 was utilized to extract spectra from the hyperspectral image of the beef samples, and the software MATLAB R2010a was utilized to the rest data processing steps.

As is known that the hyperspectral data of each sample contain rich information including space and spectral information. There is an image in each wavelength and there are many pixels in each image. When the spectra of each beef sample were extract, a region of interest (ROI) included many pixels (about 5000) was selected, then the spectra of the pixels in the ROI were extracted and the average spectrum was calculated representing the spectrum of this beef sample. At the same way, all the spectra of the beef samples were obtained.The extracted spectral data from all the beef samples were organized in a matrix (X), where the rows denote the number of samples (168 beef samples) and the columns represent the number of variables (753 wavelengths).

Abnormal samples will typically change the distribution trend of the overall data,thus affecting the accuracy of the calibration model. It is very necessary to detect and remove the abnormal samples before the calibration model is built. In this study, abnormal beef samples were detected by using the Monte Carlo sampling algorithm(MCS) [11].

Selection of the calibration and prediction sets may also affect the prediction performance of the model. In this study, 5 kinds of sample set partition methods, such as sample set partitioning based on joint X-Y distances (SPXY) [12], Kennard-Stone(KS) [13], random sampling (RS), duplex, CG, were compared and discussed.

Spectral data pretreatment may have effects on the calibration model too, so some spectral pretreatment methods were compared here. For example, mean centering(MC), autoscale (AS), multiplicative scattering correction (MSC), standard normalized variate (SNV), first derivative (FD), second derivative (SD), orthogonal signal correction (OSC) and normalization were discussed.

Hyperspectral data exhibits inter-band correlation and multi-colinearity, which will lead to data redundancy and convergence instability in the calibration model. Wavelength selection have some advantage in solving above problem, because it can effectively prevent the repetition of information or redundancies by picking out the optimum wavelengths. The competitive adaptive reweighted sampling (CARS) algorithm [14] is excellent in selecting optimum wavelengths from the full spectral range, so it was utilized to select the optimum hyperspectral wavelengths for predicting the TVB-N content in beef.

2.6 Model performance parameters

Model performance was evaluated according to the values of four parameters like the coefficient of determination (R_{cv}^2 and R_p^2), root mean square error estimated by cross validation (RMSECV), and root mean square error estimated by prediction (RMSEP). The R_{cv}^2 and R_p^2 are higher, and the RMSECV and RMSEP are lower, the performance of the corresponding model is better.

3 Results and Discussion

3.1 Spectral profiles

The reflectance spectra extracted from the hyperspectral images of the 168 samples of tested beef are shown in Figure 1.

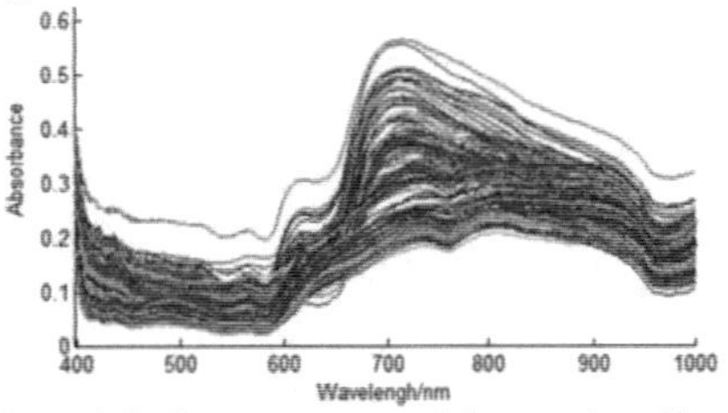

Figure1. Reflectance spectra of the samples of beef

3.2 Abnormal beef samples

The abnormal beef samples were detected by MCS algorithm, and the results are shown in Fig.2.

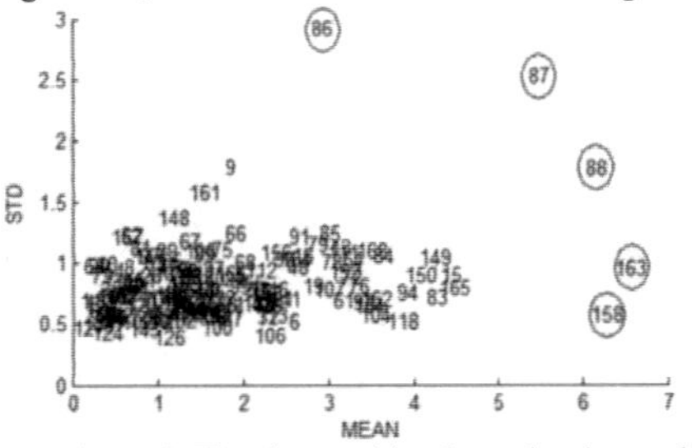

Figure 2. The abnormal beef samples detected by MCS

Figure2 indicates that sample no.86, no.87, no.88, no.158, and sample no.163 were abnormal. After deleting the five abnormal samples, 163 normal samples were obtained for the subsequent analysis.

3.3 The beef sample set partition

The beef sample set was parted by KS, SPXY, RS, duplex, and CG, respectively. At last, 123 samples were put into the calibration set ,and 40 samples were put into the test set. The statistical results of the calibration set and the test set are shown in Table 1.

Table 1 indicates that the RS and Duplex algorithm only put the minimum of TVB-N content into the calibration set, whereas KS, SPXY, and CG algorithm put the minimum and maximum of TVB-N content all into the calibration set.Table 1 also indicates that the mean value 17.6931 of the calibration set and the mean value 17.5910 of the test set obtained by CG algorithm were all closest to the mean value 17.6680 of all samples. This shows that the calibration set and the test set obtained by CG were well-distributed. So CG was choosed as the best method to part beef samples.

3.4 The full spectral range PLSR models

The results of the PLSR models which were built using the full spectral range are shown in Table 2.

Table 2 indicates that, after the spectra were preprocessed by SNV combined with AS, the PLSR model built using the full spectral range was the best, with R_{cv}^2 of 0.9124, R_p^2 of 0.8816, RMSECV of 1.5889 and RMSEP of 1.7719. Obviously, SNV combined with AS was the best spectral preprocessing method when the PLSR model for predicting the TVB-N content in beef was built using the full spectral range.

3.5 Selection of optimum wavelengths

By CARS algorithm, 54 optimum wavelengths were selected. The distribution is shown in Figure 3.

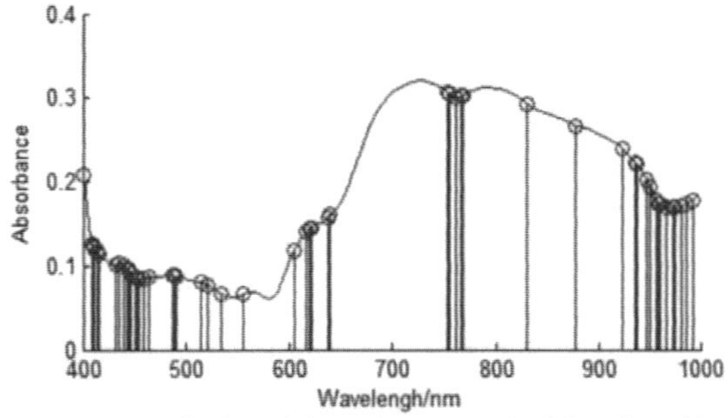

Figure 3. Distribution of the optimum wavelengths selected by CARS

3.6 PLSR models using optimum wavelengths

New PLSR models are built by using the optimum wavelengths and the results are shown in Table 3.

Table 3 shows that, after the spectra were preprocessed by OSC combined with MC, the PLSR model built using the optimum wavelengths was the best, with R_{cv}^2 of 0.9235, R_p^2 of 0.9241, RMSECV of 1.4881, and RMSEP of 1.4982, respectively. Table 2 and Table 3 also indicate that, the performance of the new PLSR model was better than that of the PLSR model built using the full spectral range.Futhermore, more than 92.8% of the wavelengths were eliminated (i.e., 54 vs. 753). All above results indicated that the CARS algorithm was efficient in selecting the beef sample optimum wavelength.

Table 1. Statistical results for the TVB-N content(mg/100g) of beef samples

Sample partition method	The calibration set			The test set		
	Range	Mean	Standard deviation	Range	Mean	Standard deviation
KS	9.80~32.4	17.4663	5.0952	10.49~30.89	18.2885	6.0168
SPXY	9.80~32.4	17.9180	5.2825	10.49~31.70	16.8993	5.4626
RS	9.80~30.8	17.1209	4.9688	11.18~32.44	19.1949	6.0264
Duplex	9.80~31.7	17.4644	5.3128	10.49~32.44	18.2942	5.3950
CG	9.80~32.4	17.6931	5.3912	10.50~30.89	17.5910	5.1968

Table 2. Results of PLSR models established using the full spectral range after different spectral pre-treatments

Spectral pretreatment method	Principal component numbers	R_{cv}^2	RMSECV	R_p^2	RMSEP
MC	20	0.8811	1.8576	0.8278	2.1369
AS	18	0.8849	1.8265	0.8340	2.0998
MSC+AS	16	0.9112	1.5997	0.8736	1.8411
SNV+AS	**17**	**0.9124**	**1.5889**	**0.8816**	**1.7719**
FD+AS	20	0.8533	2.0810	0.7802	2.4577
SD+AS	15	0.9073	1.6352	0.8439	2.0435
OSC+AS	15	0.8813	1.8540	0.8270	2.1612
Normalize+AS	19	0.9123	1.5910	0.8504	1.9857

Table 3. Results of the PLSR models established based on the selected wavelengths after different spectral pre-treatments

Spectral pretreatment method	Principal component numbers	R_{cv}^2	RMSECV	R_p^2	RMSEP
none	16	0.9008	1.6959	0.9137	1.5964
MC	16	0.9225	1.4969	0.9233	1.5010
AS	16	0.9190	1.5319	0.9097	1.6048
MSC+MC	14	0.9150	1.5679	0.9324	1.4537
SNV+MC	15	0.9165	1.5550	0.9326	1.4440
OSC+MC	**13**	**0.9235**	**1.4881**	**0.9241**	**1.4982**
Normalize+MC	15	0.9099	1.6166	0.9203	1.4871

4 Conclusions

This study aimed to investigate the aptitude of HSI technique in predicting the TVB-N content in beef. The PLSR model for predicting TVB-N content in beef was built using the full range spectra and optimized by using the spectra at the optimum wavelengths. Different data processing methods were disscused. The CG algorithm was identified as the best sample set division method for beef samples, and SNV combined with AS was determined as the best spectral preprocessing method for the PLSR model built using the full spectral range, which resulted in R_{cv}^2 of 0.9124, R_p^2 of 0.8816, RMSECV of 1.5889, and RMSEP of 1.7719, respectively. The optimum wavelengths were selected by CARS algorithm. Out of 753 wavelengths, only 54 optimum wavelengths were selected to develop a new PLSR prediction model, which performed effectively, with R_{cv}^2 of 0.9235, R_p^2 of 0.9241, RMSECV of 1.4881, and RMSEP of 1.4982, after the spectra were preprocessed by OSC combined with MC. All the results indicated that HSI system is a potential technique for rapid and non-destructive prediction of TVB-N content in beef.

Acknowledgment

Fund for this research was provided by the project 2662015QC024 supported by the Fundamental Research funds for the Central Universities.

References

1. Y.Q Lin, Z.S. Fang . *Meat Research*, **25**,62-65.(2011).
2. Z. Zhang, J. Tong, D.H. Chen, Y.B. Lan. *Journal of Bionic Engineering*, **5**, 67-73(2008).
3. X. Hong, J. Wang, Z. Hai. *Sensors and Actuators B: Chemical*, **161**,381-389(2012).
4. Z.B. Shi, Y.Y. Tong, D.H. Chen, Y. Li. *Transactions of the Chinese Society for Agricultural Machinery*, **40**,184-188(2009).
5. D.H. Noh, S.H. Chung, S.J. Choi, S.J. Hur. *Food Control*, **22**,133-136(2011).
6. F. Galgano, F. Favati, M. Bonadio, V. Lorusso, P. Romano. *Food Research International*, **42**, 1147-1152(2009).

7. G.Elmasry, D.W. Sun, P. Allen. *Food Research International*, **44**,2624-2633(2011).
8. G.Elmasry, D.W. Sun, P. Allen. *Journal of Food Engineering*, **110**,127-140(2012).
9. Y.K. Peng, J. Zhang, W. Wang, Y.Y. Li, J.H. Wu, H. Huang, X.D. Gao, W.K. Jiang. *Journal of Food Engineering*, **102**,163-169(2011).
10. National Standard of the People's Republic of China, Method for Analysis of Hygienic Standard of Meat and Meat Products, GB/T 5009.44-2003 (2003).
11. D.S. Cao, Y.Z. Liang, Q.S. Xu, H.D. Li, X. Chen. *Journal of Computational Chemistry*, **31**,592-602(2010).
12. R.K.H. Galvao, M.C.U Araujo, G.E. Jose, M.J.C. Pontes, E.C. Silva, T.C.B. Saldanha. *Talanta*, **67**,736-740(2005).
13. W. Wu, B. Walczak, D.L. Massart, S. Heuerding, F Erni, I.R. Last, K.A. Prebble. *Chemometrics & Intelligent Laboratory Systems*, **33**,35- 46(1996).
14. H.D. Li, Y.Z. Liang, Q.S. Xu, D.S. Cao. *Analytica Chimica Acta*, **648**,77-84(2009).

MATEC Web of Conferences **61**, 06008 (2016) DOI: 10.1051/matecconf/20166106008
APOP2016

16-QAM Quantum Receiver with Hybrid Structure Outperforming the Standard Quantum Limit

Yuan Zuo[1, 2], Ke Li[3] and Bing Zhu[1, 2, a]

[1] Department of Electronic Engineering and Information Science, University of Science and Technology of China, 230027 Hefei, China
[2] Key Laboratory of Electromagnetic Space Information, Chinese Academy of Sciences, 230027 Hefei, China
[3] Nanjing Research Institute of Electronic Technology, 210013 Nanjing, China

Abstract. Quantum receivers which can discriminate phase shift keying (PSK) and pulse position modulation (PPM) signals below the standard quantum limit have been proposed and some have been demonstrated experimentally. But for quadrature amplitude modulation (QAM) signals, few literatures have been reported so far. It is important to reduce the average error probability of QAM signals below the standard quantum limit (SQL), since these modulation have a high spectral efficiency. In this paper, we present a quantum receiver for 16-QAM signals discrimination with hybrid structure, which contains a homodyne receiver and a displacement receiver. By numerical simulation, we prove that the performance of the quantum receiver can outperform the SQL, and it can be improved by an optimized displacement.

1 Introduction

It is known for decades that quantum receivers can discriminate different signals below the standard quantum limit (SQL) and achieve even lower error probability limit called Helstrom bound [1]. For phase shift keying (PSK) and pulse position modulation (PPM) signals discrimination, many kinds of quantum receivers have been proposed theoretically and some of them have been demonstrated experimentally [2-11]. Dolinar receiver was proposed to discriminate binary phase shift keying (BPSK) signals. Its performance can achieve Helstrom limit theoretically [3]. Later, Bondurant proposed two type Bondurant receiver to discriminate quadrature phase shift keying (QPSK) [4]. For PPM signals, a conditionally nulling receiver was proposed by Dolinar, which is able to achieve very nearly the Helstrom limit [9].

However, few literatures have been paid attention to QAM signals discrimination. The QAM signals have a high spectral efficiency in optical communication [12]. So it is very meaningful to reduce the average error probability of QAM signals below the SQL. Inspired by Bondurant [4] and Müller [5, 11], we present an original 16-QAM quantum receiver with a hybrid structure, which contains a homodyne receiver and a displacement receiver. Numerical simulation has shown that it can outperform the SQL. What's more, its performance can be even better, if we optimize the displacement, especially when the signals are weak. By using these quantum receivers, the communication distance and capacity for 16-QAM signals are promising to be improved.

2 Schemes and Performances

2.1 Classical receiver system

In an optical communication protocol, a sender encodes information by coherent light. For 16-QAM signals, information are encoded into the amplitude and phase. The 16-QAM signals include 16 alphabet $|\alpha_{uv}\rangle$, $u,v\in\{-3,-1,1,3\}$, where

$$\alpha_{uv}=u\,|\alpha|+j\,v\,|\alpha|. \tag{1}$$

Where $j=\sqrt{-1}$ is imaginary unit. The real part and imaginary part correspond to the two quadrature amplitudes x and p respectively. It is well known that a conventional detection scheme is a heterodyne receiver, which measure the two quadrature amplitudes simultaneously. In ideal condition without any noise and interference, the probability density function is given as follows [1]:

$$P(x,p\,|\,u,v)=\frac{1}{\pi}\exp\left[-\left(x-u|\alpha|\right)^2-\left(p-v|\alpha|\right)^2\right] \tag{2}$$

The detection domain $D_{u',v'}$, where signal $|\alpha_{uv}\rangle$ is detected as $|\alpha_{u'v'}\rangle$, can be represent as

$$D_{u',v'}=\{(x,p)\,|\,D_L(u')<x\le D_U(u'),\; D_L(v')<p\le D_U(v')\}. \tag{3}$$

Where the notation D_L and D_U are defined as

[a] Corresponding author: zbing@ustc.edu.cn

$$D_L(k)=\begin{cases}-\infty, & k=-3\\ k-1, & k=-1,1,3\end{cases},$$
$$D_U(k)=\begin{cases}\infty, & k=3\\ k+1, & k=-3,-1,1\end{cases}. \tag{4}$$

Then, we can calculate the average error probability by the formula

$$P_e=1-\frac{1}{16}\sum_{u,v}\iint_{D_{u,v}}P(x,p\,|\,u,v)\,\mathrm{d}x\mathrm{d}p\,. \tag{5}$$

2.2 16-QAM Helstrom limit

In the quantum detection and estimation theory, the detectors are described by a set of positive operator-valued measure (POVM). The optimal quantum detectors can be solved by semi-definite programming (SDP) [1]. Instead of solving 16 POVM matrices, we compute only one matrix by solving the dual problem [13].

2.3 Hybrid receiver system

In order to discriminate the 16-QAM signals below SQL, we designed a hybrid structure scheme, shown in the Fig. 1 (*a*). From Fig. 1 (*a*), we can see the phase space configuration of 16-QAM signals and our receiver scheme. At first, the incoming signal is split into two equivalent portions by a beam splitter (BS) with reflect ratio $r^2=0.5$. Then one beam is directly fed into a homodyne detector (HD), whilst the other one is detected using a displacement receiver, which is controlled using a feed-forward and feedback strategy. The HD determines the *p* quadrature of the incoming signal and it is described by the POVM elements

$$\hat{\Pi}_i^{HD}=\int_{D_i}\mathrm{d}p\,|p\rangle\langle p|, i=1,2,3,4. \tag{6}$$

Where

$$D_1=\{p\,|\,p>2|r\cdot\alpha|\}, D_2=\{p\,|\,0<p\le 2|r\cdot\alpha|\},$$
$$D_3=\{p\,|\,0\ge p>-2|r\cdot\alpha|\}, D_4=\{p\,|\,p\le -2|r\cdot\alpha|\}. \tag{7}$$

For signal $|\alpha_{uv}\rangle$, the correct output corresponds to the detection domain D_i with $\mathrm{Im}(r\cdot\alpha_{uv})\in D_i$. We write this domain as $D(\alpha_{uv})$. And the probability of observing correct outcome for this signal is

$$P_c^{\mathrm{HD}}(r\cdot\alpha_{uv})=\int_{D(\alpha_{uv})}|\langle p|r\cdot\alpha_{uv}\rangle|^2\,\mathrm{d}p$$
$$=\frac{1}{2}\mathrm{erf}\left(\sqrt{2}p\right)\Big|_{p_L(D)}^{p_U(D)}. \tag{8}$$

Where $p_U(D_i)$ and $p_L(D_i)$ are the upper and lower bound for D_i .

Supposing HD yield correct results, then 16 hypotheses are reduced to only 4 hypotheses H_i (i = 1, 2, 3, 4) as shown in Fig. 1 (*b*). In order to discriminate these 4 signals, we use a displacement receiver (DR) which contains a displacement operator and an ideal single photon detector (SPD). It sequentially nulls signals 1 → 2 → 3 → 4 just like Bondurant receiver [4]. At first, the receiver choose the hypnosis H_1 and null the first signal by a displacement $D(\gamma)$. The imaginary part of γ is determined by HD detection result. If the hypnosis is correct, then after displacement $D(\gamma)$, the signal fed into SPD becomes vacuum state and no photon can be detected. So we regard the signal 1 as the correct result. Otherwise, every time a photon counting event happens, the receiver immediately changes the displacement to null the next signal and keeps counting. Suppose N is the photon count number in whole symbol interval, if $N \le 3$, we choose the hypothesis H_{N+1}; if $N > 3$, choose H_4.

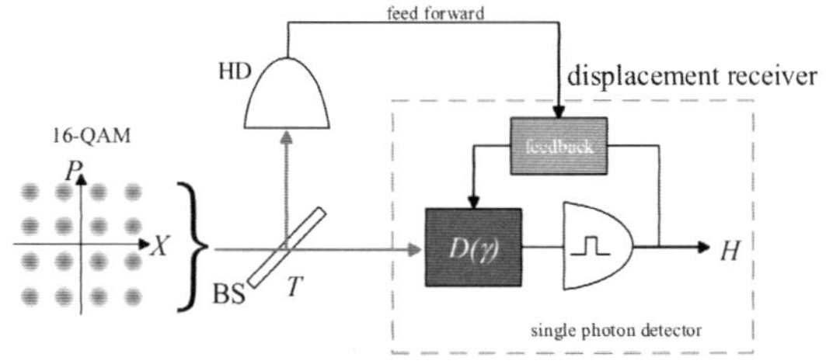

(*a*)

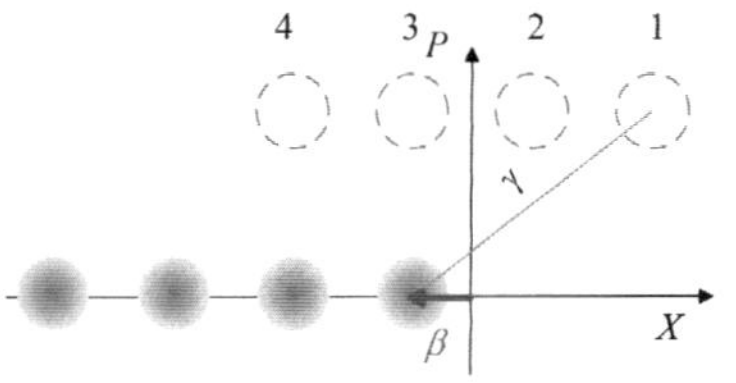

(*b*)

Figure 1. (*a*) Hybrid structure for 16-QAM signals. The incoming signal is equally divided into two parts by beam splitter (BS). One part is detected by a homodyne detector (HD) and the result is feed forward to the displacement receiver. (*b*) Displacement strategy for displacement receiver. When using the exactly nulling, choose β = 0. When using the optimal displacement, choose β by numerical optimization.

More specifically, if signal 1 is sent and no photon click event happens, then we get the correct answer with probability

$$P(1|1)=\exp\left(|\alpha_1-\gamma_1|^2\right). \tag{9}$$

If signal 2 is sent, we get the correct answer when only one photon click event happens, which occurs with probability

$$P(2|2)=\int_0^T\mathrm{d}t_1\frac{|\alpha_2-\gamma_1|^2}{T}\exp\left(-|\alpha_2-\gamma_1|^2\frac{t_1}{T}\right)$$
$$\times\exp\left(-|\alpha_2-\gamma_2|^2\frac{T-t_1}{T}\right). \tag{10}$$

Where T is signal pulse interval, and $\alpha_i\,(i=1,2,3,4)$ is the selected signal amplitude. And t_i stands for the i-th photon arrival time. If signal 3 is sent, we get the correct answer when two photon click event happen exactly, which occurs with probability

$$P(3|3)=\int_0^T dt_1\int_{t_1}^T dt_2\frac{|\alpha_3-\gamma_1|^2}{T}\exp\left(-|\alpha_3-\gamma_1|^2\frac{t_1}{T}\right)\frac{|\alpha_3-\gamma_2|^2}{T}\times\exp\left(-|\alpha_3-\gamma_2|^2\frac{t_2-t_1}{T}\right)\exp\left(-|\alpha_3-\gamma_3|^2\frac{T-t_2}{T}\right). \quad (11)$$

If signal 4 is sent, we get the correct answer when there are more than or equal to 3 photon click events happen, which occurs with probability

$$P(4|4)=\int_0^T dt_1\int_{t_1}^T dt_2\int_{t_2}^T dt_3\frac{|\alpha_4-\gamma_1|^2}{T}\exp\left(-|\alpha_4-\gamma_1|^2\frac{t_1}{T}\right)\times\frac{|\alpha_4-\gamma_2|^2}{T}\exp\left(-|\alpha_4-\gamma_2|^2\frac{t_2-t_1}{T}\right)\times\frac{|\alpha_4-\gamma_3|^2}{T}\exp\left(-|\alpha_4-\gamma_3|^2\frac{t_3-t_2}{T}\right). \quad (12)$$

Thus, for given signal α_{uv}, the correct probability for the displacement receiver is $P_c^{DR}(t\cdot\alpha_{uv})=P(j\,|\,j)$, where t is transparent ratio, $t^2=1-r^2=0.5$, and j is the index of the signal which is shown in Fig. 1(*b*). The total average probability of the hybrid receiver giving the correct outcome is

$$P_c=\frac{1}{16}\sum_{u,v}P_c^{HD}(r\cdot\alpha_{uv})P_c^{DR}(t\cdot\alpha_{uv}) \quad (13)$$

In this case, we use exact nulling strategy namely $\gamma_i=\alpha_i\,(i=1,2,3,4)$. There are at most 3 photon click events when using exact nulling strategy. Because no photon will be detected once the signal is displaced to vacuum state.

2.4 Optimal displacement

In the above, we set the exact nulling for the displacement receiver. However, exact nulling is not a good idea when signal is weak [5, 11]. By adding an additional displacement β just like generalized Kennedy receiver, namely $\gamma_i=\alpha_i+\beta\,(i=1,2,3,4)$, which is shown in Fig. 1(*b*), the error probability can be further reduced. The best value of β is given by

$$\beta^*=\underset{\beta}{\operatorname{argmax}}\{P_c(\beta)\}. \quad (14)$$

Which can be determined by numerical optimization.

2.5 Simulation and Results

We simulate our receiver under two different configurations, exact nulling (Type I) and optimal displacement (Type II). In the Type II configuration, we use numerical optimization to determine the best value of β to gain the smallest average error probability. Results are shown in Fig. 2. As in Fig. 2 (*a*), the optimal displacement decays with the average photon number increasing. When the signal is strong enough, the additional displacement β can be ignored, so the performance of the Type I receiver approaches to the performance of the type II receiver as shown in Fig. 2(*b*). In this case, type I receiver (blue) and type II receiver (green) can both outperform SQL (red). However, when the signal is weak, the performance of the type I is above the SQL, but the type II receiver can still outperform SQL within a wider range of average photon number.

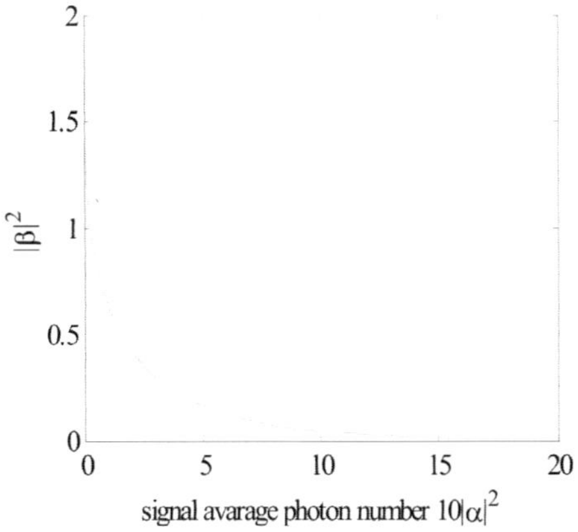

(*a*)

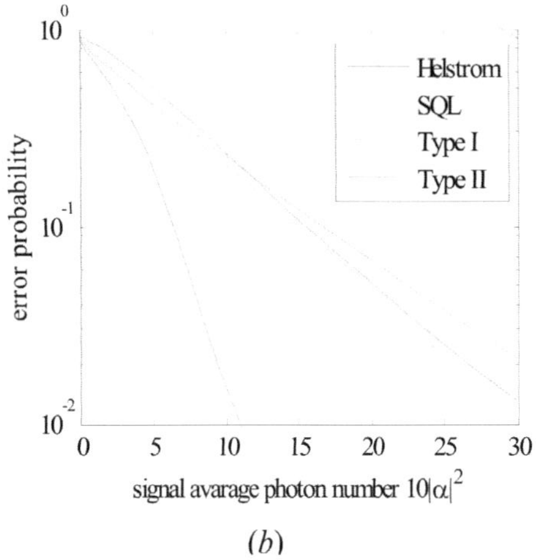

(*b*)

Figure 2. (*a*) Optimal displacement parameter $|\beta|^2$ as a function of signal energy. (*b*) Error probability for hybrid structure receiver using exact nulling (type I, blue) and optimal displacement (type II, green) compare to SQL (red) and Helstrom limit (black).

3 Conclusion

In this paper, two type receivers have been proposed for 16-QAM signals discrimination, and ran numerical

DOI: 10.1051/matecconf/20166106008

simulation to show that the hybrid structure quantum receivers can be also applied to 16-QAM signals. And the performance of these receivers outperforms the SQL in bright light regime. Finally, comparing two type receiver, we conclude that the performance can be improved by optimal displacement, especially when the signal is weak. Since these receivers can outperform the SQL, it is promising to apply these receivers to increase the distance and capacity in optical communication.

References

1. C. W. Helstrom, *Quantum detection and estimation theory*(Academic press, 1976).
2. R. S. Kennedy, MIT Research Laboratory of Electronics Quarterly Progress Report, **110**, 219 (1973).
3. S. Dolinar, MIT Research Laboratory of Electronics Quarterly Progress Report, **111**, 115 (1973).
4. R. S. Bondurant, Opt Lett, **18**, 1896 (1993).
5. C. R. Müller, M. A. Usuga, C. Wittmann, M. Takeoka, C. Marquardt, U. L. Andersen, G. Leuchs, New J Phys, **14**, 83009 (2012).
6. R. L. Cook, P. J. Martin, J. M. Geremia, Nature, **446**, 774 (2007).
7. F. E. Becerra, J. Fan, G. Baumgartner, J. Goldhar, J. T. Kosloski, A. Migdall, Nat Photonics, **7**, 147 (2013).
8. K. Li, Y. Zuo, B. Zhu, IEEE Photonic Tech L, **25**, 2182 (2013).
9. S. J. Dolinar Jr, The Telecommunications and Data Acquisition Progress Report, **42**, 72 (1982).
10. J. Chen, J. L. Habif, Z. Dutton, R. Lazarus, S. Guha, Nat Photonics, **6**, 374 (2012).
11. C. R. Müller, G. Leuchs, C. Marquardt, in CLEO: QELS_Fundamental Science, Optical Society of America, 2014, p. M3A.
12. P. J. Winzer, Lightwave Technology, Journal of, **30**, 3824 (2012).
13. Y. C. Eldar, A. Megretski, G. C. Verghese, Information Theory, IEEE Transactions on, **49**, 1007 (2003).

The research on temperature sensing properties of photonic crystal fiber based on Liquid crystal filling

Xiangzhen Zan[1,a], Liliang Qian[2], Miao Wang[3,4] and Wenbin Liu[5]

[1] *Department of Information Engineering, City College of Wenzhou University, Wenzhou city, Zhejiang province, China*
[2] *Lujiang Middle school, Hefei city, Anhui province, China*
[3] *Science and Technology on Avionics Integration Laboratory, Shanghai, China*
[4] *China National Aeronautical Radio Electronics Research Institute, Shanghai, China*
[5] *Department of Physics and Electronic Information Engineering, Wenzhou University, Wenzhou city, Zhejiang province, China*

Abstract. Based on the photonic bandgap-photonic crystal fibers(PBG-PCF) fiber core fills the namitic liquid crystal. By readjusting the temperature to change the refractive index, constitute new liquid fiber-optic temperature sensor. In this paper, we use finite element COMSOL software to simulate and analyze photonic crystal optical fiber sensitive properties. The research show that after the PBG – PCF filling the liquid crystal, its mode field distribution, effective refractive index, waveguide dispersion etc changing with temperature is so big. Therefore, the properties that the refractive index of PCF mode CF changing with temperature sensitive medium, provides the theoretical basis for designing optic fiber temperature sensors.

1 Introduction

Photonic crystal fiber (PCF) is two dimensional photonic crystals of a kind of defects with line. This new type of optical fiber will field a new revolution in Optical fiber application domain while ordinary fiber doesn't have these characteristics. According to light's transmission mechanism, PCF can be divided into two kinds: one kind similar with the traditional optical fiber, whose refractive index of fiber core is sightly bigger than refractive index of the cladding equivalent refractive index, can form total internal reflection, that we call total internal reflective photonic crystal fibers (TIR-PCF); The other kind is photonic band gap (PBG-PCF), that uses the photonic band gap (PBG) effects of light guide. When PCF is produced, the optical transmission characteristics are determined. However, in practice, due to the fact that PCF is not very sensitive to temperature, its optical properties can be adjusted according to the requirement. Different material is filled the air hole, we can change photonic crystal optical fiber transmission characteristics and sensing properties. The refractive index of liquid crystal (LC) changing with temperature and wavelength is very sensitive. By filling the LC into photonic crystal fibers' air hole, we can adjust the temperature on the transmission properties of photonic crystal fibers. Nowadays, LC filling the cladding air holes of photonic crystal fibers is much researched, but, temperature on LC filling PCF core's research is less. This paper mainly deals with the subjects about using LC filling photonic crystal fiber's core at different temperatures, which can be adjusted photonic crystal fiber optical transmission characteristics. We use all-vector finite element COMSOL software to simulate and analyze PCF transmission characteristic of temperature, on hexagonal structure photonic crystal core filling liquid crystal optical guidance. The main research covers temperature on the effective refractive index, mould field and the influence of chromatic dispersion waveguide.

[a] Corresponding author: xiangcheng2436@163.com

2 Numerical model

Maxwell equation for the transmission characteristics of photonic crystal fiber is still applicable. We can get followed by photonic crystal fiber eigenvalue equation:

$$\nabla\times(\frac{1}{\varepsilon_r}\nabla\times H)=(\frac{\omega}{c})^2\mu_r H$$
$$\nabla\times\nabla\times E=(\frac{\omega}{c})^2\varepsilon_r E \quad (1)$$

Where H, E denotes the magnetic field intensity and electric field intensity respectively, ε_r is the dielectric constant and μ_r is Magnetic conductivity, ω is the light wave angular frequency, c is the speed of light in vacuum, β is mode propagation constant. PBG-PCF model is constructed as shown in Figure1.

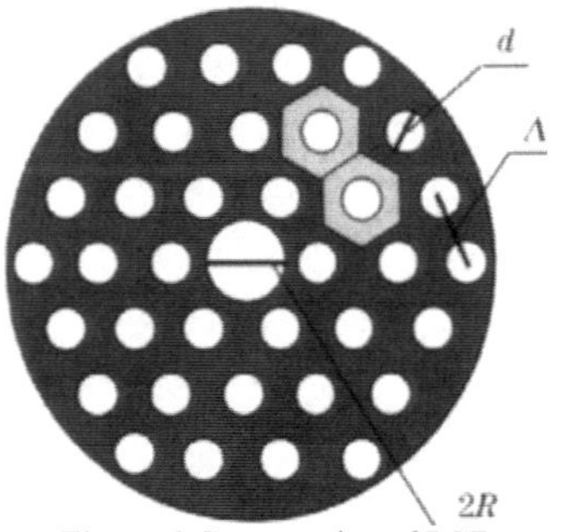

Figure 1 Cross section of PCF

Just consider the influence of temperature variations, the arrangement of liquid crystal molecules along the fiber direction is generally z axis direction, then LC in the x and y directions are ordinary polarized light, that is $n_x = n_y = n_o$, $n_e = n_z$, where n_o and n_e are refractive index of ordinary light and extraordinary light respectively, and function of temperature and wavelength.

Table 1

A=1.74	$A_0 = 1.50$
B=2.16e-4	$B_0 = 0.0078\mu m^2$
β=0.12	$C_0 = 0.0003\mu m^4$
$(\Delta n)_0 = 0.6$	$A_e = 1.864$
$T_c = 330K$	$B_e = 0.008\mu m^2$
$C_e = 0.0026\mu m^4$	

$$n_0(T) = A - BT - \frac{(\Delta n)_0}{3}(1-\frac{T}{T_C})^{\beta}, T < T_C \quad (2)$$

$$n_e(T) = A - BT + \frac{2(\Delta n)_0}{3}(1-\frac{T}{T_C})^{\beta}, T < T_C \quad (3)$$

$$n_0(\lambda) = A_0 + \frac{B_0}{\lambda^2} + \frac{C_0}{\lambda^4}, T < T_C \quad (4)$$

$$n_e(\lambda) = A_e + \frac{B_e}{\lambda^2} + \frac{C_e}{\lambda^4}, T < T_C \quad (5)$$

$$n(T,\lambda) = A - BT, T \geq T_C \quad (6)$$

3 Mode field distribution

Figure 2 describes the case that when wavelength of PCF is 1.55, $d/\Lambda = 0.5$ the mode field distribution at different temperature. We can see from the mode field distribution figure, temperature on filling the photonic crystal fiber core mode field distribution changes significantly. The reason is when fiber core filling liquid crystal, LC is sensitive to the change of temperature, in certain temperature range. Temperature increases, fiber core liquid crystal refractive index decreases, making limited decreased ability of light. So the increasing temperature makes liquid crystal filled fiber core's mode field distribution area increase. So PCF in 323K mode field distribution is greater than mode field distribution in 273K .

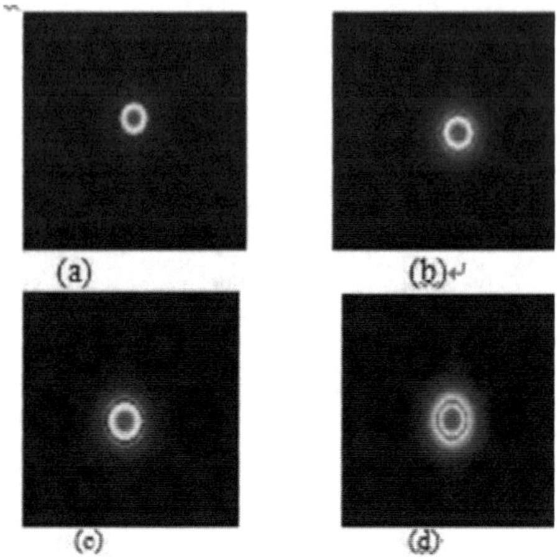

Figure 2 Mode field of PCF at 1.55 μm (a) 273K after filling LC, (b)293K after filling LC, (c)313K after filling LC, (d)323K after filling LC mode field distribution

4 The influence of temperature on the effective refractive index of guide mode

We change the PCF structure parameters: duty ratios is 0.3, 0.4, 0.5 respectively, and hole spacing is 2.3 μm . We analyze the relationship between the effective index of guide mode with wavelength and different temperature after filling LC, with core of the PCF which is filled liquid crystal .We can see from figure 3 clearly: when the core of PCF is filled LC, the effective refractive index of the guide mode decreases with the increase wavelength. The reason is the refractive index of liquid crystal decreases with the increasing wavelength, and that long wavelength light penetrates into the cladding air holes capacity is stronger than the short-wave, for short wavelength is concentrated in the fiber core. We also find that the refractive index of the guide mode increases with increasing duty ratio, at the same hole space. Due to the larger duty ratio, the equivalent refractive index is smaller. The same ability to make light of increased restrictions guide mode, also the smaller the mode field distribution is, the effective refractive index of guide mode will be. Figure 4 is when the wavelength is 1.55 μm and the PCF core is filled with LC changing with temperature, the effective index of guide mode decreases with increasing temperature. Due to the fact that liquid crystal refractive index decreases with increasing temperature, and that the core's refractive index decreases with increasing temperature, the equivalent refractive of guide mode index must decrease. In Figure 4, we still find that the refractive index of the guide mode increases with increasing duty cycle.

Figure 5 shows the relationship between the effective index and wavelength with the PCF of different temperature after filling LC, at the duty ratio of 0.5. In fact, from Figures 3 and 4, we know the relationship between them. Figure 5 can be further verified that the effective refractive index of the guide film decreases with increasing temperature, with the wavelength length increases.

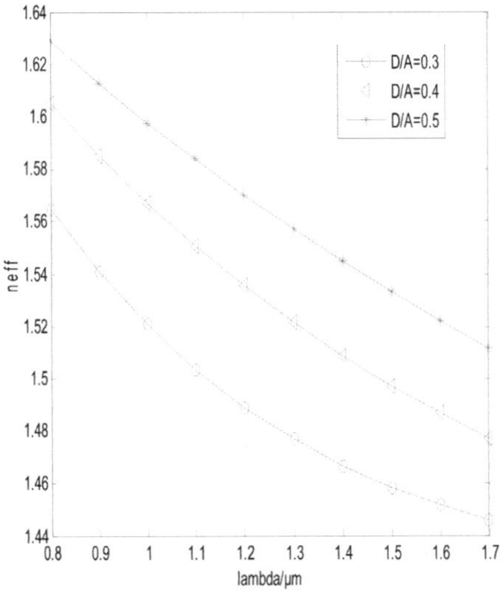

Figure 3 Relationship between the effective index and wavelength with different duty ratios of PCF after filling LC

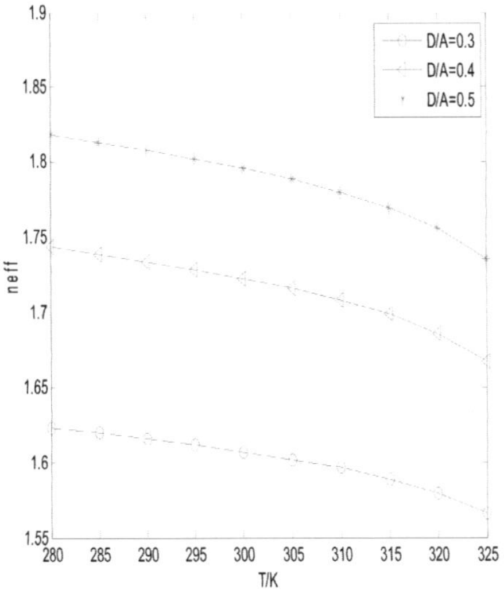

Figure 4 Relationship between the effective index and temperature with different duty ratios of PCF after filling LC

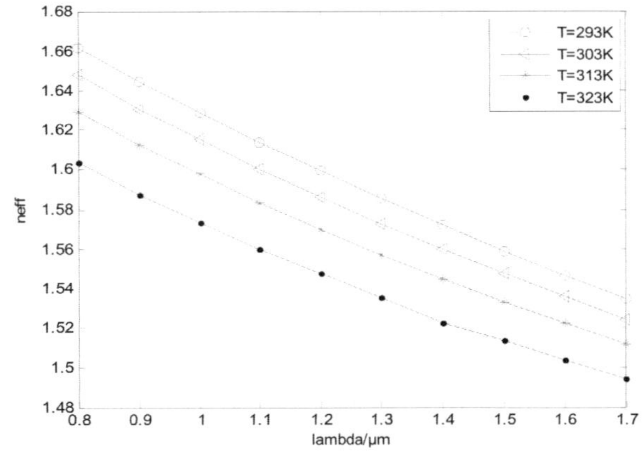

Figure 5 Relationship between the effective index and wavelength with the PCF of different temperature after filling LC

Parameters are calculated in Table 2 below.

Table 2

Temperature	A_e	$B_e(\mu m^2)$	$C_e(\mu m^4)$
293K	1.7055	0.0087	0.0028
303K	1.6933	0.0078	0.0026
313K	1.6711	0.0091	0.0025
323K	1.6395	0.0095	0.0020

5 The temperature characteristics of waveguide dispersion

The PCF dispersion, includes material dispersion (D_m) and waveguide dispersion(D_w). The different structure of the PCF, the same material dispersion, and the size is calculated by the Sellmeier formula. Waveguide dispersion is due to a pattern of the group speed of light's frequency changing with the resulting dispersion, which is defined as:

$$D_w = -\frac{\lambda}{c}\frac{d^2 n_{eff}}{d\lambda^2} \tag{7}$$

By inputting the different wavelengths in the Comol, we can obtain the relationship between the effective refractive index of guide mode and wavelength, then six fitting in Matlab, and draw function of refractive index and wavelength, the results that are based on waveguide dispersion formula are shown in Figure 6. It can be seen that the waveguide dispersion is related to temperature and wavelength. When the wavelength is concerned, the higher temperature is, the bigger is the waveguide dispersion is, which is due to the effective refractive index of the guide film decreases with increasing temperature, decreasing of the refractive of guide mode waveguide dispersion increases. When the temperature is concerned, the liquid crystal waveguide dispersion increases with wavelength increasing.

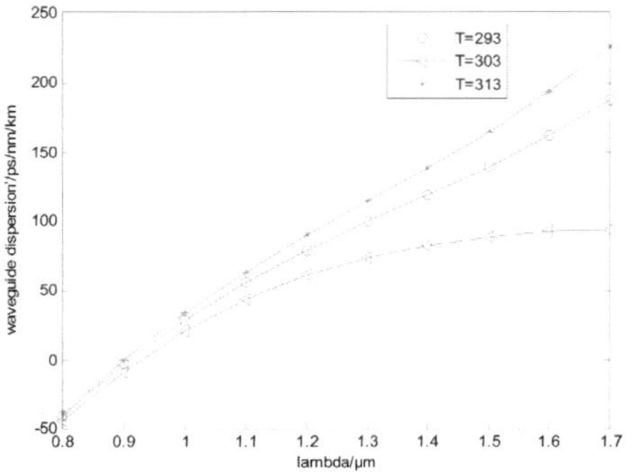

Figure 6 Relationship between the waveguide dispersion and temperature with different duty ratios of PCF after filling LC.

6 Conclusion

By using an all-vector finite element Comsol software, the temperature characteristic of PCF-PBG of hexagon structure arrangement was analyzed. Filling the air hole of the core with LC, the different structure as well as the relationship between the effective refractive index, mode field distribution and waveguide dispersion with temperature changing, were discussed. We find that when the core of photonic crystal fiber is filled with LC, the refractive index is very sensitive to temperature and wavelength, and that the transmission characteristic of PCF is very closely to temperature. Therefore, filling the medium of sensitive refractive index which can modulate optical fiber temperature sensing is feasible. These are valuable for design and application of temperature tunable PCF device.

Acknowledgement

This paper is supported by Aviation Foundation under Grant No.201510281326321

References

1. P.Russell. Science. **299(5605)**: 358-362(2003)
2. T.R.Wolinski.K.SZalliawska.S.Ertman et al. Meas. Sci.Technol. **17(5)**: 985-991(2006)
3. Liu Jie,Yang Changxi,C.Gu et al. Acta Optica Sinca. **26(10)**: 1569-1574(2006)
4. T.Kato,Y.Suetsugu,M.Nishimura. Opt.Lett, **20(22)**: 2279-2281(1995)
5. D.C.Zografopoulos,T.D.Tsiboukis. Opt Exp, **14(2)**: 914-925(2006)
6. Wen Fuzheng, Fan Wande, Li Rucheng et al.Optoelectronic Technology. **26(2)**: 133-137(2006)
7. W.Lin,S.Brugioni et al. Appl.Phys. **17(6)**: 501-505(2005)
8. N.Palka. J.Phys.France, **129**: 143-146(2005)
9. Qian Xiangzhong. Acta Optica Sinica. **28(5)**: 988-991(2008)
10. Li Xuejin. Chinese Journal of Lasers. **36(6)**: 1140-1145(2009)
11. Xu, Hua, et al. Journal of Materials Chemistry C 1.**38**: 6087-6098(2013).

A Fibre Bragg Grating Interrogation Technique Based on High Birefringence Fibre Loop Mirror and WDM

Xun Zhang[1] Shuyang Hu[1,a] and Chuan He[1]

[1]*School of applied Sciences, Beijing University of Technology, Beijing, 100124, China*

Abstract. In this paper, a fibre Bragg grating (FBG) interrogation technique based on high Birefringence fibre loop mirror (Hi-Bi FLM) and wavelength division multiplexer (WDM) is proposed and demonstrated. The approximate linear edge of the transmittance of the Hi-Bi FLM, which is a sinusoidal function of wavelength, is used to interrogate the sensing FBGs and WDMs is used to realize wavelength discrimination .Suitable for both static and dynamic sensing, this interrogation method has the advantages of all fibre design and high stability.

1 Introduction

Fibre Bragg grating has been applied broader and broader because of its many advantages. The critical aspect related to the practical use of an FBG sensor is the necessity of performing accurate measurement of the small wavelength shift associated with thermal and strain state changes. To facilitate the broad use of this class of sensors, compact, rugged, wide range and low cost interrogation systems are required. In addition, improvements in the fabrication of FBGs using phase mask techniques have reduced the cost of grating fabrication so that the interrogation unit, rather than the sensor, accounts for a large proportion of the cost of a complete sensing system .Among the often used technic such as Matched-filter interrogation technique [1] , interferometric Fourier transform technique [2] and the technic using tunable light source[3], the edge filter demodulation technic in which the FBG is interrogated by measuring the power of the output light has the advantages such as high demodulation speed, cheapness, convenience and both suitable for both static and dynamic measurement. On the other hand the optical fiber-loop mirror has been used in several applications, namely in optical communications like a terahertz optical asymmetric demultiplexer [4] or a nonlinear optical loop mirror [5]. In the optical sensor, it has been used as a Sagnac interferometer for temperature measurement [6].
High birefringence fiber loop mirror (Hi-Bi FLM) has been used as edge filter to interrogate the FBG sensors in recent years [7,8]. In 2001, Seunghwan Chung reported a FBG demodulating technic [9] using Hi-Bi FLM with the resolution of 2.12με. Zhouguang reported an edge filter demodulation method [10] using Hi-Bi FLM and demodulation range of 6nm was reached. Although this technic has many advantages such as fast demodulation speed, it can only interrogate one FBG at the same time, which limits its wider application. In this paper, an interrogation technique based on Hi-Bi FLM and WGM which can interrogate multiple FBG sensors is studied.

2 Operational principle

2.1 High-birefringnece fiber loop mirror

High-birefringence fiber loop mirror is composed of a section of high-birefringence fiber, a polarization controller and a 3 dB coupler, as shown in Figure1.The incident beam splits into two beams through the 3 dB coupler. Then two beams interfere at the output port. The transmittance function is[11] :

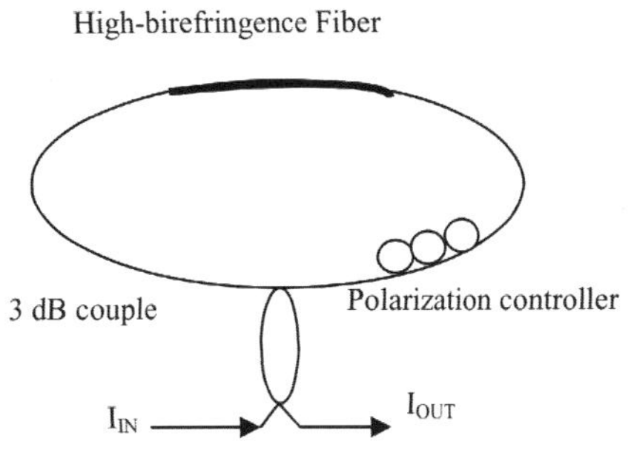

Figure1 Schematic diagram of high-birefringence fiber loop mirror

$$T = [\cos\frac{\pi LB}{\lambda}\sin\theta]^2 \qquad (1)$$

[a] Shuyang Hu: hushuyang@bjut.edu.cn

where L is the length of HB fiber, λ is wavelength, $B = \Delta n = |n_e - n_o|$, is the difference of the index along fast and slow axes, θ (is a certain constant when Hi-Bi FLM having been installed)is the angle change of the polarization of the light propagating through the fiber loop, respectively. When $\frac{\Delta\lambda}{\lambda_0} \ll 1$, we get $\frac{\pi L B}{\lambda} \approx \frac{\pi L B \lambda}{\lambda_0^2} = \beta\lambda$, where $\beta = \frac{\pi L B}{\lambda_0^2}$, $\Delta\lambda$ and λ_0 is the wavelength range and center wavelength of the source, respectively. We simplify (1) as:

$$T = \frac{1+\cos\beta\lambda}{2}\sin^2\theta \qquad (2)$$

From (2), we know that the transmission of HiBi-FLM is a sine function of λ with the period of λ_0^2/LB, which is decided by L and B. The transmission of the f HiBi-FLM we fabricated is shown in Figure 2, in which the curve A, B and c is the spectrum of light source, the light transmitting through the FLM and the transmittance, respectively.

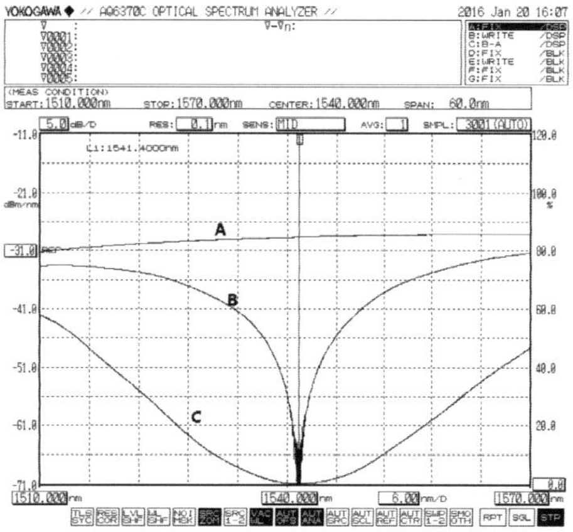

Figure2 The transmission spectrum of the Hi-Bi FLM

2.2 Interrogation principle

When the narrow band light reflecting by the FBG whose spectrum band width is narrow comparing with the period of transmission and its spectrum can be treated as δfunction is launched into HiBi- FLM, the intensity of the light transmitted through is given by

$$I_t = I_i T(\lambda_B) = I_i \frac{1+\cos\beta\lambda_B}{2}\sin^2\theta \qquad (3)$$

Where I_i and λ_B is the intensity and wavelength of reflected light of FBG, respectively. So, the transmittance of the light reflectting from the sensing FBG and passing through the Hi-Bi FLM can be given as

$$\eta = \frac{I_t}{I_i} = \frac{1+\cos\beta\lambda_B}{2}\sin^2\theta \qquad (4)$$

From (4) we know that we can interrogate the FBG by measuringη. Especially when the transmittance is in the nearly linear range the Hi-Bi FLM can be treated as a linear filter.

The Hi-Bi FLM and WDM based FBG interrogation system is carried out with the setup shown in Figure 3.The wide band light from SLED passes through coupler1 and enter the sensing FBGs. The reflecting light goes through coupler1 and is split into two beams via coupler 2. One of the beam goes through Hi-Bi FLM and WDM successively while the other beam only goes through the other WDM the same as the first. Eventually the light reflecting from different sensing FBGs is separated into different optical detector. Via measuring the corresponding beam going through FLM or not, the transmittance (η) of FBGs can be calculated out and λ_B ,the wavelength of FBGs can be determined.

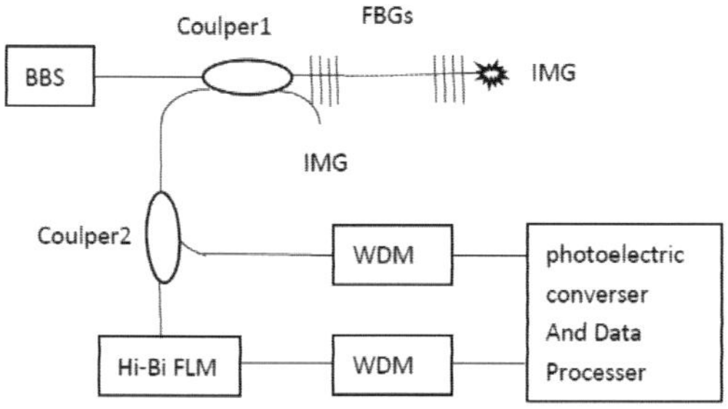

Figure 3 Schematic of the interrogation system

3 Experimental Results

In the experiment, CWDM is used to demonstrate the technique whose transmission spectrum is shown in Figure 4,in which P1 and P2 is the transmission spectrum of port1 and port3,respectively.SLED is used as light source and Hi-Bi FLM is made and the spectrum of them are both shown in Figure.1. Two FBGs with the center wavelength of 1520nm and 1550nm are used and the spectrum of their reflecting light are shown in Figure 5.

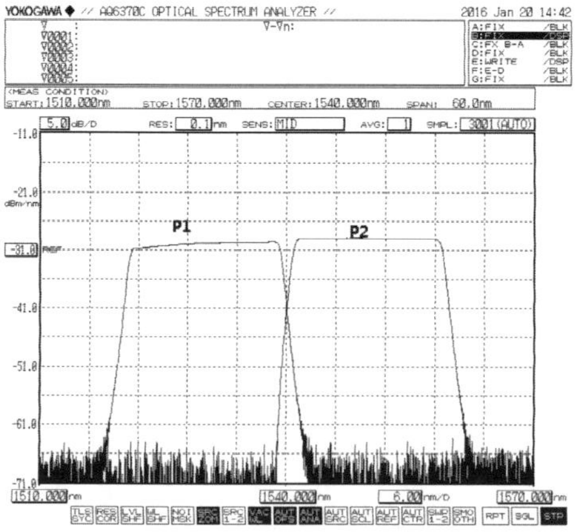

Figure 4 The spectrum of light passing through the WDM

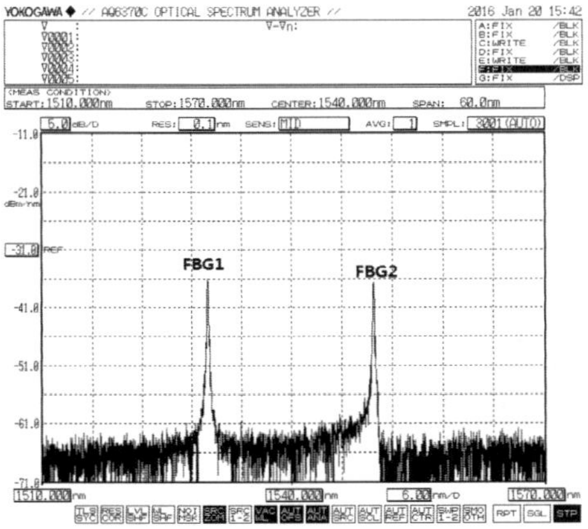

Figure 5 The reflective spectrum of the FBGs

The reflecting light of FBGs go through Hi-Bi FLM and WDM are separated into different detector and the two spectrum of FBG1 and FBG2 are shown in one picture in Figure 6. The wavelength of the FBGs can be interrogated by the value of η which can be calculated by comparing the corresponding power of the light passing through Hi-Bi FLM or not.

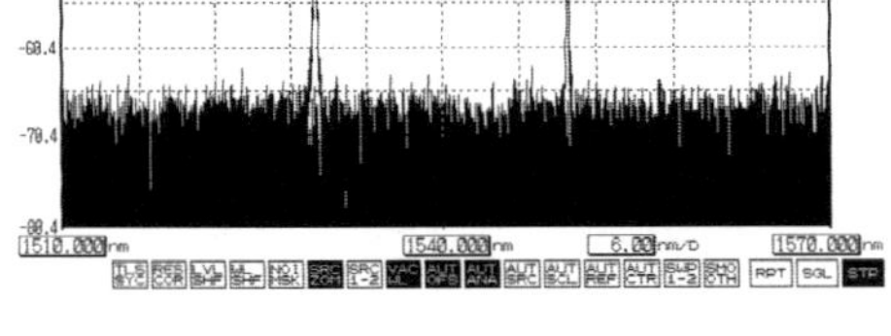

Figure.6 The reflective spectrum of the FBGs passing through WDM and Hi-Bi FLM

We adhere FBG1 to the equal deform beam to tune it which is shown in Figure 7. The beam is placed horizontal with one end is fixed while the other is free. So, we can tuning the wavelength of FBG by loading weight to the free end.

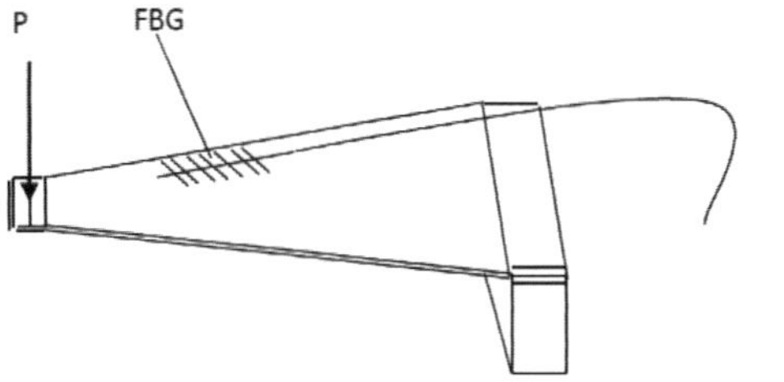

Figure 7 The schematic of wavelength tuning of FBG using equal deform beam.

By tuning the weight the transmission (η)varies with wavelength shift and the relationship between them is shown in Fig.8, with R^2=0.9938.

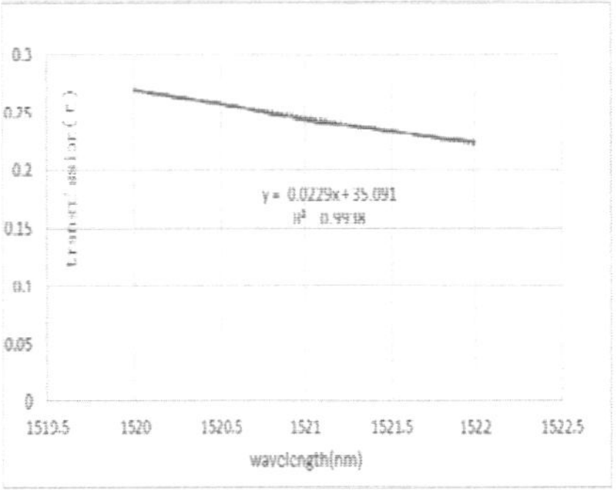

Figgure 8 Relationship between transmission and wavelength of FBG1

4 Conclusions

In this paper, a FBG interrogation technique based on Hi-Bi FLM and WDM is proposed and demonstrated. With the capability of measure multiple FBG signal at the same time based on WDM, the technique has the advantages of fast demodulation, simple structure and suitable for both static and dynamic measuring.

References

1. Davis M A , Kersey A D. Matched-filter interrogation technique for fiber Bragg grating arrays Electro. Lett, 1995, 31(10):822 ~ 825
2. Ohn M M , Huang S Y , Measures R M, et al.. Arbitrary strain profile measurement within fiber gratings using interferometric Fourier transform technique Electron. Lett , 1997, 33(14):1242 ~ 1243
3. Br ady G , Kalli K , Webb D J, et al, Simultaneous interrogation of interferometric and Bragg grating sensors Opt .Lett ., 1995, 20(11):1340 ~ 1342

4. J.P. Sokoloff, P.R. Prucnal, I. Glesk, and M. Kane, A terahertz optical asymmetric demux (TOAD), IEEE Photon Tech Lett 5 (1993), 787 – 790.
5. N.J. Doran and D. Wood, Nonlinear-optical loop mirror, Opt Lett 13 (1988), 56 – 58.
6. A.N. Starodumov, L.A. Zenteno, D. Monzon, and E. De La Rosa, Fiber Sagnac interferometer temperature sensor, Appl Phys Lett 70 (1997),19 – 21.
7. MORTIMORE D B, Fiber loop reflectors J. Lightwave Technol., 1988, 6(7):1217-1224.
8. Zhou Guang, Zhao Qida, Liu Yange, et al., Filtering Characteristics of fiber loop mirror filter composed of multi-stage high birefringent fibers Acta Optica Sinica 2004, 24(3):341-345.
9. SEUNGHWAN CHUNG, JUNGHO KIM, BONG-AHN YU, et al., A Fiber Bragg Grating Sensor Demodulation Technique Using a Polarization Maintaining Fiber Loop Mirror IEEE Photonics Technology Letters, 2001, 3(2):1343-1346.
10. Zhou Guang, Zhao Qida,Kai Guiyun, et al., A novel edge filter demodulation method using high birefringence fiber loop mirror, Journal of Optoelectronics Laser, 2003, 14(12):1245-1249.
11. Yange Liu, Bo Liu, Xinhuan Feng, et al, High-birefringence fiber loop mirrors and their applications as sensors, Applied Optics, 44(12): 2382-2390(2005)}

MATEC Web of Conferences **61**, 06011 (2016) DOI: 10.1051/matecconf/20166106011
APOP2016

Threshold Characteristics Analysis of Dual-end-pumped Nd^{3+}-doped Gain-guided and Index-antiguided Fiber Lasers

Xiao Shen[1, 2], Wei Wei[1,a]

[1]College of Optoelectronic Engineering, Nanjing University of Posts and Telecommunications, Nanjing 210023, PR China
[2]State Key Laboratory of Transient Optics and Photonics, Xi'an Institute of Optics and Precision Mechanics, Chinese Academy of Sciences,Xian,710119,Shanxi,China

Abatract. An analytical model for dual-end-pumped Nd^{3+}-doped gain-guided and index-antiguided (GG-IAG) fiber laser is established. The corresponding analytical expression of the threshold pump power is obtained by solving the improved rate equations. Pump light propagation and threshold characteristics are both researched. Simulation results show that there are optimum values of N, a and L for the lowest P_{th}. When a=50μm, L=9cm, N=1.8*10^{20}cm^{-3}, R_{1s}=1, R_{2s}=0.75, the lowest threshold pump power P_{th} of the GG-IAG fiber laser is about 11.95W, the results of the work may be helpful for the later experiments.

1 Introduction

High power fiber lasers have many important applications in industry, defense and commercial fields. However, the major limitations to high power lasers in conventional single mode fibers are undesired optical nonlinear effects [1-4]. The most direct method to solve these problems is increasing the mode area to reduce the optical power density [5], thus large mode area single-mode fibers have recently attracted a lot of attentions, such as photonic crystal fibers [6], chirally coupled core fibers [7-8] and so on. However, the maximal obtainable mode field diameter (MFD) is still less than 100μm[3].In 2003,A.E. Siegman proposed a new type of gain-guided and index-antiguided (GG-IAG) fiber with large mode area[9-10], and the maximal MFD reported is about 400 μm[1]. In this paper, we report an analytical model for a dual-end-pumped Nd^{3+}-doped GG-IAG fiber laser, and an exact analytical expression of the threshold pump power for this fiber laser has been obtained by solving the improved rate equations (REs) with the additional leakage losses of the signal light. The effects of each parameters of the fiber on the threshold pump power are also discussed.

2 Theoretical analyses

The structure of the GG-IAG fiber is showed in fig.1. The refractive index of the fiber cladding n_2 = n_0, the refractive index difference of the core and the cladding is Δn (Δn<0), the power gain factor of the signal light is g.

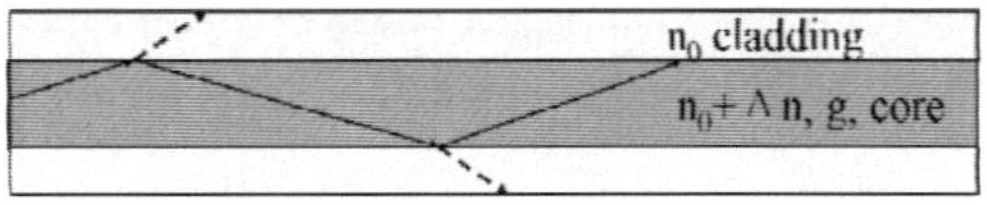

Fig.1 The schematic diagram of GG-IAG fiber

The GG-IAG fiber laser is as shown in Fig.2. Dual-end-pumped model means that pump light $P_p^+(0)$ and $P_p^-(L)$ are injected into the fiber simultaneously from both ends with $R_{1p} = R_{2p} = 0$. R_{1s} and R_{2s} are the reflectivity of the signal light. Forward and backward

[a]corresponding author: weiwei@njupt.edu.cn

pump light power are $P_p^+(z)$ and $P_p^-(z)$, respectively. Forward and backward signal light power are $P_s^+(z)$ and $P_s^-(z)$, respectively.

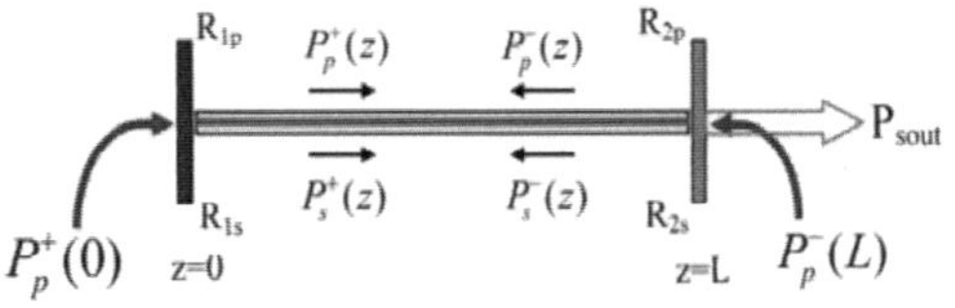

Fig.2 Schematic diagram of GG-IAG fiber laser

There are leakage losses of signal light and pump light, then according to the literature [11], a set of improved REs. of Nd^{3+} doped dual-end-pumped GG-IAG fiber laser can be expressed as formulas (1), (2) and (3):

$$\frac{N_2(z)}{N}=\frac{\dfrac{[P_P^+(z)+P_P^-(z)]\cdot\sigma_{ap}\Gamma_P}{h\nu_P A}+\dfrac{[P_s^+(z)+P_s^-(z)]\cdot\sigma_{as}\Gamma_S}{h\nu_s A}}{\dfrac{[P_P^+(z)+P_P^-(z)]\cdot\sigma_{tp}\Gamma_P}{h\nu_P A}+\dfrac{1}{\tau}+\dfrac{[P_s^+(z)+P_s^-(z)]\cdot\sigma_{ts}\Gamma_S}{h\nu_s A}} \tag{1}$$

$$\pm\frac{dP_P^\pm(z)}{dz}=\Gamma_p\left[\sigma_{tp}N_2(z)-\sigma_{ap}N\right]P_P^\pm(z)-\alpha_{tp}P_P^\pm(z) \tag{2}$$

$$\pm\frac{dP_s^\pm(z)}{dz}=\Gamma_s\left[\sigma_{ts}N_2(z)-\sigma_{as}N\right]P_s^\pm(z)-\alpha_{ts}P_s^\pm(z) \tag{3}$$

Where, N_2 *(z)* is the upper lasing level population density, N is the concentration of dopant Nd^{3+}. Γ_p and Γ_s are the pump and signal field overlaps, respectively. ν_p and ν_s are the pump and signal light frequencies, respectively. A is the cross-sectional area of the core. τ is the spontaneous lifetime of the upper-level atoms and the parameter h is Planck's constant. The parameter c is the speed of light in vacuum. $\sigma_{tp}=\sigma_{ap}+\sigma_{ep}$, σ_{ap} and σ_{ep} are absorption and emission cross sections of pump light, respectively. $\sigma_{ts}=\sigma_{as}+\sigma_{es}$, σ_{as} and σ_{es} are absorption and emission cross sections of signal light respectively. $\alpha_{tp}=\alpha_p+l_p$, $\alpha_{ts}=\alpha_s+l_s$, where α_p and α_s represent the fundamental material absorption coefficients of the pump and signal light respectively, l_p and l_s represent the additional leakage losses of pump and signal light respectively, and they are[12]:

$$l_p=g_{01}^{th}=[\lambda_p^2/(2\pi)^2a^3]\sqrt{133.8/(-2n_0^3\Delta n)} \tag{4}$$

$$l_s=g_{01}^{th}=[\lambda_s^2/(2\pi)^2a^3]\sqrt{133.8/(-2n_0^3\Delta n)} \tag{5}$$

Eq. (3) can be derived as:

$$\frac{1}{L}\int_0^L N_2(z)dz=\frac{-\dfrac{\ln(R_{1s}R_{2s})}{2L}+\Gamma_s\sigma_{as}N+\alpha_{ts}}{\Gamma_s\sigma_{ts}} \tag{6}$$

From Eq. (2), Eqs. (7) and (8) can be obtained:

$$P_p^+(z)=P_p^+(0)\cdot\exp(\beta g_{th}-\gamma)z \tag{7}$$

$$P_p^-(z)=P_p^-(L)\cdot\exp(-\beta g_{th}+\gamma)(z-L) \tag{8}$$

Where: $\beta=\Gamma_p\sigma_{tp}/\Gamma_s\sigma_{ts}$, $\gamma=\Gamma_p\sigma_{ap}N+\alpha_{tp}$

Let: $P_{s,sat}=\dfrac{h\nu_sA}{\tau\Gamma_s(\sigma_{es}+\sigma_{as})}$, $P_{p,sat}=\dfrac{h\nu_pA}{\tau\Gamma_p(\sigma_{ep}+\sigma_{ap})}$, $K=\dfrac{1}{L}\int_0^L N_2(z)dz$.

Then, Eq. (1) can be expressed as:

$$\frac{P_s^+(z)+P_s^-(z)}{P_{s,sat}}\cdot\frac{N_2(z)(\sigma_{as}+\sigma_{es})-N\sigma_{as}}{N(\sigma_{as}+\sigma_{es})}=-\frac{K}{N}+\frac{P_p^+(z)+P_p^-(z)}{P_{p,sat}}\cdot\frac{N\sigma_{ap}-N_2(z)(\sigma_{ap}+\sigma_{ep})}{N(\sigma_{ap}+\sigma_{ep})} \tag{9}$$

Combining Eqs. (2) and (3), the following equation can be derived:

$$\frac{\tau}{h\nu_sA}\left\{\frac{dp_s^+(z)}{dz}-\frac{dp_s^-(z)}{dz}+\alpha_{ts}[p_s^+(z)+p_s^-(z)]\right\}+\frac{\tau}{h\nu_pA}\left\{\frac{dp_p^+(z)}{dz}-\frac{dp_p^-(z)}{dz}+\alpha_{tp}[p_p^+(z)+p_p^-(z)]\right\}+K=0 \tag{10}$$

From Eq. (3), the following conclusion can be obtained:

$$P_s^+(z)\cdot P_s^-(z)=P_s^+(0)\cdot P_s^-(0)=P_s^+(L)\cdot P_s^-(L)=\text{Constant} \tag{11}$$

The boundary conditions of dual-end-pumped fiber laser are as followings:

$$P_s^+(0)=R_{1s}P_s^-(L) \tag{12a}$$

$$P_s^-(L)=R_{2s}P_s^+(L) \tag{12b}$$

Combining Eqs. (11) and (12a-b), after Eq. (10) is integrated, the threshold pump power can be derived as following:

DOI: 10.1051/matecconf/20166106011

$$P_{th} = P_p^+(0) + P_p^-(L) = P_{s,sat} g \frac{(N\Gamma_s \sigma_{as} + \alpha_{ls})L + \ln\frac{1}{\sqrt{R_{1s}R_{2s}}}}{\frac{v_s}{v_p} g(1-\frac{\alpha_{tp}}{\alpha}) g(1-\exp(-\alpha \cdot L))} \quad (13)$$

Where, $\alpha = N\Gamma_p \sigma_{ap} - \Gamma_p(\sigma_{ap} + \sigma_{ep})KL + \alpha_{tp}$.

3 Results and discussions

All the parameters used in the calculations are given in table1.

Table 1. The parameters of the Nd3+-doped GG–IAG fiber laser.

PRMs	value	PRMs	value
Γ_p	0.25	τ	1.9×10^{-4} s
Γ_s	0.6	σ_{ap}	0.7×10^{-20}cm^2
λ_s	1054nm	σ_{ep}	3.76×10^{-22}
λ_p	803nm	σ_{as}	7.82×10^{-22}c
L	0-15cm	σ_{es}	4.4×10^{-20}
a	40-100μm	α_p	0.016cm^{-1}
n_0	1.5734	α_s	0.016cm^{-1}
Δn	-0.0045	R_{1s}	1
N	1×10^{20}-1×10^{21}cm^{-3}	R_{2s}	0.75
h	6.6260775×10^{-34}J·s	R_{1p}, R_{2p}	0

3.1 The pump light propagation characteristics

The forward and backward pump light can be expressed as $P_p^+(z)$ and $P_p^-(z)$ respectively, and the total pump power in the fiber core is $P_p^+(z)+P_p^-(z)$. Supposing $P_p^+(0)=P_p^-(L)=10W$, then the pump light distribution in the fiber core is shown in fig.3. $P_p^+(L)$ and $P_p^-(0)$ are very low when fiber length L is 9cm, and at the middle of the fiber the total pump power is 2.835W.

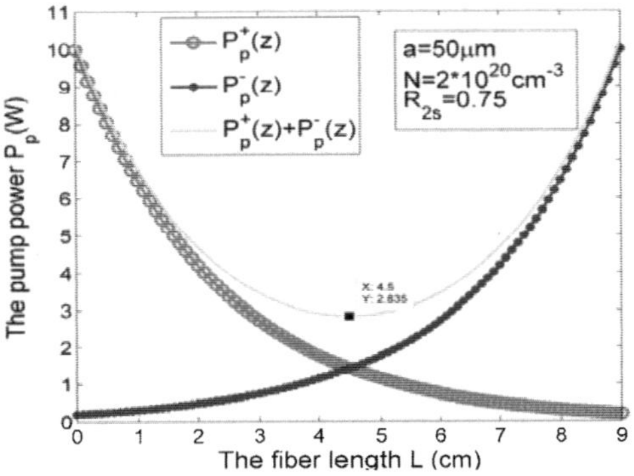

Fig.3 P_p distribution along the fiber length

Supposing that the maximal (minimal) pump power along the whole fiber is P_{pmax} (P_{pmin}), respectively. The allowed fractional variation in the pump light distribution in the fiber core should be below 61% [4] for GG-IAG fiber to maintain single mode oscillation, and which is called uniformity of pump light distribution here and can be defined as: $P_u = (P_{pmax}-P_{pmin})/P_{pmax} \times 100\%$ [4]. P_u changes process with the fiber length L is shown in Fig.4. P_u increases with the increasing of L. When P_u is 61%, the allowed maximal fiber length L is about 9.3 cm, which is shown in Fig.4.

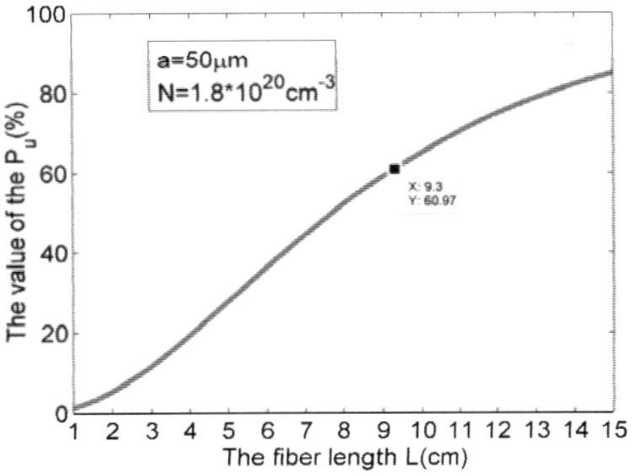

Fig.4 *Pu* changes with the fiber length

DOI: 10.1051/matecconf/20166106011

3.2 Threshold pump power characteristics

Supposing the threshold pump power is P_{th}, which changes with N is shown in Fig.5 and Fig.6. It can be found that in a certain concentration range, the threshold pump power is lower and keeps nearly invariable, but when the concentration is beyond this range, the threshold increases higher. The reason is when N is lower, it is difficult to supply enough gain to overcome the losses, and when N is too high, the pump power is absorbed fast at the two ends of the fiber, so both make the threshold power increase. In this paper, when L=9cm.a=50μm and R_{2s}=0.75, the minimum of P_{th} is about 11.95W, the corresponding concentration N is about 1.8×10^{20}cm^{-3}.

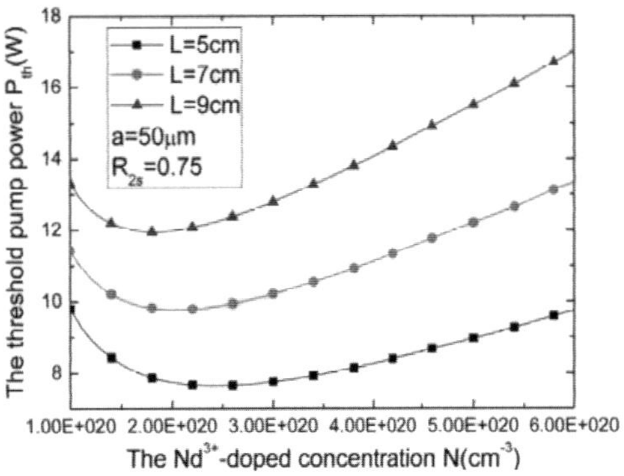

Fig.5 The relationship between P_{th} and N with different L

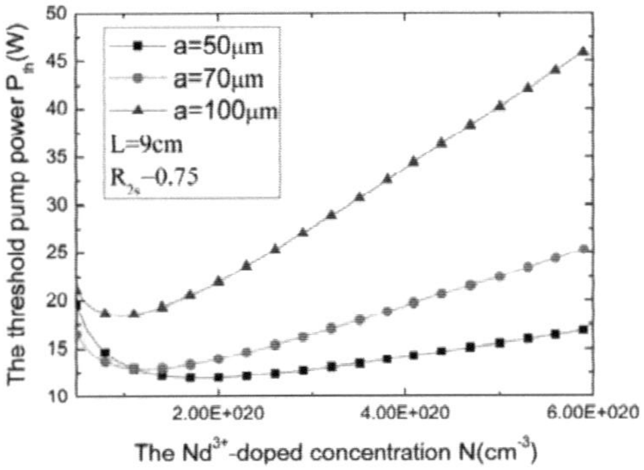

Fig.6 The relationship between P_{th} and N with different a

P_{th} changes process with L is shown in Fig.7. Fig.7 shows that the threshold pump power is lowest with the optimum fiber length, when the fiber length L is too long or too short, the P_{th} will increases correspondingly.

Because when the fiber is short, the gain is too small to satisfy the oscillation conditions, so the threshold becomes higher. When the fiber is long, the loss of the fiber becomes big, so the threshold becomes higher.

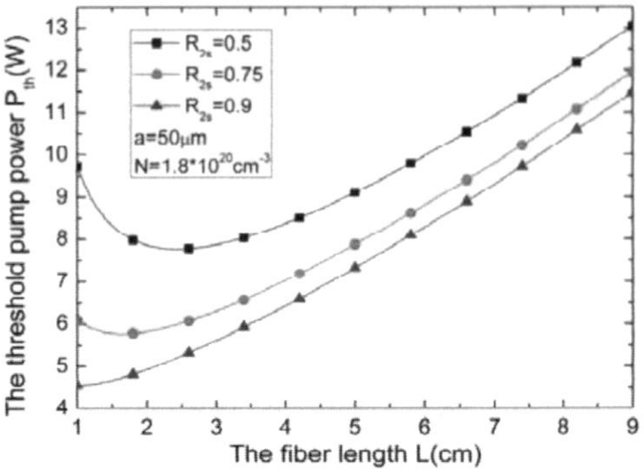

Fig.7 The relationship between P_{th} and L with different R_{2s}

P_{th} changes process with fiber core radius a is shown in Fig.8. The optimum fiber core radius a is about 50-55μm with different R_{2s} when L is 9 cm.

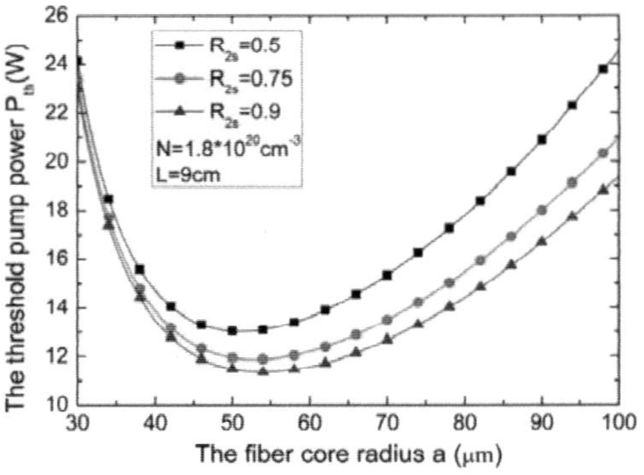

Fig.8 The relationship between P_{th} and a

4 Conclusions

A dual-end-pumped Nd^{3+}-doped GG-IAG fiber laser is analyzed in this work. Pump light propagation and threshold characteristics are both researched. Under the requirement of the uniformity of pump power along the fiber [4], the maximal fiber length is 9.3cm. The relationship between threshold pump power P_{th} and doped concentration N, fiber length L, fiber radius a is researched respectively. Simulation results show that there are optimum value of N, a and L where P_{th} is the lowest. When a=50μm, L=9cm, N=1.8*10^{20}cm^{-3}, R_{1s}=1, R_{2s}=0.75, the lowest threshold pump power P_{th} of GG-IAG fiber laser is about 11.95W, the results of the work may be helpful for the later experiments.

Acknowledgments

This work is supported by the National Natural Science Foundation of China (Nos. 61077070 and 61177086), the Jiangsu Province Universities Graduate Student Research and Innovation Program of China (No. CXZZ13_0467), the Open Research Fund of State Key Laboratory of Transient Optics and Photonics, Chinese Academy of Sciences (No. SKLST201405) and Research Fund of Nanjing University of Posts & Telecommunications (No. NY214125).

References

1. Y. Chen, T. McComb, V. Sudesh, M. Richardson, and M.Bass.Opt. Lett. **32**, 2505(2007)
2. A. E. Siegman, Y. Chen, V. Sudesh, M. C. Richardson, M.Bass, P. Foy, W. Hawkins, and J. Ballato.Appl. Phys. Lett. , **89**, 251101(2006)
3. V. Sudesh, T. McComb, Y. Chen, M. Bass, M. Richardson, J.Ballato, and A. E. Siegman. Appl. Phys. B, **90**, 369(2008)
4. W. Hageman, Y. Chen, X. Wang, L. Gao, G. U. Kim, M.Richardson and M. Bass. J. Opt. Soc. Am. B, **27**, 2451(2010)
5. H.S. Kim, T. McComb, V. Sudesh, M.C. Richardson. *Conference on Lasers and Electro-Optics/Quantum Electronics and Laser Science Conference*, **1-9**, 2581(2008)
6. P. Russell, Science, **299**, 358(2003)
7. C.H. Liu, G.Q. Chang, N.Litchinitser, D.Guertin, N.Jacobsen, K.Tankala, A.Galvanauskas. *Conference on laser & electro-optics/quantum electronics and laser science conference*, **1-5**,765 (2007)
8. A. Galvanauskas, M. C. Swan and C.H. Liu. *Conference on Lasers and Electro-Optics/Quantum Electronics and Laser Science Conference*, **1-9**,356 (2008)
9. A. E. Siegman. J. Opt. Soc. Am. A, **20**, 1617 (2003)
10. A. E. Siegman. J. Opt. Soc. Am. B, **24**, 1677 (2007)
11. I. Kelson, A. A. Hardy. **34,** 1570(1998)
12. E.Y. Zhou, B.Y. Zhao, X.R. Wang, Y.Wang, W.Wei, B.Peng. Applied Physics B, **99,** 747 (2009)

A demodulation method of Temperature-Spectrum width based on Chirped fiber grating*

Si Chen[1], Xiaogang chen[1], Zhihui Luo[1,a] and Desuo Cai[1,2]

[1]*Colleges of Science, China Three Gorges Univ. Yichang 443002, China*
[2]*Guangxi Water & Power Design Institute,Nanning,GuangXi，530023，China*

Abstract. Based on spectral reflection characteristics of chirped FBGs, a demodulation method for complex temperature fields is proposed in this paper. Relationships between the reflective spectrums and the temperatures are investigated through theoretical analyses. Testing results are acquired from a FBG sensing model which is designed according to the newly-developed method. A simulation to get the relationships between reflective spectrums and temperatures is set which shows that this demodulation method is feasible and its demodulation precision is beyond +/-1°C. The method provides a new method for analyzing complex temperature fields.

1 Introduction

FBG sensors are widely used to monitor the safety of petro chemistry industry, bridges and large equipments owing to their considerable advantage such as anti-interference, wavelength encoding and so on. The key physical parameters of the FBGs monitored are temperature. When the temperature field on FBGs changes, the grating period will change because of coaction from the temperature expansion effect, the thermo-optic effect and the photoelastic effect resulted from internal thermal strains, and arises from a corresponding shift of the central Bragg wavelength. How to detect the Bragg tiny wavelength-shift is a major technology for FBG sensor in engineering projects. Many conventional strategies such as edge-wave based on tunable laser, tunable Fabry-Derot filter or unbalance Mach-Zehnder interference have been proposed[1][2]. However, above methods assume that FBG was loaded under uniform temperature field, and the demodulator should lock wavelength-shift accurately and process optical signal. Sensing systems based on above methods become complex, because of having some loss of accuracy for measuring imperfect temperature fields. In this paper, a novel demodulation method for complex temperature fields is proposed, a spectral reflection characteristic of chirped FBGs is analyzed intensively, an FBG sensing model using professional software is designed and this demodulation scheme by numerical simulation is verified.

2 Principle of the Temperature-spectrum Width demodulation

2.1 Theories analysis

In order to make the refractive index of fiber core higher than in cladding, the metal elements such as germanium (Ge) mixed in the fiber core and the cladding is pure quartz. Therefore the fiber core possesses photosensitive properties.

Usually, the period of uniform FBG is a constant. When FBG is loaded under complex temperature field [3] [4], gratings in different positions inside it will change in a varying manner, changing it into a chirped FBG (CFBG) [5]. Compared with the bandwidth of different pitchs, the gap between their reflective spectrums is smaller, so all spectrums will overlap with each other to form a whole broadening reflective spectrum[6][7][8]. The increment of the whole reflective spectrum is $\Delta\lambda_{BW}$, which is the difference between the maximal increment $\Delta\lambda_{\max}$ and the minimum increment $\Delta\lambda_{min}$ of resonant wavelength in different pitchs [9] [10] [11]:

$$\Delta\lambda_{BW} = \Delta\lambda_{\max} - \Delta\lambda_{\min} \tag{1}$$

Given that uniform temperature field is loaded without strain resulted from external forces，taking no account of the effect of thermal expansion and other physical effects, only axial temperature changes grating pitchs and leads the shift of Bragg wavelength, so of the shift value may be given bellow[6]:

$$\Delta\lambda_B = 2\left[\frac{\partial n_{eff}}{\partial T}\Delta T + (\Delta n_{eff})_{ep} + \frac{\partial n_{eff}}{\partial a}\Delta a\right]\Lambda + 2n_{eff}\frac{\partial \Lambda}{\partial T}\Delta T \tag{2}$$

where n_{eff} is the effective index of refraction of the core of single mode fiber(SMF); ΔT is the difference of temperatures in different positions; Δa is the increment

[a] Corresponding author: zhihui_luo@126.com

of fiber core diameter; $(1/n_{eff})\dfrac{\partial n_{eff}}{\partial T}$ is the thermo-optic coefficient α_n of SMF; $(\Delta n_{eff})_{ep}$ is the photoelastic coefficient resulted from thermal expansion; $\dfrac{\partial n_{eff}}{\partial a}$ is the waveguide effect resulted from core diameter change induced by thermal expansion; $\dfrac{1}{\Lambda}\dfrac{\partial \Lambda}{\partial T}$ is the linear coefficient α_Λ of thermal expansion. Formula (2) may be simplified as follows:

$$\frac{\Delta\lambda_B}{\lambda_B \Delta T} = \frac{1}{n_{eff}}\left[n_{eff}\alpha_n + \frac{n_{eff}}{\Delta T}(\Delta n_{eff})_{ep} + \frac{\partial n_{eff}}{\partial a}\frac{\Delta a}{\Delta T} \right] + \alpha_\Lambda \qquad (3)$$

Given that SMF is an isotropy structure, obtain the increment of FBG's bandwidth resulted from both the photoelastic effect and the waveguide effect by analyzing strain sensing model[6],Formula (4) of the temperature sensibility coefficient is deduced as follows:

$$\frac{\Delta\lambda_B}{\Delta T} = \left\{ \frac{1}{n_{eff}}\left[n_{eff}\alpha_n - \frac{n_{eff}^3}{2}(P_{11}+2P_{12})\alpha_\Lambda + S_{wg}\frac{\Delta a}{\Delta T} \right] + \alpha_\Lambda \right\}\lambda_B \qquad (4)$$

where S_{wg} is the shift coefficient of Bragg wavelength by the waveguide effect. When the material of SMF is given, the increment of FBG's bandwidth is correlated with the temperature coefficient and the material coefficient. For SiO2 fiber, S_{wg} is out of account, α_n is $8.6X10^{-6}/{}^\circ C$, α_Λ and is $5.5X10^{-7}/{}^\circ C$ .When the central wavelength of FBG is 1.55um, the ratio of $\Delta\lambda_B$ to ΔT is $10.8pm/{}^\circ C$ theoretically. When a uniform FBG is loaded by different temperatures along axial direction, according to Formula (4), the increment on the whole bandwidth is proportional to the difference between increments of both ends of FBG. So Formula (5) is given bellow:

$$\Delta\lambda_{BW} = \Delta\lambda_{B2} - \Delta\lambda_{B1} \approx 10.8(\Delta T_{max} - \Delta T_{min}) \qquad (5)$$

For nonuniform temperature fields such as linear increment fields or Gaussian fields, if variations of the temperature on FBG are continuous, Formula (5) is universal. But the coefficient of 10.8 should be modified according to specific temperature fields and parameters of FBGs. The following discussion will focus on nonuniform linear field, and at the same time given that the temperature at the start point of FBG is a basic value and increases along the axial of FBGs gradually[4][5].

2. 2 the system model

According to the above discussion, there is a linear relationship between the increment of the reflective spectrum width and the temperature difference on both ends of FBG, theoretically. The key point of the demodulation of the temperature is how to measure the increment. In related papers [1] [2] some methods such as complex circuits or optical interference are discussed to detect the increment of the spectrum width, but they are difficult to implement. In order to simplify the spectrum width measuring, the sensor system is designed as Figure 1. Choosing the tunable SOA laser with the narrow spectrum as the optical source, the narrow optical pulses from the source pass through the optical circulator and launch into SMF. After arriving at the FBG, pulses are reflected within the range of the FBG reflective spectrum and the others transmit away. Reflective pulses transmit backward in the same SMF and pass through the circulator into the APD detector, and they are converted into electronical signals. The control device makes a high-speed decision according to the signal power and counts the number of qualified signals, and sends control signals to the pulse generator so that the laser is adjusted in time. At the same time, the control device shares datum with the connected computer which displays the number of qualified pulses and calculates the spectrum width by a given method. Usually, calibrating the sensor should be considered at first. Then calculating the temperature value according to Formula (5), the demodulation of the temperature is fulfilled.

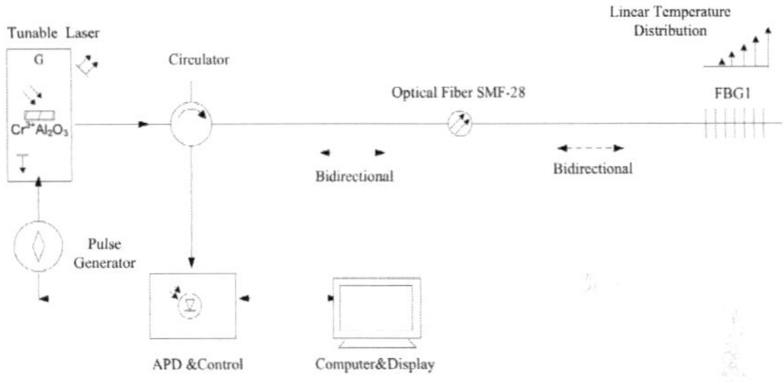

Figure 1. Schematic diagram of FBG sensing system

3 The simulation design

OptiGating is a professional software based on Coupled Mode Equations to design and simulation FBG. Design the uniform FBG using OptiGraing as follows: the central wavelength is 1550nm, the radius of the core/cladding is 4.15um/62.5um and their refractive indices are 1.46/1.45, the refraction index modulation depth is 0.0005, the period is 0.5138um, the refraction of gate region conforms with sinusoidal distribution and the total length is 4.8cm[12][13][14]. Using the embedded sensor module in OptiGrating, we set the thermo-optic coefficient α_n to be $8.6X10^{-6}/{}^\circ C$ and the linear expansion coefficient α_Λ $5.5X10^{-7}/{}^\circ C$, and the ambient temperature $25^\circ C$ .The temperature field is distributed linearly along the axial, and then a FBG model which may be dragged freely is obtained. Its physic effects are simulated more closely using the high-capacity numerical calculation. Create the sensing platform of temperature-spectrum width in OptiSystem, a professional optical communication software, as Figure 1 and transplant the FBG model from OptiGrating to build a virtual FBG sensing system as Figure 2. We set the output power of the cw laser with iterative sweeping function to 0 dbm and the bandwidth to 10MHz, and the number of sweeping steps according to specific sites. We choose G.652 SMF to transmit the optical signal and set its loss coefficient to 0.2dB/km and the length to 0.5km. After arriving at FBG, the optical signal within spectrum width was reflected and it transmits 0.5km into detector ADP. In order to collect real-time datum easily and observe directly, we use the virtual instruments such as optical power meter, optical spectrum analyzer, oscilloscope visualize and so on which are found in the OptiSystem toolbox that performs signal detecting and extracting the power and optical spectrums.

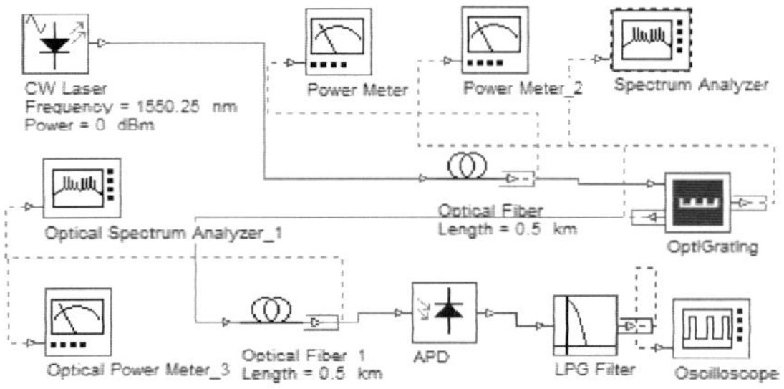

Figure 2. Model of FBG sensing system

Set parameters of the laser as follows: the tunable range of wavelength ranges from 1549.505nm to 1550.50nm with the step scanning of 0.005nm, and the linear increment of temperature is 1°C from 25°C to 35°C. Run the simulation and get the output spectrum of APD as Figure 3 (which only shows the spectrogram at 25° C and 35° C) , the number of pulses whose power are among Full Wave at Half Maximum (FWHM)of the pulse envelope is 15、17、18、20、22、24、27、29、32、35 and 37 respectively.

During simulation, the reflective spectrum broadens with the increase of temperature, and at the same time the central wavelength of FBG shifts. Moreover, both edges of spectrums fluctuate randomly as a result of laser side lobes and optical line bandwidth, and this instability will lead to some error during counting qualified output pulses. In order to make the reflective spectrums homogenized, the new technology of tan apodization method is adopted to suppress laser side lobes effect and improve the measuring accuracy. Figure 4 is reflective spectrums of apodized FBG, and all edges become steady. We run simulation and count the number of qualified pulses as follows: 11、12、13、14、16、17、18、21、23、25、27.

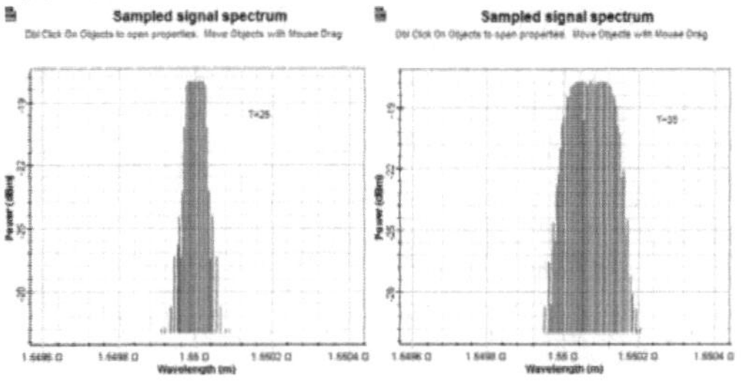

Figure 3. Spectrums of FBGs at 25°C、35°C

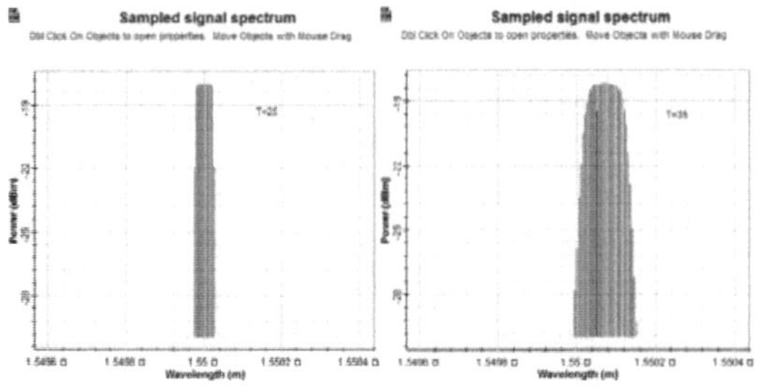

Figure 4. Spectrums of apodized FBG in 25°C、35°C

Analyzing two sets of datum, we draw curves of the temperature-pulse as Figure 5. Graients of two curves are different, one is 11.2(Curve 1) and the other is 7.91(Curve 2) respectively. Because the Curve 2 is from datum of apodized FBG, fit Curve 1 using Origin software firstly and get the linear relationship of temperature and pulse number as follows:

$$\Delta\lambda_{FWHM} \approx 11.2\Delta T \tag{6}$$

The coefficient error of simulated curves is 0.389, and the measure of its relevancy is 0.994. Because the minimum increment of temperature is set as 1°C and the counting error is +/-1, the error of spectrum width from the detector is +/-5pm and the error of the whole demodulation is less than +/-1°C. Decreasing the sweeping space between steps of the laser to 1pm, the accuracy of FBGs will be improved effectively.

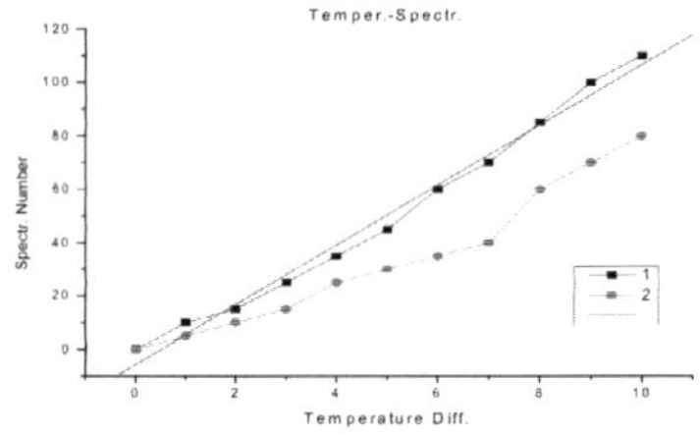

Figure 5. Curves of temperature-pulse

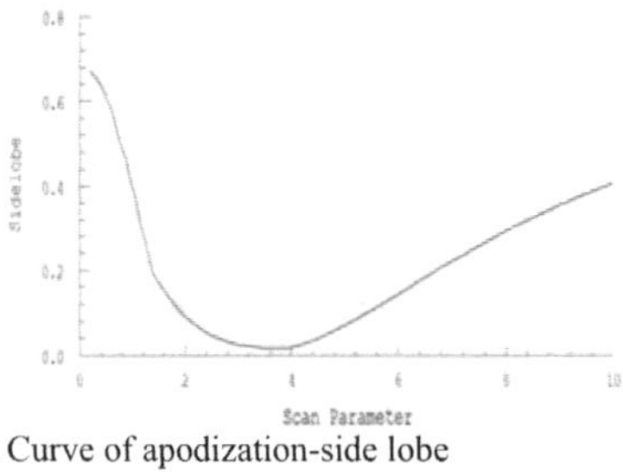

Figure 6. Curve of apodization-side lobe

4 Results and discussion

According to the above analysis, under the linearly increasing nonuniform temperature field, there is a linear relationship between the temperature and the increment of spectrum width, and the linear coefficient is 11.2 which is close to the theoretical value of 10.8, these data demonstrate the validity of the Formula. The coefficient error of 0.4 results from noises of reflective spectrums and curve fitting. Some measures such as apodizing, decreasing the space between sweeping steps and optimizing FBG are helpful to improve the demodulation accuracy. Figure 6 is the curve of apodization-side lobe which scans the apodization value from 0.1 to 10 by simulation. When the apodization value is 3.5, the suppression of the undesirable modes is optimal [15]. But nonuniform temperature field will displace the centre wavelength of FBG and weaken the adopization effect. From Figure 5, edge pulses are distinctly different within the range of 5°C and intensity detecting is reliable. On the other hand, the curve of temperature-spectrum width from the common FBG is good within a wide temperature range, but affected by the crosstalk of sidelopes, there is fluctuating or even obvious missed orders in the reflective spectrums. At this time, the error from the decision device may be serious. So it is meaningful to optimize apodization according to the measuring range and accuracy of the actual FBG system.

Spectrum width is obtained by direct intensity detecting in this demodulation, and at the same time some temperature information is collected on different points of FBG. There are obvious advantages of measuring nonuniform temperature fields. By analyzing variations of pulse intensity, the gradient will be obtained for monotonic temperature fields. For complex temperature fields, the direction change will be demodulated by comprehensive analysis of reflective spectrums [6].

5 Conclusions

In conclusion, the demodulation method of temperature-spectrum width is feasible and the measuring accuracy of temperature is better than 1°C. It provides a new method for measuring temperature fields.

Acknowledgment

This work was supported in part by Hubei Natural Science foundation (No.2015CFB436); Guangxi Province Water Power Survey and Design Institute entrusting project (SDHZ2014055)

References

1. Li Lan, Dong Xin-yong, Zhao Chun-liu, Sun Yi-ling, Jin Shang-zhong, Zhang Zai-xuan. Laser Optoelectronics Prog. **47**, 090603 (2010).
2. Z.C.Zhuo, B.S.Ham. Opt.FiberTechnol.**15**, 442(2009).
3. Q.Wu, Semenova Y.Opt. Laser Technol.**42**, 653(2010)
4. Li Bin, TAN Wei-zhong, Liu Yan, Wang Yan-hua, Ren Wen-hua, Jian Shui-sheng1, ActaOpt. Sin. **27**, 2133(2007).
5. Tsai P, Fengguo Sun, Gaozhi-xiao, Zhiyi Zhang, Rahimi S, Dayan Ban. Photonic Tech. Lett. **20**,300(2008).
6. ZHANG Zi-jia. Sci. Press China. 2009. (in Chinese)
7. Che Ya-liang, Luo Kai-bin, Du Ting-long. ACTA OPT. SINICA. **29**, 2973(2009). (in Chinese)
8. LI Zhi-quan, CAO ping, REN Xiao-li, ZHANG Jing-ru, WANG Yan-nan. J. Optoelectronics-Laser. **22**,491(2011).
9. YANG Dong-Zhou, FENG De-Jun, WANG Jing, JIANG Ming-shun, SUI Qimei.J.Optoelectronics-Laser. **20**, 1438(2009). (in Chinese)
10. LI Yong-guo, DAI Heng, YAO Kai-fang, WU an, ZHANG Dong-sheng.J. WUHAN UNIV. TECHNOL. **32**, 59(2010). (in Chinese)
11. LV Shuang-ming. Shandong Univ.2009. (In Chinese).
12. HAN Ping, ZHOU Zu-de, CHENG Song-lin.J. Optoelectronics Laser. **21**,817(2010). (in Chinese)
13. CAI Hong, WEI Zhi-fen. J.Shanxi Univ. (Nat. Sci. ED.).**34**,237(2011). (in Chinese)
14. YANG Lin, DUAN Kai-Liang, LUO Shi-long, ZHAO Wei. HIGH POWER LASER PART. BEAMS. **23**,308(2011). (in Chinese)
15. ZHANG Hao, CHENG Shu-ming. J. Changzhi Univ.**27**, 4(2010). (in Chinese)

Optical Camouflage Effect Objective Evaluation Method Research Under The Condition of Complex Backgrounds

Jianfei Qin[a], Liyong Qu , Liyan Zhu , Jianghua Hu Shuwei Song

National Key Laboratory of Electromagnetic Environment Effects and Electro-optical Engineering, PLA University of Science and Technology, Nanjing 210007, China [5]94860 PLA Troops, Nanjing 210007, China

Abstract. According to the problem that the intelligence is not high of the soccer robot, using the mechanical theory as a guide, making some mechanical analyses and calculations on the pressure and transmutation states of chip kick mechanics, and conducting optimal design too, then making the structure of chip kick mechanics more and more rationalization. Experiments show that the new soccer robot controller features a quick response and high servo rigidity, and provide a kind of method for improving and perfecting the soccer robot control system, at the same time,filling the needs of producing

1 Introduction

Objective evaluation of camouflage effect of camouflage technology has important direction, the camouflage effect evaluation models are mainly texture analysis[1,2], color histogram method [3], spectral reflection characteristics method [4], etc. Texture analysis makes the objective and accurate evaluation of texture is difficult due to texture indistinct and difficult to measure. The method of gray co-occurrence matrix can't show a comprehensive accurate performance for texture attributes (directivity, crudeness or fineness, etc). Based on Gabor wavelet texture analysis is agreed with human visual system, but its descriptive difference. The pixel position information is lost by color histogram method the different images may have the same color histogram distribution, so that the analysis result to reduce the accuracy. The method of spectral reflection characteristic can distinguish artificiality and natural background but the data obtained by the great influence from outside, in general only suitable for laboratory analysis and battlefield bad adaptability. The above methods achieve the effect of camouflage evaluation respectively from different angles, but the existence of high professional requirements for interpreters, the evaluation of a single, not accurate and evaluation result is not enough visual.

Itti proposed the concept of image saliency[3], the salient characteristics of image analysis results more intuitive but the analysis range is too wide, no specific aim. Followed by a large number of literatures on its improvement and used in various fields and good results have been achieved[6-8]. Most of these methods are only applicable to simple background and are not clear for a complicated background target significant performance. In this paper, according to the optical camouflage principle, firstly the color clustering, color space conversion, DFT conversion, filtering, normalization and a series of processing to the obtained image, get complex background saliency characteristic maps. Using image saliency characteristics on camouflage target and background to carry on the analysis, thus realize the complicated background of optical camouflage effect under the condition of the objective evaluation, by with Itti, HSV color space significance and methods, from the saliency maps look, this paper presents methods more intuitive, reasonable, this method can be used as optical camouflage effect objective evaluation to provide effective means[5].

2 Image Saliency Analysis

2.1 Color cluster

Optical camouflage stealth technology is based on patterning camouflage. The types of patterning colors are not too much for the military target generally 3~5 colors as patterning main color. According to this law, the image clusters by color k-means. K-means clustering is HCM clustering algorithm and has been applied to various fields. The n vector Xi (I = 1, 2,......, n) is divided into K groups by this algorithm, and ask each group clustering center, makes the non similarity (or distance) value function of index (or objective function) minimum, the clustering center is the color of the class in the smallest and distance between class distance the biggest characteristic, which can effectively reflect the characteristics of the color background advantages, and its advantage is not demanded to know in advance to distribution function of sample to be divided[9].

2.2 . The color space conversion

Modern research shows that the RGB color space based on the three color synthesis theories suited for color picture tube and other equipment and was more difference with human visual characteristic. In the HSV color space, the luminance component V didn't relate to color information, the hue H and saturation S describe the image color information. The ways of people feel colors are closed connected with HSV. HSV color space reflects the observation color way, which is very suitable for

[a] Corresponding author: qjfth@163.com

people with visual system to perceive color characteristics of the image processing algorithm [10]. HSV color space is directly corresponded to the human eye color visual characteristics of the three elements. Between channels are independent, the space distance is accorded with human visual characteristics. HSV color space of quantitative results can produce dimension smaller color space, more convenient to describe and control color information, help color features inquires[11] and application. In this paper, the image is converted from RGB color space to HSV color space, in order to content better the analysis and evaluation of optical camouflage effect of the background image.

2.3. Saliency map generation

Based on color clustering, the image of HSV color space decomposes by Gaussian pyramid, after using Gaussian template for image processing, gets the Gaussian pyramid structure adjacent two stage to reduce the image size ratio. In order to simulate the human eyes multi-resolution multi-scale sampling and at the same time content the professional requirements for details as high characteristics, the three component and 5 layer of use of Gaussian pyramid model are calculated $H(\sigma)$, $S(\sigma)$, $V(\sigma)$ (where $\sigma \in [1,2,...,5]$). Using the centre-around difference operator produces feature image set, defined Θ as different scale space characteristics map corresponding point of difference, three component characteristics calculation is as follows:

$$H(c,s) = |H(c)\Theta H(s)|$$
$$S(c,s) = |S(c)\Theta S(s)|$$
$$V(c,s) = |V(c)\Theta V(s)| \tag{1}$$

Where $c \in \{1,2\}$, $s = c + \delta$, $\delta \in \{2,3\}$.

Through the above calculation, each characteristic can obtain four center-around the characteristic difference map, and then to the different scale characteristic difference fuse by multi-scale. Because the front of the scale of the Gaussian pyramid structure from 1 to 5, so that choose in the middle of the scale for a saliency map scale[12], so we can give consideration to the large scale and small scale of local differences and get better effect. Before fusion to the characteristics difference for scale difference, we first need to normalize. In the Itti model defines a normalization operator $N(\cdot)$ that can inhibit differences map of the strong contrast area, and enhance influence power of the having less strong contrast the characteristics difference, in order to burst many saliency characteristic by stacking multiple saliency map[8,13].

The characteristic difference map of normalization does Fourier transform and then nature logarithm spectrum of sin and cosine, and finally to do inverse Fourier transform. Saliency map is normalized above images and linear additive, the characteristic image to merge get hue, saturation and brightness of saliency map respectively. Calculation is as follows:

$$\overline{H} = \left|\bigoplus_{c=1}^{2}\bigoplus_{s=c+2}^{c+3} N(H(c,s))\right|$$
$$\overline{S} = \left|\bigoplus_{c=1}^{2}\bigoplus_{s=c+2}^{c+3} N(S(c,s))\right|$$
$$\overline{V} = \left|\bigoplus_{c=1}^{2}\bigoplus_{s=c+2}^{c+3} N(V(c,s))\right| \tag{2}$$

Where $\oplus$ defined sum operation of characteristic image.

Entropy weight method [14] based on image of each feature weights with information variety, it is more reasonable. Synthesized the saliency of all characteristics and can get saliency map S corresponding to the input image:

$$S = \omega_1 N(\overline{H}) + \omega_2 N(\overline{S}) + \omega_3 N(\overline{V})) \tag{3}$$

Where $\sum_{n=1}^{3} \omega_n = 1$, decision index information entropy value determines the index degrees of importance.

Now, the complicated background area of camouflage target analyses the saliency, as shown in Figure 1:

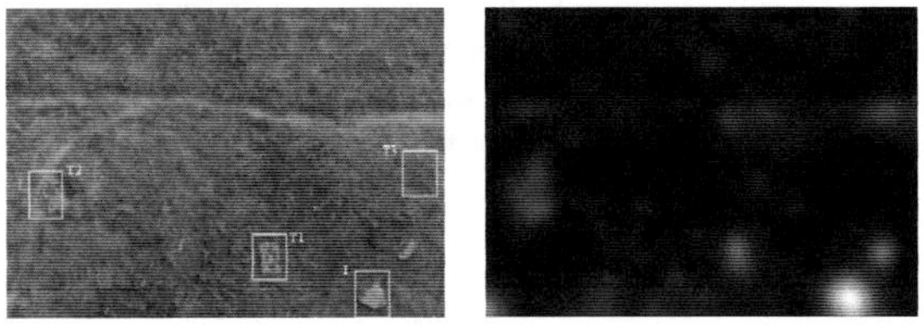

Fig.1 Background image and salience map

In complex background image (Figure 1 left) has three the patterning target, respectively in the image target 1, 2, and 3. Through the visual observation, we can distinguish road, target 1, target 2 and interference objects. While the target 3 and background fusion is better, it is difficult to distinguish with visual. Using this algorithm to generate the saliency map, the significance is biggest for interference objects. After the patterning target 1 and 2 significant is a low, but still can easily distinguish out. While target 3 significant is very low, it can't be distinguished in the saliency map (Figure 1 right).

From the above analysis shows that this calculation method can be used for analysis of optical camouflage effect significant in the complicated background conditions, the results reflect the background significant characteristic and the method of camouflage evaluation by the saliency map is feasible.

3 Simulation and Comparison Algorithm

In order to verify the validity of this algorithm, the analysis of significant optical patterning effect with different background complexity of the 30 images, and compared with Itti, HSV saliency map method. Results are consistent, choose three images as shown in Figure 2:

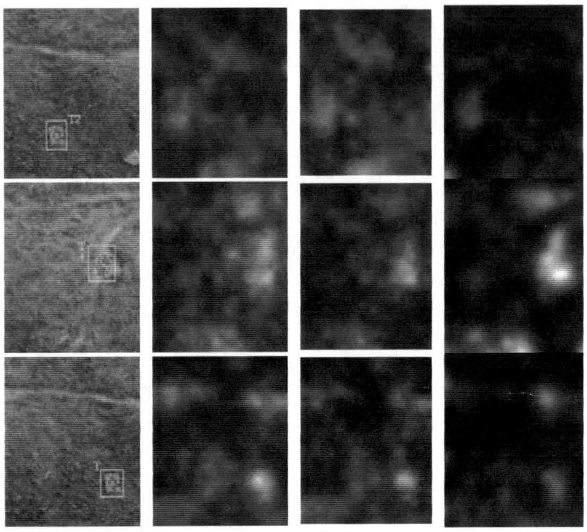

Background image | Saliency map of Itti | Saliency map of HSV | Saliency map of text

Fig.2.The comparison of the salience map

In the background image, not patterning vehicle target 1, patterning target 2, 3 and 4, and is easy to cause the background of visual attention road, bridge, bare rock etc.

From Figure 2, the Itti saliency map results from a certain extent reflects the significant features of the original, first discovered the highlighted target 1 and 3, followed by bare rock, target 2, in the saliency map, while the target 4 is not reflected. The relative Itti saliency map effect, significant area contour of HSV saliency map has been improved, but the difference is not very large. This algorithm the saliency map clearly shows the original of each significant target, such as bare rock, bridge, target 1, 2 and 3, while the target 4 is not reflected in the saliency map due to fuse good with the background.

Although Itti saliency map and HSV saliency map also correctly represent the main significant features of the complicated background, but also shows too much low significant features. It brings heavy interference for the objective evaluation of the complicated background target camouflage effect. This algorithm can effectively filter not significant information mottled in complex background, and after a series of processing makes the saliency map generated more clear, can accurately display the significant matter in the image, consistent with the visual results, convenient interpreter evaluation, non-professionals can also interpret accurately. Therefore, this algorithm can provide the effective means to target camouflage effect evaluation under complex background.

4 Conclusion

A significant feature of image can be used to camouflage effect evaluation. This paper aims at the patterning targets in the complex background, get the final saliency map by the corresponding processing, relative to other saliency map generation method, the results obtained from this method is more intuitive, accurate, clear, and can effectively filter out the low significant characteristics of background, extruded the fixation point in the image, consistent with the results and visual. Besides the interference matter, if camouflage target appears in the saliency map, then this camouflage target can easy be found by the reconnaissance equipment, camouflage effect is poor, conversely not easy be found, camouflage effect is good. Thus this algorithm results can objectively evaluate the camouflage effect of target.

A significant characteristics of image to camouflage effect evaluation is convenient, because the background of bare rock and other objects, through the saliency map in the evaluation of camouflage effect was still give the judge to bring some interference, how to effectively eliminate the interference of objects, this is a problem needed to be solved in the next step.

References

1. Xu WeiDong, Lv Xuliang, Chen Bing. A Model Based on Texture Analysis for the Performance Evaluation of Camouflage Screen Equipment, Acta Armamentarii, 2002, 23(3):229-331
2. Ling Wei, Cheng Yuhua, Gao Hongsheng, Wang Jijun. A Method of Camouflage Evaluation Based on Texture Analysis Model of Gabor Wavelet, Acta Armamentarii, 2007, 28(10):1191-1194
3. Wang Dong, Lv Xuliang, Xu Weidong. Camouflage Application Models with Pixel Frequency Analysis, Journal of PLA University of Science and Technology, 2004, 5(3):74-77
4. Xue Jingjing, Bai Tingzhu. Camouflage Assessment Based on Spectral Reflect Signature, Laser & Infrared, 2007, 37(10):1078-1084
5. Itti L,Koch C,Niebur E.A model of saliency-based visual attention for rapid scene analysis[J].IEEE Transaction on Pattern Analysis and Machine lntelligence,1998,20(11):1254-1259.
6. Dou Yan, Kong Lingfu, Wang Liufeng. A Computarional Model of Visual Attention Based on Visual Entropy, Acta Optica Sinica, 2009, 29(9):2511-2516
7. Cai Yunxiang, Qin Jianfei, Lv Zhengjian. A Model to Extract the Pattern from Background Based on Visual Attention Model, Optical Technique, 2010, 36(2):302-306
8. Itti L,Koch C. A saliency-based search mechanism for overt and covert shifts of visual attention[J]. Vision Research,2000,40(2000):1489–1506.
9. Bouman C,Sauer K. A generalized Gaussian image model of edge preserving map estimation[J]. IEEE Trans. Image Processing,1993,2(3):296-310.
10. Qin Jianfei, Hu Jianghua, Yang Gaofeng. A New Method to Determine Imitated Pattern Painting Color, Journal of PLA University of Science and Technology, 2011, 12(1):75-78
11. Xu Xu, Zhu Miaoliang, Liang Qianhui, Waheed Sajjad. A Method of Dominant Colors Eetraction and Representation for CBIR Systems, Journal computer-aided design & computer graphics, 1999,11(5) :385-388
12. Liang Chunying, Wang Guoying, Kang Jinfeng. Image Retrieval Method Based on Salient Regions, Computer Engineering, 2010, 36(5):210-214
13. Ling Yun, Xia Jun, Tu Yan. Detection of Region of

Interest and its Application in Video Image Quality Assessment, Journal of Southeast University, 2009,39(4):753-757

14. Cheng Qiyue. Structure Entropy Weight Method to Confirm the Weight of Evaluating Index, Systems Engineering-Theory & Practice, 2010, 30(7):1125-1128

Optical Digital Imitation Painting Design Based on Self-Adaptive Image Feature

Liyan Zhu[1, a], Jianfei Qin[2,b], Liyong Qu[3],Shuwei Song[2], Weidong Xu[3]

[1]National Key Laboratory of Electromagnetic Environment Effects and Electro-optical Engineering, PLA University of Science and Technology, Nanjing 210007, China
[2]94860 PLA Troops, Nanjing 210007, China
[3]National Key Laboratory of Electromagnetic Environment Effects and Electro-optical Engineering, PLA University of Science and Technology, Nanjing 210007, China

[a]zhuliyan0522@163.com, [b]qjfth@163.com

Abstract. Based on the study of existing digital imitation camouflage technology, we propose a kind of optical digital imitation camouflage design algorithm which is based on the characteristic of self-adaptive image. Picking main color feature of the background by using K-means clustering algorithm, counting the shape characteristics of each main color spots by separating layers, we generated digital camouflage pattern automatically by segmenting the background region characteristics and fill the background color image according to the statistics expected value. The simulation results show that, the digital camouflage generated automatically is blend well with the background .It keeps the background color and shape features, so has good camouflage effect. This algorithm can also provide the basic algorithm foundation for digital deformation camouflage.

Keywords: Image Processing; Digital Painting; Image Features

1 Introduction

Painting camouflage is an important part of military camouflage technique, is also the most commonly used against high technical reconnaissance means. Painting camouflage by reducing the differences in optical image between target and background, then distorting or dividing the shape of the adjacent target, in order to achieve the fusion of adjacent background and target. Digital painting is generated by digital imaging and image processing technology of multicolor block unit of camouflage. Observing near distance, we can find that digital painting unit makes the boundary between different color broken, fuzzy. This phenomenon of painting pattern makes the textural characteristics more complicated. So it is easy to simulate broken background in the nature. Observing far distance, digital units mixed to be a large painting spot because of the space color mixture. This phenomenon of digital pattern makes the target appearance divided and distorted.

At present, the spatial resolution of the human eye is often used as criterion to determine the unit size of digital painting. By analysis the human eye's discernible size in the observation distance, they can determine the pixel size of Digital painting unit[1] [2].However, as a new digital painting pattern, its design is mainly based on the optical image characteristics of background, automatically generated by computer. Its size in the image is pixel. The background image changes with the background selection, the shooting distance, angle and other factors. Resolution size of the human eye is a subjective variable. It can't reflect the spot's characteristics of the background's actual optical image. It can't adapt to the background image of different shooting conditions to play the role of a simulation of the natural background.

This paper proposes a digital imitation painting generation algorithm, which can adapt to background characteristics. It uses K-means clustering algorithm to pick the main color of background. It counts shape characteristics of spot through delaminating various spots and numbering spots. It uses the circumcircle diameter's expectations as the basis for determining the size of a digital unit. It generates digital imitation painting pattern according to the color and shape features comminuting and filling the background main-color image. It implements a design program of digital imitation painting by the Matlab programming. Compared using multiple sets of background images, the results show that the digital imitation painting generated by this algorithm can adapt to background characteristics in color and spots. It has a good camouflage effect. Its color and size can be adaptively adjusted, according to different background optical image. It can effectively avoid manual intervention in the painting design process, so as to realize the automation of painting pattern design. It provides a new idea of design digital imitation painting.

2 Algorithm Process of the Digital Imitation Painting

The spot color and shape feature of the battlefield background is the main parameter to design digital painting. By using the color and size of background as a parameter to control the process of generating digital imitation painting can realize digital imitation painting design adapted to adjacent background spot color and shape characteristics. In this paper, the digital imitation painting design method based on the characteristics of the background is based on the analysis of the background image, on the basis of spot color and shape features, we realize the automatic generation of digital imitation painting by comminuting and filling the background image. The algorithm flow chart is shown in Figure 1. According to the functions and results of each processing step in the algorithm, we can divide the algorithm into three main modules: (1) Color features analysis module, (2) Shape features analysis module, (3) Digital imitation painting generation module.

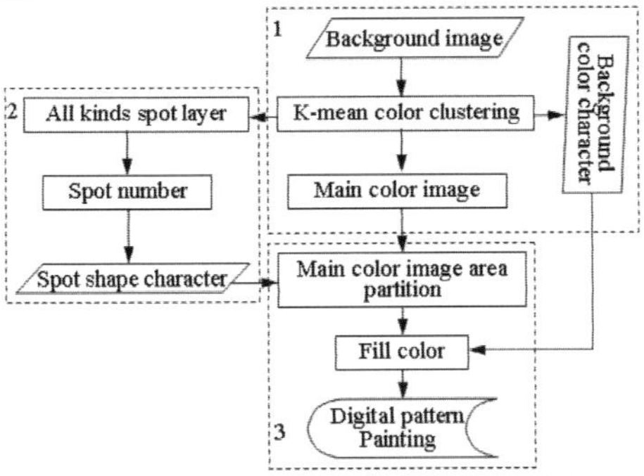

Fig.1. Flow chart of the digital imitation painting generation algorithm

3. Analysis of Characteristics of the Background Color

Natural background color is rich and colorful. In order to make the painting spot's color adapted to a variety natural backgrounds and construction conditions, we often choose 3~5 kinds of color to be the background's main color to design the painting spots [3]. Then we used $L^*a^*b^*$ color space of the K-means clustering algorithm in this paper [4].

Based on K-means color clustering results we analyze the main color feature of the background. We can statistical the chromaticity coordinates of each dominant color according to the main color image of the background, then calculate the probability of occurrence of each main color in the background, and draw the color histogram of the dominant colors background's main color image.

We collected a large number of natural backgrounds, and selected three representative background images. Including image 1 is woodland background, image 2 is the grass background image, and image 3 is desert background, as shown in Figure 2. In the algorithm research, we converted the background image size to 800 x 600 pixels to compare with the results of other studies.

(a)The background image 1 (b)The background image 2 (c)The background image 3

Fig.2. Background image

We generated background color charts of the 3 background images in Figure 2 by using K-means clustering algorithm to pick the main color of background, and then generated background color images, counted the color features of background image. We generated the main color image of background, counted the color feature of background image (red, green, blue components) as shown in Table 1.We counted the probability of occurrence of all main color in background image, as shown in Table 2.

Tab.1. Chromaticity coordinates of the main color in the background image

		Main color 1	Main color 2	Main color 3	Main color 4	Main color 5
Background 1	R	0.1216	0.5804	0.6824	0.2824	0.8275
	G	0.1608	0.6627	0.7843	0.4000	0.8667
	B	0.1020	0.3804	0.5843	0.2157	0.6745
Background 2	R	0.2353	0.5137	0.5804	0.4000	0.6784
	G	0.3333	0.6392	0.7725	0.4784	0.8314
	B	0.2157	0.4235	0.4706	03373	0.5961
Background 3	R	0.3255	0.6745	0.4667	0.8667	0.5255
	G	0.3725	0.5922	0.4706	0.7176	0.5373
	B	0.2157	0.4353	0.2941	0.5412	0.3804

Tab.2. Probability of the occurrence of the main color in each background image

	Main color 1	Main color 2	Main color 3	Main color 4	Main color 5
Background 1	0.3528	0.1665	0.1369	0.2282	0.1155
Background 2	0.1495	0.3256	0.1409	0.2824	0.1016
Background 3	0.2688	0.2015	0.2265	0.1655	0.1378

4. Analysis of Characteristics of the Background Sharp

After the main spots were separated from the

background main color in the image, the description of the spot's shape feature can be used as a basis to distinguish different spots. To describe the shape of spots, we commonly used circumscribed circle diameter, inscribed circle diameter, rectangular degree, circular degree, invariant, eccentricity, etc. Among them, the diameter of circumcircle can effectively reflect the spot size, because it isn't influenced by the shape and the direction of the spots [5] [6].In this paper, we use circumscribed circle diameter of spot as the basis of describing background spot size. To some extent, expectations of spot size in a background image reflect the shape characteristics of the background. Its value will decrease with the increase of the complexity of the background image.

We used x_i to represent the i'th size which is possible color layer in the background. We use P_i to represent the probability of occurrence of the corresponding dimension, then the context of expectations of the E in the spot size can be calculated with formula (1):

$$E = \sum_i x_i \times P_i \qquad (1)$$

If we use E_j to represent the spots size expectations of the j'th color spots layer which containing n kinds of color in background main color image, the average size of spots in the background of the color image E_{avg} was the average value of the expected value of E_j in the color spots layer, namely:

$$E_{avg} = \frac{\sum_{j=1}^{n} E_j}{n} \qquad (2)$$

Image shape analysis process is as follows, with the background image 1 as an example:

First, we stored different color spots in the corresponding layer extraction spot according to color. Then spot layer is converted into a binary image, the spot color pixel value is 1, the spot color pixel values to 0.

Second, we carried out the open-close operation of mathematical morphology on the spot layer using the circular structure of 3×3 units. Due to the large number of small spots were gathered in a main color, the same structure of the region corresponding to the original image in the background, at the same time, because the human eye space color mixing effect, large tracts of adjacent small spots is often regarded as the same. So, a piece of fine adjacent spots can be processing as large spots. Mathematical morphological closing operation can be finely formed adjacent spots connected domain. Actually opening operation calculating the image expended before corroded, it can better maintain the spot characteristics of the image, eliminating the noise and small connected in the image, which can effectively distinguish different spots.

Third, number the spots. In order to count the shape features of the background spots, we numbered and differentiated the spot layers as the basis when counting the shape feature to distinguish different spots. We numbered the spots according to the processed results, using "shuffle" color display the results of spots number.

(a) No. 1 spot image (b) The number results of No. 1 spot image (c) Spot size distribution

Fig.3. Spot shape characteristic analysis results of background 1

Fourth, the shape feature of statistical background. We calculate the circumscribed circle size according to the spot Numbers, then, count the probability statistics of each spot size, finally, draw distribution histogram of spot size, with the spot size as abscissa, the frequencies of various sizes as ordinate.

Fifth, draw distribution histogram of various spots circumscribed circle size, calculate the expectation of background probability statistics of each diameter size by the type (1, 2), as shown in Table 3.

Tab.3.the expectations distribution of background spots circumscribed circle diameter

	background 1	background 3	background 6
Average spot size	15.4076	16.7446	18.1767

5 Generation of Digital Imitation Painting

The generation digital imitation painting is filling the background main color image into a digital imitation painting by the background color and shape features, containing the following two major steps:

Step 1, the main color image division of background. According to the shape characteristics of the background image ,the main color image are divided into square area by rows and columns considering the expectations of background image spot size to be digital painting unit size.

Step 2, color filling. Select background image sub domain center pixel as characteristic pixels, and then fill the all area with the pixel color. Generate a digital imitation painting, the result is shown in Figure 4.

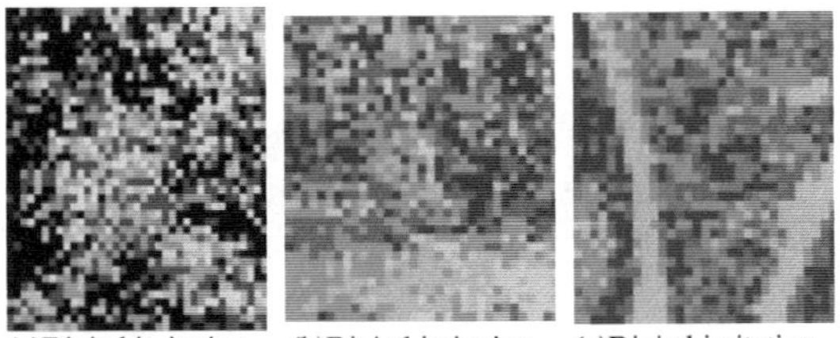

(a)Digital imitation painting 1 (b)Digital imitation painting 2 (c)Digital imitation painting 3

Fig.4. Results generated by the digital imitation painting

6 Analysis and Comparison

According to the background as shown in Figure 2, using the algorithm in this paper, probability of each main color and digital unit size distribution in a digital imitation painting are shown in Table 4 and shown in Table 5, respectively. We design digital imitation painting, using the main color which picked from the background image. Comparing Table 2 to Table 4, you can see that the main color probability in each painting pattern is consistent with its probability in the corresponding background,the maximum probability of error is 0.00. Painting pattern of the main color corresponds to the corresponding natural background; its main color appears the trend of probability is not the same. So it realizes that the digital painting color features can be self-adaptive to the background. Comparing Table 3 to Table 5, you can see that the digital painting design unit size is consistent with the corresponding background image, and digital unit size corresponds to the corresponding background image. For the woodland background image speckle is fine, the smaller digital painting unit size can simulate to the complex background details. For the monotonous desert background, the larger digital painting size can simulate to rocks and loess in the background. At the same time it reduce the increased difficulty of construction and unnecessary waste of time because of the digital spot is too small.

Tab.4. The occurrence probability of each main color in the digital imitation painting

	Main color1	Main color 2	Main color 3	Main color 4	Main color 5
Camouflage 1	0.3547	0.1623	0.1274	0.2406	0.1151
Camouflage 2	0.1508	0.3252	0.1368	0.2854	0.1027
Camouflage 3	0.2720	0.1990	0.2335	0.2335	0.1639

Tab.5 The unit size distribution of digital imitation painting

	Camouflage 1	Camouflage 2	Camouflage 3
Digital unit size	15	17	18

Literature [1] put the designed camouflage pattern into the background image and evaluates camouflage effect through the eye observation of the image designed in this paper according to the above methods. Cut the digital painting patterns and put into the background image, the camouflage effect as shown in Figure 5.

(a)Digital imitation painting 1 (b)Digital imitation painting 2 (c)Digital imitation painting 3

Fig.5. Results generated by three kinds of digital imitation painting

Known from the Figure 5, painting pattern which was designed using the digital painting design algorithm in this paper well embodies the characteristics of background details. Its color and shape features adapt to the background well. The fusion is better between painting pattern and background image. In the far distance observation, it can produce the effect of color mixing in space. So it has better camouflage effect.

7 Conclusion

Based on the study of existing digital imitation painting technology, put forward a kind of optical digital imitation painting design algorithm which is based upon the characteristic of self-adaptive image. The simulation results show that the digital painting generated automatically is blend well with the background .It keeps the background color and shape features, so has good camouflage effect. Compared to the method of digital cell size determined based on human spatial resolution, the main color and the digital unit sizes in this algorithm are determined by the background characteristics. Control conditions change as background complexity change without human intervention. So it makes the digital imitation painting design self-adaptive to the background. This algorithm can also provide the basic algorithm foundation for digital deformation painting.

References

1. Yu Jun, YANG Wu-xia, Hu Zhi-yi Research of Digital Camouflage Generation Algorithm. Opto-Electronic Engineering
2. Xu Wei-dong, Qin Jian-fei, Zhu Jin-song, Lv Xu-liang. Study on One Generating Method of Digital Painting Patterns, ICIMME, 2014.
3. YAO Yong-ping Camouflage of Army equipment[M], Beijing: Military Science Press,2006:125-160
4. Qin Jianfei, Hu Jianghua, Yang Gaofeng. A New Method to Determine Imitated Pattern Painting Color, Journal of PLA University of Science and Technology, 2011, 12(1):75-78
5. Rafael C. Gonzalez, Digital Image Processing[M].

Third Edition. Publishing House of Electronics Industry, 20012.
6. WISHART D.K-means clustering with outlier detection[C]. The 25th Annual Conference of the German Classification Society. Munich: Univisity of Munich, 2001.

 DOI: 10.1051/matecconf/20166106015

Concentrations of Nicotinamide in Plasma by RP-HPLC With Fluorescence Detection

Zhipeng Pan[1], Da Li[2], Xiaokun Yao[1], Shisheng Zhou[1] and Yongzhi Lun[1,a]

[1]*Liaoning Provincial University Key Laboratory of Biophysics, College of Medicine, Dalian University, Dalian, Liaoning 116622, P.R. China*
[2]*Department of Obstetrics and Gynecology, Shengjing Hospital of China Medical University, Shenyang, Liaoning 110004, P.R. China*

Abstract. The purpose of this study is to establish a new method for detecting nicotinamide concentration in plasma. In the experiment, the high performance liquid chromatography (HPLC) method was used, with a fluorescence detector. The nicotinamide in the plasma was first converted to N1- methylnicotinamide, then reacted with acetophenone under certain conditions to produce fluorescent derivatives for testing. The method is a kind of highly sensitive detection, of which the lower limit is 10 ng/mL, the recovery rate is between 92.75% and 105.13%, and the relative standard deviation (RSD) is between 3.76% and 4.43%. The results showed that this measurement method is accurate, sensitive and rapid. It meets the requirements of the experiment, and applies to the detection of nicotinamide concentration in plasma.

Nicotinamide (NM), namely Nike amide, vitamin B3 and vitamin PP, is a water-soluble vitamin(Figure 1) [1]. Nicotinamide is an alkylated derivative of nicotinic acid, which is also the precursor of coenzyme I (NAD) and coenzyme II (NADP) in the body, thus the pharmacokinetic observation of nicotinamide is an important tool for monitoring metabolism levels in the body [2]. Our research shows that long-term nicotinamide overload can cause insulin resistance and oxidative stress, and eventually lead to diabetes [3, 4]. Currently, the detection of nicotinamide in plasma samples is a complex method, which requires high instrument conditions and liquid chromatography - mass spectrometry [5-7]. This method is also highly time-consuming, mainly due to the very low levels of nicotinamide in the plasma. In this study, concentrations of nicotinamide were measured using the ordinary high-performance liquid chromatography method with cleverly designed, simple processing and saving supplies, and satisfactory results were achieved [8-14].

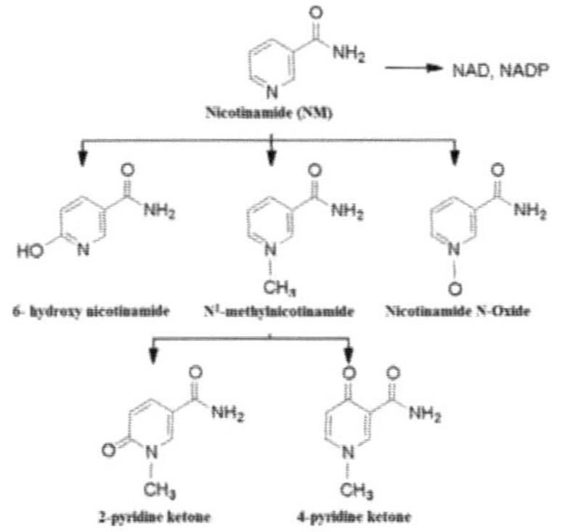

Figure 1. Vitamin PP metabolism in human

1 Methods

1.1 Instruments: High-performance liquid chromatography (Shimadzu LC-9A pump, Rheodyne 7725i injector, 20 μL sample loop, Shimadzu Co.), Waters470 fluorescence detector (Waters, USA), N2000 chromatography workstation (Zhejiang University Zhida Information Engineering Co., Ltd.).

1.1.1 Reagents: Ethyl acetate (AR), acetophenone (AR), formic acid (AR), NaCl (Beijing Jingke Hongda), 0.5M HCl; 0.5M NaH2PO4, 0.4M KOH, 20% trichloro acetic acid (TCA), Nicotinamide (NM), and N1-methylnicotinamide (NMN) (Japan Takeda Co.). N1-ethylnicotinamide (NEN) was synthesized according to the methods described by Hirayama et al. [15, 16], with a purity of 99.99%, and this was used as an internal standard.

1.2 Chromatographic conditions: The column in the experiments: ODS-C18 column (Thermo, Bellefonte, PA), 250 mm*4.6 mm (id), particle size 5 μm; mobile phase: 10 mM heptane sulfonate, 50 mM triethylamine solution (pH 3.2): acetonitrile (78:22 v/v); excitation wavelength 366 nm, emission wavelength 418 nm; flow rate 1.0 mL/min; injection volume 20 μL.

1.3 Sample preparation: 200 μL blood was input into the EP tube with 100 μL 20% TCA and then mixed by vortex, and 12000 rpm of centrifugation for 10 min; next, 200 μL supernatant was removed and placed in a new EP tube. 25 μL 0.5M HCl was added with 100 mg NaCl into the supernatant, then shocked by vortex until the water phase was fully saturated by the NaCl, which was extracted three times by 1 mL of water-saturated ethyl acetate extract, then the upper phase of ethyl acetate was

[a] Corresponding author: Yongzhi Lun:lunyz@163.com

DOI: 10.1051/matecconf/20166106015

discarded and the residual ethyl acetate was blow-dried with N_2. 25 μL 0.5M NaH_2PO_4 and 25 μL 0.4M KOH were continuously added, and these were shocked and mixed well, followed by the addition of 1 mL anhydrous ethyl acetate, which was shocked and mixed. Centrifugation was performed at 3000 rpm for 3 min, then the 800 μL upper phase was taken in another EP tube, followed by N_2 drying [17]. 250 μL iodomethane was added and the two were mixed together, then saved in dark for 24 h before being dried with N_2. Then 200 μL double distilled water was added, and extracted three times with ethyl acetate saturated with water after the shocking; the residue in the ethyl acetate was dried by N_2, at which point all of the nicotinamide in the plasma was converted into the N1-methylnicotinamide. 200 μL internal standard was added (50 ng/mL), then 200 μL of 100 mM acetophenone was added [18], and placed in the ice bath for 10 min after shocking. 400 μL of 6M NaOH was added and mixed well, then placed in the ice bath for 60 min. Then 200 μL of formic acid was added, and placed in the ice bath for 60 min. Finally, the solution was placed in the boiling water bath for 3 min to generate the fluorescent derivatives, then mixed well after cooling for machine testing.

1.3.1 Determination: using the internal standard method and analyzed by HPLC with fluorescence detection, and the N2000 data acquisition system collected the data and calculated the concentration of nicotinamide in the plasma. The nicotinamide concentration was calculated according to the following formula[Equation (1)] based on N1-methylnicotinamide determination:

$$m = 9.375\ m_0 \times M/M_0 \qquad (1)$$

Where m is the concentration of nicotinamide in the plasma, unit: ng/ml; m_0 is the mass of the N1-methylnicotinamide, unit: nanograms (ng); M_0 is the molecular weight of the N1-methylnicotinamide, molecular weight: 172.61; and M is the molecular weight of the nicotinamide, molecular weight: 122.12.

2 Results

2.1 Chromatogram of nicotinamide in plasma: Figure.2 is a blank of the chromatogram of the nicotinamide in the plasma, and Figure.3 is a marked chromatogram of nicotinamide in the plasma. Peak 1 of the chromatogram is nicotinamide, and peak 2 is the internal standard in the figure.

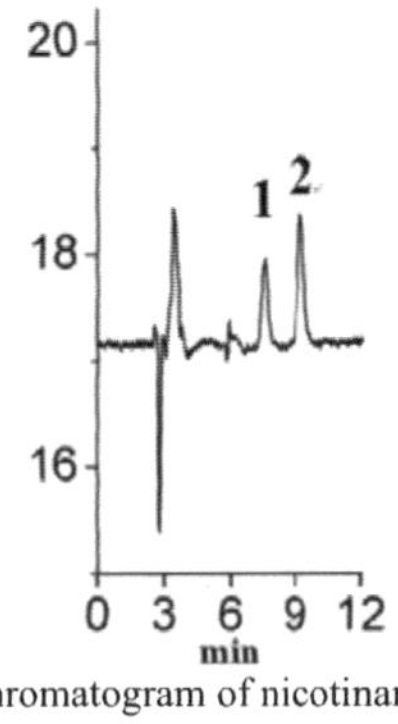

Figure.2 Chromatogram of nicotinamide in plasma

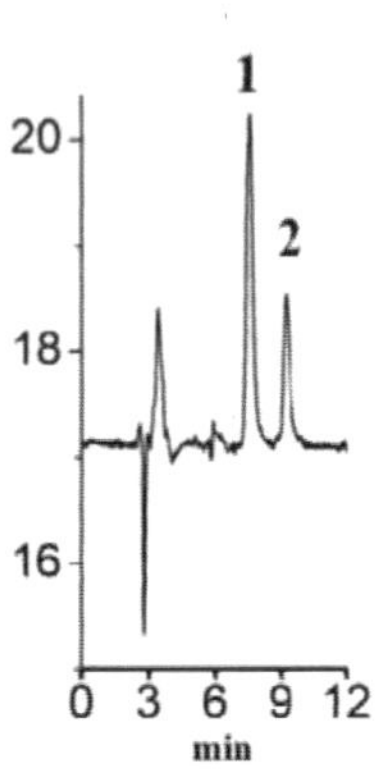

Figure.3 Marked chromatogram of nicotinamide in plasma

2.2 The minimum detection limit: the minimum detection limit for the RP-HPLC fluorescence detection method in the experimental conditions and chromatographic conditions, according to the peak response value of 3 times of the noise, the sample volume is calculated as the minimum detectable amount [19, 20]. The detection lower limit of this method is 10 ng/mL.

2.3 The recovery rate and precision of the experiment: Three plasma samples were treated using the same method, once per day for 6 days. The results are as shown in Table 1.

Table 1. Precision experiments among batches (n = 6)

Sample*	Number Measured value (ng/mL)						X±SD	RSD(%)
Sample1	158.05,	168.67,	175.58,	148.47,	154.76,	149.43	159.16±18.86	4.43
Sample2	161.40,	140.98,	154.97,	156.69,	168.46,	170.53	158.84±10.72	4.38
Sample3	141.65,	138.79,	159.85,	144.19,	133.77,	150.23	144.75±9.21	3.76

*Samples 1-3 are all plasma of normal humans.

Precision and recovery rate experiment among batches: taking 3 parts of normal human plasma, which is divided into three parts after mixing, the standard of nicotinamide was added by the concentration gradient, according to the same procedure used to deal with the samples, which were then detected. Under the same conditions, each sample was continuously injected six times, and according to the corresponding peak area to calculate the RSD, the experimental results are as shown in Table 2.

Table 2. Recoveries and precision (n = 6)

Sample	background value	marked value	measured value	recovery value	recovery rate	RSD
Sample4	153.67	100.00	246.42	92.75	92.75	4.72
Sample4	153.67	200.00	363.93	210.26	105.13	3.94
Sample4	153.67	500.00	635.57	481.90	96.38	3.27

3 Discussions

3.1 The high performance liquid chromatography with fluorescence detector was used in this paper to measure the concentration of nicotinamide in the plasma. Its operation is simple and reproducible, with a high recovery rate, and the experimental results were satisfactory compared with those of other liquid chromatography-mass spectrometry methods, showing that it provides an experimental method for the study of pharmacokinetics and dynamics of nicotinamide.

3.2 In the sample processing, it is necessary to be careful when shaking with ethyl, otherwise the ethyl may overflow, possibly causing the loss of nicotinamide, thus causing the results to decrease in accuracy. When N1-methylnicotinamide was converted into nicotinamide, the saturated ethyl acetate in the acidic conditions can be extracted to the original N1-methylnicotinamide of plasma, but this cannot be done under neutral conditions, thus it is important to pay particular attention to the pH value. The final amount of N1-methylnicotinamide was solely converted from the nicotinamide, thus the final amount of N1-methylnicotinamide is equal to the nicotinamide.

3.3 Due to the fact that the concentration levels of nicotinamide in the body differ among people, the nicotinamide content of samples 1, 2 and 3 in this experiment are different. Nicotinamide is easily oxidized in air to form oxidized nicotinamide, thus it is necessary to pay particular attention to sample storage; the samples should be stored in a -20℃ refrigerator, and processed and detected promptly after treatment.

Acknowledgments

We would like to express our thanks to the following projects for funding this study: the National Natural Science Foundation of China (No. 30471510), and Liaoning Provincial Department of Education College Project Plan (No. L2012441, 2009S005). We would also like to thank our tutor Dr. Lun for the guidance of the relevant theories, experiment technology and experimental results of this experiment, as well as Dr. Sun Wuping, Professor Guo Ming, and other friends for their assistance with the experiment.

References

1. J. Wohlrab and D. Kreft, Skin Pharmacol Physiol. **27**, 311 (2014).
2. S.J. Yang, J.M. Choi, L Kim, S.E. Park, E.J. Rhee, W.Y. Lee, K.W. Oh, S.W. Park, and C.Y. Park, J Nutr Biochem. **25**, 66 (2014).
3. SS Zhou, D Li, WP Sun, M Guo, YZ Lun, YM Zhou, FC Xiao, LX Jing, SX Sun, LB Zhang, N Luo, FN Bian, W Zou, LB Dong, ZG Zhao, SF Li, XJ Gong, ZG Yu, CB Sun, CL Zheng, DJ Jiang, and ZN Li, World J Gastroenterol.,**15**, 5674 (2009).
4. SS Zhou, D Li, YM Zhou, WP Sun, XX Liu, and YZ Lun, Acta Physiologica Sinica. **62**, 86 (2010).
5. K. Ortmayr, J. Nocon, B. Gasser, D. Mattanovich, S. Hann, and G. Koellensperger, J Sep Sci. **37**, 2185 (1024).
6. R. Lang, A. Wahl, T. Skurk, E.F. Yagar, L. Schmiech, R. Eggers, H. Hauner, and T. Hofmann, Anal Chem. **82,** 1486 (2010).
7. M. Szafarz, Lomnicka M, Sternak M, Chlopicki S, and J. Szymura-Oleksiak, J Chromatogr B Analyt Technol Biomed Life Sci. **878**, 895 (201011-12).
8. Jia Woei Wong, Kah Hay Yuen, and Kok Khiang Peh, J Chromatogr B Biomed Sci Appl. **716**, 387 (1998).
9. A. Mancinelli, S. Pace, A. Marzo, E. A. Martelli , and G. Passetti, J Chromatogr. 575, 101 (1992).
10. V.M. Kodentsova, O.A. Vrzhesinskaia, A.A. Sokol'nikov, T.G. Zaburkina, E.D. Bender, and V.B. Spirichev, Vopr Pitan. **2**, 62 (1992).
11. D. Zhong, and X. Chen, J Chromatogr B Biomed Sci Appl. **721**, 67 (1999).
12. G.L. Chan, J.E. Axelson, F.S. Abbott, C.R. Kerr, and K.M. McErlane, J Chromatogr. **417**, 295 (1987).
13. B. Alpertunga, S. Sungur, L. Ersoy, and S.Y .Manav, Arch Pharm (Weinheim). **323**, 587 (1990).
14. T. Hirayama, K. Yoshida, K. Uda, M. Nohara, and S. Fukui. Anal Biochem. **147**, 108 (1985).
15. B.R. Clark, Methods Enzymol. **66**, 5(1980).
16. B.R. Clark, R.M. Halpem, and R.A. Smith, Anal Biochem. **68**, 54 (1975).
17. C. Musfeld, J. Biollaz, N. Bélaz, U. W. Kesselring, and L.A. Decosterd, J Pharm Biomed Anal. **24**, 391 (2001).
18. L.A. Decosterd, A. Karagiannis, J.M. Roulet, N. Bélaz, M. Appenzeller, T. Buclin, P. Vogel, and J. Biollaz , J Chromatogr B Biomed Sci Appl. **703**, 25 (1997).
19. Yu H, Cai Y, Bi S, and Huang D, Se Pu. **23**, 114 (2005).
20. Wang A, Li G, You J, Ji Z. J Sep Sci. **36**, 3853 (2013).

Growth of AlGaN Film on Si (111) Substrate

Yang Zhang[1,2], Bingzhen Chen[2], Na Peng[2], Lu Zhang[2], Cuibai Yang[2], Xu Pan[2], Shun Yao[1], Zhiyong Wang[1]

[1] *Institute of Laser Engineering, Beijing University of Technology, Beijing 100022, China*

[2] *Guangdong redsolar photovoltaic technology Co.,LTD zhongshan528437,Guangdong , China*

Abstract: At present, the applications of $Al_xGa_{1-x}N$ are extensive, such as for visible-blind ultraviolet detectors, laser diodes, light emitting diodes (LEDs) and HEMTs. In this paper, $Al_{0.25}Ga_{0.75}N$ and $Al_{0.32}Ga_{0.68}N$ films have been grown on 2 in Si (111) substrates by MOCVD. The low-temperature (6 K) photoluminescence (PL) spectrum and XRD rocking curve measurements have been employed to study the crystal quality of samples, and the phonon replica peak can be observed, which indicate that the samples have better quality in a small-localized region. The surface morphology of samples was investigated by AFM and the result of wavy surface agrees with the deduction from XRD rocking curve measurements. The sheet resistance mappings have been shown, and it indicates the nonuniformity of AlGaN film on Si (111) will increase sharply as the Al content increases.

1. Introduction

GaN and related Group-III nitrides grown by metal-organic chemical-vapor deposition(MOCVD)can be applied to high power, high frequency and high-temperature electronic devices[1-3] as well as light emitting diodes (LEDs) and laser diodes. Ternary AlGaN in particular is an important one with wide direct band gap, which has a broad range of applications in ultraviolet (UV) optoelectronic devices[4-6]. Due to the limited availability of bulk GaN or AlN substrates for homoepitaxial growth, Generally GaN is grown on sapphire, SiC and Si substrates. The improvement of the power density of the device is limited by Sapphire substrate's bad thermal conduction. Meanwhile, the high cost of SiC substrates hinders the application of GaN material grown on SiC. Compared with sapphire and SiC, Si is the best alternative for its low cost, good thermal conductivity and ability to be integrated with the mature Si-based processing techniques[7, 8]. However, for the large lattice mismatch and the large coefficient of thermal-expansion (CTE) mismatch between GaN (AlN) and Si, the growth of high quality, thick crack-free AlGaN on Si substrate is a difficult task, and the cracks and high defect density seriously degrade the performance of GaN based devices on Si substrates. There are few reports on AlGaN growth on silicon substrates[9-11]. In this study, the epitaxy of AlGaN on 2 in Si (1 11) substrates was performed in a horizontal MOCVD reactor. The low-temperature (6K) photoluminescence (PL) spectrum and XRD rocking curve measurements have been employed to study the crystal quality of samples. We also investigate the surface morphology of AlGaN epilayers on Si (111) substrate by AFM. The sheet resistance mappings of samples indicate the nonuniformity of AlGaN film on Si (111) will increase sharply as the Al content increases.

2. Experiment

AlxGa1-xN epilayers were grown by MOCVD on 2-in Si (111) substrates. The Si substrates were degreased by hot H2SO4 solutions for 5 min, NH3•H2O:H2O2:H2O (1:1:5) solutions for 5 min, HCl: H2O2:H2O (1:1:5) solutions for 5 min in due order, and then etched with HF (2%) for 0.5 min to remove the surface oxide layer. This procedure results in an oxide-free, hydrogen-terminated Si surface. Trimethylaluminum (TMA), Trimethylgallium

(TMG), and ammonia were used as Al, Ga, and N sources, respectively. H2 was used as a carrier gas. After preparation, Si (111) substrates were heated under H2 ambient at 1060 ℃ for 3 min to clean its surface prior to growth. Pre-deposition of Al was used to prevent the formation of SiNx. In the initial stage of growth, a 60nm thick HT-AlN layer was grown, which is used as nucleation layer, and then 20 periods of AlN/GaN (2nm/3nm) superlattices was grown. The introduction of such superlattices aims at creating an additional compressive strain during the growth or the cooling down of the sample, in order to counterbalance the tensile strain in nitrides deposited on silicon[12]. Finally, another 30-40nm HT-AlN buffer layer was grown. Following the composite buffer layer, 500 nm Al0.25Ga0.75N epilayers and 500 nm Al0.32Ga0.68N epilayers were grown for sample A and B respectively. The samples were investigated by AFM and XRD rocking curve measurements. UV PL was performed in a home-made vacuum spectroscope. The excitation source was a frequency quadrupled Ti: sapphire laser. The sheet resistance mappings of samples were measured by LEI 1500 series contactless resistance mapping system.

3. Results and Discussion

The HR-XRD 2θ-ω scan for the sample has been measured. Using Vegard's law, the Al content in the AlxGa1-xN film can be estimated to be 25% and 32% respectively. The low-temperature (6 K) continuous-wave (cw) PL spectrum of samples is presented in Fig.1.

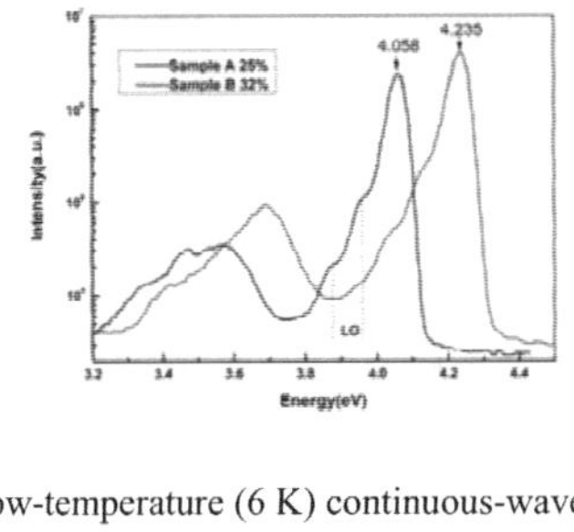

Fig.1. the low-temperature (6 K) continuous-wave(cw) PL spectrum of samples

It exhibits the main free excition emission line at 4.058 ev and 4.235 ev respectively. And the value of content deduced by band gap is in accordance with XRD's. Two weaker peaks are observed at each curve. The energy separation between the successive emission lines is about 90 mev [13-15], which corresponds to the value of the LO phonon energy in GaN. Thus, we assign the two lower energy peaks to the n=1 and n=2 phonon replicas of the main exciton transition line (n=0) in order by decreasing energy. The appearance of the phonon replica lines is a clear indicator of a strong carrier-phonon interaction in AlGaN on Si.

Under the optical microscope, many pits are observed on the surface of samples, especially in sample B. As we know, (Al)GaN epilayers on Si is a mosaic structure, several reports have indicated that the surface pits are attached to threading dislocations with screw components [16, 17] and the distribution is due to the misorientation between adjacent subgrains. We known at the same time that the full width of half maximum (FWHM) of the XRD rocking curve symmetric (0002) ω-scan is only broadened by dislocations consisting of screw components, therefore, for further studying the influence of the pits on crystal quality, the rocking curve symmetric (0002) ω-scan of our samples have been done, as shown in Fig.2 (a). A GaN epilayers sample with the same buffer structure also has been measured as a contrast. There is a visible bulge on low angle side for AlGaN samples. In combination with XRD schematic diagram, as shown in Fig.2 (b), the symmetry of curve shape is decided by the arrangement of crystal face which can satisfy a demand of Bragg reflection for X-ray.

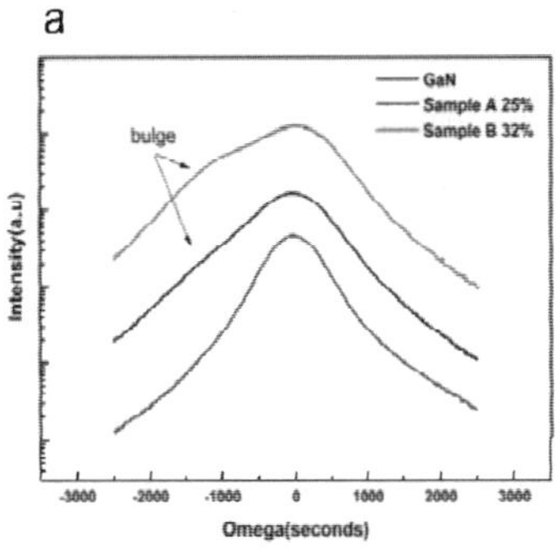

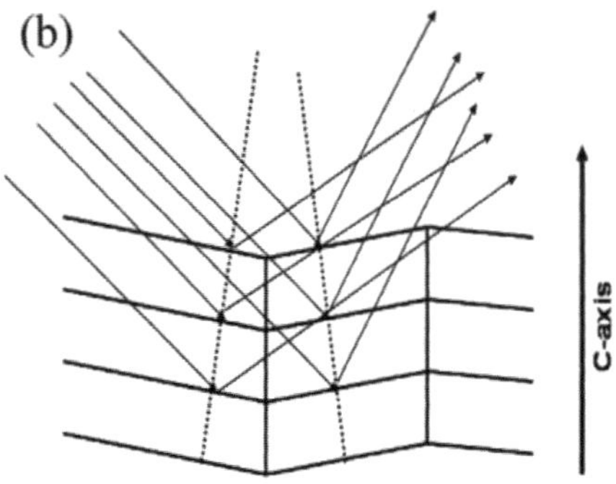

Fig.2. the rocking curve symmetric (0002) ω-scan of samples (a) and the schematic diagram of bulge appearance (b)

Thus, the result that sample B has a larger bulge implies that the film preferred orientation (001) represents a more deviation compared with sample A. The appearance of a large number of pits is a good evidence for the deviation of orientation (001) between adjacent subgrains. In fact, AFM measurements also agree with the deduction from XRD rocking curve measurements. As shown in Fig.3; the root-mean-square (rms) roughness is 3.69 nm for sample A and 5.23 nm for sample B in 5×5μm2 scan area, and pits scatter on the two sample's surface.

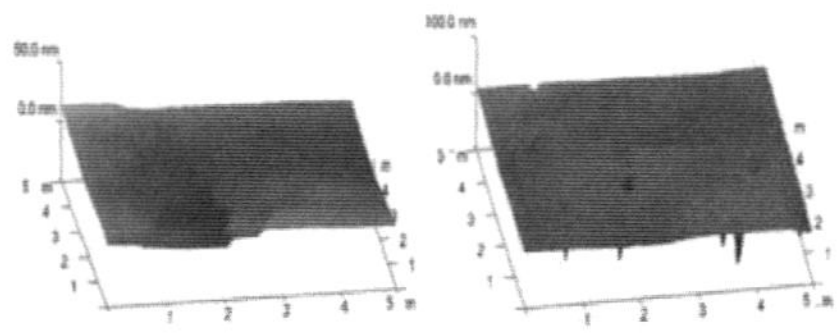

Fig.3. 3D AFM images (5×5μm2 scan) of Al0.25Ga0.75N (left) and Al0.32Ga0.68N (right)

Besides the pits, the wavy surface is also observed as shown in the set of Fig.3. The stepping curve is the white line trace in AFM measurements. The wavy surface has a low density of pits and the flat surface has a high density of pits. Taking AlGaN/Si, the big mismatch system into consideration, the probable interpretation is that the formation of the wavy surface is a way to accommodate strain energy and the formation of pits is a way to release the tensile pressure, which is all caused by CTE or lattice mismatch between AlGaN and Si. When the Al content reaches the threshold value, may be 30%, the function of accommodation will switch to stress release and led finally to the formation of pits. In addition, the sheet resistance mappings of the samples are shown in Fig.4,

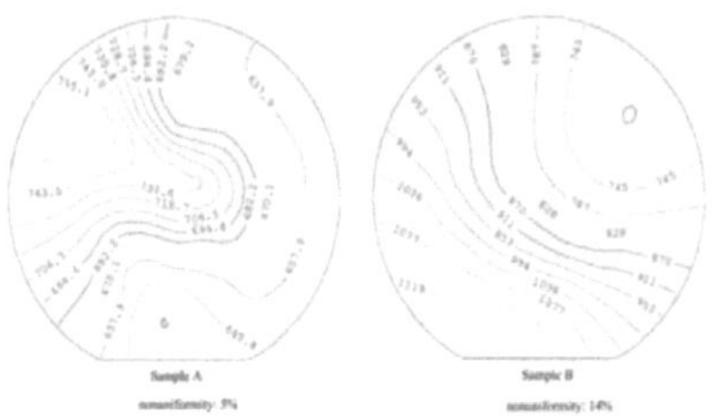

Fig.4. the sheet resistance mappings of samples

Obviously, the nonuniformity increases sharply as the Al content increases. Although the 5% nonuniformity of sample A is just passable, the value can still be reduced for preparation of devices.

4. Conclusion

AlGaN film has been grown on Si (111) substrates. The low-temperature PL spectrum shows two clear phonon replica peaks which indicate the samples have better quality in a small-localized region. The film preferred orientation (001) represents a deviation by analysis of XRD rocking curve. AFM measurements show two type of surface morphology, further analysis indicate that there may be a critical Al content value for the formation of surface pits. In addition, the wafer nonuniformity increases sharply as the Al content increases.

References

1. M.A. Khan, M.S. Shur, J.N. Kuznia, Q. Chen, J. Burm, W. Schaff, Temperature Activated Conductance in Gan/Algan Heterostructure Field-Effect Transistors Operating at Temperatures up to 300-Degrees-C, Appl Phys Lett, **66** (1995) 1083-1085.
2. S. Yoshida, J. Suzuki, Reliability of metal semiconductor field-effect transistor using GaN at high temperature, J Appl Phys, **84** (1998) 2940-2942.
3. 3.O. Aktas, Z.F. Fan, S.N. Mohammad, A.E. Botchkarev, H. Morkoc, High temperature characteristics of AlGaN/GaN modulation doped field-effect transistors, Appl Phys Lett, **69** (1996) 3872-3874.
4. V. Adivarahan, W.H. Sun, A. Chitnis, M. Shatalov, S.

Wu, H.P. Maruska, M.A. Khan, 250 nm AlGaN light-emitting diodes, Appl Phys Lett, **85** (2004) 2175-2177.

5. R. McClintock, A. Yasan, K. Mayes, D. Shiell, S.R. Darvish, P. Kung, M. Razeghi, High quantum efficiency AlGaN solar-blind p-i-n photodiodes, Appl Phys Lett, **84**(2004) 1248-1250.
6. J.Y. Duboz, N. Grandjean, F. Omnes, J.L. Reverchon, M. Mosca, Solar blind detectors based on AlGaN grown on sapphire, E-MRS 2004 Fall Meeting Symposia C and F, **2** (2005) 964-971.
7. A. Dadgar, M. Poschenrieder, J. Blasing, O. Contreras, F. Bertram, T. Riemann, A. Reiher, M. Kunze, I. Daumiller, A. Krtschil, A. Diez, A. Kaluza, A. Modlich, M. Kamp, J. Christen, F.A. Ponce, E. Kohn, A. Krost, MOVPE growth of GaN on Si(111) substrates, J Cryst Growth, **248** (2003) 556-562.
8. G.Y. Zhao, H. Ishikawa, H. Jiang, T. Egawa, T. Jimbo, M. Umeno, Optical absorption and photoluminescence studies of n-type GaN, Jpn J Appl Phys 2, **38** (1999) L993-L995.
9. S. Nikishin, G. Kipshidze, V. Kuryatkov, K. Choi, I. Gherasoiu, L.G. de Peralta, A. Zubrilov, V. Tretyakov, K. Copeland, T. Prokofyeva, M. Holtz, R. Asomoza, Y. Kudryavtsev, H. Temkin, Gas source molecular beam epitaxy of high quality AlxGa1-xN (0 <= x <= 1) on Si(111), Journal of Vacuum Science & Technology B-an International Journal Devoted to Microelectronics and Nanometer Structures-Processing Measurement and Phenomena, **19** (2001) 1409-1412.
10. M.A. Sanchez-Garcia, E. Calleja, F.B. Naranjo, F.J. Sanchez, F. Calle, E. Munoz, A.M. Sanchez, F.J. Pacheco, S.I. Molina, MBE growth of GaN and AlGaN layers on Si(111) substrates: doping effects, J Cryst Growth, **202** (1999) 415-418.
11. K. Cheng, M. Leys, J. Derluyn, K. Balachander, S. Degroote, M. Germain, G. Borghs, AlGaN-based heterostructures grown on 4 inch Si(111) by MOVPE, Physica Status Solidi C - Current Topics in Solid State Physics, Vol 5, No 6, **5** (2008) 1600-1602.
12. E. Feltin, B. Beaumont, M. Laugt, P. de Mierry, P. Vennegues, H. Lahreche, M. Leroux, P. Gibart, Stress control in GaN grown on silicon (111) by metalorganic vapor phase epitaxy, Appl Phys Lett, **79** (2001) 3230-3232.
13. W. Liu, M.F. Li, S.J. Xu, K. Uchida, K. Matsumoto, Phonon-assisted photoluminescence in wurtzite GaN epilayer, Semicond Sci Tech, **13** (1998) 769-772.
14. A. Sedhain, J. Li, J.Y. Lin, H.X. Jiang, Probing exciton-phonon interaction in AlN epilayers by photoluminescence, Appl Phys Lett, **95** (2009).
15. L. Bergman, M. Dutta, C. Balkas, R.F. Davis, J.A. Christman, D. Alexson, R.J. Nemanich, Raman analysis of the E1 and A1 quasi-longitudinal optical and quasi-transverse optical modes in wurtzite AlN, J Appl Phys, **85** (1999) 3535-3539.
16. B. Jahnen, M. Albrecht, W. Dorsch, S. Christiansen, H.P. Strunk, D. Hanser, R.F. Davis, Pinholes, dislocations and strain relaxation in InGaN, Mrs Internet J N S R, **3** (1998) art. no.-39.
17. H. Sasaki, S. Kato, T. Matsuda, Y. Sato, M. Iwami, S. Yoshida, Investigation of surface defect structure originating in dislocations in AlGaN/GaN epitaxial layer grown on a Si substrate, J Cryst Growth, **298**(2007) 305-309.

Hot Spots Detection of Operating PV Arrays through IR Thermal Image Using Method Based on Curve Fitting of Gray Histogram

Jiang Lin[1,a], Su Jianhui[1] and Li Xin[1]

[1]*Electrical Engineering and Automation School, Hefei University of Technology, Hefei, Anhui, China*

Abstract. The overall efficiency of PV arrays is affected by hot spots which should be detected and diagnosed by applying responsible monitoring techniques. The method using the IR thermal image to detect hot spots has been studied as a direct, noncontact, nondestructive technique. However, IR thermal images suffer from relatively high stochastic noise and non-uniformity clutter, so the conventional methods of image processing are not effective. The paper proposes a method to detect hotspots based on curve fitting of gray histogram. The result of MATLAB simulation proves the method proposed in the paper is effective to detect the hot spots suppressing the noise generated during the process of image acquisition.

1 Introduction

Hot spots of PV modules can be caused by poor cell matching, cracks, localized soiling (bird droppings), or shadowing [1]. A hot spot can occur when a cell (or cells) are forced into reverse bias because it (they) cannot carry the peak power current being produced by the other cells in series. Cells are supposed to be protected by the bypass diodes while the bypass diodes often fail or are never installed correctly. The performance and lifetime of PV system are affected by hot spots seriously [2].

Hot spots make the surface temperature of PV module rise abnormally so that they can be detected by infrared thermography. The detection of hot spots in PV modules through thermal image is a direct, noncontact, nondestructive technique which can not only get more accurate results, but also locate the fault point to evaluate the efficiency.

J.A. Tsanakas, etc. obtain thermal images of an operating photovoltaic array by field infrared thermographic measurements. The thermal images are analyzed by means of temperature line profiles, image histogram features and the results reveal the presence of several hot spots in the four modules of the array that are all associated with specific discontinuous cells[3]. Then, J.A. Tsanakas, etc. proposes the use of standard thermal image processing and the Canny edge detection operator as diagnostic tools for module-related faults that lead to hot-spot heating effect [4]. Wang Pei-zhen presents a new method of target feature extracting in view of low contrast of target and background, strong noise and non uniformity in the infrared image and peculiarity of the photovoltaic array. First, the subtraction operation is carried out between the two infrared images of photovoltaic array which are to be analyzed and under normal condition respectively. Then, a method of combining the two dimensional threshold and the fuzzy means algorithm is implemented [5][6].

However, there are still several key problems needed more experiments to research such as the interference of complex environment and noise. It has been found in the experiments that due to some factors such as reflecting sunlight, the thermal image of the PV arrays will appear false hot spots to cause "false alarms"; on the other hand, the emissivity of the glass on PV modules varies with the shooting angel to cause the thermal radiation quantity received by the thermal imager changed that may be conflicting with the actual variations related to hot spots and, thus, cause "false alarms"[7]. So it is necessary to suppress noise in the thermal image to accurate hot spots detection.

If the quantity of thermal image is good enough with little noise, the hot spots can be detected successfully using the method proposed by J.A. Tsanakas and Wang Peizhen etc. However, even If the images are preprocessed with Gauss filter or media filter, the hot spots can't be detected successfully from a thermal image with a large amount of noise.

In this paper, the method based on curve fitting of gray histogram is proposed in thermal image processing to suppress noise and detect hot spots accurately. Experiments and research on the interference of the IR thermal image of the PV module (noise sources in complex environment and image acquisition) are carried out.

2 Method based on curve fitting of gray histogram to detect hot spots

[a] Corresponding author: jianglin_hfut@126.com

The gray level histogram of image is a function of the gray level, which is described as the number of pixels with a gray level in the image: its horizontal coordinate is gray level, and the vertical coordinate is the frequency of the gray scale (the number of pixels in the digital image). "Peak"on histogram shows the frequency of the gray level is relative large, whereas "valley" is illustrated in that the probability of the gray level appears small.

Fig. 1 is an infrared thermal image of photovoltaic module. Fig.2 is the gray histogram of Fig.1. It can be seen from the two figures that the backgrounds, the normal working part and the hot spots of the PV module have obviously different gray levels which correspond to multiple "peaks" and "valleys" in the gray histogram obviously. As marked in the figure, the normal working part of the PV module corresponds to the highest "peak" of the gray histogram, and the hot spots correspond to lower "peak" of the gray histogram.

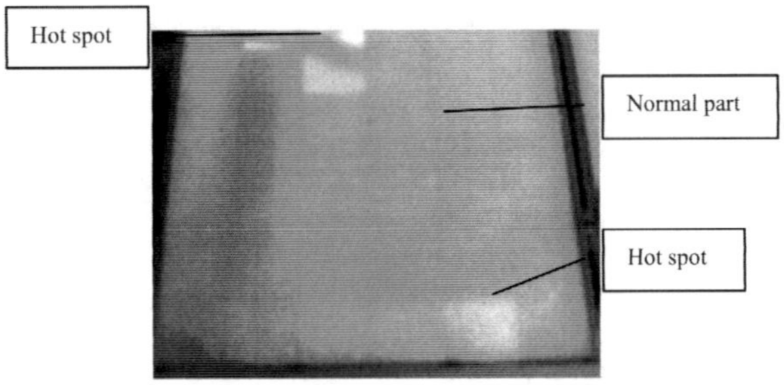

Figure 1. Infrared thermal image of PV module

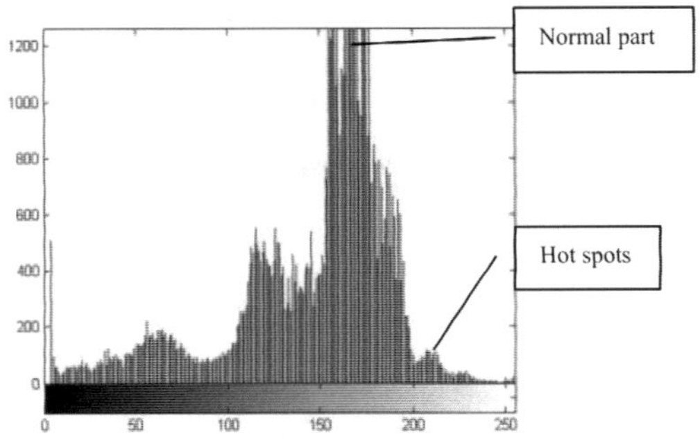

Figure 2. Gray histogram of Fig.1

Due to the influence of noise and environment interference to the infrared image, the gray histogram exists burrs and small local ups and downs, so in many cases it is difficult to detect the interest "peak" by conventional methods robustly and accurately. In order to solve this problem, this paper proposes a new image segmentation method based on curve fitting of gray histogram which is used to detect hot spots of PV modules rapidly and accurately. The method is described as follows:

Firstly, the infrared thermal image of the PV module is preprocessed. Because the thermal image is easy to be polluted by noise, the Gauss filter is used to filter the image. Gauss filter is a kind of linear smoothing filter based on the shape of Gauss's function which is effective to suppress the noise subjected to the normal distribution.

Secondly, the gray histogram of the infrared image is identified and determined. The vertical coordinate system for the histogram is set: the positive direction of the X axis is set to the increase of the gray value, the Y axis is marked by the frequency of gray level. So the gray histogram of the infrared image is established as formula (1):

$$y = p(x) \qquad x = 0,2,3K\ 255 \tag{1}$$

x: gray levels of 8 bit gray scale digital image

y: the number of pixels of a gray level

Due to the influence of noise and environment interference to the infrared image, the gray histogram exists burrs and small local ups and downs, so it is difficult to detect the interest "peak" and "valley" by conventional methods. Therefore, the least square method is used to fit the gray level histogram. The gray level histogram provides us with 256 sample points whose coordinates are seen as formula (2) :

$$(x_i, y_i), i = 0K\ 255, x_i = i, y_i = p(x_i) \tag{2}$$

According to the principle of least square method, a curve $y = \phi(x)$ is found which makes the sum of squares of deviation [formula (3)] between the curve and the sample points minimal. The histogram is fitted as a smooth curve eliminating the burrs and local fluctuations. The fitting curve of gray histogram of Fig.2 is shown in Fig.3.

$$\sum_{i=1}^{256} \delta_i^2 = \sum_{i=1}^{256} (\phi(x_i) - y_i)^2 \tag{3}$$

The result of first-order derivative and second derivative of the fitting curve function $y = \phi(x)$ can be used to determine the abscissa of the "peaks" and "valleys" in the histogram.

If $\phi'(x_{p_i}) = 0, \phi''(x_{p_i}) > 0$, $[x_{p_i}, \phi(x_{p_i})]$ is the position of "peak";

Else if $\phi'(x_{v_j}) = 0, \phi''(x_{v_j}) < 0$, $[x_{vj}, \phi(x_{v_j})]$ is the position of the "valley".

DOI: 10.1051/matecconf/20166106017

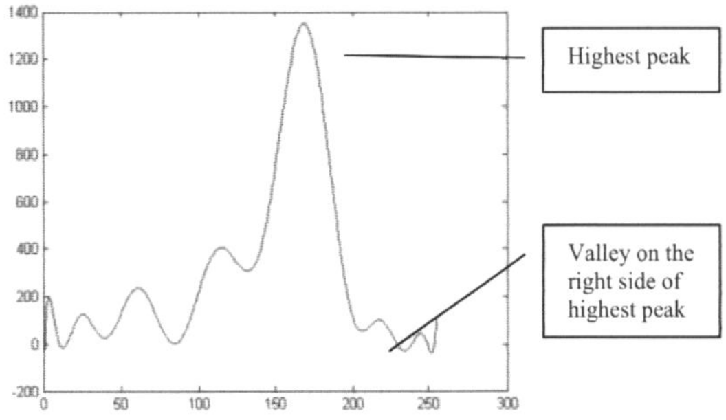

Figure 3. Fitting curve of gray histogram

Based on the characteristics of infrared thermal image of PV module , we can select gray value corresponding to the first 'valley' on the right hand side of the highest peak as the segmentation threshold, so that the hot spots can extracted .The threshold T is set according to formula(4).

$$T = \min\{x_{v_j} > x_{p_m}\} \tag{4}$$

x_{p_m} is the abscissa of the highest "peaks"

x_{vj} is the abscissa of all the "valleys"

3 Experimental result

3.1. Hardware and software

The experimental part of this study includes one portable infrared thermal imager regarding PV arrays installed on the rooftop of a building in Sungrow Power Supply Co., Ltd. located in Hefei, China. PV arrays consist of crystalline silicon modules, type JA SOLAR JAP6-60-255/3BB. The infrared thermal camera employed for the performed measurements is SAT-S80 .

Basic processing of the obtained thermal images is performed with the use of SatIr Report software. Further processing is implemented in Matlab R2014a environment.

3.2. Experimental outline

The thermographic measurements took place in the city of Hefei (north-eastern China, longitude: N31° 52′ , latitude: E117° 17′ , mean elevation: 19.05 m) at 11:30 in Nov.26th. under clear-sky conditions. The environmental conditions are given in Table 1.The inspected PV arrays are installed at a fixed angle θ ≈40 °

Table 1. Environmental conditions in experiment

Time	temperature	illuminance	Wind speed
2015/11/26 11:30	7°C	4530lux	3.4m/s

3.3 Result and discussion

In the experiment on 26 November, we collected one image of the PV module with hot spots, as shown Fig 1. It is very clear from the thermal image that some hot spots appear on the PV module which may be caused by dust or internal defects.

In the past research, the hot spots are detected by the thermal image of the PV modules, using the edge detection method based on Canny operator or threshold segmentation method or fuzzy clustering method which are effective for the thermal image in absence of noise (noise sources in the complex environment such as light, shadow or the acquisition process of thermal images). If the thermal image is good enough with little noise, the hotspots can be detected successfully using the OSTU method or Canny operator.

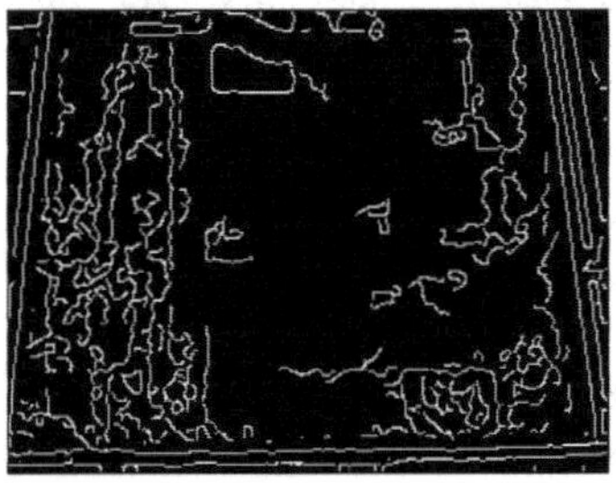

Figure 4. Canny edge detection of hot spots

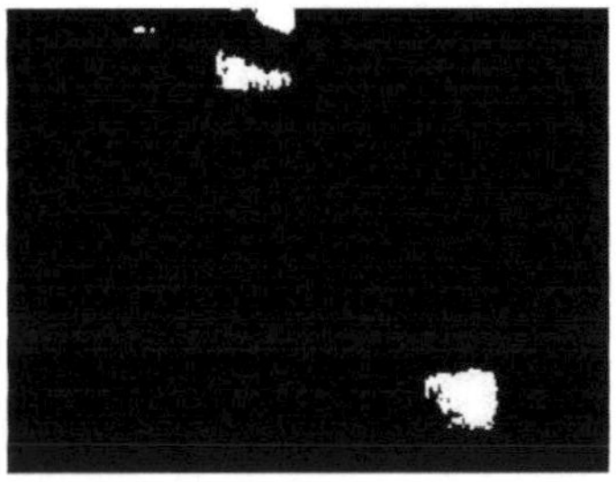

Figure 5. Hot spots detection using method of fitting curve of histogram

However, the noise in the thermal image as shown in Fig.1 is obvious: the dark areas in Fig.1 don't always mean the low temperature, instead, the emissivity of the PV modules varying with the shooting angel causes the thermal radiation quantity received by the thermal imager changed; Some dark areas in Fig. 1 are the gaps between PV modules, namely ambient temperature. The noises in Fig. 1 is so obvious that even If the images are preprocessed with conventional methods such as Gauss filter or media filter, the actual effect is poor.

Fig.1 is processed using the canny operator to result in fig.4 in which the hot spots are detected but submerged in large of noises.

The research about the infrared thermal image with noises in the complex environment is less, so the main work in the paper is to propose a method to detect hotspots based on curve fitting of gray histogram for this kind of thermal image.

Gray histogram is a function of gray level, which means the number of pixels of each gray level in the image, that is, the frequency of each gray level appears in the image.

Fig.2 is the gray histogram of Fig.1. Fig. 3 is the fitting curve of Fig.2 using the fitting technique based on the least square.

It can be seen from the histogram as shown in Fig.2 that the pixels with the highest frequency of gray level mean the normal area of PV modules. So it is realized the first valley bottom of the fitting curve after its peak is the key point which can separate hot spots from normal areas of PV modules and the abscissa of the valley bottom is the threshold for segmentation.

According to the formula (4), the segmentation threshold is determined and the hot spots are extracted successfully from the thermal image as shown in Fig.5.

The result of MATLAB simulation proves the method proposed in the paper is effective to detect the hot spots suppressing the noise generated during the process of image acquisition .

4 Conclusions

The detection of hot spots in operating PV array through IR thermal images is studied in the paper. The experiment is made on the rooftop of a building in Sungrow Power Supply Co., Ltd. in Hefei and multi thermal images of the PV modules with hotspots are collected. It is found that the thermal images suffer from relatively high stochastic noise and non-uniformity clutter caused by the complex environment and emissivity uncertainty. From the MATLAB simulation, we can see the hotspots can't be detected using the conventional image processing method. So a new method based on curve fitting of gray histogram is proposed in this study. The result of MATLAB simulation proves the new method is effective to detect the hot spots suppressing the noise generated during the process of image acquisition.

The new method proposed in this paper need to be improved. In curve fitting of gray histogram spine function may be chose to accurate the threshold. The result of segmentation of thermal images can be classified to identify the fault of PV arrays. These are further work of our research team.

References

1. John H. Wohlgemuth , Sarah R. Kurtz , "How Can We Make PV Modules Safer? " , *2012 IEEE Photovoltaic Specialists Conference*(2012)
2. Ahuja, D., and M. Tatsutani. "Sustainable Energy for Developing Countries",*A report to the Academy of Sciencesfor the DevelopingWorld (TWAS)* (2008)
3. J. A. Tsanakas, P. Botsaris, "On the Detection of Hot Spots in Operating Photovoltaic Arrays through Thermal Image Analysis and a Simulation Model", *Materials Evaluation* (2013)
4. J.A. Tsanakas, D. Chrysostomou, P.N. Botsaris and A. Gasteratos, "Fault diagnosis of photovoltaic modules through image processing and Canny edge detection on field thermographic measurements", *International Journal of Sustainable Energy*, **34**(2015)
5. Wang Peizhen ,Zheng shicheng, "Fault Diagnosis of Photovoltatic Array Based on infrared image", *Acta Energias Sinica* , **31**(2010) (in Chinese)
6. WANG Peizhen, WANG Qunjing, YANG Weihan, "A method for the feature extraction of the infrared image of the photovoltaic array", *Journal of Hefei UniversityN of Techolygy*(2004)(in Chinese)
7. Zhiyuan Zou, Yihua Hu, Bin Gao, W. L. Woo, Xinyu Zhao, "Temperature recovery from degenerated infrared image based on the principle for temperature measurement using infrared sensor", *Journal of Applied Physics* (2014)(in Chinese)

The Application based on Fourier Transform in Time Lens

Rulin Liu[1], Shuqin Guo[1,2] and Dacai Xu[1]

[1]Zhejiang Provincial Key Laboratory for Signal Processing, College of Information Engineering, Zhejiang University of Technology, Hangzhou 310023, Zhejiang, P.R. China
[2]Key Laboratory of Specialty Fiber Optics and Optical Access Networks, Shanghai University, 200072 Shanghai

Abstract. Using the similarity between the transmission of narrow band light pulses in the dispersion media and the transmission of spatial beam in paraxial diffraction, build a time lens system similar to the space lens. Space lens can achieve compression and amplification of imaging function, and it has the nature of Fourier transform in light field. Then the time lens can also realize the compression and amplification of the light pulse and its Fourier transform. Using the function of Fourier transform in the time domain of time lens system, achieved the effect of eliminating noise and removing side lobe, improved the white noise which generally exist in the communication system and the side lobe effect which will produce in the relevant processes.

1 Introduction

By comparing the propagation equation (Fraunhofer diffraction) of the paraxial diffraction in free space with the dispersion equation of the narrow band light pulse in the dispersion medium transmission, we found they are very similar. When the paraxial beam is transmitted its space frequency introduced the second-order phase, and narrow band light pulse also introduced second-order phase in time frequency. In addition, in spatial optical processing, space lens can be viewed as a spatial phase modulator, and it provides a secondary phase modulation for space position variables, so people provide a secondary phase modulation for time variables using the similar principle of diffraction and dispersion transport. Similar to the spatial imaging systems, people set up a time imaging system and call it the time lens [1-2].

We know the space lens can achieve the function of compression and amplification imaging and spatial Fourier transform on the focal length of the light field. Similar to the space lens, time lens can realize signal compression, amplification and Fourier transform. Then, it opens up a new path in multiple fields of optical signals. In recent years, people carried on the thorough research of time lens both at home and abroad. The theory of time lens continuously perfect [3-6], and it has broad application prospects such as the optical pulse compression and amplification [7], transformation of time domain, frequency domain [8], optical signal detection and process [9-10], etc..

2 The principles of Fourier transform using time lens

In the system of time lens, the similarity of the signal's dispersion transmission and the diffraction of light are based on the effective ignorance of the nonlinear effects [1], high-order dispersion effect and the loss of the fiber.

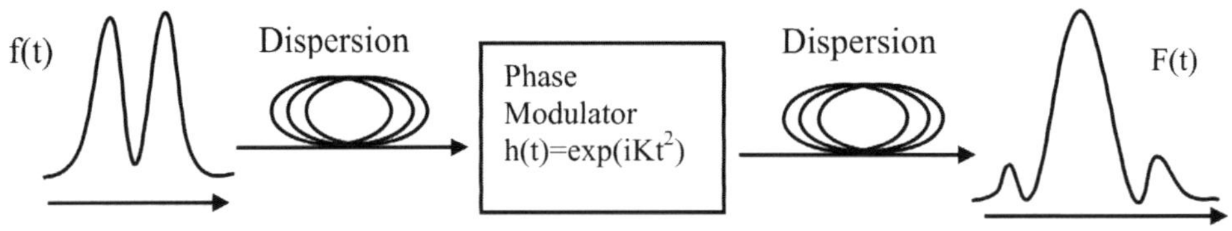

Figure 1. The structural model of time lens

Figure 1 is a structural model of the time lens which is based on the spatial lens system. The system can be part of two dispersion fiber, and the phase modulator can be used the Mach-Zehnder Interferometer (MZI).

Assuming that the input signal is f(t), its spectrum is F(w), the dispersion is equal to the length of the fiber multiplied by the second-order dispersion, phase modulator h(t)= exp(iKt2). In the frequency domain, when ignoring the nonlinear effect, the loss and the high-

This work is supported by STCSM

order dispersion of the fiber, the transfer function of the fiber or the dispersion device is

$$H_D(w) = \exp(\frac{i}{2}\beta_2 L w^2) \quad (1)$$

For the phase modulator h(t)= exp(iKt2), K is the phase shift coefficient of the phase modulator, its transfer function in spectrum domain is

$$H_M(w) = F[h(t)] = \sqrt{i\pi / K}\exp(\frac{-iw^2}{4K}) \quad (2)$$

Where the F is the Fourier transform operator. The input signal is expressed as s(t,0), after the first fiber but before the modulator, it is expressed as $s(t,L_1)$, its spectrum is $S(w,L_1)$, then,

$$S(w,L_1) = S(w,0)H_D(w) \quad (3)$$

$$s(t,L_1) = F^{-1}[S(w,L_1)] \quad (4)$$

After the modulator,

$$s(t,L_1+M) = s(t,L_1)h(t) \quad (5)$$

$$S(w,L_1+M) = F[s(t,L_1+M)] \quad (6)$$

Then through the second fiber, the final signal is

$$s(t,L_1+M+L_2) = F^{-1}[S(w,L_1+M)H_D(w)] \quad (7)$$

Finally, after finishing,

$$s(t,L_1+M+L_2) = \frac{1}{2\pi|\beta_{22}L_2|}\sqrt{\frac{i\pi}{K}}S(\frac{t}{|\beta_{22}L_2|},0)*$$
$$\exp(t^2(\frac{\beta_{21}L_2+\beta_{22}L_2}{2(\beta_{22}L_2)^2}-\frac{1}{\beta_{22}L_2})) \quad (8)$$

The result of formula 8 show that, when $\beta_{21}L_1 = \beta_{22}L_2$, the output signal is the Fourier transform waveform of the input signal, this is exactly the same with the conditions of the spatial Fourier transform. And in the process of the formula 8, it found that the coefficient K is related to second dispersion, it must meet

$$K = \frac{1}{2\beta_{22}L_2} \quad (9)$$

From the above, the time lens system which are composed of input dispersion fiber, output dispersion fiber and phase modulator can achieve the Fourier transform from the input signal in spectrum to the output signal in time domain. And about the principles and conditions for the time lens imaging, Kolner made a detailed discussion [1-2].

3 Constructing double-lens system to achieve signal shaping and de-noising

After using a single lens system to achieve the Fourier transform of the time domain, that we can not only do a series of effective treatment to the signal, it is but also need to convert the signal from the time domain to frequency domain. The relationship between Fourier transform and inverse transform are as follows.

$$F(f(t)) = F(w) \Leftrightarrow F[F(t)] = 2\pi f(-w) \quad (10)$$

From the above equation, the signal can be converted from the frequency form to the time form by the time lens. Thus we try to build a two-lens system and converse them to the original time-domain signal after the signal is processed.

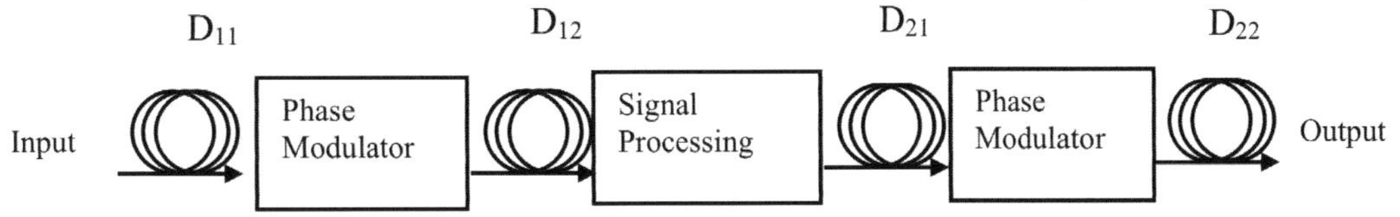

Figure 2. The structural model of double-lens

The two-lens system modal is shown as Fig 2.In the current communication system, white noise very commonly exists. However, it is difficult to eliminate white noise which throughout the frequency-domain by conventional signal processing devices, and for ultra short light pulses, especially for those picoseconds femtoseconds optical pulses, the demanding to sensitivity of the device is quite high , but reaction rates of general signal processing device is not so high. However, the narrower the ultra short pulse signals in the time domain, the spectrum will be wider, so that the time-domain waveform signals can be reduced after converting to a wide spectrum of the signal sensitivity of the device, such as using a filter to removal signal of side lobes etc., but in communication system to process the related signal will produce side lobes. Try to use double lens to convert the main lobe of the spectrum to the time domain, which effectively eliminate the ultra short pulse of side lobes.

In the time domain, Gaussian white noise obeys Gaussian distribution, but in the frequency domain, it is uniformly distributed. As a result, in the time domain Gaussian white noise signal has an effect of the disturbances of a signal, but it has little effect on the frequency spectrum of the signal. Assume the Gaussian white noise is n(t), and the original signal is s(t). Gaussian white noise is additive noise. So The input signal which contains Gaussian white noise of f(t) is n(t)+s(t), and the Fourier transform is as following :

$$F[f(t)] = \int_{\infty}^{\infty}[s(t)+n(t)]e^{-iwt}dt = F[s(t)] + F[n(t)] \quad (11)$$

In the above equation, F[n(t)] can be effectively treated as a constant, That's to say Gaussian white noise hardly changes the spectrum envelope of the original signal, thus in the time domain we can do the Fourier transformation by the time lens, after that the spectrum of the input signal envelope package will change up to the time domain, and then converted back through double-lens system. When the signal goes through the first lens system, it can be added a band-pass filter to improve noise reduction effect.

When a delayed signal s(t) produced becomes s(t-t_0), the spectrum is

$$F(s(t-t_0)) = \frac{1}{2\pi}\int_{-\infty}^{\infty} S(w)e^{iw(t-t_0)}dw = e^{-iwt_0}S(w) \quad (12)$$

Thus, the signal is shifted in time introducing a phase shift in its spectrum, it does not change the shape of the spectral envelope, and so after the two-lens system can be restored s (t).

Use MATLAB to simulate the following procedure for the double-lens system of Fourier transform functions and Fourier transform functions. On the basis of Fourier transform that a single lens satisfy two dispersion values are equal and the phase shift coefficient with the number , assume the first time provided a lens system, at this time, and dispersion values of the lens system between the second time and the first time is equal, but opposite in sign, so the last signal inversion is not appear. As shown in Figure 3.

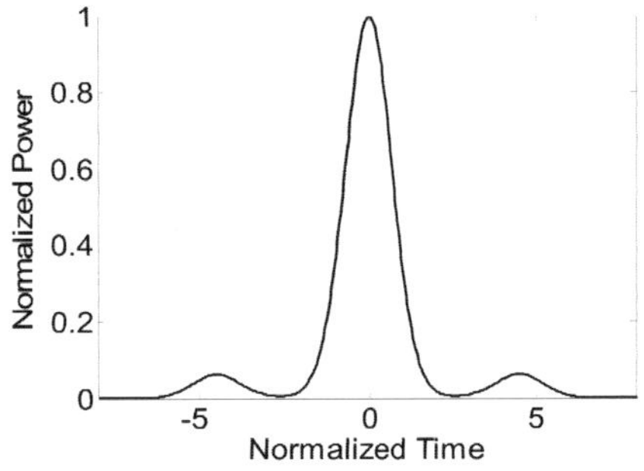

(a) The original signal in time

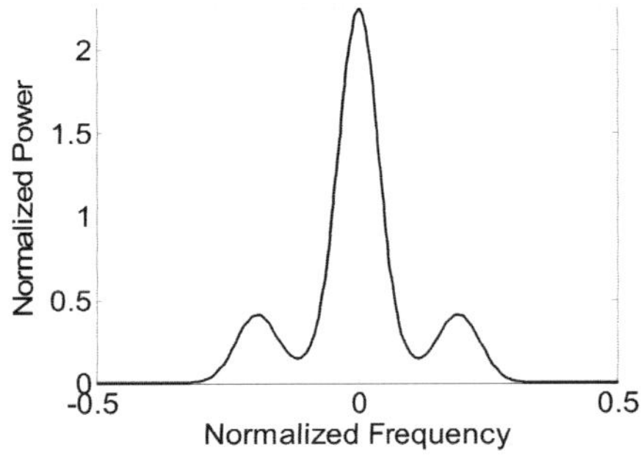

(b) The original signal in spectrum

Figure 3. The waveform and spectrum of the original

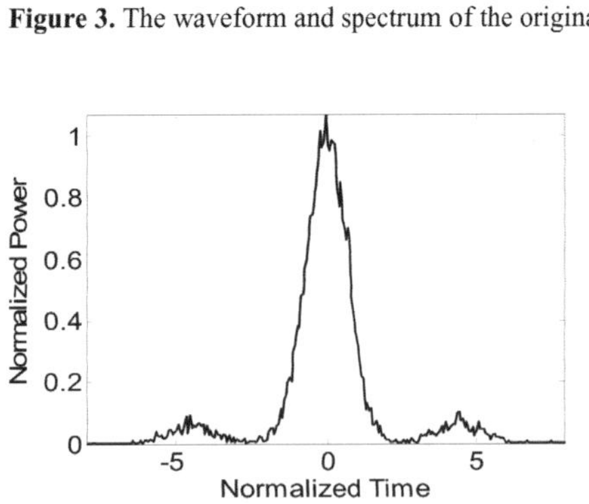

(a) The time signal with side lobe and AWGN

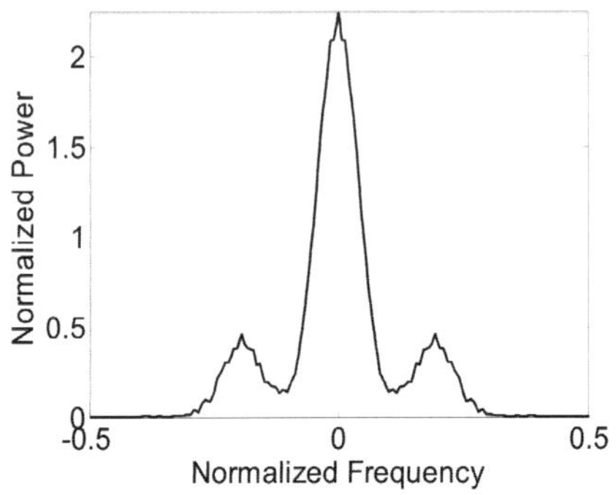

(b) The spectrum of the time signal

Figure 4. The waveform and spectrum of the input signal

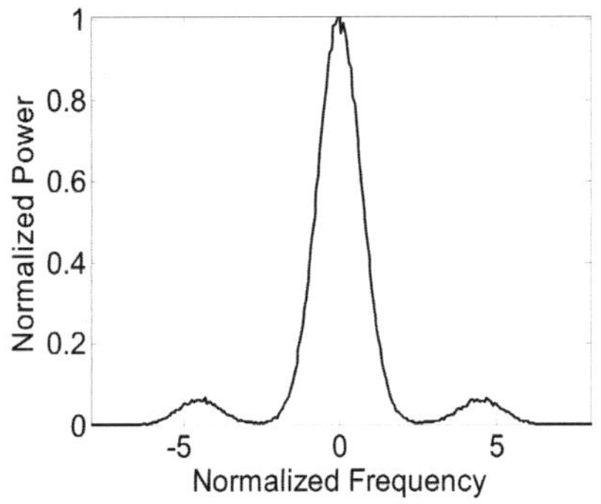

Figure 5. The spectrum of the signal after the first time lens

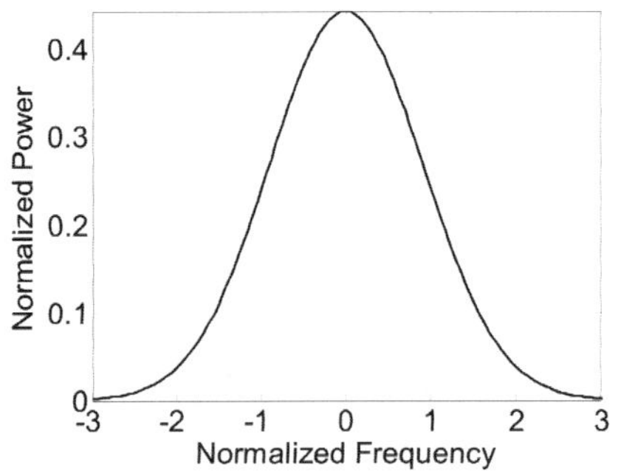

Figure 6. The spectrum after the filter

The Gaussian signal with Gaussian white noise and side lobe as an input signal of the two-lens system, after the first time through the lens into the system, the output waveform and spectrum are shown above, in order to facilitate the observation of comparison, we do the normalization process for variables.

The time-domain waveform of the output is the spectral shape belonging to the original the Fourier transform through the time lens system, which is consistent with the theoretical derivation. At the same time, the spectrum has changed to the corresponding time-domain waveform and scarcely observed original Gaussian white noise. If use an ordinary filter, the original signal side lobe cannot be effectively cut off. However, time-lens will transform the original extremely narrow time-domain waveform to a very wide frequency domain, using a common signal filtering can effectively filter in the frequency domain. Setting up a band-pass fiber in the first output of time lens system, the ideal band-pass filter, the transfer function h(t) = sinc(at), is used in order to idealized. Finally, the output

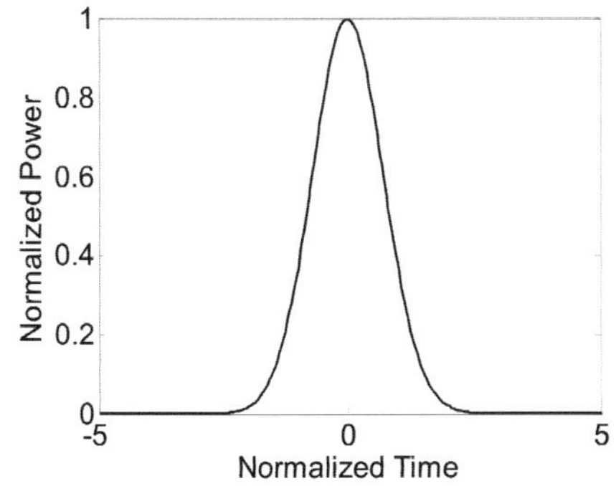

(a) The signal after the processing of double-lens

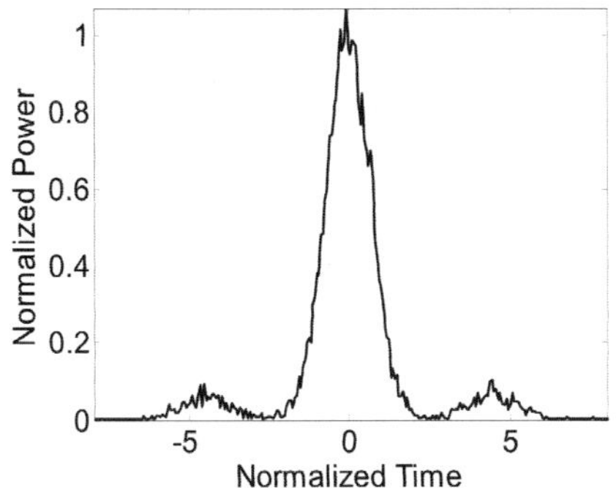

(b) The original signal side lobe and AWGN

Figure 7. The comparison between the original signal and processed signal

waveform and spectrum shown in Figure 7. Its frequency spectrum is Gaussian for a Gaussian pulse. After the noise reduction and side lobe amputated, the second time lens system is used to recovery ideal signal pulses to be obtained. The final output signal waveform shown in Figure 9, and compared with the original signal noise and side lobe. The above theoretical description and simulation visually demonstrate that the treatment effect of signal pulse, the effective realization of the noise filter, used the double-lens system by Fourier transform [10].

4 Conclusions

In this paper, we use the Fourier transform lens function of time to achieve a narrow pulse de-noising and cut side lobe. It is effective to improve the white noise witch exist in communication system commonly, and side lobe effect when we process the signal. But because it is the simulation to the signal, the experimental results will be affected by signal time window and sampling points, it can be not accurately show the relationship between narrow band time domain and frequency domain, and using the ideal band pass filter, so the actual effect of the theory can not be reached the theoretical results.

References

1. B.H. Kolner, M. Nazarathy, Opt. Lett., **14**, 630-632(1989)
2. B. H. Kolner, IEEE J. Quantum Electron., **30**, 1951 - 1963(1994)
3. A.A. Godil, Auld B A, Bloom D M, Appl. Phys. Lett, **62**, l047-1479(1993)
4. R. Salem, M.A. Foster, A.C. Turner, D.F. Geraghty, M.s Lipson and A. L. Gaeta, Opt. Lett., **33**, 1047- 1049(2008)
5. C.V. Bennett, B.H. Kolner, IEEE J. Quantum Electron., **36**, 430-437(2000)
6. T.Ng T., F. Parmigiani, M. Ibsen, Zhang ZH W, IEEE Photon. Technol. Lett., **20**, 1097-1099(2008)
7. J. A. Howe, C. Xu , Digest of the LEOS Summer Topical Meetings(2005)
8. M. T. Kauffman, W. C. Banyai, A.A. Godil and D. M Bloom, Appl. Phys. Lett., **64** , 270-272(1994)
9. A. A. Godil, B. A. Auld and D. M.Bloom., IEEE J. Quantum Electron., **30**, 827-837(1994)
10. S. Kumar, Opt. Lett, **32**, 346-348(2007)

The Illumination Model of the Valley Based on the Diffuse Reflect of Forest

Guoliang He[a], Jiashi Zhou, Cheng He and Zonghui Xie

University of Electronic Science and Technology of China, 611731 Chengdu, China

Abstract. In this paper, models are build to evaluate the impact of the forest on the valley's illumination. Based on the assumes that all the light reach the ground comes from the diffuse reflection which comes from the sun directly and from the diffuse reflection of other points, One model is build to consider the impact of time and latitude on the direction of the sunlight. So we can get the direction of the sunlight at different time and latitude through the model. Besides, this paper develops a illumination model to evaluate the intensity of illumination of the ground. Combining the models above, this paper get a complete model which can not only evaluate the overall light intensity of the valley but also convert the light intensity to the intensity of illumination. Simulation of the intensity illumination of some basic terrains and finally gives a comprehensive results which is practical and close to the common sense.

1 Introduction

It is known that in mountainous terrain, the brightness of the valley is generally lower than that of the hillside and the mountaintop which attribute to the shelter of trees and mountain peaks. Because of the various of the terrain, the latitudes, the time and the other factors, the intensity of illumination will be different from place to place.

As we all know, the intensity of illumination in the valley is not decided by the sunlight directly. That is to say, because of the diffuse reflection, the area without the sunlight is not necessarily all black. We therefore divide the influence factors into three aspects.

Firstly, we consider the impact of altitude on the intensity of illumination. Although the illumination is not influenced by the altitude directly, the altitude imposes a dramatic effect on the direction of the sunlight. We build the light directions model to contact the latitude and the intensity illumination. So the direction of sunlight finally determines the amount of incoming light.

Secondly, we consider the shelter of the forest. Here we develop a reflection model. In this model, the light from the sum contains two parts. One go through trees and the other reflect to the sky by trees. We will calculate the light reach the land to evaluate the light intensity of the valley.

Thirdly, we know that the density of trees varies with the slope of the mountain and there is also a great relationship between the diffuse reflection coefficient and the density of the tree. So we will build two models: One discuss the relationship between the diffuse reflection coefficient and the density of the tree while the other discuss the relationship between tree density and slope gradient.

Later on, we will combine the models above so as to get a complete model which can not only evaluate the overall light intensity of the valley but also convert the light intensity to the intensity of illumination.

At last, after analysis the impact of these factors, some propagation simulation of real situation can be carried out. We will use our models to simulate the intensity illumination of some basic terrains to assess the accuracy and practicality of our model. Furthermore, we can get a comprehensive simulation results which is practical and close to the common sense.

In order to establish the mathematics model easily, we introduce the following assumptions. In the paper, No consideration of the height of the tree. Generally speaking, comparing to the height of the mountain, the height of trees is so small that we treat them as a belt of trees with the same height and all kinds of terrains are evenly covered with trees. Besides, to simplify the question, we assumed that the shelter of the atmosphere has the same effect on the intensity of illumination everywhere. Namely, trees in the valley is close enough and entered no ray of light. So the ground of the valley can not receive light from the sun directly.

2 Theoretic analysis

2.1 Model Direction of Sunlight

Because the direction of the sunlight varies with the latitude and the time, firstly we have to create a coordinate system and build a model to discuss the relationship between the direction of the sunlight and the latitude as well as the local time.

Here is a glossary of specific audio terminology in our model[1].

(1)Solar time t_s,: Solar time is a reckoning of the passage of time based on the sun's position in the sky. The fundamental unit of solar time is the day. Two types

[a] Corresponding author: hegl@uestc.edu.cn

of solar time are apparent solar time (sundial time) and mean solar time (clock time).

China adopts the mean solar time of longitude 120 degrees east as the standard time for the whole country, namely "Beijing time". The latitude and longitude of Beijing is (39.9N,116E), showing that "Beijing time" is not the mean solar time of local Beijing. The longitude differ from the true mean solar time about 5 degrees, 20 minutes. The accurate mean solar time can be calculate as follows:

$$t_s = t_b + \frac{\varphi - 120}{15},$$

where t_b is the Beijing time, φ is the local latitude.

(2)Solar hour angle ω: Observing the sun from earth, the solar hour angle is an expression of time, expressed in angular measurement, usually degrees, from solar noon. At solar noon the hour angle is 0.0 degrees, with the time before solar noon expressed as negative degrees, and the local time after solar noon expressed as positive degrees. For example, at 10:30 AM local apparent time the hour angle is -22.5°(15°per hour times 1.5 hours before noon). So the solar hour angle is defined as:

$$\omega = 15(t_s - 12).$$

(3)Solar declination δ: The angle between the equator and the place on Earth where the sun is straight above you is called the "solar declination", The approximation value is given by

$$\delta = 23.45\sin\left(\frac{2\pi(284+n)}{365}\right).$$

(4)The solar elevation angle α: The solar elevation angle is the altitude of the sun, the angle between the horizon and the centre of the sun's disc. The approximate value can be calculated with the following formula:

$$\sin\alpha = \sin\varphi\cdot\sin\delta + \cos\varphi\cdot\cos\delta\cdot\cos\omega$$

where

h is the hour angle, in the local solar time,

δ is the current declination of the Sun,

φ is the local latitude.

(5)The solar azimuth angle A: The solar azimuth angle is the azimuth angle of the sun.It defines in which direction the sun is, whereas the solar zenith angle or its complementary angle solar elevation defines how high the sun is. There are several conventions for the solar azimuth, however it is traditionally defined as the angle between a line due south and the shadow cast by a vertical rod on Earth. This convention states the angle is positive if the line is east of south and negative if it is west of south The solar azimuth angle can be calculated to a good approximation with the following formula, however angles should be interpreted with care because the inverse sine, i.e. $x=\sin^{-1}(y)$, has multiple solutions, only one of which will be correct. So we have

$$\sin A = \frac{-\sin\omega\cdot\cos\delta}{\cos\alpha}.$$

From the above calculation formula, we can have a systems of equations as follows:

$$\begin{cases} t_s = t_b + \dfrac{\phi - 120}{15} \\ \omega = 15(t_s - 12) \\ \delta = 23.45\sin\left(\dfrac{2\pi(284+n)}{365}\right) \\ \sin\alpha = \sin\phi\cdot\sin\delta + \cos\phi\cdot\cos\delta\cdot\cos\omega \\ \sin A = \dfrac{-\sin\omega\cdot\cos\delta}{\cos\alpha} \end{cases} \quad (1)$$

where t_b is the local Beijing time, φ is the local latitude, ω is the solar hour angle and δ is the Solar declination at that time.

From the formulas above, we can calculate the sun's direction in a certain latitude at a certain time,

$$\overset{\mathrm{r}}{n} = (-\tan A, \frac{\varphi}{|\varphi|}, -\tan \mathrm{h} \times \sqrt{\tan^2 A + 1}). \quad (2)$$

2.2 Figure the light intensity of a certain point as a source of the diffuse reflection.

When the sunlight reach the trees, we take the layer of trees as a medium. So all the light must first pass to the tree layer, and then to the ground. As the same time, the tree layer receive sunlight and becomes a diffuse light source. Therefore, there are two parts of light that the tree layer receives, one is the light from the sun directly and the other part is the diffuse reflection from other trees(See the figure 1). Calculating the intensity of the illumination in the forest is finally the question of calculating the radiation intensity of all the diffuse light source(the tree layer).

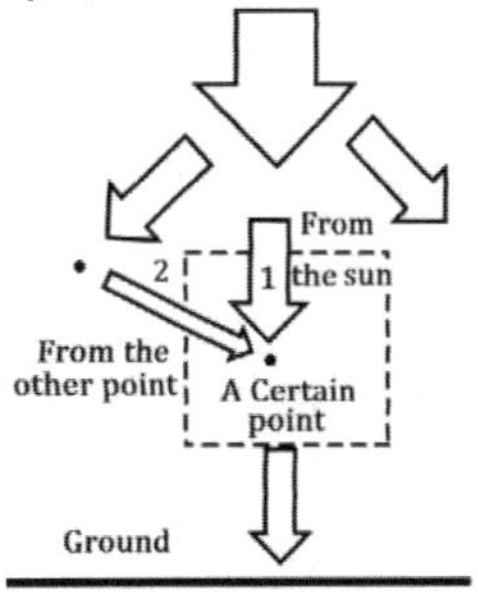

Figure 1. The light of a certain point.

Then we will discuss how to figure the light intensity of a certain point as a source of the diffuse reflection.

2.2.1 The Light From The Sun Directly Makes A Source of The Diffuse Reflection

The light intensity can be defined as the Lambert model. The intensity of diffuse reflected light is approximated to Lambert's law[2,3], which can be expressed as follows,

$$I_1 = I_0 k_d \cos\theta \quad (3)$$

where,

I_1 is the intensity of diffuse reflection of the surface,

k_d is the diffuse reflection coefficient,

I_0 is the intensity of the sunlight,

θ is the angle between the surface normal of the object and the direction of the sun.

2.2.2 The Diffuse Reflection From Other Point Makes A Source of The Diffuse Reflection

A certain point catch the light from a other point consists 3 stages.

Stage 1: The sunlight shine the other point, and the other point becomes a source of the diffuse reflection. The light intensity can be expressed as,

$$I_1 = I_0 k_d \cos\theta. \tag{4}$$

Stage 2: A certain point shined by the diffuse reflection from the other point. The luminous flux that a certain point received can be expressed as,

$$\Phi_{in} = I_1'\Omega. \tag{5}$$

Stage 3: A certain point becomes a source of the diffuse reflection. A certain point receive the diffuse reflection from the other point. These light minus the part absorbed by the leaves is the reflection luminous flux.

$$\Phi_{out} = a_l \Phi_{in}. \tag{6}$$

And the light intensity of a certain point because of the diffuse reflection can be expressed as,

$$I_2 = \frac{\Phi_{out}}{\Omega} = \frac{a_l I_1 \Omega}{4\pi}. \tag{7}$$

2.2.3 The light intensity of a certain point

The light intensity of a certain point as a source of the diffuse reflection consists two parts. one is the light from the sun directly(like formula 4) and the other part is the diffuse reflection from other trees(like formula 7).So we can figure the light intensity of a certain point as a source of the diffuse reflection[3,4].

$$I_{all} = I_1 + I_2. \tag{8}$$

2.3 The relationship of tree density, slope gradient and the diffuse reflection coefficient

In the previous analysis, we used the diffuse reflection coefficient as a constant. However, in fact, there is a great relationship between the diffuse reflection coefficient and the density of the tree, the greater the tree density, the greater the diffuse reflection coefficient. In the third part we thus build the model A to discuss the relationship between the diffuse reflection coefficient and the density of the tree. And model B to discuss the relationship between tree density and slope gradient.

2.3.1 Model A-- tree density and slope gradient

In general, the greater the slope gradient, the greater the tree density. Initially, the density of the tree is not greatly affected by the change of the slope, but it falls down sharply at a certain slope and finally tends to 0 when the slope is high enough. The relationship between the two flat at first, then violent and flat again at last, so we use the logistic function to describe its changes.

$$d_t = \frac{1}{1+e^{-b\times\cos\theta_s + c}}, \tag{9}$$

where b and c are constants, θ_s is the slope gradient.

2.3.2 Model B-- the diffuse reflection coefficient and the density of the tree

The greater the tree density, the greater the diffuse reflection coefficient, here we assume that the diffuse reflection coefficient is approximately proportional to the tree density.

So we can get a formula as follows,

$$k = d_t. \tag{10}$$

But the diffuse reflection coefficient has an upper limit instead of growing and growing with the increase of tree density. Therefore, we add a constraint factor to the formula (10)

$$k = d_t \times e^{-f\times d_t}, \tag{11}$$

where f is a constant, d_t is the trees' density and k is diffuse reflection coefficient.

Finally we chose the following parameters, and curves of A and B are shown as follows

We can find the density of the tree is an intermediate variable. And the relationship of slope gradient and the diffuse reflection coefficient also can be computed.

2.4 The Complete Model

Now we get the diffuse reflection model from formulas (4),(7),(8) and diffuse reflection coefficient model from formula (9),(11). Combining the two we can get a formula to determine the overall light intensity.

$$\begin{cases} I_1 = I_0 k_d \cos\theta \\ I_2 = \dfrac{a_l}{4\pi}\iint I_1 \times \Omega \, dS \\ I_{all} = I_1 + I_2 \end{cases}. \tag{12}$$

The overall luminous flux can be obtained by an integral.

$$\Phi = \int I_{all} d\Omega \,. \tag{13}$$

Therefore, the intensity of the illumination which means the luminous flux per unit area of the ground can be expressed as follow

$$E_{gound} = \frac{d\Phi}{dS} = \frac{\int_0^{2\pi} I_{all} \times \Omega \, d\Omega}{S} = \frac{2\pi I_{all}}{S}. \tag{14}$$

3 Numerical results

3.1 The Influence of Time and latitude

As for the impact of the latitude and time on the direction of the sunlight, we control variable and divided the the question into two parts[2,3].

Figure 2. The impact of time.

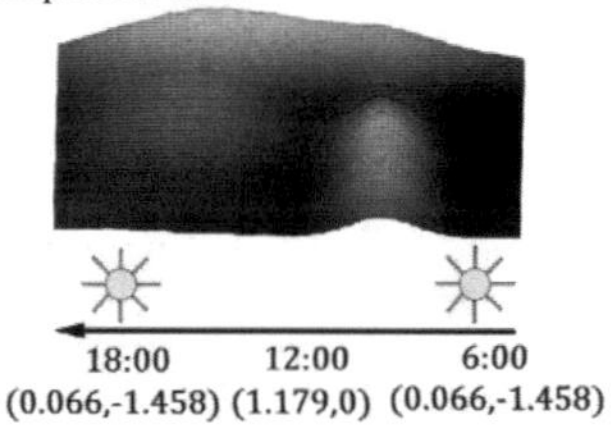

Figure 3. The impact of latitude.

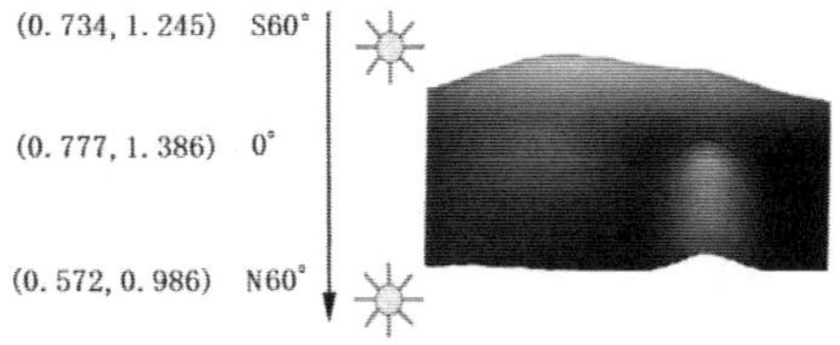

In the figure 2: A valley is shown in the centre of the figure. At a certain latitude, the sun rises at the east(the Figure 2), reach the middle of the figure at midday and finally reach the west(the Figure 2). From this figure, we can get the variations of the sun's direction at different time.

In the figure 3: A same valley is shown in the figure 3. At a certain time, the direction of the sunlight various with the latitude of the valley. The latitude of different valleys is marked in the figure, the position of the sun at midday is shown in the figure, too.

The solar elevation angle and the the solar azimuth angle are marked in the figure.

From the figure, we can get the variations of the sunlight's direction at different latitude. We will make a concrete analysis of this question after getting the result of the simulation.

3.2 Terrain generation algorithm

The actual terrain of mountainous region is complexity and variety. In order to evaluate the accuracy and practicality of our mode land take fully consideration of different situations as much as possible, we use terrain algorithm generate a random fractal terrain.

The flow-process figure of the terrain generation algorithm is shown as follows

Step 1. Generating boundaries;
Step 2. Internal expanding to generate random terrain;
Step 3. Superimposed mountain terrain(Please see figure 4).

Figure 4. Superimposed mountain terrain.

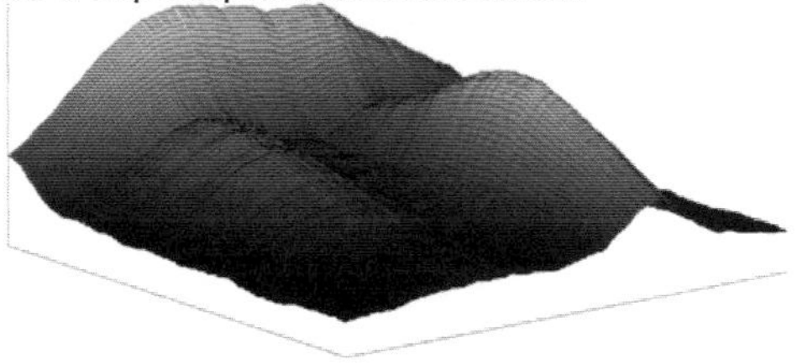

3.3 The simulation results of the complete model

Step1. Judge whether a certain point of the tree layer is under the direct light irradiation.

Contrast to the coordinates of the light rays and the actual mountain, you can know whether the light is blocked. The black line in the picture is the direction of sunlight. See the follow figure 5.

Figure 5. Judging process.

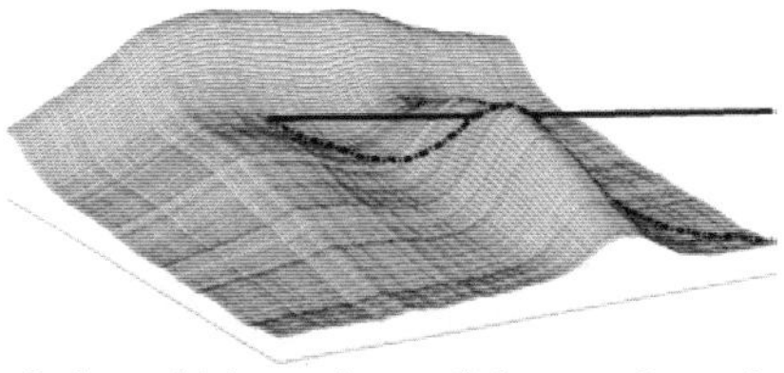

Step2. Judge which portions of the tree layer is in the direct sun.

Step3. In the first stage of diffuse reflection, if a certain point receive the sunlight directly, it can be regard as a diffuse reflection light source. The energy of the source are only comes from the sunlight directly.

Step4. In the second stage of diffuse reflection, every points of the tree layer receive light from other points more or less.

Step5. Therefore, every point of the tree layer becomes a diffuse reflection light source which contains two parts of light intensity. One comes from the sun directly and the other comes from other points of the tree layer.

Step6. Finally, the ground receive the overall light intensity from both the two stages of diffuse reflection. So the intensity of the illumination which means the luminous flux per unit area of the ground can be expressed in the following figures 6.

Figure 6. The intensity of the illumination.

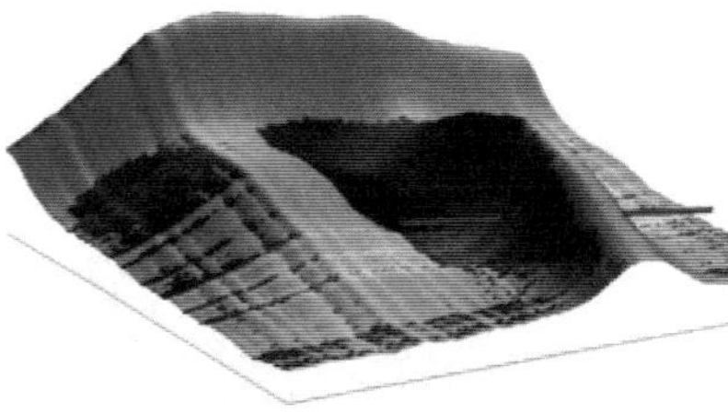

In this figure, the lighter the point seems to be, the greater the light intensity and the average luminescent of the area which receive the sunlight directly is 1.024. The average luminescent of the area which not receive the sun's irradiation is 0.287.

We can conclude that the valley is very dark because it is not exposed to the sunlight directly, but it is not without a little light. Finally, we have a comprehension simulation as follows:

Figure 7. Comprehension Simulation.

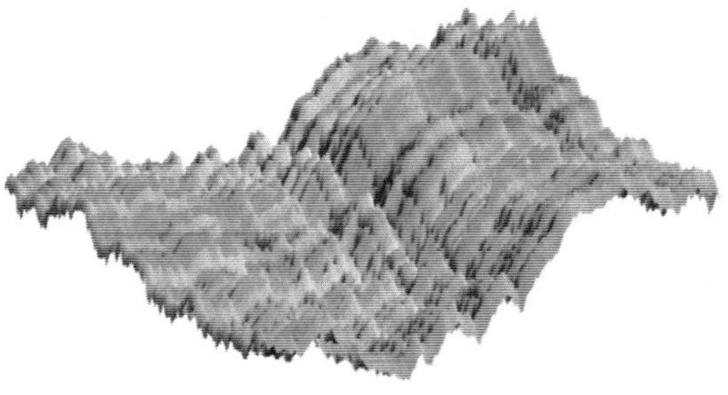

3.4 Result Analysis

The impact of the time and latitude on the direction of the sun are listed in the following figures.

Figure 8. The impact of the different times: 9:00,12:00,15:00 (the latitude is 0°).

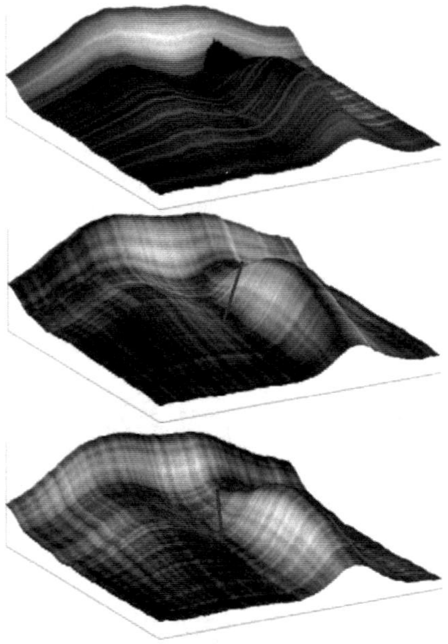

Figure 9. The impact of different latitudes: N60°,0°,S60° (the time is 12:00).

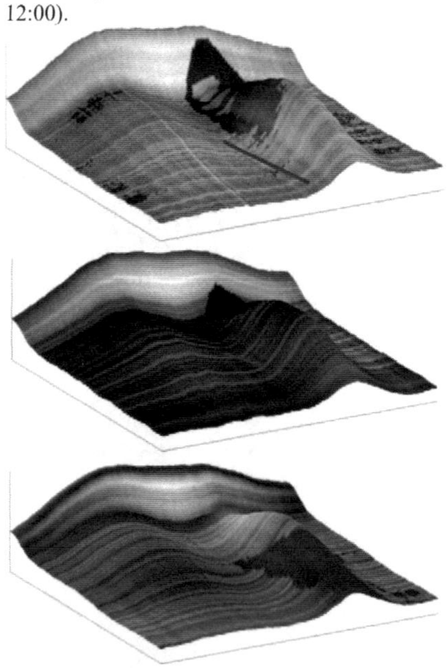

Apparently, the direction of the sun various with the time and the latitude and finally have impact on the intensity of illumination.

From figure 8 and 9, We can conclude that, the valley will be the brightest near the midday and quite dark in the early morning and at dusk. That is to say, brightness of the valley increases with the increase of the solar elevation angle.

Besides, the greater the absolute value of the latitude, the greater the light intensity of the valley. So we can know that the intensity of the illumination will be much lower near North and South Pole while the valleys close to the equator are more likely to receive adequate illumination.

4 Conclusion

In our paper, we change the question of the intensity of illumination into the effect under the diffuse reflection. This method not only simplify the question but also applicable in many conditions. Then we attempt to generate mountain randomly and study the illumination of the valley. What's more, our model can be used in many other conditions to evaluate the illumination situation. For instance, in the fields which vegetables are planted under fruit trees, there is a great relationship between the diffuse reflection of the trees and the light intensity the vegetables received. The model can be feasible to discuss whether the area is suitable for the vegetables.

However, The height of the tree is not in the consideration and we assumed that all kinds of terrains are evenly covered with trees. Moreover, most trees in the valley is not close enough to prevent the ground from the sunlight entirely. Especially in the sparse areas, large amounts of light go through trees and reach the ground directly. In this way, the real light intensity that the ground received will be much larger than our results, so does the intensity of illumination.

References

1. W.W. Zhang, X. You, G.M. Song, A Illumination model for landscape factors, Scienceof Surveying and Mapping, 34, 5 Sep.
2. C.Z. He, An Introduction to Mathematical Modelling, Chengdu, University of Science and Technology press, 1998.
3. G.Y. Ray, Mathematical Model of Lectures, Peking University Press, 1999.
4. Z.M. Zhu, H. Liu, S.M. Chen, The Design of Diffuse Reflective Free-Form Surface for Indirect Illumination With High Efficiency and Uniformity, IEEE P. J., 7, 3, 2015, 1600510.

Discussion on Reliability of Narrow-band TD-LTE Network Transmission for Distribution Automation System

CHEN Yong-tao [1], WANG Hao [1a], WANG Min [1]

[1]Guangzhou Power Supply Co.Ltd. , Guangzhou, Guangdong, 510000, China

Abstract: In Guangzhou, there is a poor communication between power control systems in equipment rooms and the 100,000 distribution terminals scattered around the city. Guangzhou Power Supply Co.Ltd. (CZPS) planned to upgrade smart power distribution grid with wireless communications technology. It built an integrated data collection platform and several small-scale test zones to collect and analyze data from a pilot. In the GZPS Communication Laboratory, to confirm the reliability and accuracy, the transmission performance indicator of DTU are compared with different RSRP by using wireless public network (GPRS) ,230MHz and 1.8 GHz wireless private network testing system.
Keywords: Power Grid Wireless Private Network; Distribution Automation; Remote Control, TD-LTE

1 Introduction

Power communication service has widely used various communication technologies. With the development of the construction of the smart grid, distribution automation system (DAS) is not just a remote control but it results into a highly reliable,efficient power system that responds rapidly to real-time events with appropriate actions.

The distribution grids have a complex network topology by feeding the power in a wide coverage area. At present, optical fiber communications are the main communication means for the access networks. However, in well-established urban areas that feature complicated terrains and tall buildings, routing of optical fibers is difficult and the costs are often very high. The power wireless private network construction attracts more and more attention. In this scenario, the LTE-based private wireless network can be used as the alternative and economic access network. It is a cost-effective solution to the last-kilometer problem.

Public networks failed to meet service requirements because of poor security, high lease costs, and insufficient service resources. The private wireless network is a solution that eased construction problems and freed wired networks from environmental restrictions. Wireless broadband private network also overcame the poor performance and limited resources of 2G/3G public networks.

1.1 Time-Division Long-Term Evolution (TD-LTE) technology

The solution uses 4G TD-LTE base stations in existing substations and office buildings. The base stations operate on a 1.8G and 230MHz industry-dedicated wireless spectrum. In addition, it deployed dedicated TD-LTE distribution terminals with Data Transformation Units (DTUs), Remote Metering Terminal, and Video Surveillance. The base stations combine data from multiple electric power terminals and send it via an existing transmission network to a host station.

1.2 Multi-level Quality of Service (QoS)

Different service applications in automatic power distribution networks work on different priorities. Private network TD-LTE system integrates radio resource management and tailored priorities for electricity services. The solution also supports multiple services while ensuring that important data, such as automatic remote service data, transmits preferentially. The QoS mechanism works with real-time online terminals and a latency minimizing solution to ensure less than 100 ms end-to-end latency. The TD-LTE solution also uses an array of end-to-end data protection methods, including 128-bit Advanced Encryption Standard (AES) and authentication encryption, to ensure high data security and reliability.

2 Performance Indicators Testing

In the Guangzhou Power Supply Co. Ltd. (GZPS) Communication Laboratory, to confirm the reliability and accuracy, the transmission performance indicator of DTU are compared with different RSRP by using wireless public network (GPRS), TD-LTE 230MHz and TD-LTE 1.8 GHz wireless private network testing system. The experiment Results is shown in Appendix I.

[a] Corresponding author: wanghao_hust@163.com

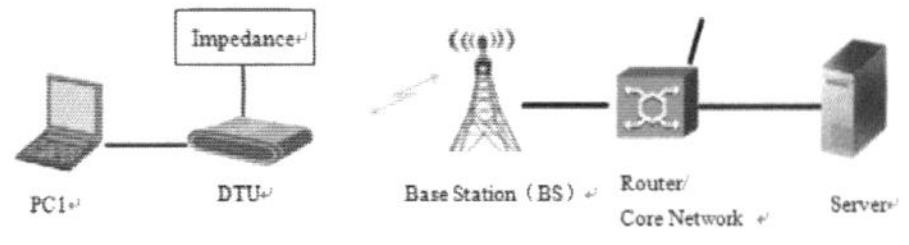

Figure 1. Testing environment.

2.1 Channel Performances

There is a positive correlation between the channel uplink /downlink speed and RSRP, its decline from large to small order is 230MHz network, 1.8GHz network, GPRS network. Besides, The packet delay and packet loss is positively related to the RSRP. The channel can remain the same under the condition of stable,

TD-LTE is the next-generation wireless broadband technology, using a variety of advanced wireless communication technologies such as Orthogonal Frequency Division Multiplexing (OFDM), Multi-Input Multi-Output (MIMO), a multi-channel smart antenna technology and so on, which effectively enhance the data throughput rate, coverage and the number of user online. GPRS is a best-effort service, implying variable throughput and latency that depend on the number of other users sharing the service concurrently.

2.2 Communication Requirements

The distribution equipment can be divided into three categories: remote measuring, remote communication, remote control(three-remote system). Currently, the transmission requirements of distribution automation system are that the largest bandwidth is not less than 64bps; packet delay is not greater than 1 second. The experiment results show that the public network GPRS and private network (TD-LTE 230MHz, TD-LTE 1.8GHz) all can satisfy the requirement of three-remote system. In addition, all the operations of switching device were controlled successfully with 1k uplink data, its purpose is to simulate the real remote signaling by loading 1k data.

2.3 Benefits of Private Wireless Network

Bandwidth of Public network GPRS is 20Kbps – 40kbps and latency is 2 second. Experimental results show that the uplink and downlink speed remain at 5KB/s,10KB/s respectively, but a gradual increase in the proportion of the retransmission packets. Compared with the TD-LTE network, the pubic GPRS is low reliability and security, its recourses shared by public and network controlled by operators.

Currently, the approved bandwidth of the 1.8GHz wireless private network is 5MHz. the testing results show that the maximum throughput of uplink and downlink are 1.95Mbps 2.27Mbps respectively. Besides, the 1.8G CPE device support multiple uplink/downlink ratio configuration, including 0(3:1),1(2:2),2(1:3) (uplink: downlink). It also offer flexible bandwidth from 5Mhz to 20Mhz. it is adapt for streaming data-intensive uplink service such as video surveillance and data collection.

TD-LTE 230 system uses Adaptation Modulation and Coding (AMC), which can offer an alternative link adaptation method that promises to raise the overall system capacity [1]. AMC provides the flexibility to match the modulation-coding scheme to the average channel conditions for each user. With AMC, the power of the transmitted signal is held constant over a frame interval, and the modulation and coding format is changed to match the current received signal quality or channel conditions. The AMC scheme of TD-LTE 230 system uses QPSK, 8-PSK and 16 and 64 QAM modulation using R=1/2 and R=3/4 Turbo code. In table 4.2, it shows that the uplink and downlink speed of LCM change with channel quality. In a system with AMC, the modulation-order or code rate will increase as the quality of channel increases (e.g. 64QAM with R=3/4 turbo codes). The maximum uplink speed can reach 45.6kbps. In order to ensure the reliability, it is typically assigned lower order modulation with lower code rates when the quality of channel decreases.

According to the State Radio Regulatory Commission frequency allocation strategy, there are 35 available frequencies for TD-LTE 230. If available frequencies are sufficiently used, the uplink and downlink speed will achieve 1.51Mbps and 289.5Kb/s respectively

2.4 The Reliability of Three Remote

As shown in the experimental results, the RSRP limit of the GPRS public network is -95dB, the RSRP limit of 1.8GHz private network is -117dB, the RSRP limit of 230 MHz private network -121dB for multi-frequency, -127dB for single frequency. The private network can tolerate lower RSRP than public network, and the channel is more stable. When RSRP = -97dB, the GPRS channel is extremely unstable, the Dial-Up networking could not complete the connection to the server. Even if successful connection, the three remote function cannot work properly. When RSRP = -120dB, the channel of private network is starting to become unstable, it will take a lot of time to connect successfully. Once the connection is established successfully, the three remote function can work well.

Compared to the GPRS public network, the TD-LTE private network is little time-delay,low packet loss rate,less retransmission rate, so the round trip time of remote control commands and shift information is expected to be much lower.

Table 1. AS the RSRP= -90dBm, a brief description of time delay are listed below

	GPRS	TD-LTE 1.8Ghz	TD-LTE 230MHz	
			LCM	CPE
Non-business state		33ms	181ms	145ms
In-business state	333.2ms	32ms	162ms	131ms

Table 2. As the channel is stable, the average time of three remote is list below.

	GPRS	TD-LTE 1.8GHz	TD-LTE 230MHz	
			LCM	CPE
Switch on	7832ms	2476ms	2808ms	2939ms
Switch off	8205ms	3141ms	2821ms	3046ms

Table 3. The measurement of GPRS network

Impedace (dB)	RSRP (dB)	Uplink speed(Kbps)	Downlink speed(Kbps)	Packet Delay (ms)	Packet Loss (%)	Packet Delay with 1k uplink	Packet Loss	Three remote test		Accuracy
								Switch-off(ms)	Switch-on(ms)	
+0dB	-65	5	10.8	335	0%	410.5	0%	7622	7575	100%
+5dB	-70	5	10.1	343.6	0%	356.9	0%	7565	8080	100%
+10dB	-75	5.2	11	367.4	0%	346.9	0%	5050	7571.5	100%
+15dB	-81	5.1	10.5	316	0%	347.7	0%	12120	7070	100%
+20dB	-85	5.1	10.1	323.4	0%	427.6	0%	7575	9090	100%
+25dB	-90	5	10.0	333.2	0%	398.2	0%	7070	8585	100%
+30dB	-95	4.9	11.3	331.4	0%	359.1	0%	8080	8585	100%
+32dB	-97	5.2	Instable	324.4	0%	387.5	0%	7576.5	9090	100%
+34dB	-98	5.2	Instable	324.4	0%	387.5	0%	Instable	Instable	Instable
+35dB	-99	5.1	Instable	1203.7	0%	514.8	0%	7095	8111.5	100%
				618.6	5%	837.2	0%			
				645.1	0%	566.3	11%			
+37dB	-101	5.2	Instable	865.4	57%/	1125.7	9%	Instable	Instable	Instable
				690.5	16%	485	2%			
				2400	40%	384.3	5%			

Table 4. The measurement of TD-LTE 1.8GHz Network

Impedace (dB)	RSRP (dB)	Uplink speed(Kbps)	Downlink speed(Kbp)	Packet Delay (ms)	BER	Packet Delay with 1k uplink	BER	Three remote test		Three remote test
								Switch-off(ms)	Switch-off(ms)	
+0dB	-92	1.95	2.27	33	0	32	0	3030	2530	100%
+5dB	-97	1.12	2.38	31	0	32	0	3015	4040	100%
+10dB	-103	0.684	2.24	34	0	34	0	2020	3030	100%
+15dB	-108	0.303	2.2	45	0	42	0	2020	3080	100%
+20dB	-112	0.143	2.49	49	0	49	0	2525	3030	100%
+25dB	-117	0.095	1.73	83	0	197	0	2525	3030	100%
+30dB	-120	0.011	1.67	228	0	Instable	Instable	2525.5	3570	100%
+32dB	-122	0.008	1.1	291	0	Instable	Instable	2559	3030	100%
+34dB	-124	0.005	0.463	594	6%	Instable	Instable	2525	3540	100%
+36dB	-126	Instable	Instable	2079	83	Instable	Instable	2020	2530	100%

Table 5. The measurement of TD-LTE 230MHz network for CPE

	Rsrp	Downlink snr	Uplinksnr	Uplink speed (Mbps)	Downlink (kbps)	Packet Delay (ms)	Packet Loss (%)	Packet Delay with 1k uplink	Packet Loss (%)	Switch-off (ms)	Switch-on(ms)
CPE (Multi-frequency)	-80	28	24.6	1.51	289.5	147	0	130	0	2631	2020
	-90	28	24.6	1.51	290.6	145	0	131	0	2525	2525
	-100	27	18.4	1.14	290.2	143	0	146	0	2525	5555
	-105	26	13.6	0.714	290.2	148	0	143	0	2610	2640
	-110	25	10.6	0.6796	288.8	141	0	139	0	3030	2518
	-115	21	4.2	0.2837	208.3	157	0	142	0	5050	4033
	-120	18	2.2	0.1098	128.4	178	3%	162	0	2525	3030
	-121	15	-1	0.0545	127.6	178	2%	167	3%	3030	3030
	-122	14	-2.2	0	126.7	287	8%	268	10%	2525	2065

Table 6. The measurement of TD-LTE 230MHz network for LCM

	Rsrp	Downlink snr	Uplinksnr	Uplink speed(Mbps)	Downlink (kbps)	Packet Delay (ms)	Packet Loss (%)	Packet Delay with 1k uplink	Packet Loss (%)	Switch-off (ms)	Switch-on(ms)
LCM (Single frequency)	-80	25	21.6	45.6	15.5	187	0	154	0	3032	3030
	-90	25	21.2	45.5	15.4	181	0	162	0	2519	2525
	-100	24	20.6	45.2	15.1	193	0	164	0	3030	3030
	-105	23	20.4	45.2	14.3	185	0	181	0	2560	2670
	-110	20	19.4	44	10.8	227	0	253	0	3535	3030
	-115	18	16.2	33.04	6.7	239	0	316	0	2525	2525
	-120	13	12.2	18	3.3	403	0	410	0	3029	3030
	-123	10	10.2	16.6	2.1	462	0	431	0	2020	2525
	-127	9	9.2	16.6	2	517	0	501	0	3030	3032

3. The Security of Network

The GPRS was originally designed to offer circuit-switched service (voice and circuit-switched data) primarily, with limited capability to offer low-bit-rate packet data service. When the amount of data is very large, there is no guarantee for the priority of key customers. Besides, the GPRS public network cannot support massive real-time online terminal, so most of the users in the standby state. The TD-LTE solution uses an array of end-to-end data protection methods, including 128-bit Advanced Encryption Standard (AES) and authentication encryption, to ensure high data security and reliability.

References

1. TSGR1#17(00)1395, Adaptive Modulation and Coding (AMC), Adhoc#24, HSDPA, 20th-24th Oct 2000
2. ZHONG Yu-Qing, WANG Hao, CHEN Bao-ren. Research on Application Principles of Communication Technologies in Smart Distribution Network, Electric Power Information Technology. 2013, 11(5):43–47.

MATEC Web of Conferences **61**, 07002 (2016) DOI: 10.1051/matecconf/20166107002
APOP2016

A fusion networking model for smart grid power distribution backbone communication network based on PTN

WANG Hao [1a], CHEN Yongtao [1], CAI Jitao [1]

[1]*Guangzhou Power Supply Co.Ltd. , Guangzhou, Guangdong, 510000, China*

Keywords: smart grid communication, communication network for distribution, TD-LTE power private wireless network, PTN

Abstract. In current communication network for distribution in Chinese power grid systems, the fiber communication backbone network for distribution and TD-LTE power private wireless backhaul network of power grid are both bearing by the SDH optical transmission network, which also carries the communication network of transformer substation and main electric. As the data traffic of the distribution communication and TD-LTE power private wireless network grow rapidly in recent years, it will have a big impact with the SDH network's bearing capacity which is mainly used for main electric communication in high security level. This paper presents a fusion networking model which use a multiple-layer PTN network as the unified bearing of the TD-LTE power private wireless backhaul network and fiber communication backbone network for distribution. Network dataflow analysis shows that this model can greatly reduce the capacity pressure of the traditional SDH network as well as ensure the reliability of the transmission of the communication network for distribution and TD-LTE power private wireless network.

1 Introduction

Communication network for distribution is a key point in smart grid. In current communication network for distribution in Chinese power grid systems, optical fiber communication are most mainstream method, including industrial switch networks and EPON[1].Most of the fiber communication networks for distribution can be divided into backbone layer and access layer. The backbone layer which is bearing by the SDH network is used to gather traffic. In recent years, TD-LTE power private wireless network grow rapidly[2]. The backhaul network of the power private wireless network is also bearing by the SDH network. As the distribution automation and measurement automation keep rapid growth of the scale, this traditional networking model will l have a big impact with the SDH network's bearing capacity which is mainly used for main electric communication in high security level. The research of a new networking model is very necessary.

2 Capacity analysis of the power distribution backbone communication network

Figure 1. shows a typical networking model of fiber backbone communication network for distribution of industrial switches. The access layer is composed of "hand in hand" ring of the access switches in the power distribution room. The backbone layer which is mainly used for main electric communication is bearing by the MSTP link of the SDH network from backbone layer switches to core router.

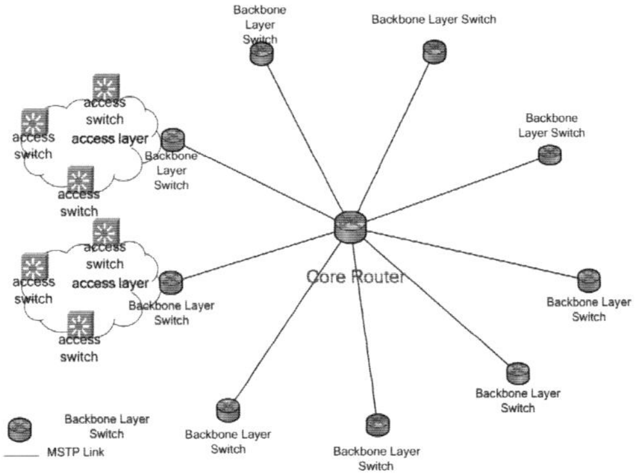

Figure 1. power distribution backbone communication network.

Table 1.shows the traffic statistics of backbone communication network for distribution. Table 1. shows that the distribution automation information is data packet which has a random data flow traffic. The network in Figure 1. configures the time slot of the MSTP link in 2*5Mbps. The MSTP link of the SDH network is mostly used for the TDM service like relay protections. When the SDH network carries the data packets, it will cause huge waste of to the time slots, and as the traffic of distribution automation information grow rapidly in recent years, it will make a big pressure to the SDH network.

Table 1. Traffic statistics of power distribution backbone communication network.

convergence point	Access switch number	Traffic(kbps)
1	92	129.536

[a] Corresponding author: wanghao_hust@163.com

convergence point	Access switch number	Traffic(kbps)
2	60	211.2
3	66	219.648
4	100	1886.72
5	196	2007.808
6	234	3032.832
7	320	4336.64
8	568	4432.384
9	462	10407.936
10	1024	11875.072

3 Capacity analysis of the TD-LTE power private wireless backhaul network

Figure 2. shows a typical networking model of power private TD-LTE wireless backhaul network. In order to save the SDH time slots,each base station have a dedicated 3*2M channel MSTP to a convergence SDH device. Up to 5 to 6 base stations have a common 2*5 MSTP channel from the convergence SDH device to the core SDH device. In this model, the convergence SDH device use the layer 2 switch function to converge the data flow, each base station in fact have only 1.66Mbps time slot capacity in average. The 1.8GHz TD-LTE base station have 42 Mbps peak data rates for 3 cells when works at 5MHz bandwidth,as the bandwidth expanding,the cell peak data rates will grow as well. So the TD-LTE wireless backhaul network in SDH channel will be a great bottle-neck to the performance of the TD-LTE power private wireless network. In addition,the TD-LTE power private wireless backhaul network carries data packets,which is also a waste of SDH time slots.

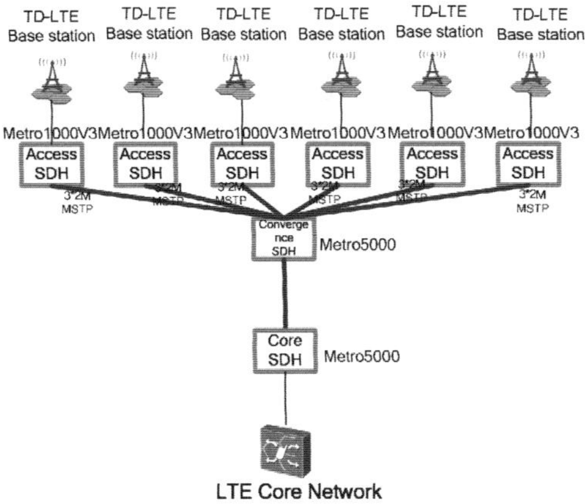

Figure 2. Typical TD-LTE power private wireless backhaul network.

4 A new fusion networking model based on PTN

4.1 Analysis of the PTN technology

PTN (Packet Transport Network) technology is be widely used in the mobile backhaul network telecom operators. This technology is designed for the bursty traffic and statistical multiplexing characteristics of data packets networks[3]. And compared to the common IP networks which are composed of switches and routers, PTN have great network management performance and Fault self-healing capability. In addition, PTN support 50ms path switching to ensure the self-healing ability of communication equal to SDH. Therefore, PTN is quite suitable for TD-LTE power private wireless backhaul network and power distribution backbone communication network, which carries data packets and have great demand in reliability and network management performance.

4.2 Network planning of the fusion networking model based on PTN

Figure 3. shows a fusion networking model for power distribution backbone communication network based on PTN. In this model, PTN network is the unified bearing of the TD-LTE power private wireless backhaul network and fiber communication backbone network for distribution. This model is divided into 3 layers, which are called PTN convergence ring, PTN backbone ring and PTN core nodes.

PTN convergence ring is the PTN GE ring which is composed of the PTN equipment in the 110kV substation. Each PTN equipment in the PTN convergence ring has two independent links to the industrial switches ring in the distribution room to converge the data flow of distribution automation and insure the reliability. Meanwhile, Each PTN equipment in the PTN convergence ring also connects to the BBU (Building Baseband Unit) of the TD-LTE power private wireless network in the 110kV substation by FE interface, as the backhaul equipment between the base station and core network.

PTN backbone ring is the PTN 10G ring which is composed of the PTN equipment in the 220kV substation. PTN backbone ring is used in converge the data flow in the PTN convergence ring and send it to the PTN core nodes. Meanwhile, each PTN equipment in the PTN backbone ring also connects to the BBU of the TD-LTE power private wireless network in the 220kV substation by FE interface, as the backhaul equipment between the base station and core network.

PTN core nodes converge the data flow in the PTN backbone ring and distribute it to the core network of the TD-LTE power private wireless backhaul network, NMS(Network Management System)of TD-LTE power private network ,master station of distribution automation system and NMS of fiber communication network for distribution. Each master station system has two independent links to the PTN core nodes, and each PTN core node has two independent links to different PTN equipment in the PTN backbone ring.

In this fusion networking model based on PTN, the distribution automation data, NMS data and the backhaul data of TD-LTE power private wireless network can be divided into different VLANs or L3 VPNs to meet the data isolation security demand.

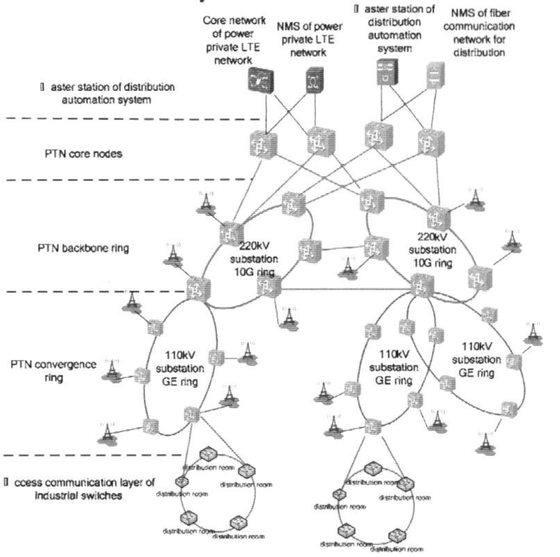

Figure 3. A fusion networking model based on PTN.

4.3 Path protection and data flow planning of the fusion networking model

In this network model, PTN convergence GE ring and PTN backbone 10G ring use the LSP APS protection. This path protection strategy uses PTN hardware to transmit the OAM message, and the switching time is less than 50ms, which has a high reliability. The key nodes PTN equipment can use TPS device protection to promote the device reliability. Moreover, VRRP, L3 VPN FRR and other protection technology such as linear protection switch, ring protection and dual-homing protection can also be used to achieve end-to-end production.

Table 2. shows the data flow analysis of the fusion networking model. Considering the TD-LTE power private wireless network uses the TDD (time division duplex),which allocate the uplink and downlink data packets in different time slots, each 230 MHz base station's backhaul flow demand is 15Mbps ,and each 1.8GHz base station's backhaul flow demand is 20Mbps.This above-mentioned bandwidths can meet the huge backhaul flow demand when the 230 MHz base station runs in 340 frequency points(25kHz per point) and 1.8GHz base station runs in 10MHz bandwidths in the future. The data flow demand of each access ring of industrial switches is 0.5Mbps*20distribution rooms=10Mbps.Each PTN convergence ring is deployed of 5 230 MHz base stations, 5 1.8GHz base stations and 20 industrial switches, the total bandwidth demand of a PTN convergence ring is 750Mbps. .Each PTN backbone ring is deployed of 10 PTN convergence rings, 5 230 MHz base stations and5 1.8GHz base stations, the total bandwidth demand of a PTN backbone ring is 7850Mbps.The PTN core nodes are 10GE.In this analysis, all the bandwidth demands are considered in max demand, actual bandwidth of the network will be less than the planning data flow. The calculation proves that the fusion networking model can meet the data flow demand of the fiber communication backbone network for distribution and TD-LTE power private wireless backhaul network in max bandwidth configuration, on the premise that using the 1+1 path protection to ensure the communication reliability.

Table 2. Data flow analysis of the fusion networking model.

	eNodeB (Mbit/s)	Access ring of industrial switches (Mbit/s)	PTN convergence ring (Mbit/s)	PTN backbone ring (Mbit/s)	PTN core nodes
data flow demands	15Mbps (230MHz base station) 20Mbps (1.8GHz base station)	0.5Mbps*20distribution rooms=10Mbps	(15Mbps*5 230 MHz base stations+20Mbps*5 1.8GHz base stations +10Mbps*20 industrial switches in an access ring)*2(LSP APS protection)=750Mbps	(375Mbps*10 PTN convergence rings +15Mbps *5 230 MHz base stations +20Mbps *5 1.8GHz base stations)* 2(LSP APS protection)=7850 Mbps	10GE

5 Conclusions

Traditional SDH/MSTP network is based on TDM technology, which is not suitable for the transmitting of the IP data packets of the fiber communication backbone network for distribution and TD-LTE power private wireless backhaul network. This paper presents a fusion networking model which use a multiple-layer PTN network as the unified bearing of the TD-LTE power private wireless backhaul network and fiber communication backbone network for distribution. The calculation result shows that this model can meet the data flow demand of the fiber communication backbone network for distribution and TD-LTE power private wireless backhaul network as well as ensure the reliability of the transmission.

References

1. ZHONG Yu-qing, WANG Hao, CHEN Bao-ren. Research on Application Principles of Communication Technologies in Smart Distribution Network, Electric Power Information Technology. 2013, 11(5):43–47.
2. WANG Hao,ZHONG Yu-qing,RONG Zhi-neng,CHEN Bao-ren.Research of Multi-frequency Hybrid Networking of TD-LTE Power Wireless Broadband System, Electric Power Information and Communication Technology. 2013, 11(8):24–30.
3. JIA Xiao-tie,LEI Xue-yi,WU Yun-feng,LI Wei-hua.PTN Provides An Ideal Information and Communication Platform for Smart Grid, Telecommunications for Electric Power System. 2010, 31(213):20–23.

MATEC Web of Conferences **61**, 07003 (2016) DOI: 10.1051/matecconf/20166107003
APOP2016

A Hybrid Networking Model for the Access Layer of the Communication Network for Distribution in Smart Grid

Wang Hao [1a], Chen Yongtao [1], Zhong Yuqing [1]

[1]Guangzhou Power Supply Co.Ltd. , Guangzhou, Guangdong, 510000, China

Abstract: The access layer in the communication network for distribution is an important link in the automation of smart distribution power grid. In current access layer of communication network for distribution in Chinese power grid systems, several communication methods like optical fiber, medium-voltage carrier communication, 1.8GHz TD-LTE power private wireless network, 230MHz TD-LTE power private wireless network, public wireless network are constructed concurrently and running simultaneously in an identical power supply area. This traditional networking model will cause repeated construction and operation and maintenance difficulties in the communication network of power grid. On the basis of giving a detailed analysis of the radio link budget of TD-LTE power private wireless network in two frequencies, this paper present a multi-communication methods hybrid networking model, which gives a clear boundary for different communication methods based on the isoline with equal signal strength of the TD-LTE power private wireless network and accomplish the optimization of communication resources for distribution.

Keywords: multi-communication methods, communication network for distribution, TD-LTE power private wireless network

1. Introduction

The access layer in the communication network for distribution showed in Figure 1. is used in gather the data flow in the distribution automation equipment such as DTU(Distribution Terminal Unit) and FTU(Feeder Terminal Unit) and send it to the backbone layer in the communication network.

In current access layer of communication network for distribution in Chinese power grid systems, several communication methods like optical fiber(industrial switch or EPON), medium-voltage carrier communication, 1.8GHz TD-LTE power private wireless network, 230MHz TD-LTE power private wireless network, public wireless network[1] are constructed concurrently and running simultaneously in an identical power supply area. This traditional networking model will cause repeated construction and operation and maintenance difficulties in the communication network of power grid. The research of a new networking model is very necessary to solve these problems.

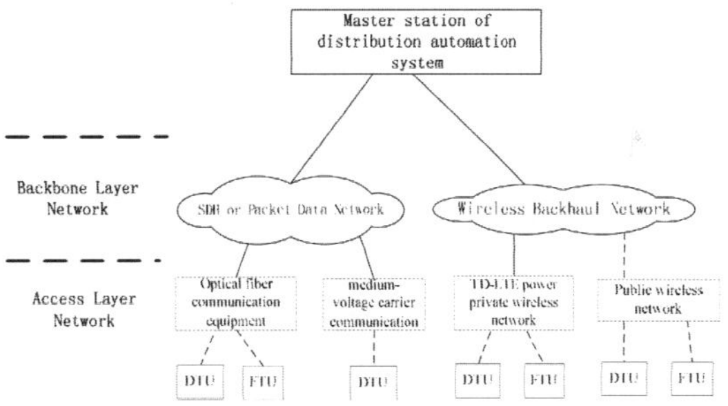

Figure 1. Several communication methods of the access layer in the communication network for distribution

2. Radio link budget of TD-LTE power private wireless network

Radio link budget is required before giving a clear boundary for wire communication and wireless communication. 1.8GHz TD-LTE power private wireless network and 230MHz TD-LTE power private wireless network is two mainstream technologies and different algorithms should be used in the radio link budget.

The Okumuram- Hata model[2] is used in the radio link budget of 230MHz TD-LTE power private wireless network. This model is established based on measured data, and is suitable for 150-1920MHz frequency. In different communication surroundings, the transmission loss can be calculated in following formula:

[a] Corresponding author: wanghao_hust@163.com

DOI: 10.1051/matecconf/20166107003

L=69.55+26.16*lg(f)-13.82*lg(ht)–[(1.1*lg(f)-0.7)*hr-1.56*lg(f)+0.8] + [44.9-6.55*lg(ht)]*lg(d) + C (1)

urban area:C=0

suburb:C= -2* [lg(f/28)]2 -5.4

rural area:C=-4.78* lg(f)* lg(f)+18.33* lg(f)-40.98

fc:working frequency(MHz);ht:height of transmitting antenna(m);hr:height of receiving antenna (m);d:coverage range(km);C:Environmental compensation factor

Table 1. Radio link budget of 230 MHz TD-LTE power private wireless network

basic parameter			
Modulation Mode	QPSK	16QAM	64QAM
Base station/Terminal transmitted power(dBm)	37/23	37/23	37/23
Base station/Terminal antenna gain(dBi)	11.5/3.5	11.5/3.5	11.5/3.5
Thermal noise density (dBm/Hz)	-174	-174	-174
Receiver noise factor(dB)	6	6	6
Service bandwidth(KHz)	25	25	25
Receiver noise power(dBm)	-124.02	-124.02	-124.02
Cable loss(dB)	1	1	1
Okumuram- Hata model parameter			
System operating frequency (MHz)	230	230	230
Antenna height of base station/terminal(m)	45/1.5	45/1.5	45/1.5
Terminal antenna correction factor(dB)	-0.04	-0.04	-0.04
Distance factor	34.07	34.07	34.07
Constant factor	108.52	108.52	108.52
urban area / suburb / rural area			
Interference Margin (dB)	20/15/15	20/15/15	20/15/15
Service SNR(dB)	4	8	22
Receiver sensitivity (dBm)	-100/-105/-105	-96/-101/-101	-82/-87/-87
Maximum allowable transmission loss(dB)	137/142/142	133/138/138	119/124/124
Shadow fading loss (dB)	8.00	8.00	8.00
maximum allowable path loss (dB)	129/134/134	125/130/130	111/116/116
Environmental compensation factor	0/-7.07/-24.35	0/-7.07/-24.35	0/-7.07/-24.35

According to the radio link budget of 230MHz TD-LTE power private wireless network in Table 1. ,the coverage performance of single 230MHz TD-LTE base station in urban area / suburb / rural area can be calculated in Table 2.

Table 2. Coverage performance of single 230 MHz TD-LTE base station

Modulation / **Environment**	**QPSK**	**16QAM**	**64QAM**
urban area (km)	4.00	3.05	1.18
suburb (km)	9.04	6.90	2.68
rural area (km)	29.04	22.17	8.61

The radio link budget of 1.8GHz TD-LTE power private wireless network can be simulated in C0ST231-Hata model [3]. The coverage performance of single 1.8GHz TD-LTE base station can be derived by the C0ST231-Hata model path loss formula:

$$L = 46.3 + 33.9 \lg f - 13.82 \lg h_t - a(h_r) + (44.9 - 6.55 \lg h_t)(\lg d)^{\gamma} + C \quad (2)$$

a(hr): correction factor of the height of receiving antenna

γ : Long distance transmission correction factor

It can be calculated from the path loss formula of Okumuram- Hata model and C0ST231-Hata model that in the same coverage environment, the 230 MHz TD-LTE base station will have a longer coverage distance in average and both the isoline with equal signal strength of 230 MHz frequency and 1.8GHz frequency will make logarithmic attenuation around the base station.

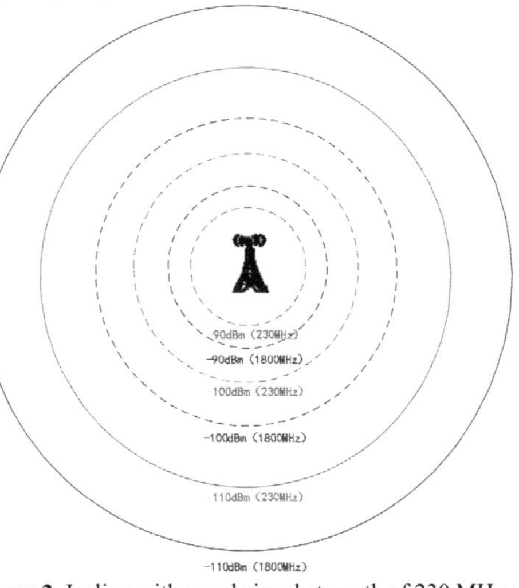

Figure 2. Isoline with equal signal strength of 230 MHz and 1.8GHz frequency

3. Hybrid networking model for the access layer of the communication network for distribution

This paper presents a hybrid networking model for the access layer of the communication network for distribution base on the signal attenuation characteristic of the 1.8GHz TD-LTE power private wireless network and 230MHz TD-LTE power private wireless network which is showed in Figure 3.

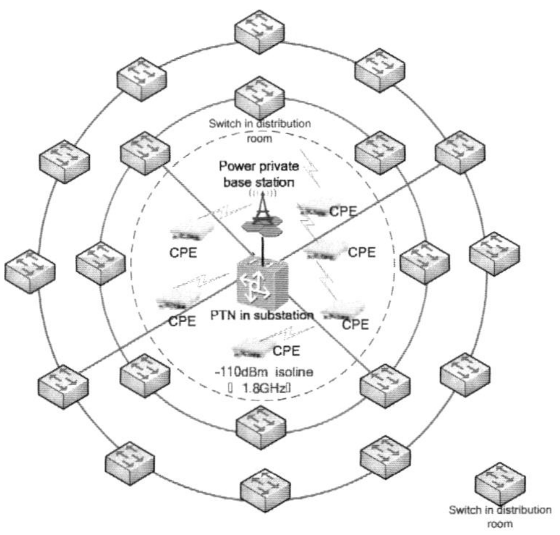

Figure 3. Hybrid networking model for the access layer of the communication network for distribution

Considering EPON network is hard to make adjustment after the construction is completed and medium-voltage carrier communication will cause non-planned power cut during operation and maintenance, this networking model choose industrial switch installed in the distribution room as the only wire communication method in the distribution communication network. The DTU in the distribution room can access the industrial switch by FE interface.

In this model, different communication methods will take the isoline with equal signal strength of the power private base station as the access boundary. Take 1.8GHz base station for example, all the DTU equipment in distribution room inside in coverage of the -110dBm isoline will access in power private wireless network by CPE(Customer Premises Equipment) terminal. Some distribution rooms in the coverage blind area of the TD-LTE power private wireless network where there is the sheltered architecture or in basement can access the public wireless network of telecom operators, and when condition allowed these distribution rooms can gradually access the TD-LTE power private wireless network by wireless repeater or other signal extension technologies. The power private base station will gather all the distribution automation data transmitted by the CPE and send it to the PTN or other backbone layer network equipment in the substation.

The industrial switches in the distribution room outside the -110dBm isoline make up several multiple hand-in-hand rings networks according to their geographic position. Each ring of industrial switches has at least two independent optical cable links to the backbone layer network. The equipment backbone layer network open up three –layer gateway to each ring of industrial switches in the access layer, and each ring of industrial switches opens up RSTP (Rapid Spanning Tree Protocol) to isolate the broadcast storm. Each industrial switch in a hand-in-hand ring has self-healing ability against single point of failure in optical cable based on RSTP

In this model, the ring networks of industrial switches are planned according to the signal test results of -110dBm isoline. The optical cables between the distribution rooms are planned after the construction of power private base station is finished according to the typical connection mode of power distribution.

4 Conclusions

TD-LTE power private wireless network has less construction cost than optical fiber communications and has higher safety characteristics than public wireless network. In the networking model presented in this paper, DTU equipment in the area around the power private base station access TD-LTE power private wireless network instead of optical fiber can make the construction model of optical cable change from plane to line, which reduces network construction cost as well as ensure reliability. Moreover, the DTU in the signal overlapping area of adjacent two power private base station access the hand-in-hand rings of industrial switches can avoid the unstable communication causing by frequent cell handover.

References

1. ZHONG Yu-qing, WANG Hao, CHEN Bao-ren. Research on Application Principles of Communication Technologies in Smart Distribution Network, Electric Power Information Technology. 2013, 11(5):43–47.
2. ZHAO Qi, LI Ying-fan, LU Jun, QI Bing. Improvement and optimization of wireless digital network in power system, Power System Protection and Control. 2009, 37(15):100–103.
3. WANG Hao,ZHONG Yu-qing,RONG Zhi-neng,CHEN Bao-ren.Research of Multi-frequency Hybrid Networking of TD-LTE Power Wireless Broadband System, Electric Power Information and Communication Technology. 2013, 11(8):24–30.

DOI: 10.1051/matecconf/20166107004

An interlacing networking model for TD-LTE power private wireless network in multiple frequency bands

WANG Hao [1a], CHEN Yongtao [1], ZHONG Yuqing [1]

[1]*Guangzhou Power Supply Co.Ltd. , Guangzhou, Guangdong, 510000, China*

Abstract: TD-LTE power private wireless network develops rapidly in recent years for the communication of distribution and sale of electricity in smart grid. The mainstream technologies of TD-LTE power private wireless network are 1.8GHz TD-LTE power private wireless network and 230MHz TD-LTE power private wireless network. These two kinds of wireless technologies are now separately constructing network in different power supply area, which will cause Co-Channel interference in the adjacent cells and signal blind zones. This paper present an interlacing networking model for TD-LTE power private wireless network in multiple frequency bands, this model can reduce the Co-Channel interference in the overlap area of the base stations and promote continuous coverage performance of TD-LTE power private wireless network.

1 Introduction

According to the frequency assignment policy in China, power grid enterprises can use 230MHz and 1.8GHz frequency bands to construct power private wireless network, and TD-LTE technology is the most mainstream wireless communication technology. The 230MHz TD-LTE power private wireless network works in 40 discrete frequency points in 223~235MHz frequency bands, each frequency point occupies 25kHz bandwidth. The 1.8GHz TD-LTE power private wireless network works in 1790~1795MHz continuous frequency[1]. These two kinds of power private wireless network are now planed and constructed separately which is showed in Figure 1. In this construction model, a power supply bureau construct the 1.8GHz TD-LTE power private wireless network and 230MHz TD-LTE power private wireless network in different power supply area and the base stations in different working frequencies have no signal overlap areas. The 1.8GHz TD-LTE power private wireless network is constructed in urban area for the larger throughput capacity and the 230MHz TD-LTE power private wireless network is constructed in suburb for the better coverage performance in low frequency. The base station location of power private wireless network is restricted in the real estate of power grid enterprises like substations or office buildings[2], which will cause the cell distance of the TD-LTE power private wireless network too big or too small. Unreasonable distribution of the TD-LTE power private wireless network will cause Co-Channel interference in the adjacent cells and signal blind zones. So the research of a new networking model to reach maximum utilization of frequency resources of TD-LTE power private wireless network is very necessary.

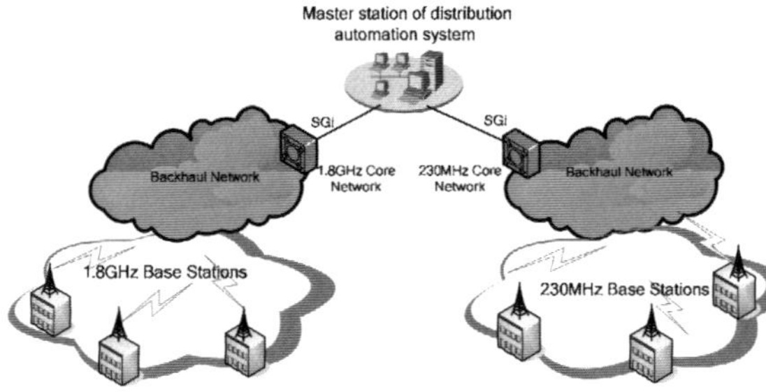

Figure 1. Current networking mode of TD-LTE power private wireless network in two frequencies.

2 Analysis of Co-Channel interference of neighbour cells in TD-LTE power private wireless network

When base stations of the TD-LTE power private wireless network in same frequency make an intensive construction in a power supply area, Co-Channel interference of neighbour cells is inevitable.

TD-LTE system support same frequency network, each cell occupy the whole frequency range. The frequency reuse factor of the same frequency network is 1, which reaches the max spectrum efficiency but will cause bad cell-edge user performance because of Co-Channel interference and has high requirements of network planning optimization. TD-LTE power private wireless network is hard to meet the network planning requirements in same frequency network because the base station location is limited.

Three different frequencies division networking can solve the problem of Co-Channel interference but

[a] Corresponding author: wanghao_hust@163.com

throughput will greatly decrease to one third of the max throughput.

FFR(Fractional Frequency Reuse), SFR(Soft Frequency Reuse) and other ICIC(Inter Cell Interference Coordination) technologies can reduce the Co-Channel interference of cell-edge users as well as ensure the throughput of cell-centre users, but frequency division in these kind of technologies will also cause throughput decrease of cell-edge users.

Figure 2. shows the using of SFR of 230MHz TD-LTE power private wireless network with omnidirectional antenna.

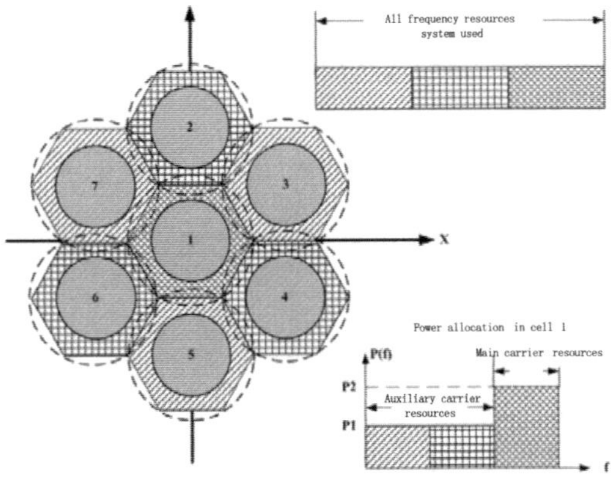

Figure 2. SFR of 230MHz TD-LTE power private wireless network with omnidirectional antenna.

Figure 3. shows 4 Neighbouring 230MHz TD-LTE power private wireless base stations in Huadu District, Guangzhou. The 230MHz TD-LTE power private wireless base stations support SFR. Table 1. shows the terminals statistics with max detach numbers in 40 days with the SFR enabled. The statistics shows that Co-Channel interference of cell-edge users can cause frequently detaches of some terminals which will influence the stability of the communication, and the SFR can not completely solve this problem.

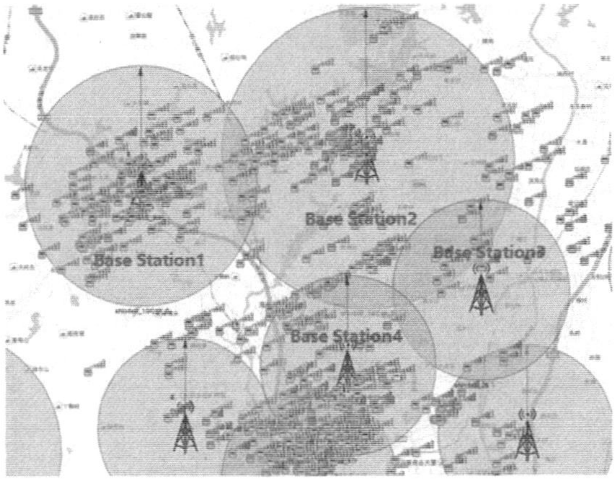

Figure 3. 4 Neighbouring 230MHz TD-LTE power private wireless base stations in Huadu District, Guangzhou.

Table 1. Terminals statistics with max detach numbers in 40 days.

Terminal serial number	Base Station	Detach numbers
1	Base Station2	854
2	Base Station4	815
3	Base Station2	780
4	Base Station4	685
5	Base Station2	661
6	Base Station4	648
7	Base Station4	601
8	Base Station2	503
9	Base Station3	450
10	Base Station4	415
11	Base Station4	404
12	Base Station2	366

3 Analysis of signal blind zones in TD-LTE power private wireless network

Figure 4. shows the emulation of the coverage capability of 1.8GHz TD-LTE power private wireless network in Yuexiu District, Guangzhou. 1.8GHz TD-LTE power private wireless network have a lager throughput than 230MHz TD-LTE power private wireless network so it is always constructed in urban area like Yuexiu District which has dense distribution room distributed. The urban area have tall buildings to shield the signal transmission and 1.8GHz frequency has less signal coverage than 230MHz frequency, so the signal blind zones in 1.8GHz TD-LTE power private wireless network appear in the blue area in Figure 4.

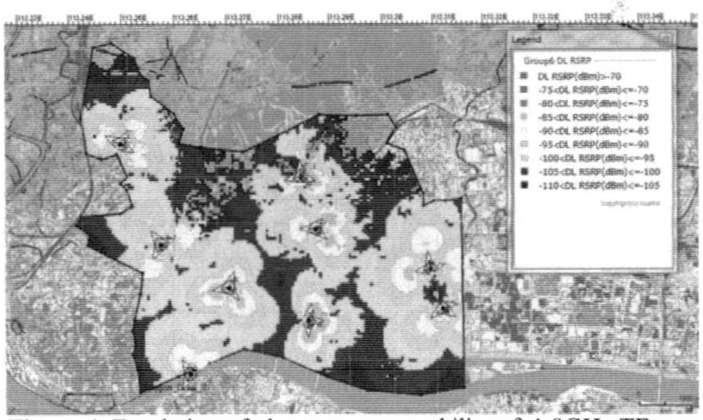

Figure 4. Emulation of the coverage capability of 1.8GHz TD-LTE power private wireless network in Yuexiu District, Guangzhou.

Figure 5. shows the emulation of the coverage capability of 230MHz TD-LTE power private wireless network in Huadu District, Guangzhou. 230MHz TD-LTE power private wireless network uses omnidirectional antenna. It is hard to adjust the downward inclination angle to reduce the signal overlap area of the neighbouring base stations to avoid Co-Channel interference. Increase the distance between the base stations to avoid Co-Channel interference will also cause signal blind zone in the yellow area in Figure 5.

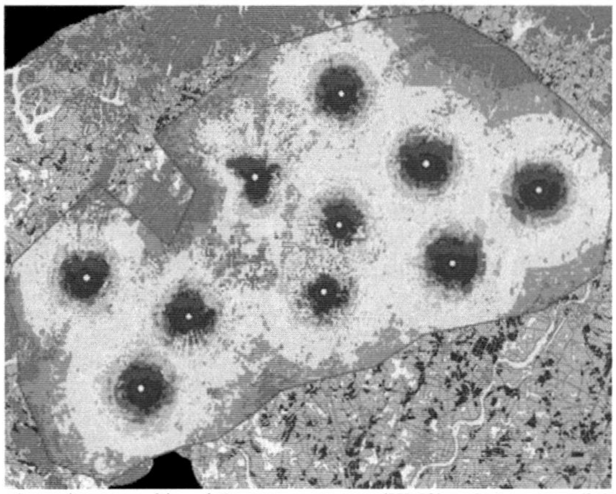

Sample	Signal Range	Area(KM2)	Percentage(%)
	X < -110.0	35.136	7.2
	-110.0 <= X < -105.0	86.464	17.7
	-105.0 <= X < -100.0	127.622	26.1
	-100.0 <= X < -95.0	108.870	22.3
	-95.0 <= X < -90.0	62.848	12.9
	-90.0 <= X < -85.0	32.486	6.6
	-85.0 <= X	35.597	7.3

Figure 5. Emulation of the coverage capability of 230MHz TD-LTE power private wireless network in Huadu District, Guangzhou.

4 An interlacing networking model for TD-LTE power private wireless network in multiple frequency bands

According to the analysis above, using a flexible networking model of interlacing the 230MHz base stations and 1.8GHz base stations can reduce the overlap area of the base stations in same frequency to reduce Co-Channel interference in neighbouring cell. Base stations in different frequencies can be very near without Co-Channel interference to reduce the signal blind zone, and the base stations can use all the carrier resources in their working frequencies to reach the maximum throughput capacity. This interlacing networking model will influence the handover performance of single frequency terminal, as the position of power distribution automation and electric measurement automation equipment is fixed, this influence can be ignored.

When the base stations in two frequencies have an interlacing rate of 1:1, the network will be too complex to operation and maintenance. Using base stations in a frequency as the main signal coverage and base stations in other frequency as auxiliary signal coverage will reach the balance of throughput capacity and maintainability.

The 1.8GHz base stations working at 5MHz bandwidth have higher throughput capacity but lower coverage performance than 230MHz base stations. The 230MHz base stations working at 1MHz discrete bandwidth have lower throughput capacity but higher coverage performance than 1.8GHz base stations. The interlacing mode should be based on the technical characteristics of the base stations in two frequencies. This paper presents two modes of the interlacing networking model. Mode1: 1.8GHz base stations as the main signal coverage and 230MHz base stations as auxiliary signal coverage in urban area. Mode2: 230MHz base stations as the main signal coverage and 1.8GHz base stations as auxiliary signal coverage in suburb area.

Mode1 is showed in Figure 6.The 230MHz base stations use omnidirectional antenna with one cell(circular regions represent) and the 1.8GHz base stations use directional antennas with 3 cells(hexagon regions represent). Distribution rooms have a dense deployment in the urban area, so the substations with high roofing and low shelter can construct the 1.8GHz base stations to ensure the throughput capacity. The remainder substations with low roofing and high shelter can construct the 230MHz base stations to make auxiliary coverage. The substations adjacent to several 1.8GHz base stations can also consider to construct a 230MHz base station to reduce Co-Channel interference in 1.8GHz frequency as well as insure the signal coverage of the power private wireless network.

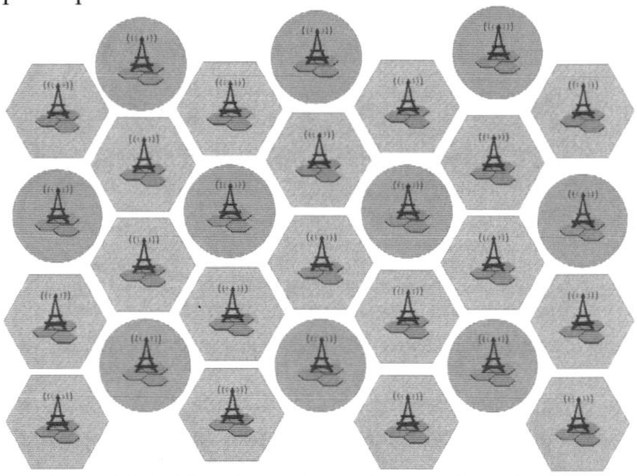

Figure 6. Mode1: 1.8GHz base stations as the main signal coverage and 230MHz base stations as auxiliary signal coverage in urban area.

Mode2 is showed in Figure 7. Distribution rooms have a sparse deployment in the suburb area, so the substations with high roofing and low shelter can construct the 230MHz base stations to ensure signal coverage. The remainder substations adjacent to several 1.8GHz base stations can also considered to construct a 1.8GHz base station as auxiliary coverage to reduce Co-Channel interference in 230MHz frequency, and the DTU(Distribution Terminal Unit) in the signal overlap area of several 230MHz base stations can access the master station[3] by 1.8GHz CPE(Customer Premises Equipment) to avoid frequent detaches caused by Co-Channel interference.

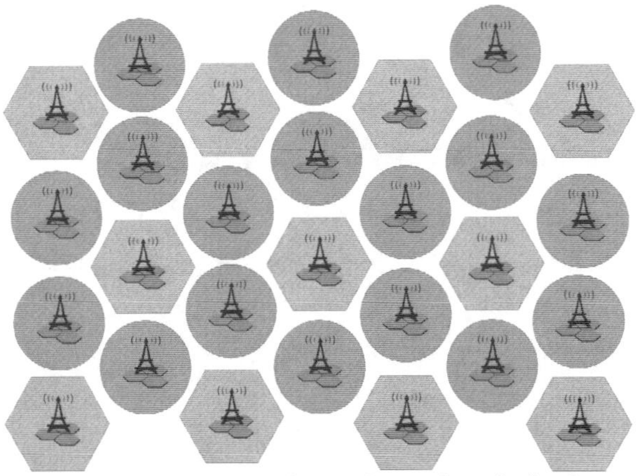

Figure 7. Mode2: 230MHz base stations as the main signal coverage and 1.8GHz base stations as auxiliary signal coverage in suburb area.

5 Conclusions

The interlacing networking model for TD-LTE power private wireless network in multiple frequency bands presented in this paper can avoid the unstable communication causing by frequent cell handover as well as reduce the signal blind zones. Moreover, develop communication terminals which can work in both 230MHz and 1.8GHz power private network can improve the handover performance of this interlacing networking model.

References

1. WANG Hao,ZHONG Yu-qing,RONG Zhi-neng,CHEN Bao-ren.Research of Multi-frequency Hybrid Networking of TD-LTE Power Wireless Broadband System, Electric Power Information and Communication Technology. 2013, 11(8):24–30.
2. LI Wen-wei, CHEN Bao-ren, WU Qian, LAI Mu-bo. Applied Research of TD-LTE Power Wireless Broadband Private Network. Telecommunications for Electric Power System.2012, 33(11):82–87.
3. ZHANG Lan. Overview of power distribution automation communication technologies. Telecommunications for Electric Power System. 2008, 29(4): 42–46.

A UWB Power Control Algorithm Based on Activity Link Protection

Xuefeng Pan[a],Jing Hu,Jiaheng Cao

Department of Computer Science,Luojia College,Wuhan University, 430064 Wuhan,China

Abstract. In the wireless local area network, the mobile terminal by a large carrier power consumption, which greatly influence the performance of mobile, in order to reduce the energy consumption of a control algorithm to establish the link. During the link establishment , the power controlled multiple access mode, using a modified maximum tolerate interference based on the (MEI) channel access algorithm for distributed power control. The simulation results show that, with the increasing distance, the link between the multiuser interference decreases, can effectively improve the efficiency of wireless local area network.
Key words:ultra wide band(UWB),wireless sensor network(WSN),match-filter(MF), energy-detection(ED)

1 Introduction

The technology of traditional radio is based on the carrier, but on the modulation signals to the carrier, the modulation is carried out on the sender and the receiver must carry on the demodulation. It increases the complexity of hardware, and larger power consumption is needed to launch a carrier, the mobile terminals mobile performance caused great influence. Under this background, the federal communications commission (FCC) definition of civil UWB[1] (the technology of Ultra wide band).The technology of UWB has the characteristics of low power consumption, high bandwidth and low complexity[2-3], and the communication technology of traditional wireless has many different places. UWB is particularly suitable for indoor and other places of dense multipath high-speed wireless access.

At present, in the research of wireless sensor network, strengthen the signal and reduce energy consumption and so on all make for correlation. Literature 3 by using the algorithm to enhance the signal coverage, literature 4 from the perspective of reducing energy consumption to improve the relevant agreements. In the link level control energy consumption by power control algorithm, can effective improve the efficiency of the network.

2 Analysis of UWB Link

2.1 Analysis of UWB signal

UWB signal is defined as the signal bandwidth is more than 500 MHz, the ratio of the signal bandwidth and center frequency between 1% ~ 25% for broadband, less than 1% for narrowband, visible all the bandwidth of the UWB significantly greater than the current technology of communication bandwidth[4-5].

Relative bandwidth of signal is defined as: the relative bandwidth=$(f_H-f_L)/[(f_H+f_L)/2]$[6].If relative bandwidth is greater than 0.25, it is a super wideband signal. Uwb baseband transmission, realizing the way is to send impulse radio signal transmit voice and image data, can send up to 1 billion per second on behalf of the pulse signal of 0 s and 1 s. The pulse signal frequency domain is very wide, range in a few Hz to several GHz, the low frequency part can realize communications through walls. Because the transmission power of the signal is very low, only the equivalent of some background noise, not to interfere with other narrowband signal[7].Because only when needed to send out the pulse wave, thus greatly reducing the power consumption.

2.2 Analysis of UWB Modulation Method

UWB transmission power of restricted by the transmit power spectral density, and thus affect the choice of modulation mode in two ways. One is for every bit of energy modulation will need to provide the best error performance. The second is the choice of modulation scheme affect the structure of the signal power spectral density, therefore may add some additional restrictions on the transmission power.

There are three kinds of UWB modulation mode: direct sequence spread spectrum UWB (DS - UWB), jump in ultra wideband (TH - UWB) and multiband UWB. Because binary reverse polarity code has better properties than the PPM code, design the TH - PAM modulation, namely jump add binary code when reverse polarity

[a] Corresponding author:panxuefeng800103@sina.com

modulation.Under the same ber requirements, binary reverse polarity signal transmission power is only half the binary orthogonal PPM modulation.

In this kind of modulation, time is divided into frames. T_f is time of each frame, and each frame is composed of N_h. It tentative $N_h = 7$.T_c is the time of each time slot, so $T_f=T_c*N_h$. A bit of information with N_s pulse, said here $N_s = 4$.A user within A frame to send only one pulse, so the user A's A bit of information in the above four frames after the T_f sent to receive completely.

Assume that the TH code of user A is {0,1,5,3}, therefore, user A in the first frame 0 time slot detection, in order to distinguish between A bit of information, here using the binary reverse polarity PAM modulation. Here the first frame detected pulse to -1.By the same token, the receiver in the first time slot in the second frame, the third time slot 5 frames, the fourth frame timeslot distinguish 3 detection, the signal respectively +1,+1 and -1.At this point, A user of A bit of information represented by A pulse test is completed, all for the {-1,+1,+1,-1}.

According to the characteristics of uwb signal, assuming there are N users in the system, so the link i the signal-to-noise ratio of formula is as follows:

$$\gamma_i = \frac{P_i g_{ii}}{R_i(\eta_i + \sigma^2 T_f \sum_{j=1, j\neq i}^{N} P_j g_{ij})} \quad (1)$$

Among them, N is the number of links.T_f is each time frame.R_i is the link rate of bits of information,and g_{ij} is link gain from the sending node j to the receiving node i.η_i is receiving nodes of the ith link I background noise power spectral density. σ^2 is the formation of the UWB pulse factor.

A bit by NS pulse, T_f is a frame of time, such the time of a bit completely sent is $T_b = T_f * N_S$.In this way, the bit rate Ri = 1 / Tb, average power $P_i= E/t= E_{frame}/T_{frame}= E_{m,i}/T_{frame}= E_{m,i}/T_f= E_{m,i}/T_f$, among them, the E_{frame} is the power of link i to send in one frame. $E_{m,i}$ is the energy of a pulse.Therefore, in order to change the power of link i,it can by changing the pulse energy or changing the T_f. The changes of T_f can be divided into Nh and Tc.

3The Algorithm of Power Control

3.1 The Main Idea

In establishing a link, multiple access method adopts an improved distributed power control based on MEI channel access algorithm.Network topology in the form of the AD - hoc network, each link between don't need to be synchronized, only need to synchronization between receiver and send points.For the upper bound of the transmission power, as UWB spectrum are unauthorized frequency band, if transmitted power of UWB without control, is likely to cause interference to other systems.So you have to consider the upper bound of the transmission power, therefore using FCC rules here.

3.2 Establish Linkment

In order to facilitate described the algorithm of power control, first introduced the concept of largest can tolerate interference (MEI).The biggest can tolerate interference is a link in order to keep normal communication QoS and is able to bear the biggest additional interference.If link i target the minimum signal-to-noise ratio is γ_i^T, ultra-wideband signal-to-noise ratio according to the formula, the calculation formula of MEI below:

$$\gamma_i^T = \frac{P_i g_{ii}}{R_i(\eta_i + \sigma^2 T_f \sum_{j=1, j\neq i}^{N} P_j g_{ij} + \sigma^2 T_f M_i)} \quad (2)$$

Each link, regularly check the interference around me, and his own gamma information such as γ_i^T and R_i.It can work out their own Mi by formula 2.

When a new link to be created, assuming that the system has N activities links in the link, each link has been worked out their own Mi, and timing will send these information through radio channel. When another new link to build, it first listening out of all the links of Mi,1≤i≤N. Suppose N+1 the link rate of signal to noise ratio and requirements are: γ_{N+1}^T and R_{N+1}.

$$\gamma_{N+1}^T = \frac{P_{N+1} g_{N+1,N+1}}{R_{N+1}(\eta_{N+1} + \sigma^2 T_f \sum_{j=1}^{N} P_j g_{N+1,j})} \quad (3)$$

In order to guarantee the minimum rate R_i to send, the power requirements is:

$$P_{N+1} = \frac{\gamma_{N+1}^T R_{N+1}(\eta_{N+1} + \sigma^2 T_f \sum_{j=1}^{N} P_j g_{N+1,j})}{g_{N+1,N+1}} \quad (4)$$

Already exists in the system N the activities of the link, the link priority is higher than the new link. When N+1 the link be established, must make the pledge that we now have normal communication link, so the article N + 1 link first calculate $\min(M_i/g_{i,N+1})$, 1≤i≤N.And considering the FCC masking power, it is concluded that the upper bound of the power of the N + 1 link limit: $P_{max}= \min(P_{FCC},\min(M_i/g_{i,N+1}))$, 1≤i≤N.

Compared with P_{N+1} Pmax values, if $P_{N+1} \leq P_{max}$, can be successfully established the new link;Otherwise unable to build new links, has reached the awaited state.After a random experiment to try to access it again.So if after a time out later still cannot be established, the N + 1 link will be the issue to top processing, or reduce the rate of demand.When the $P_{N+1}=P_{max}$, clearly link the N + 1 to send Pmax.When the $P_{N+1} \leq P_{max}$, the best is sending power $P_{opt} \in \{P_{N+1},P_{max}\}$.In reference [1] in order to calculate the Popt, MEI algorithm designed a kind of balance, but the algorithm is too complex.For the sake of simplicity, we adopted a compromise, take $P_{opt}=(P_{N+1}+P_{max})/2$.

3.3 Algorithm Description

3.3.1 The Sending Node Algorithm Description

(1)Initialize a link establishment request, and send the request information to the RX node.

(2)Calculate the node in the case of without interfering with the existing other link, can use one of the biggest power P_{max}.

(3)When it receive the reply the RX node minimal power requirements P_{min}, to (4), otherwise wait.

(4)Compare, if $P_{min} \leqslant P_{max}$, the link can be successfully established, turn to (5), or the link cannot be established, after a random time, return to(1) the initialization step.

(5)Calculate the optimal power P_{opt}, successfully established and sent to link status information to RX node.

(6)Start sending data.

3.3.2 The Receiving Node Algorithm Description

(1)Check its receive buffer, if received the TX initialization link request information, then transferred to (2), otherwise return to (1).

(2)Calculate the oneself in order to maintain the required SNR, P_{min}, and sent to the TX.

(3)When you receive the success to establish a link information, the TX to (4), otherwise it return to (1).

(4)Began to receive data.

3.4 Algorithm Implementation

UWB power control algorithm based on activity link protection according to the sending and receiving nodes of two parts. Algorithm implementation program includes two main parts: the first part is a custom message class powerMsg, definition to carry power control information. The second part is sent and received the core code of the algorithm.

3.4.1 The Definition of Class PowerMsg

Custom powerMsg classes, the main tag carries information about power control, including size, link power gain, SNR, node status information (free or busy) and so on.

```
message powerMsg
{
fields:
double txPower;
double rxWantMinpower;
double MEI;
int  sourceNodeIndex;
int  destintNodeIndex;
int  x;
int  y;
bool wantCommWithYou=false;
bool linkSucessToCreat=false;
}
```

After this MSG files by compiling, can generate the other two files: powerMsg_m.h and powerMsg_m.cc.

3.4.2 The Core Code of the Sending and Receiving Node Information

Create a module in the physical.Main processing module and the physical channel information, such as the calculation of the signal-to-noise ratio of the two nodes, compute a total interference, the receiving node calculation are receiving node MEI, calculates the maximum send power sending node, and so on.

(1)Calculate the signal-to-noise ratio of the two nodes

```
double Physic::getGain(powerMsg *pmsg)
{
 double gain;
 double dist=getDistance(pmsg);
 if(dist<35)
 {gain=1;
 }
 gain=1225/(dist*dist);
 return gain;
}
```

(2)Calculation of total interference a receive node

```
double Physic::getInterference()
{
 double totleNiose=0.0;
 for (int i=0;i<txpmsgArray.items();i++)
 {
 powerMsg *pmsg=(powerMsg *)txpmsgArray[i];
 if((pmsg->getSourceNodeIndex())!=(parentModule()
->index()))
 {
 double gains=getGain(pmsg);
 totleNiose+=(pmsg->getTxPower())*gains;
 }
 }
 return totleNiose;
}
```

(3)Calculate the MEI is receiving node

```
double Physic::getmyMEI(double tSIR)
{
 double interfere=getInterference();
 double xixi=tSIR*requireRate*shapingFactor*Tf;
 double
myMei=(initxPower*myLinkGain-tSIR*requireRate*
backnoise-xixi*interfere) /(xixi);
   return myMei;
}
```

(4) Calculate the maximum send power sending node

```
double Physic::maxPower()
{
 double thetxPower;
 double maxpower=1000.0;
 for (int i=0;i<rxpmsgArray.items();i++)
 {
 powerMsg *pmsgs=(powerMsg *)rxpmsgArray[i];
 double gain=getGain(pmsgs);
 if (((pmsgs->getMEI())/gain)<maxpower)
 maxpower=(pmsgs->getMEI())/gain;
```

```
    }
    if(maxpower>(deviceMaxpower))
    maxpower= deviceMaxpower;
    return maxpower;
}
```

4 The Simulation Experiment and Result Analysis

For UWB simulation USES is OMNeT++, it is open source component based modular open network simulation platform[9].

When the power control algorithm simulation, the TH - PAM modulation, the comprehensive binary reverse polarity jump code modulation. Binary reverse polarity required by the code under the same bit error rate, a signal transmitted power is only half the binary orthogonal PPM modulation. Test information in TH - PAM modulation modulation, time is divided into individual frames, each frame of T_f, expressed in four pulse one bit of information. A user within a frame to send only a pulse. Test information of one bit of information in four frames after the T_f sent to receive completely. Through the system after the adjustment information is converted into the corresponding binary sequence and pulse form users to send.

Will be set up to send information, combined algorithm simulation program, through the node link signal strength and after into the algorithm to optimize signal spectrum ratio on the simulation analysis of signal source.

Establish a node link simulation diagram is shown in figure 1, the blue point said mobile terminal, black has told the arrow link.

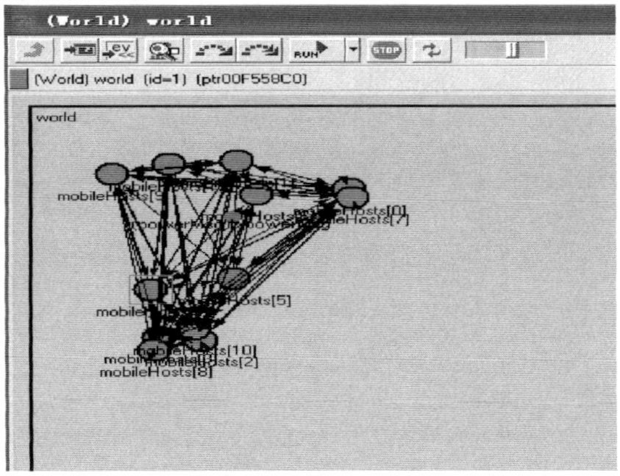

Figure 1. the Link of Node.

Set the signal frequency is 3.389 GHz, rate of 100 MHZ UWB signal.Figure 2 and figure 3 is before and after the use of power control algorithm for UWB signal contrast test figure, can see the power of UWB narrow pulse signal increased to 6.18 from -8.56 dBm dBm, gain of 14.74 dBm.Considering the actual test of the line loss, about 2-3 DBM.The actual gain 11 DBM, effectively enhance the quality of UWB signal.

Through the above simulation experiment, it can be seen that using UWB as the physical carrier, multi-user interference between each link is very small.It is very suitable for application in more than one user of the system, and complex multipath environment, so as to meet the needs of UWB short high-speed data transmission system.

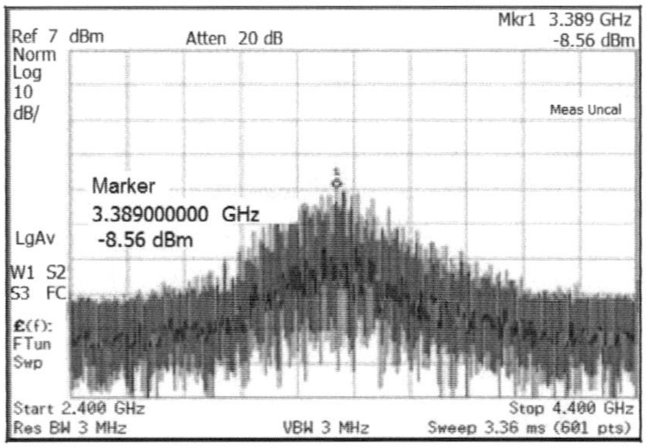

Figure 2. the Spectrum of Source Signal.

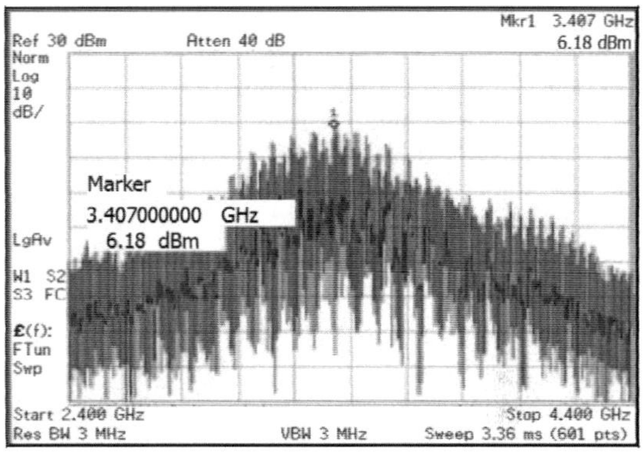

Figure 3. **S**ignal Spectrum after Using Power Control Algorithm

5 Conclusion

For mobile terminal for launch carrier consumed power and reduce the mobility of the larger problem, this paper proposes a control algorithm to build links.In establishing a link, the power control multiple access methods, for distributed power control based on MEI channel access algorithm is improved.After the improvement, reduce the energy consumption and the link between the multi-user interference is very small, the multiple users of wireless sensor network (WSN) life cycle.

Acknowledgment

The scientific research project of Education Department of Hubei Province under Grant No.B2015291. The education science planning project of Hubei Province under Grant No.2015GB156.

References

1. Ding Rui, Qian Zhi-hong, Wang Xue. UWB positioning system based on joint TOA and DOA estimation[J]. Journal of Electronics & Information Technology, 2010, 32(2): 313-317.
2. Muruganathan S D, Ma D C F, Bhasin R I, etal. A centralized energy-efficient routing protocol for wireless sensor networks. IEEE Communications Magazine, 2005, 43(3):8-13.
3. Zhang Hua, Song Zheng-xun, Shi Yun, et al. Research of TOA ranging technology based on UWB[J]. Journal of Jilin University (Information Science Edition), 2008, 26(1): 106- 110.
4. Tang Yong,Zhou Mingtian,Zhang Xin.Routing protocol for wireless sensor networks research progress[J]. Journal Of Software,2006,17 (3):410-421.
5. LIU Zhi-jun, LI Na-yuan, YANG Shao-hua. Base on energy balance LEACH-DC protocol design[J]. Computer engineering & design, 2012, 33(4): 1337-1341.
6. Shen Bo,Zhang Shiyong.wireless sensor network clustering routing protocol[J].Journal Of Software, 2006, 17(7): 1588- 1600.
7. LI Chun-xia,ZHAN Yong-zhao,WANG Liang-min. Secure Localization Scheme for Wireless Sensor Networks Based on UWB and Mobile Anchors[J]. Journal of Chinese Computer Systems,2012,33(2):347- 352.

A fast CU Size Decision Algorithm for 3D-HEVC

Yubing Wang, Yongfang Wang[a] and Yawen Shi

Shanghai University, Shanghai, China

Abstract. The emerging 3D-HEVC has achieved the highest coding efficiency but requires a very high computational complexity. To speed up the encoding process for the dependent texture views, we propose a fast CU depth range selection algorithm by jointly making use of the inter-view and temporal-spatial correlations. Firstly, adaptive correlation weights are proposed to predict coding unit(CU) depth range and skip some specific depth levels rarely used in independent view, the previous frame and neighboring CUs. Besides, a new early termination algorithm is proposed to further reduce the coding time. Experimental results demonstrate that the proposed method saves about 56% coding time on average compared to HTM with maintaining the similar video quality.

1 Introduction

With the rapid development of multimedia technology, compared to text, voice, images, video is used more and more abroad. However, the traditional 2D video can't meet people's need, and 3D video is becoming more and more popular. It is well known that the data size of video is huge, so it asks for higher requirement for transferring, storing and playing. Furthermore, 3D video uses more than one camera to shoot the same scene, which leads to increasing a large amount of information.

To improve the coding efficiency of high-definition video, the Joint Collaborative Team on Video Coding (JCT-VC) designs the standard High Efficiency Video Coding (HEVC)[1]. Subsequently, the Joint Collaborative Team on 3D Video Coding Extension Development (JCT-3V) develops HEVC based 3D video coding standard (3D-HEVC), which is for the compression of multi view video plus depth (MVD) format[2]. In 3D-HEVC, similar to HEVC, the mode decision process in HTM is performed using all the possible CU sizes , prediction modes, and coding tools(disparity-compensated prediction (DCP), inter-view motion prediction, backward view synthesis prediction (BVSP)) to find the optimal one with the least rate distortion (RD) cost using Lagrange multiplier, which leads to high computational workloads. It obstructs wide application of 3D-HEVC.

In order to reduce the complexity of video coding, much work has been done to explore the fast algorithms for H.264/AVC. Liu et al.[3] propose block partition algorithm based on image features to reduce the coding complexity. The low complexity mode prediction in [4] is proposed based on the spatial-temporal correlation. The algorithm in [5] uses SKIP mode to early terminate mode decision. Fast algorithms for HEVC inter prediction mainly aim at the process of coding unit (CU) selection which has high complexity. Method in [6] is proposed to terminate procedures of CU splitting by setting a threshold value based on the RD-cost of the CUs which have already been coded, if the RD-cost of the current CU is less than this threshold ,CU will stop splitting, and a fast CU size decision method based on coding tree pruning is proposed in [7]. Shen et al.[8] propose three early termination methods based on motion homogeneity checking, RD cost checking and SKIP mode checking to skip the procedure of motion estimation on unnecessary small CU sizes. In 3D-HEVC, the maximum depth of the related coding block in dependent view is first used to early terminate the CU splitting in [10]. A fast mode decision algorithm based on variable size CU and disparity estimation in [11] is proposed to reduce 3D-HEVC computational complexity. The texture quad-tree initialization (QTI) and depth quad-tree limitation (QTL) coding tools and their associated predictive coding (PC) algorithm by utilizing the correlations between the quad-tree of the texture and its associated depth is proposed in [12], which was adopted by the 3D-HEVC working draft. Zhang et al.[13] speed up the encoding process of dependent texture views based on inter-view correlation, which uses an early merge mode decision algorithm and an early CU splitting termination algorithm. CU depth correlation between the independent view and the dependent view is also studied to accelerate CU splitting process of dependent views in [14].

The aforementioned methods for 3D video coding only use view correlation or temporal-spatial correlation to predict the depth of current coding block, while not jointly considering these correlations. Consequently, in this paper, we comprehensively exploited the correlations between views , and spatial-temporal

[b] Corresponding author: yfw@shu.edu.cn

DOI: 10.1051/matecconf/20166107006

correlation, and then proposed a fast CU size decision algorithm based on adaptive weights called depth range selection algorithm and an early termination algorithm. Experimental results show that the proposed algorithm saves an average of 56% time compared to the original algorithm of HTM with negligible loss in compression efficiency.

The rest of the paper is organized as follows. Section 2 describes the proposed algorithm in detail. Experimental results are shown in Section 3. Conclusions are given in Section 4

2 Proposed fast coding algorithm

2.1 CU depth splitting analysis for texture views

3D-HEVC inheriting the same quad-tree structure as HEVC, which must traverse from depth level 0 to depth level 3, hugely increases the coding complexity of 3D-HEVC[15]. Zhang et.al[13] made a statistical analysis about the probability of all four splitting depth levels of dependent texture view for each test sequence. The result shows that 76.2%, 1%, 3.9% and 1.4% of treeblocks choose the depth level "0", "1", "2" and "3", respectively. It demonstrates that small depth levels are always selected, so selecting suitable CU depth range is an effective method to save coding time.

In HTM15.0, we make a statistic of the encoding time of texture video and depth video, the result of which is shown in Figure 1. Every test sequence contains three views, and every view includes texture video and depth video. Figure 1.(a) shows that the encoding time ratio is 75% for texture pictures and 25% for depth maps. Figure 1.(b) shows the encoding time ratio of three texture views, it can be seen that View0 accounts for about 17%, while View1 and View2 take up about 40% respectively, which demonstrates that the two dependent textures (View1 and View2) occupy most of the encoding time in 3D-HEVC. Accordingly, the proposed fast algorithm in this paper is only for the two dependent texture views.

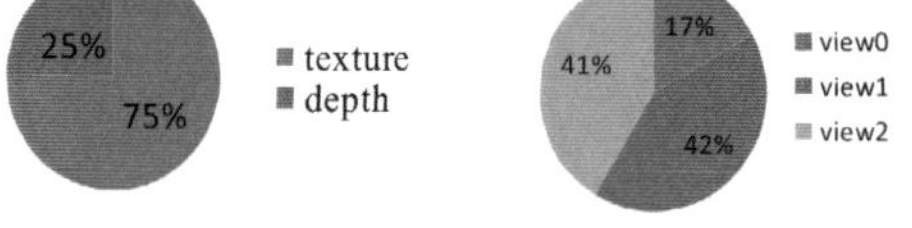

(a) (b)

Figure 1. Analysis of coding complexity:(a) Encoding time comparison between texture and depth maps;(b) Proportion of encoding time of three views

2.2 The depth range selection algorithm

Many previous fast CU decision algorithms use spatial-temporal correlations to predict the depth range of current coding block, but most of them adopt fixed weights that can't adapt to the degree of correlation, which may lead to imprecise prediction. Based on the characteristics of 3D video, we propose a depth range selection algorithm based on adaptive weights using split complexity(SC), which is defined as follows:

$$SC\_i = \begin{cases} \frac{1}{256}\cdot\sum_{j=0}^{255} d_j & depth_{max} = 0,1,2 \\ \frac{1}{128}\cdot\sum_{j=0}^{255} d_j & depth_{max} = 3 \end{cases} \quad (1)$$

Where d_j is the value of depth level, $depth_{max}$ is the maximum value of depth level. The corresponding blocks are shown in Figure 2. i indicates the type of correlated CU, for example, SC_Col is expressed as the split complexity of inter-view correlated CU. To avoid getting the same SC when depth levels are different, unlike the method used in [15], which directly averages the depth value, we adopt different definitions according to the maximum depth level.

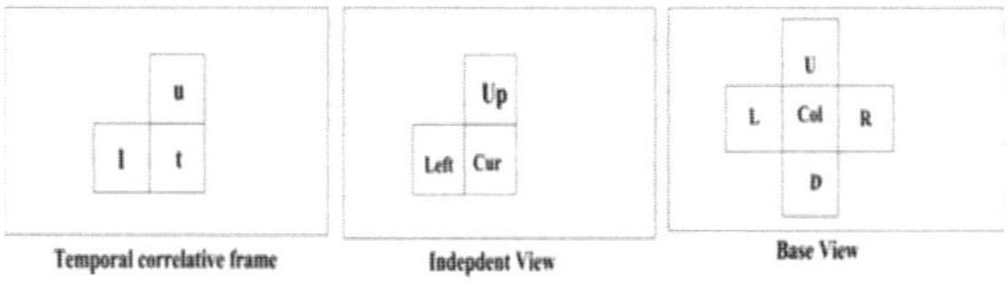

Figure 2. inter-view, spatial, and temporal correlated CUs. Cur: current CU; Left: Left CU; Up: upper CU; t: the co-located CU in the previously frame; Col: inter-view correlated CU.

Experimental result in [15] demonstrates that the spatial-temporal correlation is connected with temporal average depth error. The larger the value is, the stronger the temporal correlation is, and the weaker the spatial correlation is. We evaluate the degree of correlation by using temporal split complexity error (TSCE) and inter-view spatial split complexity error (ISCE) which are defined as follows:

$$TSCE = 1/2\cdot(|SC\_Left - SC\_l| + |SC\_Up - SC\_u|)$$
$$ISCE = 1/2\cdot(|SC\_Col - SC\_L| + |SC\_Col - SC\_U|) \quad (2)$$

The SC of current CU is predicted using spatial neighboring CU (Left, Up in Figure 2), the co-located CU (t in Figure 2) and inter-view correlated CU (Col in Figure 2) as follows.

$$SC\_pre = w_{col}\cdot SC\_col + w_l\cdot SC\_Left + w_u\cdot SC\_Up + w_t\cdot SC\_t \quad (3)$$

Where SC_pre is predicted SC of the current coding block, w_{col}, w_l, w_u, and w_t are weights, which are obtained in (4) from normalizing the W_{col}, W_l, W_u, W_t respectively in (5).

$$w_{col} = \frac{W_l\cdot W_u\cdot W_t}{W_{col}\cdot W_l\cdot W_u + W_l\cdot W_u\cdot W_t + W_{col}\cdot W_u\cdot W_t + W_{col}\cdot W_l\cdot W_t}$$
$$w_l = \frac{W_{col}\cdot W_u\cdot W_t}{W_{col}\cdot W_l\cdot W_u + W_l\cdot W_u\cdot W_t + W_{col}\cdot W_u\cdot W_t + W_{col}\cdot W_l\cdot W_t} \quad (4)$$
$$w_u = \frac{W_{col}\cdot W_l\cdot W_t}{W_{col}\cdot W_l\cdot W_u + W_l\cdot W_u\cdot W_t + W_{col}\cdot W_u\cdot W_t + W_{col}\cdot W_l\cdot W_t}$$
$$w_t = \frac{W_{col}\cdot W_l\cdot W_u}{W_{col}\cdot W_l\cdot W_u + W_l\cdot W_u\cdot W_t + W_{col}\cdot W_u\cdot W_t + W_{col}\cdot W_l\cdot W_t}$$

Where W_{col}, W_l, W_u, and W_t are calculated by jointly utilizing the temporal-spatial and inter-view correlations as follows:

$$\begin{aligned} W_{col} &= 1.2 \\ W_l &= 1 + 0.2 \cdot ISCE \\ W_u &= 1 + 0.2 \cdot ISCE \\ W_t &= 1 + 0.2 \cdot TSCE \end{aligned} \quad (5)$$

In addition, in order to take full advantage of inter-view correlation, we define a Correlated Split Complexity (CSC) in (6).

$$CSC = \max\{SC\_col, SC\_L, SC\_U, SC\_R, SC\_D\} \quad (6)$$

According to SC_pre and CSC, we can predict depth range of current LCU as follows:

If CSC<Th, DR is [0,1], otherwise do as follows:

1) When $SC\_pre \leq$ th1, DR is [0,1];

2) When th1< $SC\_pre \leq$ th2, DR is [0,2];

3) When th2< $SC\_pre \leq$ th3, DR is [1,3];

4) When SC_pre >th3, DR is [2,3].

Where Th,th1,th2,th3 are thresholds, and they are set as 3.0,1.0,2.0 ,5.0 according to extensive experiments.

To verify the effectiveness of the proposed adjusted depth range, four sequences with different activities are tested in Table 1 and Table 2. It is observed from Table I that when Th=3.5, the probability of selecting depth level "0" and "1" can be achieved about 98%. When th1, th2, th3, th4 are set as 3.0,1.0, 2.0, 5.5, we can see from Table II that the average prediction accuracy can be 92%-97%.

Table 1. Probability of selecting low depth level for small CSC.

Sequences	25	30	35	40
Ballons	98%	99%	99%	99%
Newspaper	97%	97%	98%	99%
Poznan_Street	97%	97%	98%	98%
GT_Fly	98%	99%	99%	99%
Average	97%	98%	98%	99%

Table 2. Accuracy of depth range selection

Sequences	25	30	35	40
Ballons	93%	95%	96%	98%
Newspaper	93%	94%	94%	96%
Poznan_Street	90%	93%	95%	96%
GT_Fly	93%	95%	96%	97%
Average	92%	94%	95%	97%

2.3 Early termination(ET) of CU

An early CU splitting termination is proposed to make a more accurate prediction about the splitting depth level of current CU by considering both inter-view correlation and spatial-temporal correlations.

For DR selection algorithm , the size of predicting CU is too large to predict accurately, so when early terminating the CU splitting, the size of referring CUs are 8x8, which locate in the left and top of current CU, and the maximum depth is named as Depth_neighbour. Besides, we also consider the depth level of corresponding block in forward reference frame, backward reference frame, and independent view of current CU, which are named as Depth_t$_0$, Depth_t$_1$, Depth_col respectively.

If the following two conditions are true, early CU splitting termination is performed for the current CU.

1)
$$Depth\_max = \max\{Depth\_neighbour, Depth\_col, Depth\_t_0, Depth\_t_1\}$$
$$Depth\_max \leq uidepth$$

2) Skip mode is selected as the best prediction mode for the current CU.

Where $Depth\_max$ is the maximum value of all the depth levels, $uidepth$ is the splitting depth level of the current CU.

2.4 Overall algorithm

Based on the aforementioned analysis, the proposed overall algorithm incorporates the depth range selection algorithm and the early termination method. A flowchart of the proposed algorithm is shown in Figure 3. First, calculate SC_pre and CSC; Next, predict the depth range of current CU according to CSC and SC_pre; Last, check the early termination conditions of current CU, and if the conditions are met, CU splitting ends, otherwise, the depth level increases by 1.

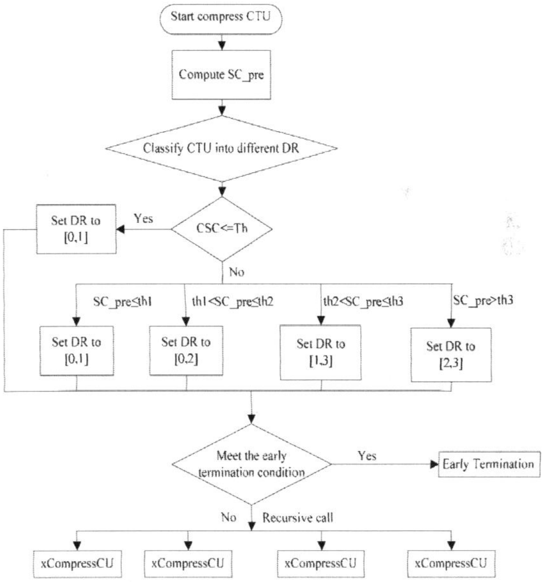

Figure 3. Flowchart of the proposed overall algorithm

3 Experimental results

To evaluate the performance of the proposed algorithm, it is implemented on HTM15.0 Test conditions are set as follows: 7 standard sequences (3 texture plus 3 depth views), four QPs (25/34, 30/39, 35/42, 40/45), QTLPC enabled, fast decision for Merge RD Cost (FDM) enabled, both fast merge mode decision and early CU determination for texture component of dependent view disabled.

Performance of the proposed algorithm is measured by BD-rate and ΔEncT. BD-rate evaluates the coding gains correspond to bitrate reductions, positive and negative values represent increments and decrements respectively. ΔEncT shows the encoding time reduction compared with HTM15.0.

we compare the performance of the proposed algorithm called "DR&ET" and the fast encoder decision algorithm for texture coding called "EMD&ECUST"[13] with HTM15.0. The experimental results are presented in Tables III–IV. The "video1" and "video2" columns show the BD-rate of the two dependent texture views. The "Video PSNR/video bitrate" column shows the BD-rate performance considering Y-PSNR of the coded texture views over the bitrate of texture data.

As can be seen from Table 3, the proposed method can reduce the entire encoding time by 56% on average. Meanwhile, the loss of BD-rate in dependent views is 0.52% and 0.57%. Therefore, the proposed DR&ET can efficiently reduce encoding time with a little loss of RD performance. The performance of "EMD&ECUST" is so good that it has been adopted to 3D-HEVC reference software, but our proposed DR&ET can reduce more time than it, and the BD-rate seen in "Video PSNR/video bitrate" merely increases about 0.5%.

Table 3. Encoding performance compared with HTM15.0

Sequence	video 1	video 2	Video PSNR/ video bitrate	ΔEncT
Balloons	0.45%	0.80%	0.39%	58%
Kendo	0.97%	0.72%	0.73 %	52%
Newspaper	0.78%	0.68%	0.55%	58%
GhostTownFly	-0.15%	-1.40%	0.41%	51%
PoznanHall2	0.85%	1.41%	1.22%	63%
PoznanStreet	0.33%	1.00%	0.64%	55%
UndoDancer	0.40%	0.74%	0.88%	52%
average	0.52%	0.57%	0.69%	56%

Table 4. Encoding performance of EMD&ECUST compared with HTM15.0

Sequence	video1	video 2	Vide PSNR/ video bitrate	ΔEncT
Balloons	-0.67%	-0.29%	-0.10%	56%
Kendo	0.11%	-0.47%	0.15%	48%
Newspaper	-0.11%	-0.37%	0.22%	52%
GhostTownFly	-1.36%	-0.73%	0.05%	46%
PoznanHall2	-0.43%	0.52%	0.60%	52%
PoznanStreet	-0.62%	-0.57%	0.12%	48%
UndoDancer	-1.07%	-0.54%	0.28%	45%
average	-0.59%	-0.24%	0.19%	48%

Figure 4 (a) and (b) illustrates an example of RD and time saving curves under four different QPs compared to HTM15.0. As shown in Fig.4, the proposed method can reduce more coding time with maintaining the RD performance than that of EMD&ECUST. In addition, according to Fig.4(b), we can see that when QP is low, DR&ET is more effective than EMD&ECUST, while as the increase of QP, the performance of both algorithms will tend to be similar. In general, the proposed algorithm is better performance than EMD&ECUST.

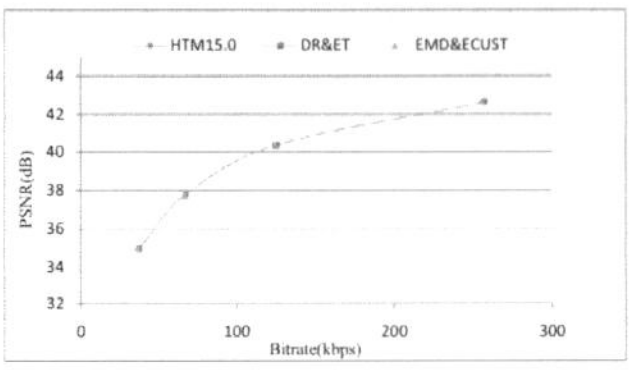

(a)

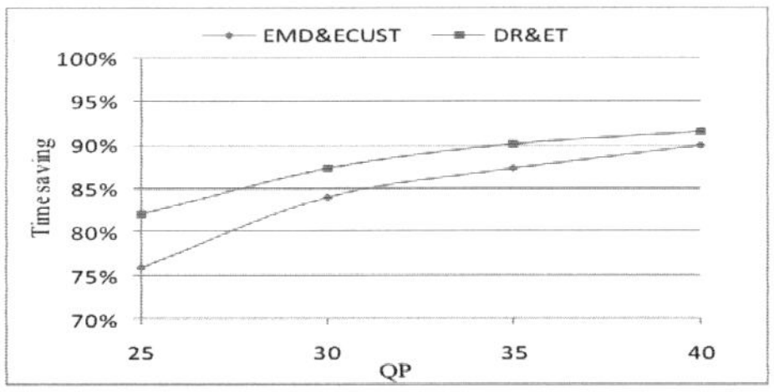

(b)

Figure 4. Experimental results of "Ballons" in video 1 under different QP settings. (a) RD curves of "Ballons" in video 1. (b) Time saving curves of "Ballons" in video 1.

4 Conclusions

In this paper, we propose a fast CU depth range selection algorithm and an early termination CU algorithm for 3D-HEVC. Experimental results show that the proposed algorithm can significantly reduce the computational complexity with negligible loss of BD-rate. And this algorithm will make 3D-HEVC more available for real-time applications.

Acknowledgments

This work was supported by Natural Science Foundation of China under Grant No. 61301113, 60972137, and Shanghai National Natural Science under Grant No.13ZR1416500.

References

1. Joint Collaborative Team–Video Coding, "Test Model under Consideration", JCTVC-A205 output document, Dresden, Germany, 15-23 April, 2010.
2. Ohm J R, Overview of 3D video coding standardization[C], International conference on 3D systems and applications, Osaka. 2013.
3. Liu Z, Shen L, Zhang Z, An efficient inter mode decision algorithm based on motion homogeneity for H. 264/AVC[J], Circuits and Systems for Video Technology, IEEE Transactions on, IEEE, 2009.
4. Zhao T, Wang H, et al, Fast mode decision based on mode adaptation[J], Circuits and Systems for Video Technology, IEEE Transactions on, IEEE ,2010.
5. Avishek Saha, Kallol Mallick, et al, SKIP prediction for fast rate distortion optimization in H.264, Consumer Electronics, IEEE Transactions on 53.3 (2007): 1153-1160.

6. Tan H L, Liu F, Tan Y H, et al, On fast coding tree block and mode decision for high-efficiency video coding (HEVC)[C], Acoustics, Speech and Signal Processing (ICASSP), 2012 IEEE International Conference on, IEEE.
7. Choi K, Jang E S, Fast coding unit decision method based on coding tree pruning for high efficiency video coding[J], Optical Engineering, 2012, 51(3): 030502-1-030502-3.
8. Shen L, Liu Z, Zhang X, et al, An effective CU size decision method for HEVC encoders[J]. Multimedia, IEEE Transactions on, IEEE, 2013.
9. Wu W, Song B, Just-noticeable-distortion-based fast coding unit size decision algorithm for high efficiency video coding[J], Electronics Letters, 2014, 50(6): 443-444.
10. Thaísa Leal da Silva, Luis A. da Silva Cruz, Inter-view prediction of coding tree depth for HEVC-based multiview video coding[C], Electronics, Circuits, and Systems (ICECS), 2013 IEEE 20th International Conference on, IEEE, 2013: 165-168.
11. Zhang Q, Li N, Gan Y, Low Complexity Mode Decision for 3D-HEVC[J], The Scientific World Journal, 2014.
12. Mora E G, Jung J, Cagnazzo M, et al, Initialization, limitation and predictive coding of the depth and texture quadtree in 3D-HEVC[J], IEEE Transactions on Circuits and Systems for Video Technology, 2014.
13. Zhang N, Zhao D, Chen Y W, et al, Fast encoder decision for texture coding in 3D-HEVC[J], Signal Processing: Image Communication, 2014, 29(9): 951-961.
14. Chi G, Jin X, Dai Q, A quad-tree and statistics based fast CU depth decision algorithm for 3D-HEVC[C], Multimedia and Expo Workshops (ICMEW), 2014 IEEE International Conference on. IEEE, 2014: 1-5.
15. Fang Shuqing, Yu Mei, Chen Fen, Jiang Gangyi, et al, Fast Adaptive Coding Unit Depth Range Selection Algorithm for High Efficiency Video Coding[J], Sensors & Transducers, 2014, 183(12): 1726-5479.

DOI: 10.1051/matecconf/20166107007

In-band Spectrum Sensing Method Based on Guard Frequency Bandwidth for Cognitive OFDM Systems

Guoqing Jia[1,a], Shanji Chen[1], Chao Chen[1], and Xiaohong Ji[1]

[1] Qinghai University for Nationalities, No.3 Middle Bayi Road, Xining, China

Abstract. A novel in-band spectrum sensing method based on guard frequency bandwidth for cognitive OFDM systems is proposed in this paper. The proposed spectrum sensing procedure is based on leakage-negligible-BW, on which power detection or other detection methods can be implemented. The advantage of this in-band sensing method is the impact to OFDM system frame structure could be minimized.

Keywords: OFDM System; Cognitive Radio; In-band Sensing; Guard Band

1 Introduction

As the development of wireless communications, there are fewer and fewer spectrum resources which can be allocated to a new implemented radio transmission system. However, the frequency utilization efficiency is very low for some wireless communication systems, such as TV broadcast system. So, cognitive radio technology[1-5] has been developed to allow unlicensed system (secondary system) to exploit spectrum of licensed system (primary system) opportunistically. Thus the spectrum allocation problem can be solved and the spectrum utilization efficiency could be promoted greatly[6-9].

OFDM (Orthogonal Frequency Division Multiplexing) is accepted as the most promising technique for next generation wireless communication system. The system BW (BandWidth) for wireless OFDM system could be as much as 100MHz, such as LTE-A (Long Term Evolution Advanced) system. In addition, there is further need for spectrum for femtocell implementation, etc. So the spectrum allocation for wireless OFDM system will be very difficult as no more available wide band spectrum can be used. As a result, it is very important for wireless OFDM system to sense and reuse the frequency resource of licensed systems which has low spectrum utilization efficiency. Figure 1 gives an illustration of wireless communication system existing in TV spectrum.

In figure 1, the procedure of spectrum sensing by wireless communication system is composed of two scenarios:

- Out-band frequency spectrum sensing: the procedure of sensing a free spectrum band of primary system by secondary system;
- In-band frequency spectrum sensing: the procedure of sensing to detect whether the primary system re-starts transmitting, so a decision will be made to switch the secondary system to another spectrum band or not.

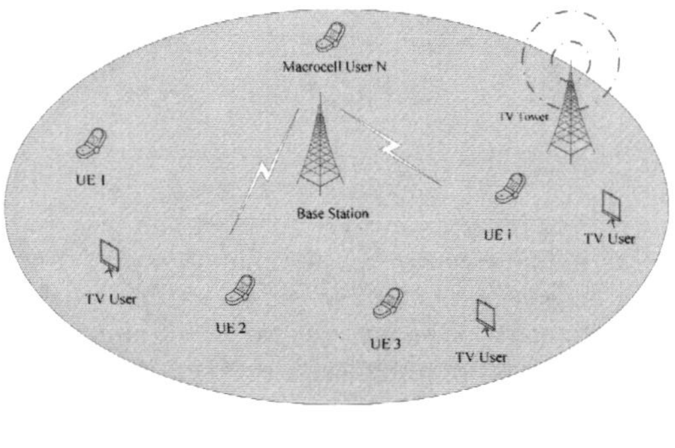

Figure 1. Wireless communication system existing in TV system

In this paper, we focus on the latter scenario in OFDM systems. That is, the OFDM system is working on a spectrum band as a secondary system, and the spectrum sensing is aiming for detect whether the primary system begins transmitting signals again.

2 Technique Analysis

2.1 OFDM System Guard Band

In OFDM system, system BW is composed by transmission BW and guard BW. Guard BW is used to prevent inter-system interference (interference from power leakage of OFDM system), and there is no signals

[a] Corresponding author: 13709759761@139.com

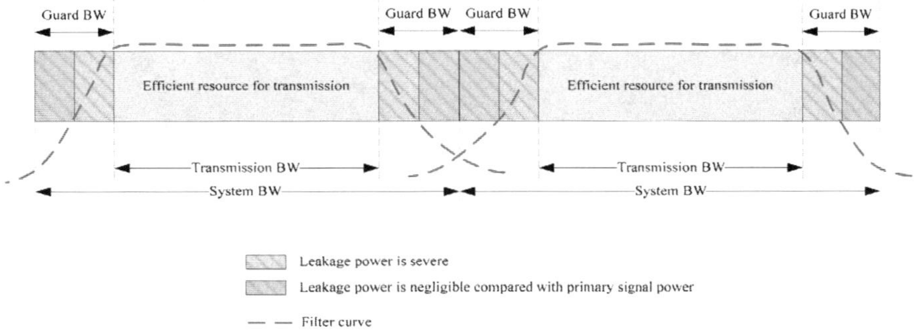

Figure 2. OFDM system guard BW illustration

on the guard BW except for leakage power from neighbor transmission BW and white noise. Figure 2 gives an illustration of how guard BW prevents the power leakage from inter OFDM systems. From figure 2, it can be considered that the guard BW is composed by two parts: the inner part and the outer part. We define the outer part of guard BW as leakage-negligible-BW here.

Under the above inner and outer parts based participation, the signals in the inner part of the guard BW will mostly be affected by the signal leakage due to non-ideal filter, while the signals in the outer part of the guard BW will mostly be influenced by the signal leakage due to inter-system interference. However, the leakage power is restricted to a low level so that there is little impact for sensing to decide whether primary signals exist or not. Further, for those users with well-designed filters, the signal leakage from transmission BW can be reduced to a very low level. As a result, we can effectively use the guard BW, specifically, the named leakage-negligible-BW, to sense the primary system.

Here we take LTE/LTE-A system for example, which is the most promising OFDM technique based cellular wireless communication system. The guard BW configurations of LTE/LTE-A system are given by Table 1.

Table 1. LTE/LTE-A system guard BW configuration

System BW	Transmission BW	Guard BW
20M	18M	2M
15M	13.5M	1.5M
10M	9M	1M
5M	4.5M	0.5M
3M	2.7M	0.3M
1.4M	1.08M	0.32M

2.2 Prior Art

One of the major problems in spectrum sensing is how to sense primary system for wireless communication systems without too much change on its frame structure or too much signaling complexity. The invention proposed here gives a spectrum sensing method with no frame structure changed and little signaling complexity increased for OFDM systems.

There are some contributions about how to sense the spectrum resource of primary system. The most of these contributions are based on time domain QP (Quiet Period), such as setting some subframes as QP [10, 11] for sensing. Contribution [10] gives a general in-band sensing method based on QP, and contribution [2] gives a QP based in-band sensing method for TD-LTE system. However, the QP based sensing methods introduce too many changes on system structure, which is illustrated as follows by taking contribution [11] for example.

The method in [11] includes a functional unit which takes in charge of the selection of Up/Down link configuration as in table 2. It is assumed here that spectrum sensing just occupies the uplink subframes, because the uplink throughput in TD-LTE networks is relatively small in contrast to the downlink communication throughput. The implement steps are as following:

Firstly, select several possible configuration modes according to the required spectrum sensing time to reach enough sensing performance. For example, if it needs 2ms to sensing spectrum to reach detection probability of 99%, mode 0, 1, 3, 6 are optional.

Then, choose certain mode from the four modes mentioned above depending on up and downlink throughput of network.

And finally, base station broadcasts a sensing instruction in the downlink subframe (0 or 5), UEs (User Equipment) sense the unlicensed bands, and report sensing results in the following uplink subframes.

Table 2. Uplink / Downlink subframe configurations for LTE TDD

Up/Down configure	Sub-frame number									
	0	1	2	3	4	5	6	7	8	9
0	D	S	U	U	U	D	S	U	U	U
1	D	S	U	U	D	D	S	U	U	D
2	D	S	U	D	D	D	S	U	D	D
3	D	S	U	U	U	D	D	D	D	D
4	D	S	U	U	D	D	D	D	D	D
5	D	S	U	D	D	D	D	D	D	D
6	D	S	U	U	U	D	S	U	U	D

As explained above, due to a whole subframe or several consecutive subframes being used as quiet periods, these contributions introduce too many changes on current LTE/LTE-A frame structure such as HARQ (Hybrid Automatic Repeat Request) timing and control signaling transmission, which bring great challenges for current frame structure and might not be attractive in standardization as well as practical implementation.

As explained above, due to a whole subframe or several consecutive subframes being used as quiet periods, these contributions introduce too many changes on current LTE/LTE-A frame structure such as HARQ (Hybrid Automatic Repeat Request) timing and control signaling transmission, which bring great challenges for current frame structure and might not be attractive in standardization as well as practical implementation.

3 Novel In-band Sensing Method

3.1 Overall Introduction

The prior inventions are based on QP, which affect the frame structure severely. Here we propose a guard BW based spectrum sensing method, by which the impact to OFDM system frame structure could be minimized.

The proposed spectrum sensing procedure is based on leakage-negligible-BW, on which power detection or other detection methods can be implemented. For the method of power detection, the leakage-negligible-BW power is very low when no primary signals exist, else could be relative high. For the other sensing method, a somewhat sensing result can also be derived. The sensing result will be reported to BS (Base Station) via uplink transmission to derive a relative high sensing performance by cooperative sensing.

UE category, which symbolizes the UE capability, is reported to BS by uplink transmission. As the power leakage will affect the sensing result on leakage-negligible-BW, those UEs whose filter is good enough will be chosen to sense. So a parameter which is able to indicate the filter performance should be added in UE category.

On one hand, the sensing result on leakage-negligible-BW has to be reported by uplink transmission; on another hand, the filter parameter added in UE category introduces some changes on system standardization. These two points are our novel in this paper.

3.2 Implementation

The procedure of proposed spectrum sensing method is:

Step 1: Adding a parameter in UE category to indicate the UE filter capability and reporting to BS;

Step 2: BS selects UEs whose filters are good enough for sensing, and notices these UEs to feed back the sensing results periodically based on leakage-negligible-BW;

Step 3: The selected UEs calculate and report the sensing results to BS periodically;

Step 4: A decision will be made by BS whether to continue reusing the spectrum of primary system or trigger a finer sensing procedure based on these reports.

3.3 Advantages and Disadvantages

For our proposed scheme, the spectrum sensing performance will be affected by the UE filter property. In order to avoid sensing performance degradation, we choose those UEs with good filter property by including filter capabilities in UE category to sense. Although the filter property has impact to spectrum sensing, it's not a critical issue. In addition, to avoid sensing performance degradation, a frequency margin can be reserved by the edge of leakage-negligible-BW.

As the guard band always exists for both downlink and uplink, the samples for sensing are time domain continuous, which guarantee the sensing performance. In addition, the proposed method has little impact to system frame structure and little signaling increased, so it's very convenient in practical implementation.

4 Conclusion

A novel in-band spectrum sensing method based on guard frequency bandwidth for cognitive OFDM systems is proposed in this paper. The proposed spectrum sensing procedure is based on leakage-negligible-BW, on which power detection or other detection methods can be implemented. The advantage of this in-band sensing method is the impact to OFDM system frame structure could be minimized. So the proposed method is a practical implementation which can be used in cognitive OFDM system.

Acknowledgment

This work was supported by the International Science & Technology Cooperation Project of Qinghai (2014-HZ-821) and Application Basic Research Project of Qinghai(2015-ZJ-721).

References

1. H. Simon. Cognitive radio: brain-empowered wireless communications. IEEE Journal on Selected Areas in Communications, **23**(2005)
2. J. Mitola. Cognitive radio: making software radios more personal. IEEE Personal Communications, **6**(1999)
3. T. Yucek, H. Arslan. A survey of spectrum sensing algorithms for cognitive radio applications. IEEE Communications Surveys & Tutorials, **11**(2009)
4. J. Mitola. Cognitive radio for flexible mobile multimedia communications. Proc. IEEE Int. Workshop MoMuC, (1999)
5. H. Urkowitz. Energy detection of unknown deterministic signals. Proc. IEEE, **55**(1967)
6. F. Digham, M. Alouini and M. Simon. On the energy detection of unknown signals over fading channels. IEEE Trans. Commun, **55** (2007)

7. T. Yucek and H. Arslan. A survey of spectrum sensing algorithms for cognitive radio applications. IEEE Commun. Surveys Tuts, **11**(2009)
8. B. Wang and K. Liu. Advances in cognitive radio networks: A survey. IEEE J. Sel. Topics Signal Process, **5**(2011)
9. E. Axell, G. Leus, E. G. Larsson and H. V. Poor. Spectrum sensing for cognitive radio: State-of-the-art and recent advances. IEEE Signal Process. Mag, **29**(2012)
10. IEEE P802.22™/D0.1, Draft Standard for Wireless Regional Area Networks Part 22: Cognitive Wireless RAN Medium Access Control (MAC) and Physical Layer (PHY) specifications: Policies and procedures for operation in the TV Bands.
11. Z. Feng, P. Zhang, et al. Cooperative cognitive technique based on LTE system. Chinese patent number:201010248004.

Performance of FSO-OFDM based on BCH code

Jiao xiao-lu,Zhao Li and Peng Kai, She Bin-hua

School of Electronic Information Engineering, Xi'an Technological University, Xi'an 710032, China

Abstract. As contrasted with the traditional OOK (on-off key) system, FSO-OFDM system can resist the atmospheric scattering and improve the spectrum utilization rate effectively. Due to the instability of the atmospheric channel, the system will be affected by various factors, and resulting in a high BER. BCH code has a good error correcting ability, particularly in the short-length and medium-length code, and its performance is close to the theoretical value. It not only can check the burst errors but also can correct the random errors. Therefore, the BCH code is applied to the system to reduce the system BER. At last, the semi-physical simulation has been conducted with MATLAB. The simulation results show that when the BER is 10-2, the performance of OFDM is superior 4dB compared with OOK. In different weather conditions (extension rain, advection fog, dust days), when the BER is 10-5, the performance of BCH (255,191) channel coding is superior 4~5dB compared with uncoded system. All in all, OFDM technology and BCH code can reduce the system BER

Key words: Free Space Orthogonal (FSO); Orthogonal Frequency Division Multiplexing (OFDM); BCH code

1 Introduction

Free space optical communication can transfer information by optical signals in the space directly. Due to the Free Space Optical communication has the advantages of low cost, flexible networking, convenient installation and dispense with frequency permission, it has become a major technology in the present information technologies[1-2]. As the channel is atmosphere channel the signals have to experience atmospheric scattering effects causing by dust, raindrop and fog drop, which result in severe fading of the power of the optical signal[3]. In references , it pointed out that the FSO-OFDM technology not only could resist the effects of light scattering, but also improve the velocity without increasing the bandwidth. BCH code has a good error correcting ability, particularly in the short-length and medium-length code, and its performance is close to the theoretical value. It not only can check the burst errors but also can correct the random errors[4-6]. Therefore, the BCH code is applied to the system to reduce the system BER.

2 The Model of the FSO-OFDM

The basic principle of FSO-OFDM is divided the channel into many sub channels through the use of multiple sub carriers. The signal through serial parallel conversion to modulate the subcarrier respectively.

2.1 The Model of OFDM

The FSO-OFDM system was formed by transmitting part and receiving part.

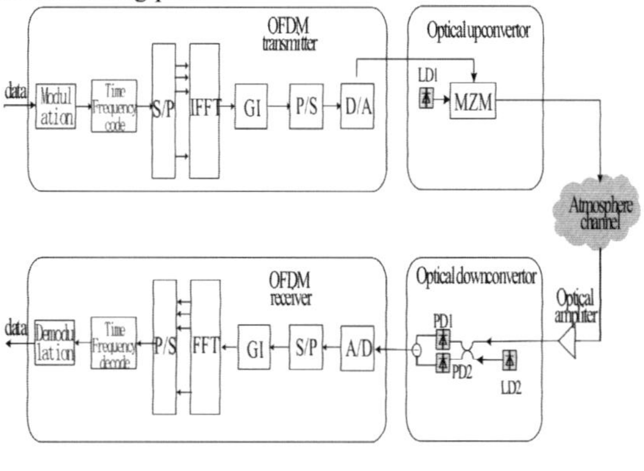

Figure 1. signal model

The baseband OFDM signal format is:

$$S(t)=\frac{1}{N}\sum_{i=1}^{\infty}\sum_{k=1}^{N}X_{i,k}e^{j2\pi f_k t} \tag{1}$$

$$f_k=(k-1)/T_s$$

in the formula : f_k represents carrier frequency. $X_{i,k}$ represents sub-carrier signal.

$$f_k=f_0+n\Delta f \tag{2}$$

in the formula : Δf represents the sub-carrier interval; f_0 represents practical emission carrier frequency. T is the symbol period. $\Delta f=\frac{1}{T}$ .Generally every N sub-carriers make up an OFDM signal.

We need spectrum shift the base-band OFDM signals to the intermediate frequency f_{LO1}, and the intermediate frequency signal can be expressed as formula (3).

$$S_{IF}(t)=\frac{1}{N}\exp(j2\pi f_{LO1}t)\sum_{i=1}^{\infty}\sum_{k=1}^{N}X_{i,k}\exp(j2\pi f_k t) \tag{3}$$

Take one OFDM symbol as an example, the real part of the signal is shown as:

$$S_{OFDM}(t)=\mathrm{Re}\left[S_{IF}(t)\right]$$
$$=\frac{1}{N}\left[\sum_{K=1}^{N}\mathrm{Re}(X_{i,k})\cos(2\pi(f_k+f_{LO1})t)-\sum_{K=1}^{N}\mathrm{Im}(X_{i,k})\sin(2\pi(f_k+f_{LO1})t)\right] \tag{4}$$

Euler's formula shows that the real part of the intermediate frequency signal has contained the phase components and the orthogonal components of the signal X_k. Therefore, the format of the ith OFDM symbol's kth sub-carrier is $S_{OFDM}(i,k)\big|_{i=1:\infty,\,k=1:N}$ 。

2.2 The Model of Atmospheric Scattering Channel

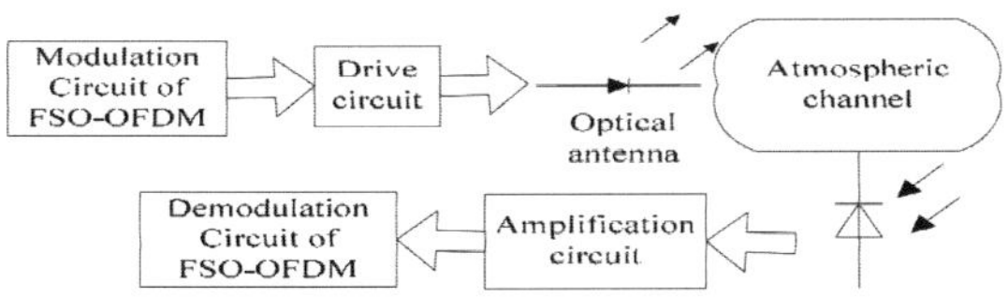

Figure 2. The channel model

FSO-OFDM communication system in this paper is a point to point communication system. the signals will be effected by atmospheric scattering caused by dust, frog, rain, etc, which will make the optical signal intensity was significantly attenuated. According to the Bougure rule, the power of the light with optical power I_0 after through the transmitter substance with h distance can be written as[1-3]:

$$I(\varepsilon)=I_0(\varepsilon)\exp\left[-\int_0^h\sigma(\varepsilon)dh\right] \tag{5}$$
$$=I_0(\varepsilon)\exp[-\sigma(\varepsilon)h]$$

in the formula : $\sigma(\varepsilon)$ represents attenuation coefficient, ε is frequency parameter.

3 The Principle of BCH Coding and Decoding

3.1 BCH coding

Let the input signal polynomial is

$$I(x)=I_0+I_1x+\Lambda+I_{k-1}x^{k-1} \tag{6}$$

The check polynomial is

$$P(x)=P_0+P_1x+\Lambda+P_{n-k-1}x^{n-k-1} \tag{7}$$

From the generator polynomial $g(x)$ we can get $P(x)$: $P(x)=X^{2t}I(x)(\mathrm{mod}(x))$

the polynomial of generational chip can be written as :

$C(x)=X^{2t}I(x)+P(x)$

3.2 BCH Decoding

Let $W(x)=r_0+r_1x+\cdots+r_{n-1}x^{n-1}$ is the receiving code, we can get the adjoin polynomial from the receiving code :

$$S(x)=s_0+s_1+\cdots+s_{2t-1}x^{2t-1} \tag{8}$$

From formula (8) we can calculate the error locator polynomial $\sigma(x)$ and error polynomial $R(x)$. We can get error pattern from the second step. Finally, we will recover the correct code $C(x)$ by subtract error pattern from receiving group.

4 Simulation Experiment

In this paper, we conduct a simulation on the system's BER on different weather condition by Monte Carlo method, and set the simulation parameters as follows: 10^6 OFDM symbols, the laser wavelength $\lambda=1.064\mu m$, the number of subcarriers $N=64$, the laser-band corresponds to the refractive index of the rain medium $m=1.178-i\times0.071$, transmission distance is 1km, adopting MPSK.

The system performance in the extend rain days which average rainfall is 25mm is shown in figure 3. After BCH coding the system's performance improved about 2.5dB than the original performance when the BER is 10^{-3} and improved about 5dB than the original performance when the BER is 10^{-5}.

The system performance under advection fog which average visibility is 600m is shown in figure 4. After BCH coding the system's performance improved about 2dB than the original performance when the BER is 10^{-3} and improved about 4.5dB than the original performance when the BER is 10^{-5}.

The system performance under dust condition which average visibility is 2500-3000m is shown in figure 5. After BCH coding the system's performance improved about 2dB than the original performance when the BER is 10^{-3} and improved about 5dB than the original performance when the BER is 10^{-5}.

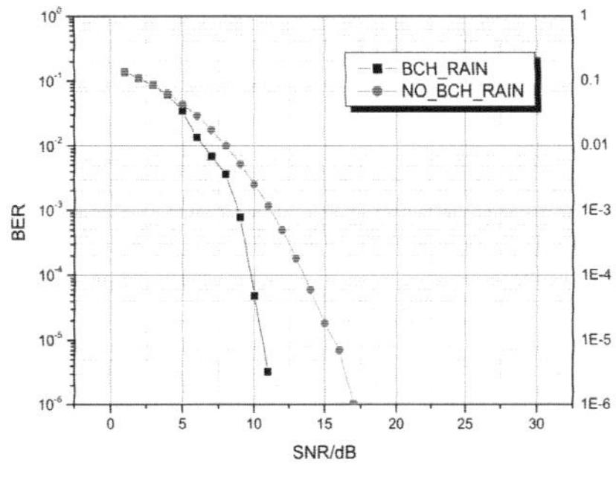

Figure 3. the BER under extend rain conditions

 DOI: 10.1051/matecconf/20166107008

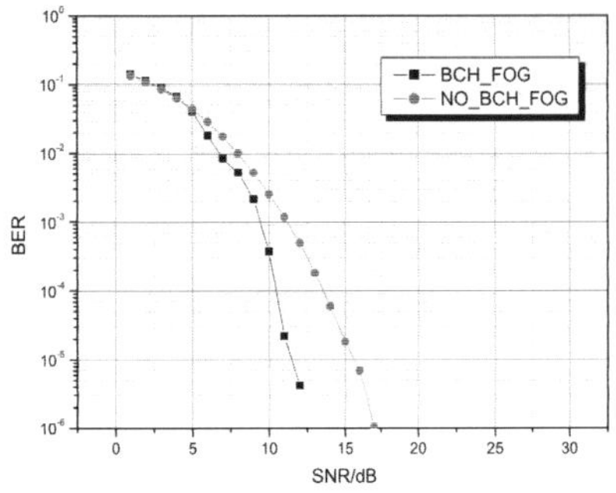

Figure 4. the BER under advection fog conditions

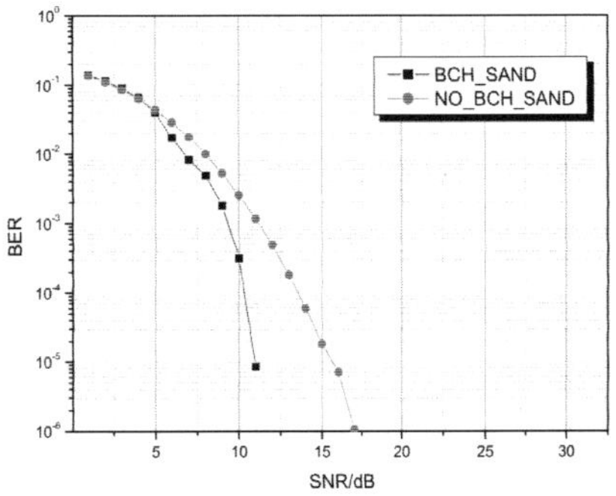

Figure 5. the BER under dust conditions

5 Conclusion

In this paper, the OFDM technique was introduced into FSO system so as to establish the FSO-OFDM system model, and the system error performance of FSO-OFDM was analyzed under the scattering channel condition. Meanwhile, according to the simulation results the performance of the FSO-OFDM system is better than the OOK system under the lower SNR condition, and the system with BCH coding technique is better than the system without channel coding technique.

References

1. Zhao Li, Ke Xi-zheng. Research and simulation of the influence of the scattering effect on FSO-OFDM system [J]. Laser Technology, 185~188, **4**,2(2010).
2. Zhao Li, Ke Xi-zheng. A FSO-OFDM Model based on Time Frequency Code [J]. Chinese Journal of Lasers, 2757~2762, **3**,10(2009).
3. Wen T, Wei J b, MA D T. Analysis of effect of multiple scattering on laser communication in light haze weather [J]. Laser Technology, 500~502, **31**,5(2007).
4. Zhang X Y, Li W, Wang H S, et al. Research on Improving BER Performance of OFDM System[J]. Computer Technology and Development, 233~237,**23**,4(2013).
5. CHOI H.LIU W.SUNG W.VLSI implement of BCH error correction for multilevel cell NAND flash memory [J].IEEE Transaction on Very large Scale Integration(VLSI)Systems,32-35,**6**,8(2008).
6. LIU X,GUO J,YAN F.Group counter-change of BCH code [J].Information Engineering and Electronic Commerce(IEEC),58-61,**1**,4(2010).

MATEC Web of Conferences **61**, 07009 (2016) DOI: 10.1051/matecconf/20166107009
APOP2016

A Joint Frequency-Domain Equalization (Fde) and Antenna Diversity Combining Method for the Multipath Fading in the Frequency-Selective Channel

Jinpeng Wang [a], Yongtao Li, Youran Li,Jierong Zhao, Nianyu Zou, Xiaoyang He

Research Institute of Photonics, Dalian Polytechnic University, 116034 Dalian, China

Abstract. In this study, a theoretical foundation is built to evaluate the downlink performance of an MC-CDMA cellular system with site diversity operation with and receive antenna diversity combining. An expression for the theoretical conditional bit error rate (BER) for the given set of channel gains is derived based on Gaussian approximation of the interference components in the composite receive signal. The local average BER is then obtained by averaging the conditional BER over the given set of channel gains using Monte-Carlo numerical method. The outage probability is measured from the numerically obtained cumulative distribution of the local average BER to determine the downlink capacity. Results from theoretical computation are compared to the results from computer simulation and discussed.

1 Introduction

The multicarrier-CDMA (MC-CDMA)[1,2] has been considered as a wireless access candidate for a wideband downlink transmission due to its robustness against the frequency-selectivity of the multipath channel and high frequency efficiency [3-5]. It is one of the new multiple access techniques proposed which based on the combination of the OFDM and CDMA. A basic MC-CDMA transmitter spreads the user's data symbol using an orthogonal spreading code in the frequency domain. Figure 1 shows the comparison between the three schemes: DS-CDMA, OFDM and MC-CDMA. The DS-CDMA spreads the user's data symbol using the spreading code in the domain whereas the OFDM transmit different user's data symbol on each subcarrier.

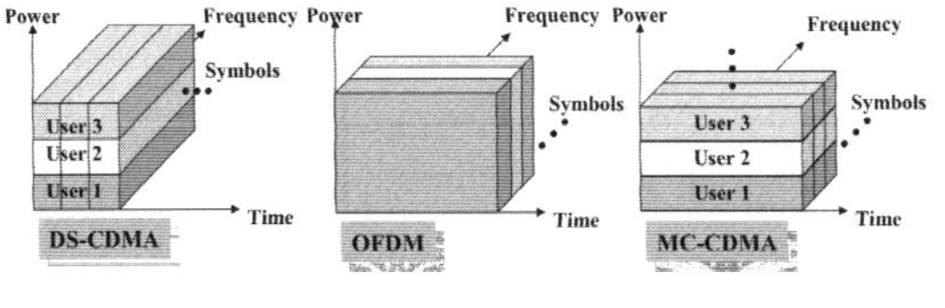

Figure 1. Comparison between DS-CDMA, OFDM and MC-CDMA.

The characteristics of MC-CDMA can be summarized as follows:

a) Transmits the same symbol in parallel through numbers of subcarriers in frequency domain.

b) Each data stream is serial-parallel converted before spreading to obtain almost frequency non-selective fading for each subcarrier (see Figure 2).

c) MC-CDMA lower the symbol rate in each subcarrier and the longer symbol duration makes it easier to quasi-synchronize the transmission.

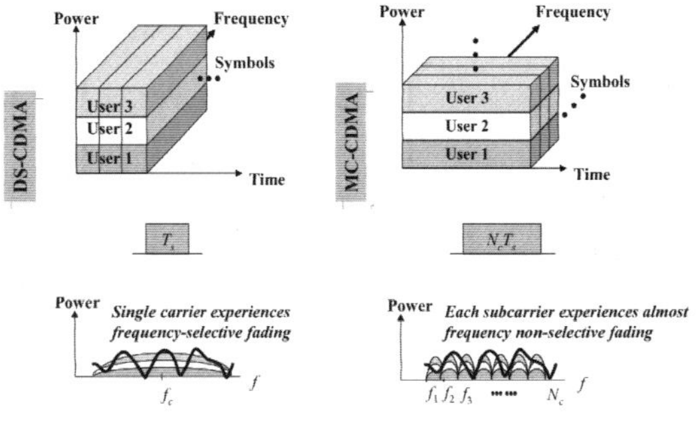

Figure2. Single-carrier CDMA (DS-CDMA) versus MC-CDMA.

OFDM should be used to mitigate the performance degradation due to frequency-selective fading while DS-CDMA is great to mitigate the interference in cellular deployment thought the use of scrambling codes [6,7]. Therefore, MC-CDMA inherits the benefits from the combination effort of CDMA and OFDM and makes it a great wireless access candidate for future cellular communications. A more detailed transmission system of a MC-CDMA cellular system is discussed in Chapter 4. In this study, joint frequency-domain equalization (FDE) and antenna diversity combining is applied to the reception of MC-CDMA cellular system, for combating ISI caused by the multipath fading in the frequency-

[a] Corresponding author: wangjp@dlpu.edu.cn

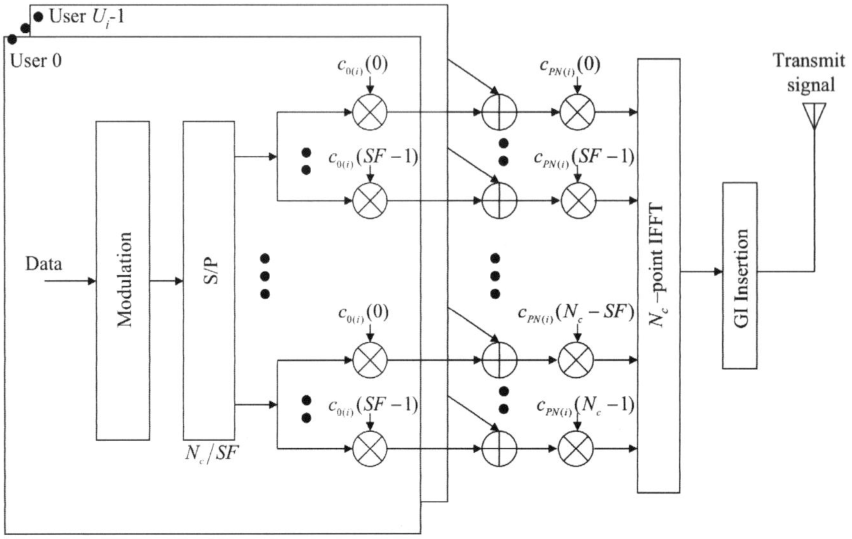

(a) Base station transmitter.

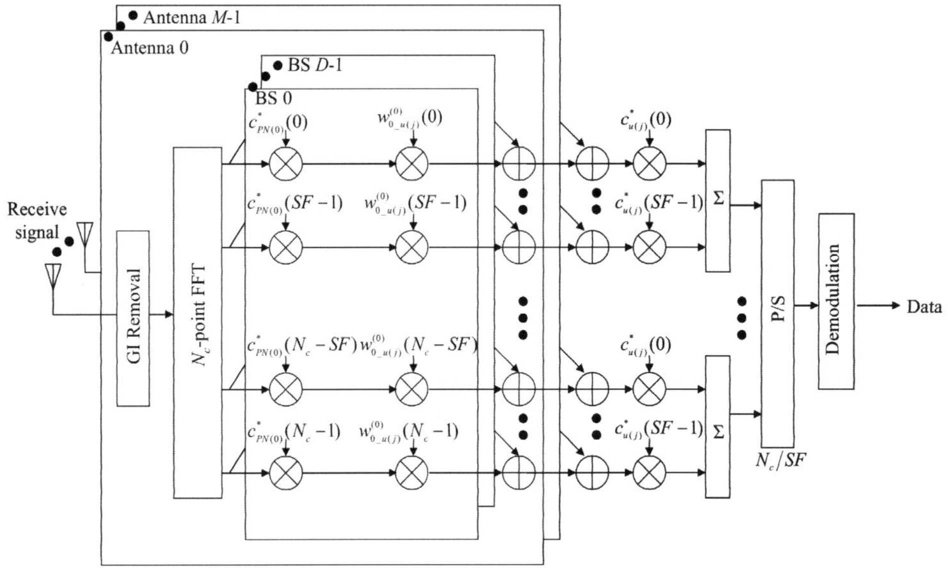

(b) Mobile station Receiver.

Figure 3. Downlink MC-CDMA transmitter/receiver with site diversity operation.

selective channel [8-11]. The FDE[12,13] and antenna diversity[14,15] are discussed in the next sections.

2 Downlink transmission model

The downlink MC-CDMA transmitter/receiver systems for site diversity operation with joint use of MMSE-FDE and receive antenna diversity is illustrated in Figure 3, where a system with Nc subcarriers and spreading factor SF is assumed.

3 Transmit signal representation

At BS i, the data modulated symbol sequence $\{d_{u(j)}(n); n=0\sim N_c/SF-1\}$ for user u of cell i is first serial-to-parallel (S/P) converted into Nc/SF parallel data sequences. Each of the S/P converter output is copied SF times and multiplied with the orthogonal spreading code $\{c_{u(j)}(k); k=0\sim SF-1\}$. All users' spread signal components at each subcarrier are combined and multiplied with the common scrambling code $\{c_{PN(i)}(k); k=0\sim N_c-1\}$ of BS i. Different scrambling codes are used in different cell sites for separating cell sites [1], besides making the resultant signal to be white-noise like. The composite transmit signal at subcarrier k is given as:

$$s_i(k)=\sqrt{\frac{2P_i}{SF}}\sum_{u(i)=0}^{U+\delta u_i-1} d_{u(i)}\left(\left\lfloor\frac{k}{SF}\right\rfloor\right)c_{u(i)}(k \bmod SF)c_{PN(i)}(k) \quad (1)$$

with Pi and U+δui defined as the transmit power and the number of active channels (users) per BS i, respectively; where δu_i denotes the number of additional channels necessary for BS i in the site diversity operation; and $|d_{u(i)}(n)|=|c_{u(i)}(k)|=|c_{PN(i)}(k)|=1$. The orthogonal spreading and the scrambling codes have the following characteristics

$$\begin{cases} \frac{1}{SF}\sum_{k=0}^{SF-1} c_{u(i)}(k)c^{*}_{u'(i)}(k)=\delta(u-u') \\ E\left[c_{PN(i)}(k)c_{PN(i)}(k')\right]=\delta(k-k') \end{cases} \quad (2)$$

However, perfect orthogonality between users cannot be achieved in a multipath fading channel, and this produces the multiple access interference (MAI). Furthermore, the inter-cellular interference (ICI) from other neighboring cells has to be considered since the scrambling codes between cells are not orthogonal.

After Nc-point IFFT, cyclic Ng-sample guard interval (GI) is inserted at the beginning of each symbol frame to mitigate the inter-symbol interference (ISI). The MC-CDMA transmit signal is then expressed as

$$s(t)=\sum_{k=0}^{N_c-1} s_i(k)\exp\left(j2\pi k\frac{t}{N_c}\right) \quad (3)$$

for $t=-N_g \sim N_c-1$.

4 Numerical results and discussion

4.1 Monte-Carlo process

The evaluation of the downlink capacity is performed by the Monte-Carlo numerical method. An MC-CDMA system with Nc=256 subcarriers is assumed. The signal is assumed to propagate through a block Rayleigh fading channel with propagation path L=16 having the exponential power delay profile with decay factor γ dB. In an interference-limited environment with negligible AWGN effect, the interference coming from the second-tier cells is relatively weak and can be neglected. Therefore each user will receive the dominant interference from the six adjacent cells' BS's. The user of interest is located in cell 0 and there are 18 surrounding cells considered as co-channel cells. The maximum number of active BS, D varies from 1 to 7 where D=1 corresponds to the no site diversity operation and D=7 corresponds to the site diversity operation with six adjacent co-channel cells.

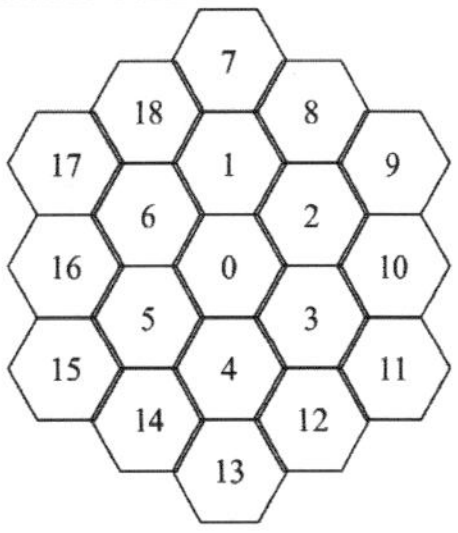

Figure 4. Cellular Structure.

4.2 Approximate analysis

The interference components have been approximated as the complex-valued Gaussian random variables for computing the conditional SINR and BER. To verify this approximation, the probability density functions (pdf's) of the MAI and ICI components obtained from the computer simulation are examined and plotted in Figure 5. A uniform power delay profile for the channel (γ =0 dB) and the number of users per cells, U=27 are assumed. Pth=4.5 dB is used for the case with site diversity operation. Only the real parts of μ MAI and μ ICI are plotted in Figure 5. The computer simulation plots show that the pdf's of the real part of μ MAI and μ ICI follow the Gaussian distribution with zero-mean. This therefore justifies the approximation used in the theoretical analysis.

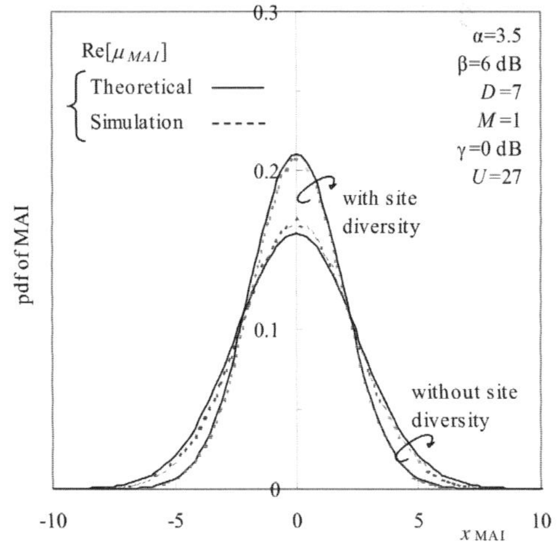

(a) MAI component.

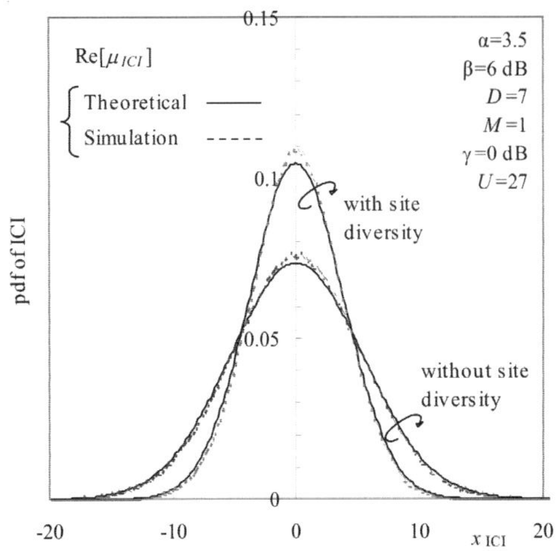

(b) ICI component.

Figure 5. Distributions of MAI and ICI components.

5 Conclusions

In this study, the theoretical treatment was developed for the downlink site diversity reception with joint use of MMSE-FDE and antenna diversity in a cellular MC-CDMA system. The expressions for the conditional SINR and BER were derived based on Gaussian approximation of the interference components; and the local average BER was numerically computed using Monte-Carlo computation method. The theoretical performance results were compared with computer simulation results to show a high degree of agreement. The pdf plots showed the real parts of MAI and ICI follow the Gaussian distribution and therefore, justifying the approximation used in the theoretical treatment.

Evidently, from both theoretical and simulation evaluations, the site diversity operation improves the MC-CDMA downlink performance. Both results showed that three (D=3) active BS's participating in site diversity operation are sufficient for achieving the maximum downlink capacity. Significant effect on the downlink capacity performance can be seen from different values of α, whereas the capacity performance is almost unaffected by different values of β. Increasing the number of receiving antenna increases the downlink capacity performance almost linearly. The larger the decay factor in the exponential power delay profile leads to a weaker frequency-selectivity of the fading channel, thus reducing the frequency-diversity effect exploited by MC-CDMA system and degrading the system overall performance.

Acknowledgements

This research was supported by Project of the National Natural Science Foundation of China (61402069), 2012 key supported project for undergraduate education by Education Dept. of Liaoning province([2012]108), 2016 Project of the Natural Science Foundation of Liaoning province(2016020027), 2012 Education reform project for undergraduate by Education Dept. of Liaoning province ([2012]130), 2014 Dalian municipal science and technology plan project（2014A11GX050）, 2014 Dalian municipal science and technology plan project (2014A11GX052).

References

1. R. Jaume, F. Josep, Sig. Pro.IEEE Trans,**62**, 1899(2014)
2. H. Juinn, T. Tsui, Wirel. Pers.Com, **65**,537 (2012)
3. Y. T. Wu, Y. S. Zhu, Ac. Astro,**115**,367(2015)
4. H. Yang, J. Ye and Y. H. Liu, Opt. Fib. Technol,**24**,127(2015)
5. R. Q. Shad, A. B. Mohammada， J. Net .Comp. Appl， **41**,459(2014)
6. Y. Bithas, S. Petros, IEEE. Tra. Com,**63**,724(2015)
7. C. Li, H. Li and X. Q. Zhang, Mic. W. Mob. Com, **26**,1843(2013)
8. S. P. Tseng, Opt. Fib. Technol, **23** ,61(2015)
9. I. H. Lee, H.knu, S. J. Lee, Com. IET, **8**,2624(2014)
10. A. L. Hai , M .Vu, IEEE Trans Com, **63**,1162(2015)
11. S. Bas, C. U. Ergen, Veh. Technol, IEEE. Trans, **62**,14(2013)
12. N. Maeda, H. Atarashi and S. Abeta ,IEICE. Trans. Com. **85**,1966(2002)
13. F. Adachi , T. Sao, IEICE Trans. Commun, **86**, 352 (2003)
14. T. Inoue, S. Takaoka,IEICETrans.Com,**88**, 298(2005)
15. U.Rehman, M.Abbasi, IET.Mic.Ant.Pro,**9**,243(2015)

MATEC Web of Conferences **61**, 07010 (2016) DOI: 10.1051/matecconf/20166107010
APOP2016

Study of the Interference Rejection Based on ATF in DSSS

Pinghua Zhang[1,2,a], Hongli Xu[1] ,Manman Peng[2] and Luyun cao[1]

[1]Hunan Information College, School of Electronic Engineering, 410200 Changsha, China
[2]Hunan University, College of Information Science and Engineering, 410082 Changsha, China

Abstract. A novel, adaptive time-frequency interference rejection for suppressing the interference in Direct Sequence Spread Spectrum communication systems is studied in this paper. For a time-localized interference, excision is performed in the time domain. When a frequency domain excision is decided, an adaptive sub-band transform is employed. The most proper sub-spectral split is defined according to the spectrum of the received signal. This joint treatment of time and frequency domains overcomes the performance limitations of the fixed transform-based excision algorithms. The paper presents the performance evaluation of the rejection technique via the analytical and computer simulation tools. It is found that the analytical performance tools match up with the computer simulation results. It is also shown that the proposed adaptive time-frequency interference rejection consistently outperforms the existing interference excision techniques in the literature.

1 Introduction

Spread spectrum technology has strong anti-interference ability and multiple access capability, and can timing and ranging precisely. So it is widely used in military and civil communication systems. At the same time, the research on the communication system of the spread spectrum communication system has become a hot spot in the research of reconnaissance, jamming and anti-jamming. Because the mechanism of spread spectrum technology is to use a relatively wide frequency band, so it is more easily to be subject to the attacks of single narrowband interference, multiple narrowband interference, broadband interference and other interferences. In many cases, the cost is too high to enhance the ability of resisting interference by improving the system's processing gain. Sometimes it is even impossible to achieve. So we need to use effective signal processing technology to suppress the interference, to ensure the accurate transmission and reception of information [1].

One of the most useful techniques in the anti-jamming technology is the transform domain filtering processing technique. Existing transform domain filters adopt the frequency domain processing technology using fixed transformation base, such as fast Fourier transform (FFT), the signal is mapped into the frequency domain, but the interference suppression performance is limited by the size and accuracy of FFT, especially when the interference is leaked into the adjacent bands due to the sub-band spectrum which like *sin(x)/x*, and also when the narrowband noise threshold is difficult to determine as the background additive Gaussian noise is too strong. There is also no need to decompose the disturbed frequency band and the free frequency band to the same resolution [2]. Later, sub-band transform interference suppression is put forward in spread spectrum communication systems, although it has a better frequency domain energy aggregation, but has the same limitations with the fixed transform base frequency domain interference processing technology, such as spectrum leakage between adjacent sub bands and fixed time-frequency resolution. Especially for time varying and non-stationary interferences, such as pulse jamming, the transform domain processing technique is invalid [3].

In this paper, a new method of anti-jamming in the transform domain is studied--adaptive time-frequency interference rejection (ATF). The proposed algorithm can generate a sub-band decomposition tree with different levels due to different input signal. In the process of the sub-band decomposition, when the intensity of the transform domain energy exceeds the time domain or a given threshold, the node will continue to decompose. Therefore, the algorithm avoids unnecessary decomposition, and can greatly reduce the calculation of addition and multiplication. Compared with a fixed structure filter, one of the biggest advantages of ATF is that it can adaptively change the hierarchical structure of the sub band filter banks, reduce the segmentation of the transform domain, and can locate the interference signal of the frequency domain more accurately. Thus it can reduce the sensitivity of interference, and it is a robust technique for suppressing interference. At the same time, the adopted adaptive sub tree structure algorithm can overcome the shortcomings of fixed spectral resolution in the existing transform domain processing techniques [4].

[a] Corresponding author: Pinghua Zhang, zhangpinghua@mail.hniu.cn

2 Direct Spread Spectrum System

Take the most common way – direct spread spectrum system (DSSS) for example, the input signal of the adaptive time-frequency (ATF) interference suppressor is [5]

$$r(k)=d_b c(k)+n(k)+j(k) \tag{1}$$

In the formula, the value of d_b is ± 1 equiprobability distributed; $c(k)$ is the spreading *sequences*, the system processing gain is L; $n(k)$ is the additive white Gaussian noise of zero mean and variance D, $j(k)$ is the interference signal of mean zero and covariance $R_j(k)$.

3 Adaptive Time-frequency interference Rejection ATF

The main idea of the interference suppression technology studied in the paper is to check the received signal $r(k)$ is suitable or not for interference suppression in the time domain. If yes, there is no need for frequency domain processing. When it is determined to eliminate the interference in frequency domain, the adaptive sub band transformation technology (also known as adaptive wavelet packet Technology), the signal is decomposed according to the frequency domain characteristics of the received signal. In order to find the optimal wavelet packet tree structure to reduce the frequency of interference, we use the energy aggregation degree to reflect the distribution of the signal power spectrum density. The energy aggregation degree of M-band is defined as

$$G=\frac{\delta_x^2}{\left[\prod_{i=1}^{M}\delta_i^2\right]^{1/M}} \tag{2}$$

Where δ_x^2 is the variance of input signal, δ_i^2 is the output variance of the *i-th* sub-band. The bigger G is, the less flat the frequency spectrum of the output signal is, and the signal energy is more focused on some sub bands [6].

The concrete steps of the adaptive tree structure algorithm are described as follows:

1) Calculate the power spectrum density of the received signal $P_{xx}(\omega)$.

2) a. Calculate the sub-band variances δ_{2l}^2 and δ_{2h}^2 of the two sub-band decomposition and the corresponding energy accumulation degree $G(2)$.

b. Calculate the sub-band variances δ_{3l}^2 , δ_{3b}^2 and δ_{3h}^2 of the three sub-band decomposition and the corresponding energy accumulation degree $G(3)$.

3) If $G(2)\leqslant T$ and $G(3)\leqslant T$, means that the signal spectrum is flat, and the decomposition will stop (T is the pre-defined threshold).

Otherwise, if $G(2)\geqslant G(3)$, two band decomposition is used , and if $G(2)< G(3)$,then three band decomposition is used.

4) Repeat step 2) and step 3) until there is no node can be decomposed.

4 Performance Analysis

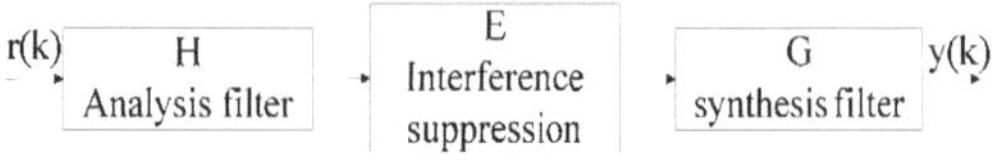

Figure 1. Diagram of interference suppression

Let $\{h_i(n)\}$ and $\{g_i(n)\}$ be the analysis and synthesis filter function, the coefficient $\{e_i\}$ is the interference weight, the sub-band with interference is multiplied by 0, the sub band without interference is multiplied by 1. Then the output of the sub-band can be written as

$$y_i(k)=[e_i(k)*h_i(k)*g_i(k)]*r(k) \tag{3}$$

The equivalent filter of the sub-band is recorded as

$$\begin{aligned} t_i(k) &\triangleq e_i(k)*h_i(k)*g_i(k) \\ &= e_i(k)*\sum_{m=0}^{N-1}h_i(m)g_i(k-m) \end{aligned} \tag{4}$$

Since $g_i(k)$ is the time reversal of $h_i(k)$, $t_i(k)$ is a linear phase FIR filter with a length $(2N-1)$. For the contaminated sub-band

$$e_i(k)=0\rightarrow t_i(k)=0 \tag{5}$$

And the output signal is

$$y(k)=\sum_{i=1}^{M}\sum_{l=0}^{2N-2}t_i(l)r(k-l) \tag{6}$$

The output variable ε can be written as

$$\begin{aligned} \varepsilon &= \sum_{k=N}^{N+L-1} y(k)c(k) \\ &= \sum_{k=N}^{N+L-1}\left[\sum_{i=1}^{M}\sum_{l=0}^{2N-2}t_i(l)r(k-l)\right]c(k) \end{aligned} \tag{7}$$

Then the mean of ε ($d_b=\pm 1$ hypothetically) is

$$E[\varepsilon]=L\sum_{i=1}^{M}t_i(0)-\frac{1}{L}\sum_{k=N}^{N+L-1}\sum_{i=1}^{M}\sum_{l=1}^{2N-2}t_i(l) \tag{8}$$

The variance of ε is

$$\begin{aligned}\operatorname{var}[\varepsilon] &= L^2\sum_{i=1}^{M}\sum_{r=1}^{M}\sum_{l=0}^{2N-2} t_i(l)t_r(l) \\ &-\left[L\sum_{i=1}^{M}t_i(0)-\frac{1}{L}\sum_{k=N}^{N+L-1}\sum_{i=1}^{M}\sum_{l=1}^{2N-2}t_i(l)\right]^2 \\ &+L\sum_{i=1}^{M}\sum_{r=1}^{M}\sum_{l=0}^{2N-2}\sum_{s=0}^{2N-2}t_i(l)t_r(s)R_j(s-l) \\ &-\frac{1}{L}\sum_{k=N}^{N+L-1}\sum_{\substack{p=N\\p\neq k}}^{N+L-1}\sum_{i=1}^{M}\sum_{r=1}^{M}\sum_{l=0}^{2N-2} \\ &\sum_{s=0}^{2N-2}t_i(l)t_r(s)R_j(k-l-p+s) \\ &+L\delta_n^2\sum_{i=1}^{M}\sum_{r=1}^{M}\sum_{l=0}^{2N-2}t_i(l)t_r(l) \\ &-\frac{1}{L}\sum_{k=N}^{N+L-1}\sum_{\substack{p=N\\p\neq k}}^{N+L-1}\sum_{i=1}^{M}\sum_{r=1}^{M}\sum_{l=0}^{2N-2} \\ &\sum_{s=0}^{2N-2}t_i(l)t_r(s)\delta_n^2\delta(k-l-p+s)\end{aligned} \tag{9}$$

While

$$E\left[\varepsilon^2\right]=E^2[\varepsilon]+\operatorname{var}[\varepsilon] \tag{10}$$

The mean value of the decision variable is the useful signal, and the variance of the decision variable is the sum of the thermal noise, the residual interference component and the noise generated by the PN sequence itself, so the output signal to noise ratio (SNR) of the system is

$$SIR_0 @ \frac{E^2[\varepsilon]}{\operatorname{var}[\varepsilon]} \tag{11}$$

Then the bit error rate (BER) of the system is

$$P_e = Q\left(\sqrt{SIR_0}\right)=Q\left(\sqrt{\frac{E^2[\varepsilon]}{\operatorname{var}[\varepsilon]}}\right) \tag{12}$$

In the equal

$$Q(x) @ \frac{1}{\sqrt{2\pi}}\int_x^{\infty} e^{-y^2/2}\,dy \tag{13}$$

5 Simulation and Performance Comparison

In computer simulation, the DSSS system using length 63 PN code to spread spectrum, BPSK modulation, first consider the single tone jamming situation, when *j(k)=Acos(* $\omega_s k+\theta$ *)* . As *SIR=-20dB*, $\omega_s=1.65$, the theoretical analysis and simulation comparison chart for bit error rate of ATF algorithm is shown in Figure 2. We can see that the theoretical analysis result is in agreement with the simulation result.

Figure 3 showing the bit error rate of the ideal case (there is no NBI), ATF algorithm and FFT algorithm respectively ($\omega_s=1.92$). Results show that, the performance of the system using the ATF algorithm has been greatly improved.

Figure 4 shows the amplitude-frequency diagram of the corresponding unequal bandwidth adaptive tree structure in direct sequence spread spectrum disturbed by single tone interference. It shows the superiority of ATF algorithm in finding the best basement, which due to the use of an adaptive tree structure algorithm can be more flexible to choose the basement which matched with the center frequency of interference.

Then consider the Gaussian interference whose time domain energy is concentrated. Figure 5 shows the BER curve when the receiver is subject to time-varying broadband Gauss interference. The bit error rate performance of the ATF algorithm and FFT algorithm is shown respectively. In this case when *SIR=-20dB*, the duty ratio of the interference is 10%, the FFT algorithm obviously lost the interference suppression ability, but ATF algorithm still has good performance in interference suppression.

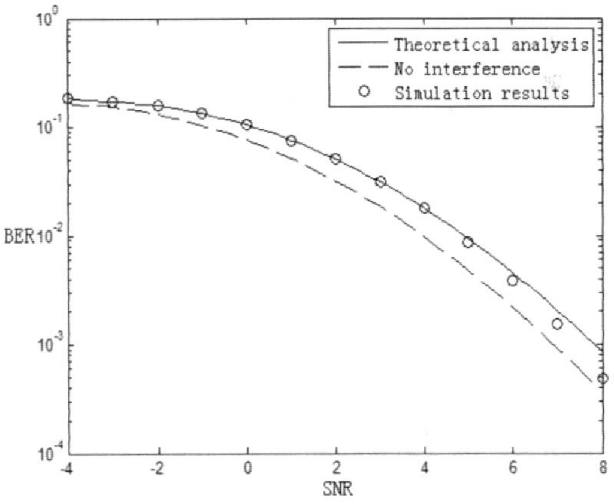

Figure 2. BER of theoretical analysis and simulation of ATF

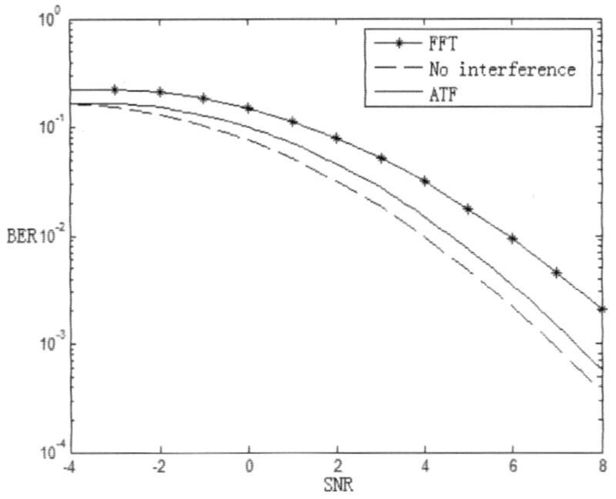

Figure 3. BER of single tone interference

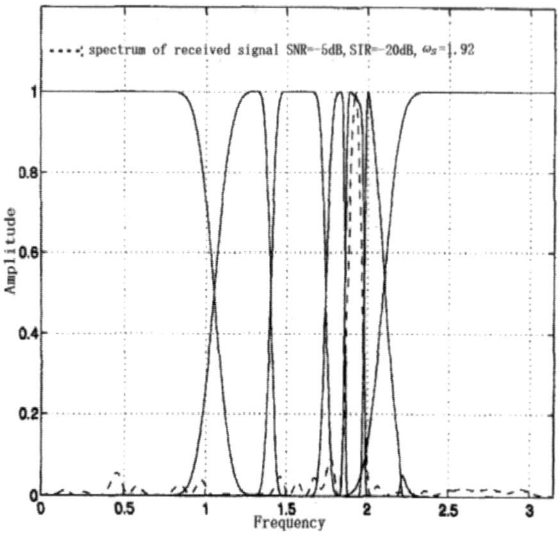

Figure 4. Structure of adaptive filter group disturbed by single tone interference

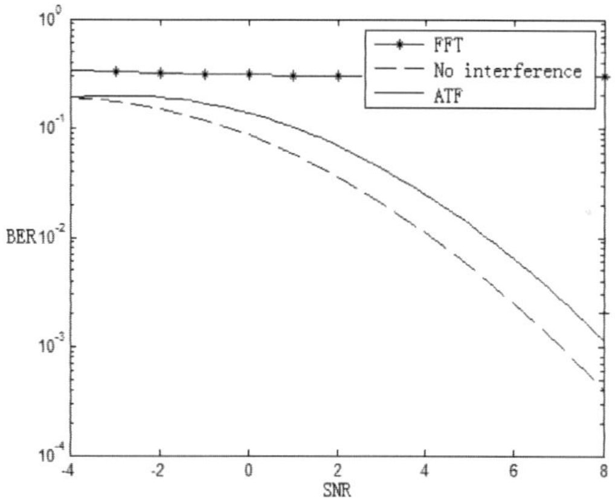

Figure 5. BER of time varying Gauss jamming

6 Conclusion

In this paper, a interference suppression technique based on adaptive time-frequency interference rejection is studied in the direct sequence spread spectrum communication system, the using of adaptive tree structure algorithm can be more flexible to select the basement matched with the center frequency of the interference. The closed form of bit error rate of the algorithm is derived. Theoretical analysis and computer simulation coincide. The results show that the algorithm is superior to the other frequency domain interference suppression techniques, and has a higher performance in bit error rate.

Acknowledgements

The scientific research project of Hunan Provincial Education Department (Number: 15C0982).

The Youth funded projects of Education Science "Twelfth Five Year Plan" in Hunan Province (Number: XJK015QZY007).

References

1. M.J.Medley,G.J.Saulnier,P.K.Da,IEEE Trans. on Commu. **45**,11 (1997)
2. S.A.Wang, J.P. An, Signal Proc. Let. IEEE, **17**,5 (2010)
3. C.Y. He, J. Guo, X.C Lu, X. Wang and F. Liu.CSNC2012, 147(2012)
4. M.V.Tazebay,A.N.Akansu,Speech and Signal Proc.Newark， **12**(1995)
5. R.L.Peterson,R.E.Aieme,D.E.Borth,*Introduction to Spread Spectrum Communication*(Publishing House of Electronic Industry,Beijing,2006)
6. L.Ping,R.Lu,2013 Inter. Conf. on QRRMS Eng. ,**2**(2013)

A Target Tracking Algorithm based on Fractional Ambiguity Function in Impulsive Noise Environment

Xiaofei Shi[1,a], Li Li[2]

[1] *Information Science and Technology College, Dalian Maritime University, Dalian, 116026, China*
[2] *Information Engineering College, Dalian University, Dalian, 116622, China*

Abstract. This paper proposes an airplane tracking algorithm based on study of the problem of interference localization. Firstly, a novel signal model to accurately estimate parameters of the airplane is proposed in impulsive noise environment. A method of instantaneous Doppler frequency estimation based on peak searching of the fractional lower-order ambiguity function based on the fractional Fourier transform (FLOS_FAF) is proposed, and a method of projection approximation subspace tracking using robust m-estimation method based on fractional lower-order ambiguity function in fractional Fourier transform domain (FF-RLM_PAST) is proposed to estimate the azimuth angle and elevation angle. As a result, the airplane tracking is achieved in bistatic radar, laying the foundations for interference localization. The correctness and effectiveness of the proposed method are verified with the computer simulation.

1 Introduction

With increase of radio stations quantity, the interfered probability of civil aviation radio frequency increases. It is key point how to make it safe operation of civil aviation radio frequency. Interference localization is realized by utilizing Doppler frequency shifts of the scattered signals from civil airplane in the receiver[1],[2]. If real-time state information of airplane is not obtained, interference localization is not realized. It is shown that accurate estimation of airplane state information is very important.

Studies and experimental measurements have shown that a class of α -stable distributions is more appropriate for modeling impulsive noise than Gaussian distribution in signal processing applications [3]. Since the stable distribution does not have finite second-order moments ($1 \le \alpha < 2$), or even first-order moment ($\alpha < 1$) due to the heavy tails, the performance of the existing parameter estimation and target tracking methods based on second-order will degrade severely.

In research background of the problem of interference localization, this paper addresses the problem of parameter estimation and target tracking in the presence of impulsive noise. With combination of the fractional ambiguity function with the fractional lower order statistics, a parameter estimation and airplane tracking algorithm is presented in this paper.

2 The Proposed Signal Model

In this paper, the airplane tracking and localization is realized by employing the principle of parameter estimation in bistatic Multiple-Input Multiple-Output radar system. Fig.1 illustrates a bistatic MIMO radar system. The considered bistatic MIMO radar is composed of M transmit antennas and N receive antennas with an interelement spacing of $\lambda/2$. D is the base line distance between the transmit reference element and the receive reference element. $\theta_t(t)$, $\varphi_t(t)$ and $\theta_r(t)$, $\varphi_r(t)$ are azimuth angle and elevation angle correspond to the transmit array and receive array, respectively. The transmitting antennas emit orthogonal waveforms $x_m(t)$ for $m=1,\ldots,M$. The received signals contain time-variant Doppler due to the three dimension motion of airplane. If the cubic phase is ignored, the performance of parameter estimation and airplane tracking will degrade. This paper presents a new signal model for the bistatic MIMO radar system. The echo of the n th receiving antenna is

$$s_n(t)=\sum_{m=1}^{M}x_m(t)\exp\left(j2\pi\left(\eta t+\mu t^2+\kappa t^3\right)\right)\cdot a_{tm}(\theta_t,\varphi_t)a_{rn}(\theta_r,\varphi_r)+w_n(t) \quad (1)$$

where η, μ, and κ are Doppler frequency parameters. $a_{tm}(\theta_t,\varphi_t)=\exp\left(-j\pi(m-1)\sin\theta_t(t)\cos\varphi_t(t)\right)$ is the mth element of the transmitter steering vector, $a_{rn}(\theta_r,\varphi_r)=\exp(-j(\pi(n-1)+2\pi D/\lambda)\sin\theta_r(\mathrm{t})\cos\varphi_r(\mathrm{t}))$ is the steering vector of the nth element of the receiver.

[a] Corresponding author: Shi_xiao_fei_57@163.com

The noise $w_n(t)$ is a sequence of i.i.d isotropic complex $S\alpha S$ random variable with $1<\alpha\le 2$.

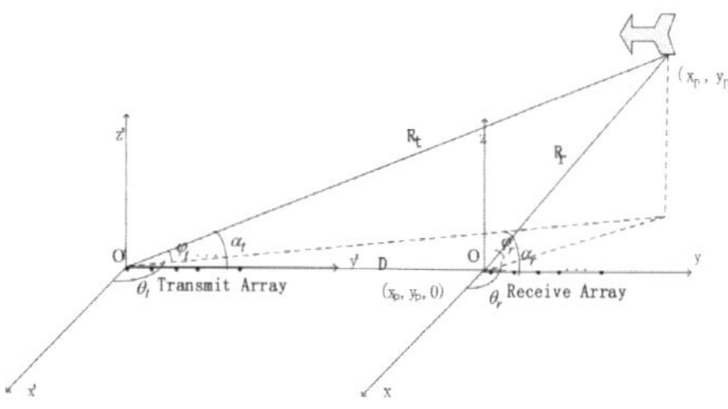

Figure 1. Bistatic MIMO radar system

Since the transmitted waves are orthogonal with each other and orthonormal, there exist two conditions, such as $\langle x_q, x_k\rangle = 0, q\ne k$ and $\|x_q\|^2=1$, for $k=1,\dots,Q$ and $A_t(\theta,\varphi)=\left[a_{t1}(\theta,\varphi),\dots,a_{tM}(\theta,\varphi)\right]^T$ $q=1,\dots,Q$. At each receiving antenna, these orthogonal waveforms can be extracted by Q matched filters. The output of the mth matched filter at the n th receive antenna can be expressed as

$$s_{mn}(t)=D(\eta,\mu,\kappa)a_{tm}(\theta_t,\varphi_t)a_{rn}(\theta_r,\varphi_r)+w_n(t) \quad (2)$$

Where $D(\eta,\mu,\kappa)=\exp\left(j2\pi\left(\eta t+\mu t^2+\kappa t^3\right)\right)$. The vector of all output of M matched filter can be expressed as

$$\boldsymbol{S}(t)=\boldsymbol{A}D(\eta,\mu,\kappa)+w_n(t) \quad (3)$$

where $\boldsymbol{A}=\boldsymbol{A}_t(\theta,\varphi)\otimes\boldsymbol{A}_r(\theta,\varphi)$, $\otimes$ is Kronecker product, and $\boldsymbol{A}_r(\theta,\varphi)=\left[a_{r1}(\theta,\varphi),\dots,a_{rN}(\theta,\varphi)\right]$.

3 Fractional ambiguity function based on fractional lower-order statistics

The ambiguity function is a useful tool for describing the ability of a waveform to simultaneously estimate the range and range-rate (speed) of targets in active (correlation-based) radar and sonar systems [4]. In recent years, a new time-frequency analysis tool, the Fractional Fourier transform, attracts increasingly more attention in signal processing and is widely applied to detection, parameter estimation and direction-of-arrival estimation of the linear frequency modulation signal.

With the development of the fractional Fourier transform (FRFT), an important relationship between the FRFT representation of the narrowband ambiguity function and its transform under the rotation operator is studied. In [5], a new ambiguity function theory based on the FRFT is presented to solve the problem of parameter estimation.

3.1 Fractional Ambiguity Function

Assume that the cubic phase signal $r(t)$ is modeled as

$$r(t)=b_0\exp\left(j2\pi\left(a_0+a_1t+a_2t^2+a_3t^3\right)\right)+w(t) \quad (4)$$

where b_0 is the signal amplitude, and $a_i, i=0,1,2,3$ are signal phase factors, the amplitude and phase factors are real and unknown . Instantaneous autocorrelation function $R_r(t,\tau)$ of the signal $r(t)$ is defined by

$$\begin{aligned}R_r(t,\tau)&=r\left(t+\tfrac{\tau}{2}\right)r^*\left(t-\tfrac{\tau}{2}\right)+R_w(t,\tau)\\&=b_0^2\exp\left(j2\pi\left(3a_3\tau t^2+2a_2\tau t+a_1\tau+a_3\tau^3/4\right)\right)\\&\quad+R_w(t,\tau)\end{aligned} \quad (5)$$

where τ denote delay, $R_w(t,\tau)$ is treated as random interference.

For a given delay τ, $R_r(t,\tau)$ has the characteristics of linear frequency modulation signal, which contains noise. According to the definition of FRFT in [6], fractional ambiguity function (FAF) of the signal $r(t)$, namely the FRFT of (5), $FAF_s(\rho,m,\tau)$ can be written as

$$FAF_r(\rho,m,\tau)=\int_{-\infty}^{+\infty}R_r(t,\tau)K_\rho(t,m)dt \quad (6)$$

where ρ is the rotation angle and m is the frequency in FRFT domain, $K_\rho(t,m)$ can be expressed as

$$K_\rho(t,m)=\begin{cases}A_\rho\exp\left(j\pi t^2\cot\rho-2j\pi mt\csc\rho\right)\times\\ \quad\exp\left(j\pi m^2\cot\rho\right), & \rho\ne n\pi\\ \delta(t-m), & \rho=2n\pi\\ \delta(t+m), & \rho=(2n+1)\pi\end{cases} \quad (7)$$

where $A_\rho=\sqrt{(1-j\cot\rho)}$

According to (6) and (7), we can see that

$$\begin{aligned}FAF_r(\rho,m,\tau)&=b_0^2A_\rho\exp\left(j\pi\left(m^2\cot\rho+2a_1\tau+\frac{a_3\tau^3}{12}\right)\right)g\\&\int_{-T/2}^{+T/2}\exp\left(j\pi\left(a_3\tau+\cot\rho\right)t^2\right)g\\&\exp\left(j\pi\left(2a_2\tau-2m\csc\rho\right)\right)dt+W(\rho,m,\tau)\end{aligned} \quad (8)$$

where $W(\rho,m,\tau)$ denotes the FAF of noise. By searching the peak of $FAF_r(\rho,m,\tau)$, we can obtain the following expressions

$$\left.\begin{aligned}&(\rho_0,m_0)=\arg\max_{\rho,m}\left[FAF_r(\rho,m,\tau)\right]\\&\mathrm{FAF}_r(\rho_0,m_0,\tau)=b_0^2TA_{\rho_0}\exp\left(j\pi\left(m^2\cot\rho_0+2a_1\tau+\frac{a_3\tau^3}{12}\right)\right)\\&\cot\rho_0=-a_3\tau\\&m_0=a_2\tau\sin\rho_0\end{aligned}\right\} \quad (9)$$

3.2 FLOS_FAF

Since the stable distribution does not have finite second-order moments ($1\le\alpha<2$), or even first-order moment ($\alpha<1$) due to the heavy tails, the performance of the existing signal selective parameters estimation methods based on second-order statistics, such as Instantaneous autocorrelation function, will degrade severely[7], [8]. Thus, this paper proposes a novel method fractional ambiguity function based on fractional lower-order statistics.

The fractional lower-order instantaneous autocorrelation function $R_r^{(p)}(t,\tau)$ of the signal $r(t)$ is defined by

$$R_r^{(p)}(t,\tau)=\left[r\left(t+\tfrac{\tau}{2}\right)\right]^{\langle p/2\rangle}\left[r^*\left(t-\tfrac{\tau}{2}\right)\right]^{\langle p/2\rangle} \quad (10)$$

where p is the order of the fractional lower order, and $1<p<\alpha\le 2$. According to (5) and (6), the FLOS_FAF of the signal $r(t)$ can be written as

$$FAF_r^{(p)}(\rho,m,\tau)=\int_{-\infty}^{+\infty}R_r^{(p)}(t,\tau)K_\rho(t,m)dt \quad (11)$$

4 Parameters Estimation based on FLOS_FAF

4.1 Doppler Instantaneous Estimation

According to (9)- (11), the FLOS_FAF of $s_{mn}(t)$ can be written as

$$FAF_{s,mn}^{(p)}(\rho,m,\tau)=\int_{-\infty}^{+\infty}R_{s,mn}^{(p)}(t,\tau)K_p(t,m)dt \quad (12)$$

Where $R_{s,mn}^{(p)}(t,\tau)=\left[s_{mn}\left(t+\tfrac{\tau}{2}\right)\right]^{\langle p/2\rangle}\left[s_{mn}^*\left(t+\tfrac{\tau}{2}\right)\right]^{\langle p/2\rangle}$.

From the relations between (ρ,m) and (a_2,a_3), we can get the estimation values of the DFCR and the DFR by finding the peak value of $FAF_{s,mn}^{(p)}(\rho,m,\tau)$

$$\left.\begin{aligned}\hat{\kappa}&=-\cot\rho_0/6\tau\\ \hat{\mu}&=m_0/2\tau\sin\rho_0\end{aligned}\right\} \quad (13)$$

Let

$$y_1(t)=s_{mn}(t)\exp\left(-j2\pi\hat{\kappa}_i t^3/6\right)\cdot\exp\left(-j2\pi\left(\hat{\mu}_i/2\right)t^2\right) \quad (14)$$

The fractional autocorrelation function of $y_1(t)$ is $R_{y_1}^{(p)}(t)=E\left[\left(y_1(t)\right)^{\langle p/2\rangle}\left(y_1^*(t)\right)^{\langle p/2\rangle}\right]$. The Fourier transform of $R_{y_1}^{(p)}(t)$ is $\hat{R}_{y_1}^{(p)}(f)$. Therefore, the IDF η can be estimated by

$$\eta=\arg\max_f\left\{\hat{R}_{y_1}^{(p)}(f)\right\} \quad (15)$$

4.2 Azimuth and Elevation Angles Dynamic Recursive Estimation

Bin Yang [9] proposed the method of projection approximation subspace tracking (PAST) based on least squares estimation, which is sensitive to impulse noise and make the performance of PAST algorithm degraded. On the base of the study of the method of PAST, this paper propose a novel RLM_PAST method based on FLOS_FCAF (FF_RLM_PAST)

According to (8) and (15), we can define $z(t)$ as

$$z(t)=\exp\left(j2\pi\left(\eta t+\mu t^2+\kappa t^3\right)\right) \quad (16)$$

The fractional cross-ambiguity functions $FCAF_{sz,mn}(\rho,m,\tau)$ between $s_{mn}(t)$ and $z(t)$ can be expressed as

$$FCAF_{sz,mn}(\rho,m,\tau)=\int_{-\infty}^{+\infty}R_{sz,mn}(t,\tau)K_\rho(t,m)dt \quad (17)$$

Where $R_{sz,mn}(t,\tau)=s_{mn}\left(t+\tfrac{\tau}{2}\right)z^*\left(t-\tfrac{\tau}{2}\right)$, $W(\rho,m,\tau)$ denotes the FAF of the noise. The $FCAF_{sz,mn}(\rho,m,\tau)$ arises the peak value when both ρ and m meet the conditions shown in (13). The peak value of the $FCAF_{yz,l}(\rho,m,\tau)$ is

$$\begin{aligned}&\mathrm{FCAF}_{sz,mn}(\rho_0,m_0,\tau)=TA_{\rho_0}a_{tm}(\theta_t,\varphi_t)\cdot\\ &a_{rn}(\theta_r,\varphi_r)\exp\left(j\pi\left(m^2\cot\rho_0+2\eta\tau+\frac{\kappa\tau^3}{2}\right)\right)\end{aligned} \quad (18)$$

According to (11) to (18), we can obtain the expression of FLOS_FCAF as

$$\begin{aligned}&FCAF_{sz,mn}^{(p)}(\rho_0,m_0,\tau)=a_{tm}(\theta_t,\varphi_t)a_{rn}(\theta_r,\varphi_r)\cdot\\ &FAF_{sz,mn}^{(p)}(\rho_0,m_0,\tau)+W^{(p)}(\rho_0,m_0)\end{aligned} \quad (19)$$

The output of all receiving antenna based on FLOS_FCAF can be expressed as

$$\boldsymbol{FCAF}=\left[\mathrm{FCAF}_{sz,11}^{(p)},\ldots,\mathrm{FCAF}_{sz,1N}^{(p)};\ldots;\ldots\mathrm{FCAF}_{sz,MN}^{(p)}\right] \quad (20)$$

Both receive subarrays $\boldsymbol{R}_1$ and $\boldsymbol{R}_2$ constructed in this paper can be expressed by

$$\boldsymbol{R}_1=\left[\boldsymbol{FCAF}_{11}\ \ \mathrm{L}\ \ \boldsymbol{FCAF}_{1N}\right]=\boldsymbol{A}_r(\theta,\varphi)D+\boldsymbol{N}_1 \quad (21)$$

$$\boldsymbol{R}_2=\left[\boldsymbol{FCAF}_{11}\ \ \mathrm{L}\ \ \boldsymbol{FCAF}_{M1}\right]^{\mathrm{T}}=\boldsymbol{A}_t(\theta,\varphi)D+\boldsymbol{N}_2 \quad (22)$$

Next, the proposed method of FF_RLM_PAST is introduced in detail by taking the subarray $\boldsymbol{R}_1$ as an example.

The matrix $\boldsymbol{W}(t)$ denotes the signal subspace of observation vector $\boldsymbol{R}_1$. This signal subspace is estimated by the minimization of a cost function $J_{\mathrm{FF\_RLM\_PAST}}(\boldsymbol{W}(t))$ in the FF_RLM_PAST algorithm.

$$J_{\mathrm{FF\_RLM\_PAST}}(\boldsymbol{W}(t))=\sum_{n=1}^{t}\beta^{t-n}\rho\left(\left\|\boldsymbol{R}_1(n)-\boldsymbol{W}(t)\boldsymbol{y}(n)\right\|_2^2\right) \quad (26)$$

where $0<\beta<1$ denotes the forgetting factor. $\boldsymbol{y}(n)=\boldsymbol{W}^H(t-1)\boldsymbol{R}_1(n)$ is projection approximation vector. $\rho(\cdot)$ is M-estimation function, whose derivative is denoted as $\varphi(\cdot)=\rho'(\cdot)$. Reference [11] indicated that evaluation function of Cauchy distribution function can obtain good preference in alpha-stable distribution noise with different characteristic exponent when parameter estimation is studied by employing M-estimation method in alpha-stable distribution noise.

We expect $J_{\mathrm{FF\_RLM\_PAST}}'(\boldsymbol{W}(t))$ to be a good approximation for $J_{\mathrm{FF\_RLM\_PAST}}(\boldsymbol{W}(t))$ and the matrix $\boldsymbol{W}(t)$ minimizing $J_{\mathrm{FF\_RLM\_PAST}}'(\boldsymbol{W}(t))$ to be a good estimate for the signal subspace of $\boldsymbol{C}(t)$. $J_{\mathrm{FF\_RLM\_PAST}}'(\boldsymbol{W}(t))$ is minimized if

$$\boldsymbol{W}(t)=\boldsymbol{C}_{\boldsymbol{R}_1\boldsymbol{y}}(t)\boldsymbol{C}_{\boldsymbol{yy}}^{-1}(t) \quad (27)$$

$$C_{R_1y}(t)=\sum_{n=1}^{t}\beta^{t-n}\varphi\left(\|e(n)\|_2^2\right)R_1(n)y^H(n) \quad (28)$$

$$C_{yy}(t)=\sum_{n=1}^{t}\beta^{t-n}\varphi\left(\|e(n)\|_2^2\right)y(n)y^H(n) \quad (29)$$

$$e(t)=R_1(t)-W(t-1)y(t) \quad (30)$$

We use RLS algorithm for updating $W(t)$ without derivations. We apply $W(t)$ to the spatial spectrum of 2D-MUSIC algorithm,

$$P(\theta_r,\varphi_r)=\frac{1}{A_r^H(\theta_r,\varphi_r)WW^HA_r(\theta_r,\varphi_r)} \quad (31)$$

Similarly, the algorithm of FF_RLM_PAST is applied to observation vector R_2. We can obtain as

$$P(\theta_t,\varphi_t)=\frac{1}{A_t^H(\theta_t,\varphi_t)WW^HA_t(\theta_t,\varphi_t)} \quad (32)$$

Searching spectral peak of $P(\theta_r,\varphi_r)$ and $P(\theta_t,\varphi_t)$, we can realize the azimuth angle $\theta_r(t)$, $\theta_t(t)$ and elevation angle $\varphi_r(t)$, $\varphi_t(t)$ dynamical recursive estimation.

5 Simulation results

The considered bistatic MIMO radar is composed of $M=6$ transmit antennas and $N=8$ receive antennas with an interelement spacing of $0.5\mathrm{m}$. The base line distance between the transmit reference element and the receive reference element is $D=5\mathrm{Km}$. Since the alpha-stable process has infinite variance for $\alpha<2$, we use a generalized signal-to-noise ratio (GSNR) measure defined as the ratio of the signal power over the implusive noise dispersion γ, $\mathrm{GSNR}=10\log_{10}\left(\mathrm{E}\left\{|r(t)|^2\right\}/\gamma\right)$. According to this choice of GSNR metric, the $S\alpha S$ noise samples are power scaled by the dispersion parameter γ. The number of Monte Carlo iterations is 500 in all simulations.

Simulation 1: Time series t

In this simulation, the noise is modeled as the α-stable noise with $\alpha=1.4$, $\mathrm{GSNR}=12\mathrm{dB}$. Fig.2 plots the estimation performance of Doppler. From Fig.2, we can find the curve of estimated Doppler is more close to the curve of true Doppler. Therefore, the proposed method has good estimation performance about three Doppler parameters.

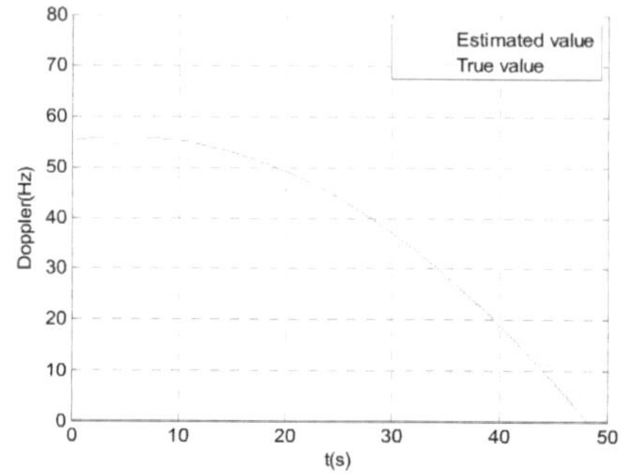

Figure 2. Estimation performance of Doppler

Fig.3 show curves of azimuth angle and elevation angle dynamical recursive estimators with time t. From Fig.3, we can find the curves of estimated azimuth angle and elevation angle are more close to the curves of true value. From Fig.2 and Fig.3, we can find that the proposed method has good estimation performance.

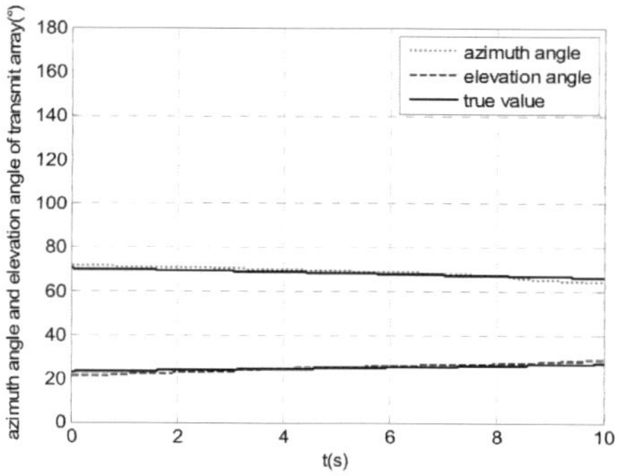

(a)

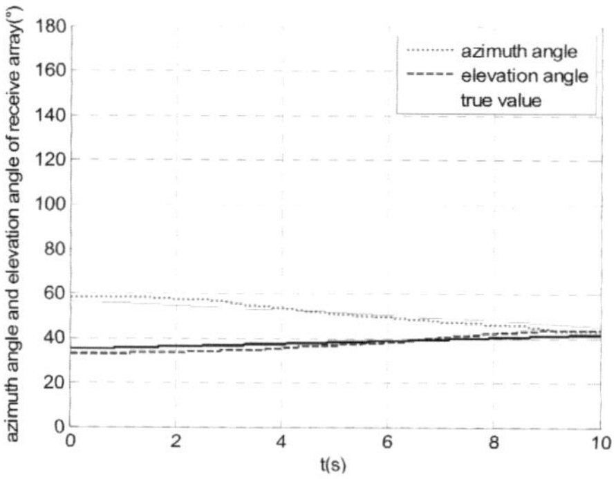

(b)

Figure 3. estimation performance of angle and airplane location

Simulation 2: Characteristic exponent α .

In this simulation, the generalized signal to noise ratio is set as $\mathrm{GSNR}=12\mathrm{dB}$. Let root mean-squared errors (RMSE) based on average is defined by $\mathrm{RMSE2}=\frac{1}{N_s}\sqrt{\sum_{t=1}^{t=N_s}\left(\hat{\xi}(t)-\xi(t)\right)^2}$, where $\hat{\xi}$ is estimator. Fig. 4 plots the RMSE of the proposed method versus various values of characteristic exponent α. The smaller the characteristic exponent α is, the stronger the pulse characteristic is. From Fig.4, we can find that the RMSE of the parameters decreases with the increase of the characteristic exponent α . Therefore, the proposed method can suppress the interference of impulse noise, and has good performance. From these figures, we also find that the estimation performance of the Doppler frequency parameters affects the estimation performances of the azimuth angle and elevation angle.

Simulation 3: Generalized signal-noise-ratio GSNR

In this simulation, the noise is modeled as α -stable noise with $\alpha = 1.4$. Fig.5 show the RMSE of angle and Doppler frequency for various values of the GSNR . From Fig.5, we can find that the proposed method still has better performance in the condition of the lower SNR.

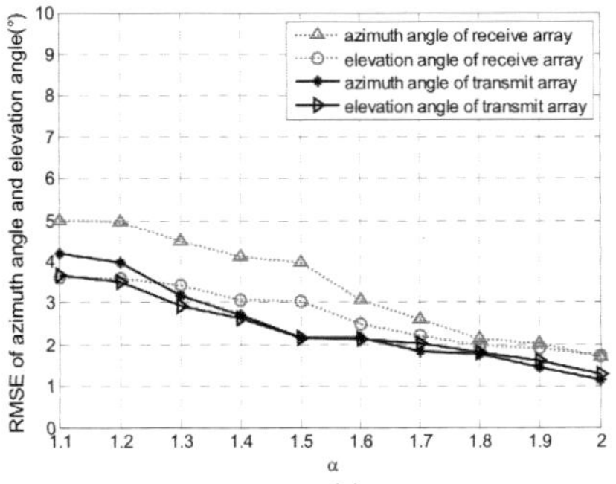

(a)

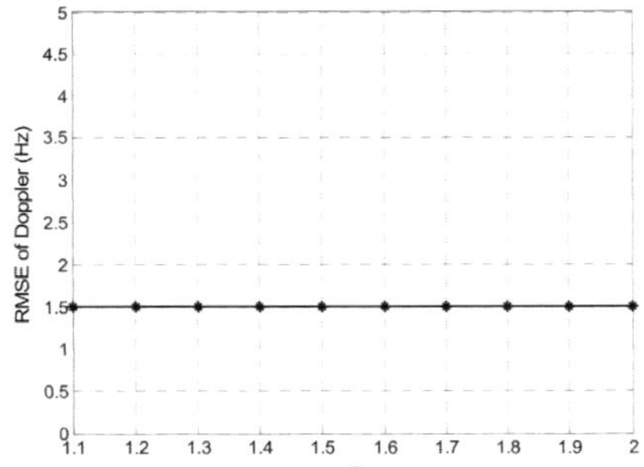

(b)

Figure 4. Root mean square error as a function of the characteristic exponent α

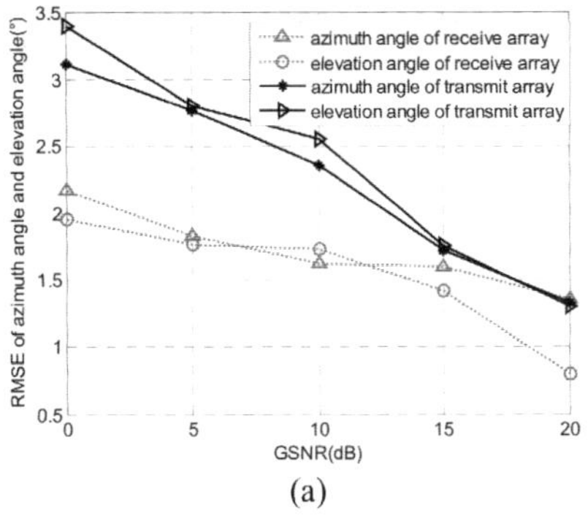

(a)

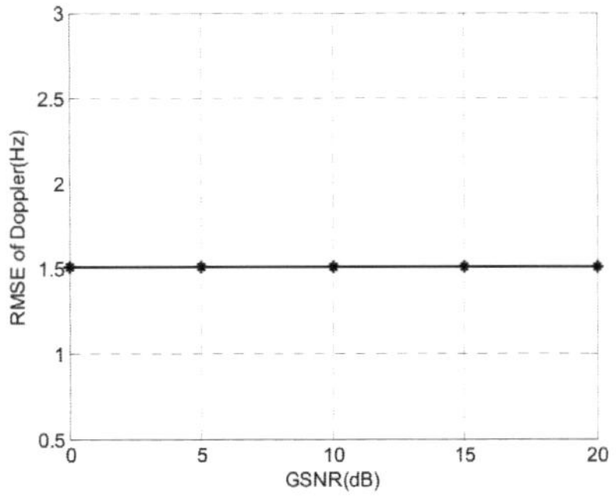

(b)

Figure 5. Root mean square error as a function of the GSNR

6 Conclusion

This paper proposes an airplane tracking algorithm based on study of the problem of interference localization. Firstly, a novel signal model is proposed in impulsive noise environment. A method of instantaneous Doppler frequency estimation based on peak searching of the FLOS_FAF is proposed, and FF-RLM_PAST algorithm is proposed to dynamical recursive estimate the azimuth angle and elevation angle. As a result, the airplane tracking is achieved in bistatic radar, laying the foundations for interference localization. The correctness and effectiveness of the proposed method are verified with the computer simulation.

Acknowledgments

This work was partly supported by the National Science Foundation of China under Grants 61401055 and the fund of Key Laboratory of marine management technology in State Oceanic Administration People's Republic of China under Grant 201504.

References

1. H. Tan, J. Li, "Emitter locating algorithm by scattered signals of civil airplanes," *Radio communications Technology*, vol. 35, pp. 40-42, Feb 2009.(in Chinese)
2. T.S. Qiu, N. Xia, J.C. Li, "An interference localization algorithm based on Gaussian approximation particle filtering with stable distribution noise," *Signal Processing*, vol. 28, pp. 1248-1253, Sep 2012. (in Chinese)
3. X. Li, W.D. Jiang, Z. W. Zhuang, "MIMO radar sensitivity analysis of antenna position for direction finding," *IEEE Transaction on Signal Processing*, vol. 60, pp. 5201-5216, Oct 2012.
4. A.A. Gorji, R. Tharmarasa, T. Kirubarajan, "Widely Separated MIMO versus Multistatic Radars for Target Localization and Tracking," *IEEE*

Transactions on Aerospace and Electronic Systems, vol.49, pp:2179-2194, April 2013.

5. Q. Qu, M.L. Jin, J.M. KIM, "FRFT based parameter Estimation of the Quadratic FM signal," *Chinese Journal of Electronics*, vol.19, pp: 463-467, Mar 2010.
6. S.H. Liu, S. Tao, R. Tao, Y.D. Zhang, G. Zhang; F. Zhang, Y. Wang, "Sparse Discrete Fractional Fourier Transform and Its Applications," *IEEE Transactions on Signal Processing*, vol. 62, pp: 6582 – 6595, Dec 2014.
7. J. Park, G. Shevlyakov, K. Kim, "Maximin distributed detection in the presence of impulsive alpha-stable noise," *IEEE Transactions on Wireless Communications*, vol. 10, pp.1687-1691, June 2011.
8. C.L. Nikias, M. Shao, "Signal Processing with Alpha-stable Distribution and Application. New York: John Wiley & Sons, Inc, 1995.
9. Y. Bin, "Projection approximation subspace tracking," *IEEE Transaction on Signal Processing*, vol. 43, pp. 95-107, Jan 1995.
10. S. Li, T.S. Qiu, "A robust past algorithm in alpha stable noise environment," *Acta Electronica Sinica*, vol.37, pp. 519-522, Mar 2009.

MATEC Web of Conferences **61**, 07012 (2016) DOI: 10.1051/matecconf/20166107012
APOP2016

Design of the Acoustic Signal Receiving Unit of Acoustic Telemetry While Drilling

Li Zhigang, Ge Shitong, Fu Zhongyao

School of Petroleum Engineering, China University of Petroleum (East China), Qingdao, Shandong, China, 266580

Abstract: Signal receiving unit is one of the core units of the acoustic telemetry system. A new type of acoustic signal receiving unit is designed to solve problems of the existing devices. The unit is a short joint in whole. It not only can receive all the acoustic signals transmitted along the drill string, without losing any signal, but will not bring additional vibration and interference. In addition, the structure of the amplitude transformer is designed, which can amplify the signal amplitude and improve the receiving efficiency. The design of the wireless communication module makes the whole device can be used in normal drilling process when the drill string is rotating. So, it does not interfere with the normal drilling operation.

1. Introduction

Acoustic Telemetry While Drilling is a new technology for the downhole information acquisition and transmission while drilling. The technology transmits information between surface and downhole using acoustic wave as the carrier and drill string as signal channel. Since entering 21th Century, the technology has made a considerable development. At present, this technology has been commercialized in North America, and has achieved good results, especially in the unbalanced drilling and air drilling [1].

Signal receiving unit is one of the core units of the acoustic telemetry system. Its function is to receive the acoustic wave (signal carrier) along the drill pipe, and then extract the useful information (data) from the carrier to achieve the purpose of signal transmission.

The acoustic signal receiving device which is currently in use is generally used in the form of a clamp [2]. These devices use a clamping structure fastening the sound wave receiving sensor and related equipment on the Kelly or the lower drill pipe under the top drive drilling device. They mainly have the following shortcomings.

The measured signal is the vibration and acoustic signals of the clamping structure and drill string, rather than the vibration and acoustic signals of the drill string. Therefore, the received signal must be different from the transmitting signal without the clamping structure. The form of the clamping structure and the degree of fastening are the main factors that affect the fidelity and reliability of the signal.

As a result of this measurement method, which was completed by an external clamping structure, some high frequency and weak acoustic signal will be lost or filtered.

Because the clamping structure is not a part of the drill string, it is a subsidiary structure, so it will come with the additional vibration and noise, which can interfere with the measurement and reception of the acoustic signal.

2. Overall structure

For shortcomings of the existing equipment, this paper designs a new type of acoustic signal receiving

Corresponding author;lee18198@126.com

 DOI: 10.1051/matecconf/20166107012

unit, as shown in Figure 1. The unit is composed of three parts, the external protection sleeve 2, the acoustic signal receiving module 4 and the lower joint 12, and the three is connected with the screw thread. The lower joint is connected with the external protective sleeve through the screw thread 10. The acoustic signal receiving module connects the lower joint by the screw thread 8. The unit is a short joint in whole, when used in connection in the drill string, and is a part of the drill string. The whole unit is connected with the other lower part by the external screw thread 13, and the upper part by means of an internal thread 1.

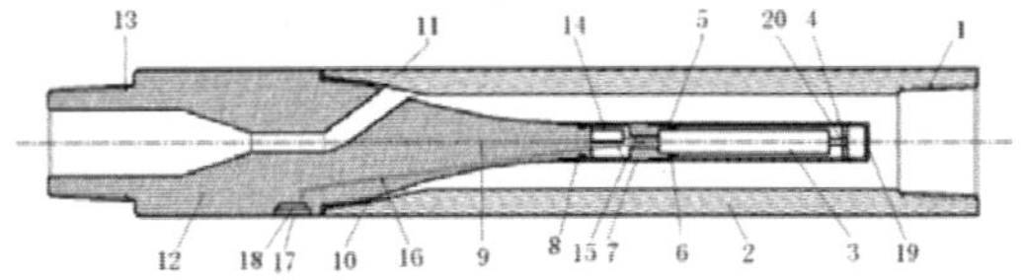

Fig.1 The overall structure of the acoustic signal receiving device

1-internal thread,2- external protection sleeve,3-circuit board,4- acoustic signal receiving module,5-screw thread,6-signal processing module,7-transducer sleeve,8-screw thread,9-amplitude transformer,10- screw thread,11-hole,12-lower joint,13-external screw thread,14-signal receiving transducer,15-cable,16-slender hole,17-cover,18-slot,19-cap,20-communication interface.

The part above the thread 10 of the lower joint is made to be an amplitude transformer structure. There are three holes along the circumferential direction (120 degrees apart) in it. On the outer wall of the lower joint, a slot 18 is provided for placing the wireless communication module. The slot was covered by the nonmetallic cover 17. The acoustic wave receiving transducer 14 is installed on the upper end face of the amplitude transformer. There is a slender hole 16 going through the lower joint between the upper end face of the amplitude transformer 9 and the slot 18. Electrical cable can be put into the hole and connect the wireless communication module and the acoustic wave signal receiving module.

The acoustic signal receiving module 4 is composed of the sensor sleeve 7 and the signal processing module 6. The sensor sleeve and the signal processing module are connected through the screw 5. There is a bracket in the sleeve of the signal processing module. The acoustic signal receiving and process circuit 3 can be mounted on it. The circuit is connected with the sensor and the wireless communication module by electric cable 15. In the signal processing module, the upper part of the sleeve 6 is provided with a communication interface 20, which can be connected with other upper equipment through the interface. The upper part of the sleeve 6 is provided with a thread. When there is other top structure, the sleeve can be connected through the thread; no other equipment, capped and sealed by cover 19.

Principle: With the tight connection of the lower joint and the lower drill string by the screw thread, the receiving unit can transfer the acoustic wave coming from the lower string while connect the drill string and itself. The sound receiving sensor mounted on the amplitude transformer structure receives the acoustic signal and converts it into an electrical signal. The signal is received and processed by the signal receiving and processing module. The processed signal and data can be stored in the memory of the acoustic signal receiving module, also can be transmitted to the wireless communication module in the outer wall of the lower joint by the cable placed in the inner hole of the lower joint. Furthermore, the wireless communication module transmits the signal to other external communication equipment or surface signal receiving and processing equipment. Due to the existence of the wireless communication module, the signal can be transmitted through the wireless way to other devices on the rig floor or ground, rather than to receive signals through a cable connection. Therefore, the unit can be used in normal drilling or while drilling, in which the drill string is rotating, and it does not interfere with the normal drilling operation.

3. The structure of amplitude transformer

In order to improve the receiving efficiency of the acoustic wave signal, the amplitude transformer structure is used in the front end of the acoustic signal receiving sensor. The amplitude transformer structure can condense the sonic energy from the larger back-end area to the smaller front-end surface where the sensor is mounted. Assuming the area of the back section is a, the area of the front end section is b, the ratio of amplitude transforming will be λ=a/b (λ>1) [3]. The sonic energy is magnifying in λ times by the amplitude transformer structure. The signal also is amplified.

In the unit described here, the size of the front end, connected with the lower drill string, is restricted by the size of the drilling tools. So, the cross-sectional area is fixed. The back end where the sensor will be mounted on is restricted by the size of the sensor and processing circuit, the section area is also fixed, at least lager than the size of the sensor. The works can be done to improve the signal receiving efficiency is to optimize the size and shape of the amplitude transformer structure through researching the relationship between the length of the amplitude transformer and the effect of the amplification. The amplitude transformer was designed as shown in Fig.2.

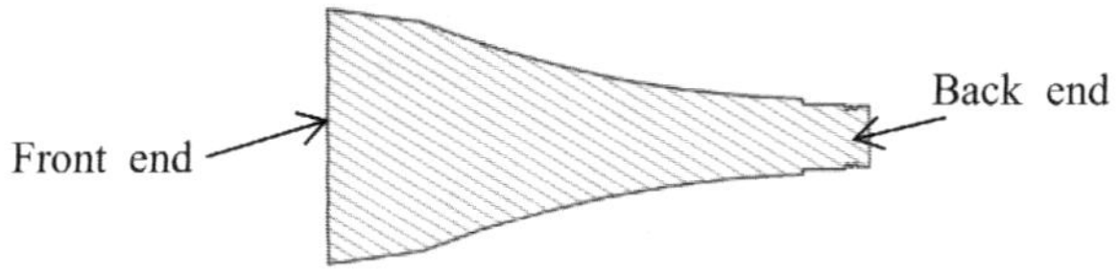

Fig.2 Longitudinal profile of the amplitude transformer

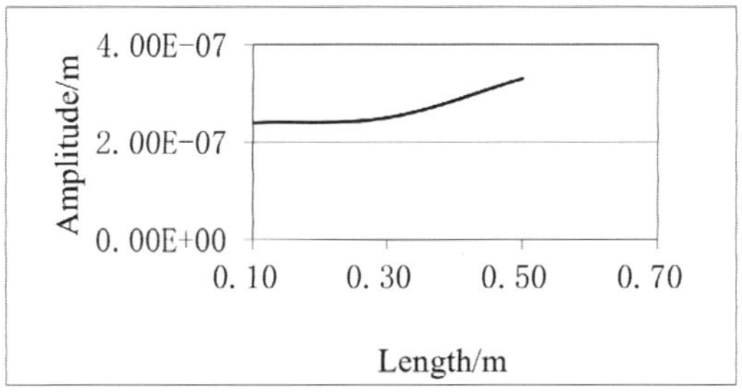

Fig.3 relationship of the length of amplitude transformer and receiving signal amplitude

Through building the three-dimensional solid finite element model, the change of the amplitude of acoustic signal is calculated in the case of different length (0.1-0.5m) of the amplitude transformer using finite element method by software ANSYS. In the calculation, a transient excitation signal is applied to the front end of the amplitude transformer, and the vibration amplitude is taken at the middle point of the back end face. The calculated result is shown in Fig.3. As shown in the figure, the amplitude of receiving signal grows up obviously when the length is bigger than 0.3m in the unit's sizes. But the length will be too long and influences the unit's strength and performance when the length is larger than 0.5m.

4. Conclusion

To sum up, the unit has the following advantages:

(1)The unit is a short joint, connected to the drill string, can not only completely received all signal transmitted from down hole, but does not produce additional interference, ensure the fidelity and anti-jamming performance. Also, threaded connections are more reliable than clamping form, no prominent structure, does not affect the normal drilling operation.

(2)The acoustic signal receiving module does not participate in the whole structural deformation, so as to ensure the strength. At the same time, it can be disassembled, easy to update and maintain.

(3)The structure of the amplitude transformer is designed to amplify the amplitude of the received signal and improve the receiving efficiency.

(4)The design of the wireless communication module makes the unit can be used while drilling, and does not interfere with the normal drilling operation.

In short, the unit can complete the reception of the acoustic signal efficiently, but also is convenient for the installation, operation and use, does not affect the normal drilling operations, do not increase the labor intensity of the drilling site.

Acknowledgements

The paper is supported by "the Fundamental Research Funds for the Central Universities" (15CX05034A) .

References

1. Reeves Michael, Smith Douglas, and etc. Unique acoustic telemetry network with distributed pressure measurement nodes enables accurate real-time analysis of sweep effectiveness [J]. Proceedings - SPE Annual Technical Conference and Exhibition, v (2) , p1690-1709, 2013.
2. Neff J M, Camwell P L. Field-Test Results of an Acoustic MWD System [R]. SPE/IADC 105021. 2007.
3. Lin Zhongmao. Principle and Design of Ultrasonic Amplitude Transformer[M]. Science Press, 1987:55-97.

MATEC Web of Conferences **61**, 07013 (2016) DOI: 10.1051/matecconf/20166107013
APOP2016

Performance Analysis of OFDM 60GHz System and SC-FDE 60GHz System

Xueyan Han[1], Jingjing Wang[1,a] , Wei Shi[1,b] and Xinli Dong[2]

[1]School of Information Science & Technology, Qingdao University of Science & Technology, Qingdao, Shandong Province, China
[2]Qingdao Branch, China United Network Communications Corporation, China

Abstract. In this paper, the performance of 60GHz wireless communication system with SC and OFDM is studied, the models of OFDM 60GHz system and SC 60GHz frequency domain equalization (SC-FDE) system are established, and the bit error rate (BER) performance of OFDM 60GHz system and SC-FDE 60GHz system in 802.15.3c channels is compared. The simulation results show that SC-FDE 60GHz system has a slight advantage over OFDM system in line-of-sight (LOS) channels, while OFDM 60GHz system has a slight advantage over SC-FDE system in non-line-of-sight (NLOS) channels. For 60GHz system, OFDM 60GHz system has a slight advantage over SC-FDE system in overcoming multipath fading, but the performance of both is close whether in the LOS or NLOS case.

1 Introduction

Currently, the high-speed interface, such as high definition multimedia interface (HDMI), Gigabit wireless local area networks(WLAN) and wireless personal area networks, have a more and more urgent requirement for Gbps-class wireless transmission technology. However, the current popular wireless high-speed communications technology, such as Ultra Broadband (UWB) wireless communication technology, generally only provides a data transfer rate of hundreds of megabits per second. 60GHz wireless communication technology has emerged. Because of several GHz wideband spectrum, up to 10W transmit power, and low-cost CMOS devices[1][2][3], it is the first choice in short distance wireless communication technologies.

At present, the study of 60GHz wireless communication physical layer for domestic and foreign research institutions and scholars mostly focus on SC-FDE and OFDM modulation. In the case of gigabit-speed transmission, the time-domain equalizer is complex. To solve this problem, [4-6] proposed SC-FDE modulation. SC-FDE can be viewed as OFDM linear precoding in special cases. The output of the receiver equalizer is inserted to compensate for linear precoding brought by the FFT matrix at the transmitter. This is the SC-FDE system. SC-FDE system has a low complexity of frequency-domain multipath channel equalizer, which is similar to OFDM. However, in the actual wireless systems, both show some different characteristics. Because of IFFT precoding at the transmitter, SC-FDE has less PAPR than OFDM. In addition, the computational complexity required for both the transmitter and receiver are different, in OFDM systems, both the transmitter and receiver should execute FFT or IFFT operations. In the SC-FDE system, the transmitter does not need to execute FFT / IFFT operations, while at the receiver, it need to perform two times of FFT / IFFT operations.

Because of OFDM system with high spectrum efficiency, which can help to achieve high-speed communications, OFDM and SC-FDE are two powerful candidate technologies that can realize gigabit transmission throughput. In 802.15.3c [7] and 802.11ad [8], both standards are chosen as the physical layer transmission modulation.802.15.3c standard divides 60GHz band into four channels band. The standard defines three physical layer technology, namely High Speed Interface (HSI) physical layer, Audio and Visual (AV) physical layer and SC physical layer, the first two kinds are based on OFDM modulation, SC physical layer uses SC-FDE modulation. The paper uses HSI and SC over 802.15.3c standards for an example, the performance of OFDM 60GHz and SC-FDE system are studied and simulated.

The remaining part of the paper is organized as follows: Section 2 analyze the 60GHz system with OFDM modulation, and conducts Monte Carlo simulations in LOS and NLOS scenes over 802.15.3c channel model; In Section 3, the 60GHz system with SC-FDE is analyzed, and compares bit error rate (BER) performance with 60GHz system with OFDM modulation; Concluding remarks are given in Section 4.

* Corresponding author: [a]*kathy1003@163.com* & [b]*shiwei6670595@126.com*

2. The 60ghz System With Ofdm Modulation

2.1.The 60GHz system model based on OFDM modulation

The 60GHz system model with OFDM modulation [9] is shown in Figure 1. The specific process is as follows: after LDPC channel coding, interleaving and constellation mapping, the frequency domain signal d(n) obtain the data symbols transmitted in N subcarriers, denoted as X (k) (k = 0, 1, 2, ..., N-1). After IFFT, the time domain signal x (n), n = 0, 1, 2 ..., N-1 is obtained.

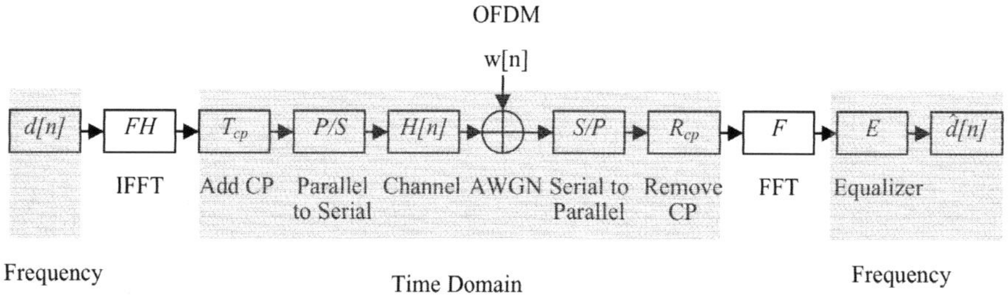

Figure 1. OFDM 60GHz system model.

$$x(n) = IFFT\{X(k)\} = \frac{1}{N}\sum_{k=0}^{N-1} X(k)\exp(j2\pi kn/N) \quad (1)$$

Then the Guard Interval (GI) is added to x(n)[10]. GI length, denoted by TGI, is greater than the maximum delay spread of the wireless channel. The original symbol is used to cycle and expand by GI generally as the Cyclic Prefix, and by this way Inter-Carrier Interference (ICI) is suppressed, Inter-Symbol Interference (ISI) is eliminated, while the linear convolution of the channel and signal is transferred to cyclic convolution. After adding CP, the signal sequence is denoted as $x_f(n)$

$$x_f(n) = \begin{cases} x(N+n) & n = -N_{GI}, -N_{GI}+1, \ldots, -1 \\ x(n) & n = 0, 1, \ldots, N-1 \end{cases} \quad (2)$$

where NGI is the length of CP.

The signal $x_f(n)$ transmitted by the antenna, goes through 60GHz channel, then the received signal $y_f(n)$ is obtained. $y_f(n)$ is given as

$$y_f(n) = x_f(n) * h(n) + w(n) \quad (3)$$

where h(n) is the sampling sequence of actual channel impulse response, w (n) is the sample sequence of additiv e white Gaussian noise with zero mean, the sampling rate is same with that of input data X (k). By removing CP fro m the received signal $y_f(n)$, $y(n)$ is obtained,

$$y(n) = y_f(n), \quad n = 0, 1, \ldots, N-1 \quad (4)$$

After FFT transform, we can obtain

$$Y(k) = FFT\{y(n)\} = \sum_{n=0}^{N-1} y(n)\cdot\exp(-j2\pi kn/N), \quad K = 0, 1, \ldots, N-1 \quad (5)$$

$H(k) = FFT\{h(n)\}$ is the frequency domain representation of h(n), $W(k) = FFT\{W(n)\}$ is the frequency domain representation of w(n), then we obtain

$$Y(k) = X(k)H(k) + W(k), \quad k = 0, 1, \ldots, N-1 \quad (6)$$

After equalizing, data is through modulation constellation de-mapping, de-interleaving, channel coding translation, and finally get the decision data $\hat{d}(n)$.

2.2. The performance simulation of OFDM 60GHz system

This section chooses HSI physical layer with 802.15.3c channel to study OFDM 60GHz system. The number of subcarriers is 512,low density parity check (LDPC) is used, BPSK, QPSK, and 16-QAM modulation modes can be chosen, 12 kinds of transmission rates of modulation and coding scheme (MCS) are provided. Table 1 lists the parameters of modulation and coding schemes.

LOS CM1.3 and NLOS CM2.3 channel are selected to simulate, and investigate bit error rate, package error rate of OFDM 60GHz system with different modulation and coding schemes. Each package contains 1Kbit information. MMSE equalization is used.

Table 1. HSI physical layer parameters of different modulation and coding modes.

MCS number	Data rate (Mb/s)	Modulation	Sprea-ding factor	Coding rate
1	1540	QPSK	1	1/2
2	2310	QPSK	1	3/4
3	2695	QPSK	1	7/8
4	3080	16-QAM	1	1/2
5	4620	16-QAM	1	3/4
6	5390	16-QAM	1	7/8
7	5775	64-QAM	1	5/8

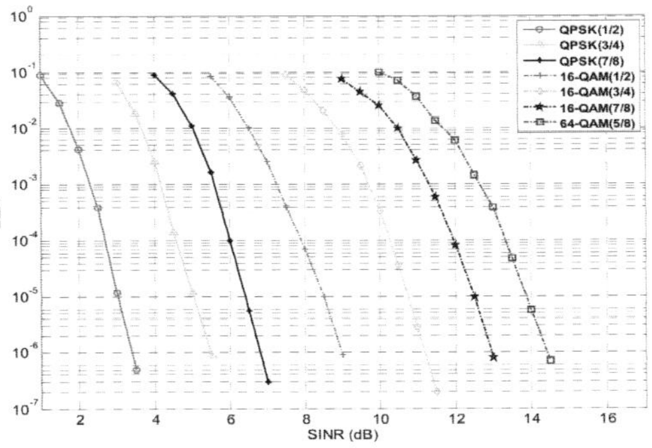

a) The BER performance simulation

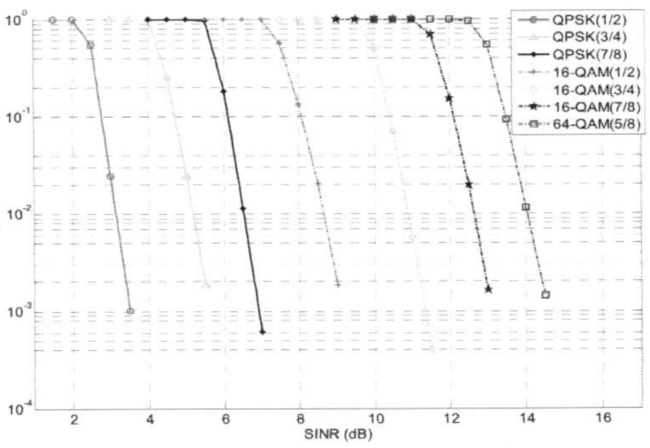

b) The PER performance simulation

Figure 2. The BER and PER performance simulation under 802.15.3c CM1.3 channel

From Figure 2 and 3, under different physical layers MCS schemes, i.e., transmission rates are 1540Mb/s, 2310Mb/s, 2695Mb/s, 3080Mb/s, 4620Mb/s, 5390Mb/s, 5775Mb/s, the SINR required for the reliable transmission with bit error rate of 10^{-6} under LOS CM1.3 channel of order are 3.3dB, 5.4dB, 6.8dB, 9.0 dB, 11.1dB, 12.9dB and 14.4dB. The SINR required for the reliable transmission with bit error rate of 10^{-6} under NLOS CM2.3 channel are 5.1dB, 7.2dB, 9.1dB, 12.6dB, 15.2dB, 17.1dB and 18.7dB. As can be seen from the simulation, in the same modulation scheme, as the coding rate increases, the bit error rate performance will deteriorate correspondingly. In order to obtain higher transmission rates, a higher SINR requirements. In whatever rate transmission, with the same SINR, the package rate and bit error rate performance of OFDM 60GHz system under LOS channel has better performance than that of OFDM 60GHz system under NLOS channel. Comparing Figure2 (a) and Figure 3 (a), we can find that, with the increasing of coding rate and modulation order, to obtain the same error rate, the SINR difference between NLOS channel and LOS channel is growing.

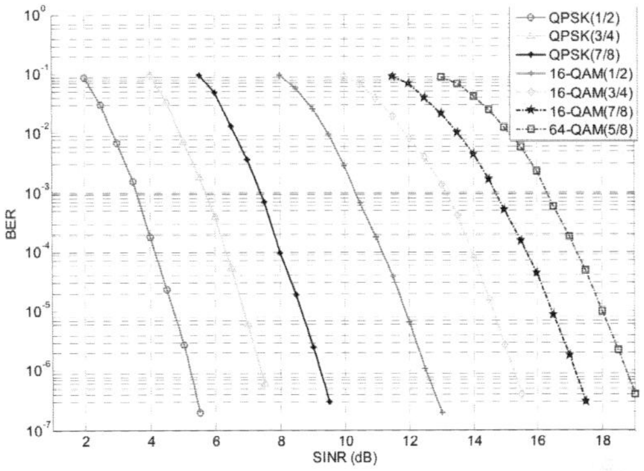

a) The BER performance simulation

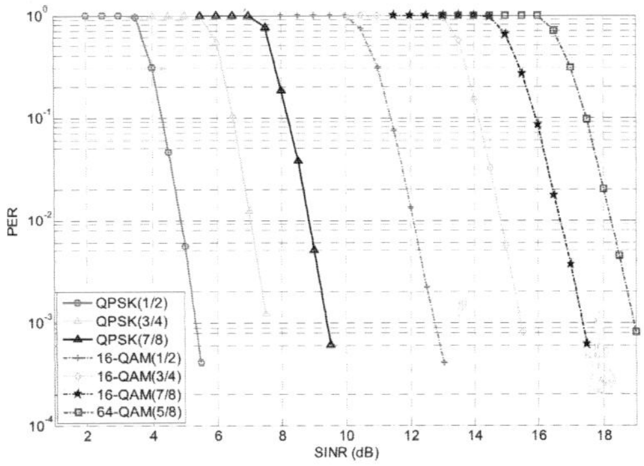

b) The PER performance simulation

Figure 3. The BER and PER performance simulation under 802.15.3c CM2.3 channel.

3. The 60ghz System With Sc-Fde Modulation

3.1. The 60GHz system model based on SC-FDE modulation

The 60GHz system model employing SC-FDE[11] is depicted in Figure 4 which moves the IFFT module, in transmitting end of OFDM system, to the receiving end. The same with the OFDM 60GHz system is that its equilibrium is also happened in frequency domain, so it is similar to the structure of the OFDM 60GHz system model.

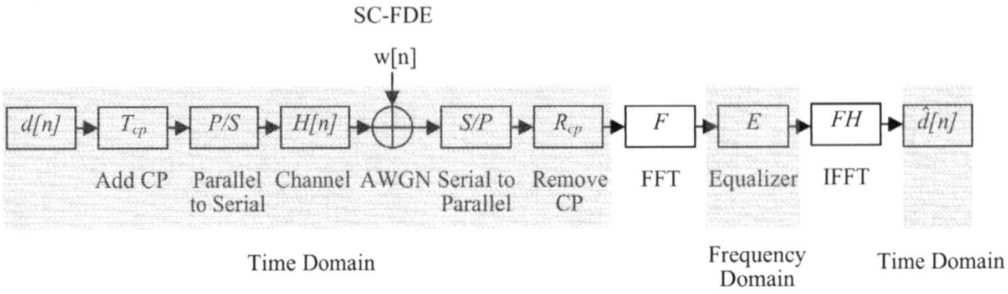

Figure 4. SC-FDE 60GHz system model.

3.2. The performance simulation and comparison of SC-FDE 60GHz and OFDM system

The study is conducted under 802.15.3c SC physical layer. When the pilot length is 64, the data transmission rate is1.54Gb/s - 4.62Gb/s. The sub-block length is 512. Low Density Parity Check Code (LDPC) and Reed-Solomon code (RS) are used. BPSK, QPSK, and 16-QAM modulation modes can be chosen. The constellation mapping takes $\pi/2$ phase shifting scheme. The modulation and coding scheme (MCS) can provide 13 kinds of transmission rate. Table 2 shows the parameters selection using the LDPC code.

Table 2. Parameter selection of modulation and coding scheme under SC physical layer.

MCS Number	Data transmission rate /(Mb/s) when pilot length=0	Data transmission rate/(Mb/s) when pilot length=64	Modulation scheme	Spreading factor	Coding rate
7	1760	1540	$\pi/2$ QPSK	1	LDPC(672,336) R=1/2
8	2640	2310	$\pi/2$ QPSK	1	LDPC(672,504) R=3/4
9	3080	2695	$\pi/2$ QPSK	1	LDPC(672,588) R=7/8
13	5280	4620	$\pi/2$-16QAM	1	LDPC(672,504) R=3/4

From Table 1 and Table 2, we can obtain that, when the pilot length is 64, the data transmission rates of SC MCS7、8、9、13 are the same with that of HSI MCS1、2、3、5.Using the MCS in Table 1 and Table 2, under the same data transmission rates, we compare the BER performance of SC-FDE and OFDM over 802.15.3c channels.

From Figure 5(a), we can see that the performance of SC-FDE 60GHz system is close to that of OFDM 60GHz system in LOS channels. With he increasing of coding rate and modulation order, the performance of SC-FDE 60GHz system has advantage over that of OFDM 60GHz system. This can be observed when SNR is large, but the advantage is small. When BER is 10^{-6} even with 16QAM high order modulation and high coding rate, the gap between them is less than 1dB. SC-FDE 60GHz system has a slight advantage in the high-speed transmission in LOS channels, but the performance of OFDM 60GHz system can be compared with that of SC-FDE system.

Figure 5(b) shows that, the performance of OFDM 60GHz system is better than that of SC-FDE 60GHz system in NLOS channels when SNR is low and QPSK (1/2LDPC) is used, and with the increasement of SNR, the performance of SC-FDE 60GHz system is slightly better than that of OFDM 60GHz system. When QPSK (3/4 LDPC) and QPSK (7/8 LDPC) are used, the performance of SC-FDE system is close to that of OFDM system. When 16QAM (3/4 LDPC) is used, the performance of OFDM system has a slight advantage over that of SC-FDE system, and the gap between them is less than 0.5dB when BER is 10-6. By comparing Figure 5, we can obtain that, SC-FDE has a slight advantage over OFDM 60GHz system in LOS channels, OFDM 60GHz system has a slight advantage over SC-FDE 60GHz system in NLOS channels. It shows that OFDM 60GHz system has a slight advantage in overcoming multipath fading, but the performance of both is close whether in the LOS or NLOS case.

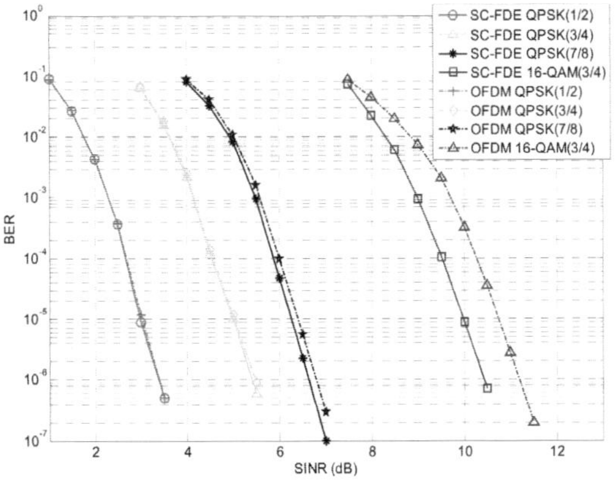

a)The BER performance simulation in CM1.3

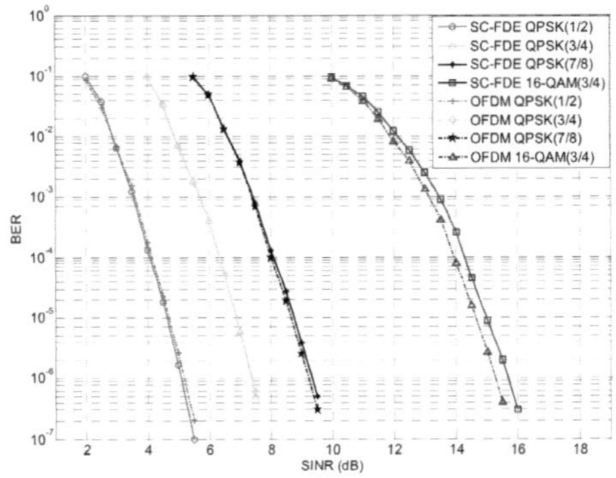

b)The BER performance simulation in CM2.3

Figure 5. The BER performance comparison of SC-FDE 60GHz system and OFDM 60GHz system under 802.15.3c CM1.3 and CM2.3 channel.

4. Conclusions

The system models of OFDM 60GHz and SC-FDE 60GHz are established respectively, and their principle are elaborated. Physical layer communication scheme under 802.15.3c standard is analyzed, and the BER performance of OFDM 60GHz system and SC-FDE system in 802.15.3c channels is compared and simulated respectively based on 802.15.3c standard and channel model. The simulation results show that, SC-FDE has a slight advantage over OFDM in LOS channels, while OFDM has a slight advantage over SC-FDE in NLOS channels. For 60GHz system, OFDM has a slight advantage over SC-FDE in overcoming multipath fading, but the performance of both is close whether in the LOS or NLOS case.

ACKNOWLEDGEMENTS

The authors would like to thank the referees and editors for providing very helpful comments and suggestions.

References

1. S.K. Yong, and C.C., Chong. An overview of multi-gigabit wireless through millimeter wave technology: Potentials and technical challenges [J], EURASIP Journal on Wireless Communications and Networking (2007), Vol.2007, pp.1~10.
2. R.C. Daniels, and R.W. Heath, 60 GHz wireless communications: emerging requirements and design recommendations [J], IEEE Vehicular Technology Magazine (2007), Vol.2, No.3, pp.41-50.
3. N. Guo, R.C. Qiu, S.S. Mo, and K. Takahashi. 60 GHz millimeter-wave radio: Principle, technology, and new results [J], EURASIP Journal on Wireless Communications and Networking (2007), Vol.2007, pp.1~8.
4. H. Sawada, Y.Shoji, C. S. Choi, K. Sato, R. Funada, H. Harada, and H. Ogawa, LOS office channel model based on TSV model. IEEE 802.15-06-0377-00- 00-3c, September, 2006.
5. T.Pollock, Office 60GHz channel measurements and model. IEEE 802.15-06-0316-00 -00-3c, July, 2006.
6. N. Moraitis, and P. Constantinou, Measurements and characterization of wideband indoor radio channel at 60GHz [J], IEEE Transactions on Wireless Communications (2006), Vol.5, No.4, pp:880-889.
7. IEEE standards 802.15.3c™ Part 15.3: wireless Medium Access Control (MAC) and PHYsical (PHY) layer specifications for high rate Wireless Personal Area Networks (WPANs) amendment 2: millimeter-wave-based alternative physical layer extension[S]. New York, USA, IEEE Computer Society, 2009.
8. IEEE P802.11.ad/D9.0, Part 11: wireless LAN Medium Access Control (MAC) and Physical Layer (PHY) Specifications. IEEE, July, 2012.
9. A. Bourdoux, J. Nsenga, W.V.Thillo, P.Wambacq, and L.V. der Perre, Gbits/s radios @60GHz:To OFDM or not to OFDM?2008 IEEE 10th International Symposium on Spread Spectrum Techniques and Applications, 2008, pp.560-565
10. R.V. Nee, and R. Prasad, OFDM for wireless Multimedia Communications, Artech House, Boston, 2000.
11. F. Horlin, and A. Bourdoux, Digital Front-End Compensation for Emerging Wireless Systems, John Wiley& Sons, Ltd (2007).

The Design of Bulk Carrier Cargo Holds State Integrated Monitoring System

Gao Ru-jiang, Wang Dan-li, Zou Chun-ming*, Xu Yan-min, Chen Min, Jin Cheng

School of Navigation, Wuhan University of Technology, Wuhan, 430063, China
Corresponding author: Zou Chun-ming

Abstract: Most ship cargo hold Internal uses artificial watch or is unattended. Therefore, it is impossible to know the appropriate information of the cargo holds Internal timely and accurately. Cargo damage and ship accidents occurred frequently. Automation remote processing and monitoring alarm system for the bulk carrier is an important part of the marine automation. The system plays a significant role to guarantee the navigation safety for bulk carriers. The paper introduced the important parts of the integrated monitoring system, structural design, hardware configuration, various modules communication transmission and various data processing software design were included. Based on embedded development, the real time information including the cargo hold internal humidity temperature, oxygen concentration, smoke gas concentration, cold water well level and other data were collected, as well the hatch closed state was detected and the cargo hold internal real time video information was collected. Then the real-time communication between the control display and acquisition modules were assured. By adjusting the corresponding buttons on the bridge according to the monitoring information, so that the cargo hold always in a safe environment, so as to avoid cargo hold accidents.

1 Introduction

With the rapid development of industry, science and technology, the automation[1] of ship equipment and systems continues to increase. Ship centralized monitoring and alarm systems[错误!未找到引用源。] as an important part of the ship automation, it is necessary to guarantee the safe navigation of the ship. In the world shipping market, bulk shipping accounted for about 30 percent of total global volume of bulk cargo shipping[错误!未找到引用源。] have large carrying capacity, types of cargo transportation complex characteristics, especially the ship transports grains , feed and cargoes, It is very important to monitor the cargo hold temperature, humidity, oxygen concentration, cold water well level and other status information, and also very necessary to get the overall status of the cargo hold and cabin hatch real-time video information. Currently, in the whole bulk shipping industry, having little research on the status of the cargo cover and cabin hatch monitoring, and now most marine detectors are using foreign products including cabin leak detectors, leak detectors, water tightness detectors, and these products are only covered diagnosis for cargo covers but no cabin condition monitoring, and can not effectively predict the alarm, even can not realize remote manual or automatic processing. Therefore, for bulk cargo ship status monitoring study still in the relatively backward level. There are many accidents including cargo damage, wrecks and other accidents have been happened just because of having not comprehensive and timely understanding the cargo status information. Such as the "ZHE Chang-xing 6266" cargo ship aground accident on the September 7, 2014. It was due to careless crew forgot to close the inlet valves while loading coal, then a large salt water poured into the cabin, causing ship rapidly sunk. If the cargo status could be got in advance, then the accident would be avoided and reduced the loss of the ship and cargo. Thus, with the rapid development of marine transport economy, the importance of studying bulk cargo hold state comprehensive monitoring has become increasingly prominent.

2 System design ideas

With the large-scale promotion and application of computers embedded development, ship automation system equipment is also rising, but the ship automation on liquid cargo tanks错误!未找到引用源。 and ship engine room错误!未找到引用源。 are applied more than bulk cargo ships. By bring ship tanks and cabin condition monitoring into the ship's cargo hold, using the STM32 错误!未找到引用源。 and DSP 错误!未找到引用源。 series of microcontrollers to develop, gathering the cargo temperature and humidity degrees, oxygen concentration, fire and smoke gas concentration, cold wells level and other status information, and by way of wired communications of 485MHZ and 2.4G wireless communications to achieve real-time data transmission. At last, the monitoring and early warning and automatic processing are achieved by the remote video information on the bridge.

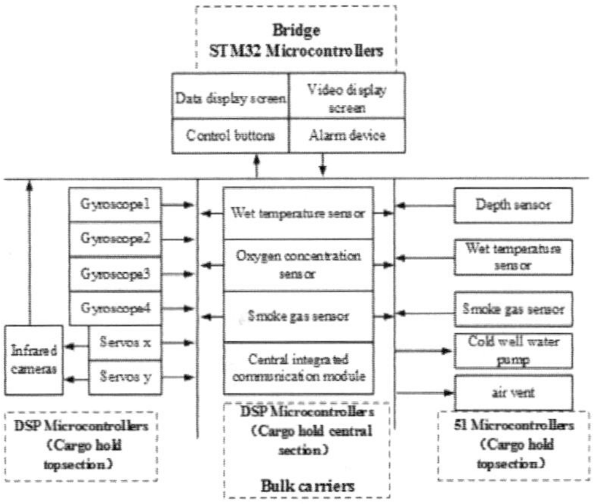

Figure 1. System Diagram

Seen from Figure 1, the system is composed of bridge controls display section, cabin top acquisition control section, central communication acquisition, side acquisition control section. System is based on STM32, DSP and 51 Microcontrollers to develop, by reasonably distributing cabin sensors including temperature sensor, humidity sensor, oxygen concentration sensor and other sensors to make up an embedded unit. The embedded modules are constituted based on data acquisition and processing and Microcontrollers performance computation. Each module between Lower Machine of the inner cabin uses 2.4G wireless communication method for data transmission, the central module belongs to the host part of the cargo is responsible for two-way data transmission of the top module and the side module and the control display communication section of the Position Machine. The RS-485 错误!未找到引用源。 serial bus standard communication mode was used on the cargo central module and bridge control display section, the standard mode uses balanced transmit and differential receive to suppress common mode interference. By using the principle of the three communications, the monitoring and control for real-time cargo was achieved well on the whole system. System top module, side and central module and display control module, see Figs. 2 and 3.

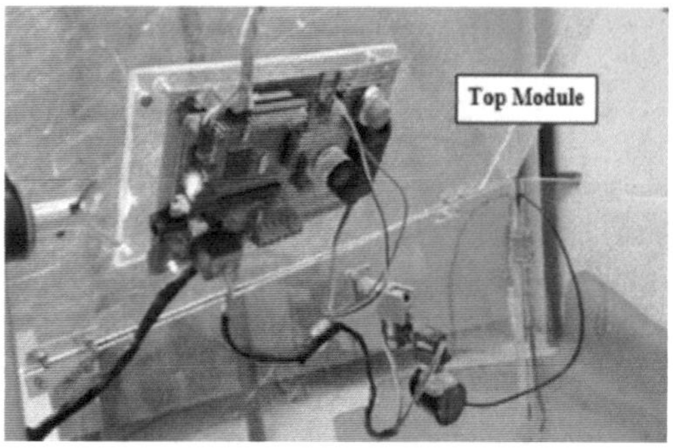

Figure 2. system top module

Figure 3. system side and central modules and display control module

3 System research methods

This paper uses system modularity step by step independent research method[错误!未找到引用源。]. System is made up bridge control display section, cargo hold central monitoring section, cargo hold top detection section, cargo hold side section. The system is divided into hardware acquisition and transmission part and software calculation aspects. Here's what the work process to achieve hardware system development and PC software calculation process.

The hardware designs mainly related to the system power supply, pins link of communication module, anti-jamming circuit. PCA82C250 is 5V power supply chip, by using the CPLD to make level conversion for CANTX and CANRX, then 5V will be converted to 3.3V accepted by DSP. Since the ships are used in the engine room automation systems, the anti-jamming capability was enhanced.

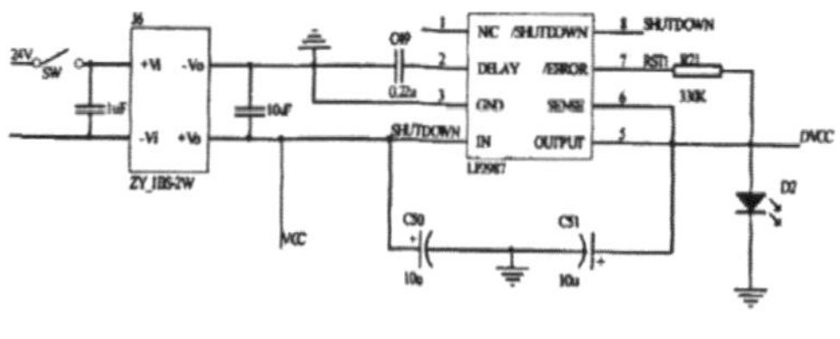

Figure 4. Power systems

The system uses a standard RS-485 serial bus communication mode, shown in Figure 4. The voltage and impedance are defined within the transmission processing, but the software protocols not. Different from RS232, RS485 features include:

(1) RS-485 electrical characteristics: logic "1" means a voltage difference between the two lines + (2 ~ 6) V, logical "0" means the voltage difference between the two lines - (2 ~ 6) V. The interface signal level lower than RS-232-C, therefore, it is not easy to damage in the chip interface circuit, and the level compatible with TTL level, can be easily connected with TTL circuits.

(2) The maximum data transfer rate of RS-485 is up to 10Mbps.

(3) RS-485 interface is a combination of balanced drivers and differential receivers, enhanced resistance to common mode interference, which has a good noise immunity.

(4) The maximum standard transmission distance of RS-485 interface is 4000 feet (about 1219 meters), in fact, it can be up to 3000 meters, in addition, RS-232-C interface on the bus only allowed to connect just one transceiver, that is a single station ability. However, RS-485 interface on the bus is allowed to connect up to 128 transceivers. That is a multi-station capability, so users can take advantage of a single RS-485 interface to easily set up the device network. RS-485 can be constituted distributed systems by network. The "nodes" RS-485 is mainly according to the "receiver input impedance".

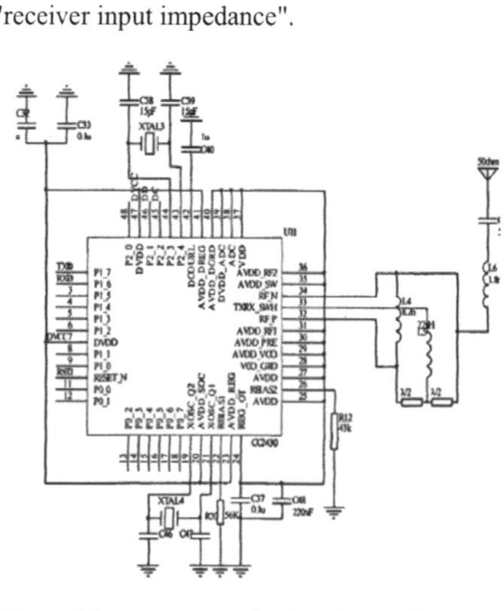

Figure 5. System communication pins configuration

System simultaneously uses the temperature and humidity sensors, oxygen concentration sensors, fire and gas sensors, ultrasonic sensors, gyroscopes. For the sensor data acquisition and processing affect the microcontroller CPU resources. Therefore, the sensor should be reasonable pins configuration. see Figs 5 and 6.

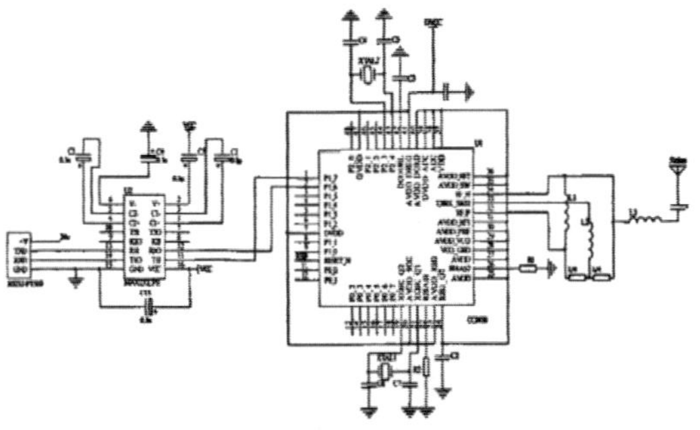

Figure 6. Sensor nodes configuration diagram

To sum up, the system power supply module, communication module, sensor configuration and anti-jamming circuit design are meet system requirements for real-time data acquisition and transmission, the system can be very good for a variety of real-time status information within the cargo hold monitoring and control. Stable and reliable hardware foundation provides a guarantee for the executing operation of software system.

4 System software design and implementation

Software design should be based on existing hardware resources and faced to applications requirements, to maximize improve the system functions. The design of the software system meets the requirements that the system collects many kinds of sensor data, real-time transmission requirement is high. The design of the software system makes monitoring to the network communication function as the center, both meet stable and reliable monitoring to the level of the cargo hold cold water wells, temperature and humidity, oxygen concentration and liquid fire gas concentration and other important parameters, but also to the achievement for smart combination of wired control network and sensor networks, as a result, both reliability and flexibility in two areas complement each other.

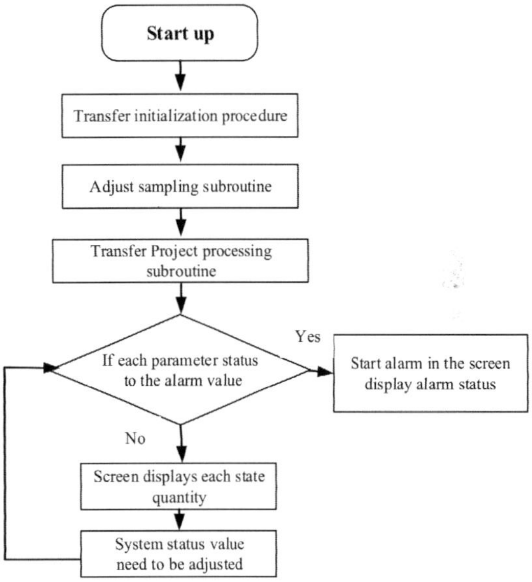

Figure 7. Software work flow chart

The monitoring software of control display section is a major component over the control layer. Control display section uses STM32F105VC model to develop, the display section in two modes, namely monitoring data display mode and monitor data setting mode. Different modes via a button on the control panel to switch. The system will display real-time status information when it is in monitoring data display mode. When there is abnormal situation occurs, the abnormal data will be marked, the operator can easily find the abnormal state quantity. When it is in monitoring data setting mode, it can set alarm threshold value for each state data by manually operating buttons, as shown in Figure 8.

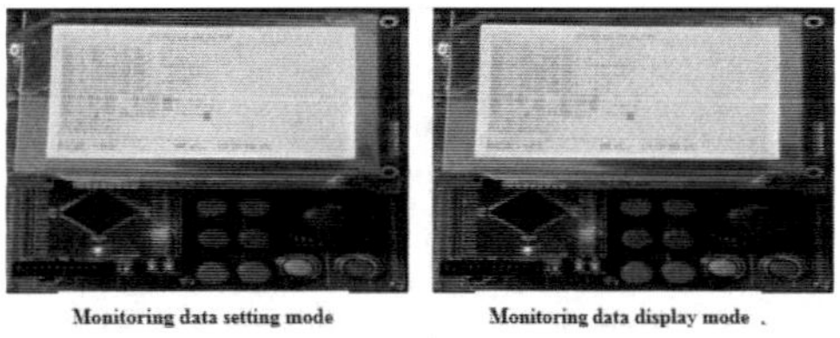

Figure 8. Two modes of Position Machine

The software of system control display module can easily get transmission data, and clearly displayed on the screen. The module uses 3.3V power supply, LCD high-brightness display, acrylic plastic spring button to ensure work reliability. In the Control Panel section, the cold well water pump switch within cargo hold and exhaust fan of the ventilation system can be manually controlled. By controlling cabin infrared camera shooting angles with joystick, the full range of real-time video cabin can be viewed well.

From the above analysis, the bulk cargo hold state monitoring system can connect and transport various of real-time information cabin. The data information is transmitted to bridge display section via RS-485 wired communication. Control display section has monitoring data display mode and monitoring data setting mode, operator can understand the real-time cabin temperature and humidity, oxygen concentration, the hatch closed, cold water well level and other status information, and can watch real-time video information cabin, and can set data alarm value in the environment of different goods to automatically monitor, the operator can manually or automatically control all cabin ventilation system exhaust fan and cold water well pumps. The entire system played a significant role in ensuring the safety of navigation.

5 System prospects

The main scope of application of the system is in the ship bulk cargo transportation, the logistics warehousing and management and other relevant aspects can also be widely applied. In the bulk cargo transport logistics warehousing will has a wide range of applications, and also corresponding industries has a big application market. Application to bulk shipping will greatly improve the safety of shipping process. It will bring great economic benefits in the ship operation safety and reduction of goods loss aspects.

References

1. Li Hong, Liu Yun, Lv Jian, Chen Yajie, Li Wenjie, Xu Bailing. Research and Application of the ship's engine room automation system generation software J diesel, 04 (2010): 21-23 + 33.
2. Zhang shengtao. Study on the modern ship monitoring and alarm systems D Instructor: Sun Jianbo. Dalian Maritime University, (2010).
3. Wang Yunlong, Ji Zhuoshang, Lin Yan. Bulk carrier status and the development trend J Ship Engineering, 01 (2006): 58-61.
4. Wu Wenbin. Design and Implementation of tanker cargo hold monitoring system software J Ship and Ocean Engineering, 04 (2010): 82-83 + 87.
5. Zhao Xiaobian. Application of liquid level gauging in the monitoring of ship loading and unloading safety aspects D. instructor: Du Jiali. Dalian Maritime University, (2013).
6. Cao Yujiao. Study on ship engine room automatic monitoring system J. China Water Transport (next fortnightly), 12(2010): 107-109.
7. Zong Yang, Wang Jianhua. Design of engine room automation monitoring system based on wireless Ethernet J East China Shipbuilding Institute (Natural Science), 01 (2004): 27-31.
8. Wang Lijie, Xu Dekai. Research on hydraulic disc brakes intelligent monitoring system based STM32 MCU J Manufacturing Automation, 01 (2013: 48-49 + 58.
9. Huang Peng, Yang Tianyi, Shi Weiren, Zhang Yuantao, Yin Lingling. Research on Marine Engine Room Automation CAN bus communication system based on DSP J Ship Engineering, 02 (2010): 51-54.
10. Qin Bingbing, Li Huawen, Qin Qin. The achievement of RS-485 serial bus and communication J mine warfare and ship protection, 02(2011): 37-41.
11. Yang Meiwen. Design and development of sluice monitoring system based on RS-485 bus D Instructor: Xu Zhixiang, Dalian University of Technology, (2006).
12. Qiu Yonghong, Wang Tingchang, Fan Jianhua, Guo Ming. Study on modular design standard system based on wireless communication system J PLA University of Technology (Natural Science),02 (2003): 11-16.

MATEC Web of Conferences **61**, 07015 (2016) DOI: 10.1051/matecconf/20166107015
APOP2016

Data Mining Based on Cloud-Computing Technology

Ren Ying[1], Lv Hong[1], Li Hua-wei[2], Zhou Li-jun[1] , Wang Li-na[1]

[1]Naval Aeronautical and Astronautical University,Yantai 264000,China;

[2]Shan dong Business Institute,Shandong Yantai 264001,China)

Abstract. There are performance bottlenecks and scalability problems when traditional data-mining system is used in cloud computing. In this paper, we present a data-mining platform based on cloud computing. Compared with a traditional data mining system, this platform is highly scalable, has massive data processing capacities , is service-oriented, and has low hardware cost. This platform can support the design and applications of a wide range of distributed data-mining systems.

1 Submitting the manuscript

With the rapid development of mobile Internet and the Internet of things, huge amounts of data are produced in every minute. Data has penetrated into each field of industry and business functions. In the age of big data,if we want to excavate implicit useful information from nonholonomic, massive,noise and random data,wo must improve the efficiency data mining algorithm. Cloud computing is to provide dynamic resource,virtualization and high available computing platform.Cloud computing into data mining can solve the efficiency problem of massive data mining.

2 Data mining and cloud computing technology

Data mining[1-3] is the technology finding the valuable information from large amounts of data by the analysis of the data.The data mining process base on cloud computing technology is basic consistent with the traditional data mining ,which are made up of data preparation,data mining,evaluation results of three stage and explain the composition.With the information age development resulting from the "big" data,data mining tasks will bring forth the new through the old,to emphasis on a large database of efficient and scalable data mining technology.

Cloud computing[4-5]distributed the tasks to a large number of computers resources pool,so that all applications can access computing power, storage space and information service according to the needs of. At the same time, cloud computing is the development of distributed computing and grid computing, parallel computing[6].Cloud computing usually consists of the following 3 levels of service: Saas,Iaas,Paas. Service model as shown in Figure 1.

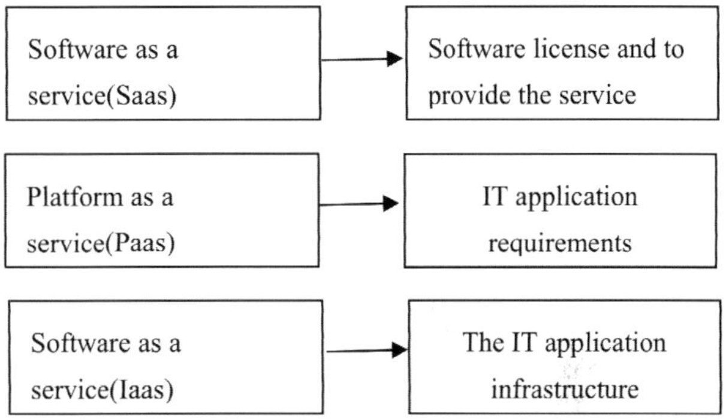

Figure 1. The service model of cloud computing

Cloud computing technology requirements to put compute nodes and storage nodes together.Task scheduling assigns and executes tasks on the preservation of equipment corresponding input file blocks as far as possible. This method makes the most of the parallel tasks to read the input data on the local machine, reducing the network data flow effectively[7].

Distributed computing is one of the effective means to solve the mass data mining tasks and improve the mass data mining[8]. Cloud computing platform provides a distributed file storage and parallel computing ability,which is a good solution to the distributed memory contained in distributed computing and parallel computing in two levels of content[9]. A good framework for the construction of cloud computing data mining platform core support ability of distributed file system and distributed parallel computing[10].

The popular distributed file system has Google file system (GFS), distributed file system(HDFS),the file system(KFS),which can effectively solve the problem of massive data storage.

3 Cloud based data mining algorithm

The algorithm of data mining is the soul,only the most efficient data mining algorithm in order to better accomplish the task of data mining. But because of a variety of data mining algorithms, there are also many data types, the requirements of different types of data mining algorithm is not the same.The most commonly used data mining algorithm has the following categories.

3.1. Classification algorithm

The main purpose of the classification algorithm is based on the existing data sets for mining to find the other data, and analysis of existing data sets and the discovery of new data, and then find the principle of data classification. This principle can be used to classify the data after adding. Classification algorithms are suitable for relational data consisting of tuple.

3.2 Cluster analysis

The main purpose of cluster analysis is find meaningful data distribution pattern of new from the potential data.The process is that the existing data is not specified grouping rules in advance,while is divided into different groups to miningn accordance with the data itself characteristics. Cluster analysis is also used for relational data consisting of tuple.

3.3 Association rules

The main purpose of the association rules is to find the interesting association or correlation between sets of items in large amounts of data. Association rules for data type is relatively more,mainly is suitable for the transaction type, transaction type and relation type data.Association rule best suited for handling the variable type is Boolean and numerical type.

Parallel mining algorithm is one of the key technologies which can effectively use the basic ability provided by loud computing platform.The general process of parallel data mining algorithms such as shown in figure 2.

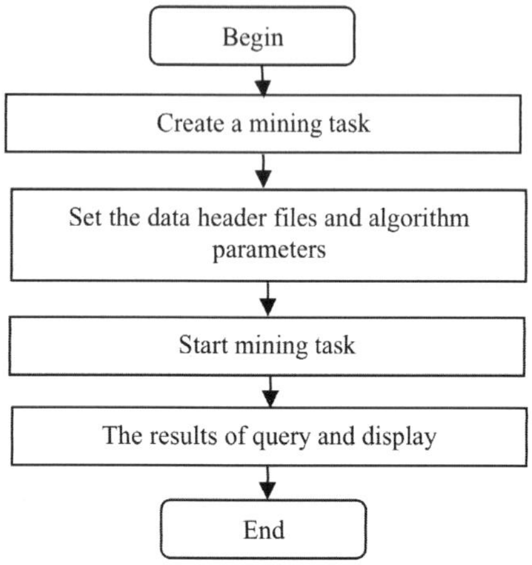

Fig 2 Parallel data mining algorithms perform general process

4 Based on the data mining system based on Cloud Computing

The data mining system based on cloud computing is built on the "cloud", transparent providing interface services for a variety of terminal users,providing an open interface for program based on for the development of the system.The user can use indirect the various service by calling the open interface provided by system.The user does not need to know how the system is to achieve, no need to worry about the computing and storage capacity of the system,only need to select the appropriate algorithm to process data,and ultimately to the way the task for system deployment area executed to get the data mining results[11].

The data mining system based on cloud computing can adopts on-demand pay way.Enterprises or individuals for a service can be directly through this platform to obtain,which do not have to buy expensive software.The data are mostly stored in storage cloud after the arrival of the cloud Era so that the mining tool based cloud computing platform based data become possible.

5 Cloud computing system architecture based on Data Mining

Cloud is a computational model based on the Internet, public participation, whose computational resources include computing power, storage capacity expansion and is virtualization, and is to provide services to the user. With the massive data increasing, diversification, and personalized data mining to strong demand,the traditional centralized data mining methods cannot adapt. Cloud computing become efficient way to solve the problem of massive data mining because of its huge storage capacity and computing ability of elastic changes.

Hadoop cloud computing framework is an open source distributed system architecture widely used[12]. Users can easily build private cloud platform. Because not needing to understand the development of distributed applications distributed the underlying details of the case, the user can make full use of the ability of cluster computing and high-speed storage.

The current cloud computing data analysis and processing widely use distributed development framework for dealing with similar MapReduce. It can execute in parallel massive data collection and analysis tasks in a large number of PC machine.This model can highly abstraction the complex operation in large -scale cluster parallel computing on process of to two functions: Map and Reduce[12].

In the stage of Map,the Map/Reduce framework will split the input data into a large number of data segments,and each data fragment is assigned to a Map task. Each Map task will be to its assigned to Key-Value to calculate, to generate an intermediate result,then all intermediate results with the same Key value of the Value pass to the Reduce function after the calculation.

In the stage of Reduce,each Reduce task take the two tuples Key-Value as input.Two tuple will call the Reduce function to merge with Value, and set the formation of a smaller Value,and each Reduce function call has only 0 or 1 value output. Each stage of the task execution are supporting fault tolerance.If one or more nodes appear error in the calculation of the process will be automatically re allocation of tasks to other nodes.

This paper designs the data mining system based on cloud computing technology,the overall structure as shown in Figure 3.The nodes in the system are divided into two categories:MainCtrlNode and WorkNode. MainCtrlNode in the system consists of NameNode, data warehouse,JobTracker,SecondaryNameNode,data mining algorithms library.WorkerNode consists of Task-Tracker,DataNode,which is responsible for actual storage and computational work.NameNode manages file system metadata,which is the main server of distributed file system and implement open,closed,operation,rename of the file system.DataNode is responsible for handling customer read and write requests,to store the actual data,in accordance with the NameNode command, performs the data block copy,delete,create work.We apply data mining to be used in the data set to uploaded to the data warehouse,NameNode will automatically block files and data redundancy storage to each DataNode. SecondaryNameNode assisted NameNode processing image files and transaction log.

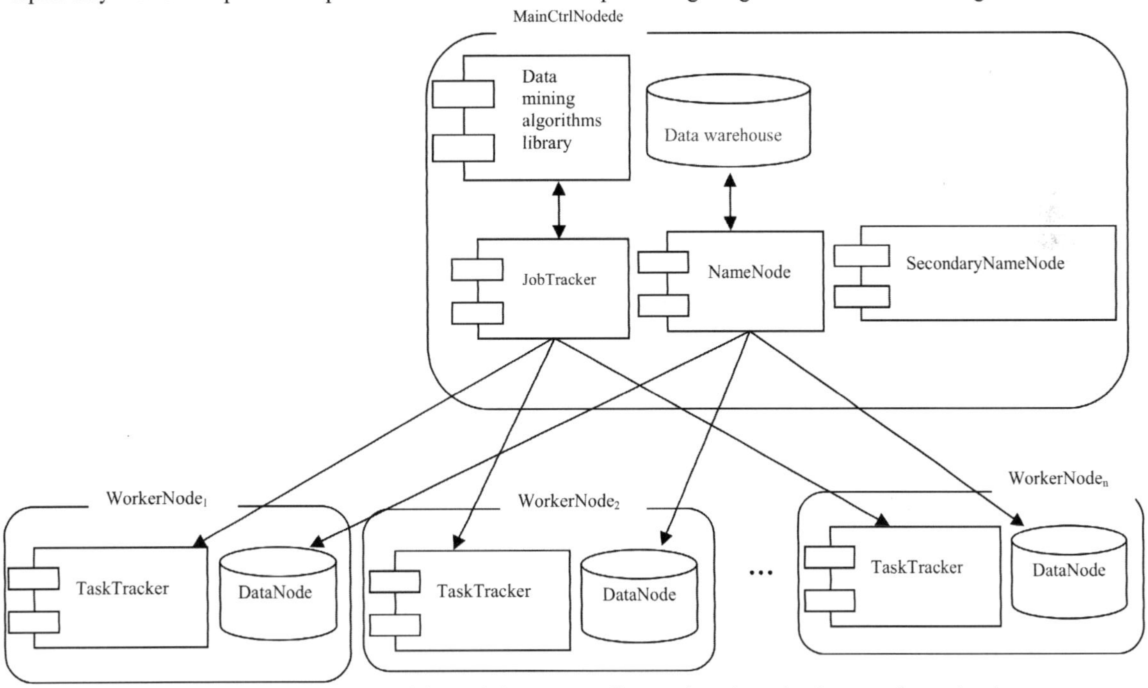

Fig 3 The overall architecture of data mining system diagram based on cloud computing technology

This paper designs the data mining system based on cloud computing technology,the overall structure as shown in Figure 3.The nodes in the system are divided Into two categories: MainCtrlNode and WorkNode. MainCtrlNode in the system consists of NameNode,data warehouse,JobTracker,SecondaryNameNode,data mining algorithms library.WorkerNode consists of Task-Tracker,DataNode,which is responsible for actual storage and computational work.NameNode manages file system metadata,which is the main server of distributed file system and implement open,closed,operation,rename of the file system. DataNode is responsible for handling customer read and write requests,to store the actual data,in accordance with the NameNode command, performs the data block copy,delete,create work.We apply data mining to be used in the data set to uploaded to the data warehouse,NameNode will automatically block files and data redundancy storage to each DataNode. SecondaryNameNode assisted NameNode processing image files and transaction log.

6 Conclusion

Massive data information and powerful computing and data processing capabilities of cloud computing provide powerful support for data mining. Through the analysis of the data mining and the cloud computing technology, this paper proposes the architecture of data mining platform based on cloud computing,for enterprise and individual user data mining task provides a good solution.

References

1. J Han,M Kamber. Data mining concepts andtechniques[M].Third Edition.San Francisco, CA,USA:Morgan Kaufmann Publishers,2012.
2. Shao feng-jing,Yu zhong-qing. Principle and algorithm of data mining[M].Beijing: Science Press,2009.
3. Shang Lin,Luo Bin. A data mining system based on Data Warehouse Framework[J]. Application Research of computers,2000,17(9):63-65.
4. Yang Yong,Dong zhen-jiang,Lu Ping. With the characteristics of cloud computing service delivery platform and its key technology research[J]. ZTE Communications,2011,17(5):55-57.
5. Wu zhu-hua. The analysis of the core technology of cloud computing[M].Bei Jing: People's Posts and Telecommunications Press,2011.
6. Mell P,Grance T. The NIST Definition of Cloud Computing ［R］.Gaithersburg,MD: National Institute of Standards andTechnology,2011.
7. Zhang jian-xun,Gu zhi-ming,Zheng chao. Review on research progress of cloud computing. 2010,27(2)：429-433.
8. Chen Quan,Deng qian-ni. Cloud computing and its key technology[J]. The computer applications,2009,29(9):2562-2567.
9. Li jian-jiang,Cui jian,Wang pin. MapReduce parallel programming model of review[J]. Chinese Journal of Electronics,2011(11):2635-2642.
10. Wang yi-jie,Sun wei-dong,Zhou Song. The key technology of distributed storage in cloud computing environment[J]. Journal of software,2012,23(4):962.
11. Wang Cong,Wang cui-rong,Wang xing-wei. The design of data center network architecture for Cloud Computing[J]. Research and development of computer,2012,49(2):286-293.
12. Hang He,Yi xiao-dong,Li shan-shan. Realization and evaluation of massive data processing platform for high performance computer[J]. Research and development of computer,2012,49:357-361.

Study of New Materials Design based on Hadoop

Wu Jun[1,2,a], Huang Zhixiong[1,b]

[1]School of Materials Science and Engineering, Wuhan University of Technology, Wuhan 430070, China

[2] School of Computer Science, Hubei University of Technology, Wuhan, 430068, China

[a] wujun@whut.edu.cn, [b] lavazza@foxmail.com

Abstract. With the rapid development of information technology, the scientific research shows that the data mining and other information technology could be used in the design of new materials. It is explicit that Intelligent Materials research focuses on using physical and chemical principles combined with computer techniques such as Big Data, Cloud computing and Intelligent modeling and simulation to solve chemical problems. In this paper, based on the cluster based outlier algorithm as the main body, this paper discusses the definition New Materials research In the Hadoop cloud platform, and the parallel processing of Map-Reduce model. The performance this model of new material was established by using the method of Map-Reduction provided the basis for the performance optimization.

Keywords: Materials Science, New Materials Design, Computational Methods, Data Mining, Hadoop

1 Introduction

In recent years, the scientific research shows that the data mining and other information technology could be used in the design of new materials. On one hand this new research method would help to make the new materials much more intelligent and useful than the traditional method, on the other hand it can get more ideal material with less experiment, and achieve the result with half the effort. From the first International Conference on Computer Aided Design of New Materials in 1990 to the Intelligent Materials developed in Big data time, it is explicit that Intelligent Materials research focuses on using physical and chemical principles combined with computer techniques such as Big Data, Cloud computing and Intelligent modeling and simulation to solve chemical problems[1-4].

Nowadays there are some international journals about this interdisciplinary research, such as Modeling and Simulation in Materials Science and Engineering, Computational Materials Science and so on. Nongnuch Artrith and Alexander Urbanb[5] have worked about Machine learning interpolation of atomic potential energy surfaces enables the nearly automatic construction of highly accurate atomic interaction potentials. Mansouri Iman and Ozbakkaloglu Togay[6] studies the ability of artificial neural network (ANN), adaptive neuro fuzzy inference system (ANFIS), multivariate adaptive regression splines (MARS) and M5 Model Tree (M5Tree) techniques to predict ultimate conditions FRP-confined concrete. Performances of the proposed models are also compared with those of the existing conventional and evolutionary algorithm models, which indicate that the proposed ANN, ANFIS, MARS and M5Tree models exhibit improved accuracy over the existing models. Cabaleiro Manuel, Riveiro Belen, and etc[7] have researched how to represent an important lack of material in the structural member. All the changes in the cross section of the beam must be considered in any kind of strength calculation and the protocols for the structural health monitoring. The results obtained demonstrate that the algorithm proposed is adequate for the automatic analysis of the geometrical properties in the section of timber beams with a lack of material or irregular section.

2 Computer Aided Material Design

Computer Aided Material Design is divided into two parts: First - Principle Calculations and Data Mining.

In the 1980s with the new material exploration and research when First - Principle Calculations method played an important role, it is important to discover those new materials, such as High temperature superconducting material, Super hard material, Nano material, Artificial low dimensional quantum structure materials. The first principle calculation is also called as Calculated based on the design of quantum theory. And its basic method has the solid quantum theory and the quantum chemistry theory. Especially suitable for the calculation and design of materials for atomic scale, Nano scale engineering, materials for many devices, and materials for electronic devices. Main research task in the aspect of material surface and interface would be as following work: reveal occur in the physical connotation of material surface and interface phenomena; how to use first principles method to calculate, design of surface and interface of physical chemical and dynamical processes. At present, the most powerful theoretical method is molecular dynamics simulation and Monte Carlo simulation. The key of the technique lies in the accurate calculation of the interaction potential between atoms.

However Data Mining is so different as First-Principle Calculations which be viewed as inductive method. Data mining is mainly defined as ''a non-trivial process of identifying valid, novel, potentially useful, and ultimately understandable patterns in data'', or ''the analysis of observational datasets to find unsuspected relationships and to summarize the data in novel ways that are both understandable and useful to the data owner''. As such, data mining and knowledge discovery are typically considered knowledge intensive tasks. Thus, knowledge plays a crucial role in Intelligent Materials research areas[8-10]. Through the data mining method would sum up the law and the method to obtain the required materials. In this paper, based on the Hadoop cloud platform, The performance prediction model of new material was established by using the method of data reduction and support vector machine or artificial neural network, which provided the basis for the performance optimization.

3 Data Mining and Hadoop

Calculation and design from the theory, although it can provide many useful clues, but the complexity of the actual material makes people tend to give priority to the induction method. Has been widely used in computational materials design, artificial neural network in data mining, pattern recognition, genetic algorithm, regression analysis and other algorithm, in recent years, the international has formed a group of computer software based on these methods. On the basis of experimental, using neural network method, the material mechanical properties and alloy composition are used as input of the network, material of other alloy composition and heat treatment temperature as the output of the network established mathematical model which can reflect the inherent law of the experimental data, in order to realize the optimization design of materials. The data mining method can also be used in the formulation design and control, forming, glazing, drying, sintering process and parameter control in ceramic industry, so as to achieve the goal of controlling and improving the quality of products.

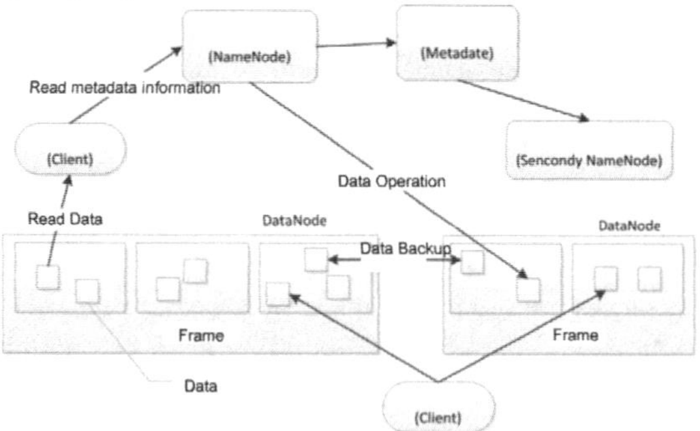

Fig1 Hadoop Architecture

Hadoop distributed file system (HDFS) is the bottom Hadoop file system, is designed by Google open source implementations of distributed file system GFS. HDFS Hadoop as one of the core components, and Hadoop HDFS data in blocks of storage in the cluster files. HDFS by superior fault tolerance, which can run on a large number of common storage hardware, to meet the data storage requirements, and has very strong extensibility, HDFS architecture figure. HDFS is a file system that can provide massive data storage and high throughput data access to the application, at the same time, she and the common file system, can be very convenient to achieve the file operation. Thanks to design architecture of HDFS master-slave (Master/slave), a HDFS cluster contains a node name (name node) and a series of data nodes (data node), which acts as a master role name node mainly responsible for the HDFS file system management, and accept the request sent by the client. As data node is slave role, is the main function of the data file is stored in HDFS by a file cut into one or more data block, the data block is storage in one or more data node, and the generated logical file storage structure on the name node nodes. HDFS file system in order to achieve high fault tolerance of the system, will be a number of copies of each data block to be stored in a number of different data nodes.

4 Map-Reduce Model and Experiment

Map-Reduce Hadoop as an open source implementation of Map-Reduce Google, is a high reliability, high fault tolerance capability of parallel computing software framework. Map-Reduce based applications can be run in large clusters of parallel processing large data sets. Simplifies concurrent programming model provides application programming interface (API) to the user through the map reduce, make not familiar with parallel computing users can easily develop map reduce application, and can reduce the amount of repeated work, map reduce execution flow chart is as follows Fig 2.

Fig 2 Map-Reduce structural work

Map-Reduce mainly consists of two core operations: mapping (Map) and protocol (Reduce). The mapping (Map) method is used to get a set of key value pairs (key/value) mapping into a new set of key value pairs (key/value). Protocol (Reduce) is used to ensure that all the keys to share the same key group. By running the Map-Reduce computing framework and Hadoop distributed file system (HDFS) on the same node, the cluster can use the network bandwidth to achieve efficient task scheduling. In the construction of the cipher text index structure should be considering the characteristics of HDFS file system, establishes the cipher text index compression and for small files for constructing reasonable with large files and the common keyword composition large inverted list for a reasonable segmentation, make full use of the characteristics of the Hadoop cluster computing, to achieve safe and efficient indexing, storage and query.

Acknowledgments

The research is supported by the National Science Foundation (NSF) under Grants (No.61502155, Hubei Natural Science Foundation under Grants (No. 2014CFB590）, Natural Science Foundation of Hubei University of Technology under Grant (No. BSQD13039), Wuhan University of Technology Hubei Key Laboratory of Transportation Internet of Things Foundation under Grants (No.2015III015-A03).

References

1. Rong Yu,Qi Zhang,Qian Zhan. Softest elastic mode governs materials hardness. Chinese Science Bulletin. 2014(15)
2. A.L. Ivanovskii. Mechanical and electronic properties of diborides of transition 3 d –5 d metals from first principles: Toward search of novel ultra-incompressible and superhard materials. Progress in Materials Science . 2011 (1)
3. Minghai Yao,Miao Qi,Jinsong Li,Jun Kong. A novel classification method based on the ensemble learning and feature selection for aluminophosphate structural prediction. Microporous and Mesoporous Materials . 2014
4. Qunyi Wei,Xiaodong Peng,Xiangguo Liu,Weidong Xie. Materials informatics and study on its further development. Chinese Science Bulletin . 2006 (4)

DOI: 10.1051/matecconf/20166107016

5. Nongnuch Artrith, Alexander Urbanb. An implementation of artificial neural-network potentials for atomistic materials simulations: Performance for TiO2. Computational Materials Science.2016(114): 135-150.
6. Mansouri Iman and Ozbakkaloglu Togay. Predicting behavior of FRP-confined concrete using neuro fuzzy, neural network, multivariate adaptive regression splines and M5 model tree techniques. Materials and Structures/Materiaux et Constructions, p 1-16, January 5, 2016
7. Cabaleiro Manuel, Riveiro Belen. Algorithm for the analysis of the geometric properties of cross-sections of timber beams with lack of material from LIDAR data. Materials and Structures/Materiaux et Constructions, 1-14, December 30, 2015
8. U.M. Fayyad, G. Piatetsky-Shapiro, P. Smyth, Advances in knowledge discovery and data mining, in: American Association for Artificial Intelligence, Menlo Park, CA, USA, 1996, pp. 1–34.
9. D. Hand, H. Mannila, P. Smyth, Principles of Data Mining, MIT Press, 2001.
10. C. Bizer, T. Heath, T. Berners-Lee, Linked data—the story so far, Int. J. Semant.Web Inform. Syst.5 (3) (2009) 1–22.
11. Amazon Elastic MapReduce, http://aws.amazon.com/elasticmapreduce/(2013).
12. The Apache Hadoop Framework, http://hadoop.apache.org (2013).
13. Aridhi, S.,d'Orazio, L. ,Maddouri,M.,MephuNguifo,E. Density-based data partitioning strategy to approximate large-scale sub graph mining. Inf.Syst.2015vol.48, 213–223.
14. Kang, U.,Faloutsos,C.,2013.Biggraphmining:algorithmsanddiscoveries.SIGKDD Explor.Newsvol.14(2),29–36
15. Aridhi S, Lacomme P, Ren L, et al. A MapReduce-based approach for shortest path problem in large-scale networks. Engineering Applications of Artificial Intelligence, 2015, 41:151-165.

MATEC Web of Conferences **61**, 07017 (2016) DOI: 10.1051/matecconf/20166107017
APOP2016

Fast Adaptive Beamforming with Smart Antenna for Radio Frequency Repeater

Chaoqun Wang[1,a], Guohua Wei[1] and Teng Wang[1]

[1] Research Institute of Radar Technology, School of Information and Electronics, Beijing Institute of Technology, Beijing 100081, China

Abstract. We present a fast adaptive beamforming null algorithm with smart antenna for Radio Frequency Repeater (RFR). The smart antenna system is realized by a Direction Of Arrival (DOA) Estimator, whose output is used by an adaptive beamforming algorithm to shape a suitable radiation pattern of the equivalent antenna; so that the co-channel interference due to retransmitting antenna can be reduced. The proposed adaptive beamforming algorithm, which has been proved by formulaic analysis and simulation, has a lower computation complexity yet better performance.

1 Introduction

In a wireless communication system, radio frequency repeater (RFR) is widely used to extend the network coverage due to easy setting-up, less expensive and without increasing the number of base station [1]. Block diagram of the repeater is shown in Fig. 1. As it depicted, RFR plays a transit role of signal enhancement; its basic function is a RF signal power amplifier. But RFR have a problem of signal feedback where the output of the retransmitting antenna is fed back to the benefactor antenna due to the high gain of the repeater. The feedback interference is amplified in the repeater in company with desired signals. Feedback and amplification process is repeated again and again, the repeater cannot work with the accumulation of the interference signal, finally [2, 3].

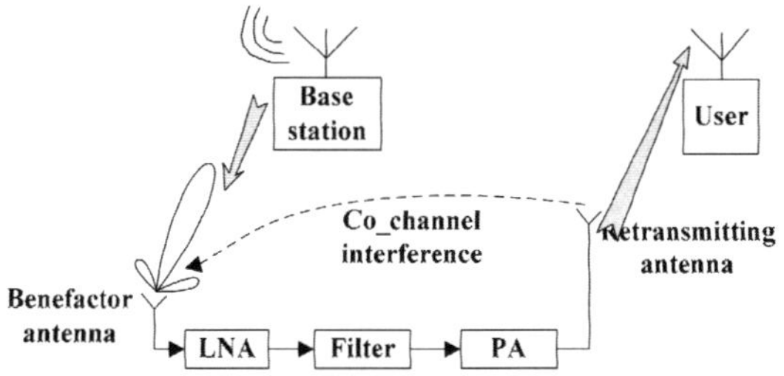

Figure 1. A radio frequency repeater with feedback interference signals.

There are many kinds of ways to solve this problem. The major method is to increase the isolation of benefactor antenna and retransmission antenna or reduce the gain of the repeater. But in fact, the former method usually increases the whole system cost greatly, and the latter can reduce the gain. So these two methods can't fundamentally solve the problem [4]. The second method is ICS (interference cancellation system) techniques. However, it requires lots of hardware complexity and computational delay [5-9]. To solve these problems, spatially adaptive filtering for RFR is studied, that is, with a certain shape of beam through the signal in the direction of the need to suppress the interference signal, which is also known as beamforming. In this paper, the beamforming algorithm based on minimum variance distortionless response (MVDR) criterion is studied. The traditional adaptive beamforming algorithm needs to construct covariance matrix of received signal and do matrix inversion directly, which is computation-intensive and time-consuming [10-12]. Formulaic analysis and simulation demonstrate that the proposed adaptive beamforming algorithm has a lower computation complexity yet better performance.

2 Adaptive BEAMFORMING

2.1. Smart antenna system

The smart antenna system needs to differentiate the desired signal from the co_channel interferences and normally requires either the knowledge of a reference signal (or training signal), or the direction of the desired signal source. There exists a range of schemes to estimate the direction of sources with conflicting demands of accuracy and processing power. Similarly, there are many methods and algorithms to update the array weights, each with its speed of convergence and required processing time. The basic functional diagram of an adaptive beamformer is shown in Fig. 2.

[a] Corresponding author: wcqan@163.com

2.2. Adaptive beamforming

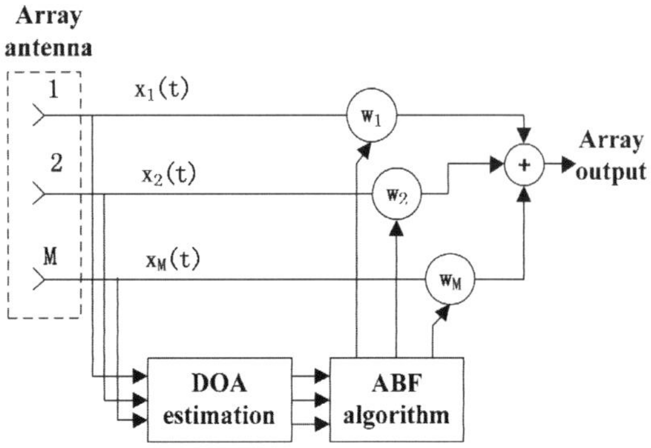

Figure 2. Functional block diagram of smart antenna system.

Adaptive beamforming is a commonly employed technique that enables system operation in an interference environment by adaptively modifying the system's antenna pattern so that nulls are generated in the angular locations of the interference sources. This approach is applicable to scenarios where multiple antenna elements are individually weighted to produce a desired directivity pattern. A typical method of forming the adaptive weights is via the MVDR algorithm, which implements a single linear constraint that maintains unit gain in the desired direction.

Considering a receiving antenna array with M array elements, the received signal is

$$\boldsymbol{X}(t) = \boldsymbol{a}(\theta_d,\varphi_d)\, d(t) + \boldsymbol{a}(\theta_e,\varphi_e)\, e(t) + n(t) \quad (1)$$

where $\boldsymbol{X}(t) = [x_1(t), x_2(t),\ldots, x_M(t)]^T$ is the complex vector of received signal; $d(t)$, $e(t)$ and $n(t)$ are the desired signal, interference and noise components, respectively. Here $\boldsymbol{a}(\theta_d,\varphi_d)$ and $\boldsymbol{a}(\theta_e,\varphi_e)$ are the desired signal steering vector and the interference steering vector, respectively.

The traditional optimal weight vector of MVDR is

$$W^H = \boldsymbol{a}^H(\theta_d,\varphi_d) R_{xx}^{-1} [\boldsymbol{a}^H(\theta_d,\varphi_d) R_{xx}^{-1} \boldsymbol{a}(\theta_d,\varphi_d)]^{-1}. \quad (2)$$

R_{xx} is the covariance matrix of received signal, whose size is M × M. For large antenna array, the calculation quantity is considerable.

For RFR, the interference signal is a direct feedback from retransmission antenna to receiving antenna, so $e(t)$ can be expressed as

$$e(t) = d(t) \times h(t) \quad (3)$$

where, $h(t)$ is the system function of the interference system, which can be equivalent to an enlarged and time-delayed system. For the benefactor antenna array whose maximum size of the physical aperture is 1 meter, the far field requirement is more than 5 meters when the signal wavelength is 800MHz. In the same repeater, the distance between the benefactor antenna and the retransmitting antenna can reach tens of meters, so the interferences can be regarded as plane waves.

The optimal weight vector of MVDR for RFR is the solution of the following optimization problem:

$$\min W^H Q W \text{ subject to } W^H \boldsymbol{a}(\theta_d,\varphi_d) = 1 \quad (4)$$

where

$$N_c(t) = \boldsymbol{a}(\theta_e,\varphi_e)\, e(t) + n(t) \quad (5)$$

$$Q = \mathrm{E}[N_c(t) N_c^H(t)]. \quad (6)$$

Here $Nc(t)$ is the combined noise (interference signal and system thermal noise) and Q is the covariance matrix of the combined noise.

The optimization problem in (4) has an elegant closed-formsolution given by:

$$W^H = \boldsymbol{a}^H(\theta_d,\varphi_d) Q^{-1} [\boldsymbol{a}^H(\theta_d,\varphi_d) Q^{-1} \boldsymbol{a}(\theta_d,\varphi_d)]^{-1}. \quad (7)$$

The covariance matrix of the combined noise is

$$Q = \boldsymbol{a}(\theta_e,\varphi_e) P_e \boldsymbol{a}^H(\theta_e,\varphi_e) + \sigma_n^2 I \quad (8)$$

where $P_e = \mathrm{E}[e(t)\, e^H(t)]$ is the covariance matrix of interference, σ_n represents the power of noise, I is an identity matrix.

According to the matrix inversion lemma, the inverse of covariance matrix can be written as

$$Q^{-1} = \{I - \boldsymbol{a}(\theta_e,\varphi_e) [I + P_e/\sigma_n^2 \boldsymbol{a}^H(\theta_e,\varphi_e) \boldsymbol{a}(\theta_e,\varphi_e)]^{-1} \times P_e/\sigma_n^2 \boldsymbol{a}^H(\theta_e,\varphi_e)\}/\sigma_n^2\}. \quad (9)$$

Substituting Q^{-1} into (7), we can obtain the weight vector:

$$W^H = \Gamma\, \boldsymbol{a}^H(\theta_d,\varphi_d) [I - \boldsymbol{a}(\theta_e,\varphi_e) (1/INR + r_{ee})^{-1} \boldsymbol{a}^H(\theta_e,\varphi_e)]. \quad (10)$$

where

$$\Gamma = \boldsymbol{a}^H(\theta_d,\varphi_d) [I - \boldsymbol{a}(\theta_e,\varphi_e) (1/INR + r_{ee})^{-1} \boldsymbol{a}^H(\theta_e,\varphi_e)] \times \boldsymbol{a}(\theta_d,\varphi_d) \quad (11)$$

$$INR = P_e/\sigma_n^2 \quad (12)$$

$$r_{ee} = \boldsymbol{a}^H(\theta_e,\varphi_e) \boldsymbol{a}(\theta_e,\varphi_e). \quad (13)$$

Here Γ is the normalized coefficient, INR is the input interference to noise radio.

For large INR, (10) and (11) can be simplified as:

$$W^H = \Gamma [\boldsymbol{a}^H(\theta_d,\varphi_d) - r_{de} r_{ee}^{-1} \boldsymbol{a}^H(\theta_e,\varphi_e)]. \quad (14)$$

$$\Gamma = \boldsymbol{a}^H(\theta_d,\varphi_d) [I - \boldsymbol{a}(\theta_e,\varphi_e) r_{ee}^{-1} \boldsymbol{a}^H(\theta_e,\varphi_e)] \boldsymbol{a}(\theta_d,\varphi_d) \quad (15)$$

where $r_{de} = \boldsymbol{a}^H(\theta_d,\varphi_d) \boldsymbol{a}(\theta_e,\varphi_e)$ is the spatial correlation coefficient between desired signal and interference. As $\boldsymbol{a}(\theta_e,\varphi_e)$ is a M × J (J is the number of retransmitting antenna) matrix, r_{ee} is a J × J matrix.

Two reasons are given to explain why the calculation of the proposed adaptive beamforming algorithm is faster than the traditional:

1) The matrix size of Q and r_{ee} is M × M and J × J,

respectively. For large receiving antenna array, the latter is significantly smaller than the former when performing matrix inversion.

2) Since the position of retransmitted antenna relatived to benefactor antenna is known for RFR, the arrival of the interference signal is known. Therefore, r^{-1}_{ee} can be calculated and stored in advance.

3 Simulation Results and Analysis

In this section, a series of numerical examples will be presented to validate the efficiency of the proposed method. In a RFR system, there are a relatively large number of interferers compared to antenna elements. Normally, to simplify the analysis, only the strongest interferers are considered. For our simulation, we consider the case of one or two interferers only. We employ uniformly circular array (UCA) with eight antenna elements having a spacing of λ, where λ denotes the signal wavelength. The geometry of UCA is shown in Fig. 3. Simulations are done for values of interference azimuths (-30°, 50°) and interference pitch angle (-35°, 10°). The azimuth and pitch angle of desired signal are 10°and 0°, respectively. The power of the input PN sequence is -29.12 dB, and the interference signal is 13.78 dB with the thermal noise is about -70 dB, that is, the input signal-to-interference-plus-noise ratio (SINR) is -39.48 dB.

The desired PN sequence is shown in Fig. 4 and parts signals received by the antenna array are showed in Fig. 5. This figure shows that the desired sequence is completely submerged by the interference signal, and then the receiving system cannot distinguish the useful signal. Under this circumstance, adaptive beamforming is necessary. Fig. 6 shows the comparison of error values of the output signal and the desired sequence between proposed algorithm and conventional MVDR algorithm. As we can see, these two methods can filter out the interference and recovery the desired signal, but the error values using the proposed algorithm is smaller than the conventional algorithm. The average error of the proposed algorithm is -44.47 dB, which is about 9 dB less than traditional algorithm.

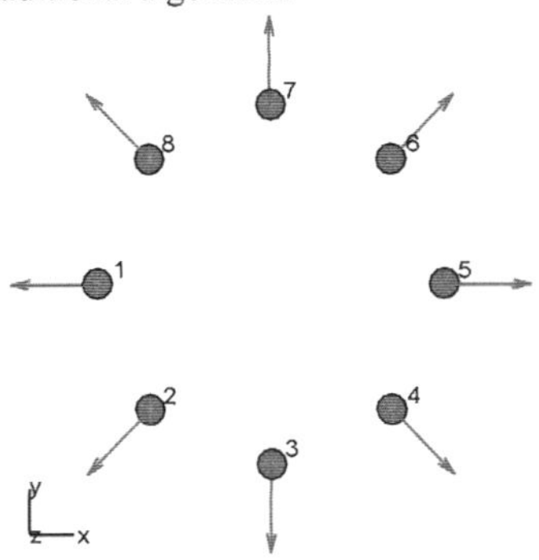

Figure 3. Geometry of UCA with eight antenna elements.

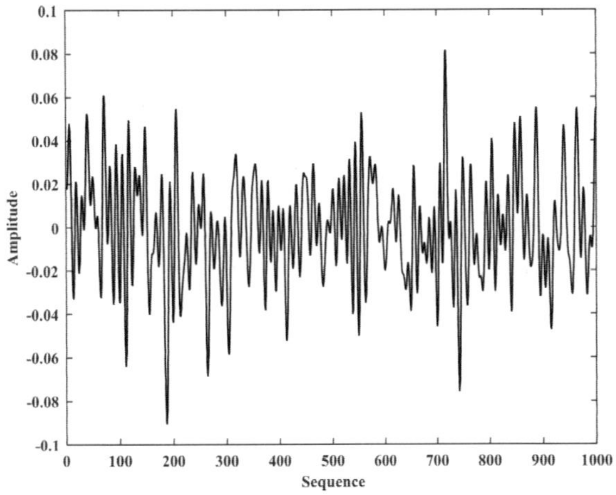

Figure 4. Desired PN sequence.

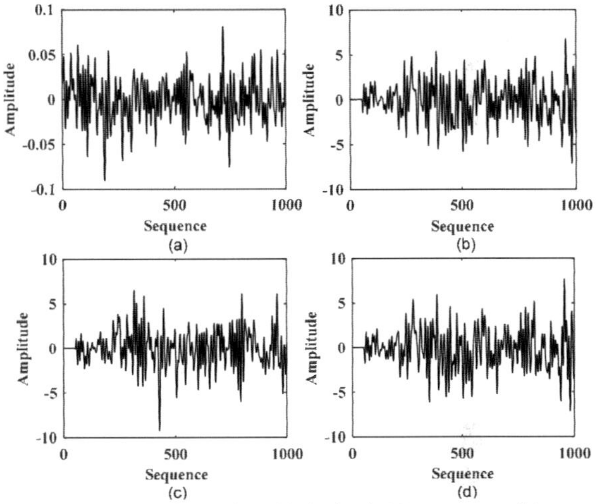

Figure 5. Comparisons for: (a) desired PN sequence, (b) signal received by the first antenna, (c) signal received by the second antenna and (d) signal received by the third antenna.

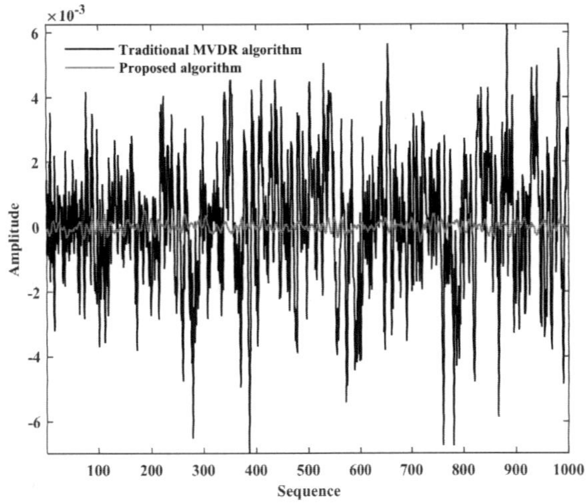

Figure 6. Comparisons of error values for: (a) traditional MVDR algorithm and (b) proposed algorithm.

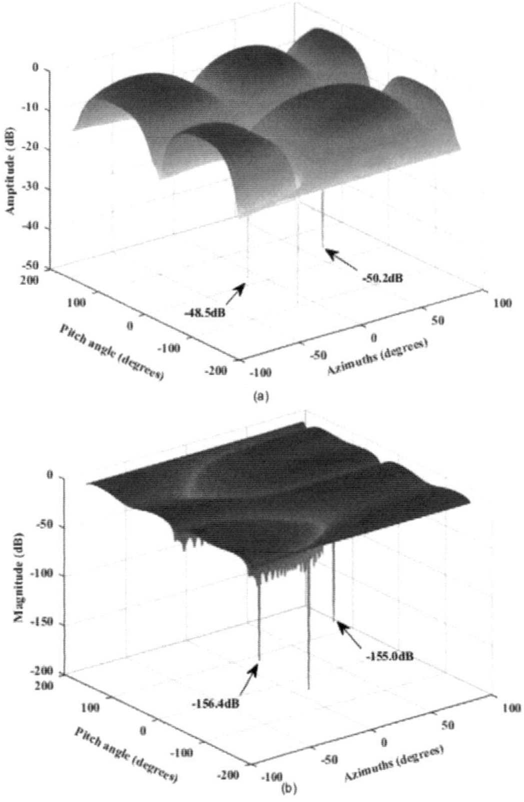

Figure 7. Antenna pattern formed by: (a) traditional MVDR algorithm, (b) proposed algorithm.

Fig. 7 compares the antenna pattern formed by traditional MVDR algorithm and proposed algorithm. There are two deeper nulls achieving -156.4 dB and -155.0 dB in the interference direction, as shown in Fig. 7(b). However, the nulls in Fig. 7(a) merely achieve -48.5dB and -50.2 dB. The output SINR of the proposed algorithm is 46.28 dB that is 10 dB more than traditional MVDR algorithm. In summary, the proposed method has better performance.

In order to further validate the efficiency of the proposed method, different direction of arrival for desired signal is studied as shown in Fig. 8. As we can see, the main beam of antenna pattern is moving with the direction of arrival. At the same time, two deep nulls is still maintaining in the direction of interference (70° and 110°).

4 Conclusion

In this paper, a fast adaptive beamforming algorithm based on MVDR criterion is studied for RFR. The analytic expression of the weight demonstrates the proposed adaptive beamforming algorithm is faster than the traditional, so it is more suitable for real-time application. Simulation results show that, in the ideal case, the nulls of the beam which are given by (15) can be completely zero, meaning that the interference direction can be fully zeroed. Meanwhile, this proposed algorithm can obtain better performance, compared with the traditional MVDR algorithm.

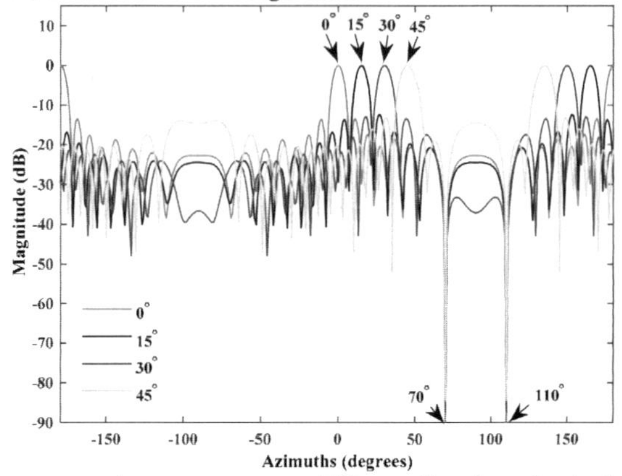

Figure 8. Antenna pattern at different direction of arrival for desired signal.

References

1. K. Salehian, M. Guillet, B. Caron, and A. Kennedy, "On-channel repeater for digital television broadcasting service," *IEEE Trans. Broadcast.*, vol. 48, no. 2, pp. 97–102, Jun. 2002.
2. S. J. Kim *et al.*, "Adaptive feedback interference cancellation system (AF-ICS)," in *IEEE MTT-S Dig.*, Jun. 2003, pp. 627–630.
3. E. Ahmed and A. M. Eltawil, "All-digital self-interference cancellation technique for full-duplex systems," *IEEE Trans. Wireless Commun.*, vol. 14, no. 7, pp. 3519–3532, July. 2015.
4. J. Wei, L. Ren, and J. Chen, "Study on repeater with adaptive interference cancellation," in *Proc. 8th WiCOM. Net. Mobile Computing*, 2012, pp. 1-4.
5. R. N. Braithwaite and S. Carichner, "Adaptive echo cancellation for an on-frequency RF repeater using a weighted power spectrum," in *Proc. 10th European Conf. Wireless Technol.*, Oct. 2007, pp. 82-85.
6. M. Lee, B. Keum, and Y. Serk Shim, "A scheme to compensate for the input multipath channel effect in an interference cancellation repeater for mobile communication systems," in Proceedings ATC, 2008, pp. 92-96.
7. T. Riihonen, S. Werner and R. Wichman, "Optimized gain control for single-frequency relaying with loop interference," *IEEE Trans. Wireless Commun.*, vol. 8, no. 6, pp. 2801–2806, Jun. 2009.
8. E. Habets, J. Benesty, I. Cohen, S. Gannot, and J. Dmochowski, "New insights into the MVDR beamformer in room acoustics," *IEEE Trans. Audio, Speech, Lang. Process.*, vol. 18, no. 1, pp. 158–170, Jan. 2010.
9. J. Y. Choi *et al.*, "Interference cancellation techniques for digital on-channel repeaters in T-DMB system," *IEEE Trans. Broadcast.*, vol. 57, no. 1, pp. 46–56, Mar. 2011.
10. S. I. Park, Y. J. Lee, H. M. Kim, and H.N. Kim, "Equalisation digital on-channel repeater with a

feedback interference canceller for the advanced television systems committee terrestrial digital television system," *IEEE Trans. Commu.*, vol. 7, no. 16, pp. 1769–1776, Aug. 2013.

11. C. Pan, J. D. Chen, and J. Benesty, "Performance study of the MVDR beamformer as a function of the source incidence angle," *IEEE/ACM Trans. Audio, Speech, Lang. Process.*, vol. 22, no. 1, pp. 67–79, Jan. 2014.
12. R. R. Qian, M. Sellathurai, and D. Wilcox, "A study on MVDR beamforming applied to an ESPAR antenna," *IEEE Signal Process. Lett.*, vol. 22, no. 1, pp. 1070-9908, Jan.2015.
13. J. J. Blanz, A. Papathanassiou, M. Haardt, I. Furio, and P.W. Baier, "Smart antennas for combined DOA and joint channel estimation in time-slotted CDMA mobile radio systems with joint detection," *IEEE Trans. VT.*, vol. 49, no. 2, pp. 293- 306, Mar. 2000.
14. S. Hwang, S. Burintramart, T. K. Sarkar, and S.R. Best, "Direction of arrival (DOA) estimation using electrically small tuned dipole antennas," *IEEE Trans. Antennas Propagat.*, vol. 54, no. 11, pp. 3292–3301, Nov. 2006.
15. J. P. González and B. Rodríguez, "Interference rejection degradation in function of the DOA in a beamforming system," *IEEE Latin America Trans.*, vol. 13, no. 1, pp. 48-53, Jan. 2005.

A solution of High-precision WLAN positioning based on TDOA and PTP

Junjie Zhou [1,a], Yue Ke[2] , Kan Yu[1] and Di Tian[1]

[1]Wenhua College, Huazhong University of Science and Technology, Wuhan, China
[2]School of Computer, Hubei University of Technology, Wuhan, China

Abstract. For indoor WLAN positioning it is difficult to achieve decimeter or centimeter-level precision questions, to IEEE1588v2 (PTP) precision network clock synchronization technology and the arrival time difference principle (TDOA), combined with non line of sight (NLOS) error mitigation techniques ranging research and design a set based on the existing WLAN (IEEE802.11x series of standards) device with high precision real-time location solutions. First, build a high-performance software-defined radio network communications platform, for accurate measurement of the radio signal arrival time and is calculated based on TDOA location. Secondly, the use of high-precision clock IEEE1588v2 standard equipment PTPGrand-2100 as a network clock source, through the integrated application of multi-hop latency compensation and asymmetric compensation algorithm to ensure nanosecond clock synchronization between the various WLAN access points (AP) accuracy(Experiments show that 1 nanosecond time synchronization error is equivalent to 10 to 30 cm of positioning error). Finally, based NLOS error becomes stability principle, filtered NLOS error and pre-positioned before the data, eliminate the influence of the greatest degree of NLOS error, the time domain error control in wireless ranging 5ns or less to achieve decimeter or more accurately estimate the position. The case can be widely used in mobile Internet, logistics management, mineral exploration, health and other needs of the precise location services (LBS) in other industry sectors.

Keywords: High-precision positioning; PTP (Precision Time Protocol);Time Difference of Arrival principle; WLAN; Non-line-ranging error

1 Introduction

With the advent of the "Internet of Things" era, the wireless location technology as a key technology, has caused the industry wide attention and domestic and international research. The wireless location technology has a very broad application prospects in many areas of military defense, traffic management, disaster relief, environmental hazards, automation workshop and logistics management. The wireless location services have broad market prospects, by 2020 in China the location-based services (LBS) market scale will reach 396.2 billion yuan[1]. Vigorously promote the rapid development of optoelectronic information industry, with the needs of social development and economic construction, more and more industries implementation of sophisticated management. The demand for high-accuracy wireless location technology will show rapid growth, particularly for wide-area decimeter accuracy real-time location, or urban areas of real-time meter level accuracy positioning technology, in the next few years.

The level of positioning accuracy and the size of positioning range of the wireless positioning technology has a direct influence on the performance of the positioning system[2]. As WLAN popularity, based on WLAN (IEEE802.11a / b) wireless location technology is the most common, is the lowest-cost way of positioning. However, the existing WLAN positioning system does not meet these requirements emerging decimeter or centimeter-level high-precision positioning applications. The existing WLAN positioning technology is mainly based on signal strength (RSSI) positioning. This kind of technology is greatly influenced by the signal propagation environment. Its general accuracy is not high and the error is usually more than 20 meters[3], which greatly exceeds the needs of the practical application of many high precision positioning. In the field of wireless precise positioning, the current domestic companies are generally used in the most simple RSSI positioning. So in the technology research and development, far behind the foreign counterparts. The Ekahau company of Finland through the acquisition of signal fingerprint way to get the industry's leading WLAN positioning accuracy, but it is only in the ideal situation to improve the positioning accuracy to 5 meters or so. The UK's Ubisense and Germany's Nanotron can provide decimeter or even centimeter level of high-precision wireless positioning solution[4]. But the two company's solution is not based on common WLAN signals, but based on industrial wireless communication

[a] Corresponding author: zjj0718@126.com
Fund Project: Hubei Provincial Department of Education 2015 annual science and technology research projects (No.B2015193); 2014 Hubei Provincial Higher Teaching research project (No. 2014424); 2014 Hubei Provincial Colleges and Universities outstanding young scientific and technological team plans funded project (No.T201431);2014 Wuhan Office of Information Industry wireless city construction of demonstration projects (No.GCJH2014005).

technology, such as UWB (ultra wideband) and Spread Spectrum Chirp (linear frequency modulation spread spectrum), which means that the two companies of the product can not be used together with ordinary wireless terminals. On the other hand, the current commercial WLAN devices or chips do not support the high accuracy of the signal arrival time record.

2 Comparative Analysis of WLAN positioning method

2.1 Overview of WLAN positioning technology

Wireless local area network (WLAN) based on IEEE802.11 protocol, based on the radio channel as the computer local area network transmission medium. Which is the combination of computer network and wireless communication technology. It uses wireless multiple access channel as the transmission medium, and provides the function of the traditional wired LAN . The WLAN can truly achieve users at any time, anywhere, broadbandInternetaccess[5].Mobile user demands for im mediacy of information and in-place more and more intense, which gives the location-based services WLAN systems provide a broad space for development. WLAN positioning system technologies include GPS / Compass satellite positioning, based TOA / TDOA / AOA triangulation, the signal strength (RSSI) positioning technology[5]. Among them, the signal strength positioning technology mainly includes two kinds of methods, signal strength fingerprint and signal propagation model.

2.2 Based on time/Angle measurement positioning

WLAN networking in general will have a lot of high density low power wireless devices, if all wireless devices did not achieve network clock synchronization, distance is also very difficult to achieve using TDOA estimation. Although it is possible by measuring the arrival time (TOA) to estimate the distance, but indoor environment in user space shorter, AP coverage is generally not more than 100 meters, the radio transmission delay can be neglected, and there is a more serious and diffraction and other non-linear propagation diffraction and multipath component pieces of the same user signal is very close in time, the need for improved resolution of the device to distinguish[6]. Therefore, accurate TOA and TDOA estimation, needs the help of more advanceddigital signal processing techn iques to achieve. In addition, many obstacles influencing signal transmission, the wireless signal reflection and scattering. Amplitude components of different paths, phase, time of arrival and the angle of incidence varies, so that the received composite signal has a serious distortion in both amplitude and phase. Therefore, the angle of arrival (AOA) is not suitable for indoor wireless positioning.

2.3 Based on signal strength positioning technology

Approximate perception, triangulation and scenario analysis are three basic principles to solve the position. Location Based on WLAN signal strength, based on the three principles of the specific positioning technology were the strongest base method, propagation modeling, and location fingerprinting method [7]. WLAN network environment based on the principles and architecture positioning system signal strength can refer to Figure 1.

The simplest method is to locate a wireless access terminal for data communication hotspots position, approximately as the estimated location of the wireless terminal, namely the strongest base station method. The precision of this method is limited by the coverage of AP, and it is difficult to pinpoint. The fingerprint positioning is in WLAN network environment under the condition of unchanged, according to different space a location change trend of wireless signal strength consistency rule the same time, based on scenario analysis principle, by matching the measured signals and the fingerprint similarity between to locate the user terminal. Propagation model positioning is based on the received signal strength varies with propagation distance spatial variability law (That is the signal strength path loss propagation) to determine the physical distance between the user terminal and the WLAN hotspot, which use the triangulation and calculation principles completion positioning. Table 1 is a typical example of the WLAN terminal's scan or listener's return to the AP and its related properties. However, based on the signal strength (RSSI) positioning technology is greatly affected by the signal propagation environment, the general accuracy is not high, the error is typically greater than 20 meters, far beyond the need for high precision positioning of many practical applications.

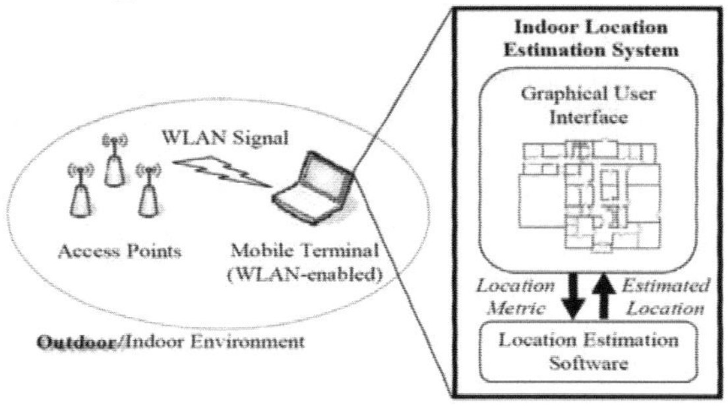

Figure 1. WLAN based on the principle architecture of signal intensity positioning system.

Table 1. AP and its attribute list sample returned by the terminal scan.

SSID	MAC	RSSI	Encrypted?
Capital City WiFi	00:0f:34:ab:0c:e0	-66 dB	No
First Security Trust	00:0f:f7:0c:e9:c0	-78 dB	Yes
First Security Trust	00:0f:f7:0c:4f:03	-105 dB	Yes

2.4 WLAN high accuracy positioning scheme b ased on TDOA and PTP

2.4.1 The time difference of arrival (TDOA) location technology

Time difference of arrival (TDOA) positioning, also known as hyperbolic positioning, and is measured by processing the data of three or more measuring stations acquired signals the arrival time measurement data of radiation source to locate[8]. In the two-dimensional plane, the radiation source signal arrival time difference between the two measuring station provides a pair of two stations focus hyperbolic. If you use three or four stations to form a two curve, then two of the two pairs of points of intersection, and then use the other auxiliary information to remove the ambiguity, which can determine the location of the radiation source [8]. In the three-dimensional space, at least three pairs of double-curved form to produce an intersection. Hyperbolic form geometric principles is: distance to two fixed points difference locus constant moving point is that two fixed points of hyperbolic focus on two hyperbolic plane intersecting point can be determined.

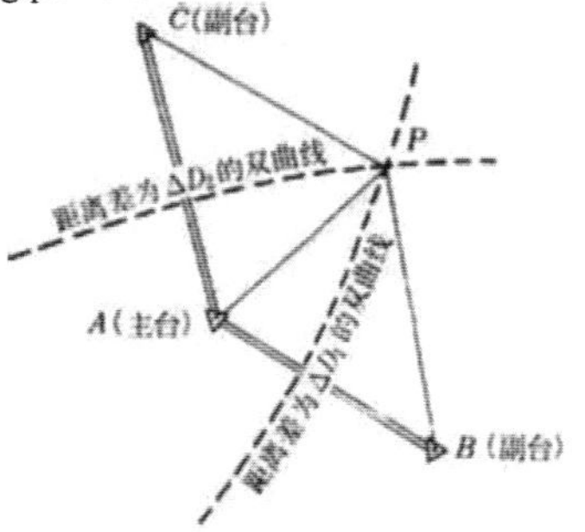

Figure 2. TDOA arrival time difference (hyperbolic) positioning principle

The figure for the position of P point measured point, A (main station), B (slave station), C (slave station) is 3 reference platform. If the measured distance between BP and AP difference ΔD1, CP and AP distance difference is ΔD2, and AC distance, AB distance can be made ΔD1, ΔD2 draw two hyperbolic on the chart are, the intersection P is being measuring the position of the point.

TDOA positioning application in WLAN (IEEE 802.11X) case is known propagation speed, the radio wave propagation time is proportional to the distance and it spread. But this method needs to pay attention to several problems: (a) radio waves (in the non line of sight NLOS environment) of the transmission characteristics; the general solution is to increase the number of measurements, the measured values obtained statistically significant; (b) clock precision. (c) clock synchronization. We must ensure the synchronization of clocks between participation with a positioning reference point.

Meanwhile, TDOA signal arrival time record with high accuracy requirements, industrial realization is difficult, needs the help of more advanced digital signal processing techniques. Especially WLAN based positioning precision arrival time difference, not only between the wireless access point (AP) to achieve the level of accuracy of time synchronization nanosecond (1 ns time synchronization error corresponding to 10 to 30 cm of positioning error), further recording error signal arrival time is also needed to control within a few nanoseconds. But now the market is relatively time synchronization accuracy between the only viable AP microsecond or millisecond, and no more mature programs can take advantage of WLAN positioning precision TDOA positioning method. GPS and Compass module can provide nanosecond synchronization accuracy, but wireless AP is usually installed indoors, the position of the blind spot below the satellite signals, so this clock synchronization methods can not meet the requirements of most of the WLAN positioning system.

2.4.2 IEEE 1588v2 (PTP) precise network clock synchronization technology

IEEE 1588v2 [9], with reference to international standards "IEEE Standard for a Precision Clock Synchronization Protocol for Networked Measurement and Control Systems", IEEE Std 1588TM-2008, March 2008. Also known as PTP (Precision Time Protocol), it is a new generation (2008 to complete the agreement for Standardization) clock synchronization technology. Its main feature is applicable to a variety of timing topology, bringing all nodes within a range of IP or Ethernet (such as WLAN AP) synchronized to a common reference clock source (such as GPS / Compass) and achieve sub-microsecond (us) synchronization accuracy. PTP clock synchronization network bandwidth is small, reusable AP's existing WLAN network connection. For this reason, we chose to have independent intellectual property rights IEEE1588v2 nanosecond precision clock device PTPGrand-2100 [11], the time source as WLAN network time; and providing PTPWatch network clock monitor, as the network location server and WLAN among TDOA-AP access station and positioning terminal precision clock synchronization monitoring tool when interacting. IEEE1588v2 Master clock device PTP Grand-2100 as shown in Figure 3.

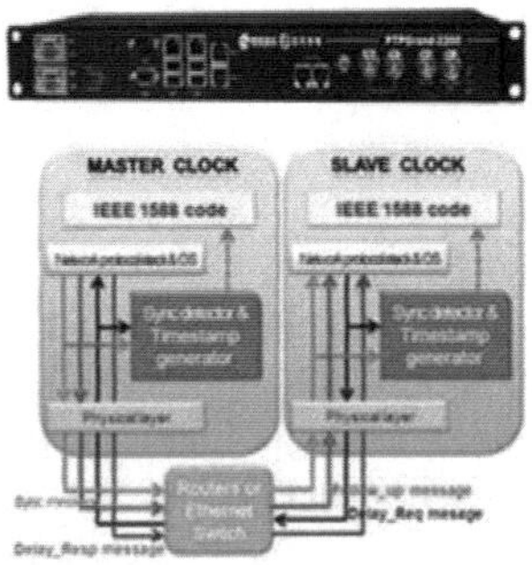

Figure 3. Nanosecond master clock device PTP Grand-2100 and Synchronization software IEEE1588v2 protocol stack.

PTPGrand-2100 embedded IEEE 1588v2 precision network clock protocol stack software (Protocol Stack). The protocol stack to achieve full IEEE 1588v2 functionality, including IEEE1588-2008 protocol engine,

 DOI: 10.1051/matecconf/20166107018

the best master clock competition (BMC) algorithm, software or hardware time stamp, as well as various clock status and mode of network communication and message standards defined management mechanism [11]. The stack either the primary clock, the clock can also be made, depending on the results of BMC algorithm competition or user settings. When an external clock reference source, PTPStack-1000 can be better than +/- 50ns (hardware timestamp) or +/- 10us (software timestamp) precision lock the reference source [10]. In summary, in order to address indoor WLAN positioning is difficult to achieve decimeter or centimeter precision problem, the research group studied and proposed to reach the time difference principle (TDOA) and IEEE1588v2 (PTP) precise network clock synchronization technology, combined with NLOS ranging error elimination technology, research and design based on a set of high-precision conventional WLAN (IEEE802.11x series of standards) devices (decimeter) RTLS solutions.

3 The architecture of the high-precision WLAN positioning system

3.1 Framework design of the positioning system

In this design, firstly, we have to TDOA (time difference of arrival) positioning principle as the theoretical basis, construction of a high-performance software-defined radio network communications platform, locating the physical layer to achieve a wireless signal arrival time accurate measurement and signal time difference of arrival (TDOA) calculation. Next, referring to IEEE1588v2 (PTP) international standards, through the integrated application of multi-hop delay compensation and asymmetric compensation algorithm to ensure nanosecond-level time synchronization accuracy
(one nanoseconds between WLAN access points (AP) synchronization error equivalent to 10 to 30 cm positioning error [12]).Finally, based NLOS error becomes stability principle, through NLOS error filtering and positioning before the data pre-processing, the greatest degree of eliminating the effects of NLOS error, the time domain error control in wireless ranging 5ns or less to achieve sub meter level or more accurate location estimate. Accordingly, we propose a design of high-precision positioning system WLAN solutions, as shown in Figure4.

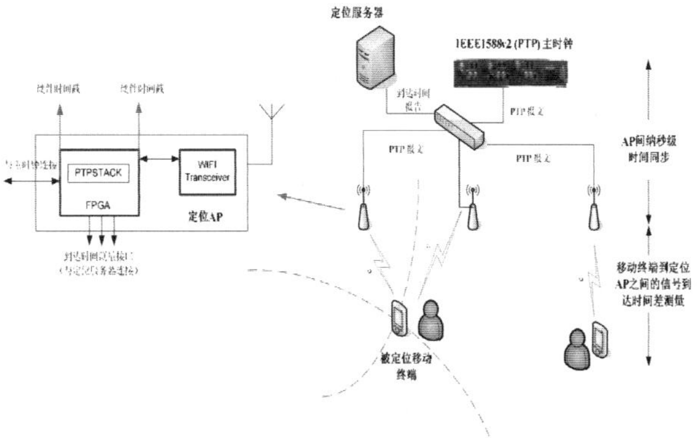

Figure 4. Based on TDOA and the PTP high-precision WLAN positioning system architecture

In the composition of the entire system, including the IEEE1588V2 master clock, TDOA WLAN access station (TDOA-AP), locate the server and the terminal equipment is located. Exploded above figure precision WLAN positioning system, various devices to work together in a unified clock pace to achieve their respective functions.

IEEE1588v2 master clock: using Rubidium time-keeping (GPS optional), by IEEE1588v2 protocol stack and each TDOA WLAN access point (TDOA-AP) interact to synchronize the time between each AP.

TDOA-AP: Each TDOA-AP in addition to traditional wireless network systems, but also added a programmable module (FPGA). FPGA module embedded 1588v2 precision proprietary module, supports upstream (master clock direction) of the PTP clock synchronization and local clock calibration process packets 1588 hardware timestamp direction from the master clock, and the positioning signals from the mobile terminal direction arrival time stamp, and other functions.

Positioning terminal: Each mobile terminal under the control of the software, the timing to send positioning signals to the AP. Develop positioning terminal to be solved mainly low power consumption, small size and other engineering challenges.

Location server: This server through the network switching device connected to one or more wireless positioning stations, and its main function is to load the system software to summarize the positioning signals provided by TDOA-AP arrival time and the arrival time difference of position calculation and data processing.

3.2 Key technical analysis and implementation

To design the high-precision WLAN positioning system based on TDOA and the PTP, the key technical innovations and difficulties mainly reflected in the measurement of high-precision positioning system WLAN physical layer radio signal arrival time, precision clock synchronization between IEEE1588v2 standards-based AP, data pre-processing NLOS error (NLOS) and the elimination of pre-positioning and other three aspects. Specific solutions and implementation as follows.

 DOI: 10.1051/matecconf/20166107018

3.2.1 high-precision measurement of the physical layer signal arrival time

In order to minimize the error signal arrival time record and improve TDOA location accuracy, we intend through a software-defined radio architecture, a WLAN AP system 802.11MAC, baseband, IF / RF conversion, RF and power amplifier chip modules together. shown in Figure 5, based on a high-speed programmable gate array (FPGA) and multicore processors DSP development board, through proprietary wireless signal recognition algorithm, implemented on the development board data bus to detect the WiFi signal, stamped. Signal arrival time stamp of playing point (Time-Stamping Point) is in the data path 802.11 M AC and baseband modules, avoiding the system processing error in the MAC layer and above the protocol stack, thereby positioning arrival signals exact time recording and real-time processing.

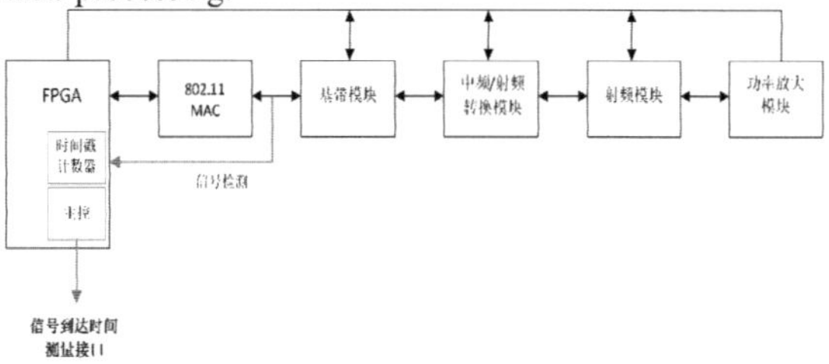

Figure 5. Key Techniques 1: physical layer based on software radio positioning signal measurement time of arrival

3.2.2 Precision clock synchronization based on IEEE 1588v2 between the AP

Based on IEEE 1588v2 precision clock synchronization protocol, we intend to pass a series of technical innovations, including network traffic prediction, multi-hop delay compensation and asymmetric compensation algorithm to achieve the following single-hop 80ns clock synchronization accuracy. This means that the time to use a switch to connect the AP, time recording error AP devices can be controlled in a single measurement 80ns or less. As shown in Figure 6, WLAN AP equipment in the work if the load from the master clock, in theory, we calculate the average transmission network information exchange process produces delays. After the adoption of multiple measurements, we will be modeling error distribution (probability distribution to obtain the mean, variance, etc.), the final error control in 5ns or less, so that the error can be controlled in TDOA positioning one meter or less.

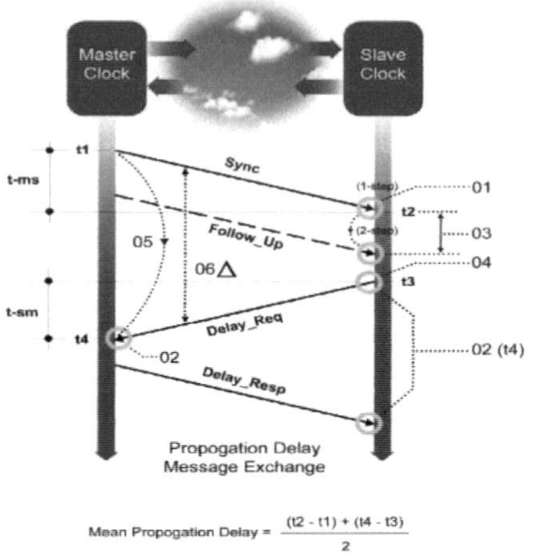

Figure 6. VLAN propagation delay information exchange schematic

3.2.3 NLOS error cancellation technology

The presence of a non-line of sight (NLOS) signal transmission between the AP and the positioning signal source is positioned, or ranging TDOA-based location accuracy will be affected. Through theoretical study, using fingerprints and TDOA hybrid positioning signal, and non-line of sight (NLOS) error based on the principle of variable stability NLOS error by filtration and pre-positioned before the data, we can maximize the impact of the elimination of NLOS error achieve decimeter positioning accuracy of less TODA. As shown in Figure 7, based on the principle of variable stability NLOS error when filtered by methods ranging error NLOS geometric analysis[12].

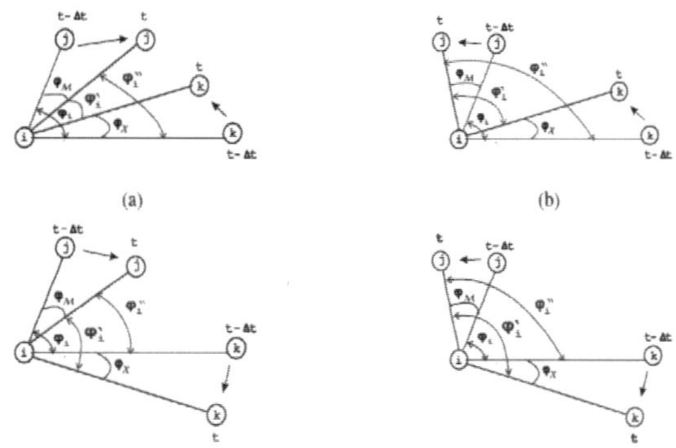

Figure 7. The NLOS error mitigation Geometric Analysis

In addition, high-precision positioning system WLAN AP also relates to positioning design prototyping, among IEEE1588v2 of TDOA-AP nanoscale clock synchronization algorithm NLOS error filtering algorithms, design and implementation of energy saving algorithms, and embedded hardware and software system integration, commissioning and optimization of the design part.

4 Typical applications for solutions and indicators

With the continuous development of the electronic information technology and the continuous improvement of the degree of social modernization, precise positioning technology, especially based on the precise positioning technology in recent years has been rapid development.In the next few years, there is a need for high precision positioning technology products in the province and the whole country. This kind of positioning technology has already entered the social life and industrial production.The system uses the most common WiFi signal (rather than the other industrial wireless technology), and can be widely for marketing produce huge economic and social benefits.The following table (Table 2) are listed in real time with high precision positioning industry in seven examples of the application and its specific accuracy requirements.

High precision TDOA positioning scheme based on PTP and WLAN is applied to a certain industrial production enterprise, as shown in Figure 8.The positioning system is realized by the method of 10 cm level accuracy positioning of the human, vehicle and object in a factory. By a finite number of TDOA-AP laid at the factory, real-time accurately positioning workers, forklift trucks, cargo and other micro tag location, people, vehicles and position information recorded in a few nanoseconds to almost zero delay transmit to the control center to a location server. Factory to achieve precise control on lean production and convenient management personnel for regional security control, on-the-job monitoring, vehicle trajectory in real time monitoring.

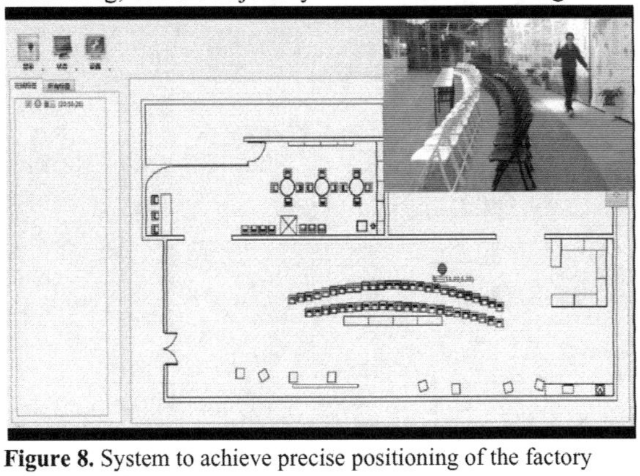

Figure 8. System to achieve precise positioning of the factory workers renderings

This case is a typical application of industrial production of human, vehicle, and material objects such as hybrid location and category management, truly precise, efficient, real-time construction of the Internet of things in the factory. After the completion of the system and repeated testing, the positioning parameters are as follows:

1.WLAN positioning signal: IEEE 802.11X;

2.Time recording error between AP (AP via hop between switches), a single measurement 12 ~ 50ns, the average of multiple measurements of 25ns;

3. real-time positioning accuracy: positioning station (TDOA-AP) and positioning the terminal for the next Line of sight (LOS) conditions mean 9 cm, the positioning station (TDOA-AP) and the positioning terminal is under NLOS conditions mean of 21 cm;

4. positioning terminal standby / charge cycles: 168 h.

Table 2. High-precision positioning Applications in different industries.

Industry	Main applications	Application description	Positioning accuracy
Healthcare	Medical equipment management ; patient activity tracking	WLAN positioning in the medical industry not only can be used to track the placement of medical devices but also can be convenient for medical personnel to real-time tracking of key patients	< 1 meter (instrument management) < 3 meters (patient tracking)
Mining exploration	Miner localization	The main use of the coal mine positioning system is to track and manage the activities of the underground workers. When the accident occurred, the location information by the miners, rescue workers can quickly determine the specific location of underground personnel and related conditions	< 5 meters
Traffic	Satellite navigation aids	The existing vehicle or pedestrian navigation system (such as GPS/) in the city center due to poor satellite visibility, often can not work effectively. You can achieve seamless precision navigation under various environments by mixing Compass / WLAN positioning system.	< 5 meters
Industrial production	Parts management ; Production Line Monitoring	By positioning the WLAN to realize automatic management of production parts, production tools and production processes, to increase productivity and shorten the production cycle.	< 1 meter
Logistics management.	Accurate material tracking and warehouse management	The automatic recording and tracking of the material location is realized by WLAN, which can realize the automation of warehouse management. According to the size of the tracking of the material on the positioning accuracy requirements are also different .	< 1 meters
Internet or mobile Internet	Location-based services (LBS) and precision marketing	The position and trajectory of the user to enter the mall is obtained by WLAN. Use this information, accurate marketing system can speculate the consumer behavior and interest, and accordingly targeted advertising push, accurate to buy, discount, new listing and other information targeted to push to the user.	< 5 meters
Government	Crowd flow management	Through outdoor AP crossing the smartphone passive positioning, to achieve the position of the crowd, the monitoring of the flow route, and the real-time management of the distribution of people.	< 5 meters

5 Conclusion

With the rapid development of the internet of things related industries, various sectors in the field of business need more and more requirements on wireless location, especially the need to WLAN indoor positioning precision LBS location services. Based on TDOA and PTP high - precision WLAN positioning system, built on a software radio network communications platform, complete WiFi wireless signal arrival time accurate measurement, based on the principle of realization of the arrival time difference TDOA ranging positioning calculation; and innovation proposed based IEEE1588v2 (PTP) standards, protection of WLAN access points (AP) between the nanosecond level time synchronization accuracy; and finally, the system through NLOS error filtering and positioning before the data preprocessing, eliminate the influence of NLOS error, the time domain error control in wireless ranging 5ns hereinafter, in order to achieve higher accuracy decimeter or location estimate.

The case can be widely used in mobile Internet, logistics management, mineral exploration, health and other needs of the precise location services (LBS) in other industry sectors. The system will be integrated into a multi-hop delay compensation, compensation for asymmetric algorithms such integrated application, high-speed positioning solver engine optimization, positioning and other open GIS Engine API interface, and gradually improve the system positioning accuracy,system integrators and end users to meet dimensional positioning business needs.

References

1. CCID Consulting, China location services (GPS / LBS) Market Competition Analysis Report [R / OL] [2014-07-14] . http //www.ccidconsulting.com/scyx/
2. Xu Jingwei. LBS system based on WLAN wireless location technology [D]. Wuhan: Huazhong University of Science and Technology master's degree thesis, 2011: 22-24
3. Wu Z, et al. Location Estimation via Support Vector Regression [J]. IEEE Transactions on Mobile Computing, 2012,6 (3)
4. Yan Xiaotian, warm bright students, Wuwei. WLAN positioning technology [EB / OL]. http://labs.chinamobile.com/mblog/712208_82775. 2012,3
5. Anthony LaMarca and Eyal de Lara Location Systems An Introduction to the Technology Behind Location Awarenes [J]. Morgan and Claypool, 2008,8 (3): 25-28
6. MC Su, DY Huang, and SC Lin A regression-mode-based approach to indoor location estimation [J] Journal of Engineering and Applied Sciences, 2008,3 (4):307-311
7. Huang Yi, Hu. Based on TDOA measurements of high-precision positioning technology research. Aiqun high-tech communications, Vol. 22, No. 1 .2012: 8-13
8. IEEE.IEEE Standard for a Precision Clock Synchronization Protocol for Networked Measurement and Control Systems [S] .IEEE1588v2, March 2008: 125-128.
9. Zhuoqun Li, et al Range-based Relative Velocity Estimators for Networked Mobile Devices [J] .IEEE Transactions on Vehicular Technology, vol.58, issue 4, pp 2095 -2099, May 2009
10. Jingang Huang, Junjie Zhou. Software Process Improvement Based On CMM [C] .2013 Asia-Pacific Computational Intelligence and Information Technology Conference .PP594-598.Dec 2013
11. Study of things for clock synchronization architecture [J] Bo and other Computer Engineering and Design 2011,32 (5): 16-21
12. Wang Yuan, Lin Jinguo. Discuss several common indoor positioning technology [J]. Chinese instruments 2011,2 (5): 23-26

AUTHOR INDEX

AUTHOR INDEX

AUTHOR INDEX

AUTHOR INDEX